FRESHWATER NEMATODES

Ecology and Taxonomy

DEDICATION

Over the past decade more than 130 students (including me), mainly from developing countries have been trained in the field of nematology at Ghent University, Belgium, in collaboration with and through the financial support of the government of Belgium.

This book is dedicated to the **Postgraduate International Nematology Course (PINC), Ghent University, all the professors, scientific and technical staff** and especially to two professors, **Dr. August Coomans** and **Dr. Etienne Geraert**. These prominent nematologists initiated the programme, spearheaded its progress over the past decade, guided the research work of numerous students and continue to support this unique international programme that aims at promoting the science of nematology on a global scale.

Eyualem-Abebe

Freshwater Nematodes
Ecology and Taxonomy

Edited by

Eyualem-Abebe

University of New Hampshire
USA

I. Andrássy

ELTE Allatrendszertani Tanszek
Hungary

and

W. Traunspurger

Universitaet Bielefeld
Germany

CABI Publishing

CABI Publishing is a division of CAB International

CABI Publishing
CAB International
Wallingford
Oxfordshire OX10 8DE
UK

Tel: +44 (0)1491 832111
Fax: +44 (0)1491 833508
E-mail: cabi@cabi.org
Website: www.cabi-publishing.org

CABI Publishing
875 Massachusetts Avenue
7th Floor
Cambridge, MA 02139
USA

Tel: +1 617 395 4056
Fax: +1 617 354 6875
E-mail: cabi-nao@cabi.org

A catalogue record for this book is available from the British Library, London, UK.

A catalogue record for this book is available from the Library of Congress, Washington, DC.

Library of Congress Cataloging-in-Publication Data

Freshwater nematodes: ecology and taxonomy/edited by A. Eyualem, I. Andrássy, W. Traunspurger.
p. cm.
Includes bibliographical references and index.
ISBN-13: 978-0-85199-009-5 (alk. paper)
ISBN-10: 0-85199-009-6 (alk. paper)
1. Freshwater nematodes. I. Abebe, E. (Eyualem) II. Andrássy, I. (István) III. Traunspurger, W. (Walter) IV. Title.

QL391.N4F74 2006
592′.57176--dc22 2005010266

ISBN 0 85199 009 6
978 0 85199 009 5

Typeset by SPI Publisher Services, Pondicherry, India
Printed and bound in the UK by Biddles Ltd, King's Lynn

Table of Contents

Contributors ix

Preface xi

Foreword xvii
P.A.A. Loof

Foreword xix
D.H. Wall

PART I: Ecology 1

1 **Introduction: Summary of Present Knowledge and Research Addressing the Ecology and Taxonomy of Freshwater Nematodes** 3
Paul De Ley, Wilfrida Decraemer and Eyualem-Abebe

2 **Techniques for Processing Freshwater Nematodes** 31
Mike Hodda and Eyualem-Abebe

3 **Composition and Distribution of Free-living Freshwater Nematodes: Global and Local Perspectives** 46
Walter Traunspurger, Iris C. Michiels and Eyualem-Abebe

4 **Dynamics of Freshwater Nematodes: Abundance, Biomass and Diversity** 77
Eyualem-Abebe, Walter Traunspurger and Iris C. Michiels

5 **Production of Freshwater Nematodes** 94
Matthias Bergtold and Walter Traunspurger

6 **Feeding Ecology of Free-living Benthic Nematodes** 105
Tom Moens, Matthias Bergtold and Walter Traunspurger

7 **Patterns in the Size Structure of Freshwater Nematode Communities: the Cases of Lakes Königssee and Brunnsee, Germany** 132
Walter Traunspurger and Matthias Bergtold

8 **Freshwater Nematodes in Environmental Science** 144
Sebastian Höss, Walter Traunspurger and Aldo Zullini

9 **Nematodes in Lotic Systems** 163
Mike Hodda

10 **Nematodes from Extreme Freshwater Habitats** 179
Mike Hodda, Amelia Ocaña and Walter Traunspurger

11 **Computation and Application of Nematode Community Indices: General Guidelines** 211
Deborah A. Neher and Brian J. Darby

PART II: Taxonomy 223

12 **Order Enoplida** 225
Nic Smol and August Coomans

13 **Order Triplonchida** 293
Aldo Zullini

14 **Dorylaimida Part I: Superfamilies Belondiroidea, Nygolaimoidea and Tylencholaimoidea** 326
Reyes Peña-Santiago

15 **Dorylaimida Part II: Superfamily Dorylaimoidea** 392
Maria Teresa Vinciguerra

16 **Order Mononchida** 468
Aldo Zullini and Vlada Peneva

17 **Orders Chromadorida, Desmodorida and Desmoscolecida** 497
Wilfrida Decraemer and Nic Smol

18 **Order Monhysterida** 574
August Coomans and Eyualem-Abebe

19 Order Araeolaimida 604
Agnes Muthumbi and Ann Vanreusel

20 Order Plectida 611
Oleksandr Holovachov and Paul De Ley

21 Order Rhabditida: Suborder Tylenchina 648
Wim Bert and Gaëtan Borgonie

22 Order Rhabditida: Suborder Rhabditina 696
Joaquín Abolafia

Appendix 722
Oleksandr Holovachov and Paul De Ley

Index 736

Contributors

Joaquín Abolafia, *Departamento de Biología Animal, Biología Vegetal y Ecología, Universidad de Jaén, Campus "Las Lagunillas" s/n. 23071-Jaén, Spain. Email: abolafia@ujaen.es*

Matthias Bergtold, *Department of Animal Ecology, University of Bielefeld, Morgenbreede 45, 33615 Bielefeld, Germany. Email: m.bergtold@uni-bielefeld.de*

Wim Bert, *Department of Biology, Ghent University, Ledeganckstraat 35, 9000 Ghent, Belgium. Email: Wim.bert@ugent.be*

Gaëtan Borgonie, *Department of Biology, Ghent University, Ledeganckstraat 35, 9000 Ghent, Belgium. Email: gaetan.borgonie@ugent.be*

August Coomans, *Department of Biology, Ghent University, Ledeganckstraat 35, B-9000 Ghent, Belgium. Email: August.coomans@ugent.be*

Brian J. Darby, *Department of Plant and Soil Science, University of Vermont, Burlington, Vermont 05405, USA. Email: Brian.Darby@uvm.edu*

Wilfrida Decraemer, *Royal Belgian Institute of Natural Sciences, Vautierstraat 29, B-1000 Brussels, Belgium, and Department of Biology, Ghent University, Ledeganckstraat 35, B-9000 Ghent, Belgium. Email: Wilfrida.Decraemer@naturalsciences.be*

Paul De Ley, *Department of Nematology, University of California, Riverside, CA 92521, USA. Email: paul.deley@ucr.edu*

Eyualem-Abebe, *Hubbard Centre for Genome Studies and Department of Zoology, University of New Hampshire, 35 Colovos Rd, Gregg Hall, Durham, NH 03824, USA. Email: Eyualem@unh.edu*

Mike Hodda, *Nematode Biosystematics & Ecology, CSIRO Entomology, GPO Box 1700, Canberra, ACT 2601, Australia. Email: Mike.Hodda@csiro.au*

Oleksandr Holovachov, *Department of Zoology, Ivan Franko National University of Lviv, Hrushevsky str. 4, Lviv 79005, Ukraine. Email: zoomus@franko.lviv.ua*

Sebastian Höss, *Ecossa, Thierschstr. 43, 80538 München, Germany. Email: hoess@ecossa.de*

Iris C. Michiels, *Department of Animal Ecology, University of Bielefeld, Morgenbreede 45, 33615 Bielefeld, Germany. Email: iris.michiels@uni-bielefeld.de*

Tom Moens, *Marine Biology Section, Biology Department, Ghent University, Krijgslaan 281/S8, B-9000 Ghent, Belgium. Email: tom.moens@UGent.be*

Agnes Muthumbi, *Department of Zoology, University of Nairobi, PO Box 30197 Nairobi, Kenya. Email: amuthumbi@uonbi.ac.ke*

Deborah A. Neher, *Department of Plant and Soil Science, University of Vermont Burlington, Vermont 05405, USA. Email: Deborah.Neher@uvm.edu*

Amelia Ocaña, *Department of Animal Biology and Ecology, Faculty of Science, University of Granada, 18071 Granada, Spain. Email: amelia@ugr.es*

Reyes Peña-Santiago, *Departamento de Biología Animal, Biología Vegetal y Ecología, Universidad de Jaén, Campus 'Las Lagunillas' s/n, 23071-Jaén, Spain. Email: rpena@ujaen.es*

Vlada Peneva, *Central Laboratory of General Ecology, 2 Gagarin Street, 1113 Sofia, Bulgaria. Email: vpeneva@ecolab.bas.bg*

Nic Smol, *Department of Biology, Ghent University, Ledeganckstraat 35, B-9000 Ghent, Belgium. E-mail: Nic.smol@ugent.be*

Walter Traunspurger, *Department of Animal Ecology, University of Bielefeld, Morgenbreede 45, 33615 Bielefeld, Germany. Email: traunspurger@uni-bielefeld.de*

Ann Vanreusel, *Marine Biology Section, Ghent University, Krijgslaan 281-S8 B-9000, Ghent, Belgium. Email: ann.vanreusel@UGent.be*

Maria Teresa Vinciguerra, *Dipartimento di Biologia Animale, Università di Catania, Via Androne 81, 95124 Catania, Italy. Email: vincimar@unict.it*

Aldo Zullini, *Dipartimento di Biotecnologie & Bioscienze, Università di Milano-Bicocca, Piazza della Scienza 2, 20126 Milan, Italy. Email: aldo.zullini@unimib.it*

Preface

Over a decade ago, as a graduate student working on freshwater nematodes and having experienced the difficulty of acquiring up-to-date and singularly comprehensive material, I wrote in my thesis '…*the absence of an organized and comprehensive reference material in this group (freshwater nematodes) has made the task of identification cumbersome. Most of the literature is widely distributed in space and time, and to make it easily accessible it needs, at least, to be compiled together*…' This statement still holds true. Over these past few years, I have received repeated requests from budding researchers for a current and relatively inclusive reference work, and this underlined the widespread need for such a resource. This book, though delayed by a decade since I first thought about it, is an attempt to address the prevailing demand by bringing together in a single volume nearly all the available information on the ecology and taxonomy of nematodes of freshwater bodies.

Nematodes are ubiquitous, the most diverse and numerically dominant metazoans in freshwater habitats, and these properties bestow exceptional significance to their role in the environment. An array of functional roles has been attributed to them: they are grazers on bacteria and primary producers, regulators of decomposition, predators, prey for other animals and closely associated symbionts of bacteria and other organisms. They also serve as potential indicators of pollution and general environmental disturbance. Despite all these claims, their importance is very much masked by their small size. In general, their invisibility to the naked eye has played a pernicious role against a better understanding of their importance. Current research on nematode biology generally follows this unfortunate trend, leaving freshwater forms as the least studied: parasitic nematodes are better studied than (>) free-living; terrestrial > aquatic; and marine > freshwater forms.

Discussion on nematode ecology is conspicuously lacking in most limnological treatises. In spite of some recent works on taxonomy and ecology of this group, the nematofauna of the vast majority of the world's inland water bodies still remains largely unknown. Moreover, currently there is a growing awareness of the loss of biological diversity and environmental degradation. There is an increasing interest among geneticists, genomic researchers, ecologists and phylogeneticists in knowing more about organisms that can be good model systems for revealing life's basic biological processes. Also, ecologists are hybridizing modern genomic tools with classical experimentation to unravel ecological processes and interactions. Several nematode species have been demonstrated to be useful models.

The fact that one can find a dozen or more nematode species in a spoonful of sediment is interesting and even challenging *vis-à-vis* the currently little understood relationship between biodiversity and ecosystem functioning. Seen in the light of ongoing genetic research on model nematodes, it is plausible to assume that in the future other nematode species too may play a central role as model systems in understanding ecological principles. Besides, the enhancement of future genetic research on these animals necessitates our better understanding of their biology and biological diversity. On the other hand, this growing interest in nematodes as model systems can also have a boosting effect on nematological research in that recent developments in genomic tools can help us understand hitherto little understood ecological processes and mechanisms, e.g. by providing new ways of quantifying the various roles of nematodes in the sediment and facilitating their inclusion in benthic food web modelling; something that is currently missing from trophic studies.

This book aspires specifically to bring this largely ignored group of free-living nematodes to the forefront. The year 2003 was celebrated as the 'International Year of Freshwater', giving much needed recognition to the importance of freshwater resources. Nevertheless, informed decisions about this habitat can be made only if we understand its individual components as well as the whole network of interactions within and outside the system. Recognizing the importance of freshwater bodies but leaving out their most numerous and most diverse metazoan inhabitants, i.e. nematodes, indeed do not go together. I believe that future freshwater research will need to emphatically include these largely ignored animals. Consequently, notwithstanding the paucity of information on some aspects of freshwater nematology, I strongly believe that the information contained in this book will help fill an overdue gap in the literature. I also believe it will help enhance research on freshwater nematodes and other fields of biology by facilitating the accessibility of a first comprehensive synthesis of the widely scattered available literature.

In the various chapters of this book, internationally recognized experts review the distribution, abundance, biomass, diversity, production and size structure of freshwater nematodes, as well as their role in environmental studies, and their adaptations to extreme environments. More general chapters such as those on techniques, classification of feeding habits, and the application of indices address issues with broader implications and will be useful to any ecological research on nematodes. One important group that is not reviewed in depth is the Order Mermithida, my efforts to involve experts in mermithid taxonomy and ecology were not successful, for reasons beyond my control. At the eleventh hour, Drs Holovachov and De Ley managed to provide us with a summary of this group (see Appendix); I am grateful to both.

One of the many challenges I encountered during the course of the work on this book was the decision on the appropriate depth and breadth of the taxonomic component. Finding the right balance between producing an up-to-date reference, which can be of use to a wide group of users, and an all-inclusive book that may require a longer time commitment from contributors, was key in shaping the taxonomic content. I discussed this issue with the co-editors and many contributors extensively, not least during a face-to-face meeting of all prospective contributors in 2002 during the Fourth International Congress of Nematologists in Tenerife. All participants agreed that ideally a book that would enable the identification of any freshwater nematode to the species level, by providing descriptions of all known freshwater species, would be the most complete and most useful material to produce. However, it would also require a significantly longer time than it took us to produce this current one (4 years). In the end, despite the need for species-level descriptions, feasibility dictated that the contributors and I limit the primary purpose of this book to be a starting point for all researchers on freshwater nematodes. To this effect, each taxonomic chapter in the book provides information that enables a genus-level identification, references to important reviews, as well as a list of species reported from freshwater bodies and available ecological information. This first step clearly forebodes that more encyclopaedic completeness is expected from freshwater nematologists in the future.

Another important decision I made was what classification scheme to follow for the presentation of the book; the reasons for such a decision are discussed in depth in the introductory chapter. The choice to follow this most recent classification scheme was openly discussed among the various authors. Seen in light of the different classifications entertained in the nematological literature, I neither expected everyone to agree on a single scheme nor was I naïve enough to assume that this issue could be addressed in depth to the satisfaction of all involved. Notwithstanding these problems, all contributors agreed to use the suggested scheme to the taxonomic category level of order and they used their choice of preference below that level. Although some authors do support the used classification scheme, readers should take note of the fact that authorship in no way implies a *de facto* acceptance of the classification scheme used in the book.

Contributors provided taxonomic diagnosis to categories at the genus level and above, and I have attempted to maintain relative uniformity with regard to content and depth of information provided for these categories. For each genus, a taxonomic diagnosis is provided, accompanied by figures illustrating generic characters and a complete list of those species reported from inland water bodies. However, in some chapters, a full and complete list of species is also provided. The advantages of providing such a complete list of species had to be deferred in some chapters for practical reasons; presenting the information on every species of a group with such a huge diversity in just a single volume would have far exceeded the intended size of the book.

With regard to the style of presentation, I have chosen not to enforce a uniform style of presentation and have left the choice to the contributors to follow their preferences. For example, some contributors have provided dichotomous keys to simplify identification at various levels of taxonomic category in addition to species lists. At the species level, each species name is accompanied by the type of environment/water body, as well as a list of geographical localities it has been reported from, with the relevant references. However, in this accompanying information too, I have encouraged variations in the styles of presentation and the balance between the different authors' emphases on morphology *versus* ecological information.

By providing an overall review of the ecology and taxonomy at genus and higher level taxon category, this book is intended to provide a useful reference to a broad user community: students, beginners and established researchers in the field of freshwater nematology, benthologists, invertebrate biologists, limnologists, ecologists, microbiologists, soil biologists and other researchers whose area of study is closely tied to nematodes but whose interest and attempts to include them in their research were hampered by the inaccessibility of the literature. Despite the fact that fairly detailed microscopic studies are required for nematode identification, line drawings that accompany generic diagnoses, tabulated summaries of characters and character states, and dichotomous keys in the book are intended to help ease the identification process. Be that as it may, nematological research needs to attract a much larger number of students in the next decade to offset current and recent decline of student interest in the field. As I mentioned in my introductory paragraph, for students and budding researchers, finding their way in nematological literature is time-consuming, frustrating and therefore not an easy stumbling block to overcome; it is my sincere hope that this book will reduce this hurdle by providing a starting point of reference and stimulate students enough to plant the seeds of interest in working in the field of freshwater nematology.

Many people have generously extended their assistance to me in various ways from the beginning till the completion of this book. I am deeply indebted to William Kelley Thomas and Thomas Kocher, co-directors of the Hubbard Center for Genome Studies, University of New Hampshire, USA, for allowing me to use the facilities of the centre, especially W.K. Thomas for his encouragement and for providing the necessary time and resources I needed for the editorial work. I also thank members of Thomas' Environmental Genomics Lab: Krystalynne Morris, Darren Bauer, Jobriah Anderson, Rick Roy and Laurie Szilagyi for supporting my editorial work in multiple ways. I thank Mark Blaxter, Institute of Cell, Animal and Population Biology, University of Edinburgh, UK and members of the Blaxter Nematode Genomics Lab for their support during the early phase of this project.

This book took its final form as a result of extended exchanges of ideas with and unreserved input from various co-authors. I thank August Coomans, Ghent University, Belgium, Paul De Ley, University of California, Riverside, USA, Wilfrida Decraemer, Royal Belgian Institute of Natural Sciences, Belgium, Maria Teresa Vinciguerra, Università di Catania, Italy, Reyes Peña Santiago, Universidad de Jaén, Spain, Aldo Zullini, Università di Milano-Bicocca, Italy, and Nic Smol, Ghent University, Belgium, for their many constructive suggestions. I especially thank the first three for their continuous encouragement and support, and for the long and frequent discussions that shaped the content and presentation of this book. All co-authors have carefully read the various versions of their manuscripts and accommodated the suggestions from the editors for change; I thank them all.

The two co-editors, Professor Drs Walter Traunspurger and István Andrássy, generously accepted the responsibility to be co-editors despite their other pressing commitments, and played a central role in enhancing the quality of the manuscripts submitted for the ecological and taxonomic parts, respectively. Both were supportive of the idea of the book from its inception and continued to provide constructive suggestions to me and the other co-authors until its completion. My journey with both from the beginning to the completion of this book has been both enjoyable and highly educational; I am deeply indebted to both. Nevertheless, despite the help I received from these co-editors and other co-authors, the final decision (and any blame that might derive therefrom) as to the depth of contents of the different chapters in the book was my own and mine alone.

One person who missed the most while I was grappling with the organization and editorial work over the past 3 years was my son, Eyob. I thank him for being extremely patient with me and, hopefully, he will benefit from his patience in the long run.

Eyualem-Abebe

Foreword

In this hectic time new possibilities open up several directions for new research. New scientific results are, with increasing speed, being distributed along electronic channels. Nevertheless books retain their right to exist. They bring together scattered information, may weave it into a general pattern, give more or less complete reviews on literature on any subject, provide syntheses; in short they serve as monuments which cause scientists to sit down and reflect. Thus they act as a necessary counterweight against the ever-increasing stream of new data.

Ecology and taxonomy always have been close allies. In order to understand the position of any animal group in a biocoenosis, it is an absolute prerequisite that the identities of the taxa involved have been established firmly. Nematology is in the happy position that morphology and taxonomy of free-living and plant-parasitic nematodes have developed strongly during the last 50–60 years (T. and J.B. Goodey, Chitwood, Steiner, Thorne and the next generation, consisting of so many investigators that naming some would do others an injustice). Another factor is the rapidly growing information technology providing databases which bring together many facts and enable these to be ordered into a consistent whole. In this way nematode communities can be analysed and their relationships to ecological conditions detected. This in turn might lead to nematode communities being used for characterization of soils and sediments.

What is needed, therefore, is a combination of ecology and taxonomy, and this is what this book gives. Many outstanding ecologists and taxonomists have contributed, and discussed relevant subjects. Also the three editors are well-known authorities in their fields and they are to be congratulated on producing a book of this quality. I am confident that it will act as a potent stimulus for both branches of science.

P.A.A. Loof
Laboratory of Nematology,
Wageningen University,
The Netherlands

Foreword

Freshwater sediments teem with nematodes. Scientists who delve into the mud of ponds, lakes, streams and rivers, find a teeming, fascinating world of microscopic nematodes, one of the most dominant invertebrate groups on earth. Often the scientists are asked by colleagues and non-scientists alike, 'why do you study these creatures?' and 'how do nematodes contribute to ecosystems?' The answers will vary, but for all intrigued by the shapes, sizes, biology and natural histories of nematodes living amongst the diverse freshwater habitats this book is an obvious next step for addressing these questions. Throughout the chapters experts present current scientific information on freshwater nematodes: the methods for extraction from sediments; the taxonomy of the freshwater species; food web relationships; and nematode ecology and biogeography. Additionally, these experts extend their understanding of the animals and their interactions to applications for environmental science. In a way, these chapters also explain why we study nematodes and determine their role in ecosystems: we are captivated by the vast diversity found in tiny representative samplings of the world's freshwater sediments and we are amazed at their many abilities and interactions in the regulation of ecosystem processes, above and below the surface. Learning about nematodes in sediments helps explain how the wealth of biodiversity in freshwater ecosystems operates and how it will need to be sustained in the future.

The themes of this book, ecology and morphology, are at the core of research for current global issues addressing the loss of biodiversity and resulting changes in ecosystem functioning. Biodiversity and ecosystem functioning were considered as separate research disciplines 15 years ago, but that has changed as we have recognized the connections between species and ecosystem processes. Biodiversity, the variability in life across many scales, developed along lines of research exploring morphological and genetic differences between species, their phylogenetic relationships and biogeographical distributions. Taxonomists thus provided the foundation on nematode identities, associations, types of freshwater habitats and biogeography. Ecologists concentrated their efforts on understanding nematode natural history, food web structure and processes of decomposition, primary production and transfer of nutrients, energy and materials. The result was a clearer understanding of the role of nematodes and nematode trophic groups in critical ecosystem processes. Today, our knowledge of nematode biodiversity and ecosystem functioning are becoming intertwined because we need to know the taxa and habitats within freshwater sediments that are most vulnerable to global changes.

There is global scientific consensus that the biodiversity within Earth's freshwaters and sediments, the ponds, lakes, streams, rivers and groundwater, is degrading, with confounding and potentially devastating impacts for humans. The benefits or services that are provided by the biodiversity within freshwater ecosystems, such as food, clean drinking water, economic livelihood (tourism, fishing, water sports) and the aesthetic and cultural benefits, are becoming recognized by policy makers and the public. With this recognition has been an increased appreciation by scientists for the poorly known invertebrates, such as nematodes, and greater attention to the discovery of needed information on whether hot spots of biodiversity exist on global or local scales, whether there are key species for an ecosystem process, such as predaceous nematode taxa, or how an invasive species may change the rate of carbon mineralization. Nematodes are a major player in biodiversity worldwide. There is an urgency to improve our understanding of the species and how nematodes contribute to and regulate the operation of freshwater ecosystems if we are to meet the challenge of sustaining these ecosystems for the future. This is the reason why this volume is needed and timely.

Diana H. Wall
Natural Resource Ecology Laboratory
Colorado State University

I Ecology

1 Introduction: Summary of Present Knowledge and Research Addressing the Ecology and Taxonomy of Freshwater Nematodes

PAUL DE LEY,[1] W. DECRAEMER[2] AND EYUALEM-ABEBE[3]

[1]*Department of Nematology, University of California, Riverside, CA 92521, USA;* [2]*Royal Belgian Institute of Natural Sciences, Vautierstraat 29, B-1000 Brussels, Belgium, and Department of Biology, Ghent University, Ledeganckstraat 35, B-9000 Ghent, Belgium;* [3]*Hubbard Centre for Genome Studies and Department of Zoology, University of New Hampshire, 35 Colovos Rd, Gregg Hall, Durham, NH 03824, USA*

Ecological Studies on Freshwater Nematodes

Research on the role and importance of nematodes in freshwater ecosystems has been limited in extent and primarily descriptive in nature, as reflected in the first part of this book. Most of what we know of freshwater nematode ecology is based on research in European freshwater bodies only. To make matters worse, a significant proportion of these studies lacks a temporal dimension: temporal analyses have only recently become available in the literature, adding a new dimension to a prior track record composed almost exclusively of taxonomic research on freshwater nematodes. As a result, present knowledge on the subject includes a relatively extensive species inventory, but only the beginning of an understanding of their interactions with each other, with other organisms or with their environment.

Discussions on the global distribution or endemicity of nematode species are still largely premature mainly because of the glaring incompleteness of global nematological surveys (Chapter 3). The scarcity of competent taxonomists, as well as the lack of agreement among experts about the correct identities of certain species, are indicated as important reasons. Easily communicable species delimitation criteria are suggested as helpful, and supplemental methods are recommended to help resolve some of the issues with species identification, thereby also providing better prospects of addressing biogeographical questions. Genus identification is overall much less contentious, and most freshwater nematode genera seem to have a cosmopolitan (or at least an extensive) distribution. This may reflect large-scale dispersal, not only owing to the small size of nematodes but also due to various adaptations enhancing nematode phoresis and survival, e.g. through various mechanisms for quiescence or anabiosis.

Comparison of freshwater nematode communities shows that similarity in community composition at the species level is higher at small scales (among cores within a lake) or large scales (between wide geographical areas) than at an intermediate scale (between lakes within a geographical area). Understanding aspects of community ecology without getting trapped in protracted taxonomic research is an issue that can be addressed by focusing on size structure of communities (Chapter 7). Size structure of freshwater zoobenthos seems to be influenced by e.g. the feeding habits of fishes and differs greatly from that in the marine environment.

Using the species richness as a unit of diversity, the highest number previously reported from an entire lake was only 152 species found in a study of 12,000 individuals (Chapter 3). Despite the limited data, this low number of species may actually indicate that freshwater sediments harbour significantly lower nematode species diversity compared to marine habitats (Eyualem *et al.*, 2004). Furthermore, abundance and biomass also tend to be lower in freshwater sediments compared to the marine environment. Paradoxically, although springs are relatively stable and have a higher ionic content than true freshwater bodies, the general trend in such habitats appears to be one where higher abundance and species richness occur in those springs with lower ionic content (Chapter 10).

Empirical data show that most freshwater nematode species and individuals remain close to the surface of the sediment, although they do migrate vertically to some extent in sediments. Furthermore, nematode communities change both temporally and spatially (Chapter 4). Despite their numerical dominance, inclusion of nematodes in productivity studies of freshwater habitats has not become a focal point for either limnologists or other benthologists, probably due to the relatively low biomass of these animals compared to other freshwater meio- and macrofauna. Empirical evidence now shows us that nematode production is indeed modest (Chapter 5). Nevertheless, nematodes play a critical ecological role in the decomposition of organic matter and serve as an important link in benthic food chains.

Traditionally, aquatic nematodes have been classified into feeding groups based on presumed associations between buccal morphology and types of consumed food (Wieser, 1953). Despite its wide use among nematologists, however, solid supporting evidence was hard to come by and as a result this approach lost ground over the years. Alternative proposals of feeding habit classifications are summarized in Chapter 6; the most recent version subdivides feeding habits of both aquatic and terrestrial free-living nematodes into five major types (i.e. plant feeders, hyphal feeders, substrate ingesters, carnivores and unicellular eukaryote feeders). These classifications ignore dissolved organic matter as a potential source of food for nematodes, an omission stressed in Chapter 6.

Nematodes possess a number of traits that make them especially useful for environmental studies as a tool to monitor change: these include high species richness, abundance, pervasiveness and tolerance, close association with the sediment and interstitial water, short generation time and large body of accumulated knowledge on model nematode systems (Chapter 8). For example, single species tests are reported to be more sensitive than the chironomid test, which uses macroinvertebrates. Comparative research on nematode communities within a uniform environment that has both impacted and undisturbed areas would contribute critically to our knowledge of natural variations in communities, and would thereby help refine their usefulness in environmental monitoring.

Nematodes can also provide a useful model system for fundamental ecological studies on the interactions between biodiversity and ecosystem functioning (Moens *et al.*, 2004), although this aspect has not yet received the attention it warrants. The contributors to this book emphasize the lack of paradigms for several aspects of freshwater nematode studies. We must stress the need for future systematic and comprehensive research, e.g. on lotic nematodes; on nematodes inhabiting hydric soils that are persistently saturated with water or ephemerally flooded for protracted periods of time; on comparisons of nematode communities responding to human impact with those under relatively undisturbed conditions; on temporal patterns and effects; and on controlled experiments quantifying their tolerance to various key environmental factors.

Previous Work on Freshwater Nematodes

Soil and Freshwater Nematodes by T. Goodey (1951) and J.B. Goodey (1963, second edition) is the first book to provide a comprehensive account of free-living nematodes, including freshwater taxa. Also essential to biodiversity studies of freshwater nematofauna are the two volumes of the *Bremerhaven Checklist of Aquatic Nematodes* by Gerlach and Riemann (1973, 1974). However, a comprehensive reference work specifically devoted to the freshwater nematofauna has not been available before this book, despite the fact that nematodes are dependent on the presence of a water film for their ecological and metabolic activities. Worldwide, general limnological contributions and series either omit nematodes altogether, or discuss the phylum only summarily. Important recent exceptions are: (i) for North America, the *Ecology and Classification of North American Freshwater Invertebrates* by Thorps and Covich (2001) and *Pennak's Freshwater Invertebrates of the United States* by Smith (2001), respectively, with a dichotomous key to genus level and to family level of freshwater nematode taxa; and (ii) for Europe, the contributions on nematodes by Andrássy (1979) in *Limnofauna Europaea* with a checklist of the animals inhabiting European inland waters with accounts of their distribution and ecology, as well as the book by Andrássy (1984) within the series *Bestimmungsbücher zur Bodenfauna Europas* (= *Identification Guides of Soil Fauna from Europe*), and the two taxonomic books by Loof (1999, 2001), respectively, on Dorylaimida and Tylenchida within the series *Süsswasserfauna von Mitteleuropa* (= *Freshwater Fauna from Central Europe*). Andrássy (1984) provided a review with identification keys to species level of free-living terrestrial and freshwater nematodes belonging to the orders Monhysterida, Desmoscolecida, Araeolaimida, Chromadorida and Rhabditida.

Data on freshwater nematodes have a patchy geographic distribution and are related to the presence of specialists in the field. Furthermore, several important contributions focus on national faunas and are written in languages other than English. For specific European countries, important books are available on the ecology and taxonomy of freshwater nematodes from Russia and adjacent countries (Tsalolikhin, 1980, 1985; Gagarin, 1981; all written in Russian), while Bongers (1988) provided an illustrated book (in Dutch) with identification keys to species level for the free-living nematofauna of the Netherlands and Zullini (1982) did the same for Italy (in Italian).

Apart from books, several major articles appeared during the last decennium. A compendium of our knowledge of the free-living nematofauna of ancient lakes by Decraemer and Coomans (1994) appeared within a book on speciation in ancient lakes. Important contributions for Africa are, for example, the review article by Jacobs (1984) with a checklist of the free-living inland aquatic fauna, several articles by Eyualem and Coomans (e.g. 1996a,b,c) and Eyualem (2000, 2002) on aquatic nematodes from Ethiopia and a checklist of free-living nematodes from freshwater habitats in Southern Africa (Heyns, 2002). For South America, several recent taxonomic articles deal with the nematofauna from nature reserves in Costa Rica (Zullini *et al.*, 2002; Esquivel and Arias, 2004) and with freshwater nematodes from the Galápagos (Eyualem and Coomans, 1995); Doucet and Doucet (1999) provided an updated checklist of soil and freshwater nematodes from continental Argentina. For Asia, recent taxonomic articles became available on freshwater nematodes from various locales: Tian, Shan Mour (Lemzina, 1989), Eurasia (Gagarin, 2000), Singapore and Japan (Tsalolikhin, 2001), and high mountain lakes in the Pamir and Himalayas (Tsalolikhin, 1998), as well as Vietnam (Gagarin *et al.*, 2003).

Changing Perspectives in Nematode Classification

Nematode taxonomy has at times had a turbulent history, not only as a result of wider developments in animal systematics, but also because of the unique skills and perspectives of the relatively few nematologists who have produced comprehensive classifications. Most of these classifications were based on relatively few morphological characters derived primarily from light microscopy. Although phylogenetic assumptions and deductions were often built into the underlying hypotheses, such assumptions were rigorously spelled out for the first time only in the 'natural classification' system of Lorenzen (1981, 1994), which aimed to define and classify monophyletic taxa *sensu* Hennig (1950). Lorenzen's classification was very much built by individual effort; as a specialist in free-living aquatic nematodes he largely omitted parasites as well as some of the most diverse groups of free-living nematodes (e.g. those belonging to Rhabditida *sensu* De Ley and Blaxter, 2002).

The advent of molecular techniques has brought substantial changes to the strengths and limitations of nematode systematics. These methods can easily be applied in many different settings and often generate a flood of data that can easily overwhelm even the most dedicated solitary specialist. Comprehensive molecular analyses also require the exchange of identified specimens, as well as primary data, to ensure reasonably representative coverage of nematode diversity: the usefulness of molecular phylogenetic analyses in nematodes may depend more on the particular selection of taxa than on mere production of data from additional genes (Ferris *et al.*, 2004). Similarly, a broad array of computational tools is required for meaningful analysis of these primary data, which again creates conditions where collaborative strategies are more productive than individual effort.

These changes in approach were embodied by the analyses of Blaxter *et al.* (1998), based on small subunit (SSU) rDNA sequences of 'only' 53 nematode species selected across the whole phylum. Since then, the number of nematode SSU rDNA sequences available in public databases has mushroomed to cover more than 600 species. The basic structure of the overall topology obtained by Blaxter *et al.* (1998) appears to hold up well, although it has also become increasingly clear that certain parts of the nematode tree cannot be resolved by SSU data alone. No single gene is likely to provide a completely accurate picture across all timescales of evolution, especially for such an ancient and diverse group as nematodes. Modern phylogenetic analysis employs a diverse array of algorithms and parameters, which combine with the inevitable occurrence in any dataset of some conflicts or uninformative parts. Both factors typically result in an outcome where multiple different but 'equally good' trees are produced, which means that the consensus tree of all analyses will inevitably retain some unresolved areas. This is both a weakness and a strength: on the one hand it suggests that we may never be able to completely reconstruct the entire evolutionary history of organisms with absolute confidence, but on the other hand it allows us to pinpoint and quantify uncertainties in trees where and when these do occur, and this with much greater objectivity and accuracy than previously possible with intuitive phylogenies.

The classification proposed by Chitwood (1937, 1958) has had a strong influence on nematologists and zoology textbooks, especially in the Western Hemisphere. However, SSU analyses support some substantially different hypotheses of taxonomic relationships, many of which were actually previously proposed by different authors in various alternative morphological systems, which usually did not catch the eye of the public quite as successfully as that of Chitwood (1937, 1958). Two conclusions stand out in particular from SSU phylogenies. First, some of the most important diagnostic characters of major groups have arisen repeatedly during evolution, and similarities in these characters have repeatedly misled previous analyses into inferring closer relationships between taxa with homoplastic morphologies. Second, one of the two traditional classes (Secernentea) actually arose from a major evolutionary radiation nested deep within the other 'class' (Adenophorea). This implies that the name Adenophorea should actually cover the entire phylum, thereby becoming a redundant and inappropriately ranked synonym of Nematoda, while Secernentea cannot be retained at the rank of class unless all nematode orders formerly classified in Adenophorea are upgraded to classes as well. In an effort to translate these and other implications of SSU rDNA phylogenies into a relatively balanced and consistent Linnaean classification, De Ley and Blaxter (2002, 2004) proposed a system that is based primarily on the phylogenetic backbone provided by SSU sequence analyses, but which also incorporates morphological, ontogenetic and biological characters to help resolve some of the more ambiguous regions of SSU trees. The resulting phylogenetic classification combines different elements from different previous systems, and introduces some new features with respect to the employed nomenclatorial ranks within the phylum, such as the usage of infraorder level taxa to resolve certain clades with deeply nested parasitic taxa. For the sake of convenience, we will follow the nomenclature of this system here to outline the major features of SSU-based nematode phylogeny.

The Backbone of the Nematode Tree

The SSU sequences published to date distinguish between three early nematode lineages, roughly corresponding to the previously recognized subclasses Chromadoria, Enoplia and Dorylaimia (Pearse, 1942; Inglis, 1983). The exact order of appearance of these three lineages is as yet unresolved, however (Fig. 1.1). It seems likely that Enoplia appeared first, and SSU sequence data suggest they could even be paraphyletic to one or both of the other two lineages. In other words, for all we know Dorylaimia and/or Chromadoria could have originated from within Enoplia which would mean that the subclass Enoplia in its present form would no longer be acceptable as a formal taxon. On the other hand, SSU data also allow for the possibility that Dorylaimia diverged first. This is in our opinion not the most likely scenario, but it is nevertheless an especially intriguing possibility in the context of this book: because of the general absence of Dorylaimia from marine habitats, a 'Dorylaimia first' topology would imply that the ancestor of all nematodes was perhaps a freshwater organism, rather than a marine one as more commonly assumed (see discussion in De Ley and Blaxter, 2004).

In summary, SSU phylogenies neither contradict nor confirm that Dorylaimia and Enoplia are joined as sister groups in a monophyletic clade, separate from Chromadoria. They are therefore treated here in separate chapters. Most morphological systems (e.g. De Coninck, 1965) specifically assumed a closer relationship between the former two, and on that basis Inglis (1983) proposed the classes Chromadorea and Enoplea. In the absence of contradictory evidence, these two classes were retained in the system of De Ley and Blaxter (Fig. 1.1) with little or no significant change to the composition of Enoplea, retaining both the subclasses Enoplia and Dorylaimia. However, in their new system Inglis' class Rhabditea (= class Secernentea in the system of Chitwood, 1958) was subsumed into Chromadorea as the redefined and greatly expanded order Rhabditida. Furthermore, they only accepted the subclass Chromadoria within Chromadorea: the previously proposed subclasses Diplogasteria, Monhysteria, Rhabditia and Spiruria are not retained because they are polyphyletic (in the former case) or are imbalanced in ranking compared to their respective sister clades (in the latter three cases). The resulting chromadorian monophylum *sensu* De Ley and Blaxter (2002) includes a number of major clades that have arisen in a series of successive divergences. These clades are currently classified as separate orders and most of them are predominantly aquatic, including several that are common in freshwater sediments (especially the orders Monhysterida and Plectida). However, the redefined order Rhabditida (formerly classified as the class Secernentea or Rhabditea) represents a hugely successful radiation of predominantly terrestrial and parasitic nematodes, emerging at the crown of Chromadoria as sister group to the order Plectida.

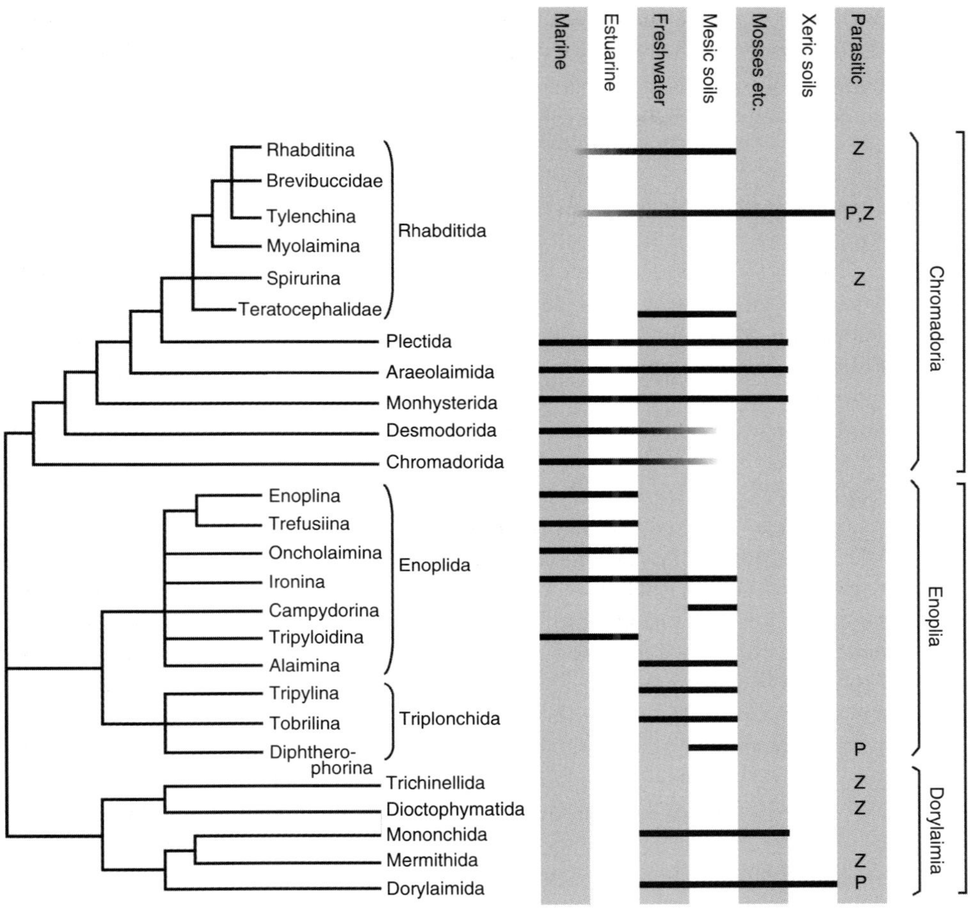

Fig. 1.1 Overview of the phylogenetic relationships (primarily based on small subunit rDNA), habitat range and higher classification of major nematode groups. Modified from De Ley and Blaxter (2004).

Speculations on the Evolutionary Ecology of Early Freshwater Nematodes

The phylum Nematoda occurs in an incredibly wide spectrum of ecological habitats and natural histories, ranging from e.g. deep-sea sediments to arid deserts, or from interstitial bacterivores to obligate parasites with multiple intermediate hosts. Several of its constituent major orders cover a large subset of this ecological spectrum, but interestingly none of them is found to span the full ecological range of the entire phylum (Fig. 1.1). The Rhabditida seem to come closest to covering all habitats, although their occurrence in marine habitats is marginal at best, being restricted to just a handful of sublittoral species (e.g. *Rhabditis marina*). This suggests that the evolution of ecological adaptations within each nematode taxon was constrained by innate limitations on the rates of change in genes and ecophysiology, by competitive exclusion from habitats previously colonized by other nematode taxa, or both.

Although excluded from marine habitats, Dorylaimia are taxonomically and ecologically diverse in many other respects. The subclass includes unusual and highly derived animal parasites such as mermithids and trichinellids, suggesting even greater diversity existed in the past but was lost to extinction. Among the surviving orders, Dorylaimida and Mononchida are common in freshwater environments and moist soils, but also in habitats subject to frequent desiccation (such as mosses). Dorylaimida are especially species rich and have also successfully adapted to xeric and cryogenic environments. The present diversity and ecological range of Dorylaimia both suggest that this lineage could very well have been the first among nematodes to pervasively colonize freshwater and terrestrial habitats. Perhaps Dorylaimia long preceded freshwater Chromadoria, such as monhysterids and plectids, although we can only infer approximate chronological sequences of events from the SSU phylogenies and not actual dates or timescales.

SSU phylogenies subdivide Enoplia into several clades, but do not provide clear resolution of their reciprocal phylogenetic relationships. Enoplia are often found in freshwater sediments and/or moist soils (Fig. 1.1) and at least one enoplian clade comprises both freshwater and marine taxa. Since Enoplia are especially prevalent in marine habitats, the mosaic of their current osmotic requirements suggests that early enoplians were perhaps characterized by greater osmotic tolerance than early Dorylaimia. Because of the lack of resolution in the molecular phylogenies, it is unclear whether some of the enoplian freshwater lineages arose earlier or later than their dorylaimian counterparts. They do appear to have radiated less extensively in terms of parasitic or terrestrial adaptations: there are (with the possible exception of a few enigmatic taxa of uncertain position) no surviving lineages that parasitize animals nor are any Enoplia known to have adapted to terrestrial environments that are subject to extreme temperatures and/or frequent dehydration. However, in SSU phylogenies one enoplian order stands out as having radiated extensively in freshwater and terrestrial habitats, the Triplonchida include not only plant parasitic trichodorids, but also morphologically very disparate free-living nematodes: tobrilids and prismatolaimids. A highly plesiomorphic pattern of embryogenesis was recently discovered in *Tobrilus diversipapillatus* (Schierenberg, 2005), hinting at the possibility that Triplonchida could be a very ancient clade. As with Dorylaimia, a basal origin of Triplonchida would be another scenario implying that the last common ancestor of Nematoda could have been a freshwater organism.

Phylogenetic relationships within Chromadoria appear to resolve better with SSU sequences, suggesting that the order Chromadorida arose first, while subsequent radiations gave rise successively to Desmodorida, Monhysterida, Araeolaimida, Plectida and Rhabditida (Fig. 1.1). Chromadoria have diversified to such an extent that they have succeeded in occupying greater habitat breadth than all of Enoplia and much of Dorylaimia. This diversification is reflected morphologically in a much wider range of amphideal, cuticular, pharyngeal, genital and caudal shapes and structures. The cuticle is a structure of fundamental importance for nematode movement, body shape and protection. Although its ultrastructure has been used to analyse deep relationships (Maggenti, 1979), important features appear to have repeatedly and independently arisen multiple times within Chromadoria, or to have disappeared secondarily. Examples of such features are cortical radial striae (found in Monhysterida and Chromadorida), median struts (found in Monhysterida, Chromadorida and certain clades within Rhabditida) and spiral fibre layers (in Monhysterida, Chromadorida and a different subset of Rhabditida). As an example of a highly homoplastic character suite, the cuticle is only of limited value as a source of data for resolving deeper nematode relationships (Decraemer *et al.*, 2003). Nevertheless, basal radial striae are characteristic for the mainly terrestrial Rhabditida, including dauers and infective juveniles of animal as well as plant parasitic members of the order. This character is apparently a synapomorphy for the order, and may play an important role in reducing the permeability of the cuticle and/or dependence on the presence of water. It could therefore well represent a key adaptation that opened up a broad spectrum of terrestrial habitats as well as conferring Rhabditida with greater ability to bypass the immune systems of other animals and repeatedly evolve into various types of parasitic lifestyles.

The pharynx of several chromadorian orders is characterized by the presence of one or more spherical bulbs, which also appear to allow for greater functional specialization and subdivision of the remainder of the pharynx (Fürst von Lieven, 2003). The pumping action of the bulb(s) presumably results in more effective ingestion of food compared to a simple cylindrical pharynx, and this may in turn allow for more compact design of the entire nematode body. Free-living Chromadoria are on average smaller than Enoplia and Dorylaimia, reflecting a greater preponderance in the latter two classes of comparatively large predators and omnivores. Instead, many Chromadoria are bacterivorous, and some groups have evolved towards very short generation times and specialized adaptations for efficient dispersal, two sets of properties that clearly allowed them to become highly successful in colonizing and exploiting eutrophic environments. Like Dorylaimia, Chromadoria have flourished in habitats subject to frequent episodes of rapid de- and rehydration (such as mosses, lichens, etc.) as well as in extremely xeric and/or cryogenic soils where most interstitial organisms are presumably dormant for much of the year.

An exception to the trend for compact pharynx design are the Monhysterida, most of which have a relatively simple cylinder. However, monhysterids exemplify a different recurring feature in Chromadoria: the ability to actively swim through free water by means of bursts of rapid oscillations of the body. We have also observed similar swimming behaviour in some panagrolaims and diplogasterids as well as in *Tobrilus* (the latter being the only example known to us outside of Chromadoria).

Introduction to Classification Listing

As noted above, DNA sequencing and genomics are bringing substantial change to nematode taxonomy. This change is not unusual in itself, as the introduction of new character and data suites has always heralded significant revisions in classification throughout the entire history of biological systematics. What is new is that, perhaps for the first time, vast amounts of new character data are being produced at rates far exceeding our ability to thoroughly analyse their implications. At the same time, we cannot meaningfully discuss nematode ecology or identification without a taxonomic framework that takes into account an updated picture of relationships between nematode groups. For the purposes of structuring the contents of this book, we have therefore opted to use the system of De Ley and Blaxter (2002, 2004) as an overall taxonomic scaffolding for the respective chapters contributed by different authors. We have employed this system because it is partly derived from recent molecular phylogenies and because it largely covers phylum-wide diversity. We do not claim or consider the system in question to be definitive or flawless, only that it provides an appropriate starting point at a time when even more drastic changes may well lie just around the corner. The willingness of the other authors to contribute does not necessarily imply their agreement with any and all aspects of the system in question; they were left free to make different classification choices within the orders discussed in their respective chapters. Two families listed in Plectida by De Ley and Blaxter (2002) are here considered *incertae sedis* (see Appendix).

Freshwater Habitat as a Category: the Issue of Clarity *versus* Convenience

The taxonomic classification below provides a complete general framework to family level, as proposed by De Ley and Blaxter (2004). Within each family, we list only those nematode genera that were reported from inland water bodies, since the answer to the question 'Which nematodes are considered denizens of freshwater habitats?' is neither direct nor simple, for a number of reasons. First, all nematodes are essentially dependent on water whether they are inhabitants of wet terrestrial environments, temporary pools or dry soil, and they remain active only as long as they are covered by at least a thin film of water. Therefore, we recognize that the very theme of this book, i.e. grouping based on habitat, is in some respects artificial as it is not based on a phylogenetic criterion. Second, what has been attempted in this book is to include all nematodes reported from inland water bodies. Nematodes excluded from this book are those considered marine or estuarine. The latter encompasses a broad and dynamic environmental range, even with regard to salt content, and that leaves a rather elastic demarcation between estuarine *versus* freshwater habitats. Different chapters attempt to include all nematodes reported from inland water bodies and these may not always be strictly freshwater, depending on the preferred definition. Third, even within these inland water bodies, the separation between nematodes that inhabit wet terrestrial habitats *versus* those that prefer sediments covered with water is somewhat indistinct, and no less so between intermittent *versus* permanent presence of water. Consequently, many contributors in this book have opted to provide an account of the range of relevant environments for each taxon, instead of simply categorizing their habitats as freshwater. The reader should therefore keep in mind that the following list must not be taken too literally, in view of the lack of a clear demarcation between different habitats and their definitions.

Classification

PHYLUM NEMATODA Potts, 1932

Incertae sedis:

Family Bastianidae De Coninck, 1935

Family Rhabdolaimidae Chitwood, 1951

ORDER BENTHIMERMITHIDA Tchesunov, 1995

Family Benthimermithidae Petter, 1980

ORDER RHAPTOTHYREIDA Tchesunov, 1995

Family Rhaptothyreidae Hope & Murphy, 1969

CLASS ENOPLEA Inglis, 1983

SUBCLASS ENOPLIA Pearse, 1942

ORDER ENOPLIDA Filipjev, 1929

Incertae sedis: Family Andrassyidae Tchesunov & Gagarin, 1999

Suborder Enoplina Chitwood & Chitwood, 1937

Superfamily Enoploidea Dujardin, 1845

Family Enoplidae Dujardin, 1845

Family Thoracostomopsidae Filipjev, 1927

Subfamily Enoplolaiminae De Coninck, 1965

Genus *Enoploides* Ssaweljev, 1912

Genus *Mesacanthion* Filipjev, 1927

Family Anoplostomatidae Gerlach & Riemann, 1974

Subfamily Anoplostomatinae Gerlach & Riemann, 1974

Genus *Anoplostoma* Bütschli, 1874

Family Phanodermatidae Filipjev, 1927

Family Anticomidae Filipjev, 1918

Suborder Trefusiina Siddiqi, 1983

Superfamily Trefusioidea Gerlach, 1966

Family Simpliconematidae Blome & Schrage, 1985

Family Trefusiidae Gerlach, 1966

Family Laurathonematidae Gerlach, 1953

Family Xenellidae De Coninck, 1965

Suborder Oncholaimina De Coninck, 1965

Superfamily Oncholaimoidea Filipjev, 1916

Familiy Oncholaimidae Filipjev, 1916

Subfamily Oncholaimellinae De Coninck, 1965

Genus *Viscosia* de Man, 1890

Subfamily Adoncholaiminae Gerlach & Riemann, 1974

Genus *Adoncholaimus* Filipjev, 1918

Subfamily Oncholaiminae Filipjev, 1916

Genus *Oncholaimus* Dujardin, 1845

Genus *Pseudoncholaimus* Kreis, 1932

Family Enchelidiidae Filipjev, 1918

Genus *Calyptronema* Marion, 1870

Genus *Eurystomina* Filipjev, 1921
Genus *Polygastrophora* de Man, 1922
Suborder Ironina Siddiqi, 1983
Superfamily Ironoidea De Man, 1876
Family Ironidae De Man, 1876
Subfamily Ironinae De Man, 1876
Genus *Ironus* Bastian, 1865
Family Leptosomatidae Filipjev, 1916
Family Oxystominidae Chitwood, 1935
Subfamily Oxystomininae Chitwood, 1953
Genus *Oxystomina* Filipjev, 1921
Genus *Thalassoalaimus* De Man, 1893
Subfamily Halalaiminae De Coninck, 1965
Genus *Halalaimus* De Man, 1888
Suborder Tripyloidina De Coninck, 1965
Superfamily Tripyloidoidea Filipjev, 1928
Family Tripyloididae Filipjev, 1928
Genus *Tripyloides* De Man, 1886
Suborder Campydorina Jairajpuri, 1983
Superfamily Campydoroidea Jairajpuri *et al.*, 1976
Family Campydoridae (Thorne, 1935) Clark, 1961
Suborder Alaimina Clark, 1961
Superfamily Alaimoidea Micoletzky, 1922
Family Alaimidae Micoletzky, 1922
Subfamily Alaiminae Micoletzky, 1922
Genus *Alaimus* De Man, 1880
Subfamily Amphidelinae Andrássy, 2002
Genus *Amphidelus* Thorne, 1939
Genus *Paramphidelus* Andrássy, 1977
Genus *Scleramphidelus* Clausi & Vinciguerra, 1995

ORDER TRIPLONCHIDA Cobb, 1920

Suborder Diphtherophorina Micoletzky, 1922
Superfamily Diphtherophoroidea Micoletzky, 1922
Family Diphtherophoridae Micoletzky, 1922
Genus *Diphtherophora* De Man, 1880
Family Trichodoridae Thorne, 1935
Genus *Trichodorus* Cobb, 1913
Genus *Paratrichodorus* Siddiqi, 1974
Suborder Tobrilina Tsalolikhin, 1976
Superfamily Prismatolaimoidea Micoletzky, 1922
Family Prismatolaimidae Micoletzky, 1922
Genus *Prismatolaimus* De Man, 1880
Superfamily Tobriloidea De Coninck, 1965
Family Triodontolaimidae De Coninck, 1965
Family Rhabdodemaniidae Filipjev, 1934
Family Pandolaimidae Belogurov, 1980

Family Tobrilidae De Coninck, 1965
Subfamily Tobrilinae De Coninck, 1965
Tribe Paratrilobini Tsalolikhin, 2001
Genus *Paratrilobus* Micoletzky, 1922
Genus *Mesotobrilus* Tsalolikhin, 1981
Genus *Quasibrilus* Tsalolikhin, 1976
Genus *Kurikania* Tsalolikhin, 1976
Genus *Lamuania* Tsalolikhin, 1976
Genus *Tobrilus* Andrássy, 1959
Genus *Eutobrilus* Tsalolikhin, 1981
Genus *Asperotobrilus* Shoshin, 1991
Genus *Setsalia* Shoshina, 2003
Subfamily Neotobrilinae Tsalolikhin, 2001
Tribe Neotobrilini Tsalolikhin, 1981
Genus *Neotobrilus* Tsalolikhin, 1981
Genus *Semitobrilus* Tsalolikhin, 1981
Genus *Brevitobrilus* Tsalolikhin, 1981
Tribe Epitobrilini Tsalolikhin, 2001
Genus *Epitobrilus* Tsalolikhin, 1981
Suborder Tripylina Andrássy, 1974
Superfamily Tripyloidea de Man, 1876
Family Tripylidae de Man, 1876
Genus *Tripyla* Bastian, 1865
Genus *Trischistoma* Cobb, 1913
Genus *Tobrilia* Andrássy, 1967
Genus *Abunema* Khera, 1971
Family Onchulidae Andrássy, 1964
Genus *Onchulus* Cobb, 1920
Genus *Caprionchulus* Swart & Heyns, 1993
Genus *Kinonchulus* Riemann, 1972
Genus *Limonchulus* Andrássy, 1963
Genus *Pseudonchulus* Altherr, 1972
Genus *Stenonchulus* W. Schneider, 1940

SUBCLASS DORYLAIMIA Inglis, 1983
ORDER DORYLAIMIDA Pearse, 1942
Suborder Dorylaimina Pearse, 1936
Superfamily Belondiroidea Thorne, 1939
Family Belondiridae Thorne, 1939
Subfamily Belondirinae Thorne, 1939
Genus *Axonchium* Cobb, 1920
Genus *Belondira* Thorne, 1939
Subfamily Dorylaimellinae Jairajpuri, 1964
Genus *Dorylaimellus* Cobb, 1913
Subfamily Swangeriinae Jairajpuri, 1964
Genus *Falcihasta* Clark, 1964
Genus *Lindseyus* Ferris & Ferris, 1973

Genus *Oxydirus* Thorne, 1939
Genus *Paraoxydirus* Jairajpuri & Ahmad, 1979
Superfamily Tylencholaimoidea Filipjev, 1934
Family Tylencholaimidae Filipjev, 1934
Subfamily Tylencholaiminae Filipjev, 1934
Genus *Chitwoodius* Furstenberg & Heyns, 1966
Genus *Discomyctus* Thorne, 1939
Genus *Tantunema* Siddiqi, 1982
Genus *Tylencholaimus* De Man, 1876
Subfamily Vanderlindiinae Siddiqi, 1969
Subfamily Metadorylaminae Andrássy, 1976
Family Leptonchidae Thorne, 1935
Subfamily Belonenchinae Thorne, 1964
Genus *Basirotyleptus* Jairajpuri, 1964
Subfamily Leptonchinae Thorne, 1935
Genus *Funaria* van der Linde, 1938
Genus *Leptonchus* Cobb, 1920
Genus *Proleptonchoides* Ferris, Goseco & Kumar, 1979
Genus *Proleptonchus* Lordello, 1955
Subfamily Scalpenchinae n. subfam.
Subfamily Tyleptinae Jairajpuri, 1964
Genus *Tyleptus* Thorne, 1939
Subfamily Xiphinemellinae Jairajpuri, 1964
Genus *Xiphinemella* Loos, 1950
Family Tylencholaimellidae Jairajpuri, 1964
Subfamily Athernematinae Ahmad & Jairajpuri, 1978
Genus *Athernema* Ahmad & Jairajpuri, 1978
Subfamily Tylencholaimellinae Jairajpuri, 1964
Genus *Doryllium* Cobb, 1920
Genus *Oostenbrinkella* Jairajpuri, 1965
Genus *Tylencholaimellus* Cobb in M.V. Cobb, 1915
Family Mydonomidae Thorne, 1964
Subfamily Calolaiminae Goseco, Ferris & Ferris, 1976
Genus *Calolaimus* Timm, 1964
Subfamily Mydonominae Thorne, 1964
Genus *Dorylaimoides* Thorne & Swanger, 1936
Genus *Morasia* Baqri & Jairajpuri, 1969
Family Aulolaimoididae Jairajpuri, 1964
Genus *Adenolaimus* Andrássy, 1973
Genus *Aulolaimoides* Micoletzky, 1915
Suborder Nygolaimina Ahmad & Jairajpuri, 1979
Superfamily Nygolaimoidea Thorne, 1935
Family Aetholaimidae Jairajpuri, 1965
Subfamily Aetholaiminae Jairajpuri, 1965
Genus *Aetholaimus* Williams, 1962
Family Nygellidae Andrássy, 1958
Subfamily Nygellinae Andrássy, 1958

Genus *Nygellus* Thorne, 1939
Family Nygolaimellidae Clark, 1961
Subfamily Nygolaimellinae Clark, 1961
Genus *Nygolaimellus* Loos, 1949
Genus *Nygolaimium* Thorne, 1930
Family Nygolaimidae Thorne, 1935
Subfamily Nygolaiminae Thorne, 1935
Genus *Aquatides* Heyns, 1968
Genus *Clavicauda* Heyns, 1968
Genus *Clavicaudoides* Heyns, 1968
Genus *Feroxides* Heyns, 1968
Genus *Laevides* Heyns, 1968
Genus *Nygolaimus* Cobb, 1913
Genus *Paranygolaimus* Heyns, 1968
Genus *Paravulvus* Heyns, 1968
Genus *Solididens* Heyns, 1968
Superfamily Dorylaimoidea De Man, 1876
Family Crateronematidae Siddiqi, 1969
Family Lordellonematidae Siddiqi, 1969
Family Longidoridae Thorne, 1935
Family Thornenematidae Siddiqi, 1969
Family Dorylaimidae De Man, 1876
Subfamily Prodorylaiminae Andrássy, 1969
Genus *Protodorylaimus* Andrássy, 1988
Genus *Prodorylaimus* Andrássy, 1959
Genus *Prodorylaimium* Andrássy, 1969
Subfamily Dorylaiminae De Man, 1876
Genus *Dorylaimus* Dujardin, 1845
Genus *Ischiodorylaimus* Andrássy, 1969
Subfamily Laimydorinae Andrássy, 1969
Genus *Laimydorus* Siddiqi, 1969
Genus *Calodorylaimus* Andrássy, 1969
Genus *Crocodorylaimus* Andrássy, 1988
Genus *Idiodorylaimus* Andrássy, 1969
Genus *Chrysodorus* Jiménez-Guirado & Cadenas, 1985
Genus *Mesodorylaimus* Andrássy, 1959
Subfamily Afrodorylaiminae Andrássy, 1969
Genus *Afrodorylaimus* Andrássy, 1964
Genus *Drepanodorylaimus* Jairajpuri, 1966
Genus *Apodorylaimus* Andrássy, 1988
Genus *Paradorylaimus* Andrássy, 1969
Family Actinolaimidae Thorne, 1939
Genus *Paractinolaimus* Meyl, 1957
Genus *Paractinolaimoides* Khan, Ahmad & Jairajpuri, 1994
Genus *Westindicus* Thorne, 1967
Genus *Egtitus* Thorne, 1967
Genus *Scleroactinolaimus* Ahmad, Khan & Ahmad, 1992

Genus *Afractinolaimus* Andrássy, 1970
Genus *Neoactinolaimus* Thorne, 1967
Genus *Mactinolaimus* Andrássy, 1970
Genus *Metactinolaimus* Meyl, 1957
Genus *Stopractinca* Khan, Ahmad & Jairajpuri, 1994
Genus *Actinolaimus* Cobb, 1913
Genus *Parastomachoglossa* Coomans & Loof, 1986
Genus *Actinca* Andrássy, 1964
Genus *Brasilaimus* Lordello & Zamith, 1957
Genus *Afractinca* Vinciguerra & Clausi, 2000
Genus *Practinocephalus* Andrássy, 1974
Family Qudsianematidae Jairajpuri, 1963
Subfamily Chrysonematinae Siddiqi, 1969
Genus *Chrysonema* Thorne, 1929
Subfamily Discolaiminae Siddiqi, 1969
Genus *Discolaimus* Cobb, 1913
Genus *Discolaimoides* Heyns, 1963
Subfamily Qudsianematinae Jairajpuri, 1969
Genus *Labronema* Thorne, 1939
Genus *Eudorylaimus* Andrássy, 1959
Genus *Ecumenicus* Thorne, 1974
Genus *Takamangai* Yeates, 1974
Genus *Pachydorylaimus* Siddiqi, 1983
Genus *Labronemella* Andrássy, 1985
Genus *Microdorylaimus* Andrássy, 1986
Genus *Allodorylaimus* Andrássy, 1986
Genus *Epidorylaimus* Andrássy, 1986
Genus *Gopalus* Khan, Jairajpuri & Ahmad, 1988
Genus *Crassogula* Andrássy, 1991
Genus *Boreolaimus* Andrássy, 1998
Subfamily Arctidorylaiminae Mulvey & Anderson, 1979
Genus *Arctidorylaimuys* Mulvey & Anderson, 1979
Family Aporcelaimidae Heyns, 1965
Subfamily Aporcelaiminae Heyns, 1965
Genus *Aporcelaimus* Thorne & Swanger, 1936
Genus *Aporcelaimellus* Heyns, 1965
Genus *Makatinus* Heyns, 1965
Genus *Epacrolaimus* Andrássy, 2000
Subfamily Sectonematinae Siddiqi, 1969
Genus *Sectonema* Thorne, 1930
Family Nordiidae Jairajpuri & A.H. Siddiqi, 1964
Subfamily Nordiinae Jairajpuri & A.H. Siddiqi, 1964
Genus *Longidorella* Thorne, 1939
Subfamily Pungentinae Siddiqi, 1969
Genus *Pungentus* Thorne & Swanger, 1936
Genus *Enchodelus* Thorne, 1939
Genus *Lenonchium* Siddiqi, 1963

Genus *Rhyssocolpus* Andrássy, 1971
Genus *Dorydorella* Andrássy, 1987
Family Thorniidae De Coninck, 1965
Subfamily Thorniinae De Coninck, 1965
Genus *Thornia* Meyl, 1954
Genus *Nygolaimoides* Meyl in Andrássy, 1960
Subfamily Thorneellinae Andrássy, 1987
Genus *Thorneella* Andrássy, 1960

ORDER MONONCHIDA Jairajpuri, 1969
Suborder Bathyodontina Coomans & Loof, 1970
Superfamily Cryptonchoidea Chitwood, 1937
Family Bathyodontidae Clark, 1961
Family Cryptonchidae Chitwood, 1937
Genus *Cryptonchus* Cobb, 1913
Superfamily Mononchuloidea De Coninck, 1965
Family Mononchulidae De Coninck, 1962
Genera *Mononchulus* Cobb, 1918
Genus *Oionchus* Cobb, 1913
Suborder Mononchina Kirjanova & Krall, 1969
Superfamily Anatonchoidea Jairajpuri, 1969
Family Anatonchidae Jairajpuri, 1969
Subfamily Iotonchinae Jairajpuri, 1969
Genera *Jensenonchus* Jairajpuri & Khan, 1982
Genus *Iotonchus* Cobb, 1916
Genus *Prionchuloides* Mulvey, 1963
Subfamily Anatonchinae Jairajpuri, 1969
Genera *Anatonchus* Cobb, 1916
Genus *Miconchus* Andrássy, 1958
Genus *Promiconchus* Jairajpuri & Khan, 1982
Superfamily Mononchoidea Chitwood, 1937
Family Mononchidae Chitwood, 1937
Subfamily Mononchinae Filipjev, 1934
Genus *Mononchus* Bastian, 1865
Genus *Paramononchus* Mulvey, 1978
Genus *Coomansus* Jairajpuri & Khan, 1977
Genus *Clarkus* Jairajpuri, 1970
Genus *Prionchulus* Cobb, 1916
Genus *Judonchulus* Andrássy, 1958
Subfamily Cobbonchinae Jairajpuri, 1969
Genus *Cobbonchus* Andrássy, 1958
Genus *Comiconchus* Jairajpuri & Khan, 1982
Family Mylonchulidae Jairajpuri, 1969
Subfamily Mylonchulinae Jairajpuri, 1969
Genus *Mylonchulus* Cobb, 1916
Genus *Polyonchulus* Mulvey & Jensen, 1967
Genus *Granonchulus* Andrássy, 1958

ORDER ISOLAIMIDA Cobb, 1920

Superfamily Isolaimoidea Timm, 1969

Family Isolaimiidae Timm, 1969

ORDER DIOCTOPHYMATIDA Baylis & Daubney, 1926

Suborder Dioctophymatina Skrjabin, 1927

Family Dioctophymatidae Castellani & Chalmers, 1910

Family Soboliphymatidae Petrov, 1930

ORDER MUSPICEIDA Bain & Chabaud, 1959

Suborder Muspiceina Bain & Chabaud, 1959

Family Muspiceidae Bain & Chabaud, 1959

Family Robertdollfusiidae Chabaud & Campana, 1950

ORDER MARIMERMITHIDA Rubtzov, 1980

Family Marimermithidae Rubtzov & Platonova, 1974

ORDER MERMITHIDA Hyman, 1951

Suborder Mermithina Andrássy, 1974

Superfamily Mermithoidea Braun, 1883

Family Mermithidae Braun, 1883

Family Tetradonematidae Cobb, 1919

ORDER TRICHINELLIDA Hall, 1916

Superfamily Trichinelloidea Ward, 1907

Family Anatrichosomatidae Yamaguti, 1961

Family Capillariidae Neveu-Lemaire 1936

Family Cystoopsidae Skrjabin, 1923

Family Trichinellidae Ward, 1907

Family Trichosomoididae Hall, 1916

Family Trichuridae Railliet, 1915

CLASS CHROMADOREA
SUBCLASS CHROMODORIA

ORDER CHROMADORIDA Chitwood, 1933

Suborder Chromadorina Filipjev, 1929

Superfamily Chromadoroidea Filipjev, 1917

Family Chromadoridae Filipjev, 1917

Subfamily Spilipherinae Filipjev, 1918

Subfamily Chromadorinae Filipjev, 1917

Genus *Chromadorina* Filipjev, 1918

Genus *Prochromadora* Filipjev, 1922

Genus *Punctodora* Filipjev, 1919

Subfamily Euchromadorinae Gerlach & Riemann, 1973

Subfamily Harpagonchinae Platonova & Potin, 1972

Subfamily Hypodontolaiminae De Coninck, 1965

Genus *Chromadorita* Filipjev, 1922

Genus *Dichromadora* Kreis, 1929

Genus *Neochromadora* Micoletzky, 1924

Family Ethmolaimidae Filipjev & Schuurmans Stekhoven, 1941
Genus *Ethmolaimus* De Man, 1880
Family Neotonchidae Wieser & Hopper, 1966
Family Achromadoridae Gerlach & Riemann, 1973
Subfamily Achromadorinae Gerlach & Riemann, 1973
Genus *Achromadora* Cobb, 1913
Subfamily Kreisonematinae Khera, 1969
Family Cyatholaimidae Filipjev, 1918
Subfamily Nyctonematinae Bussau, 1993
Subfamily Pomponematinae Gerlach & Riemann, 1973
Subfamily Paracanthonchinae De Coninck, 1965
Genus *Paracyatholaimus* Micoletzky, 1922
Subfamily Xenocyatholaiminae Gerlach & Riemann, 1973
Subfamily Cyatholaiminae Filipjev, 1918
Family Selachinematidae Cobb, 1915

ORDER DESMODORIDA De Coninck, 1965
Suborder Desmodorina De Coninck, 1965
Superfamily Desmodoroidea Filipjev, 1922
Family Desmodoridae Filipjev, 1922
Subfamily Desmodorinae Filipjev, 1922
Genus *Desmodorella* Cobb, 1933
Genus *Sibayinema* Swart & Heyns, 1991
Subfamily Spiriniinae Gerlach & Murphy, 1965
Subfamily Pseudonchinae Gerlach & Riemann, 1973
Subfamily Stilbonematinae Cobb, 1936
Subfamily Molgolaiminae Jensen, 1978
Subfamily Prodesmodorinae Lorenzen, 1981
Genus *Prodesmodora* Micoletzky, 1923
Family Epsilonematidae Steiner, 1927
Family Draconematidae Filipjev, 1918
Superfamily Microlaimoidea Micoletzky, 1922
Family Microlaimidae Micoletzky, 1922
Genus *Microlaimus* De Man, 1980
Family Aponchiidae Gerlach, 1963
Family Monoposthiidae Filipjev, 1934
Genus *Nudora* Cobb, 1920

ORDER DESMOSCOLECIDA Filipjev, 1929
Superfamily Desmoscolecoidea Shipley, 1896
Family Desmoscolecidae Shipley, 1896
Subfamily Desmoscolecinae Shipley, 1896
Tribe Desmoscolecini Shipley, 1896
Genus *Desmoscolex* Claparède, 1863
Subgenus *Desmoscolex* Claparède, 1863
Genus *Desmolorenzenia* (Freudenhammer, 1975) Decraemer, 1984

Tribe Greeffiellini Decraemer, 1985
Subfamily Tricominae Lorenzen, 1969
Genus *Tricoma* Cobb, 1894
Family Meyliidae De Coninck, 1965
Family Cyartonematidae Tchesunov, 1990

ORDER MONHYSTERIDA Filipjev, 1929

Suborder Monhysterina De Coninck & Schuurmans Stekhoven, 1933
Superfamily Monhysteroidea De Man, 1876
Family Monhysteridae De Man, 1876
Genus *Anguimonhystera* Andrássy, 1981
Genus *Diplolaimella* Allgén, 1929
Genus *Diplolaimelloides* Meyl, 1954
Genus *Eumonhystera* Andrássy, 1981
Genus *Gammarinema* Kinne & Gerlach, 1953
Genus *Geomonhystera* Andrássy, 1981
Genus *Monhystera* Bastian, 1865
Genus *Monhystrella* Cobb, 1918
Genus *Sinanema* Andrássy, 1960
Genus *Thalassomonhystera* Jacobs, 1987
Genus *Tridentulus* Eyualem & Coomans, 1995
Superfamily Sphaerolaimoidea Filipjev, 1918
Family Sphaerolaimidae Filipjev, 1918
Genus *incertae sedis Hofmaenneria* Gerlach & Meyl, 1957
Family Xyalidae Chitwood, 1951
Genus *Daptonema* Cobb, 1920
Genus *Theristus* Bastian, 1865
Suborder Linhomoeina Andrássy, 1974
Superfamily Siphonolaimoidea Filipjev, 1918
Family Siphonolaimidae Filipjev, 1918
Family Linhomoeidae Filipjev, 1922
Genus *Terschellingia* De Man, 1888
Family Fusivermidae Tchesunov, 1996

ORDER ARAEOLAIMIDA De Coninck & Schuurmans Stekhoven, 1933

Superfamily Axonolaimoidea Filipjev, 1918
Family Axonolaimidae Filipjev, 1918
Family Comesomatidae Filipjev, 1918
Family Coninckiidae Lorenzen, 1981
Family Diplopeltidae Filipjev, 1918
Subfamily Diplopeltinae Filipjev, 1918
Subfamily Cylindrolaiminae Micoletzky, 1921
Genus *Cylindrolaimus* De Man, 1880

ORDER PLECTIDA Malakhov, 1982

Family *incertae sedis* Ohridiidae Andrássy, 1976
Genus *Domorganus* Goodey, 1947

Superfamily Leptolaimoidea Örley, 1880
Family Leptolaimidae Örley, 1880
Genus *Paraplectonema* Strand, 1934
Family Aphanolaimidae Chitwood, 1936
Subfamily Aphanolaiminae Chitwood, 1936
Genus *Aphanolaimus* De Man, 1880
Genus *Paraphanolaimus* Micoletzky, 1923
Genus *Aphanonchus* Coomans & Raski, 1991
Subfamily Anonchinae Andrássy, 1973
Genus *Anonchus* Cobb, 1913
Family Rhadinematidae Lorenzen, 1981
Family Aegialoalaimidae Lorenzen, 1981
Family Diplopeltoididae Tchesunov, 1990
Family Paramicrolaimidae Lorenzen, 1981
Family Ohridiidae Andrássy, 1976
Family Odontolaimidae Gerlach & Riemann, 1974
Superfamily Plectoidea Örley, 1880
Family Chronogastridae Gagarin, 1975
Genus *Chronogaster* Cobb, 1913
Family Plectidae Örley, 1880
Subfamily Plectinae Örley, 1880
Genus *Plectus* Bastian, 1865
Subfamily Anaplectinae Zell, 1993 grad. n.
Genus *Anaplectus* De Coninck & Schuurmans Stekhoven, 1933
Subfamily Wilsonematinae Chitwood, 1951
Genus *Tylocephalus* Crossman, 1933
Family Metateratocephalidae Eroshenko, 1973
Genus *Metateratocephalus* Eroshenko, 1973
Genus *Euteratocephalus* Andrássy, 1968
Superfamily Ceramonematoidea Cobb, 1933
Family Tarvaiidae Lorenzen, 1981
Family Ceramonematidae Cobb, 1933
Family Tubolaimoididae Lorenzen, 1981
Superfamily Haliplectoidea Chitwood, 1951
Family Peresianidae Vitiello & De Coninck, 1968
Family Haliplectidae Chitwood, 1951
Family Aulolaimidae Jairajpuri & Hooper, 1968

ORDER RHABDITIDA Chitwood, 1933
Incertae sedis:
Family Teratocephalidae Andrássy, 1958
Family Chambersiellidae Thorne, 1937
Family Brevibuccidae Paramonov, 1956
Suborder Spirurina Railliet & Henry, 1915
Incertae sedis:
Superfamily Dracunculoidea Cameron, 1934
Family Dracunculidae Leiper, 1912
Family Philometridae Baylis & Daubney, 1926

Family Phlyctainophoridae Roman, 1965
Family Skrjabillanidae Schigin & Schigina, 1958
Family Anguillicolidae Yamaguti, 1935
Family Guyanemidae Petter, 1975
Family Micropleuridae Travassos, 1960

INFRAORDER GNATHOSTOMATOMORPHA De Ley & Blaxter, 2002
Superfamily Gnathostomatoidea Railliet, 1895
Family Gnathostomatidae Railliet, 1895, emend. Nicoll, 1927

INFRAORDER OXYURIDOMORPHA De Ley & Blaxter, 2002 (after Railliet, 1916)
Superfamily Thelastomatoidea Travassos, 1929
Family Thelastomatidae Travassos, 1929
Family Travassosinematidae Rao, 1958
Family Hystrignathidae Travassos, 1929
Family Protrelloididae Chitwood, 1932
Superfamily Oxyuroidea Cobbold, 1864
Family Oxyuridae Cobbold, 1864
Family Pharyngodonidae Travassos, 1919
Family Heteroxynematidae Skrjabin & Shikhobalova, 1948

INFRAORDER RHIGONEMATOMORPHA De Ley & Blaxter, 2002
Superfamily Rhigonematoidea Artigas, 1930
Family Rhigonematidae Artigas, 1930
Family Ichthyocephalidae Travassos & Kloss, 1958
Superfamily Ransomnematoidea Travassos, 1930
Family Ransomnematidae Travassos, 1930
Family Carnoyidae Travassos & Kloss, 1960
Family Hethidae Travassos & Kloss, 1960

INFRAORDER SPIRUROMORPHA De Ley & Blaxter, 2002 (after Railliet, 1914)
Superfamily Camallanoidea Travassos, 1920
Family Camallanidae Railliet & Henry, 1915
Superfamily Physalopteroidea Leiper, 1908
Family Physalopteridae Leiper, 1908
Superfamily Rictularioidea Railliet, 1916
Family Rictulariidae Railliet, 1916
Superfamily Thelazioidea Skrjabin, 1915
Family Thelaziidae Skrjabin, 1915
Family Rhabdochonidae Skrjabin, 1946
Family Pneumospiruridae Wu & Hu, 1938
Superfamily Spiruroidea Railliet & Henry, 1915
Family Gongylonematidae Sobolev, 1949
Family Spiruridae Railliet & Henry, 1915
Family Spirocercidae Chitwood & Wehr, 1932
Family Hartertiidae Quentin, 1970
Superfamily Habronematoidea Ivaschkin, 1961
Family Hedruridae Railliet, 1916
Family Habronematidae Ivaschkin, 1961
Family Tetrameridae Travassos, 1914
Family Cystidicolidae Skrjabin, 1946

Superfamily Acuarioidea Railliet, Henry & Sisoff, 1912
Family Acuariidae Railliet, Henry & Sisoff, 1912
Superfamily Filarioidea Chabaud & Anderson, 1959
Family Filariidae Chabaud & Anderson, 1959
Family Onchocercidae Leiper, 1911
Superfamily Aproctoidea Skrjabin & Schikhobalova, 1945
Family Aproctidae Skrjabin & Schikhobalova, 1945
Family Desmidocercidae Cram, 1927
Superfamily Diplotriaenoidea Anderson 1958
Family Diplotriaenidae Anderson 1958
Family Oswaldofilariidae Chabaud & Choquet, 1953
INFRAORDER ASCARIDOMORPHA De Ley & Blaxter, 2002 (after Skrjabin, 1915)
Superfamily Ascaridoidea Baird, 1853
Family Heterocheilidae Railliet & Henry, 1915
Family Ascarididae Baird, 1853
Family Raphidascarididae Hartwich, 1954
Family Anisakidae Skrjabin & Karokhin, 1945
Superfamily Cosmocercoidea Travassos, 1925
Family Cosmocercidae Travassos, 1925
Family Atractidae Travassos, 1919
Family Kathlaniidae Travassos, 1918
Superfamily Heterakoidea Railliet & Henry, 1912
Family Heterakidae Railliet & Henry, 1912
Family Aspidoderidae Skrjabin & Schikhobalova, 1947
Family Ascaridiidae Travassos, 1919
Superfamily Subuluroidea Yorke & Maplestone, 1926
Family Subuluridae Yorke & Maplestone, 1926
Family Maupasinidae Inglis, 1959
Superfamily Seuratoidea Railliet, 1906
Family Seuratidae Railliet, 1906
Family Cucullanidae Cobbold, 1864
Family Quimperiidae Baylis, 1930
Family Chitwoodchabaudiidae Puylaert, 1970
Family Schneidernematidae Freitas, 1956
Suborder Myolaimina Inglis, 1983
Superfamily Myolaimoidea Andrássy, 1958
Family Myolaimidae Andrássy, 1958
Suborder Tylenchina Thorne, 1949
INFRAORDER PANAGROLAIMOMORPHA De Ley & Blaxter, 2002
Superfamily Panagrolaimoidea Thorne, 1937
Family Panagrolaimidae Thorne, 1937
Genus *Panagrolaimus* Fuchs, 1930
Superfamily Strongyloidoidea Chitwood & McIntosh, 1934
Family Steinernematidae Filipjev, 1934
Family Strongyloididae Chitwood & McIntosh, 1934

Family Rhabdiasidae Railliet, 1916
Family Alloionematidae Chitwood & McIntosh, 1934
INFRAORDER CEPHALOBOMORPHA De Ley & Blaxter, 2002
Superfamily Cephaloboidea Filipjev, 1934
Family Cephalobidae Filipjev, 1934
Family Elaphonematidae Heyns, 1962
Family Osstellidae Heyns, 1962
Family Alirhabditidae Suryawanshi, 1971
Family Bicirronematidae Andrássy, 1978
INFRAORDER TYLENCHOMORPHA De Ley & Blaxter, 2002
Superfamily Aphelenchoidea Fuchs, 1937
Family Aphelenchidae Fuchs, 1937
Family Aphelenchoididae Skarbilovich, 1947
Genus *Aphelenchoides* Fischer, 1894
Superfamily Criconematoidea Taylor 1936
Family Criconematidae Taylor 1936
Family Hemicycliophoridae Skarbilovich, 1959
Genus *Hemicycliophora* De Man, 1921
Family Tylenchulidae Skarbilovich, 1947
Superfamily Sphaerularioidea Lubbock, 1861
Family Anguinidae Nicoll, 1935
Genus *Ditylenchus* Filipjev, 1936
Family Sphaerulariidae Lubbock, 1861
Family Neotylenchidae Thorne, 1941
Family Iotonchidae Goodey, 1935
Superfamily Tylenchoidea Örley, 1880
Family Hoplolaimidae Filipjev, 1934
Genus *Hoplolaimus* von Daday, 1905
Genus *Rotylenchus* Filipjev, 1936
Genus *Helicotylenchus* Steiner, 1945
Family Meloidogynidae Skarbilovich, 1959
Family Tylenchidae Örley, 1880
Genus *Tylenchus* Bastian, 1865
Genus *Filenchus* Andrássy, 1954
Genus *Malenchus* Andrássy, 1968
Genus *Aglenchus* Andrássy, 1954
Genus *Coslenchus* Siddiqi, 1978
Genus *Lelenchus* Andrássy, 1954
Genus *Eutylenchus* Cobb, 1913
Genus *Psilenchus* De Man, 1921
Family Belonolaimidae Whitehead, 1959
Genus *Tylenchorhynchus* Cobb, 1913
Genus *Nagelus* Thorne & Malek, 1968
Family Pratylenchidae Thorne, 1949
Genus *Hirschmanniella* Luc & Goodey, 1964
Genus *Pratylenchus* Filipjev, 1936

Family Dolichodoridae Chitwood, 1950
Genus *Dolichodorus* Cobb, 1914
Superfamily Myenchoidea Pereira, 1931
Family Myenchidae Pereira, 1931
INFRAORDER DRILONEMATOMORPHA De Ley & Blaxter, 2002
Superfamily Drilonematoidea Pierantoni, 1916
Family Drilonematidae Pierantoni, 1916
Family Ungellidae Chitwood, 1950
Family Homungellidae Timm, 1966
Family Pharyngonematidae Chitwood, 1950
Family Creagrocercidae Baylis, 1943
Suborder Rhabditina Chitwood, 1933
INFRAORDER BUNONEMATOMORPHA De Ley & Blaxter, 2002
Superfamily Bunonematoidea Micoletzky, 1922
Family Bunonematidae Micoletzky, 1922
Genus *Craspedonema* Richters, 1908
Genus *Rhodolaimus* Fuchs, 1930
Family Pterygorhabditidae Goodey, 1963
INFRAORDER DIPLOGASTEROMORPHA De Ley & Blaxter, 2002
Superfamily Cylindrocorporoidea Goodey, 1939
Family Cylindrocorporidae Goodey, 1939
Superfamily Odontopharyngoidea Micoletzky, 1922
Family Odontopharyngidae Micoletzky, 1922
Superfamily Diplogasteroidea Micoletzky, 1922
Family Pseudodiplogasteroididae Körner, 1954
Family Diplogasteroididae Filipjev & Schuurmans Stekhoven, 1941
Genus *Diplogasteroides* De Man, 1912
Genus *Rhabditidoides* Rahm, 1928
Family Diplogasteridae Micoletzky, 1922
Genus *Acrostichus* Rahm, 1928
Genus *Butlerius* Goodey, 1929
Genus *Diplogaster* Schultze in Carus, 1857
Genus *Diplogastrellus* Paramonov, 1952
Genus *Paroigolaimella* Paramonov, 1952
Family Neodiplogasteridae Paramonov, 1952
Genus *Koerneria* Meyl, 1960
Genus *Fictor* Paramonov, 1952
Genus *Mononchoides* Rahm, 1928
Genus *Neodiplogaster* Cobb, 1924
Genus *Oigolaimella* Paramonov, 1952
Family Mehdinematidae Farooqui, 1967
Family Cephalobiidae Filipjev, 1934
INFRAORDER RHABDITOMORPHA De Ley & Blaxter, 2002
Incertae sedis:
Family Carabonematidae Stammer & Wachek, 1952
Family Agfidae Dougherty, 1955

Superfamily Mesorhabditoidea Andrássy, 1976
Family Mesorhabditidae Andrássy, 1976
Genus *Mesorhabditis* Osche, 1952 (Dougherty, 1953)
Family Peloderidae Andrássy, 1976
Genus *Pelodera* Schneider, 1866
Superfamily Rhabditoidea Örley, 1880
Family Diploscapteridae Micoletzky, 1922
Family Rhabditidae Örley, 1880
Genus *Caenorhabditis* (Osche, 1952) Dougherty, 1953
Genus *Cephaloboides* (Rahm, 1928) Massey, 1974
Genus *Cruznema* Artigas, 1927
Genus *Diploscapter* Cobb, 1913
Genus *Oscheius* Andrássy, 1976
Genus *Pellioditis* (Dougherty, 1953) Timm, 1960
Genus *Poikilolaimus* Fuchs, 1930
Genus *Protorhabditis* (Osche, 1952) Dougherty, 1953
Genus *Prodontorhabditis* Timm, 1961
Genus *Rhabditella* (Cobb, 1929) Chitwood, 1933
Genus *Rhabditis* Dujardin, 1845
Genus *Rhabditoides* Goodey, 1929 (*incertae sedis*)
Superfamily Strongyloidea Baird, 1853
Family Heterorhabditidae Poinar, 1975
Family Strongylidae Baird, 1853
Family Ancylostomatidae Looss, 1905
Family Trichostrongylidae Witenberg, 1925
Family Metastrongylidae Leiper, 1912
Family Diaphanocephalidae Travassos,1920
Family Heligmosomidae Cram, 1927
Family Molineidae Durette-Desset & Chabaud, 1977

References

Andrássy, I. (1979) Nematoda. In: Illies, J. (ed.) *Limnofauna Europaea*. Eine Zusammenstellung aller die europäischen Binnengewässer bewohnenden mehrzelligen Tierarten mit Angaben über ihre Verbreitung und Ökologie. Gustav Fischer Verlag, Stuttgart, Germany, pp. 98–117.

Andrássy, I. (1984) *Klasse Nematoda (Ordnungen Monhysterida, Desmoscolecida, Araeolaimida, Chromadorida, Rhabditida)*. Gustav Fischer Verlag, Stuttgart, Germany, 509 pp.

Blaxter, M.L., De Ley, P., Garey, J.R., Liu, L.X., Scheldeman, P., Vierstraete, A., Vanfleteren, J.R., Mackey, L.Y., Dorris, M., Frisse, L.M., Vida, J.T. and Thomas, W.K. (1998) A molecular evolutionary framework for the phylum Nematoda. *Nature* 392, 71–75.

Bongers, T. (1988) *De nematoden van Nederland.* Stichting Uitgeverij Koninklijke Nederlandse Natuurhistorische Vereniging, Utrecht, Netherlands (Natuurhistorische bibliotheek nr 46), 408 pp.

Chitwood, B.G. (1937) A revised classification of the Nematoda. In: Anon (ed.) *Papers on Helminthology, 30 Year Jubileum K.J. Skrjabin.* All-Union Lenin Academy of Agricultural Sciences, Moscow, pp. 67–79.

Chitwood, B.G. (1958) The designation of official names for higher taxa of invertebrates. *Bulletin of Zoological Nomenclature* 15, 860–895.

De Coninck, L.A.P. (1965) Systématique des nématodes. In: Grassé, P.P. (ed.) *Traité de Zoologie: Anatomie, Systématique, Biologie*, 4th edn. Masson et Cie., Paris, pp. 586–731.
Decraemer, W. and Coomans, A. (1994) A compendium of our knowledge of the free-living nematofauna of ancient lakes. In: Martens, K., Goddeeris, B. and Coulter, G. (eds). *Speciation in Ancient Lakes. Archiv für Hydrobiologie Beiheft 44 Ergebnisse der Limnologie*. E. Schweizerbart'sche Verlagsbuchhandlung, Stuttgart, Germany, pp. 173–181.
Decraemer, W., Karanastasi, E., Brown, D. and Backeljau, T. (2003) Review of the ultrastructure of the nematode body cuticle and its phylogenetic interpretation. *Biological Reviews* 78, 465–510.
De Ley, P. and Blaxter, M.L. (2002) Systematic position and phylogeny. In: D.L. Lee (ed.) *The Biology of Nematodes*. Taylor & Francis, London, pp. 1–30.
De Ley, P. and Blaxter, M. (2004) A new system for Nematoda: combining morphological characters with molecular trees, and translating clades into ranks and taxa. *Nematology Monographs and Perspectives* 2, 633–653.
Doucet, M.E. and Doucet, M.M.A. de (1999) Supplement to the list of soil and freshwater nematodes recorded in continental Argentina. *Nematology* 1, 217–220.
Esquivel, A. and Arias, M. (2004) Nematode survey in Costa Rican conservation areas. *Nematology Monographs and Perspectives* 2, 463–468.
Eyualem, A. (2000) Free-living Limnic Nematofauna of Ethiopia: the species reported from sites out of the rift valley. *SINET, Ethiopian Journal of Science* 23, 273–297.
Eyualem, A. (2002) Free-living aquatic nematodes of the Ethiopian Rift Valley. In: Tudorancea, C. and Taylor, W.D. (eds) *Ethiopian Rift Valley Lakes*. Backhuys Publishers, Leiden, Netherlands, pp. 143–156.
Eyualem, A. and Coomans, A. (1995) Freshwater nematodes of the Galápagos. *Hydrobiologia* 299, 1–51.
Eyualem, A. and Coomans, A. (1996a) Aquatic nematodes from Ethiopia I. The genus *Monhystera* Bastian, 1865 (Monhysteridae: Nematoda) with the description of four new species. *Hydrobiologia* 324, 1–51.
Eyualem, A. and Coomans, A. (1996b) Aquatic nematodes from Ethiopia II. The genus *Monhystrella* Cobb, 1918 (Monhysteridae: Nematoda) with the description of six new species. *Hydrobiologia* 324, 53–77.
Eyualem, A. and Coomans, A. (1996c) Aquatic nematodes from Ethiopia III. The genus *Eumonhystera* Andrássy, 1981 (Monhysteridae: Nematoda) with the description of *E. geraerti* n. sp. *Hydrobiologia* 324, 79–97.
Eyualem, A., Grizzle, R.E., Hope, D. and Thomas, W.K. (2004) Nematode diversity in the Gulf of Maine, USA, and a Web-accessible relational database. *Journal of Marine Biological Association of the United Kingdom* 84, 1159–1167.
Ferris, V.R., Sabo, A., Baldwin, J.G., Mundo-Ocampo, M., Inserra, R.N. and Sharma, S. (2004) Phylogenetic relationships among selected Heteroderoidea based on 18S and ITS ribosomal DNA. *Journal of Nematology* 36, 202–206.
Fürst von Lieven, A. (2003) Functional morphology and evolutionary origin of the three-part pharynx in nematodes. *Zoology* 106, 183–201.
Gagarin, V.G. (1981) *Freshwater Nematodes of the European Part*. Nauka, Leningrad, Russia, 249 pp.
Gagarin, V.G. (1991) Seven new species of freshwater nematodes. *Zoologicheskyii Zhurnal* 8, 20–27.
Gagarin, V.G. (2000) New species of free-living freshwater nematodes from Eurasia. *Zoosystematica Rossica* 9, 11–17.
Gagarin, V.G., Nguyen Vu Thank and Nguyen Dinh Tu (2003) Three new species of free-living nematodes from freshwater bodies of Vietnam (Nematoda : Araeolaimida). *Zoosystematica Rossica* 12, 7–14.
Gerlach, S. and Riemann, F. (1973) The Bremerhaven Checklist of Aquatic Nematodes. *Veröffentlichungen des Instituts für Meeresforschung in Bremerhaven*, Suppl. 4, part 1, 404 pp.
Gerlach, S. and Riemann, F. (1974) The Bremerhaven Checklist of Aquatic Nematodes. *Veröffentlichungen des Instituts für Meeresforschung in Bremerhaven*, Suppl. 4, part 2, 734 pp.
Goodey, J.B. (1963) *Soil and Freshwater Nematodes*, 2nd edn. Methuen, London, 544 pp.
Goodey, T. (1951) *Soil and Freshwater Nematodes*. Methuen, London, 390 pp.

Hennig, W. (1950) *Grundlage einer Theorie der phylogenetischen Systematik.* Deutscher Zentralverlag, Berlin, 370 pp.

Heyns, J. (2002) Checklist of free-living nematodes recorded from freshwater habitats in Southern Africa. *Water S A* 28, 449–456.

Inglis, W.G. (1983) An outline classification of the phylum Nematoda. *Australian Journal of Zoology* 31, 243–255.

Jacobs, L. (1984) The free-living inland aquatic nematodes of Africa – a review. *Hydrobiologia* 113, 259–291.

Lemzina, L.V. (1989) *Free-living Nematodes of Lakes Issyk-Kul and Son-Kul.* Ilim, Frunze, Kyrgyzstan, 143 pp.

Loof, P.A.A. (1999) Nematoda, Adenophorea (Dorylaimida). In: Schwoerbel, J. and Brauer, A. (eds). *Süßwasserfauna von Mitteleuropa*, 21 Bde., Bd.4/2-1. Spektrum Akademischer Verlag, Heidelberg, Berlin, 274 pp.

Loof, P.A.A. (2001) Nematoda, Secernentea (Tylenchida, Aphelenchida). In: Schwoerbel, J. and Peter Z.P. (eds), *Süßwasserfauna von Mitteleuropa*, 21 Bde., Bd.4/1-1. Spektrum Akademischer Verlag, Heidelberg, Berlin, pp. 240.

Lorenzen, S. (1981) Entwurf eines phylogenetischen Systems der freilebenden Nematoden. *Veröffentlichungen des Institut für Meeresforschungen Bremerhaven*, Suppl. 7, 472 pp.

Lorenzen, S. (1994) *The Phylogenetic Systematics of Freeliving Nematodes.* The Ray Society, London, 383 pp.

Maggenti, A.R. (1979) The role of cuticular strata nomenclature in the systematics of Nemata. *Journal of Nematology* 11, 94–98.

Moens, T., Yeates, G.W. and De Ley, P. (2004) Carbon and energy sources for nematodes. *Nematology Monographs and Perspectives* 2, 529–545.

Pearse, A.S. (1942) *An Introduction to Parasitology.* Charles C. Thomas, Springfield, Illinois, 375 pp.

Schierenberg, E. (2005) Unusual cleavage and gastrulation in a freshwater nematode: developmental and phylogenetic implications. *Development, Genes and Evolution* 215, 103–108.

Smith, D.G. (2001) *Pennak's Freshwater Invertebrates of the United States: Porifera to Crustacea*, 4th edn. John Wiley & Sons, New York, pp. 191–215.

Thorps, J.H. and Covich, A.P. (2001) *Ecology and Classification of North American Freshwater Invertebrates*, 2nd edn. Academic Press, San Diego, California, pp. 255–295.

Tsalolikhin, S.J. (1980) *Free-living nematodes of Lake Baikal. (Svobodnozhivuschchi nematody Baikala).* Nauka, Novosibirsk, Russia, 120 pp.

Tsalolikhin, S.J. (1985) *Nematodes of Fresh and Brackish Waters of Mongolia.* Nauka Publications, Leningrad, Russia, 115 pp.

Tsalolikhin, S.J. (1998) On nematode fauna of high mountain lakes in the Pamir and Himalayas. *Zoologicheskii Zhurnal* 77, 1058–1063.

Tsalolikhin, S.J. (2001) Some species of freshwater nematodes from Singapore and Japan. *Zoosystematica Rossica* 10, 231–239.

Wieser, W. (1953) Die Beziehung zwischen Mundhöhlengestalt, Ernährungsweise und Vorkommen bei freilebenden marinen Nematoden. *Arkiv für Zoologie* 4, 439–484.

Zullini, A. (1982) Nematodi (Nematoda). *Guide per il riconoscimento delle specie animali delle acque interne Italiane* 17, 1–117.

Zullini, A., Loof, P.A.A. and Bongers, T. (2002) Free-living nematodes from nature reserves in Costa Rica. 2. Mononchina. *Nematology* 4, 1–23.

2 Techniques for Processing Freshwater Nematodes

M. HODDA[1] AND EYUALEM-ABEBE[2]

[1]*Nematode Biosystematics and Ecology, CSIRO Entomology, GPO Box 1700, Canberra ACT 2601, Australia;* [2]*Hubbard Centre for Genome Studies and Department of Zoology, University of New Hampshire, 35 Colovos Rd, Gregg Hall, Durham, NH 03824, USA*

Introduction

General techniques for processing nematodes are presented in many basic nematological textbooks (McIntyre and Warwick, 1984; Hooper, 1986a–e; Pfannkuche and Thiel, 1988; Fortuner, 1991; Traunspurger, 2002). Most of these techniques are described with reference to either terrestrial or marine nematodes. There is little to be gained by repeating the basic techniques, so this chapter presents discussions of the variations of the basic techniques demanded by work on freshwater nematodes.

Fixation

Most workers recommend fixation immediately after collection (Hooper, 1986c; Fortuner, 1991; Traunspurger, 2002). Immediate fixation produces specimens in the best possible condition for detailed morphological study and identification. However, fixing a sample immediately precludes some extraction options. Immediate fixation precludes the use of any method that relies on the active movement of the nematodes to leave the substrate and collect together (see below). Fixing a sample immediately also precludes the possibility of relaxing nematodes if they adhere to the substrate. Adherence is more likely in sediments of larger grain size and where there are higher water velocities or frequent sediment suspension.

Where nematodes are likely to adhere to sediment particles or agglomerations, there are two options. One is to relax the nematodes. Distilled water, gentle heat and addition of $MgCl_2$ have all been used (Bousseau, 1957; Uhlig *et al.*, 1973; Hooper, 1986c; Fortuner, 1991; Traunspurger, 2002). Relaxing the nematodes aids later identification because many nematodes have characteristic relaxed postures, which are mostly not tightly coiled. Without relaxation, most nematodes will coil tightly on fixation, rendering later morphological studies difficult or impossible (Jacobs, 1984).

The use of stains, such as Rose Bengal, has also been suggested as a means of finding nematodes which remain in the bulk substrate after extraction procedures, but extensive tests with marine nematodes have shown that the efficiency of locating stained nematodes by visual sorting is very low (Hulings and Gray, 1971).

Active methods of nematode extraction may alleviate some of the potential inaccuracies where nematodes adhere to substrate, but may require modifications to individual circumstances. They are also generally less labour intensive than passive methods, but also take longer to obtain results. They are at best semi-quantitative. Active methods include baiting and Whitehead trays or Baermann Funnels.

Extraction: active methods

Baiting

In estuarine mudflats fish, algae and other organic matter will attract nematodes in much the same way that *Galleria* larvae attract entomophilic nematodes and known host plants attract plant-parasitic nematodes in terrestrial soils (Meyers and Hopper, 1966; Bedding and Akhurst, 1975; Gerlach, 1977; Norton and Niblack, 1991). The method has not been greatly used in studies of freshwater aquatic nematodes.

Whitehead trays and Baermann Funnel

These two methods are variations on a single theme of saturation of substrate suspended by a nematode-permeable filter above a water reservoir (Figs 2.1 and 2.2). They are among the simplest methods available, and can be used on a wide variety of materials and substrates. The main disadvantages are great variability in extraction efficiency; lack of quantifiable results; very poor efficiency for some groups of nematodes; and time until results are obtained (generally a minimum of 3 days). Advantages are low labour and equipment requirements, and the fact that many species will hatch if present as eggs or revive if cryptobiotic so identifiable stages may be present which were not present in the initial sample. Many authorities have maintained freshwater nematodes on Whitehead trays for lengthy periods, which produced a steady stream of specimens in good condition for a period of weeks. Baermann funnels are more prone to deoxygenation of the nematode suspension than Whitehead trays, and so the nematodes need to be drawn off regularly.

Baermann funnel methods are frequently listed in general methodological recommendations, e.g. those for biological testing of freshwater by the American Public Health Association (1998). For specialist studies, they are seldom recommended.

Direct observation

This method is notoriously unreliable for all but the clearest samples. It is the simplest method and relies on nothing more than spotting nematodes in a sediment under some form of magnification. The nematodes can then be removed by pipette, hairbrush, probe or other fine implement. There are several variations for each of these implements. In practice, this method relies on the movement of nematodes or

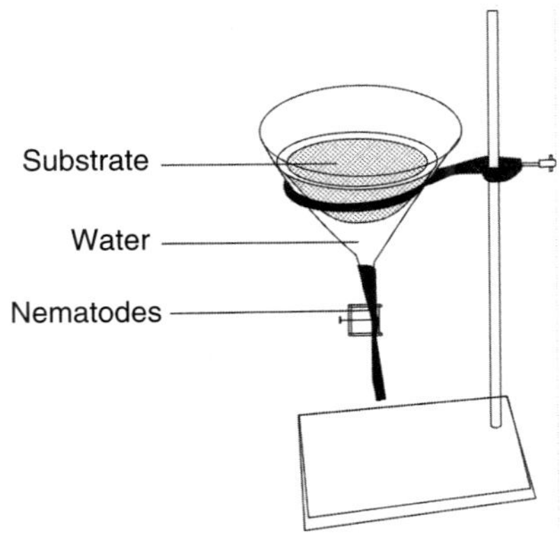

Fig. 2.1 Baermann Funnel extraction apparatus.

staining to aid recognition. Thus there is a strong bias to large, active nematodes, and the method is not suitable after fixation unless the nematodes are processed to a non-toxic medium: any fixative containing formaldehyde is unsuitable. For very large nematodes, such as adult mermithids, the method is quite efficient and commonly used. To improve visibility of nematodes, sediment can be greatly diluted. Allowing settling after mixing also improves visibility.

Extraction: passive methods

Sieving

Sieving is the simplest method. In sieving, sieves are selected such that most particulate matter is allowed through the sieves while many nematodes are retained on the sieve. Basically the finer the mesh openings, the more nematodes are retained, but also more substrate is retained. Sieves between 20 and 75 µm are commonly used

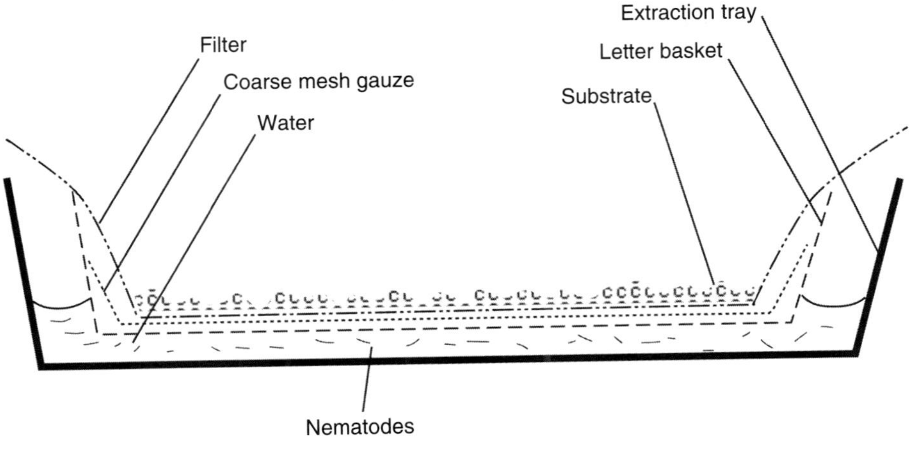

Fig. 2.2 Whitehead Tray extraction apparatus.

and can give satisfactory results where substrates are suitable, but many nematodes are lost through even the finest sieve practicable (Seinhorst, 1956; Hooper, 1986a). There may be considerable biases in the size and shape of nematodes lost through sieves: small straight nematodes are lost relatively easily and large coiled nematodes much less easily (Hooper, 1986a).

Decanting

Decanting is also a very simple method. Nematodes are generally less dense than mineral particulate substrates, so if a solution of sediment is suspended in water, the mineral sediment particles will often sink faster than the nematodes. This phenomenon is a result of Stokes' Law, which states that the settling velocity is directly proportional to the difference in density, but proportional to the square of the diameter of the particles. Thus large heavy particles, such as sand grains, sink much faster than less dense, smaller diameter nematodes. Decanting relies on pouring the water and still suspended nematodes away after the heavier soil particles have sunk, but before the nematodes sink. If this supernatant containing nematodes and small particles is poured through a fine mesh sieve, the nematodes will be retained and the small particles pass through. Without suspension before pouring through the sieve, the sieve would retain many larger substrate particles as well as nematodes, and little would be gained in terms of extracting nematodes.

The method is very simple and quick, requiring little more than a flask or measuring cylinder, but can be very inefficient, both in the proportion of nematodes extracted and in the amount of extraneous material removed. There can also be serious biases in the type of nematodes extracted. Because of Stokes' Law, large nematodes sink much faster than small ones, and so are less likely to be in the supernatant, which is sieved. The method is also very inefficient at separating nematodes from substrate, if the substrate contains a lot of material of similar size or density to nematodes, which settles at a similar rate. Many freshwater aquatic (and marine) sediments contain organic matter and particulate aggregations which fall into this category, and so the method is mostly used on coarse mineral substrates with little particulate organic debris. On this sort of substrate, the method has proved reasonably efficient (Hodda, 1981; Giere, 1993).

A more sophisticated version of decantation known as the 'Seinhorst 2-flask method' can also be used (Fig. 2.3).

Settling

Settling methods rely on the opposite process to decanting: a sample is suspended in excess water and the nematodes allowed to settle, leaving extraneous matter still suspended so that it can be carefully drawn off with the water. This method can be adequate for very large nematodes or very fine particulate substrates, such as clays, provided a dispersing agent is used to break up any aggregates. As with decanting, in the frequent cases where the size of nematodes is similar to that of the substrate, this class of methods is not effective. It can be effective for separating entomophilic nematodes, which are often large, from other material. However, the method is not often used for freshwater or marine sediments because of the amount of fine material of size and density similar to the nematodes present, which renders the procedure ineffective.

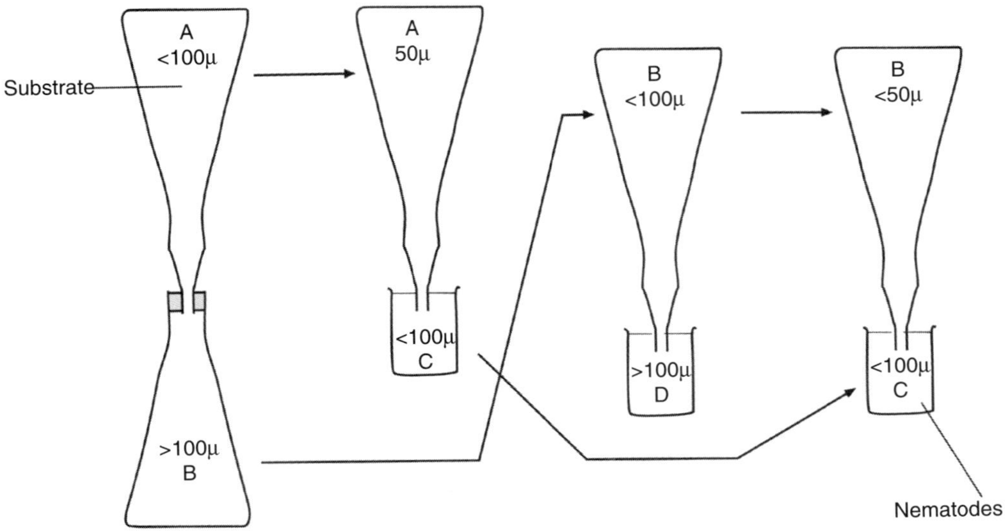

Fig. 2.3 Seinhorst 2-flask method for nematode extraction. All flasks filled with water.

Elutriation

Elutriation involves balancing the sinking rate of nematodes with an upward flow of water, most of the substrate sinking either more rapidly or more slowly, and thus being washed out of the water column. There are a number of commercial apparatuses available, often optimized for particular situations or types of nematodes. With the right samples these systems can extract nematodes very rapidly, but they often require extensive testing, particularly for different substrate types. Depending on the parameters used, some nematodes are extracted more efficiently than others.

Examples of elutriators include those designed by Oostenbrink (1962), Seinhorst (1956, 1962), Trudgill *et al.* (1973), Byrd *et al.* (1972, 1976) and Winfield *et al.* (1987). None of these variants has been tested on freshwater sediments, but Oostenbrink elutriators have been tested on both terrestrial soils and marine sediments and were found to be effective (Fricke, 1979; Viglierchio and Schmitt, 1983b). Hence, they should be effective in extracting nematodes from freshwater substrates too (Fig. 2.4).

Differential flotation

Flotation involves placing a sample into several solutions of different density such that material of different density to nematodes is removed either by flotation to the top or precipitation to the bottom. The solutions used normally have densities of 1.00 (water) and between 1.15 and 1.30. Few nematodes have densities less than 1.00 (but note that the few known 'planktonic' or swimming nematodes are close to this figure), and few nematodes have densities above 1.30. Little mineral material falls within the range of nematodes, although much organic matter does.

Water is used for the solution of density 1.00. Several different solutions are used for the higher density solution. Sugar is the most easily obtainable, but a lot of

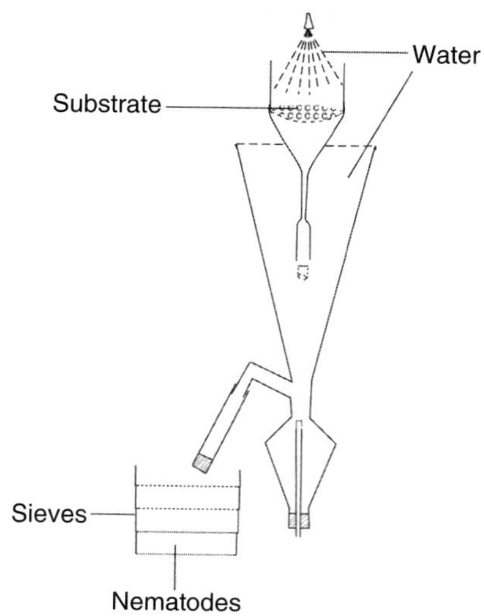

Fig. 2.4 Oostenbrink elutriator extraction apparatus.

sugar is required (approximately 450 g/l of solution) and the resultant solution is very viscous, which lowers extraction efficiency. Common salt, NaCl, can be used and is also easily obtainable. It is less viscous than sugar, but very much more osmotic, which can severely damage nematodes extracted and render morphological studies difficult. $MgSO_4$ is more expensive, but gives superior rates of extraction and better specimens than both NaCl and sugar. The best solution is a colloidal silica solution termed LUDOX® that has low viscosity and osmolarity.

Flotation can be performed by simply suspending a sample first in one solution then in the other for several days. Frequently a number of modifications are used. Centrifugation speeds the separation of the sunken from the floating material and minimizes the time in solutions damaging to the nematodes. Many freshwater and terrestrial nematodes are more susceptible to damage than most marine nematodes, and nematodes from relatively mobile environments, such as surf beaches, are often particularly resistant to damage. Thus centrifugation is mostly recommended for freshwater or semi-aquatic nematode extractions by this method. A powder of fine clay, Kaolin, can also be added to form a barrier between the material that has sunk and the floating material. Using this method and centrifugation to facilitate extraction, nematodes can be isolated in less than 1 h.

The flotation method is relatively free of bias in efficiency for different types of nematodes, and has the highest efficiency for most nematodes (around 70% for a single extraction). Efficiencies of around 95% can be obtained if the procedure is repeated three times, and virtually 100% of nematodes are extracted by between three and ten repetitions (Heip, 1974; Hodda and Bloemers, 1995). Higher efficiencies have been obtained for one or a few target species in trials with particular substrates (Viglierchio and Yamashita, 1983).

The method was originally developed for soils, but has also proven effective in marine sediments and a variety of freshwater substrates (De Jonge and Bouwman,

1977; Nichols, 1979; Nicholas and Hodda, 1982, 1999; Hodda, 1984; Hodda and Nicholas, 1985).

Combinations of methods

These methods can be combined to achieve better results. Decanting and sieving are frequently used together, as are sieving and elutriation, or flotation and sieving. The many variations and their rationales are reviewed by Hooper (1986a,b), McIntyre and Warwick (1984) and Viglierchio and Schmitt (1983a).

Comparisons of methods

Many of the methods listed above have been compared for particular combinations of substrates, nematodes and operators (Viglierchio and Schmitt, 1983a,b; Viglierchio and Yamashita, 1983; McIntyre and Warwick, 1984). These comparisons are mostly on samples from marine sediments or terrestrial soils. Enough of these studies have been done to draw general conclusions, which will almost certainly apply to freshwater sediments, which have not been tested. Unfortunately, the general conclusion is that the substrate, nematodes and operators all affect the extraction efficiency (e.g. Goodell, 1982; McIntyre and Warwick, 1984; McSorley, 1987; Hooper, 1990; Hodda and Bloemers, 1995). For maximum accuracy, testing of the method is needed for each situation.

Even the figures for extraction efficiencies are only comparative and not verifiably absolute: they can only be compared with the value obtained from another extraction method, not with an absolute or 'true' value. Seeding artificial or defaunated substrate with known numbers of nematodes, repeated extraction of the same sample and hand sorting have all been used. All have disadvantages and these have been discussed at considerable length (Green and Parrott, 1967; Kermarrec and Berge, 1972; Goodell, 1982; Peterson, 1982; McSorley, 1987; Hodda and Bloemers, 1995). Hand sorting nematodes from anything other than a very small sample is unreliable and fewer nematodes may be obtained this way than by other methods (Hulings and Gray, 1971; Higgins and Thiel, 1988). Differential flotation in LUDOX® (if available) is currently the preferred method of many nematologists, and the details of the method are presented at the end of this chapter.

Counting

Before counting nematodes, it is important to obtain the nematodes to be counted in a condition in which they can be counted. Nematodes may be so abundant that many lie on top of one another, making counting of individuals difficult or impossible. Alternatively, nematodes may be so rare that a large area must be searched to find each individual, resulting in wasted time and operator error through boredom. Dilution or concentration are used to avoid these problems and maximize efficiency of the counting operation. Both are used after extraction of nematodes from substrate.

The aim of dilution or concentration is to obtain nematodes spaced apart and in a single layer so that they can be counted most easily. In dilution a known volume of the same liquid in which the nematode sample is lying is added to the known volume containing the nematodes. In concentration, a known volume of liquid is drawn off. This can be achieved in several ways. A common way, if the nematode sample is in water or another non-viscous liquid less dense than nematodes, is to allow the nematodes to sink to the bottom of a tall, narrow container, such as a measuring cylinder, then very carefully suck excess liquid off the top with a pipette. A few minutes is sufficient to allow nematodes to sink. Another way of concentrating nematodes, also useful if the nematodes are in water or in a non-viscous liquid, is to pour the liquid through a sieve that will retain the nematodes, then wash the nematodes off the sieve with a smaller amount of liquid. This is essentially the same technique as using sieving to extract nematodes from soil (see section on 'Extraction' for details). The final volume of liquid containing nematodes can be adjusted by adding more liquid as necessary. Another method of concentration, useful for fixed nematodes if time is not limiting, is evaporation. Especially under warm conditions of low humidity, large flat dishes containing nematodes in liquid will reduce in volume over a period of a week or so. The advantage of this method is that it requires no labour, but care is needed that the sample does not dry completely (a little glycerol in the liquid will prevent this). If the nematodes are in a dangerous fixative, the evaporation should be carried out in a fume hood.

The amount of dilution or concentration needed must be determined individually for each sample. Having obtained a sample of nematodes at a suitable concentration, there is a limit to the number of nematodes that it is practical to count. The number of nematodes in a sample can be controlled by sample size, as discussed above, but frequently it is not possible to do this, so that subsampling is necessary. Whether the aim is detection, diagnosis or quantification, the accuracy required and whether a single known species or multiple unknown species are the targets will determine the extent to which subsampling can be used. In practice, the following prescriptions have proved to give good results. A single subsample of up to 1000 nematodes, or more than 10% of the total number, can be used, or multiple subsamples of smaller numbers or percentages.

Once the nematode sample is in appropriate condition for counting, there are a number of devices that can be used. All rely on having a good dissecting microscope and some means of dividing the nematode sample into areas that can be counted under one field of view. Various types of counting slides are available that place a known volume of liquid in a thin layer under a grid. There are also various types of counting dishes, which in basic form are Petri dishes with radii, concentric rings, grids or parallel lines to divide the area. It is possible to buy many types of these counting devices, but many can also be constructed easily.

Automated counting systems have been constructed, but few have proved practical. The main problems have been distinguishing nematodes from contamination, cost and the amount of set-up time required.

Identification

With experience, of both nematodes generally and particular situations, some nematodes can be tentatively identified to some degree under magnifications lower than 100×, generally the upper limit of most dissecting microscopes. Particularly in unfixed samples, nematodes with stylets or large buccal cavities can be distinguished, allowing crude trophic classifications. Likewise, the size and shape of nematodes can be used for tentative identification if the range of taxa present is known and/or the accuracy of identifications or counts is not required to be high. Identifications obtained at low magnifications are notoriously unreliable: there are often misidentifications, trophic categories are difficult to determine in groups like dorylaimids, and unusual occurrences will be missed if one is assuming that all the nematodes in a particular sample are the same as those already known from a particular site or habitat. For higher accuracy, specimens must be mounted on slides for study under high-power compound microscopes.

Preparation for microscopic examination

Nematodes can be examined either live or fixed. There are advantages to each, with different characters more easily visible in each state. Examination in both live and fixed states is often desirable, but frequently only one state is viewed because of time or resource constraints.

Some form of relaxation is desirable. Many species form characteristic postures in this condition.

Immobilization

If specimens are to be examined alive then immobilization is necessary. Heat is the most common means to achieve this. Most nematodes will become inactive at around 45°C with gentle heating in a small amount of water using a hotplate, alcohol burner or Bunsen flame. Cold can also immobilize nematodes. There is much greater variation in the success of using cold rather than heat for different species, but it can be effective especially for nematodes from warm climates.

Viscous media can also be used for immobilization. Depending on the species various concentrations of agar solutions, glycerol and gelatine have been used. Chemical narcotics can also be used. Sodium azide (0.01 M), CO_2 (from cylinder or dry ice), propylene phenoxetol (0.5–1.0%) and dichlorodiethyl ether have all been used on different nematodes, and there are undoubtedly other potential agents.

Fixation

Many recipes for fixatives contain alcohol, generally ethanol. These are no longer recommended for specimens requiring morphological examination because alcohol causes distortion.

Hot 2–5% formaldehyde (about 60°C), then rapid cooling with cold formaldehyde, gives very satisfactory results for most types of nematodes. If formaldehyde concentrations at the upper end of the range are used, the solution can be used on relatively large volumes of nematodes or bulk, unextracted soil if a passive method of extraction can be used. The best long-term results are obtained if the nematodes are allowed to remain in this fixative for at least 2 weeks. If this time is allowed, no cutting or manipulation of the nematodes to enhance penetration of the fixative should be necessary.

Mounting

Mounting nematodes on microscope slides always takes longer per nematode than most other stages of sampling and enumeration (with the exception of identification). Hence, subsampling is commonly necessary before mounting. This must be undertaken with great care to avoid artefacts. Nematodes are subject to Stokes'

Law, with the largest settling much faster than the smallest. There are also differences in density between different taxa, and even between different life stages of the same species (Coolen and D'Herde, 1972, 1977; Gooris and D'Herde, 1972; Freckman *et al.*, 1975, 1977; Hendrickx *et al.*, 1976; Coolen, 1979; Decraemer *et al.*, 1979). In practice this may mean that there are differences in the number and types of nematodes in different sections of a counting dish or slide. The only solutions to this problem are continuous agitation of the solution if drawing measured aliquots, or selection of a number of randomly located areas of a dish. The former has been achieved by bubbling air through the liquid.

If resources permit, mounting of entire samples is desirable, and sample sizes can be chosen accordingly. However, this option requires either more resources than are normally available, or prior knowledge of the likely abundance, also often not available. Most studies have relied on a subsample of either 100 nematodes or 10% of the total number. The former has the advantage of a known, definite amount of time required per sample, whilst the latter may be quicker if some samples have fewer nematodes.

For diversity studies, the most common approach has been to identify a certain number of randomly selected nematodes. Common numbers are 100 (Hodda, 1984) or 200 (Vincx, 1996). A test of 18 species of estuarine nematodes produced excellent correlation between estimates from 100 nematodes and the actual number in the sample (Hodda, 1984) (Fig. 2.5). Alternatively, a given proportion of a sample can be identified.

A novel method for non-experts is the 'sequential comparison technique' (Moore *et al.*, 1987).

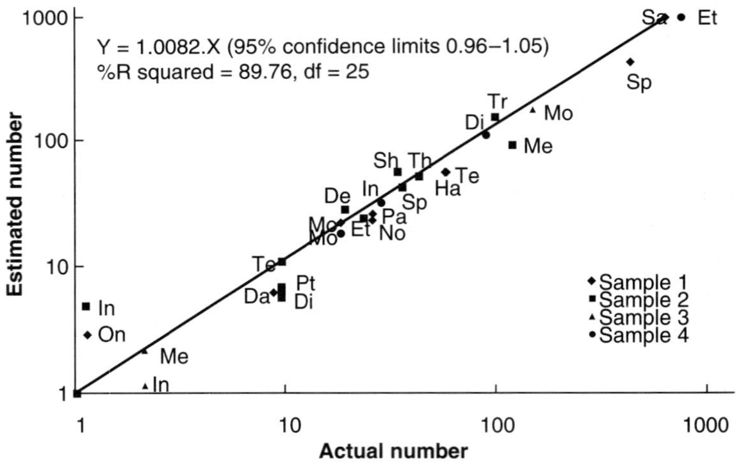

Fig. 2.5 Correlation between actual and estimated numbers of 18 species of nematodes obtained by subsampling 100 nematodes from four samples containing totals of 200–1000 nematodes (Hodda, 1984).
Da = *Daptonema* sp., De = *Desmodora cazca*, Di = *Diplolaimelloides* sp., Et = *Ethmolaimus* sp., Ha = *Haliplectus* sp., In = Low abundance sp., Me = *Metalinhomoeus* sp., Mo = *Monhystera* sp., No = *Nordia* sp., On = *Oncholaimus* sp., Pa = *Parodontophora* sp., Pt = *Ptycholaimellus* sp., Sa = *Sabatieria pulchra*, Sh = *Sphaerolaimus* sp., Sp = *Spirinia* sp., Te = *Terschellingia longicaudata*, Th = *Theristus interstitialis*, Tr = *Tripyloides* sp.

Nematodes are mostly mounted on standard microscope slides for identification. Round cover slips, rather than square or rectangular, are recommended if the slides are to be permanent or long term because there is less chance of air bubbles, and they can be sealed more easily. Also used are slides consisting of two cover slips held by thick cardboard in the centre of a metal frame (Cobb Slides). However, Cobb Slides are expensive and time consuming to make, and advances in optics since the time of Cobb (early 1900s) have made them largely unnecessary.

Nematodes can be mounted in water for examination at all but the highest magnifications. At very high magnifications, such as those required for identification to species, water mounts result in a poor view due to refraction of light passing through a series of media with different refractive indices (air, glass, water, nematode, water, glass, air, then back to glass in the objective lens). Oil immersion lenses reduce the number of different media, but mounting the nematodes in a material with the same refractive index as glass is essential for good microscopy. Lactophenol and glycerol are commonly used. Processing to lactophenol is quicker, but precautions need to be taken, and glycerol is better for permanent mounts. Immersion oil has also been used successfully, but only for rapid observation because specimens collapse after a short period.

Cover slips should be supported in some way, especially for large nematodes. Small glass beads, fine wire or wax can be used to support the cover slip. Transferring nematodes from fixative to mounting medium can be achieved in several ways, differing in the duration of the overall process and the amount of operator time required during this period. All methods have been shown to produce specimens in good condition if care is taken. The fastest is processing to lactophenol (Franklin and Goodey, 1949) (but note that this is generally undesirable on the grounds of safety and durability of slides, see above). This process takes only a few minutes. Next fastest, in terms of overall duration, is a method involving transfer through five solutions with progressively decreasing concentrations of phenol, lactate, water and formaldehyde, and an increasing concentration of glycerol (Baker, 1953). This procedure should take about an hour, but there is a lot of manual handling of nematodes. Next fastest is a method involving adding solutions, first of water, ethanol and glycerol, then of glycerol and ethanol in a saturated atmosphere of ethanol (Seinhorst, 1959). This method requires about a day, but only a small amount of time is required by the operator. The slowest method uses just the evaporation of glycerol and water. This method takes about 2 weeks, but almost no attention is required during this time.

Once nematodes have been transferred to the final mounting medium, they can be transferred to slides by pipette, fine probe, a sharpened bamboo skewer or a hair attached to a handle. For ecological studies, up to 100 small nematodes have been mounted on one slide. Ten or fewer is more usual. However, for taxonomic work, one or a few nematodes putatively of the same species are mounted together.

References

American Public Health Association (1998) Nematological examination: collection and processing techniques for nematodes. In: Clesceri, L.S., Greenberg, A.E. and Eaton, A.D. (eds) *Standard Methods for the Examination of Water and Wastewater*, 20th edn. APHA, Washington, DC, pp.10–76.

Baker, A.D. (1953) Rapid method for mounting nematodes in glycerine. *Canadian Entomologist* 85, 77–78.

Bedding, R.A. and Akhurst, R.J. (1975) A simple technique for the detection of insect parasitic rhabditid nematodes in soil. *Nematologica* 21, 109–110.

Bouisseau, J.-P. (1957) Techniques pour l'étude quantitative de la faune interstitielle des sables. *Comptes Rendus du Congrès des Sociétés Savantes de Paris et des Départements* 1957, 117–119.

Byrd, D.W., Ferris, H. and Nusbaum, C.J. (1972) A method for estimating numbers of eggs of *Meloidogyne* spp. in soil. *Journal of Nematology* 4, 266–269.

Byrd, D.W., Barker, K.R., Ferris, H., Nusbaum, C.J., Griffin, W.E., Small, R.H. and Stone, C.A. (1976) Two semiautomatic elutriators for extracting nematodes and certain fungi from soil. *Journal of Nematology* 8, 206–212.

Coolen, W.A. (1979) Methods for extraction of *Meloidogyne* spp. and other nematodes from roots and soil. In: Lamberti, F. and Taylor, C.E. (eds) *Root-knot Nematodes* (Meloidogyne *species*): *Systematics, Biology and Control.* Academic Press, London, pp. 317–329.

Coolen, W.A. and D'Herde, C.J. (1972) *A Method for the Quantitative Extraction of Nematodes from Plant Tissue.* State Agricultural Research Centre, Ghent, Belgium.

Coolen, W.A. and D'Herde, C.J. (1977) Extraction de *Longidorus* et *Xiphinema* spp. du sol par centrifugation en utilisant du silice colloidal. *Nematologia Mediterranea* 5, 195–206.

De Jonge, V.N. and Bouwman, L.A. (1977) A simple density separation technique for quantitative isolation of meiobenthos using colloidal silica Ludox-TM. *Marine Biology* 42, 143–148.

Decraemer, W., Coolen, W.A. and Hendrickx, G.J. (1979) Evaluation of extraction methods for a survey of Trichodoridae in fields of seed potatoes. *Nematologica* 25, 494–495.

Fortuner, R. (1991) Methods for collection and preparation of nematodes. Part 1. Field sampling and preparation of nematodes for optic microscopy. In: Nickle, W.R. (ed.) *Manual of Agricultural Nematology*. Marcel Dekker, New York, pp. 75–87.

Franklin, M.T. and Goodey, J.B. (1949) A cotton-blue-lactophenol technique for mounting plant-parasitic nematodes. *Journal of Helminthology* 23, 175–178.

Freckman, D.W., Mankau, R. and Ferris, H. (1975) Nematode community structure in desert soils: nematode recovery. *Journal of Nematology* 7, 343–346.

Freckman, D.W., Kaplan, D.T. and Van Gundy, S.D. (1977) A comparison of techniques for extraction and study of anhydrobiotic nematodes from dry soil. *Journal of Nematology* 9, 176–181.

Fricke, A.H. (1979) Meiofauna extraction efficiency by a modified Oostenbrink apparatus. *Helgolander Wissenschaftliche Meeresuntersuchungen* 32, 436–443.

Gerlach, S.A. (1977) Attraction to decaying organisms as a possible cause for patchy distributions of nematodes in a Bermuda Beach. *Ophelia* 16, 151–165.

Giere, O. (1993) *Meiobenthology: The Microscopic Fauna in Aquatic Sediments.* Springer-Verlag, Berlin.

Goodell, P.B. (1982) Soil sampling and processing for detection and quantification of nematode populations for ecological studies. In: Freckman, D.W. (ed.) *Nematodes in Soil Ecosystems.* University of Texas Press, Austin, Texas.

Gooris, J. and D'Herde, C.J. (1972) *A Method for the Quantitative Extraction of Eggs and Second Stage Juveniles of* Meloidogyne *spp. from soil.* Publications of the State Nematology and Entomology Research Station, Merelbeke, Belgium.

Green, C.D. and Parrott, D.M. (1967) A re-appraisal of extraction of *Heterodera* cysts from large volumes of soil. *Nematologica* 12, 601–609.

Heip, C. (1974) A rapid method to estimate nematode density. *Nematologica* 20, 266–268.

Hendrickx, G.J., Coolen, W.A. and Moermans, R. (1976) Influence de la nature et de la densité des liquides de séparation sur les nématodes pendant le processus de centrifugation-flottation. *Nematologia Mediterreanea* 4, 31–40.

Higgins, R.P. and Thiel, H. (1988) *Introduction to the Study of Meiofauna.* Smithsonian Institution Press, Washington, DC.

Hodda, M. (1981) Behavioural adaptions of nematodes in an exposed sandy beach. Honours thesis, Australian National University, Canberra.

Hodda, M. (1984) Population processes in estuarine littoral nematodes. MSc thesis, Australian National University, Canberra.

Hodda, M. and Bloemers, G.F. (1995) A method for extraction of nematodes from a tropical forest soil. *Pedobiologia* 39, 331–343.

Hodda, M. and Nicholas, W.L. (1985) Meiofauna associated with mangroves in the Hunter River estuary and Fullerton Cove. *Australian Journal of Marine & Freshwater Research* 36, 41–50.

Hooper, D.J. (1986a) Extraction of free-living stages from soil. In: Southey, J.F. (ed.) *Laboratory Methods for Work with Plant and Soil Nematodes*, 6th edn. HMSO, London, pp. 5–30.

Hooper, D.J. (1986b) Extraction of nematodes from plant material. In: Southey, J.F. (ed.) *Laboratory Methods for Work with Plant and Soil Nematodes*, 6th edn. HMSO, London, pp. 51–58.

Hooper, D.J. (1986c) Handling, fixing, staining, and mounting nematodes. In: Southey, J.F. (ed.) *Laboratory Methods for Work with Plant and Soil Nematodes*, 6th edn. HMSO, London, pp. 59–80.

Hooper, D.J. (1986d) Preserving and staining nematodes in plant tissues. In: Southey, J.F. (ed.) *Laboratory Methods for Work with Plant and Soil Nematodes*, 6th edn. HMSO, London, pp. 81–85.

Hooper, D.J. (1986e) Culturing nematodes and related experimental techniques. In: Southey, J.F. (ed.) *Laboratory Methods for Work with Plant and Soil Nematodes*, 6th edn. HMSO, London, pp. 133–157.

Hooper, D.J. (1990) Extraction and processing of plant and soil nematodes. In: Luc, M., Sikora, R.A. and Bridge, J. (eds) *Plant-parasitic Nematodes in Subtropical and Tropical Agriculture.* CAB International, Wallingford, UK.

Hulings, N.C. and Gray, J.S. (1971) A manual for the study of meiofauna. *Smithsonian Contributions to Zoology* 78, 1–83.

Jacobs, L.J. (1984) The free-living inland aquatic nematodes of Africa – a review. *Hydrobiologia* 113, 259–291.

Kermarrec, A. and Berge, J.B. (1972) Méthodes d'estimation des populations de nématodes des sols et des racines. *Bulletin de Société Écologique* 3, 151–169.

McIntyre, A.D. and Warwick, R.M. (1984) Meiofauna techniques. In: Holme, N.A. and McIntyre, A.D. (eds) *Methods for the Study of Marine Benthos.* Blackwell Scientific, Oxford, UK, pp. 217–244.

McSorley, R. (1987) Extraction of nematodes and sampling methods. In: Brown, R.H. and Kerry, B.R. (eds) *Principles and Practice of Nematode Control in Crops.* Academic Press, Sydney, Australia.

Meyers, S.P. and Hopper, B.E. (1966) Attraction of the marine nematode *Metoncholaimus* sp. to fungal substrates. *Bulletin of Marine Science* 16, 142–150.

Moore, C.G., Mathieson, S., Mills, J.L. and Bett, B.J. (1987) Estimation of meiobenthic nematode diversity by non-specialists. *Marine Pollution Bulletin* 18, 646–649.

Nicholas, W.L. and Hodda, M. (1982) Quantitative extraction of nematodes from soil and marine mud by centrifugal flotation in colloidal silica Ludox. *Australasian Nematology Newsletter* 5, 1–5.

Nicholas, W.L. and Hodda, M. (1999) Free-living nematodes of a temperate, high-energy sandy beach: faunal composition and variation over space and time. *Hydrobiologia* 394, 113–127.

Nichols, J.A. (1979) A simple flotation technique for separating meiobenthic nematodes from fine-grained sediments. *Transactions of the American Microscopical Society* 98, 127–130.

Norton, D.C. and Niblack, T.L. (1991) Biology and ecology of nematodes. In: Nickle, W.R. (ed.) *Manual of Agricultural Nematology.* Marcel Dekker, New York, pp. 47–72.

Oostenbrink, M. (1962) Estimating nematode populations by some selected methods. In: Sasser, J.N. and Jenkins, W.R. (eds) *Nematology.* University of North Carolina Press, Chapel Hill, North Carolina, pp. 85–102.

Peterson, H. (1982) Basis for the evaluation of the density and biomass data. *Oikos* 39, 295–305.

Pfannkuche, O. and Thiel, H. (1988) Sample processing. In: Higgins, R.P. and Thiel. H. (eds) *Introduction to the Study of Meiofauna.* Smithsonian Institution Press, Washington, DC, pp. 134–145.

Seinhorst, J.W. (1956) The quantitative extraction of nematodes from soil. *Nematologica* 1, 249–267.

Seinhorst, J.W. (1959) A rapid method for the transfer of nematodes from fixative to anhydrous glycerine. *Nematologica* 4, 67–69.

Seinhorst, J.W. (1962) Modification of the elutriation method for extracting nematodes from soil. *Nematologica* 8, 117–128.
Traunspurger, W. (2002) Nematoda. In: Rundle, S.D., Robertson, A.L. and Schmidt-Araya, J.M. (eds) *Freshwater Meiofauna: Biology and Ecology*. Backhuys Publishers, Leiden, The Netherlands, pp. 63–104.
Trudgill, D.L., Evans, K. and Faulkner, G. (1973) A fluidising column extracting nematodes from soil. *Nematologica* 18, 469–475.
Uhlig, G., Thiel, H. and Gray, J.S. (1973) The quantitative separation of meiofauna. *Helgolander wissenschaftliche Meeresuntersuchungen* 25, 178–185.
Viglierchio, D.R. and Schmitt, R.V. (1983a) On the methodology of nematode extraction from field samples: Baermann funnel modifications. *Journal of Nematology* 15, 438–444.
Viglierchio, D.R. and Schmitt, R.V. (1983b) On the methodology of nematode extraction from field samples: comparison of methods for soil extraction. *Journal of Nematology* 15, 450–454.
Viglierchio, D.R. and Yamashita, T.T. (1983) On the methodology of nematode extraction from field samples: density flotation techniques. *Journal of Nematology* 15, 444–450.
Vincx, M. (1996) Meiofauna in marine and freshwater sediments. In: Hall, G.S. (ed.) *Methods for the Study of Organismal Diversity in Soils and Sediments*. CAB International, Wallingford, UK, pp. 187–195.
Winfield, A.L., Enfield, M.A. and Foreman, J.H. (1987) A column elutriator for extracting cyst nematodes and other small invertebrates from soil samples. *Annals of Applied Biology* 111, 223–231.

Differential Flotation Technique for Freshwater Nematodes

1. Mix up an aqueous solution of specific gravity 1.15. Sugar, $MgSO_4$ or LUDOX® may be used. LUDOX® is recommended and must be diluted before use.

2. Break up sample and wash into as many centrifuge tubes or bottles as is necessary, each a maximum of one-third full of material.

3. Fill tube with water to an equal volume for all tubes (to balance centrifuge).

4. Add Kaolin if necessary. Kaolin (clay) may or may not be needed in freshwater sediments. It is used to separate and settle the sediment from floating organic matter in suspension. The more clay in the sediment, the less Kaolin is needed; the more organic matter, the more needed. If needed, Kaolin has to be added before any centrifugation. A rate of about a spatula/30 ml centrifuge tube or about a teaspoon/100 ml of sediment is generally appropriate.

5. Cap tube or bottle and thoroughly suspend contents.

6. Centrifuge until pellet formed. Often 400× gravity (*g*) for 5 min is used. However, 300*g* for 10 min or other centrifugation power and time combinations in between have been used. When using solutions with high osmotic potential such as sugar, the time of centrifugation should be kept to a minimum as these solutions cause nematodes to collapse. Excessive centrifuging will make step 9 more difficult.

7. Discard supernatant (containing mostly organic material) without disturbing pellet.

8. Add solution of high density (sugar, $MgSO_4$ or LUDOX®). The maximum ratio of solution to residue is desirable.

9. Cap and resuspend pellet. Mix well. This step is key for aquatic nematodes because of their tendency to stick to sediment material by their tail tip. However, care should be taken as shaking a lot may also damage nematodes if sediment contains a lot of sand.

10. Centrifuge until pellet is formed. Often 300*g* for 5 min is sufficient.

11. Pour supernatant (containing nematodes) through fine sieve (38 μm), retaining filtrate and leaving pellet in tube or bottle. Rinse very well with water. Rinsing is strongly recommended because LUDOX® and formaldehyde can react to form a gel which is difficult to wash from the nematode surface. Once formed, the gel on the surface of nematodes will make studying morphological details difficult.

12. Wash residue on sieve into appropriate receptacle.

13. Repeat step 12 using retained filtrate to increase the percentage recovery of nematodes.

3 Composition and Distribution of Free-living Freshwater Nematodes: Global and Local Perspectives

W. TRAUNSPURGER,[1] I.C. MICHIELS[1] AND EYUALEM-ABEBE[2]

[1]*Department of Animal Ecology, University of Bielefeld, Morgenbreede 45, 33615 Bielefeld, Germany;* [2]*Hubbard Centre for Genome Studies and Department of Zoology, University of New Hampshire, 35 Colovos Rd, Gregg Hall, Durham, NH 03824, USA*

Global Species Composition

There is growing evidence that free-living limno-nematodes, previously largely neglected, play important roles in ecosystem function. Free-living nematodes are present in virtually all types of limnetic sediments, including those subject to hot, acidic and anoxic conditions that exclude many other benthic invertebrates. However, despite their almost ubiquitous distribution and their diverse ecological roles, the taxonomic composition, global and local distribution and functional roles of nematodes are largely unknown for many habitats, especially for limnetic waters. In contrast, one of the relatively better-studied aspects of limnetic habitats is their nematode species composition. A large number of papers have been published that deal with the taxonomic description of species from lakes, rivers, ponds and hot springs both from temperate and tropical regions. The taxonomic aspect of nematode research still continues to dominate, as there are more papers published on nematode taxonomy than on nematode ecology from these habitats (e.g. Eyualem-Abebe and Coomans, 1995, 1996a–e, 1997a,b; Tsalolikhin, 1996, 1998; Doucet and de Doucet, 1999; Wu and Liang, 2000; Eyualem-Abebe *et al.*, 2001; Wanless and Hunter, 2001; Yun Liang *et al.*, 2002; Zullini *et al.*, 2002; Holovachov *et al.*, 2003).

Despite these continuing efforts, most nematological surveys of limnetic habitats, especially those from the tropics, remain largely incomplete mainly because most studies were momentary. Another fact that exacerbates the problem is the incomparability of the data resulting from these studies in terms of abundance, biomass or diversity because of differences in the number of samples collected and in the techniques employed for sample collection and extraction.

At a larger scale various authors have reviewed nematodes reported from limnetic habitats. Limno-nematodes from Africa were reviewed by Jacobs (1984), those from Europe by Andrássy (1978), from North America by Cobb (1914) and Esser and Buckingham (1987), from the former Soviet Union by Tsalolikhin (1988) and those from China by Wu and Liang (2000). Gerlach and Riemann (1973/74) published an invaluable work in which they listed all species that were reported from all aquatic habitats up to that point. Andrássy (1984, 1992, 1999) also provided a census of free-living genera and subgenera. These reviews are extremely useful at the genus level, but at the species level they are largely outdated and therefore cannot be relied upon completely. Most other reviews, however, remain focused within the limit of the borders of a single nation or are focused on a single nematode group (Andrássy, 1981; Coomans, 1989; Eyualem-Abebe, 2000; Traunspurger, 2003). In spite of these, apart from the above mentioned incomparability, research in this line is hindered by another major drawback: published information is widespread in time and space and currently there is no single important, accessible and compiled work that provides a readily available reference for researchers in this field.

Discussions about the global distribution or endemicity of nematode species are still largely premature mainly because of the lack or incompleteness of nematological surveys from the southern hemisphere (Jacobs, 1984; Decraemer and Coomans, 1994). The causes for this are a dearth of competent taxonomists and the lack of agreement on the identity of some species. Andrássy (1978) listed 160 genera and 605 species, including tylenchids, from European limnetic habitats. Traunspurger (2003) reported 122 genera and at least 207 species from aquatic habitats of Germany alone. The number of freshwater nematodes of Europe is hitherto estimated to be about 2500 species (http://www.faunaeur.org). The number of genera reported from African waters is far less (112 genera and 302 species) compared to Europe (Jacobs, 1984). A recent review on nematodes from Ethiopia mentioned a total species number of 89 (Eyualem-Abebe, 2002). Eyualem-Abebe's (2000) other review on limno-nematodes from a part of that country also listed three families which were not included in Jacobs' (1984) review.

Do nematodes have a ubiquitous dispersal?

Most limnetic genera are claimed to be either ubiquitous or to have a widespread distribution (Andrássy, 1978; Jacobs, 1984). Jacobs (1984) cautiously added that not only most free-living genera, but also a large number of species, could be cosmopolitan. His caution is that our knowledge of the distribution of many species is based on 'doubtful identification', and is liable to change. Generally, a number of these cosmopolitans may in fact have acquired special reproductive strategies: parthenogenesis, drought- (and ingestion-) resistant stages, rapid maturation upon hatching and short generation times. These strategies contribute to maximal species dispersal, and therefore, cosmopolitanism seems to be a logical consequence of these strategies (Jacobs, 1984). In a study of a remote limnetic habitat on the Galápagos archipelago, Eyualem-Abebe and Coomans (1995) concluded that 10 out of 18 species were cosmopolitan and the remaining six were widely distributed in the southern hemisphere (two were new records). They argued that the most likely hypothesis to explain the presence of freshwater nematodes on the Galápagos was through passive and very occasional transport by birds.

As a general rule, those species with higher local abundances tend to be more widely distributed. Therefore, a positive interspecific relation between local abundance and regional distribution among taxonomically related taxa was suggested (Lawton, 1993; Gaston, 1996). Density-dependent habitat selection is only one of a number of competing hypotheses that have been put forward to explain the interspecific relation of local abundance and range. Gaston *et al.* (1997) provided a comprehensive review of each of the hypotheses.

Finlay (2002) stated that the abundance of individuals in microbial species (e.g. protozoa) is so large that geographical barriers rarely restrict dispersal. This ubiquitous dispersal requires an alternative view of the scale and dynamic of biodiversity at the microbial level, wherein global species numbers are relatively low and local species richness is always sufficient to drive ecosystem functions. One obvious explanation is that they are simply so abundant that continuous large-scale dispersal sustains their global distribution. Finlay (2002) stated that there should be some size range where ubiquitous dispersal becomes less likely and where species are more likely to be geographically restricted. The organism size range where this change occurs is generally believed to be between 1 and 10 mm. He concluded that free-living eukaryotes, all having a body size of less than 2 mm, are probably sufficiently abundant to have worldwide distributions. For freshwater nematodes we exemplarily show the abundance of different size ranges of body length in the littoral of 11 selected lakes in Germany (Fig. 3.1). There is a strong decline (r^2 = 0.66–0.89) in freshwater nematode abundance with increasing size in most investigated lakes (exception: Lake Lautersee; dominant nematode species is *Prodesmodora circulata* with 57% abundance and a body length of 785 μm) and at least 80% of all nematodes in these lakes have a body length less than 2 mm. Moreover, only seven of 90 species in the littoral of Lake Königssee have a body length of more than 2 mm.

Variability in species compositions between areas, lakes within an area and within lakes

As discovered in the Galápagos study, most dominant species in alpine lakes are also cosmopolitan (Michiels and Traunspurger, 2005). When species compositions of two geographically different areas were compared (Tannheimer and Karwendel Alps), 38% of the 156 identified species were found in both areas. In contrast, nearby lakes within those two alpine areas showed a low similarity of species composition. In the Tannheimer Alps with the Lakes Vilslache, Traunsalpsee, Vilsalpsee and Rehbach, 14% of 130 species were present in >50% of all cores. In the Karwendel Alps 9% of 85 species were found in >50% of all cores sampled in the Lakes Soiernsee, Fereinsalm, Wildensee, Schmalsee, Ferchensee and Lautersee. The similarity among cores within lakes was far greater and ranged between 19 and 41% (Table 3.1). Therefore, on small (within lakes) and large (between areas) scales species composition resemblance is higher than on an intermediate (within areas) scale.

Most species occurred in several alpine lakes, but only three species (*Eumonhystera filiformis*, *E. longicaudatula* and *Tripyla glomerans*) were found in all lakes. The dominant species, *Chromadorina bioculata*, *Tobrilus stefanskii* and *P. circulata* are reported only from different lakes within Europe. *Rhabdolaimus aquaticus*,

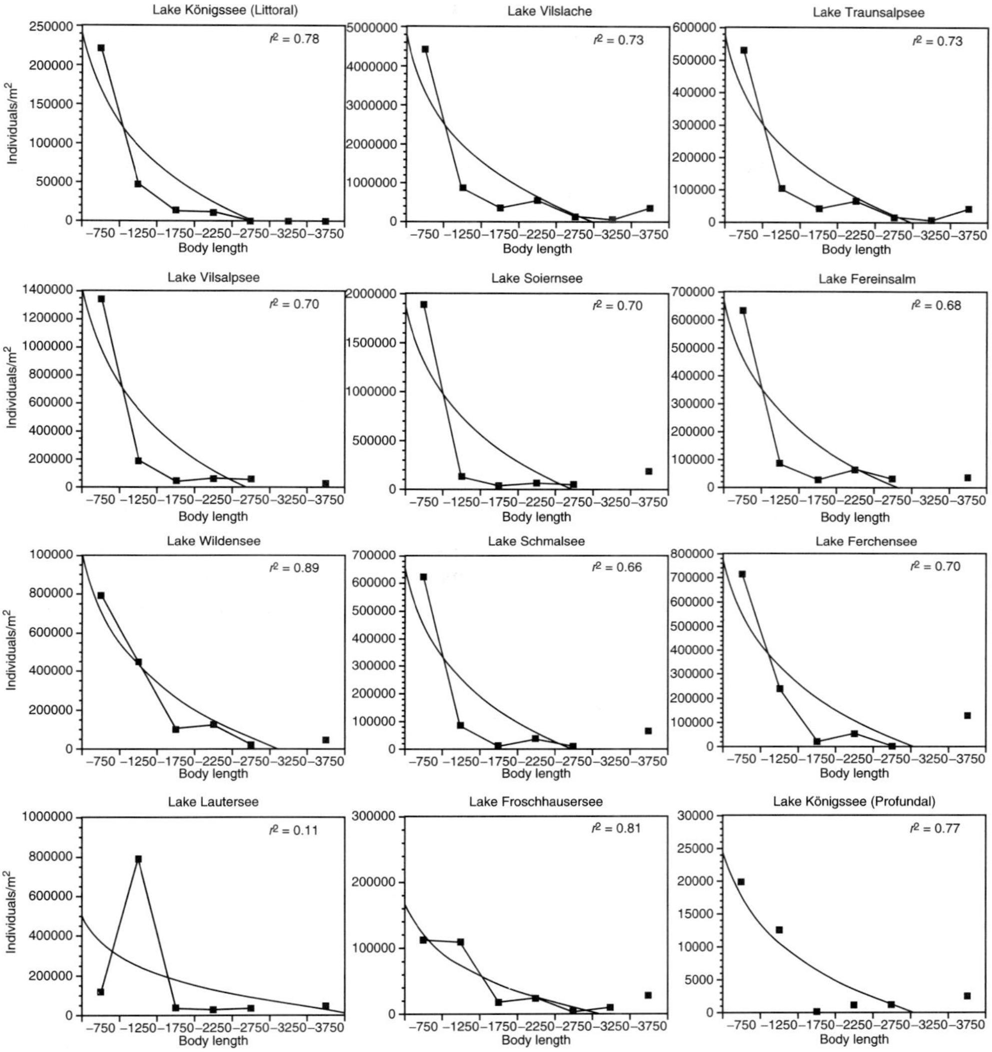

Fig. 3.1 Abundance (individuals/m^2) of nematodes with different body lengths (μm) in the littoral of 11 selected lakes and in the profundal of Lake Königssee in Germany.

Crocodorylaimus (= *Laimydorus*) *flavomaculatus* and *Plectus rhizophilus* are found in Africa as well. *E. filiformis*, *E. pseudobulbosa*, *Monhystera paludicola*, *Ethmolaimus pratensis* and *Rhabdolaimus terrestris* seem to be cosmopolitan, partially dominating freshwater habitats on at least three continents (Prejs, 1977b; Jacobs, 1984; Eyualem-Abebe and Coomans, 1995; Traunspurger, 1996a,b; Wu and Liang, 1999; Eyualem-Abebe *et al.*, 2001).

This finding is in contrast to the few studies on terrestrial habitats (e.g. Ettema, 1998). Similarity of species composition among soil cores taken within the same ecosystem was often low on local scales. Price and Siddiqi (1994) found only nine

Table 3.1 Absolute and relative (%) number of species found in five or more replicates out of ten in selected alpine lakes.

	Absolute number of species in >50% of all replicates	Relative number of species in >50% of all replicates
Tannheimer Alps		
Vilslache	23	31.51
Traunsalpsee	10	19.23
Vilsalpsee	14	38.89
Lake Rehbach	22	30.56
Karwendel Alps		
Soiernsee	7	21.21
Lake Fereinsalm	11	22.92
Wildensee	9	21.43
Schmalsee	12	41.38
Ferchensee	8	34.78
Lautersee	9	25.71
Froschhausersee	16	39.02

(6% of total) species to be present in >50% of the cores collected. These data suggest that nematode diversity has a significant local, soil core-scale component and a patchy distribution at the ecosystem scale. Though the possibility of a widespread distribution of a species within a similar type of habitat in different geographical regions may be a common phenomenon, some of the reports of a single species from a wide range of environments as well as geographical regions are alarming. For example, *Dorylaimus stagnalis* was reported within Africa from bottom sediment of freshwater, brackish water, terrestrial habitats and thermal springs, and was also reported to be planktonic (Jacobs, 1984). The doubt needs to be addressed through refinement of our taxonomy and Jacobs' (1984) caution deserves heed from both taxonomists and ecologists.

Which species?

The impact of correct identification on the resolution power of ecological studies cannot be overemphasized. Pennak (1988) wrote, '. . . nematodes are perhaps the most highly adaptable organisms from ecological and physiological stand point. The same species, for example may be found from the tropics to the sub-arctic, from warm springs to cold alpine lakes, and on many types of substrates'. Some species undoubtedly could be ubiquitous in their geographical distribution (Jacobs, 1984), but the biological flexibility of such a wide range of environmental variation may be too much for the genetic flexibility of a single species to cope with. If ubiquity is a general phenomenon in limno-nematodes, we need to confirm it using additional methods to morphology. Molecular identification has recently been introduced as a supplemental tool for nematode identification to the species level and beyond, especially in plant-parasitic nematodes (see Powers *et al.*, 1997 and references therein). Its use also has been expanded to soil (De Ley *et al.*, 1999; Floyd *et al.*, 2002;

Eyualem-Abebe and Blaxter, 2003) and marine (Blaxter *et al.*, 2002) nematode identification in diversity studies, but not in limno-nematology, so far. Therefore, it is overdue that limno-nematology embraces newly developed methods of discovering species in addition to those already used, so that we can have a readily transferable source of information for comparative purposes in cases where other methods are considered insufficient. One important aspect could be the comparison of cosmopolitan species with molecular techniques to ensure that these species are not a species complex. This is not in any way suggesting that one method is better than the other, instead it is a suggestion to use more than one method to address these unresolved issues.

Local Species Richness and Composition

The total nematode species numbers reported from different lakes with some characteristics of the lakes is given in Table 3.2. This table is not complete by any standard mainly due to the lack of information on many lakes, and only studies where the mesh size used is <45 μm are included. Moreover, the table shows that there are only a few studies of the ecology of freshwater nematodes in aquatic habitats from Asia, Africa, South America and Australia.

Looking at the relatively recent data, the highest species number (152 species) hitherto found from a single limnetic habitat is from Lake Obersee, a shallow eutrophic lowland lake in northern Germany. From this lake sediment samples were analysed from a homogeneous muddy area of approximately 2 m^2 for 3 years (Michiels and Traunspurger, 2004). One hundred and sixteen species were reported from Lake Königssee, a small (surface area 5.2 km^2) but deep (mean depth 98 m) lake in southern Germany (Traunspurger, 1991). The littoral zone (1–10 m) of this lake contains 90 species (Traunspurger, 1996a), the littoriprofundal 71 species and the profundal zone 60 species (Traunspurger, 1996b). The lake with the third highest species richness is the alpine Lake Vilslache with 75 species. A salty lake from Kyrghyzstan, Issyk-Kyl, was reported to have 74 species, most of which were not typical of freshwater habitats (Lemzina, 1995). Other species-rich lakes are Lakes Rehbach (68 species), Spitzingsee (67 species), Hopfensee (63 species), Sulzbergersee (55 species) and Baikal (53 species); (Shoshin, 1999; Traunspurger, 2001; Michiels and Traunspurger, 2005). Prejs (1977a) reported 52 species from the eutrophic Lake Mikolajskie and Wu and Liang (1999) identified 51 species in Lake Biandantang. A compilation of nematofauna of ancient lakes by Decraemer and Coomans (1994) showed those lakes having moderate species numbers similar to those reported for Lake Balaton (Biró, 1973), Neusiedlersee (Schiemer, 1978) and Piburger See (Pehofer, 1989) (Table 3.2).

Is nematode species richness related to trophic status?

At a relatively local scale (spatial resolution and extent) there is a marked tendency for a general dome-shaped relationship between species richness and available energy, with species richness increasing from low to moderate levels of energy and

Table 3.2 Abundance, biomass and number of nematode species in some selected lakes with remarks on sediment depth and sampling method including mesh size. (From Traunspurger, 2002).

Habitat	Sediment depth*	Sampling method (mesh size)	Abundance** (Individuals/10 cm^2)	Biomass (WW mg/m^2)	Species number	Reference(s)
Oligotrophic lakes						
Char Lake, Polska	4–5	Kajak corer, 45 μm	L: 250–600	n.a.	–	Prejs (1977a)
	–	–	LP: 25	n.a.	21	–
	–	–	P: 500–715	*c.*600	–	–
Mirror Lake, USA	8	Corer, unsieved	L: 750–1300	lake-wide mean 150	20	Strayer (1985)
	–	–	P: 300–900	–	–	–
Grünsee, Germany	5	Kajak corer, 40 μm	L: 10–26 (17.4 ± 7)	6.7–17.5	17	Traunspurger (1991)
Schwarzensee, Germany	5	Kajak corer, 40 μm	L: 28–81 (57.5 ± 22.8)	31.7–91.8	32	Traunspurger (1991)
Lake Königssee, Germany, hard substratum	0.5–2	Modified suction apparatus	L:1.3–65.3 LP: 0.2–1.2	0.3–12 0.1–5.2	29	Traunspurger (1992)
Erdfallsee, Germany	5	Core-sampler, 40 μm	L: 42 ± 24	n.a.	21	Traunspurger and Weischer (1993)
Lake Königssee, Germany	10	Kajak corer, 40 μm	L: 96–569 (296)	17.9–165	–	Traunspurger (1996a,b)
	–	–	LP: 43–66 (59)	9.2–127	116	–
	–	–	P: 25–50 (38)	3–69	–	–
Lake Lustsee, Germany	5	Corer, 35 μm	L: 4.1–83.5 (42.8)	6.1	47	Traunspurger (2001)
	–	–	P: 7.8–75.0 (40.8)	6.3	–	–
Lake Froschhausersee, Germany	5	Corer, 35 μm	L: 1.0–84.9 (18.2)	6.7	29	Traunspurger (2001)
	–	–	P: 0.3–2.3 (1.3)	0.3	–	–

Lake Vilslache, Austria	4	Corer, 35 µm	L: 3465–11397	1657–5347	75	Michiels and Traunspurger (2005)
	–	–	(6703 ± 2719)	(2829 ± 1097)	–	–
Traunsalpsee, Austria	4	Corer, 35 µm	L: 279–2245	40–6132	48	Michiels and Traunspurger (2005)
	–	–	(1003 ± 725)	(1486 ± 1959)	–	–
Vilsalpsee, Austria	4	Corer, 35 µm	L: 90–3119	41–935	33	Michiels and Traunspurger (2005)
	–	–	(1658 ± 1173)	(451 ± 326)	–	–
Lake Rehbach, Austria	4	Corer, 35 µm	L: 190–1433	140–1434	68	Michiels and Traunspurger (2005)
	–	–	(666 ± 445)	(463 ± 369)	–	–
Soiernsee, Germany	4	Corer, 35 µm	L: 362–5416	242–1890	31	Michiels and Traunspurger (2005)
	–	–	(2190 ± 1683)	(863 ± 529)	–	–
Lake Fereinsalm, Germany	4	Corer, 35 µm	L: 580–1382	191–851	47	Michiels and Traunspurger (2005)
	–	–	(864 ± 243)	(401 ± 229)	–	–
Wildensee, Germany	4	Corer, 35 µm	L: 17–298	3–171	38	Michiels and Traunspurger (2005)
	–	–	(148 ± 81)	(74 ± 50)	–	–
Schmalsee, Germany	4	Corer, 35 µm	L: 565–1138	152–346	28	Michiels and Traunspurger (2005)
	–	–	(801 ± 183)	(250 ± 72)	–	–
Ferchensee, Germany	4	Corer, 35 µm	L: 75–2527	9–1671	24	Michiels and Traunspurger (2005)
	–	–	(1011 ± 690)	(245 ± 503)	–	–
Lautersee, Germany	4	Corer, 35 µm	L: 106–2957	42–795	34	Michiels and Traunspurger (2005)
	–	–	(974 ± 794)	(361 ± 253)	–	–
Froschhausersee, Germany	4	Corer, 35 µm	L: 124–465	23–262	39	Michiels and Traunspurger (2005)
	–	–	(265 ± 134)	(97 ± 67)	–	–

Continued

Table 3.2 *Continued.* Abundance, biomass and number of nematode species in some selected lakes with remarks on sediment depth and sampling method including mesh size. (From Traunspurger, 2002.)

Habitat	Sediment depth*	Sampling method (mesh size)	Abundance** (Individuals/10 cm²)	Biomass (WW mg/m²)	Species number	Reference(s)
Brunnsee, Germany	4	Corer, 35 µm	P: 157–746	227–505	41	Bergtold and Traunspurger (2004)
	–	–	(385 ± 16840)	–	–	–
Mesotrophic lakes						
Zarnowieckic Lake, Polska	4–5	Kajak corer, 45 µm	L: 90	n.a.	16	Prejs (1977a)
	–	–	LP: 10	n.a.	–	–
	–	–	P: 4–8	*c.*32	–	–
Lake Starnberger See, Germany	5	Kajak corer, 35 µm	L: 107–3464 (537 ± 731)	n.a.	>70	W. Traunspurger (unpublished data)
Lake Spitzingsee, Germany	5	Corer, 35 µm	L: 1.0–153.8 (31.6)	17.1	67	Traunspurger (2001)
	–	–	P: 3.3–32.0 (12.8)	16.2	–	–
Lake Sulzberger See, Germany	5	Corer, 35 µm	L: 6–150.6 (30.1)	14.0	55	Traunspurger (2001)
	–	–	P: 3.0–6.1 (4.2)	3.3	–	–
Eutrophic lakes						
Neusiedlersee, Austria	14	Corer, 50 µm	L: 78–618	36–144	26	Schiemer *et al.* (1969)
Mikolajskie Lake, Polska	4	Pneumatic sampler and glass tube, 45 µm	L: 10–380	n.a.	39	Prejs (1970)
	–	–	LP: 10–80	n.a.	27	–
	–	–	P: 2–90	n.a.	13	–

Lake Balaton, Hungary	3	Craib's bottom-dredge, net number 25	L: 10–109	1–61	31	Biró (1973)
Mikolajskie Lake, Polska	4–5	Kajak corer, 45 μm	L: 70–538	n.a.	52	Prejs (1977a)
		–	–	LP: 5.5	n.a.	– –
		–	–	P: 6–40	*c.*80	– –
Funtensee, Germany	5	Kajak corer, 40 μm	L: 15–70 (53.1 ± 23.8)	16.5–78.2	32	Traunspurger (1991)
Großes Heiliges Meer, Germany	5	Core-sampler, 40 μm	L: 58 ± 21	n.a.	22	Traunspurger Weischer (1993)
Lake Höllerer See, Austria	5	Kajak corer, 35 μm	L: 1.8–92.4 (27.3 ± 25.1)	n.a.	26	Hehl (1996)
	–	–	P: 0.4–13.1 (5.0 ± 3.3)	–	–	–
Lake Hopfensee, Germany	5	Corer, 35 μm	L: 0.9–176.5 (14.2)	28.3	63	Traunspurger (2001)
	–	–	P: 20.0	38.1	–	–
Obersee, Germany	4	Corer, 35 μm	L: 145–3714 (978 ± 758)	96–735	152	Michiels and Traunspurger (2004)
Lake Tana, Ethiopia	10	Cores	L: 91.0–504.7	81.6–1316	40	Eyualem-Abebe *et al.* (2001)

Note: L = Littoral, LP = Littoriprofundal, P = Profundal; WW = Wetweight; n.a. = not available; *Tabled values are distances from the sediment surface, in cm; **Range and/or mean value; some mean values are given with standard deviation.

then declining again towards high levels of energy when a sufficient range of energy values is sampled (Gaston, 2000; for review see Mittelbach *et al.*, 2001). In contrast, the number of coexisting species seems to be independent of the trophic status of their habitats (Fig. 3.2). High species richness was found in the oligotrophic Lake Königssee (116 species, about 45,000 identified individuals), but higher species richness in the littoral of the highly eutrophic Lake Obersee (152 species, about 12,000 identified individuals). In eutrophic lakes between 22 and 152 species were identified, in mesotrophic lakes between 16 and 70 species, whereas in oligotrophic lakes species richness ranged between 17 and 116 species.

Hyporheic and phreatic nematodes, particularly their quantitative composition, are not well documented mainly due to inherent technical and methodological problems. Table 3.3 shows nematode abundance and species richness in different lotic habitats (see also the reviews of Traunspurger, 2000, 2002). It is not uncommon to find more than 50 nematode species (e.g. Eder, 1983; Beier and Traunspurger, 2003a). High species numbers were found in Elbe (Germany) by Riemann (1966) with 175 species and in Körsch (Germany) with 113 species by Beier and Traunspurger (2003b).

Species richness, i.e. the number of species, is currently the most widely used diversity measure. Relative species abundance in a community is another factor that affects diversity (Hurlbert, 1971). It is measured with a standardized index of species abundance (evenness) that is typically on a scale ranging from near 0 (which indicates low evenness or high single-species dominance) to 1 (indicating equal abundance of all species or maximum evenness) (Routledge, 1980; Stirling and Wilsey, 2001). Finally, proportional abundance indices such as the Shannon–Wiener index sum species weighted by their relative abundance (Magurran, 1988).

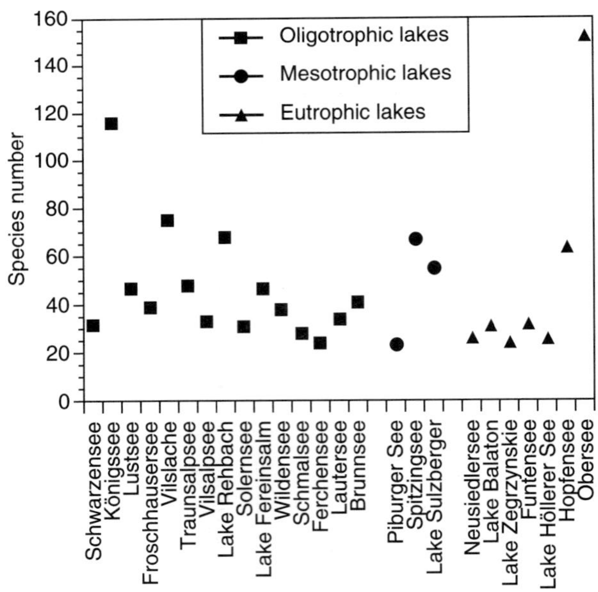

Fig. 3.2 Species number of nematodes in lakes with different productivity (*n*>750 individuals for all lakes).

Table 3.3 Species number and abundance of nematodes in some selected rivers with remarks on sediment depth, substrate and sampling method. (From Traunspurger, 2000).

Habitat	Sediment depth (cm)	Substrate	Sampling method	Species number	Abundance	Reference(s)
Lotic waters of Regnitz (Germany)	–	a) Sand and mineral mud	n.a.	a) 11	n.a.	Hirschmann (1952)
	–	b) Sand and mud with plants	–	b) 19	–	–
	–	c) Sand and mud	–	c) 23	–	–
	–	d) Sand and mud with vegetation	–	d) 26	–	–
	–	e) Sludge with sandy parts	–	e) 30	–	–
Elbe estuary (Germany)	0–10	Fine and coarse sand	Grabs	175	n.a.	Riemann (1966)
Danube (Austria)	0–100	Phreatic; all types	Bou-Rouch, 50–80 μm mesh size	7	Up to 80% of meiobenthos	Danielopol (1976)
Ohio streams, USA	0–7	Sand bars	Suction syringe	n.a.	0.2–15.2 individuals/cm^2	Hummon *et al.* (1978)
Danube (Austria)	0, 35, 70, and 105	Bed of gravel	Modified Bou-Rouch	51	n.a.	Eder (1983)
Mur (Austria)	0–5, 35 and 70	Sediment and hyporheic zone	Modified corer and Bou-Rouch, 80 μm mesh size	23	n.a.	Kirchengast (1984)
Stream in Texas (USA)	0–30	Sandy	Corer (3.85 cm); sieve 124 μm	3	4.3 individuals/10 cm^2; 0.3 individuals/ml	Whitman and Clark (1984)
South Platte River, Colorado (USA)	0–50	Gravel bar, phreatic	Bou-Rouch, 48 μm mesh size	13	n.a.	Pennak and Ward (1986)

Continued

Table 3.3 *Continued.* Species number and abundance of nematodes in some selected rivers with remarks on sediment depth, substrate and sampling method. (From Traunspurger, 2000).

Habitat	Sediment depth (cm)	Substrate	Sampling method	Species number	Abundance	Reference(s)
White Clay Creek, Pennsylvania (USA)	0–5	Rocks and sediment	Sieving and centrifugation, 30 μm sieve	n.a.	0.9–8.9 individuals/ cm^2 (fine sediment) 0.2–1.8 individuals/ cm^2 (coarse sediment)	Bott and Kaplan (1989)
Goose Creek, Virginia (USA)	0–50	Median grain size: 1000 μm; long and sandy channels	Cores: (3.3 cm dia.); 44 μm sieve; decantation	n.a.	10–160 individuals/ 10 cm^2	Palmer (1990)
Mississippi (USA)	Pool: 0–3 and 5 cm, cores 60 cm deep	Sandy and silt–sand	Ekman grab without sieving or 50 μm sieve	23	Sandy: 77–94 individuals/cm^2; silt–sand: 8–23 individuals/cm^2	Anderson (1992)
Danube (Austria)	0–70	Sandy or coarse	Freeze corer	n.a.	0.1–9 individuals/cm^2	Tockner and Bretschko (1996)
Rivers in the region of Münster (Germany)	n.a.	n.a.	250 g sediment, centrifugation	≈150 65 aquatic forms	Sandy sediment: 20 – >20000 ind/kg	Niemann *et al.* (1996)

Oberer Seebach (Austria)	0–40	Gravel	Stand-pipe traps, 30 μm mesh size	45	0–300 individuals/l; data estimated from graphs	Schmid-Araya and Schmid (1995), Schmid-Araya (1997)
14 sites in eastern USA	10, 40, and 70	All types, especially gravel bar	Bou Rouch, Sieve 100 μm	n.a.	105 (0–1157) individuals/3 l	Strayer *et al.* (1997)
Breitenbach (Germany)	0–5	Sandy	Cores: 3 cm diameter	>100	Mean: 25 individuals/ml (range:1–230)	Christl (unpublished)
Necker (Switzerland)	0–34	Coarse and sandy	Artificial cages; 35 μm mesh size	41	386–4137 individuals/ 10 cm^2; 3–38 individuals/ml	Eisenmann *et al.* (1998), Traunspurger (unpublished)
Green River, Utah (USA)	0–3	Coarse and finer sand	Core, 63 μm mesh size	n.a.	12–3600 individuals/ 100 cm^2	Jordan *et al.* (1999)
Colorado River, Utah (USA)	0–3	Coarse and finer sand	Core, 63 μm mesh size	n.a.	26–2140 individuals/ 100 cm^2	Jordan *et al.* (1999)
Krähenbach (Germany)	0–5	Sand	Core, 35 μm mesh size	71	28–126 individuals/ 10 cm^2	Beier and Traunspurger (2003a)
Körsch (Germany)	0–5	Stones, pebbles, gravel	Core, 35 μm mesh size	113	37–1205 individuals/ 10 cm^2	Beier and Traunspurger (2003b)

n.a. = not available.

Table 3.4 gives some examples of species richness relations, evenness (J') and Shannon–Wiener index (H') of some lakes and rivers, and the variation between different water depths within Lake Königssee. In alpine lakes, species diversity varied as much as between environmentally very differently structured freshwater habitats. The highest diversity and evenness indices were found in Lakes Vilslache and Rehbach which contained the highest species numbers, and various species contributed significantly to the whole nematode abundance. In all other lakes the abundance of nematode species was not equally distributed and less nematode species coexisted. Only one to four dominant species constituted up to 75% of the total density and the evenness index ranged between 0.51 and 0.68. Within Lake Königssee, nematode species numbers were surprisingly constant (about 45–55 species) among the investigated water depths over 1 year (Traunspurger, 1991, 1996a,b). In lotic habitats nematode species numbers are similar to those of lakes but diversity (H' and J') generally appears to be higher compared to lakes (Table 3.4).

One problem of comparing species richness of different habitats, among others, is the sample size and the number of identified individuals. It is important that the number of individuals identified to species level is large enough to represent abundant as well as rare species. In 11 alpine lakes the species richness was analysed with a sample size of 1000 individuals per lake (Michiels and Traunspurger, 2005). To do this, we analysed rarefaction curves (Lambshead *et al.*, 1983) where the number of identified individuals is plotted against the cumulative number of new species found per new sample. An optimistic and a conservative ranking of the replicate samples of these lakes were used: in the optimistic estimation the sample, which contained most species, was plotted first followed by the samples with the highest number of new species at each case. The conservative estimation started with the sample of the lowest species number, followed by samples with the lowest number of new species at a time. The optimistic rarefaction curves indicated that about 530 individual nematodes needed to be identified to species level to achieve a 95% estimate overview per lake. Conservative estimates required about 880 nematodes. Therefore, 880 identified individuals per habitat would be a sufficient representative for properly analysing community characteristics and ecological questions for these lakes (Fig. 3.3).

Factors Controlling Species Composition

Abiotic factors

In general, a discussion on the possible sources of variation in genus and species numbers and composition of the nematofauna in different lakes is, at present, premature for two reasons. First, information on factors playing a role in the distribution of free-living nematodes in inland water bodies is still rare. The best one can do at the moment is to assume that those factors (e.g. grain size, oxygen, temperature, food availability) influencing the distribution of the comparatively well-studied marine nematodes also affect nematodes in inland water bodies. Second, a large proportion of studies done on the nematofauna of inland water bodies does not go far beyond collecting a few samples for taxonomic purposes from a relatively small proportion of lakes and rivers. Moreover, detailed descriptions of study sites and measurements of a

Table 3.4 Species richness, Shannon–Wiener Index H' and evenness J' in selected lakes of Germany and selected rivers.

	Individuals identified	Species richness	H'	J'	Reference
LAKES					
Lake Vilslache	993	75	3.61	0.84	Michiels and Traunspurger (2005)
Lake Traunsalpsee	896	48	2.32	0.60	Michiels and Traunspurger (2005)
Lake Vilsalpsee	989	33	2.39	0.68	Michiels and Traunspurger (2005)
Lake Rehbach	991	68	3.24	0.76	Michiels and Traunspurger (2005)
Lake Soiernsee	997	31	2.12	0.62	Michiels and Traunspurger (2005)
Lake Fereinsalm	1003	47	2.16	0.57	Michiels and Traunspurger (2005)
Lake Wildensee	707	38	1.96	0.54	Michiels and Traunspurger (2005)
Lake Schmalsee	1001	28	1.69	0.51	Michiels and Traunspurger (2005)
Lake Ferchensee	915	24	2.15	0.67	Michiels and Traunspurger (2005)
Lake Lautersee	953	34	1.88	0.53	Michiels and Traunspurger (2005)
Lake Froschhausersee	892	39	2.80	0.76	Michiels and Traunspurger (2005)
Lake Königssee					
1 m	5340	51	2.10	0.37	Traunspurger (1996a)
2 m	10636	49	1.02	0.18	Traunspurger (1996a)
5 m	17530	48	1.23	0.22	Traunspurger (1996a)
10 m	2956	49	1.95	0.35	Traunspurger (1996a)
15 m	2039	51	2.05	0.36	Traunspurger (1996b)
20 m	1316	44	2.06	0.38	Traunspurger (1996b)
30 m	1982	59	2.57	0.44	Traunspurger (1996b)
60 m	1548	46	2.50	0.45	Traunspurger (1996b)
120 m	1151	54	2.34	0.41	Traunspurger (1996b)
190 m	765	31	1.92	0.39	Traunspurger (1996b)
RIVERS					
Seveso, site 1	854	55	3.97	0.53	Zullini (1976)
Seveso, site 2	798	57	2.99	0.58	Zullini (1976)
Seveso, site 3	572	28	2.92	0.62	Zullini (1976)
Seveso, site 4	1651	37	3.53	0.68	Zullini (1976)
Seveso, site 5	1133	24	2.52	0.55	Zullini (1976)
Carrega Wood, site 1	1201	37	2.5	0.49	Zullini and Ricci (1980)
Carrega Wood, site 2	1203	41	2.9	0.59	Zullini and Ricci (1980)
Mur	1453	23	2.37	0.52	Kirchengast (1984)

Continued

Table 3.4 *Continued.* Species richness, Shannon–Wiener Index H' and evenness J' in selected lakes of Germany and selected rivers.

	Individuals identified	Species richness	H'	J'	Reference
Danube	7518	51	4.20	0.74	Eder (1983)
Rhine (genera)	8032	40	2.99	0.56	Bongers and Van de Haar (1990)
Mississippi	n.a.	23	3.71	0.82	Anderson (1992)
Krähenbach	1207	71	3.96	0.65	Beier and Traunspurger (2003a)
Körsch	12042	113	3.28	0.81	Beier and Traunspurger (2003b)

minimum standard of abiotic parameters in taxonomic studies on nematodes are needed for comparative purposes and, in general, for a better scientific standard. The generally high variability of meiobenthos in freshwater habitats with respect to space and time (Traunspurger, 1996a; Pennak, 1988; Michiels and Traunspurger, 2004, 2005) also elucidates the problem of drawing conclusions from comparisons.

In this regard the contributions of Prejs and Lazarek (1988), Tudorancea and Zullini (1989), Ocaña (1991) and Ocaña and Morales (1992) stand out. In these studies the potential factors that may control the distribution of species within these lakes were pH (Prejs and Lazarek, 1988), conductivity (Tudorancea and Zullini, 1989) and water chemistry (Ocaña, 1991; Ocaña and Morales, 1992).

Prejs (1977a,b) tried to explain the variation she observed in the number of nematode species in a group of lakes using the level of primary productivity. She concluded that oligotrophic lakes contain more species than eutrophic ones.

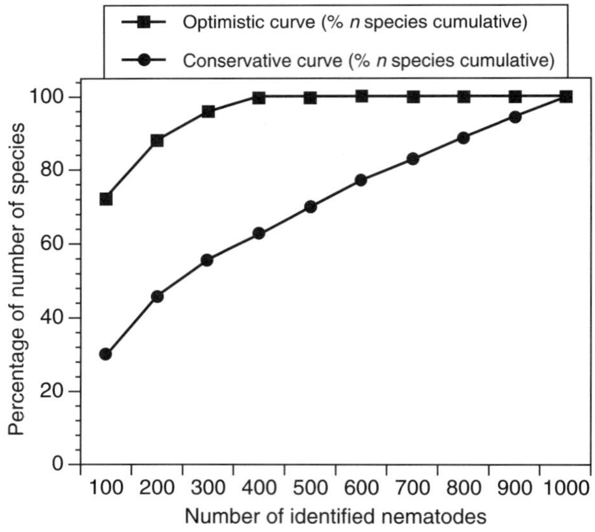

Fig. 3.3 Optimistic and conservative ranking of the percentage of species number of 10 replicate samples of 11 alpine lakes (sample size of each lake >1000 individuals).

Wu and Liang (1999) compared the nematofauna of two shallow lakes: Lake Houhu with a phytoplankton-dependent system and Lake Biandantang with a macrophyte-dependent system. Their study revealed that the macrophyte-dependent system maintained a significantly higher (51 species) total species number than the phytoplankton-dependent system (36 species), suggesting habitat heterogeneity as an important aspect affecting limno-nematofauna. Traunspurger (2002) showed that the relative abundance of dominant genera of freshwater nematodes is partly related to the trophic status of the lakes. The nematode community of the lakes, listed in Table 3.2, is presented in Fig. 3.4.

Sediment granulometry also plays a role, though its role is supposed to be more significant in explaining distribution mainly within a water body. Strayer *et al.* (1997) support the previous finding that oxygen, sediment granulometry and food supply are important in determining the distribution of hyporheic animals. A negative and significant correlation between nematode abundance with chloride was found in Krähenbach, a small, fine-grained, submountain carbonate stream, and a negative and significant correlation between deposit-feeders with ammonia. *Tobrilus* cf. *pellucidus* showed a significant negative correlation with nitrate in Krähenbach (Beier and Traunspurger, 2003a). In Körsch, a small, coarse-grained, submountain carbonate stream, all nematodes and suction feeders (mainly hyphal and plant feeding) showed a positive correlation with oxygen (Beier and Traunspurger, 2003b). These examples show that there are only a few studies available, which try to explain the distribution of nematodes and their species by abiotic factors. However, despite these reports, the basic factors that control nematode distribution at global or regional scales in limnetic habitats remain enigmatic.

Biotic factors

One of the very few studies on biotic factors controlling diversity and community of nematodes was carried out by Michiels *et al.* (2004). They showed that the ratio of predatory to non-predatory nematode abundance was positively correlated with nematode species number (Fig. 3.5) in limnetic habitats. The number of nematode species from 13 investigated alpine lakes increased from 24 to 75 when the proportion of predatory to non-predatory nematodes increased from 0.02 to 0.26. Though the relationship between nematode predator–prey ratio and species richness is not completely independent, species richness was not limited by the number of identified individuals per lake, and nematode abundance and species richness were not correlated in that dataset. The decisive factor that increased the number of coexisting species was predator abundance but not the number of predatory species. This example could be an indication that predation within nematode communities, among other factors, strongly prevents competitive exclusion.

Another important question is if there is competition among freshwater nematode species and consequently a shift in nematode community during the year. In Lake Königssee, competition seems to take place, e.g. between *Monhystera* and *Eumonhystera* (see Traunspurger, 2002). Generally, *Monhystera* dominates during winter while *Eumonhystera* reaches high abundances during the summer. Reanalysing the spatial distribution of the dominant species of *Eumonhystera* in Lake Königssee

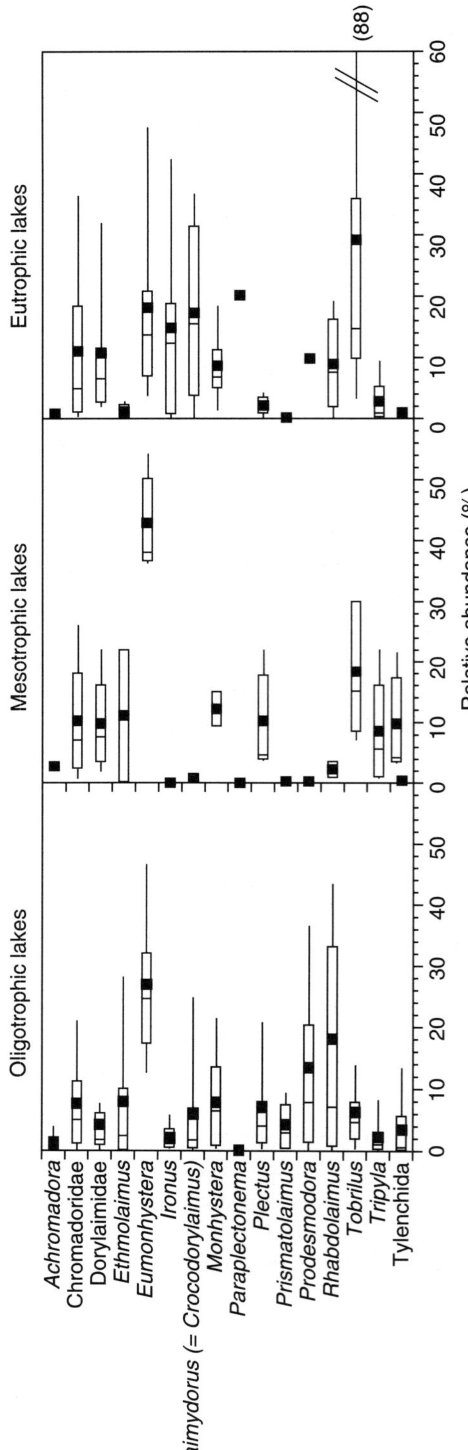

Fig. 3.4 Relative abundance of dominant nematode species in lakes with different productivity.

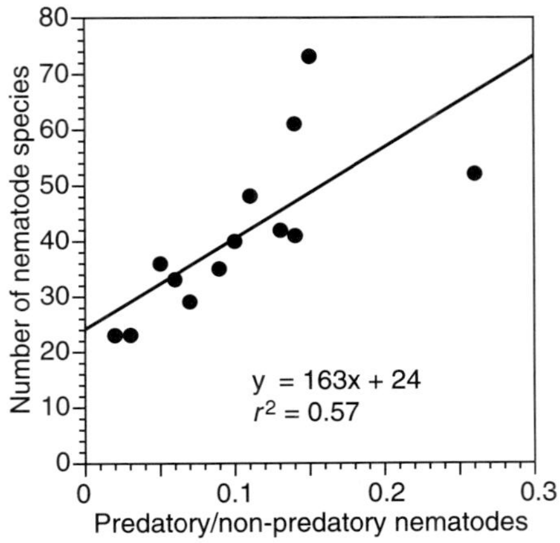

Fig. 3.5 Influence of predator–prey ratios on nematode species richness.

(Traunspurger, 1991) gives some insight into possible competition and/or coexistence among species of the same genus. The mean abundance of the dominant *Eumonhystera* species varied in the ten investigated water depths of Lake Königssee: *E. filiformis*, *E. similis* and *E. simplex* dominated at 1 and 5 m water depths, while *E. longicaudatula* dominated at 10 m (Fig. 3.6).

We know only a little about which biotic and abiotic factors regulate the distribution of nematode species. But the examples of *Eumonhystera* and *Monhystera* showed that biotic factors (e.g. population dynamics, life cycles) may also play an important role on the distribution of species.

Factors Creating Variability within a Lake

Horizontal species distributions

Recent theoretical and empirical work suggested that the study of heterogeneity, defined as variability in a process or pattern over space or time, and of its ecological consequences, requires new and more complex conceptual frameworks and empirical approaches (Kolasa *et al.*, 1991). The literature basically shows two ways of describing horizontal nematode distribution within lakes: the use of depth as a factor (Prejs, 1977b; Tudorancea and Zullini, 1989; Traunspurger, 1996a–c, 1997, 1998) and the use of statistical methods to unravel assemblages within the nematofauna of the depths considered (Eyualem-Abebe *et al.*, 2001). Most meiobenthos have a heterogeneous spatial distribution (Coull, 1988), which often trails their food source (Lee *et al.*, 1977), or is influenced by the activity of large macrofauna at the centimetre- to metre-scale (e.g. Hall *et al.*, 1994; Blome *et al.*, 1999). Other studies also showed clear variation in composition over depth. However, the picture of the available data on the relationship of species numbers with water depth in lakes is not clear-cut.

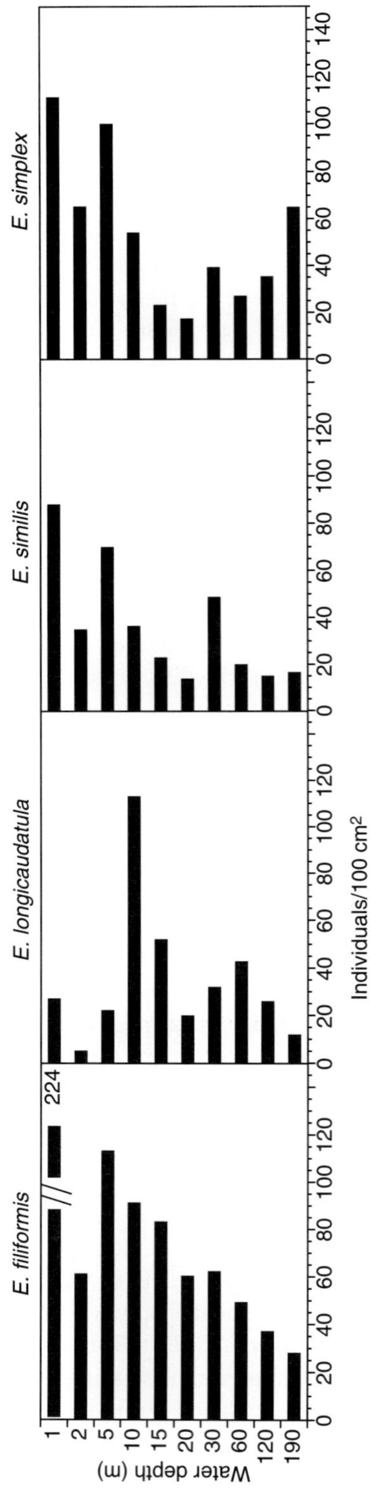

Fig. 3.6 Abundance of four dominant species of *Eumonhystera* in different water depths of Lake Königssee.

Species numbers at various depths remained constant in the littoral of Lake Königssee (Traunspurger, 1996a) and varied only slightly over water depth in the littoriprofundal and profundal zones (Traunspurger, 1996b). Nevertheless there was no 'simple' relationship of species numbers with water depth in these latter zones. Eyualem-Abebe *et al.* (2001) also reported similar inconclusive results with respect to depth-specific communities at some sites in Lake Tana, Ethiopia.

In some cases the complete absence of nematodes below a certain water depth was found. In Lake Höllerer See, a lake in Austria with a monimolimnion, no oxygen was measured below the chemocline (9–12 m water depth throughout the year). The distribution of nematode species reflects this picture: at least 21 species were identified at 1 m water depth ($n = 300$ individuals), 13 species at 6 m ($n = 167$) but only four species (*Monhystera* sp., *Tobrilus gracilis*, *Tobrilus* sp. and *Tripyla cf. glomerans*) at 8 m ($n = 128$). Only one specimen (*T. gracilis*), possibly drifted, was found in the anoxic zone below 10 m water depth. For Lake Höllerer See it can be stated that there does not exist a permanent nematode community in the anoxic monimolimnion (Hehl, 1996). The literature, on the other hand, clearly demonstrates the existence of nematode communities associated with specific parts or depths of lakes (Schiemer, 1978; Traunspurger, 1995, 1996a,b, 1997, 1998; Eyualem-Abebe *et al.*, 2001) (Fig. 3.7).

In his attempt to explain horizontal distribution patterns in Neusiedlersee, Schiemer (1978) shed some light on one of the fundamental factors that affect nematode distribution, i.e. mixing in lakes. He recognized five zones through a 'horizontal distribution pattern correlated with the wind influence on benthic conditions', emphasizing the role of local conditions that may affect mixing and sediment suspension. These factors may be closely related with depth, but are not controlled by water depth alone. Instead, a cocktail of potential factors interacts in the whole mixing process including wind regime, water depth and local physical conditions.

Wind regime, for instance, could itself be seasonal and should be considered as such in the specific site. Water depth *per se* is meaningless unless it is taken within the context of local conditions. It only becomes important when supplied with information on lake bathymetry, as in Por and Masry (1968) and Traunspurger (1996a). The interpretation and the factors involved would undoubtedly be different depending on lake bathymetry. For instance, let us say we want to compare nematofauna of a lake at 5 and 10 m depths. Depths 5 and 10 m become meaningful only if we know how far apart the two sites are situated horizontally and what the slope of the bottom is like. There could be a sharp drop in depth over a 0.5-m horizontal distance on the one hand or there could be a distance of a mile between these two depths. The impact of such a difference on sediment dwelling forms is understandable. Nevertheless, such minute but important clues as to the forces behind observed nematode distribution patterns are, most often, either not reported or not considered in the studies at all. Eyualem-Abebe *et al.* (2001) have demonstrated that nematode communities of two closely situated sites could be different in shallow parts of a lake protected from wind-induced mixing, while nematode communities may be similar in distantly situated shallow sites exposed to regular, strong wind-induced mixing. These authors, as well as Prejs and Bernard (1985), showed that species numbers from sites that were less prone to disturbance were higher than those from exposed sites. Similar observations have been documented from the marine habitat. In the littoral zone of an intertidal sand flat

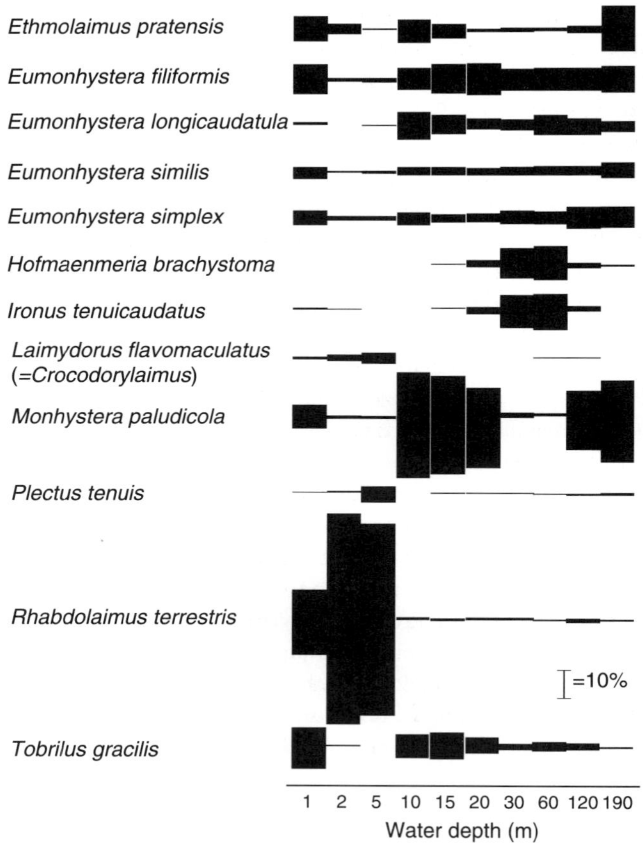

Fig. 3.7 Relative abundance (%) of dominant nematode species in different water depths of Lake Königssee.

higher diversity was recorded on a sheltered than on an exposed beach and diversity increased from a high tidal level to a low tidal level (Heip *et al.*, 1985). In addition, that beach stability controls meiofaunal biomass in the marine habitat is a long-standing notion (Gray and Rieger, 1971).

Vertical species distribution within the sediment

Though first reported from studies in marine habitats (Fenchel *et al.*, 1967; Tietjen, 1969; Bryant and Laybourn, 1974; Platt, 1977; Blome, 1983; Jensen, 1987), the general statement that nematodes tend to stay in the surface sediment layer in aquatic habitats has now become a universal fact. Various studies on limnetic habitats have assessed vertical distribution of nematodes in terms of species numbers (Holopainen and Paasivirta, 1977; Newrkla and Wijegoonawardana, 1987; Traunspurger 1996a,b; Traunspurger and Drews, 1996; Eyualem-Abebe *et al.*, 2001). The general conclusion emerging from these studies is that species numbers remain highest at or close to the surface sediment and decrease towards deeper sediment layers. For instance, of the 60 nematode species in the profundal of Lake Königssee, only two species, *E. pratensis* and *M. paludicola*, inhabited the deepest sediment layer of 10–20 cm (Traunspurger, 1996b, 1998).

Schiemer's (1978) work on the vertical distribution of limno-nematodes and environmental parameters was an important contribution to these factors. He demonstrated redox potential of the sediment, i.e. oxygen concentration, to be important in affecting vertical nematode distribution within the sediment. As yet, the existence of a truly anaerobic nematode community is still under question. Fenchel and Riedl (1970) claimed its existence, but Reise and Ax (1979) challenged this idea. These latter authors argued that meiofauna remained in close association with oxygen layers or pocket areas within the sediment (e.g. Wetzel *et al.*, 1995). Prein (1988) showed that *Pontonema vulgare* thrived on temporary anoxic conditions. However, more recent reports have demonstrated the survival and reproduction of nematodes in strict anoxic conditions (Riess *et al.*, 1999).

Irregular anoxia, even at the surface of the sediment, however, is most likely to happen in warmer, tropical waters. Oxygen is taken up fast for the decomposition of the ample organic matter in these habitats (Zinabu and Taylor, 1989). Temperate and arctic lakes that freeze for a long part of the year could also face oxygen shortage. Consequently, facultative anaerobic forms may be common in these habitats. Schiemer and Duncan (1973) demonstrated that some species (*T. gracilis*) could withstand anoxic conditions. Supporting evidence comes from studies of the two ancient lakes Baikal and Kenneret. In the case of Lake Baikal, the dominant forms are tobrilids (Tsalolikhin, 1975; Shoshin, 1999). Tobrilids and monhysterids are the two major groups that are claimed to be adapted to such an environment (Jacobs and Heyns, 1990). Moreover, without recognizing the existence of anaerobic or facultative anaerobic forms, it would indeed be difficult to convincingly explain the occurrence of nematode communities in the deepest parts of all other ancient lakes (e.g. Baikal, Tanganyika, Victoria, Malawi, Titicaca, Biwa and Ohrid). A case study is Por and Masry's (1968) ecological survey on the monomictic Lake Kinneret (Tiberias). Here, they demonstrated the adaptation of *Eudorylaimus andrassyi* to survive the anaerobic situation, reproduce and increase its abundance in the deeper, central part of the lake for 8 months every year. They also supported their observation by experimental data, in that this species survived anoxic conditions for a period of 6 months. These documented cases show us the role that oxygen plays in the distribution of nematodes in the sediment vertically as well as horizontally. Further studies of the nematofauna of previously unexplored environments, such as the deepest parts of some ancient lakes, may reveal interesting data in this regard.

Conclusions

Freshwater nematodes, because of their high abundances and species richness, are an ideal group of organisms to test ecological hypotheses and to describe diversity of an ecosystem. There is a strong decrease in freshwater nematode abundance with increasing size in most investigated lakes. If the ubiquity of small-sized species is a general phenomenon in freshwater nematodes, we need to confirm it using additional methods like molecular identification. On small (within lakes) and on large (between areas) scales, species compositions resemble each other more than on intermediate scales (within regions). The number of species and quantitative composition of hyporheic and phreatic nematodes are little known and further studies are necessary.

Information on factors playing a role in distribution of free-living nematodes in inland water bodies is still scanty but the listed examples should encourage ecologists and nematologists to investigate potential factors.

References

Anderson, R.V. (1992) Free-living nematode associations in pool 19, Mississippi river. *Journal of Freshwater Ecology* 7, 243–250.

Andrássy, I. (1978) Nematoda. A checklist of the animals inhabiting European inland waters with accounts of their distribution and ecology. In: Illies, J. (ed.) *Limnofauna Europaea.* Gustav Fischer Verlag, Stuttgart, Germany, pp. 98–117.

Andrássy, I. (1981) Revision of the order Monhysterida (Nematoda) inhabiting soil and inland waters. *Opuscula Zoologica (Budapest)* 17/18, 13–47.

Andrássy, I. (1984) *Klasse Nematoda: Ordnungen Monhysterida, Desmoscolecida, Araeolaimida, Chromadorida, Rhabditida.* Fischer, Stuttgart, Germany, 509 pp.

Andrássy, I. (1992) A short census of free-living nematodes. *Fundamental and Applied Nematology* 15, 187–188.

Andrássy, I. (1999) A census of genera and subgenera of free-living nematodes. *Journal of Nematode Morphology and Systematics* 2, 45–68.

Beier, S. and Traunspurger, W. (2003a) Seasonal distribution of free-living nematodes in the stream Krähenbach: a fine-grained submountain carbonate stream in Southwest Germany. *Nematology* 5, 113–136.

Beier, S. and Traunspurger, W. (2003b) Seasonal distribution of free-living nematodes in the stream Körsch: a coarse-grained submountain carbonate stream in Southwest Germany. *Nematology* 5, 481–504.

Bergtold, M. and Traunspurger, W. (2004) The benthic community in the profundal of Lake Brunnsee: seasonal and spatial patterns. *Archiv für Hydrobiologie* 160, 527–554.

Biró, K. (1973) Nematodes of Lake Balaton. 4. Seasonal qualitative and quantitative changes. *Annales Instituti Biologici (Tihany) Hungaricae Academiae Scientiarum* 40, 135–158.

Blaxter, M., Iredale, I., Roche, R., Floyd, R. and Eyualem-Abebe, A. (2002) Molecular surveys of marine nematode diversity. *Nematology* 4, 185.

Blome, D. (1983) Ökologie der Nematoda eines Sandstrandes der Nordseeinsel Sylt. *Mikrofauna des Meeresbodens* 88, 1–76.

Blome, D., Schleier, U. and von Bernem, K.-H. (1999) Analysis of the small scale spatial patterns of free-living marine nematodes from tidal flats in the East Frisian Wadden Sea. *Marine Biology* 133, 717–726.

Bongers, T. and Van de Haar, J. (1990) On the potential of basing an ecological typology of aquatic sediments on the nematode fauna: an example from the River Rhine. *Hydrobiological Bulletin* 24, 37–45.

Bott, T.L. and Kaplan, L.A. (1989) Densities of benthic protozoa and nematodes in a Piedmont stream. *Journal of the North American Benthological Society* 8, 187–196.

Bryant, V.M.T. and Laybourn, J.E.M. (1974) The vertical distribution of Ciliophora and Nematoda in the sediments of Loch Leven, Kinross. *Proceedings of the Royal Society of Edinburgh* 17, 265–273.

Cobb, N.A. (1914) The North American free-living fresh-water nematodes. *Transactions of American Microscopical Society* 33, 69–134.

Coomans, A. (1989) Overzicht van de vrijlevende nematofauna van België (Nematoda) (The free-living nematofauna of Belgium: a review). In: Wouters, K. and Baert, L. (eds) *Proceedings of the Symposium 'Invertebrates of Belgium'*, Brussels, Belgium, pp. 43–56.

Coull, B.C. (1988) Ecology of the marine meiofauna. In: Higgins, R.P. and Thiel, H. (eds) *Introduction to the Study of Meiofauna*. Smithsonian Institution Press, Washington, DC, pp. 18–38.

Danielopol, D.L. (1976) The distribution of the fauna in the interstitial habitats of riverine sediments of the Danube and the Piesting (Austria). *International Journal of Speleology* 8, 23–51.

Decraemer, W. and Coomans, A. (1994) A compendium of our knowledge of the free-living nematofauna of ancient lakes. *Archiv für Hydrobiologie* 44, 173–181.

De Ley, P., Felix, M.-A., Frisse, L.M., Nadler, S.A., Sternberg, P.W. and Thomas, W.K. (1999) Molecular and morphological characterization of two reproductively isolated species with mirror-image anatomy (Nematoda: Cephalobidae). *Nematology* 1, 591–612.

Doucet, M.E. and de Doucet, M.M.A. (1999) Supplement to the list of soil and freshwater nematodes recorded in continental Argentina. *Nematology* 1, 217–220.

Eder, R. (1983) Nematoden aus dem Interstitial der Donube bei Fischamend (Niederösterreich). *Archiv für Hydrobiologie* Suppl. 68, 100–113.

Eisenmann, H., Traunspurger, W. and Meyer, E. (1998) Community structure of selected micro- and meiobenthic organisms in sediment chambers from a prealpine river (Necker, Switzerland). In: Bretschko, G. and Helesic, J. (eds) *Advances in River Bottom Ecology*, Backhuys Publishers, Leiden, The Netherlands, pp. 155–162.

Esser, R.P. and Buckingham, G.R. (1987) Genera and species of free-living nematodes occupying freshwater habitats in North America. In: *Vistas on Nematology*, Society of Nematologists, Hyattsville, Maryland, pp. 477–487.

Ettema, C.H. (1998) Soil nematode diversity: species coexistence and ecosystem function. *Journal of Nematology* 30, 159–169.

Eyualem-Abebe (2000) Free-living limnic nematofauna of Ethiopia: the species reported from sites out of the rift valley. *SINET, Ethiopian Journal of Science* 23, 273–297.

Eyualem-Abebe (2002) Free-living aquatic nematodes of the Ethiopian Rift Valley. In: Tudorancea, C. and Taylor, W.D. (eds) *Ethiopian Rift Valley Lakes*, Backhuys Publishers, Leiden, The Netherlands, pp. 143–156.

Eyualem-Abebe and Blaxter, M. (2003) Comparison of biological, molecular and morphological methods of species identification in a set of cultured *Panagrolaimus* isolates. *Journal of Nematology* 35, 119–128.

Eyualem-Abebe and Coomans, A. (1995) Freshwater nematodes of the Galápagos. *Hydrobiologia* 299, 1–51.

Eyualem-Abebe and Coomans, A. (1996a) Aquatic nematodes from Ethiopia I. The genus *Monhystera* Bastian, 1865 (Monhysteridae: Nematoda) with the description of four new species. *Hydrobiologia* 324, 1–51.

Eyualem-Abebe and Coomans, A. (1996b) Aquatic nematodes from Ethiopia II. The genus *Monhystrella* Cobb, 1918 (Monhysteridae: Nematoda) with the description of six new species. *Hydrobiologia* 324, 53–77.

Eyualem-Abebe and Coomans, A. (1996c) Aquatic nematodes from Ethiopia III. The genus *Eumonhystera* Andrássy, 1981 (Monhysteridae: Nematoda) with the description of *E. geraerti* n. sp. *Hydrobiologia* 324, 79–97.

Eyualem-Abebe and Coomans, A. (1996d) Aquatic nematodes from Ethiopia V. Descriptions of *Achromadora inflata* n. sp., *Ethmolaimus zullinii* n. sp. and *Prodesmodora nurta* Zullini, 1988 (Chromadorida: Nematoda). *Hydrobiologia* 332, 27–39.

Eyualem-Abebe and Coomans, A. (1996e) Aquatic nematodes from Ethiopia VI. The genera *Chronogaster* Cobb, 1913, *Plectus* Bastian, 1865 and *Prismatolaimus* de Man, 1880 with descriptions of *C. ethiopica* n. sp. and *C. getachewi* n. sp. (Chromadorida: Nematoda). *Hydrobiologia* 332, 41–61.

Eyualem-Abebe and Coomans, A. (1997a) Aquatic nematodes from Ethiopia VIII. Enoplids, with descriptions of *Brevitobrilus fesehai* n. sp. and *B. tsalolikhini* n. sp. (Enoplida: Nematoda). *Hydrobiologia* 345, 149–184.

Eyualem-Abebe and Coomans, A. (1997b) Aquatic nematodes from Ethiopia IX. Descriptions of three dorylaimid species, *Mononchus truncatus* Bastian, 1865 and *Diploscapter coronatus* (Dorylaimida: Nematoda). *Hydrobiologia.* 253, 121–138.

Eyualem-Abebe, Mees, J. and Coomans, A. (2001) Nematode communities of Lake Tana and other inland water bodies of Ethiopia. *Hydrobiologia* 462, 41–73.

Fenchel, T. and Riedl, R.J. (1970) The sulfide system: a new biotic community underneath the oxidized layer of marine sand bottoms. *Marine Biology* 7, 255–268.

Fenchel, T., Jansson, B.-O. and von Thun, W. (1967) Vertical and horizontal distribution of the metazoan microfauna and of some physical factors in a sandy beach in the northern part of the Öresund. *Ophelia* 4, 227–243.

Finlay, B.J. (2002) Global dispersal of free-living microbial eukaryote species. *Science* 296, 1061–1063.

Floyd, R., Eyualem-Abebe, Papert, A. and Blaxter, M. (2002) Molecular barcodes for soil nematode identification. *Molecular Evolution* 11, 839–850.

Gaston, K.J. (1996) The multiple forms of interspecific abundance–distribution relationship. *Oikos* 76, 211–220.

Gaston, K.J. (2000) Global patterns in biodiversity. *Nature* 405, 220–227.

Gaston, K.J., Blackburn, T.M. and Lawton, J.H. (1997) Interspecific abundance–range size relationships: an appraisal of mechanisms. *The Journal of Animal Ecology* 66, 579–601.

Gerlach, A.S. and Riemann, F. (1973/74) The Bremerhaven checklist of aquatic nematodes. A catalogue of Nematoda Adenophorea excluding the Dorylaimida. *Veröffentlichungen des Instituts für Meeresforschung in Bremerhaven*, Suppl. 4, Heft 1+2, 736 pp.

Gray, J.S. and Rieger, R.M. (1971) A quantitative study of the meiofauna of an exposed sandy beach, at Robin Hood's Bay, Yorkshire. *Journal of the Marine Biological Association of the United Kingdom* 52, 1–19.

Hall, S.J., Raffaelli, D. and Trush, S.F. (1994) Patchiness and disturbance in shallow water benthic assemblages. In: Giller, P.S., Hildrew, A.G. and Raffaelli, D.G. (eds) *Aquatic Ecology: Scale, Pattern, and Process.* Blackwell Scientific Publications, London, pp. 333–375.

Hehl, I. (1996) Quantitative Erfassung und Verteilung des Meiobenthos dreier Seen des Alpenvorlandes (Brunnsee, Klostersee und Höllerer See). Diplomarbeit an der Ludwig-Maximilians Universität München, Fakultät Biologie.

Heip, C., Vincx, M. and Vranken, G. (1985) The ecology of marine nematodes. *Oceanography and Marine Biology: An Annual Review* 23, 399–489.

Hirschmann, H. (1952) Die Nematoden der Wassergrenze mittelfränkischer Gewässer. *Zoologische Jahrbücher (Systematik)* 81, 313–436.

Holopainen, I.J. and Paasivirta, L. (1977) Abundance and biomass of the meiozoobenthos in the oligotrophic and mesohumic lake Pääjärvi, southern Finland. *Annales Zoologici Fennici* 14, 124–134.

Holovachov, O., Zullini, A., Loof, P.A.A. and Bongers, T. (2003) Morphology and systematics of the genus *Anonchus* Cobb, 1913 (Nematoda: Leptolaimina) and reappraisal of the family Aphanolaimidae Chitwood, 1936 n. rank. *Nematology* 4, 725–758.

Hummon, W.D., Evans, W.A., Hummon, M.R., Doherty, F.G., Wainberg, R.H. and Stanley, W.S. (1978) Meiofaunal abundance in sandbars of acid mine polluted, reclaimed, and unpolluted streams in southeastern Ohio. In: Thorp, J.H. and Gibbons, J.W. (eds) *Energy and Environmental Stress in Aquatic Ecosystems*, DOE Symposium Series (CONF-771114), National Technical Information Service, Springfield, Virginia, pp. 188–203.

Hurlbert, S.H. (1971) The non-concept of species diversity: a critique and alternative parameters. *Ecology* 52, 577–586.

Jacobs, L.J. (1984) The free-living inland aquatic nematodes of Africa: a review. *Hydrobiologia* 113, 259–291.

Jacobs, L.J. and Heyns, J. (1990) An ecological strategy in the genus *Monhystera* – an hypothesis. *Revue de Nematologie* 13, 109–111.

Jensen, P. (1987) Feeding ecology of free-living aquatic nematodes. *Marine Ecology Progress Series* 35, 187–196.

Jordan, S., Shiozawa, D.K. and Schmid-Araya, J.M. (1999) Benthic invertebrates of a large, sandy river system: the Green and Colorado Rivers of Canyonlands National Park, Utah. *Archiv für Hydrobiologie* 147, 91–127.

Kirchengast, M. (1984) Faunistische Untersuchungen im hyporheischen Interstitial des Flusses Mur (Steiermark, Österreich). *Internationale Revue der Gesamten Hydrobiologie* 69, 729–746.

Kolasa, J., Allen, T.F.H. and Pickett, S.T.A. (1991) *Ecological Heterogeneity*. Springer-Verlag, New York, 332 pp.

Lambshead, P.J.D., Platt, H.M. and Shaw, K.M. (1983) The detection of differences among assemblages of marine benthic species based on an assessment of dominance and diversity. *Journal of Natural History* 17, 859–874.

Lawton, J.H. (1993) Range, population abundance and conservation. *Trends in Ecology and Evolution* 8, 409–413.

Lee, J.J., Tietjen, J.H., Mastropaolo, C. and Rubin, H. (1977) Food quality and the heterogeneous spatial distribution of Meiofauna. *Helgoländer Wissenschaftliche Meeresuntersuchungen* 30, 272–282.

Lemzina, L.V. (1995) Ecology of aquatic nematodes from Kryrghyz Republic (abstract). *First English Language International Symposium of the Russian Society of Nematologists*, 20–30 September 1995, St. Petersburg, Russia. (http://www.scri.sari.ac.uk/rjn/lemz.htm).

Magurran, A.E. (1988) *Ecological Diversity and Its Measurement.* Princeton University Press, Princeton, New Jersey, 179 pp.

Michiels, I. and Traunspurger, W. (2004) A three-year study of seasonal dynamics of a zoobenthos community in a eutrophic lake. *Nematology* 6, 655–669.

Michiels, I. and Traunspurger, W. (2005) Benthic community patterns and the composition of feeding-types and reproductive modes in nematodes. *Nematology* 7, 21–36.

Michiels, I.C., Matzak, S. and Traunspurger, W. (2004) Maintenance of biodiversity through predation in freshwater nematodes? *Nematology Monographs and Perspectives* 2, 723–737.

Mittelbach, G.G., Steiner, C.F., Scheiner, S.M., Gross, K.L., Reynolds, H.L., Waide, R.B., Willig, M.R., Dodson, S. and Gough, L. (2001) What is the observed relationship between species richness and productivity? *Ecology* 82, 2381–2396.

Newrkla, P. and Wijegoonawardana, N. (1987) Vertical distribution and abundance of benthic invertebrates in profundal sediments of Mondsee, with special reference to oligochaetes. *Hydrobiologia* 155, 227–234.

Niemann, R., Arens, M., Koczwara, K. and Sturhan, D. (1996) Untersuchungen über die Eignung von Nematoden zur Gütebewertung von Flieβgewässern. *Mitteilungen der biologischen Bundesanstalt für Land-Forstwirtschaft*, Berlin-Dahlem 317, 195–208.

Ocaña, A. (1991) A redescription of two nematode species found in hot springs. *Nematologia Mediterranea* 19, 173–175.

Ocaña, A. and Morales, R. (1992) The influence of ionic composition on the distribution of nematode species in springs of the province of Granada (Spain). *Hydrobiologia* 237, 81–92.

Palmer, M.A. (1990) Temporal and spatial dynamics of meiofauna within the hyporheic zone of Goose Greek, Virginia. *Journal of the North American Benthological Society* 9, 17–25.

Pehofer H.E. (1989) Spatial distribution of the nematode fauna and production of three nematodes (*Tobrilus gracilis*, *Monhystera stagnalis*, *Ethmolaimus pratensis*) in the profundal of Piburger See (Austria, 913 m a.s.l.). *Internationale Revue der gesamten Hydrobiologie* 74, 135–168.

Pennak, R.W. (1988) Ecology of the freshwater meiofauna. In: Higgins, R.H. and Thiel, H. (eds) *Introduction to the Study of Meiofauna.* Smithsonian Institution Press, London, pp. 211–230.

Pennak, R.W. and Ward, J.V. (1986) Interstitial faunal communities of the hyporheic and adjacent groundwater biotopes of a Colorado mountain stream. *Archiv für Hydrobiologie,* Suppl. 74, 356–396.

Platt, H.M. (1977) Vertical and horizontal distribution of free-living marine nematodes from Strangford Lough, Northern Ireland. *Cahiers de Biologie Marine* 18, 261–273.

Por, F.D. and Masry, D. (1968) Survival of a nematode and an oligochaete species in the anaerobic benthal of Lake Tiberias. *Oikos* 19, 388–391.

Powers, T.O., Todd, T.C., Burnell, A.M., Murray, P.C.B., Fleming, C.C., Szalanski, A.L., Adams, B.A. and Harris, T.S. (I997) The rDNA internal transcribed spacer region as a taxonomic marker for nematodes. *Journal of Nematology* 29, 441–450.

Prein, M. (1988) Evidence of a scavenging lifestyle in the free-living nematode *Pontonema vulgare* (Enoplida, Oncholaimidae). *Kieler Meeresforschung, Sonderheft* 6, 389–394.

Prejs, K. (1970) Some problems of the ecology of benthic nematodes (Nematoda) of Mikolajskie Lake. *Ekologia Polska* 18, 225–242.

Prejs, K. (1977a) The littoral and profundal benthic nematodes of lakes with different trophy. *Ekologia Polska* 25, 21–30.

Prejs, K. (1977b) The species diversity, numbers and biomass of benthic nematodes in central part of lakes with different trophy. *Ekologia Polska* 25, 31–44.

Prejs, K. and Bernard, B. (1985) Meiobenthos of man-made lake Zegrzynskie. *Ekologia Polska* 33, 499–509.

Prejs, K. and Lazarek, S. (1988) Benthic nematodes in acidified lakes: case of a neglected grazer. *Hydrobiologia* 169, 193–197.

Price, N.S. and Siddiqi, M.R. (1994) Rainforest nematodes with particular reference to the Korup National Park, Cameroon. *Afro-Asian Journal of Nematology* 4, 117–128.

Reise, K. and Ax, P. (1979) A meiofaunal 'thiobios' limited to the anaerobic sulfide system of marine sand does not exist. *Marine Biology* 54, 225–237.

Riemann, F. (1966) Die interstitielle Fauna im Elbe-Aestuar. Verbreitung und Systematik. *Archiv für Hydrobiologie*, Suppl. 31, 1–279.

Riess, W., Giere, O., Khols, O. and Sarbu S.M. (1999) Anoxic thermomineral cave waters and bacterial mats as habitat for freshwater nematodes. *Aquatic Microbial Ecology* 18, 157–164.

Routledge, R.D. (1980) Bias in estimating the diversity of large, uncensused communities. *Ecology* 61, 276–281.

Schiemer, F. (1978) Verteilung und Systematik der freilebenden Nematoden des Neusiedlersees. *Hydrobiologia* 58, 167–194.

Schiemer, F. and Duncan A. (1973) The oxygen consumption of a freshwater benthic nematode, *Tobrilus gracilis* (Bastian). *Oecologia* 15, 121–126.

Schiemer, F., Löffler, H. and Dollfuss, H. (1969) The benthic communities of Neusiedlersee (Austria). *Verhandlungen der Internationalen Vereinigung für Theoretische und Angewandte Limnologie* 17, 201–208.

Schmid-Araya, J.M. (1997) Temporal and spatial dynamics of meiofaunal assemblages in the hyporheic interstitial of a gravel stream. In: Gibert, J., Mathieu, J. and Fournier F. (eds) *Groundwater/Surfacewater Ecotones: Biological and Hydrological Interactions and Management Options.* Cambridge Universitiy Press, Cambridge, Unesco, New York, pp. 29–36.

Schmid-Araya, J.M. and Schmid, P.E. (1995) The invertebrate species of a gravel stream. *Jahresberichte der Biologischen Station Lunz* 15, 11–21.

Shoshin, A.V. (1999) The Baikal Lake nematode fauna: the present and prospects. Paper presented at the third symposium of English speaking nematologists in St. Petersburg, Russia, 21–27 August (http://www.zin.ru/conferences/ns/).

Stirling, G. and Wilsey, B. (2001) Empirical relationships between species richness, evenness, and proportional diversity. *American Naturalist* 158, 286–299.

Strayer, D. (1985) The benthic micrometazoans of Mirror Lake, New Hampshire. *Archiv für Hydrobiologie*, Suppl. 72, 287–426.

Strayer, D.L., May, S.E., Nielsen, P., Wollheim, W. and Hausam, S. (1997) Oxygen, organic matter, and sediment granulometry as controls on hyporheic animal communities. *Archiv für Hydrobiologie* 140, 131–144.

Tietjen, J.H. (1969) The ecology of shallow water meiofauna in two New England estuaries. *Oecologia* 2, 251–291.

Tockner, K. and Bretschko, G. (1996) Spatial distribution of particulate organic matter (POM) and benthic invertebrates in a river-floodplain transect (Danube, Austria): importance of hydrological connectivity. *Archiv für Hydrobiologie*, Suppl. 115, 11–27.

Traunspurger, W. (1991) Fischbiologie des Königssees. Nahrungsangebot und Nahrungswahl. Band I: Das Meiobenthos des Königssees. Forschungsbericht 22, Nationalpark Berchtesgaden, Berchtesgaden, 152 pp.

Traunspurger, W. (1992) A study of the free-living freshwater nematodes of hard substrates in the littoral of the oligotrophic Königssee (National Park Brchtesgaden, F.R.G.). *Spixiana* 15, 233–238.

Traunspurger, W. (1995) Autecology of *Rhabdolaimus terrestris* De Man – a dominant freshwater nematode in an oligotropic lake. *Nematologica* 41, 603–616.

Traunspurger, W. (1996a) Distribution of benthic nematodes in the littoral of an oligotrophic lake (Königssee, National Park Berchtesgaden, FRG). *Archiv für Hydrobiologie* 135, 393–412.

Traunspurger, W. (1996b) Distribution of benthic nematodes in the littoriprofundal and profundal of an oligotrophic lake (Königssee, National Park Berchtesgaden, FRG). *Archiv für Hydrobiologie* 135, 555–575.

Traunspurger, W. (1996c) Autecology of *Monhystera paludicola* De Man, 1880 – Seasonal, bathymetric and vertical distribution of a free-living nematode in an oligotrophic lake. *Internationale Revue der gesamten Hydrobiologie* 81, 199–211.

Traunspurger, W. (1997) Distribution, seasonal occurrence and vertical pattern of *Tobrilus gracilis* (Bastian, 1865) and *T. medius* (Schneider, 1916). *Nematologica* 43, 59–81.

Traunspurger, W. (1998) Distribution and sex ratio of *Ethmolaimus pratensis* De Man, 1880 (Nematoda, Chromadorida) in an oligotrophic lake. *Nematologica* 44, 391–408.

Traunspurger, W. (2000) The biology and ecology of lotic nematodes. *Freshwater Biology* 44, 29–45.

Traunspurger, W. (2001) Meiobenthos – Sedimentbewertung anhand der Nematoden. In: Bayerisches Landesamt für Wasserwirtschaft (ed) *Kleinseen in Bayern – Ökologische Bewertung von Freiwasser, Sediment, Ufer und Einzugsgebiet*, Informationsberichte des Bayerischen Landesamtes für Wasserwirtschaft, München, Heft 1/01, pp. 124–150.

Traunspurger, W. (2002) Nematoda. In: Rundle, S.D., Robertson, A.L. and Schmid-Araya, J.M. (eds) *Freshwater Meiofauna: Biology and Ecology*. Backhuys Publishers, Leiden, The Netherlands, pp. 63–104.

Traunspurger, W. (2003) Nematoda. In: Bayerisches Landesamt für Wasserwirtschaft (ed.) *Taxaliste der Gewässerorganismen Deutschlands*. Informationsberichte des Bayerischen Landesamt für Wasserwirtschaft, München, Heft, 01/03, pp. 285–290.

Traunspurger, W. and Drews, C. (1996) Vertical distribution of benthic nematodes in an oligotrophic lake: seasonality, species and age segregation. *Hydrobiologia* 331, 33–42.

Traunspurger, W. and Weischer, B. (1993) Freilebende Süßwassernematoden aus dem Uferbereich im Naturschutzgebiet 'Heiliges Meer'. *Natur und Heimat* 53, 83–91.

Tsalolikhin, S.J. (1975) New nematodes from the abyssal zone of Lake Baikal. *Zoologicheskii Zhurnal* 54, 771–775.

Tsalolikhin, S.J. (1988) A nematological miscellany. *Proceedings of the Zoological Institute of the USSR Academy of Sciences* 180, 139 .

Tsalolikhin, S.J. (1996) Review of the fauna of free living nematodes from inland waters of Ethiopia. *Zoosystematica Rossica* 4, 205–218.

Tsalolikhin, S.J. (1998) On nematode fauna of high mountain lakes in the Pamir and Himalayas. *Zoologicheskii Zhurnal* 77, 1058–1063.

Tudorancea, C. and Zullini, A. (1989) Associations and distribution of benthic nematodes in the Ethiopian Rift Valley lakes. *Hydrobiologia* 179, 81–96.

Wanless, F.R. and Hunter, R. (2001) Freshwater nematodes from Loch Ness, Scotland. Part I. The orders Tylenchida Thorne, 1949 and Rhabditida Chitwood, 1933 (Nematoda, Secernentea). *Bulletin of the Natural History Museum (Zoology Series)* 67, 1–24.

Wetzel, M.A., Jensen, P. and Giere, O. (1995) Oxygen/sulfide regime and nematode fauna associated with *Arenicola marina* burrows: new insights in the thiobios case. *Marine Biology* 124, 301–312.

Whitman, R.L. and Clark, W.J. (1984) Ecological studies of the sand-dwelling community of an East Texas Stream. *Freshwater Invertebrate Biology* 3, 59–79.

Wu, J.H. and Liang, Y.L. (1999) A comparative study of benthic nematodes in two Chinese lakes with contrasting sources of primary production. *Hydrobiologia* 411, 31–37.

Wu, J.H. and Liang, Y.L. (2000) Inland free-living nematodes in China: a historical review of taxonomic studies with descriptions of *Aphanonchus orientalis* sp. nov. (Chromadorida: Leptolaimidae) and *Daptonema limnobia* sp. nov. (Monhysterida: Xyalidae). *Annales Zoologici* 50, 307–319.

Yun Liang, P., Eyualem-Abebe, and Coomans, A. (2002) A new species of *Eumonhystera* and three species of *Monhystera* (Monhysteridae: Nematoda) from Li River, China. *Hydrobiologia* 468, 83–93.

Zinabu, G. and Taylor, W.D. (1989) Seasonality and spatial variation in abundance, biomass and activity of heterotrophic bacterioplankton in relation to some biotic variables in an Ethiopian rift-valley lake (Awassa). *Freshwater Biology* 22, 355–368.

Zullini, A. (1976) Nematodes as indicators of river pollution. *Nematologia Mediterranea* 4, 13–22.

Zullini, A. and Ricci, C. (1980) Bdelloid rotifers and nematodes in a small Italian stream. *Freshwater Biology* 10, 67–72.

Zullini, A., Loof, P.A.A. and Bongers, T. (2002) Free-living nematodes from nature reserves in Costa Rica. 3. Araeolaimida. *Nematology* 4, 709–724.

4 Dynamics of Freshwater Nematodes: Abundance, Biomass and Diversity

EYUALEM-ABEBE,[1] W. TRAUNSPURGER[2] AND I.C. MICHIELS[2]

[1]*Hubbard Centre for Genome Studies and Department of Zoology, University of New Hampshire, 35 Colovos Rd, Gregg Hall, Durham, NH 03824, USA;* [2]*Department of Animal Ecology, University of Bielefeld, Morgenbreede 45, 33615 Bielefeld, Germany*

Introduction

Earlier research on benthic forms has led to some generalizations (Lundbeck, 1936 in Brinkhurst *et al.*, 1974) with regard to basic expected horizontal distribution patterns that are characteristic of major lake productivity divisions, though admittedly these were highly idealized and focused on macroinvertebrate groups (Fig. 4.1). The first question that comes to mind, when considering this largely ignored group, i.e. nematodes, therefore is, 'do they conform to these generalizations?' The answer may not be direct for the reason that data are still incomplete or that they actually do not follow those idealized patterns. In spite of it, we present the available data on limno-nematodes and compare them to those idealized patterns in the section on 'Within-lake horizontal variation of abundance'. However, before dealing with this comparison we prefer to digress for a moment from our main focus of discussion, i.e. nematodes, to the habitats in which they live, i.e. the water bodies, to give a brief account of one of the well-documented, inherent aspects of limnetic habitats: effects of longitude and water movement on lake dynamics.

Inherent Dynamics in Limnetic Habitats: Temperate *vs* Tropics

Seasonality of the major environmental factors affecting faunistic and floristic composition, abundance, diversity and production in limnetic habitats has long been recognized and documented. Seasonality, however, functions within the context of local conditions. Global observations show that temperature is the singularly most important factor that fluctuates seasonally and in turn affects denizens of limnetic habitats through its multifaceted effects (e.g. directly by affecting the biotic component and by impacting mixing). Though temperature influences life in limnetic habitats in both temperate regions and the tropics, this, however, is more so in

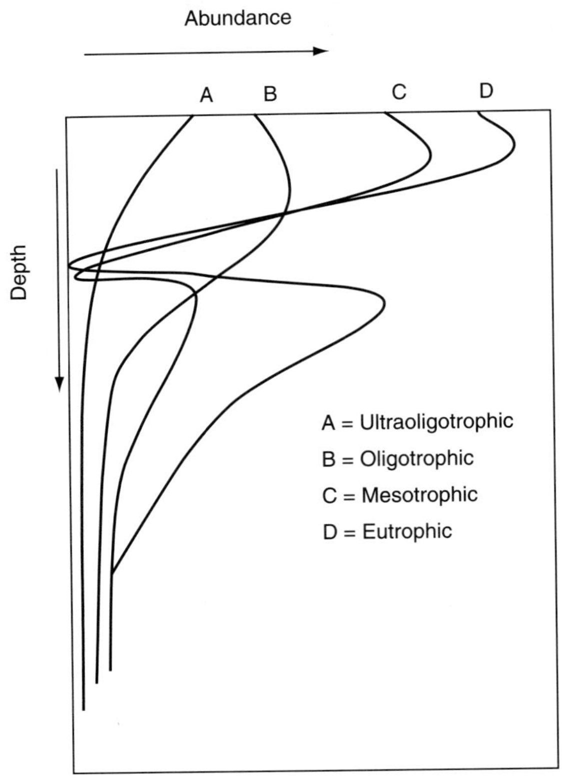

Fig. 4.1 An idealized depth distribution of benthos in relation to lake typology. (From Lundbeck, 1936 in Brinkhurts *et al.*, 1974; Courtesy of *Archiv für Hydrobiologie*).

temperate habitats than in the tropics. In the former, temperature fluctuation is clearly seasonal, but in the latter temperature fluctuates little over an annual cycle. In the tropics, temperature fluctuation is diurnal, which in some cases could be much higher than a similar diurnal or even annual fluctuation in the temperate region. This, coupled with the influence of other diurnal factors such as wind regimes that cause mixing, makes diurnal changes as important as seasonal ones in tropical limnetic habitats. Temperature, in addition to its direct effect on life in limnetic habitats, also fundamentally affects life in lakes through its effect on stratification. Local conditions such as lake morphometry and annual water budget affect temperature and density stratification, which ultimately affect nutrient cycling and the mixing regime of lakes. These mixing regimes, also known as *circulation*, are characteristic of lakes, be it in the tropics or in temperate areas. Some deep lakes may never mix at all (amictic lakes), serving as nutrient sinks, some mix once a year (monomictic), some twice a year (dimictic) and others many times a year (polymictic) depending on local conditions. Polymictic conditions are typically found in the tropics. Special cases are temperate lakes with a monimolimnion (a part of a lake that does not mix).

In the tropics, climatic factors such as rainfall and solar radiation play equally important roles. Rainfall, however, remains the most important seasonal factor in these habitats through its effect on inflow, which in turn influences water level, salinity, mixing and nutrient load.

The reason why we need to consider environmental dynamics and associated factors in limnetic habitats is primarily because they affect nematode dynamics in these habitats and nematological research has grasped little of their impact. First, the direct influence of these environmental factors and their change on nematodes needs to be understood better. Second, any seasonal change in lake primary productivity will influence all those animals at any level that may be seasonally limited by food. Third, nematodes thrive on and in the bottom sediment of these lakes and mixing moves the sediment, be it daily or once or twice annually. This undoubtedly will influence the dispersal and ultimately the spatial distribution of animals and plants in the sediment across depths. But it also may impact vertical distribution of various other important factors such as temperature, oxygen, nutrients for plankton and food for animals. These factors affect nematodes directly or indirectly.

It has been argued (Hildrew, 1992; Schmid-Araya and Schmid, 2000) that disturbance is an important factor in preserving diversity far short of the competitively excluding state, implying that environmental variability is essential for the preservation of biodiversity, i.e. intermediate disturbances maximize biodiversity (Harris, 1999). Then, how would circulation that brings about physical and chemical disturbances at a local level come into the picture of nematode diversity in limnetic habitats?

The available data on the relationship between nematode diversity and disturbance are meagre. Studies done over a 1-year period in the profundal of an oligotrophic lake where circulation of the water column is only every 5–7 years showed a total number of 60 species. A similar study in the littoral of an oligotrophic lake where circulation occurs twice a year showed a total number of nematode species of 90 and in an eutrophic lake at 1 m water depth with a monthly to weekly circulation the total number of species was 152 (Traunspurger 1996a,b; Michiels and Traunspurger, 2004; Table 4.1).

Abundance

Nematodes are the most numerous benthic organisms in limnetic habitats, therefore, their abundance is an important characteristic in terms of impact in addition to

Table 4.1 Level of disturbance and total number of nematode species in Lakes Königssee (two sites) and Obersee, Germany.

	'Low' disturbance	'Medium' disturbance	'High' disturbance
Circulation	About 5–7 years	Twice a year	Monthly to weekly
Study site	Lake Königssee (Profundal)	Lake Königssee (Littoral)	Lake Obersee
Species of nematodes (time/site dependent)	31–59	44–51	21–88
Total species	60	90	152

their diversity. Reports show that nematodes comprise as much as 40–91% of the total meiobenthos numbers in limnetic habitats and show a relatively constant relative abundance during the season (Prejs and Bernard, 1985; Prejs and Lazarek, 1988; Traunspurger, 1996a,b, 2002). However, as is true for many other aspects, making comparison of abundance values reported from different parts of the globe remains difficult. Differences in sampling scheme, including sample size and number, and extraction methodology are not uniform. Now we also know that season of sampling, sampling water depth and sediment depth considered in each core have serious impacts on the resulting abundance data. Therefore, abundance data from the literature should be treated with extreme care, as direct comparison could lead to erroneous conclusions.

In spite of this, Prejs' (1977a) report on the relationship of level of lake productivity and abundance is worth mentioning. She contends abundance in oligotrophic lakes to be much higher than in eutrophic lakes. Abundance values given by Tudorancea and Zullini (1989) for one of the most productive lakes in the world, Arenguade, and for the productive Lake Ziway, do not support her idea. These authors showed that these two productive lakes support a higher abundance than the other less productive lakes they studied in the area. A high variability of abundance within oligotrophic lakes was also reported for alpine lakes (see Chapter 3).

Traunspurger's (1996a, 2002) review of the abundance data from different lakes showed abundance of limnetic nematodes to be in the range 900 to close to 67,000 individuals/100 cm^2. These abundance values, however, are generally lower than those reported for marine habitats (up to 42×10^6 individuals/m^2) (Heip *et al.*, 1982). Seen in light of the fact that the highest abundance recorded is from a mudflat estuarine environment, productivity and level of organic matter in the sediment could be the key in impacting nematode abundance.

Temporal Variation

The variation of nematode abundance over seasons has been demonstrated for different lakes (Meschkat, 1934; Schiemer *et al.*, 1969; Prejs, 1976; Holopainen and Paasivirta, 1977; Prejs and Bernard, 1985; Tudorancea and Zullini, 1989; Traunspurger, 1996a,b; Wu and Liang, 1999; Bergtold and Traunspurger, 2004, Michiels and Traunspurger, 2004, 2005b). The pattern of seasonal variation in these lakes differed between lakes as well as water depths, but the interpretation of 'seasons' globally remains complicated. Instead of considering seasons *per se*, focus needs to be directed towards the factors within each season. In the following discussion we attempt to do this exactly although dearth of comparable data on physicochemical and biological factors hinders any meaningful analysis. For instance, the few relatively complete surveys fail to provide water or sediment temperature for the sites at dates of sampling, or for the whole annual cycle, and without such data understanding the extent of impact of these factors on nematode dynamics cannot be fully appreciated.

The lack of temperature data, for instance, probably emanates from the assumption that temperature, in defined seasons, follows a regular and known pattern. With the current trend of change at the global scale, however, these assumptions concerning temperature or other seasonal factors may be far from reality. Therefore, future discussions should direct their focus towards important factors within seasons rather than focusing on season *per se*. Generally, species composition, water depth and geographical location of lakes may determine the patterns of nematode abundance hitherto observed in these habitats. In a simplified and reductionist way three patterns can be recognized, the first two being characteristics of northern habitats while the third may explain more what we see in the south (Fig. 4.2).

The most common pattern that researchers reported is a one-peak type where abundance becomes highest in warmer months but decreases in colder months (Meschkat, 1934; Schiemer *et al.*, 1969; Prejs, 1976; Traunspurger, 1996a,b; Khodyrev, 1999). In terms of seasons, abundance peaks have been observed in spring, summer or autumn. Unlike shallower parts, deeper parts of some lakes may lack any kind of marked seasonality in abundance (Holopainen and Paasivirta, 1977), or an increase in abundance could be in the later part of the warmer part of the year, particularly in autumn as was reported for deeper parts of Königssee (Traunspurger, 1996b). Such a delayed reaction of nematofauna is expected when seen with the time at which the consequences of a higher water temperature reach there. Productivity is anticipated to increase during the spring/summer and lake water may circulate in the meantime (in mixing lakes) bringing much-needed oxygen, nutrients and plankton to the lower water layer and to the sediment. Following the higher temperature, availability of food and oxygen would make their mark on nematofaunal abundance after a certain lag phase. This, therefore, could be the reason why we observe delayed abundance peaks in deeper parts of lakes (e.g. Traunspurger, 2002).

Two peaks of abundance characterize the second pattern (Biró, 1968; Nalepa and Quigley, 1983; Traunspurger, 1996b; Wu and Liang, 1999). This pattern, in some cases, happens only at certain water depths, as at 15 and 30 m depths in the littoriprofundal zone of Königssee (Traunspurger, 1996b). In terms of seasons,

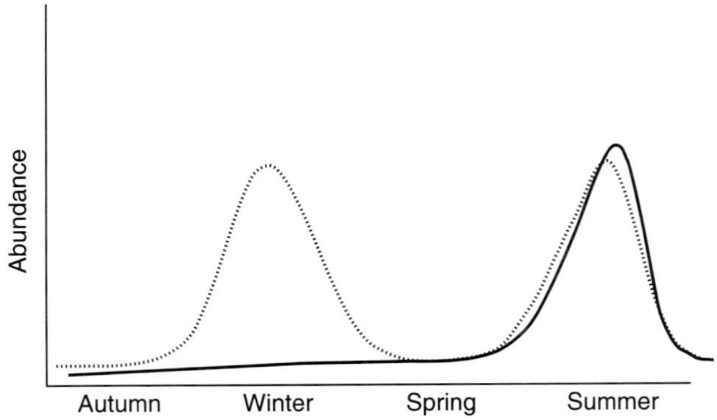

Fig. 4.2 Temporal patterns of nematode abundance in temperate lakes. Solid line characterizes the most common pattern, broken line represents the two-peak pattern.

peaks occur during both warm (summer) and cold (autumn or winter) months (Biró, 1968; Traunspurger, 1996b). Biró (1968) and Prejs and Bernard (1985) also reported a high and repeated fluctuation within a single water body in a single year. Schiemer (1975) showed that addition of nitrate to domestic waste caused an increase in nematode population and he concluded that 'nematode peaks, occurring during winter months, could be correlated with high concentration of nitrate.'

We may call the third type a site-specific pattern, that may characterize the tropics. Tudorancea and Zullini (1989), in the only seasonal study of nematode abundance in the tropical regions, reported a two-peak pattern. Abundance was highest at the end of the dry season and the beginning of a small rainy season. A second peak occurred in July, which coincided with the heavy rainy season in that region. This later peak, nevertheless, was more conspicuous for the most abundant species. As rainfall is the most important seasonal factor in the tropics and its time of occurrence varies from place to place, the months during which we expect to observe peak abundance and the number of peaks will most probably be site-specific. This, understandably, is based on limited data and therefore needs to be tested in other tropical lakes.

Though Wu and Liang (1999) demonstrated seasonal fluctuation of abundance in two Chinese lakes, Houhu and Biandantang, they failed to include any causative factor or correlation in their report. Abundance peaked only once (in January) in Lake Houhu and in Lake Biandantang it peaked twice (in August and March), but remained low during most of the year.

Despite the above generalizations, a closer look at abundance patterns shows that the pattern of seasonality at the community level (total abundance) does not always go hand in hand with the pattern of fluctuation at the species level (Biró, 1968; Tudorancea and Zullini, 1989; Traunspurger, 1996a,b; Wu and Liang, 1999). This demonstrates that species composition of a community coupled with the adaptations of the individual species ultimately draws the picture of seasonal abundance pattern in limnetic habitats. Our understanding of any observed pattern, therefore, requires detailed and purposefully gathered information on the environmental adaptations of the individual species within the context of the community.

Temporal variation in the vertical stratification of nematodes

In temperate water bodies the impact of seasonal fluctuation on temperature (with its consequences) has a pivotal impact on the vertical stratification of nematodes within the sediment. The proportion of nematodes in the upper sediment layer increases in warmer seasons while the reverse is true in colder seasons (Tietjen, 1969; Bryant and Laybourn, 1974; Hendelberg and Jensen, 1993; Traunspurger, 1996a,b). This suggests that nematodes do migrate vertically in the sediment (Traunspurger and Drews, 1996). Schratzberger *et al.* (2000) have experimentally demonstrated this ability of nematodes to migrate. This being so, Traunspurger and Drews (1996) have reported a stage-specific preference for sediment depth in a temperate lake where only adults show preference for deeper sediment layers during warmer months. But a comparable seasonal study on vertical profile from the tropics is currently lacking.

Spatial Variation

Within-lake horizontal variation of abundance

The relationship between water depth and abundance of nematodes is probably site specific, for the simple reason that water depth is only a rough and indirect indicator of factors that affect distribution and these factors may not be regulated by depth alone. All observed patterns could be grouped into four.

A number of studies in many lakes has shown a general pattern where total nematode abundance decreases as water depth increases (Prejs, 1977a,b; Schiemer, 1978; Tudorancea and Zullini, 1989; Traunspurger, 1996a,b; Eyualem-Abebe *et al.*, 2001). A completely opposite pattern was also reported in a few lakes. Here, peak abundance values were recorded in deeper waters as in the Arctic oligotrophic Lake Char (Prejs, 1977a), the tropical Lake Abaya (Tudorancea and Zullini, 1989) and temperate lakes Czarny Gasienicowy and Zielony Gasienicowy (Prejs, 1977b). The third and fourth patterns are rarely reported. One pattern is that abundance peaks are observed at both the shallowest (3 m) and deepest (7 m) sites sampled, with lower values at the middle depth, as in Mirror Lake (Strayer, 1985), and the other is that high abundance values are recorded at the middle depth considered, as in Lakes Koka and Langano (Tudorancea and Zullini, 1989).

The comparison with the general patterns recognized with regard to other benthic animals and productivity of lakes reveals that in many instances nematode horizontal distribution does not strictly follow these patterns (Fig. 4.1). Prejs (1977a), after studying 16 lakes of which seven were oligotrophic, six eutrophic, one mesotrophic and two dystrophic, reported maximum numbers of nematodes in deeper parts of oligotrophic lakes while abundance was lowest in deeper parts of eutrophic and dystrophic lakes. This goes against the idealized general pattern that nematode abundance is lowest in deeper parts of oligotrophic lakes. Other lakes that similarly do not follow these generalizations are Char, Mirror, Abaya and Abiyata, the last being a eutrophic lake. In spite of these, some lakes, like Lake Tana and Lake Königssee, do show patterns that correspond to the general pattern presented for their productivity. Both are oligotrophic where abundance is lower in deeper parts than in shallower littoral sites.

Many researchers have considered the bathymetric distribution of abundance at the species level (Biró, 1973; Prejs, 1977a; Pehofer, 1989; Traunspurger, 1995, 1996c, 1997, 1998; Eyualem-Abebe *et al.*, 2001 and references therein). But, the series of papers on Königssee by Traunspurger deserves special attention for their depth of treatment: they depicted *Tobrilus gracilis* and *Ethmolaimus pratensis* to be most abundant at 1 m depth, *Rhabdolaimus terrestris* at 2 and 5 m depths, *Monhystera paludicola* at 10 and 15 m depths and *Tobrilus medius* at 10 and 60 m depths.

Factors that govern nematode abundance over water depth in limnetic habitats are far from being fully understood. Nature of the sediment is probably the only well-recognized factor. Abundance was highest in coarse and medium sand in two productive tropical lakes, Awassa and Ziway (Tudorancea and Zullini, 1989). In another tropical lake Eyualem-Abebe *et al.* (2001) similarly indicated communities in silty sites to be less abundant than in sandy sites. Höss *et al.* (1999) have experimentally demonstrated that nematodes grow better in coarser sediments.

Other factors suggested as affecting bathymetric distribution of nematodes in limnetic habitats include decrease in primary production with increasing depth, a relatively low sedimentation rate and widely fluctuating temperature (Holopainen and Paasivirta, 1977; Traunspurger, 1996a). It is evident that the first two factors influence food availability, albeit indirectly. A case where a dense population of algae (mainly *Chara strigosa* and *C. aspera*) was correlated with nematode abundance peaks (mainly dorylaimids) in the littoral zone was reported by Traunspurger (1996a). Eyualem-Abebe *et al.* (2001), however, based on the fact that dorylaimids have a wide range of food habits in aquatic environments, questioned the validity of this statement. Currently, we lack the necessary evidence or experimental data to confirm or refute these suggestions; hence they remain as only reasonable speculations. Giving due consideration to dominance hierarchy in nematode communities (Biró, 1968; Traunspurger, 1996b), Eyualem-Abebe *et al.* (2001) speculated that dominance could be associated with niche partitioning instead of environmental tolerance.

Vertical variation of abundance within the sediment

The generalization that 'most individuals remain close to the surface of the sediment and their number decreases towards deeper sediment layers' (Fig. 4.3) is demonstrated for different waters, though the proportion may vary from site to site (Traunspurger and Drews, 1996; Eyualem-Abebe *et al.*, 2001). However, some species may show population maxima at deeper sediments (Fenchel *et al.*, 1967; Platt, 1977; Jensen, 1981; Traunspurger, 1996b; Traunspurger and Drews, 1996; Eyualem-Abebe *et al.*, 2001), indicating a possible species-specific nature of depth profile for nematodes.

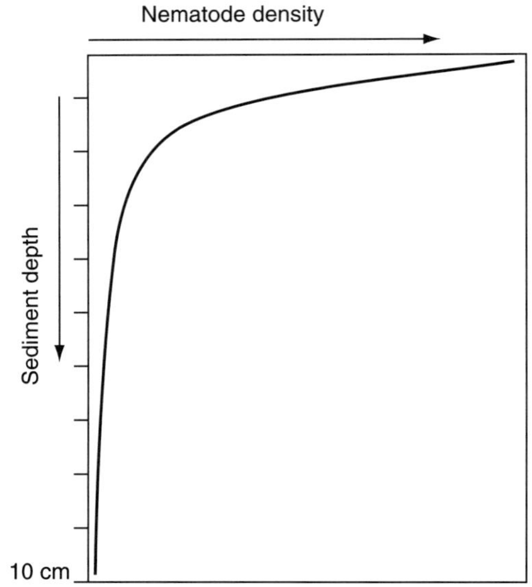

Fig. 4.3 Vertical profile of nematode abundance within lake sediment.

An important generalization that originated from the studies of Traunspurger and Drews (1996) and Eyualem-Abebe *et al.* (2001), as to the relation of water depth and vertical distribution, was that the proportion of nematodes in deeper sediment layers (below 5 cm) increases as water depth decreases (Fig. 4.4). The explanations suggested however, differ, probably due to environmental differences. The former authors suggest sediment grain size and content of organic matter as possible reasons in Königssee. 'Finer sediment inhibits the penetration of nematodes into deeper layers and that organic matter content, a likely correlate of food availability, affects the distribution of nematodes' (Traunspurger and Drews, 1996). They also agreed with Wieser's (1975) suggestion in that oxygen, temperature and alkalinity could also play a role.

Eyualem-Abebe *et al.* (2001), basically without disagreeing with Traunspurger and Drews' (1996) suggestions, gave more weight to temperature fluctuation to explain what they observed in the tropics. They maintain that in the tropics, where their study was conducted and where a marked seasonal temperature difference is lacking, inland water bodies (especially shallow lakes and rivers) are mainly affected by the rainfall regime (water level) and by the extreme diurnal fluctuation of temperature (Payne, 1986). Some nematode species in tropical waters, therefore, may inhabit the deeper sediments at shallower sites, which are exposed more to the extreme diurnal fluctuations but may concentrate only in the superficial layer in deeper and relatively protected sites, as observed in Lake Tana.

The vertical segregation of feeding types in Lake Tana was used by the same authors to argue that food availability could be one possible additional factor in determining vertical distribution of nematodes (Eyualem-Abebe *et al.*, 2001).

Traunspurger and Drews (1996) reported the occurrence of stage-specific preference at specific sediment depths; they observed vertical segregation between juveniles and adults of those species whose adults had a preference for deeper sediment layers, *vis-à-vis E. pratensis*, *Ironus tenuicaudatus*, *M. paludicola* and *T. gracilis*. These

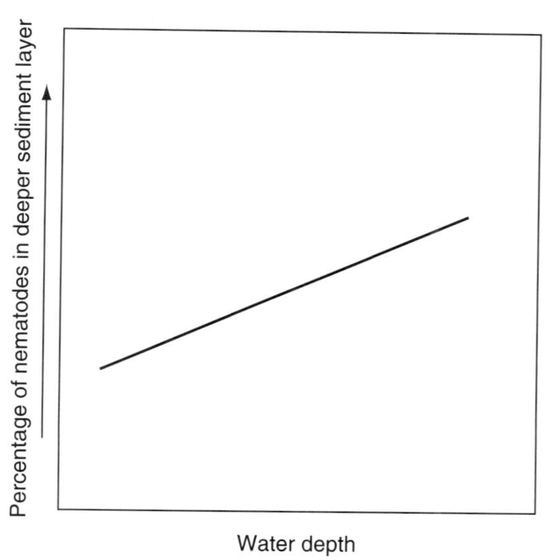

Fig. 4.4 Nematode vertical distribution within the sediment in relation to water depth.

authors argued that mating or egg laying preference, in addition to other factors (e.g. predation), could cause this inclination. In contrast to this, Eyualem-Abebe *et al.* (2001) showed the absence of any significant difference in distribution between adults and juveniles in the tropics.

Apart from the reasonable speculations pointed above, no study has unequivocally determined the reasons behind the observed vertical profile of nematodes in lakes.

Biomass

In general, meiofaunal biomass in freshwater bodies is considered low compared to total metazoan assemblage biomass (Hakenkamp and Morin, 2000; Bergtold and Traunspurger, 2005). Review of total nematode biomass from different lakes (Table 4.2) shows that it varies greatly and is lower than that in marine habitats. Within limnetic habitats, hitherto reported highest biomass values are from oligotrophic lakes while lowest ones are from eutrophic ones. But this does not imply that biomass from all eutrophic lakes is lower than that from oligotrophic lakes. Most often nematode biomass is a reflection of abundance, therefore, it follows the relationship of abundance in terms of lake productivity (see the case of Lake Arenguade above). Unfortunately comparison is hindered by the lack of biomass data for some lakes for which abundance data were available. Biomass also varies with water depth in lakes (Traunspurger, 1996a,b; Eyualem-Abebe *et al.*, 2001). Michiels and Traunspurger (2005b), in a study that spanned 11 alpine lakes, showed that biomass of nematodes far exceeded that of other benthic animals in four lakes (Lakes Vilslache, Traunsalpsee, Vilsalpsee and Soiernsee).

Seasonal studies on biomass in limnetic habitats are limited. In the well-studied Königssee, temporal variation of biomass in the littoral zone followed a pattern similar to abundance and showed variation with water depth (Traunspurger, 1996a). But, though biomass showed similar seasonal variation in the littoriprofundal and profundal zones, it did not follow the pattern of abundance (Traunspurger, 1996b). A similar discrepancy between abundance and biomass was also reported for some Polish lakes (Prejs, 1977a) and in the marine environment (Heip *et al.*, 1982). Difference in species composition was the main reason for the discrepancy, the dominance of smaller species resulted in a lower biomass. The proportion of juveniles could also potentially have an effect on biomass.

Biomass is the most important parameter to get some insight into productivity of any ecosystem. Eyualem-Abebe (2004) has provided an estimate of productivity (22 t of carbon/annum) for freshwater nematodes in a tropical setting. Bergtold and Traunspurger (Chapter 5) have dealt with the topic in depth and the reader is referred to this chapter.

Diversity

Aquatic biodiversity ensures the functioning of our lakes, wetlands and estuaries (Harris, 1999). Species richness is claimed to enhance the carbon transfer function

Table 4.2 Summary of biomass (μg wet weight/10 cm^2) reported from different lakes in the literature. Some values are transformed from published data to a uniform unit of μg/10 cm^2.

Number	Lake	Biomass	Water depth sampled (m)/lake zone	Reference
1.	Dgal Maly	2	12 and 14	Prejs (1977a)
2.	Smolak	2	4.5	Prejs (1977a)
3.	Piecek	7	8 and 10	Prejs (1977b)
4.	Balaton	1–61	–	Biró (1973)
5.	Königssee	20	Profundal (60, 120 and 190)	Traunspurger (1996b)
6.	Königssee	30	Littoriprofundal (15, 20 and 30)	Traunspurger (1996b)
7.	Pääjärvisee	36–70	Profundal	Holopainen and Paasivirta (1977)
8.	Königssee	20–170	Littoral (1, 2, 5 and 10)	Traunspurger (1996a)
9.	Pääjärvisee	28–200	Littoriprofundal	Holopainen and Paasivirta (1977)
10.	Ziway	90	0.6–6.0	Tudorancea and Zullini (1989)
11.	Tana	80–252	1.5	Eyualem-Abebe *et al.* (2001)
12.	Pääjärvisee	240	Littoral	Holopainen and Paasivirta (1977)
13.	Tana	180–1316	0.5	Eyualem-Abebe *et al.* (2001)
14.	Zielony Gasienikowy	1664	6–15	Prejs (1977a)
15.	Detritus-rich salt marsh	4600	–	Heip *et al.* (1982)

of the ecosystem (Reynolds, 1998). Also a higher number of species increases the potential for buffering processes (increased stability), and thus the possibility for trophic cascades decreases (Hildrew, 1992; Schmid-Araya and Schmid, 2000). In addition, there is a pressing need to keep extreme perturbations to a minimum in our waters. But, though the extent has a limit, depending probably on the tolerance of the organisms in question, variability seems to be an indispensable component in preserving diversity (Harris, 1999).

At a global scale, comprehensive studies on diversity of limno-nematodes so far are limited. The most commonly reported index is the number of species. Species number is often affected by the sampling effort. The problem is even further exacerbated by the inherent taxonomical confusion of the species involved. Diversity is an aspect of ecology. Generally, information on the ecology of nematodes from inland water bodies, including a relatively complete species list, is restricted to the temperate region (Biró, 1968; Zullini, 1974, 1976; Prejs, 1977a,b; Schiemer, 1978; Eder and Kirchengast, 1982; Prejs and Bernard, 1985; Prejs, 1986, 1987; Traunspurger, 1992, 1996a,b; Beier and Traunspurger, 2003a,b; Michiels and Traunspurger, 2004, 2005a,b). Because of the highly variable nature of inland water bodies (Pennak, 1988), which are principally influenced by specific local conditions, patterns and generalizations from these nematological studies have not yet emerged. In the tropics, the only studies on the ecology of nematodes are those of Tudorancea and Zullini (1989) and Eyualem-Abebe *et al.* (2001). Consequently, nematode diversity studies in limnetic habitats are notably far behind similar studies in the marine habitat. Although nematode diversity in the lentic environment has been a topic of research for more than a quarter of a century, there remains a lot with respect to our understanding of diversity *per se*, and its implication *vis-à-vis* maintenance of ecosystem function and resilience in these habitats.

Nematode diversity in limnetic habitats, as seen from the number of species, is close to that of some estuaries but much lower than reported for the deep-sea habitat (Tietjen, 1976; Heip *et al.*, 1982; Soetaert *et al.*, 1991). Species number in nematodes stretches up to 228 (Boag and Yeates, 1998) in soil samples, surpasses 400 along marine depth gradients (Levin *et al.*, 2001) and ranges up to 152 per lake in freshwater ecosystems (Michiels and Traunspurger, 2004).

One of the earliest and most extensive diversity reports on limnetic habitats is that of Prejs (1977b). In a study that spanned a Canadian, a Norwegian and 15 Polish lakes, she not only documented nematode diversity using two indices, i.e. species number and Shannon index, but also attempted to correlate level of lake primary productivity with nematode diversity. Her results showed nematode diversity to have a negative relationship with lake productivity (Fig. 4.5). The study of 11 alpine lakes by Michiels and Traunspurger (2005a) confirmed Prejs' (1997b) conclusion; they documented a high-diversity index in eight oligotrophic lakes compared to the rest of the 11 alpine lakes they studied. From the tropics, Tudorancea and Zullini (1989) studied nematode diversity of 11 rift valley lakes in Ethiopia, but their contribution was claimed to be 'pioneer but incomplete' (Eyualem-Abebe *et al.*, 2001) because of sampling methodology. Perhaps expectedly, the diversity they reported was low, comparable to the values Prejs (1977b) reported for the few eutrophic lakes she studied. Primary productivity in the Ethiopian rift valley lakes is high by any standard (Zinabu and Taylor, 1989, and references therein). But the inappropriate extraction method used by Tudorancea and Zullini (1989) makes it difficult to draw a conclusion on the relationship of productivity and nematode diversity. For instance, in a recent study, using a single sample from each lake, Eyualem-Abebe *et al.* (2001) described three previously unreported species from Lake Ziway and two from Lake Abijata.

In the marine environment, in a study of five European estuaries, nematode diversity was positively related to salinity, and this factor combined with sediment characteristics explained the observed nematode community better than latitudinal difference (Soetaert *et al.*, 1995).

Diversity along latitudinal gradient

At a global scale, latitudinal gradient in declining biodiversity from the tropics to the poles has been uncovered for other groups of invertebrates. The explanation for this trend is the association of diversity with solar energy–temperature gradients. Few studies have assessed diversity of nematodes at this scale in the marine and terrestrial habitats. For the marine, an assessment of the deep North Atlantic revealed that diversity, as measured by species count, shows a positive gradient between 13° and 56°N (Lambshead *et al.*, 2000). Similarly, Boag and Yeates (1998), using literature data, showed that species richness of the terrestrial habitat was greatest between 30° and 40°N, but warned that their results were not conclusive enough to support a 'humped back' theory of species richness. Currently, an assessment of diversity of nematodes at this scale does not exist for limnetic habitats. But, looking at the number of species reported from limnetic habitats (Chapter 3, this book), it would seem

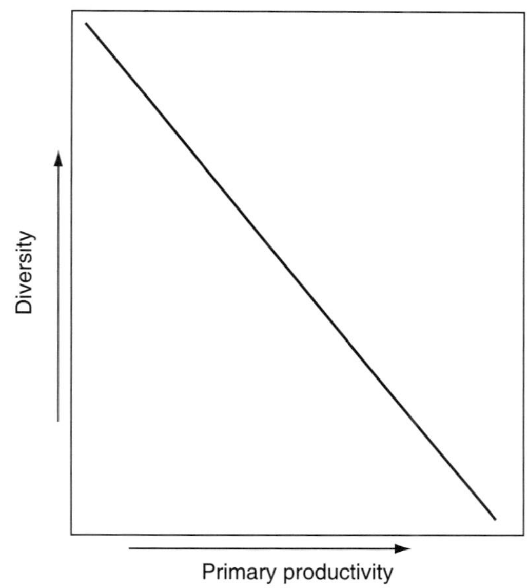

Fig. 4.5 Commonly observed relationship between nematode diversity and lake primary productivity.

that with a few exceptions, such as the highland Lake Tana, nematode diversity could be higher in higher latitudes than it is in the tropics. This is against the commonly observed trend. A note of caution, however, is that sampling intensity and the amount of research effort could also bias any observed pattern.

In the case of Lake Tana, higher diversity may be an indication that altitude with local factors modifies latitudinal gradients. But, as this is not based on statistical inferences and no attempt has been made to take sampling effort into consideration in this review, the trend from the data should be treated as no more than an impression. We hope future studies will consider this aspect to arrive at a more meaningful conclusion.

Temporal variation of diversity

There are few studies that focus on the temporal aspects of diversity in limnetic habitats. Nematode species richness, diversity, evenness and dominance in an eutrophic lake over a 3-year period were analysed based on 12,000 individuals identified to species level (Michiels and Traunspurger, 2004, 2005b). The lowest richness was found in late summer and increased in winter to reach the highest values in spring. Then the number of coexisting species decreased in summer and autumn. A similar pattern was found for species diversity and evenness. In winter 28–53 species colonized the study area, while in summer and autumn it was colonized by only 1–19 species.

Concluding Remarks

From this review it is obvious that despite the piecemeal efforts of individual researchers to study nematodes in inland water bodies and understand their abundance, diversity and ecological contribution in terms of biomass and productivity, no systematic global or regional study has been done in a meaningful way as yet. Consequently, our knowledge of their diversity, temporal dynamics and the significance of their ecological role remains largely incomplete.

The relationship of abundance and productivity is variable and currently it is not possible to make meaningful generalizations. Temporal patterns of abundance are recognized: one- and two-peak annual abundances are typical of temperate lakes whereas in the tropics abundance appears to be site-specific due to the influence of rainfall. Limnetic nematodes show temporal vertical stratification in that they prefer the upper sediment layer in warmer seasons.

Generally, abundance decreases as water depth increases. However, the relationship between abundance and water depth is neither simple nor direct. Water depth, on the other hand, has a direct bearing on the vertical sediment depth preference of nematodes: the proportion of nematodes in deeper sediment layers increases as water depth increases.

Biomass is reported in limnetic habitats to be much lower than in marine habitats and followed the patterns of abundance temporally and spatially.

Species number is the most common unit of reported diversity, and it is comparable with some estuaries but is much lower than in the deep sea. Primary productivity can be a rough indicator of nematode diversity in that oligotrophic lakes tend to have higher diversity than eutrophic lakes. Apart from productivity, little is known about the factors that determine nematode diversity in limnetic habitats. It is interesting to note that nematode diversity in limnetic habitats is higher in the temperate regions than in the tropics.

References

Beier, S. and Traunspurger, W. (2003a) Seasonal distribution of free-living nematodes in the stream Krähenbach: a fine-grained submountain carbonate stream in Southwest Germany. *Nematology* 5, 113–136.

Beier, S. and Traunspurger, W. (2003b) Seasonal distribution of free-living nematodes in the stream Körsch: a coarse-grained submountain carbonate stream in Southwest Germany. *Nematology* 5, 481–504.

Bergtold, M. and Traunspurger, W. (2004) The benthic community in the profundal of Lake Brunnsee: seasonal and spatial patterns. *Archiv für Hydrobiologie* 160, 527–554.

Bergtold, M. and Traunspurger, W. (2005) Benthic production by micro-, meio-, and macrobenthos in the profundal of an oligotrophic lake. *Journal of the North American Benthological Society* 24, 321–329.

Biró, K. (1968) The nematodes of Lake Balaton. II. The nematodes of the open water mud in the Keszthely Bay. *Annals Institute Biologie Hungary Academiae Scientiarum* 35, 109–116.

Biró, K. (1973) Nematodes of Lake Balaton. 4. Seasonal qualitative and quantitative changes. *Annales Instituti Biologici (Tihany) Hungarica Academiae Scientiarum* 40, 135–158.

Boag, B. and Yeates, G.W. (1998) Soil nematode biodiversity in terrestrial ecosystems. *Biodiversity and Conservation* 7, 617–630.

Brinkhurst, R.O., Boltt, R.E., Johnson, M.G., Mozley, S. and Tyler, A.V. (1974) *The Benthos of Lakes*. St. Martin's Press, New York, 190 pp.

Bryant, V.M.T. and Laybourn, J.E.M. (1974) The vertical distribution of Ciliophora and Nematoda in the sediments of Loch Leven, Kinross. *Proceedings of The Royal Society of Edinburgh* 17, 265–273.

Eder, R. and Kirchengast, M. (1982) The nematode-fauna (Nemathelminths, Nematoda) of a polluted part of the river Mur (Styria, Austria). *Nematologica Meditteranieae* 10, 127–134.

Eyualem-Abebe (2004) Status quo of limno-nematology, how close are we to understanding the ecology of free-living nematodes in inland water bodies. In: Cook, R. and Hunt, D.J. (eds) *Proceedings of the Fourth International Congress of Nematology*, 8–13 June 2002, Tenerife, Spain. Nematology Monographs and Perspectives 2, Brill, Leiden-Boston, pp. 739–759.

Eyualem-Abebe, Mees, J. and Coomans, A. (2001) Ecology of nematodes from L. Tana and other inland water bodies of Ethiopia. *Hydrobiologia* 462, 41–73.

Fenchel, T., Jansson, B.-O. and von Thun, W. (1967) Vertical and horizontal distribution of the metazoan microfauna and of some physical factors in a sandy beach in the northern part of the Öresund. *Ophelia* 4, 227–243.

Hakenkamp, C. and Morin, A. (2000) The importance of meiofauna to lotic systems. *Freshwater Biology* 44, 165–175.

Harris, G. (1999) This is not the end of limnology (or of science): the world may well be a lot simpler than we think. *Freshwater Biology* 42, 689–706.

Heip, C., Vincx, M., Smol, N. and Vranken, G. (1982) The systematics and ecology of free-living marine nematodes. *Helminthological Abstracts Series B, Plant Nematology* 51, 1–31.

Hendelberg, M. and Jensen, P. (1993) Vertical distribution of the nematode fauna in a coastal sediment influenced by seasonal hypoxia in the bottom water. *Ophelia* 37, 83–94.

Hildrew, A.G. (1992) Food webs and species interaction. In: Calow, P. and Petts, G. (eds) *Rivers Handbook*, vol. 1. Blackwell Scientific, Oxford, UK, pp. 309–330.

Holopainen, I. and Paasivirta, L. (1977) Abundance and biomass of the meiozoobenthos in the oligotrophic and mesohumic lake Pääjärvi, southern Finland. *Annales Zoologici Fennici* 14, 124–134.

Höss, S., Haitzer, M., Traunspurger, W. and Steinberg, C.E.W. (1999) Growth and fertility of *Caenorhabditis elegans* (Nematoda) in unpolluted freshwater sediments: response to particle size distribution and organic content. *Environmental Toxicology and Chemistry* 18, 2921–2925.

Jensen, P. (1981) Species distribution and a microhabitat theory for marine mud dwelling Comesomatidae (Nematoda) in European waters. *Cahiers de Biologie Marine* 22, 231–241.

Khodyrev, N.N. (1999) Freshwater nematodes in the south taiga belt (in Russian). *Third Congress of English Speaking Nematologists*, Pittsburgh, Russia. Available at: http://www.zin.ru/conferences/ns/abstracts.htm

Lambshead, J., Teitjen, J.H. Ferrero, T. and Jensen, P. (2000) Latitudinal diversity gradients in the deep sea with special reference to North Atlantic nematodes. *Marine Ecology Progress Series* 194, 159–167.

Levin, L.A., Etter, R.J., Rex, M.A., Gooday, A.J., Smith, C.R., Pineda, J., Stuart, C.T., Hessler, R.R. and Pawson, D.L. (2001) Environmental influences on regional deep sea species diversity. *Annual Review of Ecology and Systematics* 32, 51–93.

Lundbeck, J. (1936) Untersuchungen über die Bodenbesidlung der Alpenrandseen. *Archiv für Hydrobiologie Supplement* 10, 208–358.

Meschkat, A. (1934) Der Bewuchs in der Röhriehten des Platetensees. *Archiv für Hydrobiologie* 27, 436–517.

Michiels, I.C and Traunspurger, W. (2004) A three year study of seasonal dynamics of a zoobenthos community in a eutrophic lake. *Nematology* 6, 655–669.

Michiels, I.C. and Traunspurger, W. (2005a) Benthic community patterns and the composition of feeding types and reproductive modes in freshwater nematodes. *Nematology* (in press).

Michiels, I.C. and Traunspurger, W. (2005b) Long-term seasonal variation of biodiversity and assemblage structure in freshwater nematodes. *Archiv für Hydrobiologie* (in press).

Nalepa, T.F. and Quigley, M.A. (1983) Abundance and biomass of the meiobenthos in nearshore Lake Michigan with comparisons to the macrobenthos. *Journal of Great Lakes Research* 9, 530–547.

Payne, A.I. (1986) *The Ecology of Tropical Lakes and Rivers.* John Wiley & Sons, Chichester, UK, 301 pp.

Pehofer, H.E. (1989) Spatial distribution of the nematode fauna and production of three nematodes (*Tobrilus gracilis*, *Monhystera stagnalis*, *Ethmolaimus pratensis*) in the profundal of Piburger See (Austria 913m asl). *Internationale Revue der gesamten Hydrobiologie* 74, 135–168.

Pennak, R.W. (1988) Ecology of the freshwater meiofauna. In: Higgins, R. H. and Thiel, H (eds) *Introduction to the Study of Meiofauna.* Smithsonian Institution Press, London, pp. 39–60.

Platt, H.M. (1977) Vertical and horizontal distribution of free-living marine nematodes from Strangford Lough, Northern Ireland. *Cahiers de Biologie Marine* 18, 261–273.

Prejs, K. (1976) Bottom fauna. In: Pieczynska, E. (ed.) *Selected Problems of Lake Littoral Ecology*, University of Warsaw Press, Warsaw, pp. 145–153.

Prejs, K. (1977a) The species diversity, numbers and biomass of benthic nematodes in central part of lakes with different trophy. *Ekologia Polska* 25, 31–44.

Prejs, K. (1977b) The littoral and profundal benthic of lakes with different trophy. *Ekologia Polska* 25, 21–30.

Prejs, K. (1986) Occurrence of stylet-bearing nematodes associated with aquatic vascular plants. *Ekologia Polska* 34, 185–192.

Prejs, K. (1987) A field and laboratory study of the relation between some stylet-bearing nematodes and some aquatic vascular plants. *Archiv für Hydrobiologie* 2, 237–258.

Prejs, K. and Bernard, B. (1985) Meiobenthos of man-made Lake Zegrezynskie. *Ekologia Polska* 33, 499–509.

Prejs, K. and Lazarek, S. (1988) Benthic nematodes in acidified lakes: case of a neglected grazer. *Hydrobiologia* 169, 193–197.

Reynolds, C.S. (1998) The state of freshwater ecology. *Freshwater Biology* 39, 741–753.

Schiemer, F. (1975) Nematoda. In: C.R. Curds and H.A. Hawkes (eds) *Ecological Aspects of Used-water Treatment.* Academic Press, New York, pp. 269–288.

Schiemer, F. (1978) Verteilung und Systematik der Freilebenden Nematoden des Neusiedlersees. *Hydrobiologia* 58, 167–194.

Schiemer, F. (1985) Bioenergetic niche differentiation of aquatic invertebrates. *Verhandlungen Internationale Vereinigung für Theoretische und Angewandte Limnologie* 22, 3014–4018.

Schiemer, F., Loffler, H. and Dollfuss, H. (1969) The benthic communities of Neusiedlersees (Austria). *Verhandlungen Internationale Vereinigung für Theoretische und Angewandte Limnologie* 17, 201–208.

Schmid-Araya, J.M. and Schmid, P.E. (2000) Trophic relationships, integrating meiofauna into a realistic benthic web. *Freshwater Biology* 44, 149–163.

Schratzberger, M., Rees, H.L. and Boyd, S.E. (2000) Effects of simulated deposition of dredged material on structure of nematode assemblages – the role of burial. *Marine Biology* 136, 519–530.

Soetaert, K., Heip, C. and Vincx, M. (1991) Diversity of nematode assemblages along a Mediterranean deep-sea transect. *Marine Ecology Progress Series* 75, 275–282.

Soetaert, K., Vincx, M., Wittoeck, J. and Tulkens, M. (1995) Meiobenthic distribution and nematode community structure in five European estuaries. *Hydrobiologia* 311, 185–206.

Strayer, D. (1985) The benthic micrometazoans of Mirror Lake, New Hampshire. *Archiv für Hydrobiologie* Suppl. 72, 287–426.

Tietjen, J.H. (1969) The ecology of shallow water meiofauna in two New England estuaries. *Oecologia* 2, 251–291.

Tietjen, J.H. (1976) Distribution and species diversity of deep-sea nematodes of North Carolina. *Deep Sea Research* 23, 755–768.

Traunspurger, W. (1992) A study of the free-living freshwater nematodes of hard substrates in the littoral of the oligotrophic Königssee (National Park Berchtegaden, F.R.G.). *Spixiana* 15, 233–238.

Traunspurger, W. (1995) Autoecology of *Rhabdolaimus terrestris* de Man – a dominant freshwater nematode in an oligotrophic lake. *Nematologica* 41, 603–616.

Traunspurger, W. (1996a) Distribution of benthic nematodes in the littoral of an oligotrophic lake (Königssee, National Park Berchtegaden, F.R.G.). *Archiv für Hydrobiologie* 135, 393–412.

Traunspurger, W. (1996b) Distribution of benthic nematodes in the littori-profundal and profundal of an oligotrophic lake (Königssee, National Park Berchtegaden, F.R.G.). *Archiv für Hydrobiologie* 135, 557–575.

Traunspurger, W. (1996c) Autecology of *Monhystera paludicola* de Man, 1880 – seasonal, bathymetric and vertical distribution of a free-living nematode in an oligotrophic lake. *Internationale Revue gesamten Hydrobiologie* 81, 199–211.

Traunspurger, W. (1997) Distribution, seasonal occurrence, and vertical patterns of *Tobrilus gracilis* and *Tobrilus medius* (Nematoda). *Nematologica* 43, 59–81.

Traunspurger, W. (1998) Distribution and sex ratio of *Ethmolaimus pratensis* de Man, 1880 (Nematoda, Chromadorida) in an oligotrophic lake. *Nematologica* 44, 59–81.

Traunspurger, W. (2002) Nematoda. In: Rundle, S.D., Robertson, A.L. and Schmid-Araya, J.M. (eds) *Freshwater Meiofauna: Biology and Ecology*. Backhuys Publishers, Leiden, The Netherlands, pp. 63–104.

Traunspurger, W. and Drews, C. (1996) Vertical distribution of benthic nematodes in an oligotrophic lake: seasonality, species and age segregation. *Hydrobiologia* 331, 33–42.

Tudorancea, C. and Zullini, A. (1989) Associations and distribution of benthic nematodes in the Ethiopian Rift Valley lakes. *Hydrobiologia* 179, 81–96.

Wieser, W. (1975) The meiofauna as a toll in the study of habitat heterogeneity: ecophysiological aspects. A review. *Cahiers de Biologie Marine* 16, 647–670.

Wu, J. and Liang, Y. (1999) A comparative study of benthic nematodes in two Chinese lakes with contrasting sources of primary production. *Hydrobiologia* 411, 31–37.

Zinabu, G. and Taylor, W.D. (1989) Seasonality and spatial variation in abundance, biomass and activity of heterotrophic bacterioplankton in relation to some biotic variables in an Ethiopian rift-valley lake (Awassa). *Freshwater Biology* 22, 355–368.

Zullini, A. (1974) The nematological population of the Po River. *Bollettino di Zoologia* 41, 183–210.

Zullini, A. (1976) Nematodes as indicators of river pollution. *Nematologia Mediterranea* 4, 13–22.

Zullini, A. (1989) The ecology of the Lambro River. *Rivista di Idrobiologia* 27, 39–58.

5 Production of Freshwater Nematodes

M. BERGTOLD AND W. TRAUNSPURGER

Department of Animal Ecology, University of Bielefeld, Morgenbreede 45, 33615 Bielefeld, Germany

Production estimates are essential for studying the transfer of energy and material in natural ecosystems, but so far there is only a poor understanding of how much carbon is produced in freshwater sediments. Production estimates are available for bacteria (e.g. Sander and Kalff, 1993), protozoa (e.g. Finlay, 1978; Laybourn-Parry, *et al.*, 2000), macrofauna (e.g. Lindegaard, 1992) and meio- and macrofauna (Strayer and Likens, 1986; Kurashov, 2002). However, the simultaneous measurement of organisms covering a wide size range is scarce in freshwater sediments (Bergtold and Traunspurger, 2005). This may be attributable to the following reasons: (i) counting small organisms in sediments is time consuming; (ii) various methods are necessary to measure production of bacteria, protozoa, meio- and macrobenthos; and (iii) production values obtained can be imprecise for continuously reproducing animals (see 'Measuring Nematode Production' section). Nevertheless, production estimates provide valuable information and gain relevance if they are compared among different study sites, between different groups of organisms or between different species within a taxon. In this chapter, we focus on nematode production, briefly describing a method for estimation, and comparing different study sites and nematode species in terms of production. We also evaluate nematode production in comparison to other benthic organisms in freshwater sediments.

Measuring Nematode Production

Secondary production of a population depends on its biomass and rate of biomass turnover. For many aquatic macroinvertebrates it is possible to measure production directly because all individuals of the population grow in cohorts (Waters, 1977; Rigler and Downing, 1984; Benke, 1993). Nematodes and most other meiobenthic organisms have short generation times and reproduce continuously, lacking discrete cohorts. In this case, growth rate can either be calculated from biomass increase over time (e.g. Anderson *et al.*, 1998) or by using indirect methods. Allometric relationships of body mass, production or production/biomass (P/B) ratios have been

  Freshwater Nematodes: Ecology and Taxonomy (eds Eyualem-Abebe *et al.*)

used to estimate meiobenthic production (Banse and Mosher, 1980; Plante and Downing, 1989; Morin and Dumont, 1994). Vranken *et al.* (1986) developed a method for nematodes by measuring the duration of egg-to-egg development of 12 marine nematode species. A regression equation ($R^2 = 0.88$) explained the duration of egg-to-egg development with body mass (which can be estimated by the formula of Andrássy, 1956) and temperature, with the following equation:

$$\log T = 2.202 - 0.0461t + 0.627 \log W$$

where T is the generation time, t the temperature and W the body weight in wet weight (WW).
Assuming that biomass turnover per generation is 3, ratios of yearly P/B ratios may be calculated with the following equation:

$$P/B = 365/T \times 3$$

where P is the production, B the biomass and T the generation time in days.
P/B ratios may be calculated for shorter periods of time with the equation:

$$1/T \times D \times 3$$

where D is the duration of the period.

Multiplying biomass with P/B gives an estimate of production during this period. Hence nematode production can be calculated if temperature, biomass of the population and body mass of the adults are known.

Another approach to estimate production was published by Banse and Mosher (1980), who used body mass to predict P/B ratios, with a modification (production estimate should be divided by 4) for meiobenthic organisms. This equation has been previously used to estimate benthic production in freshwater systems (Strayer and Likens, 1986; Goedkoop and Johnson, 1996; Bergtold and Traunspurger, 2005). The relation of body mass and P/B ratio calculated after Banse and Mosher (1980) and Vranken *et al.* (1986) for a nematode community in the profundal of Lake Brunnsee is shown in Fig. 5.1. Estimates from the equation of Banse and Mosher (1980) lay within the two lines obtained with the equation of Vranken *et al.* (1986) for the selected temperature regime (5–20°C). Slope and elevation of the lines differ. Therefore, the magnitude of deviation of the production estimates obtained with the two methods is dependent both on temperature and species composition at the study site. In the following, the method of Vranken *et al.* (1986) is applied to estimate production and generation time for freshwater nematodes.

P/B ratios calculated with the method of Vranken *et al.* (1986) for freshwater nematodes growing at 10°C are in the range of 187 for the nematode *Eumonhystera simplex* (wet weight 0.03 μg) and 2.3 for *Tripyla glomerans* (wet weight 12.0 μg), respectively. Therefore, nematode P/B may differ considerably from the often-cited value of 9 (Gerlach, 1971), which is used when species composition is unknown. There is growing evidence in the literature from other groups of organisms that meiobenthic production is higher than previously thought. Anderson *et al.* (1998) examined the growth rate of three meiobenthic crustaceans (ostracods and chydorids) under

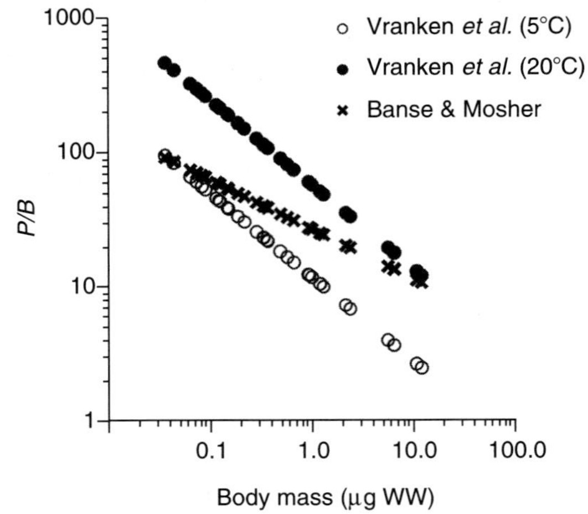

Fig. 5.1 Relation of *P/B* ratio and body mass calculated after Vranken *et al.* (1986) and Banse and Mosher (1980) (for explanation, see text). Each point represents a species of the Lake Brunnsee nematode community.

natural food conditions (floodplain swamp) and temperature regimes (10–25°C) and found *P/B* ratios ranging from 15 to 36 and generation times lying between 19.5 and 43.8 days. These values are in accordance with the literature for crustacean growth rates obtained in laboratory experiments (Anderson *et al.*, 1998, Tables 3 and 5). A further step to obtain more realistic growth rate estimates was done by Hauer and Benke (1991) who studied growth rate in field mesocosms for rapidly reproducing chironomids. With this method, Benke (1993) calculated *P/B* ratios of chironomids in the range 158–258.

A major point of criticism is the questionable transferability of generation times realized under laboratory conditions to the field situation. The production estimates derived from general equations obtained under laboratory conditions may give realistic results under favourable environmental conditions. These conditions may not be met throughout the whole year and therefore the production estimates have to be interpreted with caution. Also, the regression is based on life histories of species that have been cultured in the laboratory and many of these species show an 'opportunistic' strategy. A comparison of observations from the field could help to evaluate the validity of the equation for calculating nematode generation times for freshwater systems. Bretschko (1973) estimated generation time of some nematode species from field data in Lake Vorderer Finstertaler See (Austria). This lake is a high-altitude lake (2237 m.a.s.) with water temperature never exceeding 12°C in the littoral and 5°C in the profundal zone. Bretschko (1973) estimated the number of generations per year by plotting the frequency of weight classes of each nematode species throughout the year of the study. The author assumes that a peak in frequency of the smaller weight classes corresponds to a new generation. However, this peak could simply resemble favourable environmental conditions rather than discrete cohorts, though it could give an indication of the minimum number of generations per year. For *Tobrilus grandipapillatus* (WW ~ 4.3 μg, data from Bongers 1987), which dominated the profundal of the lake, Bretschko (1973) estimated 3 generations/year, which is higher than the value of 1.5 generations/year (5°C) calculated after Vranken *et al.* (1986). For the smaller species *Monhystera stagnalis* (WW ~ 0.9 μg, data from Traunspurger, 1991) 2 generations/year were estimated for Lake Finstertaler See in the deeper zone (4 generations/year after Vranken). The large nematode *Ironus tenuicaudatus* (WW ~ 10.5 μg, data from Traunspurger, 1991) was numerous in the littoral zone and Bretschko estimated 4 generations/year. This value is higher than the value of 1.9 generations/year (12°C) obtained by the equation from Vranken *et al.* (1986). Traunspurger (1996) recorded the occurrence of gravid females of the nematode *Monhystera paludicola* (WW ~ 0.9 μg) in Lake Königssee. These data suggest that reproduction takes place twice a year in the littoral and once a year in the deeper part of the lake (Vranken *et al.*, 1986 would suggest 5 generations/year at 5°C). *Tobrilus gracilis* (WW ~ 7.1 μg) had a high portion of gravid females twice a year in the littoral and once a year in the profundal (Traunspurger, 1997), which is in accordance with 1.2–2.5 generations/year (5–12°C) calculated after Vranken *et al.* (1986). Autecological field studies on *Ethmolaimus pratensis* (WW ~ 1.2 μg) (Bretschko, 1973; Pehofer, 1989; Traunspurger, 1998) suggest that generation time is about 4 and 6–8 months in the littoral and profundal zones, respectively. Generation time calculated after Vranken *et al.* (1986) ranges between 2 and 3 months (at 5–12°C).

Nematode Production

Nematode production may vary considerably between lakes and study sites. In the profundal of Lake Brunnsee nematodes produced 3.13 g WW/m^2/year and had an annual *P*/*B* of ~ 9. This relatively low *P*/*B* ratio is caused by the dominance of the large nematode *T. gracilis*. In Lake Königssee nematode production increased from 1 to 5 m water depth and then decreased in the deeper parts of the lake (Table 5.1). Production ranged from 0.11 g to 3.20 g WW/m^2/year in the profundal zone (190 m) and littoral zone (5 m), respectively (Table 5.1). The highest *P*/*B* ratio was observed in the littoral zone of Lake Königssee (*P*/*B* = 37). *P*/*B* ratios in the profundal of Lake Brunnsee and Lake Königssee were comparable, though production was higher in the profundal of Lake Brunnsee (Table 5.1). For marine sediments production values were given by Vranken *et al.* (1986) based on a study in muddy sediments of the Belgian coast (Vincx and Heip, 1984). Nematode production was 22.2 g WW/m/year and annual *P*/*B* was 20. Production estimates based on the study of Kito (1982) were 9.1 g WW/m^2/year with an annual *P*/*B* of 58 (Table 5.1). Also, nematode production may vary considerably throughout the year, among species and with water depth. In Lake Brunnsee most of the species had highest production values during summer and another peak in autumn (Fig. 5.2). Low production during September was observed for all nematode species in the profundal of Lake Brunnsee. In this environment *T. gracilis* had the highest production followed by *Eumonhystera filiformis*.

In Lake Königssee nematode production was high during summer and autumn in the littoral zone (Fig. 5.3A). This zone was dominated by *Rhabdolaimus terrestris* and

Table 5.1 Nematode production (g WW/m²/year) and yearly *P*/*B* ratios in marine and freshwater habitats.

Study site	Habitat (depth in m)	Production (g WW/m²/ year)	*P*/*B* ratio	Reference
Lake Königssee	Littoral (1)	1.42	16	This study
Lake Königssee	Littoral (2)	2.44	37	This study
Lake Königssee	Littoral (5)	3.20	28	This study
Lake Königssee	Littoral (10)	0.55	10	This study
Lake Königssee	Littoriprofundal (15)	0.38	11	This study
Lake Königssee	Littoriprofundal (20)	0.21	10	This study
Lake Königssee	Littoriprofundal (30)	0.26	6	This study
Lake Königssee	Profundal (60)	0.19	5	This study
Lake Königssee	Profundal (120)	0.15	9	This study
Lake Königssee	Profundal (190)	0.11	13	This study
Lake Brunnsee	Profundal (18)	3.13	9	This study
Marine (Belgium)	Littoral	22.2	20	Vranken *et al.* (1986)
Marine (Japan)	Aufwuchs	9.1	58	Vranken *et al.* (1986)

R. aquaticus. At 5 m water depth the highest yearly production value was observed (Table 5.1). The highest monthly production value was found at 2 m in October (Fig. 5.3A). Lowest production in the littoral with comparatively low summer values was calculated for the nematode community at 10 m water depth. Nematode production showed pronounced seasonal fluctuation with two peaks during February and October at 15 m water depth (Fig. 5.3B). At this depth and at 20 m *M. paludicola* dominated the nematode community. In the profundal nematode production showed moderate fluctuations during the study and increased from February to autumn (Fig. 5.3C).

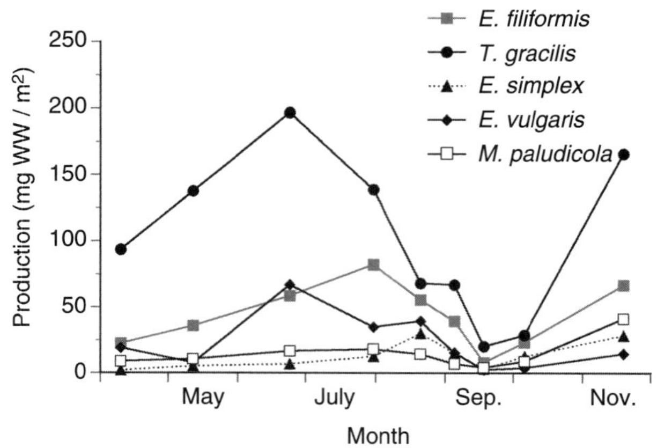

Fig. 5.2 Production (mg WW/m²) of dominant nematode species (genera *Eumonhystera*, *Tobrilus* and *Monhystera*) in the profundal of Lake Brunnsee.

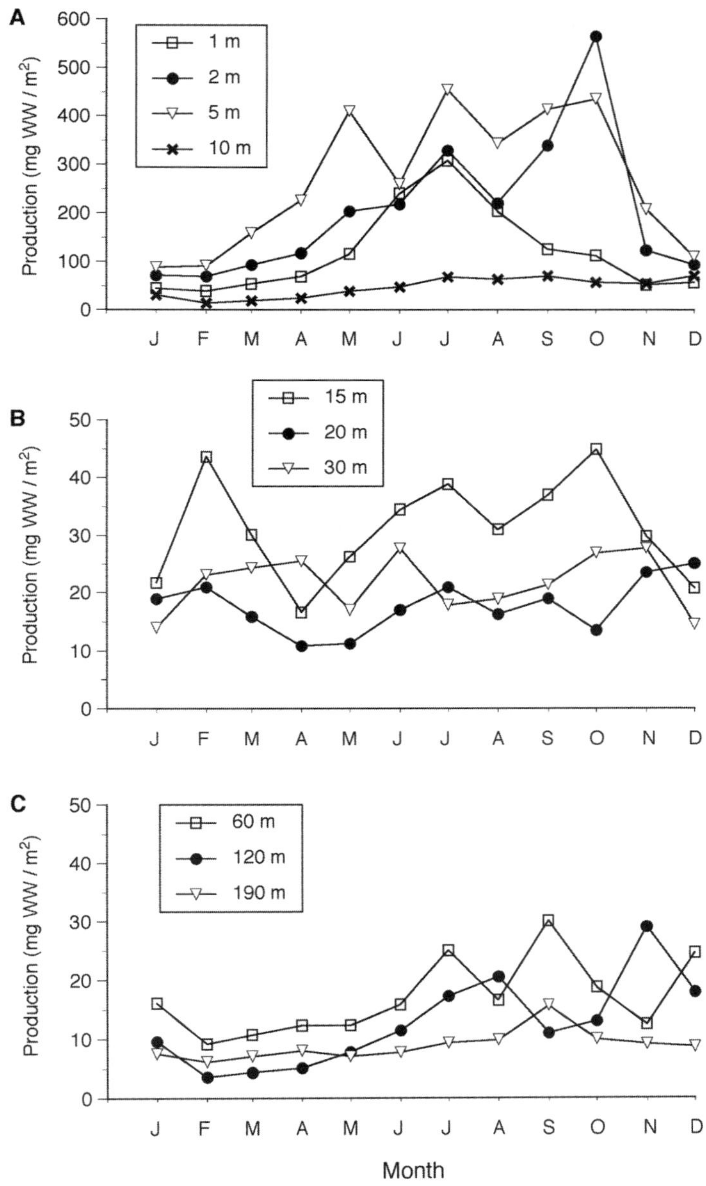

Fig. 5.3 Nematode production (mg WW/m^2) in the littoral (A), littoriprofundal (B) and profundal (C) zones of Lake Königssee.

Figure 5.4 shows nematode production among the different layers. The highest area of production was the uppermost 0.5 cm (30%), but the deepest layer (2–4 cm) showed similar high values (27%). On the other hand abundance was 46% and 18% in the uppermost and deepest layers, respectively. It is worthwhile to note that larger nematodes preferred the deeper part of the sediment in Lake Brunnsee: 39% of the

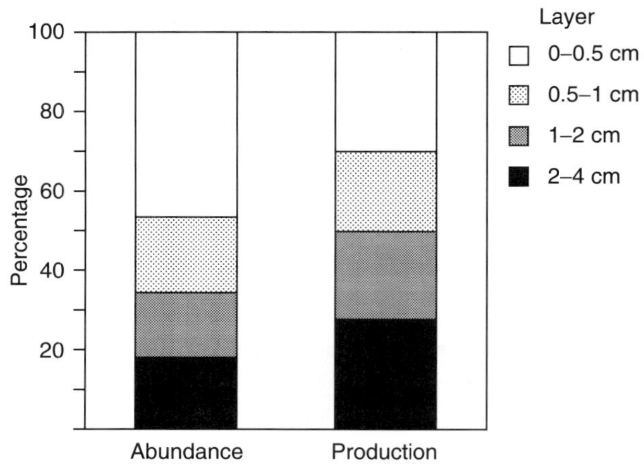

Fig. 5.4 Portion (%) of nematode abundance and production in the different sediment layers in Lake Brunnsee sediment.

individuals of *T. gracilis* inhabited the deepest investigated sediment layer in Lake Brunnsee (Bergtold, 2001) and this nematode contributed substantially to nematode production (Fig. 5.2).

Nematode Production in Relation to Other Benthic Organisms

In many freshwater and marine sediments nematodes are the most abundant representatives of the metazoans and an important component of the benthic community (Heip *et al.*, 1985; Traunspurger, 2002). However, larger organisms (macrobenthos) dominate the sediment in terms of biomass. Hakenkamp *et al.* (2002) showed for some lotic systems: Yealm River, UK (Ramsay *et al.*, 1997), 21 streams in the Outaouais region, Canada (Bourassa and Morin, 1995) and for Mirror Lake, USA (Strayer, 1985), that total biomass generally increases with body mass. Hakenkamp *et al.* (2002) concluded that the contribution of smaller sized organisms to overall biomass was higher in lentic than in lotic systems, though variance is vast and the number of studies are few. Variance of the data may be explained by several factors including grain size of the sediment, predation by fish and food availability. Strayer (1991) suggested that the smaller sized organisms of the meiofauna should predominate in food-poor environments. There is evidence from the literature supporting this hypothesis (Anderson and de Henau, 1980; Särkkä, 1995).

The share of benthic production among organisms of different size classes has been a topic of only a few studies (e.g. Strayer and Likens, 1986; Schwinghamer *et al.*, 1986; Bergtold and Traunspurger, 2005). Since the mass-specific metabolic rate of an individual decreases as body mass increases, more biomass of larger organisms compared to smaller organisms is necessary to have equal total production. For freshwater systems metabolic activity of macrobenthos in comparison to meiobenthos varied among different sites: it was equal (assimilation, lake-wide mean; Mirror Lake (Strayer and Likens, 1986)), twofold higher (assimilation, near shore zone; Lake Michigan (Nalepa and Quigley, 1983)) and 1.3 times higher (production, lake-wide mean; Lake Pääjärvi (Holopainen and Paasivirta, 1977)). In streams, contribution of meiobenthos to overall production varied between 0.7% and 52% (Hakenkamp *et al.*, 2002 and literature cited therein). The ratio of meiobenthic to macrobenthic production varied widely between selected Latgalian lakes (Latvia), and may surpass macrobenthic production by a factor of 14 (Kurashov, 2002). So far only one study provides an estimate of bacterial, protozoan and metazoan production within one lake (profundal of Lake Brunnsee; Bergtold and Traunspurger, 2005). Strayer and Likens (1986) estimated the total output (g C) of Mirror Lake. Respiration was calculated for macrobenthos and micro-meiobenthos (lake-wide mean, protozoans not measured). In both lakes benthic bacteria dominated metabolic activity (Table 5.2). Protozoan production in Lake Brunnsee contributed 22 g/m^2/year, a substantial amount, to the overall benthic production. Meiobenthic metabolic activity was higher in Mirror Lake compared to Lake Brunnsee (protozoa not included in the former) and was only a minor portion in Lake Brunnsee (Table 5.2). Lake Brunnsee was numerically dominated by nematodes (77% of overall metazoan abundance) but their contribution in terms of biomass was low (about 4% to overall benthic biomass). Nematodes contributed about 2% to overall and about 62% to meiobenthic production in Lake Brunnsee.

Conclusions

Calculating nematode production integrates both biomass and body size of the organisms. The resulting production estimate can potentially be reached under laboratory conditions. Therefore the transferability of these estimates to the field situation is limited. However, the comparison of production estimates between different habitats or groups of organisms may give valuable information. Nematodes are numerically important in sediments, though their biomass and production estimates may be of only minor relevance (see 'Nematode Production in Relation to Other Benthic Organisms' section). However, nematodes may contribute substantially to

Table 5.2 Production (g C/m^2/year) of different groups of organisms in Lake Brunnsee and percentage of production (*P*) (Lake Brunnsee) and respiration (*R*) (Mirror Lake). (Production calculated after Banse and Mosher, 1980 and respiration after Banse (1982)). Data of Mirror Lake are recalculated after Strayer and Likens (1986).

		Percentage (%)	
Organisms	*P* (g C/m^2/year)	Lake Brunnsee (*P*)	Mirror Lake (*R*)
Bacteria	27	47.1	59.0
Protozoa	22	38.4	
Meiobenthos	1.6	2.8	27.3
Macrobenthos	6.7	11.7	13.7
Total	57.3	–	–

ecosystem function. There is evidence that nematodes stimulate mineralization of organic matter (Findlay and Tenore, 1982; Alkemade *et al.*, 1992) and influence bacterial activity in sediments (Traunspurger *et al.*, 1997). Being part of the meiobenthic community nematodes may enhance microbial processes by bioturbation and feeding (Aller and Aller, 1992; Montagna, 1995). Therefore, the moderate portion of meiobenthic and more specifically nematode production may not reflect the importance of that group for the benthic community. Additional studies are necessary to evaluate the relation among these different groups of organisms and their importance for the benthic ecosystem.

References

Alkemade, R., Wielemaker, A. and Hemminga, M.A. (1992) Stimulation of decomposition of *Spartina anglica* leaves by the bacterivorous marine nematode *Diplolaimelloides bruciei* (Monhysteridae). *Journal of Experimental Marine Biology and Ecology* 159, 267–278.

Aller, R.C. and Aller, J.Y. (1992) Meiofauna and solute transport in marine muds. *Limnology and Oceanography* 37, 1018–1033.

Anderson, D.H., Darring, S. and Benke, A.C. (1998) Growth of crustacean meiofauna in a forested floodplain swamp: implication for biomass turnover. *Journal of the North American Benthological Society* 17, 21–36.

Anderson, R.S. and de Henau, A.M. (1980) An assessment of the meiobenthos from nine mountain lakes in Western Canada. *Hydrobiologia* 70, 257–264.

Andrássy, I. (1956) Die Rauminhalts- und Gewichtsbestimmung der Fadenwürmer (Nematoden) *Acta Zoologica Hungarica* 2, 1–15.

Banse, K. (1982) Mass-scaled rates of respiration and intrinsic growth in very small invertebrates. *Marine Ecology Progress Series* 9, 281–297.

Banse, K. and Mosher, S. (1980) Adult body mass and annual production/biomass relationships of field populations. *Ecological Monographs* 50, 355–379.

Benke, A.C. (1993) Concept and patterns of invertebrate production in running waters. *Verhandlungen der internationalen Vereinigung für theoretische und angewandte Limnologie* 25, 15–38.

Bergtold, M. (2001) Untersuchungen benthischer Organismen im Profundal unter besonderer Berücksichtigung der Sedimentation. Dissertation, Fakultät Biologie der Universität Bielefeld, Bielefeld, Germany, pp. 1–153.

Bergtold, M. and Traunspurger, W. (2005) Benthic production by micro-, meio-, and macrobenthos in the profundal of an oligotrophic lake. *Journal of the North American Benthological Society* 24, 321–329.

Bongers, T. (1987) *De Nematoden van Nederland.* Bibliotheekintgave KNNV, No. 49, Pirola, Schoorl, The Netherlands, pp. 1–407.

Bourassa, N. and Morin, A. (1995) Relationship between size structure of invertebrate assemblages and trophy and substrate composition. *Journal of the North American Benthological Society* 14, 393–403.

Bretschko, G. (1973) Benthos production of a high mountain lake: Nematoda. *Verhandlungen der internationalen Vereinigung für theoretische und angewandte Limnologie* 18, 1421–1428.

Findlay, S. and Tenore, K.R. (1982) Effect of a free living marine nematode (*Diplolaimella chitwoodi*) on detrital carbon mineralization. *Marine Ecology Progress Series* 8, 161–166.

Finlay, B.J. (1978) Community production and respiration by ciliated protozoa in the benthos of a small eutrophic loch. *Freshwater Biology* 8, 327–341.

Gerlach, S.A. (1971) On the importance of marine meiofauna for the benthos communities. *Oecologia* 6, 176–190.

Goedkoop, W. and Johnson, R.K. (1996) Pelagic benthic coupling: profundal benthic community response to spring diatom deposition in mesotrophic Lake Erken. *Limnology and Oceanography* 41, 636–647.

Hakenkamp, C.C., Morin, A. and Strayer, D.L. (2002) The functional importance of freshwater meiofauna. In: Rundle, S.D., Robertson, A.L. and Schmid-Araya, J.M. (eds) *Freshwater Meiofauna.* Backhuys Publishers, Leiden, The Netherlands, pp. 321–335.

Hauer, F.R. and Benke, A.C. (1991) Rapid growth of snag-dwelling chironomids in a blackwater river: the influence of temperature and discharge. *Journal of the North American Benthological Society* 10, 154–164.

Heip, C., Vincx, M. and Vranken, G. (1985) The ecology of marine nematodes. *Oceanography and Marine Biology: Annual Review* 23, 399–489.

Holopainen, I.J. and Paasivirta, L. (1977) Abundance and biomass of the meiozoobenthos in the oligotrophic and mesohumic Lake Pääjärvi, southern Finland. *Annals Zoologica Fennici* 14, 124–134.

Kito, K. (1982) Phytal marine nematode assemblages on *Sargassum confusum* Agardh, with reference to the structure and seasonal fluctuations. *Journal of the Faculty of Science Hokkaido University* 23 (Series Zoology), 143–161.

Kurashov, E.A. (2002) The role of meiobenthos in lake ecosystems. *Aquatic Ecology* 36, 447–463.

Laybourn-Parry, J., Bell, E.M. and Roberts, E.C. (2000) Protozoan growth rates in Antarctic lakes. *Polar Biology* 23, 445–451.

Lindegaard, C. (1992) Zoobenthos ecology of Thingvallavatn: vertical distribution, abundance, population dynamics and production. *Oikos* 64, 257–304.

Montagna, P.A. (1995) Rates of metazoan meiofaunal microbivory: a review. *Vie et Milieu* 45, 1–9.

Morin, A. and Dumont, P. (1994) A simple model to estimate growth rate of lotic insect larvae and its value for estimating population and community production. *Journal of the North American Benthological Society* 13, 357–367.

Nalepa, T.F. and Quigley, M.A. (1983) Abundance and biomass of the meiobenthos in nearshore Lake Michigan with comparison to the macrobenthos. *Journal of the Great Lakes Research* 9, 530–547.

Pehofer, H.E. (1989) Spatial distribution of the nematode fauna and production of three nematodes (*Tobrilus gracilis, Monhystera stagnalis, Ethmolaimus pratensis*) in the profundal of Piburger See (Austria, 913 m). *Internationale Revue der gesamten Hydrobiologie* 74, 135–168.

Plante, C. and Downing, J.A. (1989) Production of freshwater invertebrate populations in lakes. *Canadian Journal of Fisheries and Aquatic Sciences* 46, 1489–1498.

Ramsay, P.M., Rundle, S.D., Atrill, M.J., Uttley, M.G., Williams, P.R., Eslmere, P.S. and Abada, A. (1997) A rapid method for estimating biomass size spectra of benthic metazoan communities. *Canadian Journal of Fisheries and Aquatic Sciences* 54, 1716–1724.

Rigler, F.H. and Downing, J.A. (1984) The calculation of secondary production. In: Downing, J.A. and Rigler, F.H. (eds) *A Manual on Methods for the Assessment of Secondary Productivity in Freshwater,* 2nd edn. Blackwell Scientific Publication, Oxford, UK, pp. 19–58.

Sander, B.C. and Kalff, J. (1993) Factors controlling bacterial production in marine and freshwater sediments. *Microbial Ecology* 26, 79–99.

Särkkä, J. (1995) Profundal meiofauna in two large lakes: influence of pollution and bathymetric differences. *Archiv für Hydrobiologie* 132, 453–493.

Schwinghamer, P., Hargrave, B.T., Peer, D. and Hawkins, C.M. (1986) Partitioning of production and respiration among size groups of organisms in an intertidal benthic community. *Marine Ecology Progress Series* 31, 131–142.

Strayer, D. (1985) The benthic micrometazoans of Mirror Lake, New Hampshire. *Archiv für Hydrobiologie* Suppl. 72, 287–426.

Strayer, D. (1991) Perspectives on the size structure of lacustrine zoobenthos, its causes, and its consequences. *Journal of the North American Benthological Society* 10, 210–221.

Strayer, D. and Likens G.E. (1986) An energy budget for the zoobenthos of Mirror Lake, New Hampshire. *Ecology* 67, 303–313.

Traunspurger, W. (1991) *Fischbiologie des Königssees. Nahrungsangebot und Nahrungswahl. Band I: Das Meiobenthos des Königssees.* Forschungsbericht 22, Nationalpark Berchtesgaden, Berchtesgaden, Germany, pp. 1–152.

Traunspurger, W. (1996) Autecology of *Monhystera paludicola* De Man, 1880 – Seasonal, bathymetric and vertical distribution of a free-living nematode in an oligotrophic lake. *Internationale Revue der gesamten Hydrobiologie* 81, 199–211.

Traunspurger, W. (1997) Distribution, seasonal occurrence and vertical pattern of *Tobrilus gracilis* (Bastian, 1865) and *T. medius* (Schneider, 1916). *Nematologica* 43, 59–81.

Traunspurger, W. (1998) Distribution and sex ratio of *Ethmolaimus pratensis* De Man, (Nematoda, Chromadorida) in an oligotrophic lake. *Nematologica* 44, 391–408.

Traunspurger, W. (2002) Nematoda. In: Rundle, S.D., Robertson, A.L. and Schmid-Araya J.M. (eds) *Freshwater Meiofauna: Biology and Ecology*, Backhuys Publishers, Leiden, The Netherlands, pp. 65–104.

Traunspurger, W., Bergtold, M. and Goedkoop, W. (1997) The effects of nematodes on bacterial activity and abundance in a freshwater sediment. *Oecologia* 112, 118–122.

Vinxc, M. and Heip, C. (1984) Composition and seasonal fluctuation of the nematode community in a silty station off the Belgian Coast. Concerted Action Oceanography, Report to the Ministry of Scientific Policy, Brussels, Belgium, pp. 87–100.

Vranken, G., Herman, P.M.J., Vincx, M. and Heip, C. (1986) A re-evaluation of marine nematode productivity. *Hydrobiologia* 135, 193–196.

Waters, T.F. (1977) Secondary production in inland waters. *Advances in Ecological Research* 10, 91–164.

6 Feeding Ecology of Free-living Benthic Nematodes

T. Moens,[1] W. Traunspurger[2] and M. Bergtold[2]

[1]*Marine Biology Section, Biology Department, Ghent University, Krijgslaan 281/S8, B-9000 Ghent, Belgium;* [2]*Department of Animal Ecology, Universität Bielefeld, Morgenbreede 45, D-33615 Bielefeld, Germany*

Introduction

The trophic position and functioning of free-living nematodes in aquatic sediments are still poorly understood in spite of their ubiquity and abundance. As a taxon, nematodes feed on bacteria, microalgae, detritus, protozoan and metazoan prey, and may even benefit from dissolved organic matter (DOM). The quantitative importance of any of these sources in the nutrition of nematodes is, however, largely unknown. Vice versa, nematodes may affect the functioning of their prey populations and hence also benthic fluxes of carbon and energy.

The high species diversity of free-living nematodes is often attributed to a high level of niche specialization with food as an important driving factor. Observations on the feeding behaviour of aquatic nematodes are, however, scant and often anecdotal, and many other approaches to document aspects of nematode feeding ecology suffer from inherent methodological and/or interpretational problems. Feeding types of nematodes are usually inferred from buccal morphology, but there are many unresolved questions on food sources and feeding rates, which affect the accuracy with which nematodes can be included in benthic food web models (Moens *et al.*, 2004).

All nematodes are heterotrophs, ultimately relying on energy fixed through primary production. Nematodes that directly utilize living autotrophic biomass are 'grazers' – although this term is also frequently applied to feeding on (heterotrophic) bacteria – and belong to 'grazing food webs', whereas food webs based on dead organic matter may be referred to as 'detritus food webs' (Petersen and Luxton, 1982). Grazing and detritus food webs are rarely well separated (Ruess and Ferris, 2004).

For heterotrophic organisms, the intake of energy and material are closely coupled. These flows can be divided into a number of separate processes. For the purpose of clarity, we first define some of the main processes relevant to a heterotroph's carbon and energy budget (Van der Meer *et al.*, 2005).

- *Consumption or ingestion*, total uptake of energy or mass;
- *Absorption*, that part of the ingestion that crosses the gut wall[1];
- *Defaecation*, that part of the ingestion that is not absorbed, but leaves the gut as faeces;
- *Growth*, that part of the absorption that is incorporated in the body tissue of the organism;
- *Reproduction*, that part of the absorption that is released as reproductive bodies;
- *Excretion*, that part of the absorption that is released out of the body in the form of urine, or other exudates (with the exception of reproductive bodies);
- *Respiration*, that part of the absorption that is released in association with the oxidation of organic compounds, and thus causes a net loss of CO_2.

This chapter aims to give an overview of feeding habits and food sources of aquatic nematodes, and touches upon issues of food recognition, feeding selectivity, environmental constraints on feeding and complex, indirect trophic interactions with microbial prey. In addition, we give a brief overview of methods for quantifying feeding rates and link aquatic nematodes to higher trophic levels. While this contribution differs from most other monographs dealing with nematode feeding in its strong emphasis on aquatic nematodes, we recommend that the interested reader seeks additional information and insight in many of the topical papers cited here and in other recent work such as Bilgrami and Gaugler (2004).

Feeding Type Classifications and the Placement of Aquatic Nematodes

There is a long history of linking buccal morphology to (presumed) feeding habits in nematology. This has resulted in a number of largely morphology-based feeding type classifications for free-living nematodes. Separate schemes have been proposed for soil and marine nematodes. Yeates *et al.* (1993) have proposed a highly cited comprehensive grouping of soil nematodes by their feeding habits, listing documented or presumed food sources by genus. In marine- and brackish-water nematology, the feeding type classification by Wieser (1953), which is nearly exclusively based on stoma morphology, is still the most frequently used one. Freshwater nematologists draw on either or both of the above schemes, or work with their own trophic types. The classification by Traunspurger (1997) presents such a modification of both terrestrial and marine schemes. Later, we briefly describe and comment on existing feeding type classifications. We also strongly support the use of a unified trophic type classification for terrestrial and aquatic nematodes (Moens *et al.*, 2004). Aquatic nematodes fit well into the scheme of Yeates *et al.* (1993) as modified by Moens *et al.* (2004), and (dis)similarities in the organization of nematode communities from different ecological settings will be much more obvious when common

[1]The term 'assimilation' is often used but equally often poorly defined. Assimilation has been used as a synonym of absorption, as absorption minus excretion, etc. We follow the definition of Penry (1998) where assimilation relates to anabolic processes, i.e. the incorporation of absorbed products into animal tissue. Absorbed products that are not assimilated are used for catabolic processes (with end products being voided through respiration, excretion and defaecation).

trophic groupings are used. Moreover, nematodes may have flexible feeding strategies and this trophic versatility is better accounted for by schemes where nematodes can be assigned to more than one trophic type, as is the case for the schemes of Yeates *et al.* (1993) and Moens *et al.* (2004).

In the following, it is also important to note that all feeding type classifications group nematodes on the basis of food source and/or feeding habit, and while the two are often strongly linked they are not necessarily synonymous. For instance, epistrate feeders and substrate ingesters are grouped on the basis of feeding habit, while bacterial feeders are grouped on the basis of food source. Epistrate feeders and substrate ingesters may actually derive part of their nutrition from bacteria. Hence, assignment to feeding categories still remains prone to some subjectivity. Insofar as groupings are based on food source, nematode feeding types are α-guilds or resource guilds (Wilson, 1999).

Furthermore, although it is clear that feeding habits are linked to stoma morphology, presumptive feeding strategies, which are not backed by empirical evidence, may be misleading. Very similar stoma structures may be used for very different feeding behaviours, as exemplified by the range of feeding strategies in aphelenchid nematodes. On the reverse, different morphologies may serve very similar feeding behaviours (Moens *et al.*, 2004). For example, stomatostylets or odontostyles may be used to pierce microalgae, but *Pareudiplogaster pararmatus* (Romeyn *et al.*, 1983) and several Chromadoridae (Jensen, 1987; Moens and Vincx, 1997) do so with a partly eversible tooth.

The following is a brief overview of the feeding type classification by Yeates *et al.* (1993) and of the modifications proposed by Moens *et al.* (2004).

1. *Plant feeders* comprise nematodes with a tylenchoid stomatostylet or a dorylaimoid odontostyle, mouth structures that are used to feed on vascular plant tissue and fluids. In addition, some plant feeders pierce algae, lichens or mosses. The group of plant feeders is further subdivided according to feeding specializations (Yeates *et al.*, 1993; Bongers and Bongers, 1998).
2. *Hyphal feeders* pierce fungal hyphae with a stomatostylet or odontostyle. They include both obligate hyphal feeders and the alternate life cycles of some parasites of invertebrates. Feeding on yeasts or fungal spores may be included here if cells are pierced, not ingested.
3. *Bacterial feeders* feed on prokaryotes and usually ingest bacterial cells whole. Especially in species with a broad stoma, bacteria may not be the only food (see Moens *et al.*, 2004, for examples).
4. *Substrate ingesters* ingest an organic substrate together with its associated microflora and microfauna. It is not clear whether the substrate, the microflora or both are digested. Examples are the consumption of agar covered with a bacterial lawn by *Diplenteron colobocercus* (Yeates, 1998), and the ingestion of sediment particles by *Metoncholaimus scissus* (Meyers and Hopper, 1966). The term non-selective deposit-feeders, often used in marine nematology (Wieser, 1953), suggests a more or less random ingestion of suitably sized and shaped particles including substrate, and covers more or less the same behaviour. Many xyalids and comesomatids in marine and estuarine soft sediments have been considered 'non-selective deposit-feeders', but observations on the feeding behaviour of selected species suggest a

substantial selectivity where mainly unicellular pro- and eukaryotes are ingested and concomitant uptake of sediment particles is limited (Moens and Vincx, 1997). Hence, we suggest that nematodes be given only secondary classification as substrate ingesters, with a primary classification under one of the other feeding groups.

5. Many nematodes are 'carnivores', feeding on a variety of invertebrate metazoa by means of buccal adaptations such as teeth, stylet and mandibles. Ingestion of nematode prey by other nematodes with 'unarmed' buccal cavities (e.g. *Daptonema*) has also been observed, but is probably rather occasional (Moens and Vincx, 1997). Predation may be obligate in some (e.g. *Seinura*, *Labronema*, *Enoploides*, *Sphaerolaimus*) but facultative in others (e.g. many Oncholaimidae, Mononchidae). For example, *Mononchus tunbridgensis* can be cultured on bacteria as well as on prey nematodes (Yeates, 1969). Some predator nematodes may switch between protozoan (feeding group 6) and metazoan prey depending on availability (Hamels *et al.*, 2001a). Scavenging of dead metazoans is frequent, e.g. in oncholaimid nematodes (Jensen, 1987; Moens and Vincx, 1997), and also comes under this feeding category.

6. *Unicellular eukaryote feeders* feed on microalgae, fungal spores, yeasts, flagellates, ciliates and/or other protozoa. This group comprises both nematodes with 'unarmed' but fairly spacious mouth openings and so called 'epistrate-feeders', nematodes with (relatively small) teeth and/or teeth-like structures. The former typically ingest cells whole, while the latter 'pierce' (e.g. Romeyn and Bouwman, 1983; Nehring, 1992a; Moens and Vincx, 1997) or 'crack' (Jensen, 1982; Moens and Vincx, 1997) cells and suck up their contents. Size and shape of food particles greatly affect feeding success. Apparent substrate ingestion (category 4) by deposit-feeding nematodes (with unarmed stoma) may be incidental to feeding on unicellular organisms.

Yeates *et al.* (1993) additionally list two other feeding groups: free-living stages of animal parasites and omnivores. The former are of little importance in aquatic systems, and may display feeding behaviours covered by the former feeding types or, alternatively, are simply inactive and hence rather irrelevant for studies on the trophic ecology of nematodes. Some dorylaimid and most oncholaimid nematodes have often been listed as typical omnivores. However, since omnivory refers to organisms feeding at more than one trophic level, the term equally applies to a variety of nematodes covered by feeding groups 3–6. Moreover, nematodes may exhibit trophic level switching during development (see Yeates, 1987a,b; Hellwig-Armonies *et al.*, 1991), a phenomenon known as life-history omnivory. Hence, omnivory is probably the rule rather than the exception in free-living nematodes, and the nematodes listed as 'omnivores' by Yeates *et al.* (1993) should in fact be allocated to one or more of the other feeding groups.

The most commonly used feeding type classification for marine nematodes (Wieser, 1953) distinguishes four trophic types (Fig. 6.1). A primary subdivision splits between nematodes without (group 1) and with (group 2) a 'buccal armature' (teeth, onchia, jaws, etc.); a secondary subdivision discriminates within each primary group mainly on the basis of size and shape of the buccal cavity and its 'armature' (teeth and other sclerotized structures). This results in the following groupings:

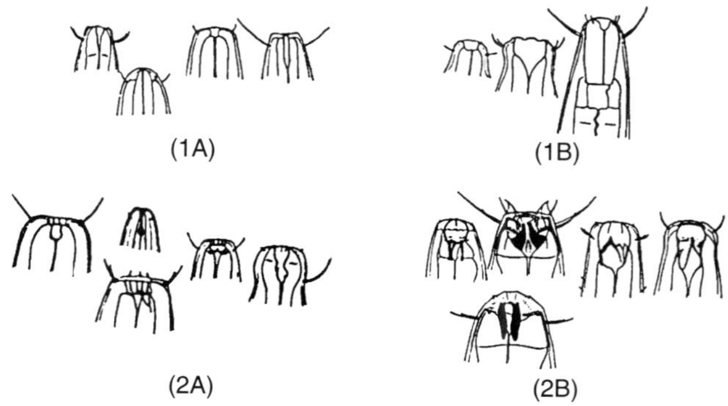

Fig. 6.1 Schematic representation of Wieser's (1953) feeding type classification for free-living marine nematodes. 1A: selective deposit-feeders, 1B: non-selective deposit-feeders, 2A: epistrate-feeders and 2B: omnivores/predators. (Redrawn after Wieser, 1953.)

(1A) *Selective deposit-feeders* have minute to small unarmed buccal cavities, allowing only particles in the bacterial size range to be ingested. There is no empirical evidence that substantiates a (more) selective feeding strategy (than in other trophic types) (Jensen, 1987), but the basic idea is that the size of the buccal cavity invalidates a non-selective feeding behaviour because it would be energetically unfavourable (Wieser, 1953). Selective deposit-feeders also tend to dominate in different environments from non-selective deposit-feeders (Heip *et al.*, 1985; Moens and Vincx, 1997).

(1B) *Non-selective deposit-feeders* also have unarmed buccal cavities, yet these are more spacious than in the previous group. These nematodes ingest bacteria as well as larger-sized protozoan and occasionally even metazoan organisms. Observations on some representatives of this group suggest that particle selection is largely a matter of particle size, shape and associated characteristics, such that particles that fit into the nematode's mouth are readily ingested (Nehring, 1992b; Moens and Vincx, 1997). However, this does not imply that food selection is only or even in a major part determined by those 'physical' particle properties, as (chemical) food recognition from a distance may be important (see living in an information-rich context: food recognition section).

(2A) *Epistrate-feeders* possess a tooth, teeth, onchia, denticles or similar structures in the buccal cavity and use these for scraping off particles from a substrate and/or to puncture cells before feeding on their contents. Food organisms can be pierced by a partly everted tooth, as in *Pareudiplogaster*, *Dichromadora*, etc. or cracked after partial intake into the mouth, as in *Hypodontolaimus*, *Chromadorita*, etc. (Jensen, 1982; Romeyn and Bouwman, 1983; Romeyn *et al.*, 1983; Moens and Vincx, 1997). Diatoms, other unicellular microalgae and filamentous green and blue-green algae have been reported as food for epistrate-feeders, but bacteria and other groups of organisms may be equally important (Moens and Vincx, 1997). Several chromadorid genera like *Hypodontolaimus* and *Ptycholaimellus* were listed as predators/omnivores (2B) in Wieser's (1953) original classification because of their large tooth and pronounced muscular pharynx, but the few available observations suggest they are in fact epistrate-feeders (Nehring, 1992a; Moens and Vincx, 1997).

(2B) *Predators/omnivores* typically have a more pronounced buccal armature and/or pharyngeal musculature than the epistrate-feeders. Many feed as predators on other metazoans, or on meta- as well as protozoans (Moens and Vincx, 1997; Moens *et al.*, 1999c, 2000; Hamels *et al.*, 2001a); however, their feeding behaviour is poorly known and other feeding strategies may prevail (Jensen, 1987; Moens and Vincx, 1997; Moens *et al.*, 1999c).

Modifications of Wieser's feeding type classification have been proposed by Jensen (1987) and Moens and Vincx (1997) on the basis of live observations and autecological information.

A recent and well-used scheme for freshwater nematodes is that by Traunspurger (1997). It recognizes four feeding types based on morphological characteristics (Fig. 6.2).

1. *Deposit feeders or swallowers*, as in the above schemes, generally have no teeth and feed on bacteria and unicellular eukaryotes that are swallowed whole.
2. *Epistrate feeders* (tear-and-swallow feeders) have a small tooth, which is used to feed on bacteria, unicellular eukaryotes, diatoms and other microalgae.
3. *Chewers* typically have a voluminous, sclerotized buccal cavity with one or more teeth and denticles. They feed on unicellular eukaryotes, nematodes, rotifers, enchytraeids and tardigrades.
4. *Suction feeders* are characterized by the presence of a stylet and are supposedly omnivorous; potential food items include algae, plants, fungi and animals. Species with particularly delicate stylets may feed mainly on plant epidermal cells and root hairs.

A detailed overview of trends in the relative abundances of nematode feeding types in different freshwater habitats is beyond the scope of this chapter. It is clear, however, that important differences exist between different freshwater environments as well as within habitats, e.g. with season, bathymetry and depth in the sediment (e.g. Traunspurger, 1997). As an example of differences between habitats, Table 6.1 gives the relative distribution of feeding types in three German lakes, spanning a gradient from oligo- to eutrophic. The data shown use the feeding type classification

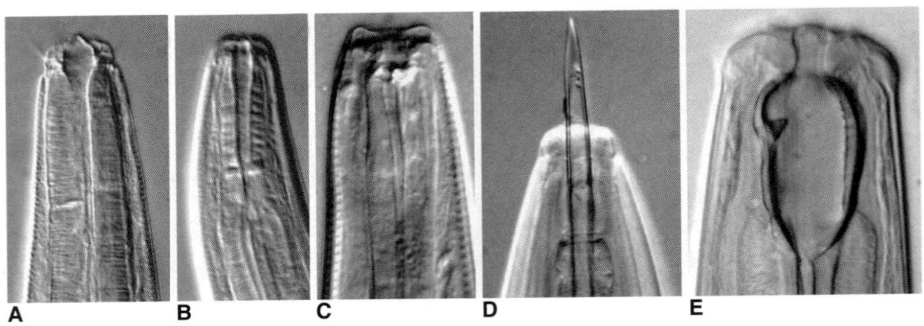

Fig. 6.2 Illustration of the feeding groups proposed by Traunspurger (1997). A and B: deposit feeders (A: *Daptonema*; B: *Plectus*); C: epistrate feeders (*Punctodora*); D: suction feeders (*Dorylaimus*); and E: chewers (*Prionchulus*).

Table 6.1 Relative distribution (annual means and ranges) of feeding types of freshwater nematodes in selected lakes.

Feeding type	Lake Königssee (oligotrophic) Littoral (n = 36,462)	Lake Königssee (oligotrophic) Profundal (n = 3464)	Lake Brunnsee (oligo-mesotrophic) Profundal (n = 5881)	Lake Obersee (eutrophic) Littoral (n = 11,806)
Deposit feeder (mainly bacteria)	86.1% (78.2–90.5)	62.4% (51.9–73.3)	72.3% (59.5–81.8)	83.9% (66.0–91.1)
Epistrate feeder (mainly algae)	4.8% (2.8–7.0)	21.9% (12.0–32.0)	2.1% (1.2–4.6)	1.0% (0.5–3.7)
Chewer (predator and omnivores)	5.1% (1.7–9.6)	13.7% (8.2–21.8)	25.3% (16.5–39.3)	12.8% (7.6–29.9)
Suction feeder (plants, fungi and omnivores)	4.0% (1.8–7.0)	2.0% (0.3–4.5)	0.3% (0.1–0.7)	2.3% (0.3–10.4)

by Traunspurger (1997), but the outcome is nearly identical when using the scheme by Yeates *et al.* (1993).

Particulate *versus* Dissolved Food

Nematode feeding type classifications inherently imply that nematodes feed on particulate organic matter (POM). However, dissolved organic carbon (DOC) in aquatic environments represents one of the largest active organic carbon reservoirs in the biosphere, the amount of DOC in aquatic systems about equalling that of CO_2 in the atmosphere (Farrington, 1992). Concentrations of DOM in aquatic systems typically surpass by far those of POM, and DOC concentrations are in general higher in freshwater than in marine environments (Wetzel, 2001). DOM may be separated into two categories: non-humic and humic (Aiken *et al.*, 1985; Thurman, 1985). Non-humic substances include carbohydrates, proteins, peptides, amino acids, fats, waxes, pigments and other low-molecular weight compounds. These substances are generally labile, easily degradable and exhibit rapid flux rates. Humic substances, by contrast, are recalcitrant to degradation and have high(er) molecular weight.

Heterotrophic bacteria are considered the major consumers of DOM in aquatic habitats (e.g. Pomeroy, 1974). However, direct use of DOM is also common among aquatic invertebrates in both marine (e.g. Stephens and Schinske, 1961; Tempel and Westheide, 1980) and freshwater environments (e.g. Hipp *et al.*, 1986). Montagna (1984a) used radiolabelled glucose to show that meiofauna, including nematodes, from intertidal sediments can incorporate DOM. Glucose was also absorbed by the marine nematodes *Pontonema vulgare* and *Adoncholaimus thalassophygas* in laboratory trials (Chia and Warwick, 1969; Lopez *et al.*, 1979). The latter species is also capable of utilizing acetate, a product of microbial fermentation processes (Riemann *et al.*, 1990). Acetate is also utilized by *Caenorhabditis briggsae* and *Turbatrix aceti* as supplemental food (Rothstein and Mayoh, 1966; Rothstein, 1970). Acetate and glucose are used for amino acid synthesis in some nematodes (*Ditylenchus triformis, Meloidogyne incognita, C. briggsae, Aphelenchoides rutgersi*) (summarized in Nicholas, 1984). Uptake of dissolved compounds such as acetate by nematodes may have interesting implications for trophic transfer. Some nematodes can use acetate for *de novo* synthesis of polyunsaturated fatty acids (PUFAs) (Rothstein and Götz, 1968; Rothstein, 1970; Bolla, 1980), which most bacteria do not produce. Nematodes could thus represent an alternative to algae-derived PUFAs that are essential for larger (macrobenthic) detritivores (Phillips, 1984).

Effects of DOM on nematode growth and/or reproduction have been shown. Addition of DOM influenced reproduction of *Caenorhabditis elegans* in laboratory experiments: negatively with one type of DOM, which was attributed to a reduction in bacterial activity, but positively with other DOM sources (Höss *et al.*, 2001). However, a major methodological difficulty in assessing the potential contribution of DOM to nematode nutrition is separating direct from bacterial-mediated uptake processes. The most telling evidence for DOM utilization, then, is provided by the axenic cultivation of several bacterial- and fungal-feeding nematodes. *C. elegans* can be sustained indefinitely axenically, although it typically has metabolic and reproductive rates at least one-third lower than when fed bacterial food (Johnson, 1985).

DOM may be ingested directly by nematodes, but DOM uptake may also be facilitated by adsorption to (bacterial) cells, exudates (Höss *et al.*, 2001), exopolymeric secretions from bacteria and/or microalgae (Decho and Lopez, 1992), and even to mucus secreted by nematodes themselves (Riemann and Schrage, 1978). It thus seems likely that DOM contributes to the nutrition of nematodes, but its significance remains to be established.

Feeding Selectivity and Intraguild Diversity

Trophic groupings tend to consider all members of one group as potential consumers of the entire resource class. High nematode species diversity, by contrast, is often attributed to a high level of niche specialization with food as a major driving factor. Clearly, feeding selectivity and flexibility may cause intraguild resource partitioning in nematodes. The range of differently sized and shaped buccal cavities, lips, appendages, etc., among bacterivorous nematodes, suggests, for instance, a variety of strategies for obtaining bacterial food and of specializations for parts of, rather than the entire, 'bacterial resource'. Rhabditidae and Panagrolaimidae feed continuously when sufficient food is available and obtain bacteria mostly from biofilms or 'suspensions' of free-living cells in interstitial water. Cephalobidae and Wilsonematidae (Plectidae) feed more intermittently, also at lower food availability, and combine complex cephalic structures with active sweeping motions during feeding, suggesting they may be more specialized in collecting bacteria attached to, or associated with a substrate (Moens *et al.*, 2004). Differences in digestive efficiency may also affect resource partitioning between species, since biochemical and ultrastructural differences exist between intestinal cells from, for example, different rhabditid nematodes (Borgonie *et al.*, 1995a,b). Furthermore, not all bacteria are suitable food for bacterial-feeding nematodes (Venette and Ferris, 1998), and even closely related nematode species may show different food preferences (Moens *et al.*,

1999a). Unfortunately, feeding selectivity and flexibility in free-living nematodes are generally poorly understood. At the same time, the issue is pivotal to the debate about functional implications of diversity, and of species redundancy within trophic levels. Species are functionally redundant if they share the same function or trophic level and do so with similar efficiency (Lawton, 1994). Mikola and Setälä (1998) demonstrated that bacterial-feeding nematodes perform species- rather than guild-specific functions in soil food webs, implying that functional redundancy is low. This is further supported by the observation that selective grazing by monhysterid nematodes results in shifts in bacterial community composition, which may typically differ even among congeneric nematodes (De Mesel *et al.*, 2004). Functional redundancy does, however, occur and may be facilitated by patchy population dynamics, as exemplified in a study on *Chronogaster* in a riparian wetland (Ettema *et al.*, 2000). Nematode species diversity thus provides an intriguing system for fundamental ecological studies on the biodiversity–ecosystem functioning relationship (Moens *et al.*, 2004).

Environmental Constraints on Nematode Feeding

Nematode feeding activity is affected by a variety of environmental factors. *Substrate characteristics* such as sediment texture and pore size may affect activity (and 'efficiency') of both consumer and resource. Elliott *et al.* (1980) observed soil texture-dependent trophic-level switching in *Mesodiplogaster* from bacterivory to predation on amoebae. Foraging efficiency of the estuarine/marine predaceous nematode *Enoploides longispiculosus* on nematode prey is strongly affected by even minor shifts in sediment silt content, grain size and water content, effectively limiting this predator's success on intertidal flats to a narrow range of sediment types and to a time-window excluding a large part of low-tide exposure (Gallucci *et al.*, in press). The authors ascribe this entirely to sediment effects on the predator, but the activity of prey organisms may also be affected by changes in sediment characteristics.

In addition to substrate characteristics, a variety of other *environmental conditions* may impact nematode feeding activity. Temperature effects on nematode (feeding) activity have, for instance, been documented. Temperature may not only affect a nematode's feeding and metabolic rate, but also its behaviour and positioning in soils and biofilms (see 'Living in an Information-rich Context: Food Recognition section).

Another crucial constraint on nematode feeding rates is their capacity to react to changes in resource availability and quality. Unfortunately, nematode *functional responses* have been poorly documented, apart from several studies on laboratory-cultivated bacterial-feeding species from terrestrial, freshwater and estuarine environments. Many bacterial-feeding nematodes need high threshold densities of bacteria to feed upon. At lower food availability, feeding is negligible, while at food concentrations above threshold, feeding rates often increase linearly or unimodally with increasing availability (Nicholas *et al.*, 1973; Schiemer, 1983; Venette and Ferris, 1998; Moens and Vincx, 2000). Much less is known about functional responses of other than bacterial-feeding nematodes. Radiotracer experiments on a marine tidal flat nematode community indicated a limited diatom-density dependence of grazing on microphytobenthos (Montagna *et al.*, 1995), but this community-averaged value may mask a variety of underlying species-specific responses. Some terrestrial and freshwater predaceous nematodes show linearly increasing predation rates with increasing prey availability, apparently without satiation (Bilgrami *et al.*, 1983, 1984; Khan *et al.*, 1991, 1995). *E. longispiculosus*, by contrast, exhibits a fairly sigmoidal prey-density dependence of feeding rate, maximal rates already being attained at relatively low prey densities (Moens *et al.*, 2000).

Interestingly, ingestion (consumption) and absorption rates may show different food-density dependent patterns, and so do absorption and respiration rates (Schiemer, 1985, 1987). For instance, a comparison of ingestion (Moens *et al.*, 1996) and absorption rates (Moens and Vincx, 2000) as a function of bacterial (food) density in *Pellioditis marina* indicates that absorption efficiencies are food-density-dependent. Next to food quantity, quality may also impact these energy budget parameters.

Foraging involves a variety of trade-offs, e.g. between energy expenditure on feeding (ingestion and digestion) and on searching for the optimal feeding conditions. Other environmental variables aside, a nematode in a patchy environment almost constantly faces two options: maximizing energy gains by moving to a better feeding location or maximizing energy gains by foraging within a given patch. The choice between these options will depend on the nematode's functional response to a given type of food, and on the presence, suitability and detectability of alternative food sources.

Living in an Information-rich Context: Food Recognition

Nematodes are capable of extracting many different types of information from and about their immediate environment. Many authors have studied the response of nematodes to different stimuli, especially their taxis towards food and food-related factors (Huettel, 1986; Perry, 1996), and their positioning in soils in response to temperature, which in turn may be a food-related behaviour (Dusenbery, 1989, 1996). Most information on chemical orientation in nematodes stems from terrestrial species. *C. elegans* can distinguish between olfactory (volatiles or odour compounds) and taste (water-soluble or dissolved compounds) clues, which are linked to separate receptor sets (Bargmann and Mori, 1997). However, the chemical sense of orientation in aquatic nematodes has not been well investigated. *A. thalassophygas* moves up gradients of (dissolved) CO_2, which may guide nematodes to spots where intense microbial activity and mineralization take place (Riemann and Schrage, 1988). More recently, the terrestrial and freshwater nematode *Bursilla monhystera*, and other nematodes isolated from the littoral of a freshwater lake, were shown to exhibit clear chemotaxis to volatile compounds produced by cyanobacterial mats. In contrast to that of *C. elegans*, the response of *B. monhystera* and *Plectus* sp. was not elicited by a single compound but rather by mixtures of different volatiles (Höckelmann *et al.*, 2004). To our knowledge, these are the only studies specifically addressing the chemical background of food finding in aquatic nematodes. Whether chemical recognition is so well developed that it can explain the often highly species-specific preferences observed in multiple food-choice experiments with nematodes (e.g. Grewal and Wright, 1992; Moens *et al.*, 1999a) remains to be established. Moens *et al.* (1999a) showed, for instance, that four coexisting species of Monhysteridae all respond differently to the presence of bacteria in their immediate environment. Such differences in response relate not only to the bacterial strains offered, but also to density, age, growth conditions and activity of the bacteria. Recent experiments indicate that such preferences are much less pronounced when mixtures of bacterial strains typical of different microhabitats frequented by these nematodes are offered as food instead of individual strains. In fact, the few preferences for bacterial mixtures that were observed in such experiments related poorly to the *in situ* microhabitat preferences of these nematodes (T. Moens *et al.*, unpublished data).

There have been some interesting studies on prey recognition by predaceous nematodes. Some mononchid predators apparently locate living prey from a distance (Bilgrami and Jairajpuri, 1988), whereas *Aporcelaimellus nivalis* depends largely on chance encounters with its prey (Khan *et al.*, 1991). The latter species, however, does perceive injured or dead prey from a distance (Khan *et al.*, 1991), as do several other predaceous nematodes (Bilgrami and Jairajpuri, 1988; Bilgrami *et al.*, 2001).

Structural heterogeneity, imposed by the substrate in which nematodes live, may further complicate their response to chemical cues emanating from food items (Anderson *et al.*, 1997a,b). While such responses are probably impossible to model in their full complexity, their implications for food web models are potentially very significant. Moreover, one can easily envisage situations where different stimuli or cues yield conflicting information. Consider, for example, a nematode grazing microphytobenthos on a tidal flat; its grazing activity may be largely restricted to (or optimal during) a few hours during ebb tide, and may be influenced over that period by a variety of biotic and abiotic factors. As an example, Fig. 6.3A shows the evolution of temperature at the sediment surface of a tidal flat station in the Schelde estuary during an ebb tide in September 1997, and compares it to the respiration rate of an abundant unicellular eukaryote feeder, *Praeacanthonchus punctatus*, as a function of temperature. The implications of the observed patterns are clear: while the sampling site was exposed for approximately 5 h, the surface activity of this grazer nematode was impaired by temperature for more than half of that time. These nematodes may escape unfavourable temperatures at the sediment surface by penetrating deeper down (Fig. 6.3B). In doing so, they end up in an environment which may offer less favourable feeding conditions in addition to being hypoxic or even anoxic. It is not difficult to envisage similar 'conflicts' in freshwater (particularly in littoral habitats) and terrestrial environments.

Many other factors may fluctuate episodically in aquatic sediments. Questions concerning the impact of unfavourable episodes on the performance/fitness of nematodes during and after are crucial to the extrapolation of data from short-term measurements, which are usually performed under conditions considered to be 'normal' in field experiments or 'close to optimal' in laboratory trials. Future work should therefore focus on hierarchies between different types of stimulus and on 'time budgets' of nematodes.

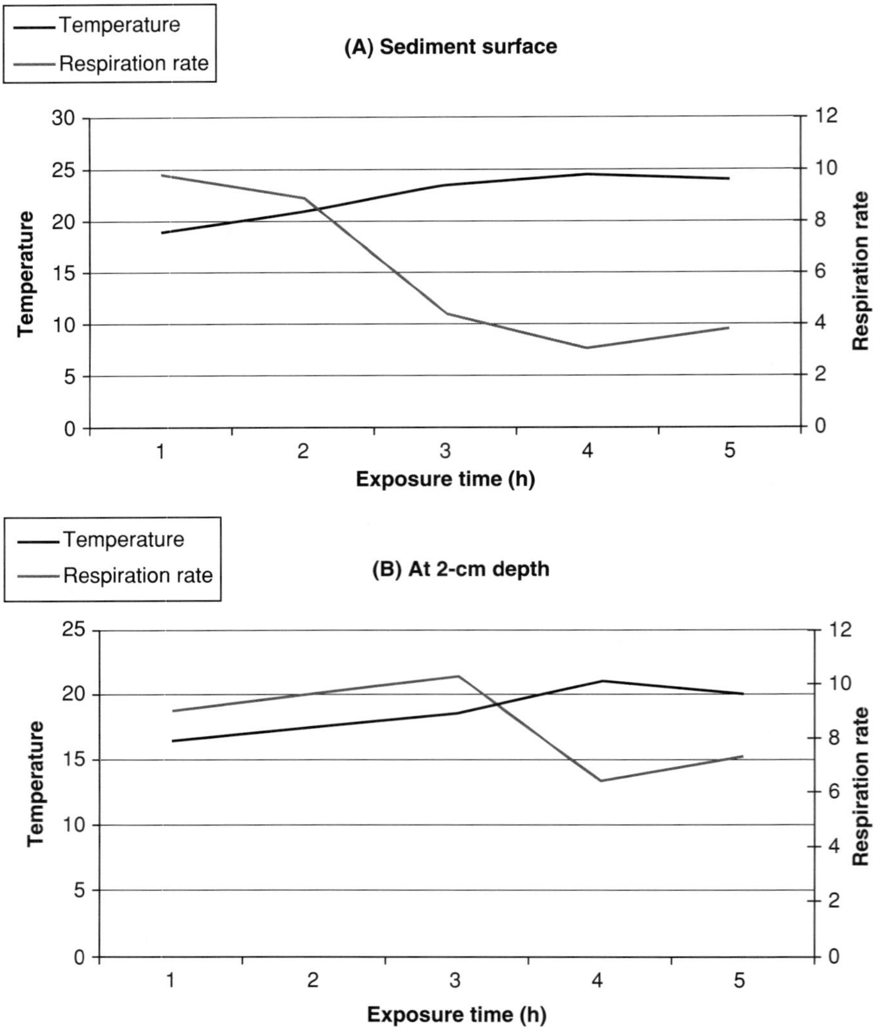

Fig. 6.3 Patterns of respiration of the deposit-feeding estuarine nematode *Praeacanthonchus punctatus* and of temperature during a 5-h low-tide exposure on an intertidal flat. A: simulation for an animal dwelling at the sediment surface and B: simulation for an animal dwelling at a depth of 2 cm.

Complex Interactions between Nematodes and their (Microbial) Resources

Nematodes are not merely passively dependent on the availability of resources. Many studies have documented how nematodes, especially bacterial-feeding species, may contribute to nutrient recycling and organic matter turnover, thus affecting both the availability of labile organic matter and the secondary production of microorganisms on which the nematodes feed (Abrams and Mitchell, 1980; Findlay and Tenore, 1982; Ingham *et al.*, 1985; De Mesel *et al.*, 2003). Bioturbation by nematodes may also increase fluxes of oxygen and nutrients (Cullen, 1973; Aller and Aller, 1992; Alkemade *et al.*, 1992). In addition, several more delicate interactions between nematodes and microbiota have been documented, especially in aquatic settings. Among the most significant are: (i) the production of copious mucous secretions, trapping sediment particles – and thereby contributing to sediment structure and stability – and perhaps DOM (see 'Particulate *versus* Dissolved Food' section), and in some cases apparently stimulating growth of microbiota (Riemann and Schrage, 1978; Warwick, 1981; Nehring *et al.*, 1990); (ii) effects – through (selective) grazing, through founder effects of mucous production or through both – on bacterial community composition and diversity (De Mesel *et al.*, 2004; Moens *et al.*, 2005); and (iii) the existence of intricate nematode–symbiotic bacteria relations, which have been particularly well documented in some groups of marine nematodes (Polz *et al.*, 1994; Nussbaumer *et al.*, 2004; Ott *et al.*, 2004). Riemann and Helmke (2002) demonstrated enzymatic activity in mucous trails of the nematode *A. thalassophygas* and proposed that both nematodes and microbiota attracted to the mucous contribute to a common exoenzyme pool and benefit from dissolved nutrients released by this enzymatic activity (the 'enzyme-sharing' concept). The general validity of these observations, and their relevance for benthic systems, remain to be established.

Measuring Feeding Rates

Methods that assess feeding rates of nematodes can be roughly divided into four categories: (i) direct observations; (ii) gut content analyses; (iii) measurements of (decreases in) food density; and (iv) tracer techniques. The latter, in turn, can be subdivided into radioactive, stable-isotope and fluorescent tracer techniques. Below, we briefly touch upon some applications as well as problems inherent to these approaches. For a more detailed overview and discussion, we refer to Somerfield *et al.* (2005) and to the topical papers cited.

Direct observations are restricted to laboratory conditions and artificial media and are most feasible when food particles are not too small compared to the consumer. The information that can be obtained from observations is qualitative rather than quantitative. An exception is the counting of pharyngeal pumping rates to quantify food consumption by rhabditid nematodes (Mapes, 1965; Moens *et al.*, 1996). Rhabditids are bacterivores that possess a valve apparatus in the pharyngeal metacorpus, the movements of which are fairly easy to observe. Other nematodes, however, may not possess such features and accurate counts of pharyngeal pulsation rates are only possible when the organism moves slowly and in a plane. The volume of medium ingested per pulsation can be estimated from the volume dilation of the pharynx (De Soyza, 1973; Woombs and Laybourn-Parry, 1984), so that when food particles are evenly distributed through the medium, food consumption can be calculated.

Gut content analysis may also be useful in generating qualitative diet information. Caution is, however, necessary when interpreting such observations, since gut content informs about substrate ingestion rather than about its utilization. As an example, Moens *et al.* (1999c) explained the presence of diatoms in the gut of *E. longispiculosus*, a predator of other nematodes and ciliates, by reference to the gut contents of its prey rather than by direct grazing. It is also quite common that food passes relatively unaltered through the guts of nematodes. Pigment analyses of gut contents using high-performance liquid chromatography (HPLC) are an alternative type of gut content analysis, which serve specifically for the detection of microalgal grazing. To our knowledge, this technique has not so far been applied to nematodes but to other meio-sized organisms (Santos *et al.*, 1995; Souza-Santos *et al.*, 1995; Buffan-Dubau *et al.*, 1996; Buffan-Dubau and Carman, 2000). The sample size needed for reproducible measurements mainly depends on the targeted pigment(s). If information on gut passage times is available, gut contents may be translated into consumption rates (Santos *et al.*, 1995; Souza-Santos *et al.*, 1995; Buffan-Dubau and Carman, 2000), but gut passage times in nematodes are poorly known (except for *C. elegans* (Thomas, 1989; Avery and Thomas, 1997)), and several commonly used fixation procedures to kill and preserve nematodes before gut content analysis tend to induce partial voidance of gut contents (Moens *et al.*, 1999b).

Measurement of (decreases in) food density is a fairly straightforward method of measuring food consumption in laboratory microcosm experiments with consumer organisms, provided adequate controls without consumers are included in the experimental design. This approach has, for instance, been used to quantify bacterivory in *C. briggsae* (Nicholas *et al.*, 1973), or predation rates of the predatory *E. longispiculosus* on other nematodes and ciliates (Moens *et al.*, 2000; Hamels *et al.*, 2001a). If prey organisms do not multiply over the time course of the incubation, ingestion rate can be readily calculated from the rate of disappearance of prey organisms. If prey items do multiply, as would be the case for bacteria and protozoans over timescales of more than 1 h, prey growth rate in the absence of predators also has to be determined. To allow extrapolation to a field situation, incubations should use prey and predator densities that closely match field densities. Prey density often has a pronounced influence on predation rate (functional response; see 'Environmental Constraints on Nematode Feeding' section), and predators may interfere with one another, thus reducing the *per capita* predation rate (Hamels *et al.*, 2001a).

An extension to this approach, potentially yielding feeding rates relevant to field conditions, is the comparison of changes in food density in micro- or mesocosms where consumer densities are manipulated *versus* experimental units with natural consumer densities. This approach has, for instance, been applied to estimate rates of bacterivory, herbivory and predation on protozoans in stream meiobenthic communities (Borchardt and Bott, 1995; Bott and Borchardt, 1999), to assess meiobenthos impacts on ciliates in a sandy tidal flat (Epstein and Gallagher, 1992) and to evaluate bacterial grazing by nematodes (Traunspurger *et al.*, 1997). Here too, potential effects of interference among consumers need to be considered.

Tracer techniques encompass approaches where tracer is added as free label (inorganic or organic) or, alternatively, as prelabelled food particles. Experiments with additions of prelabelled food particles under controlled conditions are particularly useful for studying food selectivity, but less so in determining absolute grazing rates pertinent to field situations. Most experiments in the literature on aquatic meiofauna have been performed to determine (community) grazing rates on bacteria or microalgae after direct addition of 'radioactive label' ($NaH^{14}CO_3$ for microalgae and [methyl-^{3}H]-thymidine for bacteria) to sediment samples. The latter tracer is not incorporated by a variety of bacteria (Johnstone and Jones, 1989), hence rates calculated from trials using thymidine do not accurately reflect grazing on total bacterial standing stock. ^{3}H-leucine may yield better yet still not entirely satisfactory results. Grazing rates in experiments with direct addition of free label can be calculated from the equation:

$$G = 2M/mt$$

where G is the grazing rate, M the amount of label entering grazers via feeding on bacteria or microalgae, m the amount of label in the bacteria or microalgae and t the incubation time (Daro, 1978; Montagna, 1984b; Montagna and Bauer, 1988). G is expressed in units per hour.

These calculations hold provided: (i) label uptake in the grazed compartment is linear and label recycling from this compartment does not occur; and (ii) label uptake in the grazer compartment is hyperbolic with time (Daro, 1978; Montagna, 1984b; Montagna and Bauer, 1988). In many studies, the time intervals used in grazing experiments are simply too large to properly assess these requirements and in many cases, either or both assumptions are probably violated (Moens *et al.*, 1999b). Moreover, since free label may enter grazers through non-feeding pathways, extensive control sets are a prerequisite for grazing experiments (Montagna, 1983; Montagna and Bauer, 1988).

Stable isotopes provide a powerful alternative to radioactive tracers. Consumer-stable isotope ratios tend to be very similar to those of their food sources, the offset between both (fractionation) typically being very small (on average ≤ 1‰) for carbon, but more substantial for nitrogen (3–4‰). As such, stable nitrogen isotopes can be used to discriminate between trophic levels (Rau *et al.*, 1983; Vander Zanden and Rasmussen, 2001; Post, 2002). The power of natural stable isotope ratios of meiofauna to provide information on diet depends in large part on whether the isotope ratios of different potential resources are sufficiently distinct. Moreover, natural isotope ratios usually do not allow assessment of selectivity among, for example, different diatom species. In addition to natural stable isotope analyses, enrichment experiments using either prelabelled food or free tracer (e.g. $NaH^{13}CO_3$ for labelling microphytobenthos) can be designed. Consumption and/or absorption rates can then be calculated from mass balance of the added tracer (for instance ^{13}C) (see e.g. Middelburg *et al.*, 2000; Moens *et al.*, 2002).

Most mass spectrometric analyses require in the order of 10 μg of an element (mostly carbon or nitrogen) for reliable measurements, which corresponds to *c.* 100 μg wet weight for carbon and *c.* five times more for nitrogen (but note recent analytical improvements that may reduce biomass requirements to as little as 1 or 2 μg N and/or C (Carman and Fry, 2002)). This obviously bears upon the feasibility of isotopic analyses on nematodes: carbon isotopic analyses at the community level and at the level of trophic guilds or dominant genera/species may well be feasible in many instances, but available biomass can be too limited for nitrogen analysis beyond the community level or for isotopic analyses on less abundant genera/species.

Fluorescently labelled food (bacteria or diatoms) or food analogues (similarly sized but inert microbeads) are routinely used in studies on protozoan grazing (Sanders *et al.*, 1989; Epstein and Shiaris, 1992; Sherr and Sherr, 1993), but have only rarely been applied to studies on nematode feeding (but see Epstein and Shiaris, 1992; Borchardt and Bott, 1995; Epstein, 1997). Sediment may also be stained in its entirety, so that both free-living and attached microbiota are fluorescently labelled, which may partly overcome the problem of consumer selectivity for or against prelabelled food particles (Starink *et al.*, 1994; Hamels *et al.*, 2001b), but note that this procedure kills sediment microbiota, and selection against dead microbiota may occur in microbivores. Quantification of ingested particles in nematode guts may be difficult and hampered by autofluorescence of consumer organisms (strongly aggravated by certain commonly used fixatives such as glutaraldehyde).

Caveats

Commonly used chemical fixatives such as formaldehyde, glutaraldehyde and ethanol cause substantial label loss from nematodes through leakage of low-molecular weight metabolites and voidance of gut contents upon addition of the fixative (Moens *et al.*, 1999b). Cooling samples on ice before fixation with ice-cold formaldehyde prohibits egestion, but losses through leakage are more difficult to account for.

Inappropriate incubation times (in the order of hours) will probably confound consumption and absorption and underestimate the former (Moens *et al.*, 1999b). In order to measure consumption, incubation intervals should be kept as short as possible.

Very few studies have hitherto addressed feeding rhythms in nematodes. Daily feeding rates are often calculated from experiments that last a few hours or less. The above-mentioned examples of *P. punctatus* (as affected by temperature upon low-tide exposure) and *E. longispiculosus* (as affected by sediment water content upon low-tide exposure) suggest that environmental factors may episodically constrain feeding activity in intertidal areas (as also noted for harpacticoid copepods from tidal flats (Decho, 1988; Souza-Santos *et al.*, 1995; Buffan-Dubau and Carman, 2000)), but there are few – if any – similar examples from other benthic habitats. Clearly, more research is needed on this aspect.

Finally, many experiments aimed at assessing meiofaunal ingestion rates have used starved animals. When doing so, one should always bear in mind that the responses of starved animals to food (often in excess) can substantially differ from normal behaviour. If experiments are designed to determine consumption rates under (near) natural conditions starvation should be avoided.

Nematodes as Food for Higher Trophic Levels

It is still an open question whether and to what extent freshwater nematodes are either a self-contained 'energy sink' or, alternatively, facilitate the transfer of primary food resources up the size spectrum. In the marine environment, the importance of nematodes as both producers and consumers in food webs is relatively well established (Gee, 1989; Montagna, 1995) and their role as a trophic link between primary producers/consumers and larger fauna in benthic habitats is well documented (reviewed in Coull, 1999). A significant transfer of energy from meiofauna directly to polychaetes, prawns, juvenile fishes and birds has been demonstrated (e.g. Gee, 1989; Coull, 1990, 1999; Feller and Coull, 1995; Colombini *et al.*, 1996), despite the fact that their dietary contribution may be underestimated by gut content analysis because nematodes are rapidly digested (Feller and Coull, 1995). Evidence that predation on nematodes constitutes a significant trophic link in freshwater systems is limited, although there are some examples of fishes and invertebrates feeding on the meio-size class of benthic fauna to which nematodes belong (Schmid-Araya and Schmid, 2000; Schmid and Schmid-Araya, 2002). Nematodes have been reported as prey of leeches (Schönborn, 1985; Toman and Dall, 1997), planarians (Beier *et al.*, 2004), microturbellarians and chironomids (Crisp and Lloyd, 1954; Schwank, 1981), and young dragonfly larvae respond to the presence of nematodes (Rowe, 1994).

Related to the question of energy transfer through nematodes is that of whether nematode densities and community composition are top–down controlled by predators. In estuarine and marine habitats, meiofaunal prey populations tend to be large compared to predatory removal by invertebrates and fish, which makes a profound top–down effect on nematode populations less likely (Coull, 1999). However, many aquatic nematodes may have partially or entirely predaceous feeding strategies (Wieser, 1953; Moens and Vincx, 1997) so that 'internal' predation, alone or in concert with predation by macrofauna, may significantly impact nematode density and community composition (Moens *et al.*, 1999c, 2000; Gallucci *et al.*, 2005). Such top–down impacts on aquatic nematode communities are strongly affected by habitat characteristics such as sediment texture and structure. Beier *et al.* (2004), for instance, showed that sand and fine gravel provide nematodes with refuges where they are effectively protected against predation by the planarian *Dugesia gonocephala*, whereas high predation rates were observed in coarse gravel. Similarly, sediment characteristics affect the predation efficiency of the predatory nematode *E. longispiculosus* (Gallucci *et al.*, 2005).

Table 6.2 provides an overview of nematode responses to the presence of macrofauna in enclosure and exclosure experiments. It is clear that generalizations on the effects caused by predators are difficult, even more so since nematodes may migrate to deeper sediment layers when exposed to predation at or near the sediment surface (e.g. Reise, 1979; Fitzhugh and Fleeger, 1985; Coull *et al.*, 1989; Hedqvist-Johnson and Andre, 1991).

In conclusion, a variety of predators can consume large numbers of nematodes under favourable conditions. Sediment, however, is a physically complex microhabitat and offers refuge from predation. Invertebrate predators, which do not consume sediment-dwelling meiofauna, may exert substantial predation pressure on meiofauna living as epifauna on plant stems (Gregg and Fleeger, 1998) and perhaps other substrates. At the moment, there is little information about the densities of freshwater nematodes on aufwuchs and hard substrates (Traunspurger, 1992), and we can only speculate about the intensity of predation on nematodes in such habitats.

Table 6.2 Microcosm studies and field studies using enclosures and exclosures to investigate the effect of the presence and absence of macroinvertebrate organisms (predator/disturber) on the density, vertical distribution and community structure of the nematode fauna.

Experimental design	Habitat	Sediment	Predator/ disturber	Density	Vertical distribution	Community structure	References
Marine habitats							
Field flow-through enclosure	Salt marsh	Mud + sand	Shrimp	Reduction	–	No effect	Bell and Coull (1978)
Field enclosure	Tidal flat, Germany	Sand	Macrofauna	No, small effect	–	–	Reise (1979)
Field exclosure; microcosm, enclosure	Salt marsh, South Carolina	Mud	Macrofauna	No effect	Effect	–	Bell (1980)
Field enclosure	Salt marsh, Rhode Island	Mud	Crab	Increase	No effect	–	Hoffman *et al.* (1984)
Enclosure	Salt marsh, Georgia	Mud + sand	Shrimp	Reduction	–	–	Kneib (1985)
Field exclosure	Mangrove, Australia	Mud	Crab	Increase	–	–	Dye and Lasiak (1986)
Field exclosure	Subtidal flat, Scotland	Mud	Macrofauna	No effect	–	No effect	Ólafsson and Moore (1990, 1992)
Laboratory microcosm, enclosure	Baltic proper	Mud	Amphipods	Reduction	–	Effect	Ólafsson and Elmgren (1991)
Field exclosure	Tidal flat, Australia	–	Crab	Increase	–	–	Dittmann (1993)
Field enclosure	Estuary, England	Mud + sand	Bivalves, polychaetes	No effect	–	–	Kennedy (1993)

Field flow-through enclosure	Non-tidal flat, Sweden	Sand	Prawn	No effect	–	–	Nilsson *et. al.* (1993)
Laboratory microcosm, enclosure	Baltic proper	Mud	Bivalves	No effect	No effect	No effect	Ólafsson *et al.* (1993)
Field exclosure	Mangrove, Kenya	–	Macrofauna	Increase	No effect	Effect	Schrijvers *et al.* (1995)
Laboratory microcosm, enclosure	Fjord, Norway	Mud	Urchin	No effect	–	Effect	Austen and Widdicombe (1998)
Field enclosure	Salt marsh, USA	Silty clay	Shrimp	No effect	–	–	Gregg and Fleeger (1998)
Laboratory microcosm, enclosure	Estuary, England	Sand	Crab	Reduction	–	Effect	Schratzberger and Warwick (1999)
Laboratory microcosm, enclosure	Estuary, England	Mud	Crab	No effect	–	Effect	Schratzberger and Warwick (1999)
Laboratory microcosm, enclosure	Estuary, Canada	Sand	Polychaetes	Reduction	Effect	Effect	Tita *et al.* (2000)
Freshwater habitats							
Field exclosure	Stream, Costa Rica	Sand	Macrofauna	No effect	–	Effect	Duft *et al.* (2002)
Laboratory microcosm enclosure	Lake, Germany –	Sand	Triclads, leech	No effect	Effect	No effect	Beier *et al.* (2004)

References

Abrams, B.I. and Mitchell, M.J. (1980) Role of bacterial–nematode interactions in heterotrophic systems with emphasis on sewage sludge decomposition. *Oikos* 35, 404–410.

Aiken, G.R., McKnight, D.M., Wershaw, R.L. and McCarthy, P. (1985) *Humic Substances in Soil, Sediment, and Water – Geochemistry, Isolation, and Characterization.* John Wiley & Sons, New York.

Alkemade, R., Wielemaker, A., de Jong, S.A. and Sandee, A.J.J. (1992) Experimental evidence for the role of bioturbation by the marine nematode *Diplolaimella dievengatensis* in stimulating the mineralization of *Spartina anglica* detritus. *Marine Ecology Progress Series* 90, 149–155.

Aller, R.C. and Aller, J.Y. (1992) Meiofauna and solute transport in marine muds. *Limnology and Oceanography* 37, 1018–1033.

Anderson, A.R.A., Sleeman, B.D., Young, I.M. and Griffiths, B.S. (1997a) Nematode movement along a chemical gradient in a structurally heterogeneous environment. 2. Theory. *Fundamental and Applied Nematology* 20, 165–172.

Anderson, A.R.A., Young, I.M., Sleeman, B.D., Griffiths, B.S. and Robertson, W.M. (1997b) Nematode movement along a chemical gradient in a structurally heterogeneous environment. 1. Experiment. *Fundamental and Applied Nematology* 20, 157–163.

Austen, M.C. and Widdicombe, S. (1998) Experimental evidence of effects of the heart urchin *Brissopsis lyrifera* on associated subtidal meiobenthic nematode communities. *Journal of Experimental Marine Biology and Ecology* 222, 219–238.

Avery, L. and Thomas, J.H. (1997) Feeding and defecation. In: Riddle, D.L., Blumenthal, T., Meyer, B.J. and Priess, J.R. (eds) *C. Elegans* II. Cold Spring Harbor Laboratory Press, New York, pp. 679–716.

Bargmann, C.I., and Mori, I. (1997) Chemotaxis and thermotaxis. In: D.L. Riddle, T. Blumenthal, B.J. Meyer and J.R. Priess (eds) *C. Elegans* II. Cold Spring Harbor Laboratory Press, New York, pp. 717–737.

Beier, S., Bolley, M. and Traunspurger, W. (2004) Predator–prey interactions between *Dugesia gonocephala* and free-living nematodes. *Freshwater Biology* 49, 77–86.

Bell, S.S. (1980) Meiofauna–macrofauna interactions in a high salt marsh habitat. *Ecological Monographs* 50, 487–505.

Bell, S.S. and Coull, B.C. (1978) Field evidence that shrimp predation regulates meiofauna. *Oecologia* 35, 141–148.

Bilgrami, A.L. and Gaugler, R. (2004) Chapter 4: Feeding behaviour. In: Gaugler, R. and Bilgrami, A.L. (eds) *Nematode Behaviour.* CAB International, Wallingford, UK, pp. 91–126.

Bilgrami, A.L. and Jairajpuri, M.S. (1988) Aggregation of *Mononchoides longicaudatus* and *Mononchoides fortidens* (Nematoda, Diplogasterida) at feeding sites. *Nematologica* 34, 119–121.

Bilgrami, A.L., Ahmad, I. and Jairajpuri, M.S. (1983) Some factors influencing predation by *Mononchus aquaticus. Revue de Nématologie* 6, 325–326.

Bilgrami, A.L., Ahmad, I. and Jairajpuri, M.S. (1984) Predatory behaviour of *Aquatides thornei* (Nygolaimina: Nematoda). *Nematologica* 30, 457–462.

Bilgrami, A.L., Pervez, R., Yoshiga, T. and Kondo, E. (2001) Attraction and aggregation behaviour of predatory nematodes, *Mesodorylaimus bastiani* and *Aquatides thornei* (Nematoda: Dorylaimida). *Applied Entomology and Zoology* 36, 243–249.

Bolla, R. (1980) Nematode energy metabolism. In: Zuckermann, B.M. (ed.) *Nematodes as Biological Models*, vol. 2. Academic Press, New York, pp. 165–192.

Bongers, T. and Bongers, M. (1998) Functional diversity of nematodes. *Applied Soil Ecology* 10, 239–251.

Borchardt, M.A. and Bott, T.L. (1995) Meiofauna grazing of bacteria and algae in a Piedmont stream. *Journal of the North American Benthological Society* 14, 278–298.

Borgonie, G., Claeys, M., De Waele, D. and Coomans, A. (1995a) *In vivo* and *in vitro* characterization of the intestine of fifteen bacteriophagous nematodes (Nematoda: Rhabditida). *Fundamental and Applied Nematology* 18, 115–122.

Borgonie, G., Claeys, M., De Waele, D. and Coomans, A. (1995b) Ultrastructure of the intestine of the bacteriophagous nematodes *Caenorhabditis elegans*, *Panagrolaimus superbus* and *Acrobeloides maximus* (Nematoda: Rhabditida). *Fundamental and Applied Nematology* 18, 123–133.

Bott, T.L. and Borchardt, M.A. (1999) Grazing of protozoa, bacteria, and diatoms by meiofauna in lotic epibenthic communities. *Journal of the North American Benthological Society* 18, 499–513.

Buffan-Dubau, E. and Carman, K.R. (2000) Diel feeding behavior of meiofauna and their relationships with microalgal resources. *Limnology and Oceanography* 45, 381–395.

Buffan-Dubau, E., deWit, R. and Castel, J. (1996) Feeding selectvity of the harpacticoid copepod *Canuella perplexa* in benthic muddy environments demonstrated by HPLC analyses of chlorine and carotenoid pigments. *Marine Ecology Progress Series* 137, 71–82.

Carman, K.R. and Fry, B. (2002) Small-sample methods for $\delta^{13}C$ and $\delta^{15}N$ analysis of the diets of marsh meiofaunal species using natural-abundance and tracer-addition isotope techniques. *Marine Ecology Progress Series* 240, 85–92.

Chia, F.S. and Warwick, R.M. (1969) Assimilation of labelled glucose from seawater by marine nematodes. *Nature* 224, 720–721.

Colombini, I., Berti, R., Nocita, A. and Chelazzi, L. (1996) Foraging strategy of the mudskipper *Periophthalmus sobrinus* Eggert in a Kenyan mangrove. *Journal of Experimental Marine Biology and Ecology* 197, 219–235.

Coull, B.C. (1990) Are members of the meiofauna food for higher trophic levels? *Transactions of the American Microscopical Society* 109, 233–246.

Coull, B.C. (1999) Role of meiofauna in estuarine soft-bottom habitats. *Australian Journal of Ecology* 24, 327–343.

Coull, B.C., Palmer, M.A. and Myers, P.E. (1989) Controls on the vertical distribution of meiobenthos in mud: field and flume studies with juvenile fish. *Marine Ecology Progress Series* 55, 133–139.

Crisp, G. and Lloyd, L. (1954) The community offsets in a patch of woodland mud. *Transactions of the Royal Entomological Society of London*, 105, 269–276.

Cullen, D.J. (1973) Bioturbation of superficial marine sediments by interstitial meiobenthos. *Nature* 242, 323–324.

Daro, M.H. (1978) A simplified ^{14}C method for grazing measurements on natural planktonic populations. *Helgoländer wissenschaftliche Meeresuntersuchungen* 31, 241–248.

Decho, A.W. (1988) How do harpacticoid grazing rates differ over a tidal cycle? Field verification using chlorophyll-pigment analyses. *Marine Ecology Progress Series* 45, 263–270.

Decho, A.W. and Lopez, G.R. (1992) Exopolymer microenvironments of microbial flora: multiple and interactive effects on trophic relationships. *Limnology and Oceanography* 38, 1633–1645.

De Mesel, I., Derycke, S., Swings, J., Vincx, M. and Moens, T. (2003) Influence of bacterivorous nematodes on the decomposition of cordgrass. *Journal of Experimental Marine Biology and Ecology* 296, 227–242.

De Mesel, I., Derycke, S., Moens, T., Van der Gucht, K., Vincx, M. and Swings, J. (2004) Top–down impact of bacterivorous nematodes on the bacterial community structure: a microcosm study. *Environmental Microbiology* 6, 733–744.

De Soyza, K. (1973) Energetics of *Aphelenchus avenae* in monoxenic culture. *Proceedings of the Helminthological Society of Washington* 40, 1–10.

Dittmann, S. (1993) Impact of foraging soldiercrabs (Decapoda: Mictyridae) on meiofauna in a tropical tidal flat. *Revista de Biologia Tropical* 41, 627–637.

Duft, M., Fittkau, K. and Traunspurger, W. (2002) Colonization of exclosures in a Costa Rican stream: effects of macrobenthos on meiobenthos and the nematode community. *Journal of Freshwater Ecology* 17, 531–541.

Dusenbery, D.B. (1989) A simple animal can use a complex stimulus pattern to find a location – nematode thermotaxis in soil. *Biological Cybernetics* 60, 431–437.

Dusenbery, D.B. (1996) Information is where you find it. *Biological Bulletin* 191, 124–128.

Dye, A.H. and Lasiak, T.A. (1986) Microbenthos, meiobenthos and fiddler crabs: trophic interactions in a tropical mangrove sediment. *Marine Ecology Progress Series* 32, 259–264.

Elliott, E.T., Anderson, R.V., Coleman, D.C. and Cole, C.V. (1980) Habitable pore space and microbial trophic interactions. *Oikos* 35, 327–335.

Epstein, S.S. (1997) Microbial food webs in marine sediments. 1. Trophic interactions and grazing rates in two tidal flat communities. *Microbial Ecology* 34, 188–198.

Epstein, S.S. and Gallagher, E.D. (1992) Evidence for facilitation and inhibition of ciliate population growth by meiofauna and macrofauna on a temperate zone sandflat. *Journal of Experimental Marine Biology and Ecology* 155, 27–39.

Epstein, S.S. and Shiaris, M.P. (1992) Rates of microbenthic and meiobenthic bacterivory in a temperate muddy tidal flat community. *Applied and Environmental Microbiology* 58, 2426–2431.

Ettema, C.H., Rathburn, S.L. and Coleman, D.C. (2000) On spatiotemporal patchiness and the coexistence of five species of *Chronogaster* (Nematoda: Chronogasteridae) in a riparian wetland. *Oecologia* 125, 444–452.

Farrington, J. (1992) Overview and key recommendations. Marine organic chemistry workshop, January 1990. *Marine Chemistry* 39, 5–9.

Feller, R.J. and Coull, B.C. (1995) Non-selective ingestion of meiobenthos by juvenile spot (*Leiostomus xanthurus*) (Pisces) and their daily ration. *Vie Milieu* 45, 49–59.

Findlay S. and Tenore, K.R. (1982) Effect of a free-living marine nematode (*Diplolaimella chitwoodi*) on detrital carbon mineralization. *Marine Ecology Progress Series* 8, 161–166.

Fitzhugh, G.R. and Fleeger, J.W. (1985) Goby (Pisces: Gobiidae) interactions with meiofauna and small macrofauna. *Bulletin of Marine Science* 36, 436–444.

Gallucci, F., Steyaert, M. and Moens, T. (2005) Can field distributions of marine predacious nematodes be explained by sediment constraints on their foraging success? *Marine Ecology Progress Series* (in press).

Gee, J.M. (1989) An ecological and economic review of meiofauna as food for fish. *Zoological Journal of the Linnaean Society* 96, 243–261.

Gregg, C.S. and Fleeger, J.W. (1998) Grass shrimp *Palaemonetes pugio* predation on sediment- and stem-swelling meiofauna: field and laboratory experiments. *Marine Ecology Progress Series* 175, 77–86.

Grewal, P.S. and Wright, D.J. (1992) Migration of *Caenorhabditis elegans* (Nematoda: Rhabditidae) larvae towards bacteria and the nature of the bacterial stimulus. *Fundamental and Applied Nematology* 15, 159–166.

Hamels, I., Moens, T., Muylaert, K. and Vyverman, W. (2001a) Trophic interactions between ciliates and nematodes from an intertidal flat. *Aquatic Microbial Ecology* 26, 61–72.

Hamels, I., Muylaert, K., Casteleyn, G. and Vyverman, W. (2001b) Uncoupling of bacterial production and flagellate grazing in aquatic sediments: a case study from an intertidal flat. *Aquatic Microbial Ecology* 25, 31–42.

Hedqvist-Johnson, K. and Andre, C. (1991) The impact of the brown shrimp, *Crangon crangon* L., on soft-bottom meiofauna: an experimental approach. *Ophelia* 34, 41–49.

Heip, C., Vincx, M. and Vranken, G. (1985) The ecology of marine nematodes. *Oceanography and Marine Biology: An Annual Review* 23, 399–489.

Hellwig-Armonies, M., Armonies, W. and Lorenzen, S. (1991) The diet of *Enoplus brevis* (Nematoda) in a supralittoral salt marsh of the North Sea. *Helgoländer Meeresuntersuchungen* 45, 357–372.

Hipp, E., Bickel, U., Mustafa, T. and Hoffmann, K.H. (1986) Integumentary uptake of acetate and propionate (VFA) by *Tubifex* sp., a freshwater polychaete. II. Role of VFA as nutritional resources and effects on anaerobiosis. *Journal of Experimental Zoology* 240, 299–308.

Höckelmann, C., Moens, T. and Jüttner, F. (2004) Odor compounds from cyanobacterial biofilms acting as attractants and repellents for free-living nematodes. *Limnology and Oceanography* 49, 1809–1819.

Hoffman, J.A., Katz, J. and Bertness, M.D. (1984) Fiddler crab deposit-feeding and meiofaunal abundance in salt marsh habitats. *Journal of Experimental Marine Biology and Ecology* 82, 161–174.

Höss, S., Bergtold, M., Haitzer, M., Traunspurger, W. and Steinberg, C.E.W. (2001) Refractory dissolved organic matter can influence the reproduction of *Caenorhabditis elegans* (Nematoda). *Freshwater Biology* 46, 1–10.

Huettel, R.N. (1986) Chemical communicators in nematodes. *Journal of Nematology* 18, 3–8.

Ingham, R.E., Trofymow, J.A., Ingham, E.R. and Coleman, D.C. (1985) Interactions of bacteria, fungi, and their nematode grazers: effects on nutrient cycling and plant growth. *Ecological Monographs* 55, 119–140.

Jensen, P. (1982) Diatom-feeding behaviour of the free-living marine nematode *Chromadorita tenuis*. *Nematologica* 28, 71–76.

Jensen, P. (1987) Feeding ecology of free-living aquatic nematodes. *Marine Ecology Progress Series* 35, 187–196.

Johnson, T.E. (1985) Aging in *Caenorhabditis elegans*: Update 1984. In: Rothstein, M. (ed.) *Review of Biological Research in Aging*, vol. 2. Alan R. Liss , New York, pp. 45–60.

Johnstone, B.H. and Jones, R.D. (1989) A study on the lack of [methyl-^{3}H]thymidine uptake and incorporation by chemolithotrophic bacteria. *Microbial Ecology* 18, 73–77.

Kennedy, A.D. (1993) Minimal predation upon meiofauna by endobenthic macrofauna in the Exe Estuary, south west England. *Marine Biology* 117, 311–319.

Khan, Z., Bilgrami, A.L. and Jairajpuri, M.S. (1991) Observations on the predation ability of *Aporcelaimellus nivalis* (Altherr, 1952) Heyns, 1966 (Nematoda, Dorylaimida). *Nematologica* 37, 333–342.

Khan, Z., Bilgrami, A.L. and Jairajpuri, M.S. (1995) A comparative study on the predation by *Allodorylaimus americanus* and *Discolaimus silvicolus* (Nematoda: Dorylaimida) on different species of plant parasitic nematodes. *Fundamental and Applied Nematology* 18, 99–108.

Kneib, R.T. (1985) Predation and disturbance by grass shrimp, *Palaemonetes pugio* Holthuis, in soft-substratum benthic invertebrate assemblages. *Journal of Experimental Marine Biology and Ecology* 93, 91–102.

Lawton, J.H. (1994) What do species do in ecosystems? *Oikos* 71, 367–374.

Lopez, G., Riemann, F. and Schrage, M. (1979) Feeding biology of the brackish water oncholaimid nematode *Adoncholaimus thalassophygas*. *Marine Biology* 54, 311–318.

Mapes, C.J. (1965) Structure and function of the nematode pharynx. 2. Pumping in *Panagrellus*, *Aplectana* and *Rhabditis*. *Parasitology* 55, 583–594.

Meyers, S.P. and Hopper, B.E. (1966) Attraction of the marine nematode, *Metoncholaimus* sp., to fungal substrates. *Bulletin of Marine Science* 16, 142–150.

Middelburg, J.J., Barranguet, C., Boschker, H.T.S., Herman, P.M.J., Moens, T. and Heip, C.H.R. (2000) The fate of intertidal microphytobenthos carbon: an *in situ* ^{13}C-labeling study. *Limnology and Oceanography* 45, 1224–1234.

Mikola, J. and Setälä, H. (1998) Relating species diversity to ecosystem functioning: mechanistic backgrounds and experimental approach with a decomposer food web. *Oikos* 83, 180–194.

Moens, T. and Vincx, M. (1997) Observations on the feeding ecology of estuarine nematodes. *Journal of the Marine Biological Association of the United Kingdom* 77, 211–227.

Moens, T. and Vincx, M. (2000) Temperature, salinity and food thresholds in two brackish-water bacterivorous nematode species: assessing niches from food absorption and respiration experiments. *Journal of Experimental Marine Biology and Ecology* 243, 137–154.

Moens, T., Vierstraete, A. and Vincx, M. (1996) Life strategies in two bacterivorous marine nematodes: preliminary results. *P.S.Z.N. I Marine Ecology* 17, 509–518.

Moens, T., Verbeeck, L., De Maeyer, A., Swings, J. and Vincx, M. (1999a) Selective attraction of marine bacterivorous nematodes to their bacterial food. *Marine Ecology Progress Series* 176, 165–178.

Moens, T., Verbeeck, L. and Vincx, M. (1999b) Preservation- and incubation-time induced bias in tracer-aided grazing studies with meiofauna. *Marine Biology* 133, 69–77.

Moens, T., Verbeeck, L. and Vincx, M. (1999c) Feeding biology of a predatory and a facultatively predatory marine nematode (*Enoploides longispiculosus* and *Adoncholaimus fuscus*). *Marine Biology* 134, 585–593.

Moens, T., Herman, P.M.J., Verbeeck, L., Steyaert, M. and Vincx, M. (2000) Predation rates and prey selectivity in two predacious estuarine nematodes. *Marine Ecology Progress Series* 205, 185–193.

Moens, T., Luyten, C., Middelburg, J.J., Herman, P.M.J. and Vincx, M. (2002) Tracing organic matter sources of estuarine tidal flat nematodes with stable carbon isotopes. *Marine Ecology Progress Series* 234, 127–137.

Moens, T., Yeates, G.W. and De Ley, P. (2004) Use of carbon and energy sources by nematodes. *Nematology Monographs and Perspectives* 2, 529–545.

Moens, T., dos Santos, G.A.P., Thompson, F., Swings, J., Fonsêca-Genevois, V., Vincx, M. and DeMesel, I. (2005) Do nematode mucus secretions affect bacterial growth? *Aquatic Microbial Ecology* 40, 77–83.

Montagna, P.A. (1983) Live controls for radioisotope tracer food chain experiments using meiofauna. *Marine Ecology Progress Series* 12, 43–46.

Montagna, P.A. (1984a) Competition for dissolved glucose between meiobenthos and sediment microbes. *Journal of Experimental Marine Biology and Ecology* 76, 177–190.

Montagna, P.A. (1984b) *In situ* measurement of meiobenthic grazing rates on sediment bacteria and edaphic diatoms. *Marine Ecology Progress Series* 18, 119–130.

Montagna, P.A. (1995) Rates of metazoan meiofaunal microbivory: a review. *Vie Milieu* 45, 1–9.

Montagna, P.A. and Bauer, J.E. (1988) Partitioning radiolabeled thymidine uptake by bacteria and meiofauna using metabolic blocks and poisons in benthic feeding studies. *Marine Biology* 98, 101–110.

Montagna, P.A., Blanchard, G.F. and Dinet, A. (1995) Effect of production and biomass of intertidal microphytobenthos on meiofaunal grazing rates. *Journal of Experimental Marine Biology and Ecology* 185, 149–165.

Nehring, S. (1992a) Die Vegetarier unter den freilebenden Nematoden. 1. Die Aufwuchsfresser. *Mikrokosmos* 81, 135–140.

Nehring, S. (1992b) Die Vegetarier unter den freilebenden Nematoden. 2. Die Depositfresser. *Mikrokosmos* 81, 260–266.

Nehring, S., Jensen, P. and Lorenzen, S. (1990) Tube-dwelling nematodes: tube construction and possible ecological effects on sediment–water interfaces. *Marine Ecology Progress Series* 64, 123–128.

Nicholas, W.L. (1984) *The Biology of Free-living Nematodes*, 2nd edn. Clarendon Press, Oxford, UK.

Nicholas, W.L., Grassia, A. and Viswanathan, S. (1973) The efficiency with which *Caenorhabditis briggsae* (Rhabditinae) feeds on the bacterium *Escherichia coli*. *Nematologica* 19, 411–420.

Nilsson, P., Sundbäck, K. and Jönsson, B. (1993) Effect of the brown shrimp *Crangon crangon* L. on endobenthic macrofauna, meiofauna and meiofauna grazing rates. *Netherlands Journal of Sea Research* 31, 95–106.

Nussbaumer, A.D., Bright, M., Baranyi, C., Beisser, C.J. and Ott, J.A. (2004) Attachment mechanism in a highly specific association between ectosymbiotic bacteria and marine nematodes. *Aquatic Microbial Ecology* 34, 239–246.

Ólafsson, E. and Elmgren, R. (1991) Effects of biological disturbance by benthic amphipods *Monoporeia affinis* on meiobenthic community structure: a laboratory approach. *Marine Ecology Progress Series* 74, 99–107.

Ólafsson, E. and Moore, C.G. (1990) Control of meiobenthic abundance by macroepifauna in a subtidal muddy habitat. *Marine Ecology Progress Series* 65, 241–249.

Ólafsson, E. and Moore, C.G. (1992) Effects of macroepifauna on developing nematode and harpacticoid assemblages in a subtidal muddy habitat. *Marine Ecology Progress Series* 84, 161–171.

Ólafsson, E., Elmgren, R. and Papakosta, O. (1993) Effects of the deposit-feeding bivalve *Macoma balthica* on meiobenthos. *Oecologia* 93, 457–462.

Ott, J., Bright, M. and Bulgheresi, S. (2004) Symbioses between marine nematodes and sulfur-oxidizing chemoautotrophic bacteria. *Symbiosis* 36, 103–126.

Penry, D.L. (1998) Applications of efficiency measurements in bioaccumulation studies: definitions, clarifications, and a critique of methods. *Environmental Toxicology and Chemistry* 17, 1633–1639.

Perry, R.N. (1996) Chemoreception in plant parasitic nematodes. *Annual Review of Phytopathology* 34, 181–199.

Petersen, H. and Luxton, M. (1982) A comparative analysis of soil fauna populations and their roles in decomposition processes. *Oikos* 39, 287–388.

Phillips, N.W. (1984) Role of different microbes and substrates as potential suppliers of specific, essential nutrients to marine detritivores. *Bulletin of Marine Science* 35, 283–298.

Polz, M.F., Distel, D.L., Zarda, B., Amann, R., Felbeck, H., Ott, J.A. and Cavanaugh, C.M. (1994) Phylogenetic analysis of a highly specific association between ectosymbiotic, sulfur-oxidizing bacteria and a marine nematode. *Applied and Environmental Microbiology* 60, 4461–4467.

Pomeroy, L.R. (1974) The oceans food web, a changing paradigm. *Bioscience* 24, 499–504.

Post, D.M. (2002) Using stable isotopes to estimate trophic position: models, methods and assumptions. *Ecology* 83, 703–718.

Rau, G.H., Mearns, A.J., Young, D.R., Olson, R.J., Schafer, H.A. and Kaplan, I.R. (1983) Animal $^{13}C/^{12}C$ correlates with trophic level in pelagic food webs. *Ecology* 64, 1314–1318.

Reise, K. (1979) Moderate predation on meiofauna by the macrobenthos of the Wadden Sea. *Helgoländer wissenschaftliche Meeresuntersuchungen* 32, 453–465.

Riemann, F. and Helmke, E. (2002) Symbiotic relations of sediment-agglutinating nematodes and bacteria in detrital habitats: the enzyme-sharing concept. *P.S.Z.N. I Marine Ecology* 23, 93–113.

Riemann, F. and Schrage, M. (1978) The mucus-trap hypothesis on feeding of aquatic nematodes and implications for biodegradation and sediment texture. *Oecologia* 34, 75–88.

Riemann, F. and Schrage, M. (1988) Carbon dioxide as an attractant for the free-living marine nematode *Adoncholaimus thalassophygas*. *Marine Biology* 98, 81–85.

Riemann, F., Ernst, W. and Ernst, R. (1990) Acetate uptake from ambient water by the free-living marine nematode *Adoncholaimus thalassophygas*. *Marine Biology* 104, 453–457.

Romeyn, K. and Bouwman, L.A. (1983) Food selection and consumption by estuarine nematodes. *Hydrobiological Bulletin* 17, 103–109.

Romeyn, K., Bouwman, L.A. and Admiraal, W. (1983) Ecology and cultivation of the herbivorous brackish-water nematode *Eudiplogaster pararmatus*. *Marine Ecology Progress Series* 12, 145–153.

Rothstein, M. (1970) Nematode biochemistry XI. *International Journal of Biochemistry* 1, 422–428.

Rothstein, M. and Götz, T. (1968) Biosynthesis of fatty acids in the free-living nematode, *Turbatrix aceti*. *Archives of Biochemistry and Biophysics* 126, 131–140.

Rothstein, M. and Mayoh, H. (1966) Nematode biochemistry – VIII. *Compendium of Biochemistry and Physiology* 17, 1181–1188.

Rowe, R.J. (1994) Predatory behaviour and predatory versatility in young larvae of the dragonfly *Xanthocnemis zealandica* (Odonata, Coenagrionidae). *New Zealand Journal of Zoology* 21, 151–166.

Ruess, L. and Ferris, H. (2004) Decomposition pathways and successional changes. *Nematology Monographs and Perspectives* 2, 547–556.

Sanders, R.W., Porter, K.G., Bennett, S.J. and DeBiase, A.E. (1989) Seasonal patterns of bacterivory by flagellates, ciliates, rotifers, and cladocerans in a freshwater planktonic community. *Limnology and Oceanography* 34, 673–687.

Santos, P.J.P., Castel, J. and Souza-Santos, L.P. (1995) Microphytobenthic patches and their influence on meiofaunal distribution. *Cahiers de Biologie Marine* 36, 133–139.

Schiemer, F. (1983) Food dependence and energetics of free-living nematodes. III. Comparative aspects of special consideration of two bacterivorous species *Caenorhabditis briggsae* and *Plectus palustris*. *Oikos* 41, 32–43.

Schiemer, F. (1985) Bioenergetic niche differentiation of aquatic invertebrates. *Verhandlungen der Internationale Vereinigung für Theoretische und Angewandte Limnologie* 22, 3014–3018.

Schiemer, F. (1987) Nematoda. In: Vernberg, J.F. and Pandian, T.J. (eds) *Animal Energetics*, vol. 1. Academic Press, New York, pp. 185–215.

Schmid, P.E. and Schmid-Araya, J.M. (2002) Trophic relationships in temporary and permanent freshwater meiofauna. In: Rundle, S.D., Robertson, A.L. and Schmid-Araya, J.M. (eds) *Freshwater Meiofauna*. Backhuys Publishers, Leiden, The Netherlands, pp. 295–319.

Schmid-Araya, J.M. and Schmid, P.E. (2000) Trophic relationships: integrating meiofauna into a realistic benthic food web. *Freshwater Biology* 44, 149–163.

Schönborn, W. (1985) Die ökologische Rolle von *Erpobdella octoculata* (l.) (Hirudinea: Erpobdellidae) in einem abwasserbelasteten Fluβ (Saale). *Zoogische Jahrbücher (Systematik)* 112, 477–494.

Schratzberger, M. and Warwick, R.M. (1999) Impact of predation and sediment disturbance by *Carcinus maenas* (L.) on free-living nematode community structure. *Journal of Experimental Marine Biology and Ecology* 235, 255–271.

Schrijvers, J., Okondo, J., Steyaert, M. and Vincx, M. (1995) Influence of epibenthos on meiobenthos of the *Ceriops tagal* mangrove sediment at Gazi Bay, Kenya. *Marine Ecology Progress Series* 128, 247–259.

Schwank, P. (1981) Turbellarien, Oligochaeten und Archianneliden des Breitenbachs und anderer oberhessischer Mittelgebirgsbäche. II. Die Systematik und Autökologie der einzelnen Arten. *Archiv für Hydrobiologie*, Suppl. (Monogr. Beiträge) 62, 86–147.

Sherr, E.B. and Sherr, B.F. (1993) Protistan grazing rates via uptake of fluorescently labeled prey. In: Kemp, P.F., Sherr, B.F., Sherr, E.B. and Cole, J.J. (eds) *Handbook of Methods in Aquatic Microbial Ecology*. Lewis Publishers, Boca Raton, Florida, pp. 695–701.

Somerfield, P., Warwick, R.M. and Moens, T. (2005) Meiofauna techniques, chapter 6. In: McIntyre, A. and Eleftheriou, A. (eds) *Methods for the Study of Marine Benthos*, 3rd edn. Blackwell Science, Oxford, UK, pp. 229–272.

Souza-Santos, L.P., Castel, J. and Santos, P.J.P. (1995) Feeding rate cycle of the epibenthic harpacticoid copepod *Harpacticus flexus*: laboratory experiments using fecal pellet counts. *Vie Milieu* 45, 75–83.

Starink, M., Krylova, I.N., Bär-Gilissen, N.J., Bak, R.P.M. and Cappenberg, T.E. (1994) Rates of benthic protozoan grazing on free and attached sediment bacteria measured with fluorescently stained sediment. *Applied and Environmental Microbiology* 60, 2259–2264.

Stephens, G.C. and Schinske, R.A. (1961) Uptake of amino acids by marine invertebrates. *Limnology and Oceanography* 6, 175–181.

Tempel, D. and Westheide W. (1980) Uptake and incorporation of dissolved amino acids by interstitial Turbellaria and Polychaeta and their dependence on temperature and salinity. *Marine Ecology Progress Series* 41, 41–50.

Thomas, J.H. (1989) Genetic analysis of defecation in *Caenorhabditis elegans*. *Genetics* 124, 855–872.

Thurman, E.M. (1985) *Organic Geochemistry of Natural Waters*. Martinus Nijhof/Dr. W. Junk Publishers, Dordrecht, The Netherlands.

Tita, G., Desrosiers, G., Vincx, M. and Nozais, C. (2000) Predation and sediment disturbance effects of the intertidal polychaete *Nereis virens* (Sars) on associated meiofaunal assemblages. *Journal of Experimental Marine Biology and Ecology* 243, 261–282.

Toman, M.J. and Dall, P.C. (1997) The diet of *Erpobdella octoculata* (Hirudinea: Erpobdellidae) in two Danish lowland streams. *Archiv für Hydrobiologie* 140, 549–563.

Traunspurger, W. (1992) A study of the free-living freshwater nematodes of hard substrates in the littoral of the oligotrophic Königssee (National Park Berchtesgaden, F.R.G.). *Spixiana* 15, 233–238.

Traunspurger, W. (1997) Bathymetric, seasonal and vertical distribution of feeding-types of nematodes in an oligotrophic lake. *Vie Milieu* 47, 1–7.

Traunspurger, W., Bergtold, M. and Goedkoop, W. (1997) The effects of nematodes on bacterial activity and abundance in a freshwater sediment. *Oecologia* 112, 118–122.

Van der Meer, J., Heip, C.H.R., Herman, P.M.J., Moens, T. and Van Oevelen, D. (2005) Flow of energy and matter in marine benthic animal populations, chapter 8. In: McIntyre, A. and Eleftheriou, A. (eds) *Methods for the Study of Marine Benthos*, 3rd edn. Blackwell Science, Oxford, UK, pp. 326–408.

Vander Zanden, M.J. and Rasmussen, J.B. (2001) Variation in $\delta^{15}N$ and $\delta^{13}C$ trophic fractionation: implications for aquatic food web studies. *Limnology and Oceanography* 46, 2061–2066.

Venette, R.C. and Ferris, H. (1998) Influence of bacterial type and density on population growth of bacterial-feeding nematodes. *Soil Biology and Biochemistry* 30, 949–960.

Warwick, R.M. (1981) Survival strategies of meiofauna. In: Jones, N.V. and Wolff, W.J. (eds) *Feeding and Survival Strategies of Estuarine Organisms*. Plenum Press, New York, pp. 39–52.

Wetzel, R.G. (2001) *Limnology. Lake and River Ecosystems*, 3rd edn. Academic Press, San Diego, California.

Wieser, W. (1953) Die Beziehung zwishen Mundhöhlengestalt, Ernährungsweise und Vorkommen bei freilebenden marinen Nematoden. *Arkiv für Zoologie* 4, 439–484.

Wilson, J.B. (1999) Guilds, functional types and ecological groups. *Oikos* 86, 507–522.

Woombs, M. and Laybourn-Parry, J. (1984) Feeding biology of *Diplogasteritus nudicapitatus* and *Rhabditis curvicaudata* (Nematoda) related to food concentration and temperature, in sewage treatment plants. *Oecologia* 64, 163–167.

Yeates, G.W. (1969) Predation by *Mononchoides potohikus* (Nematoda: Diplogasteridae) in laboratory culture. *Nematologica* 15, 1–9.

Yeates, G.W. (1987a) Nematode feeding and activity – the importance of development stages. *Biology and Fertility of Soils* 3, 143–146.

Yeates, G.W. (1987b) Significance of developmental stages in the coexistence of 3 species of Mononchoidea (Nematoda) in a pasture soil. *Biology and Fertility of Soils* 5, 225–229.

Yeates, G.W. (1998) Feeding in free-living soil nematodes: a functional approach. In: Perry, R.N. and Wright, D.J. (eds) *The Physiology and Biochemistry of Free-living and Plant-parasitic Nematodes*. CAB International, Wallingford, UK, pp. 245–269.

Yeates, G.W., Bongers, T., de Goede, R.G.M., Freckman, D.W. and Georgieva, S.S. (1993) Feeding habits in soil nematode families and genera – an outline for soil ecologists. *Journal of Nematology* 25, 315–331.

7 Patterns in the Size Structure of Freshwater Nematode Communities: the Cases of Lakes Königssee and Brunnsee, Germany

W. Traunspurger and M. Bergtold

Department of Animal Ecology, University of Bielefeld, Morgenbreede 45, 33615 Bielefeld, Germany

Population, community and ecosystem ecologists have been interested in the patterns and implications of organismal size distribution. It is well known that rates of many physiological processes are related to body size (Peters, 1983), which suggests that rates of such processes as energy use, respiration, production or nutrient cycling might be inferred from knowledge of the body sizes of the organisms in a community (e.g. Sheldon *et al.*, 1977; Gerlach *et al.*, 1985). A size-based approach has also been used widely in studies of predator–prey interactions. Fish predation on zooplankton is strongly size-selective, so that the effects of fish on a zooplankton assemblage could be shown clearly by using body size, rather than taxonomy, as a descriptor of the zooplankton community (e.g. Brooks and Dodson, 1965). The size structure of planktonic communities has received much attention from ecologists interested in an alternative to the traditional taxonomic description and has been used to compare communities from different environments (e.g. Sheldon *et al.*, 1972; Strayer, 1991, 1994; Cyr and Pace, 1993; Gaedke, 1993; Duplisea, 2000).

In the early 1980s, benthic ecologists started to use a size-based approach. Most studies are limited to meio- and macroinvertebrates (Warwick, 1984; Strayer, 1986; Hanson *et al.*, 1989; Hanson, 1990; Morin and Nadon, 1991; Poff *et al.*, 1993; Rasmussen, 1993) and rarely extend to protozoa and algae (Schwinghamer, 1981; Cattaneo, 1987, 1993; Schmid *et al.*, 2000). This may be caused by the fact that size analyses are more time consuming for benthic animals because organisms have to be separated from detritus and sediment before measurement. Some peculiarities of benthic size distributions, like the marked differences between habitats (Strayer, 1991) and the low biomasses in the smaller size classes (Morin and Nadon, 1991) could simply reflect these methodological problems; some size classes could be shared by invertebrates and the large algae that can develop in attached communities (Cattaneo, 1993).

Body Size and Abundance

Damuth (1981, 1987) showed that, across species covering a wide range of body masses, population density increases as body size decreases. However, several studies have shown that the density–body size relation has a peaked or polygonal pattern, with intermediate-sized species having the highest density (Siemann *et al.*, 1996). The ability to predict and understand why many species are rare while others are extremely abundant is not only critical for ecological theories, but also has important implications for conservation biology (Blackburn and Gaston, 1994). Body size (W), an easily measurable attribute of animal species, has been suggested as a potentially good predictor of ecological density (D) (Peters, 1983; Peters and Wassenberg, 1983; Carrascal and Telleria, 1991). Average population density appeared to be related to body size as $D = W^{-0.75}$, using ordinary least squares regression. Since individual metabolic requirements scale with body mass to the power of 0.75, Damuth (1981, 1987) suggested that the −0.75 exponent between density and body size was the reflection of macroevolutionary processes acting to keep population energy use of all animal species at similar levels, independent of body size. This became known as the 'energetic equivalence rule' (Damuth, 1991). However, later studies raised several problems with this pattern and with its putative explanation (Navarrete and Menge, 1997; Arneberg and Anderson, 2003). One of the main criticisms is that the relationship found in literature compendia could be greatly affected by sampling bias against small and rare species, which are usually not well represented in ecological studies.

Our ability to discern the shape of body size–abundance relationships and to test whether these shapes are consistent with hypothesized constraints is limited by the paucity of suitable data sets that span a wide range of body sizes (Strayer, 1994). In an extensive study covering a wide size range and many different taxa, Schmid *et al.* (2000) found a strong correlation with a negative slope in lotic environments. The slope of the regression varied widely among the taxonomic groups, leading the authors to the conclusion that a single ecological model cannot describe the pattern. A numerically dominant organism group may influence the abundance–size relationship within the studied community. Nematodes are very abundant in many benthic habitats and in this chapter we want to provide information on the size structure of this organism group in the littoral, littoriprofundal and profundal zones of Lake Königssee and the profundal zone of Lake Brunnsee (Traunspurger, 1991, 1996a,b; Bergtold and Traunspurger, 2004). Nematodes ranged in length from about 0.2 to 4.0 mm (0.01–32 µg wet weight (WW) per individual) and include about 130 species for both lakes. Species number varied in Lake Königssee from 31 species in 190 m water depth to 59 species in 30 m water depth. In the profundal of Lake Brunnsee 43 species were determined. Body size of nematode species was directly measured and population densities were quantified. Body mass (µg WW) was calculated after Andrássy (1956). Data are available from ten different water depths of Lake Königssee (littoral: 1, 2, 5, 10 m; littoriprofundal: 15, 20, 30 m; profundal: 60, 120, 190 m) over 1 year (monthly sampled) and from one water depth (profundal: 17 m) over 9 months of Lake Brunnsee. Three different life stages of each species (females, males, juveniles) were distinguished and each live stage was measured separately (density, body mass). In total, 45,263 individuals from Lake

Königssee and 5881 from Lake Brunnsee were analysed. Furthermore, individual nematodes (juveniles stage 1–3, juveniles stage 4, adults) were classified according to their WW into size classes of 0.05 μg width (which nearly corresponds to lowest body mass) and densities within these classes were summed up. We are aware of the fact that this arbitrary selection of the size classes has implications for the slope of the regression (Blackburn *et al.*, 1992).

The relationship of density and body mass (accuracy is 0.01 μg WW) of Lake Königssee and Lake Brunnsee nematodes is shown in Fig. 7.1. Summing up biomass in size classes (0.05 μg WW) population density is negatively correlated with body mass in Lake Königssee, both in the littoral, littoriprofundal and profundal zones and also for each water depth (Figs 7.2 and 7.3). The slope of the regression in the littoral zone is −1.13, and in the littoriprofundal and profundal −0.69 and for all the data (whole Lake Königssee) −0.97 (Table 7.1).

In general, small individuals (species) were more abundant than large individuals (species) and this is supported by strong correlations (littoral: r^2 = 0.75; littoriprofundal: r^2 = 0.67; profundal: r^2 = 0.69). The slopes show differences between water depths and range from −0.49 to −1.19 (Fig. 7.3). However, below 10 m water depth the slope is < −0.75, while in the littoral (1, 2 and 5 m) the slope ranges between −0.8 and −1.19 (r^2 between 0.43 and 0.80).

In Lake Brunnsee only weak correlations were observed (Fig. 7.4; Table 7.1). The slope for the whole study period was −0.66. No significant relation was found in spring, while in summer and autumn the slopes were −0.63 and −0.55, respectively.

The clear correlation between body mass and population density of nematodes in Lake Königssee contrasts with Lake Brunnsee and most other community-level analyses, which have reported no or weak correlations between body size and population density (e.g. Blackburn *et al.*, 1993). Possibly the large data set (about 45,000 individuals), the high number of species (about 120), the differentiation of three life stages and the introduction of size classes could be the reasons for this finding. In the zoobenthos study of Mirror Lake (Strayer, 1994), body mass and density were loosely correlated within higher taxonomic groups. In the three cases for which

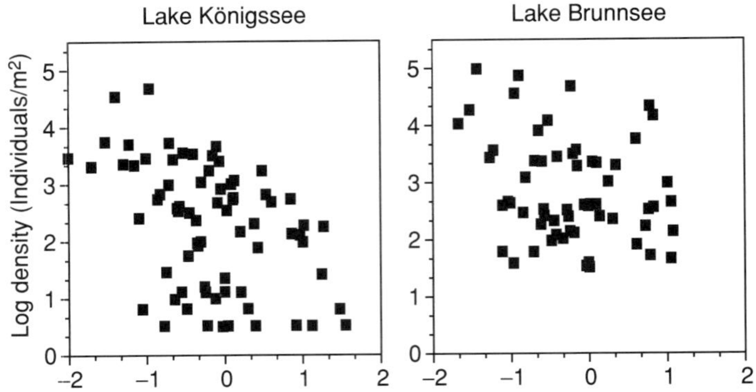

Fig. 7.1 Density of nematodes against nematode body mass in Lake Königssee (whole lake) and Lake Brunnsee (profundal).

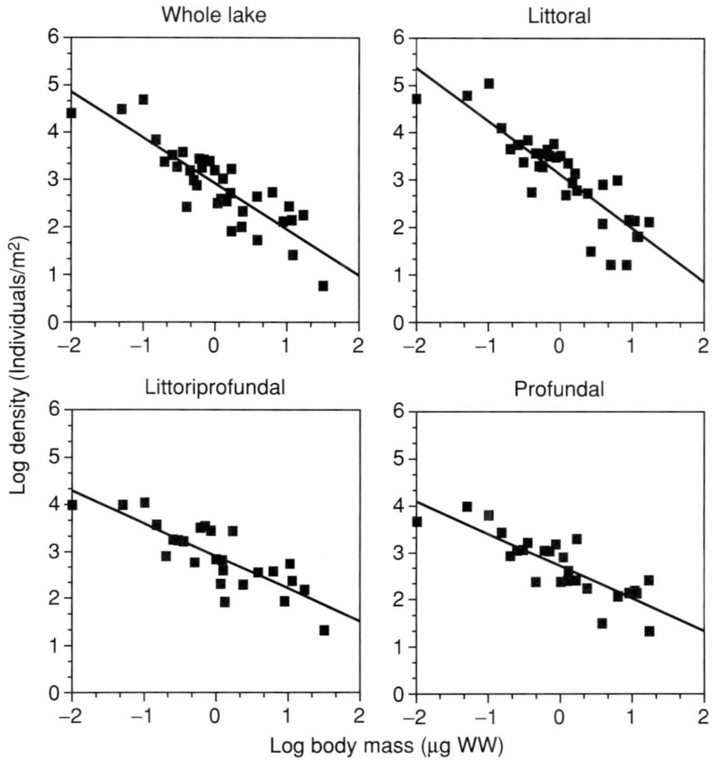

Fig. 7.2 Density of nematodes against nematode body mass in Lake Königssee (whole lake; littoral = 1–10 m; littoriprofundal = 15–30 m; profundal = 60–190 m). Nematodes were clustered in size classes of 0.05 µg wet weight (WW).

these correlations were marginally significant, the regression slope was negative. For nematodes, the authors found a weak correlation ($r^2 = 0.2$; $P = 0.06$) with a reduced major axis (RMA) slope of −1.21. Schmid *et al.* (2000) found a relatively strong correlation in the stream Oberer Seebach for nematodes with a slope of −0.9. Analysing all organisms, the authors found a slope in the range of −0.702 to −0.794.

Benthic Biomass Size Spectra

Ecologists in the pelagic zone have indeed used approaches based on body size widely and successfully over the last 30 years. Biomass size spectra are constructed by allocating all organisms (particles) into size classes according to their individual body mass (size) and computing the total biomass per size class. Consequently, biomass size distributions provide a holistic ecosystem description that facilitates ecosystem comparisons in time and space. Being a non-taxonomic approach, it permits community analysis at a general level and thus provides an alternative and complementary perspective to taxonomic analysis (Rasmussen, 1993). Biomass size

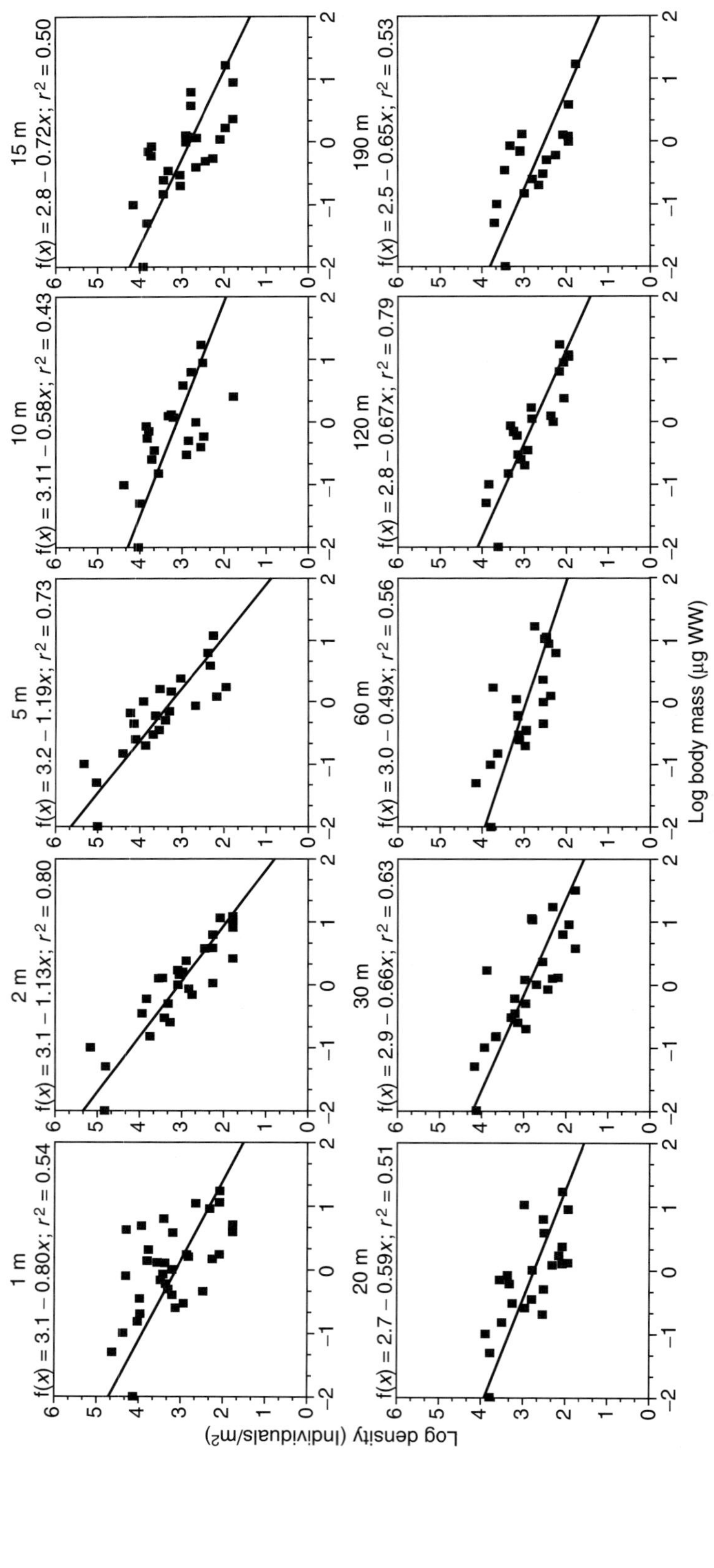

Fig. 7.3 Density of nematodes against nematode body mass at different water depths in Lake Königssee. Nematodes were clustered in size classes of 0.05 µg wet weight (WW).

Table 7.1 Linear regression (l.r.) and reduced major axis (RMA) analysis of log density and log body mass: slopes, r^2, and statistical significance of the relationship (*P* value).

Lake	Slope (l.r.)	r^2	*P*	Slope (RMA)	r^2
Königssee					
Whole lake	−0.97	0.74	0.001	−1.12	0.74
Littoral	−1.13	0.75	0.001	−1.31	0.75
Littoriprofundal	−0.69	0.67	0.001	−0.85	0.67
Profundal	−0.69	0.69	0.001	−0.83	0.69
Brunnsee					
Whole year	−0.66	0.19	0.01	−1.50	0.19
Spring	−0.13	0.01	0.70	−1.35	0.001
Summer	−0.63	0.21	0.01	−1.37	0.21
Autumn	−0.55	0.16	0.05	−1.36	0.16

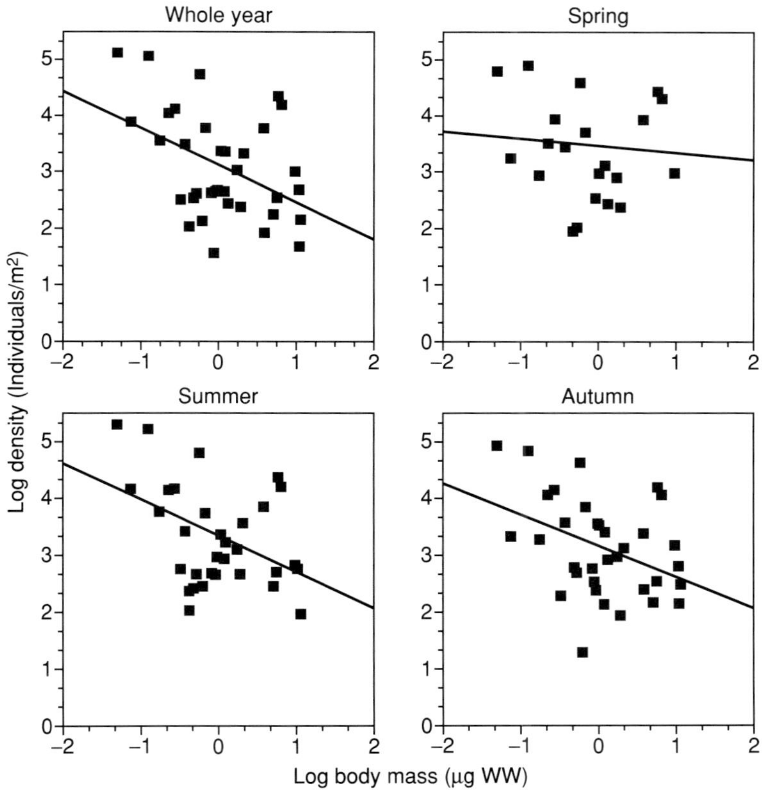

Fig. 7.4 Density of nematodes against nematode body mass at different seasons in Lake Brunnsee. Nematodes were clustered in size classes of 0.05 μg wet weight (WW).

distributions may also provide a tool for ecosystem analysis owing to the close relationships between body mass and metabolic processes and between body mass and related ecological properties of the organisms, especially in pelagic ecosystems (e.g. seasonal variability, type of nutrition) (Gaedke, 1993). The size-based approach to pelagic ecology owes its appeal to a few simplifying features such as automated measurement (e.g. electronic particle counter), its empirical simplicity and its theoretical appeal that is facilitated by the consistent predator–prey size ratios of pelagic food webs (Kerr, 1974; Borgmann, 1987).

The first biomass size spectra were obtained by Sheldon *et al.* (1972) in open oceans and comprised a restricted size range of the pelagic community based on measurements of particulate material with an electronic particle counter. Total biomass tended to be approximately uniformly distributed over logarithmically equally spaced size classes in the pelagic zone of some large aquatic environments (Sheldon *et al.*, 1972, 1977), which makes parameter estimation very simple. That is, the biomass of algae within a particular size class (logarithmic interval) is roughly the same as the biomass of algae in any other size class, which in turn is the same as the biomass of zooplankton or fish in a size class. Hence, a continuum of organisms exists with respect to body mass, which fills the potential niches defined by body mass and related parameters (e.g. generation times, turn-over rates) within the size range of plankton organisms.

The success of size-based research on the plankton led benthic ecologists to wonder about the size structure of benthic communities. Several points apply to benthic communities as well as to pelagic ones: benthic communities are rich and complex, often taxonomically difficult to describe, and they span a wide size range (about 10 orders of magnitude within the animal community). But Strayer (1991) has correctly pointed out that none of the appealing features in pelagic communities apply to benthic communities. Automated measurement is not and likely never will be possible.

Published benthic biomass size-spectra (BBSS) of infaunal eukaryotes in coastal marine systems are often not uniform (Schwinghamer, 1981; Gerlach *et al.*, 1985; Warwick and Joint, 1987). The first person to describe the size structure of an entire benthic community was Schwinghamer (1981), who took samples from several marine sites in Nova Scotia and found remarkable trimodal size spectra. Schwinghamer (1981) suggests that although benthic communities do not have constant amounts of material in logarithmically increasing size classes they do show predictable patterns. This trimodality of BBSS has been considered as a reflection of sediment grain size. The three peaks represent microbiota (e.g. bacteria and algae), meiofauna (e.g. nematodes) and macrofauna (e.g. annelids, clams). Bacteria utilize the surface of small particles, meiofauna utilize the pore space and macrofauna, the sediment–water interface. This sediment living space argument provides an intuitively satisfying explanation for BBSS trimodality because of its purely physical and external (to the organisms) forcing mechanisms; however, living organisms are not passive particles and have evolved means to overcome some of their bondage to physics and thus may defy simple physical explanations for their patterns. Warwick (1984) showed that the size structure of the marine zoobenthos was relatively unresponsive to sediment characteristics and confirmed that zoobenthic size spectra are bimodal or trimodal (with bacteria) in marine sediments. Therefore, several workers (e.g. Warwick, 1984; Strayer, 1991) suggest that the apparently constant size-spectra patterns are caused by differences in life-history strategy.

However, there have been several empirical studies in both marine and freshwater systems showing exceptions to bimodality (*sensu* Schwinghamer, 1981) lentic freshwater (Strayer, 1986), lotic freshwater (Bourassa and Morin, 1995), brackish water (Duplisea and Drgas, 1999) and marine (Edgar, 1994; Duplisea, 2000). These exceptions to bimodality indicate that a small set of physical constraints, such as sediment grain size, cannot be used to predict the size-spectra pattern in all benthic systems.

When biomass distribution across the logarithmic size categories is not constant, but follows a size trend, the slope parameter of the normalized size distribution (NSD) can be used to assess the degree of departure from the uniform biomass or 'steady-state' distribution (Sprules and Munawar, 1986). NSD plots are obtained by plotting the log (biomass/body mass interval) against the upper bound of the body mass interval (log) (Sprules and Munawar, 1986). A slope of −1 indicates a tendency towards uniform biomass, slopes shallower than −1 indicate a trend of biomass increasing with size and slopes steeper than −1 indicate a decreasing trend. Only a few data are available on the size structure of freshwater zoobenthos (e.g. Hanson, 1990; Strayer, 1991; Rasmussen, 1993).

In the following, we want to show the biomass–body size distribution of nematodes in Lake Königssee and Lake Brunnsee. Nematodes were classified according to their biomass in log 2 intervals, from 0.016 to 320 μg. Biomass in those size classes varied by 3.5 orders of magnitude over the entire range. To compare with other studies we estimated the slope parameter of the NSD (Fig. 7.5A). The slopes showed a high correlation for the whole lake and for the littoral, littoriprofundal and profundal zones and ranged between −0.69 and −1.01. In Lake Brunnsee the slope of the NSD plots ranged between −1.16 and −0.48 (Fig. 7.5B; Table 7.2). The slope is decreasing from spring to autumn, indicating that the smaller size classes are becoming more important during the year. In fact, Lake Brunnsee profundal sediment is dominated by the large nematode *Tobrilus gracilis* during springtime. In summer and autumn smaller species have been abundant.

Direct comparisons between size-spectrum parameters for planktonic and benthic organisms are complicated by several factors, e.g. differences in community structure and dimensionality. In Lake Königssee and in the study of Rasmussen (1993) the slope of the NSD plot became steeper (more negative) with increasing lake productivity (in Lake Königssee profundal to littoral), a pattern opposite to that described by Sprules and Munawar (1986) for lake plankton communities. Thus, there seems to be no general pattern relating the slope of the NSD (i.e. the biomass trends across the size spectrum) to trophic status that is valid for both pelagic and benthic systems. On the other hand benthic studies are scarce and a general statement is currently not possible. Rasmussen's (1993) analysis of the seasonal average size spectra of littoral macroinvertebrates from 53 sites in 11 lakes (NSD plots ranging from −0.6 to −1.7) provided some support for the general importance of plant community structure in determining size spectra variation. Hanson (1990) showed that the slope of the NSD was significantly different for communities in *Chara* beds (−0.89, i.e. biomass increasing with size) and those in *Potamogeton* beds (−1.11, i.e. biomass decreasing with size) within the same lake.

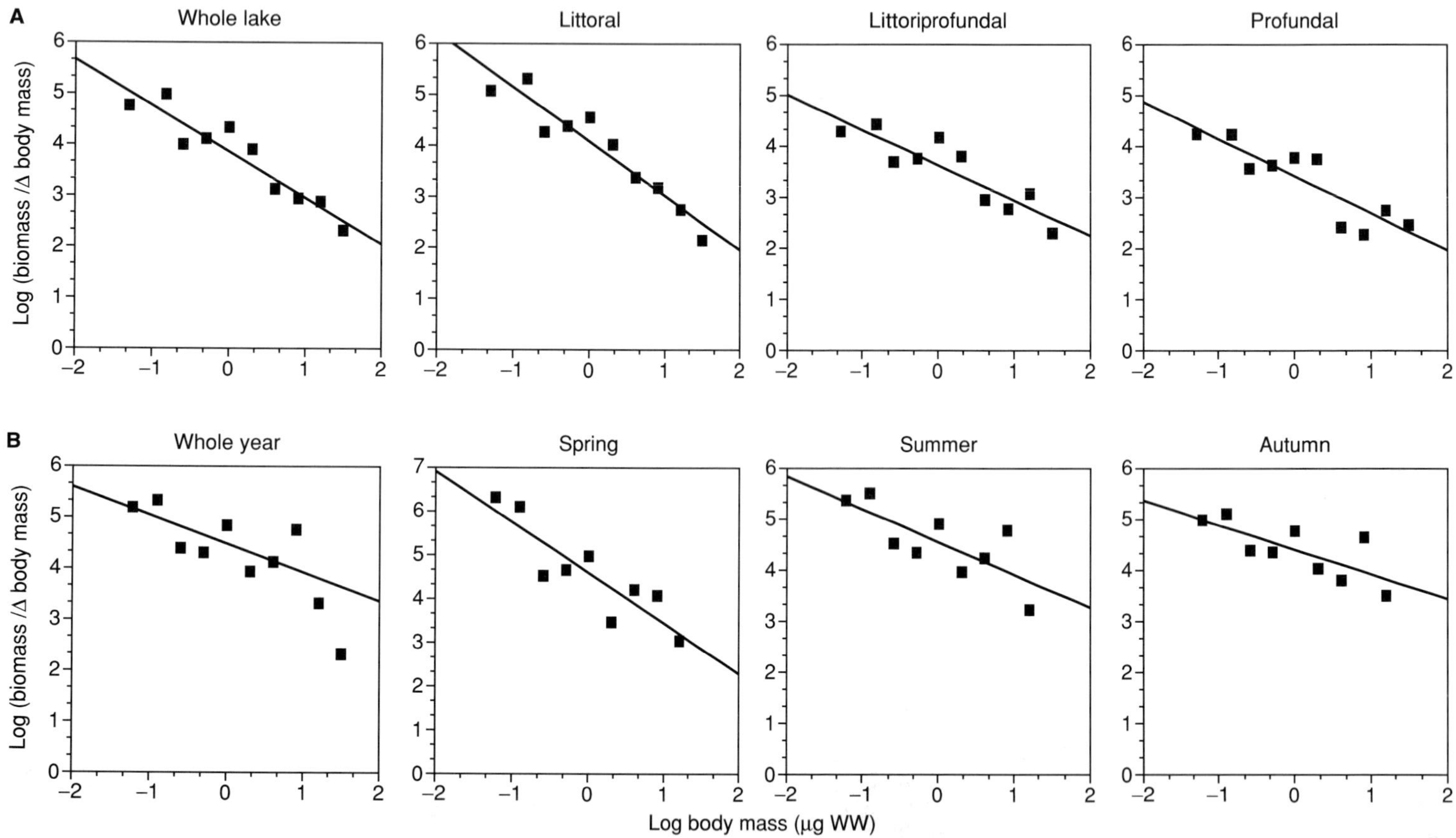

Fig. 7.5 Normalized size distribution (NSD) plots of Lake Königssee (A) at different water depths (littoral = 1–10 m; littoriprofundal = 15–30 m; profundal = 60–190 m) and of Lake Brunnsee (B) at different seasons.

Table 7.2 Linear regression (l.r.) of log (biomass/Δ body mass) and log body mass (NSD plots): slopes, r^2 and statistical significance of the relationship (*P*-value).

Lake	Slope (l.r.)	r^2	*P*
Königssee			
Whole lake	–0.91	0.89	0.001
Littoral	–1.01	0.90	0.001
Littoriprofundal	–0.69	0.79	0.001
Profundal	–0.73	0.77	0.001
Brunnsee			
Whole year	–0.56	0.53	0.05
Spring	–1.16	0.77	0.01
Summer	–0.65	0.57	0.05
Autumn	–0.48	0.55	0.05

Conclusions

More than ten years ago Strayer (1991) reviewed the size structure of the freshwater zoobenthos, and his major points are still valid: (i) apparently, the size structures of marine and freshwater zoobenthic communities differ greatly; (ii) this difference may have an evolutionary origin because animals with life history traits of Warwick's (1984) marine macrofaunal groups are scarce in freshwaters; and (iii) lakes may differ in their zoobenthic size structures, in response at least to differences in water chemistry. He concluded that a body size-based approach would not be sufficient by itself in many areas of zoobenthic ecology. By comparison with the relatively monotonous zooplankton, the lacustrine zoobenthos is an extraordinarily rich community, typically containing several hundred species per lake. The potential of the size structure approach to describe benthic communities can only be evaluated if many different systems and groups are investigated.

References

Andrássy, I. (1956) Die Rauminhalts- und Gewichtsbestimmung der Fadenwürmer (Nematoden). *Acta Zoologica* 2, 1–15.

Arneberg, P. and Anderson, J. (2003) The energetic equivalence rule rejected because of a potentially common sampling error: evidence from carabid beetles. *Oikos* 101, 367–375.

Bergtold, M. and Traunspurger, W. (2004) The benthic community in the profundal of Lake Brunnsee: seasonal and spatial patterns. *Archiv für Hydrobiologie* 160, 527–554.

Blackburn, T.M. and Gaston, K.J. (1994) Animal body size distributions: patterns, mechanisms and implications. *Trends in Ecology and Evolution* 9, 471–474.

Blackburn, T.M., Lawton, J.H. and Perry, J.N. (1992) A method of estimating the slope of upper bounds of plots of body size and abundance in natural animal assemblages. *Oikos* 65, 107–112.

Blackburn, T.M., Brown, V.K., Doube, B.M., Greenwood, J.J.D, Lawton, J.H. and Stork, N.E. (1993) The relationship between abundance and body size in natural animal assemblages. *Journal of Animal Ecology* 62, 519–528.

Borgmann, U. (1987) Models on the slope of, and the biomass flow up, the biomass size spectrum. *Canadian Journal of Fisheries and Aquatic* Sciences 44 (Suppl. 2), 136–140.

Bourassa, N. and Morin, A. (1995) Relationships between size structure of invertebrate assemblages and trophy and substrate composition in streams. *Journal of North American Benthological Society* 14, 393–403.

Brooks, J.L. and Dodson, S.L. (1965) Predation, body size, and composition of plankton. *Science* 150, 28–35.

Carrascal, L.M. and Telleria, J.L. (1991) Bird size and density: a regional approach. *American Naturalist* 138, 777–784.

Cattaneo, A. (1987) Size distribution in periphyton. *Canadian Journal of Fisheries and Aquatic Sci*ences 44, 2025–2028.

Cattaneo, A. (1993) Size spectra of benthic communities in Laurentian streams. *Canadian Journal of Fisheries and Aquatic Sci*ences 50, 2659–2666.

Cyr, H. and Pace, M.L. (1993) Allometric theory: extrapolations from individuals to communities. *Ecology* 74, 1234–1245.

Damuth, J. (1981) Population density and body size in mammals. *Nature* 290, 699–700.

Damuth, J. (1987) Interspecific allometry of population density in mammals and other animals: the independence of body mass and population energy-use. *Biological Journal of the Linnean Society* 31, 193–246.

Damuth, J. (1991) Of size and abundance. *Nature* 351, 268–269.

Duplisea, D.E. (2000) Benthic organism biomass size-spectra in the Baltic Sea in relation to the sediment environment. *Limnology and Oceanography* 45, 558–568.

Duplisea, D.E. and Drgas, A. (1999) Sensitivity of a metazoan benthic biomass size spectrum to differences in sediment granulometry. *Marine Ecology Progress Series* 177, 73–81.

Edgar, G.J. (1994) Observations on the size-structure of macrofaunal assemblages. *Journal of Experimental Marine Biology and Ecology* 176, 227–243.

Gaedke, U. (1993) Ecosystem analysis based on biomass size distributions: a case study of a plankton community in a large lake. *Limnology and Oceanography* 38, 112–117.

Gerlach, S.A., Hahn, A.E. and Schrage, M. (1985) Size spectra of benthic biomass and metabolism. *Marine Ecology Progress Series* 26, 161–173.

Hanson, J.M. (1990) Macroinvertebrate size-distributions of two contrasting freshwater macrophyte communities. *Freshwater Biology* 24, 481–491.

Hanson, J.M., Prepas, E.E. and MacKay, W.C. (1989) Size distribution of the macroinvertebrate community in a freshwater lake. *Canadian Journal of Fisheries and Aquatic Sciences* 46, 1510–1519.

Kerr, S.R. (1974) Theory of size distribution in ecological communities. *Journal of Fisheries Research Board Canada* 31, 1859–1862.

Morin, A. and Nadon, D. (1991) Size distribution of epilithic lotic invertebrates and implications for community metabolism. *Journal of the North American Benthological Society* 10, 300–308.

Navarrete, S.A. and Menge, B.A. (1997) The body size–population density relationship in tropical rocky intertidal communities. *Journal of Animal Ecology* 66, 557–566.

Peters, R.H. (1983) *The Ecological Implications of Body Size.* Cambridge University Press, Cambridge, UK, 329 pp.

Peters, R.H. and Wassenberg, K. (1983) The effect of body size on animal abundance. *Oecologia* 60, 89–96.

Poff, N.L., Palmer, M.A., Angermeier, P.L., Vadas, R.L., Jr., Hakenkamp, C.C., Bely, A., Arensburger, P. and Martin, A.P. (1993) Size structure of the metazoan community in a Piedmont stream. *Oecologia* 95, 202–209.

Rasmussen, J.B. (1993) Patterns in the size structure of littoral zone macroinvertebrate communities. *Canadian Journal of Fisheries and Aquatic Sciences* 50, 2192–2207.

Schmid, P.E., Tokeshi, M. and Schmid-Araya, J.M. (2000) Relation between population density and body size in stream communities. *Science* 289, 1557–1560.

Schwinghamer, P. (1981) Characteristic size distributions of integral benthic communities. *Canadian Journal of Fisheries and Aquatic Sciences* 38, 1255–1263.

Sheldon, R.W., Prahash, A. and Sutcliffe, W.H. (1972) The size distribution of particles in the ocean. *Limnology and Oceanography* 17, 327–339.

Sheldon, R.W., Sutcliffe, W.H., Jr. and Paranjape, M.A. (1977) Structure of pelagic food chain and relationship between plankton and fish production. *Journal of the Fisheries Research Board of Canada* 34, 2344–2353.

Siemann, E., Tilman, D. and Haarstad, J. (1996) Insect species diversity, abundance and body size relationships. *Nature* 380, 704–706.

Sprules, W.G. and Munawar, M. (1986) Plankton size spectra in relation to system productivity, size, and perturbation. *Canadian Journal of Fisheries and Aquatic Sciences* 43, 1789–1794.

Strayer, D. (1986) The size structure of a lacustrine zoobenthic community. *Oecologia* 69, 513–516.

Strayer, D. (1991) Perspectives on the size structure of lacustrine zoobenthos, its causes, and its consequences. *Journal of the North American Benthological Society* 10, 210–221.

Strayer, D. (1994) Body size and abundance of benthic animals in Mirror Lake, New Hampshire. *Freshwater Biology* 32, 83–90.

Traunspurger, W. (1991) Fischbiologie des Königssees. Nahrungsangebot und Nahrungswahl. Band I: Das Meiobenthos des Königssees. Forschungsbericht 22, Nationalpark Berchtesgaden, Berchtesgaden, 152 pp.

Traunspurger, W. (1996a) Distribution of benthic nematodes in the littoral of an oligotrophic lake (Königssee, National Park Berchtesgaden, FRG). *Archiv für Hydrobiologie* 135, 393–412.

Traunspurger, W. (1996b) Distribution of benthic nematodes in the littoriprofundal and profundal of an oligotrophic lake (Königssee, National Park Berchtesgaden, FRG). *Archiv für Hydrobiologie* 135, 555–575.

Warwick, R.M. (1984) Species size distributions in marine benthic communities. *Oecologia* 61, 32–41.

Warwick, R.M. and Joint, L.R. (1987) The size distribution of organisms in the Celtic Sea: from bacteria to metazoa. *Oecologia* 73, 185–191.

8 Freshwater Nematodes in Environmental Science

S. Höss,[1] W. Traunspurger[2] and A. Zullini[3]

[1]Ecossa, Thierschstr. 43, 80538 München, Germany; [2]Department of Animal Ecology, University of Bielefeld, Morgenbreede 45, 33615 Bielefeld, Germany; [3]Dipartimento di Biotecnologie e Bioscienze, Università di Milano-Bicocca, Piazza della Scienza 2, 20126 Milano, Italy

The use of freshwater nematodes in the field of ecotoxicology and applied ecology is reviewed. After a general introduction to quality assessment of freshwater habitats, we try to elucidate the suitability of nematodes in this field of science and application. With examples of studies using nematodes as test organisms for single-species tests, as well as studies investigating nematode communities in unpolluted and polluted habitats, we highlight different fields of application and various methods for assessing the quality of ecosystems by focusing on studies performed in freshwater environments.

Quality Assessment of Freshwater Ecosystems

Anthropogenic pollution represents a massive stress factor for aquatic ecosystems. By affecting the living organisms, pollutants strongly influence the functions of an ecosystem, such as nutrient cycling. Moreover, pollutants may accumulate in food webs, thus also affecting human health via contaminated food. Recognizing the risk of pollution of water resources, scientists and regulators have developed new methods for assessing ecotoxicological risks for aquatic ecosystems (American Society for Testing and Materials, 2003). An important regulative basis for the protection of aquatic ecosystems is given by the European Union Water Framework Directive (EU-WFD), aiming to prevent further deterioration as well as the protection and enhancement of the status of aquatic ecosystems (European Community, 2000). Besides the technological progress in chemical analysis, biological methods allowed for the estimation of the bioavailability and toxicity of pollutants for the biota. Since the late 1960s, these biological methods have improved enormously, considering various habitats (pelagial, benthal), organisms (bacteria, plants, animals), trophic levels (decomposers, producers, consumers), organization levels (from molecules to food webs) and types of effects (specific receptors). The broad range of contaminants that is released into rivers and lakes via municipal and industrial emissions requires a variety of ecotoxicological approaches:

- Many contaminants are scarcely soluble in water, subsequently being adsorbed by particles and accumulated in sediments. The risk of sediment-bound pollutants has to be assessed with bioassays using benthic organisms rather than organisms living in the water column (American Society for Testing and Materials, 1990).
- Toxicants that were specifically designed for certain pests (e.g. bactericides, fungicides, algaecides, insecticides, nematicides) can be best assessed with the target organisms. The toxicity of mixtures of different pesticides that are present in many aquatic systems can only be covered with a test battery, using various types of test organisms.
- Many compounds that were released during previous decades were considered harmless, until very specific effects on aquatic organisms could be related to these compounds. One example is endocrine disruption (Sumpter, 1998) that can arise from various types of human applications (pharmaceuticals, detergents) affecting the hormonal regulation of aquatic invertebrates and vertebrates (Depledge and Billinghurst, 1999). Only specific bioassays can estimate the ecological risk of compounds with specific effects.
- To estimate the effect of pollution on ecosystem functioning, it is not sufficient to use single-species tests. Complex experimental units such as micro- and mesocosms as well as field studies are required to study pollution-induced changes in population, community or food web structures.

Suitability of Nematodes for Ecotoxicological Studies

Besides the ecological relevance of nematodes for lentic and lotic habitats that are described elsewhere in this book, there are advantages in using nematodes for pollution studies in the laboratory and the field over commonly used macrobenthic organisms (Bongers and Ferris, 1999). An overview of environmental studies on nematodes in terrestrial and marine habitats can be found elsewhere (e.g. Kennedy and Jacoby, 1999).

High abundance

Nematodes represent the most abundant metazoans in soils and sediments (e.g. Yeates, 1981). They can reach densities in freshwater habitats of up to 11.4 million/m^2 (Michiels and Traunspurger, 2005). Due to these high densities, statistically valid sampling can be achieved more easily than with macrofauna, even with small, easily processed samples.

Species richness

Compared to macrofauna, the numbers of meiofaunal species belonging to a single taxon in a given habitat can be one order of magnitude higher. Because of this diversity, meiofauna covers a broad range of physiological and feeding types, and thus provides a balanced assessment of effects of prevailing conditions on food webs and community processes.

Pervasiveness and tolerance

Nematodes are found in all environments examined so far. These include such extremes as hot, volcanic springs, anoxic sediments, sea ice and polluted sediments. This is possible because some nematodes tolerate a wide range of different environmental stresses. However, the group also includes sensitive, stress-intolerant species. Thus changes throughout a wide spectrum of stress conditions can be assessed for a large number of different ecosystems.

Limited mobility

Nematode communities, because of limited mobility, are continuously exposed to harmful materials that enter their environment. Although nematodes are able to move relatively fast within the sediment (e.g. Croll and Zullini, 1972), community structure is generally more directly related to the physico-chemical conditions of the habitat sampled than is that of macrofaunal communities.

Generation time

Effects of short-term as well as long-term influences become apparent in meiofaunal communities, as their life cycles cover a broad range of time spreads (from as little as 6 days to over 2 years). Most species have a generation time of about 1–3 months (Traunspurger, 2002).

Accrued knowledge

Since the late 1960s, nematodes have been used as model organisms in the laboratory (Brenner, 1974). The nematode *Caenorhabditis elegans* is one of the most investigated multicellular organisms (Riddle *et al.*, 1997). Scientists in the field of genetics, developmental biology and medical sciences exploited the advantages of *C. elegans* as a study object: simple genetic structure, transparent body, eutely (959 somatic cells), short generation time (3 days) and easy culturing. Studies with *C. elegans* led to substantial progress in genome (The *C. elegans* Sequencing Consortium, 1998) and medical research. The many years of experience with and extensive research on *C. elegans* in the laboratory could also be used in the field of environmental science.

Single-species Tests

Nematodes as test organisms

For several decades, nematodes have been used as test organisms for laboratory bioassays (e.g. Boroditsky and Samoiloff, 1973). Various nematode species were used to assess the potential toxicity of various compounds in aqueous media (Haight *et al.*, 1982), on agar (Popham and Webster, 1979), as well as in more complex matrices such as sediments and soils (Donkin and Dusenbery, 1993). Mainly free-living, bacterivorous nematodes were chosen as test organisms. Besides the marine *Monhystera disjuncta* (Vranken and Heip, 1986), species of the genus *Panagrellus*

(Samoiloff *et al.*, 1980) and *Plectus acuminatus* (Kammenga *et al.*, 1998) have been used in ecotoxicological studies. However, most studies used the soil-dwelling species *C. elegans*. Recently, methods have been standardized for the assessment of waste water (Hitchcock *et al.*, 1997), sediment (Traunspurger *et al.*, 1997) and soil (Freeman *et al.*, 2000). For these purposes, a variety of toxicity parameters were studied for *C. elegans*: lethality (Williams and Dusenbery, 1990a), growth (Traunspurger *et al.*, 1997), reproduction (Traunspurger *et al.*, 1997) and behaviour (Williams and Dusenbery, 1990b). Also, *C. elegans* was used as a bioindicator for studying pollution-induced gene expression (Guven *et al.*, 1994), as a test organism for bioconcentration studies (Haitzer *et al.*, 1999) and for screening mutagenic substances (Lew *et al.*, 1983).

Although *C. elegans* only rarely appears in freshwater habitats (Hirschmann, 1952; Zullini, 1988), this species lives in aquatic biofilms and has been used for toxicity studies in freshwater sediments and waste water. The following are some examples of the use of *C. elegans* as a test organism in single-species bioassays.

Testing of chemical substances

Toxicity tests with *C. elegans* can be performed in aqueous media, such as waste water or sediment pore-water, as well as in whole sediment. In aqueous media the nematodes are exposed to dissolved pollutants by uptake via the body surface. However, in whole sediment, *C. elegans* is exposed to contaminants via all possible uptake routes, including pore-water, particle contact and particle ingestion. Cadmium (Cd) toxicity was higher in whole sediment exposure compared to exposure in pore-water that was extracted from the respective sediments (Höss *et al.*, 2001) (Fig. 8.1). This phenomenon, however, was dependent on the organic content in the sediment. Cadmium that was bound to organic particles might have become available after digestion in the gut of the nematodes. The uptake of particle-bound contaminants via dietary ingestion has been reported for various benthic organisms (e.g. Reinfelder and Fisher, 1991).

Different types of contaminants have shown various effects on *C. elegans*. Hood *et al.* (2000) showed the presence of oestrogen receptors in *C. elegans* and *Panagrellus redivivus*, and could show that pesticides, such as toxaphen and dieldrin, as well as known endocrine disrupters such as nonylphenol, were able to interact with oestrogen receptors in *P. redivivus*. Custodia *et al.* (2002) showed that *C. elegans* might be used as a laboratory and field model for screening endocrine disruption by studying gene expression and vitellogenin responses to vertebrate steroids. The authors found an increased vitellogenin synthesis in the presence of oestrogens and testosterone, and lower vitellogenin levels in progesterone-treated cultures. Studies on the reproductive effects of hormones and hormone-like substances on *C. elegans* showed dose-dependent stimulation of the reproduction of the nematodes (Höss *et al.*, 2002; Weltje *et al.*, 2003), which might be a consequence of endocrine disruption in the nematodes.

The idea that movement behaviour may be preferentially sensitive to neurotoxicants has led several researchers to use locomotion in the examination of chemicals known to be neurotoxic. Anderson *et al.* (2004) studied the behaviour of *C. elegans* by computer tracking in a 4-h assay and could show that movement was more sensitive to neurotoxic compounds than to substances not believed to be neurotoxic. Using behaviour as a toxicity endpoint, *C. elegans* was suggested as a model for mammalian neurotoxicity (Cole *et al.*, 2004).

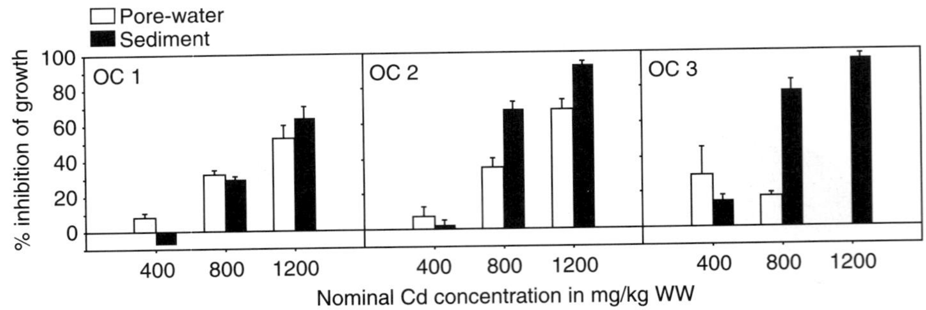

Fig. 8.1 Inhibition (%) of growth of *Caenorhabditis elegans* in pore-water (white bars) and sediment (black bars) with different organic contents (OC) and cadmium (Cd) concentrations. OC 1–3 = increasing organic content; error bars = standard deviation (n = 6). Reprinted with permission from Höss *et al.* (2001). Copyright SETAC, Pesacola, Florida.

Testing of environmental samples

Besides studies on the toxicity of known single or mixed substances, *C. elegans* has been used as a test organism to assess the toxicity of very complex material, such as freshwater sediments. Duft (2004) investigated 206 sediment samples taken from 12 large German rivers using the nematode sediment test in order to classify the ecological status of the various sampling sites. Sublethal toxicity parameters, such as growth, fertility and reproduction, were used to rank the samples according to their toxicity. According to the EU-WFD (European Community, 2000) sediments were ranked in five different classes of ecological status, ranging from very good (I) to bad (V). Only 22% of the samples could be assigned to classes I and II, the highest proportion could be found in class III and 44% ended up in classes IV and V (Fig. 8.2). Compared to the results of the commonly used sediment test with *Chironomus riparius*, which was applied to the same river sediments (Tillmann, 2003), the nematode test was found to be more sensitive (Fig. 8.2). Besides the toxicity data, Duft (2004) also compared some test criteria, such as effort in cultivation, material and test preparation, test duration, required space, flexibility and experience. In most criteria, the nematode test was considered to be equal or even preferable compared to the chironomid test.

While Duft (2004) classified sediment quality according to levels of inhibition of the toxicity parameters in the original sediments, Höss and Krebs (2003) diluted toxic sediments to get information about dose–response in these sediments. Toxicity ranking of the investigated sediments based on inhibition of nematode fertility in original sediments did not correspond to the toxicity ranking based on dose–response curves (pT-values[1]; Table 8.1; Krebs, 2000). Moreover, the degree of toxicity reduction by dilution was strongly dependent on the type of diluting sediment and original sediment (Fig. 8.3), which complicates the interpretation of sediment toxicity. However, using standardized dilution methods, dose–response curves might provide useful information for risk assessment of sediments (e.g. dredged material).

[1]pT-value = Negative $\log_2$ of the first step in a dilution series with factor 2 that had no toxic effect.

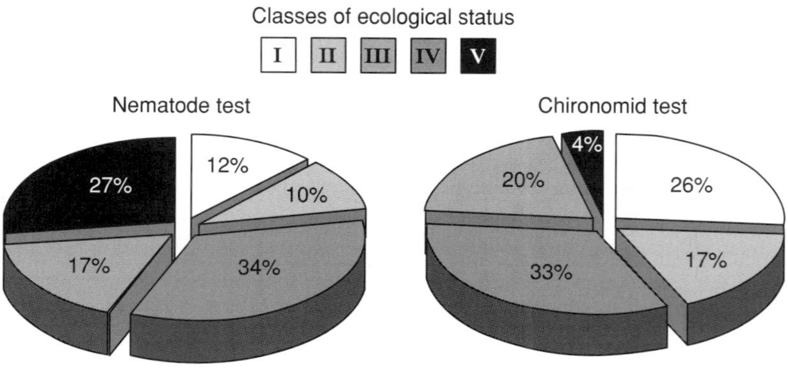

Fig. 8.2 Ranking of 206 sediment samples in five different classes of ecological status, revealed with the nematode test (left) and the chironomid test (right) (from I = very good to V = bad status). Adapted from Duft (2004) with kind permission of Cuvillier Verlag and Dr Martina Duft.

Community Level Assessment

Assessment of *in situ* benthic communities or assemblages is an ecologically more relevant approach than single-species bioassays, as it integrates biotic and abiotic interactions of the benthos, including pollution. But, this approach also contains more imponderables. Information about the ecology of nematodes in unpolluted aquatic environments is absolutely essential for evaluating pollution-induced changes in community structure. Ferris and Ferris (1979) reviewed knowledge about the ecology of nematodes in both 'normal' and polluted habitats, emphasizing the importance of obtaining baseline data on natural variability within habitats. Since various factors, regardless of sediment pollution, are able to influence nematode communities, it is

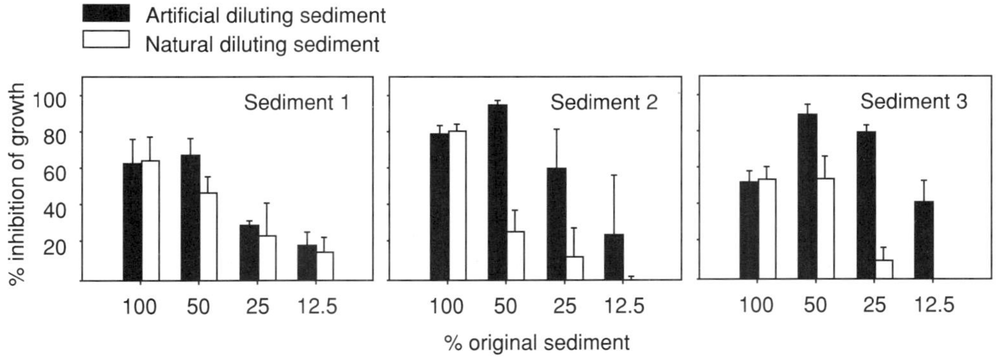

Fig. 8.3 Toxicity of three different sediments (Sediment 1, Sediment 2, Sediment 3) that were diluted with two different diluting sediments in three dilution steps (100% = original sediment; 50% = 1 : 2 dilution; 25% = 1 : 4 dilution; 12.5% = 1 : 8 dilution).

Table 8.1 Toxicity ranking of ten different sediments (A–J), based on percent inhibition of fertility in original sediments and pT-values, calculated from dilution series with natural diluting sediments (NS) and artificial diluting sediments (AS)

		pT-value	
Sample	% Inhibition in original sediments	NS	AS
A	56.1	2	2
B	100.0	1	>3
C	35.4	3	3
D	30.3	2	>3
E	69.8	1	2
F	70.3	1	3
G	100.0	1	2
H	100.0	2	>3
I	61.4	3	2
J	92.0	2	3

pT-value, negative $\log_2$ of the first step in a dilution series with factor 2 that had no toxic effect.

nearly impossible to demonstrate an unequivocal cause–effect relationship with a single factor. Food availability, particle size and salinity were found to have considerable influence on marine nematode community structure (e.g. Tietjen, 1980; Yeates and Coleman, 1982; Jensen, 1984; Vanreusel, 1991; Soetaert *et al.*, 1995) and, therefore, these factors are able to mask effects of pollutants. Moreover, the response of communities is dependent on the environmental conditions that the communities normally experience (Schratzberger and Warwick, 1999).

Thus, it is not possible to define a universally valid baseline for nematode communities, which evokes one of the major problems in pollution assessments: the negative control. Relative changes within communities, as implied by the term itself, can only be measured by comparing one set of data to another. Usually, the community of a polluted site is compared to that of an unpolluted site, the control. An ideal control site comprises 'natural' sediment properties comparable to the polluted site except for contamination, allowing any changes in community structure to be traced back to the pollution. These ideal conditions are rarely found in the field. However, suitable statistical techniques, combined with knowledge gathered from laboratory studies (model ecosystems) can help to accurately interpret community data in biomonitoring studies.

Characterization of nematode communities

There are various methods to characterize the structure of nematode communities (Neher and Darby, Chapter 11 this book; Table 8.2). The methods differ considerably in terms of their explanatory power for ecological indication. There are indices, such as the Shannon–Wiener index of diversity (*H*'; Shannon and Weaver, 1949), the evenness index *J*' ($= H/H_{max}$; Pielou, 1969) or the maturity index (MI; Bongers,

Table 8.2 Methods for analysing nematode communities.

Measures	Information	References	Examples
Univariate methods			
Species richness: S	Number of species in a defined sampling unit	Magurran (1988)	Den Besten *et al.*, (2000)
Shannon–Wiener Index (H') and Evenness (J')	Distribution of species abundances: gives more weight to rare species; higher index indicates higher diversity	Shannon and Weaver (1949); Pielou (1969)	Höss *et al.* (2004a,b)
k-dominance curves	Distribution of species abundances: plotting percentage of cumulative abundance (*k*-dominance) against species rank (*k*) of two nematode assemblages A and B; B is more diverse than A if the curve is everywhere below or touching that of A	Lambshead *et al.* (1983)	Danovaro *et al.* (1995)
Feeding types	Distribution of feeding types	Wieser (1953); Jensen (1987)	Korthals *et al.* (1996b)
Maturity index (MI)	Distribution of different life-history strategies of nematodes: based on colonizers to persisters; scale of 1–5; a low index indicates disturbance	Bongers (1990)	Korthals *et al.* (1996b)
% Secernentea or ratio of Secernentea/ Adenophorea (S/A)	High proportion of nematodes belonging to the subclass Secernentea indicates pollution	Zullini (1976)	Zullini (1988); Beier and Traunspurger (2001)
Egg index	(log egg size)–(log expected egg size): negative egg index indicates organic pollution	Zullini and Pagani (1989)	Zullini and Pagani (1989)
Multivariate methods			
Cluster analysis	Clustering based on similarities of taxa composition	Clarke (1993)	Trett *et al.* (2000)
Multidimensional scaling (MDS) ordination	Biplots with data points based on similarities of taxa composition; distance of data points reflects similarity of taxa composition	Clarke (1993)	Austen and McEvoy (1997)
Canonical correspondance analysis	Taxa composition can be related to environmental factors	Ter Braak (1994)	Fiscus and Neher (2002)
Principle response curves (PRC)	Redundancy analysis based on taxa composition including the factor time; useful for experimental studies	Van den Brink and Ter Braak (1999)	Höss *et al.* (2004a,b)

1990), reducing community data to single figures, that already contain ecological interpretation of the data, which facilitates interpretation, but also means loss of information. On the other hand there are multivariate methods, dealing with very complex data such as species composition. However, a sophisticated graphical analysis, such as principle response curves (PRC; Van den Brink and Ter Braak, 1999), is required to more easily interpret multivariate analysis.

There are some indices that are especially used for evaluating nematode communities, such as percentage of Secernentea, egg index, maturity indices or feeding type composition. The following is a description of nematode indices in more detail.

Percentage of Secernentea

The possibility of using nematode community parameters in ecological indices was examined by Tsalolikhin (1976) who, analysing data from literature, proposed that the relationship between the two subclasses Adenophorea and Secernentea may be used. Also, Zullini (1976) published data showing that, in river nematode communities, the percentage of Secernentea is a good pollution indicator. In particular, this percentage parallels fairly well the dissolved organic matter content (measured as DOC) in water. Among Secernentea, the Rhabditoidea were more tolerant to pollution than Diplogasteroidea. Water sediments with >20% Secernentea (Tylenchida excluded) are considered polluted; moreover, very heavily polluted sediments contain almost 100% Secernentea. Subsequent research confirmed these data (Eder and Kirchengast, 1982). The percentage of Secernentea or, in contemporary systematics terms (De Ley and Blaxter, 2002), the percentage of Rhabditida, is easy to assess because it requires the distinction of only two groups of nematodes. But its value has not yet been ascertained in soils and it cannot be applied in brackish waters and in 'normal' lakes where only a small number of Rhabditida are present (Prejs, 1977).

Life history strategies

Ecological indices are based not only on field and laboratory observations and experiments, but also on theoretical considerations. Thus, the assessment of an organism into the *r*–*K* strategists gradient is a preliminary step to refine and check the ecological indexes. A reproductive strategy is called *r*-type when it takes advantage of the sheer number of the progeny: it is typical of a *colonizer* (*c*) *species*, i.e. opportunists with high reproductive rate living in disturbed or polluted environments. Such types of habitats usually offer irregularly favourable living conditions such as living or dead exploitable organic matter, new, ripe for colonization, spaces or good climatic periods. For example, heaps of litter, dung or dead animals, are extreme instances of a wealthy and ephemeral resource for many colonizer organisms: bacteria, fungi, protozoa, nematodes (especially rhabditids) and insect larvae and adults (especially dipterans and beetles). All of them exhibit a very rapid and episodic demographic increase typical of *r*-strategist organisms. At the opposite point of the gradient, *K*-strategists (*persister* (*p*) *species*) exhibit a low reproductive rate and usually occupy stable, complex, mature, unstressed and unpolluted habitats. They deposit fewer eggs than their colonizer counterparts, but their eggs are relatively large (more rich in yolk) and the progeny will be more developed and with higher survival probability. There are several approaches to condense this ecological information in the form of indices to assess the degree of disturbance in ecosystems. Here, we explain one approach that was developed for freshwater nematode communities, the egg index. A description of the Maturity Indices, primarily developed for soil nematode communities, is given in detail in Chapter 11 (Neher and Darby) of this book.

Egg index

Often *r*-strategist nematodes are smaller than *K*-strategists and therefore lay smaller eggs, but the absolute egg dimension is not, *per se*, very significant. However, the deviation of the egg dimension from the expected value is more important. The expected value is the average ratio between the egg volume (y) and the female volume (x). From 463 examined soil and freshwater nematode species, this correlation was found as follows (Zullini and Pagani, 1989):

$$\log y = 2.39 + 0.58 \log x \quad (r = 0.81, \text{ significance } 99.9\%)$$

Deviations from the expected volume (*egg index*) are calculated as (log egg volume) – (log expected egg volume). These values are negative when eggs are smaller than the expected value, as happens for *r*-strategists/*c species* (frequent in organic polluted habitats), whereas numbers are positive for *K*-strategists/*p species*. Among Rhabditina, the lowest egg index pertains to *Diploscapter coronatus* (value −4.39), whereas Mononchida have positive egg indexes. Predators such as *Seinura*, *Ironus* and *Tripyla* have average egg indexes of 0.36, 0.64 and 1.00, respectively (Zullini and Pagani, 1989).

Feeding types

(see also Chapter 11 this book, Neher and Darby). The functional diversity of nematodes (feeding type diversity) offers an additional option for ecological indication (Wieser, 1953; Jensen, 1987; Yeates *et al.*, 1993). Pollution-induced changes of feeding type composition reflect on one hand the varying sensitivities of different taxa, which partly overlaps with the information of other indices, such as the MI and percentage of Secernentea. On the other hand secondary food web effects can be recognized, as changes in quality or quantity of food sources due to pollution also influence the composition of nematode feeding types.

Model ecosystems

In controlled experiments, natural sediments usually are transferred into experimental containers, to obtain small and well-defined representatives of ecosystems. These model ecosystems (micro- or mesocosms, depending on their size) are then manipulated experimentally for a certain period of time. Such experiments can also be regarded as community-level bioassays. Nematodes are suitable organisms for microcosm experiments: due to the relatively short generation time of most nematode species, community changes can be measured over the timescale at which such experiments can be realistically maintained (several months). Also, there are no problems of recruitment in microcosm systems since nematodes have direct benthic development (unlike macrobenthic groups where many species have planktonic larval recruitment).

In a microcosm study, Höss *et al.* (2004b) investigated the influence of nonylphenol on freshwater nematode communities, which is known to have toxic, (Staples *et al.*, 1998) and especially endocrine, effects on invertebrates (Baldwin *et al.*, 1997). Nonylphenol was applied to seven microcosms over a period of 6 weeks, reaching maximal concentrations ranging from 0.30 to 3.37 mg/kg dry weight in the sediment. Nematode community structure in these treatments was compared to four controls over a period of 15 weeks. Species composition was analysed using the PRC multivariate method. The PRC analysis showed a nonylphenol-induced change in species composition over a period of 7 weeks, from the end of the application until the end of the experiment (Fig. 8.4). In the highest dosed treatment, deposit-feeding species, classified as colonizers (*Eumonhystera*) increased their dominance, while epistrate feeders and chewers (*Prodesmodora, Tobrilus*) decreased in relative abundance compared to the control. Also, the composition of feeding types and the MI were affected at the highest dosed treatment. However, these effects occurred only within the last 3 weeks of the study. Nematode abundance and diversity indices were not affected by nonylphenol throughout the entire experiment.

Field studies

In field studies, the investigator can neither control sediment properties nor the quantity and quality of contaminants. Hence, it is much more difficult to recognize pollution-induced changes of community structure. Investigations of communities in polluted sediments require control sites to which polluted sites can be compared. As mentioned above, the sediment at the control site should have similar properties, but without contamination. Of course, these criteria can only be roughly achieved in field studies. In principle, there are two possibilities for how control sites are separated from the polluted sites: temporally or spatially. The former case occurs if a site where data on benthic communities are available is affected by an acute pollution event, so that the community structures before and after the accident can be compared. In addition, the progression of recovery can be monitored over time. In the latter case, a site that is remote from the source of pollution (e.g. sewage effluent) is selected as control.

Den Besten *et al.* (2000) investigated the success of sediment remediation at two pilot sites from the rivers Rhine and Meuse. The remediation consisted of partial excavation of contaminated sediments followed by application of a clean layer of sandy material on the top of the remaining contaminated sediment. As one part of the biological assessment, nematode communities were studied at remediated and non-remediated sites. The study showed that changes in community structure (relative abundance of specific families and genera) could be useful to indicate changes in sediment quality. At the Rhine river (Nieuwe Merwede) distinct differences occurred between the remediated and non-remediated sites 2 years after the remediation. While in the remediated site 24 taxa were found, only 11 taxa were present in the non-remediated sites. Moreover, differences in species composition occurred with a more complete trophic structure in the remediated site (omnivorous, carnivorous and fungivorous species were present). Species belonging to the family Tripylidae were present in all samples of the remediated site, but not in samples

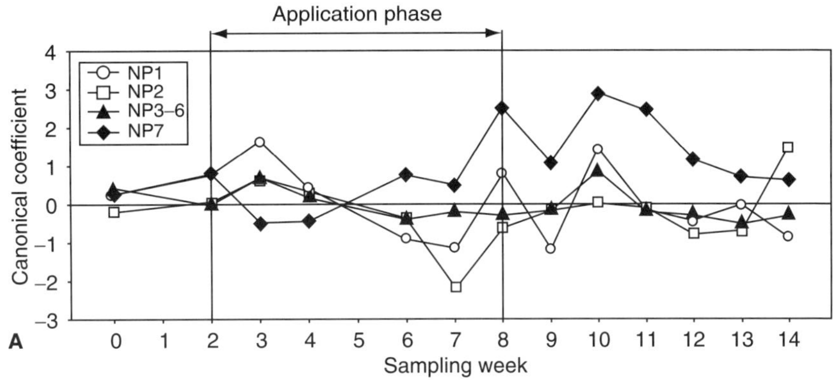

Species	Species score
Chromadorina bioculata	0.88
Eumonhystera dispar	0.63
Rhabdaimus sp.	0.16
Achromodora sp.	0.14
Prodesmodora sp.	–0.16
Aphanolaimus sp.	–0.18
Tobrilus gracilis	–0.38
Monhystera paludicola	–0.53

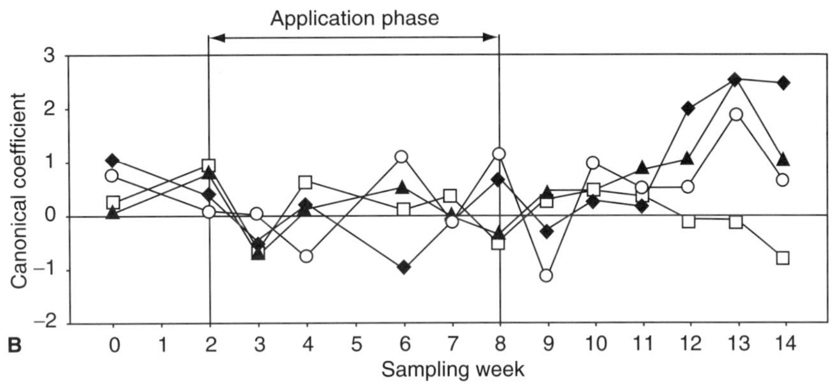

Species	Species score
Eumonhystera simplex	0.51
Rhabdolaimus sp.	–0.12
Tobrilus gracilis	–0.12
Monhystera paludicola	–0.12
Chromadorina bioculata	–0.14
Aphanolaimus sp.	–0.26
Eumonhystera dispar	–0.27
Eumonhystera similis	–0.33
Prodesmodora sp.	–0.77

Fig. 8.4 Principle response curves (PRC) and species scores calculated from the species composition of nematode communities in microcosms dosed with nonylphenol (NP1, NP2, NP3–6, NP7); A = first explanatory variable (22.1%; Eigenvalue: 0.06), B = second explanatory variable (15.4%; Eigenvalue: 0.04); for controls and NP3–6: $n = 4$, for NP1, NP2 and NP7 $n = 1$. Reprinted with permission from Höss *et al.* (2004b). Copyright SETAC, Pesacola, Florida.

from the non-remediated sites. The differences in species composition were also reflected in the MI, showing values of 2.5 to 2.8 in the remediated and 2.1 to 2.6 in the non-remediated sites, respectively. For the site of the Meuse river (Spijkerboor) remediation-mediated changes in nematode community structure were different compared to the sites of the river Rhine. At the Meuse no differences in number of taxa and MI could be observed 2 years after remediation. However, clear differences in taxa composition were observed. Relative abundances of genera belonging to the family Halaphanolaimidae (*Aphanolaimus* and *Paraphanolaimus*, deposit feeders, *c–p* 3) decreased considerably in the remediated site, while these genera remained abundant in the non-remediated sites.

In the study of Höss *et al.* (2004a) various sites of three German river catchments with varying degrees of contamination were investigated in terms of their nematode community structure. Composition of genera and feeding types, as well as the MI, were examined in relation to quality and quantity of sediment contamination. Genus-level composition showed clear differences between sites with high and low heavy metal contamination (Fig. 8.5). Deposit-feeding genera, such as *Monhystera* and *Daptonema*, as well as suction feeders, such as *Dorylaimus*, were more abundant at sites with low to medium metal contamination, whereas predatory and omnivorous genera, such as *Mononchus* and *Tobrilus*, were found more abundantly in sites with medium to high contamination. This difference in taxonomic composition is also reflected in the composition of feeding types with a significant correlation of predators and omnivores with some heavy metals (e.g. zinc). Nematode groups with high *c–p* values were more abundant in the highly contaminated sites, which is contradictory to the theory of the MI. However, community structure could have been influenced by secondary food web effects, which might have masked effects on the MI.

Beier and Traunspurger (2001) studied nematode communities of two small German streams comparing sediments from two 'unpolluted' sites (with relatively low anthropogenic impact) with two organically polluted sites (highly influenced by sewage effluents). In contrast to the latter examples, here benthic communities were exposed to a continuous release of organic contaminants. The authors found no substantial difference in the distribution of feeding types (as classified by

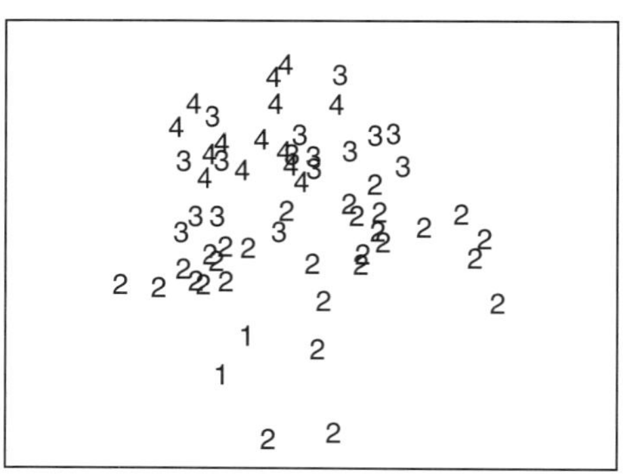

Fig. 8.5 Multidimensional scaling (MDS) ordination based on a Bray–Curtis similarity matrix calculated from nematode genera composition (relative abundance; square root transformed) of 61 different sediment samples; ranking of heavy metal contamination: 1 = low; 2 = medium; 3 = high; 4 = very high.

Traunspurger, 1997) of polluted and 'unpolluted' sites. All communities were clearly dominated by deposit feeders, with relative abundances ranging from 70% to 75%. To uncover changes within the deposit feeders, taxonomic identification to family level was required. At the polluted sites, the deposit-feeding Monhysteridae and Plectidae were replaced by deposit-feeding Diplogasteridae, Diploscapteridae and Rhabditidae. The change in the relative abundance within the dominant nematode families led to a distinct shift in the ratio of Secernentea to Adenophorea (S/A ratio). Beier and Traunspurger (2001) found a distinctly higher S/A ratio at the polluted compared to the 'unpolluted' sites (Fig. 8.6a). In addition, the MI (including *c–p* 1 nematodes) showed a clear decrease towards the polluted sites, indicating a more disturbed nematode community at the polluted than at the 'unpolluted' sites (Fig. 8.6b). This decrease is mainly driven by a higher relative abundance of nematodes belonging to the *c–p* 1 group ('colonizers'), rather than by a distinction of 'persisters' (*c–p* 3–5). When the *c–p* 1 group was excluded from MI calculations, the differences between the various sites disappeared (Fig. 8.6b). The additional use of the MI2–5 can provide supplementary information on the nature of the underlying changes in the habitat. In this case it indicated that the disturbance of the nematode community is related to eutrophication, rather than to inorganic pollution (De Goede *et al.*, 1993).

The above examples show that the response of freshwater nematode communities to pollution is variable, depending strongly on the investigated habitat (lotic, lentic, small rivers, large rivers) and on the type of pollution (organic, metals). Structural changes of nematode communities, such as shifts in species or genus composition, seem to be most sensitive to pollution. Due to the high diversity of nematode communities in most habitats, changes in species or genus composition can be very complex. Therefore, multivariate methods, such as PRC, multidimensional scaling (MDS) ordination, cluster analysis or correspondence analysis are required to examine changes in community structure. Also in studies of marine and terrestrial habitats species composition has proved to be a valuable parameter to detect the influence of chemical stress on nematode communities (e.g. Somerfield *et al.*, 1994).

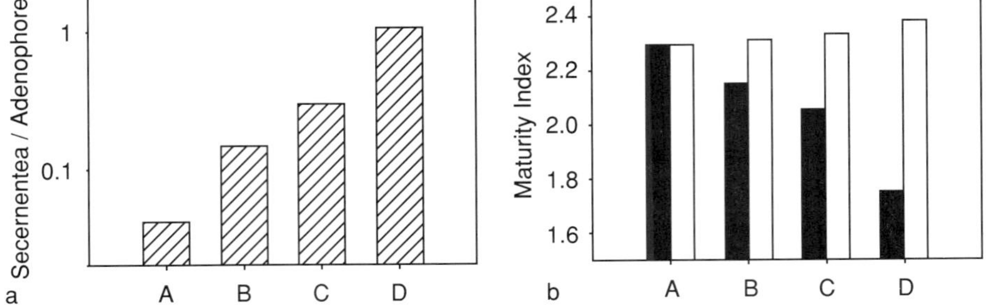

Fig. 8.6 (a) Ratios of Secernentea to Adenophorea and (b) Maturity Indices (MI, MI2–5) for nematode communities from sediments of two unpolluted (A, B) and two polluted (C, D) sites in small streams. Data from Beier and Traunspurger (2001).

Structural changes of communities can be a consequence of functional alterations in the habitat, such as food supply. Pollution-induced effects on the food web might also influence the composition of nematode feeding types that can be acquired much more easily than species composition by studying the buccal cavities of the nematodes. Effects of contaminants on the feeding type composition have been shown for terrestrial systems (Korthals *et al.*, 1996b).

The use of indices (Shannon–Wiener: *H'* and *J'*, MI) for estimating ecological quality is desired by regulators, as complex information is condensed in a single figure, which facilitates decision making. Moreover, indices, such as MI, try to provide an ecological interpretation of complex community data, including sensitivity to pollution. For several reasons, however, indices should always be used with caution. Condensing information always means loss of information. Though the Shannon–Wiener index is based on species data, intrinsic ecological information gets lost. The MI attempts to include autecological information of various nematode taxa, by ranking them in a *c–p* scale according to their life-history strategies. There are two major constraints that make the general applicability of the MI doubtful. First, the *c–p* scaling of nematode taxa is based on the taxonomic level of families. Differences between various genera of one family are not considered. Studies on freshwater nematode communities show that this taxonomic resolution was not satisfactory to describe contaminant-induced changes in community structures, as e.g. two genera of the family Monhysteridae, *Eumonhystera* and *Monhystera*, showed different preferences for sediments with various degrees of pollution (Höss *et al.*, 2004a,b). Second, problems with the allocation of taxa to *c–p* groups due to a lack of knowledge and experience (i.e. in freshwater habitats) complicate the interpretation, especially when comparing different studies or habitats (Yeates, 2002). However, accruing knowledge on the ecology of nematodes and increasing experience with sensitivities of certain nematode taxa to habitat properties (including pollution) might diminish imponderables and improve the usefulness of the MI for estimating disturbance in freshwater sediments.

Conclusions and Perspective

Nematodes have great potential as bioindicators for estimating quality of freshwater environments. Attributes such as high abundance, high diversity and short generation times make them suitable for single-species bioassays as well as for community studies using experimental or field designs. The nematode *C. elegans* is used as a test organism for assessing waste water and freshwater sediments, integrating all kinds of exposure routes in sediments and offering a variety of toxicity end points, such as mortality, growth, reproduction and movement behaviour.

Environmental studies on freshwater nematode communities are scarce. However, a few examples show that it is worthwhile to intensify applied research on the relation of nematode community structure and contaminants in freshwater environments, as has been done for terrestrial and marine ecosystems. Examples in this review include: (i) effects of single substances in microcosms; (ii) remediation success of polluted river sediments; (iii) comparisons of unpolluted and polluted sediments of large rivers; and (iv) influence of sewage effluents in small rivers. Indices and multivariate methods were used to detect changes in nematode-community structure, with species and genus composition being most sensitive to pollution. In some cases the MI was found to be related to pollution.

It is obvious that there is still a need for further research on nematode ecology in unpolluted and polluted habitats. This is particularly true for freshwater habitats. In contrast to marine and terrestrial systems, freshwater nematodes so far have rarely been taken into consideration for ecological (pollution) studies. It is necessary to put more effort into the investigations of nematode communities in freshwater model ecosystems to increase the understanding of the interactions of nematode communities with pollutants in different types of sediments. Further investigation will undoubtedly enable researchers to use this ubiquitous group of animals in future biomonitoring programmes.

References

American Society for Testing and Materials (1990) *Standard Guide for Conducting Sediment Toxicity Tests with Freshwater Invertebrates.* ASTM Standard E1383. Philadelphia, Pennsylvania.

American Society for Testing and Materials (2003) *Annual Book of ASTM Standards.* American Society for Testing and Materials, West Conshohocken, Pennsylvania.

Anderson, G.L., Cole, R.D. and Williams, P.L. (2004) Assessing behavioral toxicity with *Caenorhabditis elegans. Environmental Toxicology and Chemistry* 23, 1235–1240.

Austen, M.C. and McEvoy, A.J. (1997) The use of offshore meiobenthic communities in laboratory microcosm experiments: response to heavy metal contamination. *Journal of Experimental Marine Biology and Ecology* 211, 247–261.

Baldwin, W.S., Graham, S.E., Shea, D. and LeBlanc, G.A. (1997) Metabolic androgenization of female *Daphnia magna* by the xenoestrogen 4-nonylphenol. *Environmental Toxicology and Chemistry* 16, 1905–1911.

Beier, S. and Traunspurger, W. (2001) The meiofauna community of two small German streams as indicators of pollution. *Journal of Aquatic Ecosystem Health* 8, 387–405.

Bongers, T. (1990) The maturity index: an ecological measure of environmental disturbance based on nematode species composition. *Oecologia* 83, 14–19.

Bongers, T. and Ferris, H. (1999) Nematode community structure as a biomonitor in environmental monitoring. *Trends in Ecology and Evolution* 14, 224–228.

Boroditsky, J.M. and Samoiloff, M.R. (1973) Effects of growth inhibitors on development of the reproductive system of the free-living nematode *Panagrellus redivivus* (Cephalobidae). *Canadian Journal of Zoology* 51, 483–492.

Brenner, S. (1974) The genetics of *Caenorhabditis elegans. Genetics* 77, 71–94.

The *C. elegans* Sequencing Consortium (1998) Genome sequence of the nematode *C. elegans*: a platform for investigating biology. *Science* 282, 2012–2018.

Clarke, K.R. (1993) Non-parametric multivariate analyses of changes in community structure. *Australian Journal of Ecology* 18, 117–143.

Cole, R.D., Anderson, G.L. and Williams, P.L. (2004) The nematode *Caenorhabditis elegans* as a model of organophosphate-induced mammalian neurotoxicity. *Toxicology and Applied Pharmacology* 194, 248–256.

Croll, N.A. and Zullini, A. (1972) Observations on the bionomics of the freshwater nematode *Chromadorina bioculata. Journal of Nematology* 4, 256–260.

Custodia, N., Won, S.J., Novillo, A., Wieland, M., Li, C. and Callard, I.P. (2002) *Caenorhabditis elegans* as an environmental monitor using DNA microarray analysis. *Annals of the New York Academy of Science* 948, 32–42.

Danovaro, R., Fabiano, M. and Vincx, M. (1995) Meiofauna response to the *Agip Abruzzo* oil spill in subtidal sediments of the Ligurian Sea. *Marine Pollution Bulletin* 30, 133–145.

De Goede, R.G.M., Bongers, T. and Ettema, C.H. (1993) Graphical presentation and interpretation of nematode community structure: c–p triangles. *Mededelingen Faculteit Landbouwwetenschappen Universiteit Gent* 58, 743–750.

De Ley, P. and Blaxter, M.L. (2002) Systematic position and phylogeny. In: Lee, D.L. (ed.) *The Biology of Nematodes.* Taylor & Francis, London, pp. 1–30.

Den Besten, P.J., Postma, J.F., Wegener, J.W.M., Keidel, H., Klink, A., Mol, J. and van de Guchte, C. (2000) Biological and chemical monitoring after pilot remediations in the delta of the rivers Rhine and Meuse. *Aquatic Ecosystem Health and Management* 3, 317–334.

Depledge, M.H. and Billinghurst, Z. (1999) Ecological significance of endocrine disruption in marine invertebrates. *Marine Pollution Bulletin* 39, 32–38.

Donkin, S.G. and Dusenbery, D.B. (1993) A soil toxicity test using the nematode *C. elegans* and an effective method of recovery. *Archives of Environmental Contamination and Toxicology* 25, 145–151.

Duft, M. (2004) Ökologische Sedimentbewertung groβer Fliessgewässer mit Nematoden und Gastropoden – vom Biotest zum Freiland. Cuvillier Verlag, Göttingen, Germany.

Eder, R. and Kirchengast, M. (1982) The nematode fauna (Nemathelminthes, Nematoda) of a polluted part of the river Mur (Styria, Austria). *Nematologia Mediterranea* 10, 127–134.

European Community (2000) Directive 2000/60/EC of the European Parliament and the council of 23 October 2000 establishing a framework for community action in the field of water policy. *Official Journal* L327, 1–73.

Ferris, V.R. and Ferris, J.M. (1979) Threadworms (Nematoda). In: Hart, C.W. and Fuller, S.L.H. (eds) *Pollution Ecology of Estuarine Invertebrates.* Academic Press, London, UK, pp. 1–33.

Fiscus, D.A. and Neher, D.A. (2002) Distinguishing sensitivity of free-living soil nematode genera to physical and chemical disturbances. *Ecological Applications* 12, 565–575.

Freeman, M.N., Peredney, C.L. and Williams, P.L. (2000) A soil bioassay using the nematode *Caenorhabditis elegans.* In: Henshel, D.S., Black, M.C., and Harrass, M.C. (eds) *Environmental Toxicology and Risk Assessment: Standardization of Biomarkers for Endocrine Disruption and Environmental Assessment.* American Society for Testing and Materials, West Conshohocken, Pennsylvania, pp. 305–318.

Guven, K., Duce, J.A. and de Pomerai, D.I. (1994) Evaluation of a stress-inducible transgenic nematode strain for rapid aquatic toxicity testing. *Aquatic Toxicology* 29, 119–137.

Haight, M., Mudry, T. and Pasternak, J. (1982) Toxicity of seven heavy metals on *Panagrellus silusiae*: the efficacy of the free-living nematode as an *in vivo* toxicological bioassay. *Nematologica* 28, 1–11.

Haitzer, M., Burnison, B.K., Höss, S., Traunspurger, W. and Steinberg, C.E.W. (1999) Effects of quantity, quality, and contact time of dissolved organic matter on bioconcentration of Benzo[a]pyrene in the nematode *Caenorhabditis elegans. Environmental Toxicology and Chemistry* 18, 459–465.

Hirschmann, H. (1952) Die Nematoden der Wassergrenze mittelfränkischer Gewässer. *Zoologische Jahrbücher (Systematik)* 81, 313–436.

Hitchcock, D.R., Black, M.C. and Williams, P.L. (1997) Investigations into using the nematode *Caenorhabditis elegans* for municipal and industrial wastewater toxicity testing. *Archives of Environmental Contamination and Toxicology* 33, 252–260.

Hood, T.E., Calabrese, E.J. and Zuckerman, B.M. (2000) Detection of an oestrogen receptor in two nematode species and inhibition of binding and development by environmental chemicals. *Ecotoxicology and Environmental Safety* 47, 74–81.

Höss, S. and Krebs, F. (2003) Dilution of toxic sediments with unpolluted artificial and natural sediment – effects on *Caenorhabditis elegans* (Nematoda) in a whole sediment bioassay, *13th Annual Meeting of SETAC Europe.* Society of Environmental Toxicology and Chemistry, Brussels, Belgium, p. 148.

Höss, S., Henschel, T., Haitzer, M., Traunspurger, W. and Steinberg, C.E.W. (2001) Toxicity of cadmium to *Caenorhabditis elegans* (Nematoda) in whole sediment and porewater – the ambiguous role of organic matter. *Environmental Toxicology and Chemistry* 20, 2794–2801.

Höss, S., Jüttner, I., Traunspurger, W., Pfister, G., Schramm, K.-W. and Steinberg, C.E.W. (2002) 4-Nonylphenol can enhance the reproduction of *Caenorhabditis elegans* (Nematoda). *Environmental Pollution* 120, 169–172.

Höss, S., Heininger, P., Claus, C., Pelzer, J. and Traunspurger, W. (2004a) Nematode communities in large German rivers and the relation to sediment contamination, *14th Annual Meeting of SETAC Europe*. Society of Environmental Toxicology and Chemistry, Brussels, Belgium, pp. 51–52.

Höss, S., Traunspurger, W., Severin, G.F., Jüttner, I., Pfister, G. and Schramm, K.-W. (2004b) Influence of 4-nonylphenol on the structure of nematode communities in freshwater microcosms. *Environmental Toxicology and Chemistry* 23, 1268–1275.

Jensen, P. (1984) Ecology of benthic and epiphytic nematodes in brackish water. *Hydrobiologia* 108, 201–217.

Jensen, P. (1987) Feeding ecology of free-living aquatic nematodes. *Marine Ecology Progress Series* 35, 187–196.

Kammenga, J.E., Arts, M.S.J. and Oude-Breuil, W.J.M. (1998) HSP60 as a potential biomarker of toxic stress in the nematode *Plectus acuminatus*. *Archives of Environmental Contamination and Toxicology* 34, 253–258.

Kennedy, A.D. and Jacoby, C.A. (1999) Biological indicators of marine environmental health: meiofauna – a neglected benthic component? *Environmental Monitoring and Assessment* 54, 47–68.

Korthals, G.W., Van de Ende, A., Van Megen, H., Lexmond, T.M., Kammenga, J.E. and Bongers, T. (1996) Short-term effects of cadmium, copper, nickel, and zinc on soil nematodes from different feeding and life-history strategy groups. *Applied Soil Ecology* 4, 107–117.

Krebs, F. (2000) Ecotoxicological assessment of dredged material from federal waterways by the pT-value method. *Hydrologie und Wasserbewirtschaftung* 44, 301–307.

Lambshead, P.J.D., Platt, H.M. and Shaw, K.M. (1983) The detection of differences among assemblages of marine benthic species based on an assessment of dominance and diversity. *Journal of Natural History* 17, 859–874.

Lew, K.K., Nichols, D.G. and Kolber, A.W. (1983) *In vivo* assay to screen for mutagens/carcinogens in the nematode *C. elegans*. In: Kolber, A.R., de Woskin, R.S. and Hughes, T.J. (eds) *In Vitro Toxicity Testing of Environmental Agents: A Survey of Test Systems*. Plenum Publishing, New York, pp. 139–150.

Magurran, A.E. (1988) *Ecological Diversity and its Management*. Croom Helm, London.

Pielou, E.C. (1969) *Introduction to Mathematical Ecology*. John Wiley & Sons, New York.

Popham, J.D. and Webster, J.M. (1979) Cadmium toxicity in the free-living nematode, *Caenorhabditis elegans*. *Environmental Research* 20, 183–191.

Prejs, K. (1977) The littoral and profundal benthic nematodes of lakes with different trophy. *Ekologia Polska* 25, 21–30.

Reinfelder, J.R. and Fisher, N.S. (1991) The assimilation of elements ingested by marine copepods. *Science* 251, 794–798.

Riddle, D.L., Blumenthal, T., Meyer, B.J. and Priess, J.R. (1997) *C. elegans II*. Cold Spring Harbor Laboratory Press, New York.

Samoiloff, M.R., Schulz, S., Jordan, Y., Denich, K. and Arnott, E. (1980) A rapid simple long-term toxicity assay for aquatic contaminants using the nematode *Panagrellus redivivus*. *Canadian Journal of Fisheries and Aquatic Science* 37, 1167–1174.

Schratzberger, M. and Warwick, R.M. (1999) Differential effects of various types of disturbances on the structure of nematode assemblages: an experimental approach. *Marine Ecology Progress Series* 181, 227–236.

Shannon, C.E. and Weaver, W. (1949) *The Mathematical Theory of Communication*. University of Illinois Press, Champaign, Illinois.

Soetaert, K., Vincx, M., Wittoek, J. and Tulkens, M. (1995) Meiobenthic distribution and nematode community structure in five European estuaries. *Hydrobiologia* 311, 185–206.

Somerfield, P.J., Gee, J.M. and Warwick, R.M. (1994) Soft sediment meiofaunal community structure in relation to a long-term heavy metal gradient in the Fal Estuary system. *Marine Ecology Progress Series* 105, 79–88.

Staples, C.A., Weeks, J., Hall, J.F. and Naylor, C.G. (1998) Evaluation of aquatic toxicity and bioaccumulation of C8- and C9-alkylphenol ethoxylates. *Environmental Toxicology and Chemistry* 17, 2470–2480.

Sumpter, J.P. (1998) Xenoendocrine disruptors: environmental impacts. *Toxicology Letters* 102/103, 337–342.

Ter Braak, C.J.F. (1994) Canonical community ordination. Part 1. Basic theory and linear methods. *Ecoscience* 1, 127–140.

Tietjen, J.H. (1980) Population structure and species distribution of the free-living nematodes inhabiting sands of the New York Bight apex. *Estuarine and Coastal Marine Science* 10, 61–73.

Tillmann, M. (2003) Sedimenttoxikologische Untersuchungen mit Gastropoden und Insekten unter besonderer Berücksichtigung endokrin wirksamer Substanzen. PhD thesis, J.-W. Goethe Universität, Frankfurt/Main, Germany.

Traunspurger, W. (1997) Bathymetric, seasonal and vertical distribution of feeding types of nematodes in an oligotrophic lake. *Vie Milieu* 47, 1–7.

Traunspurger, W. (2002) Nematoda. In: Rundle, S.D., Robertson, A. and Schmid-Araya, J. (eds) *Freshwater Meiofauna: Biology and Ecology*. Backhuys Publishers, Leiden, The Netherlands, pp. 63–104.

Traunspurger, W., Haitzer, M., Höss, S., Beier, S., Ahlf, W. and Steinberg, C.E.W. (1997) Ecotoxicological assessment of aquatic sediments with *Caenorhabditis elegans* (Nematoda) – a method for testing in liquid medium and whole sediment samples. *Environmental Toxicology and Chemistry* 16, 245–250.

Trett, M.W., Calvo-Urbano, B., Forster, S.J., Hutchinson, J.D., Feil, R.L., Trett, S.P. and Best, J.G. (2000) Terrestrial meiofauna and contaminated land assessment. *Environmental Science and Technology* 34, 1594–1602.

Tsalolikhin, S.J. (1976) Free-living nematodes as indicators of polluted freshwaters. In: Skarlato, O.A. (ed.) *Methods of Biological Analysis of Freshwaters*. Akademia Nauk, Leningrad, Russia, pp. 118–122.

Van den Brink, P.J. and Ter Braak, C.J.F. (1999) Principal response curves: analysis of time dependent multivariate responses of biological community to stress. *Environmental Toxicology and Chemistry* 18, 138–148.

Vanreusel, A. (1991) Ecology of free-living marine nematodes in the Voordelta (southern bight of the North Sea). 2. Habitat preferences of the dominant species. *Nematologica* 37, 343–359.

Vranken, G. and Heip, C. (1986) Toxicity of copper, mercury and lead to a marine nematode. *Marine Pollution Bulletin* 17, 453–457.

Weltje, L., Höss, S., van Doormalen, J., Markert, B. and Oelmann, J. (2003) Endocrine disruption in the nematode *Caenorhabditis elegans, 13th Annual Meeting of SETAC Europe*. Society of Environmental Toxicology and Chemistry, Brussels, Belgium, p. 183.

Wieser, W. (1953) Die Beziehung zwischen Mundhöhlengestalt, Ernährungsweise und Vorkommen bei freilebenden marinen Nematoden. *Arkiv for Zoologi* 2, 439–484.

Williams, P.L. and Dusenbery, D.B. (1990a) Aquatic toxicity testing using the nematode *Caenorhabditis elegans*. *Environmental Toxicology and Chemistry* 9, 1285–1290.

Williams, P.L. and Dusenbery, D.B. (1990b) A promising indicator of neurobehavioral toxicity using the nematode *Caenorhabditis elegans* and computer tracking. *Toxicology and Industrial Health* 6, 425–440.

Yeates, G.W. (1981) Nematode populations in relation to soil environmental factors: a review. *Pedobiologia* 22, 312–338.

Yeates, G.W. (2002) Nematodes as soil indicators: functional and biodiversity aspects. *Biology and Fertility of Soils* 37, 199–210.

Yeates, G.W. and Coleman, D.C. (1982) Role of nematodes in decomposition. In: Freckman, D.W. (ed.) *Nematodes in Soil Ecosystems*. University of Texas Press, Austin, Texas, pp. 55–80.

Yeates, G.W., Bongers, T., De Goede, R.G.M., Freckman, D.W. and Georgieva, S.S. (1993) Feeding habits in soil nematode families and genera – an outline for soil ecologists. *Journal of Nematology* 25, 315–331.

Zullini, A. (1976) Nematodes as indicators of river pollution. *Nematologia Mediterranea* 4, 13–22.

Zullini, A. (1988) The ecology of the Lambro River. *Rivista di Idrobiologia* 27, 39–58.

Zullini, A. and Pagani, M. (1989) The ecological meaning of relative egg size in soil and freshwater nematodes. *Nematologica* 35, 90–96.

9 Nemataodes in Lotic Systems

M. Hodda

Nematode Biosystematics & Ecology, CSIRO Entomology, GPO Box 1700, Canberra, ACT 2601, Australia

Introduction

Where freshwaters are running, the habitat is known as lotic (as compared with still freshwater habitats, which are known as lentic). In lotic habitats, water, nutrients and organic matter are constantly introduced from direct rainfall, runoff, subsurface flows and groundwater. In running waters impacted by humans, there may also be outlet pipes and the like introducing either natural or anthropogenic chemical compounds or particulate material. The substances washed in are also washed away downstream after residing in the lotic system for a time. The movement of water can also wash organisms downstream, making the ability to either stay in the same place or to recolonize important. Another consequence of the flow of water is that stream beds may change frequently.

The chemical, physical and biological characteristics of lotic systems can change, but oxygen, salinity or temperature are generally not limiting at any time. Changes in the nematode fauna due to stream channels shifting are considered the province of terrestrial nematology and so are not discussed in this book. The impact of water flows and the physical changes in the sediment associated with them are undoubtedly the major influence on nematodes in lotic systems.

Many of the studies of nematodes from lotic systems have used different collecting or extraction methods. This limits the comparisons and generalizations possible between different studies.

Nematodes in the Water Column

Nematodes in lotic systems are found in both the sediment and the water column. In permanent stream channels, most are found in the sediment, but large numbers have been found in fast-moving waters (Mott *et al.*,1981; Jacobs, 1984). Fewer nematodes were found in water moving more slowly (Mott *et al.*, 1981; Jacobs, 1984). Irrigation water may have many nematodes: 18 of 24 samples of irrigation water in Tenerife contained nematodes (Grimm, 1978). Runoff water flowing overland or in temporary channels is also particularly rich in nematodes (Cadet *et al.*, 2003; Villenave *et al.*, 2003). Water in permanent channels but flowing only intermittently also seems to be rich in nematodes (Smith *et al.*, 2001a,b). Disturbance of the vegetation and exposure of bare soil may increase the abundance of nematodes in the water columns of small streams for up to 2.5 years after the disturbance (Smith *et al.*, 2001b).

Many nematodes suspended in flowing waters are the same as those in the substrate below. However, the suspended nematodes are not a random assortment of the soil fauna. In Senegal, plant-parasitic nematodes were generally less abundant in the water than in the underlying soil, except for the genus *Scutellonema*, which was very abundant in the runoff (Cadet *et al.*, 2003). This may be an important part of the survival strategy of this genus (Cadet *et al.*, 2003).

In the Sahel, the situation was the reverse of that in Senegal (Villenave *et al.*, 2003). Many plant-parasitic nematodes, such as *Ditylenchus*, aphelenchids and Tylenchidae, were more abundant in runoff water than in soil, along with a few microbivorous taxa, such as Rhabditidae, Rhabdolaimidae, *Pseudacrobeles* and *Chiloplacus*. No plant-feeding nematodes were under-represented in the runoff water, but at least 14 microbivorous taxa were (Villenave *et al.*, 2003).

Plant-parasitic nematodes are frequently found in water samples from rivers in the USA (American Public Health Association, 1998). All the nematodes found suspended have been regarded as basically soil or sediment nematodes, which have been temporarily suspended (Cadet and Albergel, 1999). However, because some taxa seem to be suspended much more than others, and perhaps even rely on transport in the water column for dispersal, they may be regarded as at least temporary inhabitants of the water column. There seems little evidence that speculation on the existence of substantial numbers of genuinely planktonic nematodes (Jacobs, 1984) will prove founded.

Nematodes in the Substrate

There have been a large number of studies of nematodes from the substrates under flowing waters, but comparisons are complicated by the different types of data recorded as well as by different methodologies (as discussed in the previous section). Too many taxa have been found to list separately here, but the main parameters of the studies are listed in Table 9.1. A species list for Europe is available (Andrássy, 1978a).

Many studies on taxa other than nematodes differentiate between a ‘hyporheic’ zone immediately below flowing waters and a ‘phreatic’ (sometimes ‘stygal’) zone deeper in sediments and at the stream sides and which merges with groundwaters, but physiographic connections and faunal interdependencies make this delineation equivocal and closely similar (Danielopol, 1989, 1991). In nematodes, no clear boundary between hyporheic and groundwater species has been observed. In a study of the boundary of the ‘hyporheic’ zone in Ontario, Canada, only one species occurred purely in the groundwater (a species of *Mononchus*) (Williams, 1993). Most species spanned the boundary.

Table 9.1 Population parameters for nematodes from lotic habitats. Studies with most similar methods are grouped together. Abundance estimated for studies where other units were used, such as number per litre.

Sediment type	Stream type	Sampling and extraction method	Mesh (μm)	Sample depth (cm)	River (R) or location	Abundance (number/10 cm^2)	Number of species	Reference(s)
Fine sand	Small stream (Piedmont)	Sieving, flotation and centrifugation	30	–	Appalachians, USA	10–90	–	Bott and Kaplan (1989)
Coarse sand	Small stream (Piedmont)	Sieving, flotation and centrifugation	30	–	Appalachians, USA	0–20	–	Bott and Kaplan (1989)
Coarse sand	Stream	Centrifugation and sieving	44	50	Virginia, USA	10–160	–	Palmer (1990)
Sand	Small river	Core, centrifugal flotation	38	5	Murrumbidgee R, Australia	–	17	Hodda (1999)
Mud	Large river	Core, centrifugal flotation	38	5	Darling R, Australia	–	8	Hodda (1999)
Mud	Large river	Core, centrifugal flotation	38	5	Murray R, Australia	–	10	Hodda (1999)
Silt	Small stream	Core, centrifugal flotation	38	5	Barrington Tops, Australia	–	3	Hodda (1999)
Mud	Large river	Core, centrifugal flotation	–	10	R Abbay, Ethiopia	–	10	Eyualem-Abebe *et al.* (2001)
Fine sand	Large river	Core, centrifugal flotation	–	10	R Abbay, Ethiopia	–	8	Eyualem-Abebe *et al.* (2001)
Fine sand	Large river	Core, centrifugal flotation	–	10	R Abbay, Ethiopia	–	8	Eyualem-Abebe *et al.* (2001)

Continued

Table 9.1 *Continued.* Population parameters for nematodes from lotic habitats. Studies with most similar methods are grouped together. Abundance estimated for studies where other units were used, such as number per litre.

Sediment type	Stream type	Sampling and extraction method	Mesh (μm)	Sample depth (cm)	River (R) or location	Abundance (number/ 10 cm²)	Number of species	Reference(s)
Fine sand	Large river	Core, centrifugal flotation	–	10	R Gelda, Ethiopia	–	4	Eyualem-Abebe *et al.* (2001)
Mud	Small rivers and brooks	Centrifugation	–	–	Munster, Germany	–	150	Niemann *et al.* (1996)
Gravel	Streams	Bou-Rouch	100	–	Eastern USA	0–386	–	Strayer *et al.* (1997)
Gravel	Large river	Bou-Rouch	50–80	100	Danube R, Austria	–	7	Danielopol (1976)
Gravel	Large river	Modified Bou-Rouch	–	135	Danube R, Austria	–	51	Eder (1983)
Gravel	Small river bar	Bou-Rouch	48	50	South Platte R, Colorado, USA	–	13	Pennak and Ward (1986)
Gravel	Large river	Modified corer, Bou-Rouch	–	70	R Mur, Austria	–	23	Kirchengast (1984)
Sand	Large river	Grab with or without sieving	50	60	Mississippi R, USA	770–940	23	Anderson (1992)
Silty sand	Large river	Grab with or without sieving	50	60	Mississippi R, USA	80–230	23	Anderson (1992)
Mud	Slow flowing river	Grab	250	30	North Canadian R, Oklahoma, USA	9–92	–	Bass and Walker (1992)
Sand	Slow flowing river	Grab	–	10	Elbe estuary, Germany	–	175	Riemann (1966)

Sand	Small stream riffle	Core	–	30	Texas, USA	–	3	Whitman and Clark (1984)
Sand	Stream	Core	–	5	Munster, Germany	–	100	Christl (unpublished); Traunspurger (2002)
Sand	River	Core	63	3	Green R, Utah, USA	12–3600	–	Jordan *et al.* (1999)
Sand	River	Core	63	3	Colorado R, USA	26–2140	–	Jordan *et al.* (1999)
Sand	Small stream	Core	35	5	Krahenbach, Germany	28–126	46	Beier and Traunspurger (2001)
Gravel	Small stream	Core	35	5	Korsch, Germany	37–1205	94	Beier and Traunspurger (2001)
Sand	Small stream	Core	30	1	Appalachians, USA	18–492	–	Borchardt and Bott (1995)
Sand	Sand bars, small streams	Suction syringe	–	–	Ohio R, USA	2–152	–	Hummon *et al.* (1978)
Sand	Large river	Freeze corer	–	70	R Danube, Austria	1–90	–	Tockner and Bretschko (1996)
Gravel	Small stream	Stand pipe traps	30	40	Oberer Seebach, Austria	–	45	Schmid-Araya and Schmid (1995); Schmid-Araya (1997); Traunspurger (2002)

Continued

Table 9.1 *Continued.* Population parameters for nematodes from lotic habitats. Studies with most similar methods are grouped together. Abundance estimated for studies where other units were used, such as number per litre.

Sediment type	Stream type	Sampling and extraction method	Mesh (μm)	Sample depth (cm)	River (R) or location	Abundance (number/ 10 cm^2)	Number of species	Reference(s)
Sand	Small river	Cage	35	34	Necker R, Germany	400–4100	41	Eisenmann *et al.* (1998)
–	Small river	Karaman–Chappuis	–	–	Skycomish R, Washington, USA	–	2	Altherr and Debouttеville (1972)
–	Small river	Karaman–Chappuis	–	–	Wenatchee R, Washington, USA	–	5–14	Altherr and Debouttеville (1972)
–	Small creek	Karaman–Chappuis	–	–	Boulder/Sth and Nth St Vrain Ck/ Fall R, Colorado, USA	–	1–10	Altherr and Debouttеville (1972)
–	Small river	Karaman–Chappuis	–	–	Woods Hole, Massachusetts, USA	–	10–12	Altherr and Debouttеville (1972)
Mud	Large river	Core, elutritor and flotation	–	25	Rhine, Netherlands	5–1770	40	Bongers and Van Der Haar (1990)
Gravel	Large river	–	–	–	R Inn, Austria	–	25	Stefanski (1916)
Gravel	Small river	–	–	–	R Still, Austria	–	11	Stefanski (1916)
–	Large river	–	–	–	R Adige, Italy	–	20	Andrássy (1959)

–	Large river	–	–	–	R Danube, Hungary	–	27	Andrássy (1962)
–	Small river	–	–	–	Chiavenna R, Italy	–	19	Zullini (1974)
–	Large river	–	–	–	R Po, Italy	–	10–23	Zullini (1974)
–	Small river	–	–	–	R Seveso (unpolluted), Italy	–	55–57	Zullini (1976)
–	Small river	–	–	–	R Seveso (polluted), Italy	–	24–37	Zullini (1976)
Gravel	Small brook	–	–	–	Himalayas	–	4	Andrássy (1978b)
Gravel	Small brook	–	–	–	Himalayas	–	1	Andrássy (1978b)
Gravel	Small brook	–	–	–	Himalayas	–	5	Andrássy (1978b)
Gravel	Glacial stream	–	–	–	Himalayas	–	1	Andrássy (1978b)
Mud	Large river	–	–	–	Indiana, USA	–	6	Callahan *et al.* (1979)
Sand	Large river	–	–	–	Regnitz, Germany	–	11	Hirschmann (1952)
Sand and mud with plants	Large river	–	–	–	Regnitz, Germany	–	19–26	Hirschmann (1952)
Sand and mud	Large river	–	–	–	Regnitz, Germany	–	23	Hirschmann (1952)
Mixed sediment	Large river	–	–	–	Regnitz, Germany	–	30	Hirschmann (1952)
Silt/mud	Stream	–	–	–	Carrega Wood, Italy	–	37–41	Zullini and Ricci (1980)

Diversity

The number of nematode species found in the substrates under flowing waters varies from 0 to over 150 (Table 9.1). This variation is partly related to differences in sampling intensity, systematic expertise and the methods used. Part of the variation is also related to habitat. Sandy sediments seem to have more species than either fine-grained muds or coarse gravels (Table 9.1). The sediment grain size has a direct effect on the interstitial space between the grains in which nematodes live, with intermediate grain sizes providing spaces optimal for most nematodes to move (Nicholas, 1984). In coarse gravels the interstitial spaces are so large that all but the largest nematodes would need to swim. In muddy sediments the spaces are so small that all but the smallest nematodes must physically push the sediment particles aside. Thus it is quite likely that sediments of intermediate grain sizes, such as sands, offer the best conditions for most nematode species, as is the case in terrestrial soils (Nicholas, 1984).

In aquatic sediments, unlike in terrestrial soils, the grain size is related to other features of the environment, particularly the current velocity and pattern of erosion. Hence, the number of species in sandy sediments may be related to one or more factors.

Few of the species found in lotic systems seem restricted to this habitat: most are also found frequently in other habitats. In small streams in Germany, only about a third of the species were restricted to the streams, with most species found frequently in surrounding terrestrial habitats (Niemann *et al.*, 1996). In the lower reaches of rivers, many species were also found in adjacent estuarine habitats (Riemann, 1966).

There seems little gradient in diversity between the deeper sediment and that near the surface. In a small stream, the number of species in the upper 2 cm of sediment was 59, compared to 49 species deeper in the sediment (Beier and Traunspurger, 2003).

Most species seem to have very patchy distributions. In a very extensive survey of 600 samples from seven different streams, about half the species occurred in fewer than 20 samples (Niemann *et al.*, 1996). By contrast, a few species occurred in many samples from similar habitats. On an even wider scale, in a study of the nematodes from the Murray–Darling River Basin, which covers one-seventh of the Australian continent or over 1 million km^2, several species were common in many of the samples (Hodda, 1999). More than half of the species were found in fewer than 5% of the samples (Hodda, 1999). Thus, diversity on a scale larger than the individual sample (often termed α diversity) of lotic nematodes is typically very high.

Species distributions

Only a few nematode species from lotic systems occur sufficiently frequently to make any generalizations about distribution. As noted above, most species have very patchy distributions.

Genera ubiquitous in lotic systems include *Tobrilus* and *Eutobrilus* (Triplonchida), *Monhystera* and *Eumonhystera* (both Monhysterida). Species of these genera were found throughout the extensive Murray–Darling Basin of

Australia, as well as in a much smaller catchment in southern Germany, the River Danube in Austria and the lower Rhine in the Netherlands (Eder, 1983; Bongers and Van Der Haar, 1990; Hodda, 1999; Beier and Traunspurger, 2003). While these genera seem to show few differences in horizontal distribution, they sometimes show distinct patterns of vertical distribution. In the Danube River, the triplonchids were mostly found deeper in the sediment, while the monhysterids were mostly found within the sediment near the surface (Eder, 1983). In other places no vertical zonation was observed, but it is possible that the surface layer was thinner and so was undetectable with the sampling depths used (Bongers and Van Der Haar, 1990; Beier and Traunspurger, 2003).

At the family level, there is some consistency in distributions, which is associated with the grain size of the sediment: Tylenchidae are frequently associated with coarser sediments, and Tobrilidae are associated with finer sediments (Beier and Traunspurger, 2003).

More generalizations are possible regarding the trophic status of nematodes from lotic freshwaters because the patchiness of individual species distributions is evened out. In small, clear alpine streams, many nematodes are plant-feeding and associated with algae or submerged bryophytes (Jacobs, 1984; Suren, 1991; Poff *et al.*, 1993). In many other streams, particularly those with fine sediments, nematodes feeding by ingesting the small microbes and organic particles in the water (deposit feeders) are dominant (Zullini, 1974, 1976; Zullini and Ricci, 1980; Bongers and Van Der Haar, 1990; Ocana and Picazo, 1991; Traunspurger, 2002; Beier and Traunspurger, 2003).

By contrast, other streams had a fauna dominated by omnivores or predators (Zullini, 1974; Eder and Kirchengast, 1982; Eder, 1983; Strayer and O'Donnell, 1992). In some of these studies, a relatively large 100 μm mesh sieve was used to extract nematodes, and the results may have been biased towards large nematodes such as predators.

Water velocity and depth have also been related to changes in species composition on a local scale (Filipjev, 1930; Hirschmann, 1952). However, sediment grain size is related to these factors (see section on 'Diversity').

Water composition has also been related to changes in species composition in a number of places, such as India, Sicily, Germany and Austria (Colomba and Vinciguerra, 1980; Venkateswarlu and Das, 1980a,b, 1982a,b; Eder and Kirchengast, 1982). The concentration of carbonate, calcium and magnesium ions (often referred to in combination as 'water hardness') or chloride ions is cited as the most important (Colomba and Vinciguerra, 1980; Venkateswarlu and Das, 1980a,b; Eder and Kirchengast, 1982; Beier and Traunspurger, 2003).

The level of organic enrichment of sediments has been frequently related to the species composition of nematodes in lotic systems (Zullini, 1976; Eder and Kirchengast, 1982; Bongers and Van Der Haar, 1990; Niemann *et al.*, 1996; Bazzanti, 2000; Beier and Traunspurger, 2003). Some species seem to be affected, while some do not seem to be affected (Zullini, 1974, 1976; Niemann *et al.*, 1996; Beier and Traunspurger, 2003). Furthermore, different species are mostly found at different levels of organic enrichment (Table 9.2). There have been a number of attempts to codify which taxa are affected, be they species, genera, families or other groups, and how they are affected by organic enrichment (summarized by Bongers, 1990; Bongers and Ferris, 1999). Increasing proportions of the Orders Rhabditida and Diplogasterida with increasing organic enrichment are frequently observed (e.g. Zullini, 1976; Ocana and Picazo, 1991; Beier and Traunspurger, 2003). However, many of the attempts to generalize responses of nematodes to organic enrichment at lower taxonomic levels have had limited geographic validity (e.g. Niemann *et al.*, 1996; Beier and Traunspurger, 2001).

Abundance

Many studies have considered only total nematode abundance in lotic systems (Table 9.1). This is a simple parameter of nematode populations to measure, but interpretation is often difficult. Total abundance is enormously variable, with reported values spanning over three orders of magnitude (Table 9.1). As with other parameters of nematode populations, part of the variability is associated with methodological differences and part with ecological differences in the nematode populations themselves (Chapter 2). The methodological differences, however, make comparisons between different studies extremely tenuous.

Studies using the same techniques to estimate total abundance of nematodes have been limited to a relatively narrow range of grain sizes within the sand to fine gravel range. Within this narrow range, there was little consistency of results. Sometimes abundance was higher in finer sandy sediments (Callahan *et al.*, 1979; Bott and Kaplan, 1989), in coarser sands (Anderson, 1992) and in fine gravels (Beier and Traunspurger, 2001). Nutrient status may be more important than sediment grain size in some situations (Callahan *et al.*, 1979). In other situations where flow rates are variable, flow rate may be the most important influence on total nematode abundance (Boulton and Stanley, 1995).

Seasonal patterns of abundance were equally inconsistent. In the south-eastern USA, maximum total abundance was during spring (Palmer, 1990). Further north in the USA, there were no differences between the seasons (Bott and Kaplan, 1989; Strayer and O'Donnell, 1992). In Switzerland, farther north again in latitude, maximum abundance was consistent in spring and summer (Eisenmann *et al.*, 1998).

The most consistent patterns in total nematode abundance may be in their relationship to the abundance of other phyla. In the least stable or favourable conditions, nematodes seem consistently more abundant relative to other organisms. In sediments which become anoxic or are subject to highly variable flows, the total abundance of nematodes is relatively greater compared to oligochaetes and chironomids than in sediments subject to less severe or variable conditions (Palmer, 1990; Ward and Voelz, 1990; Giere, 1993; Wolz and Shiozawa, 1995; Hakencamp and Morin, 2000).

Adaptations

A major problem for the organisms in lotic habitats is removal, preferentially downstream, from a particular location. Associated with this is the need to colonize new substrate, which is itself coming from upstream. Artificial substrates have been used to study these processes.

Table 9.2 Occurrence of different species of nematodes at various levels of organic enrichment (Niemann *et al.*, 1996) in 600 samples from 15 small rivers and streams around Munster, Germany.

Limited to least organically enriched	Most abundant in least organically enriched	Eurytopic	Most abundant in organically enriched	Limited to organically enriched
Mononchus	*Fictor*	*Plectus* two spp.	*Pelodera*	*Tobrilus*
–	*Monhystera*	*Mesorhabditis*	*Curviditis*	*Diploscapter*
–	*Plectus*	*Eumonhystera*	*Diplogaster*	*Paroigolaimella*
–	*Tripyla*	*Aphanolaimus*	*Aphanolaimus*	*Mononchoides*
–	*Eumonhystera*	*Eudorylaimus*	*Diplogasteritus*	*Oigolaimella*
–	*Anaplectus*	*Panagrolaimus*	*Cuticularia*	*Rhabditoides*
–	*Mononchus*	*Tobrilus* two spp.	*Chromadorita*	–
–	–	*Dorylaimus*	–	–

In a small permanent stream, nematodes (and oligochaetes) colonized equally from upstream and downstream (Williams and Hynes, 1976). In a temporary stream, colonization was more upstream when water first returned after a dry period, then was not directional (Williams and Hynes, 1976). Dispersal was in the water column. Another mechanism is recolonization from refugia (small locations sheltered from the force of the water current (Giere, 1993)).

Specific morphological adaptations to lotic environments include long anterior and somatic setae, pigment spots or ocelli, well-developed amphideal foveae and caudal glands and long filiform tails (Gerlach, 1953; Wieser, 1959; Warwick, 1971; Riemann, 1974). These adaptations are similar to those found in nematodes from other high-flux fluid environments such as high-energy beaches (Hodda, 1986; Nicholas and Hodda, 1999).

Importance of Nematodes in Lotic Food Webs

The importance of nematodes in lotic food webs has been investigated in several different ways: theoretical modelling, direct observations of gut contents and experimental manipulation of microcosms. These three approaches together indicate that nematodes are an important component of lotic systems, despite having a small biomass relative to larger insect larvae and smaller unicellular organisms (Giere, 1993).

In models of lotic systems, the presence of nematodes and other meiofauna interacting with macrofauna makes a large difference to the model predictions, particularly in terms of macrofaunal diversity and system resilience (Hildrew, 1992; Strong, 1992).

Observations of gut contents have shown that interactions with macrofauna do occur, although the size and importance of the interactions remain to be quantified. Some nematodes in lotic systems prey on diets including rotifers and small oligochaetes (Schmid-Araya and Schmid, 1995). Different species have slightly different prey. However, the proportion of these prey in the total diet remains unknown, as is the range of other possible foods. The range of foods may be quite broad (Yeates *et al.*, 1993). At the other end of the trophic chain, nematodes were themselves the prey of some Diptera from lotic habitats (Schmid-Araya and Schmid, 2000). However, the proportion of nematodes in the total diet is unknown.

The nematodes found associated with bryophytes and algae are almost certainly feeding on these plants (see section on 'Species distribution'). There is also no reason to believe that the observations of nematodes feeding on aquatic vascular plants in other freshwater aquatic systems do not apply to lotic habitats.

Microcosm experiments have demonstrated the importance of nematode interactions with the microbial component of lotic systems. When stream sediment bacteria were incubated with microbivorous or fungivorous nematodes, activity was much higher than in controls with nematodes absent (Traunspurger *et al.*, 1997). This parallels the results of similar experiments on terrestrial soils (Ingham *et al.*, 1985; Verhoef and Brussaard, 1990; Bouwman *et al.*, 1994; Griffiths, 1994; Ekschmitt and Griffiths, 1998; Bardgett *et al.*, 1999; Ekschmitt *et al.*, 1999; Sonnemann *et al.*, 1999) and marine sediments (Findlay and Tenore, 1982).

Another series of microcosm studies investigated feeding by the aquatic triclad platyhelminth *Dugesia gonocephala* on nematodes (Beier *et al.*, 2004). In this series of experiments, predation on nematodes was demonstrated, but there was no effect on the nematode composition or distribution, and predation only occurred in coarse gravel.

Conclusions

Nematodes in lotic habitats can be abundant, diverse and an important part of the ecosystem, yet they have been studied infrequently relative to nematodes from other freshwater habitats and relative to other organisms from lotic habitats. Methodological difficulties have undoubtedly contributed to this situation. The many different solutions to these difficulties have, in turn, led to different studies being partially comparable at best, and this has led to a lack of general paradigms about the nematode fauna of these habitats.

Generalizations, which can be made about nematodes from lotic habitats, are that they are highly variable in species composition, diversity and abundance, and influenced by the conditions of the sediment as well as the waters above. Many of the nematodes in lotic habitats are apparently occasional, even accidental, occupants of either the sediment or the water column, but the time in the lotic environment can be an important part of the survival mechanisms of these otherwise terrestrial taxa. More permanent inhabitants may be shared with groundwater.

Whether limited to lotic habitats or not, the nematodes found in these habitats have adapted to the particular variabilities and challenges of the habitat, including strong and sometimes highly variable water flows, organic enrichment, changes in sediment granulometry, physical characteristics and chemistry. They may also be significant for the other organisms through their interactions with the macrofaunal, microbial and phytal components of the system. They are deserving of much more study to clarify the many apparently conflicting results or observations, as well as the high variability in their environment and their responses to it.

References

Altherr, E. and Deboutteville, D. (1972) Freshwater interstitial nematodes from the States of Washington, Colorado and Massachusetts, USA, collected by Cl. Delamare Deboutteville. *Annales de Speleologie* 27(4), 683–760.

American Public Health Association (1998) Nematological examination: introduction. In: Clesceri, L.S., Greenberg, A.E. and Eaton, A.D. (eds) *Standard Methods for the Examination of Water and Wastewater*, 20th edn. APHA, Washington DC, pp. 10-74–10-76.

Anderson, R.V. (1992) Free-living nematode associations in pool 19, Mississippi River. *Journal of Freshwater Ecology* 7, 243–250.

Andrássy, I. (1959) Nematoden aus dem Psammon des Adige-Flusses. I. *Memorie Museo civico di storia naturale di Verona* 7, 163–181.

Andrássy, I. (1962) Nematoden aus dem Ufergrunwasser der Donube von Bratislava bis Budapest. *Archiv fuer Hydrobiologie Supplement* 27, 91–117.

Andrássy, I. (1978a) Nematoda. A checklist of the animals inhabiting European inland waters with accounts of their distrbution and ecology. In: Illies, J. (ed.) *Limnofauna Europaea*, 2nd edn. Gustav Fischer Verlag, Stuttgart, Germany, pp. 98–117.

Andrássy, I. (1978b) Freshwater nematodes from the Himalayas (Nepal). *Opuscula Zoologica Budapest* 15, 3–21.

Bardgett, R.D., Cook, R., Yeates, G.W. and Denton, C.S. (1999) The influence of nematodes on below-ground processes in grassland ecosystems. *Plant and Soil* 212, 22–33.

Bass, D. and Walker, V. (1992) A preliminary report of invertebrates from hyporheic sediments of the North Canadian River. *Proceedings of the Oklahoma Academy of Science* 72, 3–4.

Bazzanti, M. (2000) Macrobenthic nematodes as biological indicators in a Mediterranean lowland river in central Italy: a case study. *Archiv für Hydrobiologie* 148, 59–70.

Beier, S. and Traunspurger, W. (2001) The meiofauna community of two small German streams as indicator of pollution. *Journal of Aquatic Ecosystem Stress and Recovery* 8, 387–405.

Beier, S. and Traunspurger, W. (2003) Seasonal distribution of free living nematodes in the Krahenbach, a fine grained submountain carbonate stream in southwest Germany. *Nematology* 5, 113–136.

Beier, S., Bolley, M. and Traunspurger, W. (2004) Predator–prey interactions between *Dugesia gonocephala* and free-living nematodes. *Freshwater Biology* 49, 77–86.

Bongers, T. (1990) The maturity index: an ecological measure of environmental disturbance based on nematode species composition. *Oecologia* 83, 14–19.

Bongers, T. and Ferris, H. (1999) Nematode community structure as a bioindicator in environmental monitoring. *Trends in Ecology & Evolution* 14, 224–228.

Bongers, T. and Van Der Haar, J. (1990) On the potential of basing an ecological typology of aquatic sediments on the nematode fauna: an example from the river Rhine. *Hydrobiological Bulletin* 24, 37–45.

Borchardt, M.A. and Bott, T.L. (1995) Meiofaunal grazing of bacteria and algae in a Piedmont stream. *Journal of the North American Benthological Society* 14, 278–298.

Bott, T.L. and Kaplan, L.A. (1989) Densities of benthic protozoa and nematodes in a Piedmont stream. *Journal of the North American Benthological Society* 8, 187–196.

Boulton, A.J. and Stanley, E.H. (1995) Hyporheic processes during flooding and drying in a Sonoran Desert stream. 2. Faunal dynamics. *Archiv für Hydrobiologie* 134, 27–52.

Bouwman, L.A., Bloem, J., van den Boogert, P.H.J.F., Hoenderboom, G.H.J. and DeRuiter, P.C. (1994) Short-term and long-term effects of bacterivorous nematodes and nematophagous fungi on carbon and nitrogen mineralisation in microcosms. *Biology and Fertility of Soils* 17, 249–256.

Cadet, P. and Albergel, J. (1999) Passive transport of phytoparasitic nematodes by runoff water in the Sudano-Sahelian climatic area. *Journal of Hydrology* 214, 91–102.

Cadet, P., Planchon, O., Esteves, M. and Lapetite, J.M. (2003) Experimental study of the selective transport of nematodes by runoff water in the Sudano Sahelian area. *Applied Soil Ecology* 19, 223–236.

Callahan, C.A., Ferris, V.R. and Ferris, J.M. (1979) The ordination of aquatic nematode communities as affected by stream water quality. *Statistical Ecology Series* 11, 101–116.

Colomba, G. and Vinciguerra, M.T. (1980) Nematodi d'acqua dolce della Sicilia. 1. Nematodi dell'Anapo. *Animalia (Catania)* 6, 89–120.

Danielopol, D.L. (1976) The distribution of the fauna in the interstitial habitats of riverine sediments of the Danube and the Piesting (Austria). *International Journal of Speleology* 8, 23–51.

Danielopol, D.L. (1989) Groundwater fauna associated with riverine aquifers. *Journal of the North American Benthological Society* 8, 18–35.

Danielopol, D.L. (1991) Spatial distribution and dispersal of interstitial Crustacea in alluvial sediments of a backwater of the Danube at Vienna. *Stygologia* 6, 97–110.

Eder, R. (1983) Nematoden aus dem Interstitial der Donube bei Fischamend (Niederosterreich). *Archiv für Hydrobiologie Supplement* 68, 100–113.

Eder, R. and Kirchengast, M. (1982) The nematode-fauna (Nemathelminthes, Nematoda) of a polluted part of the River Mur (Styria, Austria). *Nematologia Mediterranea* 10, 127–134.

Eisenmann, H., Traunspurger, W. and Meyer, E.I. (1998) Community structure of selected micro- and meiobenthic organisms in sediment chambers from a prealpine river (Necker, Switzerland). In: Bretschko, G. and Helesic, J. (eds) *Advances in River Bottom Ecology*. Backhuys Publishers, Leiden, The Netherlands.

Ekschmitt, K. and Griffiths, B.S. (1998) Soil biodiversity and its implications for ecosystem functioning in a heterogeneous and variable environment. *Applied Soil Ecology* 10, 201–215.

Ekschmitt, K., Bakonyi, G., Bongers, M., Bongers, T., Boström, S., Dogan, H., Harrison, A., Kallimanis, A., Nagy, P., O'Donnell, A.G., Sohlenius, B., Stamou, G.P. and Wolters, V. (1999) Effects of the nematofauna on microbial energy and matter transformation rates in European grassland soils. *Plant and Soil* 212, 45–61.

Eyualem-Abebe, Mees, J. and Coomans, A. (2001) Nematode communities of Lake Tana and other inland water bodies of Ethiopia. *Hydrobiologia* 462, 41–73.

Filipjev, I.N. (1930) Les nematodes libres de la baie de la Neva et de l'extremité orientale du Golfe de Finlande. *Archiv fuer Hydrobiologie* 21, 1–64.

Findlay, S. and Tenore, K.R. (1982) Effect of a free-living marine nematode (*Diplolaimella chitwoodi*) on detrital carbon mineralization. *Marine Ecology Progress Series* 8, 161–166.

Gerlach, S.A. (1953) Die biozönotische Gleiderung der Nematodenfauna an den deutschen Küsten. *Zietschrift fuer Morphologie und Ökologie der Tiere* 41, 411–512.

Giere, O. (1993) *Meiobenthology: The Microscopic Fauna in Aquatic Sediments*. Springer-Verlag, Berlin.

Griffiths, B.S. (1994) Microbial-feeding nematodes and protozoa in soil: their effects on microbial activity and nitrogen mineralization in decomposition hotspots and the rhizosphere. *Plant and Soil* 164, 25–33.

Grimm, R. (1978) Die Fauna des Bewasserungssystems der Insel Teneriffa unter besonderer Berucksichtigung der Naididae (Oligochaeta). *Zoologischer Anzeiger* 201, 143–150.

Hakencamp, C.C. and Morin, A. (2000) The importance of meiofauna to lotic ecosystem functioning. *Freshwater Biology* 44, 165–175.

Hildrew, A.G. (1992) Food webs and species interactions. In: Calow, P. and Petts, G.E. (eds) *The Rivers Handbook: Hydrological and Ecological Principles*, vol. 1. Blackwell Scientific Publications, Oxford, UK, pp. 309–330.

Hirschmann, H. (1952) Die Nematoden der Wassergrenze mittelfränkischer Gewässer. *Zoologischer Jahrbucher (Systematik)* 81, 313–436.

Hodda, M. (1986) Sandworms. *Australian Natural History* 21, 520–523.

Hodda, M. (1999) Nematodes of Australia: free-living, insect- and plant-associated. Available at: http://www.ento.csiro.au/science/nematodes/nematodes

Hummon, W.D., Evans, W.A., Hummon, M.R., Doherty, F.G., Wainberg, R.H., Stanley, W.S., Thorp, J.H. and Gibbons, J.W. (1978) In: Thorp, J.H. and Gibbons, J.W. (eds) *Energy and Environmental Stress in Aquatic Systems*. Department of Energy Symposium Series No. 48, Technical Information Center, US Department of Energy, Oak Ridge, Tennessee, pp. 188–203.

Ingham, R.E., Trofymow, J.A., Ingham, E.R. and Coleman, D.C. (1985) Interactions of bacteria, fungi, and their nematode grazers: effects of nutrient cycling and plant growth. *Ecological Monographs* 55, 119–140.

Jacobs, L.J. (1984) The free-living inland aquatic nematodes of Africa – a review. *Hydrobiologia* 113, 259–291.

Jordan, S., Shiozawa, D.K. and Schmid-Araya, J.M. (1999) Benthic invertebrates of a large, sandy river system: the Green and Colorado Rivers of Canyonlands National Park, Utah. *Archiv für Hydrobiologie* 147, 91–127.

Kirchengast, M. (1984) Faunistische Untersuchungen im hyporheischen Interstitial des Flusses Mur (Steiermark, Osterreich). *International Revue de Gesamten Hydrobiologie* 69, 729–746.

Mott, J.B., Mulamoottil, G. and Harrison, A.D. (1981) A 13-month survey of nematodes at three water treatment plants in southern Ontario, Canada. *Water Research* 15, 729–738.

Nicholas, W.L. (1984) *The Biology of Free-living Nematodes.* Clarendon Press, Oxford, UK.

Nicholas, W.L. and Hodda, M. (1999) Free-living nematodes of a temperate, high-energy sandy beach: faunal composition and variation over space and time. *Hydrobiologia* 394, 113–127.

Niemann, R., Arens, M., Koczwara, K. and Sturhan, D. (1996) Studies on the potential of nematodes for evaluating the quality of flowing waters. *Mitteilungen aus der Biologischen Bundesanstalt für Land- und Forstwirtschaft Berlin-Dahlem* 317, 195–208.

Ocana, A. and Picazo, J.S. (1991) Study on nematode species encountered in the Monachil River (Granada, Spain): response to organic pollution. *Internationale Vereinigung für Theoretische und Angewandte Limnologie Verhandlungen* 24, 2729–2737.

Palmer, M.A. (1990) Temporal and spatial dynamics of meiofauna within the hyporheic zone of Goose Creek, Virginia. *Journal of the North American Benthological Society* 9, 17–25.

Pennak, R.W. and Ward, J.V. (1986) Interstitial faunal communities of the hyporheic and adjacent groundwater biotopes of a Colorado mountain stream. *Archiv für Hydrobiologie, Supplement* 74, 356–396.

Poff, N.L., Palmer, M.A., Angermeier, P.L., Vadas, R.L. Jr., Hakenkamp, C.C., Bely, A., Arensburger, P. and Martin, A.P. (1993) Size structure of the metazoan community in a Piedmont stream. *Oecologia (Berlin)* 95, 202–209.

Riemann, F. (1966) Die interstitielle Fauna im Elbe-Aestuar. Verbreitung und Systematik. *Archiv für Hydrobiologie Supplement* 31, 1–279.

Riemann, F. (1974) On hemisessile nematode with flagelliform tails living in marine soft bottoms and on microtubes found in deep sediments. *Mikrofauna Meeresboden* 40, 1–15.

Schmid-Araya, J.M. (1997) Temporal and spatial dynamics of meiofaunal assemblages in the hyporheic interstitial of a gravel stream. In: Gibert, J., Mathieu, J. and Fournier, F. (eds) *Groundwater/Surfacewater Ecotones: Biological and Hydrological Interactions and Management Options.* International Hydrology Series, Cambridge University Press, Cambridge, UK, pp. 29–36.

Schmid-Araya, J.M. and Schmid, P.E. (1995) The invertebrate species of a gravel stream. *Jahresberichte der Biologischen Station Lunz* 15, 11–21.

Schmid-Araya, J.M. and Schmid, P.E. (2000) Trophic relationships: integrating meiofauna into a realistic benthic food web. *Freshwater Biology* 44, 149–163.

Smith, F., Brown, A.V. and Pope, M. (2001a) Meiofauna in intermittent streams differ among watersheds subjected to five methods of timber harvest. *Hydrobiologia* 464, 1–8.

Smith, F., Brown, A.V. and Pope, M. (2001b) Benthic meiofauna responses to five forest harvest methods. *Hydrobiologia* 464, 9–15.

Sonnemann, I., Dogan, H., Klein, A., Pieper, B., Ekschmitt, K. and Wolters, V. (1999) Response of soil microflora to changes in nematode abundance – evidence for large scale effects in grassland soil. *Journal of Plant Nutrition and Soil Science* 162, 385–391.

Stefanski, W. (1916) Die frielebende Nematoden des Inn, ihre Verbreitung und Systematik. *Zoologischer Anzeiger* 46, 363–384.

Strayer, D.L. and O'Donnell, E.B. (1992) The hyporheic nematode community of some streams in southeastern New York State, USA. *Stygologia* 7, 143–148.

Strayer, D.L., May, S.E., Nielsen, P., Wollheim, W. and Hausam, S. (1997) Oxygen, organic matter, and sediment granulometry as controls on hyporheic animal communities. *Archiv für Hydrobiologie* 140, 131–144.

Strong, D.R. (1992) Are trophic cascades all wet? Differentiation and donor-control in speciose ecosystems. *Ecology* 73, 747–754.

Suren, A.M. (1991) Bryophytes as invertebrate habitat in two New Zealand alpine streams. *Freshwater Biology* 26, 399–418.

Tockner, K. and Bretschko, G. (1996) Spatial distribution of particulate organic matter (POM) and benthic invertebrates in a river–floodplain transect (Danube, Austria): importance of hydrological connectivity. *Archiv für Hydrobiologie Supplement* 115, 11–27.

Traunspurger, W. (2002) Nematoda. In: Rundle, S.D., Robertson, A.L. and Schmid-Araya, J.M. (eds) *Freshwater Meiofauna: Biology and Ecology*. Backhuys Publishers, Leiden, The Netherlands, pp. 63–104.

Traunspurger, W., Bergtold, M. and Goedkoop, W. (1997) The effects of nematodes on bacterial activity and abundance in a freshwater sediment. *Oecologia (Berlin)* 112, 118–122.

Venkateswarlu, G. and Das, V.M. (1980a) Studies on the correlation of benthic nematode populations with hardness of water. *Proceedings of the Indian Academy of Parasitology* 1, 129–130.

Venkateswarlu, G. and Das, V.M. (1980b) Studies on the correlation of benthic nematode populations with chlorides in water. *Proceedings of the Indian Academy of Parasitology* 1, 133–134.

Venkateswarlu, G. and Das, V.M. (1982a) Studies on the correlation of benthic nematode populations with dissolved oxygen content of water. *Indian Journal of Nematology* 11, 235–238.

Venkateswarlu, G. and Das, V.M. (1982b) Studies on the correlation of benthic nematode populations with nitrate content of water. *Proceedings of the Indian Academy of Parasitology* 3, 48–49.

Verhoef, H.A. and Brussaard, L. (1990) Decomposition and nitrogen mineralisation in natural and agro-ecosystems: the contribution of soil animals. *Biogeochemistry* 11, 175–211.

Villenave, C., Cadet, P., Planchon, O., Esteve, M. and Lapetite, J.M. (2003) Transport of free living nematodes by runoff water in a Sudano Sahelian area. *Applied Soil Ecology* 23, 85–91.

Ward, J.V. and Voelz, N.J. (1990) Gradient analysis of interstitial meiofauna along a longitudinal stream profile. *Stygologia* 5, 93–99.

Warwick, R.M. (1971) Nematode associations in the Exe estuary. *Journal of the Marine Biological Association of the UK* 51, 439–454.

Whitman, R.L. and Clark, W.J. (1984) Ecological studies of the sand-dwelling community of an east Texas stream. *Freshwater Invertebrate Biology* 3, 59–79.

Wieser, W. (1959) Free-living marine nematodes. IV. General part. *Acta Universatitis Lund (N.F.)* 55, 1–111.

Williams, D.D. (1993) Changes in freshwater meiofauna communities along the groundwater-hyporheic water ecotone. *Transactions of the American Microscopical Society* 112, 181–194.

Williams, D.D. and Hynes, H.B.N. (1976) The recolonization mechanisms of stream benthos. *Oikos* 27, 265–272.

Wolz, E.R. and Shiozawa, D.K. (1995) Soft sediment benthic macroinvertebrate communities of the Green River at the Ouray National Wildlife Refuge, Uintah County, Utah. *Great Basin Naturalist* 55, 213–224.

Yeates, G.W., Bongers, T., De Goede, R.G.M., Freckman, D.W. and Georgieva, S.S. (1993) Feeding habits in soil nematode families and genera – an outline for soil ecologists. *Journal of Nematology* 25, 315–331.

Zullini, A. (1974) The nematological population of the Po River. *Bollettino di Zoologia* 41, 183–210.

Zullini, A. (1976) Nematodes as indicators of river pollution. *Nematologica Mediterranea* 4, 13–22.

Zullini, A. and Ricci, C. (1980) Bdelloid rotifers and nematodes in a small Italian stream. *Freshwater Biology* 10, 67–72.

10 Nematodes from Extreme Freshwater Habitats

M. Hodda,[1] A. Ocaña[2] and W. Traunspurger[3]

[1]Nematode Biosystematics and Ecology, CSIRO Entomology, GPO Box 1700, Canberra, ACT 2601, Australia; [2]Department of Animal Biology and Ecology, Faculty of Science, University of Granada, 18071 Granada, Spain; [3]Department of Animal Ecology, University of Bielefeld, Morgenbreede 45, 33615 Bielefeld, Germany

Introduction

This chapter discusses the ecology and biogeography of nematodes from freshwater environments, which are extreme in terms of temperature, chemical composition, variability or isolation. The composition of nematode faunas from freshwater pools in bromeliads or tree hollows, hot or mineral springs, pools and bogs in polar regions, seasonal lakes or pools, fresh groundwaters and caves will be described and compared. The nematode faunas from these extreme habitats will also be compared with those from more typical freshwater environments. Nematodes with evolutionary affinities to freshwaters but found in estuarine sediments will also be discussed. Conversely, nematodes from freshwaters with evolutionary affinities to otherwise marine taxa will also be considered. Common themes in the adaptations of all these nematodes to such unusual environments will then be synthesized.

The emphasis will be on broad ecological patterns rather than on detailed species interactions with the various freshwater environments. Thus, the discussion will focus on genera or higher taxa rather than species. For more detail on the species involved, see the taxonomic chapters.

Springs

Among the 'extreme' freshwater habitats for nematodes, springs are some of the most studied. Southern Europe is particularly well studied (Granada in Spain: Ocaña, 1991a–c, 1993; Ocaña and Morales, 1992; Germany and Austria: Pax and Soos, 1943; Paetzold, 1958; Andrássy, 1978; Schiemer, 1978; Seiml-Buchinger and Traunspurger, 2005; the Italian Isle of Ischia: Meyl, 1953a,b, 1954; and Yugoslavia: Schneider, 1940). There have also been studies in Central Africa

(De Coninck, 1935), the USA (Hoeppli, 1926), Kyrghyzstan (Gagarin and Lemzina, 1992) and China and Taiwan (Hoeppli and Chu, 1932).

Springs are unique freshwater habitats because their physical and chemical environments are usually nearly constant. However, there are great variations between different springs in chemical composition, velocity of water and temperature (Botosaneanu, 1998). There are also great variations between different springs in how long they have existed at a particular place, their remoteness from other springs and the surrounding habitat. This is in marked contrast to the situation in most lakes, rivers, marine and terrestrial environments where the physical and chemical environment is more internally variable. Because of these ecological characteristics, springs are regarded as particularly important for ecological, biogeographic, evolutionary and genetic studies (Hynes, 1970; Odum, 1971). This level of interest is vastly out of proportion to the size and number of springs relative to other freshwater habitats for nematodes.

Springs have been used as model ecosystems in very influential studies of productivity, energy flow and trophic relationships (Odum, 1957; Teal, 1957; Minckley, 1963; Minshall, 1967; Tilly, 1968; Iversen, 1988). In these studies nematodes were considered at most in terms of gross trophic groups, and often not considered at all. The broad, but relatively shallow, approach of these studies was combined with inefficient techniques for sampling and enumerating nematodes, so the conclusions about the ecological importance and trophic roles of nematodes have a large associated uncertainty.

Furthermore, springs have also been used as controls or reference points for comparisons involving aquatic habitats polluted to some extent because they include some of the few habitats unimpacted by human activities (Ocaña and Picazo, 1991).

Nematodes seem to represent a larger proportion of the total number of metazoan animals in springs, both in terms of numbers and biomass, as conditions become more extreme. This is not necessarily because nematodes become more abundant, but because nematodes are, as a phylum, one of the most resistant multicellular organisms, and other organisms disappear first. This phenomenon has also been noted in relation to pollution (Chapter 8). Nevertheless, nematodes do not occur where conditions become too extreme: at high temperatures above 43°C and at high ionic concentrations, particularly chloride ions above 7600 meq/l and total ions expressed in conductivity data, above 17,000 μS/cm (Ocaña, 1991a,b).

The nematode genera which have been found in springs are listed in Table 10.1. Abundance, diversity and distribution of individual species are all highly variable. For example, in an extensive study by a single author using comparable methods, the five most frequent species were found in 65%, 55%, 53%, 45% and 39% of springs. Most species were found in less than 10% of the springs sampled, corresponding to three springs or less of 38 studied (Ocaña, 1991a). Many species in the orders Dorylaimida, Rhabditida, Tylenchida and Aphelenchida have been recorded only once and are considered accidental inhabitants (Ocaña *et al.*, 1986).

Hot or thermal springs

Hot or thermal springs are among the most studied types of springs. Entire faunas have been studied in Ruwenzori, USA, China and Taiwan, Italy, Germany, Spain

Table 10.1 Number of species in nematode genera found in springs in Granada, Spain. (From Ocaña, 1991a,b, 1993; Ocaña and Morales, 1992.)

Order	Genus	Fresh springs	Hot springs	Low or variable O_2 springs	High salt springs	High SO_4 springs	Eurytopic	Sporadic, probably accidental
Monhysterida	*Daptonema*	–	–	–	1	–	–	–
	Eumonhystera	–	3	–	–	1	2	–
	Monhystera	–	2	–	1	–	–	–
	Monhystrella	–	2	1	1	–	1	–
Araeolaimida	*Aphanolaimus*	–	–	–	–	–	1	1
	Chronogaster	–	2	–	–	–	1	–
	Cylindrolaimus	–	–	–	–	–	–	1
	Paraplectonema	–	–	–	1	-	–	–
	Plectus	–	3	3	–	1	1	–
	Rhabdolaimus	–	1	–	–	–	–	–
Triplonchida	*Prismatolaimus*	–	1	1	–	–	1	–
	Tobrilus	1	2	–	–	–	1	1
	Tripyla	–	–	–	–	–	–	1
Rhabditida	*Rhabditis* s.l.	–	1	1	–	–	–	–
	Other species	–	–	–	–	–	–	32
Chromadorida	*Achromadora*	–	1	–	–	1	–	1
	Chromadorita	–	–	–	1	–	–	–
	Ethmolaimus	1	–	–	–	1	–	–
	Paracyatholaimus	–	1	–	1	–	–	–
	Prodesmodora	–	–	–	1	–	–	–
	Udonchus	–	1	–	–	–	–	–
Enoplida	*Alaimus*	1	–	–	–	1	–	–
	Ironus	–	–	–	1	1	1	–
	Odontolaimus	–	1	–	–	–	–	–
	Trischistoma	–	1	–	–	–	–	–
Dorylaimida	–	–	–	–	–	–	–	24
Mononchida	–	–	–	–	–	–	–	2
Tylenchida	–	–	–	–	–	–	–	8
Aphelenchida	–	–	–	–	–	–	–	2
Total		3	22	6	8	6	9	73

and Indonesia (Hoeppli, 1926; Hoeppli and Chu, 1932; De Coninck, 1935; Schneider, 1937; Pax and Soos, 1943; Meyl, 1953a,b, 1954; Ocaña, 1991b; Eyualem-Abebe *et al.* 2001; W. Traunspurger, unpublished data). In addition there have been species descriptions from Spain and Austria (Schiemer, 1978; Ocaña, 1991c).

The occurrence of nematode species in hot springs is mostly highly variable (Ocaña, 1991a,b; W. Traunspurger, unpublished data). However, at least two species are found in hot springs all over the world: *Rhabditis terrestris* (Ocaña, 1991a in Spain; Schiemer, 1978 in Austria; and Schneider, 1937 in Sumatra, Java and Bali) and *Udonchus tenuicaudatus* (De Coninck, 1935 in Ruwenzori, Democratic Republic of Congo; Meyl, 1953a,b, 1954 in Italy; Ocaña, 1991b in Spain; and Paetzold, 1958 in Germany). Even these species are not found in all hot springs, and chance seems to be the most important influence on the species in any particular hot spring. Other genera commonly found are listed in Table 10.1.

In hot springs, total nematode abundance and species richness are also highly variable, perhaps associated with the particular species of both nematodes and other organisms present in a particular spring (Fig. 10.1). In general, the trend is for total nematode abundance to decrease where water temperature is about 25°C and higher. However, in one very hot spring (over 40°C) very high abundance has been found, which consisted almost entirely of *R. terrestris*. Species richness shows a similar trend to abundance: within the orders Monhysterida, Araeolaimida, Chromadorida and Enoplida, eight species were found in springs above 40°C, 11 species were found at 30–40°C and more species were found at 20–30°C (Ocaña, 1991b).

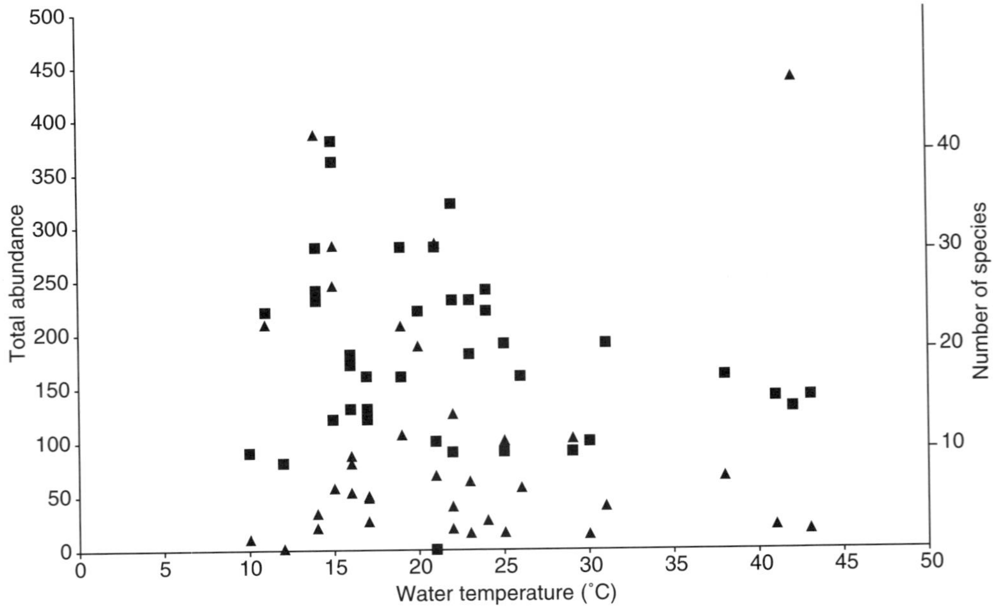

Fig. 10.1 Total abundance and species richness of nematodes in hot springs and temperature (data from Ocaña 1991b). Triangles represent abundance, squares represent species richness.

Low-oxygen springs

The characteristics of the nematode faunas in springs with different levels of oxygen potential in the waters are less variable than in springs with varying temperatures. This is perhaps related to the lower productivity and diversity of other organisms in springs with low oxygen availability: there are fewer opportunities for nematodes to colonize successfully. The number of species found in low-oxygen springs is small compared with that in other springs, and the total abundance of nematodes is low.

In springs where the oxygen availability is always low, nematode species of the order Rhabditida are often most abundant (Ocaña, 1993). Rhabditids are thought to be some of the nematodes with the most opportunistic ecological strategy: they are easily dispersed, rapidly increase population size under favourable conditions because of high fecundity and short generation times and they are very resistant to adverse conditions (Zullini, 1976; Bongers, 1988; Zullini and Pagani, 1989).

Where the amount of oxygen was often low but more variable the species are different: one species each of the monhysterid genera *Eumonhystera* and *Monhystrella*, together with several species of the genus *Plectus*, and a single species of the genus *Prismatolaimus* (Ocaña, 1993). Monhysterida are known to possess physiological adaptations to low oxygen environments (see section on 'Adaptations to Extreme Freshwater Environments') and the genus *Plectus* shares many of the ecological characteristics of Rhabditida. The genus *Prismatolaimus* occurs in very diverse habitats and must be tolerant of a wide range of conditions.

Mineral springs

Mineral springs are springs containing significantly more ions than is usual in freshwater. The concentrations of particular ions may vary considerably among springs and this, in turn, affects many other aspects of their chemistry and biology. Among the most common ions present are chloride, sulphate, carbonate or bicarbonate, calcium, magnesium, sodium, potassium and iron.

As in other types of springs, total abundance and number of species present are very variable and very poorly correlated with any environmental parameters. The general trends are towards higher abundance and species richness in springs with lower ionic content (Ocaña, 1991a; Ocaña and Morales, 1992) (Fig. 10.2).

Overall, about the same numbers of species have been found in mineral springs as in hot springs (Table 10.1). Most species present were also extremely sporadic, but a few species occurred frequently in a wide range of conditions. Many of the species which occurred frequently were also those found to be tolerant to fluctuating oxygen conditions, viz., species of the genera *Eumonhystera*, *Monhystrella*, *Chronogaster*, *Plectus*, *Prismatolaimus*, *Ironus*, *Aphanolaimus* and *Tobrilus* s.l. Most of the 31 species occurring more sporadically were not limited to any particular ionic composition or range; however, a few species occurred only in springs with a high concentration of particular ions (Table 10.1). Of these genera, some were found mainly where chloride was the predominant anion (*Monhystera*, *Monhystrella*, *Paracyatholaimus* and *Chromadorita*), and some where sulphate was predominant (*Ironus* and *Paraplectonema*). Those species found mainly where carbonate or bicarbonate were predominant ions

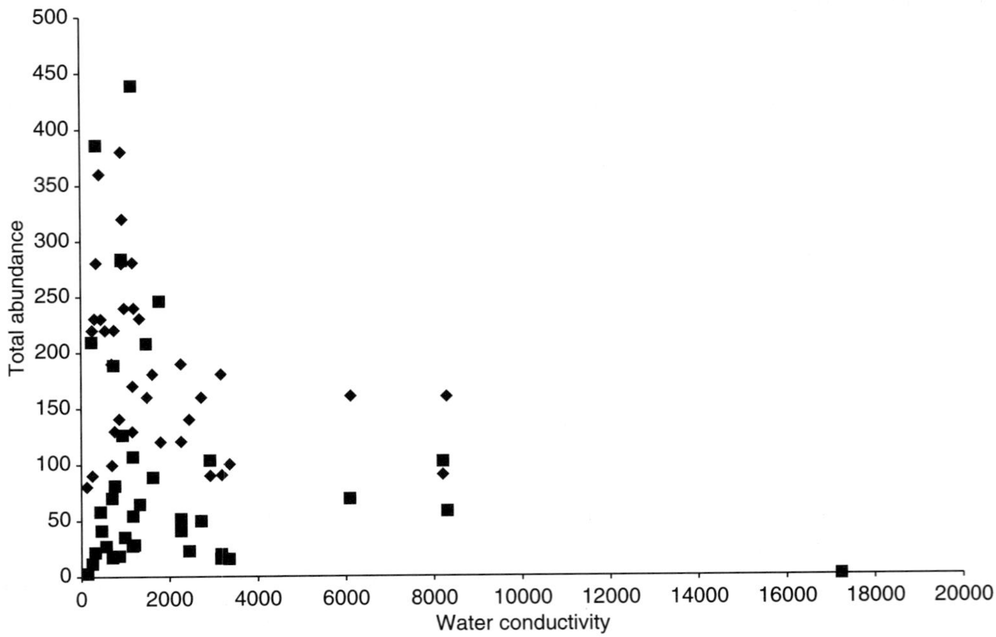

Fig. 10.2 Total abundance and species richness of nematodes in springs of different water conductivity (data from Ocaña, 1991a). Diamonds represent abundances, squares represent species richness.

were only found in springs of relatively low ionic concentration hence they are not listed in Table 10.1 (*Eumonhystera*, *Chronogaster*, *Plectus*, *Tobrilus* s.l. and *Achromadora*) (Ocaña and Morales, 1992). These last-mentioned species are also present in many rivers and streams (Ocaña and Picazo, 1991).

Freshwater springs

Of the nematodes which have been found in springs, few seem restricted to springs with low concentrations of ions, moderate to high oxygen and near-ambient temperature. One of these parameters may be limiting, and sometimes two, but rarely all three. For example, *U. tenuicaudatus* was restricted to springs with low ionic concentrations, but was widespread in hot springs. *Monhystrella lepidura* was restricted to springs with moderate oxygen, but present in hot springs and those with high ionic concentration. These observations strongly suggest that many, if not most, of the species found in springs are eurytopic to some extent. Strictly, freshwater springs generally support the largest numbers of species and highest abundances of any category of springs (Ocaña, 1993). Many of the species from this type of spring occur in other habitats, in addition to other types of springs (Gerlach and Riemann, 1974; Ocaña and Picazo, 1991; Seiml-Buchinger and Traunspurger, 2005).

Polar Freshwaters

Polar regions in both northern and southern hemispheres contain many freshwater habitats. Freshwaters in polar regions are, by definition, frozen for much of the year, but are abundant during the brief summer thaws. Freshwater nematodes have been found in many of these small pools and bogs.

Sub-polar islands

Extensive studies have been made on the sub-Arctic and sub-Antarctic islands of Spitzbergen (78°N) and Macquarie Island (54°S) (Bunt, 1954; Loof, 1971; Nunn, 1993; Marchant and Lillywhite, 1994; Hodda and Keith, unpublished data). In all of these studies freshwater aquatic habitats were poorly represented, but they nevertheless give some indication of freshwater nematodes in the sub-polar islands.

A total of 15 species were recorded from a small pool on Spitzbergen, including two which were also found in European spring waters, as well as in other habitats (*Monhystera stagnalis* and *Eumonhystera vulgaris*) (Loof, 1971; Table 10.2). Sampling in a similar habitat many years later produced a similar ecological suite of species in a site with fine sediment (Nunn, 1993) (Table 10.2; Fig. 10.3). Many of the same genera were present in the later study, although no species were identified by name. The fauna consisted of about equal proportions of hyphal feeders, microbivores and omnivores. A similar site with coarse sediment had many more microbial-feeding nematodes, and also predatory nematodes, which were not found in the earlier study.

High nematode abundance was found in a small stream sampled on Macquarie Island three times during 1952, but the species were not identified (Bunt, 1954). Single samples from several streams in 1998 produced 11 species with a similar trophic composition to the samples from Spitzbergen (Table 10.2; Fig. 10.3). Many of the genera on Macquarie Island were the same as those on Spitzbergen. The similarity in faunal composition at the generic level is remarkable given the geographic isolation of both islands, because chance events may be expected to influence composition strongly, as in spring waters (see section on 'Springs').

The freshwater aquatic nematodes of both sub-polar islands are separated from the nearest freshwater habitat by large distances of sea water. The results from Spitzbergen may be explained by the presence of land bridges during recent glaciation and lower sea levels. Alternatively, nematodes may be deposited with snow, as has been observed on high mountains in southern Europe and northern Africa (Duval *et al.*, 1999). Unlike Spitzbergen, Macquarie Island has never been connected to a continental land mass, and is very remote from freshwaters even on other islands, much more so than Spitzbergen. Human transport may be involved, since at least one species found on Macquarie Island, *Enoploides stewarti*, is elsewhere found only in southern Australia (Nicholas, 1993; Marchant and Lillywhite, 1994; Nicholas and Marples, 1995; Hodda and Keith, unpublished data). Many early visitors to Macquarie Island departed from southern Australia.

Table 10.2 Percentage abundances of nematode genera found in sub-polar freshwaters.

Order	Genera	Loof (1971)	Nunn (1993) (fine sediments)	Nunn (1993) (coarse sediments)	Hodda and Keith (unpublished data)
Enoplida	*Enoploides*	–	–	–	5
Monhysterida	*Eumonhystera*	20	–	5	30
	Monhystera	4	–	–	–
Araeolaimida	*Plectus*	8	–	22	10
Triplonchida	*Prismatolaimus*	1	–	2	2
	Tripyla	4	–	–	–
Dorylaimida	*Dorylaimus*	–	2	–	–
	Eudorylaimus s.l.	25	29	19	23
	Prodorylaimus	1	–	–	–
Mononchida	*Mononchus*	–	7	4	5
Rhabditida	*Rhabditis* s.l.	–	20	12	5
	Heterocephalobus	12	4	26	–
	Eucephalobus	4	5	3	15
	Chiloplacus	1	–	5	–
Tylenchida	*Tylenchus* s.l.	–	7	–	3
	Neotylenchidae	4	–	–	–
	Ditylenchus	1	–	–	–
Aphelenchida	*Aphelenchoides*	18	27	–	2

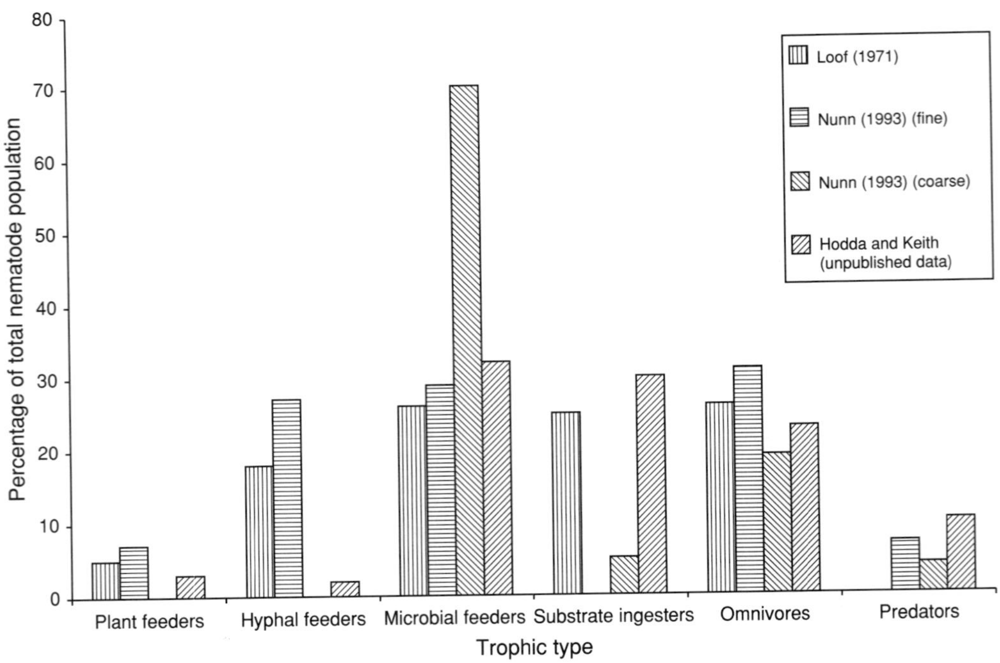

Fig. 10.3 Trophic composition of nematode faunas from freshwaters on two sub-polar islands, Spitzbergen and Macquarie Island.

Polar freshwaters

Nematodes from polar freshwaters have been recorded from continental Antarctica and islands south of the Antarctic convergence which are considered 'polar' (King George Island (62°S), Adelaide Island (67°S), Alexander Island (70°S) and Ross Island (77°S)).

Only a very few species of nematodes are known from these polar freshwaters: they represent very simplified freshwater nematode communities (Table 10.3). The genera are a subset of those found in the less extreme and less isolated sub-polar islands. However, many of the species seem to be endemic to the continent, like the terrestrial nematodes from Antarctica (Maslen, 1979). The trophic structures of these polar freshwater nematode communities seem remarkably similar to those of sub-polar freshwaters. This may seem unusual as many of the species are also found in 'terrestrial' polar habitats (Treonis *et al.*, 1999; Sinclair and Sjursen, 2001). The relative abundances of species may vary between terrestrial and freshwater aquatic habitats. Either the nematode species living in the extreme polar environment are extreme physiological and ecological generalists, the limited number of species

Table 10.3 Freshwater aquatic nematode taxa from polar regions.

Location	Habitat	Genera (number of species)	Reference(s)
King George Is.	Water puddle	*Eudorylaimus* (5)	Tsalolikhin (1989)
		Eumonhystera (1)	Tsalolikhin (1989)
	Lake	*Rhabditis* s.l (1)	Tsalolikhin (1989)
Antarctica	Nunataks	*Panagrolaimus* (1)	Swart and Harris (1996)
		Rotylenchus (1)	Van den Berg and Harris (1996)
		Plectus s.l. (1)	Heyns (1995)
		Chiloplacoides (1)	Heyns (1994)
		Eudorylaimus (1)	Heyns (1993)
Adelaide Is.	Pool	*Monhystera* (1)	Dartnall (1980)
MacMurdo	Stream	*Panagrolaimus* (1)	Treonis *et al.* (1999)
		Eudorylaimus (1)	Treonis *et al.* (1999)
		Plectus (1)	Treonis *et al.* (1999)
	Small pool with algae	*Scottnema* (1)[a]	Andrássy (1998)
		Plectus s.l. (1)	Andrássy (1998)
	Mire	*Eudorylaimus* (1)	Andrássy (1998)
Dry Valleys	Lake	*Monhystera* s.l.	Timm (1971); Wharton and Brown (1989)
Bunger Hills	Deep lakes	*Plectus* s.l. (1)	Andrássy (1998); Kirjanova (1958)
Alexander Is.	Pond	*Mesodorylaimus* (1)[b]	Maslen (1982)
Ross Is.	Intermittent stream	*Panagrolaimus davidi*	Sinclair and Sjursen (2001)

[a]In a very extensive survey of the same area Treonis *et al.* (1999) found this species in drier areas only, not in streams.
[b]Found in 45 of 50 samples (Maslen, 1982).

allows species to occupy vacant niches they would normally be excluded from or the polar habitats are so variable in time and space that only highly vagile species able to colonize a variety of habitats can survive.

Estuarine Sediments

For the purposes of this chapter, estuarine is defined as having salinity more than 0.1%. Estuarine nematodes with close taxonomic affinities to freshwater forms rather than marine forms are very much in the minority. Extensive studies of estuarine sediments on the eastern and northern coast of Australia have revealed many species from genera typically associated with freshwater or terrestrial substrates which were living in sediments with distinct marine influences (Table 10.4). Some of these species have been found in estuaries only once, and may have been washed accidentally downstream. Others have been found frequently enough to be classed as occupants. Many of these nematode species are associated with flowering plants, which have evolved adaptations to tidal flats bordering estuaries. Mangroves or salt marsh vegetation are commonly found on these estuarine tidal flats.

Many of the nematodes with freshwater affiliations are most abundant at the upper edge of the intertidal zone where the marine influence is often weak (Nicholas *et al.*, 1992). However, other species with freshwater affinities occur primarily in salt marshes where hypersaline conditions occur at least periodically (Furstenberg and De Wet, 1983; Wouts *et al.*, 1995; Wouts and Sturhan, 1996; Baldwin *et al.*, 1997).

Some of the freshwater nematodes in estuaries are plant-root feeders, associated with vascular plants (e.g. species of *Tylenchus*, *Helicotylenchus*, *Meloidogyne*, *Heterodera*, *Hirschmanniella*) (Table 10.4). Species from all these genera, except the first, cause severe damage to plants in many situations, but the limited testing conducted has shown that they seldom affect the plants in the estuarine situation (Dormann and Van Der Wal, 2001). The related genus *Halenchus* is the only tylenchid completely absent from freshwaters. It feeds exclusively on seaweeds (Siddiqi, 2000).

Other large groups of nematodes with freshwater affinities found in estuaries are omnivorous Dorylaimida and two genera of the microbivorous Tobrilidae, a family typical of freshwater sediments. At least one species of Tobrilidae (*Eutobrilus heptapapillatus*) is also very widespread in freshwater rivers and lakes upstream of the estuary where it was found (Nicholas *et al.*, 1992; Hodda, 1999). It has also been found in salt lakes in the same river system (Hodda, 1999). This distribution suggests tolerance of a wide range of conditions, making the invasion of estuarine sediments from freshwaters potentially feasible. Dorylaimida are very diverse in all terrestrial and freshwater aquatic habitats where they seem to have the ability to colonize in almost any situation (Jairajpuri and Ahmad, 1992).

The normally freshwater and terrestrial genus *Aphelenchoides* has also been found in estuaries (Bongers and van der Haar, 1990). This genus is very diverse ecologically, including plant-parasitic, entomophilic and mycetophagous species, and so has enough plasticity to colonize the estuarine environment.

Table 10.4 Freshwater nematode genera from estuaries.

Place	Habitat	Genus or species	Reference(s)	Notes
Europe	Seaweed	*Halenchus*	De Man (1892); Davide (1980); Garrad (1978)	–
Australia	Sublittoral	*Eutobrilus heptapapillatus*	Nicholas *et al.* (1992)	–
Australia	Sublittoral	*Miconchus*	Nicholas *et al.* (1992)	–
Australia	Sublittoral	*Tylenchida*	Nicholas *et al.* (1992)	Found once only
Australia	Sublittoral	*Rhabditis*	Nicholas *et al* (1992)	Found once only
Australia	Sublittoral	*Mononchus*	Nicholas *et al.* (1992)	Found once only
Australia	Sublittoral	*Dorylaimidae*	Nicholas *et al.* (1992)	Found four times
Australia	Mangroves	*Enchodelus*	Nicholas and Stewart (1984a)	–
Australia	Mangroves	*Proleptonchus*	Hodda and Nicholas (1985)	–
Australia	Mangroves	*Criconemella avicenniae*	Nicholas and Stewart (1984a)	–
Australia	Mangroves	*Labronema*	Hodda and Nicholas (1985)	–
Australia	Mangroves	*Tylenchus*	Hodda and Nicholas (1985)	–
Brunei	Salt marsh	*Meloidogyne mersa*	Siddiqi and Booth (1991)	–
New Zealand	Salt marsh	*Heterodera litoralis*	Wouts and Sturhan (1996)	–
Netherlands	Salt marsh	*Heterodera spinicaudata*	Wouts *et al.* (1995)	–
South Africa	Salt marsh	*Helicotylenchus*	Furstenberg and De Wet (1983)	–
South Africa	Salt marsh	*Tylenchus*	Furstenberg and De Wet (1983)	–
Netherlands	Salt marsh	*Pratylenchoides*	Bor and S'Jacob (1966)	–
Mexico	Salt marsh	*Cactodera salina*	Baldwin *et al.* (1997)	–
China	Beach	*Tobrilus*	Wu *et al.* (2002)	–
Russia	–	*Paratrilobus*	Tsalolikhin (1981b)	–
Texas, USA	Marine sediment	*Hirschmaniella mexicana*	Sher (1968)	–
Florida, USA	Marine plant	*Hirschmaniella marina*	Sher (1968)	–

Seasonal and Temporary Freshwaters

Seasonal and temporary freshwaters have been studied in Africa, North America and Australia. They seem to have a faunal composition distinct from most other freshwaters, with a mixture of terrestrial, aquatic and specialized groups (Table 10.5). Different groups seem to predominate in different regions, but this may be related to the differing patterns of inundation. Unfortunately, the hypothesis cannot be tested using the data available because the studies in the different regions have emphases on different aspects of the nematode fauna, so they are not directly comparable.

In the Sahara, many small freshwater bodies have been sampled for nematodes (Goossens, 1976; Grootaert, 1976). Many of these water bodies were highly intermittent or seasonal, with periods of little or no water alternating with periods of inundation. Abundance data were limited in both these studies, but the small water bodies contained a surprisingly large number of species: up to 23. Most of the species were regarded as typically freshwater and cosmopolitan (Goossens, 1976; Grootaert, 1976). A few species were distinctively Ethiopian in distribution (Goossens, 1976; Coomans and Jacobs, 1983).

Table 10.5 Nematodes from seasonal or intermittent freshwater aquatic habitats.

Genus	Sahara (Grootaert, 1976)	Sahara (Goossens, 1976)	South-east Australia, less intermittent (Hodda, 1999)	South-east Australia, more intermittent (Hodda, 1999)
Alaimus s.l.		X		
Oncholaimus	X			
Diplolaimelloides	X			
Diplolaimella	X			
Eumonhystera			X	
Monhystera		X		
Monhystrella	X			
Plectus	X		X	X
Achromadora	X		X	
Paracyatholaimus	X			
Ironus		X		
Tobrilus	X			
Mononchus	X		X	
Actinolaimus			X	
Afrodorylaimus			X	X
Apocelaimellus			X	X
Dorylaimus	X		X	
Laimydorus		X	X	
Mesodorylaimus			X	X
Neoactinolaimus			X	
Paractinolaimus			X	
Rhabditis s.l.				X
Tylenchorhynchus				X

X = genus found in study listed.

In North America, data on species composition are rare, but abundance data are available, including not only different localities, but also time sequences covering dry and wet periods (Boulton and Stanley, 1995; Leeper and Taylor, 1998). In two different locations in Arizona and North Carolina, total abundance of nematodes showed a small initial increase upon wetting, but then increased further as the temporary surface water dried (Boulton and Stanley, 1995; Leeper and Taylor, 1998). The fauna in these intermittent waters may have been largely ecological opportunists, rather than specialized aquatic ones.

In Australia, several intermittent lakes and pools of various sizes have been studied, with both abundance and species composition recorded but not in a time sequence as with the North American reports (Hodda, 1999; W.L. Nicholas, personal communication). In the most frequently filled lakes, the fauna consisted of similar genera to those in the Sahara (Table 10.5). In the least frequently filled lakes, many of the species were the same as those found in surrounding arid areas (Table 10.5). Abundance was high, but the nematodes present may have been active only when water was present. In all intermittent lakes studied, most species seemed to be widespread.

By contrast, in isolated, but relatively constant freshwaters in the Australian desert, there seem to be many fewer species, but the species present are from typical freshwater genera and often endemic (Nicholas and Hodda, 2000). Endemic species have also been reported from oases in the Sahara (Coomans and Heyns, 1983).

Bromeliads and Tree Hollows

Bromeliads are epiphytic monocotyledonous plants with large upright leaves, especially common in Central and South American rainforests. Up to 8 l of water may be retained in the leaf axils, supporting a distinctive fauna and flora. This unique freshwater habitat has been much studied by biogeographers, and includes many nematode species (Table 10.6).

The nematode genera found in the water retained by bromeliads include genera typical of other freshwaters (*Tobrilus* s.l. and *Trischistoma*), species often found in freshwaters but not confined to them (*Mesodorylaimus* and other Dorylaimida, *Mononchus* and other Mononchida, *Plectus*, *Chronogaster* and *Ironus*), terrestrial genera associated with plants or fungi (*Aphelenchoides*, *Aphelenchus* and *Helicotylenchus*) and genera probably opportunistic in ecological characteristics (*Caenorhabditis*, *Prismatolaimus*, *Plectus* and *Cephalobus*). Most of these genera are either microbivorous or omnivorous.

Some temperate trees retain water in hollows just as the bromeliads do in their leaf axils. Nematodes have been found in this tree-hollow water, but the species were not identified (Sota, 1996).

Groundwaters

Groundwater nematodes have been studied most commonly in association with caves or flowing waters, and are discussed under these headings.

Table 10.6 Nematodes found in bromeliad phytotelmata.

Genus	Number of species	Jacobs (1984) ($n = 1$)	Zullini (1977) ($n = 3$)	Traunspurger (unpublished data) ($n = 1$)
Achromadora	2	1	–	1
Actinca	2	1	–	–
Actinolaimus	1	1	–	–
Afractinolaimus	1	1	–	–
Aphelenchoides	1	1	–	1
Aphelenchus	1	1	–	–
Aporcelaimus	1	–	–	1
Bunonema	1	–	–	1
Caenorhabditis	1	–	1	–
Cephalobus	1	1	–	1
Chronogaster	1	1	–	–
Chrysonema	2	1	–	–
Clarkus	1	–	2	–
Cylindrolaimus	1	1	–	–
Diphtherophora	1	1	–	–
Dorylaimus	4	1	–	1
Eudorylaimus	2	–	1	1
Euteratocephalus	1	1	–	–
Helicotylenchus	1	–	–	–
Ironus	1	1	–	–
Ischiodorylaimus	3	1	–	–
Laimydorus	1	1	–	–
Mactinolaimus	1	1	–	–
Mesodorylaimus	2	1	1	1
Miconchus	1	1	–	–
Monhystera	4	1	–	1
Mononchus	2	1	–	–
Mylonchulus	2[a]	1	–	–
Neoactinolaimus	2	1	1	–
Oionchus	1	1	–	–
Oncholaimus	1	1	–	–
Panagrolaimus	1	–	–	1
Paractinolaimus	1	–	–	1
Plectus s.l.	2	–	3	1
Prismatolaimus	2	1	1	1
Prodesmodora	1	–	–	1
Prodorylaimus	1	–	–	1
Rhabditis	1	–	–	1
Teratocephalus	2	–	–	2
Tobrilia	1	1	–	–
Tobrilus s.l.	4	1	–	–
Trischistoma	2	–	2	–
Tylencholaimus	1	–	1	–
Tylenchus	–	–	–	1
Wilsonema	–	–	–	1

[a]One species found by Zullini *et al.* (2002) included in total.
Note: n = number of samples in which found.

Caves

The freshwaters of caves have some of the habitat characteristics of springs, such as very constant temperature, and some of the characteristics of seasonal or intermittent freshwaters, such as highly variable flow rates.

Cavernicolous nematodes have been studied relatively frequently relative to the number of species known and the amount of habitat. As with many other unusual freshwater habitats, the fauna seems to consist of a few species found only in this or very similar habitats together with accidental occupants and species with wide environmental tolerances.

In a cave containing guano deposits in Mexico five species of nematodes were found; one omnivorous, one microbivorous and three predatory, (Zullini, 1977) (Table 10.7). All were from genera found in a range of terrestrial or aquatic habitats, and may have been accidental occupants. In less extensive surveys, species of the common freshwater aquatic genera *Dorylaimus* and *Mesodorylaimus* were found in caves in North America (Chitwood, 1938; Poinar and Sarbu, 1994). Species from a range of genera were found in caves in Cuba (Andrássy, 1973).

Table 10.7 Cavernicolous nematodes.

Genus	Reference(s)
Monhystrella	Cayrol (1973)
Amphidelus	Zullini (1977)
Cylindrolaimus	Andrássy (1959)
Halalaimus	Schneider (1940)
Thallassoalaimus	Schneider (1940)
Anaplectus	Cayrol (1973)
Plectus	Cayrol (1973)
Desmoscolex	Stammer (1935)
Monochromadora	Cayrol (1973)
Stenonchulus	Schneider (1940)
Tobrilus	Cayrol (1973)
Miconchus	Cayrol (1973)
Mononchus	Cayrol (1973); Zullini (1977)
Mylonchulus	Schneider (1940); Zullini (1977)
Aporcelaimellus	Zullini (1977)
Dorylaimus	Chitwood (1938)
Mesodorylaimus	Cayrol (1973); Poinar and Sarbu (1994); Zullini (1977)
Mermis	Schuurmans Stekhoven (1950)
Acrobeloides	Cayrol (1973)
Cephalobus	Cayrol (1973); Andrássy (1967)
Cervidellus	Cayrol (1973)
Chronogaster	Poinar and Sarbu (1994)
Myolaimus	Andrássy (1959)
Rhabditis s.l.	Cayrol (1973)
Criconema	Schneider (1940)
Criconemoides	Cayrol (1973); Schneider (1940)
Rotylenchus	Cayrol (1973)
Tylenchus	Cayrol (1973)

In the European Alps, seven species were found altogether in two caves during dry periods (Cayrol, 1973). Four species were from humid terrestrial habitats and three from still waters. After rain, 14 species were present, but only three were the same as during the dry periods. Species from humid terrestrial habitats and still waters were complemented by two species of generalized aquatic predators and three plant parasites. The plant-parasitic nematodes are certainly accidental inhabitants because caves have no autotrophic production. Many of the other species may have been accidental occupants also.

In another cave system eight species were found (Schneider, 1940). A few genera were the same as those from Mexico, and some were probably accidental occupants, being from terrestrial plant-parasitic genera such as *Criconema* and *Criconemoides* (Schneider, 1940). At least one species may be a true cavernicolous nematode, *Stenonchulus troglodytes*, which has now been found in caves and similar habitats elsewhere (Schneider, 1940; Eder, 1975).

Romanian caves contain what may be other true cavernicolous nematodes: *Chronogaster troglodytes* was found frequently in fungal mats in a cave system relying on autochthonous primary production from thermal sulphur springs (Poinar and Sarbu, 1994). This species has crystalloids, which may be associated with a detoxifying mechanism for the wastes of sulphur metabolism (see section on 'Adaptations to Extreme Freshwater Environments'). A species of the entomophilic genus *Mermis* may also be truly cavernicolous (Schuurmans Stekhoven, 1950). All species of this genus are obligate parasites of insects as juveniles, and the host may be one of the several specialized cavernicolous insects known from cave systems only. Alternatively, it may have been washed into the cave accidentally, like so many other nematode taxa.

The opportunist genera *Protorhabditis* and *Panagrolaimus* were also found in the Romanian cave systems (Poinar and Sarbu, 1994).

In Africa, no nematodes were found in the few caves studied (Jacobs, 1984). This is further evidence that most nematodes in caves are accidental introductions. Many nematodes found in caves are in a quiescent state (Andrássy, 1965).

Marine Nematodes in Freshwater Environments

Just as a few species from 'freshwater' taxa have been found in estuarine or marine habitats (see section on 'Estuarine Sediments'), a few nematodes from 'marine' taxa have also been found in freshwaters (Table 10.8). Some genera have several species found frequently in freshwaters, while others appear to have just one species occurring in freshwaters. There are also many single records of marine nematodes in freshwaters, which undoubtedly represent accidental occurrences.

The genus *Oncholaimus* (Enoplida: Oncholaimidae) is abundant and diverse in marine and estuarine sediments, but has evolved several species in freshwater habitats (Table 10.8). There are few other freshwater forms in this order or family. The number of species in freshwaters suggests that there is a genuine trend, which may be related to the morphological and ecological similarities between *Oncholaimus* and the common freshwater nematodes of the order Mononchida. Both are large omnivorous predators with a barrel-shaped buccal cavity armed with teeth. Perhaps the ecological niche or niches occupied by these taxa make it easy to survive in either environment.

The genus *Theristus* (Monhysterida: Monhysteridae) is another that is very abundant and diverse in marine and estuarine sediments, but has several species which have been found in freshwaters (Table 10.8). While some of these records probably represent accidental occurrences, the number of records and species involved suggests that this too may represent either an ecological type or evolutionary lineage, which is particularly able to transfer between the marine and freshwater biotopes. The genus *Theristus* is overwhelmingly marine, but the family and order contain both many freshwater and marine taxa. In this respect, *Theristus* is different to *Oncholaimus*, and the Monhysterida seem particularly adept at crossing the marine/freshwater boundary, because of the physiological, ecological or evolutionary characteristics of the lineage. The genus is thought to be selective deposit-feeders.

A third 'marine' genus, which seems to have genuine freshwater species, is *Paracyatholaimus* (Chromadorida: Cyatholaimidae) (Table 10.7). This genus is different trophically to *Oncholaimus* and *Theristus*, being an epistrate-feeder. There are a few closely related genera, which are entirely freshwater or terrestrial (e.g. *Achromadora*), so this genus may represent a third evolutionary crossing of the marine/freshwater boundary.

Many of the other marine taxa, which have been found in freshwaters, are estuarine and may have been washed into freshwaters accidentally. This process does seem to have occurred frequently enough to have been observed, so it is perhaps significant that few evolutionary lineages other than the three noted above seem to have radiated in freshwater environments. The discussion in the preceeding paragraphs on freshwater nematodes in marine environments illustrates that there is continual interchange of species between the two environments. Which species are doing this is of evolutionary and ecological interest. So too is the fact that the flow of species seems of similar magnitude in either direction. Marine nematodes are much more diverse at higher taxonomic levels than freshwater or terrestrial nematodes and represent the biotope in which nematodes originally evolved. On this basis a greater flow of taxa from marine to freshwater may be expected over evolutionary time.

Over short, ecological timescales, distinct changes in nematode faunas have been observed in habitats being changed from marine to freshwater, as for example when land is reclaimed from the sea. Most of the taxa changed from 'marine' to 'freshwater' or 'terrestrial' within a few years as the land lost all marine influence (Table 10.8). During the transition, within about 1 year after dykes were built, some of the 'marine' fauna were still present. Only one marine species seemed to survive the transition to a fully freshwater environment: a species of *Neochromadora*. Like *Paracyatholaimus*, this genus is in the Order Chromadorida, but these two genera are not closely related (different families) (Lorenzen, 1994).

Ecologically, the reclamation of estuarine lands involved a transition from a fauna dominated by bacterial feeders and predators, to one dominated by plant feeders and omnivores (Wu *et al.*, 2002). There was little change in the proportion of nematodes thought to be ecological opportunists versus those characteristic of stable conditions.

Table 10.8 Numbers of nematode species found in freshwaters from genera mostly found in marine habitats.

Genus	Number of species found exclusively in freshwater	Number of species found in marine and freshwaters	Number of species known (Gerlach and Riemann, 1974)	Freshwater habitat	Survived transition to freshwater over 1 year (Wu *et al.*, 2002)	Survived transition to freshwater soybean field over 7 years (Wu *et al.*, 2002)	Reference(s)
Adoncholaimus	–	1	21	Cave, Yugoslavia	–	–	Andrássy (1973)
Anoplostoma	–	1	18	Various	–	–	Gerlach and Riemann (1974)
Cytolaimium	1	–	3	Lake, Colombia	–	–	Riemann (1975)
Halalaimus	1	–	59	Cave, Yugoslavia	–	–	Andrássy (1973)
Mesacanthion	2	–	40	Lake, Colombia	–	–	Riemann (1975)
Oncholaimus	6	1	74	Waterfall, Java;	Yes[a]	–	Schneider (1937)
				Moss, subantarctic, Campbell Island;	–	–	Allgen (1929)
				Insect, The Netherlands;	–	–	Fuchs (1937)
				Lake, Colombia;	–	–	Riemann (1975)
				Crater lakes, New Guinea	–	–	Nicholas and Stewart (1984b)
				Oasis, Algerian Sahara	–	–	Coomans and Heyns (1983)
Oncholaimellus	–	1	12	Land reclaimed from river, China	Yes[b]	–	Wu *et al.* (2002)
Parodontophora	1	–	11	River, China	Yes	–	Wu *et al* (2000)
Polygastrophora	–	1[a]	14	Land reclaimed from river, China	Yes	–	Wu *et al.* (2002)
Thalassoalaimus	1	–	23	Cave, Yugoslavia	–	–	Andrássy (1973)
Thalassogenus	1	–	1	Cave, New Guinea	–	–	Andrássy (1973)
Viscosia	–	1[a,c]	76	–	Yes	–	Micoletzky (1925); Gerlach (1957)

Desmoscolex s.l.	–	5[c]	87	Guatemala	–	–	Thames (1966)
Microlaimus	–	1	68	Cave, Yugoslavia	–	–	Andrássy (1973)
Chromadorina	–	1	32	Cave, Cuba	–	–	Andrássy (1973)
Chromadorita	–	1	31	Cave, Cuba	–	–	Andrássy (1973)
Cyatholaimus	–	1	75	Land reclaimed from river, China	Yes	–	Wu *et al.* (2002)
Dichromadora	2	–	19	Various streams, rivers and lakes	–	–	Hodda (1999)
Neochromadora	–	1	27	Land reclaimed from river, China	Yes	Yes	Wu *et al.* (2002)
Paracyat-holaimus	3	–	25	Lake, Colombia two species	–	–	Riemann (1975)
Desmolaimus	1	–	21	Cave, Cuba	Yes	–	Andrássy (1973)
Diplolaimella	2	–	12	Lake, Colombia Rivers, Australia	–	–	Riemann (1975) Hodda (1999)
Diplolaimelloides	–	2	7	Land reclaimed from river, China	Yes	–	Wu *et al.* (2002)
Terschellingia	–	1[c]	29	Cave, Cuba	–	–	Andrássy (1973)
Theristus s.l.	10	2[d]	*c.* 200	Lake, Colombia plus others	Yes	–	Schiemer (1984); Riemann (1975)
Axonolaimus	1	–	41	Cave, Yugoslavia	–	–	Andrássy (1973)
Caribplectus	–	1[e]	2	Lake, Colombia	–	–	Riemann (1975)
Haliplectus	1[e]	–	15	Caves, Cuba	–	–	Andrássy (1973)
Setoplectus	–	1	2	Lake, Colombia	–	–	Riemann (1975)
Tripyloides	2	–	14	Lake, Colombia	Yes	–	Riemann (1975)

[a]Estuarine, sometimes found in freshwater part.
[b]Higher abundance than in estuarine.
[c]Accidental occurrence?
[d]One species estuarine, one found in completely fresh or marine.
[e]Freshwater species also found in marine.

Terrestrial Nematodes in Freshwater Environments

The records of terrestrial nematodes in freshwater environments are almost all plant parasites associated with aquatic plants (Table 10.9). The only other records are of entomophilic nematodes associated with aquatic insects. The paucity of other records is possibly more a reflection of the type of sampling than the actual situation.

Many plant-parasitic nematode genera associated with freshwater aquatic plants have been found (Table 10.9). Most were found only a few times, and possibly represent either nematodes transported to the freshwater plants accidentally, very occasional inhabitants or nematodes which occur in freshwater plants only briefly and temporarily. A few genera occur frequently enough to be considered genuine inhabitants.

Table 10.9 Terrestrial nematodes found in freshwaters. (Tabulated from Mulvey and Anderson, 1980; Ebsary and Eveleigh, 1983; Prejs, 1986; Gerber and Smart, 1987a; Revilla *et al.*, 1991; Hodda, 2003.)

Genus (alphabetic order)	Number of species in genus[a]	Number of species found in freshwaters[b]	Number of records
Anguina	11	0	1
Aphelenchoides	164	11	53
Aphelenchus	26	1	5
Basiria	57	3	3
Belonolaimus	5	1	6
Boleodorus	26	1	1
Cacopaurus	1	0	6
Calolaimus	6	0	1
Caloosia	10	3	3
Chrysonemoides	3	1	6
Criconema	85	2	3
Criconemella	8	1	2
Criconemoides	19	11	47
Ditylenchus	57	4	8
Dolichodera	1	1	1
Dolichodorus	16	4	28
Helicotylenchus	185	11	56
Hemicriconemoides	44	4	16
Hemicycliophora	119	11	20
Heterodera	65	9	18
Hirschmaniella	34	14	90
Hoplolaimus	32	5	16
Longidorella	22	1	2
Longidorus	69	1	1
Macroposthonia	112	0	1
Meloidoderita	9	1	4
Meloidogyne	80	13	78
Nanidorus	6	1	1
Paralongidorus	34	2	2
Paratylenchus	121	5	10
Peltamigratus	23	2	5
Pratylenchus	90	12	46

Table 10.9 *Continued.* Terrestrial nematodes found in freshwaters. (Tabulated from Mulvey and Anderson, 1980; Ebsary and Eveleigh, 1983; Prejs, 1986; Gerber and Smart, 1987a; Revilla *et al.*, 1991; Hodda, 2003.)

Genus (alphabetic order)	Number of species in genus[a]	Number of species found in freshwaters[b]	Number of records
Psilenchus	20	1	2
Quinisulcius	15	1	2
Radopholus	20	1	7
Rotylenchulus	11	2	8
Rotylenchus	81	1	5
Sarisodera	1	1	1
Scutellonema	45	3	8
Seinura	45	4	4
Telotylenchus	17	0	1
Trichodorus	34	4	25
Trichotylenchus	4	1	1
Trophotylenchulus	14	1	4
Tylenchorhynchus	105	11	27
Tylenchulus	4	1	4
Tylenchus s.l.	148	9	31
Uliginotylenchus	7	2	2
Verutus	2	1	1
Xiphidorus	7	1	3
Xiphinema	138	9	22
Xiphinemella	9	1	1

[a]Number of species recognized currently (Hodda, 2003; Jairajpuri and Ahmad, 1992; Siddiqi, 2000). Not necessarily the same as the number recognized by the authors or at the time of the publication.
[b]Includes only species given specific epithets, and excludes species cited as 'sp.'.

The genus *Hirschmanniella* is a genuine freshwater inhabitant from the overwhelmingly terrestrial family Pratylenchidae. The genus contains about 35 species, almost all associated with aquatic plants. At least one species is a major economic pest of rice (*Hirschmanniella oryzae*), and one other species can cause damage to aquatic plants (Gerber and Smart, 1987b). One species is associated with a marine plant (see section on 'Estuarine Sediments'). Most species in the genus are notable in being relatively thin and moderately large. It is tempting to speculate that the body form is associated with the evolutionary success of this group of endoparasites, by facilitating movement through very wet substrates and up damp plant stems.

The genus *Xiphinema* is another large, relatively thin nematode commonly associated with aquatic plants, but the genus as a whole is most common in fully terrestrial plants. The entomophilic Mermithidae are another group of nematodes commonly associated with aquatic or semiaquatic plants, which share the long, thin body form. Unlike *Hirschmanniella* but like *Xiphinema*, mermithids are not associated only with aquatic plants.

Two other terrestrial genera commonly associated with aquatic plants also share a relatively thin body: *Ditylenchus* and *Tylenchus*. Both are considerably smaller than *Hirshmanniella*, *Xiphinema* and the mermithids. Both are more common in fully terrestrial plants than in aquatic or semiaquatic ones, but have been recorded frequently enough to be considered genuinely aquatic (Table 10.9).

Other 'terrestrial' genera commonly associated with aquatic plants have notably short, thick bodies and are known for sluggish movement, not proficiency in swimming in water films like the genera mentioned in the preceding paragraph. Examples are the genera *Criconemoides*, *Hemicycliophora*, *Hemicriconemoides*, *Paratylenchus* (all Tylenchida: Criconematina), *Hoplolaimus*, *Scutellonema* (Tylenchida: Hoplolaimidae) and *Trichodorus* (Triplonchida: Trichodoridae).

Other genera commonly reported on aquatic plants include *Dolichodorus*, *Helicotylenchus*, *Pratylenchus* and *Aphelenchoides*. These genera are neither particularly large nor small, thin nor thick, active nor sluggish. Species of sedentary endoparasites from the genera *Meloidogyne* and *Heterodera* are also commonly reported on aquatic plants. *Meloidogyne* is far more common than *Heterodera*, and indeed rivals *Hirshmanniella* as the most frequently reported on aquatic plants (Table 10.9).

Overall then, the 'terrestrial' nematode genera associated with aquatic plants seem to represent a broad selection of the genera associated with fully terrestrial plants, except for the fully aquatic *Hirshmanniella*. The genera seem to fall into three groups: (i) those found very frequently, which seem best able to adapt to aquatic environments (e.g. *Aphelenchoides* and *Meloidogyne*); (ii) those found less frequently and presumably adapt less well (e.g. *Dolichodorus* and *Xiphinema*); and (iii) those which only appear infrequently, perhaps accidentally (e.g. *Anguina* and *Basiria*). Into which group a particular genus will fall does not seem related to either body form, size, site of attack on the plant or the level of activity. Perhaps there is a physiological feature associated with ability to attack freshwater plants, or an ecological characteristic associated with opportunity to attack the plants which has yet to be discovered.

Inland Hypersaline Waters

This section deals with hypersaline waters not in contact with the open sea. They are included because they contain nematode species with a wide range of habitat affinities, including freshwater. There seem to be few nematodes which are adapted to these extreme habitats, but the observations that some nematodes occur in even these habitats lends weight to the 'nematologists' creed' first enunciated by Cobb (1914) that nematodes occur in every conceivable habitat where free water occurs. A situation of rapidly increasing salinity is also compared with a much more gradual increase.

Of the nematodes from inland saline waters, some are otherwise freshwater nematodes which have evolved to the conditions, others are marine species.

Inland saline lakes in Australia are mostly filled very intermittently, but when filled there is a huge burst of biological activity. Nematode genera found in these lakes include *Monhystera*, *Prodesmodora*, *Mesodorylaimus* and a plectid (M. Hodda, unpublished data). The occurrence of these species is very sporadic, and they are quiescent most of the time, making species identifications very difficult.

In Africa, only a single species has been reported from soda lakes, also in the genus *Mesodorylaimus* (Tudorancea and Zullini, 1989). No nematodes were found in a hypersaline lake. In another study, they were found in salt streams, springs and water holes in Namibia (Procter, 1982). Unfortunately, the species and genera were not identified.

In North America, an artificial inland hypersaline lake, the Salton Sea, has a fauna of marine nematodes (Warwick *et al.*, 2002). This lake is barely 100 years old, and has been increasing in salinity rapidly. The nematodes are thought to have been introduced from nearby marine habitats on the Pacific coast or Sea of Cortez through human influence.

In Asia, a natural freshwater lake in a closed drainage basin, which is becoming saline much more slowly, provides a contrast to the situation of rapid salinity increase in the Salton Sea. Lake Issyk-kul' in Kyrghyzstan is thought to be about 6000 years old, and currently has a salinity of about 6 ppt (below sea water, but well above most freshwaters). In Lake Issyk-kul' the nematode fauna is part freshwater and part marine (Tsalolikhin, 1979). Here the local freshwater species, including species of *Tobrilus* s.l., are seemingly adapting to the slow increase in salinity. Other genera (*Leptolaimus*, *Monhystera* and *Pseudoncholaimus*) with closer evolutionary links to marine environments may be increasing in abundance as the lake becomes more saline. These genera may be relicts of the marine conditions existing long in the geological past, because the lake is now isolated from the sea (Tsalolikhin, 1979).

Adaptations to Extreme Freshwater Environments

Nematodes have adapted to a wide range of freshwater environments, and some generalizations can be drawn from the particular groups which have adapted to these environments and how. This section discusses some of these generalizations.

Low oxygen is a common physiological stress in aquatic habitats, and has been studied in some detail. Genera of Monhysterida and Tobrilidae (Triplonchida) seem particularly common in freshwater habitats where oxygen is limited or absent (Nuss, 1984; Jacobs, 1987). Some Tobrilidae have crystalline inclusions in the body, which are thought to be part of a detoxification mechanism for hydrogen sulphide, the generally toxic end product of respiration using sulphur rather than oxygen as the final electron acceptor (Nuss, 1984; Nuss and Trimkowski, 1984). However, the mechanism for the crystalloids detoxifying the sulphides remains controversial (Nicholas *et al.*, 1987). The crystalloids may not be a mechanism for storing wastes at all (Bird *et al.*, 1991).

Anaerobic metabolism may be important for Monhysterida in low-oxygen freshwater habitats, too. Monhysterida with crystalloid inclusions occur in the same habitat as the Tobrilidae (M. Hodda, personal observation). At least some species of Monhysterida from estuarine habitats can metabolize anaerobically (Bryant *et al.*, 1983). In an extensive test of many estuarine species, including Monhysterida and several other orders but not Tobrilidae, most species survived anaerobic sediments by either anaerobic metabolism, facultative anaerobic metabolism or quiescence (Bryant *et al.*, 1983). A species of *Eudorylaimus* (Dorylaimida) lives in oxygen-free sediments of Lake Tiberias (Israel) for 8 months each year and can survive in a sealed jar with anoxic sediment for 6 months (Por and Masry, 1968). The ability to survive under anaerobic conditions may thus be quite widespread among nematodes, although different mechanisms may be involved (Schiemer and Duncan, 1974;

Bryant *et al.*, 1983). Why Tobrilidae and Monhysterida seem particularly prominent in freshwater habitats where oxygen is low is thus still problematic.

High salinity and high temperature are less common physiological stresses for aquatic nematodes than low oxygen. In springs, the same two genera were most able to tolerate high salinity and temperature: *Udonchus* and *Rhabdolaimus* (Ocaña, 1991a,b). However, in intermittent lakes subject to high salinity and temperature the genus *Mesodorylaimus* seems the only genus common to the few studies completed (Tudorancea and Zullini, 1989; M. Hodda, unpublished data). These genera are not the 'freshwater' taxa which have crossed to marine or estuarine habitats. Perhaps there are no genera particularly adapted physiologically to these conditions.

There is considerable evidence from studies of nematodes and industrial pollution that many freshwater nematodes can adapt physiologically to environmental challenges (e.g. Samoiloff *et al.*, 1980, 1983; Mutwakil *et al.*, 1997). Indeed, this ability is the basis of several methods proposed for pollution monitoring (Zullini, 1976; Millward and Grant, 1995, 2000). This ability would be a very powerful adaptive feature for 'extreme' freshwater environments.

Isolation from similar habitats, variability in time and physical disturbance are ecological challenges for freshwater nematodes, and are perhaps more important than the physiological challenges outlined above. The common adaptations to these challenges are quiescence of some form, specialized means of food storage and dispersal mechanisms.

The ability to enter a quiescent state with no water loss in response to lack of water (anhydrobiosis) seems common in freshwater nematodes from temporary ponds (Wharton, 1986; Womersley and Ching, 1989). It also seems common in nematodes from polar regions (Pickup and Rothery, 1991; Wharton and Barclay, 1993; Wharton, 2004). The ability to enter anhydrobiosis may be one of the most important and widespread adaptations in evolutionary terms amongst nematodes, although it may remain unexpressed in many species from relatively stable habitats. Genera such as *Tobrilus*, which often live in stable or deep lakes, have seldom been reported cryptobiotic. More importantly, it may not limit which taxa are found where conditions become more extreme, and even hostile.

Food storage may be another mechanism used by freshwater nematodes to cope with variable environments. The crystalline inclusions thought by some to be a detoxifying mechanism for sulphides (see section on 'Adaptations to Extreme Freshwater Environments') may be for food storage instead (Bird *et al.*, 1991). The group commonly known as mermithids (with an insect-parasitic juvenile and a free-living but non-feeding adult) have evolved an oesophagus which functions as a food storage organ in the adult. The adults, commonly associated with freshwaters, are also relatively large for nematodes, which enables a long period of survival without a new host for the juveniles.

Parthenogenetic reproduction may also be an important mechanism for coping with intermittent or widely separated patches of favourable habitat (Nicholas, 1975). Ecological theory predicts that asexual reproduction will be most common in the most spatially and temporally variable environments (Townsend *et al.*, 2003). This prediction conforms well to the large proportions of species without males which have been repeatedly observed in the most variable habitats (Grootaert, 1976; Wharton, 1986; Ocaña, 1991a).

Some genera may be especially non-specialized in their ecological requirements, e.g. *Chronogaster* (Heyns and Coomans, 1980; Poinar and Sarbu, 1994). In addition to the species found in caves and hot springs, species in the genus occur in many other habitats throughout the world, including marine and terrestrial (e.g. De Man, 1921; Gerlach, 1956; Maggenti *et al.*, 1983). Indeed, a large proportion of the genera associated with 'extreme' freshwater habitats are thought to be omnivorous, non-specialized microbial feeders or general predators, all of which probably feed on a wider range of organisms than many other nematodes.

Many species associated with extreme habitats are small (about 500 μm body length). This is especially so in hot springs (Ocaña, 1991b) and may be associated with faster life cycles (Nicholas, 1975).

The few genera with specialized feeding habits found in the extreme freshwater habitats (e.g. the root knot nematodes (*Meloidogyne* spp.), the cyst nematodes (*Heterodera* spp.) and *Cactodera*) all modify their environment extensively to create conditions most suitable for themselves. All induce galls of some form on their plant hosts.

Rhabditida and Mermithida are other taxa frequently associated with extreme freshwater habitats. Their success may be associated with highly effective dispersal between habitats often widely separated in space and time. These orders are often associated with insects, enabling transmission between patches of favourable habitat separated by considerable distances. Many other methods for dispersal have also been observed, including in dust and birds' feet, but few of these seem specific to freshwater nematodes. Transmission by flowing water, especially during periods of heavy flows and floods, is an often neglected means of dispersal, but has been frequently observed. More investigation is warranted.

Conclusions

There are several groups which stand out as occurring in many of the extreme habitats discussed in this chapter: the closely related *Monhystera*, *Eumonhystera* and *Monhystrella* (Monhysterida), the genera or subgenera listed as *Plectus* s.l. (Plectida), *Rhabditis* s.l. (Rhabditida) and the closely related *Mesodorylaimus* and *Eudorylaimus* (Dorylaimida). All these taxa seem to have more than one of the adaptations discussed in the previous sections. All are ecological generalists: the monhysterids, *Plectus* s.l. and *Rhabditis* s.l., are microbivorous, but do not have any specialized structures in their stoma; the dorylaimids are known to be highly omnivorous (Yeates *et al.*, 1993). The rhabditids have highly developed forms of quiescence, food storage and dispersal (Bird and Bird, 1991). The monhysterids and *Plectus* s.l. seem to have considerable physiological plasticity. The dorylaimids are very ubiquitous in freshwater and terrestrial soils, and so have had many opportunities to colonize freshwater habitats not readily colonized by other taxa. They must also have very great abilities for quiescence from their abundance in completely dry desert lakes (i.e. lakes that remain dry but are filled with water once every few years).

Other genera which are prominent in multiple extreme freshwater habitats, but less so than the above, include *Chronogaster* (Plectida, related to *Plectus*), *Tobrilus* (Triplonchida) and *Mononchus* s.l. (Mononchida). These taxa are similar in many ways to those discussed in the last paragraph: the former two are general microbivores and the latter an omnivore (Yeates *et al.*, 1993). *Chronogaster* seems to have great physiological plasticity (Heyns and Coomans, 1980; Poinar and Sarbu, 1994). *Tobrilus* s.l. seems very ubiquitous in freshwaters and possesses considerable morphological plasticity (a large number of similar taxa have been described, which may also represent a few variable taxa e.g. Tsalolikhin, 1981a, 2001).

Nematodes from 'extreme' freshwater habitats demonstrate the ability to cope with some apparently hostile conditions: highly variable, or extremes of, salinity, oxygen or moisture, and isolation in space and time. The taxa which have managed to adapt to these conditions seem to be a combination of three groups of nematodes. First are taxa having particular attributes, which make them able to adapt to these conditions, coming from clusters of closely related genera such as *Monhystera*, *Eumonhystera*, *Monhystrella* (Monhysterida), *Plectus* s.l., *Chronogaster* (Plectida), *Rhabditis* s.l. (Rhabditida), *Mesodorylaimus*, *Eudorylaimus* (Dorylaimida), *Tobrilus* s.l. (Triplonchida) and *Mononchus* s.l. (Mononchida). Second are taxa, which have perhaps accidentally arrived in an extreme habitat and have adapted to a largely vacant niche, coming from a wide range of taxonomic groups, including some normally considered 'marine' or 'terrestrial'. Third is a large group of nematodes found accidentally or opportunistically in the various habitats, but which have not adapted and are only temporary residents.

References

Allgen, C. (1929) Uber einige freilebende Moosnematoden. *Nytt magasin for naturvidenskapene* 67, 211–230.

Andrássy, I. (1959) Nematoden aus der Tropfsteinhoehle "Baradla" bei Aggletek (Ungarn), nebst einer Uebersicht der bisher aus Hoehlen bekannten freilebenden Nematoden-Arten. *Acta Zoologica Hungarica* 4, 253–277.

Andrássy, I. (1965) Nematodes and their role in caves. In: Rapoport, E.H. (ed.) *Progressos en biologia del suelo: Actas del primer coloquio latino americo de boilogia del suelo en la Universidad Nacional del Sur Bahia Blanca (Argentina).* UNESCO, Montevideo, Uruguay, pp. 303–312.

Andrássy, I. (1967) Die Unterfamilie Cephalobina (Nematoda: Cephalobidae) und ihre Arten. *Acta Zoologica Hungarica* 13, 1–37.

Andrássy, I. (1973) Nematodes from coastal and cave biotopes in Cuba. *Acta Zoologica Academiae Scientiarum Hungaricae* 19, 233–270.

Andrássy, I. (1978) Nematoda. A checklist of the animals inhabiting European inland waters with accounts of their distribution and ecology. In: Illies, J. (ed.) *Limnofauna Europaea*, 2nd edn. Gustav Fischer Verlag, Stuttgart, Germany, pp. 98–117.

Andrássy, I. (1998) Nematodes in the sixth continent. *Journal of Nematode Morphology and Systematics* 1, 107–186.

Baldwin, J.G., Mundo-Ocampo, M. and McClure, M.A. (1997) *Cactodera salina* n. sp. from the estuary plant, *Salicornia bigelovii*, in Sonora, Mexico. *Journal of Nematology* 29, 465–473.

Bird, A.F. and Bird, J. (1991) *The Structure of Nematodes.* Academic Press, San Diego, California.

Bird, A.F., McClure, S.G. and Nicholas, W.L. (1991) Observations on crystalloid bodies in the pseudocoelom of *Eutobrilus heptapapillatus*. *Journal of Nematology* 23, 39–47.

Bongers, T. (1988) The maturity index: an ecological measure of environmental disturbance based on nematode species composition. *Oecologia* 83, 14–19.

Bongers, T. and van der Haar, J. (1990) On the potential of basing an ecological typology of aquatic sediments on the nematode fauna: an example from the River Rhine. *Hydrobiological Bulletin* 24, 37–45.

Bor, N.A. and S'Jacob, J.J. (1966) *Pratylenchoides maritimus*, a new nematode species from the Boscplaat, Terschelling. *Nematologica* 12, 462–466.

Botosaneanu, L. (1998) *Studies in Crenobiology: The Biology of Springs and Springbrooks*. Backhuys, Leiden, The Netherlands.

Boulton, A.J. and Stanley, E.H. (1995) Hyporheic processes during flooding and drying in a Sonoran Desert stream. II. Faunal dynamics. *Archiv für Hydrobiologie* 134, 27–52.

Bryant, C., Nicholas, W.L. and Hodda, M. (1983) Does anaerobic metabolism facilitate the survival of free-living nematodes? *Report to Australian Marine Sciences and Technologies Scheme*, 56 pp.

Bunt, J.S. (1954) Soil-inhabiting nematodes from Macquarie Island. *Australian Journal of Zoology* 2, 264–274.

Cayrol, J.C. (1973) Fluctuations observées dans les populations de nématodes cavernicoles en fonction de la pluviométrie. Comptes Rendus du 96e Congrès National des Sociétés Savantes, Toulouse 1971. Section des Sciences, t. III. *Biologie Générale et Animale*, pp, 157–164.

Chitwood, B.G. (1938) Some nematodes from the caves of Yucatan. In: Pearse, A.S. (ed.) *Fauna of the Caves of Yucatan*. Carnegie Institute of Washington, Washington, DC, pp. 51–56.

Cobb, N.A. (1914) *Nematodes and Their Relationships*. US Department of Agriculture Yearbook 1914, pp. 457–490.

Coomans, A. and Heyns, J. (1983) *Oncholaimus sahariensis* sp.n. (Nematoda) from the Algerian Sahara. *Hydrobiologia* 107, 193–201.

Coomans, A. and Jacobs, L.J. (1983) *Halalaimus algeriensis* n. sp. (Nematoda) from the Sahara. *Hydrobiologia* 102, 39–44.

Dartnall, H.J.G. (1980) Freshwater biology at Rothera Point, Adelaide Island: 1. General description of the pools and the fauna. *British Antarctic Survey Bulletin* 50, 51–54.

Davide, R.G. (1980) Occurrence and observations on *Halenchus fucicola* found occurring in galls of *Ascophyllum nodosum* on the south coast of Ireland. *Scientific Proceedings of the Royal Dublin Society Series* A 6, 385–396.

De Coninck, L.A. (1935) Contribution à la connaissance de Nématodes libres du Congo Belge. I. Les Nématodes libres des marais de la Nyamuamba (Ruwenzori) et des sources chaudes du Mont Banze (Lac Kivu). *Revue Zoologique et Botanique Africain* 26, 249–326.

De Man, J.G. (1892) Ueber eine neue, in Gallen einer Meeresalge lebende Art der Gattung Tylenchus Bast. Festschrift zum 70 Geburtstage Rudolf Leuckarts, Leipzig, Germany.

De Man, J.G. (1921) Nouvelles recherches sur les Nématodes libres terricoles de la Hollande. *Capita Zoologica* 1, 1–62.

Dormann, C.F. and Van Der Wal, R. (2001) Impact of soil nematodes on salt-marsh plants: a pilot experiment. *Journal of Coastal Research* 17, 383–388.

Duval, B., Duval, E. and Hoham, R.W. (1999) Snow algae of the Sierra Nevada, Spain, and High Atlas mountains of Morocco. *International Microbiology* 2, 39–42.

Ebsary, B.A. and Eveleigh, E.S. (1983) *Meloidogyne aquatilis* n. sp (Nematoda: Meloidogynidae) from *Spartina pectinata* with a key to the Canadian species of *Meloidogyne* (root knot nematode). *Journal of Nematology* 15, 349–353.

Eder, R. (1975) Zwei neue Funde von *Stenonchulus troglodytes* Schneider 1940 (Onchulidae, Nematoda). *Carinthia* II 165, 291–294.

Eyualem-Abebe, Mees, J. and Coomans, A. (2001) Nematode communities of Lake Tana and other inland water bodies of Ethiopia. *Hydrobiologia* 462, 41–73.

Fuchs, A.G. (1937) Neue parasitische und halbparasitische Nematoden bei Borkenkaefern und einige andere Nematoden. I. Teil. *Zoologischer Jahrbuecher Abteilung fuer Systematik Oekologie und Geographie der Tiere* 70, 291–380.

Furstenberg, J.P. and De Wet, A.G. (1983) Effect of salt marsh vegetation on *Helicotylenchus* and *Tylenchus* species of Swartkops Estuary. *Nematological Society of Southern Africa Newsletter* 16, 5–6.

Gagarin, V.G. and Lemzina, L.V. (1992) The nematode fauna of mineral springs in Kyrghyzstan. *Biologiya Vnutrennikh Vod Informatsionnyi Byulleten* 94, 51–55.

Garrad, L.S. (1978) Radley college records of the intertidal marine fauna. *Calf of Man Bird Observatory Annual Report* 1976–1977, 109–122.

Gerber, K. and Smart, G.C. Jr (1987a) Plant-parasitic nematodes associated with aquatic vascular plants. In: Veech, J.A. and Dickson, D.W. (eds) *Vistas on Nematology: a Commemoration of the Twenty-fifth Anniversary of the Society of Nematologists*. Society of Nematologists, Hyattsville, Maryland, pp. 488–501.

Gerber, K. and Smart, G.C. Jr (1987b) Effect of *Hirschmanniella caudacrena* on the submersed aquatic plants *Ceratophyllum demersum* and *Hydrilla verticillata*. *Journal of Nematology* 19, 447–453.

Gerlach, S.A. (1956) Brasilianische Meeres-Nematoden. I. *Boletim do Instituto Oceanographico* (Sao Paulo) 5, 3–69.

Gerlach, S.A. (1957) Zwei Meeres-Nematoden aus dem Nicaragua-See. *Zoologischer Anzeiger* 159, 251–254.

Gerlach, S.A. and Riemann, F. (1974) The Bremerhaven checklist of aquatic nematodes. *Veroeffentlichungen des Instituts fuer Meeresforschung in Bremerhaven Supplement* 4, 1–736.

Goossens, A. (1976) A study of the aquatic nematodes of the Adrar mountains in Mauretania. *Biologisch Jaarboek Dodonaea* 44, 178–190.

Grootaert, P. (1976) A note on the aquatic nematodes from Rio de Oro (North-western Sahara). *Bulletin de l'Institut Français d'Afrique Noire, A.* 38 (4), 832–835.

Heyns, J. (1993) *Eudorylaimus nudicaudatus* sp. n. from Antarctica (Nematoda: Dorylaimoidea). *South African Journal of Antarctic Research* 23, 33–36.

Heyns, J. (1994) *Chiloplacoides antarcticus* n. gen., n. sp. from western Dronning Maud Land, Antarctica (Nematoda: Cephalobidae). *Fundamental and Applied Nematology* 17, 333–338.

Heyns, J. (1995) *Plectus antarcticus* de Man, 1904 from western Dronning Maud Land, Antarctica. *Nematologica* 41, 1–6.

Heyns, J. and Coomans, A. (1980) Freshwater nematodes from South Africa. 5. *Chronogaster* Cobb, 1913. *Nematologica* 26(2), 187–208.

Hodda, M. (1999) Nematodes of Australia: free-living, insect- and plant-associated. Available at: http:/www.ento.csiro.au/science/nematodes/nematodes

Hodda, M. (2003) Checklist of Aphelenchida. Available at: http:/www.ento.csiro.au/science/nematodes/checklist_dec2003.rtf

Hodda, M. and Nicholas, W.L. (1985) Meiofauna associated with mangroves in the Hunter River estuary and Fullerton Cove. *Australian Journal of Marine & Freshwater Research* 36, 41–50.

Hoeppli, R.J.C. (1926) Studies of free-living nematodes from the thermal waters of Yellowstone Park. *Transactions of the American Microscopical Society* 45, 234–255.

Hoeppli, R. and Chu, H.J. (1932) Free-living nematodes from hot springs in China and Formosa. *Hongkong Naturalist Supplement* 1, 15–28.

Hynes, H.B.N. (1970) *The Ecology of Running Waters*. University of Toronto Press, Toronto, Ontario.

Iversen, T.M. (1988) Secondary production and trophic relationships in a spring invertebrate community. *Limnology and Oceanography* 33, 582–592.

Jacobs, L.J. (1984) The free-living inland aquatic nematodes of Africa – a review. *Hydrobiologia* 113, 259–291.

Jacobs, L.J. (1987) *A Checklist of the Monhysteridae (Nematoda, Monhysterida)*. Rand Africans University, Johannesburg.

Jairajpuri, S. and Ahmad, W. (1992) *Dorylaimida: Free-living, Predaceous and Plant-parasitic Nematodes*. Oxford & IBH Publishing, New Delhi.

Kirjanova, E.S. (1958) Antarctic specimens of freshwater nematodes of the genus *Plectus* Bastian (Nematoda: Plectidae). *Information Bulletin of the Soviet Antarctic Expedition* 3, 163–165 [English summary – Russian version pp. 101–103].

Leeper, D.A. and Taylor, B.E. (1998) Abundance, biomass and production of aquatic invertebrates in Rainbow Bay, a temporary wetland in South Carolina, USA. *Archiv für Hydrobiologie* 143, 335–362.

Loof, P.A.A. (1971) Free-living and plant-parasitic nematodes from Spitzbergen, collected by Mr H. van Rossen. *Mededelingen Landbouwhogeschool Wageningen* 71, 1–86.

Lorenzen, S. (1994) *Phylogenetic Systematics of Freeliving Nematodes.* Ray Society, London.

Maggenti, A.R., Raski, D.J., Koshy, P.K. and Sosamma, V.K. (1983) A new species of *Chronogaster* Cobb, 1913 (Nemata: Plectidae) with an amended diagnosis of the genus and discussion of cuticular ornamentation. *Revue de Nematologie* 6, 257–263.

Marchant, R. and Lillywhite, P. (1994) Survey of stream invertebrate communities on Macquarie Island. *Australian Journal of Marine & Freshwater Research* 45, 471–481.

Maslen, N.R. (1979) Additions to the nematode fauna of the Antarctic region with keys to taxa. *British Antarctic Survey Bulletin* 49, 207–229.

Maslen, N.R. (1982) An unidentified nematode-trapping fungus from a pond on Alexander Island. *British Antarctic Survey Bulletin* 51, 285–287.

Meyl, H. (1953a) Beitraege zur Kenntnis der Nematodenfauna vulkanisch erhitzer Biotope. I. Die terrikolen Nematoden im Bereich von Fumarolen auf der Insel Ischia. *Zeitschrift für Morphologie und Oekologie der Tiere* 42, 67–116.

Meyl, H. (1953b) Beitraege zur Kenntnis der Nematodenfauna vulkanisch erhitzer Biotope. II. Die in Thermalgewaessern der Insel Ischia vorkommenden Nematoden. *Zeitschrift für Morphologie und Oekologie der Tiere* 42, 159–208.

Meyl, H. (1954) Beitraege zur Kenntnis der Nematodenfauna vulkanisch erhitzer Biotope. III. Mitteilung. Nematoden aus der Mischungszone strandnaher, heisser Suesswasserquellen mit dem Meerwasser auf der Insel Ischia. *Zeitschrift für Morphologie und Oekologie der Tiere* 42, 421–448.

Micoletzky, H. (1925) Die freilebenden Suesswasser- und Moornematoden Daenemarks. *Kongliga Svenska Vetenskaps-Akademiens Handlingar* 10, 55–310.

Millward, R.N. and Grant, A. (1995) Assessing the impact of copper on nematode communities from a chronically metal-enriched estuary using pollution-induced community tolerance. *Marine Pollution Bulletin* 30, 701–706.

Millward, R.N. and Grant, A. (2000) Pollution-induced tolerance to copper of nematode communities in the severely contaminated Restronguet creek and adjacent estuaries, Cornwall, United Kingdom. *Environmental Toxicology and Chemistry* 19, 454–461.

Minckley, W.L. (1963) The ecology of a spring stream, Doe Run, Meade County, Kentucky. *Wildlife Monographs* 11, 1–124.

Minshall, G.W. (1967) Role of allochthonous detritus in the trophic structure of a woodland springbrook community. *Ecology* 48, 139–149.

Mulvey, R.H. and Anderson, R.V. (1980) Description and relationships of a new root-knot nematode, *Meloidogyne sewelli* n. sp. (Nematoda: Meloidogynidae) from Canada and a new host record for the genus. *Canadian Journal of Zoology* 58, 1551–1556.

Mutwakil, M.H.A.Z., Reader, J.P. and Holdich, D.M. (1997) Use of stress-inducible transgenic nematodes as biomarkers of heavy metal pollution in water samples from an English river system. *Archives of Environmental Contamination and Toxicology* 32, 146–153.

Nicholas, W.L. (1975) *The Biology of Free-living Nematodes.* Clarendon Press, Oxford, UK.

Nicholas, W.L. (1993) Two new species of nematode (Nematoda: Enoplida: Thoracostomopsidae) from Lake Alexandrina, South Australia. *Transactions of the Royal Society of South Australia* 117, 163–170.

Nicholas, W.L. and Hodda, M. (2000) *Dorylaimus baylyi* sp. nov. (Dorylaimidae, Dorylaimida) a new nematode collected from freshwater in a rock-hole in the Northern Territory. *Transactions of the Royal Society of South Australia* 124, 163–168.

Nicholas, W.L. and Marples, T.G. (1995) Observations on the geographical distribution and morphometrics of *Enoploides stewarti* Nicholas 1993 (Nematoda; Thoracostomopsidae). *Bulletin van het Koninklijk Belgisch Instituut voor Wetenschappen* 65, 41–51.

Nicholas, W.L. and Stewart, A. (1984a) *Criconemella avicenniae* n.sp. (Nematoda: Criconematidae) and *Enchodelus coomansi* n.sp. (Nematoda: Nordiidae) associated with the roots of the mangrove *Avicennia marina* (Forsk.) Vierh. *Nematologica* 30, 429–436.

Nicholas, W.L. and Stewart, A. (1984b) *Oncholaimus balli* n. sp. (Nematoda: Oncholaimidae) from a volcanic crater lake in Papua New Guinea, with observations on the colonisation of the lake by nematodes. *Nematologica* 30, 1–10.

Nicholas, W.L., Goodchild, D.J. and Stewart, A. (1987) The mineral composition of intracellular inclusions in nematodes from thiobiotic mangrove mud-flats. *Nematologica* 33, 167–179.

Nicholas, W.L., Bird, A.F., Beech, T.A. and Stewart, A. (1992) The nematode fauna of the Murray River estuary, South Australia; the effects of the barrages across its mouth. *Hydrobiologia* 234, 87–101.

Nunn, R.A. (1993) *Trophic Ecology of Nematodes from Spitzbergen.* Honours Thesis, Brunel University, London.

Nuss, B. (1984) Ultrastructural and ecophysiological studies of crystalloid inclusions in the muscles of a sulphide-tolerant limnetic nematode (*Tobrilus gracilis*). *Veroeffentlichungen des Instituts für Meeresforschung in Bremerhaven* 20, 3–15.

Nuss, B. and Trimkowski, V. (1984) Physical microanalyses of crystalloid inclusions in *Tobrilus gracilis* (Nematoda, Enoplida). *Veroeffentlichungen des Instituts für Meeresforschung in Bremerhaven* 20, 17–27.

Ocaña, A. (1991a) Relationship between nematode species and the physico-chemical characteristics of spring waters. 1. Conductivity. *Nematologia Mediterranea* 19, 19–24.

Ocaña, A. (1991b) Relationship between nematode species and the physico-chemical characteristics of spring waters. 2. Temperature. *Nematologia Mediterranea* 19, 25–28.

Ocaña, A. (1991c) A redescription of two nematode species found in hot springs. *Nematologia Mediterranea* 19, 173–175.

Ocaña, A. (1993) Relationship between nematode species and physico-chemical characteristics of spring water. 3: pH, oxygen and iron. *Nematologia Mediterranea* 20, 181–186.

Ocaña, A. and Morales, R. (1992) The influence of ionic composition on the distribution of nematode species in springs of the province of Granada (Spain). *Hydrobiologia* 237, 81–92.

Ocaña, A. and Picazo, J.S. (1991) Study on nematode species encountered in the Monachil River (Granada, Spain): response to organic pollution. *Internationale Vereinigung fuer Theoretische und Angewandte Limnologie Verhandlungen* 24, 2729–2737.

Ocaña, A., Peña, R. and Morales, R. (1986) Distribucion de los nematodos del Orden Dorylaimida en determinados manantiales minero-medicinales de la provincia de Granada. *Limnetica* 2, 189–195.

Odum, E.P. (1971) *Fundamentals of Ecology*, 3rd edn. W.B. Saunders, Philadelphia.

Odum, H.T. (1957) Trophic structure and productivity of Silver Springs, Florida. *Ecological Monographs* 27, 55–112.

Paetzold, D. (1958) Beitraege zur Nematodenfauna mitteldeutscher Salzstellen im Raum von Halle. *Wissenschaftliche Zeitschrift Martin-Luther-Universitaet Halle-Wittenburg* (*Mathematik-naturw Reihe*) 8, 17–48.

Pax, F. and Soos, A. (1943) Die Nematoden der deutschen Schwefelquellen und Thermen. *Archiv fuer Hydrobiologie* 40, 123–183.

Pickup, J. and Rothery, P. (1991) Water-loss and anhydrobiotic survival in nematodes of Antarctic fell-fields. *Oikos* 61, 379–388.

Poinar, G.O. Jr. and Sarbu, S.M. (1994) *Chronogaster troglodytes* sp. n. (Nemata: Chronogasteridae) from Movile Cave, with a review of cavernicolous nematodes. *Fundamental and Applied Nematology* 17, 231–237.

Por, F.D. and Masry, D. (1968) Survival of a nematode and an oligochaete species in the anaerobic benthal of Lake Tiberias. *Oikos* 19, 388–391.

Prejs, K. (1986) Occurrence of stylet-bearing nematodes associated with aquatic vascular plants. *Ekologia Polska* 34, 185–192.

Procter, D.L.C. (1982) Free-living soil nematodes of the Namib desert. *Bulletin of the Desert Ecological Research Unit* October 1982, 6–7.

Revilla, E.P., Sastroutomo, S.S. and Rahim, M.A.A. (1991) Survey on aquarium plants of quarantine importance and their associated nematodes. *BIOTROP Special Publication* 40, 205–215.

Riemann, F. (1975) Immigration of marine nematodes into freshwater. Investigations in Colombia compared with non-tropical regions. *Internationale Revue der Gesamten Hydrobiologie* 60, 393–407.

Samoiloff, M.R., Schulz, S. and Jordan, Y. (1980) A rapid simple long-term toxicity assay for aquatic contaminants using the nematode *Panagrellus redivivus. Canadian Journal of Fisheries and Aquatic Sciences* 37, 1167–1174.

Samoiloff, M.R., Pulak, R.A. and Birkholz, D.A. (1983) The nematode bioassay for toxicity in biological samples or sediments from contaminated aquatic ecosystems. *Proceedings of the Third Biennial Plains Aquatic Research Conference,* Montana State University, Bozeman, Montana, pp. 15–23.

Schiemer, F. (1978) Verteilung und Systematik der freilebenden Nematoden des Neusiedlersees. *Hydrobiologia* 58, 167–194.

Schiemer, F. (1984) *Theristus franzbergeri* n.sp., a groundwater nematode of marine origin from the Danube. *Archiv für Hydrobiologie* 101, 259–263.

Schiemer, F. and Duncan, A. (1974) The oxygen consumption of a freshwater benthic nematode, *Tobrilus gracilis* (Bastian). *Oecologia* 15, 121–126.

Schneider, W. (1937) Freilebende Nematoden der deutschen limnologischen Sundaexpedition nach Sumatra, Java und Bali. *Archiv für Hydrobiologie Supplement* 15, 30–108.

Schneider, W. (1940) Neue freilebende Nematoden aus Hoehlen und Brunnen. I. Nematoden aus jugoslawischen Hoehlen. *Zoologischer Anzeiger* 132, 84–94.

Schuurmans Stekhoven, J.H. (1950) Études biospeleologiques 33. Nématodes des grottes et des eaux souterraines de Roumanie. *Bulletin de l'Institut Royal des Sciences Naturelles de Belgique* 26, 1–3.

Seiml-Buchinger, R. and Traunspurger, W. (2005) Nematoda. In: Gerecke, R. and Franz, H. (eds) *Quellen im Nationalpark Berchtesgaden*, Nationalpark Berchtesgaden, Forschungsbericht, Berchtesgaden, Germany.

Sher, S.A. (1968) Revision of the genus *Hirschmanniella* Luc and Goodey, 1963 (Nematoda: Tylenchoidea). *Nematologica* 14, 243–275.

Siddiqi, M.R. (2000) *Tylenchida Parasites of Plants and Insects*, 2nd edn. CAB International, Wallingford, UK.

Siddiqi, M.R. and Booth, W. (1991) *Meloidogyne* (Hypsoperine) *mersa* sp. n. (Nematoda; Tylenchina) attacking *Sonneratia alba* trees in mangrove forest in Brunei Darussalam. *Afro-Asian Journal of Nematology* 1, 212–220.

Sinclair, B.J. and Sjursen, H. (2001) Terrestrial invertebrate abundance across a habitat transect in Keble Valley, Ross Island, Antarctica. *Pedobiologia* 45, 134–145.

Sota, T. (1996) Effects of capacity on resource input and the aquatic metazoan community structure in phytotelmata. *Researches on Population Ecology* 38, 65–73.

Stammer, H.J. (1935) *Desmoscolex aquaedulcis* n.sp. der erste suesswasserbewohnende Desmoscolecidae aus einer slovenischen Hoehle (Nematoda). *Zoologischer Anzeiger* 109, 108–126.

Swart, A. and Harris, J.M. (1996) *Panagrolaimus magnivulvatus* Bostrom, 1995 in nest material, algae and soils from inland nunataks in western Dronning Maud Land, Antarctica. *South African Journal of Zoology* 31, 15–22.

Teal, J.H. (1957) Community metabolism in a temperate cold spring. *Ecological Monographs* 27, 283–302.

Thames, W.H. (1966) A soil-borne *Desmoscolex* from Guatemala. *Nematologica* 12, 100.

Tilly, L.J. (1968) The structure and dynamics of Cone Spring. *Ecological Monographs* 38, 169–197.

Timm, R.W. (1971) Antarctic soil and freshwater nematodes from the McMurdo Sound region. *Proceedings of the Helminthological Society of Washington* 38, 42–52.

Townsend, C.R., Begon, M. and Harper, J.L. (2003) *Essentials of Ecology*, 2nd edn. Blackwell, Malden, Massachusetts.

Treonis, A.M., Wall, D.H. and Virginia, R.A. (1999) Invertebrate biodiversity in Antarctic dry valley soils and sediments. *Ecosystems* 2, 482–492.

Tsalolikhin, S.Y. (1979) The origins of the nematode fauna in Lake Issyk-Kul'. *Zoologicheskii Zhurnal* 58, 629–636.

Tsalolikhin, S.Y. (1981a) A revision of the genus *Tobrilus* (Nematoda, Tobrilidae). *Zoologicheskii Zhurnal* 60, 1302–1313.

Tsalolikhin, S.Y. (1981b) On the genus *Paratrilobus* (Nematoda, Tobrilidae). *Zoologicheskii Zhurnal* 60, 1094–1098.

Tsalolikhin S.Y. (1989) Rare and new nematode species from Antarctica. *Trudy Zoologicheskogo Instituta, Akademiya Nauk SSSR* 194, 96–101.

Tsalolikhin, S.Y. (2001) Synopsis of the system of the family Tobrilidae (Nematoda: Enoplida). *Russian Journal of Nematology* 9, 19–24.

Tudorancea, C. and Zullini, A. (1989) Associations and distribution of benthic nematodes in the Ethiopian rift valley lakes. *Hydrobiologia* 179, 81–96.

Van den Berg, E. and Harris, J.M. (1996) *Rotylenchus capensis* Van den Berg and Heyns (Tylenchida: Hoplolaimidae) in soils of isolated inland nunataks in western Dronning Maud Land, Antarctica. *African Plant Protection* 2, 19–24.

Warwick, R.M., Dexter, D.M. and Kuperman, B. (2002) Free-living nematodes from the Salton Sea. *Hydrobiologia* 473, 121–128.

Wharton, D.A. (1986) *A Functional Biology of Nematodes.* Croom Helm, Sydney, Australia.

Wharton, D.A. (2004) Survival strategies. In: Gaugler, R. and Bilgrami, A.L. (eds) *Nematode Behaviour.* CAB International, Wallingford, UK, pp. 371–399.

Wharton, D.A. and Barclay, S. (1993) Anhydrobiosis in the free-living Antarctic nematode *Panagrolaimus davidi* (Nematoda: Rhabditida). *Fundamental and Applied Nematology* 16, 17–22.

Wharton, D.A. and Brown, I.M. (1989) A survey of terrestrial nematodes from the McMurdo Sound region, Antarctica. *New Zealand Journal of Zoology* 16, 467–470.

Womersley, C. and Ching, C. (1989) Natural dehydration regimes as a prerequisite for the successful induction of anhydrobiosis in the nematode *Rotylenchulus reniformis*. *Journal of Experimental Biology* 143, 359–372.

Wouts, W.M. and Sturhan, D. (1996) *Heterodera litoralis* sp. n. (Nematoda: Heteroderidae) from austral glasswort, *Sarcocornia quinqueflora*, in New Zealand. *Nematologica* 42, 62–70.

Wouts, W.M., Schoemaker, A. and Sturhan, D. (1995) *Heterodera spinicauda* sp. n. (Nematoda: Heteroderidae) from mudflats in the Netherlands, with a key to the species of the *H. avenae* group. *Nematologica* 41, 575–583.

Wu, J.H., Somerfield, P.J. and Austen, M.C. (2000) The free-living nematode genus *Parodontophora* Timm 1963 (Nematoda: Axonolaimidae) is not exclusively marine: *Parodontophora limnophila* sp. nov. from freshwater in China. *Hydrobiologia* 431, 205–210.

Wu, J.H., Fu, C., Chen, S. and Chen, J. (2002) Soil faunal response to land use: effect of estuarine tideland reclamation on nematode communities. *Applied Soil Ecology* 21, 131–147.

Yeates, G.W., Bongers, T., De Goede, R.G.M., Freckman, D.W. and Georgieva, S.S. (1993) Feeding habits in soil nematode families and genera – an outline for soil ecologists. *Journal of Nematology* 25, 315–331.

Zullini, A. (1976) Nematodes as indicators of river pollution. *Nematologica Mediterranea* 4, 13–22.

Zullini, A. (1977) Some freshwater nematodes of Southern Mexico and Guatemala. Subterranean fauna of México. Part III. *Quaderno, Accademia Nazionale dei Lincei* 171, 75–85.

Zullini, A. and Pagani, M. (1989) The ecological meaning of relative egg size in soil and fresh-water nematodes. *Nematologica* 35, 90–96.

Zullini, A, Loof, P.A.A. and Bongers, T. (2002) Free-living nematodes from nature reserves in Costa Rica. 2. Mononchina. *Nematology* 4, 1–23.

11 Computation and Application of Nematode Community Indices: General Guidelines

D.A. NEHER AND B.J. DARBY

Department of Plant and Soil Science, University of Vermont, Burlington, Vermont 05405, USA

Community indices, such as maturity or diversity, condense information regarding the structure and composition of communities into a single metric. Soil health and quality can be inferred from such indices by assuming that communities with different structure and composition function differently. Thus, these indices can be instrumental in monitoring soil and sediment quality as well as assessing ecosystem sustainability and biodiversity.

Historical Perspective

In the 1980s, interest increased for using nematode communities as indicators for environmental monitoring of terrestrial communities (Freckman, 1988; Bongers, 1990). Initially, simple indices of abundance and proportions or ratios of nematodes by trophic group were proposed. Subsequently, diversity indices were employed and a maturity index (MI) was developed for terrestrial nematodes (Yeates, 1970, 1984; Bongers, 1990). Later, the application of the MI was extended successfully to marine and brackish sediments (Bongers *et al.*, 1991). Since then, others have applied the concept to freshwater systems (Beier and Traunspurger, 2003; Höss *et al.*, 2004).

Maturity Indices

Maturity indices are used as a measure of the ecological successional status of a soil community. They are based on the principle that different taxa have contrasting sensitivities to stress or disruption of the successional sequence because of their life-history characteristics. Bongers originally proposed separate indices for free-living (MI) and plant-parasitic (PPI) nematodes. The index is represented by a colonizer–persister (*c–p*) value that ranges from a colonizer (c–p = 1) to a persister (c–p = 5) with the index values representing life-history characteristics associated with *r*- and *K*-selection, respectively. Those with a *c–p* = 1 are *r*-selected or colonizers, with short generation times, large population fluctuations and high fecundity. Those with a *c–p* = 5 are *K*-selected or persisters, produce few offspring and generally appear later in succession (Bongers and Bongers, 1998; Bongers and Ferris, 1999). Small and large *c–p* weights correspond with taxa relatively tolerant and sensitive to ecological disturbance, respectively (Table 11.1).

Maturity Indices are computed as a weighted mean frequency,

$$\mathrm{MI} = \sum \frac{v_i \times f_i}{n}$$

where v_i = *c–p* value assigned to family, f_i = frequency of family *i* in sample, *n* = total number of individuals in a sample. MI may be viewed as a measure of disturbance, with smaller values being indicative of a more disturbed environment and larger values characteristic of a less disturbed environment. The MI decreases with increasing microbial activity and pollution-induced stress when opportunistic nematodes rapidly become dominant. The PPI may (Neher and Campbell, 1994) or may not (Bongers *et al.*, 1997) correlate positively with MI.

Both stress and enrichment result in MI decrease; stress decreases numbers of sensitive species and enrichment increases abundance of opportunists. To differentiate decreases of MI caused by enrichment and pollution from environmental stress, it is better to omit *c–p* = 1 (Bongers and Korthals, 1993; Bongers *et al.*, 1995; Korthals *et al.*, 1996). Opportunistic taxa (*c–p* = 1) are re-evaluated because they are considered enrichment opportunists and their population densities increase rapidly in response to addition of nutrients to soil and may not necessarily reflect long-term changes in soil ecological condition. Those with *c–p* values between 2 and 5 are more stable temporally and may provide relatively long-term information about environmental conditions. Therefore, it is recommended that marine Oncholaimidae be omitted because they accumulate to generate mono-populations under anoxic enriched conditions known to result in fish dying. Furthermore, dauer larvae of Rhabditidae, Diplogasteridae and Panagrolaimidae are not included because they are non-feeding or inactive stages at the time of sampling.

Yeates (1994) and Wasilewska (1994) proposed a modification of the index based on merging free-living and plant-parasitic nematodes in a soil community (ΣMI). Neher and Campbell (1996) suggested a ΣMI25, which represents a combination of the concepts proposed by Bongers *et al.* (1995) and Yeates (1994). In each

Table 11.1 Colonizer–persister (*c–p*) value assignments for selected nematode families (Bongers and Bongers, 1998).

c–p value	Family
1	Rhabditidae, Diplogasteridae (s.l.), Panagrolaimidae, Bunonematidae
2	Cephalobidae, Plectidae, Monhysteridae, Aphelenchoididae
3	Teratocephalidae, Chromadoridae, Diphtherophoridae, Prismatolaimidae
4	Alaimidae, Mononchidae, Leptonchidae, Qudsianematidae, Dorylaimidae
5	Aporcelaimidae, Actinolaimidae, Thornenematidae, Belondiridae

case, the same basic calculation is performed except different combinations of feeding groups (Yeates *et al.*, 1993) and *c–p* groups are included. The difference in these two more comprehensive indices is that they include and exclude *c–p* = 1 nematodes, respectively. As no distinction between free-living and plant-parasitic nematodes is necessary these indices solve a disagreement on whether to include (Bongers, 1990) or exclude (Neher and Campbell, 1996) Tylenchidae in the PPI. The controversy is based on whether Tylenchidae are considered fungal or root feeders.

Diversity

Diversity has been equated solely with numbers of taxa, and the popular press has perpetuated this misconception. Rather, a more appropriate indication of diversity integrates numbers of taxa ('species richness') and equitability among taxa ('species evenness') (Hurlbert, 1971). Two assumptions of most diversity indices include: (i) an index limited to one taxonomic group; and (ii) all species are equal (Cousins, 1991). Good (1953) outlined a generalized diversity index that incorporates richness and evenness into a single value that generally increases with both richness and evenness:

$$\mathrm{H}(\alpha,\beta) = \sum_{i=1}^{S} p_i^{\alpha} \left\{-\ln(p_i)\right\}^{\beta}$$

where p_i is the relative abundance of taxon i, S is the total number of species present and α and β define structural attributes of the algorithm. Good's generalized diversity index demonstrates the mathematical relation of several commonly used diversity, dominance and evenness indices (Table 11.2). Namely, Shannon's diversity index can be interpreted as a variant of Good's diversity index using values of 1 and 1 for α and β, respectively (or H(1,1)). Simpson's dominance index can be interpreted as H(2,0). Rényi (1961) defined a notation of this generalized equation, which allows a user to put in different coefficients for species richness and dominance concentration (followed by Hill, 1973), and Baczkowski *et al.* (1997) discuss optimal bounds of α and β for ecological applications. Most other indices constitute logarithmic, exponential, reciprocal, complementary or relative transformations of the Shannon or Simpson indices. In addition to the main indices listed in Table 11.2, numerous alterations have been reported as evenness indices, including $(1-D)/(1-1/S)$, $(1-D)/S$, and $(-\ln D)/(\ln S)$ by Smith and Wilson (1996), and $(1/D-1)/(e^{H'}-1)$ by Alatalo (1981) ('D' is Simpson's dominance index as in Table 11.2).

These variations on a theme demonstrate unique sensitivities to changes in various structural attributes of an abundance distribution within a community. For example, first, the Camargo diversity index may be more sensitive for assessing structural alterations in aquatic communities than the Shannon (H') and MacArthur indices, which are sensitive to the number of taxa present and the whole spectrum of taxon proportions (Camargo, 1992). Second, the Camargo index typically increases with the addition of subordinate species (those defined as having a relative abundance less than $1/S$) more than most indices, valuing rare species. Beisel *et al.* (1996) argues that such sensitivity to rare species is an undesirable property of a diversity index and favours the Simpson and McIntosh indices that are more sensitive to changes in dominant taxa. Camargo (1997) rebuts with the

Table 11.2 Common indices used to characterize the distribution of abundance within a community.

Name	Equation*	Application	Reference
Diversity indices			
Shannon's diversity	$H' = -\Sigma (p_i \ln p_i)$	This widely used and versatile index can be applied for both large and small sample sizes. The Shannon index is generally more influenced by rare species than the Simpson index.	Shannon and Weaver (1949)
Hill's diversity	$N_1 = \exp [-\Sigma (p_i \ln p_i)] = \exp (H')$	An exponential form of Shannon's H′, the value of this index can be interpreted as the number of abundant taxa (Ludwig and Reynolds, 1988).	Hill (1973)
Brillouin's diversity	$H = \frac{1}{N} \log \frac{N!}{\prod N_i!}$	Use only on fully censused communities because it is a true statistic and, thus, free from statistical error.	Brillouin (1962)
Camargo's diversity	$d = \sum \left[P_i - \left(\frac{1}{S} \right) \right]$	Estimates the (structural) asymmetry in relative abundance between dominant and subordinate species, not necessarily differences between dominant species or between subordinate species.	Camargo (1992)
Margalef's diversity	$D_{Marg} = \frac{(S-1)}{\ln (N)}$	Though simple to calculate, this index is unaffected by evenness or dominance and is sensitive only to species richness and sample size. Thus, its use	Margalef (1958)

		should be restricted to comparing species richness among large communities.	
Dominance indices			
Simpson's dominance (infinite community)	$D=\Sigma p_i^2$	Probability that two randomly chosen individuals of an infinite community belong to the same class, thus inversely related to diversity. It is often reported as 1–*D*, but see Hurlbert's PIE.	Simpson (1949)
Hill's reciprocal of *D*	$N_2 = (\Sigma p_i^2) = 1/D$	The reciprocal of Simpson's *D*, the value of this index can be interpreted as the number of very abundant taxa (Ludwig and Reynolds, 1988)	Hill (1973)
Simpson's dominance (finite community)	$\lambda = \frac{\sum n_i(_i - 1)}{N(N-1)}$	Similar to Simpson's *D*, but corrected for finite communities. Mathematically, it is usually more appropriate in ecological studies than Simpson's *D*, but is used less often.	Simpson (1949)
Probability of interspecific encounter (PIE)	$PIE = \left(\frac{N}{N-1}\right)\left(1 - \sum p_i^2\right)$	Simpson's dominance index *D* converted to a diversity index and corrected for finite communities	Hurlbert (1971)
McIntosh dominance	$D1 = \frac{\left(N - \sqrt{\sum n_i^2}\right)}{\left(N - \sqrt{N}\right)} \approx \sqrt{\sum n_i^2}$	Recommended by Beisel *et al.* (1996) as the most relevant dominance index since most sensitive to variations on dominant taxa and not highly sensitive to variations on rare or medium taxa.	McIntosh (1967)

Continued

Table 11.2 *Continued.* Common indices used to characterize the distribution of abundance within a community.

Name	Equation*	Application	Reference
Evenness indices			
Brillouin's maximum diversity	$H_{max} = \frac{1}{N} \ln \frac{N!}{\left(\frac{N}{S}\right)^{S-1} \left[\left(\frac{N}{S}\right)+1\right]^{r}}$	Represents maximum possible evenness of a sample of *N* individuals and *S* species such that *N* = *S*[*N*/*S*] + *r* where [*N*/*S*] is the integer portion of *N*/*S* and *r* is the remainder.	Brillouin (1962)
Brillouin's minimum evenness	$H_{min} = \frac{1}{N} \ln \frac{N!}{(N-S+1)}$	Represents minimum possible evenness of a sample of *N* individuals and *S* species.	Brillouin (1962)
Brillouin's evenness	$J = \frac{H}{H_{max}}$ or $J' = \frac{H'}{\ln S}$	Use J for samples (and J′ for collections) to determine the evenness portion of diversity; J or J′ represent observed and maximum diversity, respectively.	Pielou (1966)
Brillouin's relative evenness	$V = \frac{H - H_{min}}{H_{max} - H_{min}}$	Unlike J and J′, V is not influenced by species richness (S)	Hurlbert (1971)
Heip's evenness	$E_{Heip} = \frac{\left(e^{H'} - 1\right)}{(S-1)}$	Hypothesized by Beisel *et al.* (2003) to be more sensitive to variations in rare species richness and/or abundance.	Heip (1974)

(Dis) Similarity indices			
Morisita's similarity	$C_H = \frac{2\sum(p_{ij}\,p_{ik})}{\sum p_{ij}^2 + \sum p_{ik}^2}$	Quantitative data; represents the degree of overlap between segment j and all segments combined (k). It ranges from 0 (no similarity in community structure) to 1 (complete similarity) and is often expressed as a percentage (C_H * 100).	Morisita (1959)
Bray–Curtis dissimilarity	$BC_{ij} = \sum \frac{\lvert n_{ik} - n_{jk} \rvert}{(n_{ik} + n_{jk})}$	Quantitative data; an alternative to Morisita's similarity.	Bray and Curtis (1957)
Jaccard similarity	$S_J = 100^* \frac{c}{(a+b+c)}$	Binary data (presence/absence); represents the percent of taxa present that are similar to both groups.	Jaccard (1912)
Sørensen similarity	$CC = 100^* = \frac{2c}{(a+b+2c)}$	Represents the percent similarity of each group with respect to taxa present.	Sørensen (1948)

*p_i represents the proportion of the *i*-th taxa in a sample, or n_i the number, with N individuals and S total species.

†Quantitative data: $p_{ij\ or\ ik}$ represents the proportion of the *i*-th taxa in sample j or k, or n_i the number; binary data: a the number of taxa unique to group A, b the number of taxa unique to group B, and c the number of taxa common to both group A and group B.

argument that the 'conviction that rare taxa should not contribute to the response of a dominance index is absurd'. In fact, the Camargo index is one of few diversity indices developed from an *a priori* ecological definition that dominance is the appropriation of potential niche space of certain subordinate species by other dominant species (McNaughton and Wolf, 1970).

The debate defining 'dominance' and 'diversity' raises a second question, that of which taxa to include in an index. Ideally a summary statistic such as a diversity index could relate the abundance structure of an entire community. For a freshwater nematode community, this would include all the bacterivores, algivores, herbivores, predators and omnivores. However, if the McNaughton and Wolf (1970) definition of dominance is accepted as a complement to diversity, then logically only one trophic group should be included in any diversity or dominance index. This creates a practical problem because the specific feeding habits of most nematode species are diverse, changing or unknown. In fact, first, the effect of each species on ecosystem processes has not been determined (Chapin *et al.*, 1992). Second, the reduction of diversity indices to specific functional groups entirely contradicts the intended use of indices, i.e. to summarize complex and varied community data into a single useful datum (Beisel, 1997).

Not all indices are contested so vigorously. Although the sensitivities of respective indices are still unclear, some generalization can be made from the literature. Typically, Shannon's (H') index is sensitive to rare taxa and Simpson's (λ) index weights common taxa (Boyle *et al.*, 1990). Hill's family of diversity numbers are easy to interpret ecologically because the indices define units as taxa (Peet, 1974) but they are not necessarily superior from a statistical perspective (Heip *et al.*, 1988; Ludwig and Reynolds, 1988). Hill's diversity numbers N0, N1 and N2 are defined as numbers of all taxa, abundant taxa and very abundant taxa, respectively (Ludwig and Reynolds, 1988). N1 equates with an antilog of a Shannon index (eH') and N2 equals the reciprocal of a Simpson index ($1/\lambda$).

Diversity is linked artificially to the taxonomic resolution an investigator employs. Even though diversity is most often equated with species, it can be applied at various taxonomic levels of resolution, such as genotype, genus, family and trophic group. For free-living nematodes, it is more common to apply diversity indices to taxonomic levels above species because species identifications based on morphology are difficult. Appropriate caution must be taken when applying indices at the family or trophic group levels. Unfortunately, ambiguity in trophic classification of nematodes usually occurs because it is inferred by morphology rather than actual experiments on feeding preferences (Yeates *et al.*, 1993). Furthermore, feeding-habit groupings may be ambiguous and/or not mutually exclusive in some cases. For example, abundant populations of *Aphelenchoides*, *Tylenchus*, *Tylencholaimus* and *Ditylenchus* can be classified as 'plant-/fungal-feeding' nematodes (Sohlenius *et al.*, 1977), or some 'predaceous' *Mesodorylaimus* sp. can grow and reproduce by feeding on bacteria (Russell, 1986). *Tylenchus* spp. are often considered fungivores in ecological studies, but the basis of the judgement is dubious because several species feed and reproduce on roots. In other cases, some species may always be placed in one category and may have developmental stages or generations that fit in another category.

As diversity indices are based on relative abundances of community species they are insensitive to taxonomic differences between species. Alternatively, biodiversity indices are based on taxonomic relations among species and ignore species abundances. However, in many environmental planning and protection programmes, there is interest in both species composition and relative distribution. Therefore, new combined indices have been proposed such as the quadratic entropy index (Q), which incorporates relative abundances of both species and a measure for the pairwise taxonomic differences between species (Izsak and Papp, 2000; Ricotta, 2002). Unfortunately, these new combined indices violate part of the mathematical properties of an ecological diversity index, so they are dubbed 'weak diversity indices'.

Critical to valid interpretation of diversity indices are appropriate sampling and statistical techniques. Generally, stratified- or simple-stage cluster sampling are touted as generating less bias in diversity estimates than simple random sampling (Gimaret-Carpentier *et al.*, 1998; Mayoral, 1998; di Battista, 2002). Commonly, diversity indices are analysed statistically with traditional ANOVA procedures. However, care must be taken to ensure that assumptions of normality and equal variances are met, especially for small sample sizes. Sometimes, distributions can be transformed to resemble a Gaussian distribution by application of log and log-normal scales (Hill *et al.*, 2003). Rogers and Hsu (2001) propose an asymptotically correct method for diversity indices with unequal variances, when sample sizes are equal, and transformations cannot remedy the situation.

Trophic Group Ratios

Ratios of trophic groups have been proposed to describe the relative balance of positive to negative impacts of nematodes on primary productivity or stage of decomposition. Wasilewska (1989) proposed a ratio that computes the sum of fungivores and bacterivores divided by plant parasites. A ratio greater than 1 suggests that the positive impacts of nematodes outweigh the negative impacts on plant productivity. Two forms of decomposition pathway indices have been proposed, differing in the denominator. Statistically, the form fungivores divided by the sum of fungivores and bacterivores [$F/(F+B)$] is considered mathematically more stable than simply dividing fungivores by bacterivores (F/B). Because each variation gives a contrasting result, it is critical that the user defines the ratio employed in the results of a report. Typically, individual or ratios of trophic groups have not withstood the level of statistical rigour that maturity and trophic diversity indices do in their ability to differentiate the ecological condition of soils on a large geographic scale (Neher *et al.*, 1995; Neher and Campbell, 1996).

Multivariate Approaches

Canonical Correspondence Analysis (CCA) is useful to compare suites of taxon data with suites of environmental variables. Environmental variables can include treatment classes as nominal 0 or 1 variables or chemical properties or pollutants as continuous variables. CANOCO (ter Braak and Smilauer, 2002) and PRIMER-E (http://www.pml.ac.uk/primer/index.htm) software are simple tools to perform these procedures. In CANOCO, abundances are transformed as log $(x + 1)$ before analysis. Transformations are unnecessary in PRIMER because the scaling is non-metric multi-dimensional scaling. CCA results are displayed graphically with biplots. In CCA biplots, each vector for an environmental variable defines an axis, and site or genera scores can be projected on to that axis (Jongman *et al.*, 1995). An indication of relative importance of a vector is its length; the angle indicates correlation with other vectors and CCA axes. Eigen values for CCA axes indicate the importance of the axes in explaining relationships in the genera–environment data matrices. Unfortunately, CCA analyses are restricted to illustrating one instance at a time.

Principal response curves (PRC) is a multivariate method for the analysis of repeated measurement design. PRC is based on redundancy analysis (RDA); each experimental unit and sampling times and unit by time interactions are treated as dummy explanatory variables. The result is a diagram showing the sampling periods on the *x*-axis and the first principal component of the variance explained by treatment on the *y*-axis. For illustrative purposes, undisturbed condition was treated as a 'control', representing a zero baseline, and 'disturbed' of the same experimental unit as the 'treatment' to focus on the 'differences' between the two states of condition through time. Monte Carlo permutation tests permuting whole time series are applied to compute statistical significance. van den Brink *et al.* (2003) provide a review of the analytical procedure and detailed instruction is provided in the manual of CANOCO Version 4.5 software (ter Braak and Smilauer, 2002).

References

Alatalo, R.V. (1981) Problems in the measurement of evenness in ecology. *Oikos* 37, 199–204.

Baczkowski, A.J., Joanes, D.N. and Shamia, G.M. (1997) Properties of a generalized diversity index. *Journal of Theoretical Biology* 188, 207–213.

Beier, S. and Traunspurger, W. (2003) Seasonal distribution of free-living nematodes in the Krahenbach, a fine-grained submountain carbonate stream in southwest Germany. *Nematology* 5, 113–136.

Beisel, J.N. (1997) Assessing changes in community structure by dominance indices: a comparative analysis – response. *Journal of Freshwater Ecology* 12, 641–646.

Beisel, J.N., Thomas, S., Ussegliopolatera, P. and Moreteau, J.C. (1996) Assessing changes in community structure by dominance indices: a comparative analysis. *Journal of Freshwater Ecology* 11, 291–299.

Beisel, J.N., Usseglio-Polatera, P., Bachmann, V. and Moreteau, J.C. (2003) A comparative analysis of evenness index sensitivity. *International Review of Hydrobiology* 88, 3–15.

Bongers, T. (1990) The maturity index: an ecological measure of environmental disturbance based on nematode species composition. *Oecologia* 83, 14–19.

Bongers, T. and Bongers, M. (1998) Functional diversity of nematodes. *Applied Soil Ecology* 10, 239–251.

Bongers, T. and Ferris, H. (1999) Nematode community structure as a bioindicator in environmental monitoring. *Trends in Evolution and Ecology* 14, 224–228.

Bongers, T. and Korthals, G. (1993) The maturity index, an instrument to monitor changes in the nematode community structure. *Proceedings of the International Symposium on Crop Protection* 45, 80.

Bongers, T., Alkemade, R. and Yeates, G.W. (1991) Interpretation of disturbance-induced maturity decrease in marine nematode assemblages by means of the Maturity Index. *Marine Ecology Progress Series* 76, 135–142.

Bongers, T., de Goede, R.G.M., Korthals, G. and Yeates, G.W. (1995) Proposed changes of c–p classification for nematodes. *Russian Journal of Nematology* 3, 61–62.

Bongers, T., van der Meulen, H. and Korthals, G. (1997) Inverse relationship between the nematode maturity index and plant parasite index under enriched nutrient conditions. *Applied Soil Ecology* 6, 195–199.

Boyle, T.P., Smillie, G.M., Anderson, J.C. and Beeson, D.R. (1990) A sensitivity analysis of nine diversity and seven similarity indexes. *Research Journal of the Water Pollution Control Federation* 62, 749–762.

Bray, J.R. and Curtis, J.T. (1957) An ordination of the upland forest communities of southern Wisconsin. *Ecological Monographs* 27, 325–349.

Brillouin, L. (1962) *Science and Information Theory*, 2nd edn. Academic Press, New York.

Camargo, J.A. (1992) New diversity index for assessing structural alterations in aquatic communities. *Bulletin of Environmental Contamination and Toxicology* 48, 428–434.

Camargo, J.A. (1997) Assessing changes in community structure by dominance indices: a comparative analysis – rebuttal. *Journal of Freshwater Ecology* 12, 637–641.

Chapin, F.S., Schulze, E.D. and Mooney, H.A. (1992) Biodiversity and ecosystem processes. *Trends in Ecology and Evolution* 7, 107–108.

Cousins, S.H. (1991) Species diversity measurement: choosing the right index. *Trends in Ecology and Evolution* 6, 190–192.

di Battista, T. (2002) Diversity index estimation by adaptive sampling. *Environmetrics* 13, 209–214.

Freckman, D.W. (1988) Bacterivorous nematodes and organic-matter decomposition. *Agriculture, Ecosystems, and Environment* 24, 195–217.

Gimaret-Carpentier, C., Pelissier, R., Pascal, J.P. and Houllier, F. (1998) Sampling strategies for the assessment of tree species diversity. *Journal of Vegetation Science* 9, 161–172.

Good, I.J. (1953) The population frequencies of species and the estimation of population parameters. *Biometrika* 40, 237–264.

Heip, C. (1974) A new index measuring evenness. *Journal of the Marine Biological Association* 54, 555–557.

Heip, C., Herman, P.M.J. and Soetaert, K. (1988) Data processing, evaluation, and analysis. In: Higgins, R.P. and Thiel, H. (eds) *Introduction to the Study of Meiofauna.* Smithsonian Institution Press, Washington, DC, pp. 197–231.

Hill, M.O. (1973) Diversity and evenness: a unifying notation and its consequences. *Ecology* 54, 427–432.

Hill, T.C.J., Walsh, K.A., Harris, J.A. and Moffett, B.F. (2003) Using ecological diversity measures with bacterial communities. *FEMS Microbiology Ecology* 43, 1–11.

Höss, S., Traunspurger, W., Severin, G.E., Jüttner, I., Pfister, G. and Schramm, K.W. (2004) Influence of 4-nonylphenol on the structure of nematode communities in freshwater microcosms. *Environmental Toxicology and Chemistry* 23, 1268–1275.

Hurlbert, S.H. (1971) The nonconcept of species diversity: a critique and alternative parameters. *Ecology* 52, 577–586.

Izsak, J. and Papp, L. (2000) A link between ecological diversity indices and measures of biodiversity. *Ecological Modelling* 130, 151–156.

Jaccard, P. (1912) The distribution of flora in the alpine zone. *The New Phytologist* 11, 37–50.

Jongman, R.H.G., ter Braak, C.J.F. and van Tongeren, O.F.R. (1995) *Data Analysis in Community and Landscape Ecology.* Cambridge University Press, Cambridge, UK.

Korthals, G.W., de Goede, R.G.M., Kammenga, J.E. and Bongers, T. (1996) The Maturity Index as an instrument for risk assessment of soil pollution. In: van Straalen, N.M. and Krivolutski, D.A. (eds) *Bioindicator Systems for Soil Pollution.* Kluwer Academic, Dordrecht, The Netherlands, pp. 85–93.

Ludwig, J.A. and Reynolds, J.F. (1988) *Statistical Ecology: A Primer on Methods and Computing.* John Wiley & Sons, New York.

Margalef, R. (1958) Information theory in ecology. *General Systems* 3, 36–71.

Mayoral, M.M. (1998) Rényi's entropy as an index of diversity in simple-stage cluster sampling. *Information Sciences* 105, 101–114.

McIntosh, R.P. (1967) An index of diversity and the relation of certain concepts to diversity. *Ecology* 48, 392–404.

McNaughton, S.J. and Wolf, L.L. (1970) Dominance and the niche in ecological systems. *Science* 167, 131–139.

Morisita, M. (1959) Measuring of interspecific association and similarity between communities. *Memoirs of the Faculty of Science, Kyushu University, Japan, Series E. (Biol.)* 3, 65–80.

Neher, D.A. and Campbell, C.L. (1994) Nematode communities and microbial biomass in soils with annual and perennial crops. *Applied Soil Ecology* 1, 17–28.

Neher, D.A. and Campbell, C.L. (1996) Sampling for regional monitoring of nematode communities in agricultural soils. *Journal of Nematology* 28, 196–208.

Neher, D.A., Peck, S.L., Rawlings, J.O. and Campbell, C.L. (1995) Measures of nematode community structure for an agroecosystem monitoring program and sources of variability among and within agricultural fields. *Plant and Soil* 170, 167–181.

Peet, R.K. (1974) The measurement of species diversity. *Annual Review of Ecology and Systematics* 5, 285–307.

Pielou, E.C. (1966) Species-diversity and pattern-diversity in the study of ecological succession. *Journal of Theoretical Biology* 10, 370–383.

Rényi, A. (1961) On measures of entropy and information. *Proceedings of the Fourth Berkeley Symposium on Mathematical Statistics and Probability* 1, 547–561.

Ricotta, C. (2002) Bridging the gap between ecological diversity indices and measures of biodiversity with Shannon's entropy: comment to Izsak and Papp. *Ecological Modelling* 152, 1–3.

Rogers, J.A. and Hsu, J.C. (2001) Multiple comparisons of biodiversity. *Biometrical Journal* 43, 617–625.

Russell, C.C. (1986) The feeding habits of a species of *Mesodorylaimus*. *Journal of Nematology* 18, 641.

Shannon, C.E. and Weaver, W. (1949) *The Mathematical Theory of Communication.* University of Illinois, Urbana, Illinois.

Simpson, E.H. (1949) Measurement of diversity. *Nature* 163, 688.

Smith, B. and Wilson, J.B. (1996) A consumer's guide to evenness indices. *Oikos* 76, 70–82.

Sohlenius, B., Persson, H. and Magnusson, C. (1977) Distribution of root and soil nematodes in a young Scots pine stand in central Sweden. *Ecological Bulletins (Stockholm)* 25, 340–347.

Sørensen, J. (1948) A method of establishing groups of equal amplitude in plant sociology based on similarity of species content. *Danske Videnskabers Selskabs Biologiske Skrifter* 5, 1–34.

ter Braak C.J.F. and Smilauer, P. (2002) *CANOCO Reference Manual and CanoDraw for Windows User's Guide, Software for Canonical Community Ordination* (version 4.5), Biometris, Wageningen and Èeské Budìjovice, The Netherlands, 500 pp.

van den Brink, P.J., van den Brink, N.W. and ter Braak, C.J.F. (2003) Multivariate analysis of ecotoxicological data using ordination: demonstrations of utility on the basis of various examples. *Australasian Journal of Ecotoxicology* 9, 141–156.

Wasilewska, L. (1989) The role of nematodes in agroecosystems. *Zeszyty Problemowe Postepów Nauk Rolniczych* 358, 7–16.

Wasilewska, L. (1994) The effect of age of meadows on succession and diversity in soil nematode communities. *Pedobiologia* 38, 1–11.

Yeates, G.W. (1970) The diversity of soil nematode faunas. *Pedobiologia* 10, 104–107.

Yeates, G.W. (1984) Variation in soil nematode diversity under pasture with soil and year. *Soil Biology and Biochemistry* 16, 95–102.

Yeates, G.W. (1994) Modification and qualification of the nematode Maturity Index. *Pedobiologia* 38, 97–101.

Yeates, G.W., Bongers T., de Goede, R.G.M., Freckman, D.W. and Georgieva, S.S. (1993) Feeding habits in soil nematode families and genera – an outline for soil ecologists. *Journal of Nematology* 25, 315–331.

II Taxonomy

12 Order Enoplida

N. Smol and A. Coomans

Department of Biology, Ghent University, Ledeganckstraat 35, B-9000 Ghent, Belgium

Introduction

Enoplida are among the largest free-living aquatic nematodes. Adults are generally larger than 3 mm (except Alaimina and Campydorina) and the largest species occur within the Leptosomatidae, a family with some species commensal with marine sponges and anemones, and regarded as a link to the parasitic Marimermithida.

Most representatives of the Enoplida are true marine species, some of them occur in brackish water and up to now relatively few have been found in freshwater. The recent inclusion of the Alaimina and Campydorina increases the importance of terrestrial species.

Morphology

Cuticle

The cuticle varies from thin (e.g. in *Oncholaimus*) to very thick (Leptosomatidae) as observed with the light microscope and generally has a smooth or weakly striated appearance; rarely punctations can be observed (e.g. in *Enoplus*).

The cuticle at the head end of many enoplids is different from the rest of the body. Anteriorly the cuticle seems to split (Fig. 12.1A) and its inner layer (endocuticle) is attached to the stoma wall or to the upper end of the pharynx forming an *endocupola* or internal skeleton of the head. There is also a thickening of the median layer (mesocuticle), which becomes filled with a fluid and is called the *cephalic ventricle*. It may act as a shock absorber and in combination with the solid *endocupola* and *exocupola* (exocuticle) it creates an effective system for protecting internal tissues of the head end. In many cases the muscles of the anterior end of the pharynx attach to the body cuticle, resulting in a thickening of the cuticle, forming the *cephalic capsule* (or *exocupola*) (Fig. 12.1B).

The posterior end of the cephalic capsule is frequently divided into lobes by a series of incisions; these incisions represent the external expression of the internal foramina in the *endocupola* through which the nerves pass. Punctations on the

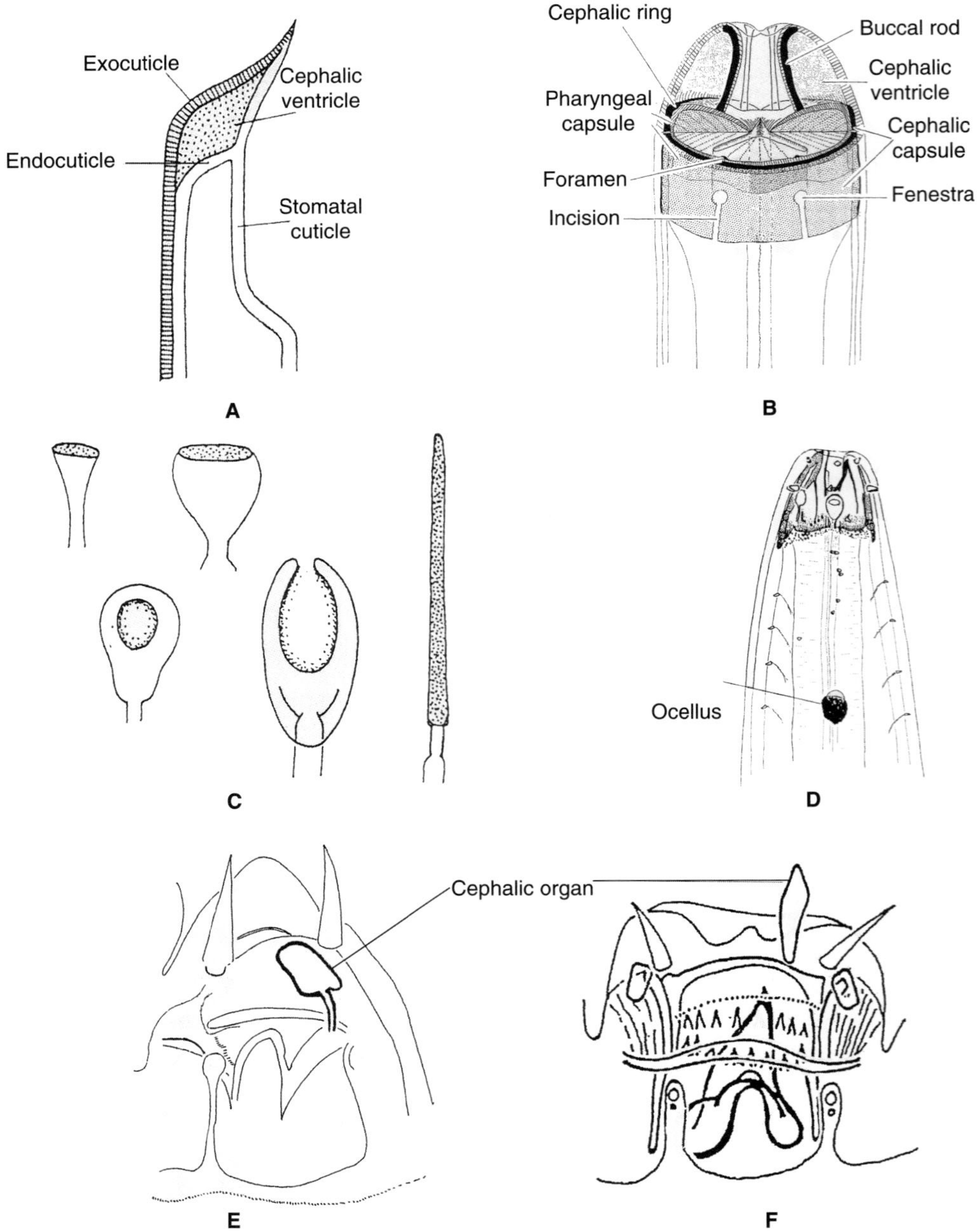

Fig. 12.1 Morphological features of Enoplida. A: Schematic drawing of the split cuticle in *Pontonema* (after Malakhov, 1994); B: generalized Enoplida (Enoplina) head from the dorsal side (after Inglis, 1964); C: non-spiral amphids in Enoplida (pocket-shaped, oval and elongate forms) (after Lorenzen, 1981, 1994); D: anterior end of *Thoracostoma* with well-developed ocellus and cephalic capsule, the posterior edge of which is granulated (after Platt and Warwick, 1983); E: cephalic organ in *Mesacanthion* (after Inglis, 1964); F: cephalic organ in *Oxyonchus* (after Inglis, 1964).

cephalic capsule appear to be the reflection of the attachments of the pharynx to the body cuticle.

The structure of the Enoplida head is extensively described by Inglis (1964).

Sensory organs

Anterior sensilla. The inner labial sensilla are mostly papilliform, rarely setiform; the outer labial and cephalic sensilla are generally setiform (exception Campydorina and Alaimina) and can be jointed in a few families; mostly the six outer labial setae are larger than the four cephalic setae. Additional sub-cephalic setae are often observed in the Enoplina.

Amphids (Fig. 12.1C) are typically non-spiral with a transverse slit-like or oval aperture and a large, pocket-shaped (also called stirrup-shaped or cyathiform) fovea. Occasionally the aperture is a longitudinal groove (*Halalaimus*) or indistinct (*Rhabdodemania*); in the latter case a subcuticular sinusoidal structure can be observed as far back as the nerve ring. In cases where spiral amphids occur they are either ventrally spiral (Tripyloididae) or dorsally spiral (e.g. *Rhabdocoma*). Amphids have not been observed in a number of Enoplidae. In all species the amphids lie posterior to the labial region and behind the outer labial sensilla.

Sexual dimorphism in the shape of the amphids occurs in *Chaetonema* and in some genera of the Enchelidiidae; in these cases the amphids of the males differ in appearance from those of females and juveniles.

Ocelli (Fig. 12.1D) or eyespots are a pair of pigmented areas (orange, red, purple, brown or black, depending on the species) (Croll, 1966; Siddiqi and Viglierchio, 1970a,b; Burr and Burr, 1975; Burr and Webster, 1975; Bollerup and Burr, 1979) positioned laterally in or partly outside the pharynx. These pigments are concentrated in cup-shaped eyespots or localized in specialized regions of pharyngeal muscle cells and in some species a distinct lens-like structure is visible.

Cephalic organ (Fig. 12.1E and F). Within the Enoplina an extra pair of sense organs = cephalic slits (cf. de Man, 1886) or cephalic organs (cf. Filipjev, 1927) are positioned lateroventrally between the circles of the anterior sensilla. They appear to be universal among the members of the Enoplidae, Thoracostomopsidae and Ironidae and show marked variation from genus to genus. In a few cases the cephalic slit is modified and forms a cirrus (= projecting club-shaped organ), such as in *Oxyonchus dentatus.* Wieser (1953a) speculated that the cirri might function as a sort of sense organ (chemical or mechanical?) connected with the predatory activity of these species because they are provided with the most powerful buccal armature and should be regarded as the most efficient predators.

Metanemes (Fig. 12.2A–C). The metanemes are spindle-shaped proprioceptors (Hope and Gardiner, 1982), which may act as stretch receptor organs (Lorenzen, 1978, 1981) sensitive to the dorsoventral bending of the nematode, positioned in the lateral epidermal chords from the pharyngeal region to the caudal region. With the light microscope metanemes can be seen as serially arranged, subcuticular, thread-like structures consisting of a median scapulus (sclerotized wall and associated with a cell) and an anterior and posterior filament. In *Enoplus*, *Deontostoma* and *Oxyonchus* metanemes can even be observed at low magnification.

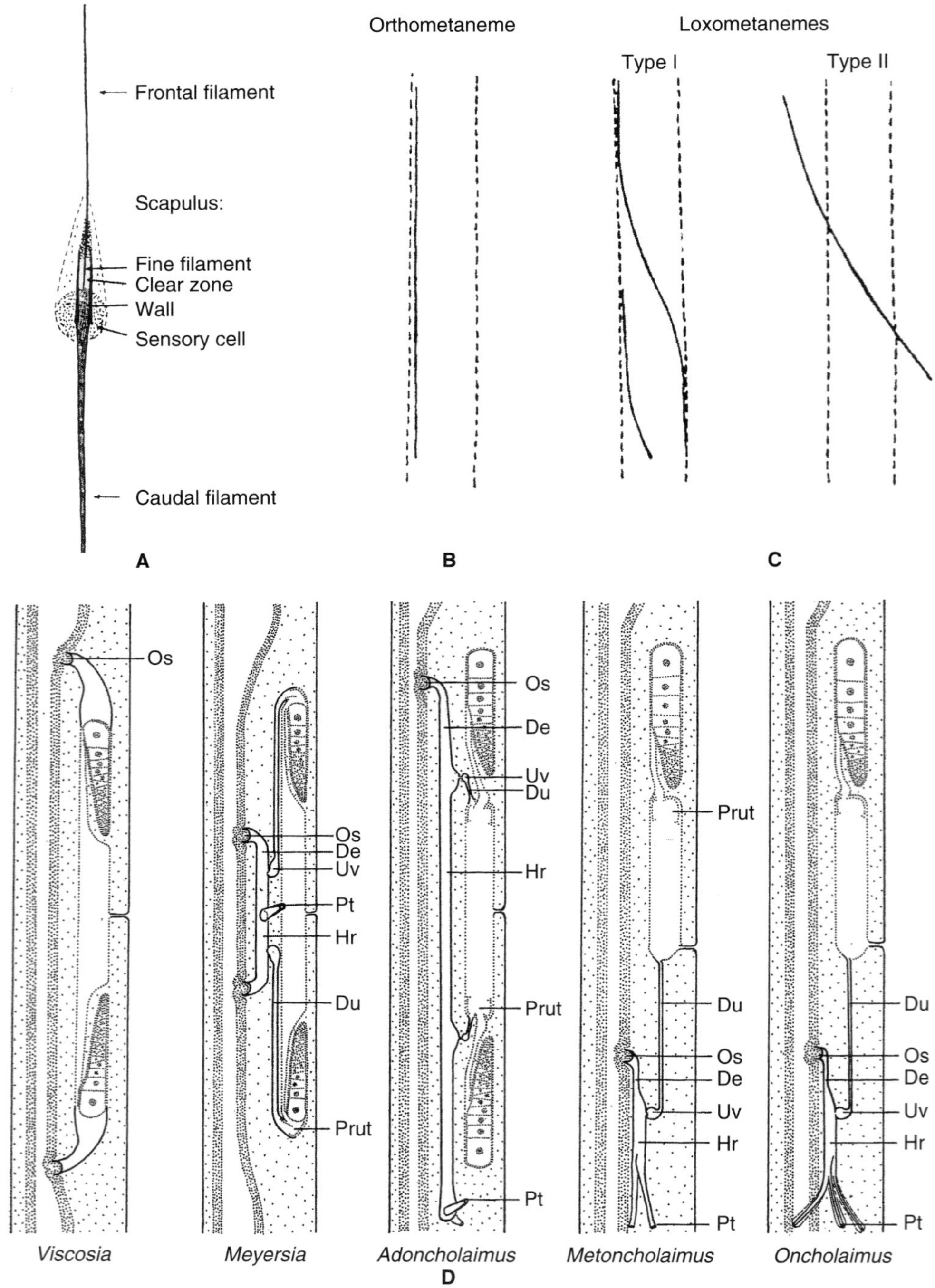

Fig. 12.2 Metanemes and demanian system. A: Detailed structure of a metaneme as seen under light microscope (after Lorenzen, 1994); B and C: different types of metanemes (after Lorenzen, 1994); and D: scheme of the demanian system in *Viscosia, Meyersia, Adoncholaimus, Metoncholaimus* and *Oncholaimus*. De = ductus entericus, Du = ductus uterinus, Hr = main duct, Os = osmosium, Prut = pre-uterus, Pt = terminal pore, Uv = uvette (after Rachor, 1969).

Metanemes run either parallel (*orthometanemes*) or obliquely (*loxometanemes*) to the longitudinal axis of the body. Within the loxometanemes two types can be distinguished: *type I* where the frontal and caudal filaments are strictly confined within the region of the lateral epidermal chord; and *type II* where the frontal filament and often also the caudal filament penetrate into the thin intrachordal epidermis. Orthometanemes always occur on the dorsal and often also at the ventral edge of the epidermal chord: respectively referred to as dorsolateral and ventrolateral orthometanemes. Hope and Gardiner (1982) have shown that a cilium is situated in the core of each spindle: the anterior component is a process from a modified single ciliated epidermal cell, and the posterior component is a dendrite of a nerve cell. They are not convinced that these receptors respond to stretch and compressions as suggested by Lorenzen (1978), however they agree that they are likely to monitor bending of the nematode during locomotion.

Somatic sensilla. Somatic setae are frequently present in Enoplida (except in Alaimina and Campydorina) and are more numerous in the cervical and tail region. Genital sense organs are widespread in the male enoplids (see reproductive system).

Glandular system

Epidermal glands of varying structure are often observed. They consist of unicellular glands situated in the epidermal chords, the ducts of which show unusual differentiation in the Enoplidae and Phanodermatidae.

The secretory–excretory (S–E) *system* is usually present, consisting of a ventral gland cell positioned in the pharyngeal region or opposite the anterior part of the intestine, a duct, an ampulla and an S–E pore, the latter mostly positioned anterior to the nerve ring.

Caudal secretory glands are typical for the enoplids (except Campydorina and Alaimina). These glands consist of mostly three (rarely two or more than three) unicellular structures with a large nucleus and are mostly situated in the tail and open to the exterior through a terminal (rarely subterminal) valve, known as the spinneret. In most representatives of the Enoplina and Oncholaimina the glands extend far into the pre-caudal body region. The secretion of these glands serves for attaching the nematode temporarily to the substratum.

Digestive system

Enoplids have either three or six lips. The buccal cavity varies considerably in size and shape and is a valuable taxonomic characteristic, reflecting the diverse feeding habits and trophic positions of the species. The buccal cavity can be absent, small or large, with or without strong sclerotized walls and with or without armature. The armature varies from small denticles to large teeth (which occasionally can be protruded) or mandibles (claws). A toothed buccal cavity is found in many enoplids, often three teeth are present and each of these is positioned on one of the three buccal walls: one dorsal and two ventrosublateral. In the Oncholaimina the terms onchium/onchia are often used in the literature, but for the sake of uniformity we use the terms tooth/teeth.

Most enoplids have a cylindrical or conoid pharynx, which occasionally has single (*Syringolaimus*) or multiple (*Belbolla, Polygastrophora*) terminal bulbs. In the Alaimina the anterior part of the pharynx is slender and the posterior third is enlarged. The pharyngo-intestinal valve is generally imbedded within the anterior intestinal cells; rarely cardiac gland cells are present.

Generally three to five pharyngeal glands have been described (except seven glands in Alaimina, see below), which open near or in the buccal cavity, often through teeth. However, in a number of species the openings of the pharyngeal glands have not yet been studied in detail.

Rectal glands may be present or absent; if present three are observed in females and six in males; they are, however, often overlooked in descriptions.

Pre-rectum absent.

Reproductive system

Most enoplids are amphimictic species, some are parthenogenetic. Commonly, females are didelphic–amphidelphic with antidromously reflexed (outstretched in *Cytolaimium exile*) ovaries; monodelphic females occur regularly, mostly prodelphic, rarely opisthodelphic (Oxystominidae, Alaimidae). Males are commonly diorchic with opposed testes. Males have a variety of secondary sexual organs: they may occur as a single tubular medioventral supplement, pre- and/or post-cloacal papillae or setae, positioned on a medioventral line or circumcloacal, or as winged supplements.

The *demanian system* (Fig. 12.2D), unique to the Oncholaimidae, is a connection between the reproductive and digestive systems. It has been described in different genera of the Oncholaimidae.[1] Unique within Nematoda is traumatic insemination (Maertens and Coomans, 1979; Chabaud *et al.*, 1983; Coomans *et al.*, 1988) referring to copulation not occurring through the vulva, but through puncturing of the cuticle followed by the formation of terminal ducts giving connection with the blind ending of a part of the demanian system. There is a high diversity in the structure (and function) of the demanian system varying from simple in *Viscosia* to highly developed in *Adoncholaimus* and *Oncholaimus,* or it can be totally absent. The system consists of a duct connecting the ovary with the intestine through a specialized part in the intestinal wall (*osmosium*). The uterus is connected via a uterine duct with the main duct through the uvette, the main duct with the intestine through the *osmosium* and one or more teminal ducts connecting the main duct with the cuticle pore(s).

In *Viscosia* (and *Oncholaimellus*) the demanian system consists of a sac-like structure at the reflexion point of the ovary, which is connected to the intestine by the *osmosium.*

In *Adoncholaimus* a short *ductus uterinus* connects the uterus with the main duct through the uvette and a long dorsal main duct ends in the region of the cloaca;

---

[1] de Man (1884, 1889); zur Strassen (1894); Stewart (1906); Cobb (1920, 1932); De Coninck and Schuurmans-Stekhoven (1933); Kreis (1934); Gerlach (1965); Rachor (1969); Heyns and Coomans (1977); Maertens and Coomans (1979); Calcoen and Dekegel (1979, 1980); Chabaud et al. (1983); Coomans and Heyns (1983, 1986); Coomans et al. (1988).

there are two or more terminal ducts (surrounded by many epidermal gland cells) and terminal pores.

In *Oncholaimus* with prodelphic females, the demanian system is post-vulvar, with only one *osmosium* and uvette present. The terminal duct can end blindly in virgin females; number of terminal pores and ducts is variable; glands are present near the posterior part of the main duct.

The following hypotheses are formulated for the function of the demanian system depending on the species under consideration: (i) a *receptaculum seminis* where spermatozoa are kept alive by secretions from the *osmosium*; (ii) a secretory organ producing substances that serve as a sex attractant or to protect deposited eggs; (iii) increasing the turgor pressure facilitating egg deposition and transfer of sperm to the intestine; and (iv) increasing the genetic diversity by elimination of superfluous sperm coupled with an additional food source for the female digesting this sperm.

Classification

The following classification is based on De Ley and Blaxter (2004) from order down to family level and on Lorenzen (1981, 1994) from family down to genus level, supplemented with more recent classifications at genus level. Genera which occur in freshwater are printed in bold.

According to Lorenzen the monophyly (= holophyly) of the order Enoplida is established by the presence of metanemes as a derived character. The metanemes have not been found outside the Enoplida and this character was used by him as a systematic criterion (Lorenzen, 1981, 1994).

According to Lorenzen (1981, 1994) the order Enoplida consists of two suborders: the extensive and monophyletic Enoplina, and the relatively small, paraphyletic Tripyloidina. Members of the Trefusiina were included in the Enoplia as the paraphyletic order Trefusiida (representing the non-monophyletic remains of the two holophyletic orders Enoplida and Dorylaimida).

Recent molecular data (Rusin *et al.*, 2001) have revealed a close relationship between *Trefusia zostericola* and Enoplida. Therefore, Trefusiida have been lowered to the rank of suborder Trefusiina. Another controversial position was that of the enigmatic genus *Campydora* Cobb, 1920. In the past it was mostly considered to belong to Dorylaimida, except by Siddiqi (1983) who regarded it as belonging to Enoplida. Molecular data (Mullin *et al.*, 2003) confirmed that *Campydora* belongs to Enoplida. Likewise the position of Alaimina inside Enoplida has been confirmed by molecular data (unpublished data by J. Vanfleteren and P. De Ley, see De Ley and Blaxter, 2002).

The order Enoplida has seven suborders: Enoplina, Trefusiina, Oncholaimina, Ironina, Tripyloidina, Campydorina and Alaimina.

For details and references about the localities where species were found before 1974 we refer to Gerlach and Riemann (1974); below are added the details from thereafter.

Jacobs (1984) gives a review of the inland aquatic nematodes of Africa and Andrássy (1978) of Europe.

Order ENOPLIDA Filipjev, 1929

Diagnosis. Amphids typically non-spiral with amphidial fovea pocket-like, rarely spiral. Cuticle light microscope mostly smooth or weakly striated. Anterior sensilla: six inner labial papillae or short setae + six outer labial setae + four cephalic setae in two or three circles, the setae can be jointed. Labial region often divided into three lips. Pharynx attached to the wall of the buccal cavity and/or to the body wall, cylindrical, sometimes slightly widening at the base (only in *Syringolaimus* and *Campydora* a posterior muscular bulb). Female reproductive system typically didelphic, amphidelphic with antidromously reflexed ovaries (outstretched ovaries only in *Cytolaimium exile*). Male reproductive system diorchic with opposed testes. Caudal glands when present open through a common pore.

The dorsal and two ventrosublateral pharyngeal glands open in each of the pharyngeal sectors either in or directly posterior to the buccal cavity (but in Alaimina described as opening posterior to the nerve ring).

Metanemes present or absent (the monophyly of Enoplida based on the presence of metanemes no longer holds as proved by molecular analysis).

Mainly marine taxa, only a few freshwater taxa.

An overview of diagnostic characters of suborders is given in Table 12.1 and the head ends of typical genera are represented in Fig. 12.3A–G.

Suborder ENOPLINA Chitwood & Chitwood, 1937 (Figs 12.3A, 12.4A–I and 12.5A–I)

Diagnosis. Enoplida. Metanemes present. Anterior sensilla in two or three circles: six inner labial papillae or setae, six outer labial papillae or setae and four cephalic papillae or setae, the outer labial and cephalic sensilla mostly together in one circle. Amphidial aperture a transverse slit or oval, fovea pocket-shaped. Buccal cavity of variable shape, mostly well developed and armed and completely embedded in pharyngeal tissue. Pharynx cylindrical, slightly enlarged posteriorly; outer wall sometimes crenate. Pharynx musculature inserts on the body wall in the region of the buccal cavity; inner layer of body wall differentiated into a cephalic capsule. The dorsal and two ventrosublateral pharyngeal glands open in the stomatal region. Female reproductive system mostly didelphic–amphidelphic with antidromously reflexed ovaries. Caudal glands generally extending pre-caudally (exceptions: *Enoplus*, Anticomidae, Halalaiminae, partly Ironidae).

Only one superfamily Enoploidea.

Superfamily ENOPLOIDEA Dujardin, 1845

Diagnosis. Three types of metanemes: orthometanemes and loxometanemes of type I and II present, mostly positioned dorsolaterally, less frequently also ventrolaterally. Outer pharyngeal wall smooth or crenate by alternate sequence of muscular and glandular areas. When a secretory–excretory gland is present it is confined to the pharyngeal region. Anterior and posterior gonads positioned at the left side of the intestine. Males with a copulatory supplement consisting of a single ventral pre-

Tabel 12.1 An overview of the diagnostic characters of the suborders of Enoplida.

Suborders	Enoplina	Trefusiina	Oncholaimina	Ironina	Tripyloidina	Campydorina	Alaimina
Metanemes	Present	Absent	Present	Present	Present	Absent	Absent
Anterior sensilla	Papillae and setae 6 + 10 or 6 + 6 + 4	Papillae and setae 6 + 6 + 4, rarely 6 + 10	Papillae and setae 6 + 10	Papillae and setae 6 + 10	Papillae and setae 6 + 10	Only papillae 6 + 10[a]	Only papillae 6 + 10
Amphid (fovea)	Pocket	Pocket, rarely spiral	Pocket	Pocket	Spiral	Pocket	Pocket or pore
Cephalic capsule	Present	Absent	Absent	Absent	Absent	Absent	Absent
Buccal cavity	Mandibles, or mandibles + teeth or spear	Conical or barrel-shaped, sometimes minute, toothless	Large, prismatic, heavily sclerotized walls, mostly three teeth, sometimes only one or more teeth	Elongated, 3–5 movable teeth anteriorly or poorly developed and toothless or with tooth-like structure anteriorly	One or more portions; teeth common, dorsal tooth may be dominant	Mural tooth in dorsal position	Vestigial, unarmed
Pharynx	Cylindrical, conoid	Cylindrical, conoid	Cylindrical, conoid	Cylindrical	Cylindrical, conoid	Cylindrical with elongated triquetrous lumen in terminal bulb	Slender with enlarged posterior part

Continued

Tabel 12.1 *Continued.* An overview of the diagnostic characters of the suborders of Enoplida.

Suborders	Enoplina	Trefusiina	Oncholaimina	Ironina	Tripyloidina	Campydorina	Alaimina
Secretory–excretory system	In pharyngeal region	Absent, but sometimes a secretory–excretory pore	In post-pharyngeal region	Limited to neck region or extension tube backwards	Present or absent	Pore and duct cuticularized	Only secretory–excretory pore observed, small, often obscure
Female gonads	Two	Two, vulva can merge with anus	One or two, demanian system present or absent	Mostly two	Two	Two, mostly asymmetrical	Two or one
Male gonads	Two	Usually two	Two	Two	One	One	One
Male supplements	One tube pre-cloacal	Present or absent	Pericloacal papillae or setae	Setiform	Present or absent, papilliform	Papilliform, leptoderan alae	Papilliform
Caudal glands	Present, extending pre-caudally	Present	Present, extending pre-caudally	Present, extending pre-caudally or confined to the tail	Present, confined to the tail	Absent	Absent
Habitat	Marine, one genus also in freshwater	Marine	Marine, brackish and two genera in freshwater	Marine, one genus in freshwater	Marine, rarely penetrating in freshwater	Terrestrial	Terrestrial, some genera in freshwater

[a]Cobb (1920) noted 6 ± 12, which is probably a mistake.

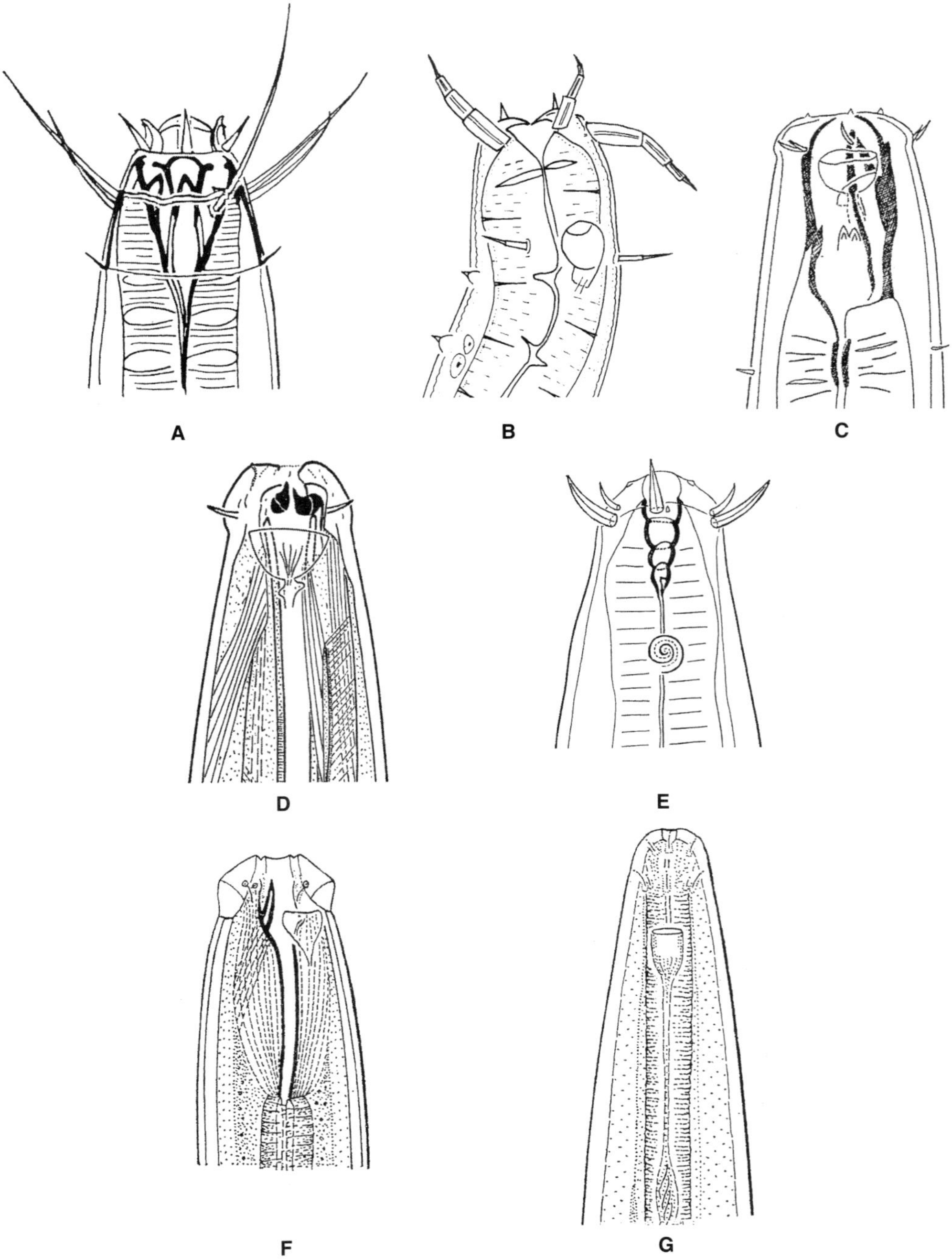

Fig. 12.3 Overview of the suborders of Enoplida represented by the head end of typical genera. A: Enoplina (after Wieser, 1953a); B: Trefusiina (after Vincx and Vanreusel, 1989); C: Oncholaimina (after Smol and Sharma, 1984); D: Ironina (after Van Der Heiden, 1974); E: Tripyloidina (after Schuurmans-Stekhoven and De Coninck, 1933); F: Campydorina (after Thorne, 1939); and G: Alaimina (after Thorne, 1939).

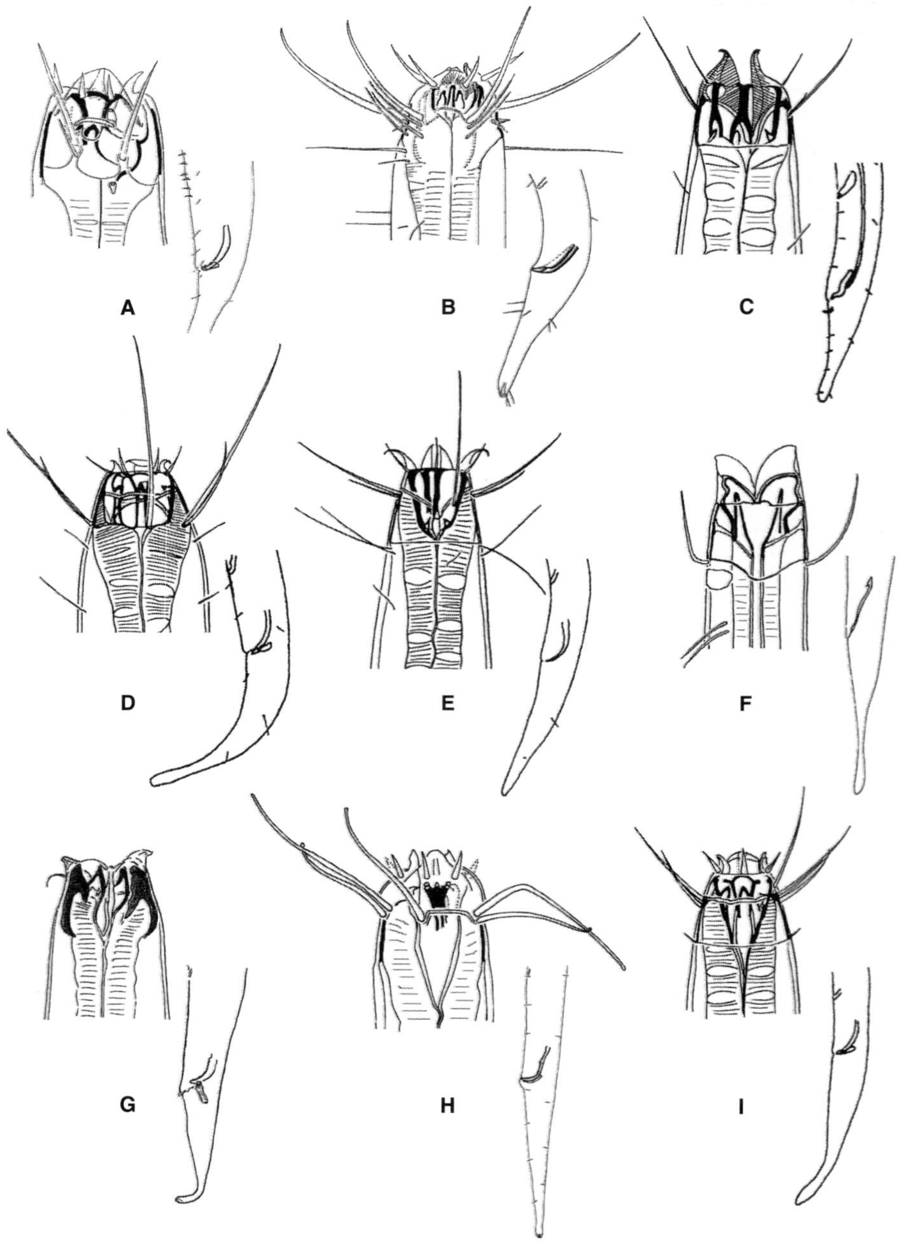

Fig. 12.4 Genera of the Enoplina, Enoplidae and Enoplolaiminae; each genus is represented by the head end and the tail. A: *Africanthion* (after Inglis, 1964); B: *Cryptenoplus* (after Riemann, 1966); C: *Enoploides* (after Wieser, 1953a); D: *Enoplolaimus* (after Wieser, 1953a); E: *Epacanthion* (after Wieser, 1953a); F: *Fenestrolaimus* (after Filipjev, 1927); G: *Filipjevia* (after Kreis, 1928); H: *Fleuronema* (after Greenslade and Nicholas, 1991); and I: *Mesacanthion* (after Wieser, 1953a).

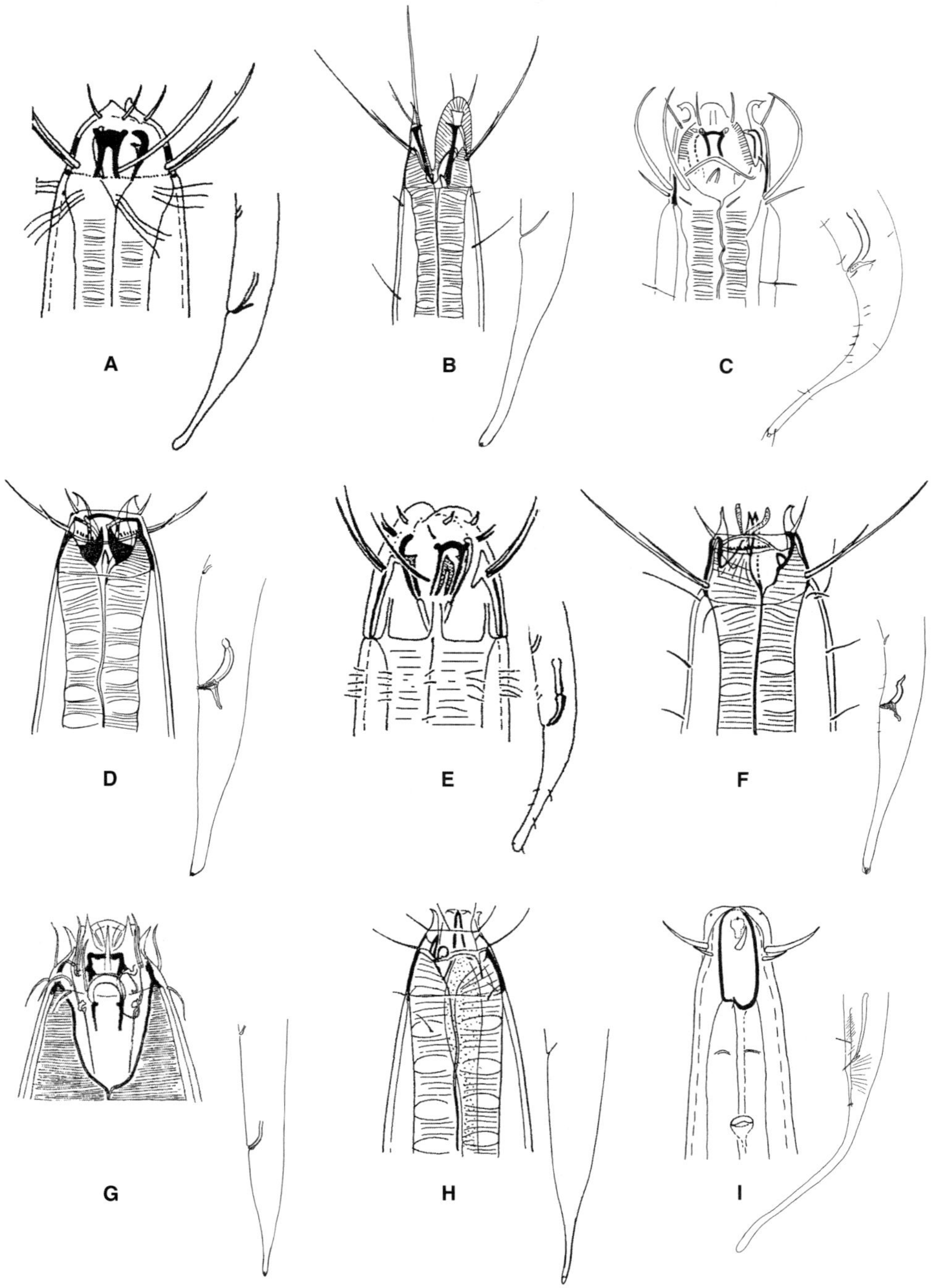

Fig. 12.5 Genera of the Enoplina (continued): Enoplidae, Enoplolaiminae (A–H) and Anoplostomatidae, Anoplostomatinae (I). A: *Mesacanthoides* (after Platt and Warwick, 1983); B: *Metenoploides* (after Wieser, 1953a); C: *Okranema* (after Greenslade and Nicholas, 1991); D: *Oxyonchus* (after Wieser, 1953a); E: *Paramesacanthion* (after Platt and Warwick, 1983); F: *Parasaveljevia* (after Wieser, 1953a); G: *Parenoplus* (after Filipjev, 1927); H: *Saveljevia* (after Wieser, 1953a); and I: *Anoplostoma* (after Platt and Warwick, 1983).

cloacal tubule (rarely absent and nine tubules in *Epacanthion multipapillatum*). The Enoploidea are marine, only isolated species occur in freshwater.

The Enoploidea consist of five families: Enoplidae, Thoracostomopsidae, Anoplostomatidae, Phanodermatidae and Anticomidae (Lorenzen, 1981, 1994). Molecular analysis corroborates the position of Anoplostomatidae within the Enoploidea and suggests a close relationship to the Enoplidae (Pegova *et al.*, 2004).

Freshwater species only occur within the Thoracostomopsidae and Anoplostomatidae.

Key to families

1. Buccal cavity large and spacious 2
   Buccal cavity small 4
2. Buccal cavity toothless . . . . .Anoplostomatidae
   Buccal cavity with mandibles, teeth or spear 3
3. Buccal cavity with only mandibles . . . . .Enoplidae
   Buccal cavity with mandibles + teeth or a spear . . . . .Thoracostomopsidae
4. Grouped lateral cervical setae present, secretory–excretory pore and duct prominent, spicules short . . . . .Anticomidae
   No such cervical setae present, spicules long . . . . .Phanodermatidae

Family ENOPLIDAE Dujardin, 1845

Diagnosis. Cuticle smooth or with oblique rows of fine dots (*Enoplus brevis* and *E. michaelseni*). Large epidermal glands with outlets of different shape. Dorsolateral and ventrolateral loxometanemes type II in adults, juveniles only dorsolateral loxometanemes type I; caudal filament sometimes present (no metanemes in *Enoplus schulzi*). Lips low. Inner labial sensilla papilliform, outer labial and cephalic sensilla setiform and in one circle. Amphids either at level of or posterior to cephalic capsule. Cephalic organs (cephalic slits) look like introverted cirrus and positioned ventrofrontally to the lateral cephalic setae. Eyespots present or absent. Anterior pharyngeal muscles insert on to cephalic capsule. Well-developed cephalic capsule. Buccal cavity only with three mandibles, no teeth. Dorsal and two ventrosublateral pharyngeal glands open at the base of the buccal cavity. Outer wall of pharynx smooth. Ventral gland lies constantly to the left of the pharynx. Female reproductive system didelphic–amphidelphic with antidromously reflexed ovaries. Males diorchic with opposed testes, rarely monorchic (one anterior testis in *E. schulzi*). One large pre-cloacal tubiform supplement. Caudal glands completely in the tail in females and juveniles, whereas in males they penetrate into the pre-caudal region. Marine.

Type and only genus: *Enoplus* Dujardin, 1845

For revision see Wieser (1953a).

Family THORACOSTOMOPSIDAE Filipjev, 1927

Diagnosis. Lips high. Only dorsolateral orthometanemes with a robust scapulus but no caudal filament. Inner labial sensilla robust and setiform (papilliform only in *Fenestrolaimus*), outer labial and cephalic setae robust and long. Epidermal glands with particularly well-differentiated outlet. Inner layer of cuticle forms a cephalic capsule on to which pharyngeal muscles are attached. Cephalic organs often present and of variable shape, situated frontally or ventrofrontally to the lateral setae. Amphids small and situated posterior to the cephalic capsule or absent. Spacious buccal cavity with three mandibles and three teeth (one dorsal and two ventrosublateral) or with one long eversible spear. Female reproductive system didelphic–amphidelphic with antidromously reflexed ovaries (a single posterior ovary in *Mesacanthion monhystera* only). Caudal glands penetrate into the pre-caudal region.

The family Thoracostomopsidae consists of the subfamilies Thoracostomopsinae, Trileptiinae and Enoplolaiminae. Only subfamily Enoplolaiminae has representatives in freshwater.

Key to subfamilies

1. Buccal cavity with a long eversible spear .... Thoracostomopsinae
2. Buccal cavity with three teeth of equal size situated well anteriorly, mandibles small or absent .... .Trileptiinae
3. Buccal cavity with three mandibles and three teeth .... .Enoplolaiminae

Subfamily Thoracostomopsinae Filipjev, 1927

Diagnosis. Thoracostomopsidae. Buccal cavity with a long eversible spear made up of the elements of the three buccal cavity sectors. Three pharyngeal glands open dorsally and ventrosublaterally at the jointed part of the spear. Marine.

Type (and only) genus: *Thoracostomopsis* Ditlevsen, 1918

For revision see Inglis (1964).

Subfamily Trileptiinae Gerlach & Riemann, 1974

Diagnosis. Thoracostomopsidae. The three teeth of equal size are situated well anterior into the buccal cavity. Pharyngeal glands open through the teeth. Mandibles small or absent. Only two caudal glands, which penetrate into the pre-caudal region. Marine.

Type and only genus: *Trileptium* Cobb, 1933

Subfamily Enoplolaiminae De Coninck, 1965

Diagnosis. Thoracostomopsidae. Lips high. Buccal cavity always with three mandibles and three teeth. One mandible and one tooth together form a unit, which can be moved back and forth by specialized pharyngeal muscles, whereby the frontal section of the unit is moved in line with the body axis. A pharyngeal gland opens through each tooth. All three teeth have the same length or the dorsal tooth is distinctly smaller than the two ventrosublateral teeth; the two ventrosublateral teeth always are equal in length. The genera are differentiated by the degree of development of the mandible–teeth complex.

This subfamily needs to be revised, as the complex structure of the buccal cavity and the head as well as the cephalic organs have been insufficiently described and misunderstood by some authors and many species need to be redescribed. Until now only species of the genera *Enoploides* and *Mesacanthion* have been found in freshwater.

Type genus *Enoplolaimus* de Man, 1893

Enoplolaiminae consist of 18 genera, including the genus *Hyptiolaimus,* of which the type species was considered as *species inquirendum* (*sp. inq.*) by Wieser (1953a).

Key to genera (excluding *Hyptiolaimus*)

The genera can be distinguished from each other by the following characters: the relative size of the mandibular teeth and their length with regard to the anterior mandibular rim (or bar), the shape and degree of sclerotization of the mandibles, the position of the second circle of anterior sensilla (although this character is variable within one genus), the position of the sub-cephalic setae, the structure of the lips and the shape of the spicules and gubernaculum.

1. Mandibles reduced, teeth unequal . . . . *Saveljevia*
   Mandibles existing of two longitudinal bars united by a thin sheath . . . . *Epacanthion*
   Mandibles solid 2
   Mandibles arch-shaped (two longitudinal rods anteriorly united by a bar) 6
2. Mandibular teeth equal 3
   Mandibular teeth unequal 5
3. Lips striated . . . . *Enoploides*
   Lips unstriated 4
4. Mandibles extremely long, teeth reduced in size . . . . *Metenoploides*
   Mandibles and teeth of normal size . . . . *Mesacanthoides*
5. Dorsal tooth reduced, mandibles with long posterior apophysis . . . . *Filipjevia*
   Dorsal tooth absent, mandibles anteriorly with central claw in addition to lateral claws . . . . *Fleuronema*
6. Teeth absent, lips plicate . . . . *Parenoplus*
   Teeth present and equal in length 7
   Teeth present and unequal in length 11
7. Teeth arch-shaped, extending anterior to the mandibular bar . . . . *Cryptenoplus*
   Teeth shorter than the mandibles 8
8. Teeth with broad base and spine-shaped, anterior bar of mandibles forming three large anteriorly directed bows . . . . *Fenestrolaimus*
   Teeth not so and anterior bar of usual shape 9
9. Cuticle very thick with distinct 'shoulder',[2] outer labial and cephalic setae at posterior edge of cephalic capsule . . . . *Okranema*
   Cuticle of same thickness all over the body 10
10. Outer labial and cephalic setae positioned anterior to cephalic capsule, spicules in two articulating parts . . . . *Paramesacanthion*
    Outer labial and cephalic setae positioned at middle or in anterior half of cephalic capsule . . . . *Mesacanthion*
    Outer labial and cephalic setae positioned at posterior edge of cephalic capsule . . . . *Enoplolaimus*
11. Teeth slightly unequal and far posterior, males with a ventral file of stout setiform supplements . . . . *Africanthion*

---

[2] A thickening of the cuticle below the head.

Dorsal tooth small, two large ventrosublateral teeth extend anterior to mandibular bar 12

12. Mandibular plate with denticles at inner surface . . . . *Oxyonchus*
Mandibular plate with or without denticles . . . . *Parasaveljevia*

Genus *Africanthion* Inglis, 1964 (Fig. 12.4A)
Diagnosis. Enoplolaiminae. Mandibles with lateral processes (claws) very well developed and mandibular walls fairly narrow in optical section; mandibular plate thin; teeth slightly unequal: dorsal smaller than ventrosublateral; teeth lying far posterior to mandibles. Cephalic setae arising from the middle of cephalic capsule. Male spicules short and stout; gubernaculum small and complex; pre-cloacal supplement replaced by a file of stout, short setae. Marine.

Type and only species: *A. nudum* Inglis, 1964

Genus *Cryptenoplus* Riemann, 1966 (Fig. 12.4B)
Diagnosis. Enoplolaiminae. Three high striated lips. Mandibles existing of two strongly sclerotized lateral rods with claws anteriorly and a weakly sclerotized part situated interlabial. Teeth modified into narrow arch-shaped tooth-like structures, which rise above the mandibles. Sub-cephalic setae present. Marine.

Type and only species: *C. gerlachi* Riemann, 1966

Genus ***Enoploides*** Ssaweljev, 1912 (Fig. 12. 4C)
Diagnosis. Lips high, striated. Buccal cavity with well-developed solid mandibles with claw-like anterior; mandibles not extremely slender (ratio length/width <6); teeth shorter than the mandibles; spicules usually long. Mostly marine, two freshwater species.

Type species *E. typicus* Ssaweljev, 1912

For revision see Wieser (1953a) and Wieser and Hopper (1967).

Species recorded from freshwater habitats:

E. fluviatilis Micoletzky, 1923 (Volga river, USSR) (Filipjev, 1928, Oka River, USSR; Gagarin, 1981, Caspian Sea, Rivers Volga, Dniepr, Don; Gagarin, 2001, River Volga, water reservoirs in Kiev and Kremenchug, Russia).
E. stewarti Nicholas, 1993 (Lake Alexandrina, South Australia) (Marchant and Lillywhite, 1994, freshwater streams in Macquarie island, sub-Antarctica).

Genus *Enoplolaimus* de Man, 1893 (Fig. 12.4D)
Diagnosis. Enoplolaiminae. Cuticle smooth or striated and punctated. Buccal cavity with mandibles with claw-like anterior; mandibles arch-shaped, consisting of two pieces, which are united by an anterior bar only. Teeth shorter than the mandibles. Outer labial and cephalic setae situated at posterior end of cephalic capsule. Marine and two brackish water species: *E. balgensis* Skwarra, 1921(Baltic) and *E. afflongicaudatus* (Southern, 1914) found by Galhano (1970) in a mixo-oligohaline estuary in Portugal.

Type species *E. vulgaris* de Man, 1893

For revision see Wieser (1953a) and Hopper (1962).

Genus *Epacanthion* Wieser, 1953 (Fig. 12.4E)
Diagnosis. Enoplolaiminae. Cuticle usually smooth. Head broadly wedge- or cone-shaped. Lips high, mostly striated. Inner labial setae long and inserted at the base of lip flaps; outer labial and cephalic setae situated at middle or anterior end of cephalic capsule. Cervical setae often present and can be numerous in males, which exhibit sexual dimorphism. Mandibles consisting of two plate-shaped columns (usually long and parallel) separated by a thin sheet of cuticle (space between columns not solid) and only connected anteriorly by a bar[3] (an intermediate stage between *Enoploides* and *Mesacanthion*); mandibular teeth small with gland opening at tip. Pharynx relatively long and cylindrical; cardia pyriform. Females didelphic–amphidelphic with reflexed[4] ovaries at left side of the intestine. Males diorchic with both testes at left side of the intestine. Spicules mostly long ($\geq$ 2.5 anal diameters long) or short; gubernaculum without apophysis present or absent. Pre-anal supplement present or absent. Three caudal glands, cells pre-caudally. Tail narrowly conical or attenuated. Marine.

Type species *E. buetschlii* (Southern, 1914)

For revision see Greenslade and Nicholas (1991).

Genus *Fenestrolaimus* Filipjev, 1927 (Fig. 12.4F)
Diagnosis. Enoplolaiminae. Body strongly attenuated anteriorly. Cuticle smooth. Tail conical. Amphid pocket-shaped with round aperture. Mandibles arch-shaped, the edges of the arch thinly sclerotized. Three teeth broad at the base of the buccal cavity and strongly attenuating anteriorly to become spine-shaped. Marine.

Type species *F. insulaealbae* Filipjev, 1927

Genus *Filipjevia* Kreis, 1928 (Fig. 12.4G)
Diagnosis. Enoplolaiminae. Lips high, all anterior sensilla setose. Mandibles solid with posterior apophyses and with teeth. Dorsal tooth reduced. Two large ventro-sublateral teeth not surpassing the mandibles. Marine.

Type and only species: *F. macrolabiata* Kreis, 1928

Genus *Fleuronema* Greenslade & Nicholas, 1991 (Fig. 12.4H)
Diagnosis. Enoplolaiminae. Large specimens. Cuticle thick and smooth. Head blunt but apically rounded, bulbous, with distinct external groove at the base of head capsule. Lip flaps low and finely striated, lips low. Mandibles solid, anterior bar with

---

[3] In contrast to Greenslade and Nicholas (1991) we accept the presence of a solid connection between the mandibular columns because Wieser (1953a) erected the new genus to accommodate the new species *E. microdentatum* of which the two mandibular plates are anteriorly united by a bar. He designated *E. bütschlii* as type species; if, later, the type species does not fit the genus diagnosis, the type species should be changed. Moreover we doubt the existence of outstretched ovaries within the family Thoracostomopsidae.
[4] Outstretched ovaries according to Greenslade and Nicholas (1991).

central claw in addition to the two lateral claws, each with an apical lateral projection; ventrosublateral teeth large, blunt and slanting, reaching as far as cephalic arch, dorsal tooth absent. Inner labial setae short and stout, outer labial and cephalic setae long and slightly clavate, with surface ornamented by meshwork, inserted just above the base of head capsule. Posterior margin of head capsule indistinct. Pharynx long and cylindrical; cardia pyriform; nerve ring about one-third of pharynx length. Female gonad on left of intestine, outstretched. Spicules heavy and encircled by a distinct ridge, which divides spicules into half; gubernaculum and pre-cloacal supplement present. Three caudal glands with cell pre-caudally. Tail conical with rounded tip. Marine.

Type species *F. dorca* Greenslade & Nicholas, 1991

Genus ***Mesacanthion*** Filipjev, 1927 (Fig. 12.4I)
Diagnosis. Enoplolaiminae. Outer labial and cephalic setae situated at middle or anterior end of cephalic capsule. Mandibles well developed, provided with claws, arch-shaped, consisting of two rod-like columns anteriorly united by a curved bar. Teeth shorter than mandibles. Spicules mostly short, if long (*M. diplechma*) then gubernaculum with caudal apophysis. One freshwater species and one brackish water species: *M. longispiculum* Gerlach, 1954 (described as *M. cf. longispiculum* and found in oligohaline lagoon in Madagascar by Gerlach, 1958).

Type species *M. luciferum* (Filipjev, 1927)

For revision see Wieser (1953a).

Species recorded from freshwater habitat:

M. alexandrinum Nicholas, 1993 (Lake Alexandrina, South Australia).

Genus *Mesacanthoides* Wieser, 1953 (Fig. 12.5A)
Diagnosis. Enoplolaiminae. Transition between *Mesacanthion* and *Enoploides*. Lips not striated. Mandibles solid with claws. Teeth shorter than mandibles. Marine.

Type species *M. sculptilis* Wieser, 1953

For revision see Wieser and Hopper (1967).

Genus *Metenoploides* Wieser, 1953 (Fig. 12.5B)
Diagnosis. Enoplolaiminae. Lips very high and deeply cut. Mandibles solid, extremely long, exceeding at least ten times the corresponding width at the middle of the shaft. Teeth reduced in size. Outer labial setae two head diameters long or more. Marine, two species, only females and juveniles known.

Type species *M. alatus* Wieser, 1953

Genus *Okranema* Greenslade & Nicholas, 1991 (Fig. 12.5C)
Diagnosis. Enoplolaiminae. Body short, broad, slightly hook-shaped when preserved. Cuticle very thick and annulated, distinct shoulder at the base of head capsule where the cuticle sharply increases in thickness. Head square. Lip flaps well developed

and with fine ridges forming striations along their edges; cephalic setae inserted just above the base of head capsule; short sub-cephalic setae. Mandibular teeth S- or hook-shaped; mandibular columns distinct, arch anteriorly between teeth; labial setae fairly long but stout. Pharynx cylindrical; cardia pyriform. Female gonad to left of intestine. Spicules with thick walls, about as long as cloacal diameter; S-shaped gubernaculum; pre-cloacal supplement absent. Caudal glands pre-caudally. Tail extremely short and broad. Marine.

Type species *O. eileenae* Greenslade & Nicholas, 1991

Genus *Oxyonchus* Filipjev, 1927 (Fig. 12.5D)
Diagnosis. Enoplolaiminae. Lips high or low. Cuticle smooth or striated. Cephalic organ present or absent. Mandibles well developed, arch-shaped, rods connected by broad transverse bar with claws, denticles can be present at the inner surface of the mandibular plate. Teeth unequal; two large ventrosublateral teeth that extend to the anterior end of the mandibles. Dorsal tooth small. Spicules short (1–2 cloacal body diameters). Pre-cloacal supplement present or absent. Gubernaculum with or without apophysis. Marine.

Type species *O. hamatus* (Steiner, 1916) Filipjev, 1927

For revision see Keppner (1988) and Nicholas (2004).

Genus *Paramesacanthion* Wieser, 1953 (Fig. 12.5E)
Diagnosis. Enoplolaiminae. Outer labial and cephalic setae in front of anterior end of cephalic capsule, about the level of anterior end of mandibles. Sub-cephalic setae at middle of cephalic capsule. Mandibles arch-shaped and with claws, consisting of two pieces united by an anterior bar only. Teeth shorter than mandibles. Spicules consisting of two portions, a distal and a proximal one articulating with each other. Males with or without supplement. Sexual dimorphism in the pilosity of the head. Marine.

Type species *P. klugei* (Filipjev, 1927) Wieser, 1953

For revision see Warwick (1970) and Boucher (1970).

Genus *Parasaveljevia* Wieser, 1953 (Fig. 12.5F)
Diagnosis. Enoplolaiminae. Lips high. All anterior sensilla setose. Similar to *Saveljevia* but with well-developed mandibles; dorsal tooth small or reduced, two large ventrosublateral teeth surpassing anterior end of mandibles. Cirri-shaped cephalic organs present or absent. Denticles on mandibular plates present or absent. Marine.

Type species *P. clavicauda* (Filipjev, 1925b)

Wieser (1953a) described two *Parasaveljevia* species with cirrus-shaped cephalic organs and denticles on the mandibular plates, stating that the presence of cirrus-shaped cephalic organs and denticles on the mandibular plate point towards a close relationship with *Oxyonchus*. This renders the distinction between the two genera unclear.

Genus *Parenoplus* Filipjev, 1927 (Fig. 12.5G)
Diagnosis. Enoplolaiminae. Lips high and plicate. Inner and outer labial and cephalic sensilla setiform. Cephalic capsule short. Mandibles arch-shaped without claws and teeth (visible in juveniles only). Marine.

Type species *P. edentatus* Filipjev, 1927

For revision see Wieser (1953a).

Only two species known. There seems to be a discrepancy between the type species and *P. serratus* described by Wieser: in *P. edentatus* the mandibles are arch-shaped and in *P. serratus* the mandibles are solid. Furthermore, there is a lapsus in the size of the cephalic setae of *P. edentatus* as mentioned by Wieser (1953a): according to Filipjev (1927) the six outer labial setae are 25 μm and the four cephalic setae are a little shorter (so not 8 μm as mentioned by Wieser, 1953a).

Genus *Saveljevia* Filipjev, 1927 (Fig. 12.5H)

Diagnosis. Enoplolaiminae. Mandibles reduced or vestigial. Teeth unequal, two ventrosublateral ones surpassing the anterior end of mandibles, dorsal tooth small or reduced. The cephalic ring, however, can be very well developed. Marine.

Type species *S. kolaensis* Filipjev, 1925

For revision see Wieser (1953a).

Family ANOPLOSTOMATIDAE Gerlach & Riemann, 1974

Diagnosis. Enoploidea. Three lips, each with short inner and outer labial sensilla. Buccal cavity spacious, cylindrical, toothless and surrounded by pharyngeal tissue only in the posterior section. The cephalic capsule is not an insertion point of the pharynx musculature. Amphids with small aperture and relatively large fovea (at least in females and juveniles) and always situated behind the cephalic capsule. Only dorsolateral loxometanemes present, predominantly of type I and some of type II; all with caudal filament. Three pharyngeal glands open directly posterior to the buccal cavity. Outline of pharynx smooth. Of the secretory–excretory system only the pore observed. Males diorchic, the posterior testis proceeds forwards a little and is then reflexed. Female reproductive system didelphic–amphidelphic with antidromously reflexed ovaries. Gonads positioned to the left of the intestine. Three caudal glands penetrate into the pre-caudal region.

Anoplostomatidae has been moved from Oncholaimoidea to Enoploidea by Lorenzen (1981, 1994) because of the constant position of the gonads to the left side of the intestine. Anoplostomatidae consists of two subfamilies Chaetonematinae and Anoplostomatinae, only the latter having freshwater species.

For revision see Belogurov and Alekseev (1977).

Key to subfamilies

1. Males with pre-cloacal tubule . . . . Chaetonematinae
   Males with bursa . . . . Anoplostomatinae

Subfamily Anoplostomatinae Gerlach & Riemann, 1974

Diagnosis. Anoplostomatidae. Amphids similar in both sexes. Males with bursa.

Type and only genus: *Anoplostoma* Bütschli, 1874

Genus ***Anoplostoma*** Bütschli, 1874 (Fig. 12.5I)

Diagnosis. Anoplostomatinae. Body narrowing towards both ends. Six inner labial papillae, six outer labial setae and four shorter cephalic setae. Large buccal cavity cylindrical with sclerotized walls and without teeth. Amphids small and situated behind the buccal cavity; aperture a small transverse slit, fovea pocket-shaped. Secretory–excretory gland present. Cardia large. Female reproductive system didelphic–amphidelphic with reflexed ovaries. Males diorchic, with long and slender spicules, short gubernaculum and bursa.

Type species *A. viviparum* (Bastian, 1865) Bütschli, 1874

For revision see Wieser (1953a) and Chitwood (1960).

The genus comprises 13 species, of which one brackish, *A. viviparum* (Bastian, 1865), and one freshwater species. Coomans *et al.* (1985) have found two juveniles of this genus in a freshwater pool with more salts than usual on a coral island in the Solomon Islands.

Species recorded from inland waters:

A. heterurum (Cobb, 1914) Kreis, 1934 (syn. *Oncholaimellus heterurus* Cobb, 1914) (freshwater pond near Ocala, Florida, USA).

A. viviparum (Bastian, 1865) (The Netherlands; Gagarin, 2001, mineral spring, Russia).

Subfamily Chaetonematinae Gerlach & Riemann, 1974

Diagnosis. Anoplostomatidae. Amphids with extreme sexual dimorphism in their structure (formerly the amphids in the male were mistaken for the Steiner's organ cf. Lorenzen (1981, 1994). In many species the four cephalic setae are situated anterior to the six outer labial setae. The males have a pre-cloacal tubule. Marine.

Type and only genus: *Chaetonema* Filipjev, 1927

Family PHANODERMATIDAE Filipjev, 1927

Diagnosis. Enoploidea. Dorsolateral + ventrolateral orthometanemes, a few loxometanemes (type I) and caudal filament absent. Buccal cavity small with one smaller dorsal and two larger ventrosublateral teeth, which all point forwards (it is unclear whether the teeth protrude freely in the buccal cavity or whether they only represent thickenings of the wall). Cephalic capsule strong or weak. Outline of posterior part of pharynx wall crenate. The pharyngeal glands open immediately posterior to the buccal cavity. Excretory–secretory gland usually present and confined to the pharyngeal region. Females didelphic–amphidelphic with antidromously reflexed ovaries. Males diorchic with opposite testes, pre-cloacal tubule present or absent. Marine.

Family ANTICOMIDAE Filipjev, 1918

Diagnosis. Enoploidea. Metanemes variable in form. Buccal cavity small and surrounded by pharyngeal tissue. Cephalic capsule very narrow, pharyngeal musculature inserts into it. Outline of pharynx wall smooth. Pharyngeal glands open immediately posterior to the buccal cavity. Secretory–excretory system positioned left or dorsal (in *Paranticoma* according to Lorenzen (1981, 1994) to the pharynx; pore situated on a spine-like projection in *Paranticoma*. Female reproductive system didelphic–amphidelphic with reflexed ovaries. Males diorchic or a single anterior testis, most species have a pre-cloacal tubule. Caudal glands completely within the tail. Marine.

Suborder TREFUSIINA (Lorenzen, 1981) De Ley & Blaxter, 2002 (Fig. 12.3B)

Diagnosis. Enoplida. Amphids usually non-spiral; spiral in only part of the Trefusiidae (where the *canalis* lies laterally in the body). Outer labial and cephalic sensilla setiform and positioned in two well-separated circles (except in *Trefusialaimus* and Lauratonematidae). Metanemes are absent. Opening of pharyngeal glands unknown in most of the species, but three openings near buccal cavity in *Trefusialaimus*. Caudal glands when present lie completely within the tail or position unknown. All species belonging to this suborder are marine.

The suborder Trefusiina has only one superfamily Trefusioidea comprising the families Simpliconematidae, Trefusiidae, Lauratonematidae and Xenellidae.

There are no records of freshwater species among the Trefusiina. An unidentified *Trefusia* (?) species has been found in the hypersaline Salton Sea, the largest inland lake in California, USA by Warwick *et al.* (2002).

Suborder ONCHOLAIMINA De Ley & Blaxter, 2002 (Figs 12.6A–I, 12.7A–G, 12.8A–G and 12.9A–G)

De Ley and Blaxter (2002) have raised the taxon Oncholaimacea Lorenzen (1981, 1994) to suborder level. Lorenzen (1981, 1994) erected the Oncholaimacea as a different taxon next to the Enoplacea differentiating them from each other by the presence or absence of a cephalic capsule, the type of metanemes, the position of the secretory–excretory gland and the position of the gonads in reference to the intestine.

Diagnosis. Enoplida. Caudal glands penetrate into the pre-caudal region. Secretory–excretory gland extends into post-pharyngeal region.

This suborder has only one superfamily Oncholaimoidea.

Additional information is published by Belogurov (1985).

Superfamily ONCHOLAIMOIDEA Filipjev, 1916

Diagnosis. Oncholaimina. Cuticle smooth or striated (*Oncholaimoides*). Inner labial sensilla usually papilliform (setiform in *Pontonema ardens* and *Ditlevsenella* partim). Six outer labial and four cephalic setae in a single circle.

Oncholaimoidea consists of the families Oncholaimidae and Enchelidiidae.

Key to families (after De Coninck, 1965)[5]

1. Pharynx cylindrical, outline never crenate;
   male supplements (if present) papilliform or setiform . . . . Oncholaimidae
   Pharynx gradually enlarged posteriorly, sometimes crenate
   or with many bulbs; male supplements papillae or suckers . . . . . Enchelidiidae

[5] The family Thalassogeneridae is not included because it occurs in terrestrial environments only.

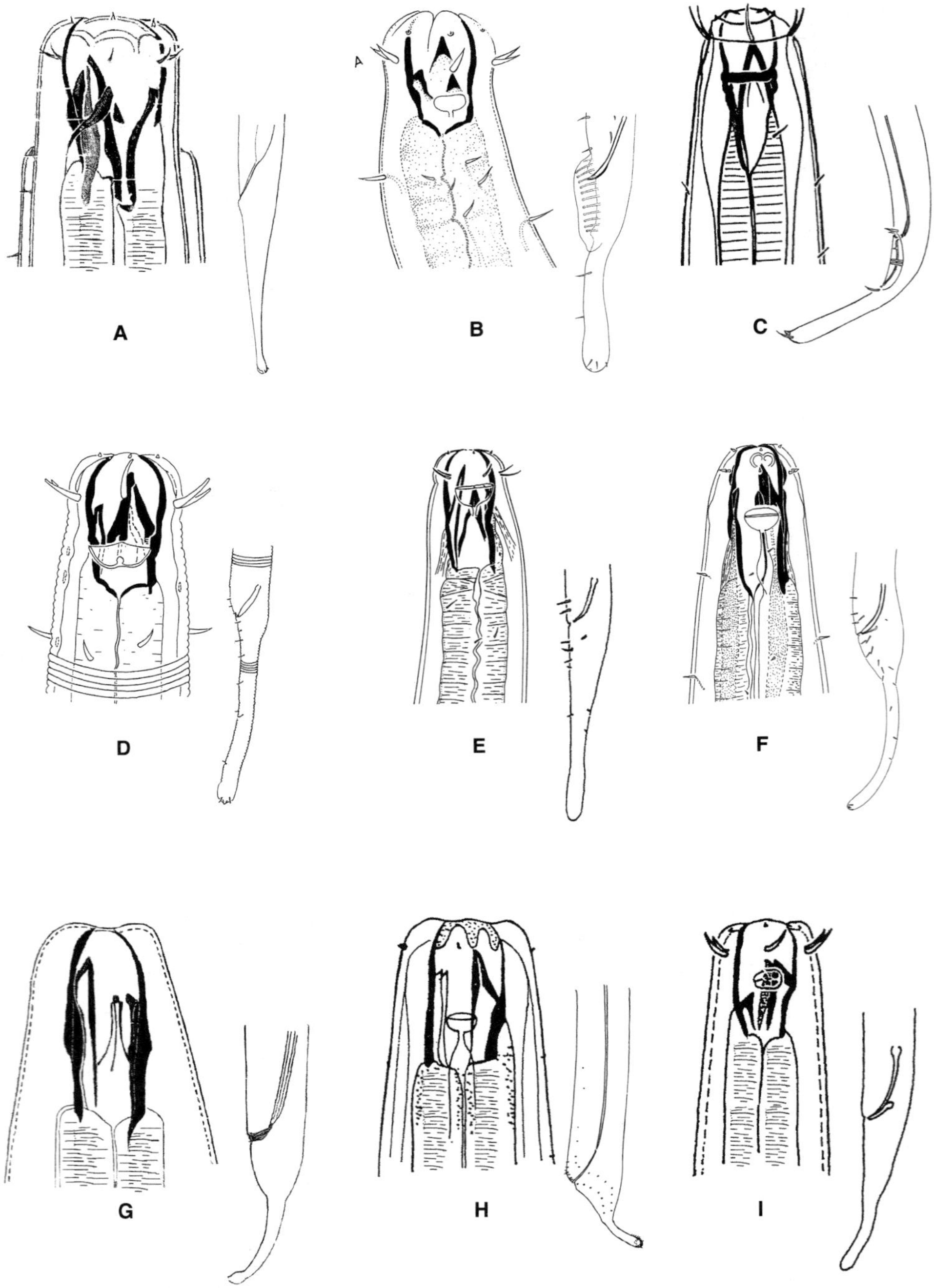

Fig. 12.6 Genera of the Oncholaimina, Oncholaimidae, Oncholaimellinae (A–E) and Adoncholaiminae (F–I); each genus is represented by the head end and the tail. A: *Cacolaimus* (after Kreis, 1934); B: *Oncholaimelloides* (after Murphy, 1966); C: *Oncholaimellus* (head after Wieser, 1953a; tail after De Coninck and Schuurmans-Stekhoven, 1933); D: *Oncholaimoides* (after Keppner and Keppner, 1989); E: *Viscosia* (after Coomans *et al.*, 1985); F: *Adoncholaimus papillatus* (after Furstenberg and Vincx, 1989); G: *Kreisoncholaimus* (after Kreis, 1934); H: *Metoncholaimoides* (after Wieser, 1953a); and I: *Meyersia* (after Platt and Warwick, 1983).

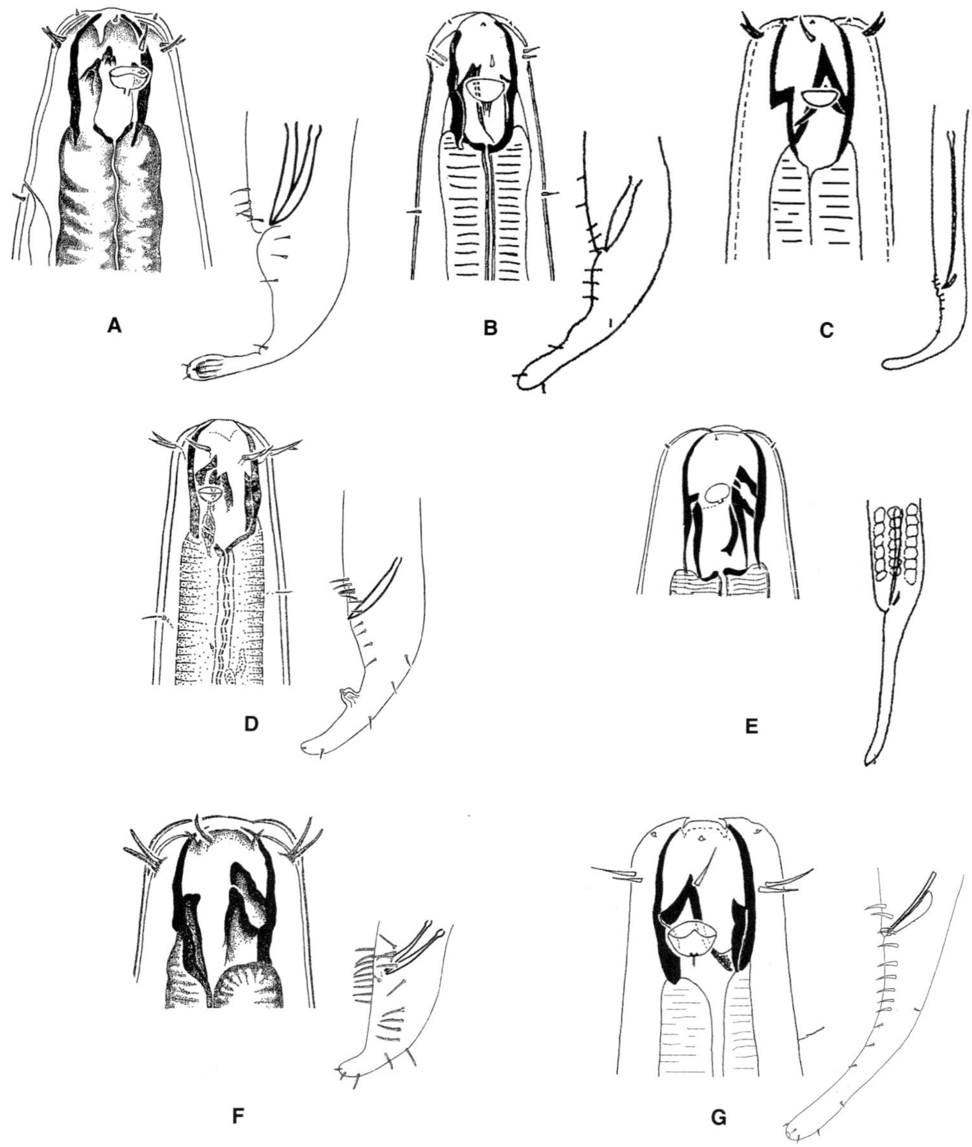

Fig. 12.7 Genera of the Oncholaimina, Oncholaimidae and Oncholaiminae; each genus is represented by the head end and the tail. A: *Fotolaimus* (after Belogurova and Belogurov, 1974); B: *Metaparoncholaimus* (after De Coninck and Shuurmans-Stekhoven, 1933); C: *Metoncholaimus* (after Platt and Warwick, 1983); D: *Oncholaimus* (after Heyns and Coomans, 1977); E: *Prooncholaimus* (head after Kreis, 1934; tail after Platt and Warwick, 1983); F: *Pseudoncholaimus venustus* (after Belogurov *et al*., 1972); and G: *Wiesoncholaimus* (after Inglis, 1964).

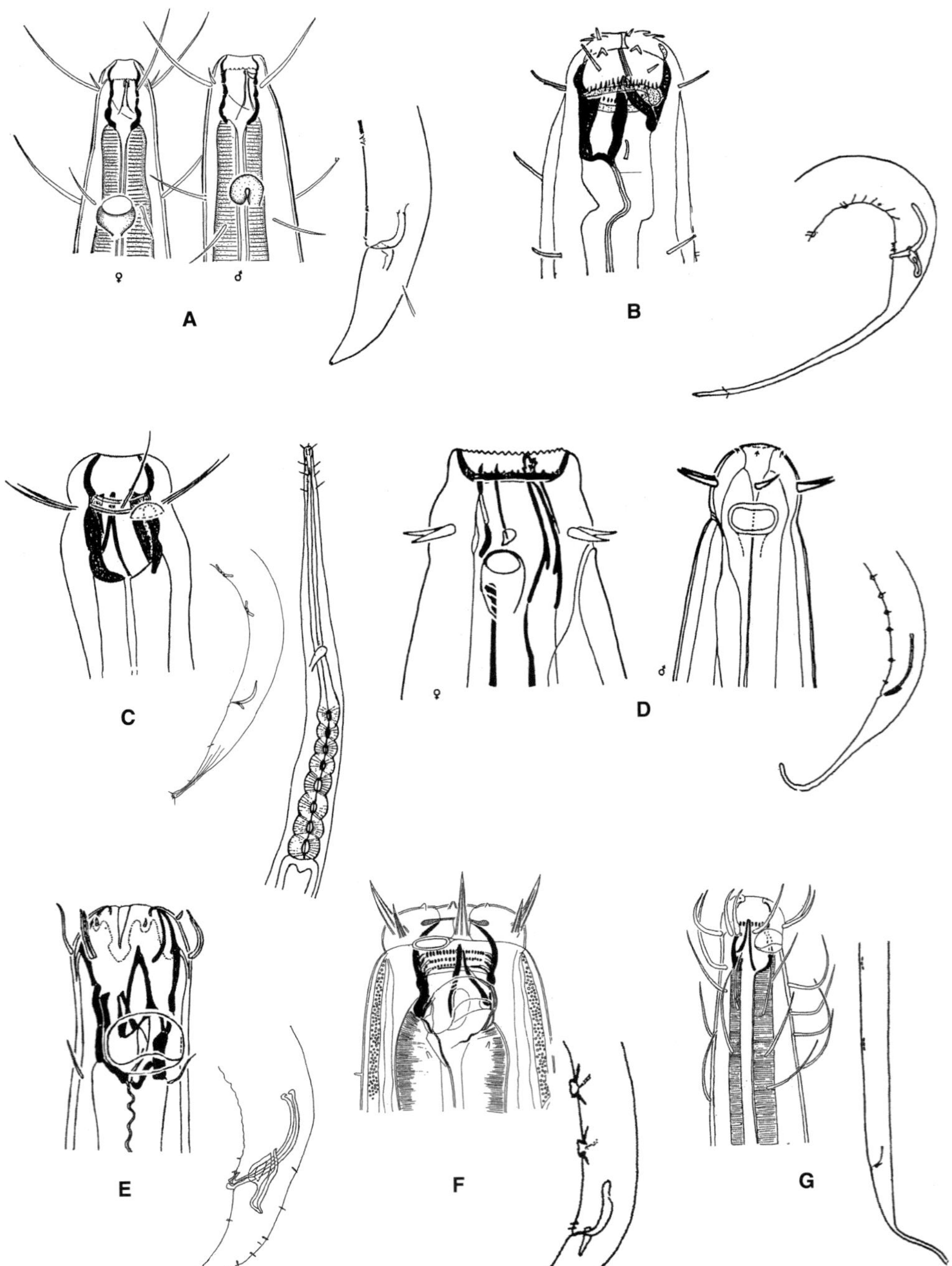

Fig. 12.8 Genera of the Oncholaimina, Enchelidiidae; each genus is represented by the head end and the tail. A: *Aronema* (after Fadeeva and Belogurov, 1988); B: *Bathyeurystomina* (after Lambshead and Platt, 1979); C: *Belbolla* (after Inglis, 1961); D: *Calyptronema* (male head after Bresslau and Schuurmans-Stekhoven, 1940; female head and tail after Lorenzen, 1969); E: *Ditlevsenella* (after Filipjev, 1927); F: *Eurystomina* (head after de Man, 1907; tail after Platt and Warwick, 1983); and G: *Ledovitia* (after Filipjev, 1927).

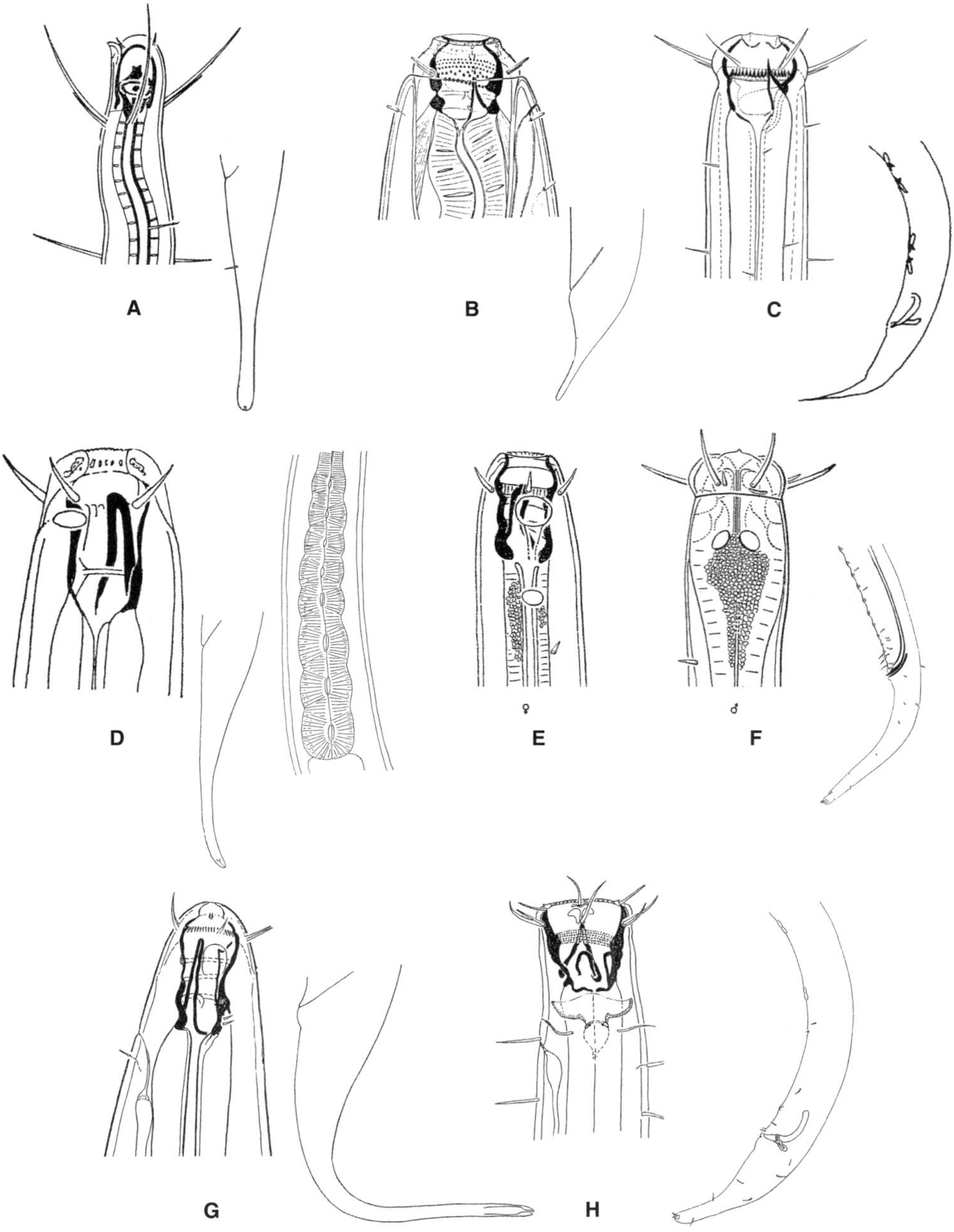

Fig. 12.9 Genera of the Oncholaimina and Enchelidiidae, continued; each genus is represented by the head end and the tail. A: *Lyranema* (after Timm, 1961); B: *Megeurystomina* (after Luc and De Coninck, 1959); C: *Pareurystomina* (head after Micoletzky, 1930; tail after Platt and Warwick, 1983); D: *Polygastrophora* (after Gerlach, 1957a); E: *Symplocostoma* head of female (after Luc and De Coninck, 1959); F: *Symplocostoma* (head of male after Luc and De Coninck, 1959; tail after Schuurmans-Stekhoven, 1950); G: *Symplocostomella* (head after Micoletzky, 1930; tail after Schuurmans-Stekhoven, 1950); H: *Thoonchus* (after Keppner, 1988).

Family ONCHOLAIMIDAE Filipjev, 1916

Diagnosis. Oncholaimoidea. Lips usually merged. Buccal cavity spacious, posteriorly surrounded by pharyngeal tissue, usually with three unequal immovable teeth. One of the two ventrosublateral teeth usually the largest, sometimes the two ventrosublateral teeth are equal in size and are bigger than the dorsal tooth, rarely all the three teeth are equal in size. Pharynx does not insert on to the body wall. Outline of pharynx smooth. Three pharyngeal glands open through teeth. Amphids generally pouch-shaped (dorsally spiral in many Enchelidiidae). Delicately built dorsolateral and ventrolateral orthometanemes with pronounced caudal filament. The secretory–excretory system distinct with gland cell to the right side of the intestine. Female reproductive system variable: either didelphic–amphidelphic or monodelphic–prodelphic or exceptionally monodelphic–opisthodelphic. The demanian system absent or present in different degrees of development. Males mostly diorchic with opposite testes. Spicules of variable shape; gubernaculum present or absent. Gonads always at the right side of the intestine.

The family Oncholaimidae has seven subfamilies.

For revision see Kreis (1934) and Belogurov and Belogurova (1978, 1989).

The number of teeth, the position of the largest tooth, the structure of the female reproductive system and the presence and development of the demanian system are important diagnostic characters. Exceptionally for aquatic (marine) nematodes both females and males are necessary for exact identification of the genus.

Key to subfamilies

1. Buccal cavity with three teeth 2
   Buccal cavity with many teeth . . . . Octonchinae
2. Teeth very small or absent 3
   Teeth distinct 4
3. Females didelphic, amphidelphic . . . . Pelagonematinae
   Females monodelphic, prodelphic . . . . Krampiinae
4. Both ventrosublateral teeth of equal size and larger than dorsal tooth . . . . Pontonematinae
   Ventrosublateral teeth unequal 5
5. Left ventrosublateral tooth largest . . . . Oncholaiminae
   Right ventrosublateral tooth largest 6
6. Demanian system simple (only short connection between ovary and intestine) . . . . Oncholaimellinae
   Demanian system highly developed . . . . Adoncholaiminae

Subfamily Oncholaimellinae De Coninck, 1965

Diagnosis. Oncholaimidae. Cuticle smooth. Buccal cavity free; right ventrosublateral tooth always larger than the other teeth, if the other teeth occur at all. Females didelphic–amphidelphic, rarely monodelphic–prodelphic (*Oncholaimelloides*). Demanian system absent or present and if present simple. Bursa present or absent.

Oncholaimellinae consist of five genera.

Type genus *Oncholaimellus* de Man, 1890

Key to genera

1. Cuticle thickened at the base of buccal cavity forming a distinct shoulder . . . . *Cacolaimus*
   Cuticle uniformly thick over the body 2
2. Females monodelphic, males with copulatory bursa . . . . *Oncholaimelloides*
   Females didelphic 3
3. Males usually with copulatory bursa . . . . *Oncholaimellus*
   Males without copulatory bursa 4
4. Cuticle transversely striated and with longitudinal ridges . . . . *Oncholaimoides*
   Cuticle smooth, without longitudinal ridges . . . . *Viscosia*

Genus *Cacolaimus* Kreis, 1932 (Fig. 12. 6A)
Diagnosis. Oncholaimellinae. Head distinctly offset by thickened cuticle at the base of buccal cavity. Buccal cavity large with thin walls, anteriorly broader than posteriorly. Right ventrosublateral tooth largest. Amphid indistinct. Tail elongated conical, swollen at the end. Only juveniles known. Marine.

Type and only species: *C. papillatus* Kreis, 1932

Genus *Oncholaimelloides* Timm, 1969 (Fig. 12.6B)
Diagnosis. Oncholaimellinae. Left[6] ventrosublateral tooth probably the largest, the right ventrosublateral and dorsal tooth equal. Females monodelphic–prodelphic. Demanian system absent. Male spicules short, broadening distally, copulatory bursa with 15 ribs. No information about the demanian system. Marine.

Type and only species: *O. vonhaffneri* (Murphy, 1966) Timm, 1969

Genus *Oncholaimellus* de Man, 1890. (Fig. 12.6C)
Diagnosis. Oncholaimellinae. Right ventrosublateral tooth large and solid. Buccal cavity divided transversely by sclerotized band. Spicules long, equal or unequal; gubernaculum short. Pre- and post-cloacal papillae. Copulatory bursa usually present, sometimes absent (or overlooked?). Demanian system similar to *Viscosia* or absent (*O. calvadosicus*). Marine.

Type species *O. calvadosicus* De Man, 1890

For revision see Timm (1967) and Keppner (1987).

Genus *Oncholaimoides* Chitwood, 1937 (Fig. 12.6D)
Diagnosis. Oncholaimellinae. Cuticle transversely striated bearing longitudinal ridges broken by striae. Six lips. Anterior sensilla in two circles: six inner labial papillae, six outer labial setae + four cephalic setae. Ampideal aperture elliptical to ovoid, moderate to large in size. Buccal cavity wide with three teeth, of which the right ventrosublateral is the largest, the other two are equal in size. Females didelphic–amphidelphic

[6] According to Timm (1969), if this is true, then this genus does not belong to this subfamily.

with antidromously reflexed ovaries; demanian system apparently absent. Males with two short straight spicules. Caudal glands and spinneret present. Marine.

Type species *O. rugosus* Chitwood, 1937

For revision see Hopper (1961a) and Keppner and Keppner (1989).

Genus ***Viscosia*** de Man, 1890 (Fig. 12.6E)
Diagnosis. Oncholaimellinae. Buccal cavity large, right ventrosublateral tooth largest. Females didelphic–amphidelphic with reflexed ovaries. Spicules short. Gubernaculum absent. Demanian system present and simple consisting of a prolongation of the ovary at the reflexed point, connecting it with the intestine through the *osmosium*. Mostly marine species, some occur in brackish water: *V. glabra* (Bastian, 1865), *V. viscosa* (Bastian, 1865) and two in freshwater.

Type species *V. viscosa* (Bastian, 1865) de Man, 1890

For revision see Wieser (1953a) and Wieser and Hopper (1967).

Species recorded from inland waters:

V. nicaraguensis Gerlach, 1957 (Lake Nicaragua, Nicaragua).
V. uipii Coomans *et al.*, 1985 (small deep freshwater pool in coral rock on Uipi Island, Solomon Islands).

Subfamily Adoncholaiminae Gerlach & Riemann, 1974
Diagnosis. Oncholaimidae. Right ventrosublateral tooth largest or both ventrosublateral teeth of equal size (*Meyersia*). Females didelphic–amphidelphic with antidromously reflexed ovaries. Demanian system always present and highest degree of development.

Adoncholaiminae consist of four genera, mostly marine; one genus is also represented in freshwater.

Type genus *Adoncholaimus* Filipjev, 1918

Key to genera
1. Two ventrosublateral teeth of equal size . . . .*Meyersia*
   Right ventrosublateral tooth largest 2
2. Anterior sensilla indistinct, demanian system without terminal duct and pore . . . .*Kreisoncholaimus*
   Anterior sensilla distinct, demanian system with terminal duct and pore(s) 3
3. Spicules ≤3 cloacal diameters . . . .*Adoncholaimus*
   Spicules ≥3 cloacal diameters . . . .*Metoncholaimoides*

Genus ***Adoncholaimus*** Filipjev, 1918 (Fig. 12.6F)
Diagnosis. Adoncholaiminae. Buccal cavity large, right ventrosublateral tooth largest. Females didelphic–amphidelphic with antidromously reflexed ovaries. Spicules long, gubernaculum present or absent. Demanian system present, parallel to the reproductive system, with long main duct and terminal duct in anal region; the main duct is anteriorly connected with the intestine through the *ductus entericus* and the *osmosium*.

Mainly marine species, some occur in brackish water: *A. aralensis* Filipjev, 1918, *A. fuscus* (Bastian, 1865), *A. islandicus* Kreis, 1963, *A. lepidus* (de Man, 1889), *A. papillatus* Kreis, 1932 and *A. thalassophygas* (de Man, 1876) and three in freshwater.

Type species *A. fuscus* (Bastian, 1865) Filipjev, 1918

For revision see Kreis (1934) and Wieser (1953a).

Species recorded from inland waters:

A. aralensis Filipjev, 1924 (Gagarin, 1993a,b, fresh and brackish water, Russia; Gagarin, 2001, Rivers Dnieper and Volga, Russia).
A. punctatus (Cobb, 1914) Filipjev, 1924 (Cobb, 1914, Nova Scotia).
A. thalassophygas (de Man, 1876) (Mol, 1984, The Netherlands; Andrássy, 1978, inland waters, Europe).

Genus *Kreisoncholaimus* Rachor, 1969 (Fig. 12.6G)
Diagnosis. Adoncholaiminae. Body strongly attenuating anteriorly. Anterior sensilla indistinct. Amphids positioned far anteriorly. Buccal cavity large, dorsal wall longer posteriorly than ventral wall, right ventrosublateral tooth largest. Females didelphic–amphidelphic with antidromously reflexed ovaries. Spicules long, gubernaculum present or absent. Demanian system present, symmetrical, the main duct runs parallel to the uterus and ends in the region of the uvette, short rounded ductus uterine connects the uvette with the uterus, *osmosium* not observed, and is connected to the vagina, no terminal duct and terminal pores but a connection between main duct and vagina. Marine.

Type and only species: *K. nudus* (Kreis, 1932) Rachor, 1969

Genus *Metoncholaimoides* Wieser, 1953 (Fig. 12.4H)
Diagnosis. Adoncholaiminae. Right ventrosublateral tooth largest. Females didelphic–amphidelphic. Spicules enormously elongated, gubernaculum absent. Demanian system well developed with uvette and moniliform terminal duct and pore, however, details unknown. Marine.

Type and only species: *M. squalus* Wieser, 1953

Genus *Meyersia* Hopper, 1967 (Fig. 12.6I)
Diagnosis. Adoncholaiminae. Similar to *Viscosia*, except ventrosublateral teeth equal, dorsal tooth smaller. Gubernaculum present. Demanian system symmetrical, the main duct runs parallel to the uterus and is anteriorly and posteriorly connected with the intestine and with both gonads, *osmosia* and uvettes two, terminal pore in region of vulva. Marine.

Type species *M. major* Hopper, 1967

Subfamily Oncholaiminae Filipjev, 1916
Diagnosis. Oncholaimidae. Left sub-ventral tooth almost always larger than the other teeth; very rarely all three teeth are the same size (*Oncholaimus keiensis*, *O. leptos*). Females monodelphic–prodelphic with antidromously reflexed ovary. Demanian system present or absent.

Oncholaiminae comprise seven genera, two of which occur also in freshwater.

Type genus *Oncholaimus* Dujardin, 1945

For revision see Belogurov and Belogurova (1975, 1977, 1978).

Key to genera

1. Both ventrosublateral teeth equal in length and larger than dorsal tooth 2
   Left ventrosublateral tooth largest 3
2. Spicules very long and slim, gubernaculum large . . . . . *Wiesoncholaimus*
   Spicules short, gubernaculum absent . . . . . *Metaparoncholaimus*
3. Demanian system absent 4
   Demanian system present 5
4. Large bubble-like cells between intestine and longitudinal chords, gubernaculum present . . . . . *Prooncholaimus*
   No such cells, gubernaculum absent . . . . . *Pseudoncholaimus*
5. Spicules short, gubernaculum absent or weakly developed 6
   Spicules long, gubernaculum present . . . . . *Metoncholaimus*
6. Demanian system with terminal ducts embedded in a sac, gubernaculum weakly developed . . . . . *Fotolaimus*
   Demanian system without such sacs, gubernaculum absent . . . . . *Oncholaimus*

Genus *Fotolaimus* Belogurova & Belogurov, 1974 (Fig. 12.7A)

Diagnosis. Oncholaiminae. Left sub-ventral tooth largest, the other two teeth equal in size. Females monodelphic–prodelphic with antidromously reflexed ovary. Demanian system present, similar to *O. domesticus* Chitwood & Chitwood, 1938. However, the caudal end of the main duct forms two sacs pierced by the terminal ducts ending in terminal pores. Males with slightly curved spicules with weak cephalation. Gubernaculum weak. Genital setae mainly in pre-cloacal region. Marine.

Type species *F. marinus* Belogurova & Belogurov, 1974

Genus *Metaparoncholaimus* De Coninck & Schuurmans-Stekhoven, 1933 (Fig. 12.7B)

Diagnosis. Oncholaiminae. Both ventrosublateral teeth equal and larger than dorsal tooth. Females monodelphic–prodelphic with antidromously reflexed ovary. Demanian system well developed with terminal ducts ending in pre-anal region. Spicules short, pericloacal setae and pre-cloacal papillae present. Marine.

Type species *M. campylocercus* De Man, 1876

For revision see Kreis (1934).

Genus *Metoncholaimus* Filipjev, 1918 (Fig. 12.7C)

Diagnosis. Oncholaiminae. Spicules usually long, gubernaculum usually present. Demanian system well developed with single uvette and double moniliform terminal ducts. Marine.

Type species *M. demani* (Zur Strassen, 1894) Filipjev, 1918

For revision see Chitwood (1960) and Wieser and Hopper (1967).

Genus ***Oncholaimus*** Dujardin, 1845 (Fig. 12.7D)
Diagnosis. Oncholaiminae. Left ventrosublateral tooth largest. Females monodelphic–prodelphic with antidromously reflexed ovary. Demanian system well developed, terminal ducts and pores present in variable number or absent in virgin females. Males diorchic. Spicules short, gubernaculum absent. Tail short. Many are marine species, the following species occur in brackish water: *O. bajulus* Paramonov, 1937; *O. conicauda* Filipjev, 1929; *O. domesticus* Chitwood & Chitwood, 1938; *O. orientalis* Tchesunov, 1976; *O. oxyuris* Ditlevsen, 1911; *O. vanderlandi* Loof, 1973. Two terrestrial species: *O. campbelli* Allgen, 1929 (in moss at Campbell Island, Subantarctica) and *O. diversidens* Fuchs, 1938 (uncertain origin, The Netherlands). An unidentified *Oncholaimus* species has been recorded in the hypersaline Salton Sea, the largest inland lake in California, USA, by Warwick *et al.* (2002).

Type species *O. attenuatus* Dujardin, 1845

For revision see Wieser (1953a) and Chitwood (1960).

Species recorded from freshwater:

O. aquadulcis W. Schneider, 1937 (waterfall in East Java, Indonesia)
O. balli Nicholas & Stewart, 1984 (a volcanic crater lake, Papua New Guinea)
O. deconincki Heyns & Coomans, 1977 (South Africa)
O. jessicae Coomans & Heyns, 1986 (Blyde River, South Africa)
O. oxyuris Ditlevsen, 1911 (Andrássy, 1978, inland waters, Europe)
O. rhopalocercus Schuurmans-Stekhoven, 1942 (in aquaria)
O. sahariensis Coomans & Heyns, 1983 (oasis, Algeria).

Genus *Prooncholaimus* Micoletzky, 1924 (Fig. 12.7E)
Diagnosis. Oncholaminae. Cuticle smooth or with fine striations. Large bubble-like cells in body cavity between intestine and longitudinal chords. Left ventrosublateral tooth largest, the other two equal. Demanian system absent. Spicules slightly curved. Gubernaculum present. Marine.

Type species *P. megastoma* (Eberth, 1863) Micoletzky, 1924

For revision see Wieser and Hopper (1967).

Genus ***Pseudoncholaimus*** Kreis, 1932 (Fig. 12.7F)
Diagnosis. Oncholaminae. Large and medium-sized nematodes. Cuticle smooth. Buccal cavity well developed with left ventrosublateral tooth largest, its tip reaching the level of the second circle of sensilla. Dorsal and right ventrosublateral tooth considerably smaller and equal. Secretory–excretory gland post-pharyngeal. Females monodelphic–prodelphic. Demanian system absent. Males diorchic with opposed testes,[7] spicules short dagger-shaped, moderately cephalate, gubernaculum absent,

[7] Monorchic according to Kreis (1934).

genital papillae and setae present. Mostly marine, however, a few species occur in brackish inland waters.

Type species *P. elegans* Kreis, 1932

For revision see Belogurov *et al.* (1972, 1980) and Tsalolikhin (1982).

Species recorded from inland waters:

P. arenarius Tsalolikhin, 1979 (brackish, Issykkul Lake, Tien Shan, Kyrgyzstan)
P. charon Tsalolikhin, 1982 (brackish lakes of Mongolia).
P. issykulensis Tsalolikhin, 1979 (brackish, Issykkul Lake, Tien Shan, Kyrgyzstan)
P. neglectus Tsalolikhin, 1982 (brackish lakes of Mongolia; Gagarin, 2001, mineral spring, Russia).

Genus *Wiesoncholaimus* Inglis, 1966 (Fig. 12.7G)
Diagnosis. Oncholaminae. Cuticle smooth. Length of buccal cavity less than twice its diameter. Dorsal tooth smallest; two ventrosublateral teeth equal. Demanian system present, of oncholaimoid type. Spicules long, thin and slightly curved. Gubernaculum present. Males with a double papilliform structure on anterior lip of cloaca. Tail relatively long and stout, slightly conical. Marine.

Type and only species: *W. mawsonae* Inglis, 1966

Subfamily Krampiinae De Coninck, 1965
Diagnosis. Oncholaimidae. Cuticle smooth. Buccal cavity partly imbedded by the pharynx. Three equal weak teeth whose apices are at the level of the anterior edge of the pharyngeal collar. Secretory–excretory pore either at level of cephalic setae or slightly behind the stoma base; terminal duct and the anterior part of the ampulla sclerotized. Female reproductive system monodelphic–prodelphic with reflexed ovary. Demanian system absent. Spicules are paired, dagger-shaped. Tail short. All species marine.

Type and only genus: *Krampia* Ditlevsen, 1921

For revision see Belogurov and Dashenko (1981).

Subfamily Octonchinae De Coninck, 1965
Diagnosis. Oncholaimidae. Buccal cavity tapers posteriorly, posterior one-third surrounded by pharyngeal tissue; armed with eight teeth, one of which is conspicuous by its size. Tail conical. The only species known so far was described from a young female, hence detailed data on the reproductive system are lacking. Marine.

Type and only genus: *Octonchus* Clark, 1961

Subfamily Pelagonematinae De Coninck, 1965
Diagnosis. Oncholaimidae. Teeth small or absent. Female reproductive system usually didelphic–amphidelphic, rarely a single anterior ovary (*Curvolaimus*).

Demanian system probably absent. This subfamily has nine genera, all of which are marine.

Type genus *Pelagonema* Cobb, 1894

For revision see Belogurov (1983).

Subfamily Pontonematinae Gerlach & Riemann, 1974

Diagnosis. Oncholaimidae. Cuticle smooth. Buccal cavity about twice as long as its width. Both ventrosublateral teeth equal in length and larger than the dorsal tooth. Females didelphic–amphidelphic. Demanian system absent. Spicules strongly curved; gubernaculum with distal wings. Males have one or several supplements. Tail blunt, often short.

Pontonematinae consist of three genera, all marine.

Type genus *Pontonema* Leidy, 1855

Family ENCHELIDIIDAE Filipjev, 1918

Diagnosis. Oncholaimoidea. Buccal cavity with three unequal teeth, of which the dominant can be extended at least in *Calyptronema*, *Polygastrophora* and *Symplocostoma*. Right ventrosublateral tooth usually larger than the other (only in *Calyptronema* can either the left or right ventrosublateral tooth be the largest). Sexual dimorphisms in shape of the buccal cavity and amphid in that the organs concerned are reduced in the males in *Calyptronema*, *Polygastrophora* and *Symplocostoma*. Amphids in part non-spiral and in part dorsally spiral. Outline of pharynx crenate. Females didelphic–amphidelphic (only in *Calyptronema sabulicola* monodelphic–opisthodelphic). Demanian system always absent. Males with or without pre-anal supplements.

Type genus *Enchelidium* Ehrenberg, 1836

Since Filipjev (1934) the Encheldiidae consisted of the subfamilies Eurystomininae and Enchelidiinae. To these were added the Thoonchinae (Gerlach and Riemann, 1974), Pareurystomininae (Andrássy, 1976) and more recently Aronematinae (Fadeeva and Belogurov, 1988). Because of the lack of holophyletic characters the subfamilies Eurystomininae, Pareurystomininae, Thoonchinae as well as Belbollinae (lapsus for Belbollidae Andrássy, 1976) are synonymized with Enchelidiinae by Lorenzen (1981, 1994). As very few species occur in freshwater, further discussion of the remaining subfamilies is omitted.

Enchelidiidae comprises 15 genera. Mostly marine, a few species in brackish water and one in freshwater.

Key to genera (*Enchelidium* not included because it is regarded as a doubtful genus)

1. Pharynx posteriorly with many muscular bulbs 2
   Pharynx without several bulbs 3
2. Ocelli absent, males with two winged supplements . . . . *Belbolla*
   Ocelli present, males without such supplements . . . . *Polygastrophora*
3. Head end divided into two lateral portions, mandibles lyre-shaped . . . . *Lyranema*

Head end not so 4
4. Males with two characteristic sclerotized supplements 5
Males with papilliform supplements or without supplements 9
5. Caudal glands and spinneret present 6
Caudal glands and spinneret absent 7
6. Extremely long cervical setae present and weakly developed winged supplements . . . .*Ledovitia*
Without extremely long cervical setae and with strongly developed winged supplements . . . .*Eurystomina*
7. One band of denticles not lined into rows, only one large right ventrosublateral tooth, males with two tubiform supplements . . . .*Bathyeurystomina*
Many rows of denticles, one large and sometimes two very small teeth 8
8. Head end offset, retractile by plicate cuticle, tail not pointed . . . .*Megeurystomina*
Head end not offset, tail finely pointed . . . .*Pareurystomina*
9. Buccal cavity large and wide, oncholaimoid; ventrosublateral teeth with broad base 10
Buccal cavity narrow, ventrosublateral teeth with narrow base, spine-like 11
10. Buccal cavity with rows of denticles . . . .*Thoonchus*
Buccal cavity without rows of denticles . . . .*Ditlevsenella*
11. Buccal cavity of females divided by an anterior transverse ring of denticles in two chambers only 12
Buccal cavity of females divided by cuticularized rings into more than two parts 13
12. Cervical setae short if present, female pharynx anteriorly with widely dilated sclerotized lumen, sexual dimorphism in shape of buccal cavity and amphid . . . .*Calyptronema*
Cervical setae very long, lumen of pharynx sclerotized, but not so wide, sexual dimorphism only in shape of amphid . . . .*Aronema*
13. Transverse rings of buccal cavity solid, ocelli present . . . .*Symplocostoma*
Transverse rings consisting of denticles, ocelli absent . . . .*Symplocostomella*

Genus *Aronema* Fadeeva & Belogurov, 1988 (Fig. 12.8A)

Diagnosis. Enchelidiidae. Body strongly narrowing anteriorly. Cuticle smooth. Endocupola with short dome, stomatal ring surrounds the stoma at the level of the denticles. Ocelli absent. Anterior sensilla in two circles, six outer labial setae longer than four cephalic setae; six long cervical setae at the level of the amphids. Amphids positioned far from buccal cavity; sexual dimorphism in shape: pocket-shaped in females and dorsally spiral in males. Buccal cavity in two parts with a row of denticles separating the two parts and one long sharply pointed sub-ventral tooth, reaching the level of the denticles and provided with a guiding piece. Pharynx cylindrical and with dilated sclerotized lumen. Females didelphic–amphidelphic with antidromously reflexed ovaries. Males with bent spicules, gubernaculum with apophysis and four weak pre-cloacal supplements. Marine.

Type species *A. crinita* Fadeeva & Belogurov, 1988

Genus *Bathyeurystomina* Lambshead & Platt, 1979 (Fig. 12.8B)
Diagnosis. Enchelidiidae. Body anteriorly attenuated. Ocelli absent. Cuticle smooth. Anterior sensilla in two circles and all setiform, six outer labial setae longer than four cephalic setae. Somatic setae short. Stomatal opening surrounded by complex flap-like lips. Buccal cavity with denticles not organized into definite rows and only a large right ventrosublateral tooth present. Pharyngeal bulbs absent. Males with two sclerotized tubular pre-cloacal supplements surmounted by a distinct cap. Additional small cuticularized body situated between posterior supplement and cloaca. Gubernaculum with well-developed apophysis. Tail conico-flagellate. Caudal glands absent. Marine.

Type species *B. valeriae* Lambshead & Platt, 1979

Genus *Belbolla* (Cobb, 1920) Andrássy, 1973 (Fig. 12.8C)
Diagnosis. Enchelidiidae. Ocelli absent. Buccal cavity large and narrow divided in two parts by a sclerotized transverse band, three teeth of which one ventrosublateral is the largest. Amphids dorsally spiral. Posterior section of pharynx with series (7–10) of bulb-like muscular swellings. Males mostly with winged pre-cloacal supplements, spicules short and curved, gubernaculum absent or present with apophysis. Marine.

Type species *B. tenuidens* (Cobb, 1920)

For revision see Wieser (1953a) and Belogurov and Belogurova (1980).

Genus ***Calyptronema*** Marion, 1870 (Fig. 12.8D)
Diagnosis. Enchelidiidae. Ocelli present or absent. Sexual dimorphism in anterior end. Amphids non-spiral. Female buccal cavity divided into two unequal portions by an anterior transverse band, ring or row of denticles; other rings, which might be present are faint and concentrated at the posterior end of the buccal cavity. Three slender teeth, one ventrosublateral largest and eversible. Posterior portion of the buccal cavity large, asymmetrical, almost ventricose. Female pharynx anteriorly with widely dilated sclerotized lumen. Buccal cavity in male absent, amphid relatively larger, lumen of pharynx not dilated and sclerotized. Spicules long, arcuate, pre-cloacal papillae present. Tail short or long. Most species marine, one species occurs in brackish habitats: *C. maxweberi* (de Man, 1922).

Type species *C. paradoxum* Marion, 1870

For revision see Wieser (1953a) and Belogurov (1979).

Species recorded from inland waters:

C. maxweberi (de Man, 1922) Andrássy (1978, inland waters, Europe); Mol (1984, The Netherlands).

Genus *Ditlevsenella* Filipjev, 1927 (Fig. 12.8E)

Diagnosis. Enchelidiidae. Six high lips, deeply incised. Buccal cavity large, Oncholaiminae-like, with three teeth with broad base, right or left ventrosublateral tooth largest. Amphids large, dorsally spiral. Spicules curved, gubernaculum with apophysis. Characteristic pre-cloacal supplements absent, papillae present. Caudal glands pre-caudally. Marine.

Type species *D. murmanica* Filipjev, 1927

For revision see Hopper (1963).

We do not consider the specimen drawn as *Ditlevsenella danica* by Lorenzen (1981, p. 259) and as *Ditlevsenella aff. murmanica* by Lorenzen (1994, p. 257) as a true *Ditlevsenella* species because of the oral collar and the rows of denticles, both characters are not present in *Ditlevsenella.*

Genus *Enchelidium* Ehrenberg, 1836

Diagnosis. Enchelidiidae. A diagnosis cannot be given because the type species, as most of the other species, has been transferred to the genera *Calyptronema* or *Symplocostoma*, the remaining species are either described by Allgén or are regarded as *sp. inq*. The type species might occur in brackish water.

Type species *E. marinum* Ehrenberg, 1836

For revision see Wieser (1953a,b) who regards this as a doubtful genus.

Genus ***Eurystomina*** Filipjev, 1921 (Fig. 12.8F)

Diagnosis. Enchelidiidae. Amphids dorsally spiral. Ocelli present or absent. Buccal cavity divided into two parts by several rows of denticles, three teeth, right ventrosublateral largest and acute. Pharynx without posterior bulbs. Males with two well-developed pre-cloacal supplements cup-shaped and sclerotized, with or without winged bases. Gubernaculum with or without apophysis. Caudal glands present. Marine, one brackish species: *E. terricola* (de Man, 1907).

Type species *E. spectabilis* Filipjev, 1921

For revision see Wieser (1953a), Inglis (1962), Wieser and Hopper (1967) and Warwick (1969).

E. terricola occurs in soil soaked in brackish water and in estuarine sand subjected to seepage of freshwater from the land; it is considered a typical inhabitant of the transitional zone between marine and terrestrial environments and is characteristic of coarse, relatively dry, sandy substrates with low salinity.

Species recorded in inland waters:

E. terricola (de Man, 1907) Andrássy (1978, as *Eurystoma terricola*, inland waters, Europe); Mol (1984, The Netherlands).

Genus *Ledovitia* Filipjev, 1927 (Fig. 12.8G)

Diagnosis. Enchelidiidae. Buccal cavity similar to *Eurystomina*. Extremely long cervical setae. Male genital apparatus more weakly developed. Marine.

Type species *L. hirsuta* Filipjev, 1927

For revision see Wieser (1953a) and Belogurov *et al.* (1983).

Genus *Lyranema* Timm, 1961 (Fig. 12.9A)
Diagnosis. Enchelidiidae. Body with greatly narrowed head and neck. Anterior sensilla in two circles: six inner labial setae and six very long outer labial setae + four cephalic setae, second circle at a distance from the first. Amphids bowl-shaped. Head divided into two lateral portions. Mandibles lyre-shaped anteriorly sclerotized. Pharyngeal lumen expanded and strongly sclerotized. Posterior of pharynx abruptly expanded. Females didelphic–amphidelphic. Marine.

Type and only species: *L. speciosum* Timm, 1961

Genus *Megeurystomina* Luc & De Coninck, 1959 (Fig. 12.9B)
Diagnosis. Enchelidiidae. Long body size (10 mm). Head retractile with a cuticular fold at the base of the head, which extends as a double circular sleeve. A superficial ornamentation of two rows of very fine scales between first and second circle of anterior sensilla. Below each inner labial papilla a comma-shaped internal refringed structure. Buccal cavity barrel-shaped, walls divided into two pieces by a transverse band, five rows of denticles in the anterior part. One large right ventrosublateral tooth in the posterior part. Amphids dorsally spiral, slightly displaced dorsally. Secretory–excretory pore positioned near the head end. Cardia conical elongated. Female reproductive system didelphic–amphidelphic with antidromously reflexed ovaries. Tail short ($c' \leq 2$) conico-cylindrical. Caudal glands absent. Marine.

Type and only species: *M. combesi* Luc & De Coninck, 1959; only one female described.

Genus *Pareurystomina* Micoletzky, 1930 (Fig. 12.9C)
Diagnosis. Enchelidiidae. Amphids dorsally spiral. Buccal cavity divided into two parts by rows of denticles, one large right ventrosublateral slender acute tooth, and other teeth small or absent. Males with two well-developed pre-cloacal supplements. Tail conical with sharp tip. Caudal glands absent. Marine.

Type species *P. typica* Micoletzky, 1930

For revision see Luc and De Coninck (1959), Wieser (1959), Chitwood (1960) and Keppner (1989).

Genus ***Polygastrophora*** de Man, 1922 (Fig. 12.9D)
Diagnosis. Enchelidiidae. Ocelli present. Amphids dorsally spiral. Buccal cavity divided into several chambers by cuticularized rings. Posterior section of pharynx with series of bulb-like muscular swellings. Pre-cloacal supplements absent. Mostly marine, only one freshwater species.

Type species *P. attenuata* de Man, 1922

For revision see Wieser (1953a), Chitwood (1960) and Wieser and Hopper (1967).

Species recorded from inland waters:
P. octobulba Micoletzky, 1930 (Gerlach, 1957a, Lake Nicaragua, Nicaragua; Meyl, 1957, Lake Tanganyika, Africa).

Genus *Symplocostoma* Bastian, 1865 (Fig. 12.9E)
Diagnosis. Enchelidiidae (Enchelidiinae). Sexual dimorphism in anterior end. Ocelli present. Female buccal cavity narrow, almost cylindrical, transverse rings more or less irregularly arranged so as to give rise to more than two apparent chambers. Transverse rings solid. Pharyngeal lumen not dilated and sclerotized. Males lack a buccal cavity. Spicules long, gubernaculum present, pre-cloacal supplements papilliform. Marine.

Type species *S. longicolle* Bastian, 1865

Genus *Symplocostomella* Micoletzky, 1930 (Fig. 12.9F)
Diagnosis. Enchelidiidae (Enchelidiinae). Sexual dimorphism similar to *Symplocostoma.* Ocelli lacking. Buccal cavity narrow, almost cylindrical, transverse rings more or less irregularly arranged so as to give rise to more than two apparent chambers. A transverse row of denticles in the anterior portion of the buccal cavity present. Marine.

Type species *S. javaensis* Micoletzky, 1930

For revision see Vitiello (1970).

Genus *Thoonchus* Cobb, 1920 (Fig. 12.9G)
Diagnosis. Enchelidiidae (Enchelidiinae). Six lips. Inner labial sensilla papilliform or setiform. Buccal cavity large with heavily sclerotized walls, 1–1.5 times as long as broad. Three teeth of which right ventrosublateral tooth largest; several rows of denticles present. Amphid large and situated either at the level of buccal cavity or behind. Excretory pore anterior or posterior to buccal cavity. Spicules short and arcuate, gubernaculum with expanded corpus and with or without apophysis. One or two pre-cloacal supplement(s) and papillae present or absent. Marine.

Type species *T. ferox* Cobb, 1920

For revision see Keppner (1988).

Suborder IRONINA Siddiqi, 1983 (Figs 12.3D and 12.10A–I).

Diagnosis. Enoplida. Metanemes present. Body shape variable from small and very thin to very large (50 mm). Anterior sensilla in two or three circles. Buccal cavity narrow, tubular or funnel-shaped, sometimes even obscure or absent, elongated prismatic without teeth or having tooth-like structures or provided with three, four or five movable teeth at anterior end (the normal number is three, but the dorsal one or the ventrosublateral ones may be double). Pharyngeal glands do not open through the teeth but further back in the buccal cavity. Pharynx surrounding the buccal cavity and may insert on to the body cuticle in the region of the buccal cavity. Cephalic capsule variable. Posterior section of pharynx crenate or not. Females didelphic–amphidelphic with antidromously reflexed ovaries or monodelphic–opisthodelphic. Males diorchic

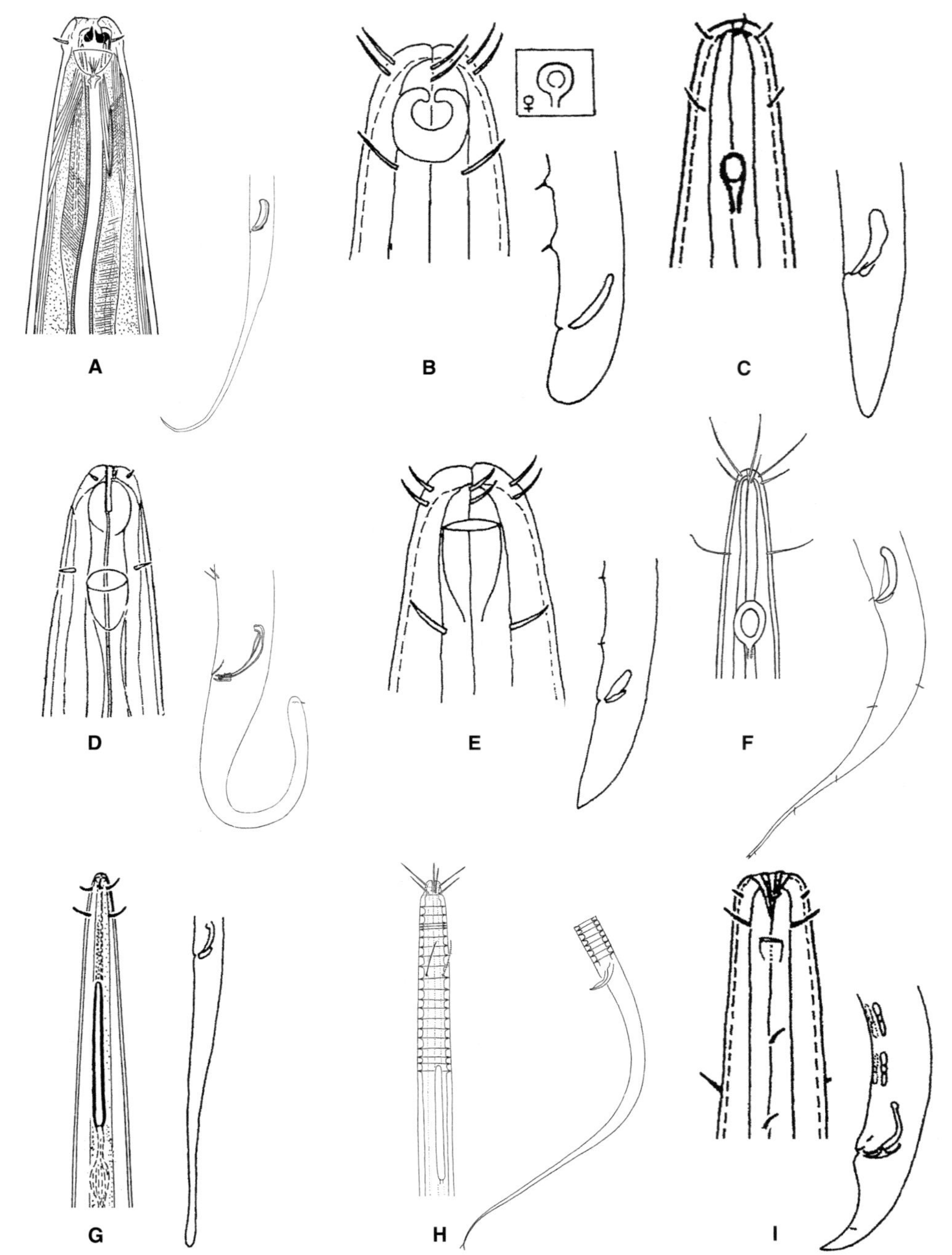

Fig. 12.10 Genera of the Ironina, Ironidae (A) and Oxystominidae (B–I); each genus is represented by the head end and the tail. A: *Ironus* (after Van Der Heiden, 1974); B: *Litinium* (after Platt and Warwick, 1983); C: *Nemanema* (after Platt and Warwick, 1983); D: *Oxystomina* (after Jensen, 1979); E: *Thalassoalaimus* (after Platt and Warwick, 1983); F: *Wieseria* (after Zhang, 1983); G: *Halalaimus* (after Coomans and Jacobs, 1983); H: *Cricohalalaimus* (after Bussau, 1993); and I: *Paroxystomina* (after Platt and Warwick, 1983).

with testes opposed or monorchic with anterior testis only; supplements present or absent. Position of caudal glands variable.

Superfamily IRONOIDEA de Man, 1876

Diagnosis. Similar to Ironina.

Ironoidea comprise three families: Ironidae, Leptosomatidae and Oxystominidae. Ironidae is a freshwater group, Leptosomatidae is a marine family and members of the Oxystominidae occur mostly in marine habitats, some occurring in brackish habitats with low salinity and in freshwater.

Key to families

1. Buccal cavity long and tubular, walls heavily sclerotized, 3–5 movable teeth at anterior edge . . . . Ironidae
   Buccal cavity absent, narrow, very small to funnel-shaped, if large, then short cylindrical with a complicated array of teeth and tooth-like structures 2
2. Small nematodes, buccal cavity very small, amphids polymorphic, pharynx anteriorly not attached to body cuticle . . . . Oxystominidae
   Large nematodes, buccal cavity very small or shortly funnel-shaped or tubiform, pharynx anteriorly attached to body cuticle, cephalic capsule variable . . . . Leptosomatidae

Family IRONIDAE de Man, 1876

Diagnosis. Ironoidea. Anterior sensilla in three circles. Several combinations are possible; all setiform (*Ironella*); all papilliform (*Trissonchulus*, *Dolicholaimus*, *Pheronus*); only inner labial papilliform (*Thalassironus*, *Conilia*, *Parironus*) or only cephalic sensilla setiform (*Ironus*). Among the setae the four cephalic setae are usually largest, only in *Thalassironus* the outer labial setae are largest. Buccal cavity elongated with three (one dorsal and two ventrosublateral, all more or less equal in length), four (two smaller dorsal and two larger ventrosublateral as in *I. ignavus*) or five (one dorsal and two pairs of ventrosublateral ones as in *Ironella*) movable teeth at the anterior edge. In juveniles the teeth for the following stage are in pharyngeal pouches behind the functional ones. Pharyngeal glands do not open through the teeth, but further back in the buccal cavity. The pharynx inserts, at least in some genera, on to the body cuticle in the buccal cavity region. Females didelphic–amphidelphic with antidromously reflexed ovaries, rarely monodelphic–opisthodelphic (*Trissonchulus oceanus* and *T. raskii*). Males diorchic with opposed testes or a single anterior testis. Gubernaculum present. A single pre-cloacal seta present in males of some species.

Two subfamilies Ironinae de Man, 1876 and Thalassironinae Andrássy, 1976.

Subfamily Ironinae de Man, 1876

Diagnosis. Ironidae. Delicately built, dorsolateral and ventrolateral orthometanemes occur in a strictly alternating sequence. Buccal cavity a long sclerotized tube with three or more eversible teeth anteriorly. Cephalic setae present. Small pharyngo-intestinal valve. Exclusively limnetic.

Type genus *Ironus* Bastian, 1865

Since Filipjev (1918) the position and composition of subfamily Ironinae has been subject to different opinions hence classification, shifting from the Dorylaimidae (based on the presence of spare teeth in juveniles) to Tripyloidea, Enoploidea and Plectoidea. For quite some time the genera *Ironella, Thalassironus, Dolicholaimus, Trissonchulus* and *Syringolaimus* were included in the Ironinae. For revision see Chitwood (1960).

However, based on the above diagnostic characters, only the genus *Ironus* fits into this subfamily according to Lorenzen (1981, 1994). Most of the above-mentioned genera are now placed within the Thalassironinae. As mainly the limnetic occurrence and the characteristics of the metanemes are used to separate the Ironinae from the Thalassironinae, a key to subfamilies is irrelevant.

Genus ***Ironus*** Bastian, 1865 (Fig. 12.10A)
Diagnosis. Ironinae. Cuticle smooth. Lip region offset by shallow constriction, three lips. Amphidial aperture slit-like, fovea cup-shaped. Buccal cavity long, tubular and sclerotized, anteriorly three strong eversible hook-like teeth, which can be bifurcated; juveniles three smaller replacement teeth a short distance behind the functional ones. Pharynx gradually swollen posteriorly. Secretory–excretory pore in vicinity of the lips. Female reproductive system didelphic–amphidelphic with reflexed ovaries. Males with strong arcuate spicules with central strengthening piece, small gubernaculum and one mid-ventral pre-cloacal seta. Tail tapering to a fine point. This genus can be considered as a typical freshwater genus, with very few species occurring in brackish water (*I. ignavus* found by Filipjev, 1929, 1930 in the Baltic Sea and *I. tenuicaudatus* found by Filipjev, 1929, 1930 and Gerlach, 1953 in Kiel Bay) and some in soil (*I. crassatus* Argo & Heyns, 1972; *I. ernsti* Argo & Heyns, 1972; and *I. lautus* Argo & Heyns, 1972) or in mosses.

Type species *I. ignavus* Bastian, 1865

For revision see Andrássy (1968), Khera (1979), Ebsary (1985), Tsalolikhin (1987) and Gagarin (1993b).

Species recorded from inland waters:

I. americanus Cobb, 1914, (Andrássy, 1978, inland waters, Scandinavia, Europe; Tsalolikhin, 1980 in Lake Baikal, Russia; Gagarin, 2001, water reservoir in Kremenchug, Russia).
I. colourus Steiner, 1919 (Andrássy, 1978, inland waters, Alps, Danube countries, Europe; Tsalolikhin, 1980, in Lake Baikal, Russia).
I. dentifurcatus Argo & Heyns, 1972 (Tsalolikhin, 1980 in Lake Baikal, Russia; Eyualem-Abebe and Coomans, 1995 in Wittmer's Well, Floreana, Galápagos).
I. elegans Colomba & Vinciguerra, 1979 (Sicily).
I. gagarini Tsalolikhin, 1987 (water store in Mingechaurskoe, Azerbaijan) (Gagarin, 2001, Lake Son Kul', Russia).
I. helveticus Daday, 1911 (Tsalolikhin, 1980 in Lake Baikal, Russia).
I. ignavus Bastian, 1865 (Andrássy, 1973, freshwater, Cuba; Zullini, 1973, Chiapas, Mexico in 650 m-deep Chorreador cave, mould fairly rich in guano and bottom mud of small water pools; and in bottom mud and water; Goossens, 1976, spring in Terjit, Mauretania; Zullini, 1977a,b, in a well, a cenote and in a brook in

Mexico; Andrássy, 1978, inland waters, Europe; Tsalolikhin, 1980, in Lake Baikal, Russia; Mol, 1984, The Netherlands; Wu and Liang, 1999, Lake Biandantang and Lake Houhu, China; Gagarin, 2001, Lake Kubenskoye, Rivers Angara and Volga, water reservoirs in Kiev, Kremenchug, Kakhov, Ribinsk,Uchinski, pond near Borok, Russia; Wu *et al.*, 2004, Lake Donghu, China).

I. intermedius Stefanski 1936 (Andrássy, 1978, inland waters, Europe).

I. longicaudatus de Man, 1884 (Zullini, 1973, Chiapas, Mexico, forest soil near the roots of a tree; and in cave with very fine yellow brown mud imbibed with water; Zullini, 1976a,b, Seveso River, Italy, relatively unpolluted water near Como, Italy; Goossens, 1976, guelta in deep canyon in El Berbara, Mauretania; Zullini, 1977a,b, in a cenote and in a brook in Mexico; Tsalolikhin, 1980 in Lake Baikal, Russia; Mol, 1984, The Netherlands; Gagarin, 2001, Lake Son Kul', water, Russia).

I. longicollis Daday, 1899 (Tsalolikhin, 1987, freshwater).

I. luci Andrássy, 1956 (Andrássy, 1973, freshwater, Cuba).

I. macramphis Schuurmans-Stekhoven & Teunissen, 1938 (Mol, 1984, The Netherlands; from P. Loof, personal communication).

I. paludicola W. Schneider, 1937 (Goossens, 1976, water pit in Tod and lakelet in Hamdoun, Mauretania; Tsalolikhin, 1996, Lakes Awash, Sago, Bale Province, Omo and Gambela, Ethiopia)

I. paramacramphis Alther, 1972 (Tsalolikhin, 1987, freshwater).

I. rotundicaudatus Kreis, 1924 (Andrássy, 1978, inland waters, Alps, Europe).

I. sphincterus Ebsary, 1985 (St. Lawrence River, Montréal, Québec, Canada) (Wu *et al.*, 2004, Lake Donghu, China).

I. tenuicaudatus de Man, 1876 (Zullini, 1976a, Seveso River, clean water near source, Como, Italy; Zullini, 1976b, experiment; Zullini and Ricci, 1980, Carrego wood stream, unpolluted, Italy; Tsalolikhin, 1980, Lake Baikal, Russia; Eder and Kirchengast, 1982, River Mur, Styria, Austria; Mol, 1984, The Netherlands; Zullini, 1988b, Lakes Zway, Abaya and Koka, Ethiopia; Tudorancea *et al.*, 1989, Lakes Zway, Abaya and Koka, Ethiopia; Gagarin, 2001, Lakes Siverskoye, Sevan, Kubenskoye, Pleshcheyevo, Rivers Angara and Volga, water reservoirs in Kiev, Kremenchug, Kakhov and Ribinsk, Russia).

I. terranovus Ebsary, 1985 (stream at Cape Spear, Newfoundland, Canada).

I. truncates Stefanski, 1916 (Andrássy, 1978, inland waters, Europe).

Ironus species were also found in Lambro River, Italy by Zullini (1988a,b) and in Murray River, Australia by Nicholas *et al.* (1992).

Subfamily Thalassironinae Andrássy, 1976

Diagnosis. Ironidae. Only dorsolateral loxometanemes of type II or no metanemes. The Thalassironinae consists of eight genera, all marine. (Some species may penetrate into brackish waters.)

Family LEPTOSOMATIDAE Filipjev, 1916

Diagnosis. Ironoidea. Large nematodes (2–50 mm). Two circles of anterior sensilla: six inner labial sensilla mostly papilliform, six outer labial + four cephalic setiform, often the cephalic setae are very short. Amphids pocket-shaped. Large number of metanemes with caudal filament: dorsolateral and ventrolateral or only dorsolateral orthometanemes and loxometanemes of type I. Many species with ocelli. Buccal cavity narrow, sometimes with tooth-like thickenings. Pharynx inserts on to the body cuticle in the region of the buccal cavity, the cephalic capsule is variable in form. Three pharyngeal glands open in the buccal cavity. Pharynx always smooth in outline. Secretory–excretory system, if present, restricted to the pharyngeal region. Female reproductive system didelphic–amphidelphic with antidromously reflexed ovaries. Males with two testes opposed. Sub-ventral or ventral pre-cloacal papillae (never tubules) often present. Caudal glands mostly present, extending into the pre-caudal region. Marine.

Type genus *Leptosomatum* Bastian, 1865

Family OXYSTOMINIDAE Chitwood, 1935

Diagnosis. Ironoidea. Body elongated and very thin at the anterior end. Cephalic sensilla is three separate circles, the second and third circle clearly separated; the inner labial sensilla papilliform or setiform, outer labial and cephalic setae very slender. Buccal cavity narrow, tubular or funnel-shaped and without teeth. Among species amphids are unusually polymorphic. Only orthometanemes with very short caudal filament present. Pharynx inserts on to the body cuticle in the region of the buccal cavity; however, the cephalic capsule is not well developed. The posterior section of the pharynx has an undulating outline. Females didelphic–amphidelphic with antidromously reflexed ovaries or monodelphic–opisthodelphic. Males diorchic with opposed testes or only one anterior testis. Position of caudal glands variable.

Three subfamilies, two of them have freshwater species.

Key to subfamilies

1. Amphidial aperture a longitudinal groove . . . . .Halalaiminae
   Amphidial aperture not a longitudinal groove 2
2. Males with two rows of sub-ventral winged supplements, females with supplements anterior and posterior to the vulva . . . . .Paroxystomininae
   Males without sub-ventral rows of winged supplements, females without supplements . . . . .Oxystomininae

Subfamily Oxystomininae Chitwood, 1953

Diagnosis. Oxystominidae. Only dorsolateral orthometanemes. Ventral gland present and confined within the pharyngeal region. Females monodelphic–opisthodelphic. Caudal glands extend into the pre-caudal region.

Type genus *Oxystomina* Filipjev, 1921

Key to genera

1. Amphids anterior to cephalic setae 2
   Amphids at the level of or posterior to cephalic setae .3
2. Amphidial aperture a transverse slit, fovea pocket-shaped . . . .*Thalassoalaimus*
   Amphidial aperture horseshoe-shaped in males and round in females . . . .*Litinium*

3. Amphidial aperture round with double contour, tail elongated conical with sharply pointed tail tip, males with one pre-cloacal spine . . . . *Wieseria*
   Amphidial aperture round to longitudinally oval, fovea as two backward projections, males with setiform supplements 4
4. Tail conico–cylindrical, tip clavate . . . . *Oxystomina*
   Tail conical . . . . *Nemanema*

Genus *Litinium* Cobb, 1920 (Fig. 12.10B)
Diagnosis. Oxystomininae. Sexual dimorphism in the shape of the amphid: in males horseshoe-shaped and in females a round aperture surrounded by a heart-shaped fovea, continuing in a distinct *canalis*. Inner and outer labial sensilla setiform, cephalic setae situated behind the amphid. Buccal cavity absent. Marine.

Type species *L. aequale* Cobb, 1920

Genus *Nemanema* Cobb, 1920 (Fig. 12.10C)
Diagnosis. Oxystomininae. Similar to *Oxystomina*, except rounded tail. Marine.

Type species *N. simplex* Cobb, 1920

For revision see Wieser (1953a) and Hope and Murphy (1972).

Genus ***Oxystomina*** Filipjev, 1921 (Fig. 12.10D)
Diagnosis. Oxystomininae. Anterior sensilla in three circles: six inner labial papillae indistinct, six outer labial setae, four cephalic setae backwardly positioned (due to elongated neck region). Buccal cavity absent. Amphidial aperture typically oval-shaped, sometimes larger in the male than in the female. Prominent oval cells (? epidermal glands) are scattered throughout the body. Secretory–excretory pore usually conspicuous and sclerotized. Tail clavate.

Type species *O. elongata* (Bütschli, 1874) Filipjev 1921

The genus *Adorus* Cobb in Thorne, 1939 (formerly placed within the Alaimidae) was considered different from *Oxystomina* by Jensen (1979) on the basis of the shape of the amphid and the reduced anterior sensilla; *Adorus* has, however, been synonymized with *Oxystomina* by Lorenzen (1981, 1994) because of the existence of *Oxystomina* species with papilliform sensilla.

Some species occur in salt marshes, mangroves: *O. tenuis* (Cobb in Thorne, 1939); brackish water habitats with low salinity: *O. astridae* (Jensen, 1979) and *O. aff. astridae* (Jensen, 1979); and in hot springs.

Species recorded from inland waters:

O. islandica (De Coninck, 1943) Wieser, 1953 (hot spring, Iceland).

Genus ***Thalassoalaimus*** de Man, 1893 (Fig. 12.10E)
Diagnosis. Oxystomininae. Inner and outer labial sensilla setiform, four cephalic setae in backward position. Amphidial aperture a transverse slit, fovea large and

pocket-shaped. Males usually have pre-cloacal papilliform supplements. Thick cuticular lining at the tail tip, called the caudal capsule.

Freshwater species:

T. aquaedulcis W. Schneider, 1940 (freshwater, cavernicolous, Yugoslavia; Andrássy, 1978, relict, inland waters, Balkan, Europe).

Genus *Wieseria* Gerlach, 1956 (Fig. 12.10F)
Diagnosis. Oxystomininae. Inner and outer labial sensilla setiform, four cephalic setae in backward position. Amphid circular with double contour. Males have a distinct pre-cloacal spine. Tail tip sharp. Marine.

Type species *W. pica* Gerlach, 1956

For revision see Vitiello (1972) and Zhang (1983).

Subfamily Halalaiminae De Coninck, 1965
Diagnosis. Oxystominidae. Amphidial aperture a longitudinal groove, fovea nearly non-existent. Two genera, one with freshwater species.

Type genus *Halalaimus* de Man, 1888

Key to genera
1. Females monodelphic, opisthodelphic . . . . *Cricohalalaimus*
   Females didelphic–amphidelphic . . . . *Halalaimus*

Genus ***Halalaimus*** de Man, 1888 (Fig. 12.7G)
Diagnosis Halalaiminae. Anterior and posterior end of body strongly attenuated. Amphid greatly elongated longitudinally. Anterior sensilla in three circles: six inner labial sensilla sometimes indistinct, papilliform or setiform; six outer labial setae and four cephalic setae. Cuticle thin from the anterior end to the level of the cephalic setae and thickened posterior to it; fine transverse striations present or absent. Alae-like structures at lateral chords present in some species. Buccal cavity absent. Pharynx long, narrow anteriorly and broader posteriorly. Female reproductive system didelphic–amphidelphic with reflexed ovaries. Males diorchic with opposed testes; pre-cloacal sensillum (seta) and/or pore present or absent; caudal alae present or absent. Egg and sperm dimorphism occurs in at least one species. Tail conical–cylindrical, tip blunt or bifurcate, cylindrical part of tail with or without transverse cuticular striations. Caudal glands present, spinneret present or undetermined. Mostly marine, some brackish and freshwater species.

Type species *H. gracilis* de Man, 1888

For revision see Keppner (1991).

Species recorded from inland waters:

H. algeriensis Coomans & Jacobs 1983 (brackish lake in Algeria)
H. dolgovi Alekseev & Linnik, 1994, 1995 (Lake Khanka, Russia)
H. durus Gagarin & Thanh, 2004 (Mekong river delta, Vietnam)
H. lineatus Gagarin & Thanh, 2004 (Mekong river delta, Vietnam)

H. luticolus Gagarin & Thanh, 2004 (Mekong river delta, Vietnam)
H. minor Gagarin & Thanh, 2004 (Mekong river delta, Vietnam)
H. stammeri W. Schneider, 1940 (in cave lake in Yugoslavia, Europe; Andrássy, 1978, relict, cave lake in Yugoslavia, Europe).

Genus *Cricohalalaimus* Bussau, 1993 (Fig. 12.7H)
Diagnosis. Halalaiminae. Cuticle thick and coarsely annulated. All anterior sensilla long, setiform and equal in length. Amphidial aperture a longitudinal groove. Sexual dimorphism in size of the amphid: shorter in females. Female reproductive system monodelphic–opisthodelphic with reflexed ovary. Tail elongated with two terminal setae. Marine.

Type and only species: *C. vagabundus* Bussau, 1993

Subfamily Paroxystomininae De Coninck, 1965
Diagnosis. Oxystominidae. Buccal cavity small, conical. Males with winged pre-cloacal supplements arranged in two sub-ventral rows. Females with supplements anterior and posterior to the vulva. Two marine genera.

Type genus *Paroxystomina* Micoletzky, 1927

Suborder TRIPYLOIDINA De Coninck, 1965 (Fig. 12.11A–F).

Diagnosis. Enoplida. Metanemes present. Three lips. Anterior sensilla in two circles, the setae of the second circle often jointed. No cephalic capsule. Amphids usually ventrally spiral, rarely dorsally spiral. Buccal cavity well developed, mostly consisting of different chambers and armed with teeth or denticles. The secretory–excretory gland, if present, extends into the post-pharyngeal region. Pharynx cylindrical. Female reproductive system didelphic–amphidelphic with reflexed ovaries. Only anterior testis present. The gonads lie ventral to the intestine. Three caudal glands confined within the tail.

The suborder Tripyloidina has only one superfamily Tripyloidoidea.

Lorenzen (1981, 1994) also put the families Tobrilidae, Tripylidae, Triodontolaimidae, Rhabdodemaniidae and Pandolaimidae within the Tripyloidina because of the presence of metanemes. However, according to De Ley and Blaxter (2004) these are put into the Triplonchida, suborder Tobrilina.

Superfamily TRIPYLOIDOIDEA Filipjev, 1928

Diagnosis. Tripyloidina. Only one family Tripyloididae.

Family TRIPYLOIDIDAE Filipjev, 1928
Diagnosis. Tripyloidoidea. Metanemes, when present, are almost exclusively ventrolateral loxometanemes of type II, whereas dorsolateral loxometanemes occur only rarely or are absent altogether. Two circles of anterior sensilla: first circle with six inner labial sensilla often setiform; second circle of six outer labial + four cephalic setae, which may be jointed. Buccal cavity consists of one, three or four portions,

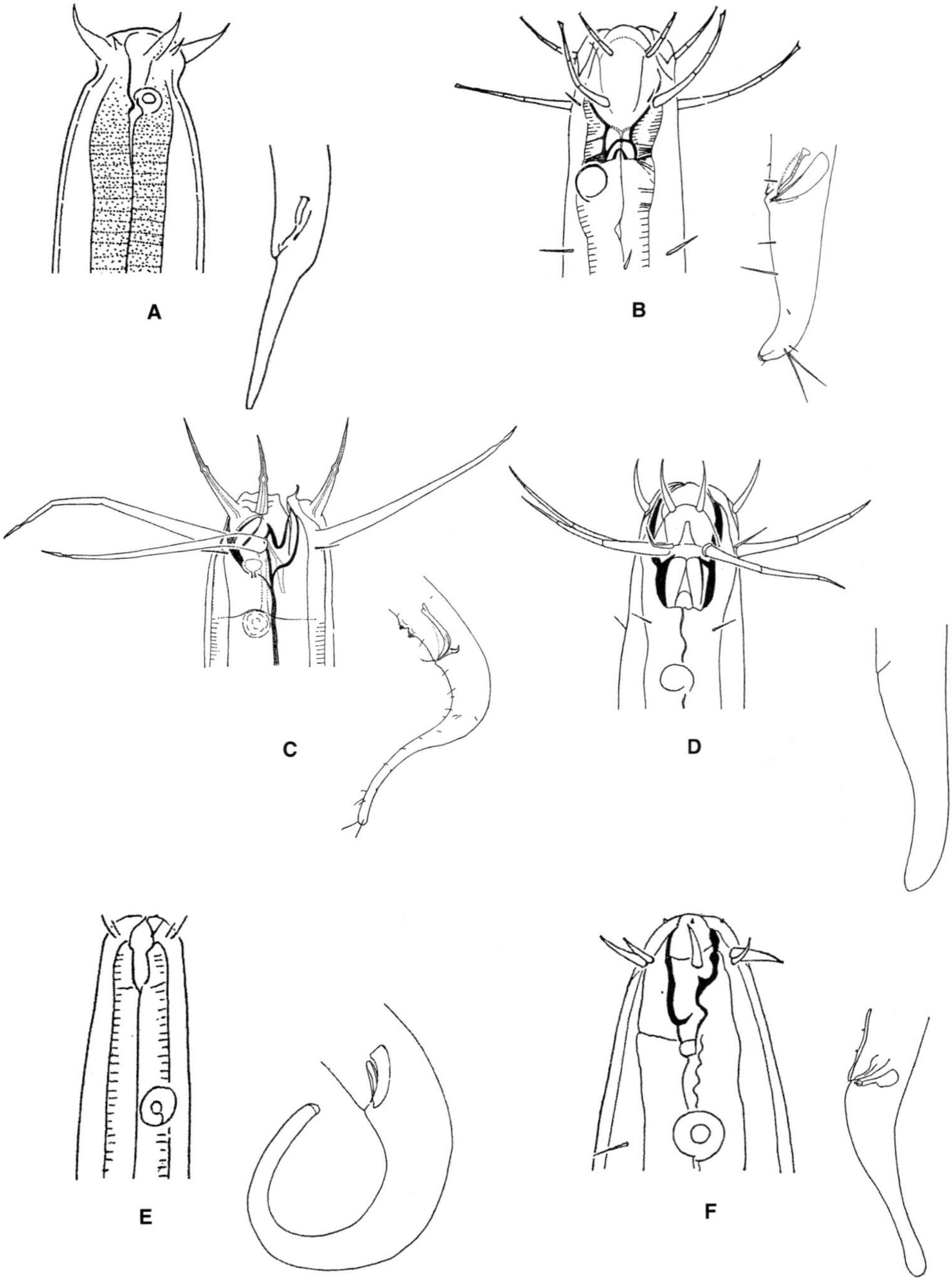

Fig. 12.11 Genera of the Tripyloidina and Tripyloididae; each genus is represented by the head end and the tail. A: *Arenasoma* (after Yeates, 1967); B: *Bathylaimus* (after Riemann, 1966); C: *Gairleanema* (after Warwick and Platt, 1973); D: *Ingenia* (after Gerlach, 1957a,b); E: *Paratripyloides* (after Schuurmans-Stekhoven, 1950); and F: *Tripyloides* (after Gerlach, 1957b).

which lie one behind the other; tooth-like projections are common, of which the dorsal tooth can be dominant.

Type genus *Tripyloides* de Man, 1886

The family consists of six genera, mostly marine, some brackish water and terrestrial species and only the type genus has freshwater species.

Key to genera

1. Buccal cavity consisting of one part only 2
   Buccal cavity consisting of more than one part 4
2. Inner labial sensilla setiform 3
   Inner labial sensilla papilliform . . . . *Arenasoma*
3. Inner labial setae with bulbous swelling . . . . *Gairleanema*
   Inner labial setae without bulbous swelling . . . . *Ingenia*
4. Buccal cavity in two parts, lips high and deeply incised . . . . *Bathylaimus*
   Buccal cavity in several parts, lips low and not deeply incised 4
5. Buccal cavity with teeth . . . . *Tripyloides*
   Buccal cavity without teeth . . . . *Paratripyloides*

Genus *Arenasoma* Yeates, 1967 (Fig. 12.11A)
Diagnosis. Tripyloididae. Cuticle with very fine transverse striations. Anterior sensilla in two circles: six inner labial papillae and six outer labial setae + four cephalic setae. Amphid unispiral. Buccal cavity elongate–conoid without teeth. Females didelphic–amphidelphic with antidromously reflexed ovaries. Males monorchic with outstretched testis, spicules short with single joint; gubernaculum without caudal apophysis; supplements absent. Caudal glands absent. Terrestrial.

Type and only species: *A. terricola* Yeates, 1967

Genus *Bathylaimus* Cobb, 1894 (Fig. 12.11B)
Diagnosis. Tripyloididae. Buccal cavity large and deep, lips deeply incised. Buccal cavity large and consisting of two parts with teeth in posterior part and cuticular projections. Males with large gubernaculum. Mostly marine species, some occur in brackish water: *B. capacosus* Hopper, 1962 and *B. longisetosus* (Allgen, 1929) Jensen, 1979 in brackish water in Finland.

Type species *B. australis* Cobb, 1894

For revision see Wieser (1956).

Genus *Gairleanema* Warwick & Platt, 1973 (Fig. 12.11C)
Diagnosis. Tripyloididae. Cuticle with indistinct transverse striations. Lips high. Anterior sensilla in two circles: six inner labial setae large and with bulbous swellings, six outer labial setae very long + four small and thin cephalic setae; labial setae jointed. Amphids dorsally spiral. Buccal cavity in one part with a pointed heavily sclerotized dorsal tooth and two less heavily sclerotized ventrosublateral teeth with rounded tip. Cardia surrounded by a sphincter. Males monorchic, testis sub-ventrally left or right of intestine. Pre-cloacal supplements mammiform. Marine.

Type and only species: *G. anagremilae* Warwick & Platt, 1973

Genus *Ingenia* Gerlach, 1957 (Fig. 12.11D)
Diagnosis. Tripyloididae. Cuticle with very fine transverse striations. Lips high. Anterior sensilla in two circles: six long inner labial setae; six very long and jointed outer labial setae + four short and thin cephalic setae. Amphids in sub-ventral position. Buccal cavity large consisting of only one part with a large sharply pointed dorsal tooth and two smaller ventrosublateral teeth. Females didelphic–amphidelphic with antidromously reflexed ovaries. Males unknown. Marine.

Type and only species: *I. mirabilis* Gerlach, 1957

Genus *Paratripyloides* Schuurmans-Stekhoven, 1950 (Fig. 12.11E)
Diagnosis. Tripyloididae. Lips low and not deeply incised. Anterior sensilla in two circles: six inner labial papillae or short setae, six outer labial setae long and mostly jointed + four cephalic setae shorter and finer. Amphids cryptospiral: circular in outline and provided with a median dot. Buccal cavity consisting of two parts, without teeth or cuticular projections. Spicules inconspicuously curved; gubernaculum large. Tail elongate, cylindrical. Marine.

Type species *P. longicauda* Schuurmans-Stekhoven, 1950

Genus ***Tripyloides*** de Man, 1886 (Fig. 12.11F)
Diagnosis. Tripyloididae. Lips low and not deeply incised. Anterior sensilla in two circles: six inner labial papillae or short setae, six outer labial setae long and mostly jointed + four cephalic setae shorter and finer. Amphids cryptospiral. Buccal cavity consisting of several parts with teeth or cuticular projections. Males sometimes with circular musculature surrounding the spicules. Mainly marine species; however, some occur in brackish and freshwater inland bodies.

Type species *T. vulgaris* de Man, 1886

For revision see Wieser (1956).

Species recorded from inland waters:

T. acherusius Gerlach, 1952 (Gerlach, 1952, syn. *Bathylaimus latisetosus* Altherr 1958, coastal groundwater, Kieler Bucht; Hussman, 1962, brackish dune water in Heligoland, Germany; Andrássy, 1978, as *B. latisetosus*, inland waters, Europe).
T. amazonicus (Gerlach, 1957) Riemann, 1970 (Gerlach, 1958, freshwater lagoon in Madagascar; Riemann, 1970, Caribbean coast, Colombia).
T. granulatus (Cobb, 1913) Wieser, 1956 (Cobb, 1913, oligohaline region, Virginia, USA).
T. marinus (Bütschli, 1874) de Man, 1886 (Riemann, 1966, oligohaline region of Elbe estuary, Germany; Andrássy, 1978, inland waters, Europe; Tchesunov, 1981, oligohaline region of the Caspian Sea).
T. pallidus Tchesunov, 1981 (oligohaline region of the Caspian Sea).

Suborder CAMPYDORINA Jairajpuri, 1983 (Fig. 12.3F)

Diagnosis. Enoplida. Cuticle apparently smooth but fine striae visible (especially in the region of tail), as well as 4–8 faint longitudinal lines. Lip region offset, lips large, mammiform. Anterior sensilla papilliform in two circles, labial papillae prominent. Amphids very small and obscure; amphidial aperture slit-like, fovea stirrup-shaped. Stoma narrow and tubular with dorsally located hollow, acute, mural tooth. Pharynx slender anteriorly, basal bulb with a distinct, elongated triquetrous chambered lumen. Secretory–excretory pore and duct present, cuticularized. Female reproductive system didelphic–amphidelphic with reflexed ovaries extending beyond vulva; asymmetrical, usually anterior branch more developed. Males monorchic, testis outstretched, spicules slightly arcuate, gubernaculum short, supplements papilliform, well-developed pericloacal (leptoderan) alae. Pre-rectum absent. Caudal glands absent.

Superfamily CAMPYDOROIDEA Jairajpuri, Ahmad & Bajaj 1976

Diagnosis. Campydorina.

Family CAMPYDORIDAE (Thorne, 1935) Clark, 1961

Diagnosis. Campydoroidea.

Type genus *Campydora* Cobb, 1920

Genus *Campydora* Cobb, 1920

Diagnosis. Campydoridae.

Type and only species: *C. demonstrans* Cobb 1920

For revision see Jairajpuri (1983), Winiszewska (2001) and Mullin *et al.* (2003).

Andrássy (1954) transferred *Desmolaimus balatonicus* Daday, 1894 from Lake Balaton, Hungary, to this genus. However, the original description and drawing by Daday (1894) does not reveal any correspondence with the genus diagnosis. *D. balatonicus* has a distinct annulated cuticle and a short funnel-shaped buccal cavity without any buccal tooth. We therefore follow Jairajpuri (1983) and reject this freshwater species as belonging to *Campydora*. *C. demonstrans*, a terrestrial species, remains the only species and the genus is regarded as terrestrial.

Suborder ALAIMINA Clark, 1961 (Figs 12.3G and 12.12A–N)

Diagnosis. Enoplida. Metanemes absent. Anterior sensilla in two circles; pits or papilliform, never setiform. Amphids with crescent apertures (*Amphidelus*), or minute, pore-like (*Alaimus*), commonly placed far back from lip region; fovea and sensillar pouch (fusus) separated by a distinct duct. Stoma vestigial, unarmed, surrounded by pharyngeal tissue. Pharynx slender, enlarged only in terminal third. Pharyngeal glands described as opening into the lumen behind the nerve ring. No pre-rectum. Females monodelphic–opisthodelphic, rarely didelphic–amphidelphic. Males monorchic with outstretched testis, spicules short and straight to longer and arcuate, gubernaculum absent, pre-cloacal supplements absent or present. Caudal glands and spinneret absent. Generally terrestrial, some genera with freshwater species.

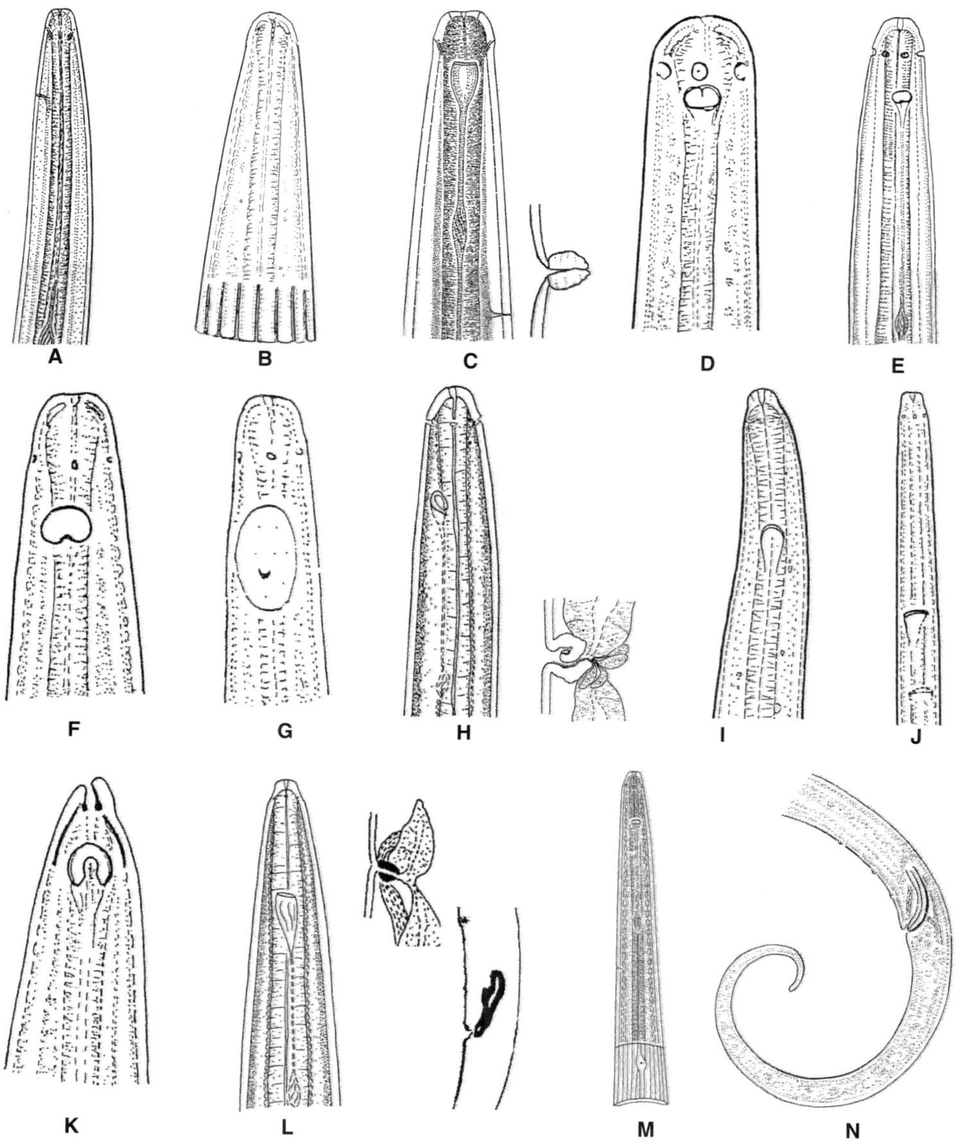

Fig. 12.12 Genera of the Alaimina, Alaimidae, Alaiminae (A, B), Amphidelinae (C–L) and Cristamphidelinae (M); each genus is represented by the head end, some with additional typical features. A: *Alaimus* (after Thorne, 1939); B: *Cosalaimus* (after Siddiqi, 1993); C: *Amphidelus* head and vagina (after Clausi and Vinciguerra, 1998); D: *Caviputa* (after Siddiqi, 1993); E: *Etamphidelus* (after Siddiqi and Vinciguerra, 1991); F: *Laxamphidelus* (after Siddiqi, 1993); G: *Megamphidelus* (after Siddiqi, 1993); H: *Metamphidelus* head and vagina (after Clausi and Vinciguerra, 1995); I: *Paramphidelus* (after Andrássy, 1977b); J: *Postamphidelus* (after Siddiqi, 1993); K: *Scleralaimus* (after Siddiqi, 1993); L: *Scleramphidelus* head, vagina and male tail (after Clausi and Vinciguerra, 1995); M: *Cristamphidelus* (after Coomans and Raski, 1988); and N: typical *Alaimina* tail (after Thorne, 1939).

The suborder Alaimina was erected by Clark (1961) and placed within the Enoplida because the species did not conform to the Tripyloidea nor to the Dorylaimoidea. They possess characters from both superfamilies as well as some special ones of their own.

The suborder Alaimina only has one superfamily Alaimoidea and one family Alaimidae

The Alaimidae (together with a residue of six other families with few species) were put by Lorenzen (1981, 1994) in the Dorylaimida because according to Clark (1961) all pharyngeal glands open posterior to the nerve ring, however, expressing the phylogenetic uncertainty of its position. Indeed, the position of these outlets needs confirmation.

Superfamily ALAIMOIDEA Micoletzky, 1922

Diagnosis. Alaimina.

Family ALAIMIDAE Micoletzky, 1922

Diagnosis. Alaimoidea. Body without cuticular appendages, cephalic or somatic setae. Cuticle non-annulated (L.M.). Head rounded or truncated. Anterior sensilla in two circles: six + (6 + 4). Buccal cavity very small and unarmed. Amphid opening pore- or slit-like, often placed far back from the lip region. Pharynx anteriorly slender, gradually widening posteriorly. Pre-rectum absent. Female gonads single or paired. Testis single. Spicules short, gubernaculum absent. Mostly few ventromedial supplements. Caudal glands and spinneret absent.

Three subfamilies and 13 genera, all of them have mainly terrestrial species; however, some occur in freshwater.

For revision see Andrássy (2002) with key to all genera of the family.

Key to subfamilies

1. Cuticle with longitudinal ridges, vagina S-shaped, spicules long . . . . Cristamphidelinae
   Cuticle usually without such ridges 2
2. Amphids with distinct fovea, with transverse slit-like aperture and long amphidial duct, labial sensilla distinct, spicules short . . . . Amphidelinae
   Amphids minute, fovea indistinct with pore-like aperture and short amphidial duct, labial sensilla obscure . . . . Alaiminae

Subfamily Alaiminae Micoletzky, 1922

Diagnosis. Alaimidae. Cuticle practically smooth or, rarely, provided with well-developed longitudinal ridges 20–30 in number. Inner labial sensilla minute and often indistinct. Amphids very small, amphidial aperture pore-like, situated very posteriorly, at one-third to one-half of pharynx length (except *A. oryzae*). Amphidial duct between fovea and sensillar pouch (fusus) quite short. Pharynx rapidly enlarging in its posterior fifth to seventh. Female reproductive system monodelphic–opisthodelphic. Spicules small with central line, shorter than anal body diameter. Two genera.

Type genus *Alaimus* de Man, 1880

Genus ***Alaimus*** de Man, 1880 (Fig. 12.12A)
Diagnosis. Alaiminae. Cuticle smooth. Lip region rounded, continuous with body contour. Anterior sensilla in two circles, inner labial sensilla mostly invisible (L.M.). Amphid apertures minute, pore-like, connected with fusus by slender and short amphidial duct. Pharynx with seven pharyngeal gland cells. Pharyngo-intestinal valve thin and disk-like. Excretory pore far back in pharyngeal region and often obscure. Females monodelphic–opisthodelphic with reflexed ovary; vulva a transverse slit, vagina thick, muscular. Testis single, outstretched. Spicules quite short and simple; gubernaculum absent; supplements consisting of a ventromedian row of three to nine papillae (exceptionally 16), adcloacal pair absent. Males known in about 50% of the species. Most species are terrestrial, some occur in freshwater or in moist soil.

Type species *A. primitivus* de Man, 1880

For revision see Siddiqi and Husain (1967) and Andrássy (2002).

Species recorded from freshwater habitats:

A. asifkhalili Mulk & Coomans, 1979 (Mount Kenya, Kenya).
A. belogurovi Alekseev, 1996 (Lake Khanka, Russia).
A. longiovatus Alekseev, 1996 (Lake Khanka, Russia).
A. meyli Andrássy, 1961 (Andrássy, 1978, inland waters, Caucasus, Europe; Mol, 1984, The Netherlands).
A. parvus Thorne, 1939 (Andrássy, 1978, inland waters, Europe; Mol (1984), The Netherlands).
A. primitivus de Man, 1880 (Andrássy, 1978, inland waters, Europe; Colomba and Vinciguerra, 1979, Sicilia; Zullini and Ricci, 1980, in Carrego wood stream, unpolluted, Italy; Mol, 1984, The Netherlands; Gagarin, 1993b, Taimir Lake, Russia; Gagarin, 1995, sewage disposal Rybinsk water reservoir, Borok, Russia; Alekseev, 1996, Frolovka River, Russia; Wu and Liang, 1999, Lake Biandantang, China; Gagarin, 2001, River Volga, water reservoirs in Kiev, Kremenchug and Kakhov, Lake Pleshcheyevo, Russia; Wu *et al.*, 2004, Lake Donghu, China).
A. thompsoni Mulk & Coomans, 1979 (Mount Kenya, Kenya).
A. wittmeri Eyualem-Abebe & Coomans 1995 (in Wittmer's Well, Floreana, Galápagos)
Alaimus sp. (Zullini & Peretti, 1986, in moss affected by lead pollution, Po River, Italy; Zullini, 1988a, in Lambro River, Italy; Nicholas *et al.*, 1992, in Murray River, Australia).

Genus *Cosalaimus* Siddiqi, 1993 (Fig. 12.12B)
Diagnosis. Alaiminae. Cuticle with 20–30 well-expressed longitudinal ridges. Anterior sensilla indistinct. Amphids indistinct. Females monodelphic–opisthodelphic, pre-vulval uterine sac absent or rudimentary. Spicules short, straight, proximally much wider. Three ventromedian supplements. Males known in one species only. Terrestrial.

Type species *C. costatus* Siddiqi, 1993

For revision see Andrássy (2002).

Subfamily Amphidelinae Andrássy, 2002

Diagnosis. Alaimidae. Cuticle smooth (L.M.), without longitudinal ridges. Outer labial sensilla distinct, circular, cup-shaped. Amphids distinct with large to very large transverse slit-like or oval, or longitudinally oval apertures (except for females of one genus), generally at 1–5 (exceptionally 6–8) labial diameters from anterior body end. Amphidial duct between fovea and sensilla long to very long. Pharynx gradually expanding. Females mono- or didelphic. Spicules massive, mostly straight, shorter to hardly longer than anal body diameter, predominantly with central line. Ten genera, mostly terrestrial.

Type genus *Amphidelus* Thorne, 1939

Genus ***Amphidelus*** Thorne, 1939 (Fig. 12.12C)

Diagnosis. Amphidelinae. Relatively large size. Head rounded, not offset. Lip region low. Amphids pocket-shaped, well developed, apertures large transverse slits, fovea broad, deeper than wide; sometimes elongate–conoid; positioned anterior to two head diameters from anterior end, mostly at one head diameter from anterior end. Pharynx relatively short, gradually expanding in its posterior part. Secretory–excretory pore anteriorly located. Females mostly didelphic–amphidelphic with reflexed ovaries (except in *A. dolichurus* and *A. pusillus*); vagina thick-walled and vulva with open lips. Males monorchic with anteriorly directed testis, spicules short with central piece, 3–5 ventral pre-cloacal papillae. Many species are found in mosses.

Type species *A. lissus* Thorne, 1939

For revision see Siddiqi and Basir (1965), Andrássy (1977b), Clausi and Vinciguerra (1998) and Andrássy (2002).

Species recorded from inland waters:

A. coronatus Andrássy, 1978 (inland waters, Danube countries, Europe).

A. dudichi Andrássy, 1957 (in sodic waters, Hungary; Andrássy, 1978, inland waters, Donau countries, Europe).

A. elegans (de Man, 1921) Thorne, 1939[8] (Andrássy, 1978, inland waters, Alps, Europe; Mol, 1984, The Netherlands; Alekseev, 1996, Frolovka River, Russia).

A. rivalis Alekseev, 1996 (Frolovka River, Russia).

A. stefanskii Juget, 1969 (in Lake Leman, Switzerland; Andrássy, 1978, inland waters, Europe).

A. tenuis Alekseev, 1996 (Lake Khanka, Russia).

A. thornei Clausi & Vinciguerra, 1998 (stream bank soil, Broad Run, Virginia, USA).

Genus *Caviputa* Siddiqi, 1993 (Fig. 12.12D)

Diagnosis. Amphidelinae. Body very slender. Outer labial sensilla large, circular, cup-like. Amphid aperture in females indistinct, in males large, circular, transversely oval

[8] Clausi and Vinciguerra (1998) consider all citations of *A. elegans* dubious, *A. elegans* was originally found in humid soil.

or kidney-shaped, located immediately behind lateral outer labial sensilla. Females didelphic–amphidelphic, anterior gonad normal or reduced. Spicules somewhat longer than anal diameter, slightly arcuate with central line. Three to four ventromedian supplements present. Males known in about 50% of the species. Terrestrial.

Type species *C. connota* Siddiqi, 1993

For revision see Andrássy (2002).

Genus *Etamphidelus* Andrássy, 1977 (Fig. 12.12E)
Diagnosis. Amphidelinae. Body slender, anteriorly narrowing strongly. Head rounded. Outer labial papillae prominent and peculiar: pits with sclerotized rim. Amphids large with oval sclerotized aperture and situated at 1–3 head diameters from anterior end. Females monodelphic–prodelphic without posterior vulval sac (one exception). Vulval musculature strongly developed at post-vulvar region; vagina thin. Spicules short, strong and arcuate with central piece; 2–4 pre-cloacal papillae. Males known in most of the species. Terrestrial.

Type species *E. japonicus* Andrássy, 1977

For revision see Coomans and Raski (1988), Siddiqi and Vinciguerra (1991) and Andrássy (2002).

Genus *Laxamphidelus* Siddiqi, 1993 (Fig. 12.12F)
Diagnosis. Amphidelinae. Outer labial sensilla as distinct pits. Amphid fovea large and roomy, aperture transversely oval to rounded, close to outer labial sensilla. Females monodelphic–opisthodelphic, no pre-vulval uterine sac. Spicules short and straight with central line. Supplements two or three. Males known in most of the species. Terrestrial.

Type species *L. laxus* Siddiqi, 1993

For revision see Andrássy (2002).

Genus *Megamphidelus* Siddiqi, 1993 (Fig. 12.12G)
Diagnosis. Amphidelinae. Outer labial sensilla minute but distinct pits. Amphids with unusually large, longitudinally oval aperture, close to head. Females monodelphic–opisthodelphic without pre-vulval branch. Spicules a little longer than anal body diameter, slightly arcuate with central line. Ventromedian supplements two. Terrestrial.

Type and only species: *M. filiformis* Siddiqi, 1993

Genus *Metamphidelus* Clausi & Vinciguerra, 1995 (Fig. 12.12H)
Diagnosis. Amphidelinae. Labial sensilla small but distinct. Amphids about two head widths from anterior with small shallow fovea and small, roundish, heavily sclerotized aperture. Females didelphic–amphidelphic, vagina thick with infolded walls. Spicules very short. Two supplements. Males known in only one species. Terrestrial.

Type species *M. brzeskii* Clausi & Vinciguerra, 1995

For revision see Andrássy (2002).

Genus ***Paramphidelus*** Andrássy, 1977 (Fig. 12.12I)
Diagnosis. Amphidelinae. Amphids positioned posteriorly at more than two head diameters from anterior end; amphidial openings crescent-like, fovea funnel- or racket-shaped. Female reproductive system typically monodelphic, mostly opisthodelphic, rarely prodelphic, with or without uterine sac, exceptionally didelphic with longer post-vulvar branch; vagina thin. Males rare. Spicules plump and shorter than cloacal diameter; 2–4 pre-cloacal supplements.

Type species *P. dolichurus* (de Man, 1876) Andrássy, 1977

For revision see Andrássy (1977b) and Andrássy (2002).

Species recorded from inland waters:

P. dolichurus (de Man, 1876) Andrássy, 1977 (Andrássy, 1978, inland waters, Europe; Eder and Kirchengast, 1982, River Mur, Styria, Austria; Mol, 1984, The Netherlands; Gagarin, 1993a,b, Taimir Lake, Russia; Eyualem-Abebe and Coomans, 1995 in Wittmer's Well, Floreana, Galapagos; Gagarin, 2001, River Volga, pond near Borok, Ribinsk water store, Russia).
P. exilis (Andrássy, 1962) Andrássy, 1977 (Andrássy, 1978, inland waters, Danube countries, Europe).
P. hortensis Zullini (1988a, Lambro River, Italy).
P. paludicola Gagarin, 1991 (Gagarin, 1993a,b, Taimir Lake, Russia).
P. propinquus (Andrássy, 1962) Andrássy, 1977 (Andrássy, 1978, inland waters, Donau countries, Europe).
P. puccinelliae Lorenzen, 1966 (brackish, salt marshes, Schleswig-Holstein, Germany).
P. uniformis (Thorne, 1939) Andrássy, 1977 (Andrássy, 1978, inland waters, Europe; Mol, 1984, The Netherlands; Gagarin, 1993a,b, Taimir Lake, Russia; Alekseev, 1996, Lake Khanka, Russia; Gagarin, 2001, Ribinsk water store, Russia).

Genus *Postamphidelus* Siddiqi, 1993 (Fig. 12.12J)
Diagnosis. Amphidelinae. Labial sensilla rather small but distinct, pit-like. Amphid aperture distinct, transverse slits, located far posterior, 6–8 labial diameters from anterior end, the two apertures not opposite to each other but separated by one corresponding body width. Pharynx very long, about one-third of body length. Females monodelphic–opisthodelphic, no pre-vulval uterine sac. Spicules short with central piece. Two supplements. Terrestrial.

Type and only species: *P. asymmetricus* Siddiqi, 1993

Genus *Scleralaimus* Siddiqi, 1993 (Fig. 12.12K)
Diagnosis. Amphidelinae. Cephalic region strongly tapering, asymmetrical, with oral opening subterminal in dorsal position. Cephalic region, stoma and fovea walls strongly sclerotized. Anterior sensilla indistinct. Amphid aperture large, bilobed, located immediately posterior to head region. Females monodelphic–opisthodelphic, without pre-vulval uterine sac. Males unknown. Terrestrial.

Type and only species: *S. serpens* Siddiqi, 1993

Genus ***Scleramphidelus*** Clausi & Vinciguerra, 1995 (Fig. 12.12L)

Diagnosis. Amphidelinae. Anterior sensilla consisting of three circles of sensorial pits. Amphids, located at 2–5 head widths from anterior end, are large, stirrup-shaped; aperture slit-like to oval with sclerotized margin; fovea wide and deep with distinct innervation. Females monodelphic–opisthodelphic with pre-vulval uterine sac; vulval lips sclerotized. Spicules strongly sclerotized, without central piece, somewhat longer than one anal diameter, 4–5 pre-cloacal supplements. Freshwater.

Type and only species: *S. lemani* (Stefanski, 1914) Clausi & Vinciguerra, 1995

Freshwater species:

S. lemani (Stefanski, 1914) Andrássy, 1977 (Stefanski, 1914, Lake Leman, Switzerland; Schiemer, 1978, in Neusiedlersee, Austria (syn. *Amphidelus lemani longicaudatus*; Andrássy, 1978, inland waters, Alps and Donau countries, Europe).

Subfamily Cristamphidelinae Andrássy, 2002

Diagnosis. Alaimidae. Cuticle with numerous fine longitudinal ridges. Labial sensilla distinct, circular. Amphid apertures transverse, either close to lip or more posterior, but never beyond one-tenth of pharyngeal length from anterior end. Amphidial duct between fovea and fusus as long as 2–4 corresponding body diameters. Pharynx gradually widening. Female gonad prodelphic, vagina S-shaped. Eggshell sculptured. Spicules distinctly longer than anal body diameter, slender, ventrally arcuate. Terrestrial.

Type and only genus: *Cristamphidelus* Siddiqi & Vinciguerra, 1991

For revision see Andrássy (2002).

Genus *Cristamphidelus* Siddiqi & Vinciguerra, 1991 (Fig. 12.12M)

Diagnosis. Alaimidae. Cuticle with 30–60 longitudinal ridges. Head continuous with body contour. Only six conspicuous outer labial sensory pits visible with light microscope. Amphidial aperture slit-like or transversely oval and rim not appearing sclerotized; fovea cup-shaped connected with fusus through a long amphidial duct. Female reproductive system monodelphic–prodelphic; distal part of uterus functioning as spermatheca; vagina narrow, thin-walled, anteriorly oblique; vulva a transverse slit; eggshell with protuberances; post-vulval sac absent; vulval dilator muscles fan-shaped. Male monorchic with long glandular vas deferens; sperm very long, elongate spindle-shaped. Spicules long, slender and arcuate; two or three medioventral supplements present. Terrestrial.

Type species *C. fuegensis* (Coomans & Raski, 1988) Siddiqi & Vinciguerra, 1991

For revision see Andrássy (2002).

References

Alekseev, V.M. (1996) A new species of freshwater nematodes, *Alaimus belogurovi* sp. n., from the Lake Khanka (Nematoda: Alaimidae). *Zoosystematica Rossica* 4, 17–18.

Alekseev, V.M. and Linnik O.V. (1995)[9] *Halalaimus dolgovi* sp.n. (Nematoda, Oxystominidae), new species of marine nematode from Lake Khanka. *Hydrobiological Journal* 31, 20–24.

Andrássy, I. (1954) Über einige von Daday beschriebene Nematodenarten. *Zoologischer Anzeiger* 152, 138–144.

Andrássy, I. (1968) Fauna Paraguayensis 2. Nematoden aus den Galeriewäldern des Acaray-Flusses. *Opuscula Zoologica Budapest* 8, 167–315.

Andrássy, I. (1973) Nematoden aus Strand- und Höhlenbiotopen von Kuba. *Acta Zoologica Academiae Scientiarum Hungaricae* 19, 233–270.

Andrássy, I. (1976) *Evolution and Systematization of Nematodes.* Pitman Publishing, London, San Francisco and Melbourne, 288 pp.

Andrássy, I. (1977a) Ergebnisse der zoologischen Forschungen von Dr. Z. Kaszab in der Mongolei. 356. Süsswasser- und Bodennematoden aus den Jahren 1967 und 1968. *Opuscula Zoologica Instituti Zoosystematici Universitatis Budapestinensis* 13, 3–24.

Andrássy, I. (1977b) Die Gattungen Amphidelus Thorne, 1939, Paramphidelus n.gen. und Etamphidelus n.gen. (Nematoda: Alaimidae). *Opuscula Zoologica Instituti Zoosystematici Universitatis Budapestinensis* 14, 3–43.

Andrássy, I. (1978) Nematoda. A checklist of the animals inhabiting European inland waters, with accounts of their distribution and ecology. In: Illies, J. (ed.) *Limnofauna Europaea.* Gustav Fisher Verlag, Stuttgart, Germany, pp. 98–117.

Andrássy, I. (2002) The genus *Cristamphidelus* Siddiqi and Vinciguerra, 1991 and a general survey of the family Alaimidae (Nematoda). *Journal of Nematode Morphology and Systematics* 4, 51–82.

Belogurov, O.I. (1979) A study of the genus *Calyptronema* Marion, 1870 (Nematoda, Enoplida). *Nauchnye Doklady Vysshei Shkoly, Biologicheskie Nauki* 5, 53–59.

Belogurov, O.I. (1983) Taxonomic and morphological problems in studies of the Pelagonematidae (Nematoda, Oncholaimina). *Zoologicheskii Zhurnal* 62, 485–492 (in Russian).

Belogurov, O.I. (1985) On the origin and stages of the evolution of the stoma in nematodes and analysis of the organization of the stoma in the suborder Oncholaimina. *Zoologicheskii Zhurnal* 64, 982–992.

Belogurov, O.I and Alekseev, V.M. (1977) Morphology of *Anoplostoma cuticularia* sp.n. (Nematoda, Enoplida) and the position of the family Anoplostomatidae Gerlach and Riemann, 1974 in the system of Nematoda. *Zoologicheskii Zhurnal* 56, 188–198 (in Russian).

Belogurov, O.I. and Belogurova, L.S (1975) Organization of the cephalic end of nematodes Oncholaiminae and in the 'cuticle splitting' phenomenon. *Biologiya Morya* 4, 24–35 (in Russian).

Belogurov, O.I. and Belogurova, L.S (1977) Systematics and evolution of Oncholaiminae (Nematoda, Oncholaimidae). I. Structure, evolution and taxonomic significance of the demanian system. *Biologiya Morya* 3, 36–47 (in Russian).

Belogurov, O.I. and Belogurova, L.S (1978) Systematics and evolution of Oncholaiminae (Nematoda). III. System of Oncholaiminae. *Biologiya Morya* 2, 22–31 (in Russian).

Belogurov, O.I. and Belogurova, L.S. (1980).Morphology of *Belbolla intarma* sp.n., diagnosis and table for species determination of the genus *Belbolla. Biologiya Morya* 4, 74–77.

Belogurov, O.I. and Belogurova, L.S. (1989) Morphology and systematics of free-living Oncholaimidae (Nematoda: Enoplida: Oncholaimina). *Asian Marine Biology* 6, 31–58.

Belogurov, O.I. and Dashenko, O.I. (1981) Krampiinae, Oncholaimidae (Nematoda) with primitive features of organization. *Nauchnye doklady vysshei shkoly. Biologicheskie nauki* 1, 44–49 (in Russian).

Belogurov, O.I., Belogurova, L.S. and Leonova, A.V. (1972) Review of the genus *Pseudoncholaimus* (Nematoda, Oncholaimidae) and notes on the monodelphic nature of nematodes and heterotopic female sex aperture. *Zoologicheskii Zhurnal* 51, 1450–1456 (in Russian).

[9] Originally published in *Zoologicheskiy Zhurnal*, (1994), 73 (4), 111–115

Belogurov, O.I., Belogurova, L.S. and Pavlyuk, O.N. (1980) Three new species of free-living nematodes of the family Oncholaimidae from the far-eastern seas of the USSR and a key to species of the genus *Pseudoncholaimus* Kreis, 1932. In: *Pribrezhnyi planktoni bentos severnoi chasti Yaponspogomorya.* USSR Acadamy of Sciences, Vladivostok, pp. 57–64 (in Russian).

Belogurov, O.I., Fadeeva, N.P. and Belogurova, L.S. (1983) Studies of nematodes of the subfamily Eurystomininae (Enoplida, Enchelidiidae) from the Far East Seas of the USSR. *Fotolaimus marinus* gen.et sp.n. (Nematoda; Oncholaimidae) from the Schikotan Island (Kuril Islands). *Zoologicheskii Zhurnal* 53, 1566–1568 (in Russian).

Belogurova, L.S. and Belogurov, O.I. (1974) New species of free-living nematodes of the family Oncholaimidae from the far-eastern seas of the USSR and a key to species of the genus *Pseudoncholaimus* Kreis, 1932. In: *Pribrezhnyi plankton i bentos severnoi chasti Yaponskogo morya.* USSR Academy of Sciences, Vladivostok, 57–64 (in Russian).

Bollerup, G. and Burr, A.H. (1979) Eyespot and other pigments in nematode esophageal muscle cells. *Canadian Journal of Zoology* 57, 1057–1069.

Boucher, G. (1970) *Paramesacanthion catellus* n.sp., nouvelle espèce d'Enoplidae (Nématode) de la vase terrigène côtière de Banyuls-sur-Mer. *Museum National d'Histoire Naturelle de Bruxelles* 42, 570–576.

Bresslau, E. and Schuurmans-Stekhoven, J.H. (1940) Marine freilebende Nematoden aus der Nordsee. *Musée Royal d'Histoire Naturelle de Belgique* 42, 1–74.

Burr, A.H. and Burr, C. (1975) The amphid of the nematode *Oncholaimus vesicarius:* Ultrastructural evidence for a dual function as chemoreceptor and photoreceptor. *Journal of Ultrastructural Research* 51, 1–15.

Burr, A.H. and Webster, J. (1975) Morphology of the eyespot and description of two pigment granules in the esophageal muscle of a marine nematode, *Oncholaimus vesicarius*. *Journal of Ultrastructural Research* 36, 621–632.

Bussau, C. (1993). Taxonomische und ökologische Untersuchungen an Nematoden des Peru-Beckens. Doctoral dissertation, Christian-Albrechts-Universität zu Kiel, Germany, 621 pp.

Calcoen, J.A. and Dekegel, D. (1979) Spermatozoa in the demanian organ of female *Adoncholaimus fuscus* (Bastian, 1865) (Nematoda). *Netherlands Journal of Zoology* 29, 142–143.

Calcoen, J.A. and Dekegel, D. (1980) Electron microscopic observations on the demanian system of female *Metoncholaimus denticulatus* Schuurmans-Stekhoven and Adam, 1931 (Nematoda). *Netherlands Journal of Zoology* 30, 516–526.

Chabaud, A., Bain, O., Hugot, J.-P., Rausch, R.L. and Rausch, V.R. (1983) Organe de de Man et insémination traumatique. *Revue de Nématologie* 6, 127–131.

Chitwood, B.G. (1960) A preliminary contribution on the marine nemas (Adenophorea) of Northern California. *Transactions of the American Microscopical Society* 79, 347–384.

Clark, W.G. (1961) A revised classification of the Order Enoplida (Nematoda). *New Zealand Journal of Science* 4, 123–150.

Clausi, M. and Vinciguerra, M.T. (1995) Description of *Metamphidelus* gen. n. and *Scleramphidelus* gen. n. (Alaimidae: Nematodea). *Nematologia Mediterranea* 23, 15–27.

Clausi, M. and Vinciguerra, M.T. (1998) Revision of the genus *Amphidelus* Thorne, 1939 (Nematoda: Alaimidae). *Journal of Nematode Morphology and Systematics* 1, 57–98.

Cobb, N.A. (1913) New nematode genera found inhabiting freshwater and non-brackish soils. *Journal of the Washington Academy of Science* 3, 432–444.

Cobb, N. A. (1914) The North American free-living fresh-water nematodes. *Transactions of the American Microscopical Society* 33, 69–119.

Cobb, N.A. (1920) One hundred new nemas (type species of 100 new genera). *Contribution to a Science of Nematology* (Baltimore) 9, 127–343.

Cobb, N.A. (1930) The demanian vessels in nemas of the genus *Oncholaimus*; with notes on four new *Oncholaimus*. *Journal of the Washington Academy of Sciences* 20, 225–241.

Cobb, N.A. (1932) *Metoncholaimus pristiuris* (zur Strassen); a nema suitable for use in courses in zoology. *Journal of the Washington Academy of Science* 22, 344–354.

Colomba, G. and Vinciguerra, M.T. (1979) Nematodi d'aqua dolce della Sicilia. *Animalia* 6, 89–120.

Coomans, A. and Heyns, J. (1983) *Oncholaimus sahariensis* sp. n. (Nematoda) from the Algerian Sahara. *Hydrobiologia* 107, 193–201.

Coomans, A. and Heyns, J. (1986) *Oncholaimus jessicae* n. sp. (Nematoda: Oncholaimidae) from freshwater in the Transvaal. *South African Journal of Zoology* 21, 197–201.

Coomans, A. and Jacobs, L.J. (1983) *Halalaimus algeriensis* n.sp. (Nematoda) from the Sahara. *Hydrobiologia* 102, 39–44.

Coomans, A.V. and Raski, D.J. (1988) Three new species of *Etamphidelus* Andrássy, 1977 (Nemata: Alaimidae) in Southern Chile. *Journal of Nematology* 20, 9–22.

Coomans, A., Vincx, M. and Decraemer W. (1985) Nematodes from a fresh-water pool on a coral island in the Solomon Islands. *Hydrobiologia* 123, 265–281.

Coomans, A., Verschuren, D. and Vanderhaeghen, R. (1988) The demanian system, traumatic insemination and reproductive strategy in *Oncholaimus oxyuris* Ditlevsen (Nematoda, Oncholaimina). *Zoologica Scripta* 17, 15–23.

Croll, N.A. (1966) The chemical nature of the pigment spots of *Enoplus communis*. *Nature* 211, 859.

Daday, E.V. (1894) New contributions towards the knowledge of the microfauna of lake Balaton (Nematoda, Rotatoria, Entomostraca). *Mathematikai és Természettudományi Értesíto* 12, 122–145 (in Hungarian).

De Coninck, L.A. (1943) Sur quelques espèces nouvelles de Nématodes libres des eaux et des terres saumâtres de l'Islande. *Biologisch Jaarboek* 10, 193–219.

De Coninck, L.A. (1965) Classe des Nématodes – Systématique des Nématodes et sous-classe des Adenophorea. In: Grassé, P. (ed.) *Traité de Zoologie*. Masson, Paris 4, 586–681.

De Coninck, L.A. and Schuurmans-Stekhoven, J.H. (1933) The free-living marine nemas of the Belgian coast. II. *Verhandelingen van het Koninklijk Belgisch Instituut voor Natuurwetenschappen* 58, 1–163.

De Ley, P. and Blaxter, M. (2002) Systematic position and phylogeny. In: Lee, D.L. (ed.) *The Biology of Nematodes*. Taylor & Francis, London and New York, pp. 1–30.

De Ley, P. and Blaxter, M. (2004) A new system for Nematoda: combining morphological characters with molecular trees and translating clades into ranks and taxa. In: Cook, R. and Hunt, D.J. (eds) *Nematology Monographs and Perspectives 2, Proceedings of the fourth International Congress of Nematology*. Brill, Leiden, The Netherlands, pp. 633–653.

de Man, J.G. (1884) Die frei in der reinen Erde und im süssen Wasser lebenden Nematoden der Niederländischen Fauna. Eine systematisch-faunistische Monographie, Leiden, The Netherlands, 206 pp.

de Man, J.G. (1886) Anatomische Untersuchungen über freilebende Nordsee-Nematoden. Paul Frohberg, Leipzig, Germany, 82 pp.

de Man, J.G. (1889) Über zwei in der feuchten Erde lebende Arten der Gattung *Oncholaimus* Duj. *Tijdschrift der Nederlandsche Dierkundige Vereeniging* 2, 162–169.

de Man, J.G. (1907) Sur quelques espèces nouvelles ou peu connues des Nématodes habitat les côtes de la Zélande. *Mémoires de la Société de Zoologie de France* 20, 33–90.

Ebsary, B.A. (1985) Two new aquatic species of *Ironus* Bastian, 1865 (Nematoda: Ironidae) from Canada. *Canadian Journal of Zoology* 63, 1368–1370.

Eder, R. and Kirchengast, M. (1982) The nematode-fauna (Nemathelminthes, Nematoda) of a polluted part of the river Mur (Styria, Austria). *Nematologia Mediterranea* 10, 127–134.

Eyualem-Abebe and Coomans, A. (1995) Freshwater nematodes of the Galápagos. *Hydrobiologia* 299, 1–51.

Fadeeva, N.P. and Belogurov, O.I. (1988) *Aronema crinita* gen. et sp. n. as a missing link in the phylogenetic tree of the family Enchelidiidae (Nematoda, Enoplida, Oncholaimina). *Zoologicheskii Zhurnal* 67, 614–617.

Filipjev, I. (1918) Free-living marine Nematodes of the Sevastopol area. *Trudy Osoboy zoologicheskoy laboratorii i Sevastopol'skoy Biologicheskoy stantsii Rossiyskoy Akademii Nauk* 4, 350 pp. (in Russian).

Filipjev, I. (1927) Les Nématodes libres des mers septentrionales appartenant à la famille des Enoplidae. *Archiv für Naturgeschichte* 91, 216 pp.

Filipjev, I. (1928) Nématodes libres du fleuve Oka. *Raboty Okskoy biologicheskoy stantsii* 5, 81–112.

Filipjev, I. (1929) Les Nématodes libres de l'extrémité orientale du Golfe de Finlande et de la baie de la Néva. *Études de la Neva* 5, 3–22.

Filipjev, I. (1930) Les Nématodes libres de la Néva et de l'extrémité orientale du Golfe de Finlande. Deuxième partie. *Archives de Hydrobiologie* 21, 1–64.

Filipjev, I. (1934) The classification of the free-living nematodes and their relation to the parasitic nematodes. *Smithsonian Miscellaneous Collections* 89, 1–63.

Furstenberg, J.P. and Vincx, M. (1989) Two oncholaimid species from a South African estuary (Nematoda, Oncholaimidae). *Hydrobiología* 183, 43–50.

Gagarin, V. (1981) Some aspects of evolution of freshwater nematodes. In: Platonova, T.A. and Tsalolikhin, S.Y. (eds) *Evolution, Taxonomy, Morphology and Ecology of Free-living Nematodes*. Academy of Sciences of the USSR, Leningrad, pp. 25–26.

Gagarin, V.G. (1993a) Free-living nematodes in the waters of Sakhalin Island. *Trudy Instituta Biologii Vnutrennikh Vod* 68, 53–77.

Gagarin, V.G. (1993b) *Free-living Nematodes from the Fresh Waters of Russia and Adjacent Countries: Order Monhysterida, Areaolaimida, Chromadorida, Enoplida, Mononchida*. Gidrometeoizdat, St. Petersburg, Russia, 352 pp.

Gagarin, V.G. (1995) Free-living nematodes recovered from sewage disposal sites at the Rybinsk water reservoir, Borok, Russia. *Russian Journal of Nematology* 3, 81–84.

Gagarin, V.G. (2001) *Free-living Nematodes of the Fresh Waters of Russia and Adjacent Countries: Fauna and the Pathways of its Formation, Ecology, Taxonomy, Phylogeny*. Nauka Publisher, Moscow, 170 pp.

Galhano, M.H. (1970) Contribuçao para o conhecimento da fauna intersticial em Portugal. *Publiçoes Instituto de Zoologia de Dr. Auguste Nobre* 110, 206.

Gerlach, S.A. (1952) Nematoden aus dem Küstengrundwasser. *Abhandlungen der mathematisch-naturwissenschaftlichen Klasse. Verlag der Akademie der Wissenschaften und der Literatur in Mainz* 6, 315–372.

Gerlach, S.A.(1953) Die biozönotische Gliederung der Nematodenfauna on der deutschen küsten. *Zeitschrift der Morphologie und Okologie der Tiere* 41, 411–512.

Gerlach, S.A. (1957a) Zwei Meeres-Nematoden aus dem Nicaragua-See. *Zoologischer Anzeiger* 159, 251–254.

Gerlach, S.A. (1957b) Die Nematodenfauna des Sandstrandes an der Küste von Mittelbrasilien (Brasilianische Meeres-nematoden IV). *Mitteilungen aus dem Zoologischen Museum in Berlin* 33, 411–459.

Gerlach, S.A. (1958) Deuxième contribution à la faune des Nématodes des eaux interstitielles littorales de Madagascar. *Mémoires de l'Institut Scientifique de Madagascar* 2, 343–365.

Gerlach, S.A. (1965) Freilebende Meeresnematoden aus der Gezeiterzone von Spitzbergen. *Verlöffenlichungen des Instituts für Meeresforschung in Bremerhaven* 9, 109–172.

Gerlach, S.A. and Riemann, F. (1974) The Bremerhaven checklist of aquatic nematodes. A catalogue of Nematoda Adenophorea excluding the Dorylaimida. Part 2. *Veröffentlichungen des Instituts für Meeresforschung in Bremerhaven* (Suppl. 4), 405–734.

Goossens, A. (1976) A study of the aquatic nematodes of Adrar Mountains in Mauritania. *Biologisch Jaarboek Dodonaea* 44, 178–190.

Greenslade, P. and Nicholas, W.L. (1991) Some Thoracostomopsidae (Nematoda: Enoplida) from Australia, including descriptions of two new genera and diagnostic keys. *Invertebrate Taxonomy* 4, 1031–1052.

Heyns, J. and Coomans, A. (1977) Freshwater nematodes from South Africa. 2. *Oncholaimus deconincki* n.sp. *Revue de Zoologie Africaine* 91, 906–912.

Hope, W.D. and Gardiner, S.L. (1982) Fine structure of a proprioceptor in the body wall of the marine nematode *Deontostoma californicum* Steiner and Albin, 1933 (Enoplida, Leptosomatidae). *Cell and Tissue Research* 225, 1–10.

Hope W.D. and Murphy D.G. (1972) A taxonomic hierarchy and checklist of the genera and higher taxa of marine Nematodes. *Smithsonian Contributions to Zoology* 137, 1–101.

Hopper, B.E. (1961a) Marine nematodes from the coast line of the Gulf of Mexico. *Canadian Journal of Zoology* 39, 183–199.

Hopper, B.E. (1961b) Marine nematodes from the coast line of the Gulf of Mexico II. *Canadian Journal of Zoology* 39, 359–365.

Hopper, B.E. (1962). Free-living marine nematodes of Rhode Island waters. *Canadian Journal of Zoology* 40, 41–52.

Hopper, B.E. (1963) Marine nematodes from the coast line of the Gulf of Mexico. III. Additional species from Gulf Shores, Alabama. *Canadian Journal of Zoology* 41, 841–863.

Hussman, S. (1962) Ökologische und verbreitungsgeschichtliche Studien über limnische Grundwassertiere aus dem küstlichen Mesopsammal der Helgoländer Düneninsel. *Archiv für Hydrobiologie* 58, 405–422.

Inglis, W.G. (1961) Two new species of free-living marine nematodes from the West coast of Scotland. *Hydrobiologia* 18, 284–294.

Inglis, W.G. (1962) Marine nematodes from Banyuls-sur-mer: with a review of the genus *Eurystomina*. *Bulletin of the British Museum (Natural History) Zoology* 8, 209–287.

Inglis, W.G. (1964) The marine Enoplida (Nematoda): a comparative study of the head. *Bulletin of the British Museum (Natural History) Zoology* 11, 263–376.

Jacobs, L. (1984) The free-living inland aquatic nematodes of Africa – a review. *Hydrobiologia* 113, 259–291.

Jairajpuri, M.S. (1983) Observations on *Campydora* (Nematoda: Dorylaimida). *Nematologia Mediterranea* 11, 33–42.

Jensen, P. (1979). Description of the aquatic nematode *Adorus astridae* sp. n., with notes on the systematic position and geographical distribution of *Adorus* species. *Annales Zoologices Fennici* 16, 84–88.

Juget, J. (1969) Description de quelques formes rares ou nouvelles de Nématodes libres du bassin du Léman. *Bulletin de la Société Vauduse des Sciences Naturelles* 70, 141–173.

Keppner, E.J. (1987) Five new and one known species of free-living marine nematodes of the family Oncholaimidae (Nematoda: Enoplida) from Northwest Florida, USA. *Transactions of the American Microscoscopical Society* 106, 214–231.

Keppner, E.J. (1988) *Thoonchus longisetosus* and *Oxyonchus striatus*, new species of free-living marine nematodes (Nematoda, Enoplida) from Northwest Florida, USA. *Proceedings of the Biological Society of Washington* 101, 183–191.

Keppner, E.J. (1989) Four new species of free-living marine nematodes in the genus *Pareurystomina* (Nematoda, Enoplida) with observations on other members of the genus. *Proceedings of the Biological Society of Washington* 102, 249–263.

Keppner, E.J. (1991). Eleven new species of free-living marine nematodes of the genus *Halalaimus* de Man, 1888 (Nematoda: Enoplida) from Florida with keys to the species. *Gulf Research Reports* 8, 333–362.

Keppner, E.J. and Keppner, L.A. (1989). A redescription of *Oncholaimoides elongatus* Hopper, 1961 (Nematoda: Enoplida) with descriptions of the other two members of the genus. *Gulf Research Reports* 8, 137–144.

Khera, S. (1979) On the genus *Ironus* with remarks on the genus *Cryptonchus*. *Indian Journal of Helminthology* 28, 34–42.

Kreis, H. (1928) Die freilebende marinen Nematoden der Spitzbergen. Expedition von F. Roemer und F. Schaudinn im Jahre 1898. *Miltteilungen aus dem Zoologischen Museum Berlin* 14, 131–197.

Kreis, H. (1934) Oncholaiminae Filipjev 1916. Eine monographische Studie. *Capita Zoologica*. 4, 271 pp.

Lambshead, P.J.D. and Platt, H. (1979) *Bathyeurystomina*, a new genus of free-living marine Nematodes (Enchelidiidae) from the Rockall Trough. *Cahiers de Biologie Marine* 20, 371–380.

Lorenzen, S. (1966) Diagnosen einiger freilebender Nematoden von der Schleswig-holsteinishen Westküste. *Veröffentlichungen des Instituts für Meeresforschung in Bremerhaven.* 10, 31–48.

Lorenzen, S. (1969) Freilebende Meeresnematoden aus dem Schlikwatt und den Salzwiesen der Nordseeküste. *Veröffentlichungen des Instituts für Meeresforschung in Bremerhaven* 11, 195–238.

Lorenzen, S. (1978) Discovery of stretch receptor organs in nematodes – structure, arrangements and functional analysis. *Zoologica Scripta* 7, 175–178.

Lorenzen, S. (1981) Bau, Anordnung und postembryonale Entwicklung von Metanemen bei Nematoden der Ordnung Enoplida. *Veröffentlichungen des Instituts für Meeresforschung in Bremerhaven* 19, 89–114.

Lorenzen, S. (1994) The phylogenetic systematics of free-living nematodes. (Translation of the 1981 German edition). *The Ray Society* 162, 1–383.

Luc, M. and De Coninck, L.A.P. (1959) Nématodes libres marins de la région de Roscoff. *Archives de Zoologie Expérimentale et Générale* 98: 103–165.

Maertens, D. and Coomans, A. (1979) The function of the demanian system and an atypical copulatory behaviour in *Oncholaimus oxyuris*. *Annales de la Société Royale Zoologique de Belgique.* 108, 83–87.

Malakhov, V.V. (1994) In: Hope, W.D. (ed.) *Nematodes. Structure, Development, Classification and Phylogeny.* Smithsonian Institution Press, Washington, DC, 286 pp.

Marchant, R. and Lillywhite, P. (1994) A survey of stream invertebrate communities on Macquarie Island. *Australian Journal of Marine and Freshwater Research* 45, 471–481.

Meyl, A.H. (1957) Freeliving nematodes: exploration hydrobiologique du Lac Tanganika (1946–1947). *Bulletin de L'Institut Royal Des Sciences Naturelles de Belgique. Biologie*, 27–51.

Micoletzky H. (1923) Freilebende Nematoden der Wolga mit besonderer Berücksichtigung der Umgebung von Saratow. *Arbeiten der Biologischen Wolga-Station* 7, 1–29.

Micoletzky, H. (1930) Freilebende marine Nematoden von den Sunda-Inseln. I. Enoplidae. (Papers from Dr. Th. Motensen's Pacific Expedition 1914–16) *Videnskabelige Meddelelser fra Dansk naturhistorisk Forening i Kobenhavn* 87, 243–339.

Mol, A.W.M. (1984) *Limnofauna Neerlandica*: Nematoda. Nieuwsbrief European Invertebrate Survey – Nederland 15, 17–25.

Mulk, M.M. and Coomans, A. (1979) Three new *Alaimus*-species (Nematoda: Alaimidae) from Mount Kenya. *Nematologica* 25, 447–457.

Mullin, P.G., Harris, T.S. and Powers T.O. (2003) Systematic status of *Campydora* Cobb, 1920 (Nematoda: Campydorina). *Nematology* 5, 699–711.

Murphy, D.G. (1966) An initial report on a collection of Chilean marine nematodes. *Mitteilungen der Hamburger Zoologisches Museum Institut* 63, 29–50.

Nicholas W.L. (1993) Two new species of nematode (Nematoda: Enoplida: Thoracostomopsidae) from Lake Alexandrina, South Australia. *Transactions of the Royal Society of South Australia* 117, 163–170.

Nicholas, W.L. (2004) *Oxyonchus longisetosus* n.sp., *Oxyonchus evelynae* n.sp. and *Oxyonchus culcitatus* Wieser, 1959 (Thoracostomopsidae: Enoplida: Nematoda) from Australian ocean beaches, together with observations on the variability of taxonomic characters and a key to species of the genus *Oxyonchus*. *Hydrobiologia* 511, 47–64.

Nicholas, W.L. and Stewart (1984) *Oncholaimus balli* n.sp. (Nematoda, Oncholaimidae) from a volcanic crater lake in Papua-New-Guinea, with observations on the colonization of the lake by nematodes. *Nematologica* 30, 1–10.

Nicholas, W.L., Bird, A.F., Beech, T.A. and Stewart, A.C. (1992) The nematode fauna of the Murray River estuary, South Australia; the effects of the barrages across its mouth. *Hydrobiologia* 234, 87–101.

Pegova, A.N., Krasnova, E. and Aleshin, V.D.V. (2004) Evidence from the small and large ribosomal RNA structure suggests that *Anoplostoma rectospiculum* Gal'tsova, 1976 (Nematoda:

Anoplostomatidae) is a member of the superfamily Enoploidea, not Oncholaimoidea. *Nematology* 6, 413–241.

Platt, H.M. and Warwick, R.M. (1983) *Free-living Marine Nematodes. Part I British Enoplids.* Cambridge University Press, Cambridge, 307 pp.

Rachor, E. (1969) Das de Mansche Organ der Oncholaimidae, eine genito-intestinale Verbindung bei Nematoden. *Zeitschrift für Morphologie der Tiere* 66, 87–166.

Riemann, F. (1966) Die interstitielle Fauna im Elbe-Aestuar. Verbreitung und Systematik. *Archiv für Hydrobiologie Supplement* 31, 1–279.

Rusin, L.Y., Aleshin, V.V., Vladychenskaya, N.S., Milyutina, I.A., Kedrova, O.S. and Petrov, N.B. (2001) Trefusiidae are a subtaxon of marine Enoplida (Nematoda): evidence from primary structure of hairpin 35 and 48 loops of SSU rRNA gene. *Molecular Biology* 35, 778–784.

Schiemer, F. (1978) Verteilung und Systematik der freilebenden Nematoden des Neusiedlersee. *Hydrobiologia* 28, 167–194.

Schneider, W. (1937) Freilebende Nematoden der Deutschen Limnologischen Sundaexpedition nach Sumatra, Java und Bali. *Archiv für Hydrobiologie Supplement* 15 (Tropische Binnengewässer) 7, 30–108.

Schneider, W. (1940) Neue freilebende Nematoden aus Höhlen und Brunnen. I. Nematoden aus jugoslawischen Höhlen. *Zoologischer Anzeiger* 132, 84–94.

Schuurmans-Stekhoven, J.H. (1942) Nematoden und Spongien. *Zoologischer Anzeiger* 137, 181–185.

Schuurmans-Stekhoven, J.H. (1950) The free-living marine nemas of the Mediterranean I. The Bay of Villefranche. *Verhandelingen van het Koninklijk Belgisch Instituut voor Natuurwetenschappen* 37, 1–220.

Schuurmans-Stekhoven, J.H. and De Coninck, L.A. (1933) Diagnoses of new Belgian marine nemas. *Bulletin du Musée Royal d'Histoires Naturelles de Belgique* 9, 1–13.

Siddiqi, M.R. (1983) Phylogenetic relationships of the soil nematode orders Dorylaimida, Mononchida, Triplonchida and Alaimida, with revised classification of the subclass Enoplia. *Pakistan Journal of Nematology* 1, 79–110.

Siddiqi, M.R. (1993) Nematodes of tropical rainforests. 1. Six new genera and eighteen new species of alaims. *Afro-Asian Journal of Nematology* 3, 60–80.

Siddiqi, M.R. and Basir, M.A. (1965) *Amphidelus sylvaticus* n. sp. and *A. candidus* n. sp. (Nematoda: Alaimidae) from India, with a key to the species of *Amphidelus*. *Nematologica* 11, 343–348.

Siddiqi, M.R. and Husain, Z. (1967) Studies on the genus *Alaimus* de Man, 1880, with description of six new species from India. *Proceedings of the Helminthological Society of Washington* 34, 158–167.

Siddiqi, M.R. and Vinciguerra, M.T. (1991) *Cristamphidelus*, a new genus of Alaimidae (Nematoda) and *Etamphidelus botrys* sp.n. from Australia. *Afro-Asian Journal of Nematology* 1, 87–90.

Siddiqui, I.A. and Viglierchо D.R. (1970a) Fine structure of photoreceptors in *Deontostoma californicum*. *Journal of Nematology* 2, 274–276.

Siddiqui, I.A. and Viglierchо D.R. (1970b). Ultrastructure of photoreceptors in the marine nematode *Deontostoma californicum*. *Journal of Ultrastructural Research* 32, 558–571.

Smol, N. and Sharma, J. (1984) Two new and three redescribed species of *Viscosia* (Nematoda, Oncholaimidae). *Hydrobiologia* 114, 123–147.

Stefanski, W. (1914) Recherches sur la faune des nématodes libres du Bassin de Léman. Thèse, Université Genève, 72 pp.

Stewart, F.H. (1906) The anatomy of *Oncholaimus vulgaris* Bast., with notes on two parasitic nematodes. *Quekett Journal of Microscopy* 50, 101–150.

Tchesunov, A. (1981) Free-living nematodes of the genus *Tripyloides* de Man, 1886 (Enoplida, Tripyloididae) from the Caspian Sea. *Byuleten Moskovskogo Obshchestva Ispytatelei Prirody, Otdel Biologicheskii* 86, 49–55 (in Russian, English summary).

Thorne, G. (1939) A monograph of the nematodes of the superfamily Dorylaimoidea. *Capita zoologica* 8, 1–261.

Timm, R.W. (1961) The marine nematodes of the Bay of Bengal. *Proceedings of the Pakistan Academy of Sciences* 1, 1–88.

Timm, R.W. (1967) Two new species of *Oncholaimellus* from the Bay of Bengal. *Proceedings of the Pakistan Academy of Sciences* 4, 19–25.

Tsalolikhin, S.Y. (1980) Free-living nematodes of Lake Baikal. *Novosibirsk, USSR; Nauka,* 120 pp (in Russian).

Tsalolikhin, S.Y. (1982) The problem of origin of the fauna of central Asian water bodies, *Pseudoncholaimus charon* sp.n. and *Pseudoncholaimus neglectus* sp. n. (Nematoda, Oncholaimidae) taken as an example. *Zoologicheskii Zhurnal* 61, 653–662.

Tsalolikhin, S.Y. (1987) Revision of the genus *Ironus* (Nematoda, Dorylaimida). *Zoologicheskii Zhurnal* 66, 662–673.

Tsalolikhin, S.Y. (1996) Review of the fauna of free-living nematodes from inland waters of Ethiopia. *Zoosystematica Rossica* 4, 205–218.

Tudorancea, C., Baxter, R.M. and Fernando, C.H. (1989) A comparative limnological study of zoobenthic associations in lakes of the Ethiopian Rift Valley. *Archiv für Hydrobiologie Supplement* 83, 121–174.

Van Der Heiden, A. (1974) The structure of the anterior feeding apparatus in members of the Ironidae (Nematoda: Enoplida). *Nematologica* 20, 419–436.

Vincx, M. and Vanreusel, A. (1989) Free-living marine nematodes from the Southern Bight of the North Sea. II. Notes on species of the Trefusiidae Gerlach, 1966. *Hydrobiologia* 175, 213–221.

Vitiello, P. (1970) Nématodes libres marins des vases profondes du Golfe d'Lion. I. Enoplida. *Téthys* 2, 139–210.

Vitiello, P. (1972) Le genre *Wieseria* Gerlach, 1956 (Nematoda, Oxystominidae). *Téthys* 4, 645–650.

Warwick, R.M. (1969) The marine nematode *Eurystomina terricola* (de Man, 1907) and its distribution on British shores. *Hydrobiologia* 34, 392–400.

Warwick, R.M. (1970) The genus *Paramesacanthion* Wieser (Nematoda, Enoplidae) off the coast of Northumberland. *Cahiers de Biologie Marine* 11, 187–194.

Warwick, R.M. and Platt, H.M. (1973) New and little-known marine nematodes from a Scottish sandy beach. *Cahiers de Biologie Marine* 14, 135–158.

Warwick, R.M., Dexter, D.M. and Kuperman, B. (2002) Free-living nematodes from the Salton Sea. *Hydrobiologia* 473, 121–128.

Wieser, W. (1953a) Free-living marine nematodes. I. Enoploidea. Chile reports 10. *Lunds Universitets Arsskrift* 49, 155 pp.

Wieser, W. (1953b) Der Sexualdimorphismus der Enchelidiidae (freilebende marine Nematoden) als taxonomisches Problem. *Zoologischer Anzeiger* 150, 152–170.

Wieser, W. (1956) Free-living marine nematodes. III. Axonolaimoidea and Monhysteroidea. Chile reports 26. *Lunds Universitets Arsskrift* 52, 1–115.

Wieser, W. (1959) Free-living nematodes and other small invertebrates of Puget Sound beaches. University of Washington Press, Seattle, 179 pp.

Wieser, W. and Hopper, B. (1967) Marine nematodes of the East Coast of North America. I. Florida. *Bulletin of the Museum of Comparative Zoology* 135, 239–344.

Winiszewska, G. (2001) Some observations on *Campydora* Cobb, 1920 (Nematoda), with description of male *Campydora demonstrans* Cobb, 1920. *Annales Zoologici* 51, 391–394.

Wu, J. and Liang, Y. (1999) A comparative study of benthic nematodes in two Chinese lakes with contrasting sources of primary production. *Hydrobiologia* 411, 31–37.

Wu, J.H., Fu, C.Z., Liang, Y.L. and Chen, J.K. (2004) Distribution of the meiofaunal community in a eutrophic lake of China. *Archiv für Hydrobiologie* 159, 555–575.

Yeates, G.W. (1967). Studies on nematodes from dune sands. 7. Monhysterida and Chromadorida. *New Zealand Journal of Science* 10, 785–801.

Zhang, Z. (1983) Three new species of free-living marine nematodes from a sublittoral station in Firemore Bay, Scotland. *Cahiers de Biologie Marine* 24, 219–229.

Zullini, A. (1973) Some soil and freshwater nematodes from Chiapas (Mexico): In: *Subterranean fauna of Mexico*, part 2, *Quaderni Accademia Nazionale dei Lincei* 171, 55–96.

Zullini, A. (1976a) Nematodes as indicators of river pollution. *Nematologia Mediterranea* 4, 13–22.

Zullini, A. (1976b) I nematodi come bioindicatori, Studio di una successione. *Accademia Nazionale dei Lincei* (*Rendiconti Scienze fisiche, matematiche, naturali*) 61, 503–507.

Zullini, A. (1977a) Some freshwater nematodes of southern Mexico and Guatemala. In: *Subterranean Fauna of Mexico*, part 3, *Quaderni Accademia Nazionale dei Lincei* 171, 75–85.

Zullini, A. (1977b) On certain moss nematodes of Central Mexico. In: *Subterranean Fauna of Mexico*, part 3, *Quaderni Accademia Nazionale dei Lincei* 171, 87–90.

Zullini, A. (1988a) The ecology of the Lambro River. *Rivista di Idrobiologia* 27, 39–57.

Zullini, A. (1988b) A new genus and five species of nematodes from Ethiopian lakes. *Revue de Nématologie* 11, 279–288.

Zullini, A. and Peretti, E. (1986) Lead pollution and moss-inhabiting nematodes of an industrial area. *Water, Air, and Soil Pollution* 27, 403–410.

Zullini, A. and Ricci, C. (1980) Bdelloid rotifers and nematodes in a small Italian stream. *Freshwater Biology* 10, 67–72.

zur Strassen, O. (1894) Über das rohrenformige Organ von *Oncholaimus*. *Zeitschrift für Wissenschaften. Zoologie* 58, 460–474.

13 Order Triplonchida

A. ZULLINI

Dipartimento di Biotecnologie & Bioscienze, Università di Milano-Bicocca, Piazza della Scienza 2, 20126 Milan, Italy

Introduction

Till recent years *Diphtherophora, Tripyla* and *Onchulus* were traditionally placed in different orders (the first in Triplonchida or in Dorylaimida, the latter two in Enoplida or in Tripylida). In fact, the families pertaining to this order possess a diversified stomal structure: a buccal spear (Diphtherophoridae, Trichodoridae), a prismatic buccal cavity (Prismatolaimidae), a large buccal cavity usually armed with teeth (Tobrilidae, Onchulidae) or a rather collapsed buccal cavity with teeth (Tripylidae).

Cobb (1920) created the order Triplonchia (amended to Triplonchida by Siddiqi in 1983) with the typical genus *Triplonchium* (now synonymized with *Tylolaimophorus* de Man, 1880). Till recent times the order Triplonchida included only two superfamilies: Diphtherophoroidea and Trichodoroidea (e.g. in Jairajpuri and Ahmad, 1992) characterized by a complex, asymmetrical buccal spear. Tsalolikhin (1983), in a completely different systematic context, considered the suborder Tobrilina, including Tobrilidae, Tripylidae, Prismatolaimidae and Onchulidae, in the order Enoplida. But earlier, Riemann (1972) stressed a relationship among *Diphterophora*, *Tripyla*, *Onchulus* and *Kinonchulus*, anticipating the possibility of gathering these and similar taxa into a higher group. Recent molecular data support this view (see 'Ecology' at the end of this chapter) and now the order Triplonchida includes a wide array of groups.

Note: In the following sections taxa with asterisks are partially or entirely freshwater.

Systematics of order Triplonchida

Phylum Nematoda

Class Enoplea

Subclass Enoplia

Order Triplonchida Cobb, 1920

1. Suborder Diphtherophorina Micoletzky, 1922

Superfamily Diphtherophoroidea Micoletzky, 1922

Family Diphtherophoridae Micoletzky, 1922
Genus *Diphtherophora* de Man, 1880
Tylolaimophorus de Man, 1880
Longibulbophora Yeates, 1967
Family Trichodoridae Thorne, 1935
Genus *Trichodorus* Cobb, 1913
Paratrichodorus Siddiqi, 1974
Monotrichodorus Andrássy, 1976
Allotrichodorus Rodriguez-M, Sher & Siddiqi, 1978
Ecuadorus Siddiqi, 2002
2. Suborder Tobrilina Tsalolikhin, 1976
Superfamily Prismatolaimoidea Micoletzky, 1922
Family Prismatolaimidae Micoletzky, 1922
Genus *Prismatolaimus* de Man, 1880
Superfamily Tobriloidea Filipjev, 1918
Family Tobrilidae Filipjev, 1918
Subfamily Tobrilinae Filipjev, 1918
Tribe Paratrilobini Tsalolikhin, 2001
Genus *Paratrilobus* Micoletzky, 1922
Mesotobrilus Tsalolikhin, 1981
Quasibrilus Tsalolikhin, 1976
Kurikania Tsalolikhin, 1976
Lamuania Tsalolikhin, 1976
Tribe Tobrilini Filipjev, 1918
Genus *Tobrilus* Andrássy, 1959
Eutobrilus Tsalolikhin, 1981
Asperotobrilus Shoshin, 1991
Setsalia Shoshina, 2003
Subfamily Neotobilinae Tsalolikhin, 2001
Tribe Neotobrilini Tsalolikhin, 1981
Genus *Neotobrilus* Tsalolikhin, 1981
Semitobrilus Tsalolikhin, 1981
Brevitobrilus Tsalolikhin, 1981
Tribe Epitobrilini Tsalolikhin, 2001
Genus *Epitobrilus* Tsalolikhin, 1981
3. Suborder Tripylina de Man, 1876
Superfamily Tripyloidea de Man, 1876
Family Tripylidae de Man, 1876
Genus *Tripyla* Bastian, 1865
Tripylella Brzeski & Winiszewska-Slipinska 1993
Tripylina Brzeski, 1963
Trischistoma Cobb, 1913
Tobrilia Andrássy, 1967
Abunema Khera, 1971
Family Onchulidae Andrássy, 1964
Subfamily Onchulinae Andrássy, 1964
Genus *Onchulus* Cobb, 1920
Stenonchulus W. Schneider, 1940

Kinonchulus Riemann, 1972
Limonchulus Andrássy, 1963
Pseudonchulinae Hodda, Bloemer & Wanless, 2005
Pseudonchulus Altherr, 1972
Caprionchulus Swart & Hayns, 1993
Tobriloides Loof, 1973

Order TRIPLONCHIDA Cobb, 1920

Diagnosis. Enoplia. Protractor muscles between the spicule heads and the body wall are absent: they are replaced by peculiar suspensor muscles forming a capsule around the spicules. These capsule-like structures surround the anterior half of each spicule and seem to squeeze out the spicules.

A clade of nematodes without protrusible spear such as Tobrilina (Tobriloidea and Prismatolaimoidea are united in it owing to robust molecular data) was found to be a sister taxon to triplonchs (Diphtherophoridae, Trichodoridae) on the basis of very robust positions of sequence analysis. Moreover, Tripylina (Tripylidae, Onchulidae) are considered close to Tobrilina owing to the strong spicule protrusion capsules. Therefore, order Triplonchida is established both on molecular basis and morphological data (De Ley and Blaxter, 2002).

Suborder DIPHTHEROPHORINA Micoletzky, 1922

Diagnosis. Triplonchida. Plant parasitic. Body short and obese. Cuticle usually loosely fitting body. Amphids post-labial, usually with large fovea and elliptical aperture. Stoma weakly cuticularized; spear complex, asymmetrical. Pharynx with a posterior pyriform or elongate bulb. Males with single testis. No pre-cloacal pair of supplements. Spicules paired and tripyloid. Gubernaculum present. No lateral guiding pieces. No protractor muscles between spicule heads and body wall: there are, instead, suspensor muscles forming a capsule around spicules. Tail short, similar in both sexes. Caudal glands and spinneret absent.

Superfamily DIPHTHEROPHOROIDEA Micoletzky, 1922

Diagnosis. Diphtherophorina. With the characters of the suborder.

Family DIPHTHEROPHORIDAE Micoletzky, 1922

Diagnosis. Diphtherophoroidea. Spear short, about equal to width of lip region, with basal swellings. Female amphidelphic with reflexed ovaries. Gubernaculum reduced. Male supplements reduced or vestigial.

Genus *Diphtherophora* de Man, 1880 (Fig. 13.1 (4))

Syn. *Chaolaimus* Cobb, 1893; *Archionchus* Cobb, 1913.

Diagnosis. Diphtherophoridae. Body less than 1 mm. Cuticle loose. Dorsal sector of stylet short, divided and well curved; spear extension with basal swelling. Spear-guiding structure arched. Pharynx a slender tube ending with a pyriform or elongate basal bulb. Secretory–excretory pore present. Female amphidelphic; vulva transverse; vagina small. Spicules slightly arcuate. Essentially soil inhabiting.

Type species: *D. communis* de Man, 1880.

Freshwater species:

**D. communis* de Man, 1880 – USA, Europe, Asia, South America (wet meadows, (fresh or brackish waters), arable lands).
**D. perplexans* (Cobb, 1913) de Coninck – Hungary (Lake Balaton).
**D. vanoyei* DeConinck, 1931 – Belgium, Austria, Italy, Kenya (wet soil around liverwort).

Genus *Tylolaimophorus* de Man, 1880 (Fig. 13.1 (B, B'))
Syn. *Triplonchium* Cobb, 1920.
Diagnosis. Diphtherophoridae. Body less than 2 mm, body contents very dense. Dorsal and ventral sectors of stylet not fused anteriorly, dorsal sector somewhat widened and pointed; base of the extension knobbed. Spear-guiding structure arched. Pharynx a slender tube ending with a pyriform basal bulb. Female amphidelphic; vulva transverse. Spicules strongly curved, with spiral protractor muscles.

Type species: *T. typicus* de Man, 1880. Soil species.

Genus *Longibulbophora* Yeates, 1967 (Fig. 13.1 (C))
Diagnosis. Diphtherophoridae. Body less than 1 mm, body contents very dense. Dorsal sector of spear with a protuberance in mid-region; small basal knobs. Posterior half of pharynx expanded, not pyriform. Spear-guiding structure V-shaped. Female amphidelphic. Spicules slightly arcuate.

Type species: *L. rotundicauda* (Paesler, 1955). Soil species.

Family TRICHODORIDAE Thorne, 1935
Diagnosis. Diphtherophoroidea. Body cigar-shaped. Anterior end rounded with lips amalgamated; anterior sensorial organs in two circlets: one with six anterior labial papillae and one with six outer labial papillae and four cephalic papillae; amphids cup-shaped located just posterior to the lip region. Stoma with stiletto-form tooth, or onchiostyle, relatively long, ventrally curved; anterior half solid, posterior half hollow; no basal knobs. Guiding ring simple, surrounds anterior end of onchiostyle. Protractor muscles distinct. Pharynx swollen with posterior glandular bulb with five gland nuclei: one small anterior ventrosublateral pair, one posterior ventrosublateral pair and a single large dorsal gland nucleus. Oviduct consisting of two finely granular cells. Male with one testis, outstretched. Supplements few, well developed. Spicules slightly curved; gubernaculum developed. Tail short, maximum length one anal body width. Ectoparasites of higher plants, some of them are vectors of plant viruses (Decraemer, 1995; 1998).

Genus *Trichodorus* Cobb, 1913 (Fig. 13.1 (D, D'))
Diagnosis. Trichodoridae. Cuticle slightly inflated after fixation. Pharyngeal bulb not overlapping intestine. Female amphidelphic; $V \approx 50–65\%$; vulva with lateral body pores; vaginal sclerotization strong. Males abundant, with strong arcuate posterior

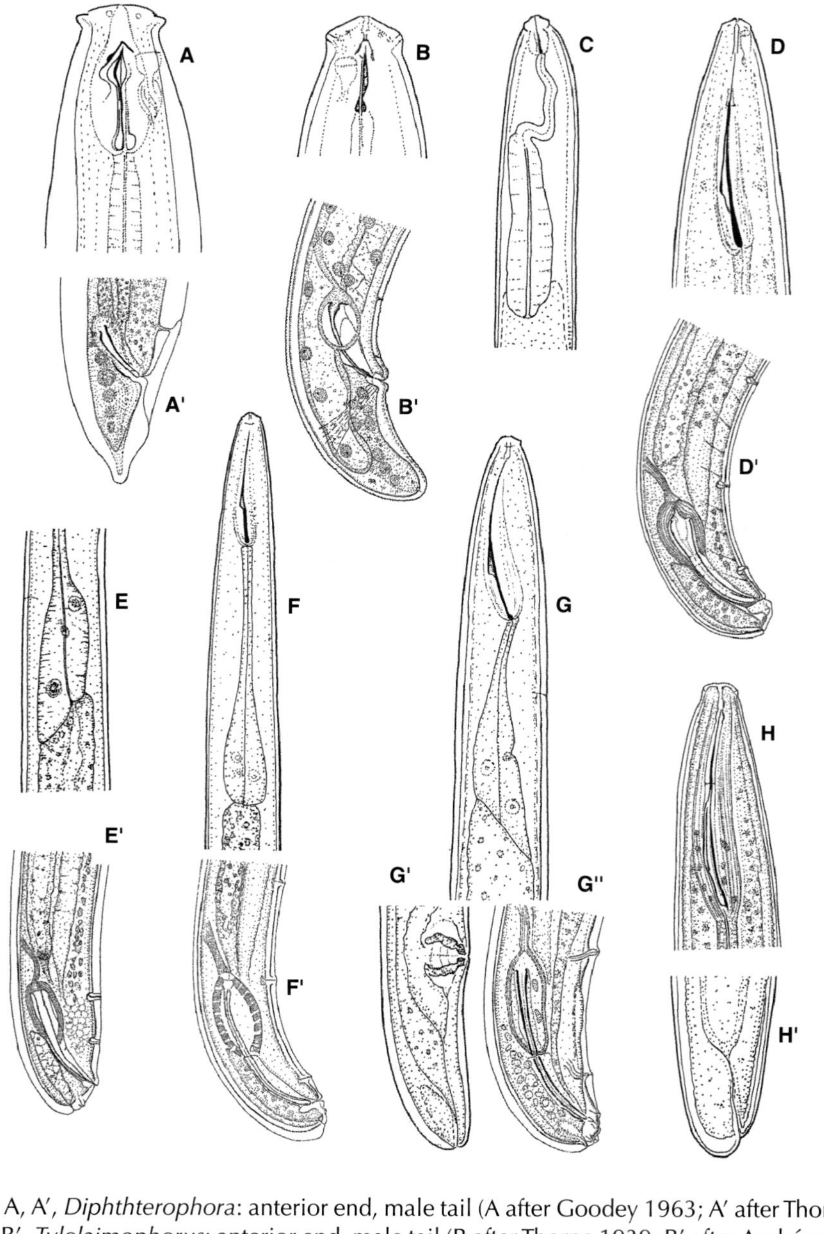

Fig. 13.1 A, A′, *Diphthterophora*: anterior end, male tail (A after Goodey 1963; A′ after Thorne 1939); B, B′, *Tylolaimophorus*: anterior end, male tail (B after Thorne 1939; B′ after Andrássy 1988); C, *Longibulbophora*: anterior end (after Yeates 1967); D, D′, *Trichodorus*: anterior end, male tail (D after Thorne 1939; D′ after Rodriguez & Bell 1978); E, E′, *Paratrichodorus*: basal pharyngeal bulb, male tail (E after Siddiqi 1963; E′ after Rodruguez & Bell 1978); F, F′, *Monotrichodorus*: anterior end, male tail (F after Allen 1957; F′ after Rodriguez, Sher & Siddiqi 1978); G, G′, G″, *Allotrichodorus*: anterior end, female tail, male tail (G, G′ after Rodriguez, Sher & Siddiqi 1978; G″ after Rashid, De Waele & Coomans 1986); H, H′, *Ecuadorus*: anterior end, female tail (after Siddiqi 2002).

region. Diagonal copulatory muscles developed. Caudal alae practically absent. Supplements three (or four) evenly spaced.

Type species: *T. obtusus* Cobb, 1913. Mostly soil species.

Freshwater species:

**T. primitivus* (de Man, 1880) Micoletzky, 1922 – Europe, Asia, North and South America (soil, wet river banks, canals, brooks).
**T. similis* Seinhorst, 1963 – Europe (soil, flooded soil, wet river banks, brooks, ponds; occasionally feeds on root tips of aquatic plants).
**T. sparsus* Szcygiel, 1968 – Europe and Asia (soil, wet soil and banks of brooks and canals).
**T. variopapillatus* Hooper, 1972 – Europe (soil, wet banks of ponds and rivers, sediment of brooks and canals).
**T. velatus* Hooper, 1972 – Europe (soil, river banks).
**T. viruliferus* Hooper, 1963 – Europe and North America (soil, wet river banks).

Genus *Paratrichodorus* Siddiqi, 1974 (Fig. 13.1 (E, E'))
Syn. *Atlantadorus* Siddiqi, 1974; *Nanidorus* Siddiqi, 1974.
Diagnosis. Trichodoridae. Cuticle very inflated after fixation. Pharyngeal bulb overlapping intestine. Female amphidelphic; V = 50–65%; no lateral body pores near vulva; vaginal sclerotization weak. Males with caudal alae.

Type species: *P. tunisiensis* (Siddiqi, 1963). Mostly soil species.

Species list:
**P. pachydermus* (Seinhorst, 1954) Siddiqi, 1974 – Europe and North America (soil, wet river banks, fresh water).
**P. teres* (Hooper, 1962) Siddiqi, 1974 – temperate regions of USA, Europe, South Africa, Australia (soil, wet river banks, canals and brooks).

Genus *Monotrichodorus* Andrássy, 1976 (Fig. 13.1 (F, F'))
Diagnosis. Trichodoridae. Cuticle not inflated after fixation. Pharyngeal bulb not overlapping intestine. Female prodelphic; V = 77–85%; vulva with lateral body pores; vaginal sclerotization strong. Spicules long and very thin.

Type species: *M. monohystera* (Allen, 1957) Andrássy, 1976. Soil species.

Genus *Allotrichodorus* Rodriguez-M, Sher & Siddiqi, 1978 (Fig. 13.1 (G, G', G''))
Diagnosis. Trichodoridae. Cuticle inflated after fixation. Pharyngeal bulb overlapping intestine. Female prodelphic; V = 80–90%; post-vulvar uterine sac present; vulva without lateral body pores; vaginal sclerotization strong. Anus subterminal. Males with caudal alae. Gubernaculum small.

Type species: *A. campanullatus* Rodriguez-M, Sher & Siddiqi, 1978. Soil species.

Genus *Ecuadorus* Siddiqi, 2002 (Fig. 13.1 (H, H'))
Diagnosis. Trichodoridae. Body less than 0.5 mm; cuticle swollen after fixation, without pores. Pharyngo-intestinal junction offset, no overlaps. Secretory–excretory pore near

base of pharynx. Female prodelphic; vulva transverse; $V \approx 60$–65%; vagina inconspicuous, less than half the body width long; vaginal sclerotizations very small. Anus subterminal. Tail hemispherical to conoid-rounded. Males unknown.

Type species: *E. equatorius* Siddiqi, 2002. Soil species.

Suborder TOBRILINA Tsalolikhin, 1976

Diagnosis. Triplonchida. Free-living. Following De Ley and Blaxter (2002), Prismatolaimoidea and Tobriloidea form a separate clade in triplonchs on the basis of small subunit (SSU) rDNA. Pharynx cylindrical with three large pre-cardiac glands. Male with at least than five supplements. Anterior end of the spicules embedded in muscle tissue in at least some *Prismatolaimus* and *Tobrilus* species.

Superfamily PRISMATOLAIMOIDEA Micoletzky, 1922

Diagnosis. Tobrilina. Stoma usually wide, prismatic; tail filiform in both sexes, ending with a mucro.

Family PRISMATOLAIMIDAE Micoletzky, 1922

Prismatolaimoidea. With the characters of the superfamily.

Genus *Prismatolaimus* de Man, 1880 (Fig. 13.2 A, A')

Syn. *Takakia* Yeates, 1967.

Diagnosis. Prismatolaimidae. Body length 0.4–1.9 mm. Cuticle thin, finely annulated with scattered setae. Head with ten cephalic setae in one circle, or in two circles very near each other: six longer setae articulate and four shorter non-articulate. Stoma usally wide, prismatic, rather cuticularized, in its base a small dorsal tooth and two sub-ventral swellings with or without rasp-like structures. Amphids transversal slits at about 2–3 cephalic diameters from anterior end. Pharynx cylindrical, with weak transverse fields. Posterior to cardia, a small organ with a dorsal pore (*organellum dorsale*) may be present (Andrássy, 2003). One or two ovaries consisting of few cells. Mature eggs one at time, 2–5 body diameters long. Male very rare, diorchic, with 11–46 small supplements in the posterior region of body, but in some species extending up to the pharyngeal region. Anterior end of spicules embedded in muscle tissue; gubernaculum thin. Tail filiform in both sexes ($c' = 8$–40); tail tip with a hook-like mucro.

Type species: *P. intermedius* (Bütschli, 1873) de Man, 1880. Soil and limnic species.

Species list:

P. afer Andrássy, 2003.

P. andrassyi Khera & Chaturvedi, 1967.

**P. andrassyanus* Coomans & Mulk, 1980 – East Africa (soil, rivers).

**P. brevicaudatus* Wu & Hoeppli, 1929 – China (mud in a brackish water canal).

P. chilensis Coomans & Raski, 1988.
**P. dolichurus* de Man, 1880 – North and South America, Europe, Africa, Asia, Australia (*Sphagnum*, ponds, springs, lakes, psammon, rivers, brooks, swampy soil, soil, non-calcareous habitats, acid waters).
P. exilis Andrássy, 2003.
**P. flagellatus* Andrássy, 2003 – Brazil (*Sphagnum* moor).
**P. hsuei* Wu & Hoeppli, 1929 – China (streams, sandy soil).
**P. intermedius* (Bütschli, 1873) de Man, 1880 – North and South America, Europe, Africa, Asia (soil, moss, rivers, lakes, bogs, aquatic plants, psammon, swampy soil, acid waters).
P. iucundus Andrássy, 2003.
**P. kenyensis* Mulk & Coomans, 1979 – Galápagos, Kenya, Spain (humid soil, lakes).
**P. leptolaimus* Andrássy, 1969 – Europe (Romania, Spain), Congo (psammon).
**P. matoni* Mulk & Coomans, 1979 – Africa (soil, lakes).
P. mulcoomus Brzeski, 1997.
P. novoporus Coomans & Raski, 1988.
**P. paraprimitivus* Turpeenniemi, 1997 – Gulf of Bothnia, Scandinavia, (0.3–0.4% salinity, sand, mud at 10–80 m).
P. parvus Milne, 1963.
P. primitivus Loof, 1971.
P. stenolaimoides Loof, 1971.
**P. tareya* Gagarin & Kuzmin, 1972 – North Siberia (humid soil, temporary waters).
P. tenuicaudatus Schuurmans Stekhoven, 1951.
**P. verrucosus* Hirschmann, 1952 – Germany, Czech Republic, Romania, Lithuania, East Siberia (subterranean waters, brooks).
**P. waipukeus* (Yeates, 1967) Andrássy, 1969 – New Zealand (sandy soil, algae from a dried pool).

Superfamily TOBRILOIDEA Filipjev, 1918

Diagnosis. Tobrilina. Body length 1–7 mm. Cuticle smooth, very finely annulated or with longitudinal ridges. Head not offset. Anterior setae: six long and four shorter. Stoma usually funnel-shaped; tail with caudal glands and spinneret, never filiform.

Family TOBRILIDAE Filipjev, 1918

Diagnosis. Tobriloidea. Stoma funnel- or cup-shaped with two teeth at its base (except *Quasibrilus*). Amphids at stoma level or slightly posteriorly. Revision by Tsalolikhin (2001b).

Subfamily Tobrilinae Filipjev, 1918

Diagnosis. Tobrilidae. Stoma with two overlapping pockets adjacent to each other, rarely absent; two teeth relatively near each other.

Tribe Paratrilobini Tsalolikhin, 2001

Diagnosis. Tobrilinae. No stomatal pockets, or pockets merged with the buccal cavity.

Genus *Paratrilobus* Micoletzky, 1922 (Fig. 13.2 (B, B'))

Syn. *Baicalobrilus* Tsalolikhin, 1976

Diagnosis. Paratrilobini. Body length 2–5 mm. Stoma big, barrel-shaped or rounded, stoma walls thick; pockets merged with the buccal cavity with two strong teeth (appearing *Miconchus*-like); amphid aperture at posterior stoma level; vulva equatorial or post-equatorial; male supplements six, large, protruding, echinulate; first and/or last supplement slightly smaller than the others, with bulbillum in top position, immediately below the supplement's cap; cap mobile and larger than the pad's ('shoulder's') thickness.

Type species: *P. grandipapilloides* Micoletzky, 1922. Limnic species.

Species list:

**P. brevis* (Tsalolikhin, 1976) Tsalolikhin, 1981 – Baikal (clay).
**P. expugnator* (Tsalolikhin, 1976) Tsalolikhin, 1981 – Baikal (clay).
**P. grandipapilloides* Micoletzky, 1922 – Northern Europe, Siberia (large lakes: Madüsee, Onega, Taimir, Baikal).
**P. ponticus* Tsalolikhin, 1981 – Estuary of Dniepr (silt bottom).
**P. rapis* Gagarin, 1991 – North Siberia (lake).

Genus *Mesotobrilus* Tsalolikhin, 1981 (Fig. 13.2 (C, C'))

Diagnosis. Paratrilobini. Body length 1.8–2.8 mm. Stoma funnel-shaped, stoma walls thin; pockets merged with the buccal cavity, with two small teeth (appearing *Tobrilus*-like); amphid aperture at anterior stoma level; vulva pre-equatorial; supplements 6–8, uniform, protruding, echinulate, with the bulbillum in low position; supplement's cap fixed and smaller than the pad's ('shoulder's') thickness.

Type species: *M. ultimus* (Tsalolikhin, 1977). Limnic species.

Species list:

**M. delicatus* Shoshin, 1988 – Baikal (fine sand).
**M. ultimus* (Tsalolikhin, 1977) Tsalolikhin, 1981 – Baikal (sand).

Genus *Quasibrilus* Tsalolikhin, 1976 (Fig. 13.2 (D, D'))

Diagnosis. Paratrilobini. Body length 3–6 mm. Stoma cup-shaped, without teeth and pockets. Amphids posterior to the stoma base. Pharynx short ($b = 6–10$); cardiac gland cells very large, round. Vagina muscular. Male supplements protruding, echinulate.

Type species: *Q. nannostomus* Tsalolikhin, 1976. Limnic species.

Species list:

**Q. nannostomus* Tsalolikhin, 1976 – Baikal (clay).
**Q. prodigiosus* (Shoshin, 1988) n. comb. – Baikal (silt).

Genus *Kurikania* Tsalolikhin, 1976 (Fig. 13.2 (E, E'))

Diagnosis. Paratrilobini. Body length 2–4 mm. Stoma very large, barrel-shaped, with thick walls, without pockets and two very minute teeth at its base. Pharyngeal musculature not encircling the stoma (= stoma free). Male supplements protruding, echinulate.

Type species: *K. sibirica* Tsalolikhin, 1976. Limnic species.

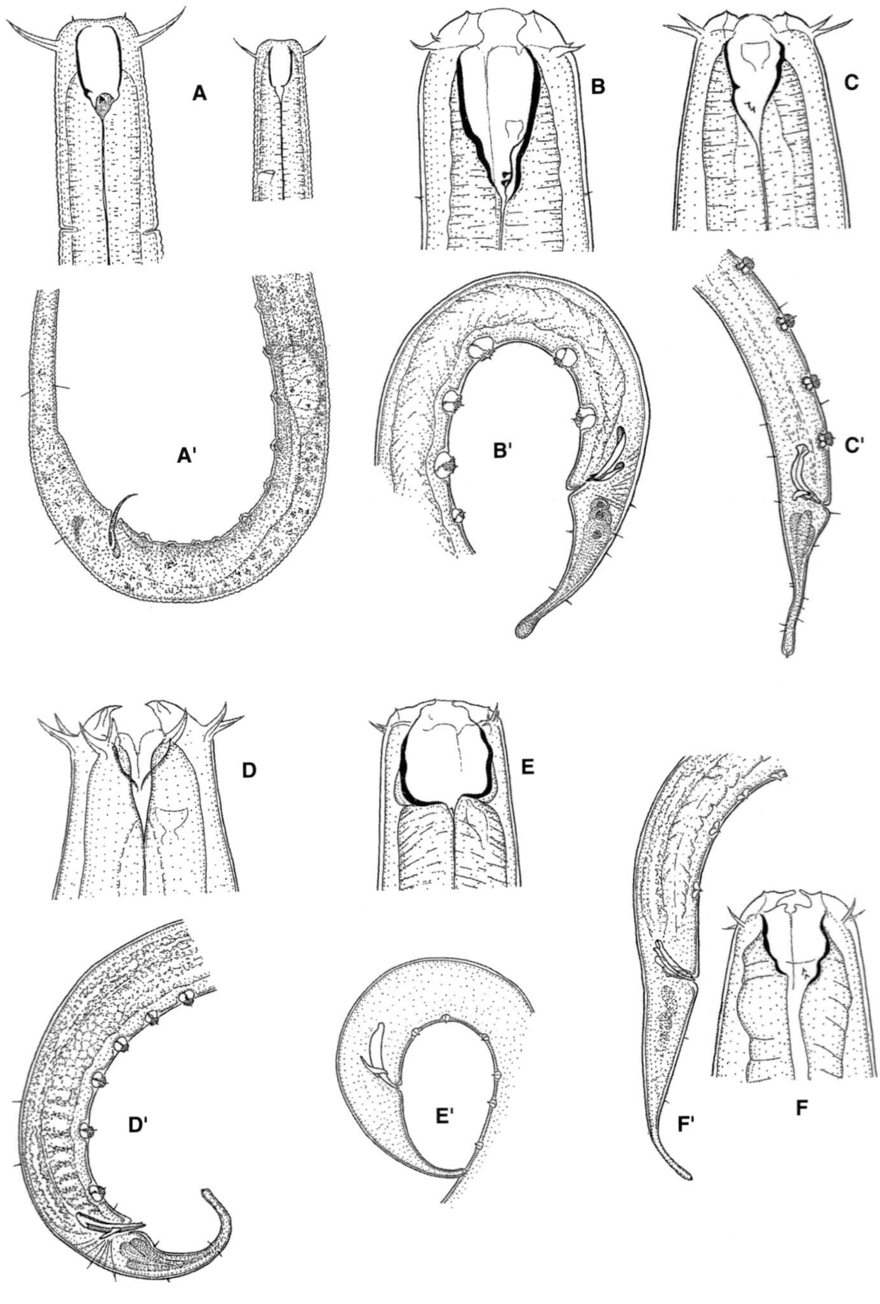

Fig. 13.2 A, A′, *Prismatolaimus*: anterior end (two species), male tail (A after Zullini 1982; A′ original); B, B′, *Paratrilobus*: anterior end, male tail (after Tsalolikhin 1981); C, C′, *Mesotobrilus*: anterior end, male tail (after Shoshin 1988); D, D′, *Quasibrilus*: anterior end, male tail (after Shoshin 1988); E, E′, *Kurikania*: anterior end, male tail (E after Shoshin 1991; E′ after Tsalolikhin 2001b); F, F′, *Lamuania*: anterior end, male tail (after Shoshin & Shoshina 2002).

Species list:
*K. *sibirica* Tsalolikhin, 1976 – Baikal (clay).
*K. *tsalolikhini* Shoshin, 1991 – Baikal (sand).

Genus *Lamuania* Tsalolikhin, 1976 (Fig. 13.2 (F, F'))
Paratrilobini. Body length 2–3 mm. Stoma very large, shaped as a wide funnel, without pockets; base of the stoma with two small teeth. Pharynx short (b = 7–9); cardiac glands very large, round. Male supplements not protruding nor echinulate.

Type species: *L. orientalis* Tsalolikhin, 1976. Limnic species.

Species list:
*L. *orientalis* Tsalolikhin, 1976 – Baikal (clay).

Tribe Tobrilini Filipjev, 1918

Tobrilinae. Stomatal pockets adjacent, overlapping; teeth near each other.

Genus *Tobrilus* Andrássy, 1959 (Fig. 13.3 (A, A'))
Syn. *Trilobus* Bastian, 1865, nec Brünnich, 1781.
Diagnosis. Tobrilini. Body length 1.0–4.5 mm. Cuticle smooth or very finely annulated. Cephalic setae length usually about one-third of head width. Stoma an anterior large buccal cavity and two overlapping adjacent pockets with one tooth, the two teeth being 0–6 μm apart. Vaginal musculature weak or not too strong. Male supplements 4–13 (often 6); small, submerged (= not protruding); distances between supplements almost equal. Spicules short (1.8–2.5% of body length). Spinneret small or absent.

Type species: *T. gracilis* (Bastian, 1865). Limnic species.

Species list:
*T. *aberrans* (W. Schneider, 1925) Andrássy, 1959: syn. *Trilobus gracilis conjungens* (Schneider, 1925) – Europe, North Baltic, Asia, North America (lakes, rivers, brackish waters).
*T. *affinis* Gagarin, 1996 – North Siberia (rivers), North America.
*T. *amabilis* Tsalolikhin, 1974 – Baikal (sand).
*T. *annettae* (Joubert & Heyns, 1979) Tsalolikhin, 1983 – Galápagos, Ethiopia, South Africa, China (brooks, rivers, pools, water reservoirs, wet soil).
*T. *bekmanae* Tsalolikhin, 1975 – Baikal (clay).
*T. *brevisetosus* (W. Schneider, 1925) Andrássy, 1959 – Central and Northern Europe to Siberia, North Baltic, Africa (lakes, rivers, brackish waters, chiefly in clay sediments).
*T. *gracilis* (Bastian, 1865) Andrássy, 1959 – Europe, doubtful data for other continents (fresh and brackish water, rhizosphere of aquatic macrophytes, psammon, subterranean waters, springs, also in polluted waters, rarely in soil and moss).
*T. *helveticus* (Hofmänner, 1914) Andrássy, 1959: syn. *Trilobus armatus* Allgén, 1925; *T. allophysoides* and *T. pseudoallophysis* in Micoletzky, 1925; *T. gracilis allophysis* var. *allophysoides* Schneider, 1925 – Europe, Asia (lakes, rivers, springs, brackish waters, clay sediments, psammon, also in polluted waters).
*T. *incognitus* Tsalolikhin, 1972 – Baikal (sand).
*T. *latens* Tsalolikhin, 1974 – Baikal (commensal of sponges).

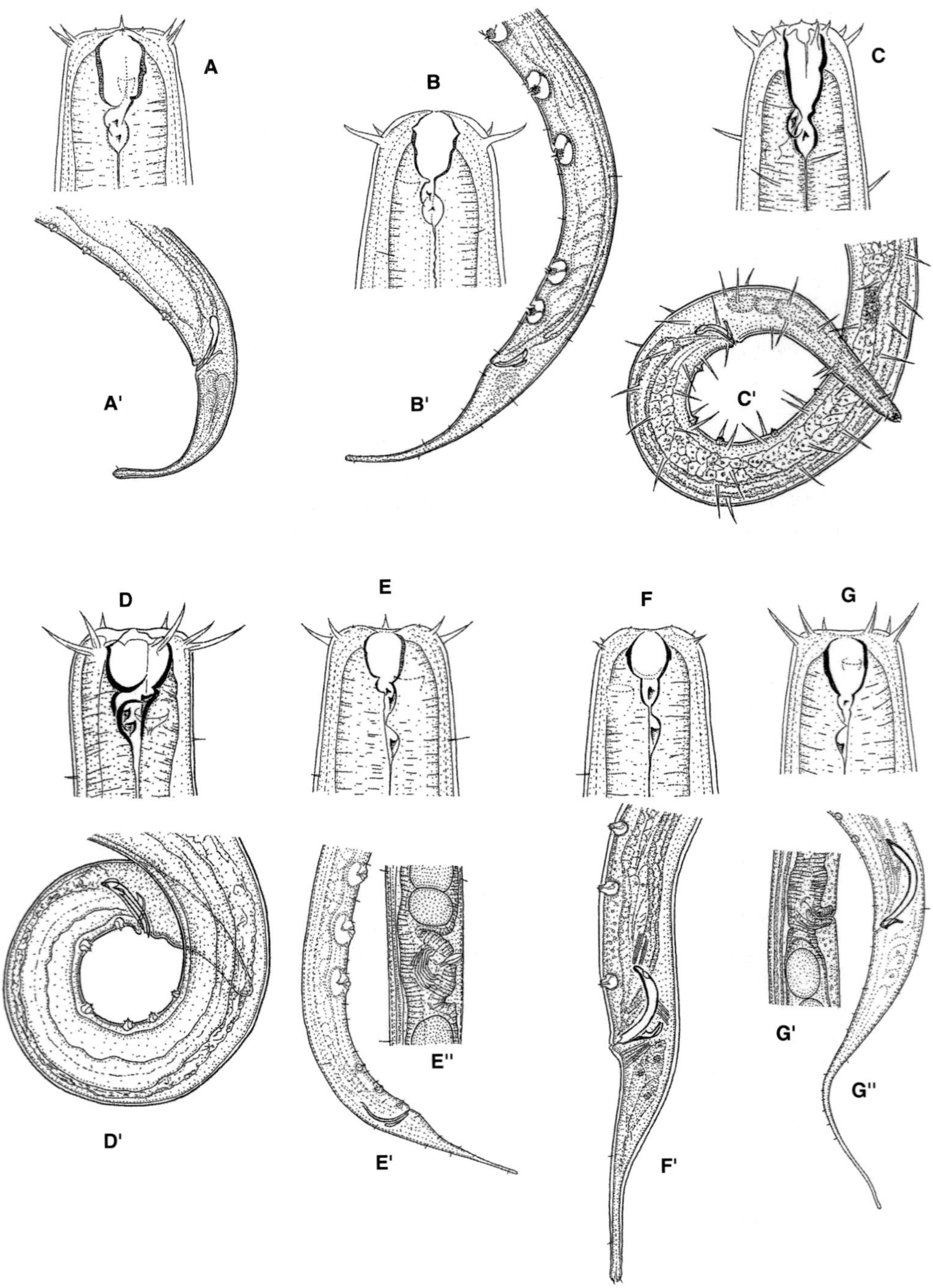

Fig. 13.3 A, A′, *Tobrilus*: anterior end, male tail (A after Zullini 1982; A′ after Shoshin & Shoshina 2003); B, B′, *Eutobrilus*: anterior end, male tail (after Zullini 1982); C, C′, *Asperotobrilus*: anterior end, male tail (C after Shoshin 1991; C′ after Shoshin 1998); D, D′, *Setsalia*: anterior end, male tail (after Shoshina 2003); E, E′, E″, *Neotobrilus*: anterior end, male tail, vulval region (after Zullini 1982); F, F′, *Brevitobrilus*: anterior end, male tail (F after Zullini 1982; F′ after Eyualem & Coomans 1997); G, G′, G″, *Semitobrilus*; anterior end, vulval region, male tail (after Zullini 1982).

*T. *macramphis* Tsalolikhin, 1977 – Baikal (clay).
*T. *minor* Gagarin & Gusakov, 1998 – Russia (lakes, rivers).
*T. *modestus* Gagarin, 1996 – North Siberia (rivers).
*T. *nepalensis* Tsalolikhin, 1983 – Nepal, Tadzhikistan (high mountain lakes).
*T. *parvus* Gagarin, 1991 – North Siberia (lakes, clay sediment).
*T. *phantasus* Tsalolikhin, 1983 – Mongolia (slightly salt lakes, sandy sediment).
*T. *tenuicaudatus* Gagarin, 1989 – Russia (lakes, clay).
*T. *tripylis* Gagarin, 1991 – North Siberia (lakes).
*T. *undophylus* Shoshin, 1988 – Baikal (coarse sand).
*T. *wesenbergi* (Micoletzky, 1925) Andrássy, 1959 – Denmark, Czech Republic, Russia (lakes).
*T. *zakopanensis* (Stefanski, 1924) Andrássy, 1959: syn. *Trilobus gracilis zakopanensis* Stefanski, 1924 – Poland, Czech Republic (lakes, moss).

Species inquirenda: *Trilobus aequiseta* W. Schneider, 1925; *T. dryophilus* Rahm, 1924; *T. gracilis intermedius* W. Schneider, 1923; *T. gracilis profundus* Micoletzky, 1914; *T. gracilis punctata* Stefanski, 1916; *T. kherai* Soni & Nama, 1981; *T. kirjanovae* Karimova, 1957; *T. lomnickii* Grochmalicki, 1911; *T. microstoma* Altherr, 1976; *T. murisieri* Altherr, 1938; *T. octiespapillatus* Linstow, 1876; *T. ornatus* Altherr, 1963.

Genus *Eutobrilus* Tsalolikhin, 1981 (Fig. 13.2 (B, B'))

Syn. *Peritobrilus* Gagarin, 1993; *Raritobrilus* Tsalolikhin, 1981

Diagnosis. Tobrilini. Body length 1.5–4.7 mm. Anterior setae long (usually 40–60% the head diameter). Stoma with two overlapping adjacent pockets; teeth 2–8 μm apart. Vaginal musculature normal. Male supplements 5–10 (often 6) protruding, echinate, with almost equal distance between them; first and last supplements usually smaller than the others. Spicules short (usually 2.0–2.5% of body length).

Type species: *E. grandipapillatus* (Brakenhoff, 1914). Limnic species.

Species list:
*E. *affectiosus* Shoshin, 1988: syn. *Eutobrilus erbaevae* Gagarin, 1991 (according to Tsalolikhin, unpublished) – Baikal, Siberia (lakes, rivers).
*E. *andrassyi* (Altherr, 1963) Tsalolikhin, 1981 – Alps, Russia (rivers).
*E. *angarensis* Gagarin, 1991 – Siberia (rivers).
*E. *anguiculus* Tsalolikhin, 1977 – Baikal (clay).
*E. *antarcticus* Tsalolikhin, 1981 – Antarctic (lakes).
*E. *arcticus* Gagarin, 1991 – North Siberia (lakes).
*E. *delamarei* (Altherr, 1963) Tsalolikhin, 1981: syn. *Eutobrilus rotundicapitatus* (Altherr, 1976) – Alps (rivers).
*E. *differtus* Shoshin, 1988 – Baikal (sand).
*E. *graciliformis* (Altherr & Delamare Deboutteville, 1972) Tsalolikhin, 1981 – North America (rivers, psammon).
*E. *grandipapillatus* (Brakenhoff, 1914) Tsalolikhin, 1981: syn. *Eutobrilus assimulatus* Shoshin, 1988 (according to Tsalolikhin, unpublished) – Europe, Baikal (lakes, rivers, sand, detritus, brackish waters).
*E. *heptapapillatus* (Joubert & Heyns, 1979) Tsalolikhin, 1981 – South Africa, South Australia (lakes, rivers).

*E. *husmanni* (Altherr, 1958) Tsalolikhin, 1981 – Europe, North Baltic (lakes, rivers, sand, detritus, brackish waters).
*E. *japonicus* Gagarin, 2000 – Japan (lake).
*E. *naumovi* Tsalolikhin, 1981 – New Guinea (brooks).
*E. *nothus* Gagarin, 1989 – Russia (lakes).
*E. *papillicaudatus* (Altherr, 1963) Tsalolikhin, 1981 – Argentina (rivers, psammon).
*E. *peregrinator* Tsalolikhin, 1983 – Baikal, Mongolia (lakes).
*E. *scallensis* (W. Schneider, 1925) n. comb. – Germany, Mongolia (lakes)
*E. *selengaensis* (Tsalolikhin, 1977) Tsalolikhin, 1981 – Selenga delta into Baikal (sand, detritus).
*E. *steineri* (Micoletzky, 1925) n. comb. – Europe, North America (psammon, lakes, slightly brackish waters).
*E. *tansaniensis* Tsalolikhin, 1983: syn. *Tobrilus brevisetosus* in Andrássy, 1968 – Tanzania (dried river bed, soil).
*E. *tumidus* Gagarin & Gusakov, 1998 – Russia (river).
*E. *unisexus* (Gagarin, 1989) n. comb. – Russia (canals).
*E. *vipriensis* (Gagarin, 1999) n. comb. – North Siberia (lakes).
Species inquirenda: *E. altherri* (Andrássy, 1959) = *Trilobus allophysoides* Altherr, 1953 (according to Tsalolikhin); *E. fortis* (Tsalolikhin, 1972); *E. strenuus* Gagarin, 1991.

Genus *Asperotobrilus* Shoshin, 1991 (Fig. 13.3 (C, C'))
Diagnosis. Tobrilini. Body length 0.6–1.6 mm. Cuticle finely but distinctly annulated, covered with numerous easily visible spine-like setae; anterior setae (unlike the other tobrilids) in two appreciably separated circles: setae of anterior circle shorter than setae of posterior circle. Cardiac glands large, rounded. Vagina muscular. Male supplements setiform; *ductus ejaculatorius* very long (>60% of the entire gonoduct). Tail conical. Spicules short (2.2–4.0% of body length).

Type species: *A. asper* Shoshin, 1991. Limnic species.

Species list:
*A. *aculeatus* Shoshin, 1998 – Baikal (coarse sand).
*A. *asper* Shoshin, 1998 – Baikal (sand).
*A. *investis* Shoshin, 1991 – Baikal (sand).

Genus *Setsalia* Shoshina, 2003 (Fig. 13.3 (D, D'))
Diagnosis. Tobrilini. Body length 3–5 mm. Buccal cavity separate from stomatal pockets by a distinct crest-like protrusion; three stomatal pockets, each armed with a well-developed tooth. Anterior part of stoma not free (as in the other tobrilids), but the muscular pharyngeal tissue reaches the anterior body cuticle. Male supplements six, protruding and echinate; spicules short (2% of body length).

Type species: *S. mirabilis* Shoshina, 2003. Limnic species.

Species list:
*S. *mirabilis* Shoshina, 2003 – Baikal (sand).

Subfamily Neotobrilinae Tsalolikhin, 2001

Diagnosis. Tobrilidae. Stomatal pockets situated longitudinally one after the other; each pocket with one tooth, both teeth being relatively distant from one another.

Tribe Neotobrilini Tsalolikhin, 1981
Neotobrilinae. Stoma with buccal cavity distinctly separated from anterior pocket.

Genus *Neotobrilus* Tsalolikhin, 1981 (Fig. 13.3 (E, E' E''))
(non-*Neotobrilus sensu* Rathore & Nama, 1989).
Diagnosis. Neotrobilini. Body length 1.2–4.3 mm. Anterior setae short or long (20–60% of head diameter). Stoma with buccal cavity distinctly separated from anterior pocket; the two pockets are separated by a narrow duct; teeth 7–20 μm apart. Vaginal musculature very strong, bulb-like and layered. Male supplements protruding, echinate, consisting of three large anterior + three small posterior ones. Spicules long and thin (about 3–5% of body length).

Type species: *N. longus* (Leidy, 1852). Limnic species.

Species list:
**N. ampiei* (Joubert & Heyns, 1979) Tsalolikhin, 1981 – South Africa (rivers).
**N. breviductus* (Loof & Riemann, 1976) Tsalolikhin, 1981: syn. *Tobrilus longus* in Riemann, 1970, in Andrássy, 1970; *partim* in Meyl, 1957 – Canada, South America, Africa (rivers, lakes).
**N. diversipapillatus* (Daday, 1905) Tsalolikhin, 1981: syn. *Tobrilus longior* Altherr, 1963; *T. longus* in Skwarra, 1922; in Riemann, 1966; in Argo & Heyns, 1973; *N. diversipapillatus occidentalis* (Daday, 1905); *N. diversipapillatus orientalis* (Argo & Heyns, 1973); *Tobrilus savaryi* Altherr, 1963 – Central and South America, Europe, Africa, Australia (rivers, brooks, also in polluted waters).
**N. hopei* (Loof & Riemann, 1976) Tsalolikhin, 1981 – North and South America (rivers).
**N. longiformis* (Loof, 1973) Tsalolikhin, 1981 – Surinam (swamp-creeks).
**N. longus* (Leidy, 1852) Tsalolikhin, 1981: syn. *Trilobus allophysis* in Hoeppli (1926); *N. filipjevi* (Ebsary, 1982); *N. longus americanus* (Leidy, 1852); *N. longus rossicus* Tsalolikhin, 1983 – Europe, North and South America, Russia, Siberia, Mongolia (rivers, pools, ditches, lakes, psammon, rhizosphere of aquatic plants).
**N. macrospiculum* (Altherr, 1963) Tsalolikhin, 1981 – Argentina (lakes, sand).
**N. nicaraguensis* (Loof & Riemann, 1976) Tsalolikhin, 1981: syn. *Trilobus longus* in Meyl 1957; *T.nicasimilis* Loof & Riemann, 1976 – Nicaragua (lakes, detritus).
**N. telekiensis* (Allgén, 1952) Tsalolikhin, 1981: syn. *Tobrilus hoehnelensis* (Allgén, 1952) – Kenya (high mountain lakes, mountain soils and moss).
**N. vicinus* (Loof, 1973) Tsalolikhin, 1981 – Surinam (canals).
Species inquirenda: *N. brzeskii* (Altherr, 1963); *N. floridensis* (Joubert & Heyns, 1979); *N. tantloyi* (Sukul, 1971)

Genus *Semitobrilus* Tsalolikhin, 1981 (Fig. 13.3 (G, G', G''))
Diagnosis. Neotrobilini. Body length 1.2–3.0 mm. Anterior setae long (30–60% of head diameter). Stoma with buccal cavity distinctly separated from anterior pocket;

the two pockets are separated by a narrow duct; teeth 6–12 μm apart. Vagina heavily muscled, usually anteriorly directed. Male supplements 6–10, small, submerged, not echinate; distance between supplements variable. Spicules long and thick (about 3.3–5.0% of body length).

Type species: *S. pellucidus* (Bastian, 1865). Limnic species.

Species list:

**S. closlongicaudatus* (Gagarin, 1971) Tsalolikhin, 2000: syn. *Trilobus longicaudatus* in Hofmänner, 1913; *T. pellucidus* in Hofmänner & Menzel, 1915; *Tobrilus longicauda* in Zullini, 1982; *T. pellucidus lemani* Gerlach & Riemann, 1974; *T. pellucidus longicaudatus* Juget, 1969 – Europe (lakes, rivers, brooks).

**S. ebsaryi* Tsalolikhin, 2000: syn. *Tobrilus longicaudatus* in Ebsary, 1982 – Canada (rivers, creeks, channels).

**S. parapellucidus* (Ebsary, 1982) Tsalolikhin, 2000 – Canada, North Siberia (rivers, brooks).

**S. pellucidus* (Bastian, 1865) Tsalolikhin, 1981: syn. *Trilobus leptosoma* de Man, 1881; *T. longicaudatus* W. Schneider, 1923 *sensu* Stefanski, 1938; *T. tenuicaudatus* Daday, 1894; *Tobrilus bayonensis*, Altherr, 1963; *T. gagarini* Ebsary, 1982; *T. lunzensis* Altherr, 1976 – North America, Europe, Asia, Africa, (lakes, rivers, brooks, springs, subterranean waters, peat-bogs, brackish waters, wet soils).

Genus *Brevitobrilus* Tsalolikhin, 1981 (Fig. 13.3 (F, F'))

Diagnosis. Neotrobilini. Body length short for Tobrilidae, usually about 1.5 mm (0.9–2.4 mm). Cuticle smooth or finely annulated, with few setae. Anterior setae short (usually 18–28% of head diameter). Stoma: buccal cavity goblet-shaped, distinctly separated from anterior pocket; the two pockets are separated by a narrow duct; teeth 8–12 μm apart. Amphids opening at the level of the anterior pocket. Vaginal musculature normal, not well developed. Male supplements six, slightly protruding with a stretched neck, not echinate; posterior supplement usually smaller than the others; distance between supplements not equal in some species. Spicules short (about 2.0–2.8% of body length).

Type species: *B. stefanskii* (Micoletzky, 1925). Limnic species.

**B. fesehai* Eyualem-Abebe & Coomans, 1997 – East Africa (lakes).

**B. findeneggi* (Schiemer, 1971) Tsalolikhin, 1981 – Austria (slightly saline lake).

**B. graciloides* (Daday, 1908) Tsalolikhin, 1981: syn. *Tobrilus africanus* Zullini, 1988; *Trilobus graciloides* Daday, 1908 – Africa (ponds, rivers, brackish waters, wet soils).

**B. granatensis* (Ocaña & Zullini, 1988) Tsalolikhin, 1992: syn. *Tobrilus graciloides* in Vinciguerra, 1972; *T. siculus* Vinciguerra & Zullini, 1991 – South Europe (rivers, warm springs, psammon).

**B. kenyensis* Tsalolikhin, 1992: syn. *Trilobus graciloides* in Allgén (1952); *Tobrilus graciloides* in Andrássy, 1964 – Africa (lakes, rivers)

**B. sardus* (Vinciguerra & Zullini) Tsalolikhin, 2002 – Sardinia (ponds).

**B. stefanskii* (Micoletzky, 1925) Tsalolikhin, 1981: not *Tobrilus stefanskii* in Joubert & Heyns, 1979; syn. *Trilobus gracilis consimilis* Schneider, 1939; *T. grandipapillatus* in Ditlevsen, 1921; *T. malayanus* W. Schneider, 1938; *T. paludicola* Filipjev, 1928; *Trilobus stefanskii paludicola* Micoletzky, 1925; *T. stefanskii stenurus* Filipjev, 1929;

T. vibratus Sukul, 1967; *Tobrilus allophysis* in Vinciguerra (1972); *B. montanus* Ocaña, Hernandez & Martin, 1996 – Europe, South Africa, Asia (rivers, lakes, clay sediments, also in polluted waters, wet soils).
B. tsalolikhini Eyualem-Abebe & Coomans, 1997 – East Africa (lakes).
Species inquirenda: *B. confusus* (Khera, 1975); *B. consimiloides* (Altherr, 1965); *B. fontinalis* (Altherr, 1976); *B. sexsetiferus* (Khera, 1975)

Tribe Epitobrilini Tsalolikhin, 2001

Diagnosis. Neotobrilinae. Stoma with buccal cavity joint with anterior pocket or not distinctly separated from it.

Genus *Epitobrilus* Tsalolikhin, 1981 (Fig. 13.4 (A, A'))
Syn. *Macrotobrilus* Tsalolikhin, 1981
Diagnosis. Epitobrilini. Body length 1.3–6.8 mm. Anterior setae long (usually 40–50% of head diameter). Buccal cavity and anterior pocket joint or almost so (i.e. the anterior tooth appears to be in the buccal cavity or just at its base); teeth distant from each other (8–20 μm apart). Male supplements echinate, protruded or semi-submerged; spicules short (1.8–2.8% of body length).

Type species: *E. meyli* Tsalolikhin, 1981. Limnic species.

Species list:
E. allophysis (Steiner, 1919) Gagarin, 1991: syn. *Trilobus gracilis allophysis* Steiner, 1919 – Europe, Asia, East Africa (rivers, lakes, brooks, springs, never in polluted waters).
E. elephas (Andrássy, 1964) n. comb. – East Africa (mountain lakes).
E. flagellatus (Andrássy, 1963) Tsalolikhin, 1981 – Argentina (wet soil, litter).
E. medius (G. Schneider, 1916) Tsalolikhin, 1991; syn. *Trilobus gracilis homophysalidis* Steiner, 1919; *T. gracilis medius* W. Schneider, 1925; *T. vistula* Pieczynska, 1960 – Europe, Asia (lakes, rivers, sand sediments, brackish waters).
E. meyli Tsalolikhin, 1981: syn. *Trilobus graciloides* in Meyl, 1957 – Lake Tanganyika.
E. sablensis (Ebsary, 1982) Tsalolikhin, 2001 – Canada (aquatic habitats).
E. setosus (Altherr, 1963) Tsalolikhin, 1981 – Argentina, East Africa (rivers, lakes).
Species inquirenda: *Trilobus parvipapillatus* Kreis, 1932.

Suborder TRIPYLINA de Man, 1876

Triplonchida. Free living. Pharynx cylindrical. Prominent spicules with protrusion capsules.

Superfamily TRIPYLOIDEA de Man, 1876

Tripylina.

Family TRIPYLIDAE de Man, 1876
Diagnosis. Tripyloidea. Lips three, with two papillae each (except *Abunema*). Outer strong labial (six) and thin cephalic (four) setae form one or two whorls. Amphids

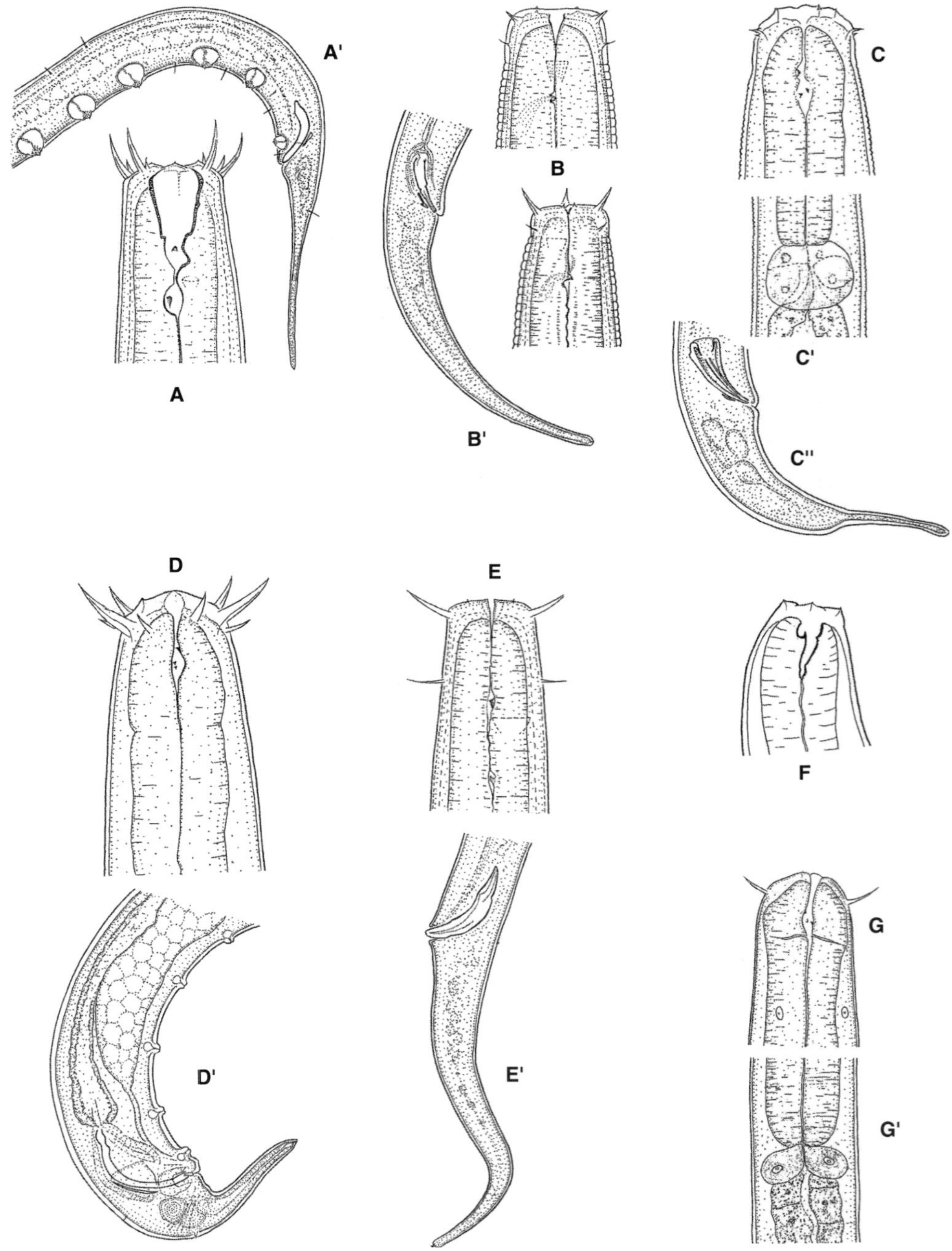

Fig. 13.4 A, A′, *Epitobrilus*: anterior end, male tail (A after Zullini 1982; A′ original); B, B′, *Tripyla*: anterior end (two species), male tail (after Zullini 1982); C, C′, C″, *Tripylella*: anterior end, cardiac region, male tail (C, C′ after Brzeski & Winiszewska-Slipinska 1993; C″ after Hernandez & Jordana 1988); D, D′, *Tripylina*: anterior end, male tail (after Brzeski & Winiszewska-Slipinska 1993); E, E′, *Trischistoma*: anterior end, male tail (after Zullini 1982); F, *Tobrilia*: anterior end (after Andrássy 1967); G, G′, *Abunema*: anterior end, cardiac region (after Khera 1971).

cup-like pockets with a slit-like aperture (except *Abunema*). Stoma narrow, encircled by pharyngeal musculature, with three teeth (often only one well visible) in one or two stomatal chambers. Pharynx a muscular tube. Cardia large. Oviduct short, without distinct lumen. Spermatheca present in bisexual species. Spicules enclosed in a muscular pouch. Tail with three glands and terminal spinneret.

Genus *Tripyla* Bastian, 1865 (Fig. 13.4 (B, B'))

Syn. *Promononchus* Micoletzky, 1923; *Paratripyla* Brzeski, 1964.

Diagnosis. Tripylidae. Body length 0.4–3.4 mm. Cuticle thick, annulated. Anterior body with six strong labial and four thin cephalic setae in two separate whorls. Stomatal chamber with two very small sub-ventral denticles and, slightly posteriorly, a larger, triangular or hook-shaped dorsal tooth. Cardiac glands large, composed of three separated cells. Female didelphic. Vagina without internal sclerotized pieces. $V \approx 45$–65%. Spicules wide, horn-shaped; numerous supplements extending from cloaca to near the head region.

Type species: *T. glomerans* Bastian, 1865. Limnic and soil species.

Species list:

**T. affinis* de Man, 1880: syn. *Tripyla tatrica* Stefanski, 1924 – Canadian and Siberian Arctic, North America, Europe, Africa, Asia (lakes, ponds, estuaries, wet soils, moss (*Sphagnum*)).

**T. aquatica* Brzeski & Winiszewska-Slipinska, 1993 – Poland (rivers).

T. crassa Alekseev & Bestalannaya, 1990.

**T. dybowskii* Tsalolikhin, 1976 – Baikal (deep sediment).

**T. elegantula* Brzeski & Winiszewska-Slipinska 1993 – Poland (flooded peat soil).

**T. filicaudata* de Man, 1880 – Europe, Siberia, Asia, North America (soil, moss, subterranean water, psammon, lakes, brooks).

**T. glomerans* Bastian, 1865: syn. *Tripyla papillata* Bütschli, 1873; *T. papillata crystallifera* Micoletzky, 1922; *Promononchus filipjevi*, Micoletzky, 1923 – Europe, North Baltic, Africa, Asia (ponds, lakes, rivers, brooks, psammon, flooded soils, brackish water, paddy fields, rare in moss).

**T. infia* Brzeski & Winiszewska-Slipinska, 1993: syn. *Tripyla filipjevi* Altherr & Delamare Deboutteville, nec *Promononchus filipjevi* Micoletzky, 1923 – North America, Africa, Europe, North Baltic, Siberia, East Asia (ponds, brooks, lakes, psammon, brooks, paddy fields, soil).

**T. italica* Tsalolikhin, 2003 – Italy (brook).

T. longicaudata Nesterov, 1979.

**T. magna* Altherr & Delamare Deboutteville, 1972 – North America, North Siberia (psammon, brooks).

T. minuta (Brzeski, 1963) Brzeski & Winiszewska-Slipinska, 1993.

T. pygmaea Micoletzky, 1922.

**T. setifera* Bütschli, 1873 – Europe, Africa, Siberia, Asia (moss, humid soil, pine bark, rivers, ponds, lakes, psammon).

**T. sibirica* Gagarin, 1993 – Siberia (river).

**T. subterranea* Tsalolikhin, 1976 – Kyrgyzstan (subterranean waters).

**T. tenuis* Brzeski, 1964 – North and Central America, Kyrgyzstan (wet soil, lakes, psammon, brackish waters).

T. terricola Brzeski & Winiszewska-Slipinska, 1993.
T. vulvata Andrássy, 1977.
Species inquirenda: *T. asiatica* Rahm, 1938; *T. bulbifera* Rahm, 1938; *T. cornuta* Skwarra, 1921; *T. crassicauda* Daday, 1899; *T. dentata* Daday, 1894; *T. filicaudata austriaca* Micoletzky, 1922; *T. filicaudata hoehnei* Rahm, 1928; *T. gigantea* Daday, 1827; *T. gracilis* Onorato de Cillis, 1917; *T. lata* Cobb, 1914; *T. minor* Cobb, 1893; *T. monohystera longicauda* Rahm, 1928; *T. punctata* Stefanski, 1916; *T. salsa* Bastian, 1865; *T. setifera triloboides* Rahm, 1932; *T. tenuicaudata* Cobb, 1893.

Genus *Tripylella* Brzeski & Winiszewska-Slipinska 1993 (Fig. 13.4 (C, C', C''))
Diagnosis. Tripylidae. Body length 0.7–1.5 mm. Cuticle thin, annulated. A single whorl of six longer and four shorter setae. Anterior stomatal chamber with a dorsal tooth, posterior chamber with two sub-ventral denticles. Cardiac glands very large, composed of six fused cells. Female didelphic. Vagina without internal sclerotized pieces. V = 47–55%. Male very rare; spicules wide, horn-shaped; with numerous supplements extending from cloaca to middle of pharynx.

Type species: *T. intermedia* (Bütschli, 1873) Brzeski & Winiszewska-Slipinska 1993. Moss and soil species.

Genus *Tripylina* Brzeski, 1963 (Fig. 13.4 (D, D'))
Diagnosis. Tripylidae. Body length 0.8–1.8 mm. Cuticle thin, not annulated, with numerous pores. A single whorl of six longer and four shorter setae. Stomatal chamber with dorsal tooth and (slightly anterior or posterior to it) two sub-ventral teeth. Female prodelphic, reflexed; no post-vulval uterine sac. Vagina with internal sclerotized pieces; $V \approx$ 56–80%. Male very rare; spicules narrow, sickle-shaped; supplements confined to pre-cloacal region. Tails of both sexes anteriorly cylindrical–conical, posteriorly a thin cylinder strongly bent.

Type species: *T. arenicola* (de Man, 1880) Brzeski, 1963. Soil and moss species.

Genus *Trischistoma* Cobb, 1913 (Fig. 13.4 (E, E'))
Diagnosis. Tripylidae. Body 0.6–1.8 mm, slender (a = 35–85), bent dorsad mainly in the posterior part. Cuticle smooth. Lips with six long setae and posteriorly four shorter and thinner subcephalic setae. Stoma denticles minute. Female prodelphic, with or without post-vulval uterine sac; vulva posterior (V = 73–83%). Tail dorsally twisted, at least in its anterior portion; c' = 4–6. Males rare or unknown.

Type species: *T. pellucidum* Cobb, 1913. Brzeski and Winiszewska-Slipinska (1993) removed *Trischistoma* from Tripylidae mainly because spicules are not surrounded by muscular pouch and because of the absence of a glandular cardia, and a peculiar body shape and posture. However, they did not propose an alternative classification, so we provisionally maintain this genus in Tripylidae.

Type species: *T. pellucidum* Cobb, 1923. Limnic and soil species.

Species list:
**T. gracile* Andrássy, 1985 – Hungary, Sicily (algae from a pool, coastal sand dune).

T. monohystera (de Man, 1880) Schuurmans Stekhoven, 1951: syn. *Tripyla monohystera* de Man, 1880, *Tripylina monohystera* de Man, 1880; *T. monohysteroides* Altherr, 1963 – North and South America, Europe, Africa, Asia (soil, sand filter beds, subterranean water, psammon, brooks, rivers, lakes).
T. pellucidum Cobb, 1913: syn. *Tripyla pellucida* (Cobb, 1913) – southern Mexico, Jamaica, Paraguay (soil, bromeliad water, inundated soil).
Species inquirenda: *T. conicaudatum* Schuurmans Stekhoven, 1951.

Genus *Tobrilia* Andrássy, 1967 (Fig. 13.4 F)
Diagnosis. Tripylidae. Body length less than 1 mm. Cuticle smooth or finely annulated, no anterior setae. Stoma a narrow funnel with one dorsal tooth and two small sub-ventral teeth; no pockets. Amphid aperture oval. Pharynx cylindrical; cardia flat, glandular. Two ovaries; vagina not cuticularized. Tail conical, elongate with spinneret. Essentially soil species.

Type species: *T. imberbis* (Andrássy, 1953). Loof (1973) considered *Tobrilia* as a genus *incertae sedis* and Tsalolikhin (1983) removed it from Tripylidae because of the lack of cephalic setae and glandular cardia, the different stoma shape and the araeolaimid amphids. However, he did not propose an alternative classification, so we provisionally maintain this genus in Tripylidae.

Type species: *T. imberbis* (Andrássy, 1953) Andrássy, 1967. Soil, rarely limnic species.

Species list:
T. imberbis (Andrássy, 1953) Andrássy 1967 – Hungary, Czech Republic, Mongolia, Tanzania (soda soil, soil of a lake bank).
T. longicaudata Andrássy, 1968 – Paraguay (irrigated soil, humid soil).

Genus *Abunema* Khera, 1971 (Fig. 13.4 (G, G'))
Diagnosis. Tripylidae. Head with six indistinct lips, each carrying a large seta. Anterior stoma funnel-shaped, posterior part with three teeth (one dorsal, two sub-ventral) about the same level. Amphids oval with a central point. Pharynx cylindrical, surrounding most of the stoma. Cardiac cells rounded with prominent nuclei. Ovary prodelphic, vulva post-equatorial. India, wet soil.

Brzeski and Winiszewska-Slipinska (1993) removed *Abunema* from Tripylidae mainly because of the six lips (instead of three) and the shape of the amphids. However, they did not propose an alternative classification, so we provisionally maintain this genus in Tripylidae.

Type species: *A. indicum* Khera, 1971. Wet soil species.

Species list:
A. indicum Khera, 1971 – India (soil from banks of a vegetated pond).

Family ONCHULIDAE Andrássy, 1964
Diagnosis. Tripyloidea. Anterior setae in two circles: six anterior longer setae + four at about 0.5–1.5 head diameter posteriad. Mouth cavity generally barrel-shaped with a large dorsal tooth and two sub-ventral swellings possessing occasionally similarly large teeth. Pharyngeal musculature areolate (= stratified in muscular bands

mingled with glandular tissue) at least in its posterior part. Cardia spheroid or elongated (one body width or somewhat more), consisting of large glandular cells. Two ovaries. Tail tip without mucro.

To this family should pertain genus *Cyathonchus* Cobb, 1933 here considered *genus inquirendum*.

Subfamily Onchulinae Andrássy, 1964.
Onchulidae. Stoma with large dorsal tooth with cytoplasmic interior and tubule opening near tip.

Genus *Onchulus* Cobb, 1920 (Fig. 13.5 (D, D'))
Diagnosis. Onchulidae. Cuticle smooth or very finely annulated. Lips with six well-developed and long setae; posteriorly, four pairs of thin cervical setae. Stoma wide, with a large dorsal, hooked tooth. Stoma base with 2–3 transverse folds. Cardia spheroid. Male supplements very small; spicules short, partially or entirely enveloped by a wide sheath. Tail filiform.

Type species: *O. longicauda* (Daday, 1889). Limnic and soil species.

Species list:
- *O. capriviensis* Swart & Furstenberg 1993 – Namibia (pond: sandy bottom).
- *O. dorsalis* Andrássy, 2001.
- *O. filicaudatus* Altherr, 1972 – Brazil (psammon near river).
- *O. gracilis* Andrássy, 2001: syn. *Onchulus longicauda* in Ali, Suryawanshi & Chisty, 1972.
- *O. longicauda* (Daday, 1899) – USA, Galápagos, Vietnam, New Guinea (soil, litter, lake).
- *O. nolli* Goffart, 1950: syn. *Onchulus fuscilabiatus* Altherr, 1965 – Europe (subterranean water, psammon near river).
- *O. oistospiculum* Allgén, 1933 – Congo (wet mud).
- *O. straticauda* (Altherr, 1972) n. comb. – Brazil (psammon near river).

Genus *Stenonchulus* W. Schneider, 1940 (Fig. 13.5 F)
Diagnosis. Onchulidae. Cuticle very finely annulated. Lips with six well-developed and long setae; posteriorly, four thin cervical setae. Anterior stoma relatively wide, posterior stoma narrow; in its middle three hooked teeth. Stoma without transverse folds. Cardia a coherent, elongated structure. Tail long.

Type species: *S. troglodytes* W. Schneider, 1940. Soil and limnic species.

Species list:
- *S. troglodytes* W. Schneider, 1940 – Europe, Azores, Madeira, Canary Islands (subterranean waters, psammon near river, humid sandy soil, beach litter, meadows).

Genus *Kinonchulus* Riemann, 1972 (Fig. 13.5 B, B')
Diagnosis. Onchulidae. Anterior end aberrant, similar to the head of a kinorhynch. Stoma with a long, acute, dorsal tooth and 12 long pricks (≈ 20 µm long) surrounding it; these structures can be protruded and retracted by movements of the lip region. Six double rows of small hooks are respectively situated between the points of attachment of one pair of the pricks and the six anterior head setae. Female

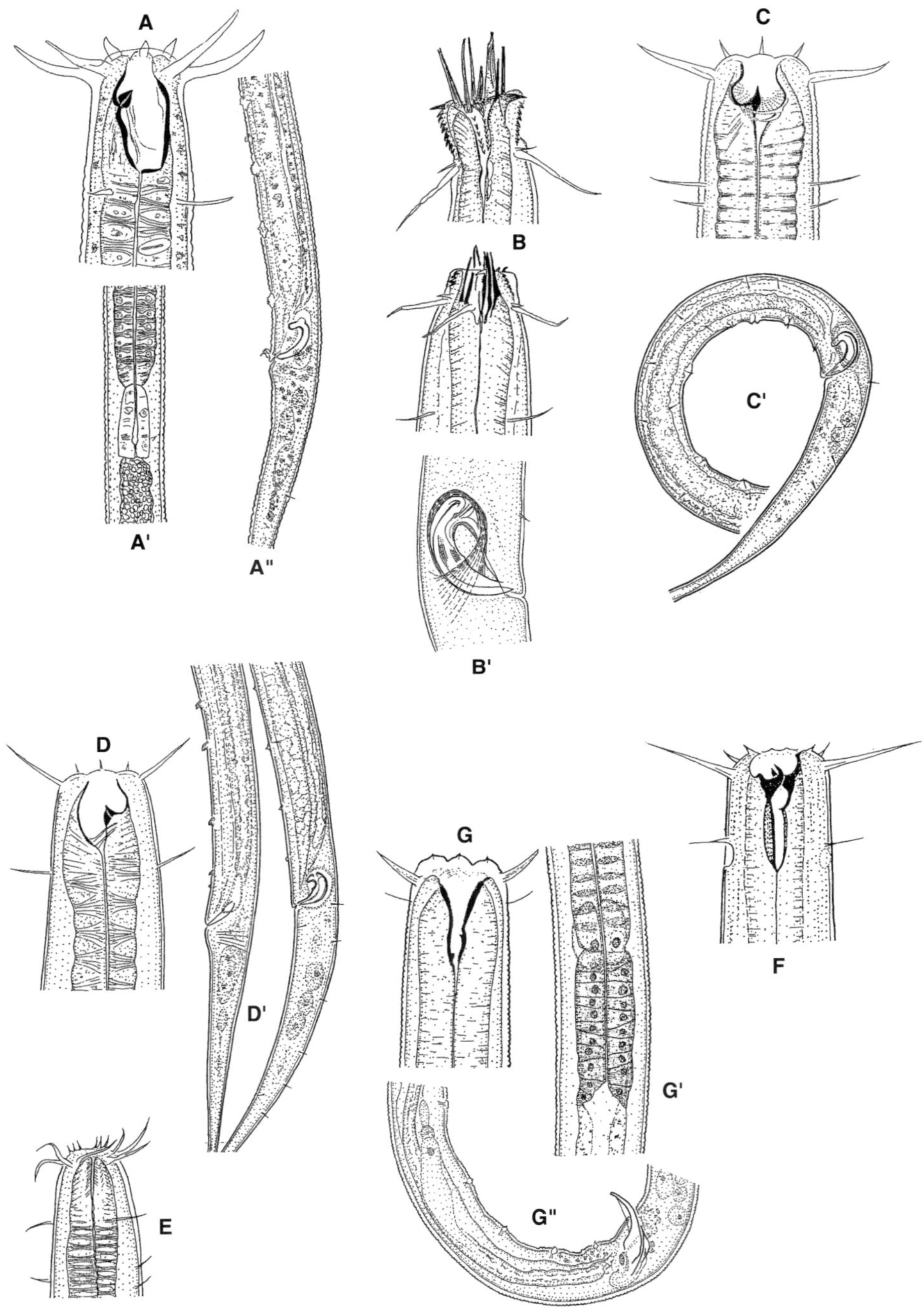

Fig. 13.5 A, A', A'', *Caprionchulus*: anterior end, cardiac region, male tail (after Swart & Heyns 1993); B, B', *Kinonchulus*: anterior end (extruded and retracted introvert), spicular apparatus (after Riemann, 1972); C, C', *Limonchulus*: anterior end, male tail (C after Andrássy 1963; C' after Swart & Furstenberg 1993); D, D', *Onchulus*: anterior end, male tail (two species) (D, D' after Andrássy 1964; D' after Swart & Furstenberg 1993); E, *Pseudonchulus*: anterior end (after Altherr 1972); F, *Stenonchulus*: anterior end (after Zullini 1982); G, G', G'', *Tobriloides*: anterior end, cardiac region, male posterior end (G, G' after Loof 1973; G'' after Swart & Heyns 1990).

amphidelphic. Male with setose pre-cloacal organs and other ventral papillae reaching the pharyngeal level, spicules short, curved, very sligthly refractive. Muscular sheath of spicules with evident spiral-like fibres. Tail long, filiform, without glands.

Type species: *K. sattleri* Riemann, 1972. Limnic species.

Species list:
**K. sattleri* Riemann, 1972 – Brazil (river psammon: fresh to brackish).

Genus *Limonchulus* Andrássy, 1963 (Fig. 13.5C, C')
Syn. *Paronchulus* Altherr, 1972.
Diagnosis. Onchulidae. Cuticle annulated. Lips with six well-developed, long setae; posteriorly four pairs of thin cervical setae. Stoma wide but short, with a large, hooked, dorsal tooth and numerous denticles forming rasp-like teeth on the stoma wall. Stoma base with two transversal folds. Cardia elongated, consisting of nine large cells. Tail long.

Type species: *L. bryophilus* Andrássy, 1963. Moss and limnic species.

Species list:
L. bryophilus Andrássy, 1963.
**L. heynsi* Swart & Furstenberg 1993 – South Africa (lake: sandy bottom).

Subfamily Pseudonchulinae Hodda, Bloemers & Wanless, 2005
Diagnosis. Onchulidae. Stoma lacking large dosal tooth or with small dorsal tooth consisting of cuticle only, without cytoplasm.

Genus *Pseudonchulus* Altherr, 1972 (Fig. 13.5E)
Diagnosis. Onchulidae. Subuticle finely annulated. Lips with six long setae and anteriorly setose papillae; posteriorly four cervical setae. Stoma thin, elongated, without teeth. Ovary opisthodelphic.

Type species: *P. inermis* Altherr, 1972. Wet soil or limnic species.

Species list:
**P. inermis* Altherr, 1972 – Brazil (psammon near river).

Genus *Caprionchulus* Swart & Heyns, 1993 (Fig. 13.5 A, A', A")
Diagnosis. Onchulidae. Body almost straight, especially in female. Cuticle smooth. Lips with six well-developed setae. Second whorl of six long setae just posterior to lips. Third whorl of four cervical setae more than halfway between the anterior end and amphids. Amphids slit-like. Stoma barrel-shaped, tapering posteriad. Dorsal tooth anterior, prominent. Cardia elongated, consisting of 12 cells. Female didelphic, vulva transverse, vagina short. Male diorchic, spicules enclosed in sheath-like tissue, rather arcuate, strongly cephalated, open at anterior end. No gubernaculum. Two adanal setae and a row of papilloid supplements. Tails filiform in both sexes.

Type species: *C. diversipapillatus* Swart & Heyns, 1993. Limnic species.

Species list:
**C. diversipapillatus* Swart & Heyns, 1993 – Namibia (pond: sandy bottom).

Genus *Tobriloides* Loof, 1973 (Fig. 13.5 G', G")

Diagnosis. Onchulidae. Genus resembling Tobrilidae (stoma structure, tail not filiform, caudal glands and spinneret), but with pharyngeal tissue and cardia typical of Onchulidae. Cuticle faintly, but distinctly, annulated. Stoma *Tobrilus*-like, but with only one pocket. Base of pharynx surrounded by four cardiac gland cells; cardia longer than body diameter. Tail elongated, conoid.

Type species: *T. choii* Loof, 1973. Soil species.

Key to genera of Triplonchida
(Asterisks refer to genera containing limnetic species)
Illustration numbers correspond to the key numbers (e.g. Tylolaimophorus is in 4')

1. Body short (rarely >1 mm) and obese (a = 15–25, rarely more); stoma with a complex, asymmetrical spear; tail short and rounded (Diphtherophorina) 2
   Body normal, slender; buccal cavity (stoma) more or less visible, without spear, but usually with tooth or teeth 9
2. Spear length about equal to lip region width (Diphtherophoridae) 3
   Spear much longer than the lip region width (Trichodoridae) 5
3. Expanded part of pharynx about one-half of total length . . . . *Longibulbophora* (Fig. 13.1 C)
   Expanded part of pharynx a short pyriform basal bulb 4
4. Anterior part of spear divided, with a strong dorsal curve . . . . *Diphtherophora** (Fig. 13.1 A)
   Anterior part of spear pointed, neither curved nor divided . . . . *Tylolaimophorus* (Fig. 13.1 B)
5. Female prodelphic; vulva at 60–78% of body length 6
   Female amphidelphic; vulva at 50–65% of body length 8
6. Body very small, less than 0.5 mm; vulva at 60–65%; female with sub-terminal anus. . . . . *Ecuadorus* (Fig. 13.1 H)
   Body usually longer than 0.5 mm; vulva at 75%; female with terminal or sub-terminal anus 7
7. Pharyngeal bulb not overlapping intestine; lateral body pores near vulva; vulva at 77–85%; posterior half of male body somewhat curved . . . . *Monotrichodorus* (Fig. 13.1 F)
   Pharyngeal bulb overlapping intestine; no lateral body pores near vulva; vulva at 80–90%; posterior half of male body not curved; female usually with terminal anus. . . . . *Allotrichodorus* (Fig. 13.1 G)
8. Pharyngeal bulb not overlapping intestine; lateral body pores near vulva; vagina usually about one-half the body width long; posterior body of male somewhat curved . . . . *Trichodorus** (Fig. 13.1 D)
   Pharyngeal bulb overlapping intestine; no lateral body pores near vulva; vagina about one-third body width long; posterior body of male not curved . . . . *Paratrichodorus** (Fig. 13.1 E)
9. Tail filiform, without caudal glands and spinneret 10
   Tail short or long, but not filiform, with caudal glands and spinneret 11

10. Stoma prismatic, in its base with a very small dorsal tooth and two sub-ventral swellings with or without rasp-like structures; tail tip with a claw-like mucro; body less than 2 mm . . . . *Prismatolaimus** (Fig. 13.2 A)
Stoma barrel- or funnel-shaped, or very narrow, usually with at least one large tooth; pharyngeal musculature areolate (= stratified muscular bands mingled with glandular tissue) in its whole length (Onchulidae, *partim*) 30
11. Pharyngeal musculature areolate (= stratified muscular bands mingled with glandular tissue) only in its posterior end; cardia longer than corresponding body width . . . . *Tobriloides* (Fig. 13.5 G)
Pharyngeal musculature normal, not areolate; cardia shorter than corresponding body width 12
12. Anterior setae absent; stoma a narrow funnel with a well-visible dorsal tooth and two small sub-ventral denticles . . . . *Tobrilia** (Fig. 13.4 F)
Anterior setae present; stoma wide or narrow 13
13. Stoma narrow 14
Stoma wide, funnel- or rarely barrel-shaped, usually with two teeth at its base (Tobrilidae) 18
14. Head with six large setae; stoma with three small teeth; amphids oval; cardiac cells large and rounded, with prominent nuclei; ovary short, prodelphic; tail conical . . . . *Abunema** (Fig. 13.4 G)
Anterior setae 10, usually 6 longer + 4 shorter, at the same level or in two circles 15
15. Body slender (a = 35–70), dorsally bent; six stronger anterior setae + four thinner, more posterior sub-labial setae; stoma with minute denticles; no cardiac glands; female gonad prodelphic . . . . *Trischistoma** (Fig. 13.4 E)
Body usually not so slender and not dorsally bent; anterior setae in one or two circles; female gonad mostly amphidelphic 16
16. Cuticle smooth; ovary prodelphic; spicules narrow, sickle-shaped; stoma with a dorsal tooth and two denticles anteriorly or posteriorly placed . . . . *Tripylina* (Fig. 13.4 D)
Cuticle annulated; two ovaries; spicules wide, horn-shaped 17
17. Cuticle thick; anterior setae in two separate circles; stoma with a dorsal tooth (triangular- or hook-shaped) and anteriorly, two tiny denticles; cardiac glands rather large; male supplements small, papillae extending up to the head . . . . *Tripyla** (Fig. 13.4 B)
Cuticle thin; anterior setae in a single circle; stoma with a dorsal triangular tooth and, posteriorly, two teeth; cardiac glands very large, their diameter being larger than the pharyngeal base; male supplements small, papillae extending halfway up to pharynx . . . . *Tripylella* (Fig. 13.4 C)
18. Stoma with overlapping pockets adjacent to each other, or pockets absent; teeth near each other, rarely absent (Tobrilinae) 19
Stoma with two pockets (each with one tooth) situated longitudinally one after the other; teeth relatively distant from each other (6–25 μm apart) (Neotobrilinae) 27

19. No stomatal pockets, or pockets merged with the buccal cavity; stoma very large, funnel- to barrel-shaped; large nematodes (2–6 mm) (Paratrilobini) 20
    Two (rarely three) overlapping stomatal pockets (each with one tooth); teeth 0–8 μm apart; small to large nematodes (Tobrilini) 24
20. Pharyngeal musculature normally encircling stoma 21
    Pharyngeal musculature not encircling stoma; stoma very large, barrel-shaped, with thick walls, and with two minute denticles at its base. Lake Baikal . . . . *Kurikania** (Fig. 13.2 E)
21. Pharynx short (b = 6–10); cardiac gland cells very large, round 22
    Pharynx normal (b = 3–6); cardiac gland cells usually elliptic and not so large 23
22. Stoma short, cup-shaped, without teeth; amphid posterior to base of stoma; supplements protruding, echinulate. Lake Baikal . . . . *Quasibrilus** (Fig. 13.2 D)
    Stoma large, roundish, a wide funnel with two small teeth at its base; amphid at the base of stoma; supplements neither protruding nor echinulate. Lake Baikal . . . . *Lamuania** (Fig. 13.2 F)
23. Stoma barrel-shaped or rounded, thick-walled; pockets merged with the buccal cavity with two strong teeth; amphid aperture posterior to stoma; male supplements six, first and/or last supplement smaller than others, with bulbillum in top position, immediately below the supplement's cap . . . . *Paratrilobus** (Fig. 13.2 B)
    Stoma funnel-shaped, thin-walled; pockets merged with the buccal cavity, with two small teeth; amphid aperture anterior to stoma; male supplements 6–8, uniform, with bulbillum in low position. Lake Baikal . . . . *Mesotobrilus** (Fig. 13.2 C)
24. Buccal cavity separated from stoma pockets by a distinct crest-like protrusion; three stoma pockets, each armed with a well-developed tooth; large nematodes. Lake Baikal . . . . *Setsalia** (Fig. 13.3 D)
    No such distinct crest between buccal cavity and pockets; only two teeth and pockets 25
25. Cuticle finely but distinctly annulated, covered with numerous well-visible spine-like setae; anterior setae in two appreciably separated circles; male supplements setiform; tail conical. Lake Baikal . . . . *Asperotobrilus** (Fig. 13.3 C)
    Cuticle smooth or very finely annulated with few and very thin setae; cephalic setae in one circle; male supplements papilliform; spicules short 26
26. Pockets overlapping and teeth near each other (0–6 μm apart); anterior setae usually short (about one-third of head width); male supplements small, almost not protruding . . . . *Tobrilus** (Fig. 13.3 A)
    Pockets overlapping or almost so; teeth fairly near each other (2–8 μm apart); cephalic setae long (usually 40–60% of head diameter); male supplements protruding . . . . *Eutobrilus** (Fig. 13.3 B)

27. Buccal cavity well separated from the anterior pocket (Neotobrilini) 28
Buccal cavity and anterior pocket joint or almost so (i.e. the anterior tooth appears to be in the buccal cavity or in its base); teeth distant from each other (8–20 μm apart); anterior setae very long (40–50% of head diameter); supplements protruded or semisubmerged. . . . .*Epitobrilus** (Fig. 13.4 A)
28. Vaginal musculature very strong; spicules long (3.3–5.0% of body length) 29
Vaginal musculature normal; spicules short (2.0–2.8% of body length) anterior setae short (usually 18–28% of head diameter); teeth far from each other (8–12 μm apart); supplements slightly protruding with a stretched neck; not (or only slightly) echinate. . . . .*Brevitobrilus** (Fig. 13.3 F)
29. Vaginal musculature very strong, bulb-like and layered; spicules long and thin; six protruding and echinate supplements, three large anterior + three small posterior; cephalic setae 25–60% of head diameter; teeth 7–20 μm apart . . . .*Neotobrilus** (Fig. 13.3 E)
Vagina heavily muscular, usually anteriorly directed; spicules long and thick; supplements small, submerged; cephalic setae 30–60% of head diameter; teeth 6–12 μm apart . . . .*Semitobrilus** (Fig. 13.3 G)
30. Aberrant anterior end with protrusible and retractile pharynx, similar to the head of kinorhynchs, with 12 narrow pricks ≈ 20 μm long and 6 double rows of small hooks . . . .*Kinonchulus** (Fig. 13.5 B)
Anterior end normal 31
31. Stoma without teeth; female gonad opisthodelphic . .*Pseudonchulus** (Fig. 13.5 E)
Stoma with one or more teeth; female gonad amphidelphic 32
32. Stoma barrel-shaped, tapering posteriorly, with an anterior dorsal tooth; cardia one body diameter long or slightly more; spicules strongly curved and cephalated. . . . .*Caprionchulus** (Fig. 13.5 A)
Stoma anteriorly widened, posteriorly narrow; teeth in the middle of the stoma; spicules almost straight or strongly curved, but not cephalated 33
33. Stoma with a large dorsal tooth; two to three transverse folds in posterior part of stoma; cardia length less or equal to body diameter 34
Stoma with three large teeth; no transverse folds in its posterior part; cardia length longer or equal to one body diameter . . . .*Stenonchulus** (Fig. 13.5 F)
34. Stoma with a large tooth and several small denticles .*Limonchulus** (Fig. 13.5 C)
Stoma with a large tooth, without small denticles . . . . . .*Onchulus** (Fig. 13.5 D)

Ecology of Triplonchida

The order Triplonchida is, from an ecological point of view, a very composite taxon. It encompasses the plant-parasitic and almost entirely terrestrial Diphtherophorina, the almost entirely freshwater Tobrilina and the mainly freshwater Tripylina.

DIPHTHEROPHORINA: Being plant parasitic, they are terrestrial nematodes. They lack setae, caudal glands, a spinneret and their tails are short in both sexes. *Diphtherophora* was occasionally found in wet soils and near liverworts (plants living only in constantly wet habitats). Trichodoridae are repeatedly found in wet or flooded soils, river banks, ponds, canals and brooks. Probably many of them are passively driven into water, but evidently they withstand submersion better than *Diphtherophora*. One species (*Trichodorus similis*) was seen to feed on root tips of aquatic plants, so it can also be considered to be a real aquatic nematode.

TOBRILINA: On the basis of molecular data (De Ley and Blaxter, 2002) this suborder comprises Prismatolaimoidea (with the only genus *Prismatolaimus*) and Tobriloidea. About one-half of *Prismatolaimus* species are aquatic and were found in a large array of aquatic habitats, including temporary waters, subterranean waters, *Sphagnum* habitats and even low-salt marine water (Gulf of Bothnia, with one-tenth of the normal marine salinity content). *Prismatolaimus* is supposed to be a bacterial feeder, but we cannot preclude the possibility that it can also eat some eukaryote cells. This genus was found in all the continents.

Tobrilidae is the only real freshwater nematode family we know. One of its smaller species, *Brevitobrilus stefanskii*, measures an average 1.5 mm and weighs 1.1 μg; one of its larger species, *Epitobrilus elephas*, measures 6.1 mm and weighs 50 μg. But a typical species, *Tobrilus gracilis*, measures 2.3 mm and weighs 5.4 μg. Tobrilidae must be considered a relatively euryhaline group, since the salinity ranges from pure freshwater to brackish water of 10–12 ppt salinity. They feed on small invertebrates and unicellular algae, mainly diatoms. According to Tsalolikhin (1983), the ancestors of Tobrilidae (like other fauna of marine origin) during ancient oceanic regressions adapted to freshwater habitats via slow desalination of large sea extents on continental shelves isolated from the open ocean. The continuous input of freshwater from rivers led first to the formation of lake–sea habitats and then to freshwater lakes and marshes with a parallel adaptation of the former marine fauna to a decreasing salt concentration. Tsalolikhin, for geological and palaeoclimatic reasons, concludes that for the tobrilids this process took place not before the Permian period. He notes that till now these nematodes have maintained some tolerance to low-salt water.

Among Tobrilidae, genus *Paratrilobus* includes large (2–5 mm) and rare species, most of them endemic to Lake Baikal (Tsalolikhin, 1980). Chaetae of oligochaetes have been found in the gut of some of *Paratrilobus*. The species of *Quasibrilus*, *Kurikania*, *Lamuania*, *Asperotobrilus* and *Setsalia* are exclusively large (2–6 mm) and endemic species of Lake Baikal, but more genera and species of this family are expected to be found in this lake alone. Each species of these genera seems to prefer a definite depth and substratum. In Lake Baikal, these two variables are well correlated: at about 1–2 m the bottom consists of coarse sand, at 2–50 m depth there is sand and fine sand, at 50–1600 m depth (profundal zone) there is silt and clay. Lake Baikal, given its great dimension and antiquity, is a hot spot for tobrilid evolution.

The genus *Tobrilus* is restricted to Europe and Asia and was found in all types of freshwater and slightly brackish water habitats; one species is a commensal of sponges in Lake Baikal. *T. gracilis* is very abundant in some lakes (it comprises more than 80% of the nematodes in the Lake of Plön and almost 40% in Lake Madü), but is rather scarce in others (about 4% in Lake Balaton and in Königssee) (Biró, 1972). In many lakes in Poland it represents about 10% of the nematode population (Prejs, 1977). But, owing to its dimensions, it represents the dominant nematode biomass even in the Königssee (German Alps) at 1 m water depth (mean yearly biomass: 52 mg/m^2); in this lake the population peaks twice a year at 1 m depth (winter and summer) and once (in early spring) in the profundal (Traunspurger, 1997). This species presents a reproductive length of time (egg-to-egg time) of 6–12 months at 4–14°C. *T. gracilis* was also one of the commonest nematodes found in Budapest tap water (up to >10 individuals/t water) (Dózsa-Farkas, 1965). This species can also tolerate polluted waters and is sometimes found in anoxic sediments.

The genus *Eutobrilus* is known from all the continents in various freshwater and slightly brackish water habitats. In a small Alpine lake *Eutobrilus grandipapillatus* is dominant at 20–28 m depth where it exhibits three generations (peaking in October, March and May); its average density in summer is 60,000 individuals/m^2, and in winter (under ice cover) 235,000/m^2 (Bretschko, 1973).

The more 'advanced' genus *Neotobrilus* is found in all continents in various freshwater habitats. It seems that this genus originated on the American continent and moved eastwards only after the Tertiary era: in fact, it is not present in the ancient (Tertiary) Lake Baikal (Tsalolikhin, 1983). Some species of this genus, such as *Neotobrilus diversipapillatus*, also occur in polluted waters (Arthington *et al.*, 1986). *Brevitobrilus* also is a 'modern' genus, with similar biogeographical peculiarities. It contains relatively small (about 1.5 mm) species, is limited to the Old World (Eurasia and Africa) and inhabits various freshwater and slightly brackish habitats. Some of its species are reported to feed on diatoms (Tsalolikhin, 2001a). *Brevitobrilus granatensis* was found also in warm springs (42–43°C) (Ocaña and Zullini, 1988). *Semitobrilus* is mainly present in Palaeoarctic regions (Eurasia, North America) in various freshwater and slightly brackish water habitats. *Epitobrilus* was found on all continents, except Australia; this genus inhabits various freshwater and slightly brackish water habitats and only one species is known from wet soil and litter. *Epitobrilus medius* in Königssee reaches high numbers at 60 m depth (mean yearly biomass: 20 mg/m^2) (Traunspurger, 1997).

Among tobrilids, *Tobrilus* is considered a primitive genus because of its small and simple supplements and their variable number; moreover, the juveniles have an *Oxystomina*-like stoma. Evolution of tobrilids leads towards a fixed and lower (six) number of differently shaped supplements, a long and strong ejaculatory canal and long and thin spicules (Tsalolikhin, 1983).

TRIPYLINA: On the basis of molecular data, the family Tripylidae was separated from the tobrilids and put in close relationship with the family Onchulidae. *Tripyla* is the richest in species and most common genus of Tripylidae and a large majority of its species are aquatic or primarily aquatic. Their length varies from 0.5 mm (*Tripyla pygmaea*) to 3.4 mm (*Tripyla magna*). This genus is more or less cosmopolitan, but almost all records refer to Europe. All kinds of freshwater habitats can host *Tripyla*. This is an omnivorous, but mainly predatory genus: residues of rotifers, tardigrades and small nematodes were found in the gut (Menzel, 1920). *Tripyla setifera* (1.2 mm) presents a reproductive time length (egg-to-egg time) of 30–40 days at 20–22°C. *Tripyla*, unlike tobrilids, is rarely dominant in lakes and in other bodies of water; it is a typical inhabitant of mesotrophic lakes (Traunspurger, 2002).

Family Onchulidae includes relatively rare and noteworthy species including some of the few exclusive, or almost exclusive, subterranean nematodes (such as *Onchulus nolli* and *Stenonchulus troglodytes*). Moreover, this family includes one of the most interesting and enigmatic nematodes: *Kinonchulus sattleri*, collected by Sattler and described by Riemann in 1972 from the psammon of a Brazilian river. This nematode has a remarkable protrusible pharynx outfitted with long pricks and rows of hooks. This feature is reminiscent of the head of kinorhynchs and of nematomorph larvae. It is interesting to note that the original description stressed the relationship existing between *Kinonchulus* and *Trischistoma* (for the stomatal structure), and with Diphtherophoroidea (for the spicular protractor muscles).

Acknowledgements

The author is deeply indebted to Dr I. Andrássy, W. Decraemer, V. Peneva, A.V. Shoshin, A.A. Shoshina and S. J. Tsalolikhin for their fruitful discussions and help and to Dr A. Coomans and T. Bongers for their useful comments.

References

Allen, M.W. (1957) A review of the nematode genus *Trichodorus* with descriptions of ten new species. *Nematologica* 2, 32–62.

Altherr, E. (1972) Contribution à la connaissance des Nématodes de l'estuaire de l'Amazone. *Amazoniana* (Kiel) 3, 141–174.

Andrássy, I. (1963) Freilebende Nematoden aus Angola, I. Einige moosbewohnende Nematoden. Museu do Dundo, Subsìdios para o estudio da biologia na Lunda. Diamang, pp. 57–79.

Andrássy, I. (1964) Onchulidae n. fam., eine neue Familie der Ordnung Enoplida (Nematoda). *Opuscula Zoologica* (Budapest) 5, 25–41.

Andrássy, I. (1967) Ergebnisse der zoologischen Forschungen von Dr. Z. Kaszab in der Mongolei. 92. Weitere Bodennematoden aus den Jahren 1964 und 1965. *Opuscula Zoologica* (Budapest) 6, 203–233.

Andrássy, I. (1988) A new species of the genus *Tylolaimophorus* (Nematoda) from Hungary. *Állattani Közlemények* 74, 183–186 (in Hungarian).

Andrássy, I. (2003) Three new species of Prismatolaimus de Man (Nematoda: Enoplida) possessing organellum dorsale, and systematics of the genus. *International Journal of Nematology* 13, 123–134.

Arthington, A.H., Yeates, G.W. and Conrick, D.L. (1986) Nematodes, including a new record of *Tobrilus diversipapillatus* in Australia, as potential indicators of sewage effluent pollution. *Australian Journal of Marine and Freshwater Research* 37, 159–166.

Biró, K. (1972) Nematodes of Lake Balaton. III. The fauna in the late-summer. *Annales Instituti Biologici Tihany* 39, 89–100.

Bretschko, G. (1973) Benthos production of a high-mountain lake: Nematoda. *Verhandlung Internationale Vereinigung Limnologie* 18, 1421–1428.

Brzeski, M.W. and Winiszewska-Slipinska, G. (1993) Taxonomy of Tripylidae (Nematoda: Enoplia). *Nematologica* 39, 12–52.

Cobb, N.A. (1920) The orders and classes of nemas. In: *Contributions to a Science of Nematology*, VIII. Waverly Press, Baltimore, Maryland, pp. 213–216.

Decraemer, W. (1995) *The Family Trichodoridae: Stubby Root and Virus Vector Nematodes.* Kluwer Academic Publishers, Dordrecht, The Netherlands, 360 pp.

Decraemer, W. (1998) A polytomous key for the identification of species of the family Trichodoridae Thorne, 1935 (Nematoda: Triplonchida). *Fundamentals of Applied Nematology* 21, 37–62.

De Ley, P. and Blaxter, M. (2002) Systematic position and phylogeny. In: Lee D.L. (ed.) *The Biology of Nematodes*. Taylor & Francis, London, pp. 1–30.

Dózsa-Farkas, K. (1965) Untersuchungen über die Fauna des Budapester Leitungswassers, mit besonderer Berücksichtigung der Nematoden. *Opuscula Zoologica* (Budapest) 5, 173–181.

Eyualem-Abebe and Coomans, A. (1997) Aquatic nematodes from Ethiopia VIII. Enoplids, with descriptions of *Brevitobrilus fesehai* n. sp. and *B. tsalolikhini* n. sp. (Enoplida: Nematoda). *Hydrobiologia* 345, 149–184.

Goodey, T. (1963) *Soil and Freshwater Nematodes*. 2nd edn. revised by J.B. Goodey, Methuen, London, 544 pp.

Hernández, M.A. and Jordana, R. (1988) Description of the male of *Paratripyla intermedia* (Bütschli, 1873) (Nematoda, Enoplida). *Miscellanea Zoologica* 12, 23–26.

Jairajpuri, M.S. and Ahmad, W. (1992) *Dorylaimida. Free-living, Predaceous and Plant-parasitic Nematodes*. E.J. Brill, Leiden, The Netherlands, 458 pp.

Khera, S. (1971) Nematodes from the banks of still and running waters. VII. Family Monhysteridae. *Nematologica* 16, 492–502.

Loof, P.A.A. (1973) *Tobriloides choii* n. gen., n. sp. (Tripylidae). *Nematologica* 19, 75–79.

Menzel, R. (1920) Über die Nahrung der freilebenden Nematoden und die Art ihrer Aufnahme. *Verhandlung Naturforschung* 31, 153–188.

Ocaña, A. and Zullini A. (1988) A new species of *Tobrilus* (Nematoda) from spring water. *Nematologica* 34, 1–5.

Prejs, K. (1977) The species diversity, numbers and biomass of benthic nematodes in central part of lakes with different trophy. *Ekologia Polska* 25, 31–44.

Rashid, F., De Waele, D. and Coomans, A. (1986) Trichodoridae (Nematoda) from Brazil. *Nematologica* 31, 289–320.

Riemann, F. (1972) *Kinonchulus sattleri* n.g. n.sp. (Enoplida, Tripyloidea), an aberrant free-living nematode from the lower Amazonas. *Veröffentlichungen des Instituts für Meeresforschung Bremerhaven* 13, 317–326.

Rodriguez, M. and Bell, A.H. (1978) Three new species of Trichodoridae (Nematoda: Diphtherophorina) with observations on the vulva in *Paratrichodorus*. *Journal of Nematology* 10, 132–141.

Rodriguez, M., Sher, S.A. and Siddiqi, M.R. (1978) Systematics of the monodelphic species of Trichodoridae (Nematoda: Diphtherophorina) with descriptions of a new genus and four new species. *Journal of Nematology* 10, 141–152.

Shoshin, A.V. (1988) New species of nematodes of family Tobrilidae from the Baikal Lake. *Proceedings of the Zoological Institute Leningrad* 180, 43–58 (in Russian).

Shoshin, A.V. (1991) Two new species of Baikal nematodes of the Enoplida order. *Zoologicheskii' Zhurnal* 70, 132–135 (in Russian).

Shoshin A.V. (1998) Two new species of Baikal nematodes of the genus *Asperotobrilus* (Nematoda: Enoplida: Tobrilidae). *Zoosystematica Rossica* 7, 223–227.

Shoshin, A.V. and Shoshina, A.A. (2002) Description of *Lamuania orientalis* (Nematoda, Enoplida, Tobrilidae) male. *Zoologicheskii Zhurnal* 81, 1271–1272 (in Russian).

Shoshin, A.V. and Shoshina, A.A. (2003) New data on a fauna of free-living nematodes of the upper Saratov reservoir. *Izvestija Samarskogo nauchnogo zentra Rossiiskoi akademii nauk* 2, 276–281 (in Russian).

Shoshina, A.A. (2003) A new trident tobrilid, *Setsalia mirabilis* gen. et sp. n., from Lake Baikal (Nematoda: Enoplida). *Zoosystematica Rossica* 12, 15–18.

Siddiqi, M.R. (1963) *Trichodorus* spp. (Nematoda: Trichodoridae) from Tunisia and Nicaragua. *Nematologica* 9, 69–75.

Siddiqi, M.R. (1983) Ecological adaptations of plant-parasitic nematodes. *Pakistan Journal of Nematology* 1, 63–77.

Siddiqi, M.R. (2002) *Ecuadorus equatorius* gen. n., sp. n. and *Nanidorus mexicanus* sp. n. (Nematoda: Trichodoridae). *International Journal of Nematology* 12, 197–202.

Swart, A. and Furstenberg, J.P. (1993) Description of two new species of the genera *Onchulus* and *Limonchulus* from Southern Africa (Nematoda: Enoplida, Onchulinae). *Phytophylactica* 25, 11–19.

Swart, A. and Heyns, J. (1990) Description of *Tobriloides loofi* n. sp. from Natal, South Africa (Nematoda: Onchulidae). *South African Journal of Zoology* 25, 138–143.

Swart, A. and Heyns, J. (1993) Description of *Caprionchulus diversipapillatus* n. gen., n. sp. from Caprivi, Namibia (Nematoda: Onchulinae). *South African Tydskrift Dierkunde* 28, 136–141.

Thorne, G. (1939) A monograph of the nematodes of the superfamily Dorylaimoidea. *Capita Zoologica* 8, 1–261.

Traunspurger, W. (1997) Distribution, seasonal occurrence and vertical pattern of *Tobrilus gracilis* (Bastian, 1865) and *T. medius* (Schneider, 1916). *Nematologica* 43, 59–81.

Traunspurger, W. (2002) Nematoda. In: Rundle, S.D., Robertson A.L. and Schmid-Araya, J.M. (eds) *Freshwater Meiofauna: Biology and Ecology*. Backhuys Publishers, Leiden, The Netherlands, pp. 63–104.

Tsalolikhin, S.J. (1980) *Free-living Nematodes of Baikal*. Nauka, Novosibirsk, Russia, 118 pp. (in Russian).

Tsalolikhin, S.J. (1981) Description of *Kurikania sibirica* male (Nematoda, Enoplida, Tobrilidae). *Zoologicheskii Zhurnal* 80, 369–370 (in Russian).

Tsalolikhin, S.J. (1983) *Nematodes of the Families Tobrilidae and Tripylidae of the World Fauna*. Nauka, Leningrad, Russia, 232 pp. (in Russian).

Tsalolikhin, S.J. (2001a) *Notes on Composition of the Genus* Brevitobrilus *Tsalolikhin, 1981 (Nematoda, Enoplida: Tobrilidae)*. Zoological Institute, St. Petersburg, Russia, pp. 25–35.

Tsalolikhin, S.J. (2001b) Synopsis of the system of the family Tobrilidae (Nematoda: Enoplida). *Russian Journal of Nematology* 9, 19–24.

Tsalolikhin, S.J. (2001c) On the genus *Paratrilobus* (Nematoda, Tobrilidae). *Zoologicheskii Zhurnal* 60, 1094–1098 (in Russian).

Yeates, G.W. (1967) Studies on nematodes from dune sands. 4. Diphtherophoroidea. *New Zealand Journal of Science* 10, 322–329.

Zullini, A. (1982) Nematodi. In: Ruffo, S. (ed.) *Guide per il riconoscimento delle specie animali delle acque interne italiane*, vol. 17. Valdonega, Verona, Italy, pp. 1–117.

14 Dorylaimida Part I: Superfamilies Belondiroidea, Nygolaimoidea and Tylencholaimoidea

R. Peña-Santiago

Departamento de Biología Animal, Biología Vegetal y Ecología, Universidad de Jaén, Campus 'Las Lagunillas' s/n, 23071-Jaén, Spain

Introduction

The order Dorylaimida Pearse, 1942 is one of the most diversified ordinal nematode taxa, and includes about 250 valid genera and 2000 species. They are very abundant in both soil and freshwater habitats and mostly free-living forms, although some of them are plant parasites; a few taxa feed on other components of the microfauna, so are considered to be carnivorous or predaceous. Several morphological features distinguish the dorylaims from other nematodes: (i) feeding apparatus armed with a protrusible odontostyle or a mural tooth; (ii) amphidial fovea usually cup-shaped, opening at the base of the lip region, and with slit-like aperture; (iii) pharynx consisting of two parts: an anterior slender portion that extends into an expanded, very muscular, basal (posterior) portion containing five unicellular glands; (iv) intestine with a peculiarly differentiated hindmost portion: the pre-rectum; and (v) males bearing paired adanal and unpaired ventromedian supplements.

Jairajpuri and Ahmad (1992) provided the most recent monography on this group.

General Morphology

General aspect

Dorylaims are relatively big nematodes, ranging from less than 0.5 mm to almost 1 cm in length, but most species are between 1 and 3 mm. The body is practically cylindrical, tapering more or less abruptly towards both extremities. Body diameter usually reaches its greatest at mid-body, and Demanian *a*-ratio provides a good perspective of the slenderness of the animals. Anterior and posterior extremities are almost always narrower than their respective adjacent body regions, but exceptions exist. After fixation, the specimens in general adopt an open 'C' shape, i.e. they

  Freshwater Nematodes: Ecology and Taxonomy (eds Eyualem-Abebe *et al.*)

appear more or less ventrad curved. Less frequently, the posterior body region is more ventrad curved than the anterior region, then resulting in a 'G' shape. In exceptional cases, e.g. some members of Nygolaimoidea, the habitus is almost straight. Obviously, as in other nematode taxa, there exists a marked sexual dimorphism because males present their posterior body region more ventrad curved than females. Dorylaims are colourless animals but their gut contents may provide greenish (species of *Enchodelus*) or brownish (frequent in some *Aporcelaimellus* species) tonalities.

Cuticle

Under light (optical) microscopy (LM)[1] the cuticle of dorylaims appears constituted of two (outer and inner) layers, but in species of the very frequent genus *Aporcelaimellus* three layers are easily identified, more distinctly at the tail level. The outer layer is relatively thin, almost always thinner than the inner layer. Species belonging to the superfamily Tylencholaimoidea are distinguished by their inner layer being irregularly outlined and loose from the outer layer; moreover, small, abundant radial refractive elements are present between both cuticle layers. The external surface of the cuticle usually bears more or less distinct fine transverse striations. Exceptionally, very distinct longitudinal ridges are observed in some members of the families Dorylaimidae and Actinolaimidae, whose number (counted in a transverse section) is of taxonomic interest.

Lateral chords

Beneath the cuticle, forming part of the body wall, the epidermis bulges into the pseudocoelom in four (ventral, dorsal and two lateral) epidermal chords, of which the lateral are especially developed and distinct, being easily observed (by transparence) under LM. They display a granular appearance with no other particular differentiation in general, but in some species a longitudinal series of large gland bodies is present throughout the body length, each one connecting to a lateral body pore. The width of the lateral chord (measured at mid-body) is frequently provided in descriptions.

Lip region

The anteriormost part of the body is termed the lip (labial) region (or head), its posterior limit lying at the level of the amphid opening, and the oral aperture being situated in the centre of its frontal face. The lip region displays the typical hexaradial symmetry and the usual nematode pattern of 6 + 6 + 4 sensilla. Sometimes the lip region is perfectly continuous with the adjacent body (Fig. 14.1B), but usually it is

[1]Cuticle structure under the transmission electron microscope (TEM) is much more complex, but its analysis is beyond the realm of this introductory discussion.

offset by depression or constriction. If the transition from lip region to adjacent body is marked off by a slight narrowing, this feature is described as offset by depression. On the other hand, the existence of a deeper narrowing separating both portions of the body is described as offset by constriction (Figs 14.1A and C). Anyway, subjectivity underlies these descriptions.

An important variability can be found affecting the general shape of the lip region. In general, it is more or less angular and offset from the adjacent body, with variably separated lips (Fig. 14.1A). However, in addition to this general pattern, other options occur. Some species present a rather rounded contour, with completely amalgamated lips, usually combined with a continuous lip region (Fig. 14.1B), e.g. members of Thornenematidae. Other members present a more or less truncated anterior margin with well-amalgamated lips, and, in general, but not necessarily, in coincidence with a continuous lip region; for instance, some species of the genera *Laimydorus* and *Prodorylaimus*. Many tylencholaimid species (as well as some belondirids) display a peculiar cap-shaped lip region (Fig. 14.1C).

Hexaradial symmetry is very well perceived in the lip region of dorylaims because six lips (two lateral, two dorsolateral and two ventrolateral), which are almost always present, with more or less angular outline, surround the oral aperture. Nevertheless, the lips may appear strongly amalgamated. It is not unusual that labial inner portions, surrounding the oral aperture, are markedly offset from the remaining part and form separated liplets. Even these inner portions may merge and become a distinct disc-like structure (Fig. 14.1C). Six inner labial, six outer labial and four cephalic button-like papillae are easily observed even under light microscopy. Every lip bears one inner and one outer labial papilla, and the lateral lips lack cephalic papillae (Figs 14.1A–C).

Amphids

In dorylaims amphids consist of three components: (i) fovea; (ii) amphidial canal; and (iii) fusus (Coomans, 1979; Fig. 14.1D). The fovea is observed just behind the depression or constriction (if these exist), beneath the body cuticle, being usually a cup-shaped cavity (but other variations exist); its aperture is a transverse slit that occupies in general one-third to two-fifths of the corresponding body diameter (Figs 14.1A–C). The fovea connects with a sensillum pouch (fusus) through the amphidial canal (*canalis amphidialis*), but these structures are only distinct in well-preserved specimens.

Feeding apparatus

The oral aperture is followed by the stoma, a cylinder that connects to the pharyngeal lining. It comprises five regions (Coomans, 1964).

The anteriormost portion is termed the *cheilostom* or *vestibulum* (some authors refer to this structure as the stoma), and extends from the oral aperture to the guiding ring. In general, it is cylindrical or a truncate cone (Figs 14.1G–J), but peculiar modifications may occur, e.g. heavily sclerotized walls (Fig. 14.1I), a flask-like shape (Fig. 14.1H), four refractive circumoral pieces or terminal thickenings present at its

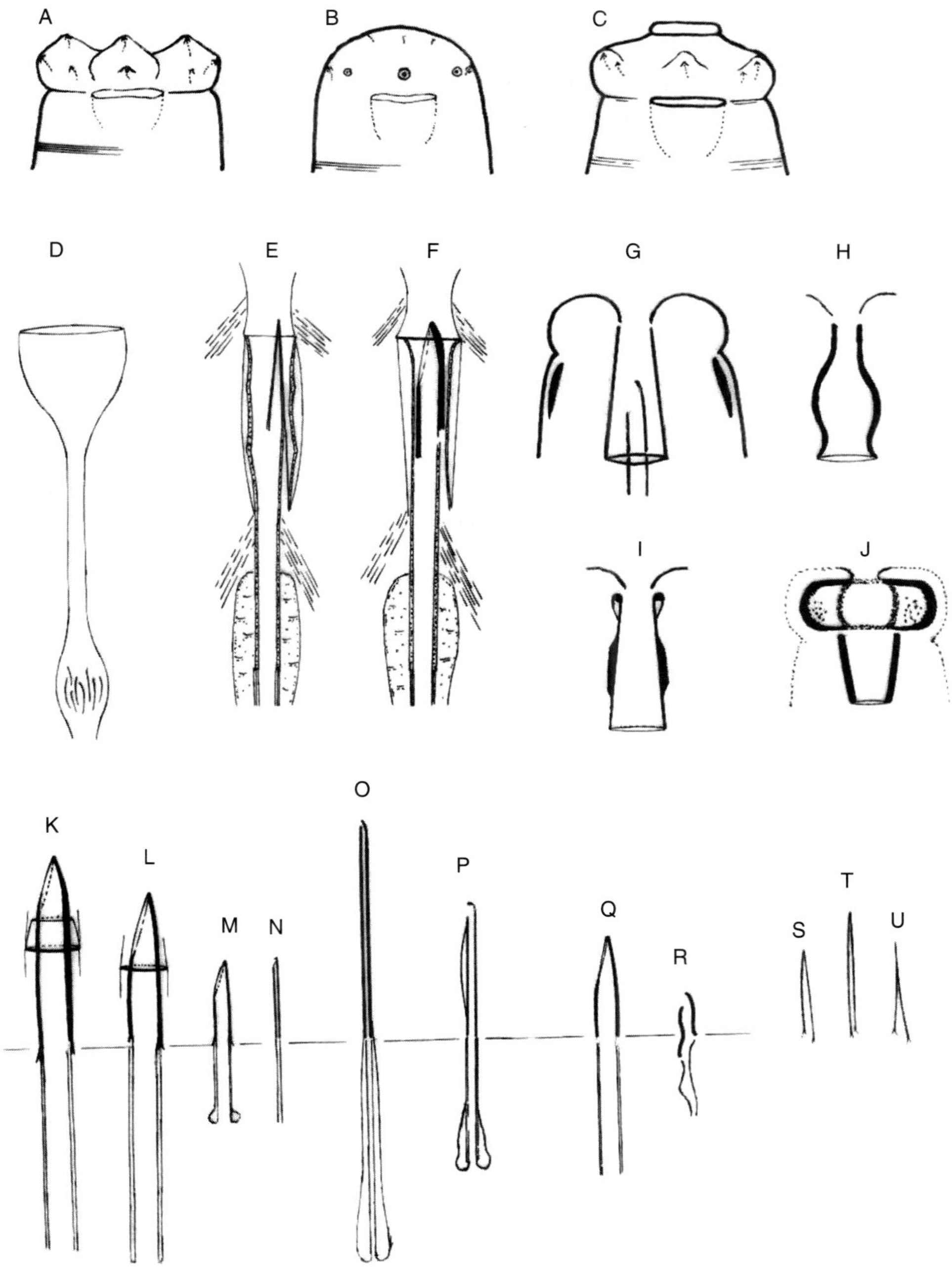

Fig. 14.1 A–C: Variations in lip region. A: Angular and offset by constriction; B: rounded and continuous; C: cap-like; D: amphid; E: stoma morphology in Nygolaimina; F: stoma morphology in Dorylaimina; G–J: variations in cheilostom; K–R: variations in odontostyle and odontophore in Dorylaimina; S–U: variations in mural tooth in Nygolaimina. (E and F: based on Coomans, 1964.)

anteriormost end (Fig. 14.1I). The case of members of the family Actinolaimidae deserves special mention in that they exhibit a very modified cheilostom and a larger and strongly sclerotized wall, consisting of two components (Fig. 14.1J): the labial portion, which is transformed into a large and wider than long basket that occupies the greater part of the corresponding body diameter, being frequently armed with a myriad of minute teeth-like structures called denticles; and the post-labial portion, which is almost as long as wide, often bearing four strong teeth (onchia).

An intermediate region, the *guiding sheath*, consists of three subregions (pro-, meso- and metastom[2]) and bears a protruding structure, either a mural tooth (Fig. 14.1E) or an axial odontostyle (Fig. 14.1F). The *guiding ring* is the anteriormost end of this region, being strongly attached to the cheilostom. It is a delicate structure, which is especially distinct in well-preserved specimens. Most dorylaims possess a simple guiding ring (Fig. 14.1L), but some genera (among others, *Dorylaimus* and *Labronema*) present a double guiding ring (Fig. 14.1K). Members of the family Aporcelaimidae present an atypical plicate guiding ring, and nygolaimid species display a very weak guiding ring (Fig. 14.1E).

Dorylaimid nematodes are typically armed with a protruding structure (stylet or spear) that shows two different models. In members of the superfamily Nygolaimoidea the three subregions (pro-, meso- and metastom) of the guiding sheath are more perceptible. In this nematode group a *mural tooth* that lacks any aperture is attached to the stomatal wall at its sub-ventral median side on a support (Figs 14.1E and S–U). In other dorylaims, guiding sheath sub-regions are difficult to observe, and a broader, symmetrical and aperture-bearing *odontostyle* (Fig. 14.1F) is present attached to the median guiding sheath in continuity with the posteriormost portion of the stoma (telostom or odontophore; see the following paragraph). The odontostyle is a tube-like structure with pointed anterior end, opening at the dorsal side of its anteriormost end. An incredible variability (Figs 14.1K–R) may be found affecting the morphology (size, shape, aperture, etc.) of the odontostyle.

The posteriormost portion of the stoma is termed the *telostom* or *odontophore.* Although it forms a functional unit with the odontostyle,[3] its origin and histology is different. It is also a tube-like structure that attaches to the odontostyle at its anterior end and connects with the pharyngeal lining (in fact, it is a specialization of the latter) at its posterior extremity. In median view under LM it is usually rod-like (Figs 14.1K, L, N and Q) and without special modifications, but basal knobs (Figs 14.1M and P) or flanges (Fig. 14.1O) are sometimes present at its base.

Pharynx

One of the most characteristic features of dorylaims is the morphology of the pharynx that extends from the odontophore base, to which it joins and

[2]Some authors (see, for instance, Jairajpuri and Ahmad, 1992) use the terms cheilostome, prostome, mesostome, metastome and telostome to refer the corresponding subregions of the stoma, instead of cheilostom, prostom, mesostom, metastom and telostom, respectively.

[3]Some authors have used the term stylet or spear to refer to the unit formed by odontostyle ± odontophore, and have described the odontophore as stylet or spear extension, or basal part of the stylet.

envelopes, to the cardia. The pharynx is typically constituted by a slender but usually muscular anterior region that extends gradually into a well muscular expanded posterior portion (Fig. 14.2A), but this basic pattern has been modified in different ways in particular taxa. In transverse section the pharyngeal lumen is triradial, delimiting three sectors, one dorsal and two ventrosublateral.

No major specialization is found in the morphology of the anterior pharyngeal region, if we accept that in some members of the superfamily Belondiroidea a small spindle-shaped swelling is sometimes present behind the junction to the odontophore base, and that in some tylencholaimoid (and other) species the anterior region is especially slender and weakly or not muscular (Fig. 14.2E). Junction between both (anterior and posterior) parts of the pharynx occurs in general by a more or less gradual widening (Figs 14.2F and G), but sometimes a distinct constriction (Fig. 14.2H), and more occasionally a short isthmus-like portion (Fig. 14.2I), separates them.

The pharyngeal posterior expansion (traditionally termed pharyngeal bulb or cylindrus) usually occupies two-fifths to one-half of the total neck length, but it is not unusual that it extends along more than half of this measurement (e.g. some representatives of the genera *Axonchium* and *Nygolaimus*), and in members of the subfamily Tylencholaimoidea a short bulb-like expanded portion occurs, occupying one-fifth to one-third of the total neck length (Fig. 14.2E). In belondirid species the pharyngeal expansion is surrounded by a spiral muscular sheath (Fig. 14.2C).

Jairajpuri and Ahmad (1992) have distinguished several basic pharyngeal patterns in dorylaims:

- In the dorylaimoid type (Fig. 14.2A) the two parts of the pharynx are muscular and similar in size, their transition being a more or less gradual (rarely abrupt) expansion.
- The presence of three large cardiac cells at the pharyngo-intestinal junction distinguishes the nygolaimoid type (Fig. 14.2B) from the dorylaimoid one.
- The members of the superfamily Belondiroidea are characterized by having a developed muscular spiral sheath involving the pharyngeal expansion: this is the belondirid type (Fig. 14.2C); both parts of the pharynx are separated by a constriction or a short isthmus-like portion.
- The tylencholaimid (or leptonchoid) type (Figs 14.2D and E) is distinguished by its relatively short pharyngeal expansion, usually occupying one-fifth to one-third of the total neck length and forming a true basal bulb.

The expanded posterior portion of the pharynx contains unicellular and mononucleate glands that are difficult to observe; however, their nuclei, nucleoli and outlets in the pharyngeal lumen are distinctly visible. Five glands (one dorsal and two ventrosublateral pairs) and their corresponding outlets are the typical pattern of dorylaims, but reduction to three gland nuclei and their outlets occurs in species belonging to the family Longidoridae and in some members of Tylencholaimoidea. Loof and Coomans (1970) and Andrássy (1998) have each proposed one formula to express the relative location of these nuclei and outlets, which are of taxonomic importance.

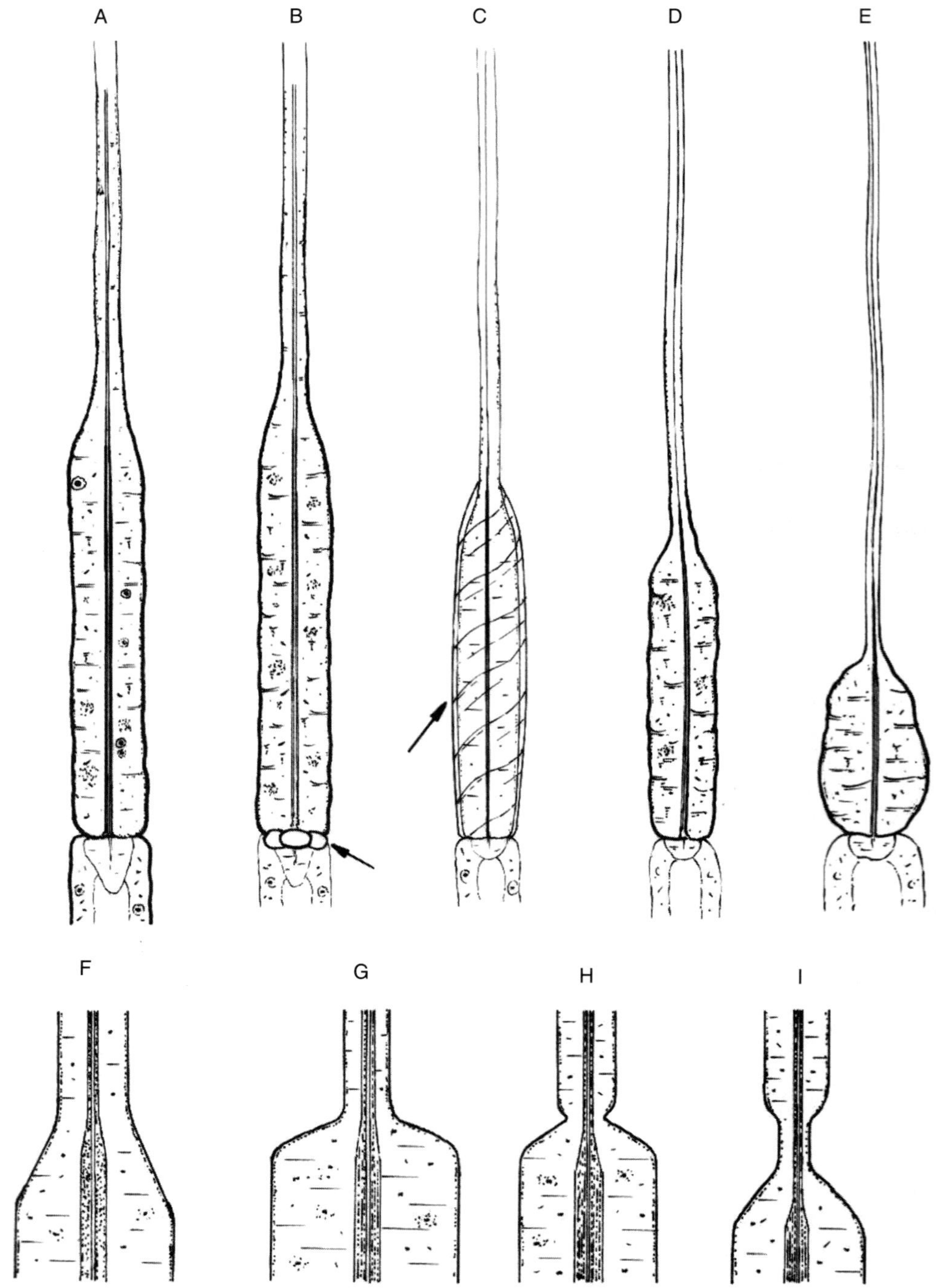

Fig. 14.2 A–E: Morphological types of pharynx. A: Dorylaimoid; B: Nygolaimoid (arrowhead pointing out the cardiac cells); C: Belondiroid (arrowhead pointing out the spiral muscular sheath); D, E: Tylencholaimoid. F–I: Junction between two main pharyngeal regions. F: Gradual expansion; G: Abrupt expansion; H: Constriction; I: Isthmus-like narrowing.

Pharyngo-intestinal junction (cardia)

The junction between the pharynx and the intestine is guarded by the cardia, a variably long conoid or conical structure that joins the pharyngeal expansion base and projects into the intestinal lumen. The cardia is a valve that controls the food flow from the pharynx to the intestine and prevents regurgitation. It is usually surrounded by intestinal tissue that embraces it at its anterior end or less frequently at its middle. Several specializations have been described concerning this basic pattern, among others a disc-like intermediate structure, three large cardiac cells, a weak ring-like structure surrounding the junction between the pharyngeal expansion base and the cardia, etc. Yeates (1972) distinguished seven types of cardia, but their analyses are beyond the contents of this contribution.

Intestine

As in many nematodes, the intestine is quite simple and tubular, but in dorylaims it appears divided into two regions: one anterior, the intestine proper, and another posterior, the pre-rectum. Almost always the anterior portion is much longer than the posterior, but exceptions exist to this rule. Pre-rectum length is an interesting diagnostic feature used in species identification.

Rectum

The hindmost part of the alimentary canal is the rectum, which is lined internally with the cuticle. In the female it is a simple tube to the anus, but in the male the ejaculatory duct opens into it to form the cloaca, which contains the copulatory apparatus. The intestine–rectum junction is provided with a valve surrounded by a sphincter.

Female reproductive (genital) system

It typically comprises two (anterior or pre-vulval and posterior or post-vulval) genital branches (didelphic condition) converging in a common vagina that opens on to the body surface in the vulva (Fig. 14.3A). One of two branches may be reduced or completely absent (monodelphic condition): if only the anterior branch is present, such condition is termed prodelphic (Fig. 14.1C); and, in the same way, if just the posterior branch is developed, such condition is named opisthodelphic (Fig. 14.1D). Each genital branch comprises reflexed ovary and genital tract, the latter formed in turn by two tubular organs: oviduct and uterus. When one of the two branches is reduced, a more or less developed uterine (pre- or post-vulval) sac is usually found (Fig. 14.1C). Coomans (1965) provided an excellent study on general morphology of genital system in dorylaims.

The *ovary* is reflexed at its junction with the oviduct, having its tip directed towards the vulva. Three regions may be distinguished: (i) germinative or multiplication zone, at the tip (distal end), consisting of several nuclei without

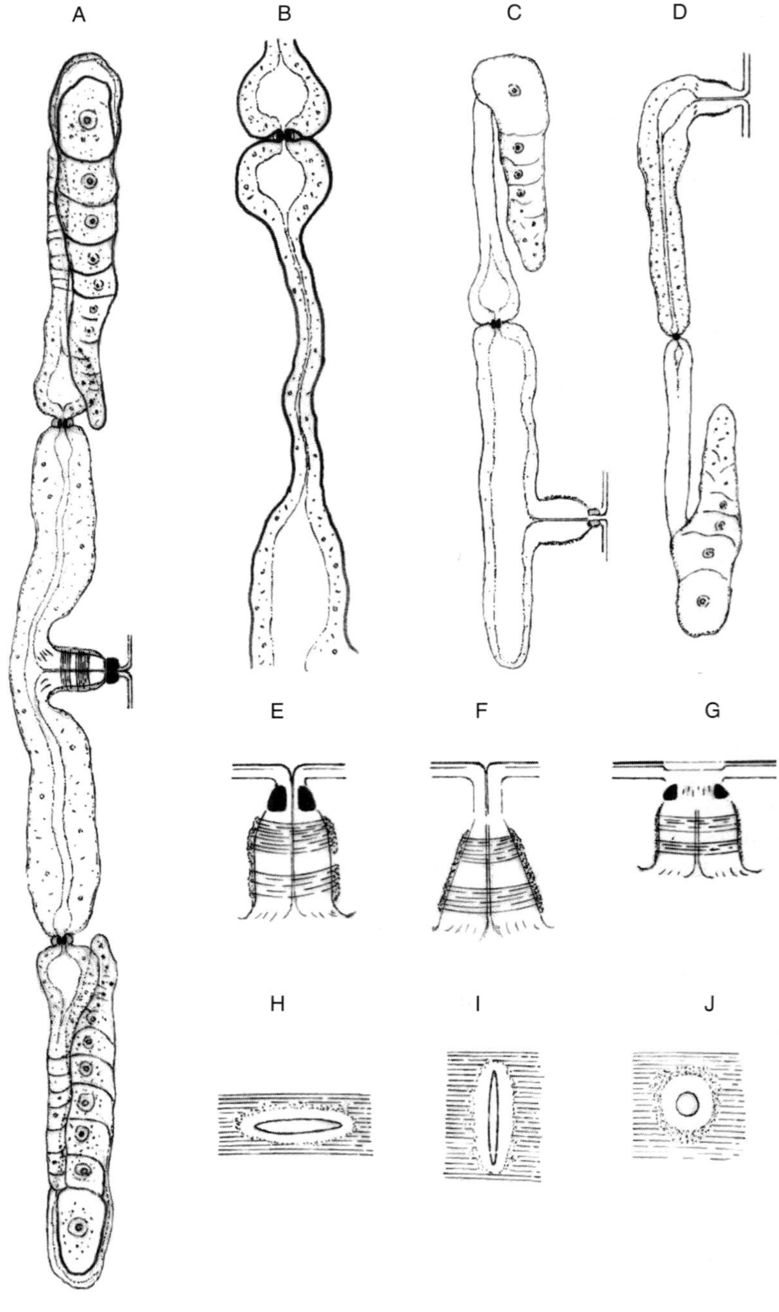

Fig. 14.3 Female genital system. A: Didelphic type; B: Uterus with three specialized regions; C: Prodelphic type with post-vulval sac; D: Opisthodelphic type with no pre-vulval sac; E–G: Vagina and vulva (lateral view); E: Vagina with the three usual parts and transverse vulva; F: Vagina lacking *pars refringens,* and transverse vulva; G: Vagina with the usual three parts and longitudinal vulva; H–J: Vulva, in frontal view; H: Transverse; I: Longitudinal; and J: Pore-like.

distinct cell walls; (ii) growth zone, where the oocytes increase in size, being arranged first in a double row and then in a single one; and (iii) ripening zone, at the proximal end of the ovary, where the oocytes accumulate yolk before passing into the oviduct. The ovary is enveloped in a thin-walled sac, which is more distinct at its proximal region. The size of the ovary may be rather short, but in general it extends to near the oviduct–uterus junction, and less frequently surpasses it.

The *oviduct* joins the ovary subterminally and consists of a more slender distal portion of high cells and indistinct lumen, and a more or less developed *pars dilatata oviductus* with wide lumen (Figs 14.3A and C), often containing spermatozoa. A marked narrowing or constriction separates oviduct and uterus, being surrounded by a variably developed *sphincter*.

The *uterus* is a tube-like organ of variable length and morphology, and usually lacks special modifications (Figs 14.3A, C and D). Nevertheless, it is not unusual to distinguish three modified regions (Fig. 14.3B): (i) a distal spherical *pars dilatata uteri* close to the sphincter, with distinct lumen, in which spermatozoa may be found; (ii) an intermediate long tubular portion with narrow lumen and associated musculature, sometimes bearing peculiar specialization (for instance, the Z-organ); and (iii) a tubular proximal part with wide lumen, sometimes with a specialized muscular region, the ovejector. Uterine eggs are frequently observed within the uterus. They are oval bodies, longer than wide, and covered by a thick shell.

The *vagina* connects both uteri to the body surface, and typically consists of three components (Figs 14.3E–G; for additional information, see De Ley *et al.*, 1998): (i) the *pars proximalis vaginae* joins the proximal part of both uteri. Under light microscopy in lateral view it appears like a translucent region. It is usually longer than wide, but a wide range of forms exists. The outer surface of its wall may be sigmoid (i.e. convergent proximally and divergent distally), practically straight, etc., it is surrounded by a more or less developed ring musculature; (ii) the *pars refringens vaginae*, an intermediate region, is characterized by the presence of a sclerotized structure, which appears as two sclerotized pieces in optical section. These are variable in shape (triangular, trapezoidal, hemispherical, linear, and so on) and size, and more or less close together. Members of the subfamily Discolaiminae and many belondirid and tylencholaimid species completely lack this part of the vagina (Fig. 14.3F); and (iii) the *pars distalis vaginae* is almost always a rather narrow and short portion that opens in the vulva. No special modification is found in this region.

The *vulva* is usually a transverse slit (Figs 14.3E, F and H), but a longitudinal slit is also frequent (Figs 14.3G and I), and even a pore-like shape (Figs 14.3J) has been described in some species. The vulva is usually located at mid-body, but anterior (see, for instance, *Tylencholaimellus* species) or posterior (some members of *Tylencholaimus*) locations are also found. Rarely the vulva is preceded by a body surface depression, thus opening at the bottom of a small cavity. Near the vulva, and in a few species the cuticle presents special modifications (wrinkles, small elevations, paravulvae, etc.; see, for instance, some *Mesodorylaimus* species), which have no relation to the genital system.

Male reproductive (genital) system

This invariably consists (Fig. 14.4D) of two gonads (testes), a condition termed diorchic, which merge in a common genital tract that connects the rectum, forming the cloaca, which opens on the body surface (in the cloacal opening). Other organs and structures are accessory elements of the genital system: copulatory apparatus (spicules, lateral guiding pieces, gubernaculum and associated musculature), copulatory muscles, glands and supplements. Some of these components deserve more attention because of their taxonomic interest.

One pair of *testes* lying opposite each other are always observed in dorylaims. Two regions are always distinguished: the germinative or multiplication zone and the maturation or differentiation zone. The proximal ends of both testes converge forming the unique common genital tract.

A long tubular structure, the *genital tract*, connects the testes to the rectum. It is distinct in well-preserved specimens, but its detailed morphology is difficult to observe, even in this material. The two testes meet at and lead to a unique *vas deferens*, often having a narrow lumen. The *ejaculatory duct* follows the *vas deferens* and leads to the rectum, thus forming the cloaca; it usually exhibits a granular aspect and no visible lumen. The rectum (cloaca) finally leads to the anus (cloacal opening) on the ventral surface of the body.

Two sclerotized structures, the *spicules* (Figs 14.4E–G), are the main components of the copulatory apparatus in dorylaimid nematodes, being lodged in special pouches of the dorsal wall of the cloaca, and consisting of two parts: head (*capitulum*) and blade (*lamina*). Their shape and size have taxonomic interest. Two small structures located at both sides of the spicules are termed the *lateral guiding pieces* or *crura* (Figs 14.4H–J). They are usually cylindrical and moderately long, with a bifurcate or more or less acute tip. In some nygolaimid species, but lacking in most dorylaims, the *gubernaculum* is a small, slender and somewhat arcuate structure; actually it is a thickening of the dorsal wall of the spicular pouch.

Dorylaimid male specimens are distinguished by having a variable number of small mammiform or papilliform, distinctly inervated organs: the *supplements* (also termed copulatory or genital papillae). Usually they appear distributed in an adanal (close to the anus) pair and a longitudinal ventromedian series that may be constituted of 1–100 supplements. Their number, arrangement and distribution have taxonomic importance. The ventromedian series is usually separated from the adanal pair by a marked distance (pre-cloacal space). In many dorylaims the ventromedian series is formed by well-separated supplements (Fig. 14.4B), but very close (contiguous) supplements (Fig. 14.4C) are also frequent; sometimes the ventromedian supplements are grouped in close-packed packages (fascicles) (Fig. 14.4A).

Tail

It is enormously variable in shape and size, both aspects having taxonomic importance (Fig. 14.5A). It ranges from very long and filiform (Figs 14.5A and B) to very short and rounded, almost hemispherical (Fig. 14.5E), but other morphologies such as elongate, conical (Fig. 14.5C), cylindrical (Fig. 14.5D), clavate (Fig. 14.5F), etc. are also frequent. The inner layer of the cuticle appears frequently thickened at the level of the tail, and more rarely a developed hyaline part exists at its terminal portion. Females and males may have similar tail morphology, but sexual dimorphism affecting this feature is found in several genera: long and filiform in females, and short and rounded in males.

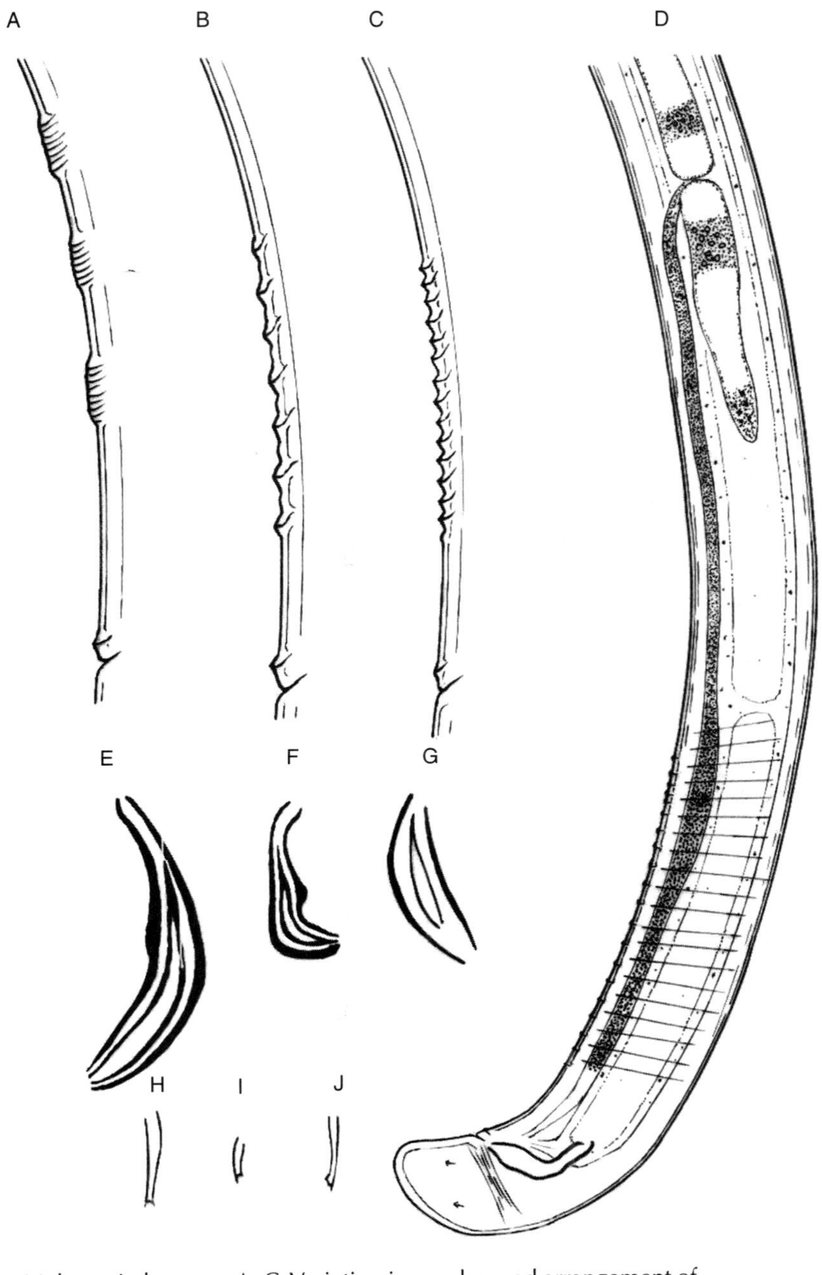

Fig. 14.4 Male genital system. A–C: Variation in number and arrangement of ventromedian supplements; A: In fascicles; B: Separated; C: Contiguous; D: General morphology; E–G: Variation in morphology of spicules; H–J: Variation in morphology of lateral guiding pieces.

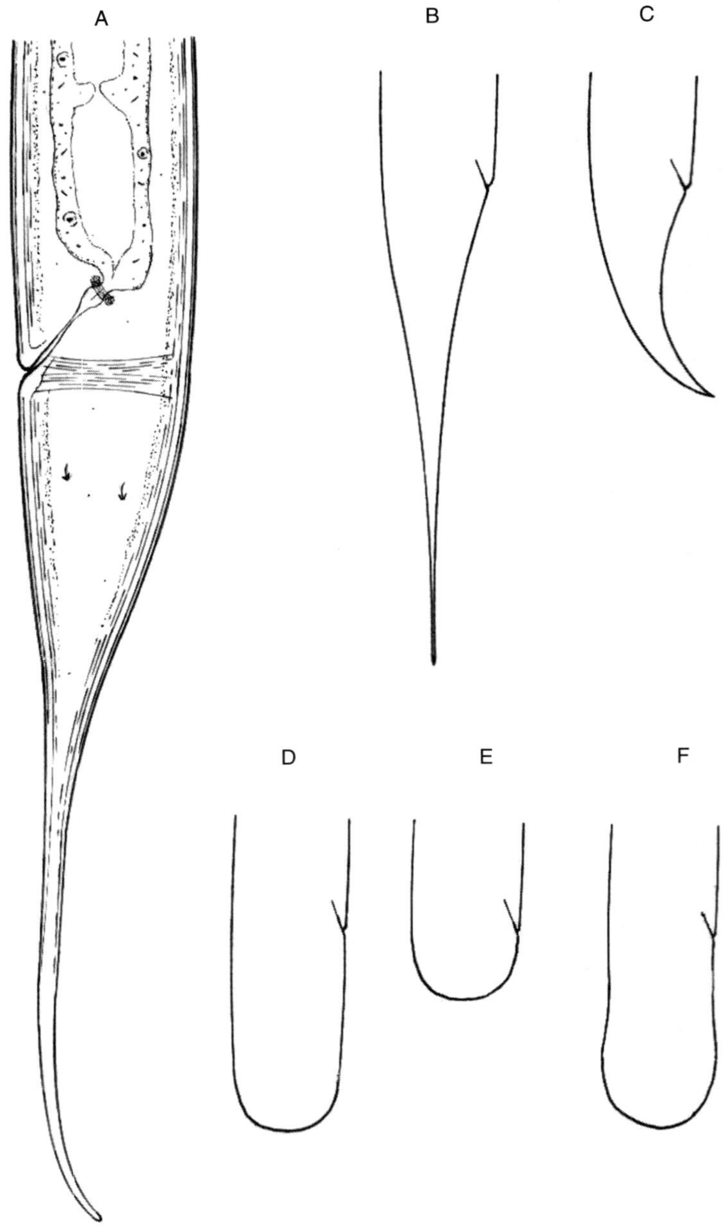

Fig. 14.5 Tail morphology. A: General morphology. B–F: Variation in shape and size. B: Elongated to filiform; C: Conical, ventrad curved; D: Cylindrical; E: Short rounded, hemispheroid; and F: Clavate.

Freshwater Dorylaimid Diversity

Nematodes are narrowly dependent on water, even in terrestrial edaphic systems, i.e. they are aquatic animals. This may be the reason many genera and species are found in both freshwater and soil, it being difficult, if not impossible, to define them as terrestrial or freshwater forms; nevertheless, there are some other genera and species that are predominantly or exclusively collected from freshwater habitats *sensu stricto*.

A compilation of dorylaimid taxa recorded from a wide range of freshwater/aquatic biotopes (rivers, lakes, streams, pools, springs, etc.) follows, but, given the copious literature on this topic – some references were not even available to the author – the list may be incomplete, although the reader will certainly find it the most representative.

The information is presented as follows: a key is provided to taxa identification, including the genera hitherto classified under Dorylaimida. Diagnoses of taxa are presented for all suprageneric taxa and also for those genera that have been recorded from freshwater habitats. For each genus a list of freshwater/aquatic species is given. And, finally, for each species are indicated the binomen, authority and geographical distribution: country, limnic habitat, locality (region, state, etc.) and the corresponding reference(s).

In general, the proposal by De Ley and Blaxter (2002) is followed for the systematics of higher taxa, and that by Jairajpuri and Ahmad (1992) for the ordination of subfamilies and genera. Nevertheless, some original proposals are also introduced.

Order DORYLAIMIDA Pearse, 1942

Nematodes of relatively large size, 0.3–10.0 mm long but very often ranging from 1 to 3 mm. Cuticle smooth, occasionally provided with longitudinal ridges. Lip region very variable in morphology, but a constant feature are the six inner labial, six outer labial and four cephalic papillae. Stoma is divided into three sections: cheilostom, guiding sheath and telostom. Cheilostom a truncate cone, sometimes modified and/or with specializations. Guiding sheath bearing a protrusible structure, either an axial odontostyle or a mural tooth. Pharynx bottle-like, i.e. consisting of a slender anterior portion, which extends into the expanded posterior (basal) portion. Pharyngeal expansion provided with five (sometimes three) glands: one dorsal and two, sometimes one, ventrosublateral pairs. Hindermost portion of intestine differentiated in pre-rectum. Female genital system didelphic or monodelphic, with reflexed ovaries. Vagina usually with three sections (*pars proximalis, pars refringens* and *pars distalis*), sometimes lacking the *pars refringens*. Vulva a transverse slit in general, more rarely a longitudinal slit or pore-like. Males bearing two testes (diorchic), copulatory spicules, lateral guiding pieces, rarely gubernaculum and ventromedian supplements. Tail very variable, similar or dissimilar in sexes: long and filiform to short and rounded. No caudal glands.

Two suborders, both including freshwater species:
Dorylaimina Pearse, 1936
Nygolaimina Ahmad & Jairajpuri, 1979

Suborder DORYLAIMINA Pearse, 1936

Diagnosis. Dorylaimida. Nematodes of variable size, usually 1–3 mm long. Cuticle with fine transverse striations, occasionally with distinct longitudinal ridges, very rarely squared. Stoma armed with axial odontostyle, very variable in size and shape; odontophore rod-like, sometimes bearing knobs or flanges at its base. Pharynx consisting of the usual two portions; basal expansion occupying about half of total neck length in general, but sometimes reduced to a bulb (less than one-fourth of total neck length) or longer. Pharyngo-intestinal junction variable in morphology. Female genital system didelphic or monodelphic. Vagina with the usual three sections, but in some taxa only two, lacking the *pars refringens*. Vulva a transverse slit, less frequently a longitudinal slit or pore-like. Male with two testes, spicules, lateral guiding pieces and ventromedian supplements. Tail similar or dissimilar in sexes, ranging from long and filiform to short and rounded.
Three superfamilies, all including freshwater species:

Belondiroidea Thorne, 1939
Dorylaimoidea de Man, 1876[4]
Tylencholaimoidea Filipjev, 1934

Superfamily BELONDIROIDEA Thorne, 1939

Diagnosis. Dorylaimina. Nematodes of variable size, less than 1 mm to more than 5 mm long. Lip region narrow, usually scarcely differentiated. Odontostyle relatively short. Odontophore rod-like, flanged at its posterior portion in one subfamily and genus. Anterior slender portion of pharynx more or less (usually weakly) muscular; basal expansion of variable length and surrounded by a distinct muscular sheath. Female genital system didelphic or monodelphic. Male with spicules very variable in size and shape, and few to numerous ventromedian supplements. Tail similar or dissimilar in sexes: long and filiform to short and rounded.

Only one family, which includes freshwater species:

Family BELONDIRIDAE Thorne, 1939 (type family)

Diagnosis. Belondiroidea. Nematodes of variable size, usually more than 1 mm long. Lip region narrow, continuous or differentiated from the adjacent body. Odontostyle short, rarely longer than the lip region width, linear or somewhat fusiform. Anterior slender portion of the pharynx more or less muscular, in general extending gradually, but sometimes separated from the basal expansion by an isthmus-like section. Pharyngeal basal expansion highly variable in size, but always surrounded by a more or less (usually well) developed spiral muscular sheath. Cardia usually enveloped by intestinal tissue, but sometimes joining the intestine only at its posterior end. Female genital system didelphic or monodelphic. Vagina with greatly variable morphology. Vulva transverse, longitudinal or pore-like. Male having dory-

[4]The superfamily Dorylaimoidea is discussed in Chapter 15 of this book.

laimid spicules widely varying in size and shape, and few to numerous spaced or contiguous ventromedian supplements. Tail similar or dissimilar in sexes: long and filiform to short and rounded.

Three subfamilies, all of them including freshwater species:

Belondirinae Thorne, 1939 (type subfamily)
Dorylaimellinae Jairajpuri, 1964
Swangeriinae Jairajpuri, 1964

Subfamily Belondirinae Thorne, 1939
Diagnosis. Belondiridae. Nematodes of medium to large size, usually more than 1 mm long. Lip region continuous or marked from the adjacent body by depression or constriction, rarely cap-like. Odontostyle short in general, linear or fusiform. Odontophore rod-like, lacking special modifications. Anterior region of the pharynx slender and more or less (occasionally not) muscular, either extending gradually into the basal expansion or separated from it by a constriction or short isthmus-like section. Pharyngeal expansion surrounded by a muscular spiral sheath with dextral, sinistral or almost straight bands. Female genital system frequently mono–opisthodelphic, but didelphic–amphidelphic in a few species. Males with spicules having highly variable morphology. Tail similar in both sexes: short, rounded or conoid, sometimes clavate, very rarely conical or digitate; inner cuticle layer often thickened.

Eleven genera, at least two of them (in bold) including freshwater species:

Amphibelondira Rahman, Jairajpuri, Ahmad & Ahmad, 1987
Anchobelondira Nair & Coomans, 1971
Axonchium Cobb, 1920
Axonchoides Thorne, 1967
Belondira Thorne, 1939 (type genus)
Belondirella Thorne, 1964
Heynsaxonchium Coomans & Nair, 1975 in Jairajpuri & Dhanachand, 1979
Immanigula Andrássy, 1991
Nimigula Andrássy, 1985
Porternema Suryawanshi, 1972
Yunqueus Thorne, 1964

Key to genera

1. Pharynx exceptionally long, more than half of total length; vulva anterior to pharyngo-intestinal junction . . . . *Nimigula*
   Pharynx not so long, one-third of total length or less; vulva posterior to pharyngo-intestinal junction 2
2. Perioral region differentiated into a small disc-like structure; cheilostom excessively short, less than half of lip region height; three cardiac glands present at the pharyngo-intestinal junction . . . . *Yunqueus*
   Perioral region lacking any small disc-like structure; cheilostom longer, more than half of lip region height; cardiac glands absent 3

3. Odontostyle long (about twice the lip region diameter) and slender, with minute aperture . . . . *Heynsaxonchium*
   Odontostyle linear or fusiform, rarely longer than the lip region diameter, and with distinct aperture 4
4. Anterior region of the pharynx more or less (usually weakly) muscular and extending gradually into the basal expansion (Fig. 14.6D) 5
   Anterior region of the pharynx slender but muscular, distinctly separated from the basal expansion by a constriction or short isthmus-like section (Fig. 14.6E) 10
5. Perioral region differentiated into six refractive liplets . . . . *Belondirella*
   Perioral region hardly differentiated 6
6. Female genital system didelphic–amphidelphic . . . . *Porternema*
   Female genital system mono–opisthodelphic 7
7. Cephalic framework completely lacking; *pars refringens vaginae* well developed . . . . *Immanigula*
   Cephalic framework more or less developed, usually with labial and post-labial sclerotization; *pars refringens vaginae* hardly differentiated 8
8. Cephalic framework present but weakly developed, with no labial nor post-labial sclerotization; pharyngeal basal expansion occupying about three-fourths of the total neck length . . . . *Axonchoides*
   Cephalic framework well developed in general, usually with labial and/ or post-labial sclerotization (Fig. 14.6B); pharyngeal basal expansion up to two-thirds of the total neck length 9
9. Female genital system didelphic–amphidelphic . . . . *Amphibelondira*
   Female genital system mono–opisthodelphic . . . . *Belondira*
10. Odonstostyle slender, linear; a pyriform thickening is present at the posterior end of the anterior slender region of the pharynx, in front of the isthmus-like section; female genital system didelphic–amphidelphic . . . . *Anchobelondira*
    Odontostyle more robust, fusiform (Fig. 14.6C); no thickening is present in front of the isthmus-like section of the pharynx; female genital system mono–opisthodelphic . . . . *Axonchium*

Freshwater genera and species:

Genus ***Axonchium*** Cobb, 1920

Diagnosis. Belondirinae (Figs 14.6C and E). Medium to large-sized nematodes, 0.9–4.3 mm long. Habitus after fixation straight or ventrad curved. Lip region of variable morphology, often cap-like with more or less separated lips, but may be also continuous with amalgamated lips. Odontostyle relatively short but robust, fusiform. Odontophore rod-like, lacking any differentiation. The anterior portion of the pharynx is slender but muscular, being separated from the basal expansion by an isthmus-like short section; pharyngeal expansion in general distinctly muscular, longer than half of total neck length, and surrounded by a strongly developed spiral muscular sheath. Female genital system mono–opisthodelphic with the anterior branch more or less reduced. Vagina well developed, with or without *pars refringens*. Vulva a transverse slit, rarely longitudinal or oval. Male with well-developed spicules, lateral guiding pieces

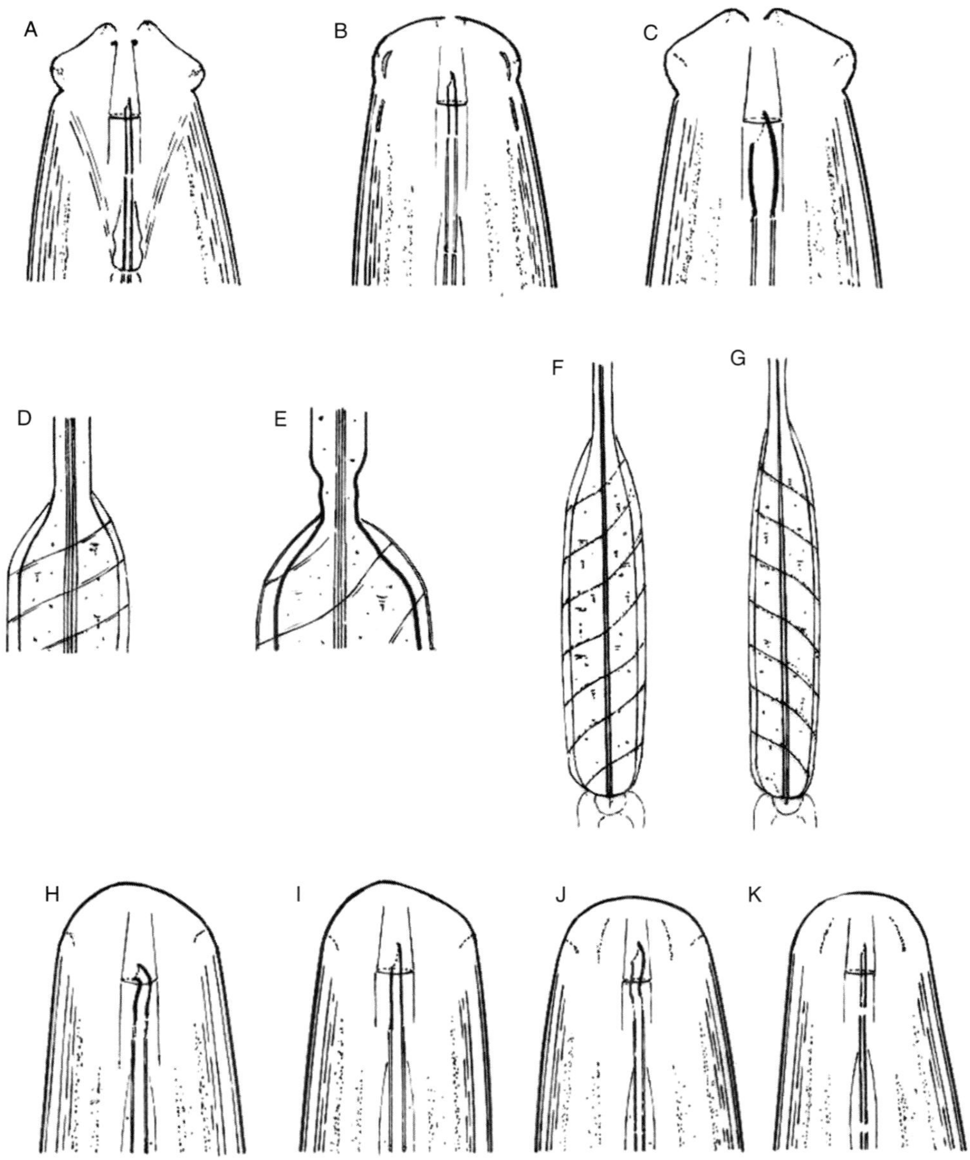

Fig. 14.6 Freshwater belondirid genera (Belondiroidea: Belondiridae). A: Subfamily Dorylaimellinae, genus *Dorylaimellus*; B–E: Subfamily Belondirinae. B, D: Genus *Belondira*; C, E: Genus *Axonchium*; F–K: Subfamily Swangeriinae; F, I: Genus *Oxydirus*; G, J: Genus *Paraoxydirus*; H: Genus *Falcihasta*; and K: Genus *Lindseyus*.

and 2–30 contiguous or spaced ventromedian supplements. Tail similar in sexes: rounded to conoid.

Remarks: Coomans and Nair (1975) provided the most recent revision of the genus, including a key to its subgenera and species. Mostly terrestrial nematodes with at least four species, among about 60, also recorded in freshwater habitats:

A. coomansi Nair, 1975 – The Netherlands (Loof, 1999).
A. coronatum (de Man, 1907) Thorne & Swanger, 1936 – Europe (Loof, 1999); Germany (subterranean water bodies/Thuringe (Altherr, 1974; Andrássy, 1978b; as *A. lissum*)).
A. nairi Altherr, 1974 – Europe (Loof, 1999); Germany (subterranean water bodies/Thuringe (Altherr, 1974)).
A. propinquum (de Man, 1921) Thorne & Swanger, 1936 – Europe (Loof, 1999); France (Moselle river basin/Lorraine (Altherr, 1963)).

Genus ***Belondira*** Thorne, 1939

Diagnosis. Belondirinae (Figs 14.6B and D). Small to medium-sized nematodes, 0.7–2.0 mm long. Habitus after fixation straight or somewhat ventrad curved. Lip region with pre- and/or post-labial sclerotizations more or less (usually well) developed. Amphidial fovea with opening occupying practically the whole body diameter. Odontostyle rather slender, and shorter than lip region width, but long and needle-like in *B. tenuidens*; odontophore rod-like, lacking any peculiar differentiation. Slender anterior portion of the pharynx more or less muscular, with an ellipsoidal swelling behind its junction to the odontophore and extending gradually into the basal expansion; this latter surrounded by a distinct spiral muscular sheath. Female genital system mono–opisthodelphic, with its anterior branch absent or reduced to a pre-vulval sac of variable length. Males bearing one or two widely separated ventromedian supplements. Tail similar in both sexes: rounded to clavate, usually with (sometimes very) thickened inner cuticle layer.

Remarks: Ferris *et al.* (1983) published the most recent revision of the genus, including descriptions and a key to its species. Mainly soil nematodes, but at least four of its almost 40 species have been recorded from freshwater habitats:

B. aquatica Ferris, Ferris & Goseco, 1983 – USA (Wilson Fork stream/Indiana (Ferris *et al.*, 1983))
B. mernyi Andrássy, 1970 – Ivory Coast (rice field/Senoufo (Andrássy, 1970a))
B. neortha Andrássy, 1968 – Paraguay (Acaray River (Andrássy, 1968b; but see Ferris *et al.*, 1983))
B. paratumicauda Dhanam & Jairajpuri, 1998 – India (rice field/Balechonur/Karnataka State (Dhanam and Jairajpuri, 1998))

Subfamily Dorylaimellinae Jairajpuri, 1964

Diagnosis. Belondiridae. Small to medium-sized nematodes. Lip region cap-like, with elevated inner region. Circumoral refractive pieces present. Odontostyle relatively short and slender. Odontophore consisting of two portions, the posterior one bearing flanges. Vulva of variable morphology. Female genital system also variable, monodelphic or didelphic. Males with moderately developed dorylaimid spicules, and ventromedian supplements arranged in pairs or separated. Tail similar in both sexes, but highly variable in size and shape; long and filiform to short and rounded. Only one genus, which includes freshwater species:

Genus ***Dorylaimellus*** Cobb, 1913

Diagnosis. Dorylaimellinae (Fig. 14.6A). Nematodes of small to medium size, whose length ranges from 0.4 to 1.8 mm. Lateral chord with abundant glandular bodies throughout its length. Lip region cap-like, usually well offset by constriction; its perioral region frequently differentiated into a disc-like structure. Anterior end of cheilostom with four more or less (usually very) distinct circumoral refractive cuticularized pieces. Odontostyle slender and relatively short, rarely longer than lip region diameter. Odontophore with its basal portion bearing three broad flanges. Anterior portion of the pharynx slender and weakly muscular, which extends gradually. Pharyngeal expansion surrounded by a more or less conspicuous spiral muscular sheath. Female genital system of variable morphology: didelphic or mono-, pro- or opisthodelphic. Vulva transverse, longitudinal or pore-like. Males with dorylaimid spicules, not specially robust, and bearing few ventromedian supplements, sometimes arranged in pairs, often widely spaced. Tail similar in both sexes, but highly variable in morphology: short, rounded to filiform.

Remarks: Siddiqi (1983) provided the most updated revision of the genus, including a key to its subgenera and species. It is the most abundant and widely distributed taxon among the belondirid taxa. Practically all of its almost 60 species have been collected in soil, but at least nine of them have been also recorded in aquatic habitats:

D. caffrae Kruger, 1965 – South Africa (Sabie River/Kruger National Park (Botha and Heyns, 1993))
D. cataracticus Andrássy, 1968 – Paraguay (Acaray River (Andrássy, 1968b))
D. egmonti Yeates & Ferris, 1984 – Hungary (moor/Bátorliget Nature Reserve (Andrássy, 1991); marsh/Fertö-Hanság National Park (Andrássy, 2002))
D. globatus Yeates, 1970 – Europe (Loof, 1999); UK (several habitats and localities (Yeates, 1970; Andrássy, 1978b))
D. heynsi Andrássy, 1968 – Paraguay (Acaray River (Andrássy, 1968b))
D. monticolus Clark, 1963 – Paraguay (Acaray River (Andrássy, 1968b))
D. muthi Dhanam & Jairajpuri, 1998 – India (rice field/Muthinakoppa (Dhanam and Jairajpuri, 1998)
D. occidentalis Thorne, 1939 – USA (stream bank/Brookings/South Dakota (Thorne, 1939))
D. parvulus Thorne, 1939 – USA (stream bank/Broad Run/Virginia (Thorne, 1939))

Subfamily Swangeriinae Jairajpuri, 1964

Diagnosis. Belondiridae. Nematodes of medium to large size, usually more than 1 mm long. Lip region continuous with the adjacent body, rounded (rarely truncated), with amalgamated lips. Cephalic framework present in some genera. Odontostyle relatively short and slender, asymmetrical in several genera. Pharyngeal basal expansion occupying one-third to one-half of total neck length, but usually rather short, fusiform and surrounded by a distinct dextral or sinistral spiral muscular sheath. Female genital system didelphic or mono–opisthodelphic. Male with dorylaimid spicules and few to numerous, contiguous or spaced, ventromedian supplements. Tail similar or dissimilar in sexes: female tail always long and filiform, male tail either long and filiform or short and conoid.

Nine genera, at least four of them (in bold) including aquatic species:

Falcihasta Clark, 1964
Lindseyus Ferris & Ferris, 1973
Oxybelondira Ahmad & Jairajpuri, 1979

Oxydirus Thorne, 1939
Paraoxybelondira Dhanam & Jairajpuri, 1998
Paraoxydirus Jairajpuri & Ahmad, 1979
Qudsiella Jairajpuri, 1967
Roqueus Thorne, 1964
Swangeria Thorne, 1939 (type genus)

Key to genera

1. Tail similar in both sexes: long and filiform 2
   Tail dissimilar in both sexes: long and filiform in female; short and conoid in male 8
2. Cheilostom basket-like, with 12 weakly sclerotized ribs . . . .*Swangeria*
   Cheilostom scarcely developed, cylindrical or a truncate cone, with no rib 3
3. Odontostyle falcate, with a distinct dorsal flexure (Fig. 14.6H) . . . .*Falcihasta*
   Odontostyle with other morphology, not falcate 4
4. Odontostyle attenuated, with very narrow lumen and minute aperture; cardia joining the intestine at its posterior end . . . .*Qudsiella*
   Odontostyle not so attenuated, with visible lumen and distinct aperture; cardia enveloped by intestinal tissue 5
5. Cephalic framework with labial sclerotizations; a fusiform swelling is present at the anterior slender region of the pharynx; female tail with its posterior half practically cylindrical and ending in a clavate or acute terminus . . . .*Oxybelondira*
   Cephalic framework lacking; fusiform swelling absent at the anterior slender region of the pharynx; female gradually long and filiform 6
6. Odontostyle symmetrical (Fig. 14.6I); pharyngeal basal expansion surrounded by a dextral spiral muscular sheath (Fig. 14.6F) . . . .*Oxydirus*
   Odontostyle asymmetrical (Fig. 14.6J); pharyngeal basal expansion surrounded by a sinistral spiral muscular sheath (Fig. 14.6G) 7
7. Large-sized nematodes (L>3 mm); perioral region not disc-like; female genital system didelphic; male with numerous contiguous ventromedian supplements . . . .*Paraoxydirus*
   Smaller nematodes (L<2 mm); perioral region disc-like; female genital system mono–opisthodelphic; male with few spaced ventromedian supplements . . . .*Paraoxybelondira*
8. Odontostyle attenuated, with narrow lumen; cephalic framework present although weakly developed (Fig. 14.6K) . . . .*Lindseyus*
   Odontostyle relatively short and robust, fusiform; cephalic framework completely absent . . . .*Roqueus*

Remarks: The genus *Duriella* Dhanam & Jairajpuri, 1998 has not been included in this subfamily because of the absence of any spiral muscular sheath surrounding the basal pharyngeal expansion, although in other features it fits the general morphology of the group.

Freshwater genera and species:

Genus ***Falcihasta*** Clark, 1964

Diagnosis. Swangeriinae (Fig. 14.6H). Nematodes of medium size, 1.1–1.8 mm long. Lip region continuous with the adjacent body, asymmetrical; lips amalgamated. Odontostyle falcate, with a distinct dorsal flexure in its ventral wall near anterior end. Pharynx consisting of a slender but muscular anterior portion extending gradually, and a very short basal pyriform bulb-like expansion that is surrounded by a spiral muscular sheath. Female genital system didelphic–amphidelphic or mono–opisthodelphic. Vulva transverse. Male bearing arcuate dorylaimid spicules and two or three spaced ventromedian supplements. Tail similar in both sexes: long and filiform, with or without very peculiar lateral cuticular expansions.

Remarks: Andrássy (1995) discussed the taxonomic position of *Falcihasta* and provided descriptions of two new and one known species. At least one of its four species has been collected in aquatic habitats:

F. palustris Clark, 1964 – New Zealand (several habitats and localities (Clark, 1964))

Genus ***Lindseyus*** Ferris & Ferris, 1973

Diagnosis. Swangeriinae (Fig. 14.6K). Very slender nematodes of large size. Lip region continuous with the adjacent body and rounded, with amalgamated lips. Odontostyle very attenuated, short and slender, with narrow lumen and small aperture. Cephalic framework present, basket-like, but weakly developed. Anterior portion of the pharynx slender but muscular, and extending gradually; basal expansion occupying about one-third of total neck length and surrounded by a distinct sinistral spiral muscular sheath. Cardia elongated, joining the intestine at its posterior portion. Female genital system didelphic–amphidelphic. Vulva transverse. Male with dorylaimid spicules and 7–10 spaced ventromedian supplements. Tail dissimilar in sexes: long and filiform in females; short and conoid in males.

Remarks: It is mainly an aquatic genus, with three of its four species having been collected in freshwater habitats:

L. costatus Ferris & Ferris, 1973 – USA (marsh/Hovey Lake/Indiana (Ferris and Ferris, 1973))

L. heterurus (Schuurmans Stekhoven & Teunissen, 1938) Coomans & Kheiri, 1986 – Ivory Coast (rice field/Baoulé (Andrássy, 1970a, as *R. africanus*); rice field (Mulk *et al.*, 1978)); South Africa (Vaal River (Andrássy, 1970b; Heyns, 2002, as *R. africanus*)); Spain (several localities (Jiménez-Guirado, 1989; Jiménez-Guirado and Murillo-Navarro, 2004))

L. indicus Dhanachand & Jairajpuri, 1980 – India (rice field/Imphal/Manipur (Dhanachand and Jairajpuri, 1980))

Genus ***Oxydirus*** Thorne, 1939

Diagnosis. Swangeriinae (Fig. 14.6F and I). Nematodes of medium size, 1.5–2.5 mm long. Lip region practically continuous with the adjacent body, rounded or weakly angular, and somewhat asymmetrical in lateral view. Odontostyle relatively short and slender, with aperture occupying about half of total length. Odontophore rod-like, without specializations. Anterior part of the pharynx slender but muscular, and extending gradually into the basal expansion that occupies about half of total neck length and is surrounded by a dextrally spiral muscular sheath. Female genital system didelphic–amphidelphic or mono–opisthodelphic. Vulva transverse. Male with dorylaimid spicules and 5–15 contiguous, more rarely spaced, ventromedian supplements. Tail similar in sexes: long and filiform.

Remarks: Ferris *et al.* (1980) published the most recent revision of the genus *Oxydirus*, including a key to its species. The genus is found in both soil and freshwater, but at least ten of its almost a dozen species have been reported (exclusively or not) in aquatic habitats:

O. amplicephalus Colomba & Vinciguerra, 1979 – Europe (Loof, 1999); Italy (Anapo river/Sicily (Colomba and Vinciguerra, 1979))

O. elongatus Altherr, 1963 – France (Moselle river basin/Lorraine (Altherr, 1963; Andrássy, 1978b; Loof, 1999))

O. gangeticus Siddiqi, 1966 – Nigeria (irrigation canal/Makarfi (Ferris *et al.*, 1980)); South Africa (Sabie river/Kruger National Park (Botha and Heyns, 1993; Heyns, 2002)); USA (irrigation canal/Beltsville, Maryland (Hechler, 1969; see Ferris *et al.*, 1980))

O. husmanni Altherr, 1972 – Brazil (Amazon river estuary (Altherr, 1977))

O. loofi Ferris, Goseco & Ferris, 1980 – Germany (stream/Kaufunger Wald/Münster (Ferris *et al.*, 1980))

O. nethus Ferris, Goseco & Ferris, 1980 – Europe (Loof, 1999)

O. oxycephaloides (de Man, 1921) Thorne, 1939 – Austria (several habitats and localities (Altherr, 1976)); Europe (Loof, 1999); Hungary (river bank/Fertö-Hanság National Park (Andrássy, 2002)); Italy (Po River (Zullini, 1975b); Anapo river/Sicily (Colomba and Vinciguerra, 1979; Zullini, 1982 – review)); Paraguay (Acaray River (Andrássy, 1968b)); Spain (several springs/Sierra Nevada/Granada (Ocaña *et al.*, 1986)); USA (several localities and habitats (Ferris *et al.*, 1980))

O. oxycephalus (de Man, 1885) Thorne, 1939 – Europe (Loof, 1999); Ghana (pool/Winneba (Andrássy, 1965))

O. tenuicaudatus Thorne, 1964 – Costa Rica (sediment/Maritza (Loof and Zullini, 2000))

O. tropicus Thorne, 1964 – Costa Rica (sediment/several localities (Loof and Zullini, 2000))

Genus ***Paraoxydirus*** Jairajpuri & Ahmad, 1979

Diagnosis. Swangeriinae (Fig. 14.6G and J). Nematodes of large size, more than 3 mm long. Lip region continuous with the adjacent body, rounded or truncated and symmetrical; with amalgamated lips; often with refractive elements present. Odontostyle relatively short and asymmetrical. Odontophore rod-like, without special modifications, but slightly sclerotized. Anterior portion of the pharynx slender, weakly muscular; basal expansion fusiform, occupying about one-third of the total neck length and surrounded by a thick sinistral spiral muscular sheath. Cardia well developed, enclosed by intestinal tissue. Female genital system didelphic–amphidelphic. Vagina with distinct *pars refringens*. Vulva transverse or pore-like. Male bearing well-developed dorylaimid spicules and 11–15 contiguous ventromedian supplements situated out of the spicule's range. Tail similar in both sexes: long and filiform.

Remarks: Ferris *et al.* (1980) provided the most recent revision of *Paraoxydirus*, with descriptions and a key to species identification. At least four of its seven species have been recorded in freshwater habitats:

P. abnormus Dhanam & Jairajpuri, 1998 – India (rice field/Chikmagahur district (Dhanam and Jairajpuri, 1998))
P. cavenessi (Ferris, Goseco & Ferris, 1980) Ferris, Goseco & Ferris, 1981 – India (rice field/Balehonnur/Karnataka State (Dhanam and Jairajpuri, 1998))
P. gigas (Jairajpuri, 1964) Jairajpuri & Ahmad, 1979 – India (rice field/Pathankot/ Punjab State (Jairajpuri and Ahmad, 1979))
P. pelinus (Ferris, Goseco & Ferris, 1980) Ferris, Goseco & Ferris, 1981 – Philippines (rice field/Bulacan (Ferris *et al.*, 1980))

Superfamily TYLENCHOLAIMOIDEA Filipjev, 1934

Diagnosis. Dorylaimina. Nematodes of small to large size, usually less than 2 mm long. Cuticle tylencholaimoid (inner layer with irregular outline and loose from the outer layer, and with abundant radial refractive elements) or dorylaimoid (with the inner layer not loose, and radial elements lacking). Lip region usually cap-like (i.e. offset from adjacent body and elevated perioral area), sometimes rounded and continuous, rarely with other morphology. Lips more or less (usually well) amalgamated, their inner region frequently forming distinct liplets or transformed into a perioral disc-like structure. Odontostyle axial, attenuate in general (with more or less distinct lumen and aperture), occasionally strong and with thickened walls. Odontophore rod-like (sometimes arcuate), more or less sclerotized, often with basal knobs or flanges. Pharynx with a slender and more or less (frequently weakly) muscular anterior portion extending gradually or abruptly. Pharyngeal basal expansion often reduced to a cylindrical or pyriform bulb occupying up to one-third of total neck length, but longer (one-third to one-half of total neck length) in one family; occasionally with a valvular chamber. Female genital system didelphic or monodelphic, with genital tract poorly differentiated in general. Vulva transverse, sometimes longitudinal or pore-like. Male with simple dorylaimid spicules and generally few spaced (very rarely contiguous) ventromedian supplements. Tail almost always similar in sexes: long and filiform to short and rounded.

Five families,[5] all of them including some freshwater species:

Aulolaimoididae Jairajpuri, 1964
Leptonchidae Thorne, 1935
Mydonomidae Thorne, 1964
Tylencholaimellidae Jairajpuri, 1964
Tylencholaimidae Filipjev, 1934 (type family)

[5]Jairapuri *et al.* (1998) proposed the new family Loofilaimidae under Tylencholaimoidea to classify a new genus and species, *Loofilaimus phadistoma*, recorded from Iran (muddy soil/Balutchestan), but serious doubts persist on the true taxonomic position of this taxon that might be a member of Dorylaimoidea (see also discussion by own authors).

Key to families

1. Pharynx consisting of three sections (Fig. 14.7A): long slender portion gradually expanding backwards, intermediate more slender portion surrounded by glandular tissue and basal expansion a pyriform bulb with well-developed valvular chamber . . . . .Aulolaimoididae
   Pharynx consisting of two sections: anterior slender portion and basal expansion, with no intermediate portion surrounded by glandular tissue 2
2. Cuticle tylencholaimoid (Fig. 14.7B): inner layer with irregular outline and loose from the outer layer, radial refractive elements present 3
   Cuticle dorylaimoid (Fig. 14.7C): inner layer with regular outline and not loose from the outer layer, radial refractive elements lacking 4
3. Odontostyle dorylaimoid, with distinct lumen and aperture (Fig. 14.7D); pharyngeal expansion cylindrical, occupying more than one-third of total neck length, generally with the usual five pharyngeal gland nuclei . . . . .Tylencholaimidae
   Odontostyle rather attenuated, with narrow lumen and minute aperture (Fig. 14.7E); pharyngeal expansion reduced to a cylindrical or pyriform basal bulb, occupying up to one-third of total neck length, often with only three pharyngeal gland nuclei .Leptonchidae
4. Odontostyle asymmetrical, with ventral arm shorter than dorsal one, and with wide aperture (Fig. 14.7F); pharyngeal bulb occupying one-fourth to one-third of total neck length . . . . .Mydonomidae
   Odontostyle symmetrical, tubular, usually with small aperture (Fig. 14.7G); pharyngeal bulb occupying one-fourth to one-fifth of total neck length . . . . .Tylencholaimellidae

Family TYLENCHOLAIMIDAE Filipjev, 1934

Diagnosis. Tylencholaimoidea. Nematodes of variable size, from less than 0.5 mm to more than 3 mm, usually less than 1 mm long. Cuticle tylencholaimoid, almost always with loose inner layer and abundant radial refractive elements. Lip region usually cap-like, but showing important variability. Lips usually amalgamated, with the perioral area frequently differentiated into a disc-like structure. Odontostyle highly variable in morphology, usually dorylaimoid, with distinct lumen and aperture, but relatively short, although sometimes long and robust. Odontophore rod-like, often bearing basal knobs, more rarely flanged. Anterior region of the pharynx more or less muscular, pharyngeal expansion occupying more than one-third of total neck length and bearing the usual five gland nuclei (except three in *Vanderlindia*). Female genital system didelphic or monodelphic. Vulva transverse, longitudinal or pore-like. Male with dorylaimid spicules and a variable number of usually spaced (more rarely contiguous) ventromedian supplements. Tail similar in sexes, but enormously variable: long and filiform to short and rounded.

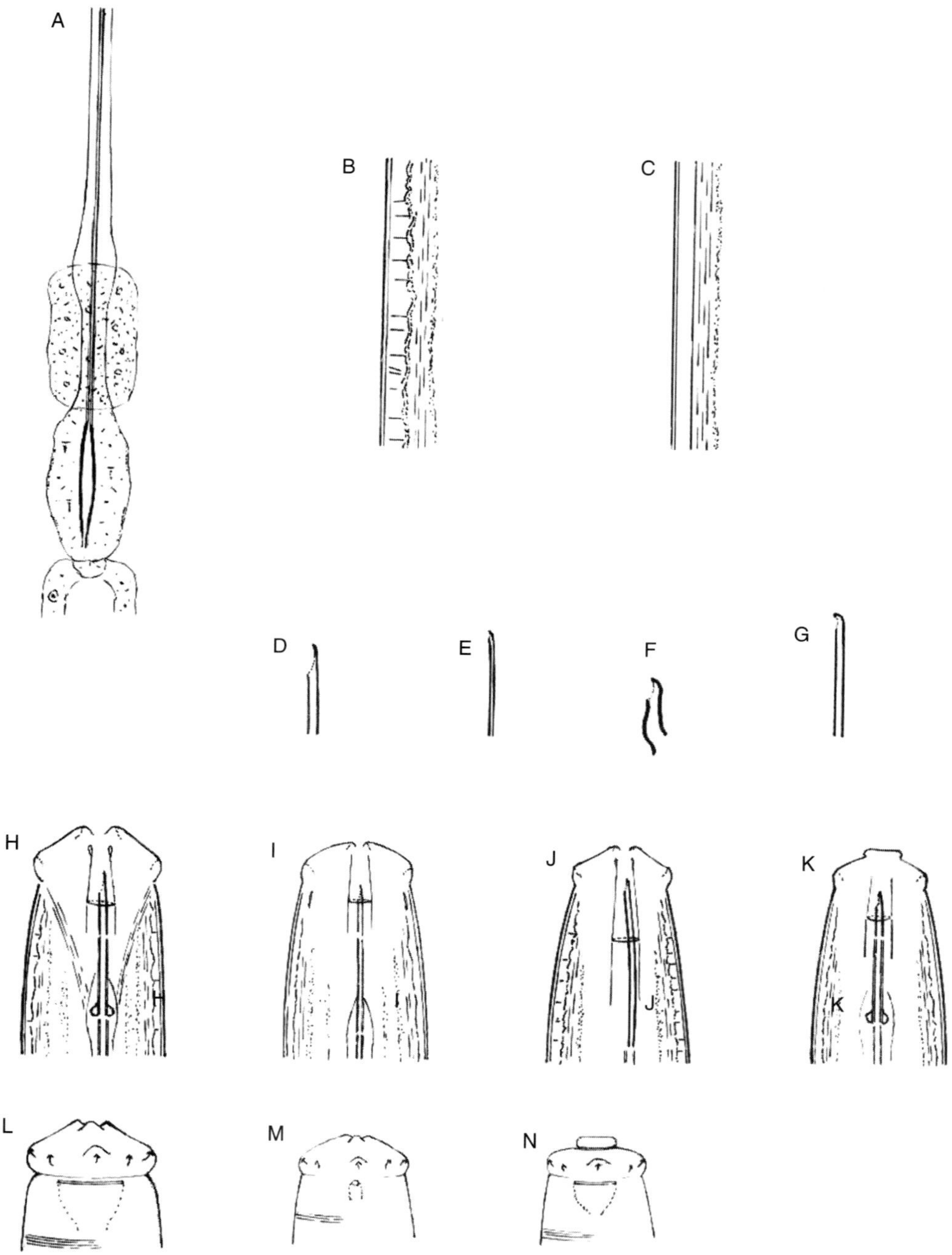

Fig. 14.7 Freshwater tylencholaimoid genera. A–G: Diagnostic characters of tylencholaimoid families; A: Pharynx in Aulolaimoididae; B: Tylencholaimoid typical cuticle, present in Tylencholaimidae and Leptonchidae; C: Dorylaimoid typical cuticle, present in Tylencholaimellidae and Mydonomidae; D–G: Typical odontostyle of the families Tylencholaimidae, Leptonchidae, Mydonomidae and Tylencholaimellidae, respectively (see also the text for variability); H–N: Genera in Tylencholamidae, Tylencholaiminae; H, L: *Tylencholaimus*; I, M: *Tantunema*; J: *Chitwoodius*; K, N: *Discomyctus*.

Three subfamilies, one of them including freshwater species:
Metadorylaiminae Andrássy, 1976
Tylencholaiminae Filipjev, 1934 (type subfamily)
Vanderlindiinae Siddiqi, 1969

Subfamily Tylencholaiminae Filipjev, 1934
Diagnosis. Tylencholaimidae. Nematodes of small to medium size, up to 2 mm long, but usually under 1 mm. Cuticle tylencholaimoid, with loose inner layer and more or less abundant and distinct radial refractive elements. Lip region variable in shape, but typically cap-like, with amalgamated lips and scarcely protruding papillae; perioral area sometimes differentiated into a disc-like structure. Odontostyle variable in length, dorylaimid, sometimes attenuate, very rarely short and massive. Odontophore rod-like, frequently with basal knobs. Pharynx consisting of a more or less muscular slender portion extending gradually or abruptly into the pharyngeal expansion, which occupies one-third to one-half of total neck length and contains the typical five pharyngeal gland nuclei. Female genital system didelphic, mono-prodelphic or mono-opisthodelphic. Vulva transverse, longitudinal or pore-like. Males with dorylaimid spicules and usually a few separated ventromedian supplements. Tail similar in sexes, but enormously variable: very long and filiform to short and hemispheroid.

Fifteen genera, at least four of them (in bold) including freshwater species:

Chitwoodielloides Ahmad & Araki, 2003
Chitwoodiellus Jiménez-Guirado & Peña-Santiago, 1992
Chitwoodius Furstenberg & Heyns, 1966
Discomyctus Thorne, 1939
Loncharionema Goseco, Ferris & Ferris, 1974
Meylonema Andrássy, 1960
Mumtazium Siddiqi, 1969
Promumtazium Siddiqi, 1982
Pseudotylencholaimus Jairajpuri & Ahmad, 1992
Rostrulium Siddiqi, 1995
Sclerolaimus Jairajpuri & Ahmad, 1992
Tantunema Siddiqi, 1982
Tsukubanema Ahmad & Araki, 2002
Tylencholaimus de Man, 1876 (type genus)
Wasimellus Bloemers & Wanless, 1996

Key to genera

1. Tail rather short (rounded, conoid or conical), always less than three times as long as the anal body diameter 2
   Tail long and filiform, more than seven times the anal body diameter 12
2. Odontostyle attenuate, with very narrow lumen; pharyngeal expansion occupying about one-third of total neck length . . . .*Sclerolaimus*
   Odontostyle more robust, with distinct lumen; pharyngeal expansion occupying more than one-third of total neck length 3

3. Odontostyle massive, with both its arms bifurcate from near half of its length, and with very small aperture; amphidial opening practically in labial position . . . . *Tsukubanema*
   Odontostyle not massive, with arms lacking any bifurcation and wider aperture; amphidial opening post-labial 4
4. Odontostyle somewhat asymmetrical, sigmoid, with aperture occupying more than one-third of its total length 5
   Odontostyle symmetrical, not sigmoid, with aperture occupying up to one-third of its total length 6
5. Odontostyle longer than lip region width, slender and somewhat sigmoid . . . . *Rostrulium*
   Odontostyle shorter than lip region width, more robust and fusiform . . . . *Promumtazium*
6. Odontostyle relatively robust, up to five times as long as wide, with wide lumen and aperture occupying one-fourth to one-third of its total length 7
   Odontostyle relatively slender, more than five times as long as wide, with narrow lumen and small aperture occupying less than one-fourth of its total length 10
7. Lip region with rounded lips, not cap-like; odontostyle strong, longer than lip region width; *pars refringens vaginae* moderately developed . . . . *Pseudotylencholaimus*
   Lip region cap-like; odontostyle not so strong, shorter than lip region width; *pars refringens vaginae* practically absent 8
8. Odontophore with more or less distinct basal knobs (Figs 14.7H and L) . . . . *Tylencholaimus*
   Odontophore without basal knobs 9
9. Amphidial fovea slit-like; post-labial sclerotization lacking . . . . *Meylonema*
   Amphidial fovea very small, pore-like; post-labial sclerotization present (Figs 14.7I and M) . . . . *Tantunema*
10. Odontostyle less than 1.5 times the lip region diameter long; junction between odontophore base and pharyngeal lining enveloped in a swollen and more consistent area; both parts of the pharynx separated by an isthmus-like portion . . . . *Chitwoodiellus*
    Odontostyle more than 1.5 times as long as the lip region diameter; junction between odontophore base and pharyngeal lining not specially differentiated; anterior portion of the pharynx extending gradually into the basal expansion 11
11. Odontostyle more robust, with distinct lumen and thickened walls (Fig. 14.7J); odontophore rod-like, lacking any specialization; anterior portion of the pharynx slender but muscular; *pars refringens vaginae* fringed; vulva longitudinal . . . . *Chitwoodius*
    Odontostyle attenuated, with narrow lumen and slender walls; odontophore with basal thickenings; anterior portion of the pharynx weakly muscular; *pars refringens vaginae* absent; vulva transverse . . . . *Chitwoodielloides*

12. Cheilostomatal and post-labial sclerotizations present; female genital system didelphic . . . . *Wasimellus*
    Cheilostomatal and post-labial sclerotizations absent; female genital system monodelphic 13
13. Odontostyle longer than lip region width, relatively robust and irregular, somewhat falcate; female genital system opisthodelphic . . . . *Mumtazium*
    Odontostyle shorter than lip region width, less robust and regular; female genital system prodelphic 14
14. Odontostyle attenuated, with very narrow lumen and obscure aperture; odontophore with basal flanges; post-uterine sac well developed . . . . *Loncharionema*
    Odontostyle not attenuated, with distinct lumen and aperture; odontophore with basal knobs (Figs 14.7K and N); post-uterine sac scarcely or not developed . . . . *Discomyctus*

Note: The genus *Tylenchodoroides* Gagarin, 2001 (syn. *Tylenchodorus* Gagarin, 2000, nec Siddiqi, 1983) is not included in the key because some doubts persist on its taxonomic position. According to the original description, both cuticle layers ('cuticle and subcuticle') are 'smooth, non-striated' and the morphology of the pharynx is very unusual within the family Tylencholaimidae.

Freshwater genera and species:

Genus ***Chitwoodius*** Furstenberg & Heyns, 1966
Diagnosis. Tylencholaiminae (Fig. 14.7J). Nematodes of medium size, 1–2 mm long. Cuticle typical tylencholaimoid, with abundant radial refractive elements. Lip region rounded or cap-like, offset by constriction or depression. Odontostyle tubular and often slightly dorsally curved, with narrow but distinct lumen and small aperture, its length being 1.6–2.1 times the lip region diameter. Odontophore rod-like, lacking any specialization. Anterior slender portion of the pharynx normal or weakly muscular. Pharyngeal basal expansion occupying about half of total neck length, distinctly muscular. Female genital system didelphic–amphidelphic. *Pars refringens vaginae* fringed or absent. Vulva longitudinal or pore-like. Male with dorylaimid spicules and 4–13 spaced ventromedian supplements. Tail similar in sexes: convex conoid to rounded.
Remarks: Peña-Santiago (2002) has provided the most recent diagnosis of this genus and a compendium of its species. *Chitwoodius* is a terrestrial taxon, but one of its seven species has been recorded in aquatic habitats:

C. undulatus Altherr, 1977 – Brazil (Amazon river estuary (Altherr, 1977))

Genus ***Discomyctus*** Thorne, 1939
Diagnosis. Tylencholaiminae (Figs 14.7K and N). Small nematodes, under 1 mm long. Lip region cap-like, with the perioral area differentiated into a disc-like structure. Odontostyle small, shorter than lip region width, and more or less robust. Odontophore provided with distinct basal knobs. Anterior portion of the pharynx slender, weakly muscular and expanding abruptly; pharyngeal basal expansion occupying about half of total neck length. Female genital system mono–prodelphic, with posterior branch completely absent. Tail similar in sexes: long and filiform.

Remarks: *Discomyctus* is a terrestrial genus, but at least two of its nine species have been recorded from freshwater habitats:

D. bisexualis Dhanam & Jairajpuri, 1999 – India (rice field/Muthinakoppa/ Karnataka State (Dhanam and Jairajpuri, 1998))
D. longicaudatus (Imamura, 1931) Thorne, 1939 – India (rice field/Margoa/ Goa (Ahmad *et al.*, 2002)); Japan (rice field/Tokyo (Imamura, 1931))

Genus ***Tantunema*** Siddiqi, 1982
Diagnosis. Tylencholaiminae (Figs 14.7I and M). Nematodes of small size, always under 1 mm long. Cuticle tylencholaimoid, but the inner layer is not very irregular and the radial refractive elements are barely distinct. Lip region cap-like, more or less differentiated from adjacent body, with amalgamated lips; small post-labial sclerotizations usually present. Opening of amphidial fovea exceptionally small, usually pore-like. Odontostyle small; odontophore rod-like, hardly thickened at the base. Anterior portion of the pharynx extending gradually into the basal expansion, which occupies about two-fifths of total neck length. Female genital system mono-opisthodelphic, with anterior branch completely absent. Vulva transverse. Female tail rounded to subcylindrical. Male unknown.
Remarks: At least one of the three species hitherto classified under this genus has been collected in freshwater habitats:

T. aquaticum Ahmad & Jairajpuri, 1983 – India (rice field/Aslali/Gujarat State (Ahmad and Jairajpuri, 1983a))

Genus ***Tylencholaimus*** de Man, 1880
Diagnosis. Tylencholaiminae (Figs 14.7H and I). Nematodes of small to medium size, usually under 1 mm long. Cuticle tylencholaimoid, with radial refractive elements present but more or less distinct. Lip region cap-like, usually offset by constriction, but sometimes by depression; lips more or less amalgamated, with their perioral area transformed or not into a disc-like structure. Cheilostom a truncate cone, often with perioral thickenings. Odontostyle somewhat shorter than lip region width; its aperture about one-third of total length. Odontophore rod-like, with more or less (generally very) distinct basal knobs. Pharynx consisting of a slender anterior part extending gradually or abruptly into the basal expansion, which occupies one-third to two-fifths of the total neck length (exception: *T. leptodorus*, less than one-third). Female genital system amphidelphic, mono-prodelphic or mono-opisthodelphic. Male with dorylaimoid spicules and few spaced (very rarely numerous and contiguous) ventromedian supplements. Tail similar in sexes: elongate–conical to hemispherical.
Remarks: Peña-Santiago and Coomans (1996) provided the most recent key to species identification. *Tylencholaimus* species are typically terrestrial, as practically all of the almost 50 species have been collected in soil, but at least 11 of them have also been recorded from aquatic habitats:

T. ibericus Peña-Santiago & Coomans, 1994 – India (rice field/Muthinakoppa/ Karnataka State (Dhanam and Jairajpuri, 1999))
T. maritus Loof & Jairajpuri, 1968 – Italy (Anapo river/Sicily (Colomba and Vinciguerra, 1979; Loof, 1999)); Mexico (water in Bromeliaceae/Chiapas (Zullini, 1977))

T. minimus de Man, 1876 – Austria (several habitats and localities (Altherr, 1976)); Denmark (several habitats and localities (Micoletzky, 1925a)); Europe (Andrássy, 1978b; Loof, 1999); Hungary (several habitats/Bátorliget Nature Reserve (Andrássy, 1991)); Italy (Po River (Zullini, 1975a)).
T. mirabilis (Bütschli, 1873) de Man, 1876 (type species) – Europe (Andrássy, 1978b)
T. obscurus Jairajpuri, 1965 – Mexico (rivulet/Chiapas (Zullini, 1977))
T. paradoxus Loof & Jairajpuri, 1968 – Europe (Loof, 1999)
T. proximus Thorne, 1939 – Europe (Loof, 1999); Spain (several springs/Sierra Nevada National Park/Granada (Ocaña *et al.*, 1986))
T. savaryi Loof & Jairajpuri, 1968 – Switzerland (Loof and Jairajpuri, 1968; Loof, 1999)
T. stecki Steiner, 1914 – Denmark (several (Micoletzky, 1925a)); Europe (Andrássy, 1978b; Loof, 1999)
T. teres Thorne, 1939 – Europe (Loof, 1999)
T. vigil Andrássy, 1959 – Romania (Andrássy, 1959, 1978b; Loof, 1999)

Subfamily Vanderlindiinae Siddiqi, 1969
Diagnosis. Tylencholaimidae. Nematodes of large size, more than 2 mm long. Cuticle tylencholaimoid, with distinct irregular inner layer outline and abundant radial refractive elements. Lip region offset by constriction. Odontostyle longer than lip region width, thick-walled and ventrad curved. Odontophore rod-like. Anterior slender portion of the pharynx more or less muscular; pharyngeal expansion well developed, occupying one-half to two-thirds of total neck length. Female genital system didelphic–amphidelphic. Vulva transverse or pore-like. Male with dorylaimid spicules and numerous ventromedian supplements. Tail similar in sexes: short and rounded.

List of genera:
Curvidorylaimus Jairajpuri & Rahman, 1983
Vanderlindia Heyns, 1964 (type genus)
No genus of this subfamily has been recorded in freshwater habitats.

Subfamily Metadorylaminae Andrássy, 1976
Diagnosis. Tylencholaimidae. Medium-sized nematodes, 2–3 mm long. Cuticle dorylaimoid, with the inner cuticle layer thicker than the outer one, lacking radial refractive elements. Lip region cap-like, with amalgamated lips. Odontostyle as long as lip region width, but robust and massive, with very thick walls; the base of the ventral wall typically furcate. Odontophore rod-like, with no differentiation. Anterior portion of the pharynx extending gradually into the basal expansion, which occupies about half of total neck length. Female genital system didelphic–amphidelphic. Vagina with three sections. Vulva transverse or longitudinal. Female tail short and rounded. Males unknown.

List of genera:
Metadorylaimus Jairajpuri & Goodey, 1966 (type genus)
Neometadorylaimus Jairajpuri & Ahmad, 1992
No genus of this subfamily has been recorded in freshwater habitats.

Family LEPTONCHIDAE Thorne, 1935
Diagnosis. Tylencholamoidea. Nematodes of small to medium size, up to 3 mm long. Cuticle typical tylencholaimoid, with loose inner cuticle layer and abundant

radial refractive elements. Lip region cap-like, usually offset by constriction, sometimes with perioral disc. Cheilostom a truncate cone, rarely with other shape. Odontostyle highly variable, in general attenuated or needle-like, rarely more robust; almost always shorter than lip region width, but occasionally longer. Odontophore rod-like, with or without basal differentiations. Anterior region of the pharynx slender and weakly muscular; pharyngeal expansion reduced to a basal bulb occupying up to one-third of total neck length but almost always about one-fourth, and in several genera with a valvulvar chamber. Female genital system didelphic or monodelphic. Vulva transverse, rarely longitudinal or pore-like. Male with dorylaimid spicules and few spaced (seldom numerous contiguous) ventromedian supplements. Tail similar in sexes: short and rounded, occasionally conical or long and filiform.

Five subfamilies, four of them (in bold) including freshwater species:
Belonenchinae Thorne, 1964
Leptonchinae Thorne, 1935 (type subfamily)
Scalpenchinae n. subfam.
Tyleptinae Jairajpuri, 1964
Xiphinemellinae Jairajpuri, 1964

Key to subfamilies

1. Odontostyle strong, asymmetrical, with aperture occupying about half of total length, scalpel-like . . . . .Scalpenchinae
   Odontostyle more slender, symmetrical, with smaller aperture, not scalpel-like 2
2. Odontostyle more robust, with distinct lumen and aperture . . . . .Tyleptinae
   Odontostyle more slender, with narrow or no lumen and small or no aperture 3
3. Odontostyle solid or needle-like, lacking lumen and aperture . . . . .Belonenchinae
   Odontostyle attenuated, with very narrow but visible lumen and minute aperture 4
4. Odontostyle longer than lip region width . . . . .Xiphinemellinae
   Odontostyle shorter than lip region width . . . . .Leptonchinae

Subfamily Belonenchinae Thorne, 1964

Diagnosis. Leptonchidae. Small nematodes, almost always under 1 mm long. Cuticle typical tylencholaimoid, usually with abundant radial refractive elements. Lip region cap-like, offset by constriction, with or without perioral disc. Cheilostom a truncate cone, occasionally bottle- or flask-shaped. Odontostyle solid or needle-like, sometimes with a very narrow lumen at its posterior half; very rarely longer than lip region width. Odontophore usually simple, but seldom with basal knobs, thickenings or flanges. Anterior portion of the pharynx weakly muscular; basal expansion a bulb with or without valvular chamber. Female genital system didelphic or monodelphic. Male with dorylaimid spicules and few spaced ventromedian supplements. Tail similar in sexes: short and rounded to conoid.

Five genera, at least one of them (in bold) including aquatic species:
Aculonchus Siddiqi, 1983

Basirotyleptus Jairajpuri, 1964 (type genus)
Glochidorella Siddiqi, 1982
Sclerostylus Goseco, Ferris & Ferris, 1981
Zetalaimus Siddiqi, 1983

Remarks: Siddiqi (1995) described the new genus *Clavigula* and classified it under Xiphinemellinae. However, its general pattern (solid odontostyle at its anterior half and shorter than lip region width, lip region cap-like, etc.) better fits that of members of Belonenchinae. In fact, *Clavigula* is very similar to *Glochidorella*, differing from this mainly in the absence of perioral disc (vs. presence), shape of pharyngeal bulb (more cylindrical vs. more pyriform) and more robust spicules. These differences should be interpreted as intrageneric variations, and then both genera are considered to be synonymous. Siddiqi (1995) mentioned that *Clavigula* has a hollow odontostyle as an additional difference, but this does not seem to be significant because in his original diagnosis it is stated: 'odontostyle needle-like, with extremely narrow lumen and solid appearing tip'; meanwhile in the original diagnosis of *Glochidorella* (Siddiqi, 1982b) it is stated: 'odontostyle straight, needle-like, without lumen for most of its length'.

Key to genera

1. Pharyngeal basal bulb having thickened cuticular inner lining and a valvular chamber (Fig. 14.8E) 2
   Pharyngeal basal bulb lacking any specialization or differentiation 4
2. Odontostyle longer than lip region width . . . .*Sclerostylus*
   Odontostyle shorter than lip region width 3
3. Odontophore simple, lacking any differentiation (Fig. 14.8A) . . . .*Basirotyleptus*
   Odontophore flanged . . . .*Aculonchus*
4. Cheilostom goblet-shaped, weakly sclerotized; odontophore with basal knobs; female genital system didelphic . . . .*Glochidorella*
   Cheilostom flask-shaped, sclerotized; odontophore flanged; female genital system mono-prodelphic . . . .*Zetalaimus*

Freshwater genera and species:

Genus ***Basirotyleptus*** Jairajpuri, 1964

Diagnosis. Belonenchinae (Figs 14.8A and E). Small nematodes, less than 1 mm long. Cuticle tylencholaimoid, with more or less abundant radial refractive elements. Lip region cap-like, offset by constriction or depression. Cheilostom a truncate cone, usually with sclerotized walls and perioral thickenings. Odontostyle solid, needle-like. Odontophore simple. Anterior portion of the pharynx slender and weakly muscular; basal expansion a pyriform bulb occupying less than one-fourth of total neck length, with valvular chamber. Female genital system didelphic or monodelphic. Vulva transverse. Male with dorylaimid spicules and one to four spaced ventromedian supplements. Tail similar in sexes: short and rounded to conoid.

Remarks: Goseco *et al.* (1974b) provided the most recent revision of the genus, including descriptions and a key to species. *Basirotyleptus* is basically a terrestrial genus, but at least three of its about 25 species have been recorded in aquatic habitats:

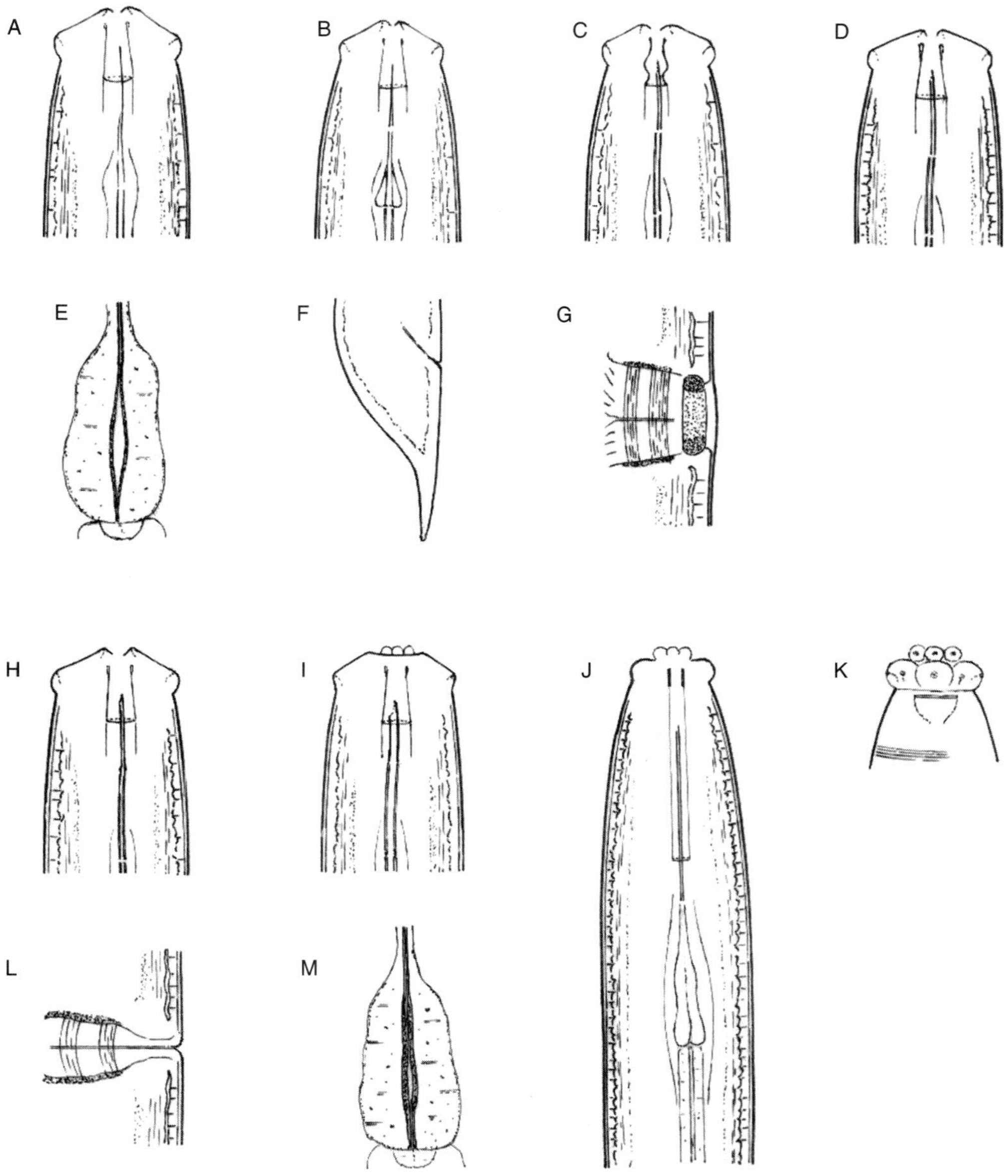

Fig. 14.8 Freshwater tylencholaimoid genera of the family Leptonchidae.
A, E: Subfamily Belonenchinae, genus *Basirotyleptus*; B–D, F–H; L: Subfamily Leptonchinae; B, F: Genus *Proleptonchoides*; C: Genus *Proleptonchus*; D, G: Genus *Funaria*; H, L: Genus *Leptonchus*; I, M: Subfamily Tyleptinae, genus *Tyleptus*; J, K: Subfamily Xiphinemellinae, genus *Xiphinemella*.

C. altherri (Heyns, 1968) Ahmad & Jairajpuri, 1982 – Europe (Loof, 1999)
C. clavicaudatus (Altherr, 1953) Ahmad & Jairajpuri, 1982 (type species) – Europe (Loof, 1999)
C. trophurus (Heyns, 1968) Ahmad & Jairajpuri, 1982 – Europe (Loof, 1999)

Genus ***Feroxides*** Heyns, 1968
Diagnosis. Nygolaiminae (Fig. 14.10I). Nematodes of large size, more than 3.0 mm long. Habitus practically straight after fixation. Cuticle particulary thick. Lip region continuous with the adjacent body. Mural tooth small, with solid anterior portion and hollow posterior (basal) portion. Cardiac cells not present at the pharyngo-intestinal junction. Female genital system didelphic. Vagina with two sections, lacking the *pars refringens*. Vulva transverse. Female tail short and rounded. Male unknown.
Only one species, collected in a freshwater habitat:

F. ferox (Thorne, 1930) Jairajpuri & Ahmad, 1992 – USA (stream bank/Alsee Mountains/Oregon (Thorne, 1930))

Genus ***Laevides*** Heyns, 1968
Diagnosis. Nygolaiminae (Fig. 14.10J)). Nematodes of medium to large size, usually over 2 mm long. Lip region continuous, slightly narrower than adjacent body. Mural tooth shorter than lip region width, relatively robust or dorylaimoid, i.e. with a dorsal anterior oblique plane resembling dorsal aperture of the odontostyle in Dorylaimina. Pharyngeal expansion surrounded by a conspicuous sheath forming basal pockets. Cardiac cells ovoid or rounded. Female genital system didelphic. Vagina with two sections, lacking *pars refringens*. Vulva transverse. Male with well-developed dorylaimid spicules, gubernaculum and 5–8 ventromedian supplements. Tail similar in sexes: short, conoid to hemispheroid, mostly somewhat clavate.
Remarks: Heyns (1968) published the most recent revision of the genus, including species descriptions, and Ahmad and Jairajpuri (1982) provided the most updated available key to species identification. At least eight *Laevides* species, among 11, have been recorded from freshwater habitats:

L. husmanni (Meyl, 1954) Ahmad & Jairajpuri, 1982 – Germany (Weser River bank/Boffzen (Meyl, 1954; Andrássy, 1978b))
L. laevis (Thorne, 1939) Thorne, 1974 (type species) – Bangladesh (rice field/Tejgaon/Dacca district (Heyns, 1968)); Russia (several habitats and localities (Gagarin, 1992)); The Netherlands (Loof, 1999)
L. loofi (Heyns, 1968) Ahmad & Jairajpuri, 1968 – Germany (Loof, 1999)
L. microdens Thorne, 1974 – USA (swamp/Brookings/South Dakota (Thorne, 1974))
L. paraaquaticus (Paetzold, 1958) Ahmad & Jairajpuri, 1982 – Germany (Andrássy, 1978b; Loof, 1999)
L. rapax (Thorne, 1939) Ahmad & Jairajpuri, 1982 – China ((Wu and Liang, 1999)); USA (stream bank/Slide Gulch/Idaho; stream bank/Alsee Mountains/Oregon (Thorne, 1939, Heyns, 1968))
L. symmetricus Khan, Jairajpuri & Ahmad, 1989 – India (rice field/Raanti/Bihar State (Khan *et al.*, 1989))
L. timmi (Heyns, 1968) Ahmad & Jairajpuri, 1982 – Bangladesh (rice field/Tejgaon/Dacca district (Heyns, 1968))

B. acus Goseco, Ferris & Ferris, 1974 – USA (several habitats and localities (Goseco *et al.*, 1974b))
B. lieberi Goseco, Ferris & Ferris, 1974 – USA (Harrison spring sediment/White Cloud/Indiana (Goseco *et al.*, 1974b))
B. minimus Jana & Baqri, 1981 – India (rice field/Teen Battimore/Darjeeling district (Jana and Baqri, 1981))

Subfamily Leptonchinae Thorne, 1935

Diagnosis. Leptonchidae. Nematodes of small to medium size, up to 3 mm long, but usually about 1 mm long. Cuticle typical tylencholaimoid, with loose and irregularly outlined inner layer and abundant radial refractive elements. Lip region almost always cap-like, with lips more or less separated and their perioral region often differentiated into a perioral disc. Cheilostom a truncate cone or flask-shaped, sometimes with sclerotized thickened walls. Odontostyle attenuate, (very rarely more robust) normally with narrow lumen and minute aperture, but variable in length. Odontophore with or without basal flanges, knobs or thickenings. Anterior region of pharynx weakly muscular; basal expansion very short, bulb-like, occupying up to one-third, usually about one-fourth of total neck length, rarely provided with a valvular chamber. Female genital system didelphic or monodelphic. Vulva transverse, rarely longitudinal or pore-like. Male with dorylaimid spicules and few spaced (rarely numerous contiguous) ventromedian supplements. Tail similar in sexes: almost always short and rounded, less frequently conical or long and filiform.

Twelve genera, at least four of them (in bold) including freshwater species:
Apoleptonchus Siddiqi, 1982
Bertzuckermania Khera, 1970
Caveonchus Siddiqi, 1982
Clavigula Siddiqi, 1995
Funaria van der Linde, 1938
Incanema Andrássy, 1997
Lawtonema Siddiqi, 1998
Leptonchus Cobb, 1920 (type genus)
Meylis Goseco, Ferris & Ferris, 1974
Paraleptonchus Dhanam & Jairajpuri, 1999
Proleptonchoides Ferris, Goseco & Kumar, 1979
Proleptonchus Lordello, 1955

Key to genera

1. Female tail long and filiform . . . . *Lawtonema*
   Female tail short and rounded to conoid 2
2. Odontophore bearing distinct basal knobs or thickenings (Fig. 14.8B) 3
   Odontophore lacking distinct basal knobs or thickenings 5
3. Female genital system mono–prodelphic; female tail conical digitate (Fig. 14.8F) . . . . *Proleptonchoides*
   Female genital system didelphic; female tail short, rounded to conoid 4

4. Odontostyle with distinct lumen throughout its length; pharyngeal bulb shorter and pyriform; vulva pore-like; female tail short and rounded . . . . *Apoleptonchus*
   Odontostyle with very narrow lumen at its anterior half, needle-like; pharyngeal bulb longer and cylindroid; vulva transverse; female tail conoid . . . . *Clavigula*
5. Cheilostom flask- or goblet-shaped (Fig. 14.8C); female genital system monodelphic 6
   Cheilostom a truncate cone; female genital system didelphic 8
6. Anterior portion of the pharynx highly convoluted, with indistinct or absent musculature; pharyngeal basal bulb occupying about one-tenth of total neck length . . . . *Paraleptonchus*
   Anterior portion of the pharynx with usual morphology; pharyngeal basal bulb occupying about one-fifth of total neck length 7
7. Perioral area not disc-like; cheilostom flask-shaped; pharyngeal bulb lacking valvular chamber; female genital system prodelphic . . . . *Proleptonchus*
   Perioral area disc-like; cheilostom goblet-shaped; pharyngeal bulb with a valvular chamber; female genital system opisthodelphic . . . . *Caveonchus*
8. Vulva longitudinal; *pars refringens vaginae* usually well differentiated (Fig. 14.8G) . . . . *Funaria*
   Vulva transverse; *pars refringens vaginae* not differentiated (Fig. 14.8M) 9
9. Lip region rounded and completely continuous with the adjacent body; amphidial fovea aperture occupying about one-fourth of lip region width; ventromedian supplements absent . . . . *Bertzuckermania*
   Lip region cap-like, more or less offset; amphidial fovea aperture occupying usually more than half of total neck length; ventromedian supplements present 10
10. A thickened short portion of the anterior slender region of the pharynx in front of the constriction and separating both pharyngeal portions; intestine surrounding the pharyngeal bulb base, forming a collar-like structure . . . . *Incanema*
    Pharyngeal constriction, if present, preceding not specially thickened short portion; intestine not surrounding the pharyngeal bulb base 11
11. Lip region distinctly offset by very deep constriction; perioral region disc-like; both parts of the pharynx separated by a narrowing or constriction; ventromedian supplements numerous (more than ten) and contiguous . . . . *Meylis*
    Lip region offset by a weak constriction; perioral region not disc-like (Fig. 14.8I); anterior portion of

the pharynx extending gradually into the basal bulb; ventromedian supplements fewer (less than ten) and widely separated . . . .*Leptonchus*

Freshwater genera and species:

Genus ***Funaria*** van der Linde, 1938

Diagnosis. Leptonchinae (Figs 14.8D and G). Nematodes of medium size, 1–2 mm long. Cuticle typical tylencholaimoid. Lip region cap-like, offset by more or less deep constriction. Cheilostom a truncate cone, its walls weakly sclerotized, with perioral thickenings. Odontostyle attenuated, with very narrow but usually distinct lumen and minute aperture, straight or somewhat arcuate and shorter than lip region width. Odontophore with slender walls, often slightly arcuate. Anterior region of the pharynx extending gradually, basal expansion a cylindrical (more rarely pyriform) bulb occupying about one-fifth of total neck length. Vulva longitudinal. *Pars refringens vaginae* well developed in general. Female genital system didelphic. Pre-rectum very long, but its junction with intestine posterior to vulva. Male with dorylaimid spicules and 4–10 spaced ventromedian supplements. Tail similar in sexes: short, rounded to conoid, with mucronate or acute tip.

Remarks: Goseco *et al.* (1974a) provided the most recent revision of the genus, and Peña-Santiago (1991) the last key to its species. Its dozen known species have been recorded in soil, but at least two of them have also been collected in freshwater habitats:

F. obtusa (Thorne, 1939) Goseco, Ferris & Ferris, 1974 – USA (several habitats and localities (Goseco *et al.*, 1974a))

F. thornei van der Linde, 1938 (type species) – USA (several habitats and localities (Goseco *et al.*, 1974a)

Genus ***Leptonchus*** Cobb, 1920

Diagnosis. Leptonchinae (Figs 14.8H and L). Nematodes of small to medium size, 0.7–1.5 mm long. Cuticle typical tylencholaimoid, with irregularly outlined inner layer and abundant radial refractive elements. Body pores arranged in two rows along margins of lateral chord. Lip region cap-like, offset by constriction. Cheilostom a truncate cone, often with developed circumoral thickenings. Odontostyle attenuated, straight, with distinct but narrow lumen and minute aperture. Odontophore usually arcuate. Pharyngeal expansion bulb-like, pyriform, occupying about one-fifth of total neck length. Female genital system didelphic. Vulva transverse. Pre-rectum very long, its junction with intestine anterior or posterior at level of vulva. Male with dorylaimid spicules and 5–7 spaced ventromedian supplements. Tail similar in sexes: short, rounded to conoid.

Remarks: Goseco *et al.* (1974a) provided the most recent revision of the genus, including descriptions and key to species. It is a terrestrial genus, but at least one of its nine species has been collected in freshwater habitats:

L. granulosus Cobb, 1920 (type species) – USA (Diermans' Lake/O'Fallon/Missouri; George Lake/New York (Goseco *et al.*, 1974a))

Genus ***Proleptonchoides*** Ferris, Goseco & Kumar, 1979

Diagnosis. Leptonchinae (Figs 14.8B and F). Nematodes of medium size, about 1.5 mm long. Cuticle tylencholaimoid, with irregular outlined inner layer and distinct

radial refractive elements. Lip region scarcely offset from adjacent body. Cheilostom almost cylindrical, thick-walled. Odontostyle attenuated, its anterior half practically solid and shorter than lip region width. Odontophore rod-like, bearing distinct basal thickenings. Anterior region of pharynx weakly muscular, basal expansion somewhat pyriform and occupying about one-fourth of total neck length; both parts of the pharynx separated by a constriction. Female genital system mono–prodelphic. Vulva transverse. Female tail conical digitate. Male unknown.
Remarks: One species of this genus, among three, was originally described from an aquatic habitat:

P. equistylus Dhanam & Jairajpuri, 1999 – India (rice field/Muthinakoppa/Karnataka State (Dhanam and Jairajpuri, 1999))

Genus ***Proleptonchus*** Lordello, 1955
Diagnosis. Leptonchinae (Fig. 14.8C). Nematodes of small to medium size, 0.8–1.9 mm long. Cuticle tylencholaimoid, with abundant radial refractive elements. Lip region cap-like, with amalgamated lips and perioral area transformed into a disc-like structure in only one species. Cheilostom cuticularized, showing a peculiar flask-like shape. Odontostyle attenuated, with very narrow but distinct lumen and minute aperture. Odontophore slender and somewhat arcuate. Anterior portion of the pharynx slender and weakly muscular; a short narrowing or constriction separates both pharyngeal regions; basal expansion occupying one-fifth to one-fourth of total neck length. Female genital system mono–prodelphic, with very long post-vulval sac. Male with dorylaimid spicules and one to six widely spaced ventromedian supplements. Tail similar in sexes: short and rounded to conoid.
Remarks: *Capilonchus* Siddiqi, 1982 is considered to be a junior synonym of *Proleptonchus*. Goseco *et al.* (1974a) provided the most recent revision of the genus, including a key to species; and Jiménez-Guirado and Arcos (1994) performed a cladistic study of its species. Most parts of the species of *Proleptonchus* have been collected in terrestrial habitats, but at least six of them were recorded from freshwater:

P. brevisaccus Goseco, Ferris & Ferris, 1974 – USA (several habitats and localities (Goseco *et al.*, 1974a))
P. bryophilus (Imamura, 1931) Goseco, Ferris & Ferris, 1974 – Japan (rice field/Imperial University Farm/Tokyo (Imamura, 1931))
P. clarus Jana & Baqri, 1981 – India (rice field/Abujhati/Burdwan district (Jana and Baqri, 1981))
P. krugeri Botha & Heyns, 1992 – South Africa (Sabie and Olifants rivers/Kruger National Park (Botha and Heyns, 1992; Heyns, 2002))
P. lacihoveyensis Goseco, Ferris & Ferris, 1974 – USA (Hovey lake/Indiana (Goseco *et al.*, 1974a))
P. saccatus (Clark, 1962) Andrássy, 1963 – Congo Republic (Congo river (Andrássy, 1968a)); Nepal (brook/Rolwaling valley (Andrássy, 1978a))

Subfamily Scalpenchinae n. subfam.
Diagnosis. Leptonchidae. Nematodes of small size, less than 1 mm long. Cuticle tylencholaimoid. Lip region practically continuous, with amalgamated lips. Odontostyle strong, asymmetrical, with large aperture. Odontophore rod-like.

Pharyngeal expansion a subcylindroid bulb. Female genital system mono–prodelphic. Vulva transverse. Male with dorylaimid spicules and spaced ventromedian supplements. Tail similar in sexes: conoid to conical.

Remarks: Siddiqi (1995) classified the genus *Scalpenchus* under the family Mydonomidae, subfamily Mydonominae. As the cuticle is tylencholaimoid, not dorylaimoid it better fits the pattern of Leptonchidae and the morphology of the odontostyle (strong, with very large aperture) does not compare with that of other members of this family, justifying the erection of a new subfamily to accommodate it.

Only one genus with one species, collected in soil:

Scalpenchus Siddiqi, 1995

Subfamily Tyleptinae Jairajpuri, 1964

Diagnosis. Leptonchidae. Small to medium-sized nematodes, 0.5–1.5 mm long. Cuticle tylencholaimoid, with more or less abundant radial refractive elements. Lip region cap-like, usually offset by constriction, with amalgamated lips. Cheilostom a truncate cone, slightly sclerotized. Odontostyle cylindrical, relatively robust, with distinct lumen and aperture. Odontophore rod-like, occasionally flanged. Anterior portion of the pharynx slender and weakly muscular; basal expansion bulb-like, occupying up to one-fourth of total neck length. Female genital system didelphic or mono–opisthodelphic. Male with dorylaimid spicules and few spaced ventromedian supplements. Tail similar in sexes: short and rounded.

Three genera, at least one of them (in bold) including freshwater species:

Gymnotyleptus Ahmad & Jairajpuri, 1982
Tyleptus Thorne, 1939 (type genus)
Utahnema Thorne, 1939

Genus ***Tyleptus*** Thorne, 1939

Diagnosis. Tyleptinae (Figs 14.8I and M). Nematodes of small to medium size, 0.5–1.2 mm long. Cuticle tylencholaimoid, with more or less distinct and abundant radial refractive elements. Lateral pores arranged in two rows. Lip region cap-like, offset. Lips amalgamated, but their inner portion differentiated into six distinct separated liplets. Odontostyle robust, typical of the subfamily. Odontophore rod-like, sometimes flanged. Pharyngeal basal expansion a pyriform bulb occupying about one-fifth of total neck length; its inner lining thickened and delimiting a more or less developed valvular chamber. Female genital system mono–opisthodelphic. Vulva transverse. Male with dorylaimid spicules and one or three spaced ventromedian supplements. Tail similar in sexes: short and rounded.

Remarks: Goseco *et al.* (1974b) provided the most recent revision of the genus, including description of its species and a key to their identification. *Tyleptus* species have been mainly collected in soil, but at least five, among eight, were also recorded from freshwater habitats:

T. amalgans Thorne, 1964 – Puerto Rico (drain ditch bank/Río Piedras (Thorne, 1964))

T. bongersi Loof & Zullini, 2000 – Costa Rica (Tempisquito River/Maritza (Loof & Zullini, 2000))

T. oryzae Dhanam & Jairajpuri, 1999 – India (rice field/Muthinakoppa/ Karnataka State (Dhanam and Jairajpuri, 1999))

T. projectus Thorne, 1939 (type species) – USA (several habitats and localities (Goseco *et al.*, 1974b))

T. striatus Heyns, 1963 – South Africa (Sabie and Olifants rivers/Kruger National Park (Botha and Heyns, 1992))

Subfamily Xiphinemellinae Jairajpuri, 1964

Diagnosis. Leptonchidae. Small to medium-sized nematodes, 0.5–3.0 mm. Cuticle typical tylencholaimoid, with abundant radial refractive elements. Lip region cap-like, usually offset by constriction. Lips amalgamated or separated, their inner portion often differentiated into a disc-like structure. Odontostyle attenuated and longer than lip region width. Odontophore flanged or bulboid at base. Anterior portion of the pharynx slender and weakly muscular; basal expansion a cylindrical bulb occupying up to one-third of total neck length; both portions of the pharynx separated by constriction or depression. Female genital system didelphic. Vulva transverse, rarely pore-like. Male with dorylaimid spicules and spaced ventromedian supplements. Tail similar in sexes: short and rounded, occasionally conical.

Two genera, one of them (in bold) including freshwater species:

Khantbhala Siddiqi, 1982

Xiphinemella Loos, 1950 (type genus)

Genus ***Xiphinemella*** Loos, 1950

Diagnosis. Xiphinemellinae (Figs 14.8J and K). Nematodes of medium to large size, 1–3 mm long. Cuticle tylencholaimoid, with very irregular outlined inner layer and abundant radial refractive elements. Lip region cap-like, offset by deep constriction. Lips more or less (often well) separated, with their perioral area usually transformed into a distinct perioral disc. (Cephalic framework present in *X. marindae.*) Cheilostom with sclerotized walls in type species. Odontostyle long and attenuated, with minute aperture. Odontophore well developed, with prominent basal flanges or thickenings. Guiding ring a truncate cone in type species, single in the others. Anterior region of the pharynx slender and weakly muscular, basal expansion a cylindroid bulb occupying about one-fourth of total neck length; both parts of the pharynx separated by a constriction. Female genital system didelphic. Vulva transverse. Male with dorylaimid spicules and spaced ventromedian supplements. Tail short and rounded in female, more conoid in male and conical in both sexes in *X. caudata.*

Remarks: Siddiqi (1982a) described the new genus *Zalophidera* differing from *Xiphinemella* mainly by the absence of a perioral disc, but such a difference does not justify the erection of a new generic taxon, and it should better interpreted as intrageneric variability; therefore *Zalophidera* is here considered to be a synonym of *Xiphinemella.* De Bruin and Heyns (1991) provided the most recent revision of the genus, including a description of its species. At least one species, among 13, has been recorded from freshwater habitats:

X. caudata Andrássy, 1970 – Ivory Coast (rice field/Senoufo (Andrássy, 1970a))

Family TYLENCHOLAIMELLIDAE Jairajpuri, 1964

Diagnosis. Tylencholaimoidea. Nematodes of small to medium size, under 1.5 mm long. Cuticle dorylaimoid, lacking radial refractive elements. Lip region usually cap-like, with amalgamated lips, more seldom continuous. Odontostyle tubular, straight and shorter than lip region width, occasionally with other morphology. Odontophore with or without specializations. Pharyngeal expansion a more or less pyriform bulb, constricted or not. Female genital system didelphic or monodelphic. Vulva transverse. Male with dorylaimid spicules and few spaced ventromedian supplements. Tail similar in sexes: long and filiform to short and rounded.

Two subfamilies, both including freshwater species:

Athernematinae Ahmad & Jairajpuri, 1978
Tylencholaimellinae Jairajpuri, 1964 (type subfamily)

Remarks: Two main features characterize this family and separate it from Tylencholaimidae and Leptonchidae: the presence of the dorylaimoid cuticle and a tubular (rarely attenuated) odontostyle.

Subfamily Athernematinae Ahmad & Jairajpuri, 1978

Diagnosis. Tylencholaimellidae. Nematodes of small size, under 1 mm long. Cuticle dorylaimoid. Lip region continuous, with amalgamated lips. Amphidial fovea strongly sclerotized. Odontostyle asymmetrical, arcuate. Odontophore rod-like, lacking any differentiation. Pharyngeal expansion a pyriform basal bulb. Intestine joining the base of cardia. Female genital system mono–opisthodelphic. Tail similar in sexes: long and filiform.

Only one genus with one freshwater species:

Genus ***Athernema*** Ahmad & Jairajpuri, 1978

Diagnosis. Athernematinae (Figs 14.9A and E). Nematodes of medium size, 1.5–1.8 mm long. Cuticle dorylaimoid, completely lacking radial refractive elements. Lip region small, rounded and continuous; lips amalgamated. Amphidial fovea bilobed, strongly sclerotized. Odontostyle asymmetrical and arcuate, with distinct lumen and small aperture. Odontophore rod-like, simple. Anterior portion of the pharynx weakly muscular, extending gradually; basal expansion a pyriform bulb occupying about one-fifth of total neck length. Cardia relatively large, with the intestine attached to its base. Female genital system mono–opisthodelphic. Vulva transverse. Male with dorylaimid spicules and few and weakly developed ventromedian supplements. Tail similar in sexes: long and filiform.

Remarks: *Athernema* is certainly an aquatic genus as its only species has been collected in freshwater habitats:

A. unicum Ahmad & Jairajpuri, 1978 (type species) – India (Hoggly river bank/Shibpur/West Bengal (Ahmad and Jairajpuri, 1978))

Subfamily Tylencholaimellinae Jairajpuri, 1964

Diagnosis. Tylencholaimellidae. Nematodes of small to medium size, 0.3–1.5 mm long. Cuticle dorylaimoid, with inner layer distinctly thicker than outer one. Lip region cap-like, more or less offset, more rarely continuous; lips usually

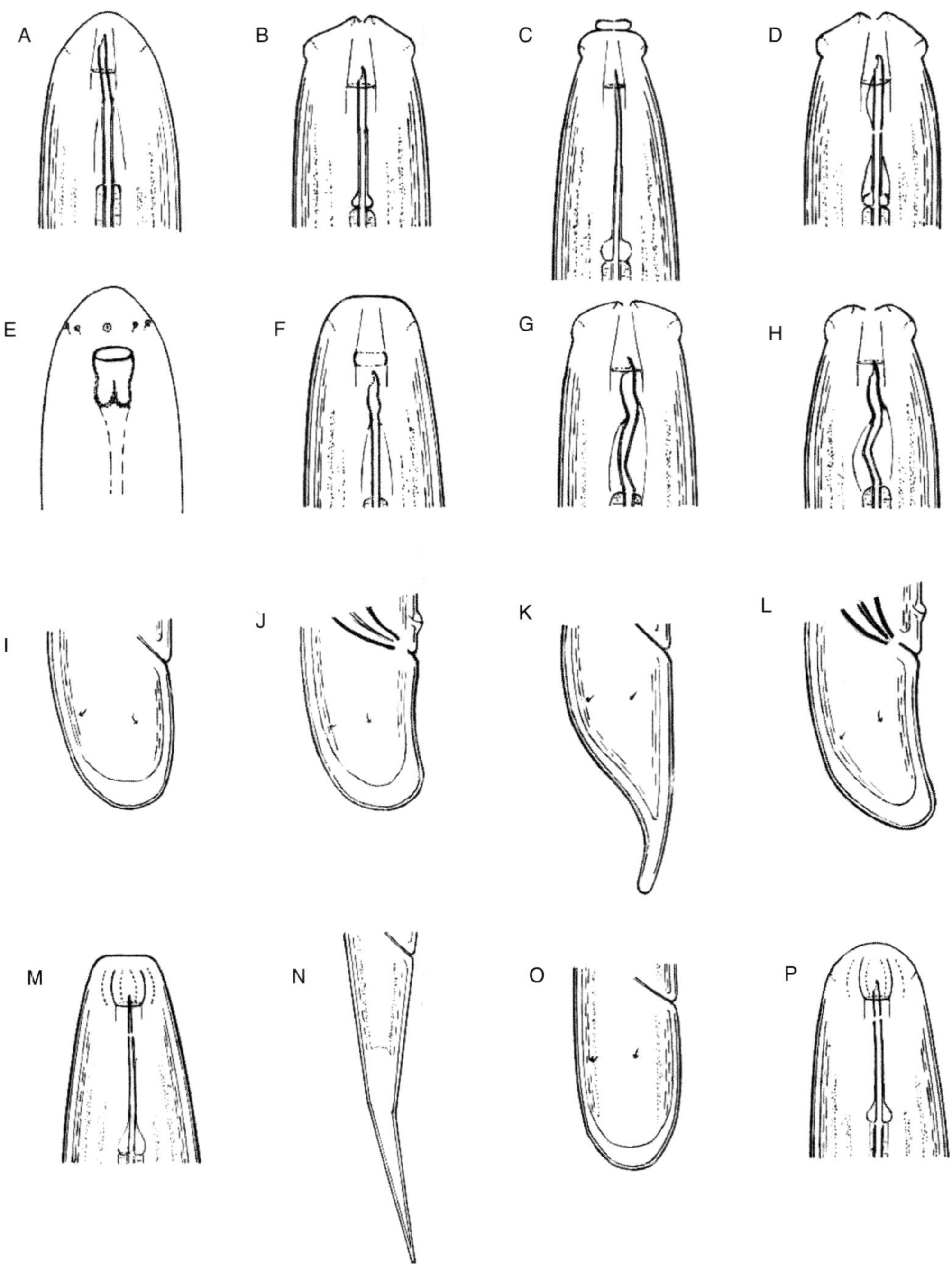

Fig. 14.9 Freshwater tylencholaimoid genera. A–F: Family Tylencholaimellidae; A, E: Subfamily Athernematinae, genus *Athernema*; B–D: Subfamily Tylencholaimellinae; B: Genus *Doryllium*; C: Genus *Oostenbrinkella*; D: Genus *Tylencholaimellus*; F–L: Family Mydonomidae; F: Subfamily Calolaiminae, genus *Calolaimus*; G–L: Subfamily Mydonominae; G, I, J: Genus *Dorylaimoides*; H, K, L: Genus *Morasia*; M–P: Family Aulolaimoididae; M, N: Genus *Aulolaimoides*; O, P: Genus *Adenolaimus*.

amalgamated, with their perioral region differentiated or not into a disc-like structure. Cephalic framework developed only in one genus. Odontostyle tubular with distinct lumen and small aperture, shorter than lip region width; occasionally more attenuated and longer than lip region width; accessory stiffening piece may be present. Odontophore with or without basal knobs. Anterior region of the pharynx weakly muscular; basal expansion a pyriform bulb, occupying about one-fifth of total neck length; both portions may be separated or not by constriction. Female genital system didelphic or monodelphic. Vulva transverse. Male with dorylaimid spicules and few spaced ventromedian supplements. Tail similar in sexes, but variable in length: long and filiform to short and hemispheroid.
Eight genera, at least three of them (in bold) including freshwater species:

Agmodorus Thorne, 1964
Dorella Jairajpuri, 1964
Doryllium Cobb, 1920
Goferus Jairajpuri & Ahmad, 1992
Margollus Peña-Santiago, Peralta & Siddiqi, 1993
Oostenbrinkella Jairajpuri, 1965
Phellonema Thorne, 1964
Tylencholaimellus Cobb in M.V. Cobb, 1915 (type genus)

Key to genera

1. Tail very long and filiform ($c' > 8$) . . . . *Oostenbrinkella*
   Tail short and rounded, rarely conoid, conical or elongated ($c' < 6$) 2
2. Odontostyle lacking any accessory stiffening piece 3
   Odontostyle bearing an accessory stiffening piece 6
3. Odontophore with distinct basal knobs 4
   Odontophore simple, lacking basal knobs 5
4. No cuticularized circumoral pieces; female genital system mono–opisthodelphic . . . . *Doryllium*
   Cuticularized circumoral pieces present; female genital system didelphic . . . . *Phellonema*
5. Anterior slender portion of the pharynx extending gradually into the basal bulb; female genital system didelphic; tail conoid to cylindroid . . . . *Goferus*
   Pharyngeal bulb separated by constriction; female genital system mono–opisthodelphic; tail elongate . . . . *Agmodorus*
6. Accessory stiffening piece ventral, arising from odontostyle base; female genital system mono–prodelphic . . . . *Dorella*
   Accessory stiffening piece dorsal and occupying most odontostyle length; female genital system mono–opisthodelphic 7
7. Cephalic framework well developed, with labial and post-labial sclerotization . . . . *Margollus*
   Cephalic framework absent or very weakly developed . . . . *Tylencholaimellus*

Freshwater genera and species:

Genus ***Doryllium*** Cobb, 1920

Diagnosis. Tylencholaimellinae (Fig. 14.9B). Small nematodes, less than 1 mm long. Lip region continuous or offset by depression or constriction; lips amalgamated, with their inner portion differentiated or not into a perioral disc. Cheilostom a truncate cone, lacking any specialization. Odontostyle relatively short, tubular, with distinct lumen and aperture. Odontophore with basal knobs or flanges. Anterior portion of the pharynx slender and weakly muscular; basal expansion a more or less pyriform bulb, separated or not from the anterior portion by a constriction. Female genital system mono–opisthodelphic, with the anterior branch absent or reduced to a pre-vulval sac. Male with dorylaimid spicules and only one ventromedian supplement. Tail similar in both sexes: short and rounded.

Remarks: Goseco *et al.* (1975) erected the new genus *Gerthus* to accommodate one new and two previously known species belonging to *Doryllium*, on the basis of absence of constriction separating the pharyngeal bulb base from the anterior slender portion of the pharynx. Such difference should be interpreted as intrageneric variation and does not justify the proposal of a new generic taxon. Then *Gerthus* is considered to be a junior synonym of *Doryllium*. Goseco *et al.* (1975) provided the most recent revision of the genus, including descriptions and key to species.

Species of *Doryllium* have mainly been collected in terrestrial habitats, but at least five of them, among ten, were also recorded from aquatic habitats:

D. aestuarii Timm, 1967 – Bangladesh (Pasar river estuary/Sunderbans (Timm, 1967))

D. cornelli (van der Linde, 1938) Jairajpuri & Siddiqi, 1963 – Argentina (rivulet/El Bolsón/Río Negro (Andrássy, 1963))

D. minor Jairajpuri, 1963 – The Netherlands (Loof, 1999)

D. uniforme Cobb, 1920 (type species) – Europe (Andrássy, 1978b; Loof, 1999); USA (estuary bank/Los Patos/California (Cobb, 1920))

D. zeelandicum (de Man, 1876) Loof, 1996 – Europe (Andrássy, 1978b; Loof, 1999); The Netherlands (several habitats and localities (de Man, 1876; Loof, 1999))

Genus ***Oostenbrinkella*** Jairajpuri, 1965

Diagnosis. Tylencholaimellinae (Fig. 14.9C). Nematodes of medium size, 1.0–1.3 mm long. Cuticle dorylaimoid. Lip region cap-like, offset by constriction. Lips amalgamated, with the perioral region differentiated into a disc-like structure. Odontostyle attenuated, with narrow but distinct lumen and small aperture, scarcely longer than lip region width, sigmoid in one species. Odontophore with developed basal knobs. Anterior portion of the pharynx weakly muscular and extending gradually; basal expansion a pyriform bulb occupying about one-fifth of total neck length. Female genital system mono–opisthodelphic. Vulva transverse. Male with dorylaimid spicules and no ventromedian supplement. Tail similar in sexes: long and filiform.

Remarks: One of the two species hitherto classified under this genus has been collected in aquatic habitats:

O. ventrostylus Dhanam & Jairajpuri, 1999 – India (rice field/Muthinakoppa/ Karnataka State (Dhanam and Jairajpuri, 1999))

Genus ***Tylencholaimellus*** Cobb in M.V. Cobb, 1915

Diagnosis. Tylencholaimellinae (Fig. 14.9D). Nematodes of small to medium size, 0.3–1.5 mm long. Cuticle with the typical dorylaimoid pattern: outer layer relatively thin and with fine transverse striations and inner layer thicker than the outer one. Lip region cap-like, in general offset by constriction, sometimes continuous; perioral region elevated and transformed into a disc-like structure in several species. Cheilostom a truncate cone, without special modifications. Odontostyle tubular, with distinct lumen and small aperture; dorsal side with an arch-like stiffening accessory piece. Odontophore rod-like, bearing basal knobs. Anterior portion of the pharynx slender and weakly muscular; basal expansion a cylindrical to pyriform bulb, occupying about one-fifth of total neck length; both pharyngeal portions separated by a more or less marked constriction. Female genital system mono-opisthodelphic, with anterior genital branch absent or reduced to an uterine blind sac of variable length. Male with dorylaimid spicules and only one ventromedian supplement. Tail similar in sexes: short and rounded to conical with rounded tip.

Remarks: Goseco *et al.* (1975) provided the most recent revision of the genus, including descriptions. Peña-Santiago and Peralta (1999a) provided a list of species, a polytomous key to their identification and a compendium of their measurements. Most *Tylencholaimellus* species inhabit mainlands, but at least 12 of them, among almost 40, have been recorded (exclusively or not) in freshwater habitats:

T. affinis (Brakenhoff, 1914) Thorne, 1939 – Europe (Andrássy, 1978b; Loof, 1999); France (peat bog/Bavois (Altherr, 1954)); USA (stream bank/Broad Run/ Virginia (Thorne, 1939))

T. amphidius Dhanam & Jairajpuri, 1999 – India (rice field/Muthinakoppa/ Karnataka State (Dhanam and Jairajpuri, 1999))

T. coronatus Thorne, 1939 – USA (stream bank/Williston/North Dakota (Thorne, 1974))

T. diplodorus Cobb in M.V. Cobb, 1915 (type species) – USA (Maple river/Douglas Lake region/Michigan (Cobb, 1915))

T. incertus Colomba & Vinciguerra, 1979 – Italy (Anapo river/Sicily (Colomba and Vinciguerra, 1979; Loof, 1999))

T. montanus Thorne, 1939 – Spain (rivulet/Despeñaperros Natural Park/Jaén (Peña-Santiago and Peralta, 1999b))

T. ozarkensis Goseco, Ferris & Ferris, 1975 – USA (sediment/White River/Bull Shoals/Arkansas (Goseco *et al.*, 1975))

T. pluvialis Siddiqi, 1965 – USA (Eal River bank/Wabash/Indiana (Goseco *et al.*, 1975))

T. polonicus Szczygiel, 1962 – Italy (Po river (Zullini, 1975b)); Nepal (Lobuche Lake (Andrássy, 1978a))

T. projectus Siddiqi, 1964 – Europe (Andrássy, 1978b, as *T. fenensis*; Loof, 1999)

T. sagittifer (de Man, 1921) Thorne, 1939 – The Netherlands (Mark River bank/Breda (de Man, 1921; Loof, 1999)

T. striatus Thorne, 1939 – Europe (Andrássy, 1978b; Loof, 1999)

Family MYDONOMIDAE Thorne, 1964

Diagnosis. Tylencholaimoidea. Medium to large nematodes, usually more than 1 mm long. Cuticle dorylaimoid, with inner layer thicker than outer one. Lip region continuous or slightly offset, sometimes cap-like. Odontostyle relatively robust, asymmetrical, with both arms of different length. Guiding ring thin or thick and refringent. Odontophore straight and simple or arcuate and more or less sclerotized. Pharyngeal expansion a cylindrical (more rarely pyriform) basal bulb occupying less than one-third of total neck length. Female genital system didelphic or monodelphic. Male with dorylaimid spicules and spaced ventromedian supplements. Tail very variable, short and rounded to long and filiform, similar or dissimilar in sexes. Two subfamilies, both including freshwater species:

Calolaiminae Goseco, Ferris & Ferris, 1976
Mydonominae Thorne, 1964 (type subfamily)

Remarks: This family is characterized by having the dorylaimoid cuticle, asymmetrical odontostyle and relatively long (up to one-third of total neck length) pharyngeal basal bulb. Jairajpuri and Ahmad (1992) considered the genus *Miranema* Thorne, 1939 to be a member of this nematode group, but the pharyngeal expansion in this genus occupies 35–42% of total neck length (cf. Goseco *et al.*, 1976, p. 42) and it should be classified under Dorylaimoidea. Siddiqi (1995) included the new genus *Scalpenchus* in Mydonomidae, but the cuticle in this taxon is tylencholaimoid and better fits the leptonchid pattern (see page 363, subfamily Scalpenchinae).

Subfamily Calolaiminae Goseco, Ferris & Ferris, 1976

Diagnosis. Mydonomidae. Nematodes of large size, more than 3 mm long. Cuticle dorylaimoid. Lip region practically continuous, with amalgamated lips. Odontostyle irregular, delicate and shorter than lip region width. Guiding ring usually sclerotized and thick. Odontophore straight. Pharyngeal expansion a cylindroid or pyriform bulb, occupying up to one-third of total neck length. Female genital system didelphic. Vulva transverse or pore-like. Tail similar in sexes: elongated to long and filiform.

Two genera, one of them (in bold) including freshwater species:

Calolaimus Timm, 1964 (type genus)
Timmus Goseco, Ferris & Ferris, 1976

Genus ***Calolaimus*** Timm, 1964

Diagnosis. Calolaiminae (Fig. 14.9F). Nematodes of large size, rarely under 4 mm long. Cuticle dorylaimoid. Lip region rounded and continuous, with amalgamated lips. Odontostyle asymmetrical and short. Guiding ring very thick and sclerotized, belt-like. Odontophore straight, sclerotized. Anterior portion of the pharynx weakly muscular and extending gradually; pharyngeal expansion a cylindrical bulb, occupying up to one-third of total neck length, and sometimes with sclerotized thick inner lining. Female genital system didelphic. Vulva transverse or pore-like. Male known in one species, with dorylaimid spicules and eight spaced ventromedian supplements. Tail similar in sexes: elongated to filiform.

Remarks: Goseco *et al.* (1976) published the most recent revision of the genus, including descriptions, and Ahmad and Jairajpuri (1983b) provided the most updated key to species identification. Four species, among six, have been recorded from freshwater habitats:

C. ditlevseni (Micoletzky, 1922) Timm, 1964 – Europe (Andrássy, 1978b; Loof, 1999); Hungary (Balaton lake (Andrássy, 2001)); Italy (rice field/Cascina Merizzo (Andrássy, 2001)); Java island (rice field/Bogor (Goseco *et al.*, 1976))
C. lenorum (Siddiqi, 1965) Siddiqi, 1966 – India (rice field/Patna City/Bihar (Siddiqi, 1965))
C. papillatus Timm, 1964 (type species) – Russia (several habitats and localities (Gagarin, 1992))
C. parapapillatus Dhanachand & Jairajpuri, 1980 – India (rice field/Khangabok/Manipur (Dhanachand and Jairajpuri, 1980))

Subfamily Mydonominae Thorne, 1964
Diagnosis. Mydonomidae. Small to medium-sized nematodes, under 2 mm long. Cuticle dorylaimoid. Lip region continuous or more or less offset by depression or constriction. Odontostyle asymmetrical, relatively robust and shorter than lip region width. Odontophore arcuate. Pharyngeal expansion a cylindrical bulb, occupying one-fourth to one-third of total neck length, occasionally enveloped in a muscular sheath. Female genital system didelphic or monodelphic. Male with dorylaimid spicules and spaced ventromedian supplements. Tail similar or dissimilar in sexes: long and filiform to short and rounded.
Three genera, at least two of them (in bold) including freshwater species:

Dorylaimoides Thorne & Swanger, 1936
Morasia Baqri & Jairjapuri, 1969
Mydonomus Thorne, 1964 (type genus)

Genus ***Dorylaimoides*** Thorne & Swanger, 1936
Diagnosis. Mydonominae (Figs 14.9G, I and J). Nematodes of medium size, almost always 1–2 mm long. Cuticle dorylaimoid: outer layer thin and with fine transverse striations, and inner layer thicker than the outer one. Odontostyle asymmetrical, ventral side shorter than the dorsal one and dorsally bent at its anterior end; dorsal side longer and somewhat sigmoid. Odontophore usually arcuate or angular, involved by the pharynx. Pharynx consisting of a slender and weakly muscular anterior portion, which extends more or less abruptly; basal expansion a cylindrical bulb occupying one-fourth to one-third of total neck length. Female genital system didelphic or mono-opisthodelphic. Vulva transverse. Male with dorylaimid spicules and 1–12 spaced ventromedian supplements, usually starting into the range of the spicules. Tail similar in sexes: long and filiform to short and rounded.
Remarks: Goseco *et al.* (1976) published the most recent revision of the genus, including descriptions of its species, and Peña-Santiago and Peralta (1997a,b) studied its general morphology and provided a compendium and key to its species. *Dorylaimoides* is basically a terrestrial genus, but at least 13 of its almost 60 species have been recorded (exclusively or not) in freshwater habitats:

D. arcuicaudatus Baqri & Jairajpuri, 1969 – India (rice field/Majehrpara/Burdwan district (Jana and Baqri, 1981))
D. ariasae Loof, 1990 – Spain (Sanabria lake/Zamora province (Loof, 1990, 1999))

D. bulbosus (Brzeski & Szczygiel, 1961) Szczygiel, 1965 – Europe (Loof, 1999)
D. elegans (de Man, 1880) Thorne & Swanger, 1936 – Europe (Andrássy, 1978b; Loof, 1999); Russia (several habitats and localities (Gagarin, 1992)); USA (several habitats and localities (Goseco *et al.*, 1976))
D. indicus Jairajpuri, 1965 – USA (river bank/Hovey lake region/Indiana; Lake of the Ozarks bank/Damsel/Missouri (Goseco *et al.*, 1976))
D. kalingus Ahmad & Jairajpuri, 1983 – India (rice field/Bhubaneshwar/Orissa State (Ahmad and Jairajpuri, 1983b))
D. limnophilus (de Man, 1880) Loof, 1964 – Europe (Loof, 1999); Hungary (lake/Boglár (Daday, 1897); Danube River/Stromkm (Andrássy, 1962; as *D. riparius*)); Spain (irrigation canal/Castellón (Gadea, 1952); pool/Zamora (Gadea, 1954); pool and peat bog/Pontevedra, Antela lagoon/Orense (Gadea, 1955); Engors-Pyrenees/Gerona (Gadea, 1956); pool/Sierra Nevada National Park/Granada (Peña-Santiago and Peralta, 1999b)); The Netherlands (marsh/unknown locality (de Man, 1880))
D. longicaudatus (Imamura, 1931) Thorne & Swanger, 1936 – Japan (rice field/Komaba/Tokyo (Imamura, 1931))
D. micoletzkyi (de Man, 1921) Thorne & Swanger, 1936 – Europe (Andrássy, 1978b; Loof, 1999); Romania (river bank/Cerna (Andrássy, 1959)); Russia (several habitats and localities (Gagarin, 1992))
D. microdentatus Altherr, 1968 (*species inquirendae*) – Germany (Stechlin Lake/Brandebourg (Altherr, 1968; Andrássy, 1978b))
D. parateres Siddiqi, 1964 – India (rice field/Tufangunj/West Bengal State (Baqri and Khera, 1979))
D. parvus Thorne & Swanger, 1926 – India (rice field/several (Jana and Baqri, 1981))
D. paulbuchneri Meyl, 1956 – Paraguay (Acaray River (Andrássy, 1968b))
D. teres Thorne & Swanger, 1936 (type species) – USA (Lake of the Ozarks bank/Damsel/Missouri (Goseco *et al.*, 1976))

Genus ***Morasia*** Baqri & Jairajpuri, 1969

Diagnosis. Mydonominae (Figs 14.9H, K and L). Nematodes of medium size, under 2 mm long. Cuticle dorylaimoid. Lip region rounded, practically continuous, with amalgamated lips. Odontostyle asymmetrical, dorsal arm longer than ventral one. Odontophore arcuate. Anterior portion of the pharynx weakly muscular and extending gradually; basal expansion a cylindroid bulb occupying about one-third of total neck length. Female genital system amphidelphic. Vulva transverse. Male with dorylaimid spicules and spaced ventromedian supplements. Tail dissimilar in sexes: elongate in female and short and rounded in male.

Remarks: Two species of this genus, among four, have been collected from freshwater habitats:

M. bengalensis Jana & Baqri, 1981 – India (rice field/Memari/Burdwan district (Jana and Baqri, 1981))
M. postmontana Loof, 1990 – The Netherlands (Loof, 1990, 1999)

Family AULOLAIMOIDIDAE Jairajpuri, 1964

Diagnosis. Tylencholaimoidea. Nematodes of small to medium size, under 2 mm long. Cuticle dorylaimoid. Lip region continuous, with amalgamated lips.

Cheilostom a truncate cone, usually surrounded by a basket-like structure bearing minute rib-like elements. Odontostyle attenuated, with narrow lumen and small or undistinct aperture. Odontophore with prominent basal thickenings. Pharynx consisting of three sections: (i) long slender portion gradually expanding backwards; (ii) more slender intermediate portion surrounded by glandular tissue; and (iii) basal expansion a pyriform bulb with well-developed valvular chamber. Female genital system didelphic or monodelphic. Male with slender dorylaimid spicules and few spaced ventromedian supplements; gubernaculum may be or may not be present. Tail similar in sexes: subcylindrical to filiform.

Four genera, at least two of them (in bold) including freshwater species:

Adenolaimus Andrássy, 1973
Aulolaimoides Micoletzky, 1915 (type genus)
Cladocephalus Swart & Heyns, 1991
Oostenbrinkia Ali, Suryawanshi & Ahmed, 1973

Genus ***Adenolaimus*** Andrássy, 1973

Diagnosis. Aulolaimoididae (Figs 14.9O and P). Nematodes of small to medium size, 0.8–1.3 mm long. Cuticle dorylaimoid. Lip region continuous, rounded or somewhat truncated, with amalgamated lips. Amphidial fovea supported by a basket-like structure. Longitudinal ribs surrounding the cheilostom. Odontostyle small, shorter than lip region width, but with distinct lumen. Odontophore longer than odontostyle, flanged. Pharynx with the three usual sections of the family. Female genital system mono–opisthodelphic. Vulva transversely oval. Male with slender dorylaimid spicules and no supplement. Tail similar in sexes: subcylindroid.

Remarks: Goseco *et al.* (1975) provided the most recent revision of the genus, including descriptions and key to species. It seems to be an aquatic taxon because its two species have been recorded in freshwater habitats:

A. ahyperoche Goseco, Ferris & Ferris, 1975 – USA (sediment/Harrison-Crawford State Forest/Indiana (Goseco *et al.*, 1975))

A. orthus (Thorne, 1939) Goseco, Ferris & Ferris, 1975 – New Guinea (brook/Mt. Wilhelm (Andrássy, 1973b; as *A. dadayi*)); USA (several habitats and localities (Goseco *et al.*, 1975))

Genus ***Aulolaimoides*** Micoletzky, 1915

Diagnosis. Aulolaimoididae (Figs 14.9M and N). Nematodes of small to medium size, 0.6–1.9 mm long. Cuticle dorylaimoid. Lip region continuous, rounded or slightly truncated, with amalgamated lips. Amphidial fovea supported by rib-like elements. Cheilostom a truncate cone, surrounded by a basket-like structure with longitudinal rib-like elements. Odontostyle attenuated, relatively long, with narrow but distinct lumen and small aperture. Odontophore with basal flanges. Pharynx consisting of the three usual sections of the family. Female genital system mono–opisthodelphic. Vulva transverse or oval. Male with slender dorylaimid spicules, 1–2 widely spaced ventromedian supplements, and gubernaculum. Tail similar in sexes: long and filiform.

Remarks: Goseco *et al.* (1975) provided the most recent revision of the genus, including descriptions and key to species. Four of the six species hitherto classified under *Aulolaimoides* have been collected in freshwater habitats:

A. altherri Andrássy, 1968 – Paraguay (Acaray River (Andrássy, 1968b))
A. elegans Micoletzky, 1915 (type species) – Europe (Andrássy, 1978b; Loof, 1999); Romania (pool/Bukowina (Micoletzky, 1915)); USA (sediment/Indian Creek and James Lake/Indiana (Goseco *et al.*, 1975))
A. lutarius Goseco, Ferris & Ferris, 1975 – USA (sediment/Tippecanoe River/Indiana (Goseco *et al.*, 1975))
A. phoxodorus Andrássy, 1964 – Hungary (Andrássy, 1964; Loof, 1999)

Suborder NYGOLAIMINA Ahmad & Jairajpuri, 1979

Diagnosis. Dorylaimida. Nematodes of small to large size. Lip region continuous or offset, usually with amalgamated lips. Cheilostom scarcely differentiated, occasionally strongly sclerotized and chamber-like. Feeding sheath provided with a mural tooth of variable morphology, but always situated on sub-ventral wall of pharynx. Pharynx consisting of a slender but muscular anterior portion and a basal expansion occupying about half of total neck length. Pharyngo-intestinal junction almost always provided with three large cardiac cells. Female genital system didelphic, rarely mono–opisthodelphic; genital tract poorly differentiated in general. Vagina typically with two sections, almost always lacking the *pars refringens.* Vulva transverse, less frequently longitudinal. Males with two testes, dorylaimid spicules, with or without gubernaculum and usually few ventromedian supplements. Tail variable in shape, always similar in sexes.
Only one superfamily:

Superfamily NYGOLAIMOIDEA Thorne, 1935

Diagnosis. Nygolaimina. Nematodes of small to large size, usually more than 1 mm long. Cuticle smooth or with fine transverse striations. Lip region in general scarcely differentiated, with amalgamated lips. Stoma armed with a mural tooth on sub-ventral wall of pharynx. Anterior slender portion of the pharynx extending gradually into the basal expansion, which occupies usually about half of total neck length. Pharyngo-intestinal junction provided with three large cardiac cells. Female genital system didelphic, rarely mono–opisthodelphic, with the genital tract poorly developed. Vagina usually lacking *pars refringens.* Vulva transverse or exceptionally longitudinal. Males with two testes, well-developed spicules, lateral guiding pieces, with or without gubernaculum and few ventromedian supplements. Tail variable in shape and size, similar in sexes.

Four families, all of them including freshwater species:

Aetholaimidae Jairajpuri, 1965
Nygellidae Andrássy, 1958
Nygolaimellidae Clark, 1961
Nygolaimidae Thorne, 1935 (type family)

Family AETHOLAIMIDAE Jairajpuri, 1965

Diagnosis. Nygolaimoidea. Nematodes of medium size, under 2 mm long. Lip region somewhat discoid and more or less offset. Cheilostom strongly sclerotized, consisting

of two, labial and post-labial, portions. Mural tooth deltoid to linear. Pharynx consisting of the usual two portions, but basal expansion surrounded by a delicate sheath, sometimes forming basal pockets. Pharyngo-intestinal junction with three well-developed cardiac cells. Female genital system didelphic. Vagina with three sections. Vulva transverse. Female tail rounded conoid to hemispheroid. Male unknown.
Only one subfamily.

Subfamily Aetholaiminae Jairajpuri, 1965
Diagnosis. Aetholaimidae. Nematodes of medium size, under 2 mm long. Lip region discoid and offset, with amalgamated lips. Cheilostom strongly sclerotized, with two, labial and post-labial, sections. Mural tooth relatively small, deltoid. Pharyngo-intestinal junction with small ovoid cardiac cells. Female genital system didelphic. Vagina with three sections, *pars refringens* present. Vulva transverse. Female tail rounded conoid to hemispheroid. Male unknown.
Only one genus, including one species collected in freshwater habitats:

Genus ***Aetholaimus*** Williams, 1962
Diagnosis. Aetholaiminae. Nematodes of medium size, usually under 2 mm long. Lip region somewhat discoid, offset by depression or constriction, with amalgamated lips. Cheilostom strongly sclerotized and consisting of two parts: (i) a wide bowl-shaped anterior (labial) portion with complex inner differentiations; and (ii) a posterior conical portion with six elongate ribs. Slender portion of the pharynx extending gradually into the basal expansion, which occupies about three-fifths of total neck length. Pharyngo-intestinal junction with three cardiac cells. Female genital system didelphic–amphidelphic. Vagina lacking *pars refringens.* Vulva transverse. Female tail rounded conoid. Male unknown.
Remarks: Coomans and Loof (1978) provided an excellent study on morphology and systematics of this genus, including descriptions and key to species. Only one species, among five, has been collected in freshwater habitats:

A. trochus Swart & Heyns, 1994 – Namibia (pond sediment/Mudumu Game Park/ Caprivi (Swart and Heyns, 1994; Heyns, 2002))

Family NYGELLIDAE Andrássy, 1958
Diagnosis. Nygolaimoidea. Nematodes of medium size, under 2 mm long. Cuticle thin and with fine transverse striations. Lip region continuous, with amalgamated lips. Mural tooth deltoid to linear. Cheilostom simple. Expanded basal portion of the pharynx surrounded by a conspicuous spiral sheath. Pharyngo-intestinal junction with three cardiac cells. Female genital system mono–opisthodelphic. Vagina with two sections, lacking the *pars refringens.* Vulva transverse. Female tail clavate. Male unknown.
Only one subfamily, including freshwater species:

Subfamily Nygellinae Andrássy, 1958
Diagnosis. Nygellidae. Nematodes of medium size, 1.0–1.5 mm long. Habitus after fixation practically straight. Cuticle with fine transverse striations. Lip region continuous with the adjacent body, with amalgamated lips. Mural tooth deltoid to linear,

and small, shorter than lip region width. Pharyngeal expansion surrounded by a distinct spiral sheath. Pharyngo-intestinal junction provided with three small ovoid cardiac cells. Female genital system mono–opisthodelphic. Vulva transverse. Female tail elongate and clavate at its posteriormost portion. Male unknown.
Only one genus, including aquatic species:

Genus ***Nygellus*** Thorne, 1939
Diagnosis. Nygellinae (Figs 14.10A and E). Medium-sized nematodes, 1.0–1.4 mm long. Habitus after fixation straight, sometimes slightly ventrad curved. Lip region rounded and continuous with the adjacent body, with closely amalgamated lips. Mural tooth deltoid to linear, shorter than lip region width, 5–7 μm long. Pharyngeal expansion surrounded by a distinct spiral muscular sheath. Pharyngo-intestinal junction with three small ovoid cells. Female genital system mono–opisthodelphic, with pre-vulval sac more or less developed. Vagina with two sections, lacking the *pars refringens.* Vulva transverse. Female tail elongate and clavate. Male unknown.
Remarks: Heyns (1968) provided the most recent revision of the genus, including descriptions and key to species. At least two species, among its four, have been collected in aquatic habitats:

N. subclavatus Timm & Ameen, 1960 – India (around water plants/Cuttack/ Orissa State (Ahmad and Jairajpuri, 1982))
N. zingli Dhanam, Jairajpuri & Sreedharan, 2002 – India (rice field/Muthinakoppa/ Karnataka State (Dhanam *et al.*, 2002))

Family NYGOLAIMELLIDAE Clark, 1961
Diagnosis. Nygolaimoidea. Nematodes of large size, more than 2 mm long. Cuticle moderately thick and with distinct lateral pores. Lip region continuous or offset, often wider than adjacent body. Cheilostom simple, sometimes with a rasp-like area. Mural tooth deltoid or dorylaimoid, located on left sub-ventral wall of pharynx. Pharyngeal basal expansion longer than half of total neck length, sometimes bibulbar and surrounded by a more or less (in general poorly) developed sheath. Pharyngo-intestinal junction provided or not with a cardiac disc, and no cardiac cells. Female genital system didelphic. Vagina with two sections, lacking the *pars refringens.* Vulva transverse. Male with dorylaimid spicules and developed ventromedian supplements, but without gubernaculum. Tail similar in sexes, short and conoid.
Only one subfamily, including freshwater species:

Subfamily Nygolaimellinae Clark, 1961
Diagnosis. Nygolaimellidae. Nematodes of large size, more than 2.5 mm long. Cuticle thick, with distinct lateral pores. Lip region distinctly offset and wider than adjacent body. Cheilostom simple, with or without rasp-like area. Mural tooth deltoid or dorylaimoid. Pharyngeal basal expansion simple or bibulbar, always occupying about two-thirds of total neck length. Pharyngo-intestinal junction with or without cardiac disc, always lacking cardiac cells. Female genital system didelphic. Vulva transverse. Male with well-developed dorylaimid spicules and ventromedian supplements. Tail similar in both sexes, short and conoid.

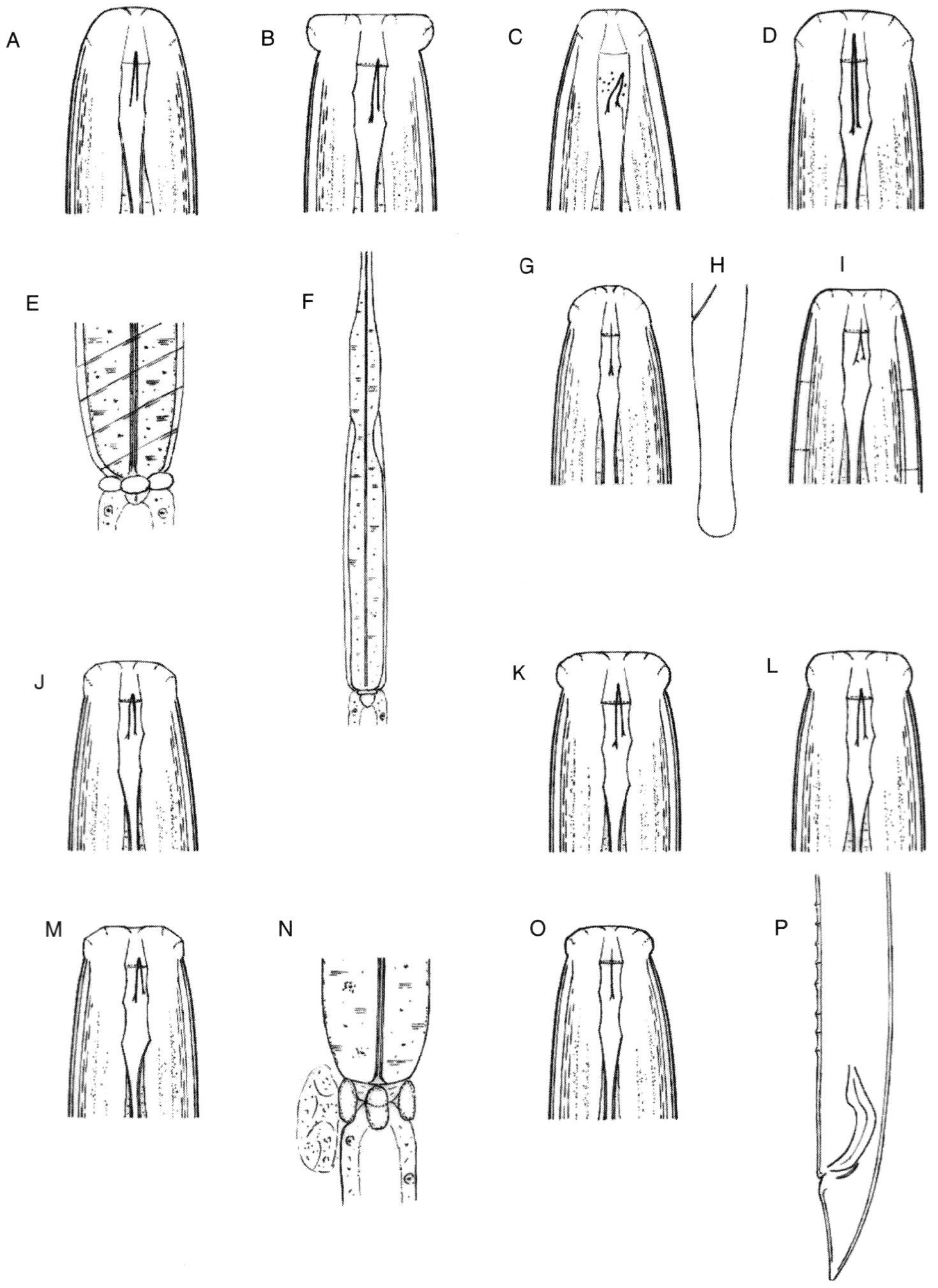

Fig. 14.10 Freshwater nygolaimid genera (Nygolaimina, Nygolaimoidea). A, E: Family Nygellidae, subfamily Nygellinae, genus *Nygellus*; B, C, F: Family Nygolaimellidae, subfamily Nygolaimellinae; B, F: Genus *Nygolaimellus*; C: Genus *Nygolaimium*; D, G–P: Family Nygolaimidae, subfamily Nygolaiminae; D: Genus *Aquatides*; G, H: Genus *Clavicauda*; I: Genus *Feroxides*; J: Genus *Laevides*; K: Genus *Nygolaimus*; L, P: Genus *Paranygolaimus*; M, N: Genus *Paravulvus*; and O: Genus *Solididens*.

Four genera, at least two of them (in bold) including freshwater species:

Aporcelaimoides Heyns, 1965
Nygolaimellus Loos, 1949 (type genus)
Nygolaimium Thorne, 1930
Scapidens Heyns, 1965

Genus ***Nygolaimellus*** Loos, 1949

Diagnosis. Nygolaimellinae (Figs 14.10B and F). Nematodes of large size, 2.5–7.0 mm long. Cuticle notably thick, especially at level of tail, with fine transverse striations, sometimes with indications of criss-cross lines overlying transverse striations. Lip region offset by constriction, hexagonal in *en face* view, and wider than adjacent body; lips fairly distinct. Mural tooth deltoid, of variable length. Pharyngeal basal expansion occupying about two-thirds of total neck length and consisting of two more dilatate sections with a narrower intermediate one, the intermediate and the posterior dilatate section surrounded by a weak sheath. Cardiac disc present at pharyngo-intestinal junction. Female genital system didelphic. Vagina with two portions, lacking *pars refringens*. Vulva transverse. Male with dorylaimid spicules, five to eight spaced ventromedian supplements, and without gubernaculum. Tail similar in sexes, convex conoid to rounded.

Remarks: Heyns (1968) provided the most recent revision of the genus, including descriptions and a key to species. Two of its six species have been recorded from aquatic habitats:

N. abnormis Loos, 1949 (type species) – Sri Lanka (stream bank/Kurunegala (Loos, 1949))

N. macmacus Heyns, 1968 – South Africa (stream bank/Mac Mac Pools/Eastern Transvaal (Heyns, 1968))

Genus ***Nygolaimium*** Thorne, 1930

Diagnosis. Nygolaimellinae (Fig. 14.10C). Nematodes of large size, more than 2 mm long. Cuticle thick. Lip region continuous or offset by constriction, with amalgamated lips. Cheilostom with a rasp-like area of minute denticles. Mural tooth deltoid but more or less arcuate, and of variable length. Pharyngeal basal expansion occupying about half of total neck length. Pharyngo-intestinal junction without cardiac cells. Female genital system didelphic. Vulva transverse. Female tail short and rounded. Male unknown.

Remarks: Heyns (1968) provided the most recent revision of the genus, including descriptions and a key to species. One of its two species was collected in freshwater habitats:

N. menzeli (Micoletzky, 1925) Heyns, 1968 – Surinam (waterfall/Brownsberg (Micoletzky, 1925b))

Family NYGOLAIMIDAE Thorne, 1935

Diagnosis. Nygolaimoidea. Nematodes of medium to large size, rarely under 1 mm long. Cuticle thin, with fine transverse striations. Lip region practically continuous with the adjacent body or offset by constriction or depression, with lips mostly amalgamated. Cheilostom simple. Mural tooth variable in shape and size. Pharyngeal expansion occupying half or more than half of total neck length, sometimes enveloped by a weak sheath. Pharyngo-intestinal junction provided with three large cardiac cells. Female genital system didelphic. Vagina usually lacking *pars refringens*. Vulva transverse, exceptionally longitudinal. Males with dorylaimid spicules, lateral guiding pieces, with or without gubernaculum and usually few and weakly developed ventromedian supplements. Tail variable in shape and size, but similar in sexes.

Only one subfamily, including freshwater species:

Subfamily Nygolaiminae Thorne, 1935

Diagnosis. Nygolaimidae. Nematodes of small to large size, 0.7–7.2 mm long. Habitus after fixation variable: practically straight to distinctly ventrad curved. Cuticle with fine transverse striations. Lip region continuous or offset, with amalgamated lips. Mural tooth of variable shape: dorylaimoid, deltoid, linear, solididentoid or acicular. Three large rounded or ovoid cells present at pharyngo-intestinal junction. Female genital system didelphic. Vagina with two sections, *pars refringens* only poorly developed in *Paravulvus*. Vulva transverse, longitudinal only in *Paravulvus*. Male with dorylaimid spicules, lateral guiding pieces, with or without gubernaculum and few ventromedian supplements. Tail short, similar in sexes, but variable in shape: short and rounded to conical or clavate.

Ten genera, practically all of them (in bold) including aquatic species:

Afronygus Heyns, 1968
Aquatides Heyns, 1968
Clavicauda Heyns, 1968
Clavicaudoides Heyns, 1968
Feroxides Heyns, 1968
Laevides Heyns, 1968
Nygolaimus Cobb, 1913 (type genus)
Paranygolaimus Heyns, 1968
Paravulvus Heyns, 1968
Solididens Heyns, 1968

Key to genera

1. Cardiac cells longitudinal, i.e. longer than wide (Fig. 14.10N); a sub-dorsal distinct cell mass is present near the cardia (Fig. 14.10N); vagina with the usual three sections, with more or less developed *pars refringens*; vulva longitudinal . . . .*Paravulvus*
   Cardiac cells rounded or transverse, i.e. wider than long; only exceptionally sub-dorsal cell mass present near the cardia; vagina with two sections, lacking *pars refringens*; vulva transverse 2
2. Mural tooth solidentoid to solid or acicular (needle-like) (Figs 14.10G and I) 3
   Mural tooth dorylaimoid, deltoid or linear 5
3. Large-sized nematodes, more than 3 mm long; cuticle specially thick (Fig. 14.10I); cardiac cells lacking at the pharyngo-intestinal junction; female tail short and rounded . . . .*Feroxides*

Small to medium-sized nematodes, usually under 2.5 mm long; cuticle of usual thickness; cardiac cells present at the pharyngo-intestinal junction 4

4. Sub-dorsal cell mass present near the cardia; female tail relatively long, cylindrical with clavate terminal portion (Fig. 14.10H) . . . . *Clavicauda*
   No sub-dorsal cell mass is present near the cardia; female tail shorter, conoid to hemispheroid . . . . *Solididens*
5. Mural tooth dorylaimoid, i.e. with distinct lumen and a dorsal anterior oblique plane resembling the aperture of a dorylaimid odontostyle (Fig. 14.10J) . . . . *Laevides*
   Mural tooth with other morphology: deltoid to linear, with narrow lumen and lacking an anterior oblique plane .6
6. Tail long, strongly ventrad curved . . . . *Afronygus*
   Tail short, conoid to hemispherical, sometimes clavate 7
7. Lip region offset by more or less deep constriction (Figs 14.10K and L) 8
   Lip region practically continuous with the adjacent body (Fig. 14.10D) 9
8. Tail conical with rounded tip; male with gubernaculum and a series of 6–7 well-developed ventromedian supplements (Fig. 14.10P) . . . . *Paranygolaimus*
   Tail rounded to conoid; male lacking gubernaculum; 0–2 poorly developed ventromedian supplements . . . . *Nygolaimus*
9. Body practically straight after fixation; mural tooth linear (Fig. 14.10D); tail convex conoid to hemispherical, never clavate; mostly freshwater species . . . . *Aquatides*
   Body ventrad curved after fixation; mural tooth deltoid to linear; tail rounded to hemispheroid, clavate; mostly edaphic species . . . . . *Clavicaudoides*

Freshwater genera and species:

Genus ***Aquatides*** Heyns, 1968

Diagnosis. Nygolaiminae (Fig. 14.10D). Nematodes of medium to large size, 1–4 mm long. Habitus almost straight after fixation. Lip region rounded or truncate, slightly offset by depression. Mural tooth linear, usually longer than lip region width. Pharyngeal expansion surrounded by a conspicuous sheath, often forming basal pockets. Cardiac cells well developed, spheroid to ovoid. Female genital system didelphic. Vulva with only two sections, lacking *pars refringens*. Vulva transverse. Male with massive arcuate spicules, well-developed gubernaculum and series of ventromedian supplements. Tail similar in sexes: convex conoid to hemispherical.

Remarks: Heyns (1968) published the most recent revision of the genus, including species descriptions, and Ahmad and Jairajpuri (1982) provided the most updated available key to species identification. *Aquatides* is practically an aquatic genus as at least eight of its ten species have been recorded from freshwater habitats:

A. aquaticus (Thorne, 1930) Thorne, 1974 (type species) – Europe (Andrássy, 1978b; Loof, 1999); Hungary (pool/Martonvásár (Andrássy, 1973a)); Russia (several habitats and localities (Gagarin, 1992)); Spain (Santa Olalla lagoon/Doñana National Park (Jiménez-Guirado, 1989)); USA (several habitats and localities (Thorne, 1930, 1974; Heyns, 1968))

A. intermedius (De Man, 1880) Ahmad & Jairajpuri, 1982 – Hungary (lake/Kenese (Daday, 1897); unknown locality (Andrássy, 1997); irrigation canal/Fertö-Hanság National Park (Andrássy, 2002)); Spain (peat bog/Pontevedra (Gadea, 1955)); The Netherlands (Loof, 1999)

A. kaburaki (Imamura, 1931) *species inquirendae* – Japan (rice field/unknown locality Tokyo (Imamura, 1931))

A. minutus Dhanam, Jairajpuri & Sreedharan, 2002 – India (rice field/Muthinakoppa/Karnataka State (Dhanam *et al.*, 2002))

A. rotundicaudatus Thorne, 1974 – USA (swamp/Brookings/South Dakota (Thorne, 1974))

A. shadini (Filipjev, 1928) Ahmad & Jairajpuri, 1982 – Russia (Oka River/Kljazjima/Baliff district and Santshur/Melenki district (Filipjev, 1928; Andrássy, 1978b))

A. smoliki Thorne, 1974 – USA (swamp/Brookings/South Dakota (Thorne, 1974))

A. thornei (Schneider, 1937) Ahmad & Jairajpuri, 1982 – India (rice field/Magdah Campus/Bihar State (Ahmad and Jairajpuri, 1982)); South Africa (Sabie and Olifants Rivers/Kruger National Park (Botha and Heyns, 1992)); Sumatra Island (several habitats/Toba Lake (Schneider, 1937))

Genus ***Clavicauda*** Heyns, 1968

Diagnosis. Nygolaiminae (Figs 14.10G and H). Nematodes of medium size, under 1.5 mm long. Habitus practically straight after fixation. Lip region rounded, offset. Mural tooth acicular, needle-like, furcate at the base and shorter than lip region width. Cardiac cells ovoid. A peculiar cell mass is present near the pharynx base. Female genital system didelphic. Vagina with two sections, lacking *pars refringens*. Vulva transverse. Female tail relatively long, cylindrical and with clavate terminal portion. Male unknown.

Remarks: One of the two species hitherto classified under this genus has been collected in aquatic habitats:

C. differentialis Dhanam, Jairajpuri & Sreedharan, 2002 – India (around aquatic grasses/Sringeri/Karnataka State (Dhanam *et al.*, 2002))

Genus ***Clavicaudoides*** Heyns, 1968

Diagnosis. Nygolaiminae. Nematodes of small to medium size, usually under 2 mm long. Cuticle provided with distinct transverse striations. Lip region rounded, continuous or offset by depression, sometimes asymmetrical. Mural tooth deltoid to linear. Pharyngeal expansion surrounded by a thin sheath. Cardiac cells ovoid to rounded. Female genital system didelphic. Vagina with two sections, lacking *pars refringens*. Vulva transverse. Males with small spicules and few rudimentary supplements, without gubernaculum. Rectum especially long, more than anal body diameter. Tail similar in sexes, clavate to hemispheroid.

Remarks: Ahmad and Jairajpuri (1982) provided the most recent available key to species identification. Three of these species, among eight, have been collected in freshwater habitats:

Genus ***Nygolaimus*** Cobb, 1913

Diagnosis. Nygolaiminae (Fig. 14.10K). Nematodes of medium to large size, 1–4 mm long. Outer cuticle with fine transverse striations. Lip region rounded, offset by more or less marked constriction. Mural tooth deltoid, variable in size: usually shorter than, more rarely equal to, lip region width. Pharyngeal expansion occupying half or more than half of total neck length, in general surrounded by a delicate (occasionally more distinct) sheath. Cardiac cells well developed, rounded to ovoid. Female genital system amphidelphic. Vagina with two sections, lacking *pars refringens*. Vulva transverse. Male with dorylaimid spicules, few weakly developed ventromedian supplements (even lacking) and without gubernaculum. Tail rounded to conoid, similar in sexes.

Remarks: Heyns (1968) published the most recent revision of the genus, including species descriptions, and Ahmad and Jairajpuri (1982) provided the most updated available key to species identification. At least five species, among about 30, have been recorded from freshwater habitats:

N. amphigonicus Thorne, 1930 – France (peat bog/Bavois (Altherr, 1954))
N. brachyuris (De Man, 1880) Thorne, 1930 – New Guinea (marsh/Lemien (Daday, 1901)); Spain (unknown locality in southern Iberian peninsula (Peña-Santiago *et al.*, 1999)); The Netherlands (unknown locality (Heyns, 1968; Loof, 1999))
N. dorotheae Heyns, 1968 – South Africa (spring/Transvaal (Heyns, 1968))
N. gracillimus Andrássy, 1968 – Congo Republic (Congo river (Andrássy, 1968a))
N. papillicaudatus Heyns, 1968 – USA (mud/Mirror lake/California (Heyns, 1968))

Genus ***Paranygolaimus*** Heyns, 1968

Diagnosis. Nygolaiminae (Figs 14.10L and P). Nematodes of large size, in general more than 2 mm long. Cuticle smooth. Lip region offset by constriction, with distinct lips. Mural tooth deltoid, about two-thirds the lip region width long. Pharyngeal expansion occupying more than half of total neck length. Cardiac cells well developed. Female genital system didelphic. Vagina with two sections, lacking *pars refringens*. Vulva transverse. Male with dorylaimid spicules, 7–8 well-developed ventromedian supplements and weak gubernaculum. Tail similar in sexes: rounded to conical with rounded terminus.

Remarks: One of the two species of the genus has been collected in aquatic habitats:

P. confusus Khan, Jairajpuri & Ahmad, 1989 – India (rice field/Raanti/Bihar State (Khan *et al.*, 1989))

Genus ***Paravulvus*** Heyns, 1968

Diagnosis. Nygolaiminae (Figs 14.10M and N). Small to medium-sized nematodes, in general under 2 mm long. Habitus after fixation ventrad curved. Cuticle with fine transverse striations. Lip region continuous or offset by depression. Mural tooth deltoid in general, sometimes solididentoid. Pharyngeal expansion with peculiarly large and distinct gland nuclei. Cardiac cells well developed, longer than wide. A distinct sub-dorsal cell mass is present near the cardia. Female genital system didelphic. Vagina with the usual three sections, but *pars refringens* more or less developed. Vulva longitudinal. 'Paravulvae' are often present near the vulva. Male with dorylaimid spicules, well-developed ventromedian supplements and slender gubernaculum. Tail similar in sexes but variable in shape, short and rounded to conical and ventrad curved.

Remarks: Heyns (1968) published the most recent revision of the genus, including species descriptions. Lazarova *et al.* (2002) provided an updated list of species and a key to their identification. *Paravulvus* is a terrestrial genus, but at least three species, among 12, have been recorded from freshwater habitats:

P. curvistilis (Stefanski, 1924) *species inquirendae* – Poland (stream/Olcyska Valley/Zakopane (Stefanski, 1924; Loof, 1999)

P. hartingii (de Man, 1880) Thorne, 1974 – Europe (Andrássy, 1978b; Loof, 1999); Hungary (several habitats/Fertö-Hanság National Park (Andrássy, 2002)); Nepal (several habitats and localities (Andrássy, 1978a)); Russia (several habitats and localities (Gagarin, 1992))

P. teres (Thorne, 1930) Thorne, 1974 – Hungary (Andrássy, 1978b)

Genus ***Solididens*** Heyns, 1968

Diagnosis. Nygolaiminae (Fig. 14.10O). Nematodes of medium to large size, usually 1–3 mm long. Habitus after fixation ventrad curved, sometimes twisted. Lip region mostly offset by constriction. Mural tooth solid, needle-like but occasionally bifurcated at the base and shorter than lip region width. Cardiac cells small, ovoid. Female genital system didelphic. Vagina with two sections, lacking *pars refringens*. Male with dorylaimid spicules, one or two weakly developed ventromedian supplements and without gubernaculum. Tail similar in sexes, conoid to hemispheroid.

Remarks: Heyns (1968) published the most recent revision of the genus, including species descriptions, and Ahmad and Jairajpuri (1982) provided the most recent available key to species identification. At least three *Solididens* species, among seven, have been collected from freshwater habitats:

S. bisexualis (Thorne, 1939) Thorne, 1974 – ? Germany (Andrássy, 1978b)

S. trichodorus (Andrássy, 1965) Ahmad & Jairajpuri, 1982 – Ghana (marsh/Winneba (Andrássy, 1965))

S. vulgaris (Thorne, 1930) Thorne, 1974 – India (rice field/Chuari Khas/Himachal Pradesh State (Ahmad and Jairajpuri, 1982))

Acknowledgements

The author thanks the editorial comments by Prof. Dr. I. Andrássy (Budapest, Hungary) and the assistance of Dr. P. Guerrero with literature written in German.

References

Ahmad, M. and Jairajpuri, M.S. (1978) *Athernema* n. gen. (Athernematinae n. subfam.), a unique genus of Leptonchidae (Nematoda: Dorylaimida). *Nematologica* 24, 445–448.

Ahmad, M. and Jairajpuri, M.S. (1982) Nygolaimina of India. *Records of the Zoological Survey of India Occasional Paper* No 34, 70 pp.

Ahmad, W. and Jairajpuri, M.S. (1983a) Three new species of Tylencholaimoidea (Nematoda: Dorylaimida) from India. *Nematologica* 29, 367–374.

Ahmad, W. and Jairajpuri, M.S. (1983b) Descriptions of new species of *Dorylaimoides* and *Calolaimus* (Dorylaimida) from India. *Revue de Nématologie* 6, 65–72.

Ahmad, W., Ahmad, A. and Shaheen, A. (2002) Observations on *Discomyctus longicaudatus* (Imamura, 1931) Thorne, 1939 (Nematoda) with first report of its male. *Annales Zoologici* 52, 347–352.

Altherr, E. (1954) Les nématodes des étangs de Bavois (Vaud). *Bulletin de la Société Vaudoise des Sciences Naturelles* 66, 3–46.

Altherr, E. (1963) Contribution à la connaissance de la faune des sables submergés en Lorraine. Nématodes. *Annales de Spéléologie* 18, 53–98.

Altherr, E. (1968) Nématodes de la nappe phréatique du réseau fluvial de la Saale (Thuringe) et psammniques du Lac Stechlin (Brandebourg du nord). *Limnologica* 6, 247–320.

Altherr, E. (1974) Nématodes de la nappe phréatique du réseau fluvial de la Saale (Thuringe). II. *Limnologica* 9, 81–132.

Altherr, E. (1976) Nématodes des eaux stygorhitrales des Alpe autrichiennes. *Revue Suisse de Zoologie* 83, 779–847.

Altherr, E. (1977) Contribution à la connaissance des Nématodes de l'estuaire de l'Amazone (2me partie). *Amazoniana* 6, 145–159.

Andrássy, I. (1959) Freilebende Nematoden aus Rumänien. *Annales Universitatis Scientiarum Budapestinensis de Rolando Eötvös nominatae Sectio Biologica* 2, 3–27.

Andrássy, I. (1962) Zwei neue Nematoden-Arten aus dem Überschwemmungsgebiet der Donau (Danubiala Hungarica XIII). *Opuscula Zoologica Budapestinensis* 4, 3–8.

Andrássy, I. (1963) The zoological results of Gy. Topál's collectings in South Argentina. *Annales Historico-Naturales Musei Nationalis Hungarici* 55, 243–273.

Andrássy, I. (1964) Neue Nematoden-Arten aus Ungarn, III. Fünf neue Arten. *Opuscula Zoologica Budapestinensis* 5, 9–23.

Andrássy, I. (1965) Erd- und Süsswasser-Nematoden aus Ghana. Klasse Adenophorea (Aphasmidia). *Opuscula Zoologica Budapestinensis* 5, 127–151.

Andrássy, I. (1968a) The scientific results of the Hungarian soil zoological expedition to the Brazzaville-Congo. 31. Nematoden aus Grundwasser. *Annales Universitatis Scientiarum Budapestinensis de Rolando Eötvös nominatae Sectio Biologica* 9–10, 3–26.

Andrássy, I. (1968b) Fauna Paraguayensis 2. Nematoden aus den Galeriewaldern des Acaray-Flusses. *Opuscula Zoologica Budapestinensis* 8, 167–315.

Andrássy, I. (1970a) Einige neue Nematodenarten aus Westafrikanischen Reisfeldern. *Annales Universitatis Scientiarum Budapestinensis Rolando Eötvös Sectio Biologica* 12, 243–254.

Andrássy, I. (1970b) Nematoden aus einigen Fluss-Systemen Südafrikas. *Opuscula Zoologica Budapestinensis* 10, 179–219.

Andrássy, I. (1973a) 100 neue Nematodenarten in der ungarischen Fauna. *Opuscula Zoologica Budapestinensis* 9, 7–48.

Andrássy, I. (1973b) Ein Meeresrelikt und einige andere bemerkenswerte Nematodenarten aus Neuguinea. *Opuscula Zoologica Budapestinensis* 12, 3–19.

Andrássy, I. (1978a) Fresh-water nematodes from the Himalayas. *Opuscula Zoologica Budapestinensis* 15, 3–21.

Andrássy, I. (1978b) Nematoda In: Illies, J. (ed.) *Limnofauna Europea.* Gustav Fischer Verlag, Stuttgart, Germany, pp. 81–117.

Andrássy, I. (1991) The free-living nematode fauna of the Bátorliget Nature Reserves. In: *The Bátorliget Nature Reserves – After Forty Years.* Académiai Kiadó, Budapest, pp. 127–197.

Andrássy, I. (1995) Tropical nematodes of rare genera (Dorylaimida). *Opuscula Zoologica Budapestinensis* 27–28, 5–24.

Andrássy, I. (1997) Nematodes in the Hévizi lake (Hungary). *Állattani Közlemények* 82, 13–27.

Andrássy, I. (1998) Once more: the oesophageal gland nuclei in the dorylaimoid nematodes. *Opuscula Zoologica Budapestinensis* 31, 165–170.

Andrássy, I. (2001) Some species of curious genera of the class Penetrantia (Nematoda). *International Journal of Nematology* 11, 43–57.

Andrássy, I. (2002) Free-living nematodes from the Fertö-Hanság National Park, Hungary. In: *The Fauna of the Fertö-Hanság National Park*. Académiai Kiadó, Budapest, pp. 21–97.

Baqri, Q.H. and Khera, S. (1979) Nematodes from West Bengal (India). IV. Three known and two new species of the genus *Dorylaimoides* Thorne & Swanger, 1936 (Leptonchidae: Dorylaimida). *Records of the Zoological Survey of India* 75, 247–254.

Botha, A. and Heyns, J. (1992) Species of *Tyleptus*, *Proleptonchus*, *Aquatides* and *Afractinolaimus* from rivers in the Kruger National Park (Nematoda: Dorylaimida). *Koedoe* 35, 43–54.

Botha, A. and Heyns, J. (1993) Species of the genera *Oxydirus*, *Dorylaimellus* (*Axodorylaimellus*), *Laimydorus* and *Rhabdolaimus* from rivers in the Kruger National Park (Nematoda: Dorylaimida and Araeolaimida). *Koedoe* 36, 49–60.

Clark, W.C. (1964) *Falcihasta palustris* n. gen. et sp., family Belondiridae (Nematoda). *New Zealand Journal of Sciences* 7, 177–180.

Cobb, M.V. (1915) Some freshwater nematodes of the Douglas Lake region of Michigan, USA *Transactions of the American Microbiological Society* 34, 21–47.

Cobb, N.A. (1920) One hundred new nemas. (Type species of 100 new genera). *Contributions to Science Nematology* 9, 217–243.

Colomba, G. and Vinciguerra, M.T. (1979) Nematodi d'acqua dolce della Sicilia. I. Nematodi dell'Anapo. *Animalia* 6, 89–120.

Coomans, A. (1964) Stoma structure in members of the Dorylaimina. *Nematologica* 9, 587–601.

Coomans, A. (1965) Structure of the female gonads in members of the Dorylaimina. *Nematologica* 10, 587–601.

Coomans, A. (1979) The anterior sensilla of nematodes. *Revue de Nématologie* 2, 259–283.

Coomans, A. and Loof, P.A.A. (1978) Observation on the subfamily Aetholaiminae Jairajpuri, 1965 (Nygolaimidae: Nematoda). *Proceedings of the Helminthological Society of Washington* 45, 82–92.

Coomans, A. and Nair, P. (1975) The genus *Axonchium* (Nematoda: Belondiridae). VI. Atypical species, keys, subgenera and conclusions. *Nematologica* 21, 296–332.

Daday, E. von. (1897) Die freilebenden Süsswasser-Nematoden Ungarns. *Zoologischen Jahrbüchern* 10, 91–134.

Daday, E. von. (1901) Mikroskopische Süsswasserthiere aus Deutsch-Neu-Guinea. *Természetrajzi Füzetek* 24, 1–16.

De Bruin, S. and Heyns, J. (1991) A review of the nematode genus *Xiphinemella* Loos, 1950. *Entomology Memoir Department of Agricultural Development Republic of South Africa* 83, 1–48.

De Ley, P. and Blaxter, M. (2002) Systematic position and phylogeny. In: Lee, D.L. (ed.) *The Biology of Nematodes*. Taylor & Francis, London, pp. 1–30 (635 pp.).

De Ley, P., Loof, P.A.A. and Coomans, A. (1998) Terrestrial nematodes from the Galápagos Archipelago. II. Redescription of *Aporcelaimellus obtusicaudatus* (Bastian, 1865) Altherr, 1968, with review of similar species and a nomenclature for the vagina in Dorylaimida (Nematoda). *Bulletin de l'Institut royal des sciences naturelles de Belgique* 63, 13–34.

de Man, J.G. (1876) Onderzoekingen over vrij in de aarde levende Nematoden. *Tijdschrift Nederlandsche dierkundige Vereeniging* 2, 78–196.

de Man, J.G. (1880) Die einheimischen, frei in der reinen Erde und im süssen Wasser lebenden Nematoden. *Tijdschrift Nederlandsche dierkundige Vereeniging* 5, 1–104.

de Man, J.G. (1921) Nouvelles récherches sur les nematodes libres terricoles de la Hollande. *Capita Zoologica* 1, 3–62.

Dhanachand, C. and Jairajpuri, M.S. (1980) Four new and one known species of Dorylaimida from Manipur, India. *Indian Journal of Nematology* 10, 152–163.

Dhanam, M. and Jairajpuri, M.S. (1998) New belondirid nematodes: two new genera, two new subgenera and seven new species from Malnad tracts of Karnataka, India. *International Journal of Nematology* 8, 205–209.

Dhanam, M. and Jairajpuri, M.S. (1999) New leptonchid nematodes: one new genus and eleven new species from Malnad Tracts of Karnataka, India. *International Journal of Nematology* 9, 1–18.

Dhanam, M., Jairajpuri, M.S. and Sreedharan, K. (2002) Four new nygolaim nematodes (Dorylaimida) from Malnad Tracts of Karnataka, India. *International Journal of Nematology* 12, 13–18.

Ferris, V.R. and Ferris, J.M. (1973) *Lindseyus costatus* gen. n., sp. n. and notes on the Roqueidae and Swangeriidae (Nematoda: Dorylaimida). *Proceedings of the Helminthological Society of Washington* 48, 163–171.

Ferris, V.R., Goseco, C.G. and Ferris, J.M. (1980) Revisions of *Oxydirus* and *Tarjanius* n. gen. in Oxydiridae, Belondiroidea (Nematoda: Dorylaimida); and *Oxydiroides* in Prodorylaimidae, Dorylaimoidea. *Purdue University Agriculture Experimental Station Research Bulletin.* No. 965, 1–29.

Ferris, V.R., Ferris, J.M. and Goseco, C.G. (1983) Revision of *Belondira* and notes on *Oxybelondira* in Belondiridae, Belondiroidea (Nematoda: Dorylaimida). *Purdue University Agriculture Experimental Station Research Bulletin* 979, 1–46.

Filipjev, I.N. (1928) Nématodes libres du fleuve Oka. *Rabot. Okskoi Biol. Stantsii* 5, 81–112.

Gadea, E. (1952) Sobre algunos nematodos libres de agua dulce de la Plana de Castellón. *Publicaciones del Instituto de Biología Aplicada* 11, 173–186.

Gadea, E. (1954) Sobre algunos nematodos muscícolas de la Sanabria. *Publicaciones del Instituto de Biología Aplicada* 17, 51–63.

Gadea, E. (1955) Nematodos dulceacuícolas de Galicia. *Publicaciones del Instituto de Biología Aplicada* 20, 77–114.

Gadea, E. (1956) Nematodos libres de los Pirineos Centrales. *Actes du Deuxième Congrés International d'Études Pyrénéennes Tome* 3, 124–146.

Gagarin, V.G. (1992) *The Free-living Freshwater Nematodes of the Soviet Union.* Hidrometeoizdat, Sankt Peterburg, Russia, 152 pp. (in Russian.).

Goseco, C.G., Ferris, V.R. and Ferris, J.M. (1974a) Revisions in Leptonchoidea (Nematoda, Dorylaimida). *Leptonchus*, *Proleptonchus*, *Funaria*, and *Meylis* n. gen. in Leptonchidae, Leptonchinae. *Purdue University Agriculture Experimental Station Research Bulletin.* No. 911, 32 pp.

Goseco, C.G., Ferris, V.R. and Ferris, J.M. (1974b) Revisions in Leptonchoidea (Nematoda: Dorylaimida). *Tyleptus* in Leptonchidae, Tyleptinae; *Basirotyleptus* in Leptonchidae, Belonenchinae; and *Loncharionema* n. gen. in Leptonchidae, Xiphinemellinae. *Purdue University Agriculture Experimental Station Research Bulletin.* No. 913, 24 pp.

Goseco, C.G., Ferris, V.R. and Ferris, J.M. (1975) Revisions in Leptonchoidea (Nematoda: Dorylaimida). *Tylencholaimellus*, *Doryllium*, *Gerthus* n. gen. and *Agmodorus* in Tylencholaimellidae; and *Aulolaimoides* and *Adenolaimus* in Aulolaimoididae. *Purdue University Agriculture Experimental Station Research Bulletin.* No. 928, 40 pp.

Goseco, C.G., Ferris, V.R. and Ferris, J.M. (1976) Revisions in Leptonchoidea (Nematoda: Dorylaimida). *Dorylaimoides* in Dorylaimoididae, Dorylaimoidinae; *Calolaimus* and *Timmus* n. gen. in Dorylaimoididae, Calolaiminae; and *Miranema* in Miranematidae. *Purdue University Agriculture Experimental Station Research Bulletin.* No. 941, 46 pp.

Hechler, H.C. (1969) Variability of the basal oesophageal bulb sheath in belondirid nematodes. *Journal of Nematology* 1, 160–165.

Heyns, J. (1968) A monographic study of the nematode families Nygolaimidae and Nygolaimellidae. *Entomology Memoirs Plant Protection Research Institute Pretoria South Africa,* No. 10, 51 pp.

Heyns, J. (2002) Checklist of free-living nematodes recorded from freshwater habitats in Southern Africa. *Water SA* 28, 449–456.

Imamura, S. (1931) Nematodes in the paddy field, with notes on their population before and after irrigation. *Journal of the College of Agriculture of the Imperial University of Tokyo* 11, 193–240.

Jairajpuri, M.S. and Ahmad, M. (1979) Some remarks on the genus *Oxydirus* Thorne, 1939 with proposal of *Paraoxydirus* n. gen. *Indian Journal of Nematology* 8, 146–148.

Jairajpuri, M.S. and Ahmad, W. (1992) *Dorylaimida. Free-living, Predaceous and Plant-parasitic Nematodes.* Oxford and IBH Publishing, New Delhi, 458 pp.

Jairajpuri, M.S., Ahmad, W. and Sturhon, D. (1998) *Lrofilaimus*, a remarkable new genus of Dorylaimida from Iran. *Nematologica* 44, 15–20.

Jana, A. and Baqri, Q.H. (1981) Nematodes from West Bengal (India). XI. Studies on the species of the superfamily Leptonchoidea (Dorylaimida). *Journal of the zoological Society of India* 33, 1–24.

Jiménez-Guirado, D. (1989) Nematodos acuáticos del Parque Nacional de Doñana. *Oxyura* 5, 83–91.

Jiménez-Guirado, D. and Arcos, S.C. (1994) A new species of *Proleptonchus* Lordello, 1955 (Nematoda: Leptonchidae) from Spain, with a cladistic study of the genus. *Nematologica* 40, 328–339.

Jiménez-Guirado, D. and Murillo-Navarro, R. (2004) Observations on *Lindseyus heterurus* (Schuurmans Stekhoven & Teunissen, 1938) Coomans & Kheiri, 1986 (Nematoda: Belondiroidea) from Spain. *Journal of Nematode Morphology and Systematics* 7, 91–96.

Khan, T.H., Jairajpuri, M.S. and Ahmad, W. (1989) Descriptions of some new and known species of dorylaim nematodes. *Nematologica* 35, 419–437.

Lazarova, S., Peneva, V. and Loof, P.A.A. (2002) *Paravulvus rhodopensis* sp. n. (Nematoda: Nygolaimidae) and three known species from Bulgaria, with notes on the taxonomy of the genus. *Nematology* 4, 505–520.

Loof, P.A.A. (1990) Systematic observations on some species of Dorylaimoididae. *Revue de Nématologie* 13, 161–170.

Loof, P.A.A. (1999) *Süsswasserfauna von Mitteleuropa 4/2–2. Nematoda, Adenophorea (Dorylaimida).* Spektrum Akademischer Verlag, Heidelberg, Germany, 249 pp.

Loof, P.A.A. and Coomans, A. (1970) On the development and location of the oesophageal gland nuclei in Dorylaimina. *Proceedings of the IX International Nematology Symposium, Warsaw, Poland,* 1967, pp. 79–161.

Loof, P.A.A. and Jairajpuri, M.S. (1968) Taxonomic studies on the genus *Tylencholaimus* De Man, 1876 (Dorylaimoidea) with a key to the species. *Nematologica* 14, 317–350.

Loof, P.A.A. and Zullini, A. (2000) Free-living nematodes from nature reserves in Costa Rica. 1. Dorylaimina. *Nematology* 2, 605–633.

Loos, C.A. (1949) Notes on free-living and plant-parasitic nematodes of Ceylon. 6. *Journal of the Zoological Society of India* 1, 30–36.

Meyl, H. (1954) *Nygolaimus husmanni* n. sp., ein neuer Nematode aus dem Grundwasser Nordwestdeutschland, sowie Bemerkungen über die bisher in Europa gefundenen Arten der Gattung *Nygolaimus* Cobb, 1913. *Zoologischer Anzeiger* 152, 127–133.

Micoletzky, H. (1915) Neue Süsswasser-Nematoden aus der Bukowina. *Mitteilungen des Naturwissenschaftlichen Vereines für Steiermark (Graz)* 445–454.

Micoletzky, H. (1925a) Die freilebenden Süsswasser- und Moornematoden Dänemarks. *Mémoires de l'Académie Royale des Sciences et des Lettres de Danemark, Copenhague* 10, 57–310.

Micoletzky, H. (1925b) Zur Kenntnis tropischer, freilebender Nematoden aus Surinam, Trinidad and Ostafrika. *Zoologischer Anzeiger* 64, 1–28.

Mulk, M.M., Coomans, A. and Baqri, Q.H. (1978) A taxonomic revision of the nematode species described by S. Stekhoven and Teunissen (1938) from National Virunga Park, Zaire Republic. III. Belondiridae. *Revue de Zoologie Africaine* 92, 711–722.

Ocaña, A., Peña-Santiago, R. and Morales, R. (1986) Distribución de los nematodos del orden Dorylaimida en determinados manantiales minero-medicinales de la provincia de Granada. *Limnética* 2, 189–195.

Peña-Santiago, R. (1991) Nematodes of the order Dorylaimida from Andalucía Oriental, Spain. The genus *Funaria* van der Linde, 1938, with description of two new species. *Nematologia Mediterranea* 19, 9–18.

Peña-Santiago, R. (2002) A new diagnosis of the genus *Chitwoodius* Furstenberg & Heyns, 1966 (Nematoda: Dorylaimida), with comments and a compendium of its species. *Nematology* 4, 441–443.

Peña-Santiago, R. and Coomans, A. (1996) Revision of the genus *Tylencholaimus* De Man, 1876. General discussion and key to the species. *Nematologica* 42, 440–454.

Peña-Santiago, R. and Peralta, M. (1997a) The genus *Dorylaimoides* Thorne & Swanger, 1936 (Nematoda: Dorylaimida). 1. Taxonomy and variability. *Fundamental and Applied Nematology* 20, 243–251.

Peña-Santiago, R. and Peralta, M. (1997b) The genus *Dorylaimoides* Thorne & Swanger, 1936 (Nematoda: Dorylaimida). 2. A compendium and key to the species. *Fundamental and Applied Nematology* 20, 253–259.

Peña-Santiago, R. and Peralta, M. (1999a) The genus *Tylencholaimellus* Cobb in M.V. Cobb, 1915 (Nematoda: Dorylaimida): general morphology, taxonomy, compendium, and polytomous key to the species. *Nematology* 1, 183–194.

Peña-Santiago, R. and Peralta, M. (1999b) La distribución de especies de doriláimidos (Nematoda: Dorylaimida) en Andalucía Oriental. I. Especies de las familias Dorylaimoididae, Tylencholaimellidae, Leptonchidae y Aulolaimoididae en relación con comunidades vegetales. In: Román-Rayo, M. (Coord.) *Educar Enseñando.* Servicio de Publicaciones, Universidad de Jaén, Jaén, Spain, pp. 305–312.

Peña-Santiago, R., Liébanas, G. and Abolafia, J. (1999) La distribución de especies de doriláimidos (Nematoda: Dorylaimida) en Andalucía Oriental. II. Especies de Dorylaimidae y *Nygolaimus* en relación con comunidades vegetales. In: Román-Rayo, M. (Coord.) *Educar Enseñando.* Servicio de Publicaciones, Universidad de Jaén, Jaén, Spain, pp. 313–318.

Schneider, W. (1937) CXIII. Freilebende Nematoden der Deutsche Limnologischen Sundaexpedition nach Sumatra, Java und Bali. *Archiv für Hydrobiologie* 15, 30–108.

Siddiqi, M.R. (1965) Seven new species of Dorylaimoidea (Nematoda) from India, with descriptions of *Lenonchium* n. gen. and *Galophinema* n. gen. *Proceedings of the Helminthological Society of Washington* 32, 81–90.

Siddiqi, M.R. (1982a) Seven new genera of dorylaimid nematodes from Colombian rain forest. *Systematic Parasitology* 4, 69–87.

Siddiqi, M.R. (1982b) Six new genera of dorylaimid nematodes. *Nematologica* 27, 397–421.

Siddiqi, M.R. (1983) Taxonomy of the subfamily Dorylaimellinae (Nematoda: Dorylaimida). *Pakistan Journal of Nematology* 1, 1–38.

Siddiqi, M.R. (1995) Nematodes of tropical rainforests. 5. Seven new genera and forty-two new species of dorylaims. *Afro-Asian Journal of Nematology* 5, 72–109.

Stefanski, W. (1924) Étude sur les nématodes muscicoles des environs de Zakopane (Massif du Tatra polonais). *Bulletin de l'Academie polonaise des Sciences et Lettres, Série Biologie Sciences Naturelles,* year 1923, 21–60.

Swart, A. and Heyns, J. (1994) Description of *Aetholaimus trochus* n. sp. and the male of *Ironus ignavus* Bastian, 1865 from Caprivi, Namibia. *South African Journal of Zoology* 29, 54–61.

Thorne, G. (1930) Predaceous nemas of the genus *Nygolaimus* and a new genus *Sectonema. Journal of Agricultural Research, U.S.D.A.* 41, 445–466.

Thorne, G. (1939) A monograph of the nematodes of the superfamily Dorylaimoidea. *Capita Zoologica* 8, 1–261.

Thorne, G. (1964) Nematodes of Puerto Rico: Belondiroidea, new superfamily, Leptonchidae Thorne, 1935, and Belonenchidae new family (Nematoda: Adenophorea, Dorylaimida). *University of Puerto Rico Agriculture Experimental Station Technical Paper* No. 39, 1–51.

Thorne, G. (1974) Nematodes of the Northern Great Plains. Part II. Dorylaimoidea in part (Nematoda: Adenophorea). *Technical Bulletin Agriculture Experimental Station South Dakota State University.* No. 41, 1–120.

Timm, R.W. (1967) Some estuarine nematodes from the Sunderbans. *Proceedings of the Pakistanian Academy of Sciences* 4, 1–13.

Wu, J. and Liang, Y. (1999) A comparative study of benthic nematode in two Chinese lakes with contrasting sources of primary production. *Hydrobiologia* 411, 31–37.

Yeates, G.W. (1970) Four Dorylaimina (Nematoda) from Wicken Fen, Cambridgshire. *Nematologica* 16, 273–283.

Yeates, G.W. (1972) The oesophago-intestinal junction in the Dorylaimoidea (Nematoda). *Journal of Natural History* 6, 343–355.

Zullini, A. (1975a) Nematodi dello psammon del Po. *Istituto Lombardo* (*Rend. Sci.*) B 109, 79–83.

Zullini, A. (1975b) Nematodi ripicoli del Po. *Istituto Lombardo* (*Rend. Sci.*) B 109, 130–142.

Zullini, A. (1977) Some freshwater nematodes of southern Mexico and Guatemala. *Quaderni dei Academia Nazionale dei Lincei* 171, 75–85.

Zullini, A. (1982) 17. Nematodi (Nematoda). In: Ruffo, S. (Coord.) *Guide per il riconoscimento delle specie animali delle acque interne italiane.* Consiglio Nazionale delle Richerche, Rome, 177 pp.

15 Dorylaimida Part II: Superfamily Dorylaimoidea

M.T. Vinciguerra

Dipartimento di Biologia Animale, Università di Catania, Via Androne 81, 95124 Catania, Italy

Superfamily DORYLAIMOIDEA de Man, 1876

Diagnosis. Dorylaimina. Cuticle generally smooth or finely transversely striated, sometimes marked with longitudinal ridges. Lip region offset or continuous with adjoining body; lips distinct or fused, bearing two circles of labial papillae. Amphidial fovea stirrup-shaped with slit-like aperture. Odontostyle axial of varying size and shape; odontophore generally rod-like, rarely provided with basal knobs (flanged). Guiding ring appearing single or double. Pharynx generally in two parts: a slender anterior and an enlarged posterior part; their relative length varying in the different taxa. A pre-rectum is always present. Female reproductive apparatus didelphic or monodelphic. Vulva transverse, longitudinal or pore-like. Male reproductive apparatus always diorchic. Spicules generally arcuate, provided with central thickening and a pair of guiding pieces. Before the cloacal aperture a pair of ventral ad-cloacal papillae is always present and, before it, a row of ventromedian supplements, varying in number and arrangement. Tail varying in shape and length, similar or different in the two sexes. The members of Dorylaimoidea are often inhabitants of terrestrial habitats but many species, and in some cases even larger taxa, can live in freshwater.
Additional references: Jairajpuri and Ahmad (1992); Thorne, 1939; Thorne and Swanger, 1936.
Note: Freshwater species of genera reported from both freshwater and terrestrial habitats are marked with an asterisk; the term abw denotes anal body widths.

Families:

DORYLAIMIDAE de Man, 1876
ACTINOLAIMIDAE Thorne, 1939
QUDSIANEMATIDAE Jairajpuri, 1965
THORNIIDAE De Coninck, 1965
THORNENEMATIDAE Siddiqi, 1969
APORCELAIMIDAE Heyns, 1965
NORDIIDAE Jairajpuri & A.H. Siddiqi, 1964
LONGIDORIDAE Thorne, 1935

CRATERONEMATIDAE Siddiqi, 1969
LORDELLONEMATIDAE Siddiqi, 1969

Key to families of DORYLAIMOIDEA

1. Cheilostome armed with four massive onchia encircling the odontostyle . . . . ACTINOLAIMIDAE
   Cheilostome without onchia 2
2. Presence of numerous body pores leading to cuticularized ducts . . . . LORDELLONEMATINAE
   Body pores, if present, indistinct 3
3. Mouth cavity cyathiform, odontostyle guiding sheath anteriorly sclerotized . . . . CRATERONEMATIDAE
   Mouth cavity tubular; odontostyle guiding sheath not sclerotized 4
4. Large size; odontostyle extremely long and thin with fine lumen and aperture . . . . LONGIDORIDAE
   Odontostyle not so long and thin or, if thin, small size 5
5. Lip region low, not offset; the two circles of labial papillae close to each other; gubernaculum present . . . . THORNIIDAE
   Lip region prominent; the two circles of labial papillae far apart; gubernaculum absent 6
6. Labial framework and mouth cavity with sclerotized pieces . . . . THORNENEMATIDAE
   Lip region without sclerotizations 7
7. Odontostyle slender, more or less long, with narrow lumen and aperture . . . . NORDIIDAE
   Odontostyle not very long, with wide lumen and aperture 8
8. Female tail elongated to filiform; male tail generally short, rarely as in females . . . . DORYLAIMIDAE
   Tail similar in both sexes, short conoid or rounded 9
9. Odontostyle aperture longer than half its length . . . . . . APORCELAIMIDAE
   Odontostyle aperture not longer than half its length . . . . QUDSIANEMATIDAE

Family CRATERONEMATIDAE Siddiqi, 1969

Diagnosis. Dorylaimoidea. Cuticle thin. Lip region off set. Lips distinct. Mouth cavity cyathiform. The guiding sheath around the distal end of odontostyle is sclerotized. Odontostyle thin, with small aperture. Pharynx enlarging near middle. Female gonads paired; vulva pore-like; vagina lacking sclerotizations. Male supplements few, spaced; spicula lacking central thickening piece. Tail elongate conoid in both sexes. The species belonging to this family are typical inhabitants of terrestrial habitats.

Family LORDELLONEMATIDAE Siddiqi, 1969

Diagnosis. Dorylaimoidea. Cuticle rather loose. Body pores conspicuous leading to prominent cuticularized ducts. Lip region offset by constriction. Odontostyle with wide lumen and aperture. Odontophore rod-like. Guiding ring single. Female reproductive apparatus amphidelphic or monodelphic. Males rare. Male supplements low, well spaced. Spicula with or without median piece. Tail short conoid to elongate conoid, similar in both sexes. The species belonging to this family are typical inhabitants of terrestrial habitats.

Family THORNENEMATIDAE Siddiqi, 1969

Diagnosis. Dorylaimoidea. Cuticle smooth. Labial framework and mouth cavity with sclerotized pieces. Odontostyle dorylaimoid; simple or double. Expanded part of pharynx about one-third its length. Second pair of sub-ventral glands and nuclei located far anterior to pharynx base. Female reproductive apparatus amphidelphic or mono–opisthodelphic. Tail similar or different in both sexes. The family includes many genera and species, which are typical inhabitants of terrestrial habitats, even though a few species have been occasionally found in wet habitats.

Family LONGIDORIDAE Thorne, 1935

Diagnosis. Dorylaimoidea. Body slender and large-sized (1–12 mm long). Lip region continuous or offset. Amphids with large fovea and slit- or pore-like aperture. Odontostyle exceedingly long and thin, with fine lumen and aperture. Odontophore rod-like or with basal flanges. Guiding ring simple or double. Anterior part of pharynx tubular, non-muscular; the muscular, expanded posterior part about one-fourth of pharynx length. Female reproductive apparatus didelphic or monodelphic. Spicula dorylaimoid. Tail generally short, conoid or rounded, similar in both sexes. The family includes many genera and species, which are typical inhabitants of terrestrial habitats, in association with plant roots, and virus vectors. Occasionally a few of the numerous known species have been found in wet habitats.

Family DORYLAIMIDAE de Man, 1876 (Andrássy, 1988; Jairajpuri & Ahmad, 1992)

Diagnosis. Dorylaimoidea. Small to large nematodes. Cuticle smooth or finely transversely striated, in a few cases marked by longitudinal ridges. Lip region off set or continuous with the adjoining body. Odontostyle straight or slightly sinuate, with rather wide lumen; its aperture about one-third its length; guiding ring simple or double. Pharynx muscular, expanded posteriorly for about half its length. Pre-rectum distinct. Female reproductive apparatus didelphic; vulva transverse or longitudinal; *pars refringens* of vagina sclerotized. Male reproductive apparatus diorchic. Spicules generally dorylaimoid, with central thickening and lateral guiding pieces, rarely simple (alaimoid) lacking central thickening and lateral pieces; a pair of ad-cloacal papillae and a row of ventromedian supplements present. Tail generally with sexual dimorphism: attenuated or filiform in female; short and rounded in male; in few cases long and filiform in both sexes. Almost all the genera of this family include freshwater or semi-freshwater species.

Subfamilies :
Dorylaiminae de Man, 1876
Prodorylaiminae Andrássy, 1969
Laimydorinae Andrássy, 1969
Afrodorylaiminae Andrássy, 1969
Amphidorylaiminae Andrássy, 1976 (no freshwater species)

Genus *incertae sedis* within the family:
Kunjudorylaimus Dhanam & Jairajpuri, 2000 (no freshwater species)

Key to subfamilies of DORYLAIMIDAE

1. Tail long, filiform in both sexes 2
   Tail almost always with sexual dimorphism 3
2. Spicules simple, alaimoid . . . . Amphidorylaiminae
   Spicules dorylaimoid . . . . Prodorylaiminae
3. Cuticle with longitudinal ridges . . . . Dorylaiminae
   Cuticle without longitudinal ridges 4
4. Pre-rectum very long; sub-medial papillae on male posterior body very small . . . . Laimydorinae
   Pre-rectum short; sub-medial papillae on male posterior body unusually strong . . . . Afrodorylaiminae

Subfamily Prodorylaiminae Andrássy, 1969

Diagnosis. Dorylaimidae. Large nematodes. Cuticle without longitudinal ridges. Labial papillae 6 + 10. Odontostyle guiding ring simple. Spicules dorylaimoid. Supplements numerous. Tail long and filiform in both sexes.

Additional reference: Andrássy (1988).

Genera:

Prodorylaimus Andrássy, 1959

Prodorylaimium Andrássy, 1969

Protodorylaimus Andrássy, 1988

Key to genera of Prodorylaiminae

1. Pre-rectum extremely long in both sexes (more than 15 anal body widths); body very slender (a to 80) . . . . *Protodorylaimus*
   Pre-rectum short in both sexes; body not so slender 2
2. Supplements not numerous, spaced . . . . *Prodorylaimium*
   Supplements numerous, contiguous . . . . *Prodorylaimus*

Genus *Protodorylaimus* Andrássy, 1988 (Figs 15.1A–C)

Diagnosis. Prodorylaiminae. Medium–large-sized nematodes. Body exceedingly slender (a = 64–81). Lip region continuous. Odontostyle short and weak; guiding ring single. Pre-rectum in both sexes extremely long (15–25 anal body widths). Female gonads paired. Vulva longitudinal. Supplements numerous, contiguous. Tail long; filiform in both sexes. One of the two described species has been found in a freshwater habitat.

Type species: *P. dalmassoi* (Loof, 1985) Andrássy, 1988 (France, Spain)*

Other species:

P. kazakhstanicus (Sagitov, 1973) Andrássy, 1988 (Kazakhstan)

Additional reference: Andrássy (1988).

Genus *Prodorylaimus* Andrássy, 1959 (Figs 15.1D–F)

Diagnosis. Prodorylaiminae. Cuticle smooth or finely transversely striated. Lip region moderately off set; lips more or less distinct. Odontostyle straight; guiding ring simple, distant from anterior end at least one lip region width. Pharynx enlarging near its middle; pre-rectum rather short (1–4 abw in females and 4–8 abw in males). Female gonads paired; vulva longitudinal or transverse; vagina with sclerotized

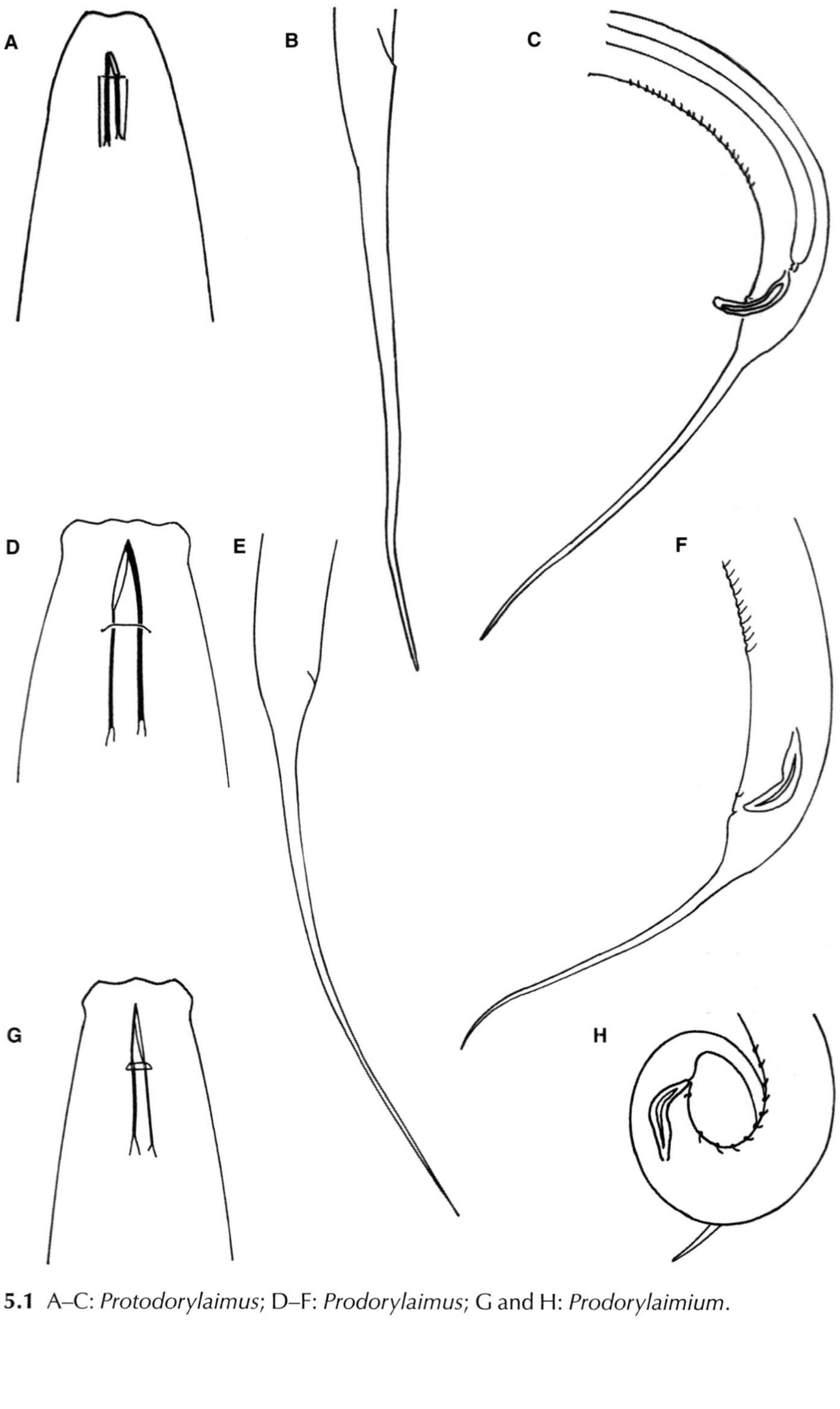

Fig. 15.1 A–C: *Protodorylaimus*; D–F: *Prodorylaimus*; G and H: *Prodorylaimium*.

pieces. Supplements numerous (13–31), contiguous. Tail elongate to filiform in both sexes. The species of this genus are frequently found in freshwater habitats.

Type species: *P. longicaudatoides* Altherr, 1968 (Germany, The Netherlands, Hungary, Spain, Russia, Kenya)

Other species:
P. andrassyi (Zullini, 1973) Loof, 1985 (Mexico)
P. brasiliensis (Meyl, 1956) Andrássy, 1959 (Brazil)
P. cantabricus Peña-Santiago, Abolafia & Jiménez-Guirado, 1998 (Spain)
P. dahli (Altherr, 1960) Andrássy, 1964 (Cameroon)
P. depressus Loof, 1973 (Surinam)
P. dolichurus (Loos, 1946) Siddiqi, 1969 (Sri Lanka)
P. donghagens Choi & Khan, 1999
P. eliavai Tsalolikhin, 1977 (Baikal Lake)
P. ensis Kleynhans, 1970 (South Africa)
P. filiarum Andrássy, 1964 (Hungary, Austria, Spain, UK, Russia, Ukraine, The Netherlands)
P. fukushimaensis Khan & Araki 2002 (India)
P. hamatus Loof, 1973 (Surinam)
P. jihuai Ahmad & Ahmad, 2001 (India)
P. kukuy Tsalolikhin, 1977 (Baikal Lake)
P. magnus Alekseev & Dolgova, 1993
P. mas Loof, 1985 (The Netherlands, Switzerland, Germany, Austria, Italy, Hungary)
P. obesus Ahmad & Jairajpuri, 1982 (India)
P. paralongicaudatus (Micoletzky, 1925) Andrássy, 1959 (Kenya, Congo Republic, Tanzania, Angola, Lithuania, Hungary)
P. paraobesus Ahmad & Shaheen, 2004
P. ranzii (Zullini, 1973) Loof, 1985 (Mexico)
P. rionensis (Gerlach, 1954) Andrássy, 1959 (Kenya, Brazil, South Africa)
P. rotundiceps Loof, 1985 (The Netherlands, Switzerland, UK, Hungary)
P. sturhani Ahmad & Ahmad, 2001 (India)
P. taebekensis Choi & Khan, 1999
P. uliginosus Loof, 1985 (The Netherlands, Switzerland, Austria, Germany, Italy)
Additional reference: Loof (1985, 2000) and Andrássy (1988).

Genus *Prodorylaimium* Andrássy, 1969 (Figs 15.1G and H)
Diagnosis. Prodorylaiminae. Cuticle smooth or finely transversely striated. Lip region moderately off set; lips more or less distinct. Odontostyle straight or slightly sinuate; guiding ring simple, distant from anterior end less than one lip region width long. Pharynx enlarging near its middle; pre-rectum 2–3 anal body widths long in females, originating at level of the anterior supplements in males. Female gonads paired; vagina with sclerotized pieces. Supplements few (6–10), spaced. Tail elongate to filiform in both sexes. The species of this genus are frequently found in freshwater habitats.

Type species: *P. brigdammense* (de Man, 1876) Andrássy, 1969 (Europe and Asia, Surinam)

Other species:
P. alpinum Andrássy, 1978 (Nepal, Korea)
P. goaense Ahmad & Jairajpuri, 1985 (India)
P. stenosoma (de Man, 1876) Andrássy, 1973 (The Netherlands)
Additional reference: Andrássy (1988).

Subfamily Dorylaiminae de Man, 1876
Diagnosis. Dorylaimidae. Generally large nematodes. Cuticle marked by longitudinal ridges. Labial papillae 6 + 10. Odontostyle robust, straight; guiding ring double. Vagina with sclerotizations. Supplements numerous, arranged in a single series or in two fascicles. Spicules dorylaimoid. Tail elongate to filiform in females, short and rounded in males. Mainly freshwater species.
Additional reference: Andrássy (1988).
Genera:
Dorylaimus Dujardin, 1845
Ischiodorylaimus Andrássy, 1969

Key to genera of Dorylaiminae
1. Supplements arranged in a continuous series . . . . *Dorylaimus*
   Supplements arranged in fascicles . . . . *Ischiodorylaimus*

Genus *Dorylaimus* Dujardin, 1845 (Figs 15.2A–C)
Diagnosis. Dorylaiminae. Cuticle thick, marked with longitudinal ridges. Lip region moderately offset; lips distinct. Odontostyle straight, robust, 2–3 times as long as lip region width; guiding ring double. Pre-rectum rather long, beginning well before the supplement series in males. Supplements numerous and contiguous. The representatives of this genus are typical freshwater nematodes, inhabiting water bodies and also wet moss and soil.

Type species: *D. stagnalis* Dujardin, 1845 (cosmopolitan)

Other species:
D. afghanicus Andrássy, 1960 (Afghanistan, Nepal, Singapore, Egypt, South Africa, Hungary)
D. alaeus Thorne, 1939 (USA, China)
D. asymphidorus Andrássy, 1969 (Hungary, Italy, Spain, Sahara, South Africa)
D. baylii Nicholas & Hodda, 2000
D. carinatus Thorne & Swanger, 1936 (USA)
D. conicus Andrássy, 1981 (Nepal)
D. crassus de Man, 1884 (Japan, China, Taiwan, USA, Chile, widespread in Europe)
D. deaconi Botha & Heyns, 1991 (South Africa)
D. fodori Andrássy, 1988 (India)
D. geraerti Baqri & Jana, 1986 (India, Ethiopia)
D. gigas Kleynhans, 1970 (South Africa)
D. helveticus Steiner, 1919 (Russia, Canada, USA, Uzbekistan, widespread in Europe)
D. innovatus Jana & Baqri, 1982 (India)
D. lineatus Altherr & Delamare-Deboutteville, 1972 (Canada, USA)
D. macroproctus Altherr, 1963 (Switzerland, Hungary)

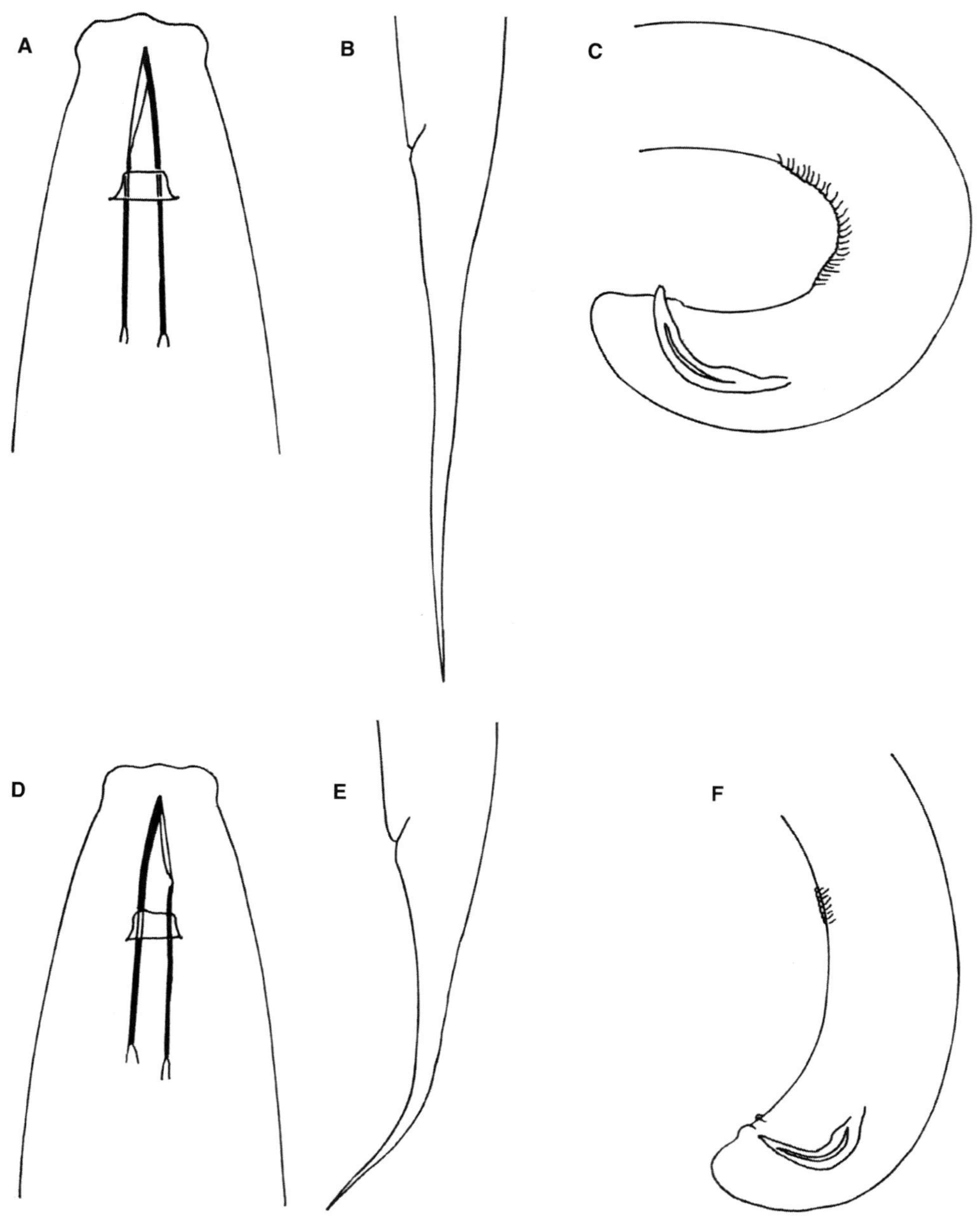

Fig. 15.2 A–C: *Dorylaimus;* D–F: *Ischiodorylaimus.*

D. macrosoma Jiménez-Guirado, 1988 (Spain)
D. montanus Stefanski, 1923 (Switzerland, Poland, Russia, Ukraine, Mongolia, USA)
D. numidicus Andrássy, 1988 (Algeria)
D. pachys Andrássy, 1970 (South Africa)
D. popus Gagarin, 1981 (Tadzhikistan)
D. siddiqii Ahmad & Jairajpuri, 1982 (India)
D. specialis Andrássy, 2003 (Alaska)

D. stekhoveni Baqri & Coomans, 1973 (Congo Republic)
D. stenus Andrássy, 1970 (South Africa)
D. stephani Andrássy, 1969 (Kenya)
D. tepidus Andrássy, 1959 (Switzerland, Hungary, Italy, Moldavia, Uzbekistan, Kenya)
D. unicus Andrássy, 1970 (South Africa)

Additional reference: Andrássy (1988).

Genus *Ischiodorylaimus* Andrássy, 1969 (Figs 15–2D–F)
Diagnosis. Dorylaiminae. Very large nematodes. Cuticle marked by longitudinal ridges. Lip region slightly offset; lips not well distinct. Odontostyle straight, 2–3 times as long as lip region width; guiding ring double. Pre-rectum long; in males beginning far before the supplements. Vulva longitudinal. Male supplements arranged in two fascicles between which there is a small row of separate supplements.
The species of this genus are typical inhabitants of freshwater habitats mostly found in Africa, but present also in Europe and Asia.

Type species: *I. gulliver* (Andrássy, 1964) Andrássy, 1969 (Kenya, South Africa)

Other species:
I. bathypyla Andrássy, 1970 (South Africa)
I. cognatus Andrássy, 1983 (Hungary)
I. loeffleri (Andrássy, 1964) Andrássy, 1969 (Kenya)
I. minimus Khan, Jairajpuri & Ahmad, 1989 (India)
I. novus Baqri & Jana, 1986 (India)
I. paraugandanus Khan & Ahmad, 1994 (India)
I. robustus Zullini, 1974 (Italy)
I. tessares Kleynhans, 1970 (South Africa)
I. ugandanus Andrássy Banage in Andrássy, 1969 (Uganda, Ethiopia)

Additional reference: Andrássy (1988).

Subfamily Laimydorinae Andrássy, 1969
Diagnosis. Dorylaimidae. Cuticle generally smooth, sometimes with annulated subcuticle without longitudinal ridges, (with longitudinal ridges in a single species). Labial papillae 6 + 10. Odontostyle straight; guiding ring simple or double. Spicules dorylaimoid. Supplements contiguous or spaced. Tail attenuated to filiform in female and short and rounded or conoid-rounded in males. Almost all the genera of Laimydorinae include truly freshwater species.
Additional reference: Andrássy (1988).

Genera:
Laimydorus Siddiqi, 1969
Calodorylaimus Andrássy, 1969
Calcaridorylaimus Andrássy, 1986 (no freshwater species)
Crocodorylaimus Andrássy, 1988
Fuscheila Siddiqi, 1982 (no freshwater species)
Halodorylaimus Andrássy, 1988 (no freshwater species)
Idiodorylaimus Andrássy, 1969

Mesodorylaimus Andrássy, 1959
Chrysodorus Jiménez-Guirado & Cadenas, 1985
Miodorylaimus Andrássy, 1986 (no freshwater species)
Namaquanema Heyns & Swart, 1995
Baladorylaimus Andrássy, 2001

Key to genera of Laimydorinae

1. Cuticle distinctly annulated . . . . . . *Idiodorylaimus*
   Cuticle smooth or finely striated 2
2. Mouth cavity with sclerotized rods . . . . . . *Fuscheila*
   Mouth cavity without sclerotized rods 3
3. Amphidial opening very small; haline species . . . . . . *Halodorylaimus*
   Amphidial opening normal 4
4. Male tail conoid, with nipped tip . . . . . . *Baladorylaimus*
   Male tail bluntly rounded 5
5. Male pre-rectum very long, beginning far before the supplements 6
   Male pre-rectum shorter, beginning at level of supplements 11
6. Body small, near 1 mm, guiding ring simple . . . . . . *Miodorylaimus*
   Body longer than 2 mm, guiding ring double 7
7. Conspicuous yellowish spots around the base of odontostyle; male tail conoid–rounded, ventrally bent . . . . . . *Crocodorylaimus*
   No yellowish spots at the odontostyle base; male tail short, bluntly rounded, straight 8
8. Odontostyle with hooked basal processes . . . . . . *Namaquanema*
   Odontostyle without hooked basal processes 9
9. Supplements in two groups with some intermediate elements . . . . . . *Calodorylaimus*
   Supplements contiguous in a continuous row 10
10. Body very slender (a = 60–92), odontostyle weak and slender . . . . . . *Chrysodorus*
    Body not so slender; odontostyle robust . . . . . . *Laimydorus*
11. Spicules with small spurs near the distal tip . . . . . . *Calcaridorylaimus*
    Spicules without spurs near the distal tip . . . . . . *Mesodorylaimus*

Genus *Laimydorus* Siddiqi, 1969 (Figs 15.3A–C)

Diagnosis. Laimydorinae. Large nematodes. Cuticle smooth. Lip region low, not or slightly off set; lips hardly separate. Odontostyle straight, not longer than twice the lip region width; guiding ring double. Pharynx expanding in or behind its middle. Pre-rectum generally very long, in males beginning before the supplements row. Supplements very numerous, contiguous. Female tail attenuated to filiform; male tail short and rounded. The species of this genus are typical inhabitants of freshwater habitats.

Type species: *L. prolificus* (Thorne & Swanger, 1936) Siddiqi, 1969 (USA)

Other species:
L. afer (Andrássy, 1964) Andrássy, 1986 (Kenya)
L. africanus Botha & Heyns, 1993 (South Africa)

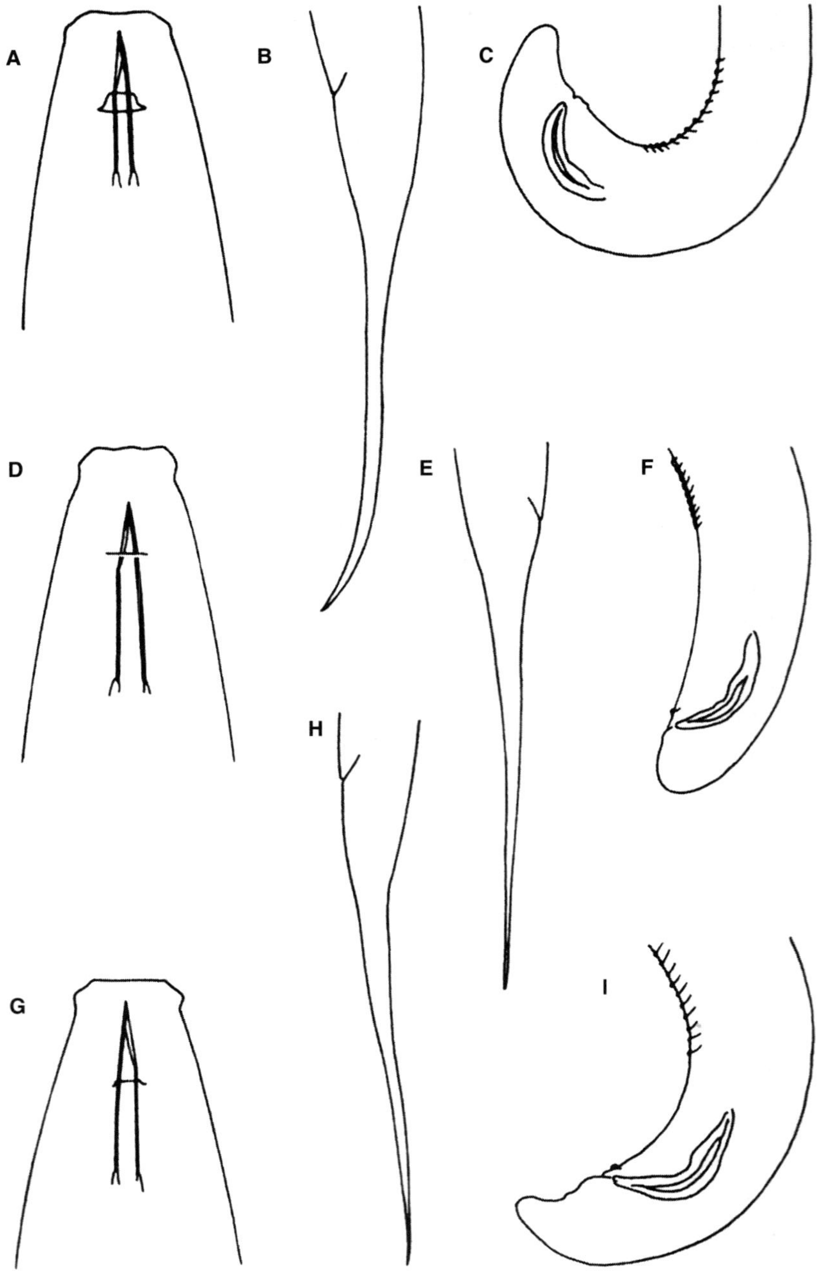

Fig. 15.3 A–C: *Laimydorus*; D–F: *Calodorylaimus*; G–I: *Crocodorylaimus*.

L. agilis (de Man, 1880) Siddiqi, 1969 (USA, Brazil, Russia, Armenia, Lithuania, Northern Russia, Uzbekistan)
L. aquatilis (Skwarra, 1921) Andrássy, 1988 (Germany)
L. baldus Baqri & Jana, 1983 (India)
L. bongersi Loof, 1996 (The Netherlands)

L. constrictus Loof, 1996 (The Netherlands)
L. coomansi Baqri, 1991 (India)
L. coroniceps Loof, 1996 (Brazil)
L. crassoides (Jägerskiöld, 1908) Siddiqi, 1969 (Switzerland)
L. cryptosperma (Loof, 1969) Baqri & Coomans, 1973 (USA)
L. distinctus Dey & Baqri, 1986 (India)
L. doryuris (Ditlevsen, 1911) Andrássy, 1986 (Denmark, Holland)
L. elephas Andrássy, 1988 (USA)
L. esquiveli Ahmad & Shaheen, 2004
L. finalis Thorne, 1975 (India, USA, Russia)
L. flevensis Loof, 1996 (The Netherlands)
L. gazella Andrássy, 1970 (Ethiopia, South Africa)
L. halophilus (Daday, 1897) Andrássy, 1969 (Hungary)
L. keilini (Lee, 1961) Andrássy, 1986 (Nigeria)
L. kherai Baqri, 1985 (India)
L. luettichaui (Meyl, 1957) Siddiqi, 1969 (Tanzania)
L. macrostylus Ahmad & Ahmad, 2002 (India)
L. mangalorensis Ahmad & Ahmad, 2002 (India)
L. massachusetsensis Loof, 1996 (USA)
L. merogaster (Steiner, 1916) Loof, 1996 (Namibia)
L. minimus Baqri, 1991 (India)
L. multialaeus (Khera, 1970) Baqri, 1985 (India)
L. olifanti Botha & Heyns, 1991 (South Africa)
L. oryzae Dey & Baqri, 1986 (India)
L. papillatus Ahmad & Ahmad, 2002 (India)
L. parabastiani (Paetzold, 1958) Siddiqi, 1969 (Europe, Russia, Uzbekistan)
L. partapurii Khaton, Sharma & Singh, 1997
L. pinguis Andrássy, 1988 (USA)
L. proximus (Thorne & Swanger, 1936) Siddiqi, 1969 (USA)
L. pseudostagnalis (Micoletzky, 1927) Siddiqi, 1969 (Italy, Spain, Poland, Russia, Uzbekistan, India, Pakistan, Japan, Ghana, Kenya, Jamaica, Cuba, Ethiopia)
L. reversus Thorne, 1974 (USA)
L. ritae Eyualem-Abebe & Coomans, 1997 (Ethiopia)
L. siddiqii Baqri & Jana, 1983 (India)
L. stenopygus (Andrássy, 1968) Siddiqi, 1969 (Congo Republic)
L. thornei Andrássy, 1969 (Brazil)
L. tropicus Ahmad & Shaheen, 2004
L. unipapillatus (Daday, 1905) Andrássy, 1969 (Paraguay)
L. uterinus Loof, 1996 (Ivory Coast)
L. vacillans Loof, 1996 (Venezuela)
L. vixamictus (Andrássy, 1962) Siddiqi, 1969 (Hungary, Moldavia)
Additional reference: Loof (1985, 2000) and Andrássy (1988).

Genus *Calodorylaimus* Andrássy, 1969 (Figs 15.3D–F)
Diagnosis. Laimydorinae. Body large. Lip region continuous or off set; lips rounded or angular. Odontostyle relatively long and slender, situated further back than

usual; guiding ring thin, situated 1, 5–2 lip region widths from anterior end. Pharynx expanding in or behind its middle. Pre-rectum long, beginning anterior to the supplements in males. Vulva transverse or longitudinal. Spicules long and slender. Supplements numerous, arranged in three groups: two contiguous rows and between them some separate elements. Tail short and rounded in male, long and filiform in female. The species of *Calodorylaimus* live in freshwater habitats.

Type species: *C. octo* Andrássy, 1969 (Ivory Coast)

Other species:
C. andrassyi Baqri & Jana, 1983 (India)
C. chassanicus (Alekseev & Naumova, 1977) Andrássy, 1988 (Eastern Russia)
C. densus Andrássy, 1988 (India)
C. gravidus (Andrássy, 1986) Andrássy, 1988 (Ethiopia)
C. indicus Ahmad & Jairajpuri, 1982 (India)
C. insignis (Gagarin, 1981) Andrássy, 1988 (Tadzhikistan)
C. mongolicus Andrássy, 1988 (Mongolia)
C. parhomalopapillatus (Schuurmans Stekhoven, 1944) Andrássy, 1988 (Congo Republic)
C. wasimi Baqri & Bohra, 2003 (India)
Additional reference: Andrássy (1988).

Genus *Crocodorylaimus* Andrássy, 1988 (Figs 15.3G–I)
Diagnosis. Laimydorinae. Cuticle smooth or finely transversely striated. Lip region continuous with the adjoining body; lips fused. Odontostyle straight, 1.3–2.3 times as long as lip region width. Guiding ring thin. Four spindle-shaped yellowish bodies present near the odontostyle basis. Pharynx expanding near middle. Pre-rectum generally long, in males beginning anterior to the supplements row. Vulva longitudinal or transverse. Spicules massive. Supplements contiguous. Female tail attenuated and long; male tail short, conoid with rounded terminus, ventrally concave. Freshwater or semi-freshwater, widespread in all continents.

Type species: *C. flavomaculatus* (Linstow, 1876) Andrássy, 1988 (Central Europe, Russia, USA, Ethiopia, Sumatra, South Africa)

Other species:
C. aequatorialis Andrássy, 1988 (Ecuador)
C. biserovi Gagarin, 1996 (Russia)
C. dadayi (Thorne & Swanger, 1936) Andrássy, 1988 (Paraguay, Colombia, Russia, Mauritania)
C. dimorphus Andrássy, 1988 (Vietnam)
C. fecundus (Cobb, 1914) Andrássy, 1988 (USA)
C. fusus Andrássy, 1992 (Hungary)
C. maior Andrássy, 1988 (USA, Hungary)
C. paraincae (Thorne, 1974) Andrássy, 1988 (USA)
C. thermalis Andrássy, 1997 (Hungary)
C. vaginatus Ahmad & Araki, 2003 (Japan)

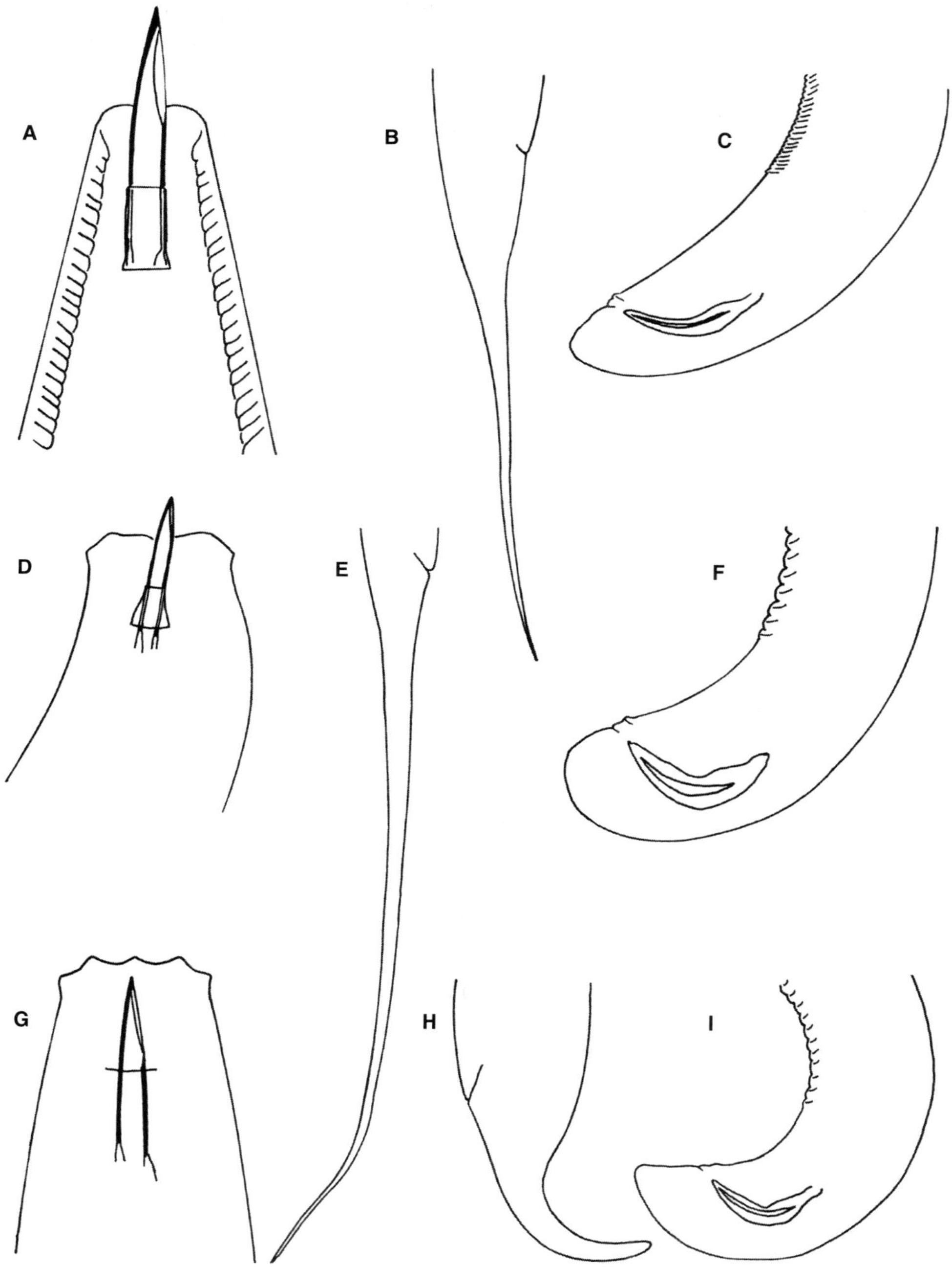

Fig. 15.4 A–C: *Idiodorylaimus;* D–F: *Chrysodorus;* G–I: *Mesodorylaimus.*

Genus *Idiodorylaimus* Andrássy, 1969 (Figs 15.4A–C)
Diagnosis. Laimydorinae. Large nematodes. Cuticle thick, superficially smooth (in one species marked with longitudinal ridges) but distinctly annulated in its inner layer. Lip region not or slightly offset; lips amalgamated. Odontostyle straight, 1.5–2 times as long as the lip region; guiding ring double. Pharynx expanding before the middle. Pre-rectum generally very long, in males beginning far before the supplements row. Vulva longitudinal; vagina with sclerotizations. Spicules large and slender. Supplements contiguous. Tail elongate, conoid to filiform in females, short and bluntly rounded in males. The species of this genus are typical inhabitants of freshwater habitats.

Type species: *I. annulatus* (Daday, 1905) Andrássy, 1969 (Paraguay, Costa Rica, Guatemala)

Other species:
I. annulatiformis Loof, 1973 (Venezuela, Surinam)
I. homalopapillatus (Kreis, 1932) Andrássy, 1969 (Paraguay)
I. kreisi Loof, 1973 (Paraguay)
I. novaezealandiae (Cobb, 1904) Andrássy, 1969 (New Zealand, USA)
I. robustus Gagarin, 1985 (Russia)
I. washingtonensis Loof, 1996 (USA)
Additional reference: Andrássy (1988).

Genus *Chrysodorus* Jiménez-Guirado & Cadenas, 1985 (Figs 15.4D–F)
Diagnosis. Laimydorinae. Body extremely slender. Cuticle thin. Lip region truncate, continuous or offset. Odontostyle weak and unusually slender; guiding ring double. Pharynx enlarging in or behind its middle; pre-rectum very long, in male originating well before the supplements. Vulva longitudinal. Supplements closely packed or contiguous. Tail elongate to filiform in female, short and rounded in male. Freshwater or terrestrial species.

Type species: *C. longicaudatus* Jiménez-Guirado & Cadenas, 1985 (Spain)*

Other species:
C. attenuatus (de Man, 1880) Andrássy, 1988 (Spain, Central Europe, Denmark, Russia, Mongolia)*
C. dhanachandi (Jairajpuri & Ahmad, 1983) Jiménez-Guirado & Cadenas, 1985 (India)
C. filiformis (Bastian, 1865) Andrássy, 1988 (New Guinea, Sumatra, Egypt, widespread in Europe)*
C. lordelloi (Meyl, 1957) Jiménez-Guirado & Cadenas, 1985 (Brazil)
Additional reference: Jiménez-Guirado and Cadenas (1985).

Genus *Mesodorylaimus* Andrássy, 1959 (Figs 15.4G–I)
Diagnosis. Laimydorinae. Cuticle smooth or finely transversely striated. Lip region not or only slightly off set from the adjoining body; lips more or less distinct; odontostyle straight; guiding ring simple, thin. Pharynx enlarging in or behind its middle; at the posterior end of intestine a tongue-like structure often projects into the lumen; pre-rectum varying in length. Vulva transverse or longitudinal; vagina

with sclerotized pieces. Supplements few or numerous, spaced or contiguous. Tail elongate in female: convex–conoid in the anterior part and then slender or digitate, varying in length (1–20 times as long as anal body width); male tail short and rounded. The species of this genus are frequently soil- or moss-inhabiting forms but many of them have been found in freshwater habitats. The latter are marked with an asterisk.

Type species: *M. mesonyctius* (Kreis, 1939) Andrássy, 1959 (Central Europe, Italy, Russia, China, India, South Africa, USA, Jamaica, Venezuela)*

Other species:

M. aberrans Loof, 1969 (Switzerland, The Netherlands, Yugoslavia, Poland, Hungary, Pakistan, Argentina)
M. adalberti Andrássy, 1963 (Argentina, Russia)
M. aduncus Andrássy, 1986 (Hungary)
M. aegypticus (Andrássy, 1958) Andrássy, 1959 (Denmark, Sweden, Russia, Egypt, South Africa, Brazil, Antarctica)*
M. aequatorialis Andrássy, 1986 (Ecuador)
M. aestuarii (Timm, 1952) Andrássy, 1959 (USA)*
M. alpestris (Thorne, 1939) Andrássy, 1959 (USA, South Africa)*
M. americanus Andrássy, 1986 (USA)
M. andrassyi Ahmad & Ahmad, 2001 (India)
M. angustus Andrássy, 1964 (Kenya)*
M. antarcticus Nedelchev & Peneva, 2000 (Antarctica)
M. arcuatus Andrássy, 1986 (Bolivia)
M. baeticus Peña-Santiago & Abolafia 2000 (Spain)
M. bainsi Basson & Heyns, 1974 (India, Ethiopia, South Africa)*
M. bastiani (Bütschli, 1873) Andrássy, 1959 (cosmopolitan)*
M. bastianoides (Meyl, 1961) Andrássy, 1967 (Germany, Hungary)*
M. bhanselii Baqri & Bohra, 2001 (India)
M. biroi (Daday, 1899) Andrássy, 1959 (Lithuania, Russia, India, Mauritius, Brazil, New Guinea)*
M. brassicus Soni & Nama, 1981 (India)
M. brevicaudatus Abolafia & Peña-Santiago, 1996 (Spain)
M. brevicylindrus Andrássy, 2002 (Hungary)
M. brevidens Thorne, 1974 (USA)
M. brevispicatus (Schuurmans Stekhoven, 1951) Andrássy, 1986 (Congo)
M. brzeskii Abolafia & Peña-Santiago, 2000 (Spain)
M. cardiacus Thorne, 1974 (USA)
M. caudatus Ahmad, 1993 (India)
M. chamoliensis Ahmad, 1995 (India)
M. chekiangesis (Wu & Hoeppli, 1929) Andrássy, 1986 (China)
M. chinensis Wu & Ahmad, 1998 (China)
M. chipevi Nedelchev & Peneva, 2000 (Antarctica)
M. clavicaudatus (Thorne & Swanger, 1936) Andrássy, 1959 (Poland, Spain, India)
M. cognatus Andrássy, 1986 (Ecuador)
M. conurus (Thorne, 1939) Goodey, 1963 (USA, Switzerland, Russia, Kenya)*
M. delicatus Lordello, 1965 (Brazil)

M. derni Loof, 1969 (Germany, The Netherlands, Russia, India, Argentina)
M. deuberti (Andrássy, 1958) Goodey, 1963 (Hungary)*
M. dolomiticus Vinciguerra, 1982 (Italy)
M. dreyeri (Ven der Linde, 1938) Goodey, 1963 (USA)
M. effilatus (Schuurmans Stekhoven & Teunissen, 1938) Andrássy, 1959
M. enigmaticus Ahmad & Ahmad, 2001 (India)
M. erdelyii Andrássy, 1965 (Ghana)*
M. exilis (Cobb, 1893) Andrássy, 1959 (Fiji, Ivory Coast, Surinam)*
M. flagellatus (Williams, 1959) Andrássy, 1960 (Mauritius)
M. flagellifer Andrássy, 2002 (Hungary)*
M. ghanae Andrássy, 1965 (Ghana)
M. globiceps Loof, 1974 (Venezuela)
M. graciosus Andrássy, 1986 (Ecuador)
M. guarani Andrássy, 1968 (Paraguay)
M. harunaglus Khaton, Sharma & Singh, 1997
M. hofmaenneri (Menzel in Hoffmänner & Menzel, 1914) Goodey, 1963 (Germany, Switzerland, Austria, Hungary, France, Spain, Yugoslavia, Russia, USA, Sumatra, Java, Latvia)*
M. ibericus Abolafia & Peña-Santiago, 2000 (Spain)
M. imperator Loof, 1975 (Antarctica)
M. importunus Basson & Heyns, 1974 (South Africa)*
M. indicus Ahmad & Ahmad, 2001 (India)
M. intermedius Dassonville & Heyns, 1984 (South Africa)*
M. intervallis (Thorne & Swanger, 1936) Andrássy, 1959 (USA, Egypt)*
M. japonicus (Cobb in Thorne & Swanger, 1936) Andrássy, 1987 (Japan, USA, Russia, Angola, Congo Republic, South Africa)*
M. johanni Basson & Heyns, 1974 (South Africa)
M. kamandeanus Baqri & Coomans, 1973 (Congo)
M. kauli Baqri & Bohra, 2001 (India)
M. keralaensis Ahmad & Ahmad, 2001 (India)
M. kittenbergeri Andrássy, 1988 (Kenya)
M. kowyni Basson & Heyns, 1974 (South Africa)*
M. lissus Thorne, 1974 (USA, China)
M. litoralis Loof, 1969 (The Netherlands, Germany, Belgium, Spain, Italy, Russia, USA)
M. longicaudatus Ahmad & Araki, 2003 (Japan)
M. loofi Ahmad, 1993 (India)
M. lopadusae Vinciguerra & La Fauci, 1978 (Italy)
M. lourdesae (Lordello, 1955) Andrássy, 1959 (Brazil)
M. luci Brzeski & Szczygiel, 1961 (Poland)*
M. macrofallus Brzeski & Szczygiel, 1961 (USA)
M. macrospiculum Zullini, 1987 (Ethiopia)
M. malacitanus Abolafia & Peña-Santiago, 2000 (Spain)
M. margaritifer Andrássy, 1986 (Hungary)*
M. margaritus Basson & Heyns, 1974 (South Africa)
M. masleni Nedelchev & Peneva, 2000
M. meridianus Andrássy, 1963 (Mexico, Argentina)

M. mexicanus Zullini, 1973 (Mexico)
M. meyli (Andrássy, 1958) Andrássy, 1959 (Germany, Czech Republic, Bulgaria, Russia, Tadzhikistan, Spain)*
M. nevadaensis Peña-Santiago & Abolafia, 2000
M. nigritulus (Schneider, 1937) Andrássy, 1959 (Sumatra)*
M. nipponi Ahmad & Araki, 2003 (Japan)
M. nodicaudatus (Dey & Baqri, 1986) Ahmad, 1993 (India)
M. novus (Dey & Baqri, 1986) Ahmad, 1993 (India)
M. nudus (Thorne, 1939) Andrássy, 1969 (USA, Italy)
M. obscurus Thorne, 1974 (USA)
M. orientalis Andrássy, 1970 (Vietnam)
M. ornativulvatus Abolafia & Peña-Santiago, 1997 (Spain)*
M. paetzoldi Altherr, 1965 (Germany)*
M. palustris Andrássy, 1991 (Hungary)*
M. parabastiani (Paetzold, 1958) Andrássy, 1988 (The Netherlands, Germany, Russia, Moldavia, Uzbekistan, Hungary)*
M. paraguayensis (Kreis, 1932) Andrássy, 1959 (Paraguay, Brazil)
M. paralitoralis Basson & Heyns, 1974 (South Africa)*
M. parapotus Ahmad & Ahmad, 2001 (India)
M. parasubtilis (Meyl, 1957) Andrássy, 1959 (Hungary, Brazil, Venezuela)
M. parasubulatus (Meyl, 1954) Andrássy, 1959 (Mongolia, Germany, Czechoslovakia, Russia)*
M. parvus Ahmad, 1995 (India)
M. paulbuchneri (Meyl, 1956) Andrássy, 1959 (Brazil)
M. pendschipentikus (Tulaganov, 1949) Andrássy, 1959 (Uzbekistan)
M. pizai Lordello, 1965 (Brazil)
M. plicatus Andrássy, 1986 (Ecuador)*
M. potus Heyns, 1963 (Spain, Russia, South Africa)*
M. procerus Andrássy, 1986 (Australia)
M. pseudobastiani Loof, 1969 (Russia, USA)
M. pseudorecurvus Abolafia & Peña-Santiago, 1996 (Spain)
M. pseudosubtilis Basson & Heyns, 1974 (South Africa)*
M. puellae Andrássy, 1963 (Argentina)
M. pulcher Andrássy, 1986
M. pusillus (Cobb, 1893) Andrássy, 1959 (Georgia, Jamaica, Australia)
M. recurvus Andrássy, 1964 (Hungary, Poland, Italy)*
M. rhenanus Altherr, 1965 (Germany)*
M. rotundolabiatus Basson & Heyns, 1974 (South Africa)
M. sanctus Basson & Heyns, 1974 (South Africa)
M. shamimi Ahmad & Araki, 2003 (Japan)
M. similibastiani Zell, 1986
M. simplex Thorne, 1974 (USA)
M. spengelii (de Man, 1812) Andrássy, 1959 (The Netherlands)
M. subtiliformis (Andrássy, 1959) Andrássy, 1959 (Hungary, Romania, Czech Republic, Japan)*
M. subtilis (Thorne & Swanger, 1936) Andrássy, 1959 (The Netherlands, Germany, Yugoslavia, Poland, Israel, Russia, India, USA, Brazil)*

M. subtiloides (Paetzold, 1958) Andrássy, 1959 (Germany)*
M. sveltus (Meyl, 1957) Andrássy, 1959 (Brazil)
M. sylphus (Thorne, 1939) Goodey, 1963 (USA)
M. szechenyii Andrássy, 1961 (Tanzania)*
M. szunyoghyi Andrássy, 1968 (Tanzania)
M. tenellus (Thorne & Swanger, 1936) Andrássy, 1959 (USA, Spain, Czech Republic)
M. thermae (Cobb in Hoeppli, 1926) Goodey, 1963 (USA)
M. tholocercus Andrássy, 1968 (Cuba, Paraguay)
M. thorneiswangerae Andrássy, 1968 (Brazil)
M. transkeyensis Basson & Heyns, 1974 (South Africa)
M. trapaefructus Andrássy, 1986 (Peru)
M. usitatoides De Bruin & Heyns, 1992
M. usitatus Basson & Heyns, 1974 (South Africa)*
M. vulneratus Andrássy, 1986 (Ecuador)
M. vulvapapillatus Bagaturia & Eliava, 1966 (Russia, Georgia)*
M. vulvastriatus Ahmad, 1993 (India)

Additional reference: Andrássy (1988).

Genus *Baladorylaimus* Andrássy, 2001 (Figs 15.5A–C)
Diagnosis. Laimydorinae. Cuticle smooth. Lip region continuous with adjoining body, lips amalgamated. Odontostyle short; guiding ring simple. Pharynx enlarging behind its middle; pre-rectum short. Reproductive apparatus amphidelphic; vulva oval; vagina with sclerotized pieces. Supplements spaced. Female tail filiform; male tail short, conoid, with terminal peg. The only species was found in wet detritus on the edge of Lake Balaton (Hungary).

Type and only species: *B. balatonicus* Andrássy, 2001

Additional reference: Andrássy (2001b).

Genus *Namaquanema* Heyns & Swart, 1993 (Figs 15.5 D–F)
Diagnosis. Laimydorinae. Lip region slightly off set; lips rather amalgamated: Odontostyle of moderate length with hooked basal processes but without sclerotized odontophore; guiding ring double. Pharynx expanding before its middle. Pre-rectum long, beginning far before the supplements in males. Vulva longitudinal, sunken. Supplements numerous, contiguous. Tail short and rounded in male, hemi-spherical but with a finger-like process dorsally directed in female. The only species was found in the soil of the bank of a stream in South Africa.

Type and only species: *N. hanki* Heyns & Swart, 1993 (South Africa)

Additional reference: Heyns and Swart (1993).

Subfamily Afrodorylaiminae Andrássy, 1969
Diagnosis. Dorylaimidae. Cuticle smooth or very finely striated. Lip region more or less set off; lips slightly separate, labial papillae 10 + 6. Amphids wide. Odontostyle straight or slightly sinuate, at least twice as long as lip region width; guiding ring simple or double. Pharynx enlarging in or behind its middle. Cardia longer than usual. Pre-rectum very short. Male supplements contiguous or spaced; sub-median papillae

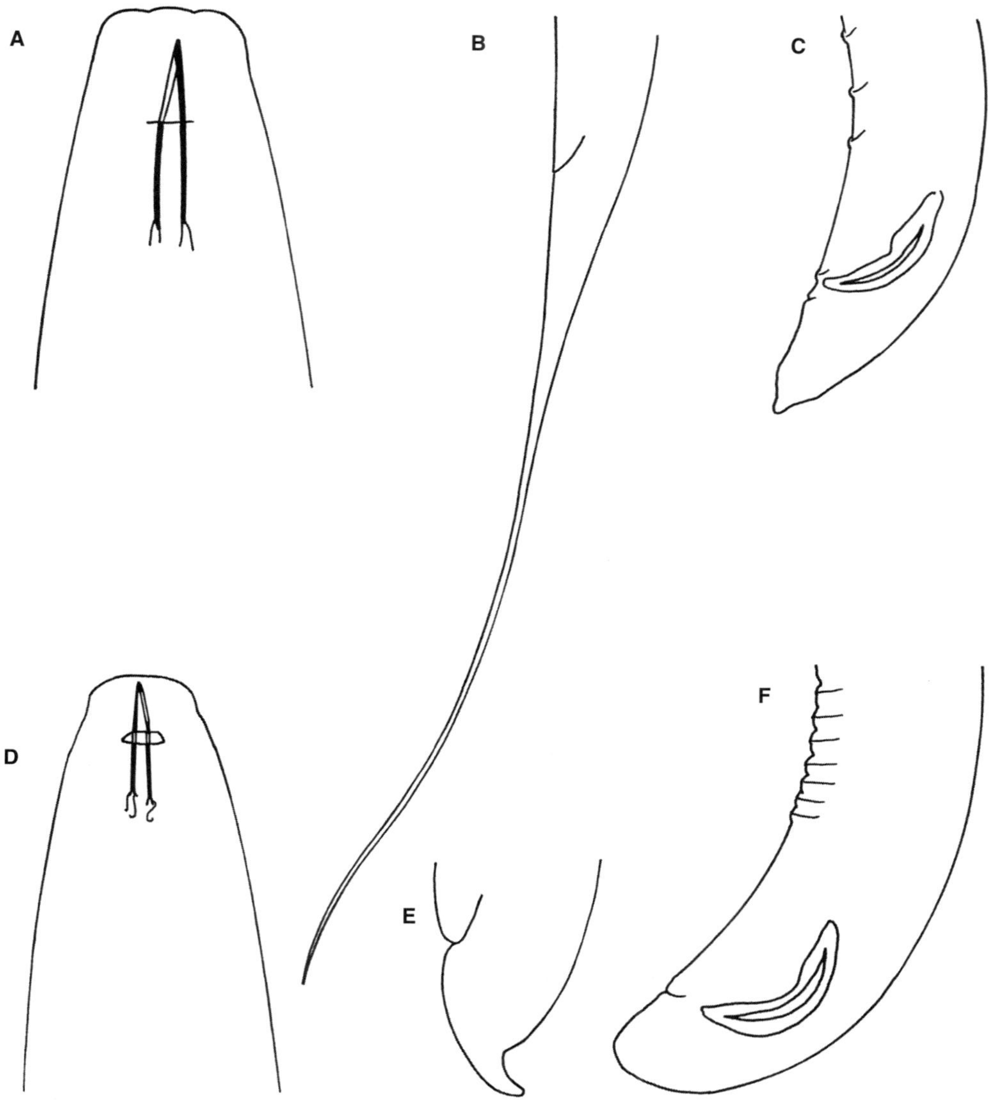

Fig. 15.5 A–C: *Baladorylaimus*; D–F; *Namaquanema*.

on posterior body very prominent. Tail long and filiform in females, variously shaped in males. Mostly freshwater species.
Additional reference: Andrássy (1969).

Genera:[1]
Afrodorylaimus Andrássy, 1964
Drepanodorylaimus Jairajpuri, 1966
Paradorylaimus Andrássy, 1969
Apodorylaimus Andrássy, 1988

[1]The attribution of these genera to Afrodorylaiminae was suggested by Andrássy (personal communication).

Key to genera of Afrodorylaiminae

1. Odontostyle sinuate, male unknown . . . . *Drepanodorylaimus*
   Odontostyle straight, male generally present 2
2. Male tail short and rounded, guiding ring double . . . . *Paradorylaimus*
   Male tail filiform or conoid, guiding ring simple 3
3. Male tail short, conoid, ventrally curved, with pointed terminus . . . . *Afrodorylaimus*
   Male tail filiform as in female . . . . *Apodorylaimus*

Genus *Afrodorylaimus* Andrássy, 1964 (Figs 15.6A–C)
Diagnosis. Afrodorylaiminae. Cuticle smooth. Lip region moderately off set. Odontostyle straight, with wide lumen, about twice as long as lip region width; guiding ring simple. Pre-rectum short. Tail long and filiform in female, short, conoid, ventrally curved, with pointed terminus in male. The species of this genus are typical inhabitants of freshwater habitats.

Type species: *Afrodorylaimus bwana* Andrássy, 1964 (Kenya, China)

Other species:
A. beaumonti (Altherr, 1952) Andrássy, 1969 (Switzerland)
A. bizane Kleynhans, 1970 (South Africa)
A. geniculatus (Andrássy, 1961) Andrássy, 1964 (Russia, India, Kenya)
A. lambsheadi Ahmad, Bloemers & Wanless, 1996 (Cameroon)
A. mediterraneus Vinciguerra & La Fauci, 1978 (Italy)
Additional reference: Andrássy (1964).

Genus *Drepanodorylaimus* Jairajpuri, 1966 (Figs 15.6D and E)
Diagnosis. Afrodorylaiminae. Cuticle smooth. Lip region moderately offset. Odontostyle about twice as long as lip region width, slightly but clearly sinuate; guiding ring simple. Pre-rectum short. Tail long and filiform in female. The species of this genus lack males and are all parthenogenetic. They are typical inhabitants of freshwater habitats.

Type species: *Drepanodorylaimus filiformis* Jairajpuri, 1966 (India, La Réunion)

Other species:
D. arganoi Zullini, 1973 (Mexico)
D. brevicaudatus Andrássy, 1970 (Vietnam)
D. brzeskii Winiszewska, 1987 (Poland)
D. filicaudatus (Daday, 1905) Andrássy, 1986 (Paraguay, Surinam)
D. flexus (Thorne & Swanger, 1936) Andrássy, 1969 (USA, Spain)
D. maqsoodi Dhanachand & Jairajpuri, 1981 (India)
D. picardi (Altherr, 1963) Monteiro, 1970 (France, Angola, Brazil)
D. renwicki (Van der Linde, 1938) Andrássy, 1969 (USA, Austria, Switzerland, Hungary, Japan)
D. szekessyi (Andrássy, 1960) Andrássy, 1969 (China, Argentina, Brazil, Paraguay)
D. williamsi (Heyns & Kruger, 1983) Andrássy, 1986 (South Africa)
Additional reference: Jairajpuri (1966).

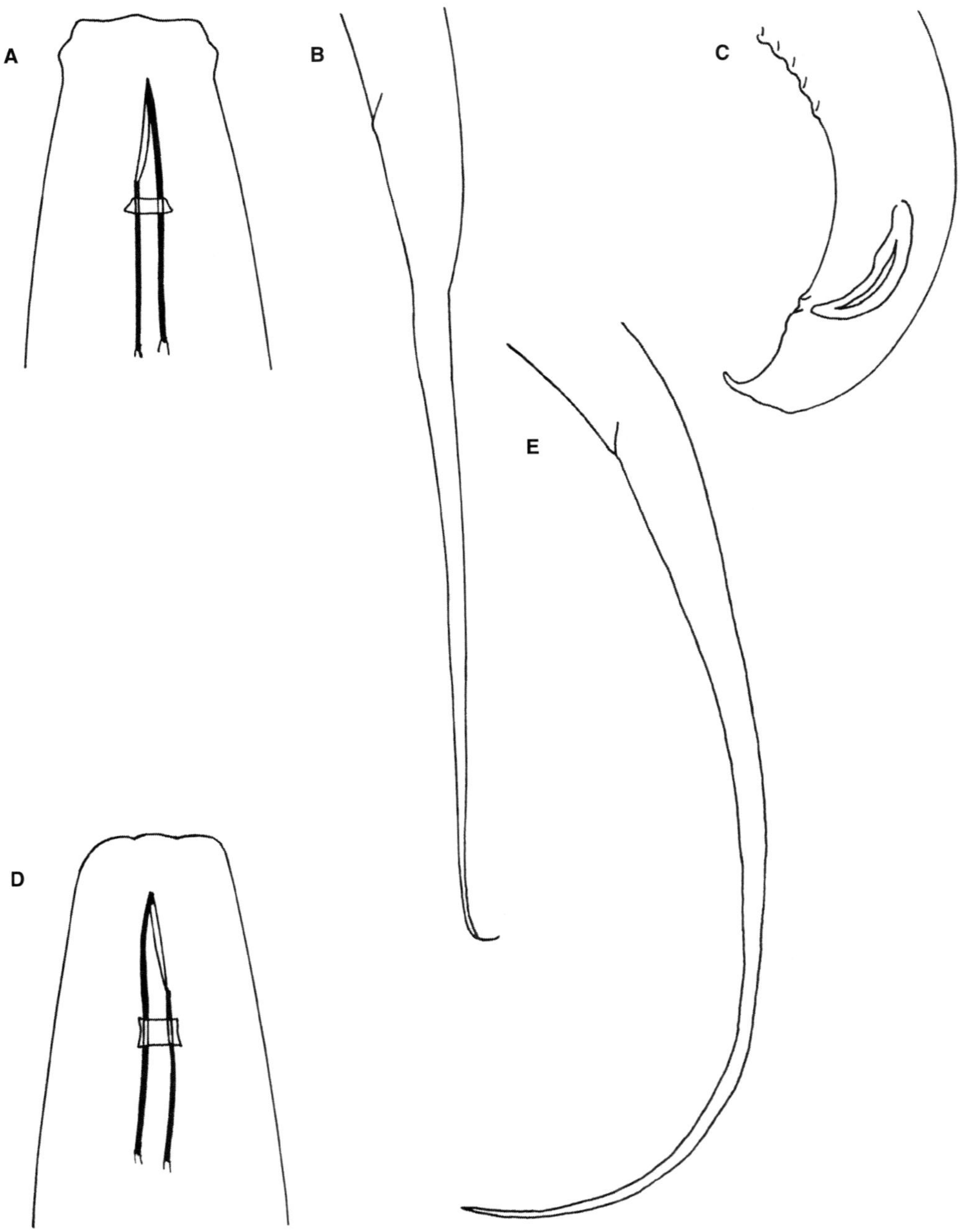

Fig. 15.6 A–C: *Afrodorylaimus;* D and E: *Drepanodorylaimus.*

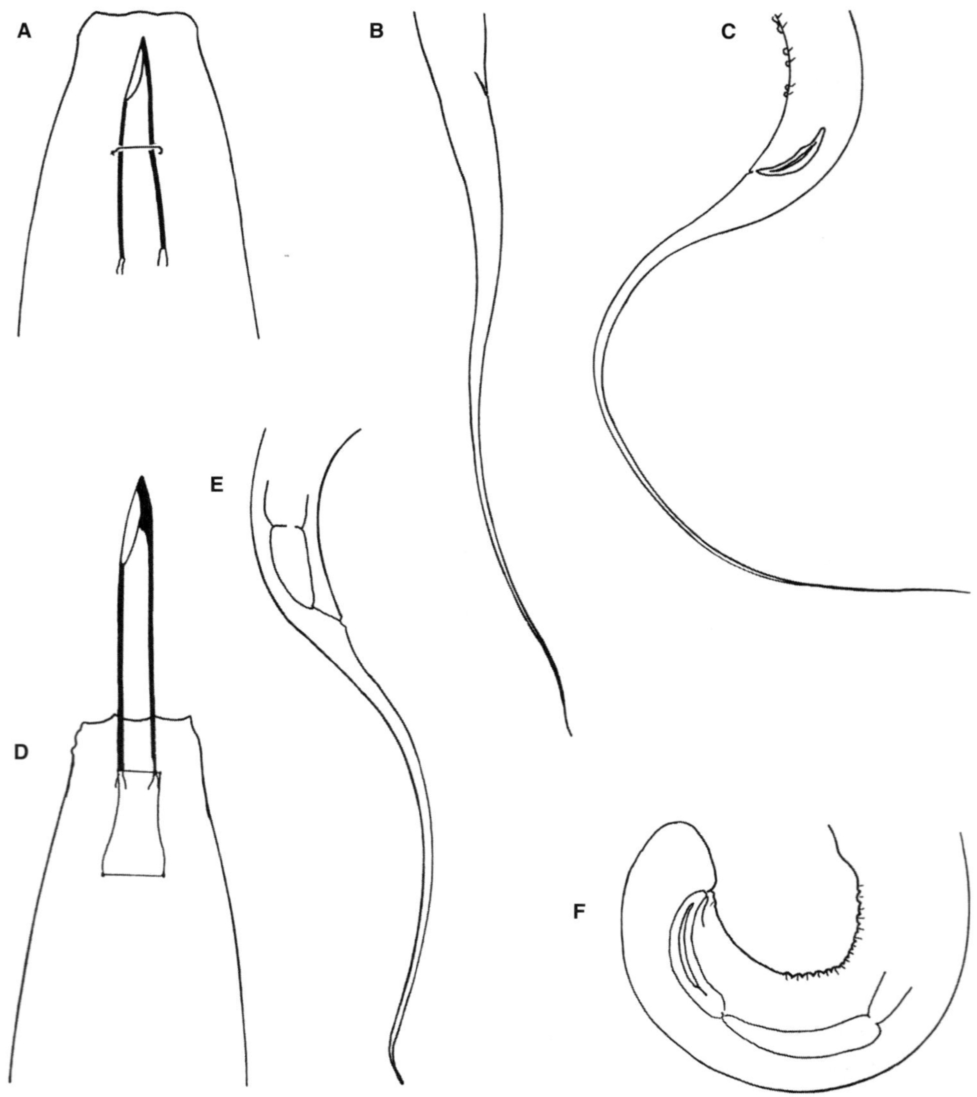

Fig. 15.7 A–C: *Apodorylaimus;* D–F: *Paradorylaimus.*

Genus *Apodorylaimus* Andrássy, 1988 (Figs 15.7A–C)

Diagnosis. Afrodorylaiminae. Cuticle smooth. Lip region moderately offset, lips slightly distinct. Odontostyle straight, with wide lumen, about twice as long as lip region width; guiding ring simple. Pre-rectum short. Vagina without sclerotizations. Spicules lacking lateral pieces and central thickening. Supplements 5–8 arranged two by two. Tail long and filiform in both sexes. Of the two described species one was found in wet soil.

Type species: *A. bini* Andrássy, 1988 (India: marshy soil)*

Other species:

A. congonensis (Andrássy, 1960) Andrássy, 1988 (Congo Republic)

Genus *Paradorylaimus* Andrássy, 1969 (Figs 15.7D–F)
Diagnosis. Afrodorylaiminae. Cuticle smooth, subcuticle often finely striated. Lip region slightly offset; lips distinct. Odontostyle straight, about twice as long as lip region width or more; guiding ring double. Pharynx enlarging in or behind its middle. Pre-rectum short, in males beginning within the range of the supplement row. Supplements 13–18, contiguous. Female tail long, filiform; male tail short and rounded. The species of *Paradorylaimus* are all freshwater.

Type species: *P. parafecundus* (De Coninck, 1935) Andrássy, 1969 (Congo Republic)

Other species:
P. conurus (Thorne, 1939) n. comb. (USA)
P. jankowskyi (Tsalolikhin, 1977) Andrássy, 1988 (Siberia)
P. wilhelmschneideri (Andrássy, 1959) Andrássy, 1969 (Sumatra)
Valid species of Dorylaimidae regarded as *incertae sedis* (according to Andrássy, 1988: insufficient descriptions, or absence of males):
Dorylaimus acris Thorne, 1939 (Central Europe, Italy, USA, Iraq)*
Dorylaimus africanus Daday, 1908
Dorylaimus angusticephalus Steiner, 1914
Dorylaimus aquaticus Eliava, 1978*
Dorylaimus atratus Linstow, 1901
Dorylaimus callosus Skwarra, 1921 (Germany)*
Dorylaimus cashmerensis Altherr & Delamare-Deboutteville, 1972 (USA)*
Dorylaimus exilicaudatus Altherr, 1953
Dorylaimus fastigatus Thorne & Swanger, 1936
Dorylaimus festivus Paesler, 1941 (Austria)*
Dorylaimus gaussi Steiner, 1916 (Denmark)*
Dorylaimus lybicus Pierantoni, 1915 (Lybia)
Dorylaimus longicaudatus Bütschli, 1874 (Croatia, Slovenia, Denmark, Germany, Kenya)*
Dorylaimus maximodorus Schuurmans Stekhoven & Teunissen, 1938
Dorylaimus nyongi Altherr, 1960 (Cameroon)*
Dorylaimus pachydermis Daday, 1908
Dorylaimus pachysoma Linstow, 1876
Dorylaimus paraagilis Altherr, 1953 (Switzerland)*
Dorylaimus polyblastus Bastian, 1865
Dorylaimus serpentinus Thorne & Swanger, 1936 (USA, Spain)*
Dorylaimus sulcatus Cobb in Cobb, 1915 (USA)*
Dorylaimus tenuicaudatus Bastian, 1865
Dorylaimus tenuissimus Stefanski, 1925
Dorylaimus tenuistriatus Schneider, 1933 (West Africa)*
Dorylaimus zograffi de Man, 1885
Laimydorus longissimicaudatus Altherr, 1977 (Brazil)*
Mesodorylaimus argentinus Altherr, 1963 (Argentina)*
Prodorylaimus uliginosus Loof, 1985 (Central Europe, Italy)*
Urolabes palustris Carter, 1859*

Family ACTINOLAIMIDAE Thorne, 1939

Diagnosis. Dorylaimoidea. Cuticle smooth or marked with longitudinal ridges or grooves; lips fused; lip region with an anterior cuticularized ring (vestibular ring) often appearing corrugated; cheilostom wide, armed with four massive onchia encircling the odontostyle, with or without denticles on the stoma walls. Gonads paired and reflexed; Vulva transverse, longitudinal or pore-like; *pars refringens* of vagina with sclerotized pieces. Male supplements arranged in a ventral series or in two or three fascicles. Tail elongate to filiform in both sexes or elongate to filiform in females and short in males (sometimes with a filiform appendix). All the representatives of this family are typical freshwater or semi-freshwater nematodes: they can be found in freshwater bodies and, more frequently, in wet soil and in wet moss. There are 18 genera in this family.

Additional references: Thorne (1967), Vinciguerra (1988), Khan and Jairajpuri (1994) and Vinciguerra and Clausi (2003).

Genera:

Trachactinolaimus Andrássy, 1963
Trachypleurosum Andrássy, 1959
Paractinolaimus Meyl, 1957
Paractinolaimoides Khan, Ahmad & Jairajpuri, 1994
Westindicus Thorne, 1967
Egtitus Thorne, 1967
Scleroactinolaimus Ahmad, Khan & Ahmad, 1992
Afractinolaimus Andrássy, 1970
Neoactinolaimus Thorne, 1967
Mactinolaimus Andrássy, 1970
Metactinolaimus Meyl, 1957
Stopractinca Khan, Ahmad & Jairajpuri, 1994
Actinolaimus Cobb, 1913
Parastomachoglossa Coomans & Loof, 1986
Actinca Andrássy, 1964
Brasilaimus Lordello & Zamith, 1957
Afractinca Vinciguerra & Clausi, 2000
Practinocephalus Andrássy, 1974

Key to genera of Actinolaimidae

1. Tail elongate in both sexes 2
   Tail elongate in female and short in male 3
2. Cheilostom walls with denticles . . . . *Trachactinolaimus*
   Cheilostom walls smooth or rugose . . . . *Trachypleurosum*
3. Cuticle without longitudinal ridges 4
   Cuticle with longitudinal ridges 13
4. Cheilostom with four sclerotized vertical plates 5
   Cheilostom without sclerotized vertical plates 6
5. Cheilostom with denticles . . . . *Paractinolaimoides*
   Cheilostom without denticles . . . . *Scleroactinolaimus*
6. Onchia basally fused . . . . *Metactinolaimus*
   Onchia distinct 7

7. Cheilostom walls strongly ribbed, basket-like . . . . *Westindicus*
   Cheilostom walls not or weakly sclerotized 8
8. Cheilostom walls with denticles 9
   Cheilostom walls without denticles 10
9. Male supplements in series . . . . *Paractinolaimus*
   Male supplements in fascicles . . . . *Afractinolaimus*
10. Male supplements in series 11
    Male supplements in fascicles 12
11. Pharynx muscular, in two parts . . . . *Egtitus*
    Pharynx tripartite, with slender anterior part . . . . *Stopractinca*
12. Onchia simple . . . . *Mactinolaimus*
    Each onchium with a small denticle behind . . . . *Neoactinolaimus*
13. Pharynx muscular, in two parts; vagina with strongly sclerotized pieces . . . . *Actinolaimus*
    Pharynx tripartite, with non-muscular anterior part, vagina with weakly or non-sclerotized pieces 4
14. Anterior hyaline part of pharynx short, middle part long, odontostyle robust, lip region low . . . . *Parastomachoglossa*
    Anterior hyaline part of pharynx long, middle part short, odontostyle slender, lip region high 15
15. Cuticle very thick, *pars refringens* of vagina thick walled, wider than *pars proximalis*, lacking sclerotizations . . . . *Brasilaimus*
    Cuticle not so thick, *pars refringens* of vagina not wider than *pars proximalis*, with weak sclerotizations 16
16. Lip region expanded, cheilostom sclerotizations heavy, male tail with filiform appendix . . . . *Practinocephalus*
    Lip region narrow, cheilostom sclerotizations weak, male tail without any appendix 17
17. Inner labial papillae external to the anterior sclerotized ring . . . . *Actinca*
    Inner labial papillae located on the anterior sclerotized ring . . . . *Afractinca*

Genus *Trachactinolaimus* Andrássy, 1963 (Figs 15.8A and B)

Diagnosis. Actinolaimidae. Cuticle smooth; lip region with vestibular ring appearing corrugated; cheilostom with denticles on the stoma walls. Vulva transverse; *pars refringens* of vagina with strongly sclerotized pieces. Male supplements arranged in a ventral series. Tail elongate to filiform in both sexes.

Type species: *Trachactinolaimus radulatus* Andrássy, 1963 (Angola)

Other species:

T. dominicus (Hunt, 1978) Vinciguerra, 1988 (Santo Domingo)

Additional reference: Coomans *et al.* (1990).

Genus *Trachypleurosum* Andrássy, 1959 (Figs 15.8C and D)

Diagnosis. Syn. *Trachypleura* Thorne, 1939 nec Jackel, 1900

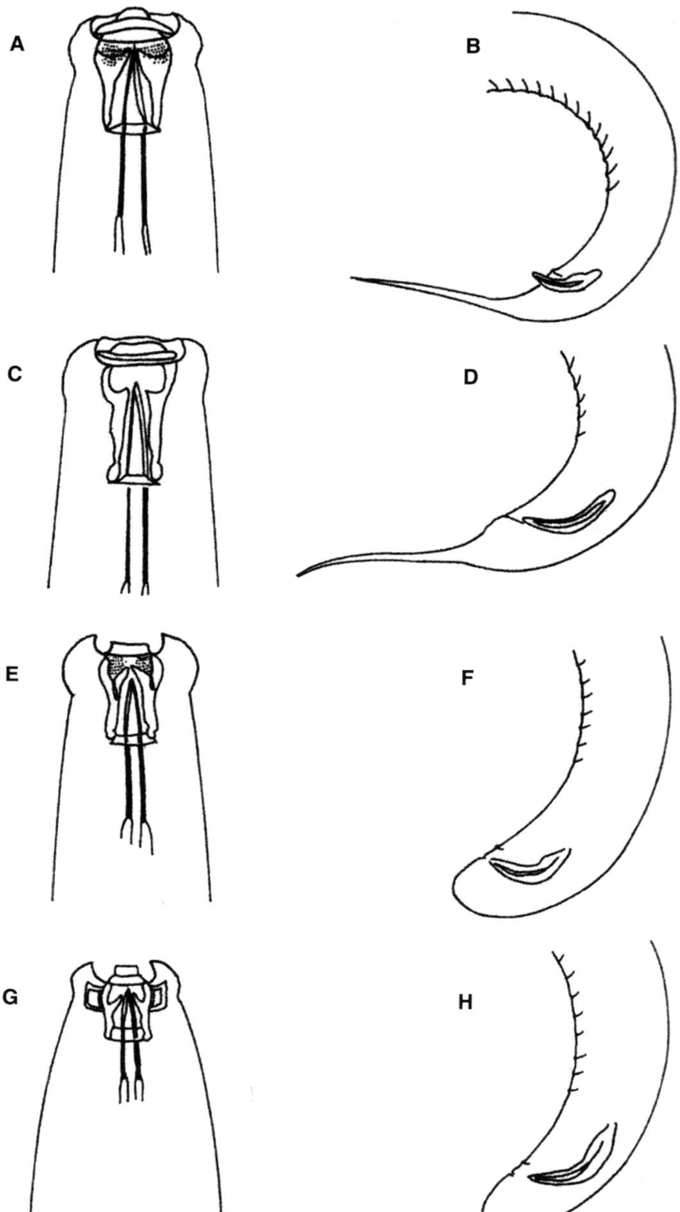

Fig. 15.8 A and B: *Trachactinolaimus*; C and D: *Trachypleurosum*; E and F: *Paractinolaimus*: G and H: *Paractinolaimoides*.

Actinolaimidae. Cuticle smooth; lip region with vestibular ring appearing corrugated; cheilostome without denticles on the stoma walls. *Pars refringens* of vagina with strongly sclerotized pieces. Male supplements arranged in a ventral series. Tail elongate to filiform in both sexes.

Type species: *T. conforme* (Schneider, 1935) Andrássy, 1959 (West Africa)

Other species:
T. belforum Bloemers, Ahmad, Wanless & Hodda, 1995 (Cameroon)
T. indicum Ahmad, Khan & Rahman, 1991 (India)
T. karnatakus Khan & Jairajpuri, 1994 (India)
T. labirynthostoma (Cobb, 1893) Andrássy, 1959 (Fiji)
T. venezolanum Coomans, Vinciguerra & Loof, 1990 (Venezuela)
Additional reference: Coomans *et al.* (1990).

Genus *Paractinolaimus* Meyl, 1957 (Figs 15.8E and F)
Diagnosis. Actinolaimidae. Cuticle smooth; lip region with vestibular ring appearing corrugated; cheilostome with denticles on the stoma walls. *Pars refringens* of vagina with strongly sclerotized pieces. Male supplements arranged in a ventral series. Tail elongate to filiform in females and short in males.

Type species: *Paractinolaimus micoletzkyi* (Schneider, 1935) Meyl, 1957

Other species:
P. acutus Khan & Park, 1999 (India)
P. aruprus Khan, Ahmad & Jairajpuri, 1994 (India)
P. baldus Thorne, 1967 (Puerto Rico)
P. chandicus Khan & Jairajpuri, 1994 (India)
P. chiki Dhanam, Jairajpuri & Khan, 1994 (India)
P. dhanachandi Khan & Jairajpuri, 1994 (India)
P. elongatus Khan & Jairajpuri, 1994 (India)
P. filipjevi (Schneider, 1935) Meyl, 1957 (West Africa)
P. girini Sukul, 1967 (India)
P. indicus Khan & Ganguly, 1988 (India)
P. intermedius Altherr, 1968 (Switzerland, Italy)
P. longidrilus Eveleigh, 1982 (Canada)
P. macrolaimus (de Man, 1880) Andrássy, 1964 (cosmopolitan)
P. microdentatus (Thorne, 1939) Meyl, 1957 (USA, Israel, South Africa, Spain)
P. occalescens (Schneider, 1937) Vinciguerra, 1988 (Sumatra, Java)
P. pachydermis Khan & Araki, 2001 (Japan)
P. parietinus Eroshenko, 1977 (Russia)
P. prodenticulatus Heyns & Argo, 1969 (South Africa)
P. proximus (Yeates, 1973) Vinciguerra & Coomans, 1991 (New Zealand)
P. rafiqi Khan & Jairajpuri, 1998 (India)
P. robustus Thorne, 1967 (Puerto Rico)
P. spanithelus Eveleigh, 1982 (Canada)
P. vulvapapillatus Khan, Ahmad & Jairajpuri, 1994 (India)
P. vigor Thorne, 1967 (Puerto Rico, St. Lucia, South Africa)
P. xosorum Heyns & Argo, 1969 (South Africa)

Genus *Paractinolaimoides* Khan, Ahmad & Jairajpuri, 1994 (Figs 15.8G and H)
Diagnosis. Actinolaimidae. Cuticle smooth; vestibular ring corrugated; cheilostom with four sclerotized plates at level of onchia and mural denticles arranged in several rows. Vulva transverse; *pars refringens* of vagina with strongly sclerotized pieces. Male supplements arranged in a ventral series. Tail elongate to filiform in females and short in males.

Type species: *P. unicus* Khan, Ahmad & Jairajpuri, 1994 (India)

Other species:
P. hawangensis Choi, Duan & Baek 1998 (Korea)
Additional reference: Khan *et al.* (1994b).

Genus *Westindicus* Thorne, 1967 (Figs 15.9A and B)
Diagnosis. Actinolaimidae. Cuticle smooth; vestibular ring corrugated; cheilostom walls strongly ribbed, basket-like, with denticles on the walls. *Pars refringens* of vagina with strongly sclerotized pieces. Male supplements arranged in a ventral series. Tail elongate to filiform in females and short in males.

Type species: *Westindicus brachycephalus* Thorne, 1967 (Puerto Rico)

Other species:
W. cheongsongensis Choi, Duan & Baek, 1998 (Korea)
W. cinctus (Cobb in Thorne, 1939) Thorne, 1937 (Jamaica)
W. keralaensis Khan, Ahmad & Jairajpuri, 1994 (India)
W. rapax Hunt, 1978 (St. Lucia)
W. senaensis Khan, Park & Choi, 1999 (Korea)

Genus *Egtitus* Thorne, 1967 (Figs 15.9C and D)
Diagnosis. Actinolaimidae. Cuticle smooth; vestibular ring corrugated; cheilostom without denticles on the stoma walls. Vulva transverse or longitudinal; *pars refringens* of vagina with strongly sclerotized pieces. Male supplements arranged in a ventral series. Tail elongate to filiform in females and short in males.

Type species: *Egtitus bryophilus* Thorne, 1967 (Puerto Rico)

Other species:
E. andhricus Khan & Jairajpuri, 1994 (India, Korea)
E. cyatholaimus (Daday, 1905) Thorne, 1967 (Paraguay)
E. elaboratus (Cobb, 1906) Thorne, 1967 (Hawaii)
E. itanagrus Khan, Ahmad & Jairajpuri, 1994 (India)
E. japonicus Khan & Araki, 2001 (Japan)
E. kazirangus Khan & Jairajpuri, 1994 (India)
E. koriensis Khan, Park & Choi, 1999 (Korea)
E. lacustris (Loof, 1973) Vinciguerra, 1988 (Surinam)
E. naunii Khan & Jairajpuri, 1994 (India)
E. neocyatholaimus (Kreis, 1936) Thorne, 1967 (India)
E. neoelaboratus (Rahman, Jairajpuri, Ahmad & Ahmad, 1987) Jairajpuri & Ahmad, 1992 (India)
E. nipponicus Khan & Araki, 2002 (Japan)

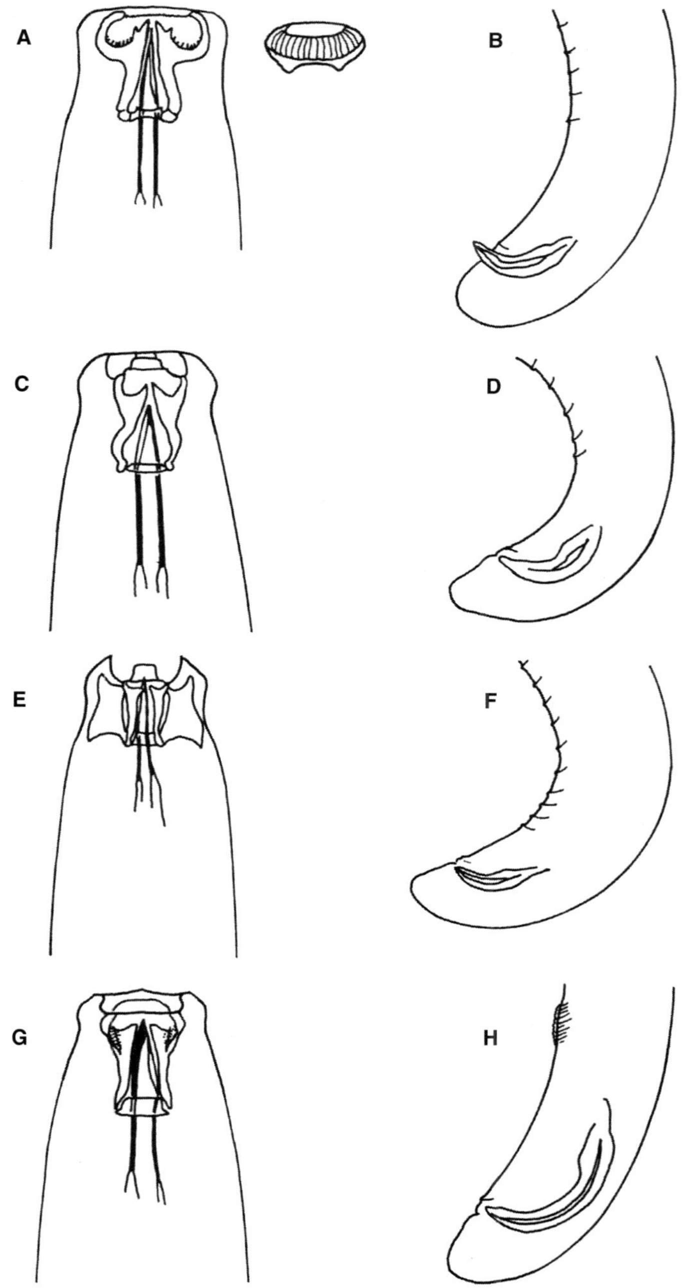

Fig. 15.9 A and B: *Westindicus;* C and D: *Egtitus;* E and F: *Scleractinolaimus;* G and H: *Afractinolaimus*.

E. nudus (Wu & Hoeppli, 1929) Thorne, 1967 (China)
E. proximus (Yeates, 1973) Vinciguerra, 1988 (New Zealand)
E. shillongensis Khan & Jairajpuri, 1994 (India)
E. siddharthi Khan & Jairajpuri, 1996 (India)
E. surinamensis (Micoletzky, 1925) Thorne, 1967 (Surinam, Guyana)
E. zealandicus (Clark, 1963) Vinciguerra & Heyns, 1984 (New Zealand)

Additional reference: Khan and Jairajpuri (1994).

Genus *Scleroactinolaimus* Ahmad, Khan & Ahmad, 1992 (Figs 15.9E and F)
Diagnosis. Actinolaimidae. Cuticle with fine transverse striae. Vestibular ring corrugated; cheilostom with four strongly sclerotized vertical plates, without denticles on the stoma walls. Gonads paired and reflexed; Vulva pore-like; *pars refringens* of vagina with strongly sclerotized pieces. Male supplements arranged in a ventral series. Tail elongate to filiform in females and short in males.

Type and only species: *S. punctatus* Ahmad, Khan & Ahmad, 1992 (India)

Genus *Afractinolaimus* Andrássy, 1970 (Figs 15.9G and H)
Diagnosis. Actinolaimidae. Cuticle smooth. Vestibular ring corrugated; cheilostom with denticles on the stoma walls. *Pars refringens* of vagina with strongly sclerotized pieces. Male supplements arranged in two fascicles. Tail elongate to filiform in females and short in males.
Additional reference: Ahmed *et al.* (1992).

Type species: *A. megaliesmontanus* (Heyns & Argo, 1969) Vinciguerra & Heyns 1984 (South Africa)

Other species:
A. capensis (Heyns & Argo, 1969) Vinciguerra & Heyns, 1984 (South Africa)
A. minor (Vinciguerra & De Francisci, 1973) Vinciguerra & Heyns, 1984 (Italy)
A. zairensis (Baqri, Coomans & Van der Heiden, 1975) Vinciguerra & Heyns 1984 (Congo, Korea)
Additional reference: Vinciguerra and Heyns (1984).

Genus *Neoactinolaimus* Thorne, 1967 (Figs 15.10A and B)
Diagnosis. Actinolaimidae. Cuticle smooth. Vestibular ring corrugated; cheilostom without denticles on the stoma walls; each onchium with a small denticle behind. *Pars refringens* of vagina with strongly sclerotized pieces. Male supplements arranged in two fascicles. Tail elongate to filiform in females and short in males.

Type species: *N. agilis* Thorne, 1967 (Puerto Rico)

Other species:

N. africanus (Filipjev, 1929) Thorne, 1967 (Congo, East Africa)
N. attenuatus Khan, Ahmad & Jairajpuri, 1994 (India)
N. barbieri Vinciguerra & Heyns, 1984 (South Africa, Seychelles)
N. brachydorus Vinciguerra & Heyns, 1984 (South Africa)
N. crassidens Heyns & Argo, 1969 (South Africa)

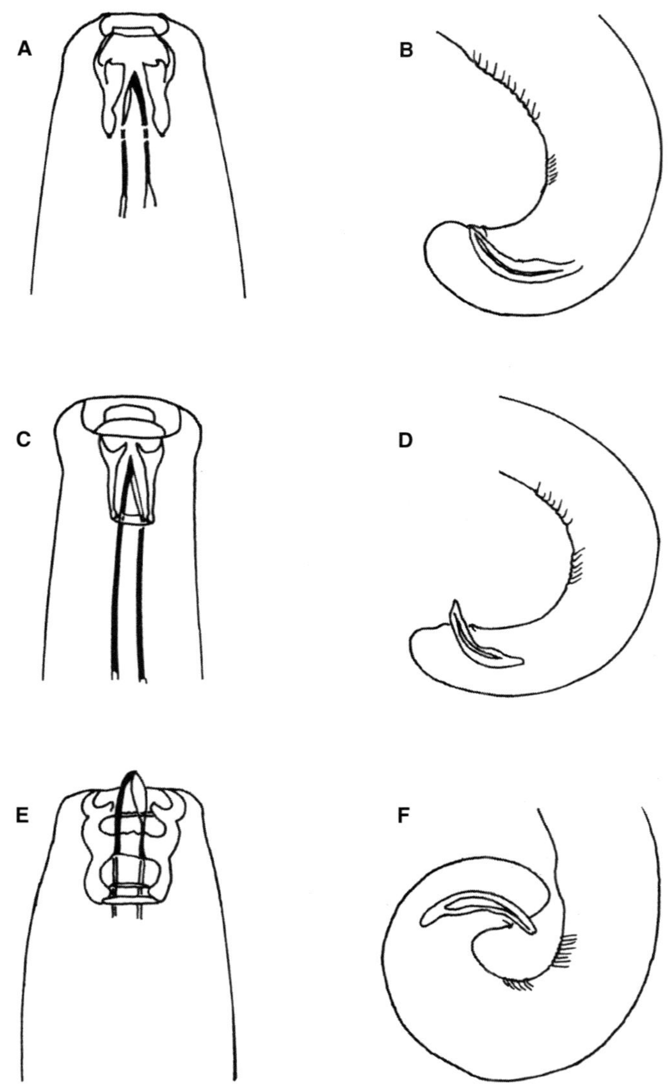

Fig. 15.10 A and B: *Neoactinolaimus*; C and D: *Mactinolaimus*; E and F: *Metactinolaimus*.

N. duplicidentatus (Andrássy, 1968) Andrássy, 1970 (Kenya, Ghana, Uganda, Tanzania, Congo, Hungary, Russia)
N. dzjubani Gagarin, 1979 (Ukraine)
N. gopeshwari Khan & Jairajpuri, 1994 (India)
N. himanicus Khan & Jairajpuri, 1994 (India)
N. imphalensis Khan & Jairajpuri, 1994 (India)
N. hintoni (Lee, 1961) Thorne, 1967 (Nigeria)
N. kosambus Khan, Ahmad & Jairajpuri, 1994 (India)
N. tepidus Andrássy, 1997 (Hungary)

N. thornei Chaturvedi & Khera, 1970 (India, South Africa)
N. vaalensis Andrássy, 1970 (South Africa)

Genus *Mactinolaimus* Andrássy, 1970 (Figs 15.10C and D)
Diagnosis. Actinolaimidae. Cuticle smooth. Vestibular ring corrugated; cheilostome without denticles on the stoma walls; onchia simple without secondary denticles. *Pars refringens* of vagina with strongly sclerotized pieces. Male supplements arranged in two fascicles. Tail elongate to filiform in females and short in males.

Type species: *M. typicus* Andrássy, 1970 (Congo)

Other species:
M. armatus (Jairajpuri, 1968) Vinciguerra, 1988 (India)
M. birketi (Altherr, 1960) Andrássy, 1970 (Cameroon)
M. chitwoodi (Moorthy, 1937) Andrássy, 1970 (India)
M. hutchinsoni (Filipjev, 1929) Andrássy, 1970 (South Africa)
M. michaelseni (Steiner, 1916) Andrássy, 1970 (South West Africa)
M. omercooperi (Filipjev, 1931) Andrássy, 1970 (Congo, Ethiopia, Sunda)
M. pooensis (Gadea, 1950) Andrássy, 1970 (North-west Equatorial Guinea)
M. tenuis (Schneider, 1935) Andrássy, 1970 (Tanganyika, West Africa)
M. transkeiensis (Heyns & Argo, 1969) Vinciguerra, 1988 (South Africa)
Additional reference: Andrássy (1970).

Genus *Metactinolaimus* Meyl, 1957 (Figs 15.10E and F)
Diagnosis. Actinolaimidae. Cuticle smooth. Vestibular ring corrugated; cheilostome with onchia fused and without denticles on the stoma walls. *Pars refringens* of vagina with strongly sclerotized pieces. Male supplements arranged in two fascicles. Tail elongate to filiform in females and short in males.

Type species: *Metactinolaimus kreisi* Meyl, 1957 (Paraguay)

Other species:
M. leloupi Meyl, 1957 (Tanganyika)
Additional reference: Meyl (1957).

Genus *Stopractinca* Khan, Ahmad & Jairajpuri, 1994 (Figs 15.11A–C)
Diagnosis. Cuticle smooth. Vestibular ring corrugated; cheilostome without denticles on the stoma walls; each onchium with an additional tooth below. Pharynx tripartite: a slender anterior part, a gradually expanding middle part and a long, enlarged posterior part. A cuticularized tongue-like piece at pharynx base. *Pars refringens* of vagina with strongly sclerotized pieces. Male supplements arranged in a series. Tail elongate to filiform in females and short in males.
Remarks: This genus was considered by Khan *et al.* (1994b) to be related to the genera *Parastomachoglossa*, *Actinca*, *Brasilaimus*, etc. (the so-called brittonems) because of the tripartite pharynx; however, the lack of longitudinal ridges in the cuticle, the low lip region, the robust odontostyle, the vagina with well-sclerotized pieces and the supplements in series show its affinity to the genera described so far.

Type species: *S. orientalis* Khan, Ahmad & Jairajpuri, 1994 (India)

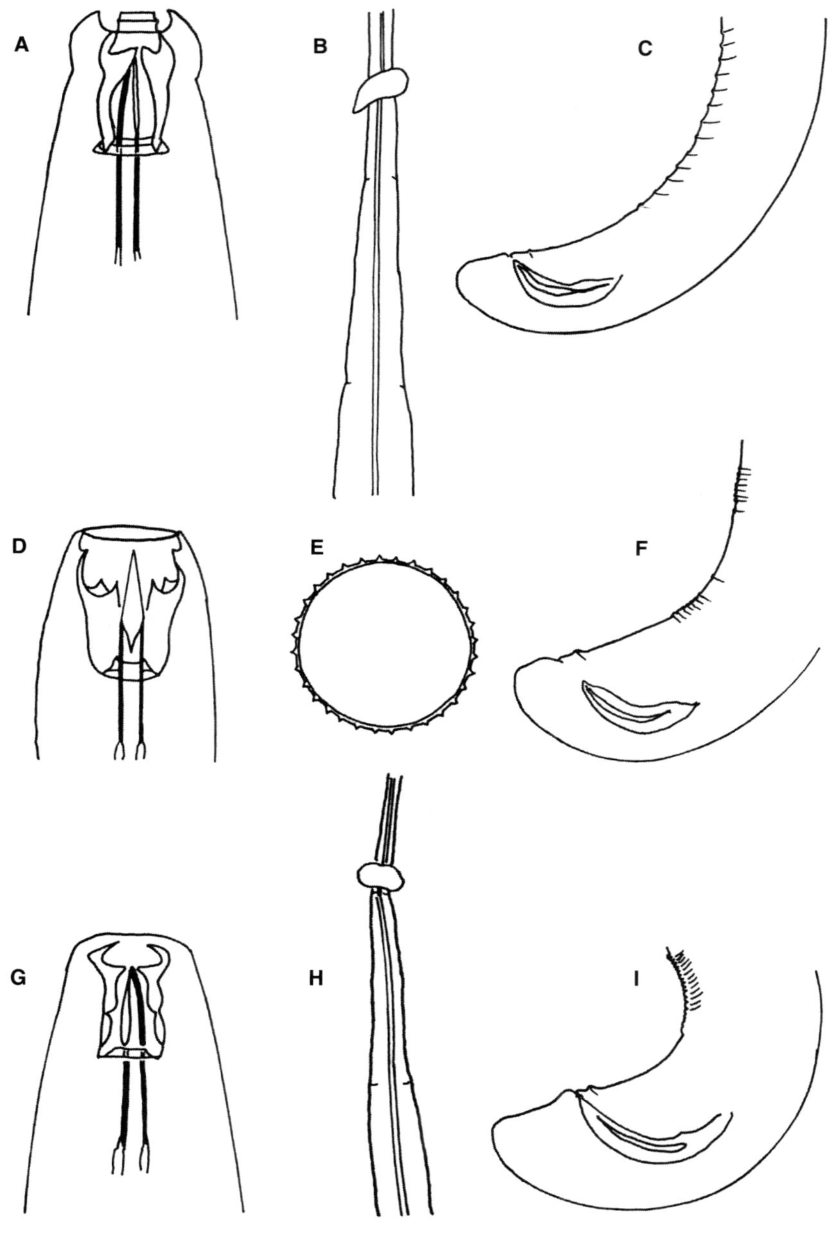

Fig. 15.11 A–C: *Stopractinca*; D–F: *Actinolaimus*; G–I: *Parastomachoglossa*.

Other species:
S. glandulus Khan, Park & Choi, 1999 (Korea)
S. kyotoensis Khan & Araki, 2001 (Japan)
S. malnadensis Dhanam, Jairajpuri & Khan, 1994 (India)
Additional reference: Khan *et al.* (1994a).

Genus *Actinolaimus* Cobb, 1913 (Figs 15.11D–F)
Diagnosis. Actinolaimidae. Cuticle with prominent cuticular ridges; lip region wide and low; vestibular ring sclerotized. Odontostyle robust. Vaginal refractive part with sclerotized pieces; male supplements in two fascicles. Tail elongate in females, short and rounded in males.

Type species: *A. tripapillatus* (Daday, 1905) Steiner, 1916 (Surinam, Paraguay, USA)

Other species:
A. chappuisi Meyl, 1957 (Tanganyika)
A. fritschi Altherr, 1972 (Amazonia)
A. palmaeri Allgen, 1933 (Congo)
A. schuurmansstekhoveni Meyl, 1957 (Kenya, Tanzania, Congo Republic)
A. straeleni Meyl, 1957 (Kenya)

Genus *Parastomachoglossa* Coomans & Loof, 1986 (Figs 15.11G–I)
Diagnosis. Actinolaimidae. Cuticle with prominent cuticular ridges; lip region wide and low; vestibular ring double and sclerotized; the six inner labial papillae external to the ring. Odontostyle robust; in between the four large onchia one to four narrow denticles arise from the same basis. Pharynx tripartite, with an anterior hyaline tube, a median muscular, gradually expanding, middle part and a posterior, evenly enlarged, muscular part; the median portion is longer or as long as the anterior one. Vulva longitudinal; vaginal refractive part with slightly sclerotized pieces; male supplements in two or three fascicles. Tail elongate in females, short and rounded in males.

Type species: *P. costata* (Schneider, 1935) Coomans & Loof, 1986 (West Africa)

Other species:
P. perplexa (Heyns & Argo, 1969) Vinciguerra & Coomans, 1988 (South Africa, Ethiopia)
P. taylori (Meyl, 1957) Coomans & Loof, 1986 (Kenya)
Additional references: Coomans and Loof (1986) and Vinciguerra and Coomans (1988).

Genus *Actinca* Andrássy, 1964 (Figs 15.12A–C)
Diagnosis. Actinolaimidae. Cuticle with conspicuous longitudinal ridges. Lips fused; lip region high; anterior end with a thin sclerotized circular ring slightly protruding, in medial position in respect of the external anterior margin; the six inner labial papillae and the four cephalic papillae are located in the same circle externally to the circular ring. Cheilostom without denticles on the walls. Pharynx tripartite, with long, non-muscular anterior part, short, muscular, expanding middle part and long, muscular, cylindrical posterior part. Vulva longitudinal; *pars refringens* of vagina with weakly sclerotized pieces; males with supplements in two fascicles. Tail elongated in females, short and round in males.

Type species: *A. gracillima* Andrássy, 1964 (Cameroon, Kenya, Uganda).

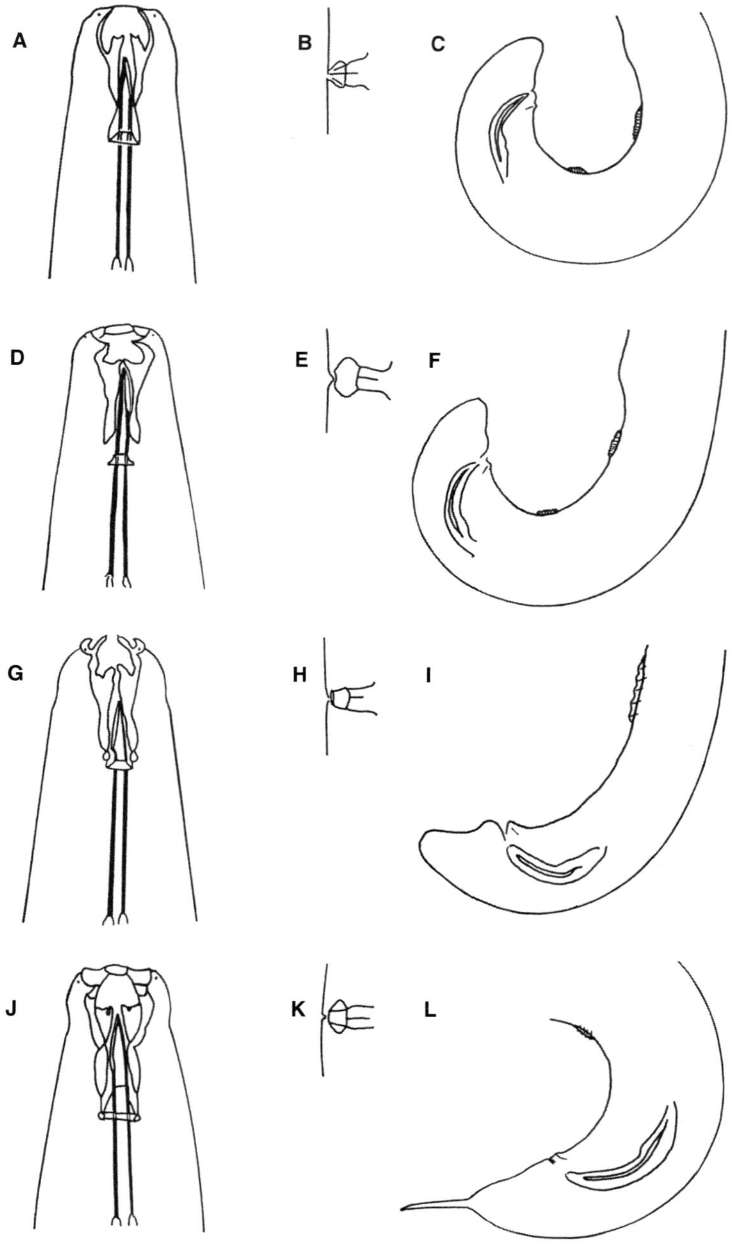

Fig. 15.12 A–C: *Actinca;* D–F: *Brasilaimus;* G–I: *Afractinca;* J–L: *Practinocephalus.*

Other species:

A. bidentata (Loof & Zullini, 2000) Vinciguerra & Clausi, 2003 (Costa Rica)
A. dicastrii Andrássy, 1968 (Paraguay)
A. intermedia Andrássy, 1968 (Tanzania)
A. memorabilis Andrássy, 1968 (Paraguay, Costa Rica)

Species inquirenda: *A. striata* (Thorne, 1939) Andrássy, 1964 (USA)

Species incertae sedis: *A. fusiformis* (Thorne, 1967) Andrássy, 1970 (Puerto Rico)

Additional reference: Vinciguerra and Clausi (2003).

Genus *Brasilaimus* Lordello & Zamith, 1957 (Figs 15.12D–F)
Syn. *Stomachoglossa* Andrássy, 1968
Diagnosis. Actinolaimidae. Cuticle with conspicuous longitudinal ridges; body anteriorly tapering; lips fused; lip region high; anterior end with a thin sclerotized circular ring slightly protruding, in medial position in respect of the external anterior margin; the six inner labial papillae and the four cephalic papillae are located in the same circle externally to the circular ring. Pharynx tripartite, with long, non-muscular anterior part, short, muscular, expanding middle part and long, muscular, cylindrical posterior part; a sclerotized tongue- or rod-like protrusion of the cuticular lining of the pharyngeal lumen is present in the cardia. *Pars refringens* of vagina without sclerotized pieces, thick-walled, wider than *pars proximalis*. Males with supplements in two fascicles. Tail elongated in females, short and round in males.

Type species: *B. subaquilus* Lordello & Zamith, 1957 (Brazil, Costa Rica)

Other species:
B. bryophilus (Hunt, 1978) Vinciguerra, Zullini & Monteiro, 1999 (St. Lucia, Costa Rica)
B. lombardoi Clausi & Vinciguerra, 2003 (Ecuador)
B. pachyderma (Andrássy, 1968) Vinciguerra, Zullini & Monteiro, 1999 (Paraguay)
B. pilatus (Andrássy, 1986) Vinciguerra, Zullini & Monteiro, 1999 (Bolivia)
B. spicatus (Thorne, 1967) Vinciguerra, Zullini & Monteiro, 1999 (Puerto Rico)
B. vinciguerrae Loof & Zullini, 2000 (Costa Rica)
Additional reference: Vinciguerra and Clausi (2003).

Genus *Afractinca* Vinciguerra & Clausi, 2000 (Figs 15.12G–I)
Diagnosis. Actinolaimidae. Cuticle with thick longitudinal ridges: lips fused; lip region high; anterior end externally bordered by a hexagonal thickening on which the six inner labial papillae and the four cephalic papillae are located. Odontostyle long and slender; cheilostom walls without denticles. Pharynx tripartite, with long, non-muscular anterior part, short, muscular, expanding middle part and long, muscular, cylindrical posterior part. Vulva longitudinal; *pars refringens* of vagina with small slightly sclerotized pieces; male supplements in two fascicles. Female tail elongate; male tail short and rounded.

Type species: *A. papillata* (Schneider, 1935) Vinciguerra & Clausi, 2000 (Ivory Coast).

Other species:
A. andrassyi Vinciguerra & Clausi, 2000 (Ivory Coast)
A. heynsi (Coomans & Vinciguerra, 1989) Vinciguerra & Clausi, 2000 (Ivory Coast)
A. irmae (De Ley & Coyne, 1997) Vinciguerra & Clausi, 2000 (Ivory Coast)
Additional reference: Vinciguerra and Clausi (2000).

Genus *Practinocephalus* Andrássy, 1974 (Figs 15.12J–L)
Syn. *Actinocephalus* Thorne, 1967
Diagnosis. Actinolaimidae. Cuticle furrowed by longitudinal ridges; lips fused; lip region high and off set from adjoining body, sometimes swollen or expanded; anterior end with a sclerotized circular ring, in medial position in respect of the external anterior margin; the six inner labial papillae and the four cephalic papillae are located in the same circle externally to the circular ring. Odontostyle long; mouth cavity walls without denticles but with heavy sclerotizations and long and thick postonchial extensions; Pharynx in three parts: the anterior part a hyaline tube, the intermediate muscular part short and the posterior part long and cylindrical. Vulva longitudinal; *pars refringens* of vagina with slightly sclerotized pieces; males with supplements in two fascicles. Female tail elongate conoid to filiform; male tail convex–conoid with a filiform appendix variable in length, shorter than female tail.

Type species: *P. bizarrus* (Thorne, 1967) Andrássy, 1974 (Puerto Rico)

Other species:
P. brzeskii Vinciguerra & Clausi, 2000 (Ecuador)
P. secundus Andrássy, 1986 (Peru)
Additional reference: Vinciguerra and Clausi (2003).

Family QUDSIANEMATIDAE Jairajpuri, 1963
Diagnosis. Dorylaimoidea: Cuticle generally smooth or finely striated; marked with longitudinal ridges in one genus. Lip region generally offset; odontostyle straight, with distinct lumen: its aperture less than 50% its length. Odontophore rod-like. Amphids stirrup-shaped. Posterior expanded part of pharynx about half its length. Pre-rectum distinct. Female genital organs generally didelphic, sometimes opisthodelphic; *pars refringens* of vagina with or without sclerotizations. Testes two; spicules dorylaimoid; a pair of ad-cloacal papillae and a row of ventromedian supplements present. Tail similar in both sexes, generally short, conoid or rounded. Terrestrial, freshwater or semi-freshwater nematodes.
Additional references: Andrássy (1990, 1991) and Jairajpuri and Ahmad (1992).

Subfamilies:
Chrysonematinae Siddiqi, 1969
Discolaiminae Siddiqi, 1969
Carcharolaiminae Thorne, 1967 (no freshwater species)
Qudsianematinae Jairajpuri, 1969
Arctidorylaiminae Mulvey & Anderson, 1979
Paraxonchiinae Dhanachand & Jairajpuri, 1981 (no freshwater species)
Genera *incertae sedis* within the family:
Kittydorylaimus Andrássy, 1999 (no freshwater species)
Kolodorylaimus Andrássy, 1999 (no freshwater species)
Cricodorylaimus Ahmad & Sturhan, 2001 (no freshwater species)

Key to subfamilies of Qudsianematidae
1. Cuticle marked with longitudinal ridges . . . . .Arctidorylaiminae
   Cuticle without longitudinal ridges 2

2. Body tapering strongly in posterior part of neck, lip region very narrow ....Paraxonchiinae
   Body regularly tapering in the neck, lip not so narrow 3
3. Vagina not sclerotized; numerous glandular cells and pores in the lateral chords 4
   Vagina with sclerotization; lateral chords without such glands 5
4. Lip region with a heavily sclerotized basket-like structure; cheilostome wall also sclerotized ....Carcharolaiminae
   Lip region very expanded, lacking such sclerotization ....Discolaiminae
5. Lips well developed, odontostyle robust with aperture about one-third its length ....Qudsianematinae
   Lips fused, odontostyle slender with very small aperture ....Chrysonematinae

Subfamily Arctidorylaiminae Mulvey & Anderson, 1979

Diagnosis. Qudsianematidae. Cuticle marked with longitudinal ridges. Lip region off set from adjoining body. Odontostyle straight, with wide lumen; its aperture shorter than half its length. Female reproductive apparatus didelphic. Tail conoid, similar in both sexes: conoid or rounded.

Remarks: Jairajpuri and Ahmad (1992) considered this peculiar taxon a subfamily of Dorylaimidae, but because of the conoid tail similar in both sexes I prefer to attribute it to Qudsianematidae.

Only genus:

Arctidorylaimus Mulvey & Anderson, 1979

Genus *Arctidorylaimus* Mulvey & Anderson, 1979 (Figs 15.13A–C)

Diagnosis. Arctidorylaiminae. Large nematodes. Lips distinct. Odontostyle aperture about one-third its length; guiding ring double. Expanded part of pharynx more than half its length. Vulva longitudinal. Males with supplements arranged in two groups; two supplements adanal. Tail conoid, ventrally curved, in both sexes. The only species was found in freshwater.

Type and only species: *A. arcticus* Mulvey & Anderson, 1979 (Canada)

Additional reference: Mulvey and Anderson (1979).

Subfamily Discolaiminae Siddiqi, 1969

Diagnosis. Qudsianematidae. Cuticle smooth with numerous glandular cells and pores. Lip region expanded, sometimes discoidal or sucker-like. Odontostyle short, with large opening; guiding ring simple. Vulva transverse; vagina not sclerotized. Tail short, conoid or rounded in both sexes. Mainly terrestrial species. Only a few species found in freshwater.

Additional reference: Andrássy (1990).

Genera:

Discolaimus Cobb, 1913

Discolaimium Thorne, 1939 (no freshwater species)

Discolaimoides Heyns, 1963

Latocephalus Patil & Khan, 1982 (no freshwater species)

Mylodiscus Thorne, 1939 (no freshwater species)

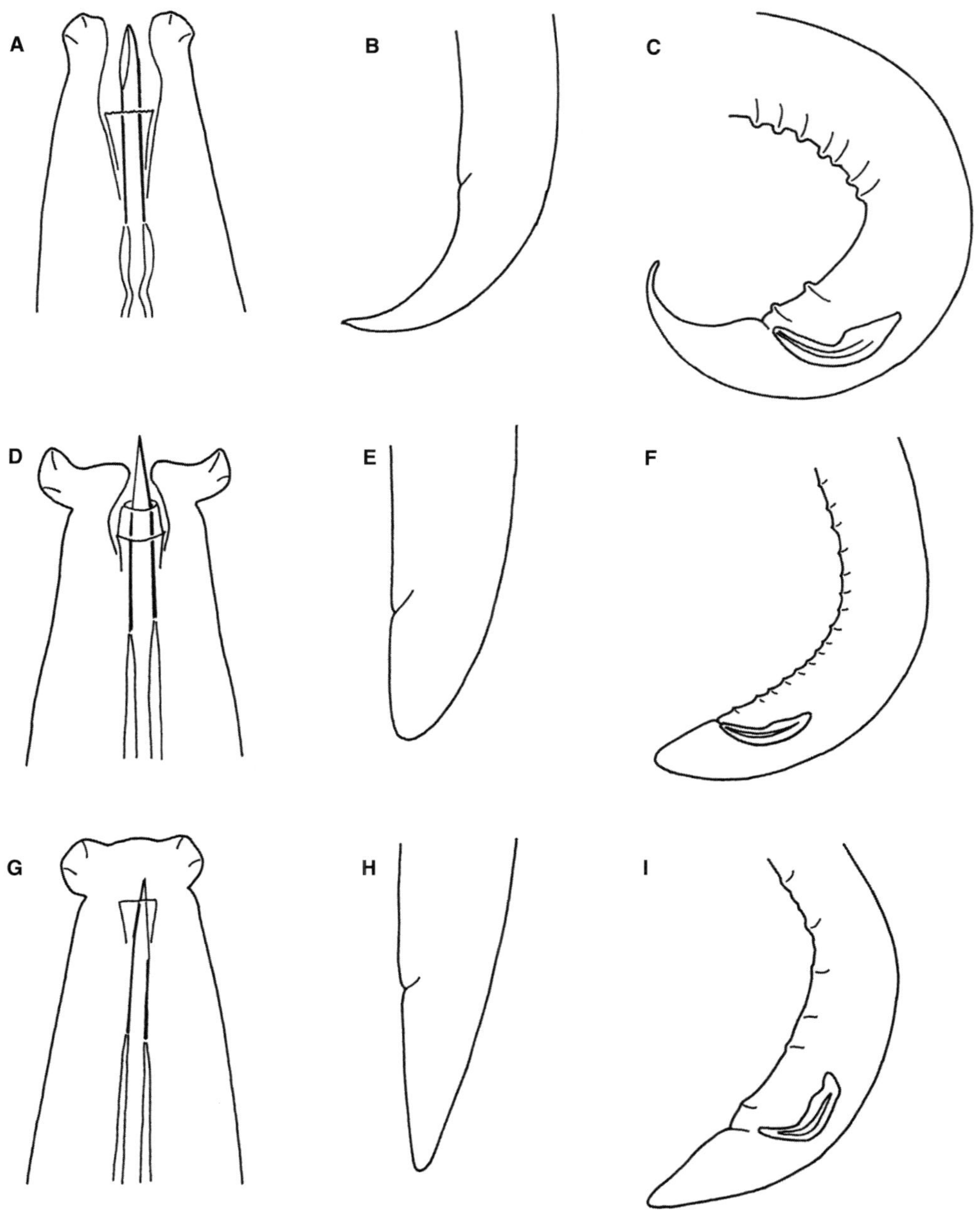

Fig. 15.13 A–C: *Arctidorylaimus*; D–F: *Discolaimus*; G–I: *Discolaimoides*.

Mylodiscoides Lordello, 1963 (no freshwater species)
Filidiscolaimus Siddiqi, 1995 (no freshwater species)

Genus *Discolaimus* Cobb, 1913 (Figs 15.13D–F)
Diagnosis. Discolaiminae. Numerous large glandular bodies in the lateral chords. Lip region very expanded, sucker-like, surrounding the sunken mouth. Odontostyle

robust, with large aperture. Female gonads paired. Tail short, rounded or bluntly conoid in both sexes. Mostly terricolous species, rarely found in freshwater habitats (marked with an asterisk in the list).

Type species: *D. texanus* Cobb, 1913 (USA, Australia, Italy, Spain, Uzbekistan, Central Europe, South Africa)

Other species:
D. acuticapitatus Furstenberg & Heyns, 1966 (South Africa)
D. affinis Loof, 1964 (Venezuela)
D. agricolus Sauer & Annels, 1986 (Australia)
D. albarossicus Merzheevskaja, 1951 (Russia, Belarus, Lithuania, Uzbekistan)
D. auritus Lordello, 1955 (Brazil)
D. bicorticus Furstenberg & Heyns, 1966 (South Africa, Bulgaria)
D. constrictus Heyns, 2001
D. discocephalus Tulaganov, 1949 (Uzbekistan)
D. elegans Sauer & Annels, 1986 (Australia)
D. gossypiorum Karimova, 1957 (Uzbekistan)
D. intermedius Heyns & Lagerway, 1965 (South Africa)
D. krugeri Furstenberg & Heyns, 1966 (South Africa)
D. labiatus Peña-Santiago, Torres, Liebanas & Abolafia, 2002 (Spain)
D. lahorensis Khan, 1998 (India)
D. laksi Khan & Laha, 1982 (India)
D. levinae Furstenberg & Heyns, 1966 (South Africa, Georgia)
D. major Thorne, 1939 (cosmopolitan)*
D. mariae Peña-Santiago, Torres, Liebanas & Abolafia, 2002 (Spain)
D. monoplanus Heyns, 1963 (South Africa, Cameroon)*
D. papillatus Khan, Ahmad & Jairajpuri, 1994 (India)
D. paramajor Coomans, 1966 (Georgia, Congo, Brazil, Italy)
D. perplexans Siddiqi, 1964 (Romania)
D. pizai Monteiro, 1970 (Brazil)
D. rotundicaudatus Khan & Laha, 1982 (India)
D. silvicolus Sauer & Annels, 1986 (Australia)
D. similis Thorne, 1939 (Italy, Uzbekistan, India, South Africa, Venezuela, USA)*
D. tenax Siddiqi, 1964 (India)
D. zicsii Andrássy, 1968 (Congo)
Additional reference: Andrássy (1990).

Genus *Discolaimoides* Heyns, 1963 (Figs 15.13G–I)
Diagnosis. Body very slender; cuticle smooth, subcuticle transversely striated. Lip region well developed, broader than adjoining body, not sucker-like. Odontostyle short, not longer than lip region width. Anterior part of pharynx very slender; dorsal pharyngeal gland far anterior to the expanded posterior part. Pre-rectum with a short caudal sack. Female gonads paired. Ventromedian supplements few, well spaced. Tail conoid or elongate conoid in both sexes. Mostly terricolous species, rarely found in freshwater habitats (marked with an asterisk in the list).

Type species: *D. bulbiferus* (Cobb, 1906) Heyns, 1963 (cosmopolitan)

Other species:

D. arcuatus (Husain & Siddiqi, 1967) Andrássy, 1990 (India)

D. arcuicaudatus (Furstenberg & Heyns, 1965) Das, Khan & Loof, 1969 (South Africa, Egypt)

D. brevicaudatus Vinciguerra & Orselli, 1998 (Italy)

D. discolaimioideus (Andrássy, 1971) Andrássy, 1991 (Italy)*

D. filiformis Das, Khan & Loof, 1969 (India, The Netherlands, Switzerland, Vietnam)

D. florealis Vinciguerra & Orselli, 1998

D. gracilis (Thorne, 1939) Andrássy, 1990 (USA, Venezuela, Italy, Uzbekistan)

D. indicus Thapa & Ganguly, 1993 (India)

D. intrastriatus (Loos, 1945) Loof, 1964 (Sri Lanka)

D. loofi Andrássy, 1990 (Hungary)

D. paraloofi Ahmad, Bloemers & Wanless, 1996 (Cameroon)

D. skrjabini (Tulaganov, 1949) Andrássy, 1990 (Uzbekistan)

D. spatilabium Khan & Laha, 1982 (India)

D. symmetricus Das, Khan & Loof, 1969 (The Netherlands, Great Britain, Hungary, Italy, Egypt)

D. tenuis (Furstenberg & Heyns, 1965) Das, Khan & Loof, 1969 (South Africa, Spain)

D. teres Khan & Laha, 1982 (India)

Additional reference: Andrássy (1990).

Subfamily Qudsianematinae Jairajpuri, 1965

Diagnosis. Qudsianematidae. Cuticle smooth, without glands in the lateral chords. Lip region off set from adjoining body but not wider than it; lips distinct; cheilostome without sclerotization. Odontostyle generally shorter than twice its lip region width; aperture shorter than half its length. Guiding ring simple or, more rarely, double. Pharynx muscular, expanded in its posterior half. Vulva transverse or longitudinal; *pars refringens* of vagina with sclerotized pieces. Female reproductive apparatus generally didelphic, mono–opisthodelphic in one genus. Tail rather short, similar in both sexes, conoid or rounded. Most species of the 14 genera are terrestrial but quite a few species have been found in freshwater habitats.

Additional reference: Andrássy (1991).

Genera:

Labronema Thorne, 1939

Eudorylaimus Andrássy, 1959

Torumanawa Yeates, 1967 (no freshwater species)

Ecumenicus Thorne, 1974

Takamangai Yeates, 1974

Pachydorylaimus Siddiqi, 1983

Labronemella Andrássy, 1985

Microdorylaimus Andrássy, 1986

Allodorylaimus Andrássy, 1986

Kallidorylaimus Andrássy, 1986 (no freshwater species)

Epidorylaimus Andrássy, 1986

Skibbenema Van Reenen & Heyns, 1986 (no freshwater species)

Gopalus Khan, Jairajpuri & Ahmad, 1988
Baqriella Ahmad & Jairajpuri, 1989 (no freshwater species)
Talanema Andrássy, 1991 (no freshwater species)
Crassogula Andrássy, 1991
Boreolaimus Andrássy, 1998
Amblydorylaimus Andrássy, 1998 (no freshwater species)
Inbionema Loof & Zullini, 2000 (no freshwater species)

Key to genera of Qudsianematinae

1. Female gonad mono–opisthodelphic . . . . *Ecumenicus*
   Female gonads didelphic 2
2. Odontostyle weakly sclerotized, irregular in shape . . . . *Amblydorylaimus*
   Odontostyle well sclerotized 3
3. Posterior part of tail appearing empty; a single anterior sub-ventral gland nucleus in the pharynx . . . . *Boreolaimus*
   Posterior part of tail never empty, one pair of anterior sub-ventral gland nuclei in the pharynx 4
4. Tail rounded or bluntly conoid, about as long as anal body width 5
   Tail conoid or convex–conoid, clearly longer than anal body width 11
5. Presence of a constriction between the thinner anterior and the expanded posterior part of pharynx . . . . *Skibbenema*
   No constriction between the two parts of pharynx 6
6. Cardia provided with three glands . . . . *Torumanawa*
   Cardia without glands 7
7. Vulva longitudinal, supplements contiguous . . . . *Labronema*
   Vulva transverse, supplements not contiguous 8
8. Odontostyle very slender, 10–15 times as long as wide; mouth sunken, surrounded by inner liplets . . . . *Labronemella*
   Odontostyle not so slender, mouth not sunken, liplets absent 9
9. Anterior part of pharynx strongly muscular; sperm very small . . . . *Crassogula*
   Anterior part of pharynx thin, weakly muscular; sperm not very small 10
10. Guiding ring single, tail subdigitate . . . . *Talanema*
    Guiding ring double, tail broadly rounded . . . . *Takamangai*
11. Tail longer than three anal body widths 12
    Tail shorter, 1–3 anal body-widths long 15
12. Odontostyle very thick and short; odontophore flanged . . . . *Pachydorylaimus*
    Odontostyle normal; odontophore rod-like 13
13. Outer margins of lips forming a flap over oral area, inner margins sunken . . . . *Baqriella*
    Lips normal 14
14. Female tail straight; a single supplement in male . . . . *Kallidorylaimus*

Female tail ventrally curved; more than four supplements in male . . . . *Epidorylaimus*

15. The most posterior supplements well anterior to the spicules . . . . *Eudorylaimus*
    The most posterior supplement in the range of the spicules or just before it 16
16. Very small size, body length less than 1 mm; pharynx relatively long, posterior expanded portion two-fifths of its length . . . . . . *Microdorylaimus*
    Larger size; pharynx expanded at about half its length 17
17. Odontostyle about as long as lip region width . . . . *Allodorylaimus*
    Odontostyle at least twice as long as lip region width 18
18. Odontostyle about twice as long as lip region width . . . . *Inbionema*
    Odontostyle more than three times as long as lip region width . . . . *Gopalus*

Genus *Labronema* Thorne, 1939 (Figs 15.14A and B)

Diagnosis. Qudsianematinae. Rather large nematodes. Cuticle thick, finely transversely striated. Lip region offset, lips well developed; six small inner liplets around mouth. Odontostyle at least as long as lip region width or longer; guiding ring double. Female gonads paired; vulva longitudinal; vagina with sclerotization. Male supplements numerous, contiguous. Tail rounded in both sexes, never longer than anal body width. Most species are terricolous but some have been found in freshwater habitats. The latter are marked by an asterisk.

Type species: *L. ferox* Thorne, 1939 (USA, Spain, Nepal)

Other species:

L. alticola (Menzel in Hofmänner & Menzel, 1914) Thorne, 1939 (Switzerland)
L. andrassyi Gagarin, 1993 (Russia)*
L. angeloi Vinciguerra & Clausi, 1994 (Italy)
L. arenicola (Altherr, 1958) Andrássy, 1986 (Germany)*
L. baqrii Khan, Jairajpuri & Ahmad, 1989 (India)
L. bathybium (Daday, 1906) Andrássy, 1962 (Switzerland)*
L. bicuticulum Furstenberg, Heyns & Swart, 1993 (Seychelles)
L. brevicauda Furstenberg, Heyns & Swart, 1993 (Madagascar)
L. carusoi Vinciguerra & Orselli, 1998 (Italy)
L. chilense Andrássy, 1967 (Chile)
L. confusum (Jana & Baqri, 1983) Andrássy, 1991 (India)
L. corii (Liebermann, 1928) Andrássy, 1960 (Czechoslovakia)*
L. deoriaensis Khan, Jairajpuri & Ahmad, 1989 (India)
L. digiturum Vinciguerra, 1984 (Italy)
L. diversum Andrássy, 2002 (Chile)
L. fimbriatum Thorne, 1939 (USA, China)
L. fluviatile Altherr, 1958 (Germany)*
L. glandosum Rahman, Jairajpuri, Ahmad & Ahmad, 1987 (India)
L. goodeyi Altherr in Altherr & Delamare-Deboutteville, 1972 (Russia, Ethiopia, USA)

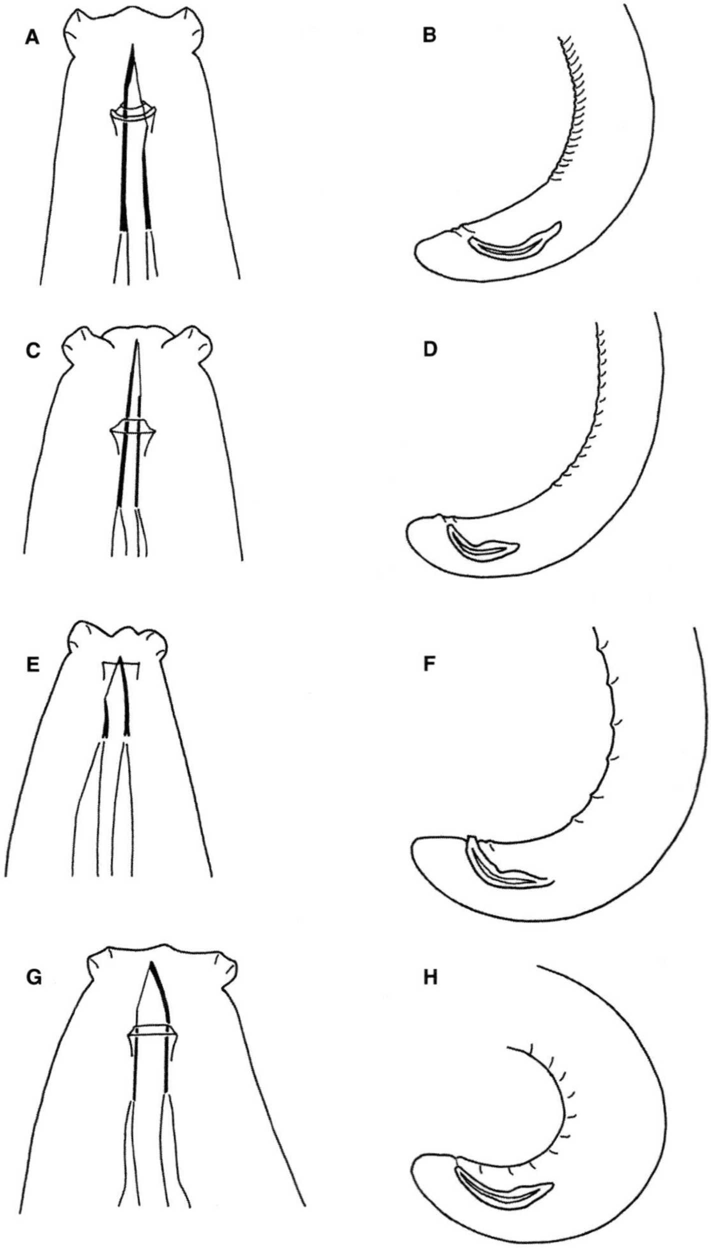

Fig. 15.14 A and B: *Labronema;* C and D: *Labronemella;* E and F: *Takamangai;* G and H: *Crassogula.*

L. ibarakiense Khan & Araki, 2002 (Japan)
L. hyalinum (Thorne & Swanger, 1936) Thorne, 1939 (Mauritius, USA)
L. khazariensis (Chesunov, 1985) Andrássy, 1991 (Caspian Sea)*
L. korandrum Choi, Khan & Choi, 2001

L. latum (Cobb, 1891) Andrássy, 1986 (Russia, Australia)
L. loeffleri Andrássy, 1978 (Nepal, Russia)*
L. magnum Altherr, 1972 (Sweden)*
L. malagasi Furstenberg, Heyns & Swart, 1993 (Madagascar)
L. mangalorense Ahmad & Ahmad, 2002 (India)
L. neopacificum Rahman, Jairajpuri, Ahmad & Ahmad, 1987 (India)
L. nepalense Ahmad & Jairajpuri, 1982 (Nepal)
L. obesum Thorne, 1974 (USA)
L. pacificum (Cobb, 1906) Thorne, 1939 (Mauritius, Jamaica, Hawaii, New Caledonia)
L. papillatum Khan, Ahmad & Jairajpuri, 1995 (India)
L. pulchrum Vinciguerra & Zullini, 1980 (Italy, Spain)
L. rapax Thorne, 1974 (USA, Canada, Spain)
L. rikia Yeates, 1967 (New Zealand)
L. seychellense Furstenberg, Heyns & Swart, 1993 (Seychelles)
L. sphinctum Mohilal & Dhanachand, 2001
L. stechlinense Altherr, 1968 (Germany, Russia, Alaska)*
L. thornei Ferris, 1968 (USA)
L. varicaudatum (Thorne, 1929) Thorne, 1939 (USA, Spain)
L. virgo Monteiro, 1970 (Brazil, Spain)
L. vulvapapillatum (Meyl, 1954) Loof & Grootaert, 1981 (The Netherlands, Belgium, Germany, Italy, Hungary, Spain)*
Additional reference: Andrássy (1991).

Genus *Labronemella* Andrássy, 1985 (Figs 15.14C and D)
Diagnosis. Qudsianematinae. Cuticle transversely striated. Lip region markedly offset, with sunk, plate-like oral field and conspicuous inner liplets. Odontostyle very slender, longer than lip region width; guiding ring double. Pharynx expanded near middle. Female gonads paired. Vulva transverse; vagina with sclerotizations. Male supplements low, closely spaced. Tail short, rounded in both sexes. Freshwater, semi-freshwater or terrestrial nematodes.

Type species: *L. labiata* Andrássy, 1985

Other species:
L. andrassyi (Baqri & Khera, 1975) Andrássy, 1985 (India)
L. czernowitziensis (Micoletzky, 1922) Andrássy, 2002 (Germany, Denmark, Austria, Romania, The Netherlands, Hungary)*
L. georgiensis Eliava & Kuchava, 2001
L. hemicaudata Baqri, 1991 (India)
L. loofi (Ahmad & Jairajpuri, 1983) Andrássy, 1985 (India)
L. octodurensis (Altherr, 1950) Andrássy, 1991 (Switzerland, Russia)
L. ruttneri (Schneider, 1937) Andrássy, 1985 (Sumatra)*
Additional reference: Andrássy (1991).

Genus *Takamangai* Yeates, 1967 (Figs 15.14E and F)
Syn. *Thonus* Thorne, 1974
Diagnosis. Qudsianematinae. Cuticle smooth. Lip region generally offset; lips separate. Odontostyle about as long or slightly longer than lip region width; guiding ring

simple, thin. Pharynx enlarged in its posterior half. Female gonads amphidelphic; vulva transverse; vagina sclerotized. Male supplements spaced; pre-cloacal space between the most posterior one and the ad-cloacal papillae present. Tail short, rounded or conoid–rounded in both sexes. Most species are terricolous but some have been found in freshwater habitats. The latter are marked by an asterisk.

Type species: *T. waenga* Yeates, 1967 (New Zealand)

Other species:
T. balda (Thorne, 1974) Andrássy, 1991 (USA)
T. brachycephala (Thorne & Swanger, 1936) Andrássy, 1991 (USA, Georgia)
T. circulifera (Loof, 1961) Andrássy, 1991 (The Netherlands, Germany, Spain, UK, Poland, Italy, USA)*
T. confusa (Thorne, 1939) Andrássy, 1991 (USA, Uzbekistan)
T. cylindrica (Thorne, 1974) Andrássy, 1991 (USA, India)
T. dogielii (Tulaganov, 1949) Andrássy, 1991 (Poland, Georgia, Uzbekistan, Kazakhstan)
T. elegans (Thorne, 1974) Andrássy, 1991 (USA)
T. eroshenkoi Andrássy, 1991 (Russia, Hungary)
T. ettersbergensis (de Man, 1885) Andrássy, 1991 (cosmopolitan)*
T. goaensis (Ahmad, 1993) n. comb. (India)
T. goldeni (Khan & Fatima, 1980) Andrássy, 1991 (Pakistan)
T. gracilis (Eroshenko, 1976) Andrássy, 1991 (Russia)
T. himala (Jairajpuri & Ahmad, 1983) Andrássy, 1991 (India, Hungary)
T. insignis (Loos, 1945) n. comb. (Sri Lanka)
T. kaszabi (Andrássy, 1959) Andrássy, 1991 (Poland, France, Yugoslavia, Italy)
T. laticollis (de Man, 1907) Andrássy, 1991 (The Netherlands, Denmark, Georgia, Congo)*
T. lauta (Andrássy, 1959) Andrássy, 1991 (Hungary, Uzbekistan)
T. major (Thorne, 1974) Andrássy, 1991
T. mediana (Eroshenko, 1976) Andrássy, 1991 (Russia)
T. minima (Steiner, 1914) Andrássy, 1991 (Switzerland, Australia)
T. nothus (Thorne & Swanger, 1936) Andrássy, 1991 (USA, Poland, Hungary, Sweden, South Africa, Czech Republic)
T. parvula (Thorne & Swanger, 1936) Andrássy, 1991 (Russia, USA)
T. pavlovskii (Tulaganov, 1949) Andrássy, 1991 (Uzbekistan)
T. porosa (Zell, 1986) Andrássy, 1991 (Germany)
T. pumila (Andrássy, 1963) Andrássy, 1991 (Hungary)
T. pusilla (Andrássy, 1985) Andrássy, 1991 (Hungary, Italy)
T. rhopalocerca (de Man, 1880) Andrássy, 1991 (cosmopolitan)*
T. saccata (Thorne, 1974) Andrássy, 1991 (Pakistan, USA)
T. steineri (Thorne & Swanger, 1936) Andrássy, 1991 (Switzerland)
T. tropica (Jana & Baqri, 1981) Andrássy, 1991 (India)
Additional reference: Andrássy (1991).

Genus *Crassogula* Andrássy, 1991 (Figs 15.14G and H)
Diagnosis. Qudsianematinae. Cuticle smooth. Lip region markedly offset; lips separate. Odontostyle slightly longer than lip region width; guiding ring double but thin.

Pharynx gradually expanded, unusually thick and muscular also in the anterior part. Female gonads paired; vulva transverse; vagina sclerotized. Male supplements numerous, minute and very close to each other. Pre-cloacal space present. Tail broadly rounded in both sexes, shorter than anal body width. The single species known so far is freshwater.

Type and only species: *C. torosa* Andrássy, 1991 (Ecuador)

Additional reference: Andrássy (1991).

Genus *Eudorylaimus* Andrássy, 1959 (Figs 15.15A and B)
Syn. *Qudsianema* Jairajpuri, 1965; *Witoldinema* Brzeski, 1960
Diagnosis. Qudsianematinae. Cuticle smooth or finely transversely striated. Lip region offset, lips distinct. Odontostyle about as long as lip region width or slightly longer, its aperture less than half its length; guiding ring simple. Pharynx expanded near middle or slightly behind. Female reproductive apparatus didelphic. Vulva transverse, rarely longitudinal; refractive part of vagina well sclerotized. Males frequent; supplements variable in number, not contiguous; pre-cloacal space present; sperm fusiform. Tail conoid, straight or ventrally bent, in both sexes. Most species are terricolous but some have been found in freshwater habitats. The latter are marked by an asterisk.

Type species: *E. carteri* (Bastian, 1865) Andrássy, 1959 (cosmopolitan)*

Other species:
E. acuticauda (de Man, 1880) Andrássy, 1959 (cosmopolitan)*
E. acutus (Thorne & Swanger, 1936) Andrássy, 1959 (USA, Georgia, Romania, Russia)
E. altherri Tjepkema, Ferris & Ferris, 1971 (USA, Romania)
E. amabilis (Jairajpuri, 1965) Siddiqi, 1966 (Botswana, India)
E. andersoni Khan, 1989
E. antarcticus (Steiner, 1916) Yeates, 1970 (Antarctica)
E. aquilonarius Tjepkema, Ferris & Ferris, 1971 (USA)
E. arcus (Thorne & Swanger, 1936) Andrássy, 1959 (USA, Poland, Italy, Russia, Uzbekistan)
E. arenarius Bussau, 1991 (Denmark)
E. badensis Zell, 1986 (Germany)*
E. bombilectus Andrássy, 1962 (Germany, Hungary, Spain, Russia, Mongolia, Uzbekistan)*
E. brevidens (Thorne & Swanger, 1936) Andrássy, 1959 (USA, Czech Republic)
E. brevis (Altherr, 1952) Andrássy, 1959 (France, Spain, Switzerland, Italy, Bulgaria, Poland, Czech Republic, Sweden, USA, Russia)*
E. bureshi (Andrássy, 1958) Andrássy, 1959 (Poland, Bulgaria, Czech Republic, France, Lithuania, Cuba)*
E. centrocercus (de Man, 1880) Andrássy, 1959 (widespread in Europe and in Asia, Congo, Ivory Coast, Kenya)*
E. chauhani (Baqri & Khera, 1975) Andrássy, 1986 (India)
E. coloradensis Loof, 1971 (USA)
E. confusus Thorne, 1974 (USA, Canada, Uzbekistan)
E. conicaudatus Thorne, 1974 (USA)

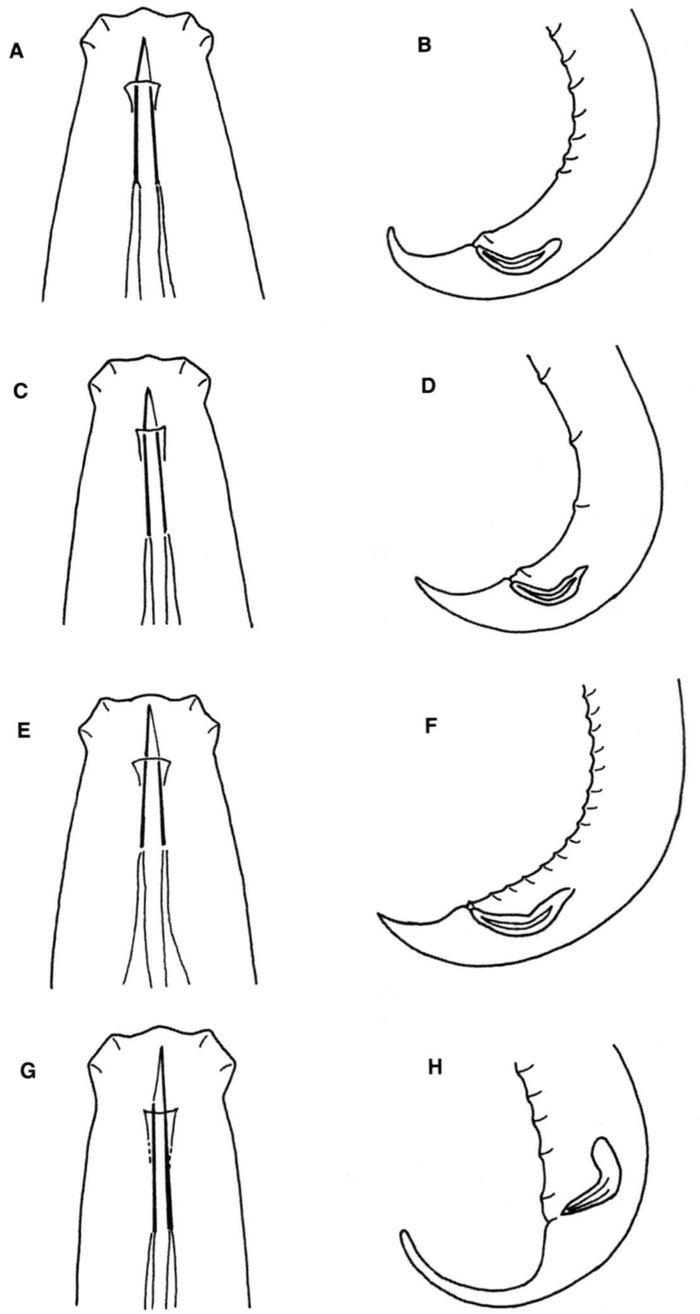

Fig. 15.15 A and B: *Eudorylaimus*; C and D: *Microdorylaimus*; E and F: *Allodorylaimus*; G and H: *Epidorylaimus*.

E. coniceps Loof, 1975 (Antarctica)
E. dermatus (Thorne, 1939) Andrássy, 1959 (USA)
E. discolaimioideus (Andrássy, 1958) Andrássy, 1959 (Bulgaria, Germany, Spain, Uzbekistan)*
E. dubius Thorne, 1974 (USA)
E. eremitus (Thorne, 1939) Andrássy, 1959 (USA, Uzbekistan)
E. eudorylaimoides (Geraert, 1962) Andrássy, 1991 (Russia, Uzbekistan, Cuba, Congo)
E. familiaris Winiszewska-Slipinska, 1987 (Poland)*
E. fransus Heyns, 1963 (South Africa)
E. franzi Andrássy, 1967 (Chile)
E. ibiti Lordello, 1965 (Brazil)
E. imitatoris Gagarin, 1982 (Mongolia)
E. indicus Soni & Nama, 1980 (India)
E. iners (Bastian, 1865) Andrássy, 1959 (Congo, Egypt, Australia, widespread in Europe and in Central Asia)*
E. isokaryon Loof, 1975 (Antarctica)
E. japonicus Khan & Araki, 2000 (Japan)
E. junctus (Cobb in Thorne & Swanger, 1936) Andrássy, 1959
E. juniperi Andrássy, 1987 (Hungary, Italy)
E. jurassicus (Altherr, 1953) Andrássy, 1959 (Switzerland, Czech Republic, Spain)
E. leuckarti (Bütschli, 1873) Andrássy, 1959 (Central Europe, Spain, France, Russia, Kazachstan, Ghana, Alaska)*
E. liangii Ahmad, Wu & Shaheen, 2002
E. lindbergi Andrássy, 1960 (Sweden, Russia, Afghanistan, Mongolia)*
E. longicardius Thorne, 1974 (USA)
E. lotharingiae Altherr, 1963 (France)*
E. magistri Andrássy, 1986 (USA)
E. maritimus (Ditlevsen, 1913) Andrássy, 1959 (Greenland)
E. maritoides Zell, 1986 (Germany)
E. maritus Andrássy, 1959 (Moldavia, Russia)
E. megadon Loof, 1971 (Spitzbergen)*
E. meridionalis Tjepkema, Ferris & Ferris, 1971 (USA, Russia)*
E. minutus (Bütschli, 1873) Andrássy, 1959 (Germany, The Netherlands, Czechoslovakia, Russia, Spain, Georgia, Uzbekistan, Kyrgyzstan, USA, Jamaica, Venezuela)
E. nitidus (Cobb in Thorne & Swanger, 1936) Andrássy, 1959 (Spain, Venezuela, USA)
E. nodus (Thorne & Swanger, 1936) Andrássy, 1959 (USA, Czechoslovakia)
E. novus Vinciguerra & Orselli, 1998 (Italy)
E. nudicaudatus Heyns, 1993 (South Africa)
E. opistohystera (Altherr, 1953) Andrássy, 1959 (Switzerland, Czech Republic, Italy, Romania, Georgia)
E. paesleri Andrássy, 1954 (Hungary)
E. parabokori Altherr, 1974 (Germany)*
E. paradiscoalimioideus Altherr, 1976 (Austria)
E. paramonovi Eliava & Bagaturia, 1968 (Georgia)
E. paucipapillatus Andrássy, 1986 (USA)
E. pectinatus Mukhina, 1970 (Russia, Belarus)
E. perspicuus (Andrássy, 1958) Andrássy, 1959 (Romania, Bulgaria)

E. productus (Thorne & Swanger, 1936) Andrássy, 1959 (USA, Uzbekistan, Spain, Central Europe)
E. pseudobokori Zell, 1986 (Germany)
E. pseudocarteri Loof, 1975 (Antarctica)
E. pussulosus Andrássy, 1991 (Hungary)
E. quadramphidius Andrássy, 1973 (Argentina)
E. retractus Thorne, 1974 (USA)
E. rugosus (Andrássy, 1957) Andrássy, 1959 (Hungary, Poland, Spain)
E. saxifragae Popovici, 1995
E. schraederi Altherr, 1974 (Germany)*
E. shirasei Kito, Shishida & Ohyama, 1996 (Japan)
E. silvaticus Brzeski, 1960 (Poland, USA, Georgia)*
E. similis (de Man, 1876) Andrássy, 1959 (Central Europe, Russia, Belarus, Kazakhstan, Georgia, Uzbekistan)*
E. sodakus Thorne, 1974 (USA)
E. solus Andrássy, 1962 (Hungary)
E. spaulli Loof, 1975 (Antarctica)
E. spongiophylus Batalova, 1983 (Siberia)
E. stefanskii (Brzeski, 1960) Andrássy, 1991 (Poland)
E. subdigitalis Tjepkema, Ferris & Ferris, 1971 (USA, Italy)
E. subjunctus Loof, 1971 (Spitzbergen, Sweden, Pakistan)
E. subulophilus Tjepkema, Ferris & Ferris, 1971 (USA, India)
E. thorneanus Andrássy, 1990 (USA)
E. truncatus (Cobb in Thorne & Swanger, 1936) Andrássy, 1959
E. turkestanicus Eliava, 1978 (Turkestan, Sweden)
E. unicus Khan & Araki, 2000 (Japan)
E. vanrosseni Loof, 1971 (Spitzbergen)
E. verrucosus Loof, 1975 (Antarctica)
E. vestibulifer (Micoletzky, 1922) Andrássy, 1959 (Austria, Czech Republic)
Additional references: Andrássy (1986, 1991).

Genus *Microdorylaimus* Andrássy, 1986 (Figs 15.15C and D)
Diagnosis. Qudsianematinae. Very small nematodes, not reaching 1 mm in length. Body rather plump. Cuticle smooth. Lip region offset; lips distinct, with prominent papillae. Odontostyle about as long as lip region width. Pharynx about one-third as long as body length, expanded in its posterior two-fifths. Female gonads paired. Vulva transverse; vagina weakly sclerotized. Males rare. Supplements spaced; no pre-cloacal space between the most posterior one and the ad-cloacal papillae. Tail conoid, straight or ventrally bent, in both sexes. Most species are terricolous but some have been found in freshwater habitats. The latter are marked by an asterisk.

Type species: *M. parvus* (de Man, 1880) Andrássy, 1986 (widespread in Europe and Asia)*

Other species:
M. angelus (Thorne, 1974) Andrássy, 1986 (USA, Spain)
M. diminutivus (Thorne & Swanger, 1936) Andrássy, 1986 (USA)

M. drepanoideus (Eroshenko, 1976) Andrássy, 1991 (Russia)
M. dubius (Thorne, 1974) Andrássy, 1991 (USA)
M. longicollis (Brzeski, 1964) Andrássy, 1986 (Poland, Hungary, Spain, Italy, Georgia)
M. minor (Cobb in Thorne & Swanger, 1936) Andrássy, 1986 (Georgia, Japan, USA, Venezuela, Brazil)
M. minusculus (Loos, 1946) Andrássy, 1986 (Sri Lanka)
M. miser (Thorne & Swanger, 1936) Andrássy, 1986 (USA, Venezuela, Hungary, The Netherlands, Italy)*
M. modestus (Altherr, 1952) Andrássy, 1986 (The Netherlands, Spain, Italy, UK, Georgia, USA)
M. modicus (Kirjanova, 1951) Andrássy, 1986 (Uzbekistan)
M. parvissimus (Bagaturia & Eliava, 1968) Andrássy, 1986 (Georgia, Canada, USA)
M. profestus (Andrássy, 1963) Andrássy, 1986 (Argentina)
M. rapsoides (Heyns & Lagerway, 1965) Andrássy, 1986 (South Africa)
M. rapsus (Heyns, 1963) Andrássy, 1986 (South Africa, India, Japan, Poland, Spain, Italy, USA)
M. thornei (Tjepkema, Ferris & Ferris, 1971) Andrássy, 1986 (USA, Spain)
Additional references: Andrássy (1986, 1991).

Genus *Allodorylaimus* Andrássy, 1986 (Figs 15.15E and F)
Diagnosis. Qudsianematinae. Cuticle smooth or finely transversely striated. Lip region offset, lips distinct. Odontostyle about as long as lip region width; guiding ring simple. Pharynx expanded in its posterior half. Female gonads paired; vulva transverse or longitudinal; vagina with sclerotizations. Male supplements spaced; no pre-cloacal space between the most posterior one and the ad-cloacal papillae. Tail conoid or convex–conoid, straight or ventrally curved, in both sexes. Most species are terricolous but some have been found in freshwater habitats.

Type species: *A. uniformis* (Thorne, 1929) Andrássy, 1986 (Austria, Russia, Germany, Georgia, USA, Nepal)*

Other species:
A. aljabaranus Quijano, Peña-Santiago & Jiménez-Guirado, 1991 (Spain)
A. allgeni (Andrássy, 1958) Andrássy, 1986 (Sweden, Romania, Bulgaria)
A. alpinus (Steiner, 1914) Andrássy, 1986 (The Netherlands, Switzerland)
A. americanus Andrássy, 1986 (USA)
A. andrassyi (Meyl, 1955) Andrássy, 1986 (The Netherlands, Germany, Russia, Israel, Spain)*
A. bokori (Andrássy, 1959) Andrássy, 1986 (Hungary, Russia, Georgia)*
A. diadematus (Cobb in Thorne & Swanger, 1936) Andrássy, 1986 (Italy, Yugoslavia, UK, Georgia, South Africa, Jamaica, Venezuela, Brazil))*
A. digiturus (Thorne, 1939) Andrássy, 1986 (Holland)
A. ferrisorum Andrássy, 1986 (USA)
A. granuliferus (Cobb, 1893) Andrássy, 1986 (USA, Spain, Switzerland, Poland, Czech Republic, Mongolia, Japan, Java, Sumatra, Fiji, Mauritius, Hawaii, Trinidad, Surinam, Venezuela, Brazil, Paraguay)

A. holdemani (Andrássy, 1959) Andrássy, 1959 (Bulgaria, Italy, Spain, Nepal)*
A. husmanni (Altherr, 1972) Andrássy, 1986 (Sweden, Hungary)*
A. infundibulicaudatus Andrássy, 1991 (Hungary)
A. irritans (Cobb in Thorne & Swanger, 1936) Andrássy, 1986 (Jamaica)
A. kosambaensis Khan, Ahmad & Jairajpuri, 1995 (India)
A. meridianus Andrássy, 1991 (Ecuador)
A. paragranuliferus Quijano, Peña-Santiago & Jiménez-Guirado, 1991 (Spain)
A. parasimilis (Kreis, 1963) Andrássy, 1986 (Iceland)
A. piracicabensis (Lordello, 1955) Andrássy, 1986 (Brazil)
A. rarus Gagarin, 1999 (Siberia)*
A. robustus (Thorne, 1974) Andrássy, 1986 (USA)
A. santosi (Meyl, 1957) Andrássy, 1986 (Brazil, USA)
A. septentrionalis (Kreis, 1963) Andrássy, 1986 (Spitzbergen, Iceland, Hungary)
A. tarkoenensis (Andrássy, 1959) Andrássy, 1986 (Hungary, Lithuania, Spain)
A. thimophilus Quijano, Peña-Santiago & Jiménez-Guirado, 1991 (Spain)
A. uniformis (Thorne, 1929) Andrássy, 1959 (Austria, Georgia, Nepal, Northern Russia)*
A. vellus Khan, Ahmad & Jairajpuri, 1995 (India)
Additional references: Andrássy (1986, 1991).

Genus *Epidorylaimus* Andrássy, 1986 (Figs 15.15G and H)
Diagnosis. Qudsianematinae. Cuticle smooth. Lip region offset; lips distinct. Odontostyle variable in length, up to 1.5 times as long as lip region width. Guiding ring simple. Pharynx enlarged in its posterior half. Female genital apparatus didelphic; vulva longitudinal or pore-like, generally pre-equatorial; vagina with sclerotization. Males rare. Supplements spaced; no pre-cloacal space between the most posterior one and the adcloacal papillae. Tail elongate conoid, ventrally curved, in both sexes. A few species of this mainly terrestrial genus have been found in freshwater habitats.

Type species: *E. lugdunensis* (de Man, 1880) Andrássy, 1986 (Central Europe, Greenland, Spitzbergen, Russia, USA, Alaska)*

Other species:
E. agilis (de Man, 1880) Andrássy, 1986 (The Netherlands, Austria, Ireland, Sweden, Spitzbergen, Russia, Armenia, Uzbekistan, Brazil, USA)*
E. angulosus (Thorne & Swanger, 1936) Andrássy, 1986 (Poland, Hungary, Romania, Serbia, Spain, USA)
E. consobrinus (de Man, 1918) Andrássy, 1986 (The Netherlands, Belgium, Czech Republic, Hungary, Norway, Spain, Russia, USA, Kenya)*
E. filicaudatus (Tjepkema, Ferris & Ferris, 1971) Andrássy, 1986 (USA, Poland, Italy)
E. humilior (Andrássy, 1959) Andrássy, 1986 (Hungary, Romania)
E. humilis (Thorne & Swanger, 1936) Andrássy, 1986 (USA, Yugoslavia, Bulgaria, Romania, Spain, Uzbekistan, Jamaica, Venezuela, New Hebrides)
E. leptosoma (Altherr, 1963) Andrássy, 1986 (Switzerland)
E. mellenbachensis (Altherr, 1974) Andrássy, 1986 (Germany)*
E. muchabbatae (Tulaganov, 1949) Andrássy, 1986 (Uzbekistan)
E. muscorum (Skwarra, 1921) Andrássy, 1986 (Germany, Spain, Romania)*
E. pseudoagilis (Altherr, 1952) Andrássy, 1986 (Switzerland, Romania, Italy, Spain)*
E. rivalis Gagarin, 1991 (Russia)*

Additional references: Andrássy (1986, 1991).

Genus *Pachydorylaimus* Siddiqi, 1983 (Figs 15.16A and B)
Diagnosis. Qudsianematinae. Body short and plump. Cuticle thick, finely striated. Lip region expanded, truncate. Amphidial aperture oval, small. Odontostyle very short and thick walled, with small aperture; odontophore with sclerotized basal flanges. Pharynx enlarging at about its middle. Female reproductive apparatus didelphic. Vulva transverse. Male supplements few, spaced. Tail elongate conoid with pointed terminus, similar in both sexes. All described species of *Pachydorylaimus* come from the South American rain forest; one of them was found in freshwater.

Type species: *P. furcatus* Siddiqi, 1983 (Colombia)

Other species:
P. aequatorialis Andrássy, 1997 (Ecuador)
P. andreasi Andrássy, 1997 (Ecuador)*
P. longicaudatus Siddiqi, 1983 (Colombia)
P. notabenus Siddiqi, 1983 (Colombia)
P. pachyvulvus Siddiqi, 1983 (Colombia)
P. schizodontus Loof & Zullini, 2000 (Costa Rica)
Additional reference: Siddiqi (1983).

Genus *Ecumenicus* Thorne, 1974 (Figs 15.16C and D)
Syn. *Indokochinema* Darekar & Khan, 1979
Diagnosis. Qudsianematinae. Cuticle smooth. Lip region moderately offset; lips distinct. Odontostyle about as long as lip region width; guiding ring simple. Pharynx enlarged in its posterior two-fifths. Female genital apparatus mono–opisthodelphic; vulva transverse, sclerotized, pre-equatorial. Male supplements six or seven, spaced. Tail conoid, ventrally curved, or convex–conoid, straight, similar in both sexes. Generally soil- or moss-inhabiting nematodes, one species also known from freshwater habitats.

Type species: *E. monohystera* (de Man, 1880) Thorne, 1974 (cosmopolitan)*

Other species:
E. conicauda (Darekar & Khan, 1979) Andrássy, 1991 (India)
E. parvus (Thorne, 1939) Andrássy, 1991 (USA, Hungary)
E. proprius Andrássy, 1991 (New Caledonia)
Additional reference: Andrássy (1991).

Genus *Boreolaimus* Andrássy, 1998 (Figs 15.16E and F)
Diagnosis. Qudsianematinae. Cuticle thin, smooth, exceptionally with some superficial striation. Lip region offset, lips distinct, angular. Buccal cavity around odontostyle unusually wide and large. Odontostyle longer than lip region width; its aperture about a third of its length. Guiding sheath long, tubular, with widened anterior ring. Pharynx muscular, gradually widening near middle; only a single anterior sub-ventral gland nucleus present. Pre-rectum with short caudal sac. Female gonads paired; vulva transverse. Male not known. Tail conoid, straight or ventrally curved, with rounded tip, 2–5 times as long as anal body width; Distal part

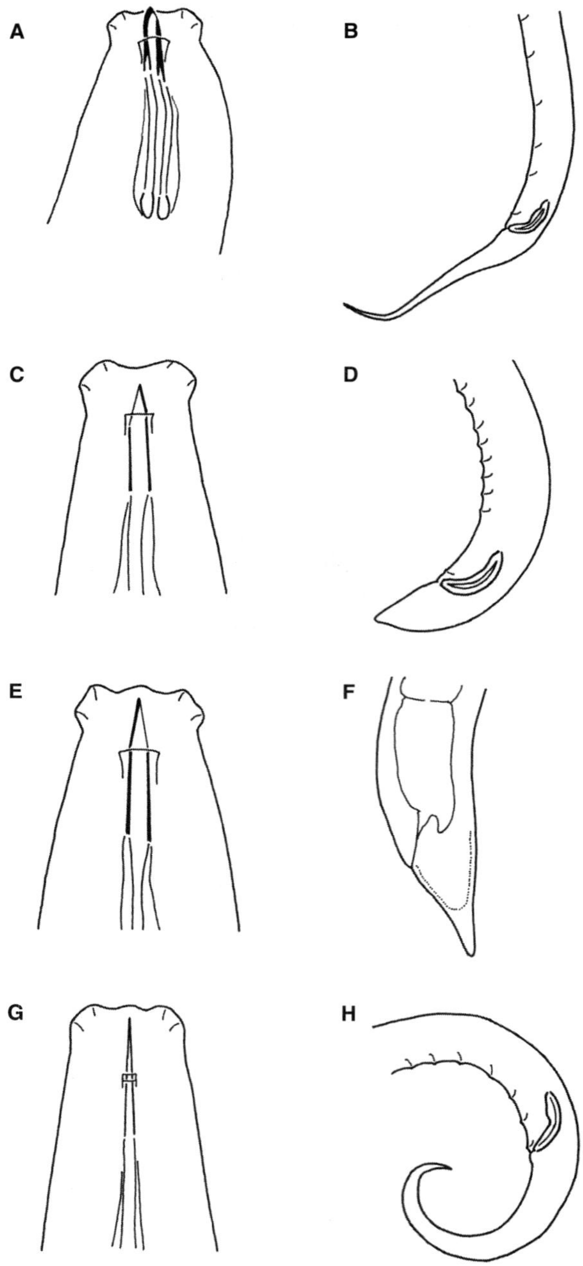

Fig. 15.16 A and B: *Pachydorylaimus*; C and D: *Ecumenicus*; E and F: *Boreodorylaimus*; G and H: *Chrysonema*.

of tail appearing 'empty'. All the species of *Boreolaimus* have been found in groundwater along river banks or lake shores in the northern part of Scandinavia, north to the Arctic Polar Circle.

Type species: *B. borealis* Andrássy, 1998 (Norway)

Other species:
B. alaskanus Andrássy, 2003 (USA)
B. arcticus Andrássy, 1998 (Norway)
B. enckelli (Andrássy, 1967) Andrássy, 1998 (Sweden, Lapland)
B. lapponicus Andrássy, 1998 (Finland, Sweden, Norway)
B. norvegicus Andrássy, 1998 (Norway)
B. septentrionalis Andrássy, 1998 (Sweden, Lapland)
Additional reference: Andrássy (1998).

Subfamily Chrysonematinae Siddiqi, 1969
Diagnosis. Qudsianematidae. Body slender, cuticle finely annulated. Lips barely distinct. Odontostyle small and slender, often needle-like; guiding ring simple. Female reproductive apparatus didelphic. Vulva transverse; vagina not sclerotized. Supplements not very numerous, spaced. Tail conoid to elongate conoid, ventrally arcuate, in both sexes. Terricolous and freshwater nematodes.
Additional reference: Andrássy (1998).
Genera:
Chrysonema Thorne, 1929
Oonaguntus Thorne, 1974 (no freshwater species)

Genus *Chrysonema* Thorne, 1929 (Figs 15.16G and H)
Syn. *Chrysonemoides* Siddiqi, 1969
Diagnosis. Chrysonematinae. Body very slender. Cuticle thin, finely annulated. Lip region barely off set, lips round, fused. Odontostyle small and thin, about as long as lip region width. Guiding ring thin. Pharynx expanding at about its middle. Vagina swollen. Male supplements not numerous, spaced; no pre-cloacal space. Tail elongate conoid in both sexes, provided with conspicuous sub-ventral papillae in males, arranged in pairs. The species of this genus are freshwater and/or terrestrial.

Type species: *C. aurum* Thorne, 1929 (USA)

Other species:
C. distinctum (Jana & Baqri, 1985) Andrássy, 1990 (India)
C. holsaticum (Schneider, 1925) Andrássy, 1990 (Central Europe, Russia, Australia)
C. limigenus (Siddiqi, 1969) Andrássy, 1990 (Poland, India, Surinam)
C. lozovense (Nesterov, 1976) Andrássy, 1990 (Moldavia)
C. maksymovi (Altherr, 1963) Andrássy, 1990 (France, Poland)
Additional reference: Andrássy (1990).

Family APORCELAIMIDAE Heyns, 1965
Diagnosis. Dorylaimoidea. Medium to large-sized nematodes. Cuticle thick, smooth or finely transversely striated, often with criss-cross lines or punctations and with

numerous body pores. Lip region clearly set off from the adjoining body; lips separate. Amphids often with sclerotized median support, fovea often duplex. Odontostyle robust, relatively short, generally axial, rarely as a mural tooth set on ventral wall of mouth cavity; its aperture longer than half its length. Guiding sheath without sclerotized fixed ring; its anterior margin widened and often plicated. Female gonads paired; vulva transverse, longitudinal or pore-like. Males with dorylaimoid spicules and spaced ventromedian supplements. Tail similar in the two sexes, generally short, conoid; long, filiform in one genus. Mostly terrestrial nematodes, but some species have been found also in freshwater habitats.
Additional references: Heyns (1965) and Jairajpuri and Ahmad (1992).

Subfamilies:
Aporcelaiminae Heyns, 1965
Sectonematinae Siddiqi, 1969
Key to subfamilies of Aporcelaimidae
1. Odontostyle axial, with wide aperture .Aporcelaiminae
   Odontostyle a mural tooth . . . . .Sectonematinae

Subfamily Aporcelaiminae Heyns, 1965
Diagnosis. Aporcelaimidae. Cuticle thick, often showing criss-cross lines or punctations on its surface. Lip region well off set. Lips distinct or amalgamated. Odontostyle with very large aperture; guiding ring plicate. Pharynx expanding gradually; dorsal pharyngeal gland nucleus far from its orifice; first pair of subventrals widely separated. Tail short conoid or filiform in both sexes. The representatives of the family are mostly soil inhabitants; four out of nine genera include some freshwater species.
Additional reference: Heyns (1965).
Genera:
Aporcelaimus Thorne & Swanger, 1936
Aporcelaimellus Heyns, 1965
Makatinus Heyns, 1965
Metaporcelaimus Lordello, 1965 (no freshwater species)
Akrotonus Thorne, 1974 (no freshwater species)
Aporcedorus Jairajpuri & Ahmad, 1983 (no freshwater species)
Epacrolaimus Andrássy, 2000
Tubixaba Monteiro & Lordello, 1980 (no freshwater species)
Aporcella Andrássy, 2002 (no freshwater species)

Key to genera of Aporcelaiminae *with freshwater species*
1. Lip region only slightly off set from the adjoining body . . . . *Makatinus*
   Lip region strongly off set 2
2. Lips provided with protruding inner liplets, vulva with wrinkled cuticle . . . . *Epacrolaimus*
   Lips without inner liplets, vulva with smooth cuticle 3
3. Very large nematodes, cuticle with criss-cross lines . . . . *Aporcelaimus*
   Medium-sized nematodes, inner layer of cuticle finely striated . . . . *Aporcelaimellus*

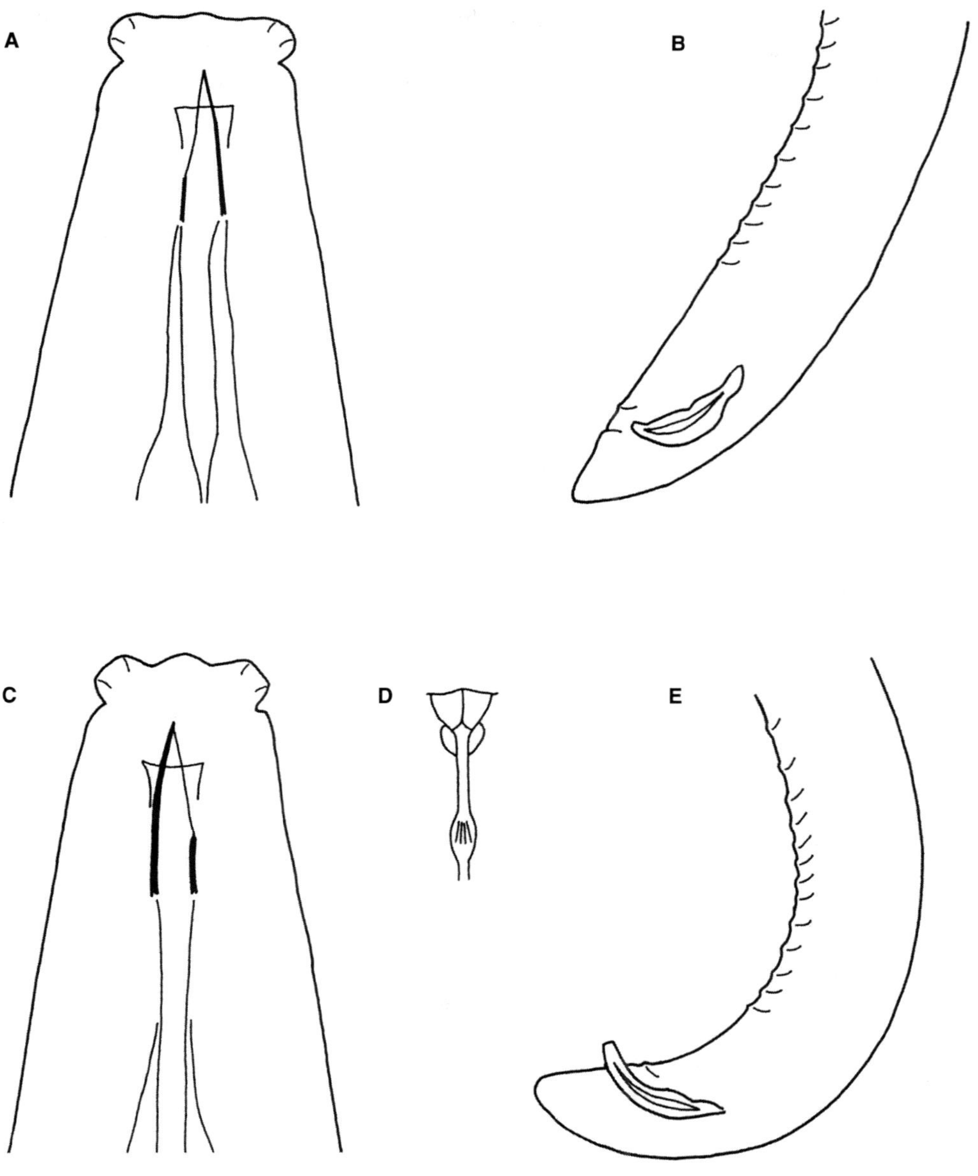

Fig. 15.17 A and B: *Aporcelaimus;* C–E: *Aporcelaimellus.*

Genus *Aporcelaimus* Thorne & Swanger, 1936 (Figs 15.17A and B)

Diagnosis. Aporcelaiminae. Large nematodes, 3–10 mm long. Body strongly tapering towards the anterior end. Cuticle thick, marked with criss-cross lines. Lip region offset by a deep constriction; lips rather amalgamated. Mouth a dorsoventral slit. Odontostyle about as long as lip region width; aperture occupying more than half its length. Amphids with fine median support. Pharynx expanding before its middle. Vulva transverse; vagina generally with sclerotized pieces. Spicules of two types: either slender, dorylaimoid or strongly swollen. Tail similar in the two sexes, short and rounded or conoid–rounded, in some cases subdigitate. The species of *Aporcelaimus* are widespread in terrestrial habitats but some species have been found in freshwater habitats. The species listed are those considered as definitely belonging to the genus by Andrássy (2001b).

Type species: *Aporcelaimus superbus* (de Man, 1880) Goodey, 1951 (widespread in Europe; Uzbekistan, Kazakhstan)*

Other species:
A. americanus Thorne & Swanger, 1936 (USA, The Netherlands, France, Poland, Hungary)
A. bestiarius Isatullaeva, 1967 (Kazakhstan, Germany, Hungary)*
A. boreus Andrássy, 2000 (Alaska, USA)
A. brzeskii Andrássy, 2000 (Alaska, USA)
A. caesar Andrássy, 2000 (Hungary)
A. cobbi Thorne, 1937 (Italy)
A. digiticaudatus Andrássy, 2001 (England, Holland)
A. eurydoris (Ditlevsen, 1911) Thorne & Swanger, 1936 (USA, Canada, widespread in Europe)*
A. femineus Andrássy, 2000 (Alaska, USA)
A. fortis Gagarin, 1992 (Russia)
A. ingens Andrássy, 2000 (USA)
A. litchi Srivastava, Rawat & Ahmad, 2000 (India)
A. macrohystera Altherr, 1974 (Germany, Norway, England, USA)*
A. pachydermus Thorne, 1937 (Gemany, Hungary, USA)*
A. paraspiralis Thorne & Swanger, 1936 (Austria, Romania)
A. polaris Andrássy, 2003 (Alaska, USA)
A. pseudospiralis Botha & Heyns, 1990 (South Africa, Sumatra)
A. sicus Gagarin, 1992 (Russia)
A. subdigiticaudatus Altherr, 1965 (Germany)*
Additional references: Heyns (1965) and Andrássy (2001a).

Genus *Aporcelaimellus* Heyns, 1965 (Figs 15.17C–E)
Diagnosis. Aporcelaiminae. Cuticle thick, with two well distinct layers (besides the thin exocuticle): the outer one hyaline, the inner one finely striated. Lip region offset by a deep constriction; lips prominent, more or less amalgamated. Odontostyle short and thick, its aperture about half its length. Amphids with or without median support; fovea sometimes duplex. Vulva transverse or pore-like. Vagina with sclerotized pieces. Spicules dorylaimoid; supplements spaced. Tail short, rounded, bluntly conoid or subdigitate, similar in the two sexes. The species of *Aporcelaimellus* are widespread in terrestrial habitats but some species have been found in freshwater habitats.

Type species: *A. obtusicaudatus* (Bastian, 1865) Heyns, 1965 (cosmopolitan)*

Other species:
A. adriani Botha & Heyns, 1990 (South Africa)
A. alius Andrássy, 2002 (Hungary)
A. amplexor (Nesterov & Lisetskaya, 1965) Heyns 1965

A. amylovorous (Thorne & Swanger, 1936) Heyns, 1965 (Hungary)
A. baqrii Ahmad & Jairajpuri, 1982
A. chauhani Baqri & Khera, 1975
A. cocophilus (Loos, 1949) Andrássy, 2001
A. conoidus Thorne, 1974
A. dubius (Thorne, 1974) Andrássy, 1986
A. duhouxi (Altherr, 1963) Baqri & Khera, 1975 (France)*
A. estonicus (Krall, 1957) Andrássy, 1990
A. faridpuriensis Singh, Sharma & Khataou, 2002 (India)
A. futeii Khan & Araki, 2002 (Japan)
A. gerlachi (Meyl, 1956) Heyns, 1965
A. glandus Botha & Heyns, 1965 (South Africa)*
A. goldeni Khan & Fatima, 1980 (Pakistan)
A. heynsi Baqri & Jairajpuri, 1968
A. hylophilus Tjepkema, Ferris & Ferris, 1971
A. index (Thorne, 1939) Andrássy, 1986
A. jairajpurii Singh, Sharma & Khataou, 2002 (India)
A. kikereensis Baqri & Coomans, 1973
A. krygeri (Ditlevsen, 1928) Heyns, 1965 (Russia, Hungary)*
A. macropunctatus (Heyns, 1967) Jiménez-Guirado, 1994
A. maitai Yeates, 1967 (New Zealand)
A. malagasi Heyns, 1966
A. medius Andrássy, 2002 (Hungary)
A. micropunctatus Botha & Heyns, 1990 (South Africa)*
A. nawabganjense Singh, Sharma & Khataou, 2002 (India)
A. nivalis (Altherr, 1952) Heyns, 1965
A. odhneri (Allgén, 1951) Andrássy, 1990
A. papillatus (Bastian, 1865) Baqri & Khera, 1975
A. paracentrocercus (De Coninck, 1935) Baqri & Coomans, 1973
A. paraconicaudatus (Meyl, 1956) Heyns, 1965
A. parapapillatus Botha & Heyns, 1990
A. productus (Thorne & Swanger, 1936) Baqri & Khera, 1975
A. propinquus (Thorne & Swanger, 1936) Tjepkema, Ferris & Ferris, 1971
A. punctatus Altherr in Altherr & Delamare-Deboutteville, 1972
A. pycnus (Thorne, 1939) Baqri & Khera, 1975 (Hungary)
A. samarcandicus (Tulaganov, 1949) Baqri & Khera, 1975 (Hungary)
A. saprophilus Gagarin & Gusakov, 2001 (Russia)*
A. seinhorsti (Meyl, 1957) Heyns, 1965
A. stilus (Kirjanova, 1951) Andrássy, 1986
A. taylori Yeates, 1967 (New Zealand, USA, Germany)*
A. tritici (Bastian, 1865) Andrássy, 1986
A. vitrinus (Thorne & Swanger, 1936) Baqri & Khera, 1975
Additional reference: Heyns (1965) and Andrássy (2002).

Genus *Makatinus* Heyns, 1965 (Figs 15.18A and B)
Diagnosis. Cuticle with two distinct layers (besides the exocuticle); the inner layer often radially striated. Lip region hardly off set; lips rather amalgamated.

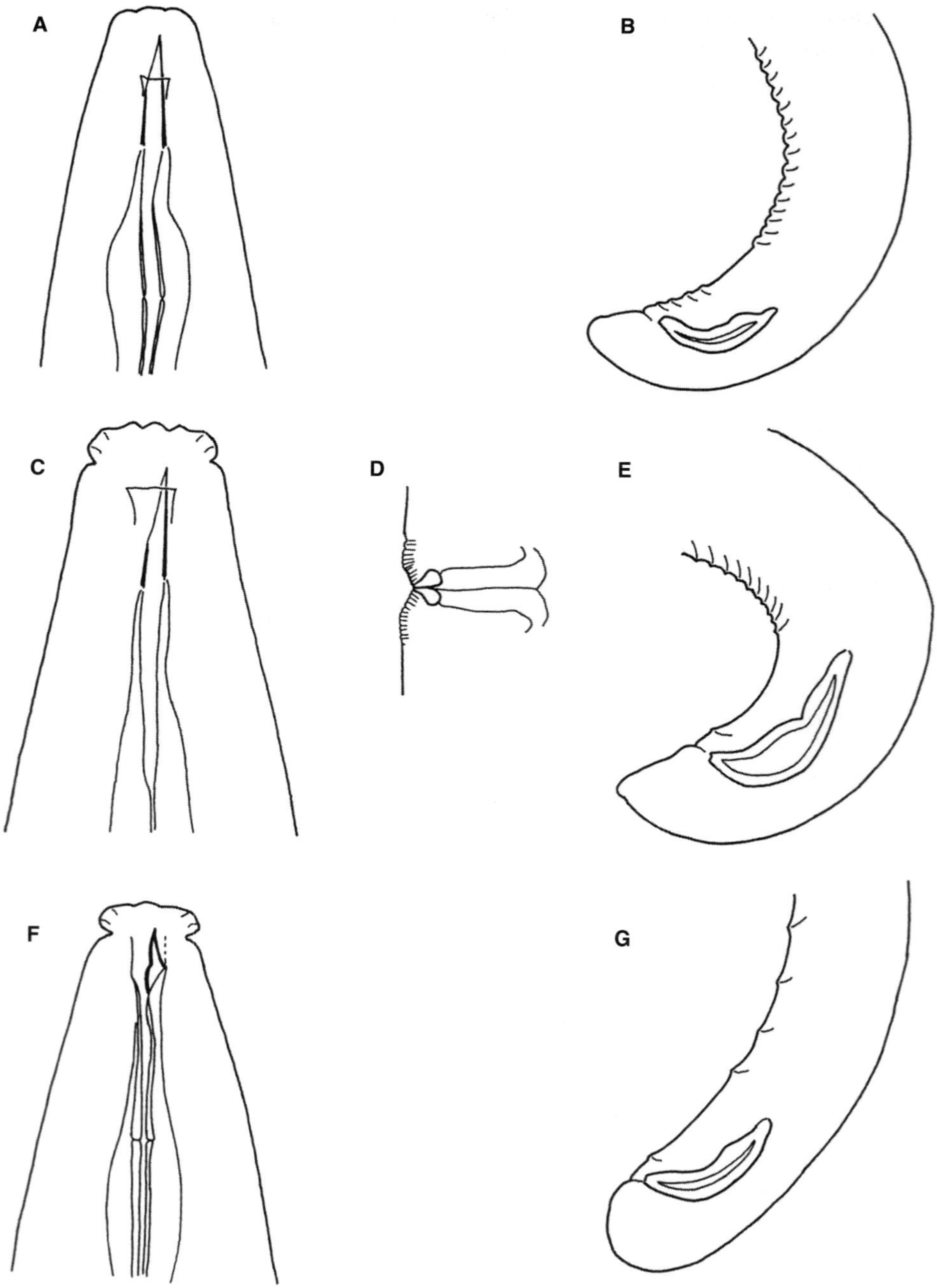

Fig. 15.18 A and B: *Makatinus*; C–E: *Epacrolaimus*; F and G: *Sectonema*.

Odontostyle aperture occupying about half its length. Amphidial fovea not divided. Cardiac disc absent. Female gonads paired. Vulva longitudinal, sunken; vagina with sclerotized pieces. Male supplements numerous, diverging into double row near anus (2–5 pairs). Tail short, convex–conoid in both sexes. Mainly terrestrial nematodes, a few species have been found in freshwater habitats.

Type species: *M. punctatus* Heyns, 1965 (South Africa)

Other species:
M. aquaticus Jiménez-Guirado, 1994 (Spain, Hungary)*
M. capensis Heyns, 1965 (South Africa)
M. crassiformis (Kreis, 1924) Andrássy, 1986 (Surinam)*
M. goodeyi Jiménez-Guirado, 1994
M. heynsi Ahmad & Ahmad, 1992 (India)
M. micropunctatus Ahmad & Ahmad, 2002 (India)
M. minor (Loos, 1945) Ahmad, 1997 (Sri Lanka)
M. simus (Andrássy, 1958) Andrássy, 2002
Additional references: Heyns (1965) and Andrássy (2002).

Genus *Epacrolaimus* Andrássy, 2000 (Figs 15.18C–E)
Diagnosis. Aporcelaimidae. Very large size (7–9 mm). Cuticle thick, marked with fine criss-cross lines. Lip region sharply set off by a deep constriction; lips distinct, each provided with a protruding inner liplet. Odontostyle with very large aperture, arched inward. Pharynx muscular, gradually expanding far before the middle. Female reproductive apparatus didelphic; vulva transverse with wrinkled cuticula; vagina with sclerotized pieces. Supplements spaced. Tail similar in both sexes, short, convex–conoid with bluntly rounded or subdigitate terminus. Of the two species known so far one has been found in both freshwater and terrestrial habitats.

Type species: *E. declinatoaculeatus* (Kreis, 1924) Andrássy, 2000 (Central Europe, South Africa, USA)*

Other species:
E. imperator Andrássy, 2000 (Alaska)
Additional reference: Andrássy (2000).

Subfamily Sectonematinae Siddiqi, 1969
Diagnosis. Aporcelaimidae. Large nematodes. Cuticle thick. Lip region offset; lips distinctly separate and prominent. Odontostyle as a mural tooth set on the ventral wall of the cheilostome. A conspicuous disc at the pharyngeal–intestinal junction. Female gonads paired.
Only genus: *Sectonema* Thorne, 1930
Additional reference: Siddiqi (1969).

Genus *Sectonema* Thorne, 1930 (Figs 15.18F and G)
Syn. *Aporcelaimoides* Heyns, 1965
Diagnosis. Sectonematinae. Mural tooth dorsally grooved and with basal projections diverging. Amphids with strong median support; fovea duplex. Vulva transverse; vagina with cuticularized pieces. Spicules dorylaimoid; supplements irregularly arranged. Tail short, conoid to hemispheroid, in both sexes.

Mainly terrestrial nematodes; some species have been found in freshwater habitats.

Type species: *S. ventrale* Thorne, 1930 (USA)

Other species:
S. amazonicum Siddiqi, 1995 (Brazil)
S. anisonchum Siddiqi, 1984 (Colombia)
S. barbatum Heyns, 1965 (USA)
S. brevicauda Heyns, 1965 (South Africa)
S. barbatoides Heyns, 1965 (The Netherlands)
S. basilgoodeyi Heyns, 1965 (England)
S. californicus (Heyns, 1965) Siddiqi, 1995 (USA)
S. conicaudatum Siddiqi, 1995 (Samoa)
S. deltatum Siddiqi, 1995 (Cameroon)
S. demani Altherr, 1965 (Germany, The Netherlands, Hungary)*
S. heynsi Altherr, 1968 (Germany)*
S. macbethi Heyns, 1965 (Venezuela, Colombia)
S. macrospiculum (Altherr, 1958) Heyns, 1965 (Germany)*
S. moderatum Siddiqi, 1995 (Cameroon)
S. mucrodens Siddiqi, 1984 (Colombia)
S. paramonovi (Eliava, 1966) Eliashvili, Aliev & Eliava, 1977
S. probulbum (Heyns, 1965) Siddiqi, 1995 (South Africa, India)
S. procta Jairajpuri & Baqri, 1966 (India)
S. pseudoventrale Heyns, 1965 (South Africa, Hawaii, The Netherlands)
S. rotundicauda Goodey, 1951 (England)
S. sica Clark, 1964 (New Zealand)
S. transsilvanicum Popovici, 1978 (Romania)
S. truxum Siddiqi, 1984 (Colombia)
Additional references: Heyns (1965) and Siddiqi (1995).

Family NORDIIDAE Jairajpuri & A.H. Siddiqi, 1964

Diagnosis. Dorylaimoidea. Cuticle smooth or very finely transversely striated. Lip region continuous with adjoining body or offset; lips distinct or fused. Odontostyle slender, attenuated, generally long, with narrow lumen and aperture; guiding ring single or double; odontophore rod-like or flanged. Pharynx muscular expanding behind its middle. Female reproductive apparatus generally didelphic, sometimes monodelphic; vulva transverse; refractive part of vagina with sclerotized pieces. Tail similar in the two sexes, short and rounded or elongate to filiform. The representatives of the family are mainly terrestrial, soil- or moss-inhabiting nematodes but several species have been found in freshwater habitats.

Remarks: Jairajpuri and Ahmad (1992) recognized five subfamilies: Cephalodorylaiminae, Nordiinae, Pungentinae, Actinolaimoidinae and Helmabiinae. The last subfamily, with the single genus *Helmabia*, is not considered to be related to the Nordiidae.

Additional reference: Jairajpuri and Ahmad (1992).

Subfamilies:
Nordiinae Jairajpuri & A.H. Siddiqi, 1964
Pungentinae Siddiqi, 1969

Cephalodorylaiminae Jairajpuri, 1967 (no freshwater species)
Actinolaimoidinae Jairajpuri & Ahmad, 1992 (no freshwater species)

Key to subfamilies of NORDIIDAE

1. Labial papillae prominent, clearly protruding above lip contour . . . . . Cephalodorylaiminae
   Labial papillae normally developed 2
2. Odontostyle not very thin, odontophore often flanged . . . . . Pungentinae
   Odontostyle very thin, odontophore never flanged 3
3. Body plump; odontostyle very thin, more than three lip region widths long . . . . . Nordiinae
   Body slender; odontostyle less than three lip region widths long . . . . . Actinolaimoidinae

Subfamily Pungentinae Siddiqi, 1969

Diagnosis. Nordiidae. Cuticle finely transversely striated. Lip region continuous or off set from body. Amphids located at level of lips or behind them. Odontostyle varying in length, but never longer than 2–3 lip region widths, slender, with narrow lumen and aperture; guiding ring single or double; odontophore with or without basal flanges. Female reproductive apparatus generally amphidelphic, rarely mono–opisthodelphic. Vulva transverse. Male supplements spaced. Tail hemispheroid to long, filiform, similar in both sexes. The majority of species of this family are typical inhabitants of terrestrial habitats (soil, moss) but some species belonging to five genera have been found in aquatic habitats and one monospecific genus, *Lanzavecchia* Zullini, 1988, has been found exclusively in freshwater.

Additional reference: Jairajpuri and Ahmad (1992).

Genera:

Pungentus Thorne & Swanger, 1936
Enchodelus Thorne, 1939
Kochinema Heyns, 1963 (no freshwater species)
Lenonchium Siddiqi, 1963
Rhyssocolpus Andrássy, 1971
Enchodorus Vinciguerra, 1976 (no freshwater species)
Dorydorella Andrássy, 1987
Lanzavecchia Zullini, 1988

Key to genera of Pungentinae

1. Tail short, rounded or conoid 2
   Tail long, filiform 7
2. Amphids very anterior, at lip level . . . . . *Kochinema*
   Amphids post-labial 3
3. Sclerotized platelets around stoma . . . . . *Pungentus*
   No sclerotized platelets around stoma 4
4. Mouth cavity with massive hyaline wall, very large size (7–8 mm) . . *Lanzavecchia*
   Mouth cavity with thin wall, relatively small size 5
5. Cuticle coarsely striated near vulva . . . . . *Rhyssocolpus*
   Cuticle near vulva smooth 6

6. Odontophore with basal flanges .... *Enchodelus*
   Odontophore simple .... *Dorydorella*
7. Odontostyle very short, odontophore flanged, supplements few, spaced .... *Enchodorus*
   Odontostyle not very short, odontophore rod-like, supplements numerous .... *Lenonchium*

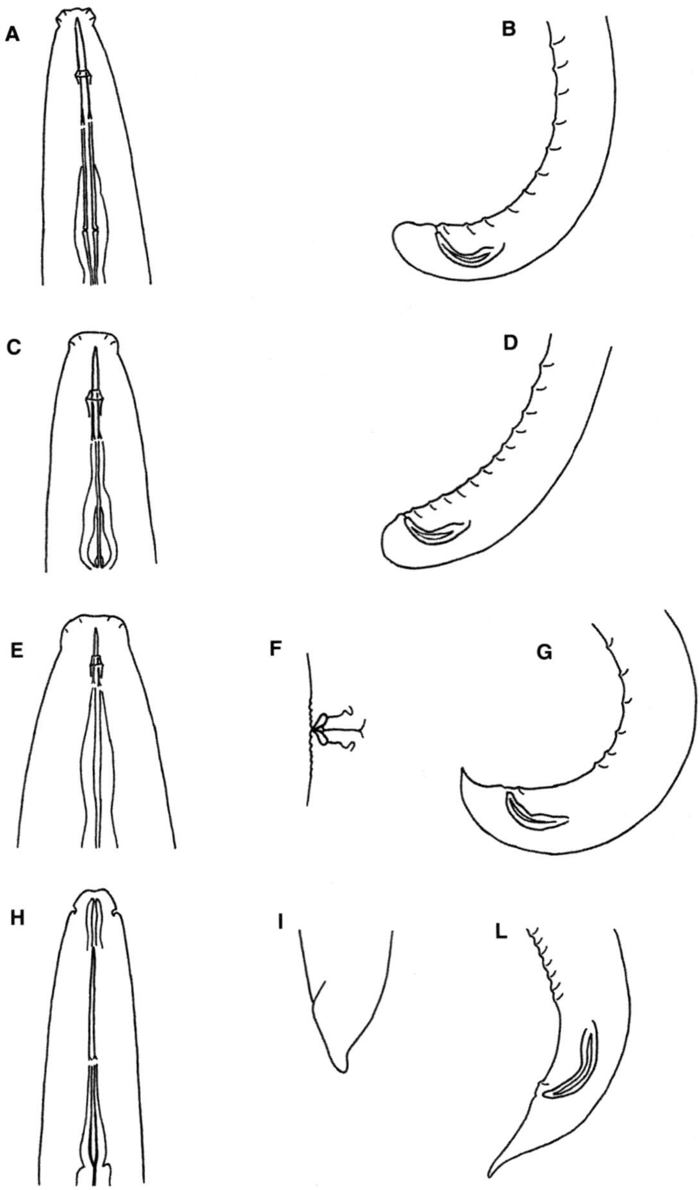

Fig. 15.19 A and B: *Pungentus;* C and D: *Enchodelus;* E–G: *Rhyssocolpus;* H–L: *Lanzavecchia.*

Genus *Pungentus* Thorne & Swanger, 1936 (Figs 15.19A and B)
Syn. *Pungentoides* Altherr, 1950; *Paramonovia* Eliava, 1973
Pungentinae. Lip region more or less set off; labial papillae prominent. Four cuticularized platelets surround the mouth. Odontostyle varying in length, sometimes slightly curved; guiding ring double; odontophore rod-like. Female reproductive apparatus generally mono–opisthodelphic, amphidelphic in a few species. Vulva transverse; vagina with sclerotized pieces. Supplements spaced. Tail short, hemispheroid, sometimes clavate or conoid, similar in both sexes. Mostly inhabitants of terrestrial habitats; some species have been found in freshwater.

Type species: *P. pungens* Thorne & Swanger, 1936 (Canada, USA, Romania)

Other species:
P. angulatus Jairajpuri & Baqri, 1966 (India, Hungary)
P. angulosus Thorne, 1939 (Canada, USA, The Netherlands, Romania, Georgia, India)
P. brevidentatus Thorne, 1939 (USA)
P. clavatus Ahmad & Jairajpuri, 1979 (Canada, Poland, India, Italy)
P. crassus Thorne, 1974 (USA)
P. engadinensis (Altherr, 1950) Altherr, 1952 (Central Europe, Italy, USA, Canada, Iraq)*
P. eroshenkoi Eliava & Eliashvili, 1990 (Russia)
P. fagi Vinciguerra & Giannetto, 1983 (Italy)
P. gracilis Eroshenko, 1976 (Russia)
P. granosus Eroshenko, 1976 (Russia)
P. juglensi Mahajan, 1972 (India)
P. macrostylus Ahmad & Sturhan, 2000 (Turkey)
P. mahunkai Andrássy, 1968 (Paraguay)
P. maorium Clark, 1963 (New Zealand)
P. marietani Altherr, 1950 (Germany, Poland, Serbia, Slovakia, Macedonia, Switzerland, Italy, Canada)
P. microdentatus Thorne, 1939 (Canada, USA, Poland, Uzbekistan)
P. monohystera Thorne & Swanger, 1936 (USA, Canada, Argentina, Hungary)
P. mosellae (Altherr, 1963) Andrássy, 1986 (France)*
P. obscurus Thorne, 1939 (USA)
P. ortocephalus Eroshenko, 1976 (Russia)
P. parapungens Gagarin, 1985 (Russia)*
P. parvus Thorne, 1939 (USA, Canada, Russia)
P. porosus Ebsary, 1983 (Canada)
P. projectus (Thorne, 1939) Andrássy, 1991
P. roposus Zell, 1987
P. silvestris (de Man, 1912) Coomans & Geraert, 1962 (Central Europe)*
P. sparsus Thorne, 1939 (USA)
P. sublatum Eroshenko, 1976 (Russia)
P. textilis (Thorne & Swanger, 1936) Thorne, 1939 (USA)
P. vesiculosus Eroshenko, 1976 (Russia)
Additional reference: Ahmad and Jairajpuri (1979).

Genus *Enchodelus* Thorne, 1939 (Figs 15.19C and D)
Diagnosis. Pungentinae. Lip region more or less offset; labial papillae prominent. Odontostyle generally long; guiding ring double; odontophore with basal flanges, rarely

rod-like. Female reproductive apparatus amphidelphic. Vulva transverse; vagina with sclerotized pieces. Supplements spaced. Tail short, hemispheroid or conoid, similar in both sexes. Mostly inhabitants of moss or soil, a few species have been found in freshwater habitats.

Type species: *E. macrodorus* (de Man, 1880) Thorne, 1939 (cosmopolitan)*

Other species:
E. altherri Vinciguerra & De Francisci 1973 (Italy)
E. analatus (Ditlevsen, 1927) Thorne, 1939 (Greenland, Spitzbergen, Romania)
E. arcticus Nesterov, 1976 (Moldavia)
E. arcuatus Thorne, 1939 (USA, Bulgaria)*
E. brevidentatus Thorne, 1939 (USA)
E. conicaudatus (Ditlevsen, 1927) Thorne, 1939 (Greenland, Spitzbergen)
E. constrictus Jairajpuri & Loof, 1969 (India)
E. coomansi Nicholas & Stewart, 1984 (Australia: brackish water)*
E. distinctus Ahmad & Jairajpuri, 1980 (India)
E. faeroensis (Ditlevsen, 1928) Thorne, 1939 (Faroe Islands)*
E. geraldi Winiszewska, 1985 (Poland)
E. groenlandicus (Ditlevsen, 1937) Thorne, 1939 (Greenland)
E. hedickei (Paesler, 1941) Andrássy, 1960 (Germany)
E. hopedoroides Altherr, 1963 (Switzerland, Georgia, Nepal)
E. hopedorus (Thorne, 1929) Thorne, 1939 (USA, Poland, Uzbekistan)
E. irregularis Altherr, 1972 (Sweden)*
E. laevis Thorne, 1939 (Korea, USA)*
E. liangi Ahmad, Wu & Shaheen, 2002 (China)
E. longidens Jairajpuri & Loof, 1968 (India)
E. lucinensis Popovici, 1968 (Romania)
E. lushani Ahmad, Wu & Shaheen, 2002 (China)
E. magnificus (Altherr, 1952) Altherr, 1963 (Switzerland, Austria)*
E. maximus Baqri & Jairajpuri, 1974 (India)
E. microdoroides Baqri & Jairajpuri, 1974 (India, Korea)
E. montanus Bagaturia, Eliava & Eliashvili, 1979 (Georgia)
E. morgensis Loof, 1989 (Germany, Austria, Switzerland, The Netherlands)*
E. nepalensis Zullini, 1973 (Nepal, China)
E. parateres Baqri & Jairajpuri, 1974 (India)
E. parvus Loof, 1971 (Spitzbergen)
E. ponorensis Popovici, 1995 (Romania)
E. satendri Baqri & Jairajpuri, 1974 (India, China)
E. saxifrage Popovici, 1995 (Romania)
E. signyensis Loof, 1975 (Signy Island: South Orkneys, UK)
E. southeyi Jairajpuri & Ahmad, 1986 (India, China)
E. striatus Thorne, 1939 (USA)
E. teres Thorne, 1939 (USA, Italy, Uzbekistan)
E. thornei Baqri & Jairajpuri, 1974 (India)
E. vestibulifer Altherr, 1952 (Switzerland)
E. vesuvianus (Cobb, 1893) Thorne, 1939 (Italy, Poland, Hungary)
E. zonatus Jairajpuri & Loof, 1968 (India, Egypt)
Additional reference: Ahmad and Jairajpuri (1980).

Genus *Rhyssocolpus* Andrássy, 1971 (Figs 15.19E–G)
Diagnosis. Pungentinae. Lip region more or less off set. Odontostyle small, about one lip region width long; guiding ring double; odontophore with basal flanges. Female reproductive apparatus amphidelphic. Vulva transverse; cuticle anterior and posterior to vulva coarsely wrinkled; vagina with sclerotized pieces. Supplements spaced. Tail short, conoid, in both sexes. The few species of this genus live in terrestrial and in freshwater or semi-freshwater habitats.

Type species: *R. vulvostriatus* (Stefanski, 1924) Andrássy, 1971 (Poland, Denmark, Italy)

Other species:
R. alleni (Brzeski, 1962) Andrássy, 1986 (Spitzbergen)
R. aquilonius Andrássy, 2003 (Alaska, USA)*
R. arcticus Ebsary, 1984 (Canada)
R. brasiliensis (Meyl, 1957) Andrássy, 1971 (Brazil)
R. fluviatilis Gagarin, 1985 (Russia)*
R. iuventutis Andrássy, 1984 (Hungary)*
R. microdorus (Schiemer, 1965) Andrássy, 1971 (Austria, Italy)*
R. paradoxus (Loof, 1975) Andrássy, 1986 (Orkney Islands, UK)
R. repis (Brzeski, 1992) n. comb. (Korea)
Additional reference: Andrássy (2003).

Genus *Lenonchium* Siddiqi, 1965 (Figs 15.20A–C)
Diagnosis. Pungentinae. Large nematodes. Cuticle with fine striae; subcuticle loose. Lip region continuous or offset; lips amalgamated. Odontostyle very slender, with narrow lumen and small aperture; guiding ring single or double. Odontophore rod-like, basal region swollen. Pharynx expanded at about its middle. Female reproductive apparatus didelphic. Vulva transverse or longitudinal. Supplement numerous, contiguous. Tail long, filiform in both sexes. Mainly terrestrial nematodes; some species have been found in freshwater.

Type species: *L. oryzae* Siddiqi, 1965 (India)

Other species:
L. asterocaudatum Choi & Jairajpuri, 1998 (Korea)
L. denticaudatum (Imamura, 1931) (Japan, The Netherlands)* (*species inquirenda*)
L. fimbricaudatum Swart & Heyns, 1991 (South Africa)*
L. longidens (Furstenberg & Heyns, 1965) Jairajpuri, 1966 (South Africa)
L. macrodorum Ahmad & Jairajpuri, 1988 (India)
Additional reference: Swart and Heyns (1991).

Genus *Dorydorella* Andrássy, 1987 (Figs 15.20D–F)
Diagnosis. Pungentinae. Lip region offset, lips prominent. Odontostyle slender, 1.3–2.0 lip region widths long. Pharynx expanded behind its middle; pre-rectum short. Vulva transverse; vagina with sclerotized pieces. Tail short, conoid, almost straight in both sexes. The few species live in terrestrial or freshwater habitats.

Type species: *D. bryophila* (de Man, 1880) Andrássy, 1986 (The Netherlands, Germany, Austria, Hungary, Sweden, Spain, France, Russia, Moldavia, Uzbekistan, Kazakhstan, USA)*

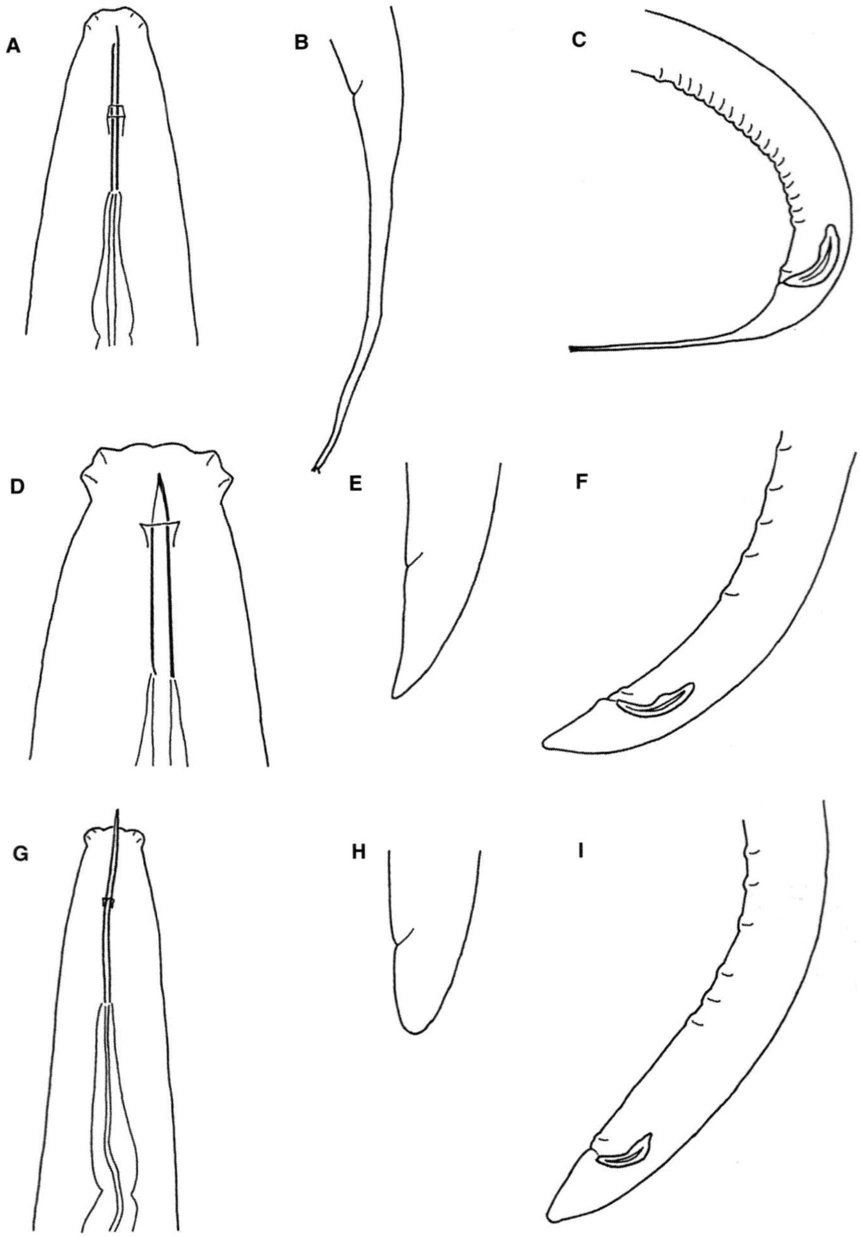

Fig. 15.20 A–C: *Lenonchium*; D–F: *Dorydorella*; G–I: *Longidorella*.

Other species:

D. pratensis (de Man, 1880) Andrássy, 1986 (cosmopolitan)*

D. tenuidens (Thorne & Swanger, 1936) Andrássy, 1987 (cosmopolitan)

Additional reference: Andrássy (1987a).

Genus *Lanzavecchia* Zullini, 1988 (Figs 15.19 H–L)

Diagnosis. Pungentinae. Very large nematodes. Cuticle smooth or faintly dotted. Lip region slightly off set. Amphid wide, stirrup-shaped. Mouth cavity surrounded by thick hyaline wall. Odontostyle more than two times as long as lip region width, narrow, slightly bent. Guiding ring simple. Odontophore with moderately developed flanges. Pharynx gradually expanding at about half its length. Female reproductive apparatus didelphic; gonads reflexed; spermatheca present. Vulva longitudinal. Supplements numerous. Tail conoid, subdigitate, in the two sexes. The single species described was found in an Ethiopian lake.

Type and only species: *L. fafner* Zullini, 1988 (Ethiopia)

Additional reference: Zullini (1988).

Subfamily Nordiinae Jairajpuri & A.H. Siddiqi, 1964

Diagnosis. Nordiidae. Body short and robust. Lip region more or less off set. Odontostyle very slender; with very narrow lumen, 3–5 times as long as lip region width. Odontophore rod-like, as long as odontostyle. Pharynx expanded behind its middle. Female reproductive apparatus didelphic or mono–opisthodelphic. Tail short, conoid, ventrally bent or sub-cylindrical, similar in the two sexes. Mainly soil inhabiting species; some of them have also been found in freshwater habitats.

Additional reference: Jairajpuri and A.H. Siddiqi (1964).

Genera:

Longidorella Thorne, 1939

Acunemella Andrássy, 2002 (no freshwater species)

Key to genera of Nordiinae

1. Posterior two-thirds of body strongly twisted, spring-like; body strongly tapering towards the anterior end .... *Acunemella*
   Body not twisted and regularly tapering .... *Longidorella*

Genus *Longidorella* Thorne, 1939 (Figs 15.20G–I)

Syn. *Nordia* Jairajpuri & A.H. Siddiqi; *Enchodorella* Khan, 1964; *Thornedia* Husain & Khan, 1965; *Saevadorella* Siddiqi, 1982

Diagnosis. Nordiinae. Lip region continuous with the adjoining body or off set from it. Odontostyle long, at least three times the lip region width, thin, sometimes slightly curved. Pharynx about a third of body length long. Female reproductive apparatus generally amphidelphic, rarely monodelphic. Tail mostly conoid, ventrally bent in both sexes. The species of *Longidorella* may be found in different terrestrial habitats; a few species have been found in freshwater.

Type species: *L. parva* Thorne, 1939 (USA, India, Japan, China, Venezuela, Ghana, Belgium, Poland, Romania, Spain, Italy, Moldavia, France)*

Other species:

L. acutis (Jairajpuri & Siddiqi, 1964) Jairajpuri & Hooper, 1969 (India)

L. arenicola Vinciguerra & Zullini, 1980 (Italy)

L. caudata Suryawanshi, 1971 (India)

L. cuspidata (Andrássy, 1964) Jairajpuri & Hooper, 1969 (Mongolia)

L. europaea Dalmasso, 1966 (France, Germany)

L. frontiniani (Dalmasso, 1966) Jairajpuri & Hooper, 1969 (France)

L. hastata (Andrássy, 1963) Jairajpuri & Hooper, 1969 (Angola)

L. longicaudata Orselli & Vinciguerra, 1999 (Italy)
L. macramphis (Altherr, 1950) Altherr, 1952 (Switzerland, Hungary, The Netherlands, Germany, Spain, UK)*
L. magna Loof, 1971 (Spitzbergen, Alaska)
L. microdora (de Man, 1880) Goodey, 1963 (The Netherlands, Austria, Italy, USA, India)*
L. morbida (Loof, 1964) Jairajpuri & Hooper, 1969 (Venezuela)
L. multipapillata (Schuurmans Stekhoven & Teunissen, 1938) Siddiqi, 1962 (Congo)
L. murithi Altherr, 1950 (Switzerland, Hungary, USA, Chile, Italy)*
L. mustafi (Husain & Khan, 1965) Jairajpuri & Hooper, 1969 (India)
L. obtusicaudata Orselli & Vinciguerra, 1999 (Italy)
L. okhlaensis (Jairajpuri & A.H. Siddiqi, 1964) Jairajpuri & Hooper, 1969 (India, USA, Brazil, Hungary)
L. opistodelphis Jairajpuri, 1964 (India)
L. parvibulbata Orselli & Vinciguerra, 1999 (Italy)
L. penetrans (Thorne & Swanger, 1936) Goodey, 1963 (USA)
L. perparvula Monteiro, 1970 (Brazil)
L. perveeni (Khan, 1964) Jairajpuri & Hooper, 1969 (India)
L. solani (Husain & Khan, 1965) Jairajpuri & Hooper, 1969 (India)
L. suviswa Patil & Khan, 1982 (India)
L. tredecima (Andrássy, 1964) Jairajpuri & Hooper, 1969 (Hungary)
L. xenura Khan & Siddiqi, 1963 (India, Venezuela)
Additional references: Jairajpuri and Hooper (1969) and Jairajpuri and Ahmad (1992).

Family THORNIIDAE De Coninck, 1965

Diagnosis. Dorylaimoidea. Small to medium-sized nematodes. Cuticle smooth, thin. Lip region low, rounded, generally continuous with the adjoining body; lips fused. The two circles of labial papillae more or less close to each other. Odontostyle small, rather weak, often narrowing at its posterior end; guiding ring simple, thin. Pharynx expanding near or behind its middle. Pre-rectum short. Both female and male gonads paired. Vagina with or without sclerotized pieces. Spicules of two types: simple, short and straight (alaimoid) or longer and arcuated (dorylaimoid). Gubernaculum small. A pair of ad-cloacal papillae is present; only one or few supplements present or none. Tail in both sexes cylindrical, broadly rounded. The representatives of this family are predominantly freshwater.
Additional reference: Andrássy (1987b).

Subfamilies
Thorniinae De Coninck, 1965
Thorneellinae Andrássy, 1987

Key to subfamilies of Thorniidae

1. The two circles of labial papillae close to each other, odontostyle narrowing posteriorly . . . . Thorniinae
   The two circles of labial papillae not so close; odontostyle cylindrical . . . . Thorneellinae

Subfamily Thorniinae De Coninck, 1965

Diagnosis. Thorniidae. The two circles of labial papillae are very close to each other. Odontostyle narrowing towards its posterior end. Spicules alaimoid or dorylaimoid. One, four or no supplements present.

Additional reference: Andrássy (1987b).

Genera:

Thornia Meyl, 1954

Nygolaimoides Meyl in Andrássy, 1960

Thorniosa Andrássy, 1996 (no freshwater species)

Sphaeroamphis Ahmad & Sturhan, 2000 (no freshwater species)

Key to genera of Thorniinae

1. Spicules dorylaimoid 2
   Spicules alaimoid; no supplement present 3
2. Amphidial aperture pore-like; only one supplement . . . . *Nygolaimoides*
   Amphidial aperture slit-like; four supplements . . . . *Sphaeroamphis*
3. Lip region offset, six inner liplets besides the lips, vagina not swollen . . . . *Thorniosa*
   Lip region continuous, liplets absent, vagina swollen . . . . *Thornia*

Genus *Thornia* Meyl, 1954 (Figs 15.21A–C)

Syn. *Timminema* Khera, 1978

Diagnosis. Thorniinae. Small to medium-sized nematodes. Lip region not off set. Pharynx expanding near or behind its middle. Vagina strongly swollen, without sclerotized pieces. Spicules alaimoid. Gubernaculum very thin; no supplement is present. Tail varying in length from shorter than anal body width to three times as long as anal body width; tail length slightly different in both sexes. Mainly freshwater nematodes, occasionally terrestrial.

Type species: *T. steatopyga* (Thorne & Swanger, 1936) Meyl, 1954 (USA, Paraguay, The Netherlands, Switzerland, Hungary, Italy, Russia)*

Other species:

T. goffarti (Meyl, 1953) Meyl, 1954 (Italy, Czechoslovakia)*

T. hirschmannae Andrássy, 1966 (Germany, Hungary)*

T. juvenilis (De Coninck, 1935) Meyl, 1954 (Congo)

T. magna (Paetzold, 1958) Andrássy, 1968 (Germany)*

T. parathermophila (Meyl, 1953) Meyl, 1954 (Italy, Russia)

T. pithecusana Meyl, 1954 (Italy)

T. propinqua (Paesler, 1941) Andrássy, 1957 (Germany, Hungary, The Netherlands, Czech Republic, France, Italy, Russia, India)*

T. rhopalocercoides (Schneider, 1937) Meyl, 1954 (Sumatra)*

T. steineri (Schneider, 1925) Andrássy, 1957 (Germany)*

T. thermophila (Meyl, 1953) Meyl, 1954 (Italy)

Additional reference: Andrássy (1987b).

Genus *Nygolaimoides* Meyl in Andrássy, 1960 (Figs 15.21D–F)

Diagnosis. Thorniinae. Small to medium-sized nematodes. Lip region not offset. Pharynx expanding behind its middle. Vagina swollen, without sclerotized pieces.

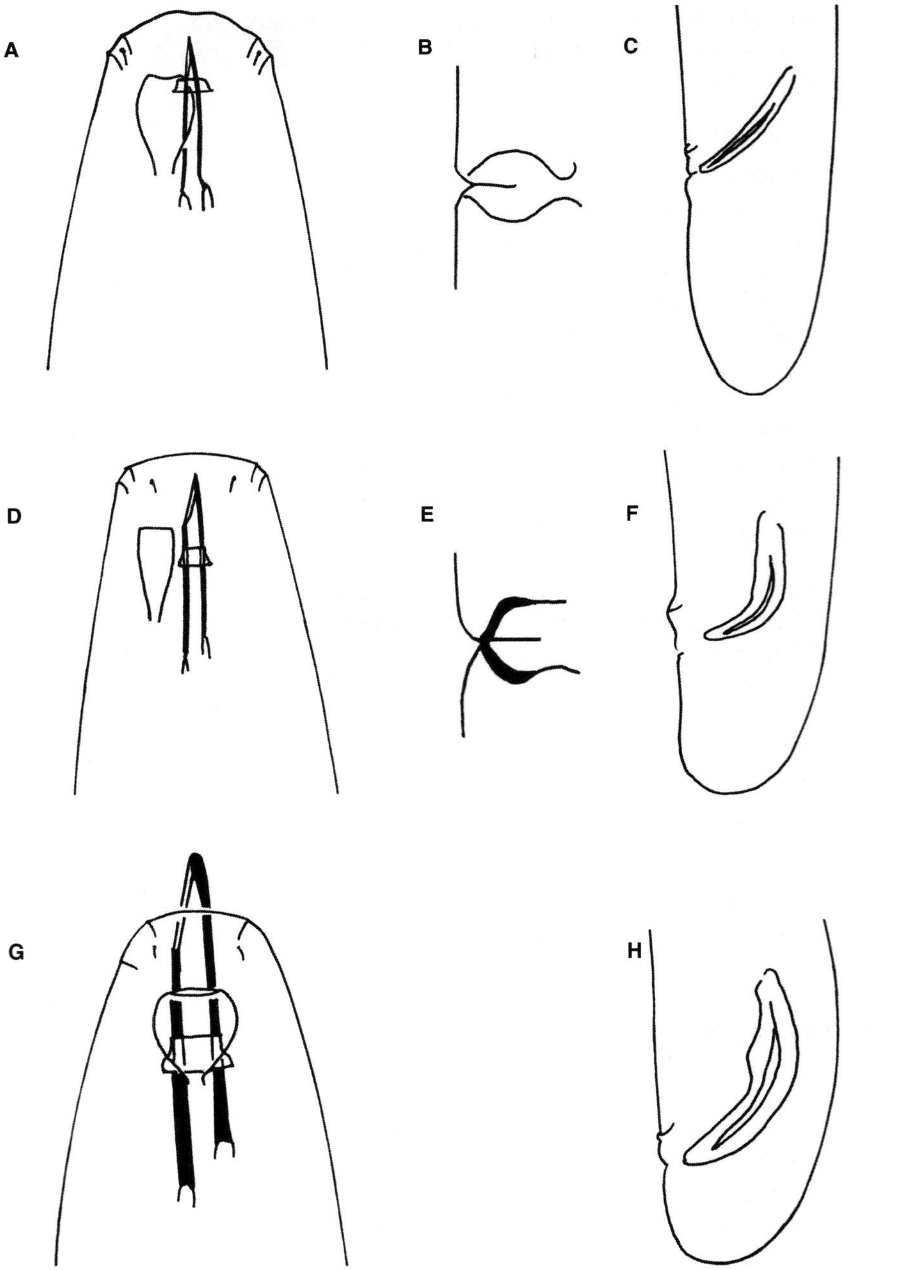

Fig. 15.21 A–C: *Thornia*; D–F: *Nygolaimoides*; G and H: *Thorneella*.

Spicules alaimoid; gubernaculum absent; one large supplement is present. Ad-cloacal supplements at different levels. Tail in both sexes shorter than anal body width. Mainly saprobic forms; few freshwater species.

Type species: *N. borborophilus* (de Man, 1876) Meyl in Andrássy, 1960 (Central Europe)*

Other species:
N. andrassyi Ahmad, Okada & Araki, 2003 (Japan)
N. fraternus Andrássy, 1987 (Hungary)
N. gubernaculifer (Andrássy, 1957) Andrássy, 1987 (Hungary, Germany)*
N. pusillus Andrássy, 1987 (Congo)
Additional reference: Andrássy (1987b).

Subfamily Thorneellinae Andrássy, 1987
Diagnosis. Thorniidae. Lip region with the two circles of labial papillae far apart. Odontostyle cylindrical. Spicules dorylaimoid. No supplements in male.
Additional reference: Andrássy (1987b).
Only genus: *Thorneella* Andrássy, 1960

Genus *Thorneella* Andrássy, 1960 (Figs 15.21G and H)
Diagnosis. Thorneellinae. With the characters of the subfamily. Cuticle smooth. Odontostyle straight, as long or slightly longer than lip region width; guiding ring thin. Pharynx enlarging at or behind its middle. Pre-rectum short. Female gonads paired. Spicules dorylaimoid; gubernaculum absent. Only the ad-cloacal pair of papillae present. Tail short and rounded in both sexes. The species of *Thorneella* live in wet soil.

Type species: *T. teres* (Thorne & Swanger, 1936) Andrássy, 1960 (USA)

Other species:
T. aculeata (Cobb in Thorne & Swanger, 1936) Andrássy, 1960 (Jamaica)
Additional reference: Andrássy (1987b).

Acknowledgements

I wish to express my gratitude to Prof. István Andrássy, University of Budapest (Hungary), who has revised the manuscript improving it with suggestions, comments and new information; to Prof. Reyes Peña-Santiago, University of Jaen (Spain), for having shared with me his bibliographical research on freshwater species; and to Drs Mirella Clausi and Lara Orselli, University of Catania (Italy) for having actively helped me to finish the manuscript in time.

References

Ahmad, M. and Jairajpuri, M.S. (1979) Nematodes of high altitude in India VIII. *Pungentus clavatus* n. sp. with a key to the species of the genus *Pungentus* (Nematoda: Dorylaimida). *Revue de Nématologie* 2, 185–189.

Ahmad, M. and Jairajpuri, M.S. (1980) A review of the genus *Enchodelus* Thorne, 1939 with descriptions of species from India. *Records of the Zoological Survey of India, Occasional Paper* 15, 42 pp.
Ahmad, W., Khan, Z. and Ahmad, I. (1992) Description of *Scleroactinolaimus punctatus* n. gen. n. sp. (Dorylaimida: Actinolaimidae) from India. *Nematologica* 38, 261–266.
Andrássy, I. (1964) Süss-wasser Nematoden aus den Grossen Gebirgsgegenden Ostafrikas. *Acta Zoologica Academia Scientiae Hungarica* 10, 1–59.
Andrássy, I. (1969) Taxonomische Übersicht der Familien Prodorylaimidae n. fam. und Dorylaimidae de Man, 1876. *Opuscula Zoologica Universitatis Budapestinensis* 9, 187–233.
Andrássy, I. (1970) Nematoden aus einigen Fluss-Systemen Sudafrikas. *Opuscula Zoologica Universitatis Budapestinensis* 10, 179–219.
Andrássy, I. (1986) The genus *Eudorylaimus* Andrássy, 1959 and the present status of its species (Nematoda: Qudsianematidae). *Opuscula Zoologica Universitatis Budapestinensis* 22, 3–42.
Andrássy, I. (1987a) The free-living nematode fauna of the Kiskunság National Park. In: Mahunka, S. (ed.) *The Fauna of the Kiskunság National Park*, vol. 2. Akadémiai Kiado, Budapest, pp. 15–46.
Andrássy, I. (1987b) The superfamily Dorylaimoidea (Nematoda) – a review. Families Thorniidae and Thornenematidae. *Acta Zoologica Academia Scientiae Hungarica* 33, 277–315.
Andrássy, I. (1988) The superfamily Dorylaimoidea (Nematoda) – a review. Family Dorylaimidae. *Opuscula Zoologica Universitatis Budapestinensis* 23, 3–63.
Andrássy, I. (1990) The superfamily Dorylaimoidea (Nematoda) – a review. Family Qudsianematidae, I, *Acta Zoologica Academia Scientiae Hungarica* 36, 163–188.
Andrássy, I. (1991) The superfamily Dorylaimoidea (Nematoda) – a review. Family Qudsianematidae, II. *Opuscula Zoologica Budapestinensis* 24, 3–55.
Andrássy, I. (1998) The genus *Boreolaimus* gen. n. and its six species (Dorylaimida: Qudsianematidae), nematodes from the European Arctic. *Fundamental and Applied Nematology* 21, 553–567.
Andrássy, I. (2000) Four large-sized species of the family Aporcelaimidae (Nematoda, Dorylaimida) with proposal of a new genus, *Epacrolaimus* gen. n. *Opuscula Zoologica Universitatis Budapestinensis* 32, 3–26.
Andrássy, I. (2001a) A taxonomic review of the genera *Aporcelaimus* Thorne and Swanger, 1936 and *Metaporcelaimus* Lordello, 1965 (Nematoda, Aporcelaimidae). *Opuscula Zoologica Universitatis Budapestinensis* 33, 7–47.
Andrássy, I. (2001b) Some species of curious genera of the Class Penetrantia (Nematoda). *International Journal of Nematology* 11, 43–57.
Andrássy, I. (2002) Free-living nematodes from the Ferto-Hanság National Park, Hungary. In: Mahunka, S. (ed.) *The Fauna of the Fertö-Hanság National Park.* Hungarian National History Museum, Budapest, pp. 21–97.
Andrássy, I. (2003) New and rare nematodes from Alaska III. Five species of the order Dorylaimida. *Journal of Nematode Morphology and Systematics* 5, 163–182.
Coomans, A. and Loof, P.A.A. (1986) Redescription of *Actinolaimus costatus* Schneider, 1935 (Nematoda: Actinolaimidae), with observations on its pharyngosomatic muscles and discussion of its taxonomic position. *Revue de Nématologie* 9, 221–231.
Coomans, A., Vinciguerra, M.T. and Loof, P.A.A. (1990) Status of the genera *Paractinolaimus* Meyl, 1957, *Trachypleurosum* Andrássy, 1959 and *Trachactinolaimus* Andrássy, 1963 (Nematoda: Actinolaimidae) with description of *Trachypleurosum venezolanum* n. sp. *Revue de Nématologie* 13, 143–154.
Heyns, J. (1965) On the morphology and taxonomy of the Aporcelaimidae, a new family of dorylaimoid nematodes. *Entomology Memoirs (South Africa)* 10, 1–51.
Heyns, J. and Swart, A. (1993) *Namaquanema hanki* n. gen., n. sp. from South Africa (Nematoda: Dorylaimoidea). *Fundamental and Applied Nematology* 16, 171–175.
Jairajpuri, M.S. (1966) *Drepanodorylaimus filiformis* n. gen., n. sp. (Nematoda: Dorylaimida) *Labdev. Journal of Science and Technology (India)* 4, 181–183.
Jairajpuri, M.S. and Ahmad, W. (1992) *Dorylaimida – Free-living, Predaceous and Plant-parasitic Nematodes.* E.J. Brill, Leiden, The Netherlands.

Jairajpuri, M.S. and Hooper, D.J. (1969) The genus *Longidorella* Thorne (Nematoda). *Nematologica* 15, 275–284.

Jairajpuri, M.S. and Siddiqi, A.H. (1964) On a new nematode genus *Nordia* (Dorylaimoidea: Nordianae n. subfam.) with remarks on the genus *Longidorella* Thorne, 1939. *Proceedings of the Helminthological Society of Washington* 31, 1–9.

Jiménez-Guirado, D. and Cadenas, M.J. (1985) Especies de *Laimydorus* Siddiqi, 1969 y descripción de *Chrysodorus longicaudatus* gen. et sp. n. (Nematoda: Dorylaimidae) de la cuenca del río Guadalquivir. *Miscelánea Zoológica* 9, 49–54.

Khan, Z. and Jairajpuri, M.S. (1994) *The Actinolaims. Predatory Soil Nematodes from India.* Section of Nematology, Department of Zoology, Aligarh Muslim University, Aligarh, India, 137 pp.

Khan, Z., Ahmad, W. and Jairajpuri, M.S. (1994a) Description of *Stopractinca orientalis* n. gen., n. sp. (Dorylaimida: Actinolaimidae) from India. *Fundamental and Applied Nematology* 17, 263–266.

Khan, Z., Ahmad, W. and Jairajpuri, M.S. (1994b) Description of *Paractinolaimoides unicus* n. sp. (Dorylaimida: Actinolaimoidea) from India. *Nematologica* 40, 494–502.

Loof, P.A.A. (1985) Taxonomic study on the genus *Prodorylaimus* Andrássy, 1959 (Nematoda: Dorylaimina). *Revue de Nématologie* 8, 193–213.

Loof, P.A.A. (2000) Dichotomous and polytomous identification keys for females of the genera *Prodorylaimus* Andrássy, 1959 and *Laimydorus* Siddiqi, 1969 (Nematoda: Dorylaimoidea). *Russian Journal of Nematology* 4, 7–28.

Meyl, A.H. (1957) Free-living nematodes. In: *Exploration Hydrobiologique du lac Tanganika (1946–47).* Institut Royal de Sciences Naturelles de Belgique 3, 27–51.

Mulvey, R.H. and Anderson, R.V. (1979) Benthic species of *Dorylaimus* Dujardin, 1845 (Nematoda: Dorylaimidae) and *Arctidorylaimus* n. gen. (Arctidorylaimidae n. fam.) from the Mackenzie and Porcupine river systems, Northwest Territories, Canada. *Canadian Journal of Zoology* 57, 743–755.

Siddiqi, M.R. (1969) *Crateronema* n. gen. (Crateronematidae n. fam.), *Poronemella* n. gen. (Lordellonematinae n. sub. fam) and *Chrysonemoides* n. gen. (Chrysonematidae n. fam.) with a revised classification of Dorylaimoidea (Nematoda). *Nematologia* 15, 18–100

Siddiqi, M.R. (1983) Four new species of *Pachylaimus* gen. n. and *Tylenchodorus tylosus* gen. n. sp. n. (Dorylaimida: Qudsianematinae) from Colombian rain forest. *Revue de Nématologie* 6, 207–215.

Siddiqi, M.R. (1995) Nematodes of tropical rainforests. 5. Seven new genera and forty-two new species of dorylaims. *Afro-Asian Journal of Nematology* 5, 72–109.

Swart, A. and Heyns, J. (1991) *Lenonchium fimbricaudatum* n. sp. from South Africa, with a key to the species of *Lenonchium* (Nematoda, Nordiidae). *Revue de Nématologie* 14, 413–418.

Thorne, G. (1939) A monograph of the nematodes of the superfamily Dorylaimoidea. *Capita Zoologica* 8, 261 pp.

Thorne, G. (1967) Nematodes of Puerto Rico: Actinolaimoidea, new superfamily with a revision of its genera and species with addenda to Belondiroidea (Nematoda, Adenophorea, Dorylaimida). *University of Puerto Rico Agricultural Experimental Station, Technical Paper* 43, 1–48.

Thorne, G. and Swanger, H.H. (1936) A monograph of the nematode genus *Dorylaimus* Dujardin, *Aporcelaimus* n. g., *Dorylaimoides* n. g. and *Pungentus* n. g. *Capita Zoologica* 6, 223 pp.

Vinciguerra, M.T. (1988) A new classification of Actinolaimoidea Thorne, 1939 using a cladistic approach. *Nematologica* 33, 251–277.

Vinciguerra, M.T. and Clausi, M. (2000) *Afractinca andrassyi* gen. n. sp. n. (Nematoda: Actinolaimidae) from Ivory Coast. *Journal of Nematode Morphology and Systematics* 2, 113–120.

Vinciguerra, M.T. and Clausi, M. (2003) Diagnostic value of morphological characters in some genera of Actinolaimidae (Nematoda: Dorylaimida) with notes on the phylogeny of the family. *Journal of Nematode Morphology and Systematics* 5, 49–59.

Vinciguerra, M.T. and Coomans, A. (1988) *Parastomachoglossa perplexa* (Heyns *et* Argo, 1969) n. comb. with a definition of the genus. *Nematologia Mediterranea* 16, 205–212.

Vinciguerra, M.T. and Heyns, J. (1984) Freshwater nematodes from South Africa. 8. New and known species of Actinolaimoidea. *South African Journal of Zoology* 19, 135–140.

16 Order Mononchida

A. ZULLINI[1] AND V. PENEVA[2]

[1]Dipartimento di Biotecnologie & Bioscienze, Università di Milano-Bicocca, Piazza della Scienza 2, 20126 Milan, Italy; [2]Central Laboratory of General Ecology, 2 Gagarin Street, 1113 Sofia, Bulgaria

Introduction

Mononchids have a strongly sclerotized and predominantly roomy stoma armed with a tooth or teeth, a practically cylindrical muscular pharynx with heavily cuticularized lumen, five pharyngeal uninucleate gland cells (one dorsal and two pairs subventral) and a gubernaculum; and precloacal supplements in males, and they lack a prerectum. The body length varies from 0.52 (*Jensenonchus ovatus*) to 7 mm (*Miconchus rex*), but usually it is >1 mm. About 25% of the species described to date can be found in freshwater (half of them being exclusively or almost exclusively freshwater). In distinguishing the genera and species, the shape and structure of the buccal cavity is of first importance. Mononchida are predators.

Systematics

Mononchida are an easily recognizable group of nematodes erected to order by Jairajpuri (1969). The first species were discovered by Dujardin (1845) (and incorrectly placed in the genera *Oncholaimus* and *Enoplus*). Twenty years later, Bastian erected the eponymous genus *Mononchus*, describing five new species. Cobb (1917) published an excellent and fundamental monograph on 'The Mononchs' with a total of 60 species; this was the basis of all the subsequent studies and revisions. Of the latter, the following are worthy of mention: Andrássy (1958, 1992a,b, 1993), Coomans and Loof (1970), Jairajpuri and Khan (1982) and De Ley and Coomans (1989). A review of freshwater Mononchida occurring in Central Europe was presented by Loof (1999).

The genus *Rahmium* Andrássy, 1973 (syn. *Stephanium* Rahm, 1938, nec Haeckel, 1887) must be considered as a *genus inquirendum* because of its poor description. It seems to be similar to *Oionchus* except for the double gonads. The genus *Tectonchus* Tsalolikhin, 1974 was not properly described, so we consider it and its four species as *genera inquirenda* and *species inquirenda*, respectively. The recently established genus *Parkellus* Jairajpuri, Tahseen & Choi, 2001 was considered different from *Coomansus* Jairajpuri & Khan, 1972. However, the distinguishing characters – larger and spacious buccal cavity (not evident from the drawing), posteriorly situated dorsal tooth and a complex gubernaculum (features occurring in some species of *Coomansus* too) – are, in our opinion, not sufficient criteria for separating these genera: consequently *Parkellus* is here regarded as a synonym of *Coomansus*. We do not support the validity of *Iotonchulus* Andrássy, 1992 and *Megaonchulus* Jairajpuri & Khan, 1982, because both genera do not differ in stomatal characters from *Mulveyellus* Siddiqi, 1984 and *Polyonchulus* Mulvey & Jensen, 1967, respectively. The differences concerning the shape and length of tail are of species value in the taxonomy of the group. Following Loof (1999), we consider *Tigronchoides* Ivanova & Dzhuraeva, 1971 and *Truxonchus* Siddiqi, 1984 as *Anatonchus*. Moreover, we subsume Andrássy's (1976) subfamily Miconchinae into Anatonchinae Jairajpuri, 1969. It is possible that molecular methods will help in clearing up the controversial opinions in the mononchid taxonomy.

Phylum Nematoda
Class Enoplea
Subclass Dorylaimia
Order Mononchida Jairajpuri, 1969
Suborder Bathyodontina Coomans & Loof, 1970
  Superfamily Cryptonchoidea Chitwood, 1937
    Family Bathyodontidae Clark, 1961
        Genus *Bathyodontus* Fielding, 1950
    Family Cryptonchidae Chitwood, 1937
        Genus *Cryptonchus* Cobb, 1913
  Superfamily Mononchuloidea de Coninck, 1965
    Family Mononchulidae de Coninck, 1962
        Genera *Mononchulus* Cobb, 1918
          Oionchus Cobb, 1913
Suborder Mononchina Kirjanova & Krall, 1969
  Superfamily Anatonchoidea Jairajpuri, 1969
    Family Anatonchidae Jairajpuri, 1969
      Subfamily Iotonchinae Jairajpuri, 1969
        Genera *Nullonchus* Siddiqi, 1984
          Caputonchus Siddiqi, 1984
          Mulveyellus Siddiqi, 1984
          Jensenonchus Jairajpuri & Khan, 1982
          Iotonchus Cobb, 1916
          Hadronchus Mulvey & Jensen, 1967
          Parahadronchus Mulvey, 1978
          Prionchulellus Mulvey & Jensen, 1967
          Prionchuloides Mulvey, 1963
          Hadronchoides Jairajpuri & Rahman, 1984
          Hadronchulus Ray & Das, 1983
      Subfamily Anatonchinae Jairajpuri, 1969
        Genera *Anatonchus* Cobb, 1916
          Micatonchus Jairajpuri, Tahseen & Choi, 2001

Miconchus Andrássy, 1958
Doronchus Andrássy, 1993
Promiconchus Jairajpuri & Khan, 1982
Crassibucca Mulvey & Jensen, 1967
Paracrassibucca Baqri & Jairajpuri, 1974
Superfamily Mononchoidea Chitwood, 1937
Family Mononchidae Chitwood, 1937
Subfamily Mononchinae Filipjev, 1934
Genera *Mononchus* Bastian, 1865
Paramononchus Mulvey, 1978
Nigronchus Siddiqi, 1984
Coomansus Jairajpuri & Khan, 1977
Clarkus Jairajpuri, 1970
Prionchulus Cobb, 1916
Actus Baqri & Jairajpuri, 1974
Sporonchulus Cobb, 1917
Judonchulus Andrássy, 1958
Subfamily Cobbonchinae Jairajpuri, 1969
Genera *Cobbonchus* Andrássy, 1958
Comiconchus Jairajpuri & Khan, 1982
Tricaenonchus Andrássy, 1996
Family Mylonchulidae Jairajpuri, 1969
Subfamily Mylonchulinae Jairajpuri, 1969
Genera *Mylonchulus* Cobb, 1916
Margaronchulus Andrássy, 1972
Oligonchulus Andrássy, 1976
Polyonchulus Mulvey & Jensen, 1967
Brachonchulus Andrássy, 1958
Granonchulus Andrássy, 1958
Crestonchulus Siddiqi & Jairajpuri, 2002

Description

Order MONONCHIDA Jairajpuri, 1969

Diagnosis. Dorylaimia. Body usually large and stout, cuticle thick and smooth (under light microscope). Cephalic papillae arranged in two or three circles. Amphids small, cup-like, with post-labial aperture. *Stoma generally large, strongly sclerotized*, in overwhelming majority of species armed with teeth or denticles. *Pharynx cylindrical, strongly muscular and with thickened lumen.* Pharyngeal gland cells uninucleate, one dorsal and two pairs ventrosublateral, their orifices posterior to nerve ring. Excretory system of renette type, excretory pore opposite or posterior to the nerve ring. Males with a series of ventromedian supplements, but without adanal supplements. Spicules paired and similar, gubernaculum and accessory pieces usually present. Caudal glands and spinneret present or absent. Mostly soil dwellers, a smaller proportion (about 25%) inhabit freshwater biotopes. Predominantly predators, but some are also bacterial feeders.

Suborder BATHYODONTINA Coomans & Loof, 1970

Diagnosis. Mononchida. Body length varying between 0.7 and 2.2 mm. Cephalic papillae arranged in two or three circles. *Stoma rather narrow, embedded almost wholly in pharyngeal tissues. One ventrosublateral tooth present.* Pharynx with five gland nuclei: the anterior one is dorsal, located about the middle of the pharynx, followed by a ventrosublateral pair and, near the posterior pharyngeal end, another ventrosublateral pair.

Superfamily CRYPTONCHOIDEA Chitwood, 1937

Diagnosis. Bathyodontina. Cephalic papillae arranged in two circles. Stoma long, cylindrical, sometimes tubular in its posterior portion, *no anterior teeth.* Cardia large.

Family CRYPTONCHIDAE Chitwood, 1937

Cryptonchoidea. *Stoma long, cylindrical.*

Genus *Cryptonchus* Cobb, 1913 (Fig. 16.1A)
Syn. *Gymnolaimus* Cobb, 1913; *Ditlevsenia* Micoletzky, 1925; *Cryptonchulus* Sukuyl, 1969
Diagnosis. Cryptonchidae. Body 0.8–2.4 mm long, slender. Lips 6, labial papillae 12. Amphids stirrup-shaped, sub-labial, up to 1.6 head region widths from anterior end. Stoma encircled by pharyngeal tissue, a long cylinder with, in its left ventrosublateral base, a small denticle. Nerve ring at 20–30% of pharyngeal length from the head end. Pharynx cylindrical, composed of a long muscular corpus and a short constricted basal region. The sequence (from anterior to posterior direction) of pharyngeal gland nuclei and orifices is: (nerve ring), DO, DN, S_1O, S_1N, S_2O, S_2N. Cardia bulboid, multinucleated. Cardiac glands absent. Tail elongate conoid to filiform, with glands and not always discernible spinneret. Female prodelphic, ovary reflexed. Vulva near mid-body or slightly behind that. Eggs (as those of related taxa) are much longer than wide. Testes two, opposed, outstretched. Spicules arcuate, without gubernaculum. A row of ventromedian supplements present.

Type species: *C. tristis* (Ditlevsen, 1911) Filipjev, 1934. In all continents except Antarctica. Freshwater or wet soil species.

Freshwater species:

C. abnormis (Allgén, 1933) Schuurmans Stekhoven, 1951 – USA (Florida), Surinam, Paraguay, Congo Republic, Ivory Coast, India, Bangladesh, Java, New Zealand (submerged soil, pond bank, water reed roots).
C. tristis (Cobb, 1893) Cobb, 1913 – Fiji Islands, West Africa (soil, river).
C. exilis (Ditlevsen, 1911) Filipjev, 1934 – North and South America, Africa, Central and Northern Europe, India, Japan, Australia (pond, river, lake, rice field, roots of water plants, brackish water, wet sandy soil).

Family BATHYODONTIDAE Clark, 1961

Cryptonchoidea. *Stoma anteriorly wider, cylindrical, posteriorly narrow, tubular.*

Genus *Bathyodontus* Fielding, 1950 (Fig. 16.1E)
Syn. *Mirolaimus* Andrássy, 1956

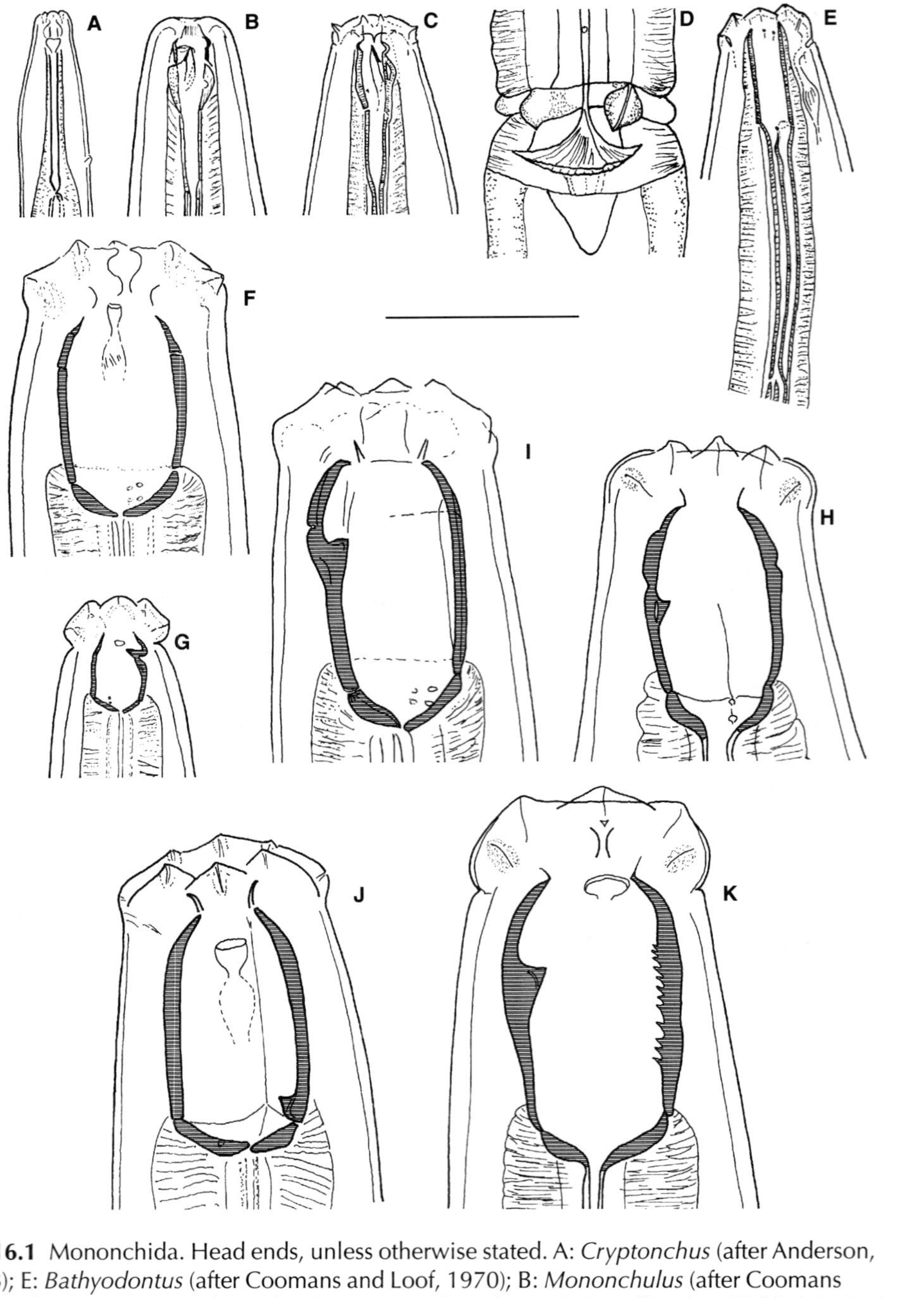

Fig. 16.1 Mononchida. Head ends, unless otherwise stated. A: *Cryptonchus* (after Anderson, 1968); E: *Bathyodontus* (after Coomans and Loof, 1970); B: *Mononchulus* (after Coomans and Loof, 1970); C: *Oionchus* (after Coomans and Loof, 1970); F: *Nullonchus* (Siddiqi, 1984a); G: *Caputonchus* (after Siddiqi, 1984b); H: *Mulveyellus* (after Mulvey and Jensen, 1967); I: *Jensenonchus* (after Peneva *et al*., 1998); J: *Iotonchus* (after Coomans and Khan, 1981); K: *Hadronchus* (after Mulvey and Jensen, 1967); D: Pharyngeal base (Anatonchoidea) (after Clark, 1960); A–D and F–K: scale bar 20 μm.

Diagnosis. Bathyodontidae. Body length 0.8–1.2 mm. Stoma long, with two cuticularized portions: anterior cylinder (15–30 μm), posterior tubular part (30–40 μm) triangular in section. Cardiac glands absent. Female didelphic (anterior branch occasionally rudimentary). Males with a ventromedian row of numerous (32–37) supplements reaching up to the neck region. Tail short, rounded, with glands and spinneret. North and South America, Europe, Africa, Asia.

Type species: *B. cylindricus* Fielding, 1950. Soil species.

Superfamily MONONCHULOIDEA de Coninck, 1965

Diagnosis. Bathyodontina. Cephalic papillae arranged in three circles. *Stoma strongly sclerotized*, with a large anterior sub-ventral tooth and 2–6 rows of transverse denticles. Spinneret conspicuous.

Family MONONCHULIDAE de Coninck, 1962

With characters of Mononchuloidea.

Genus *Mononchulus* Cobb, 1918 (Fig. 16.1B)

Diagnosis. Mononchulidae. Body 0.7–1.3 mm long, relatively plump. Stoma encircled by pharyngeal tissue. Cheilostome with thick, longitudinally striated wall. Prostome short, sclerotized, wider than cheilostome. Mesostome with sclerotized wall, with a ventrosublateral tooth-like process and *about six rows of transverse denticles*. Metastome with heavily sclerotized walls and with a rather *large grooved tooth on the right or left* (intraspecific variability) *ventrosublateral side*. Telostome elongate with triradiate walls. Amphids inconspicuous slits near the tip of the large tooth. Pharynx almost cylindrical, only slightly thickening posteriad. Nerve ring at 30–35% of pharyngeal length from the head end. The sequence of pharyngeal gland nuclei and orifices is: (nerve ring), DO, S_1O, DN + S_1N, S_2N + S_2O. Three large cardiac gland cells present. Female prodelphic, ovary reflexed, posterior genital branch rudimental. Vulva somewhat behind mid-body. Male with paired spicules, a gubernaculum and a row of ventromedian supplements. *Tail elongate*, cylindrical, with glands and a *large ventrally subterminal spinneret*.

Type species: *M. nodicaudatus* (Daday, 1901). Soil and freshwater species.

Freshwater species:

M. nodicaudatus (Daday, 1901) Schneider, 1936 – Florida, Washington, DC, Costa Rica, Surinam, Botswana (Okawango), Tanganyika, South Africa, India, Bangladesh, Indonesia, China, New Guinea (lake, pond, river, canal, peat soil, rice field, thermal water, filter beds).

Genus *Oionchus* Cobb, 1913 (Fig. 16.1C)

Syn. *Enoplochilus* Kreis, 1932

Diagnosis. Mononchulidae. Body 0.7–1.2 mm long, rather plump. Stoma encircled by pharyngeal tissues. Prostome short, sclerotized, wider than cheilostome. Mesostome sclerotized with a ventrosublateral, posteriorly placed small tooth (near the base of the large metastomal tooth) and *2–4 rows of transverse denticles*. Metastome sclerotized with a *very large grooved tooth on the right or left* (intraspecific variability) *ventrosublateral side*.

Telostome elongate with triradiate walls. Amphids near the tip of the large tooth or slightly behind it. Pharynx nearly cylindrical, slightly thickening posteriad. Nerve ring at 25–30% of pharyngeal length from the head end. The sequence of pharyngeal gland nuclei and orifices is: (nerve ring), DO, S_1O, DN + S_1N, S_2N + S_2O. Three cardiac glands. Female prodelphic, ovary reflexed, a rudimental posterior genital branch may be present. Vulva somewhat behind mid-body. *Tail short*, rounded, with glands and a *terminal* (or slightly dorsally subterminal) *spinneret.*

Type species: *O. obtusus* Cobb, 1913. Central and South America, Africa, Asia. Soil and limnic species.

Freshwater species:

O. obtusicaudatus (Kreis, 1932) Thorne, 1935 – Paraguay, Ghana, Tanzania (lake mud, riparian soil and moss, pool, swamp).
O. obtusus Cobb, 1913: syn. *O. acutangulus* Ali & Joshi, 1970; *O. setosus* Sukul, 1973 – Jamaica, Venezuela, Brazil, Andaman Islands, India (soil, riparian soil).
O. paraobtusus Jairajpuri & Khan, 1982 – India (paddy soil).
Species inquirenda: *O. didelphus* Sukul, 1971; *O. rotundicaudatus* (Kannan, 1961).

Genus *Rahmium* Andrássy, 1973 (syn. *Stephanium* Rahm, 1938 nec Haeckel, 1887). Probably it must be considered as *genus inquirendum* because of its poor description. It seems to be similar to *Oionchus* except for the double gonads.

Type species: *R. lingulatum* (Rahm, 1938) Andrássy, 1973. China. Moss species.

Suborder MONONCHINA Kirjanova & Krall, 1969

Diagnosis. Mononchida. Body length 1–7 mm. Cephalic papillae arranged in two circles. Cuticle apparently smooth, but in fact very finely transversely striated. No lateral field. *Stoma large and wide, strongly sclerotized, usually bearing teeth or denticles; its posterior quarter, or less, embedded in pharyngeal tissue.* Stoma consisting of three series of plaques (plates); six short oblique anterior plaques (cheilostome); three vertical plaques (usually divided, by a thin groove, into an anterior and a posterior part); and three other short oblique posterior plaques reaching the pharyngeal lumen. Except *Nullonchus,* dorsal tooth always present, large or small, located in anterior or posterior part of stoma, or nearly medial in position. Pharynx with five gland nuclei: the anterior one is dorsal, located near half of the pharyngeal length, followed by a ventrosublateral pair and, near the posterior pharyngeal end, another ventrosublateral pair. The sequence usually is: DO, DN, S_1O, S_1N, S_2N, S_2O. Often the two posterior structures (S_2N and S_2O) are practically at the same level. Nerve ring in the anterior third of pharynx. Females usually with two gonads. Males with two testes, 3–4 pairs of ejaculator glands and 3 pairs of rectal glands. Usually 10–25 pre-cloacal supplements.

Superfamily ANATONCHOIDEA Jairajpuri, 1969

Diagnosis. Mononchina. *Stoma broad and flattened at the base.* The three internal longitudinal folds of the pharyngeal lumen ending in three short vesicular extensions named *tubercula* (Fig. 16.1E). Distribution mainly tropical and subtropical.

Family ANATONCHIDAE Jairajpuri, 1969

Anatonchoidea. With the characters of the superfamily.

Subfamily Iotonchinae Jairajpuri, 1969

Anatonchidae. *One dorsal tooth* (lacking in one genus); ventrosublateral walls unarmed or with small teeth variously located.

Genus *Nullonchus* Siddiqi, 1984 (Fig. 16.1F)

Diagnosis. Iotonchinae. Body length 1–2 mm. *Stoma without teeth,* denticles or ribs. Tubercula small. Female prodelphic, with or without posterior uterine sac. Tail conoid arcuate with glands and spinneret.

Type species: *N. levistomus* Siddiqi, 1984. South America. Soil species of rain forests.

Genus *Caputonchus* Siddiqi, 1984 (Fig. 16.1G)

Diagnosis. Iotonchinae. Body length 0.5–0.6 mm. *Lip region sharply offset by a constriction.* Stoma relatively short. *Dorsal tooth in its anterior quarter.* Female tail short, straight, conical, without glands and spinneret.

Type species: *C. capitatus* Siddiqi, 1984. Caribbean. Soil species.

Genus *Mulveyellus* Siddiqi, 1984 (Fig. 16.1H)

Syn. *Iotonchulus* Andrássy, 1992 syn. n.

Diagnosis. Iotonchinae. Body length 0.8–3.1 mm. *Dorsal tooth from anterior third to mid-part of stoma,* or somewhat farther. Tail in both sexes *conoid arcuate* with glands and spinneret reduced or absent, or *filiform* with well-developed spinneret.

Type species: *M. jairi* (Lordello, 1959) Siddiqi, 1984 – Asia, Africa, South America, Australia–Oceania. Soil species.

Genus *Jensenonchus* Jairajpuri & Khan, 1982 (Fig. 16.1I)

Diagnosis. Iotonchinae. Body length 0.5–2.3 mm. Stoma roomy, *dorsal tooth from anterior third to midway of stoma.* On each ventrosublateral wall a small *longitudinal ridge* and/or a fine transverse rib. Pharyngeal base tuberculate. Female didelphic or prodelphic without post-vulval sac. $V \approx 60–80\%$. Spicules arcuate with bifurcate accessory pieces. Supplements 9–15. Tails of both sexes very short or conoid arcuate ($c' = 2–5$) with glands reduced, no spinneret.

Type species: *J. ovatus* (Jensen & Mulvey, 1968) Jairajpuri & Khan, 1982 – Europe, Asia, Africa, North America, Oceania. Soil and limnic species.

Freshwater species:

J. sphagni (Brzeski, 1960) Peneva, Neilson, Boag & Brown, 1998 – Czech Republic, Scotland, Poland (wet soil, *Sphagnum*).

J. vorax (Cobb, 1917) Andrássy, 1993 – East Canada, Wisconsin, New Jersey (moor, pond).

Genus *Iotonchus* Cobb, 1916 (Fig. 16.1J)

Diagnosis. Iotonchinae. Body length 0.8–6.4 mm. Stoma roomy, sometimes oblong, varying in length from 20 to 90 µm. *Dorsal tooth moderately large, basal* or suprabasal. No other armature in stoma. Female prodelphic or didelphic. $V \approx 50–80\%$. Spicules more or less arcuate with bifurcated lateral pieces. Supplements 6–22. Tails of both sexes similar, from very short and rounded to very long and filiform. Caudal glands and spinneret usually well developed, sometimes reduced or absent; spinneret often sub-terminal. In all continents, except Antarctica, especially in Africa, Asia and Australia–Oceania.

Type species: *I. gymnolaimus* (Cobb, 1893) Cobb, 1916. Soil and freshwater species.

Freshwater species:

I. bangkokensis Buangsuwon & Jensen, 1966 – Thailand (rice fields).
I. caesar Alekseev, 2001 – Siberia (Far East) (lake).
I. parabasidontus Mulvey and Jensen, 1967 – South Africa, India (soil, humid soil near water).
I. risoceiae Carvalho, 1955 – USA, Africa, South America, India, Hawaii, Hungary, Russia, Kenya, India (soil, lake, water furrows).
I. trichurus Cobb, 1917 – Central and South America, Africa, Mauritius, India, New Zealand (soil, brook).
I. vulvapapillatus Andrássy, 1964 – Nigeria, Uganda, Kenya (pond).

Genus *Hadronchus* Mulvey & Jensen, 1967 (Fig. 16.1K)
Diagnosis. Iotonchinae. Body length 1.5–2.2 mm. *Dorsal tooth in anterior half of stoma opposed by two denticulate ridges* along the entire ventrosublateral walls. Female prodelphic. Tails of both sexes conoid arcuate with glands and spinneret reduced.

Type species: *H. bisexualis* Mulvey & Jensen, 1967. Africa. Soil species.

Genus *Parahadronchus* Mulvey, 1978 (Fig. 16.2A)
Diagnosis. Iotonchinae. Body length 1.9–3.3 mm. *Both dorsal tooth and two or four longitudinal ridges* (with 2–8 denticles each) *in posterior part of stoma*. Spicules slender and arcuate. Tails of both sexes elongate to filiform.

Type species: *P. andamanicus* (Jairajpuri, 1969) Mulvey, 1978. Asia. Soil species.

Genus *Prionchulellus* Mulvey & Jensen, 1967 (Fig. 16.2B)
Diagnosis. Iotonchinae. Body length 0.9–1.4 mm. *Dorsal tooth in anterior third of stoma opposed by two denticulate ribs with an anterior hook-like loop*. Tail elongate, without glands and spinneret.

Type species: *P. cavenessi* Mulvey & Jensen, 1967. Africa. Soil species.

Genus *Prionchuloides* Mulvey, 1963 (Fig. 16.2C)
Diagnosis. Iotonchinae. Body length about 2 mm. *Dorsal tooth midway in stoma opposed by two denticulate ribs plus a clump of irregularly dispersed denticles*. Tail conoid arcuate, without glands and spinneret.

Type species: *P. micoletzkyi* (Meyl, 1954) Mulvey, 1963. Limnic and wet soil species.

Freshwater species:

P. micoletzkyi (Meyl, 1954) Mulvey, 1963; Italy, Russia, Kola Peninsula (soil, lake, water, vegetation).

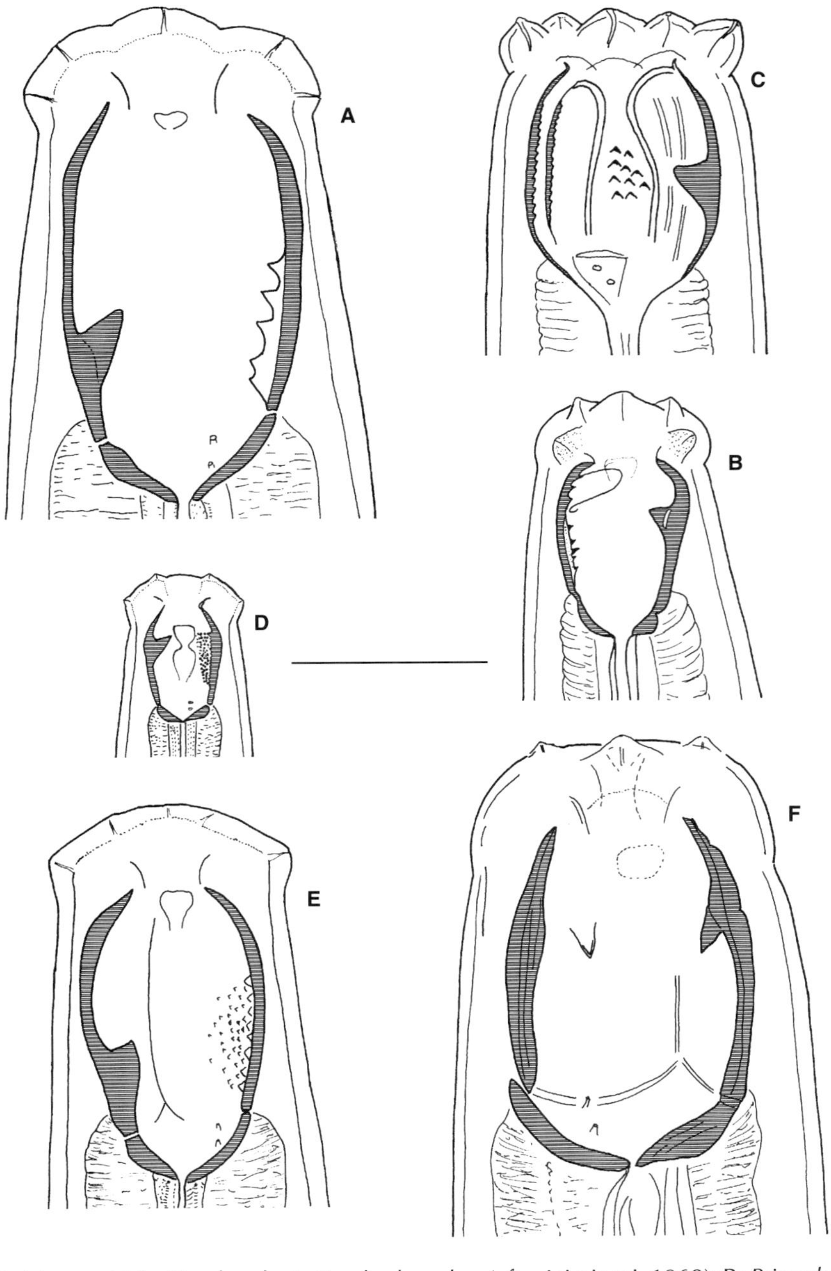

Fig. 16.2 Mononchida. Head ends. A: *Parahadronchus* (after Jairajpuri, 1969); B: *Prionchulellus* (after Mulvey and Jensen, 1967); C: *Prionchuloides* (after Meyl, 1954); D: *Hadronchoides* (after Jairajpuri and Rahman, 1984); E: *Hadronchulus* (after Ray and Das, 1982); F: *Anatonchus* (after Peneva *et al.*, 1999). Scale bar 20 μm.

Genus *Hadronchoides* Jairajpuri & Rahman, 1984 (Fig. 16.2D)
Diagnosis. Iotonchinae. Body length about 1 mm. *Dorsal tooth in anterior half of stoma, ventrosublateral walls bearing many small denticles decreasing in number posteriorly*. Tail filiform with glands and spinneret.

Type species: *H. microdenticulatus* Jairajpuri & Rhaman, 1984. Asia. Soil species.

Genus *Hadronchulus* Ray & Das, 1983 (Fig. 16.2E)
Syn. *Morenchus* Dhanachand, Renubala & Mohilal, 1991.
Diagnosis. Iotonchinae. Body length 1.8–3.2 mm. *Dorsal tooth in posterior half of stoma; ventrosublateral walls with many small denticles*. Tail of both sexes elongate conoid to filiform, with glands and spinneret.

Type species: *H. shamini* Ray & Das, 1983. Asia. Soil species.

Subfamily Anatonchinae Jairajpuri, 1969
Diagnosis. Anatonchidae. *One dorsal tooth and one or two pairs of ventrosublateral teeth*, all equal in shape or ventrosublateral teeth somewhat smaller than dorsal tooth.

Genus *Anatonchus* Cobb, 1916 (Fig. 16.2F)
Syn. *Tigronchoides* Ivanova & Dzhuraeva, 1971; *Truxonchus* Siddiqi, 1984
Diagnosis. Anatonchinae. Body length ranges from 1.5 to 6.5 mm. Stoma oblong to spheroid. *One dorsal and two ventrosublateral teeth, all retrorse*, equal in shape and position: near the base, midway or in the very anterior part of stoma. Pharyngeal base usually very clearly tuberculate. Female didelphic or prodelphic; $V \approx 50–80\%$. Small papillae near the vulva usually present. Spicules more or less massive, arcuate, with forked lateral pieces. Supplements 9 to 20. Tail of each sex similar, short conoid (rarely straight) to filiform, usually with glands and spinneret.

Type species: *A. tridentatus* (de Man, 1876) Cobb, 1916. In all continents. Soil, rarely in freshwater habitats.

Freshwater species:

A. acutus Altherr, 1974 – Iberia, Germany (river banks, psammon, soil).
A. dolichurus (Ditlevsen, 1911) Cobb, 1916 – Canada, USA (New York), Spain, UK, Alps, Central and Northern Europe, Faroe Islands (boggy soil, green turf, moss, pond, moist sand, subterranean water).
A. filicaudatus Altherr, 1971 – Carpathians (subterranean water).
A. mammillatus Altherr, 1968 – Central and Northern Europe (subterranean water).
A. monohystera Altherr, 1977 – Brazil (river).
A. tridentatus (de Man, 1876) Cobb, 1916 – Canada, USA, Mexico, France, UK, Europe, Alps, Carpathians, Turkey, Central Russia, Uzbekistan, Iran, Ghana (humid soil, pond, lake, subterranean water, brackish water).

Genus *Micatonchus* Jairajpuri, Tahseen & Choi, 2001 (Fig. 16.3A)
Diagnosis. Anatonchinae. Body length about 2 mm. *Dorsal tooth forward directed, two ventrosublateral teeth retrorse*. Tail conoid-elongate with glands and spinneret.

Type species: *M. reversus* Jairajpuri, Tahseen & Choi, 2001. Asia. Soil species.

Genus *Miconchus* Andrássy, 1958 (Fig. 16.3B)
Syn. *Miconchoides* Jairajpuri & Khan, 1982.
Diagnosis. Anatonchinae. Body length 1–7 mm. Stoma roomy, with *three basal teeth: one dorsal and two ventrosublateral, pointed forwards*. No other teeth nor ridges. Female usually didelphic; rarely prodelphic or pseudomonodelphic; $V \approx 50–80\%$. Spicules arcuate, with lateral guiding pieces. Supplements 9–24. Tail short conoid to filiform, similar or different in the two sexes. Caudal glands and spinneret well developed or reduced.

Type species: *M. digiturus* (Cobb, 1893) Andrássy, 1958. In all continents, except Antarctica. Soil or freshwater species.

Freshwater species:

M. aquaticus Khan, Ahmad & Jairajpuri, 1978 – Romania, India (rice fields).
M. crenicaudatus Gagarin, 1984 – Siberia (lake).
M. rex (Cobb, 1904) Andrássy, 1958 – New Zealand (lake).
M. schneideri (Meyl, 1957) Andrássy, 1958 – Brazil, Tanzania (lake).
M. studeri (Steiner, 1914) Andrássy, 1958 – California, Central America, UK, Spain, France, Italy, Alps, Central and Northern Europe, Mauritius Islands (soil, wet moss, wet river bank).

Genus *Doronchus* Andrássy, 1993 (Fig. 16.3C)
Diagnosis. Anatonchinae. Body length 1.8–3.6 mm. Cuticle finely annulated. Labial region off set from body. Stoma barrel-shaped, moderately roomy, armed with three basal teeth lying at same level. Dorsal tooth distinctly larger than ventrosublateral ones. Female didelphic, V ≈ 53–62%. Spicules arcuate, with forked lateral pieces. Supplements 12–16. Tails of both sexes different: in females almost straight, elongate–conoid with sharp terminus and 5–10 times as long as anal body width, in male ventrally bent, broadly conoid with bluntly rounded terminus, only 1.5–2 times as long as anal body width. Caudal glands and spinneret absent.

Type species: *D. kirikiri* (Yeates, 1967) Andrássy, 1993. New Zealand. Species of dune sands.

Genus *Promiconchus* Jairajpuri & Khan, 1982 (Fig. 16.3D)
Diagnosis. Anatonchinae. Body length 1.3–1.9 mm. Stoma barrel-shaped, roomy, with *three teeth: one dorsal tooth anterior, two ventrosublateral teeth basal*. Female prodelphic; V ≈ 60–70%. Spicules arcuate, with bifurcate lateral pieces. Supplements 5–6. Tail elongate, ventrally or first ventrally and then dorsally bent, without glands and spinneret.

Type species: *P. microdontus* (Mulvey & Jensen, 1967) Jairajpuri & Khan, 1982. Tropical America, Africa. In humid soil and freshwater.

Freshwater species:

P. conicaudatus (Altherr, 1977) Andrássy, 1993 – Brazil (river).

Genus *Crassibucca* Mulvey & Jensen, 1967 (Fig. 16.3E)
Diagnosis. Anatonchinae. Body length 1–1.5 mm. *Dorsal tooth anterior, four ventrosublateral denticles: one pair in anterior and one pair in posterior half of stoma*. Tail conoid arcuate to very long, with poorly developed glands, spinneret present or absent.

Fig. 16.3 Mononchida. Head ends, unless otherwise stated. A: *Micatonchus* (after Jairajpuri *et al.*, 2001); B: *Miconchus* (after Mulvey, 1962); C: *Doronchus* (after Yeates, 1967); D: *Promiconchus* (after Mulvey and Jensen, 1967); E: *Crassibucca* (after Mulvey and Jensen, 1967); F: *Paracrassibucca* (after Baqri and Jairajpuri, 1974); H: *Mononchus* (after Coomans and Khan, 1981); J: *Nigronchus* (after Siddiqi, 1984); I: *Paramononchus* (original); G: Pharyngeal base (Mononchoidea) (after Clark, 1960); A–F, H and I: scale bar 20 μm.

Type species: *C. penicula* Mulvey & Jensen, 1967. Africa, South America. Soil species.

Genus *Paracrassibucca* Baqri & Jairajpuri, 1974 (Fig. 16.3F)
Diagnosis. Anatonchinae. Body length 0.7–1.2 mm. *Dorsal tooth in the anterior, two pairs of ventrosublateral denticles in the posterior half of stoma,* behind each other. Tail short, arcuate, with glands and spinneret.

Type species: *P. paucidentata* (Lordello, 1970) Jairajpuri & Khan, 1982. Tropical America, Asia. Soil species.

Superfamily MONONCHOIDEA Chitwood, 1937

Diagnosis. Mononchina. *Stoma tapering at base.* The three internal longitudinal folds of the pharyngeal lumen do not form vesicular extensions (tubercula) (Fig. 16.3G). Cosmopolitan.

Family MONONCHIDAE Chitwood, 1937

Mononchoidea. Ventrosublateral *denticles,* if present, *never arranged in transverse rows.*

Subfamily Mononchinae Filipjev, 1934
Diagnosis. Mononchidae. *Dorsal tooth in anterior part of stoma* (rarely in the mid or posterior part). Ventrosublateral walls generally with *longitudinal ridges,* which are tiny and smooth or denticulated.

Genus *Mononchus* Bastian, 1865 (Fig. 16.3H)
Diagnosis. Mononchinae. Body length 1–6 mm. Stoma oval, twice or thrice as long as wide, with *one dorsal tooth, pointed forward, in anterior part;* opposite to the dorsal tooth a *thin transverse rib on each ventrosublateral wall.* Sometimes also a weak longitudinal rib in the anterior part of stoma. Female didelphic (in one species opisthodelphic); $V \approx$ 45–65%. Spicules long and slender. Supplements tubular, 10–47. Tail of both sexes short and obtuse or elongate, with evident *glands and spinneret.*

Type species: *M. truncatus* Bastian, 1865. In all continents except Antarctica, predominantly freshwater animals, rarely in moist soil.

Freshwater species:

M. angariensis Gagarin, 1984 – Siberia (river, lake).
M. aquaticus Coetzee, 1968 – Mexico, Central and South America, Iberia, UK, Northern Europe, Spain, Italy, Bulgaria, Serbia, Russia, Kyrgyzstan, Egypt, Nigeria, South Africa, Mauritius Islands, India, Sahalin Island (wet soil, canal, brook, river, mineral springs).
M. clarki Altherr, 1972 – Massachusetts (brook: psammon).
M. italicus Andrássy, 1959 – Italy (psammon).
M. laminatus Zullini, Loof & Bongers, 2002 – Costa Rica (brook).
M. loofi Winiszewska, 1998 – Poland (ditch).
M. maduei W.Schneider, 1925 – Canada, Germany: Madüsee; Russia: Onega (in deep lake); Lithuania.
M. mulveyi Andrássy, 1985 – Canada (lake).

M. niddensis Skwarra, 1921 – Canada, Alps, Central and Northern Europe, Central Russia, Baikal (lake, humid soil, sandy littoral).
M. oryzae Dhanam & Jairajpuri, 1999 – India (rice field).
M. pulcher Andrássy, 1993 – Argentina, Brazil, Chile, California, Ethiopia, Hungary (lake).
M. superbus Mulvey, 1978 – Canada N-W, Spain, Central and Northern Europe, Russia (brook, river, channel, pond).
M. tajmiris Gagarin, 1991 – Arctic Russia (deep lake).
M. truncatus Bastian, 1865 – All continents except Antarctica (brook, river, ditch, channel, pond, lake, wet soil and moss).
M. tunbridgensis Bastian, 1865 – All continents except Antarctica (ditch, brook, wet rice field, wet grass).

Genus *Paramononchus* Mulvey, 1978 (Fig. 16.3I)
Diagnosis. Mononchinae. Body length 2.5–4.0 mm. Stoma oblong, more than twice as long as wide, with one large *anterior dorsal tooth* pointed forward. Opposite to it each ventrosublateral wall armed with *a toothlet develops a thin transverse rib*; *a short longitudinal ridge is present.* Female didelphic; $V \approx 50–55\%$. Male unknown. Tail elongate, with glands and spinneret.

Type species: *P. arcticus* Mulvey, 1978. North America, Asia. Limnic species.

Freshwater species:

P. alimovi Tsalolikhin, 1990 – Arctic Russia (lake).
P. arcticus Mulvey, 1978 – Arctic Canada (river).
P. thiocrenobius (Soos, 1943) Loof, 1999 – Poland (sulphurous spring).

Genus *Nigronchus* Siddiqi, 1984 (Fig. 16.3J)
Diagnosis. Mononchinae. Body length 0.6–0.7 mm. *Lip region sharply offset by a constriction.* No ventral longitudinal ridges in buccal cavity. Female prodelphic. Tail elongate, with conspicuous glands and spinneret.

Type species: *N. parvus* Siddiqi, 1984. Africa. Soil species.

Genus *Coomansus* Jairajpuri & Khan, 1977 (Fig. 16.4A)
Syn. *Parkellus* Jairajpuri, Tahseen & Choi, 2001 syn. n.
Diagnosis. Mononchinae. Body length strongly varying, from 0.7 to 5.8 mm. Stoma oblong, usually as long as head width. *Dorsal tooth moderately large, located in anterior or posterior part of stoma, on each ventrosublateral wall opposed by a thin longitudinal ridge gradually fading anteriorly.* Posterior third of pharynx with pearl-shaped structure. Female didelphic; $V \approx 45–75\%$. Spicules relatively short; supplements 8–28. Tails in both sexes conoid, ventrally arcuate (in males shorter than in females), with rudimentary glands and without spinneret.

Type species: *C. parvus* (de Man, 1880) Jairjapuri & Khan, 1977. In all continents. In soil, rarely in freshwater.

Freshwater species:

C. fletcherensis Mulvey, 1978 – Arctic Canada (lake).

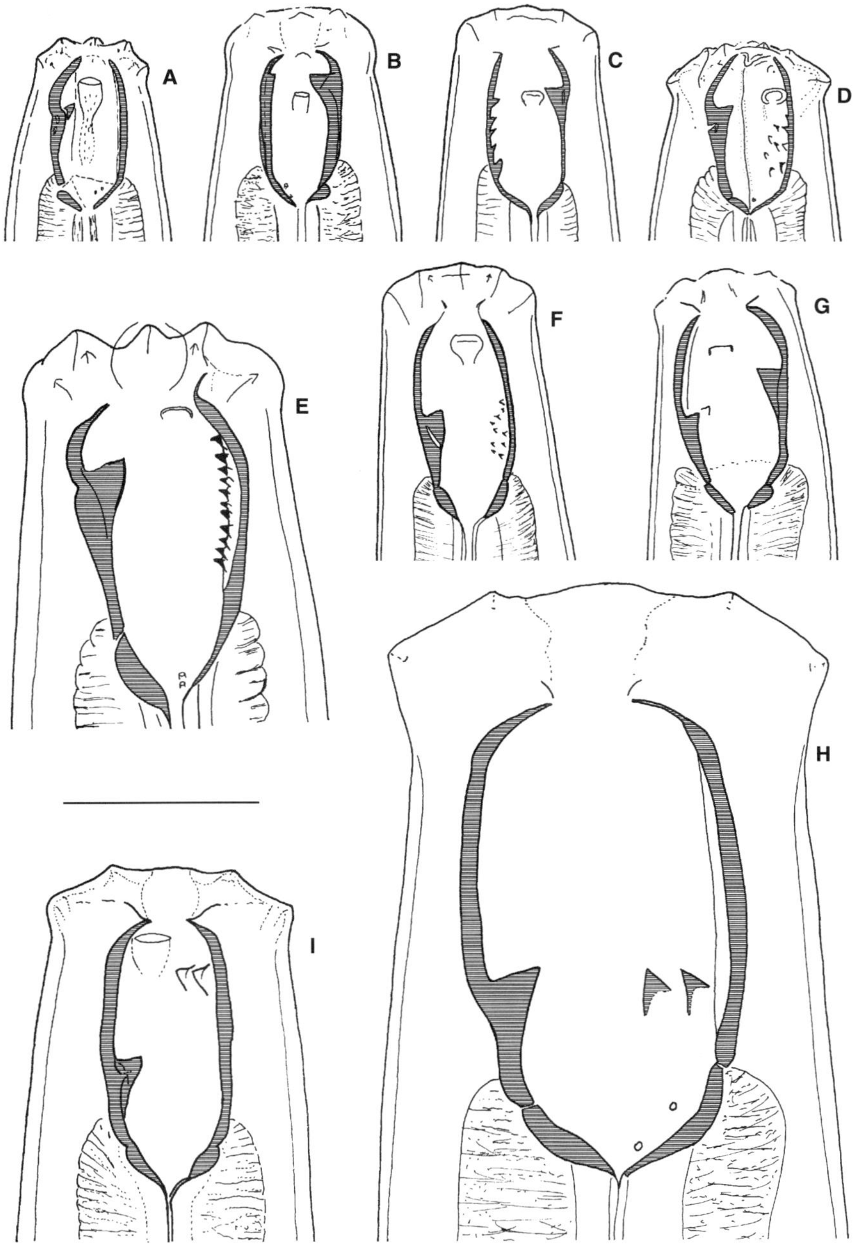

Fig. 16.4 Mononchida. Head ends. A: *Coomansus* (after Coomans and Khan, 1981); B: *Clarkus*; E: *Prionchulus* (after Mulvey, 1967); C: *Actus* (after Mulvey, 1973); D: *Sporonchulus* (after Mulvey, 1963); F: *Judonchulus* (after Mulvey, 1963); G: *Cobbonchus*; H: *Comiconchus* (original) (after Susulovsky, 2001); I: *Tricaeonchus* (after Andrássy, 1996). Scale bar 20 μm.

C. intestinus (Vinciguerra & La Rosa, 1990) Andrássy, 1993 – Argentina, Chile (peat-bog moss).
C. kastrolli (Altherr, 1958) Jairajpuri & Khan, 1977 – Central and Northern Europe (river bank, subterranen water, filter beds).
C. parvus (de Man, 1880) Jairajpuri & Khan, 1977 – North and South America, Africa, Europe, Asia, Hawaii (humid soil, sandy soil, pond, aquatic moss, river bank, ocean shore).
C. pretoriensis (Coetzee, 1968) Jairajpuri & Khan, 1977 – South Africa, India (river mud).
C. silvius (Eroshenko, 1975) Jairajpuri & Khan, 1977 – Russia (Far East), Korea.
C. simmenensis (Kreis, 1924) Loof & Winiszewska-Slipinska, 1993 – Spain, Alps, Balkans (mountain lake).
C. zschokkei (Menzel, 1913) Loof & Winiszewska-Slipinska, 1993 – UK, Europe, Central Russia, Eastern Siberia, Korea (soil, grassland, peat bog, rare in water).

Genus *Clarkus* Jairajpuri, 1970 (Fig. 16.4B)
Diagnosis. Mononchinae. Body length between 0.7 and 2.7 mm. Stoma barrel-shaped, nearly twice as long as wide. *Dorsal tooth directed forward, located in anterior half of stoma; it is opposed by a longitudinal ridge on each ventrosublateral wall abruptly ending anteriorly almost tooth-like.* Posterior third of pharynx with pearl-shaped structure. Female didelphic; $V \approx 45–70\%$. Males rare. Supplements 10–19; spicules relatively short. Tails in both sexes conoid, ventrally arcuate, with rudimentary glands and without spinneret.

Type species: *C. papillatus* (Bastian, 1865) Jairajpuri, 1970. In all continents except Antarctica. In soil, rarely in freshwater.

Freshwater species:

C. papillatus (Bastian, 1865) Jairajpuri, 1970 – all continents except Antarctica (soil, moss, spring, pond).
C. sheri (Mulvey, 1967) Jairajpuri, 1970 – North America, Galápagos, South Africa, India, Japan (soil, banks of stream, wall of dam, moss next to fountain).
C. sveltus (Altherr, 1968) Jairajpuri & Khan, 1972 – Germany, Carpathians (spring).

Genus *Prionchulus* Cobb, 1916 (Fig. 16.4E)
Diagnosis. Mononchinae. Body length 1–4 mm. Stoma oblong, large, with a *large dorsal anterior tooth* pointing forward. *Opposite to it, two ventrosublateral denticulate longitudinal ridges*. Female didelphic; $V \approx 55–70\%$. Males rare. Supplements 16–29. Tails in both sexes conoid, bent ventrad, with rudimentary glands and without spinneret.

Type species: *P. muscorum* (Dujardin, 1845) Cobb, 1916. In all continents except Antarctica. Especially in moss, but also in soil, litter and freshwater.

Freshwater species:

P. kamchaticus Susulovsky, Winiszewska & Gagarin, 2003 – Russia – Kamchatka Peninsula (lake).
P. longus (Thorne, 1929) Goodey, 1951 – North America, Italy, Nepal (soil, lake, beach sand).
P. major Gagarin, 2001 – Kamchatka Peninsula (lake).
P. muscorum (Dujardin, 1845) Cobb, 1916 – all continents except Antarctica, (moss, litter, soil, lake, subterranean water, brook, moor, thermal water).

P. punctatus Cobb, 1917 – North and South America, Europe, Asia (moss, litter, humid soil, ditch, brook, lake, beach sand).
P. spectabilis (Ditlevsen, 1912) Cobb, 1916 – Central and Northern Europe (meadow, taiga soil, brackish water, inundated marine beach).

Genus *Actus* Baqri & Jairajpuri, 1974 (Fig. 16.4C)
Diagnosis. Mononchinae. Body length 0.8–1.5 mm. *Opposite to the dorsal tooth, irregular rows of 4–6 pairs of denticles* not arranged in longitudinal ridges. Tail short, conoid, bent ventrad, with glands and spinneret.

Type species: *A. minutus* (Mulvey, 1963) Baqri & Jairajpuri, 1974. North and Central America, Oceania. Soil species.

Genus *Sporonchulus* Cobb, 1917 (Fig. 16.4D)
Diagnosis. Mononchinae. Body length 1.1–1.8 mm. *Opposite to the dorsal tooth, four irregular rows of numerous denticles.* Tail short, conoid, bent ventrad; caudal glands and spinneret normal or reduced.

Type species: *S. dentatus* Cobb, 1917. Subtropical: South America, Africa, Asia. Soil species.

Genus *Judonchulus* Andrássy, 1958 (Fig. 16.4F)
Diagnosis. Mononchinae. Body varying in length between 1.4 and 5.1 mm, slender to very slender. Stoma large, elongate with a large *dorsal tooth located midway or farther in stoma*; *opposite to it are many scattered denticles.* Pharyngeal base not tuberculate. Female didelphic; $V \approx 64\%$. Supplements 9–10. Tail elongate, almost cylindrical, with glands and spinneret.

Type species: *J. recessus* (Cobb, 1917) Andrássy, 1958. Subtropical America. In soil and freshwater.

Freshwater species:

J. magnus Altherr, 1977 – Brazil (psammon).
J. recessus (Cobb, 1917) – Florida (soil, swampy soil).

Subfamily Cobbonchinae Jairajpuri, 1969
Mononchidae. *Dorsal and each ventrosublateral wall armed with a tooth of nearly equal size and forward pointed apex.* Rarely also two denticles in basal stoma.

Genus *Cobbonchus* Andrássy, 1958 (Fig. 16.4G)
Diagnosis. Cobbonchinae. Body length 0.8–3.9 mm. Stoma oblong, a large *dorsal tooth in anterior or in mid-region of stoma,* and *two ventrosublateral teeth, somewhat smaller, slightly or distinctly posteriorly located.* Female didelphic or prodelphic; $V \approx 60–80\%$. Male supplements 6–18. Tails in both sexes short, conoid, digitate, or blunt, arcuate, with conspicuous glands and spinneret.

Type species: *C. palustris* (Cobb, 1917) Andrássy, 1958. In all continents (except Antarctica), but not certain in Europe. In freshwater, wet soil, moss.

Freshwater species:

C. abrupticaudatus (Altherr, 1960) Goodey, 1963 – Cameroon (river).
C. artemisiae Coetzee, 1968 – South Africa, India (semi-freshwater).
C. coetzeeae Andrássy, 1970 – South Africa, Central America (river bank, humid soil)
C. collaris Andrássy, 1985 – Brazil (river neuston).
C. curvicaudatus Tsalolikhin, 1988 – Brazil (pond).
C. dianae Coetzee, 1965 – South Africa (semi-freshwater).
C. mauritianus (Williams, 1958) Clark, 1960 – Mauritius, New Zealand (wet soil).
C. megalus Coetzee, 1966 – South Africa (river bank).
C. palustris (Cobb, 1917) Andrássy, 1958 – Wisconsin (swamp).
C. pounamua Clark, 1969 – New Zealand (swampy pasture, wet sand, *Sphagnum*).
C. radiatus (Cobb, 1917) Andrássy, 1985 – New Jersey (bog).
C. regulus Altherr, 1963 – Argentina (interstitial).
C. soutoi Carvalho, 1956 – Brazil (lake).

Genus *Comiconchus* Jairajpuri & Khan, 1982 (Fig. 16.4H)
Diagnosis. Cobbonchinae. Body length 3–4 mm. Stoma barrel-shaped with *three equal teeth in mid-stoma with forward pointed apex*. Female didelphic; $V \approx 60$–65%. Male not known. Tail elongate conoid, with reduced glands, no spinneret.

Type species: *C. trionchus* (Thorne, 1924) Jairajpuri & Khan, 1982. Europe, North America. In wet moss, soil and freshwater.

Freshwater species:

C. trionchus (Thorne, 1924) Jairajpuri & Khan, 1982 – North America, Volga River (moss, moist soil, river).

Genus *Tricaenonchus* Andrássy, 1996 (Fig. 16.4I)
Diagnosis. Cobbonchinae. Body length about 2.8 mm. *Dorsal tooth in posterior half, two ventrosublateral teeth in the anterior part of stoma*. Ventral contour of posterior pharynx with a pearl-like structure (small oblong or concave elements). Tail short, conoid, arcuate, without glands and spinneret.

Type species: *T. caucasius* Andrássy, 1996. Caucasus: Soil species.

Family MYLONCHULIDAE Jairajpuri, 1969
Diagnosis. Mononchoidea. Stoma strongly tapering at base, more funnel-shaped than in other families. *Ventrosublateral denticles rasp-like*, generally numerous, *arranged in transverse rows*.

Subfamily Mylonchulinae Jairajpuri, 1969
Mylonchulidae. With the characters of the family.

Genus *Mylonchulus* Cobb, 1916 (Fig. 16.5A)
Syn. *Paramylonchulus* Jairajpuri & Khan, 1982; *Pakmylonchulus* Khan & Saeed, 1987.
Diagnosis. Mylonchulidae. Body length ranged from 0.5 to 2.9 mm. Stoma goblet- or funnel-shaped, dorsal wall thicker than ventral. *Dorsal tooth very large, directed forward, with pointed apex, in anterior half of stoma; ventrosublateral rasp-like denticles arranged in rather*

regular transverse rows. Anterior and/or posterior margin of rasp-fields often with refractive arches. Often ventrosublateral denticles opposite to the base of the dorsal tooth. Female didelphic or prodelphic; $V \approx 50$–70%. Spicules slender, arcuate, with or without lateral pieces. Supplements 6–16; often some weak supplements are present anterior to the normal ones. Tails similar in both sexes, short or slightly elongated, with usually conspicuous glands and terminal or subterminal spinneret.

Type species: *M. minor* (Cobb, 1893) Cobb, 1916. In all continents except Antarctica. In soil, semi-freshwater habitat or freshwater.

Freshwater species:

M. andrassyi Loof, 1993 – Hungary, Czech Republic, Serbia, Italy, Russia, Egypt (subterranean water, psammon).

M. brachyuris (Bütschli, 1873) Cobb, 1917 – North and South America, Europe, Africa, Asia (soil, moss, subterranean water, pond, marsh, shore).

M. brevicaudatus Cobb, 1917 – North and Central America, UK, Alps, Central Europe, Spain, Northern Russia, Caucasus, Africa, Asia (soil, moor, pond, river).

M. cavensis W. Schneider, 1940 – Slovenia (pond from subterranean water).

M. gigas Gagarin, 1993 – Siberia (lake).

M. incurvus Cobb, 1917 – North and Central America, Brazil, Central and Northern Europe, France, Russia, India, Korea, Thailand, South Africa (soil, moor, spring).

M. inflatus Loof, 1993 – The Netherlands (pond, moist soil).

M. insolitus Andrássy, 1968 – Congo (psammon).

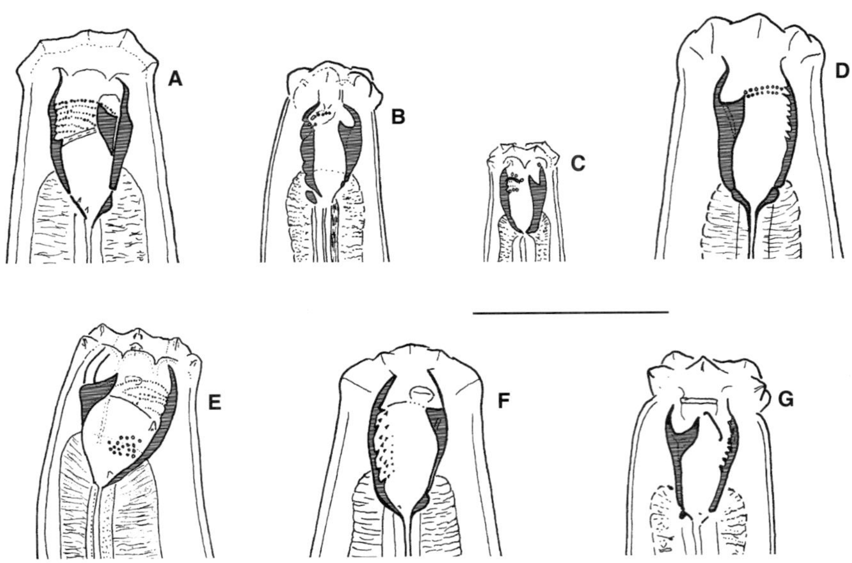

Fig. 16.5 Mononchida. Head ends. A: *Mylonchulus* (after Jairajpuri, 1970); B: *Margaronchulus* (original); C: *Oligonchulus* (after Cobb, 1917); D: *Polyonchulus* (after Mulvey and Jensen, 1967); E: *Brachonchulus* (after Micoletzky, 1925); F: *Granonchulus* (after Mulvey, 1963); G: *Crestonchulus* (after Siddiqi, 2002). Scale bar 20 μm.

M. lacustris (Cobb, 1915) Cobb, 1917 – North and South America, North and Central Europe, Central Russia, Africa, Asia (soil, psammon, lake, river bank, mineral springs, pool).
M. longus Altherr, 1972 – Northern Europe (subterranean water).
M. macrosoma Altherr, 1976 – Germany (subterranean water).
M. minor (Cobb, 1893) – North and South America, Iran, India, Nigeria, South Africa, Pakistan, Fiji Islands (soil, *Sphagnum*, well, river sand).
M. obtusicaudatus (Daday, 1899) Cobb, 1916 – Central and South America, New Guinea (soil, flooded soil, brook).
M. rotundicaudatus (Skwarra, 1921) W. Schneider, 1939 – Central and Northern Europe, Italy, Russia (brackish water, coastal dune, lake shore).
M. sigmaturellus Mulvey, 1961 – North and Central America, Belgium, Romania (soil, subterranean water, river bank).
M. sigmaturus (Cobb, 1917) – All continents except Antarctica (soil, moss, pond, river, brook, subterranean water, continental salt water).
M. vulvapapillatus Altherr, 1972 – Massachusetts, UK (pond: psammon).

Genus *Margaronchulus* Andrássy, 1972 (Fig. 16.5B)
Diagnosis. Mylonchulidae. Body length about 0.8 mm. *Ventrosublateral rasp-like denticles in a single row* (with a short transverse ridge) *or absent* (instead of those, two closely arranged refractive rings present). Tail short or filiform, with reduced glands, no spinneret.

Type species: *M. mulveyi* Andrássy, 1972. Tropical Africa. Soil species.

Genus *Oligonchulus* Andrássy, 1976 (Fig. 16.5C)
Diagnosis. Mylonchulidae. Body length about 1 mm. *A few denticles in two transverse rows*. Tail arcuate, with glands and spinneret.

Type species: *O. reversus* (Cobb, 1917). South America. Soil species.

Genus *Polyonchulus* Mulvey & Jensen, 1967 (Fig. 16.5D)
Syn. *Megaonchulus* Jairajpuri & Khan, 1982 syn. n.
Diagnosis. Mylonchulidae. Body length about 1 mm. Stoma about twice as long as wide with a *dorsal tooth in anterior half, directed forward*; *ventrosublateral rasp-like denticles in 2–7 transverse rows, two longitudinal serrate ridges also present*. No refractive arches. Pharyngeal base not tuberculate. Female didelphic or prodelphic; $V \approx 65$–70%. Spicules arcuate, without lateral pieces; six small supplements. Tails of both sexes short, conoid or subdigitate to filiform, with or without glands and spinneret.

Type species: *P. cobbi* Mulvey & Jensen, 1967. Africa. Soil and freshwater species.

Freshwater species:

P. clavicaudatus (Schuurmans Stekhoven & Teunissen, 1938) – Congo (brook).

Genus *Brachonchulus* Andrássy, 1958 (Fig. 16.5E)
Diagnosis. Mylonchulidae. Body length about 1 mm. *Denticles in two goups*: an anterior composed of 5–6 *rows* and a posterior in a *scattered* order. Tail short, conoid, with glands and spinneret.

Type species: *B. brachyuroides* (Micoletzky, 1925). South America. Soil species.

Genus *Granonchulus* Andrássy, 1958 (Fig. 16.5F)
Diagnosis. Mylonchulidae. Body length 1.0–2.2 mm. Stoma ovoid, not tapering strongly to the base. *Dorsal tooth in anterior half of stoma* pointed forward; *anteriorly a single transverse rasp-like row of denticles and posteriorly scattered denticles in longitudinal rows.* Female didelphic; $V \approx 55$–65%. Supplements 15–21. Tails in both sexes hemispheroid or conoid, with or without glands and spinneret.

Type species: *G. decurrens* (Cobb, 1917) Andrássy, 1958. North America, Europe, Africa, Asia. Soil and limnic species.

Freshwater species:

G. schulzi (Meyl, 1955) – Massachusetts, Central and Northern Europe (brackish psammon, pond).
G. subdecurrens Coetzee, 1966 – South Africa, Nigeria (soil, river banks).

Genus *Crestonchulus* Siddiqi & Jairajpuri, 2002 (Fig. 16.5G)
Diagnosis. Mylonchulidae. Body comparatively long and slender. Cephalic region sharply offset from the body by a constriction. Stoma tapering to the base. Dorsal tooth massive, two ventrosublateral longitudinal rows of denticules. *Each ventrosublateral wall provided with a sclerotized square, bracket-shaped crest at about the level of the dorsal tooth apex. Ventrosublateral teeth and transverse row of denticles absent.* Female gonad pseudo-monodelphic. Tail conoid in both sexes, caudal glands and spinneret absent.

Type species: *C. crestus* Siddiqi & Jairajpuri, 2002. Cameroon. Soil species.

Key to genera of MONONCHIDA
Note: *Asterisk indicates genera containing freshwater species.*

1. Stoma rather narrow, almost entirely embedded in pharyngeal tissues. Largest tooth ventrosublateral . . . . Bathyodontina 2
   Stoma large and wide, its posterior one-fourth, or less, embedded in pharyngeal tissues. Larger tooth typically dorsal . . . . Mononchina 6
2. Stoma strongly sclerotized with a large grooved tooth on a ventrosublateral side 3
   Stoma moderately sclerotized, without a large tooth 5
3. In anterior part of stoma about six rows of transverse denticles. Tail elongate ($c' > 2$), cylindrical, with a large (ventrally) subterminal spinneret . . . . (Fig. 16.1B) *Mononchulus**
   In anterior part of stoma 2–4 rows of transverse denticles. Tail short ($c' \approx 1$), rounded, with terminal or (dorsally) subterminal spinneret . . . . (Fig. 16.1C) *Oionchus**
4. Stoma a long cylinder (35–70 μm). Female prodelphic. Tail conical, elongate to filiform; spinneret terminal, often indiscernible . . . . (Fig. 16.1A) *Cryptonchus**
   Stoma anteriorly more spacious, cylindrical (15–30 μm), posteriorly a narrow tube (30–40 μm). Female didelphic

(anterior branch occasionally rudimentary). Tail short, rounded; spinneret terminal or sub-terminal .... (Fig. 16.1B) *Bathyodontus*

5. Stoma broad and flattened at the base; pharyngeal base with three tubercula (Anatonchoidea). .... (Fig. 16.1E) 7
   Stoma tapering at base; pharyngeal base without tubercula (Mononchoidea) .... (Fig. 16.3G) 22
6. Stoma without any tooth or denticles; tubercula small; female prodelphic .... (Fig. 16.1F) *Nullonchus*
   Stoma with one dorsal tooth and, sometimes, many denticles on the ventral wall 8
   Stoma with one dorsal tooth and 2–4 ventrosublateral teeth 16
7. Dorsal tooth faced by longitudinal denticulate ribs and scattered rasp-like denticles 9
   Dorsal tooth faced by longitudinal denticulate ribs, no rasp-like denticles 11
   Only dorsal tooth present, no ventrosublateral teeth nor rasp-like denticles 13
8. Dorsal tooth in anterior half of stoma; small denticles decreasing in number posteriorly .... (Fig. 16.2D) *Hadronchoides*
   Dorsal tooth midway or in posterior half of stoma 10
9. Dorsal tooth mid-part of stoma; few rasp-like denticles .... (Fig. 16.2C) *Prionchuloides**
   Dorsal tooth in posterior half of stoma; many rasp-like denticles .... (Fig. 16.2E) *Hadronchulus*
10. Dorsal tooth posterior .... (Fig. 16.2A) *Parahadronchus*
    Dorsal tooth anterior 12
11. Dorsal tooth faced by a hook-like loop .... (Fig. 16.2B) *Prionchulellus*
    Dorsal tooth not faced by a hook-like loop .... (Fig. 16.1K) *Hadronchus*
12. Dorsal tooth relatively small or near the base of stoma .... (Fig. 16.1J) *Iotonchus**
    Dorsal tooth at mid-part or in anterior stoma 14
13. Lip region sharply off set by constriction; dorsal tooth in the anterior quarter of stoma .... (Fig. 16.1G) *Caputonchus*
    Lip region not or slightly offset; dorsal tooth in the anterior third to midway of stoma, or somewhat posterior 15
14. A small longitudinal ridge and/or fine transverse rib on each ventrosublateral wall present . (Fig. 16.1I) *Jensenonchus**
    A ventral longitudinal ridge and transverse rib absent .... (Fig. 16.1H) *Mulveyellus*
15. Stoma with one dorsal tooth and two ventrosublateral teeth 17
    Stoma with one anterior dorsal tooth and four ventrosublateral teeth 21
16. Dorsal tooth in anterior half of stoma; ventrosublateral teeth basal .... (Fig. 16.3D) *Promiconchus**
    The three teeth lying about at the same level 18

17. Teeth in the posterior half of stoma; dorsal tooth directed forward; ventrosublateral teeth retrorse . . . . (Fig. 16.3A) *Micatonchus*
All three teeth pointing in the same direction 19
18. Teeth basal, pointed forward 20
Teeth retrorse (suprabasal, midway or anteriorly placed) . . . . (Fig. 16.2F) *Anatonchus**
19. Teeth equal in size, no sexual dimorphism in tail shape . . . . (Fig. 16.3B) *Miconchus**
Ventrosublateral teeth somewhat smaller than the dorsal one; tail in female almost straight, elongate conoid with sharp terminus, $c' = 5–10$, tail in male ventrally bent, broadly conoid with bluntly rounded terminus, $c' = 1.5–2$ . . . . (Fig. 16.3C) *Doronchus*
20. One pair of ventrosublateral teeth facing the dorsal tooth, the other pair posteriorly placed; female prodelphic . . . . (Fig. 16.3E) *Crassibucca*
Both pairs of ventrosublateral teeth in posterior half of stoma; female amphidelphic . . . . (Fig. 16.3F) *Paracrassibucca*
21. Ventrosublateral teeth, if present, never arranged in transverse rows; stoma not excessively tapered at base; dorsal tooth in anterior (usually) or in posterior half of stoma 23
Ventrosublateral denticles rasp-like, arranged in transverse rows; stoma strongly tapered at base (funnel-shaped); dorsal tooth always in anterior half of stoma 33
22. Ventrosublateral sides of stoma without ridges or denticles, or with one or two longitudinal ridges tiny and smooth or denticulated 24
Three teeth (one on each stomatal wall) of nearly equal size and forward-pointed apex 31
23. Ventral side of stoma with longitudinal rows of denticles (sometimes very tiny) or with scattered denticles 25
Ventral side of stoma without denticles 27
24. Dorsal tooth in mid-part or in lower part of stoma; opposite to it many scattered denticles; tail elongate, almost cylindrical, with glands and spinneret . . . . (Fig. 16.4F) *Judonchulus**
Dorsal tooth in anterior part of stoma 26
25. Stoma with subventral denticles (conspicuous or very small) arranged in two longitudinal regular rows; tail without spinneret . . . . (Fig. 16.4E) *Prionchulus**
Ventrosublateral walls of stoma with two irrregular rows of 4–6 pairs of denticles; tail with spinneret . . . . (Fig. 16.4C) *Actus*
Ventrosublateral walls of stoma with numerous denticles arranged in four irregular rows; tail with spinneret normal or reduced . . . . (Fig. 16.4D) *Sporonchulus*

26. Lip region sharply offset by constriction; female prodelphic;
$L < 1$ mm; tail with conspicuous spinneret .... (Fig. 16.3J) *Nigronchus*
Lip region not sharply off set 28
27. Opposite the dorsal tooth a thin, transverse ribbon each ventrosublateral wall; tail elongate cylindroid (never conical–arcuate) with distinct spinneret 29
Transverse ribs lacking; tail conoid, ventrally arcuate, without spinneret 30
28. Each transverse rib connected with a short longitudinal ridge possessing one pair of small denticles; $L = 2.5$–4 mm .... (Fig. 16.3I) *Paramononchus**
Opposite the dorsal tooth no toothlets; tail elongate cylindroid, rarely short obtuse; $L = 1$–6 mm .... (Fig. 16.3H) *Mononchus**
29. Almost entire ventrosublateral sides of stoma bearing a very thin and transparent longitudinal ridge, each gradually fading anteriorly .... (Fig. 16.4A) *Coomansus**
Almost entire ventrosublateral sides of stoma bearing a very thin and transparent longitudinal ridge, each abruptly ending anteriorly .... (Fig. 16.4B) *Clarkus**
30. Three equal teeth in mid-stoma; tail without spinneret .... (Fig. 16.4H) *Comiconchus**
Teeth otherwise arranged 32
31. Dorsal tooth in anterior or in mid-part of stoma, and two ventrosublateral teeth, somewhat smaller, somewhat posteriorly; tail with distinct spinneret (Fig. 16.4G) .... *Cobbonchus**
Dorsal tooth in posterior, two ventrosublateral teeth in anterior half of stoma; tail without spinneret (Fig. 16.4I) .... *Tricaenonchus*
32. Rasp-like denticles in two groups: in 5–6 transverse rows opposite to dorsal tooth, and scattered denticles in the base of stoma (Fig. 16.5E) .... *Brachonchulus*
Rasp-like denticles in one group opposite to dorsal tooth 34
33. Transverse rasp-like denticles absent or in one row (or one refractive ring) 35
Transversal rasp-like denticles in 2–7 rows 36
34. Ventral side of stoma with scattered denticles arranged in longitudinal rows; female didelphic (Fig. 16.5F) ... *Granonchulus**
No ventral denticles; $L < 1$ mm; female prodelphic (Fig. 16.5B) .... *Margaronchulus*
Two longitudinal serrate ridges; one square bracket-shaped crest on each ventrosublateral wall at level of the dorsal tooth apex; gonad pseudomonodelphic (Fig. 16.5G) .... *Crestonchulus*
35. Ventrosublateral denticles and refractive ring present; caudal glands and spinneret well developed (Fig. 16.5A) .... *Mylonchulus**

Ventrosublateral denticles and refractive ring absent; caudal glands and spinneret reduced or absent 37

36. Longitudinal serrate ridge present (Fig. 16.5D) . . . . *Polyonchulus**

Longitudinal serrate ridge absent (Fig. 16.5C) . . . . *Oligonchulus*

Mononchids as Predators

Only mononchs, among nematodes, have a feeding apparatus of the engulfing type (Gaugler and Bilgrami, 2004). In fact, they are regarded as predators *par excellence* within the whole Nematoda and their predatory behaviour is the basis for understanding their morphology, physiology and ecology. Cobb (1917) stressed that some mononchs swallow other nematodes that are nearly half as long as themselves: 'A similar gastronomic exploit on the part of a man would be the gulping down of a string of bologna sausages several feet in length'. Since the pharynx seldom occupies more than a quarter of the total body length, this means that the victim's body reaches the mononch's intestine before the other end has entered the mouth. Sometimes such effort can be dangerous: one ingested soil nematode (*Rhabditis*) was observed unaffected in the mononch's intestine, causing progressive harm to the motility of its predator. During the night, the *Rhabditis* bored its way through the vulva, killing the mononch (Steiner and Heinly, 1922).

Mononchs usually search for their prey 'in the midst of stygian darkness' (Cobb, 1917), so they must rely on their touch and smell. Following Thorne (1924) 'they are doubtless terrors to the myriads of minute creatures associated with them'. However, observations on the predatory behaviour of *Prionchulus punctatus* and of *Mononchus aquaticus* show that these species are not attracted towards their prey and that predation depends purely on chance encounters. Moreover, the prey species (predominantly other nematodes) seemed unable to detect the mononchs and took no evasive action (Nelmes, 1974; Bilgrami *et al.*, 1984). Small (1987) examined 212 records pertaining to 55 species of Mononchida and established that their average diet was: nematodes 75%, oligochaetes 7%, rotifers 7%, protozoa 5%, tardigrades 1% and other types of prey 5%. Mononchs, therefore, are essentially nematode-eaters. In an experiment, one individual of *Mononchus* killed 1332 nematodes during its life of 12 weeks. *Mylonchulus sigmaturus* was seen devouring other nematodes at an average rate of 1.5–2.0 individuals/h (Cohn and Mordechai, 1974). Bilgrami *et al.* (1986), examining the gut content of more than 1000 mononchid specimens (of at least 24 species, mostly soil but also freshwater inhabitants), discovered that 45% of them contained smaller nematodes in their intestine. More predators (75%) had microphagous nematodes (monhysterids, diplogastrids, rhabditids, etc.) in their gut rather than had tylenchids (45%) or dorylaims (42%). The most commonly preyed-upon nematode genera were *Pratylenchus, Hoplolaimus, Tylenchorhynchus, Tylencholaimus, Aporcelaimus, Thornenema, Iotonchus, Mylonchulus, Rhabditis, Acrobeloides* and *Chiloplacus*. The number of prey animals present in the intestine of a single predator varied from none to eight. It is interesting that about one-half of the examined mononch species show cannibalism (20% of individuals). Experimental studies on *Mylonchulus dentatus* and *M. aquaticus* showed that these two species, in the absence of other prey, feed on the younger specimens of their own species. Not only adults, but also fourth-stage juveniles, and sometimes even third-stage ones, perform cannibalism, especially on first- and second-stage juveniles (Azmi and Jairajpuri, 1979; Bilgrami and Jairajpuri, 1984).

Food preferences were detected for *M. aquaticus,* which especially preys on sluggish nematodes such as *Cephalobus* Bastian, 1865, whereas less frequently, it kills active prey such as *Prismatolaimus* de Man, 1880 (Bilgrami *et al.*, 1983). This same predator has no problem in feeding on adults of *Monhystera* Bastian, 1865, but it cannot wound nor eat adults of *Cuticularia oxycerca.* Instead, it can swallow the first stage of *C. oxycerca* (in 1–10 s), the second stage (in 30s to 2 min), the third stage (up to 3 min) and the fourth stage (up to 10 min). Another mononchid, *P. punctatus*, is less efficient as a predator on the same rhabditid (Small and Grootaert, 1983). Other nematodes resist predation owing to their thick cuticle, annulations, gelatinous matrix, speed (body undulations), vigorous escape response and toxic/unfavourable secretions (Bilgrami, 1992). Some mononchs were cultured under monoxenic conditions with bacteria, so this would theoretically suffice for their energy requirements. Nevertheless, this resource is not very effective, otherwise we should have greater populations of mononchs in nature; on the contrary, Rhabditida appear to be much more efficient as bacterial feeders. Moreover, Mononchida are *K*-strategists and require a lot of energy, especially in their later stages, and this requires animal predation (Yeates, 1987).

Interactions among different predators were studied, placing in the same agar plates three mononch species: *M. dentatus* (1.0–1.5 mm) was devoured by the other two species without being able to injure any of them; more efficient was *M. aquaticus* (1.2–1.8 mm), being able to ingest the latter species but not *Iotonchus indicus*; this latter (1.2–2.2 mm) was the strongest predator, being able to kill and devour the other two species (Bilgrami *et al.*, 1985). Mononchida can be devoured by other elements of the microfauna (e.g. tardigrades), but probably their numbers are controlled especially by parasites; in a certain field, e.g. 30% of mononchs were seen to be parasitized by sporozoans (Thorne, 1927).

The males of the mononchs, as a rule, are rare or absent in many species. A striking exception occurs in *Jensenonchus amphigonicus* (hence the name), where the males and females are about equal in number (Thorne, 1924). Cobb (1917) interpreted this fact as syngonism, when he observed extremely minute spermatozoa produced by females. The existence of males does not necessarily imply an amphimictic reproduction. In fact, observations of males of *P. punctatus* highlighted several copulations but no sperm transmission. Moreover, although active in their movements, unlike the females, they never attacked any prey (Samsoen, 1984).

References

Anderson, R.V. (1968) Emendation of *Cryptonchus* Cobb, 1913 (Nematoda: Ironidae), redescription of *C. tristis* (Ditlevsen, 1911) Filipjev, 1934, and description of the male. *Canadian Journal of Zoology* 46, 935–942.

Andrássy, I. (1958) Über das System der Mononchiden (Mononchidae Chtitwood, 1937; Nematoda). *Annales Historico-Naturales Musei Nationalis Hungarici* 50, 151–171.

Andrássy, I. (1992a) A taxonomic survey of the family Mononchidae (Nematoda). *Acta Zoologica Hungarica* 39, 13–60.

Andrássy, I. (1992b) A taxonomic survey of the family Mylonchulidae (Nematoda). *Opuscula Zoologica (Budapest)* 25, 11–35.

Andrássy, I. (1993) A taxonomic survey of the family Anatonchidae (Nematoda). *Opuscula Zoologica Budapest* 26, 9–52. (Published in 1994.)

Andrássy, I. (1996) *Tricaenonchus caucasius* gen. et sp. n., a remarkable nematode species from the Caucasus (Nematoda: Mononchidae). *Zoosystematica Rossica* 4, 19–21.

Azmi, M.I. and Jairajpuri, M.S. (1979) Effect of mechanical stimulation and crowding on *Mylonchulus dentatus*. *Nematologia Mediterranea* 7, 203–207.

Baqri, S.Z. and Jairajpuri, M.S. (1974) Studies on Mononchida. V. The mononchs of El Salvador with descriptions of two new genera, *Actus* and *Paracrassibucca*. *Nematologica* 19, 326–333.

Bilgrami, A.L. (1992) Resistance and susceptibility of prey nematodes to predation and strike rate of the predators, *Mononchus aquaticus*, *Dorylaimus stagnalis* and *Aquatides thornei*. *Fundamentals of Applied Nematology* 15, 265–270.

Bilgrami, A.L. and Jairajpuri, M.S. (1984) Cannibalism in *Mononchus aquaticus*. *Indian Journal of Nematology* 14, 202–203.

Bilgrami, A.L., Ahmad, I. and Jairajpuri, M.S. (1983) Some factors influencing predation by *Mononchus aquaticus*. *Revue de Nématologie* 6, 325–326.

Bilgrami, A.L., Ahmad, I. and Jairajpuri, M.S. (1984) Observations on the predatory behaviour of *Mononchus aquaticus*. *Nematologia Mediterranea* 12, 41–45.

Bilgrami, A.L., Ahmad, I. and Jairajpuri, M.S. (1985) Interaction between mononchs in non-sterile agar plates. *Indian Journal of Nematology* 15, 339–340.

Bilgrami, A.L., Ahmad, I. and Jairajpuri, M.S. (1986) A study of the intestinal contents of some mononchs. *Revue de Nématologie* 9, 191–194.

Clark, W.C. (1960) The oesophago-intestinal junction in the Mononchidae (Enoplida, Nematoda). *Nematologica* 5, 178–183.

Cobb, N.A. (1917) The Mononchs. A genus of free-living predatory nematodes. *Soil Science* 3, 431–486.

Cohn, E. and Mordechai, M. (1974) Experiments in suppressing citrus nematode populations by use of a marigold and a predaceous nematode. *Nematologia Mediterranea* 2, 43–53.

Coomans, A. and Khan, S.H. (1981) Mononchida from Mount Kenya. *Biologisch Jaarboek Dodonaea* 49, 64–75.

Coomans, A. and Lima, M.B. (1965) Description of *Anatonchus amiciae* n. sp. (Nematoda: Mononchidae) with observations on its juvenile stages and anatomy. *Nematologica* 11, 413–431.

Coomans, A. and Loof, P.A.A. (1970) Morhology and taxonomy of Bathyodontina (Dorylaimida). *Nematologica* 16, 180–196.

De Ley, P. and Coomans, A. (1989) A revision of the genus *Bathyodontus* Fielding, 1950 with the description of a male of *B. cylindricus* Fielding, 1950 (Nematoda: Mononchida). *Nematologica* 35, 147–164.

Dujardin, F. (1845) Histoire naturelle des helminthes ou vers intestinaux. Libraire Encyclopédique de Roret, Paris, 654 pp.

Gaugler, R. and Bilgrami, A.L. (2004) *Nematode Behaviour.* CAB International, Wallingford, UK, 419 pp.

Jairajpuri, M. S. (1969) Studies of Mononchida of India. I. The genera *Hadronchus*, *Iotonchus* and *Miconchus*, and a revised classification of Mononchida, new order. *Nematologica* 15, 557–581.

Jairajpuri, M.S. (1970) Studies of Mononchida of India. III. The genus *Mylonchulus* (Family Mylonchulidae Jairajpuri, 1969). *Nematologica* 16, 434–456.

Jairajpuri, M.S. and Khan, W.U. (1982) *Predatory Nematodes (Mononchida)*. Associated Publishing Company, New Delhi.

Jairajpuri, M.S. and Rahman, M.F. (1984) *Hadronchoides microdenticulatus* n. gen., n.sp. (Nematoda: Iotonchidae) from India. *Nematologica* 29, 266–268.

Jairajpuri, M.S., Tahseen, Q. and Choi, Y.E. (2001) *Micatonchus reversus* gen. n., sp. n. (Mononchida: Anatonchidae), a unique mononch from Korea. *International Journal of Nematology* 11, 77–80.

Loof, P.A.A. (1999). *Nematoda: Adenophorea (Dorylaimida)*. *Süßwasserfauna von Mitteleuropa* 4/2-2. Spektrum Akademisher Verlag, Heidelberg, Germany, 264 pp.

Meyl, A.H. (1954) Die bisher in Italien gefundenen freilebenden Erd- und Suesswasser-Nematoden. *Archivio Zoologico Italiano* 39, 161–264.

Micoletzky, H. (1925) Wissenschaftliche Mitteilungen. 1. Zur Kenntnis tropischer, freilebender Nematoden aus Surinam, Trinidad und Ostafrika. *Zoologischer Anzeiger* 64, 1–28.

Mulvey, R.H. (1962) The Mononchidae: a family of predaceous nematodes III. Genus *Miconchus* (Enoplida: Mononchidae). *Canadian Journal of Zoology* 40, 65–81.

Mulvey, R.H. (1963) The Mononchidae: a family of predacious nematodes. V. Genera *Sporonchulus, Granonchulus* and *Prionchuloides* n. gen. (Enoplida: Mononchidae). *Canadian* Journal of Zoology 41, 763–774.

Mulvey, R.H. (1967) The Mononchidae: a family of predaceous nematodes. VII. Genus *Prionchulus* (Nematoda: Mononchidae). *Canadian Journal of Zoology* 45, 941–953.

Mulvey, R.H. and Jensen, H.J. (1967) The Mononchidae of Nigeria. *Canadian Journal of Zoology* 45, 667–727.

Nelmes, A.J. (1974) Evaluation of the feeding behaviour of *Prionchulus punctatus* (Cobb): a nematode predator. *Journal of Animal Ecology* 43, 553–565.

Peneva, V., Neilson, R., Boag, B. and Brown, D.J.F. (1998) The morphology of *Jensenonchus sphagni* (Brzeski, 1960) n. comb. from Scotland and comments on the taxonomic status and distribution of this species. *Nematologica* 44, 255–268.

Peneva, V., Neilson, R. and Nedelchev, S. (1999). Mononchoid nematodes from oak forests in Bulgaria. 1. The subfamily Anatonchinae Jairajpuri, 1969 with descriptions of *Anatonchus genovi* sp. n. and *Tigronchoides quercus* sp. n. *Nematology* 1, 37–53.

Ray, S. and Das, S.N. (1983) *Hadronchulus shamimi* n. gen, n. sp. and *Cobbonchus artemisiae* (Mononchida: Nematoda) from Orissa, India. *Nematologica* 28, 247–252.

Samsoen, L. (1984) Experiments with the male of *Prionchulus punctatus* (Cobb, 1917) Andrássy, 1958. *Revue de Nématologie* 7, 417.

Siddiqi, M.R. (1984a) *Nullonchus*, a new genus of toothless mononchs, with comments on the phylogenetic relationship of Nullonchinae n. subfam. (Nematoda: Mononchida). *Nematologica* 30, 11–21.

Siddiqi, M.R. (1984b) Four new genera and four new species of mononchs (Nematoda). *Pakistan Journal of Nematology* 2, 1–13.

Siddiqi, M.R. (2002) *Crestonchulus crestus* gen. n., sp. n. (Nematoda: Mononchida) from Cameroon. *International Journal of Nematology* 12, 85–88.

Small, R.W. (1987) A review of the prey of predatory soil nematodes. *Pedobiologia* 30, 179–206.

Small, R.W. and Grootaert, P. (1983) Observations on the predatory abilities of some soil dwelling predatory nematodes. *Nematologica* 29, 109–118.

Steiner, G. and Heinly, H. (1922) The possibility of control of *Heterodera radicicola* and other plant-injurious nemas by means of predatory nemas, especially by *Mononchus papillatus*. *Journal of the Washington Academy of Sciences* 12, 367–386.

Susulovsky, A.S. (2001) A new species of the genus *Comiconchus* Jairajpuri & Khan from Ukraine with redescription of *C. trionchus* (Thorne) (Nematoda, Mononchida: Mononchidae). *Zoosystematica Rossica* 9, 261–266.

Thorne, G. (1924) Utah nemas of the genus Mononchus. *Transactions of the American Microscopical Society* 43, 157–171.

Thorne, T. (1927) Importance of some mononchs. *Journal of Agricultural Research* 34, 265–286.

Yeates, G.W. (1967) Studies on nematodes from dune sands. 3. Oncholaimidae, Ironidae, Alaimidae and Mononchidae. *New Zealand Journal of Science* 10, 299–321.

Yeates, G.W. (1987) Significance of developmental stages in the coexistence of three species of Mononchoidea (Nematoda) in a pasture soil. *Biology and Fertility of Soils* 5, 225–229.

17 Orders Chromadorida, Desmodorida and Desmoscolecida

W. Decraemer[1,2] and N. Smol[2]

[1]Royal Belgian Institute of Natural Sciences, Vautierstraat 29, B-1000 Brussels, Belgium; [2]Department of Biology, Ghent University, Ledeganckstraat 35, B-9000 Ghent, Belgium

Introduction

The three orders Chromadorida, Desmodorida and Desmoscolecida are treated together. Their close relationship has been discussed in several studies of high-level taxonomy. Before Lorenzen (1981), the majority of these studies were based on morphological features as markers of evolutionary relationships and resulted in conflicting versions of classification (De Ley and Blaxter, 2002). Lorenzen (1981) provided the first phylogenetic system based on cladistic principles, but again morphological characters alone appeared insufficient to resolve phylogenetic relationships. The development of molecular techniques and the identification of DNA sequences provide a much vaster amount of data, allowing a more profound analysis of phylogenetic relationships. However, because of the still-restricted number of taxa of which sequence data are available, the relationship of several taxa remains unresolved. In the most recent classification by De Ley and Blaxter (2002) the non-secernentean taxa within the Chromadoria were treated as separate orders because the Chromadorida, Desmodorida and Monhysterida do not share common ancestry, but rather represent nested series within the Chromadoria. The Desmoscolecida were also classified as a separate order in agreement with Lorenzen (1981, 1994); no sequence data for the taxa of the Desmoscolecida, except for a single unidentified species of the genus *Tricoma*, are currently available (Litvaitis *et al.*, 2000).

The Chromadorida, Desmodorida and Desmoscolecida are essentially marine nematodes. Only 2.5% (Desmodorida, Desmoscolecida) to 3.5% (Chromadorida) of valid species have been recorded from freshwater habitats and even these percentages are overestimated as the majority of the 'freshwater species' are more common in marine habitats (Fig. 17.1). Only four genera (*Achromadora*, *Punctodora*, *Ethmolaimus* and *Prodesmodora*) could be more or less considered limnetic, with 30–50% or more of their species being recorded from freshwater habitats. A possible explanation for the low representation and low speciation of chromadorids in freshwater habitats might be that, together with the monhysterids, the three orders belong to the second group of invaders. The first invaders of freshwater habitats consisted of the currently typical freshwater groups, such as the Tobrilidae, which evolved from marine enoplids (Decraemer and Coomans, 1994). Of all genera recorded from freshwater habitats, sequence data are available for the genus *Dichromadora* only (NCBI genebank).

Free-living freshwater nematodes belonging to the Chromadorida, Desmodorida and Desmoscolecida are usually small with body length (*L*) shorter than 1 mm. The main morphological characters used to distinguish these three higher taxa are body cuticle (transverse striation/annulation[1] and ornamentation), shape of the amphideal fovea, presence or absence of an offset head region described as a cephalic helmet, arrangement and position of cephalic setae and number of gonads in the male reproductive system (one testis or monorchic/two testes or diorchic).

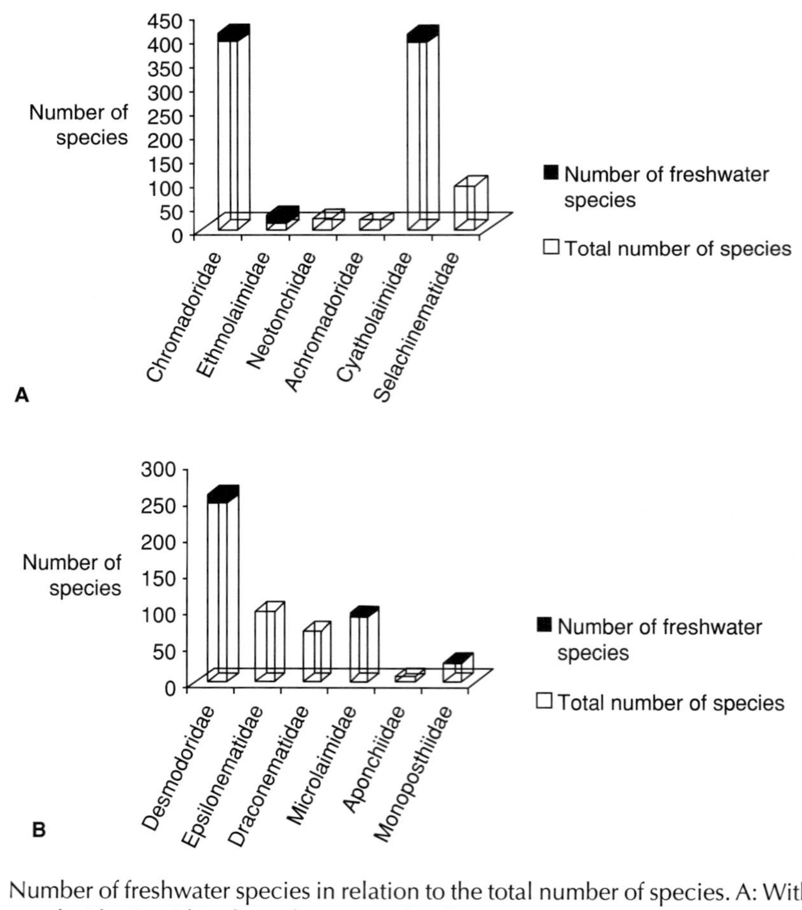

Fig. 17.1 Number of freshwater species in relation to the total number of species. A: Within the order Chromadorida; B: within the order Desmodorida.

---

[1]Transverse striae are shallow transverse grooves restricted to the cortical zone; annuli are deeper grooves involving at least two zones of the body cuticle.

Key to orders CHROMADORIDA, DESMODORIDA and DESMOSCOLECIDA.

1. Cuticle with transverse striae and punctated ornamentation. Amphideal fovea simple spirals, reniform, transverse elongate loops or multispiral. No cephalic helmet. Male diorchic . . . . .Chromadorida (Figs 17.2A–C)
2. Cuticle with transverse striae, not punctated. Brownish habitus. Amphideal fovea spiral. Cephalic helmet may be present. Male monorchic . . . . .Desmodorida (Figs 17.2D–and E)
3. Cuticle with transverse striae, often covered with desmen (= rings of secretion and foreign particles). Amphideal fovea large vesicular. Cephalic setae mainly inserted on peduncles. Male monorchic or diorchic . . . . .Desmoscolecida (Figs 17.2F–H)

Order CHROMADORIDA

Chromadorida are small (usually <1 mm) free-living nematodes, characterized by a body cuticle with transverse striae and punctated ornamentation. Chromadorids are essentially marine animals except for species of the genera *Achromadora*, *Ethmolaimus* and *Punctodora*. Within the Chromadorida about 10% of the genera have species recorded from freshwater habitats.

Classification

According to De Ley and Blaxter (2002), phylogenetic analyses based on small subunit (SSU) rDNA sequences strongly support a monophyletic clade of the chromadorids *sensu stricto* or Chromadoroidea *sensu* Lorenzen (1981, 1994). To allow a high rank for the other clades within the Chromadoria, the superfamily of Chromadoroidea of Lorenzen (1981, 1994) was raised to order level. Within the order, one suborder Chromadorina and one superfamily were recognized.

The following classification is adapted from De Ley and Blaxter (2002) and Lorenzen (1981, 1994). The classification within taxa without freshwater records is not presented.

Note: Asterisk indicates taxa reported from freshwater bodies.

Order CHROMADORIDA Chitwood, 1933

Suborder Chromadorina Filipjev, 1929
Superfamily Chromadoroidea Filipjev, 1917
Family Chromadoridae Filipjev, 1917
  Subfamily Spilipherinae Filipjev, 1918
  Subfamily Chromadorinae Filipjev, 1917

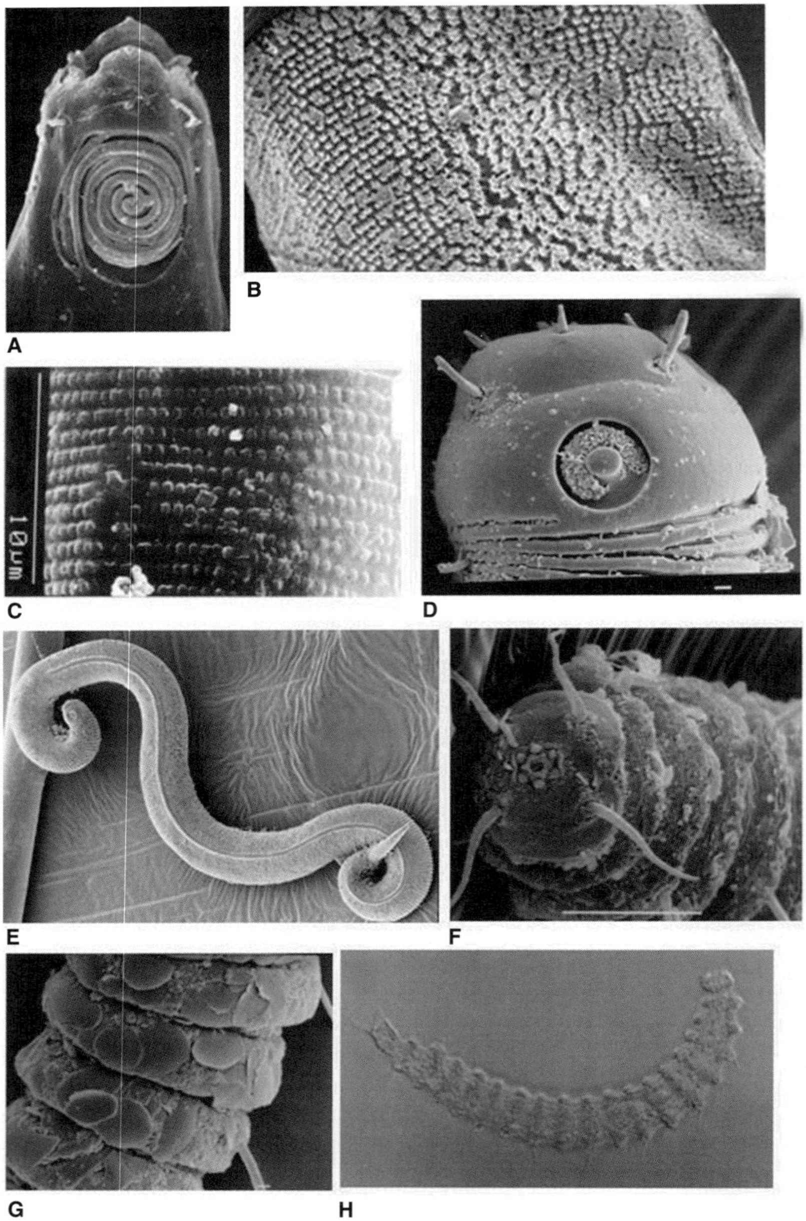

Fig. 17.2 Illustrations of representatives of the three orders. A–C: Order Chromadorida. A and B: *Cheironchus paravox* male by scanning electron microscopy (SEM), respectively of head region and detail of body cuticle (from Castillo-Fernandez and Decraemer, 1993); C. *Achromadora indica*, detail of body cuticle (from Tahseen, 2001). D and E: Order Desmodorida, *Pseudochromadora coomansi*, SEM of head region, male (D) and of total body region, female (E) (from Verschelde and Vincx, 1995). F–H: Order Desmoscolecida, *Desmoscolex parvospiculatus*; F and G: respectively SEM of front view of head region and detail of body cuticle of male; H: total view of male by light microscopy (from Decraemer, 1996). Courtesy of the *Bulletin of the Royal Belgian Institute of Natural Sciences*, Brussels (A, B, D, E); *Journal of Nematode Morphology and Systematics* (C); *Nematologica* (F).

Genus *Atrochromadora* Wieser, 1959
Chromadora Bastian, 1865
Chromadorella Filipjev, 1918
Chromadorina Filipjev, 1918*
Fusonema Kreis, 1928
Prochromadora Filipjev, 1922*
Prochromadorella Micoletzky, 1924
Punctodora Filipjev, 1919*
Timmia Hopper, 1961
Subfamily Euchromadorinae Gerlach & Riemann, 1973
Subfamily Harpagonchinae Platonova & Potin, 1972
Subfamily Hypodontolaiminae De Coninck, 1965
Genus *Chromadorissa* Filipjev, 1917
Chromadorita Filipjev, 1922*
Deltanema Kreis, 1929
Denticulella Cobb, 1933
Dichromadora Kreis, 1929*
Hypodontolaimus de Man, 1886
Innocuonema Inglis, 1969
Karkinochromadora Blome, 1982
Megodontolaimus Timm, 1969
Neochromadora Micoletzky, 1924*
Panduripharynx Timm, 1961
Parachromadorita Blome, 1974
Ptycholaimellus Cobb, 1920
Spilophorella Filipjev, 1917
Family Ethmolaimidae Filipjev & Schuurmans Stekhoven, 1941
Genus *Ethmolaimus* de Man, 1880*
Paraethmolaimus Jensen, 1994
Family Neotonchidae Wieser & Hopper, 1966
Family Achromadoridae Gerlach & Riemann, 1973
Subfamily Achromadorinae Gerlach & Riemann, 1973
Genus *Achromadora* Cobb, 1913*
Subfamily Kreisonematinae Khera, 1969
Family Cyatholaimidae Filipjev, 1918
Subfamily Nyctonematinae Bussau, 1993
Subfamily Pomponematinae Gerlach & Riemann, 1973
Subfamily Paracanthonchinae De Coninck, 1965
Genus *Acanthonchus* Cobb, 1920
Biarmifer Wieser, 1954
Paracanthonchus Micoletzky, 1924
Paracyatholaimoides Gerlach, 1953
Paracyatholaimus Micoletzky, 1922*
Subfamily Xenocyatholaiminae Gerlach & Riemann, 1973
Subfamily Cyatholaiminae Filipjev, 1918
Family Selachinematidae Cobb, 1915

Order CHROMADORIDA Chitwood, 1933

Diagnosis. Chromadorea. Cuticle with transverse striae and punctated ornamentation, showing a pattern of dots, rods or 'basket weave' and, usually, with lateral differentiation. Anterior sensilla arranged in two or three circlets at the anterior end of the body, i.e. an anterior circlet of inner labial papillae, an outer circlet of external labial sensilla (seti- or papilliform) and a third circlet of cephalic setae; the second and third circlet are often referred to in the literature as cephalic sensilla (= posterior to the lip region) – they may form a single circlet of ten sensilla. Amphideal fovea variable, simple spiral, reniform, a transverse loop or slit, or multispiral; when spiral, amphideal fovea usually located posterior to the posterior circlet of cephalic setae, but sometimes lying between the four cephalic setae and difficult to observe. Head cuticle not thickened or sclerotized and thus without cephalic helmet. Cheilostom with 12 cheilorhabdia (also referred to as folds or rugae); pharyngostom of variable size, armed with a single dorsal tooth and usually two smaller ventrosublateral teeth. Pharynx largely cylindrical, posteriorly, usually, with muscular bulb with well-cuticularized luminal wall (also described as crescentic valve). Secretory–excretory system with terminal duct cuticularized. Female reproductive system didelphic–amphidelphic with antidromously reflexed ovaries; less commonly monodelphic (Chromadoridae). Male monorchic or diorchic; spicules paired; gubernaculum present; pre-cloacal supplements usually cup-shaped, may also be setose or fine tubular. Caudal glands and spinneret present. Free-living marine, uncommon in freshwater and soil.

The diagnosis of the single suborder Chromadorina also applies to its single superfamily Chromadoroidea. Within the Chromadoroidea, six families are recognized: the Chromadoridae, Ethmolaimidae, Neotonchidae, Achromadoridae, Cyatholaimidae and Selachinematidae (Lorenzen, 1981, 1994). The family Selachinematidae was not included in the classification of the Chromadorida in De Ley and Blaxter (2002); no comment was given. Of the six families, the families Neotonchidae and Selachinematidae have not been recorded from limnetic habitats. A tabulate key to families of the Chromadoroidea is represented in Table 17.1.

Family CHROMADORIDAE Filipjev, 1917

Diagnosis. Chromadoroidea. Amphideal fovea basically a simple transverse slit-like ventrally wound spiral, sometimes obscure, located between the cephalic setae or posterior to them. Cuticular ornamentation as punctations, which may be evenly distributed and of equal size (cuticle homogeneous), or unevenly distributed, for example, enlarged in the lateral body regions or different along the body (cuticle heterogeneous). Stoma with dorsal tooth usually larger than ventrosublateral ones; teeth hollow or solid; denticles may be present. Male monorchic with anterior testis (synapomorphy); pre-cloacal supplements cup-shaped (never tubular), may be absent. Females with two antidromously reflexed ovaries, the anterior gonad right of the intestine, the posterior gonad left of the intestine (synapomorphy). Mostly marine. Of the five subfamilies, only the Chromadorinae and Hypodontolaiminae have representatives in freshwater habitats. The other subfamilies, Spilipherinae, Euchromadorinae and Harpagonchinae will not be dealt with in detail.

Table 17.1 Tabulated key to families of the Chromadoroidea.

	Chromadoridae (Figs 17.6A and B)	Ethmolaimidae (Figs 17.6C and D)	Neotonchidae (Figs 17.6E and F)	Achromadoridae (Figs 17.6G–I)	Cyatholaimidae (Figs 17.6J and K)	Selachinematidae (Figs 17.6L–N)
Arrangement anterior sensilla	Mostly on three circlets: 6 + 6 + 4	On three circlets: 6 + 6 + 4	Usually three circlets: 6 + 6 + 4	Usually two circlets: 6 + (6 + 4)	Mostly two circlets: 6 + (6 + 4)	Three circlets: 6 + 6 + 4
Outer labial sensilla (OLS) and cephalic sensilla (CS)	Setose, usually six OLS shorter than four CS (6 < 4) except in part of Euchromadorinae	Four CS; six OLS papilliform or setiform papillae (6 < 4)	Setose, four CS longer than six OLS of the second circlet (6 > 4)	Setose, six OLS longer than four CS (6 > 4)	Setose, six OLS longer than four CS (6 > 4)	Variable; sensilla of one of the three circlets may be longer than the other two
Armature stoma	Three teeth of equal size or dorsal tooth larger	Three teeth more or less of equal size	One large dorsal tooth with or without two small ventrosublateral teeth	Usually one distinct dorsal tooth; ventrosublateral teeth small or absent	Usually one distinct dorsal tooth and ventrosublateral teeth small or absent; rarely without teeth	Large stoma, unarmed or with jaws
Amphideal fovea	Transverse slit, ventrally spiral; often situated in between four CS	Ventrally unimultispiral, not wider than long; posterior to four CS	Multispiral; posterior to CS	Uni- or multispiral; well posterior to CS	Multispiral; posterior to CS	Multispiral; posterior to CS
Position gonads in relation to intestine	Gonads on opposite side of intestine (anterior gonad to the right)	Gonads on opposite side of intestine (anterior gonad to the right)	Gonads on opposite side of intestine (anterior gonad to the left)	Both gonads on same side of intestine (left or right)	On different sides of intestine	Variable

Continued

Table 17.1 *Continued.* Tabulated key to families of the Chromadoroidea.

	Chromadoridae (Figs 17.6A and B)	Ethmolaimidae (Figs 17.6C and D)	Neotonchidae (Figs 17.6E and F)	Achromadoridae (Figs 17.6G–I)	Cyatholaimidae (Figs 17.6J and K)	Selachinematidae (Figs 17.6L–N)
Number of testes; position in relation to intestine	A single anterior testis, right of intestine	Two testes, opposed, each on opposite sides (left or right) of intestine	Two testes, opposed	Two testes; males rare	Usually two, rarely one testis (either anterior or posterior gonad); on opposite sides of intestine	Usually two testes; on the same or on opposite sides of intestine
Male: presence	Common	Common	Common	Rare; parthenogenetic reproduction	Common	Common
Pre-cloacal supplements	Cup-shaped, rarely absent	Cup-shaped with external articulate flange	Cup-shaped	Absent	Pre-cloacal tubules or papillae; rarely absent	Usually cup-shaped, never tubular
Habitat	Mostly marine, few limnetic	Limnetic (freshwater, brackish, hyper-saline)	Marine	Limnetic–terrestrial	Mainly marine, rarely limnetic	Marine

The main characters used to differentiate the subfamilies within the Chromadoridae are the shape of the amphideal fovea, the stoma armature, the arrangement of the anterior sensorial organs, the cuticular ornamentation with or without differentiations and presence or absence of pre-cloacal supplements in the male (Table 17.2). The subfamily Harpagonchinae is known by three species only; all are ectoparasites of polychaetes.

Subfamily Chromadorinae Filipjev, 1917
Diagnosis. Chromadoridae. External labial papilliform sensilla and four cephalic setae in two separate circlets. Amphideal fovea transverse slit-like, difficult to observe under light microscopy.

Of the nine genera, only *Chromadorina*, *Prochromadora* and *Punctodora* have representatives in freshwater habitats. The monotypic genus *Fusonema,* based on a rather poor description and an illustration of males only, has an unclear taxonomic position. Originally classified within the Chromadoridae, the genus *Fusonema* was moved to the Monoposthiinae by Hope and Murphy (1972), probably based on the structure and ornamentation of the cuticle. Later, the genus was returned to the Chromadoridae, subgenus Chromadorinae by Lorenzen (1981, 1994), apparently based on the males being monorchic. No comments were provided by the authors. The genus is not dealt with here. The subfamily needs revision.

Key to genera of the Chromadorinae

1. Body cuticle with ornamentation of homogeneous punctation along the body, with or without lateral differentiation .2
   Body cuticle with ornamentation of heterogeneous punctation along the body, with or without lateral differentiation .6
2. No lateral differentiation of body cuticle 3
   Lateral differentiation of body cuticle present 5
3. Buccal cavity with three equal teeth; pre-cloacal supplements absent or present 4
   Buccal cavity with a large solid dorsal tooth and usually two minute ventrosublateral teeth and pre-cloacal supplements present . . . .*Prochromadora* (Figs 17.3A and B)
4. Pre-cloacal supplements when present never tubular . . . .*Chromadorina* (Figs 17.3C and D)
   Pre-cloacal supplements cup-shaped and additional one tubular . . . .*Timmia* (Figs 17.3E and F)
5. Amphideal fovea transverse slit-like; ventrosublateral wall of buccal cavity without indentation of transverse folds . . . .*Chromadora* (Figs 17.3G and H)
   Amphideal fovea transverse oval, open-looped; ventrosublateral wall of buccal cavity without indentation of transverse folds . . . . *Atrochromadora* (Figs 17.4A and B)
   Amphideal fovea transverse flattened spiral; ventrosublateral wall of buccal cavity with indentation . . . .*Punctodora* (Fig. 17.3I)

Table 17.2 Tabulated key to subfamilies of the Chromadoridae.

	Chromadorinae (Figs 17.4A and B)	Euchromadorinae (Figs 17.4a and C–E)	Harpagonchinae (Figs 17.4F and G)	Hypodontolaiminae (Figs 17.4H and I and 17.5)	Spilipherinae (Figs 17.4J and K)
Arrangement of anterior sensilla	Three circlets: 6 + 6 + 4	Two circlets: 6 + (6 + 4)	? Three circlets	Three circlets: 6 + 6 + 4	Two circlets: 6 + (6 + 4)
Outer labial sensilla (OLS) and cephalic sensilla (CS)	Only posterior four CS setose	Setose; four CS longer than six OLS	Only four CS, setose	Setose; four CS longer than six OLS	Six OLS and four CS, setose
Amphideal fovea	Transverse, more or less slit-like; usually between base of CS	Large transverse slit, obvious or obscure; posterior to CS	Unknown	Transverse, oval to oval loop-shaped; in between four CS or posterior to them	Distinctly spiral but not multispiral nor with circular outline, slit-like or single open spiral; posterior to CS
Body cuticle	Punctation homogeneous or heterogeneous; with or without lateral differentiation	Complex ornamentation, heterogeneous (rarely homogeneous); lateral differentiation or alae may be present	Homogeneous punctation without lateral differentiation	Punctation homogeneous or heterogeneous; with or without lateral differentiation	Punctation homogeneous or heterogeneous; lateral differentiation present or absent

Stoma: teeth	Solid dorsal tooth about equal or larger than ventrosublateral teeth	Usually a large solid dorsal tooth, two small ventrosublateral teeth, with or without denticles	Three motile jaws with hooks	Hollow dorsal tooth not articulating with stoma wall, with or without ventrosublateral teeth and denticles	Three solid teeth, more or less of equal size
Pre-cloacal supplements	Cup-shaped; an additional tubular supplement may be present	Absent	Present or absent	Cup-shaped or absent	Not cup-shaped, setose or absent
Number of genera recorded from freshwater	3	None	None	4	None

6. Lateral punctated differentiation of body cuticle without longitudinal rows of larger dots . . . . *Prochromadorella* (Figs 17.3J–L)
Lateral differentiation of body cuticle with two well-developed longitudinal rows of larger dots *Chromadorella* (Figs 17.3M–O)

Genus *Atrochromadora* Wieser, 1959 (Figs 17.4A and B)
Syn. pro *Chromadoropsis* Wieser, 1954 homonym to *Chromadoropsis* Filipjev, 1918.
Diagnosis. Chromadorinae. Homogeneous punctated body cuticle pattern along the body and with lateral differentiation of larger dots. Amphideal fovea transverse oval, open loop-shaped. Buccal cavity with three solid teeth, the dorsal tooth larger than the ventrosublateral teeth. Males usually with cup-shaped pre-cloacal supplements. Copulatory apparatus with ventrally bent spicules, usually cephalate and provided

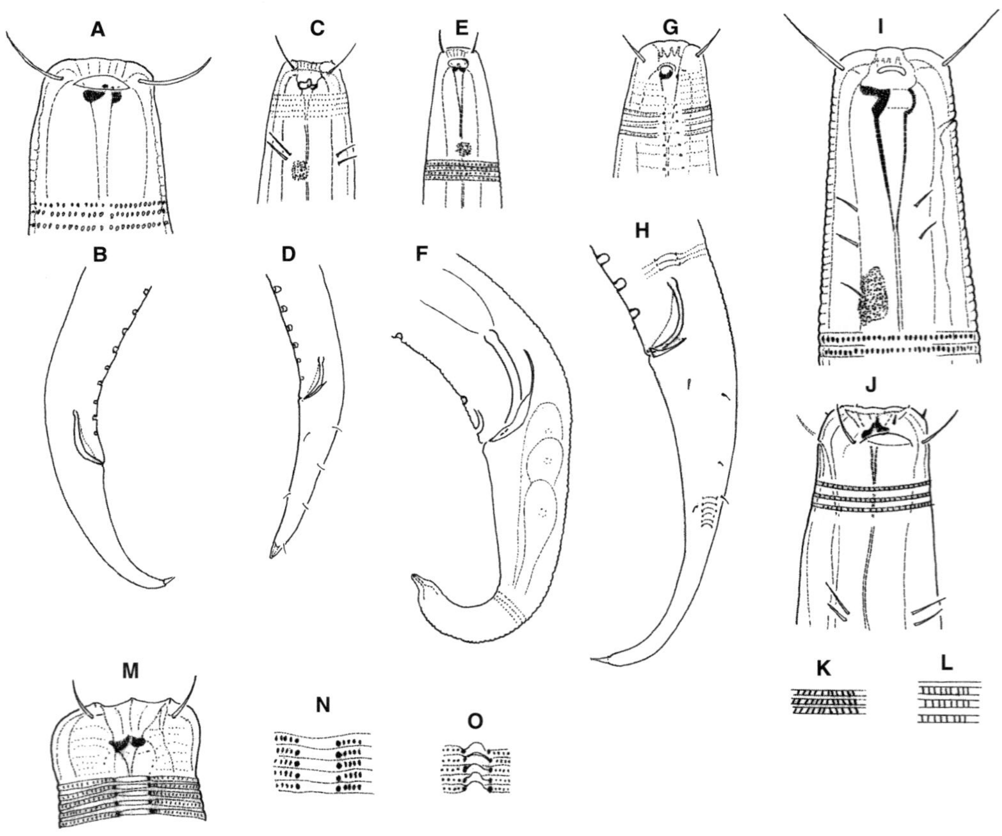

Fig. 17.3 Genera of the subfamily Chromadorinae, head and tail regions. A and B: *Prochromadora*; C and D: *Chromadorina*; E and F: *Timmia* (based on Timm, 1952); G and H: *Chromadora*; I: *Punctodora*; J–L: *Prochromadorella* (based on Wieser and Hopper, 1967); M–O: *Chromadorella* (based on Wieser and Hopper, 1967).

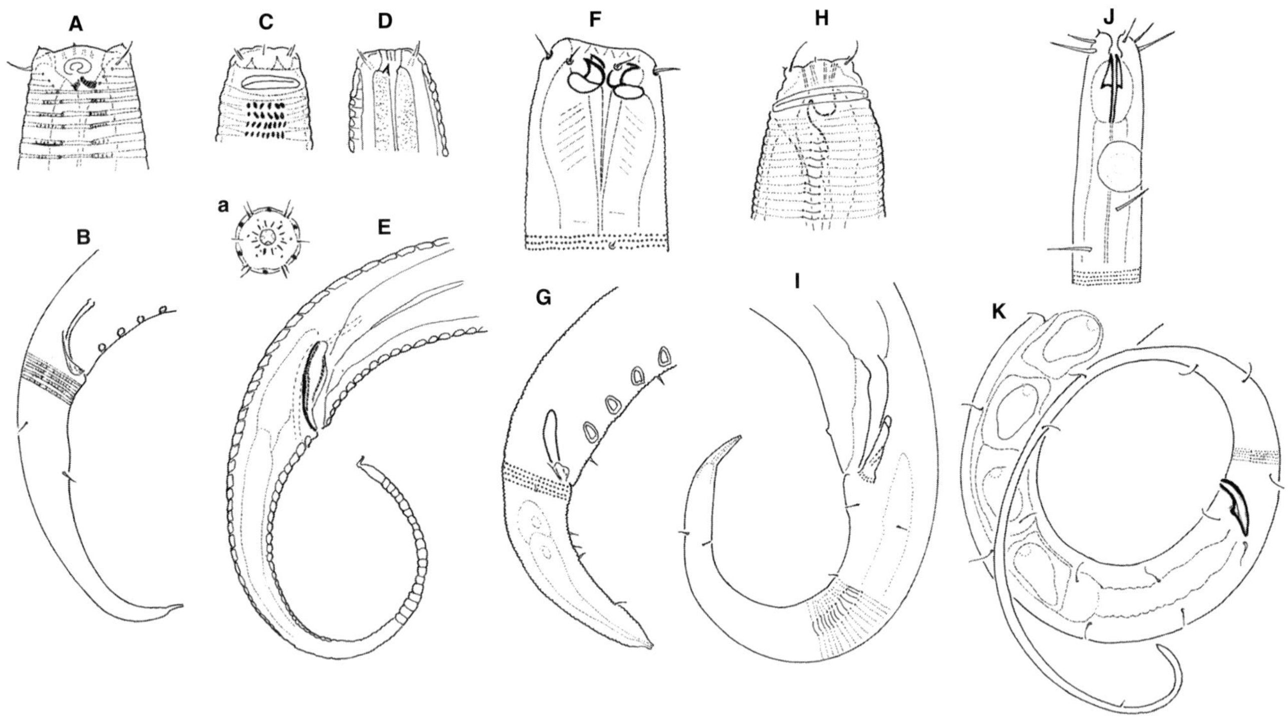

Fig. 17.4 Subfamilies of the family Chromadoridae, head and tail region respectively. Chromadorinae, A and B: *Atrochromadora*. Euchromadorinae, a and C–E: *Actinonema*, with a head in front view (based on Decraemer and Coomans, 1978). Harpagonchinae, F and G: *Harpagonchus*. Hypodontolaiminae (redrawn from Platonova and Potin, 1972), H and I: *Ptycholaimellus* (based on Decraemer and Coomans, 1978). Spilipherinae, J and K: *Acantholaimus*.

with ventral velum; gubernaculum paired proximally, distally unpaired, the distal end often enlarged, with or without denticles. Marine.

Type species: *A. parva* (de Man, 1893) Wieser, 1959.

Genus *Chromadora* Bastian, 1865 (Figs 17.3G and H)
Diagnosis. Chromadorinae. Homogeneous punctated body cuticle pattern along the body, with lateral differentiation of larger dots. Transverse slit-like amphideal fovea. Buccal cavity with three solid teeth, the dorsal tooth larger than the ventrosublateral teeth. Ocelli may be present. Males usually with cup-shaped pre-cloacal supplements. Copulatory apparatus with ventrally bent spicules, usually cephalate and provided with ventral velum; gubernaculum paired proximally, distally unpaired and distal end often enlarged, with or without denticles. *Chromadora* is very similar to *Atrochromadora* except for amphid shape; however, the differentiation is not always clear. Marine.

Type species: *C. nudicapitata* Bastian, 1865.

Genus *Chromadorella* Filipjev, 1918 (Figs 17.3M–O)
Diagnosis. Chromadorinae. Heterogeneous punctated body cuticle along the body, with lateral differentiation in longitudinal rows of larger dots. Amphideal fovea transverse oval in shape and slightly bent. Buccal cavity with three solid teeth of about equal size. Males with 5–12 (mostly 5) cup-shaped pre-cloacal supplements. *Chromadorella* resembles *Neochromadora* except for the structure of the buccal teeth. Marine.

Type species: *C. filiformis* (Bastian, 1865) Filipjev, 1918.

Genus *Chromadorina* Filipjev, 1918 (Figs 17.3C and D)
Diagnosis. Chromadorinae. Homogeneous cuticle with transverse rows of dots but without obvious lateral differentiation. Amphideal fovea when visible, transverse slit-like. Buccal cavity with three solid teeth of about equal size. Ocelli may be present. Cup-shaped pre-cloacal supplements present. The genus is largely a marine genus with a few (4) limnetic species and 26 species from marine and brackish habitats.

Type species: *C. obtusa* Filipjev, 1918

Species recorded from freshwater habitats:
C. astacicola (W.Schneider, 1932) Wieser, 1954: limnetic, in association with gills of crayfish, Germany.
C. bercziki Andrássy, 1962: in groundwater, Hungary.
C. bioculata (Schultze in Carus, 1857) Wieser, 1954: widespread in freshwater and in brackish waters of Europe, common in growth (e.g. *Cladophora*) on stones and other hard substrates, recorded from The Netherlands, Germany, Switzerland, Austria, Hungary, Czech Republic, Romania, France, Italy, Poland, Denmark, UK, Ireland, Finland, Estonia, Russia and neighbouring countries, Bulgaria (lapsus *Punctodora bioculata*) from gravel, sand, sludge in the Danube River.
C. viridis (Linstow, 1876) Wieser, 1954: in freshwater and brackish water habitats with overgrowth of aquatic vegetation; a common component of the periphyton in Russian freshwater basins.

Genus *Prochromadora* Filipjev, 1922 (Figs 17.3A and B)
Diagnosis. Chromadorinae. Cuticle with homogeneous punctation, without lateral differentiation. A single large dorsal tooth opposed by a ventrosublateral pit or at most by a small elevation of the buccal wall or small ventrosublateral teeth. Precloacal supplements present.
The genus *Prochromadora* is largely a marine genus (eight species); only one species *P. orleji* has been recorded from freshwater habitats.

Type species: *P. megodonta* Filipjev, 1922

Species recorded from freshwater habitats:
P. orleji (de Man, 1881) Filipjev, 1922: a cosmopolitan species mainly found in brackish habitats along European coasts, common in Cyanophycea sand on North Sea coast and in salt marshes on the Ost Sea coast; the species has also been recorded from the USA and from a marine salter in Shandong China. *P. orleji* is less common in marine habitats (e.g. the North Sea) and rare in freshwaters (Germany, Czech Republic, Russia, e.g. in River Neva, in overgrowths of sluices of the northern Dvina aquatic system and water bodies of the River Dnieper basin). It has been recorded from groundwater of the potamon of the River Weser. According to Bongers (1988), *P. orleji* also occurs in terrestrial habitats.

Genus *Prochromadorella* Micoletzky, 1924 (Figs 17.3J–L)
Syn. *Trichromadora* Kreis, 1929
Diagnosis. Chromadorinae. Heterogeneous punctated body cuticle along the body, without or with lateral differentiation of larger dots, the latter not arranged in longitudinal rows. Amphideal fovea transversely oval, obscure, located between the four cephalic setae. Buccal cavity with three solid teeth of about equal size or the two ventrosublateral teeth smaller. Ocelli may be present. Males with cup-shaped precloacal supplements. Marine.

Type species: *P. neapolitana* (de Man, 1876) Micoletzky, 1924.

Genus *Punctodora* Filipjev, 1919 (Fig. 17.3I)
Diagnosis. Chromadorinae. Cuticle with fine transverse striae and fine homogeneous punctation except at the level of lateral fields with some larger punctation. Amphideal fovea flattened, spiral at level of dorsal tooth. Stoma with well-developed dorsal tooth and two smaller ventrosublateral teeth visible at the anterior and posterior border of an indentation, also described as two transverse folds. Ocelli present. Secretory–excretory pore in head region. From one to 18 cup-shaped pre-anal supplements. Mainly a freshwater genus with the freshwater species also recorded from brackish habitats, more rarely in marine habitats. One species, *P. salinarum* (Linstow, 1901) Wieser, 1954, has been recorded only from inland waters with high salinity (4–21%).

Type species: *P. ratzeburgensis* (Linstow, 1876) Filipjev, 1929

Species recorded from freshwater habitats:
P. dudichi Andrássy, 1966: in freshwaters, in mud of River Danube; on submerged grass roots of Lake Balaton at Hortobágy National Park, Hungary but also in brackish waters from the Dnieper-Bug estuary (Black Sea).

P. exochopora Hopper, 1963: in freshwater habitats found in sand, fine gravel and shells from shore of Lake Ontario, Canada.

P. ohridensis W. Schneider, 1943: freshwater, in sandy sediment and stones in water near river bank, Lake Ohrid, Macedonia; Dead Sea, Israel.

P. ratzeburgensis (Linstow, 1876) Filipjev, 1929: mainly in brackish biotopes (Baltic Sea, Gulf of Bothnia–Finland, Black Sea, Kiel Bay–Germany) but also in marine habitats (France, Spain) and freshwater (Germany, Hungary, Poland, Austria, Lakes Torneträsk and Vättern in Sweden, widespread in overgrowths of aquatic vegetation in Russia and neighbouring countries).

Remark: *P. bioculata* in Stoichev (1996) from Danube River, Bulgaria (= *C. bioculata*).

Genus *Timmia* Hopper, 1961 (Figs 17.3E and F)
Syn. *Parachromadora* Timm, 1952 homonym to *Parachromadora* Micoletzky, 1913
Diagnosis. Chromadorinae. Homogeneous cuticle with transverse rows of dots but without lateral differentiation. Amphideal fovea when visible, slit-like. Buccal cavity with three teeth of about equal size. A tubular curved supplement immediately preanal in addition to the typical cup-shaped pre-cloacal supplements. *Timmia* is similar to *Chromadorina* except for the tubular pre-cloacal supplement. Marine.

Type species: *T. parva* (Timm, 1952) Gerlach & Riemann, 1973.

Subfamily Euchromadorinae Gerlach & Riemann, 1973 (Figs 17.4a, C–E)
Diagnosis. Chromadoridae. Body cuticle with complex heterogeneous ornamentation along the body (rarely homogeneous as in *Endeolophos* Boucher, 1976), often with lateral differentiation. The six outer labial and four cephalic sensilla setiform and arranged in a single circlet. Amphideal fovea transverse slit-like or oval, located posterior to the cephalic setae. Buccal cavity with large dorsal tooth, with or without denticles or smaller ventrosublateral teeth. Pharynx with or without defined terminal bulb. Gubernaculum usually with hammer- or L-shaped lateral pieces (wrongly indicated as telamon). Pre-cloacal supplements absent but a pre-cloacal differentiation of body cuticle may be present. Marine.

Type genus: *Euchromadora* de Man, 1886.

Subfamily Harpagonchinae Platonova & Potin, 1972 (Figs 17.4F and G)
Diagnosis. Chromadoridae. Body cuticle with homogeneous punctation. Buccal cavity with three strong triangular-shaped jaws, provided with big immobile hooks. Pharynx enlarged anteriorly at level of jaws and posteriorly with or without a terminal bulb. Pre-cloacal supplements present (cup-shaped) or absent. Marine, ectoparasites of Polychaeta.

Type genus: *Harpagonchus* Platonova & Potin, 1972.

Subfamily Hypodontolaiminae De Coninck, 1965 (Figs 17.4H and I and 17.5)
Diagnosis. Chromadoridae. Cuticle with transverse striae and punctated ornamentation homogeneous or heterogeneous, with or without lateral differentiation. Six outer labial and four cephalic sensilla setiform, on separate circlets. Amphideal fovea distinct or rather obscure, transverse flattened ovals, generally located

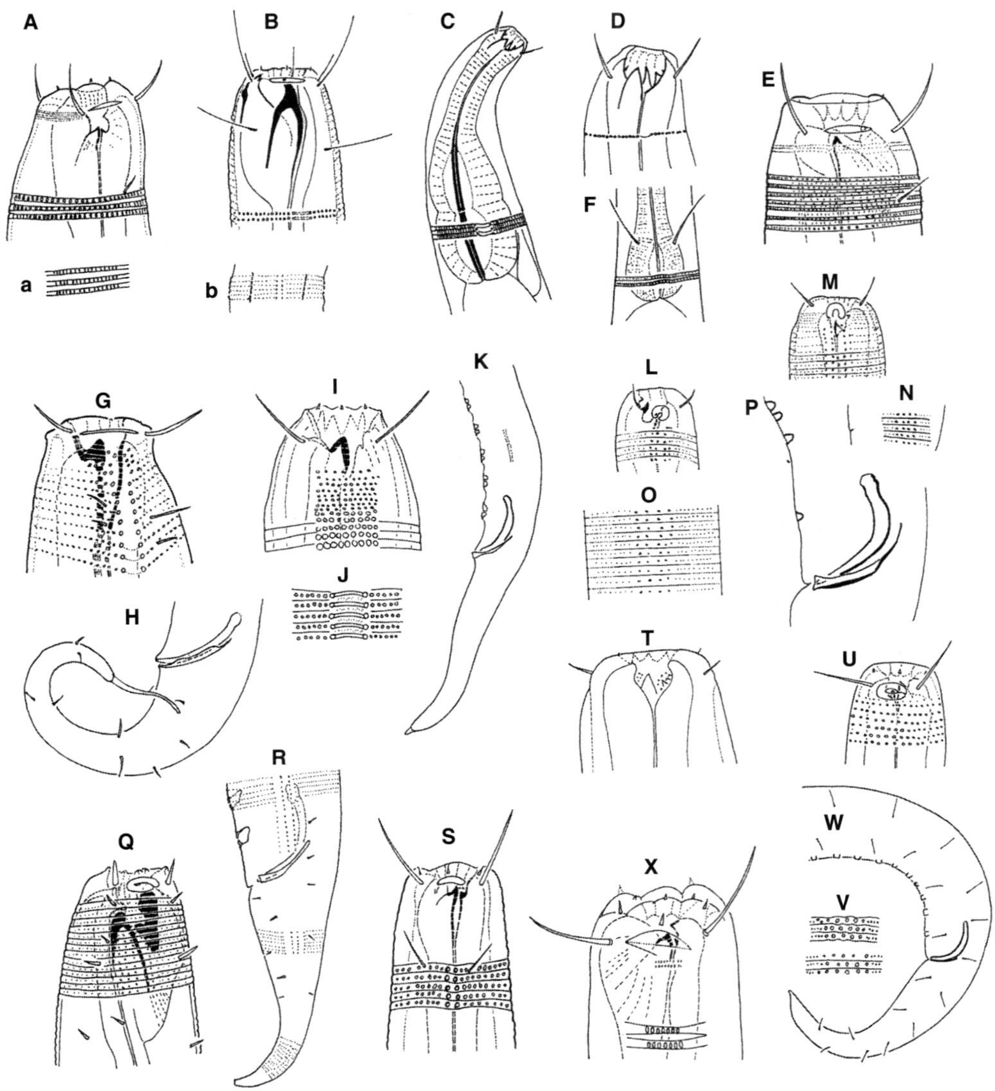

Fig. 17.5 Genera of the subfamily Hypodontolaiminae, head region, tail region and detail of body cuticle. A and a: *Innocuonema* (based on Wieser, 1959); B and b: *Megodontolaimus* (based on Timm, 1969); C and D: *Panduripharynx* (based on Timm, 1961); E and F: *Neochromadora* (based on Wieser, 1959); G and H: *Spilophorella* (based on Platt and Warwick, 1988); I–K: *Chromadorissa* (based on Filipjev, 1917); L–P: *Karkinochromadora* (based on Jensen, 1980; Blome, 1982); Q and R: *Hypodontolaimus* (based on De Coninck, 1944); S: *Dichromadora* (based on Platt and Warwick, 1988); T: *Denticulella* (based on Schulz, 1932); U–W: *Parachromadorita* (based on Blome, 1974); X: *Chromadorita* (based on Gerlach, 1952).

between the four cephalic setae. Stoma funnel-shaped, armed with immovable hollow teeth; the larger dorsal tooth may be opposed by two smaller ventrosublateral teeth, denticles may be present; anterior part of pharynx often with asymmetric muscular swelling. Males with cup-like pre-cloacal supplements, rarely absent. Gubernaculum with or without dorsal apophysis.

Of the 14 genera, only three, *Chromadorita*, *Dichromadora* and *Neochromadora*, have representatives in freshwater habitats. The genus *Deltanema*, based on a poor original description, has an unclear taxonomic position and will not be included in the key to genera.

Key to genera of the Hypodontolaiminae

1. Body cuticle with heterogeneous punctated ornamentation, with or without lateral differentiation 2
 Body cuticle with homogeneous punctated ornamentation, with or without lateral differentiation 6
2. Body cuticle without lateral differentiation . . . .*Innocuonema* (Fig. 17.5A)
 Body cuticle with lateral differentiation 3
3. Body cuticle laterally enlarged, forming alae, bordered by larger dots 4
 Body cuticle without lateral alae 5
4. Buccal cavity with large hollow dorsal tooth, two smaller ventrosublateral teeth and a solid denticle . . . .*Panduripharynx* (Figs 17.5C and D)
 Buccal cavity with different armature . . . .*Megodontolaimus* (Fig. 17.5B)
5. Lateral differentiation of body cuticle with larger dots not arranged in longitudinal rows; pharynx with single bulb; spinneret not elongated . . . . *Neochromadora* (Figs 17.5E and F)
 Lateral differentiation of body cuticle with larger dots arranged in longitudinal rows (2–3); pharynx with double end-bulb; spinneret elongated . . . . *Spilophorella* (Figs 17.5G and H)
6. Lateral differentiation in ornamentation arranged in longitudinal rows 7
 Lateral differentiation in ornamentation not arranged in longitudinal rows 11
7. Pharyngeal end-bulb strongly bipartite 8
 Pharyngeal end-bulb different 9
8. Males with 5–6 pre-cloacal supplements; lip region not offset . . . .*Chromadorissa* (Figs 17.5I–K)
 Pre-cloacal supplements usually absent; lip region offset, narrower . . . .*Ptycholaimellus* (Figs 17.4H and I)
9. Pre-cloacal supplements grouped; buccal cavity with two consecutive dorsal teeth; amphideal fovea rounded, spiral; obvious elongate

ventral gland along anterior intestine . . . .*Karkinochromadora* (Figs 17.5L–P)
Pre-cloacal supplements absent or with different arrangement; armature buccal cavity and fovea different 10
10. Anterior end of pharynx clearly swollen; S-shaped dorsal tooth with apophysis . . . .*Hypodontolaimus* (Figs 17.5Q and R)
Anterior end of pharynx not swollen; S-shaped hollow dorsal tooth without apophysis . . . .*Dichromadora* (Fig. 17.5S)
11. Amphideal fovea less obvious, a transverse slit of oval loop shape; lateral differentiation of body cuticle may be heterogeneous . . . .*Chromadorita* (Fig. 17.5X)
Buccal cavity with additional denticles next to dorsal tooth and ventrosublateral teeth 12
12. Amphideal fovea conspicuous, oval-shaped . . . .*Parachromadorita* (Figs 17.5U–W)
Amphideal fovea slit-like . . . .*Denticulella* (Fig. 17.5T)

Genus *Chromadorissa* Filipjev, 1917 (Figs 17.5I–K)
Diagnosis. Hypodontolaiminae. Cuticle with homogeneous punctated ornamentation with lateral differentiation of larger dots arranged in two longitudinal rows. Buccal cavity with large hollow dorsal tooth and small ventrosublateral teeth. Pharyngeal end-bulb bipartite. Males with 5–6 pre-cloacal supplements. *Chromadorissa* is known by three marine species only, two from the Caspian Sea (Grimma, 1876; Filipjev, 1917) and one from the Mediterranean Sea (Kreis, 1928).

Type species: *C. beklemishevi* Filipjev, 1917.

Genus *Chromadorita* Filipjev, 1922 (Fig. 17.5X)
Syn. *Allgeniella* Strand, 1934 pro *Odontonema* Filipjev, 1929 homonym to *Odontonema* Weber, 1913.
Diagnosis. Hypodontolaiminae. Cuticle ornamentation homogeneous exceptionally with slightly more pronounced punctation (may be heterogeneous) at level of lateral fields. Hollow teeth, mostly one dorsal and one or two ventrosublateral, rarely one indistinct dorsal tooth only. Amphideal fovea transverse, slit-like or oval loop-shaped. A subgroup within the *Innocuonema–Chromadorita* complex lacks pre-cloacal supplements, has a somewhat elongated angular punctation and a tail with the tip bent to left and dorsally curved. Platt and Warwick (1988) remarked that some confusion may arise between *Chromadorita* and *Innocuonema* Inglis, 1969; both genera need revision. The genus has only two species recorded from freshwater habitats but one species, *C. leuckarti*, is common in brackish and marine habitats. Twenty-seven species of the genus have been recorded, mainly from marine habitats.

Type species: *C. demaniana* Filipjev, 1922

Species recorded from freshwater habitats:
C. leuckarti (de Man, 1876) Filipjev, 1929: cosmopolitan, mainly limnetic, recorded from freshwater in Spain, France, Poland, Neusiedlersee (Hungary), Austria, Hungary and Russia, and from salt (up to 5%) inland waters of Mongolia. Common in brackish environments: Baltic Sea (Germany, Finland) and marine habitats: North Sea (The Netherlands, Belgium, Germany), Irish Sea, Black Sea, Spain, France, Italy, Denmark, Sweden, Finland, Iceland, Färöe Islands, Tunisia, USA, Brazil.
C. mirabilis Gagarin, 1993 (freshwater: Sakhalin Island (Russia)).

Genus *Deltanema* Kreis, 1929
Diagnosis. Hypodontolaiminae. Body cuticle with fine transverse striae; ornamentation not described. Amphideal fovea transverse slit-like, located anterior to a large dorsal tooth. Buccal cavity large, with well-developed dorsal tooth and two (?) smaller ventrosublateral teeth. Marine; genus originally described as close to *Chromadorina* and *Chromadorita* but distinguished from these genera by the well-developed teeth. *Deltanema* is a monotypic genus known only from females of the type population. The genus was synonymized with *Metalinhomoeus* by Hope and Murphy (1972) but recognized as valid by Lorenzen (1981, 1994). The limited description and illustrations of the type species are responsible for the unclear taxonomic status of this marine genus.

Type species: *D. parvum* Kreis, 1929.

Genus *Denticulella* Cobb, 1933 (Fig. 17.5T)
Diagnosis. Hypodontolaiminae. Body cuticle with homogeneous punctation along the body, showing lateral differentiation of larger dots not arranged in longitudinal rows and, in the region of the buccal cavity, a pattern recalling the plaiting of a basketball may be present. Amphideal fovea transverse slit-like. Buccal cavity with hollow dorsal tooth, two smaller ventrosublateral teeth and numerous additional denticles. Pharynx anteriorly with asymmetrical dorsal swelling (indicated as peristomal bulb in Dashchenko, 2002) with degree of swelling related to size of dorsal tooth, and posteriorly with end bulb. Males with cup-shaped pre-cloacal supplements.

The genus is intermediate in position between the genera *Chromadorita* and *Dichromadora*, the cuticle being laterally differentiated into larger dots, not arranged into longitudinal rows as in *Chromadorita*, and the buccal teeth similar in arrangement as in *Dichromadora*; denticles considered as typical of the genus were also observed in *Dichromadora hyalocheile* Gerlach, 1951. Wieser (1954) transferred *Dichromadora stygia* Gerlach, 1952 to *Denticulella*, but Blome (1974) transferred it to a new genus *Parachromadorita* Blome, 1974. Marine.

Type species: *D. pellucida* Cobb, 1933.

Genus *Dichromadora* Kreis, 1929 (Fig. 17.5S)
Diagnosis. Hypodontolaiminae. Cuticle ornamented with fine punctation homogeneous along the body, showing a pronounced lateral differentiation in the form of two longitudinal rows of enlarged dots bordering the lateral field without ornamentation. Amphideal fovea transverse slit-like to bean-like. Buccal cavity armed either

with a single S-shaped hollow dorsal tooth or with three teeth, the dorsal one being larger than the two ventrosublateral ones; sometimes denticles may be present. Anterior end of pharynx usually without asymmetrical enlargement, end bulb distinct. Pre-cloacal supplements in male absent or present.

Type species: *D. cephalata* (Steiner, 1916) Kreis, 1929

The genus is mainly marine with three brackish water species, e.g. *D. geophila* (de Man, 1876) Kreis, 1929, *D. tobaensis* Schneider, 1937 and *D. cephalata,* and 20 species from marine habitats. *D. geophila* has been recorded from a freshwater pond as *Spilophora canadensis* by Cobb (1914). Gerlach (1951) remarked that because of the close location of this pond to the sea, the habitat is probably brackish. The record from a mineral source on Sakhalin Island by Gagarin (2001), in his book on the fauna of freshwaters of Russia and neighbouring countries, is from a saline habitat.

Andrássy (1984) described *D. tobaensis* Schneider, 1937 as 'aquatil', based on a record by Schneider (1937) from a study of material collected by a German limnological expedition; the specimens of *D. tobaensis* from North Sumatra are apparently from a brackish environment.

Genus *Hypodontolaimus* de Man, 1886 (Figs 17.5Q and R)
Syn. *Iotadorus* Cobb, 1920.
Diagnosis. Hypodontolaiminae. Body cuticle with homogeneous punctated ornamentation with lateral differentiation of two longitudinal rows of larger dots. Buccal cavity with large S-shaped hollow dorsal tooth and a dorsal apophysis. Amphideal fovea transverse slit-like to bean-like. Pharynx dorsally swollen in buccal region to accommodate the dorsal tooth. *Hypodontolaimus* is similar to *Dichromadora* except for shape of the dorsal tooth and apophysis. Marine and brackish.

Type species: *H. inaequalis* (Bastian, 1865) de Man, 1886.

Genus *Innocuonema* Inglis, 1969 (Fig 17.5A and a)
Diagnosis. Hypodontolaiminae. Body cuticle with heterogeneous punctated ornamentation with dots larger anteriorly; lateral differentiation absent. Amphideal fovea transverse slit-like located in between the four cephalic setae. Buccal cavity with large hollow dorsal tooth and associated slight dorsal swelling of the pharynx in this region. Pre-cloacal supplements absent. Marine. Some species of *Innocuonema* are almost identical at the generic level with some *Chromadorita* species. A revision of the genus is recommended. Mainly marine.

Type species: *I. flaccidum* (Wieser, 1959) Inglis, 1969.

Genus *Karkinochromadora* Blome, 1982 (Figs 17.5L–P)
Diagnosis. Hypodontolaiminae. Body cuticle with homogeneous punctated ornamentation with heterogeneous lateral differentiation of two to four in longitudinal rows arranged larger dots. Amphideal fovea loop-shaped, located between the cephalic setae. Buccal cavity with two consecutive dorsal teeth: a small anterior and a larger posterior, and two small ventrosublateral teeth. Obvious elongated ventral gland present along anterior intestinal region; secretory–excretory pore obscure but

possibly located near the anterior end. Weak pre-cloacal supplements present with cyatholaimid arrangement, i.e. with grouped distribution. Monotypic marine genus. Originally, the *Karkinochromadora* type species was described as a *Chromadora* species (Jensen, 1980) with a single dorsal tooth; the anteriormost dorsal tooth was apparently interpreted as part of the folded stoma wall.

Type species: *K. lorenzeni* (Jensen, 1980) Blome, 1982.

Genus *Megodontolaimus* Timm, 1969 (Fig. 17.5B and b)
Diagnosis. Hypodontolaiminae. Body cuticle with prominent heterogeneous punctation along the body, i.e. with larger dots towards the extremities of the body or only anteriorly, and with lateral differentiation of larger dots at the borders of the lateral field; two or six alae are present, showing fine cross bars. Amphideal fovea transverse elliptical. Buccal cavity originally described with a large hollow ventral tooth with hooked tip and a two-pronged dorsal tooth with crescent thickening along the buccal wall. Upon revision of the type specimens from the University of California Nematode Collection (slides UCNC1080–1085), kindly put at our (W.D.) disposal by Steve Nadler; most of them being in rather bad condition, the large hollow tooth appeared inserted on the strikingly swollen dorsal sector of the anterior pharynx. Pharynx anteriorly with obvious asymmetric dorsal swelling and posteriorly with double bulb. Pre-cloacal supplements setose. Marine. The genus is based on two specimens, the type specimens of which are in very poor condition.

Type species: *M. coxbazari* Timm, 1969.

Genus *Neochromadora* Micoletzky, 1924 (Figs 17.5E and F)
Diagnosis. Hypodontolaiminae. Heterogeneous complex cuticle ornamentation with lateral differentiation visible as two or three longitudinal rows of large dots. Buccal cavity with hollow dorsal tooth and two ventrosublateral teeth. Pharynx anteriorly enlarged especially in association with the dorsal tooth and posteriorly with well-developed end-bulb. Male usually with numerous cup-shaped pre-cloacal supplements.

Type species: *N. poecilosoma* (de Man, 1893) Micoletzky, 1924.

Neochromadora is mainly a marine genus with only one species (*N. izhorica*) recorded from freshwater, one species (*N. orientalis*) described from a saline lake (Lake Issyk-Kul in Kirghizia) and 30 marine species, a few of them occasionally found in brackish waters.

Species recorded from freshwater habitats:
N. izhorica (Filipjev, 1929) Schuurmans Stekhoven, 1935: mostly in brackish habitats along European coasts, occasionally also in freshwater, e.g. in Lake Balaton in Hungary.

Genus *Panduripharynx* Timm, 1961 (Figs 17.5C and D)
Diagnosis. Hypodontolaiminae. Cuticle ornamentation complex heterogeneous; lateral differentiation with larger dots bordering the lateral field (alae). Amphideal fovea indistinct transverse slit-like. Stoma wall well cuticularized, large hollow dorsal tooth and two smaller ventrosublateral teeth, and a solid denticle projecting from

base of anterior part of stoma. Pharynx with bipartite end bulb. No pre-cloacal supplements observed. Monotypic marine genus; known from the type population only.

Type species: *P. ornatus* Timm, 1952.

Genus *Parachromadorita* Blome, 1974 (Figs 17.5U–W)
Diagnosis. Hypodontolaiminae. Body cuticle with homogeneous punctated ornamentation and lateral differentiation into larger dots, not arranged into longitudinal rows. Buccal cavity with a large hollow dorsal tooth, two smaller ventrosublateral teeth and a field of denticles. Amphideal fovea transverse oval loop-shaped. Male with cup-shaped pre-cloacal supplements. The genus is differentiated from the related genera *Chromadorita*, *Dichromadora* and *Chromadora* by the type of lateral differentiation of the cuticle ornamentation as well as the shape of the amphideal fovea in the first two genera. It is also differentiated by the shape of the amphideal fovea from the genus *Denticulella* that also possesses a field of denticles. Marine.

Type species: *P. stygia* (Gerlach, 1952) Blome, 1974.

Genus *Ptycholaimellus* Cobb, 1920 (Figs 17.4H and I)
Diagnosis. Hypodontolaiminae. Similar to *Hypodontolaimus* for the cuticular ornamentation and structure of the dorsal tooth and apophysis, but different by a double pharyngeal end bulb, labial region usually offset and narrower than the rest of the head and pre-cloacal supplements are mainly absent (except with a small papilla in *Ptycholaimellus lizardiensis* Decraemer & Coomans, 1978). Marine.

Type species: *P. carinatus* Cobb, 1920.

Genus *Spilophorella* Filipjev, 1917 (Figs 17.5G and H)
Diagnosis. Hypodontolaiminae. Cuticle with fine transverse striae and complex heterogeneous punctated ornamentation with lateral differentiation of larger dots. Amphideal fovea slit-like. Deep buccal cavity with a long hollow dorsal tooth. Pharynx with elongated double posterior bulb. Tail often with a long, pointed, smooth, offset terminal portion with spinneret.

A marine genus with rare records from brackish habitats, e.g. *S. intermedia* Gagarin & Lemzina, 1982 from Issyk Kul Lake, Kyrgyzstan and estuary of River Tim on Sakhalin Island.

Type species: *S. paradoxa* (de Man, 1888) Filipjev, 1917.

Subfamily Spilipherinae Filipjev, 1918
Diagnosis. Chromadoridae. Body cuticle with homogeneous or heterogeneous punctation, with or without lateral differentiation. Buccal cavity with three or more solid teeth with or without apophyses. Six outer labial setae and four cephalic setae usually inserted at the same level. Amphideal fovea spiral, i.e. either cryptospiral with circular outline or single open-loop spiral with at most 1.5 turns. Pharynx with subdivided end bulb. Pre-cloacal supplements setose or absent. Tail conical or elongate. Marine.

Type genus: *Spiliphera* Bastian, 1865.

Family ACHROMADORIDAE Gerlach & Riemann, 1973

Diagnosis. Chromadoroidea. Mostly small nematodes (0.5–0.6 mm long). Cuticle with transverse striae and punctated ornamentation. Six outer labial and four cephalic sensilla setose, usually arranged in a single circlet of 6 + 4, with the six outer labial setae usually longer than the four cephalic setae; the six outer labial sensilla may be papilliform and arranged on a separate circlet from four cephalic setae. Amphideal fovea spiral, well posterior to the cephalic setae. Stoma with a distinct dorsal tooth, with or without two small ventrosublateral teeth. Female reproductive system didelphic–amphidelphic with anterior and posterior gonad situated on the same side of the intestine, either both on the left or on the right side (synapomorphy). Males rare; pre-cloacal supplements absent. Reproduction mostly by parthenogenesis. A purely limnetic–terrestrial family.

Two subfamilies: Achromadorinae Gerlach & Riemann, 1973 and Kreisonematinae Khera, 1969. Previous to Lorenzen (1981, 1994), the Kreisonematinae were classified within the Leptolaimidae but the presence of cheilorhabdia, a dorsal tooth and two female gonads at the same side of the intestine indicated their relatedness to the Chromadoroidea, close to the Achromadoridae. Both subfamilies mainly differ by the shape of the amphideal fovea, i.e. spiral and located posterior to the cephalic setae in the Achromadorinae *vs.* circular and at level of posterior stoma region (no anterior sensilla described) in the Kreisonematinae.

Subfamily Achromadorinae Gerlach & Riemann, 1973

Diagnosis. Achromadoridae. Amphideal fovea spiral, located posterior to the cephalic setae.

Type genus: *Achromadora* Cobb, 1913.

Genus *Achromadora* Cobb, 1913 (Figs 17.6G and I)

Syn. Limicyatholaimus Micoletzky, 1924 (*nomen nudum*).

Diagnosis. Achromadorinae. Cuticle with fine transverse striae, ornamented with rows of fine dots. Amphideal fovea multispiral, rarely unispiral, located from level anterior part of stoma to slightly posterior to stoma. Head continuous with body, rounded, with six poorly developed lips, each with an apical papilla; six outer labial setae and four cephalic setae usually in a single row of ten setae but may also occur in two separate circlets with the outer labial sensilla of the second circlet papilliform. Stoma funnel-shaped with distinct dorsal tooth anterior to midway in cavity and one or two small ventrosublateral teeth posteriorly. Pharynx cylindrical enlarged posteriorly to a pseudobulb. Intestine with short (as long as anal body width) to long rectum. Female didelphic–amphidelphic, ovaries reflexed. Tail conoid, tapering to a bluntly rounded terminus; caudal glands and terminal spinneret present. Male rare, diorchic. Precloacal supplements usually absent or small cup-shaped when present.

The only genus within the family with real terrestrial species. Nine out of 16 valid species of the genus are terrestrial but four of them rarely occur in freshwater habitats; the other seven species are limnetic with two species recorded from thermal waters.

Type species: *Chromadora minima* Cobb, 1913, a junior synonym of *A. ruricola* (de Man, 1880) Micoletzky, 1925.

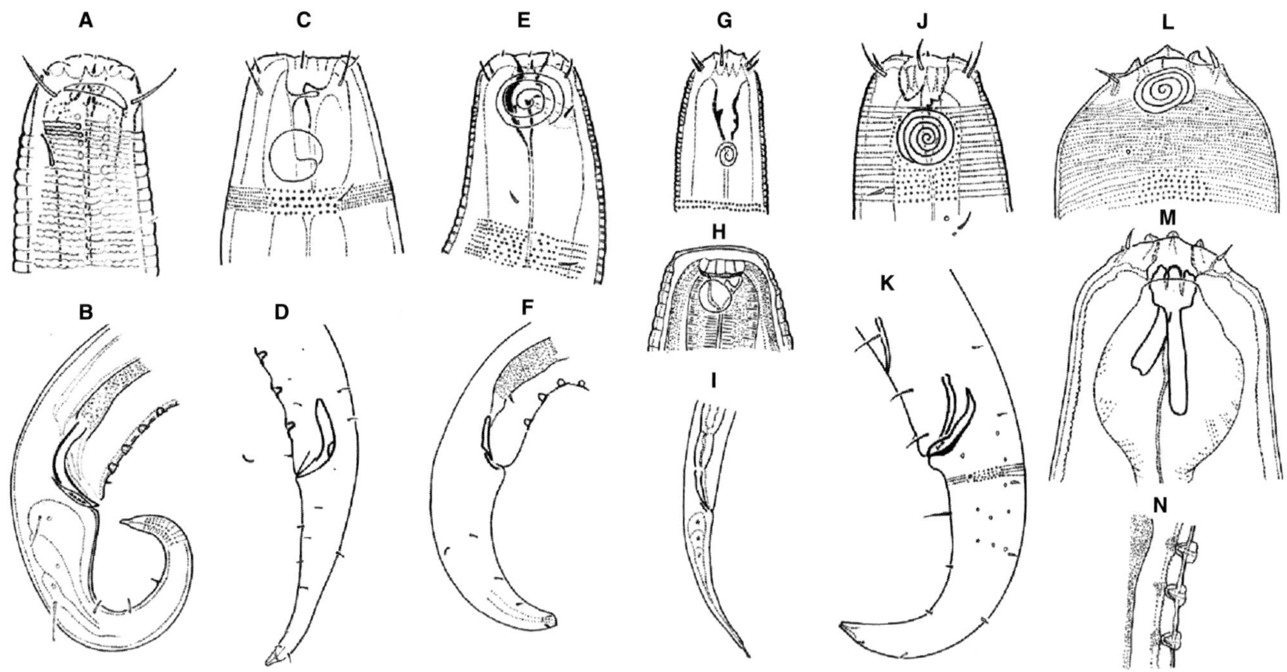

Fig. 17.6 Families of the order Chromadorida, head and tail regions. Chromadoridae, A and B: *Neochromadora*. Ethmolaimidae, C and D: *Ethmolaimus*. Neotonchidae, E and F: *Neotonchus*. Achromadoridae (based on Platt, 1982), G and I: *Achromadora*; H: *Kreisonema* (redrawn from Khera, 1969). Cyatholaimidae, J: *Paracyatholaimus*; K: *Paracanthonchus*. Selachinematidae, L–N: *Cheironchus* (based on Castillo-Fernandez and Decraemer, 1993).

Species recorded from freshwater habitats:

A. chungsani (Hoeppli & Chu, 1932) Andrássy, 1984: found in thermal water up to 48°C (China).

A. gracilis Ocaña, Hernadez & Monterrubio, 1999: from sandy sediment of high-mountain lakes in Sierra Nevada, Spain.

A. indica Tahseen, 2000: limnetic, from sewage slurry from the drains of the Aligarh Muslim University, India.

A. inflata Eyualem-Abebe & Coomans, 1996: limnetic, Lake Tana (Ethiopia).

A. longicauda Schneider, 1937: aquatic (Russia, Sumatra in Andrássy (1984)).

A. micoletzkyi (Stefanski, 1915) van der Linde, 1938: mainly terrestrial, cosmopolitan, recorded from all over Europe (The Netherlands, Germany, Switzerland, Austria, Hungary, Czech Republic, Serbia, Spain, Italy, Poland, Denmark, Sweden, Russia), South America (Mexico, Colombia, Paraguay, Galápagos), Africa (Ethiopia, Sudan), Asia (Java, Sumatra); common in wet soils (e.g. borders of rivers and pools) less common in freshwater (e.g. in high-mountain lakes in Sierra Nevada, Spain).

A. ruricola (de Man, 1880) Micoletzky, 1925: mostly terrestrial, cosmopolitan, recorded from all over Europe, USA, Canada, South America, China, Indonesia, Sumatra, Java, Japan and Zaire; common in wet humus and muck soils, in wet pastures, in a potato field (Belgium), mosses; less common in river sediments and river banks (South Africa), from sewage slurry from the drains of Aligarh Muslim University (India) and freshwater (The Netherlands, Italy).

A. sedata Gagarin, 2001: Ukraine limnetic, from mosses at bank of a brook.

A. semiarmata Altherr, 1952: mainly terrestrial, e.g. in Switzerland, West Sahara, Mauritania, Paraguay, Sichuan (China); also recorded from a freshwater habitat (Wittmer's Well, Floreana, Galápagos).

A. tenax (de Man, 1876) Kreis, 1932: mainly terrestrial, recorded from Europe and South America and common in wet pastures; also recorded from brackish water (The Netherlands).

A. terricola (de Man, 1880) Micoletzky, 1925: mainly terrestrial, cosmopolitan (Europe and North America), common in moss and humid soils, widespread in Russia where it is also recorded from freshwater habitats.

A. thermophila Lemzina & Gagarin, 1994: recorded from thermal waters in Kyrgyzstan.

Remarks: *A. granulata* (Cobb, 1913) Goodey, 1951, considered as valid in Andrássy (1984), belongs to the genus *Tripyloides* de Man, 1886 (see Wieser, 1954; Gerlach and Riemann, 1973).

There is a lot of controversy in the literature concerning the validity of the following species: *A. ambigua* Altherr, 1938; *A. dubia* (Bütschli, 1873) Micoletzky, 1925; *A. inermis* Altherr, 1952; *A. micoletzkyi*; *A. pseudomicoletzky*; and *A. tenax* (de Man, 1876) Kreis, 1932. In agreement with Mulvey (1969), Riemann (1970), Loof (1971) and Andrássy (1984) we consider *A. dubia* (Bütschli, 1873) Micoletzky, 1925 as *species inquirenda.* In agreement with Andrássy (1984) we accepted *A. micoletzkyi* and *A. tenax* as valid species (I. Andrássy, personal communication).

Subfamily Kreisonematinae Khera, 1969: a doubtful taxon (Fig. 17.6H)

Diagnosis. Achromadoridae. Cuticle with transverse striae but ornamentation obscure. Anterior sensilla not observed. Amphideal fovea large with circular outline. Stoma with well-developed cheilorhabdia, pharyngostome with a dorsal L-shaped dorsal tooth. Males unknown. Until Lorenzen (1981, 1994) the Kreisonematinae have always been classified within the Leptolaimidae. The structure and armature of the buccal cavity, a pharynx with terminal bulb and the position of the gonads in relation to the intestine justify their position with the Achromadoridae; however, the genus is poorly described and illustrated and needs revision. *Kreisonema* was considered a synonym of *Achromadora* by Andrássy (1976, 1984).

Type genus: *Kreisonema* Khera, 1969.

Family CYATHOLAIMIDAE Filipjev, 1918

Diagnosis. Chromadoroidea. Inner labial sensilla often setiform. Six outer labial sensilla and four cephalic sensilla setiform and arranged in a single circlet (except in Nyctonematinae where in separate circlets); the four cephalic setae being shorter than the outer labial setae. Amphideal fovea multispiral. Stoma with a large dorsal tooth, usually with smaller ventrosublateral teeth (without armature in Nyctonematinae). Pharynx without terminal bulb. Female didelphic, amphidelphic with anterior and posterior antidromously reflexed gonads always on different sides of the intestine. Male diorchic with both gonads on opposite sides (right or left) of the intestine, exceptionally monorchic; different types of pre-cloacal supplements or supplements absent. Mostly marine, rarely limnetic.

Five subfamilies are recognized within the family; only the subfamily Paracanthonchinae has representatives in freshwater habitats. According to Lorenzen (1994), the subfamilies are paraphyletic and need to be revised.

Key to subfamilies of the CYATHOLAIMIDAE

1. Six outer labial and four cephalic sensilla in two separate circlets; buccal cavity without teeth . . . . .Nyctonematinae (Fig. 17.7A)
   Six outer labial and four cephalic sensilla in one circlet; buccal cavity with teeth 2
2. Pre-cloacal supplements absent (exc. present and tubular in *Praeacanthonchus* or cup-shaped in *Marylynnia*); no lateral differentiation of body cuticle (exceptionally with lateral differentiation *Marylynnia*) 3
   Pre-cloacal supplements present; usually with lateral differentiation of body cuticle 4
3. Body cuticle punctated, no longitudinal ridges . . . . .Cyatholaiminae (Fig. 17.6J)
   Body cuticle with ten longitudinal ridges; punctation restricted to head region and tail . . . . .Xenocyatholaiminae (Fig. 17.7C)

4. Pre-cloacal supplements tubular (exc. sucker-like in some *Paracanthonchus* species or setose in *Paracyatholaimus*); with lateral differentiation of body cuticle (exc. in *Paracyatholaimus*); gubernaculum paired proximally . . . . .Paracanthonchinae (Figs 17.6J, 17.7B, D and E and H–M)
Pre-cloacal supplements knob-like or flattened, small; cuticle with lateral differentiation, gubernaculum paired proximally. . . . . .Pomponematinae (Figs 17.7F and G)

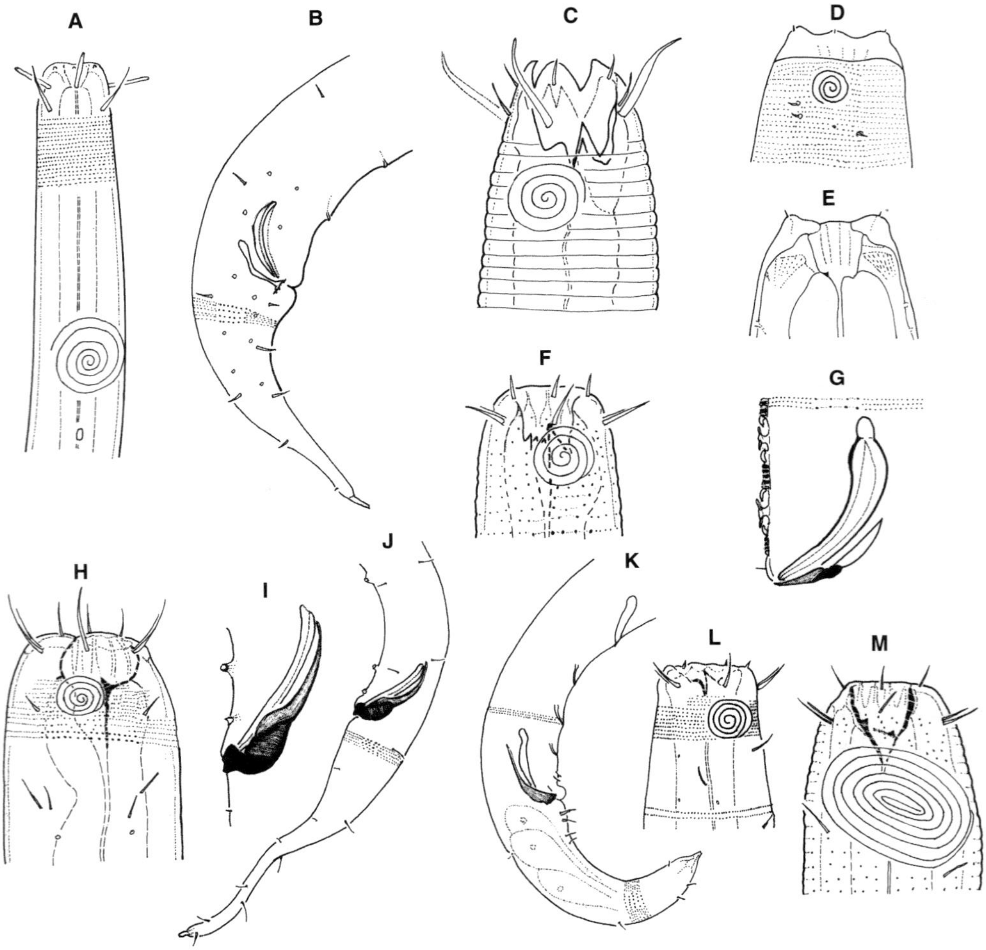

Fig. 17.7 Subfamilies and genera of the family Cyatholaimidae, head region, tail region and copulatory apparatus. Nyctonematinae, A: *Nyctonema*; Paracanthonchinae (based on Bussau, 1993), B: *Paracyatholaimus*, D and E: *Paracanthonchus*; H–J: *Biarmifer* (based on Wieser, 1959), K and L: *Acanthonchus*; M: *Paracyatholaimoides* (based on Riemann, 1966). Xenocyatholaiminae, C: *Xenocyatholaimus*. Pomponematinae, F and G: *Pomponema* (based on Platt and Warwick, 1988).

Subfamily Cyatholaiminae Filipjev, 1918
Diagnosis. Cyatholaimidae. Body cuticle with homogeneous punctation, without lateral differentiation (except in *Marylynnia*). Pre-cloacal supplements absent, except in *Praeacanthonchus* where tubular and in *Marylynnia* where cup-shaped; gubernaculum unpaired proximally, except in *Marylynnia* where paired.

Type genus: *Cyatholaimus* Bastian, 1865.

Subfamily Nyctonematinae Bussau, 1993
Diagnosis: Cyatholaimidae. Cuticle with transverse striae and fine homogeneous punctation, without lateral differentiation. Anterior sensorial organs in three circlets: six inner labial sensilla papilliform, six outer labial and four cephalic sensilla setiform; the outer labial setae thick with rounded tip and the cephalic setae fine with pointed tip. Amphideal fovea multispiral located at great distance posterior to head region. Buccal cavity without teeth; pharynx largely cylindrical. Male diorchic, without pre-cloacal supplements. Female didelphic–amphidelphic with reflexed ovaries. Tail long. Marine; animals apparently able to build sediment tubes.

Type genus: *Nyctonema* Bussau, 1993; no other genera.

Remarks: Hereby, Bussau (1993), a PhD thesis, is accepted as a valid publication because it fulfils the rules of the Code of Zoological Nomenclature, 4th edition, 1999, although it does not follow recommendation 8A, i.e. to publish new taxa by preference in recognized scientific journals. The author agreed (personal communication dated 7 August 2004) with this approach although in the past they (Bussau, 1995; Bussau and Vopel, 1999) published some of the new taxa of the thesis in other scientific journals, which could be interpreted as a disclaimer of the publication, i.e. in accordance with the 3rd edition of the code (1985). In these articles the author used the same scientific names as mentioned in the thesis, and although the thesis is not a publication, the names of the thesis are *nomina nuda* and so not available.

Subfamily Paracanthonchinae De Coninck, 1965
Diagnosis. Cyatholaimidae. Homogeneously punctated body cuticle, with or without lateral differentiation in ornamentation. Pharynx without end bulb. Pre-cloacal supplements variable, but mainly tubular; rarely absent. Gubernaculum paired proximally. Largely marine, rarely limnetic in one genus.

Type genus: *Paracanthonchus* Micoletzky, 1924.

Subfamily with five genera, only *Paracyatholaimus* has been recorded from freshwater habitats.

Key to genera of the Paracanthonchinae

1. Pre-cloacal supplements tubular 2
   Pre-cloacal supplements setose or absent 3
   Pre-cloacal supplements inconspicuous,
   cup-shaped . . . .*Biarmifer* (Figs 17.7H and I)

2. Cuticle without lateral differentiation in ornamentation; gubernaculum simple . . . . *Acanthonchus* (Figs 17.7K and L)
   Cuticle with larger dots laterally; gubernaculum proximally paired, distally dentated . . . . *Paracanthonchus* (Figs 17.6K and 17.7D and E)
3. Cephalic cuticle thickened; tightly coiled amphideal fovea showing sexual dimorphism in size . . . . *Paracyatholaimoides* (Fig. 17.7M)
   Cephalic cuticle not thickened; amphideal fovea not tightly coiled; no sexual dimorphism . . . . *Paracyatholaimus* (Figs 17.6J and 17.7B)

Genus *Acanthonchus* Cobb, 1920 (Figs 17.7K and L)
Diagnosis. Paracanthonchinae. Punctated cuticle mostly homogeneous, without lateral differentiation. Pre-cloacal supplements tubular, anteriormost larger. Gubernaculum simple. Marine.

Type species: *A. viviparous* Cobb, 1920.

Genus *Biarmifer* Wieser, 1954 (Figs 17.7H and I)
Diagnosis. Paracanthonchinae. Punctated cuticle without or with inconspicuous lateral differentiation. Buccal cavity with a large dorsal tooth. Male with inconspicuous cup-shaped pre-cloacal supplements. Gubernaculum plate-shaped, distally dilated or not.

Type species: *B. cochleatus* Wieser, 1954.

Genus *Paracanthonchus* Micoletzky, 1924 (Figs 17.6K and 17.7D and E)
Diagnosis. Paracanthonchinae. Body cuticle with transverse rows of fine dots, which laterally may be slightly larger or irregularly arranged. Buccal cavity with larger pointed dorsal tooth and smaller ventrosublateral teeth. Gubernaculum proximally paired and distally expanded and dentate. Pre-cloacal supplements fine tubules. Marine and brackish.

Type species: *P. caecus* (Bastian, 1865) Micoletzky, 1924.

Genus *Paracyatholaimoides* Gerlach, 1953 (Fig. 17.7M)
Diagnosis. Paracanthonchinae. Similar to *Paracyatholaimus* except for the thickened cephalic cuticle, the more tightly coiled amphideal fovea showing sexual dimorphism in size and sexual dimorphism in body length with the females being much smaller than the males. Marine genus.

Type species: *P. multispiralis* Gerlach, 1953.

Genus *Paracyatholaimus* Micoletzky, 1922 (Figs 17.6J and 17.7B)
Diagnosis. Paracanthonchinae. Cuticle with fine transverse striae with homogeneous ornamentation of dots, no distinct lateral differentiation but the punctations may appear somewhat larger laterally. Stoma not longer than wide; dorsal tooth stronger

than ventrosublateral teeth or denticles. Pre-cloacal supplements small cuticularized and setose, rarely absent. Gubernaculum paired, simple, hardly dilated at the distal end, rounded or notched distally but never with pronounced teeth. Tail conical, long spinneret.

Paracyatholaimus is mainly a marine genus with a few marine species occasionally recorded from brackish habitats; one marine species has also been recorded from freshwater habitats, three species occur in freshwater habitats and one species was recorded from dune sand and is considered terrestrial.

Type species: *P. dubiosus* (Bütschli, 1874) Micoletzky, 1922.

Species recorded from freshwater habitats:
P. intermedius (de Man, 1880) Filipjev, 1930: mainly marine/brackish: Baltic Sea, North Sea, Kiel Bay (Germany), Svalbard (Arctic), Iceland, English Channel, Kattegat (Northern Europe), Öresund (Scandinavia); brackish: Germany, Denmark, Poland, Gulf of Guinea, Portugal; more rarely recorded from freshwater habitats: Germany and Spain, and from maritime turf, Linga Island; rarely terrestrial (see record from clayish soil from Leiden (The Netherlands) in Loof, 1961).
Remarks: Some of the records of this species may refer to different species, e.g. *P. intermedius sensu* Gerlach (1965) has a much larger amphideal fovea.

P. lewisi Coomans, Vincx & Decraemer, 1985: found in a small freshwater pool in a coral rock on Uipi Island, Marovo Lagoon, Solomon Islands.
P. paramonovi Gagarin, 2004: freshwater, from silty sand from a lagoon–lake on Kunashir Island, Russia.
P. truncatus (Cobb, 1914) Filipjev, 1930: freshwater, Silver Springs, Florida and Rio Cordoba estuary, Colombia.

Subfamily Pomponematinae Gerlach & Riemann, 1973 (Figs 17.7F and G)
Diagnosis. Cyatholaimidae. Body cuticle punctated, with lateral differentiation in ornamentation. Pre-cloacal supplements knob-like or flattened, small; gubernaculum paired proximally.

Type genus: *Pomponema* Cobb, 1917.

Subfamily Xenocyatholaiminae Gerlach & Riemann, 1973 (Fig. 17.7C)
Diagnosis. Cyatholaimidae. Cuticle with transverse striae, only with punctated ornamentation anterior to amphideal fovea and on tail; ten longitudinal ridges present. Six outer labial setae and four cephalic setae arranged in a single circlet, the four cephalic setae being shorter than the six outer labial setae. Amphideal fovea multispiral, located at about one head width from anterior end. Buccal cavity large, with large pointed dorsal tooth. Pre-cloacal supplements absent; gubernaculum small. Marine, interstitial fauna. Monotypic genus, based on type species only.

Type genus: *Xenocyatholaimus* Gerlach, 1953.

Family ETHMOLAIMIDAE Filipjev & Schuurmans Stekhoven, 1941 (Figs 17.6C and D)
Diagnosis. Chromadoroidea. Cuticle with transverse striae, with rows of fine punctation, sometimes with lateral differentiation of fewer rows and rows of pores. Anterior

sensilla arranged in three circlets with the anterior circlet papilliform, both outer circlets setiform. Three equal buccal teeth. Amphideal fovea uni- or multispiralled. Males with ventral row of conspicuous cup-shaped (also described as knob-like) pre-cloacal supplements, each with an external articulated flange; ventral pre-cloacal spine present. Two opposed outstretched testes, both gonads on different sides of the intestine. Tail conical, usually with rounded tip.
Limnetic. One subfamily: Ethmolaiminae Filipjev & Schuurmans Stekhoven, 1941; the subfamily Neotonchinae Wieser & Hopper, 1966 (in revision of Ethmolaimidae in Platt, 1982) has been raised to family level in De Ley and Blaxter, 2002.

Subfamily Ethmolaiminae Filipjev & Schuurmans Stekhoven, 1941 (Figs 17.6C and D)
Diagnosis. Ethmolaimidae. Three large teeth, the ventrosublateral teeth being as large or almost as large as the dorsal tooth. Amphideal fovea largely unispiral, usually located at mid level of stoma.
Two genera: *Ethmolaimus* de Man, 1880 and *Paraethmolaimus* Jensen, 1994.

Genus *Ethmolaimus* de Man, 1880 (Figs 17.6C and D)
Syn. *Parachromadora* Micoletzky, 1914 pro *Triodontolaimus* Micoletzky, 1913 homonym to *Triodontolaimus* de Man, 1893; *Trichethmolaimus* Platt, 1982.
Diagnosis. Ethmolaiminae. Cuticle with fine transverse striae and transverse rows of fine homogeneous punctation between transverse striae, lateral differentiation present. Head rounded with lip region not offset. A freshwater genus with one marine species (*E. hirsutus* (Gerlach, 1956) Jensen, 1994), two brackish water species (*E. bothnicus* Jensen, 1994 and *E. hailuotoensis* Turpeenniemi, 1995) and one species from hyper-saline (*E. multipapillatus* Paramonov, 1926) habitats.

Type species: *E. pratensis* de Man, 1880.

Eyualem-Abebe and Coomans (1996) remarked that the number of species in the genus is still questionable due to taxonomic confusion and bias concerning the characteristics used for specific identification. Platt (1982, 1985) accepted the revision by Hirschman (1952) and recognized only three valid species within the genus: *E. dahli*, *E. multipapillatus* and *E. pratensis*. He did not accept as valid those species described on females only (*E. faeroeensis* and *E. caudatus*) and *E. distaphanus*, a *nomen nudum*, and he synomized *E. parapratensis* with *E. pratensis* based on the similarities in teeth and pre-cloacal supplements in the male. Platt (1982) created a new genus *Trichethmolaimus* for a single species *T. hirsutus* (Gerlach, 1956), characterized by extremely long somatic setae, a head cuticle with supporting rods and the second circlet of anterior sensilla setiform and stout. Jensen (1994) followed Platt (1982, 1985) but synonymized the genus *Trichethmolaimus* with *Ethmolaimus* as the diagnostic features were also observed in species of *Ethmolaimus* and/or are not considered of generic importance. Jensen (1994) recognized six species within the genus *Ethmolaimus*. Currently, seven freshwater species have been recorded.

Species recorded from freshwater habitats:
E. derisorius Shoshin, 1998: Lake Baikal, depth 3–4 m, very fine slightly silted sand.

E. intermedius Jensen, 1994: Atter Lake and Finstertaler Lake, Austria.

E. lanatus Shoshin, 1998: Baikal Lake, depth 3–4 m, very fine silted sand; males known only.

E. parapratensis Alekseev & Naumov in Alekseev, Naumov & Dymina, 1979: Khasan Lake, former USSR. Valid species in Andrássy (1984) but considered a synonym of *E. pratensis* by Platt (1982) because of similarity in teeth and pre-cloacal supplements in male.

E. pilosus Shoshin, 1998: Lake Baikal, depth 3–4 m, very fine slightly silted sand.

E. pratensis de Man, 1880: mostly found in freshwater but also in wet soils (e.g. wet meadow, marshland near rivers), cosmopolitan, known from all over Europe (The Netherlands, Königssee (Germany), Switzerland, Finstertaler See and Piburger See (Austria), Hungary, Czech Republic, Romania, France, Italy, Poland, Denmark, Spain, Finland, Sweden, Russia, North America (Canada, USA, e.g. Vermont, Maryland); Nepal, Indonesia, Java, China and Zaire; also recorded from a brackish habitat (salinity below 5%) in Baltic Sea together with *E. hailuotoensis*.

E. zullinii Eyualem-Abebe & Coomans, 1996: Lake Tana, Ethiopia, from surface to 1.0 cm of sediment at 0.5 m depth; females known only.

Species recorded from freshwaters but with uncertain taxonomic position:

E. foreli (Hofmänner, 1913) Hofmänner & Menzel, 1915: recorded from Switzerland, Russia. Accepted as a valid species in Andrássy (1984), Tsalolikhin (1985), Gagarin (1993) and also by Eyualem-Abebe and Coomans (1996: see differential diagnosis of *E. zullinii*). Considered a synonym of *E. pratensis* op. Micoletzky (1922), Hirschmann (1952), Gerlach and Riemann (1973), Platt (1982, 1985) and Jensen (1994).

E. gracilicaudatus N.A. Cobb in M.V. Cobb, 1915: aquatic, Faroe Islands, USA (Michigan); females known only. Considered a valid species in Hirschmann (1952), Andrássy (1984) and Eyualem-Abebe and Coomans (1996: in differential diagnosis of *E. zullinii*) but as a synonym of *E. pratensis* op. Micoletzky (1922) in Gerlach and Riemann (1973).

E. revaliensis (Schneider, 1906) Steiner, 1913: freshwater, Switzerland, Romania, Poland, former USSR, e.g. in Lake Baikal (Shoshin, 1998). Accepted as a valid species by Andrássy (1984) but considered a synonym of *E. pratensis* op. Hirschmann (1952), Gerlach and Riemann (1973), Platt (1982, 1985), Tsalolikhin (1985), Gagarin (1993) and Jensen (1994).

Remarks (on *E. pratensis*): The type species *E. pratensis* was originally described from a female specimen by de Man (1880) from a wet meadow near Leiden. Hitherto, it is the only species of the genus known from the Netherlands. The species appears to be common in freshwater and sandy river banks.

The majority of the records of *E. pratensis* and of species synonymized with *E. pratensis* do not include descriptions and illustrations and no voucher specimens are available. The earliest description of a male (with illustration) is as *Chromadora revaliensis* by Schneider (1906), followed by Hofmänner (1913) as *Chromadora foreli* and by Hofmänner and Menzel (1915, i.e. of *E. revaliensis*), Hirschmann (1952 of *E. pratensis*), and in recent revisions of the genus by Platt (1982) and Jensen (1994).

The revision of the genus *Ethmolaimus* by Hirschmann (1952) included a detailed study of *E. pratensis* based on a large number of specimens from different localities in Germany. This resulted in an unusually wide range of the morphometric data, e.g. body length in female ranges between 387 and 1123 μm, number of pre-anal supplements in male between 9 and 17 (9 and 15 in German populations), and in morphological features, e.g. position of amphideal fovea from posterior stoma region to just behind it and punctated ornamentation of the body cuticle from weak to strong and irregular within a single population. Consequently, Hirschmann (1952) found it impossible to subdivide '*E. pratensis*' into 'varieties' or species. So, the author synonymized all previously described species of the genus with *E. pratensis,* except for *E. multipapillatus* with 21–24 supplements and absence of a gubernaculum in male and short cephalic setae in both sexes.

Jensen (1994) recognized an evolutionary trend within the genus towards a reduction of cephalic sense organs with the original condition possessing six outer labial and four cephalic setiform sensilla (*E. hirsutus*, *E. intermedius*, *E. pratensis*), towards only the third crown setiform (*E. bothnicus*, *E. parapratensis*) and towards most evolved types with all three circlets papilliform (*E. multipapillatus*). However, the position of *E. pratensis* seems biased as most illustrations only show the third crown setiform, the second crown at most with raised papillae.

Genus *Paraethmolaimus* Jensen, 1994

Diagnosis. Ethmolaiminae. Mainly characterized by the offset lip region and papilloid anterior (labial/cephalic) sense organs. Eyualem-Abebe and Coomans (1996) questioned the validity of the genus because papilloid sensilla also occur in species of the closely related genus *Ethmolaimus,* from which it only differs by the offset lip region. All species are from brackish/marine biotopes.

Type species: *P. appendixocaudatus* Jensen, 1994.

Family NEOTONCHIDAE Wieser & Hopper, 1966 (Figs 17.6E and F)

Diagnosis. Chromadoroidea. Punctated cuticle with dorsolateral sub-cephalic setae and with or without sub-lateral pore-setae. Anterior sensilla arranged in three circlets, only the four cephalic sensilla are setiform. Amphideal fovea multispiral (with more than three turns). Buccal cavity with large dorsal tooth, with or without two small ventrosublateral teeth. Two testes, opposed. Pre-cloacal supplements cup-shaped. Marine.

Type genus: *Neotonchus* Cobb, 1933.

Family SELACHINEMATIDAE Cobb, 1915 (Figs 17.6L–N)

Diagnosis. Chromadoroidea. Body cuticle punctated. Anterior sensilla arranged in three circlets, one of them may be longer than the other two; sensilla may be jointed. Amphideal fovea multispiral. Buccal cavity spacious either without armature or with jaws. Pharynx with posterior end-bulb and in part also widened anteriorly. Males usually with two testes, both gonads on same or on different sides of the intestine; cup-shaped pre-cloacal supplements, never tubular. Female didelphic–amphidelphic with ovaries reflexed. Marine.

Type genus: *Selachinema* Cobb, 1914, a junior synonym of *Demonema* Cobb, 1894.

Key to freshwater species of the Chromadorinae

1. Amphideal fovea a transverse, slit-like or unispiral, wider than long, usually located between the four cephalic setae 2
   Amphideal fovea uni- or multispiralled, not wider than long, variable location 11
2. Dorsal tooth hollow; six outer labial and four cephalic sensilla setose; amphideal fovea transverse, usually oval loop-shaped (Hypodontolaiminae) . . . . . . *Neochromadora izhorica*
   Dorsal tooth solid; only four cephalic sensilla setose; amphideal fovea transverse slit-like (Chromadorinae) 3
3. Body cuticle with homogeneous punctation without lateral differentiation 4
   Body cuticle with homogeneous punctation with lateral differentiation 7
4. Numerous pre-cloacal supplements in male 5
   Pre-cloacal supplements absent 6
5. Ocelli absent; amphideal fovea transverse oval; 11–12 pre-cloacal supplements . . . . . . *Chromadorina bercziki*
   Ocelli present; amphideal fovea obscure; 14–16 pre-cloacal supplements . . . . . . *Chromadorina viridis*
6. Ocelli absent. Body length = 900–1000 μm; *c'* (female) = 9; spicule length = 28–30 μm . . . . . . *Chromadorina astacicola*
   Ocelli present; body length smaller (= 500–700 μm); *c'* (female) = 5–7; spicule length = 21–23 μm . . . . . . *Chromadorina bioculata*
7. Amphideal fovea slit-like . . . . . . *Prochromadora orleji*
   Amphideal fovea clearly spiral 8
8. Male with less than five pre-cloacal supplements . . . . . . *Punctodora ratzeburgensis*
   Male with more supplements (>10) 9
9. Spicule with blunt distal end, gubernaculum without thickened distal portion; punctation in lateral field enlarged, irregularly arranged; vulva with slightly protruding lips, *V* = 47% or more 10
   Spicule with pointed distal end; *SP* = 11–14; *L* = 660–1000 μm; punctation variable; *V* = 40–47% . . . . . . *Punctodora ohridensis*
10. Secretory–excretory tube protruding; large dorsal tooth opposite groove bordered with denticle-like protrusions; *SP* = 15–16; *L* = 600–750 μm; *V* = 47–50.8% . . . . . . *Punctodora exochopora*
    Secretory–excretory tube not protruding; large tooth opposite groove without marked

denticle-like protrusions; *SP* = 14–18 (mostly 16) ; L = 600–900 µm; V = 47–50% . . . . *Punctodora dudichi*

11. Six outer labial and four cephalic setae on separate circlets, outer labial setae shorter than four cephalic setae; pre-cloacal supplements cup-shaped 12
    Six outer labial and four cephalic setae mostly on a single circlet, outer labial setae longer than four cephalic setae; pre-cloacal supplements not cup-shaped 18
12. Numerous long somatic setae present 13
    Somatic setae shorter and less numerous 14
13. Cuticle with large homogeneous punctation; cephalic setae 15 µm; spicules not strongly bent . . . . *Ethmolaimus pilosus*
    Cuticle with linear punctation; cephalic setae longer (25 µm); spicules finer and more bent . . . . *Ethmolaimus lanatus*
14. Amphideal fovea with anterior position at level of buccal teeth . . . . *Ethmolaimus intermedius*
    Amphideal fovea more posteriorly, at level of mid-stoma or base of stoma 15
15. Secretory–excretory pore far anteriorly, close to cephalic setae . . . . *Ethmolaimus derisorius*
    Secretory–excretory pore more posteriorly, shortly posterior to nerve ring 16
16. Anterior end widened, marked off from body; ventrosublateral teeth/denticles inconspicuous . . . . *Ethmolaimus zullinii*
    Anterior end continuous with body; ventrosublateral teeth well developed 17
17. Outer labial sensilla papilloid; stoma one-fifth of pharynx long; 12 pre-cloacal supplements . . . . *Ethmolaimus parapratensis*
    Outer labial sensilla setiform; stoma shorter (one-sixth to one-seventh of pharynx); 9–17 pre-cloacal supplements . . . . *Ethmolaimus pratensis*
18. Males common. Female gonads on opposite sides of intestine; pre-cloacal supplements as tubules or papilla, rarely absent; mainly marine 19
    Males rare. Female gonads on same side of intestine; pre-cloacal supplements absent; mainly limnetic 22
19. Body length shorter than 1 mm; two prominent ventrosublateral teeth . . . . *Paracyatholaimus lewisi*
    Body length over 1 mm; ventrosublateral teeth variable in number and size 20

20. Male with >10 pre-cloacal supplements; spicules 38–39 μm long; amphideal fovea one-third of corresponding body width . . . . *Paracyatholaimus paramonovi*
    Male with less pre-cloacal supplements; spicules shorter; amphideal fovea equally wide or smaller 21
21. Four pre-cloacal supplements; cephalic setae 50% of head width long; amphideal fovea one-third of corresponding body width . . . . *Paracyatholaimus truncatus*
    Five to nine pre-cloacal supplements; cephalic setae 33% of head width; amphideal fovea smaller (20%) . . . . *Paracyatholaimus intermedius*
22. Body length mean value around 1 mm (800–1300 μm) . . . . *Achromadora terricola*
    Body length shorter (300–800 μm) 23
23. Amphideal fovea unispiral . . . . *Achromadora inflata*
    Amphideal fovea multispiral 24
24. Rectum short, one anal body diameter long 25
    Rectum longer than one anal body diameter 26
25. Thermophile; L = 310–360 μm . . . . *Achromadora thermophila*
    Not thermophile, mainly terrestrial; L = 350–380 μm and 530–710 μm (syn. *A. ambigua*) from different localities . . . . *Achromadora ruricola*
26. Amphideal fovea at level of stoma, i.e. at 50% of labial head diameter from anterior end . . . . *Achromadora semiarmata*
    Amphideal fovea more posteriorly, i.e. at more than one head diameter from anterior end 27
27. Tail long (c' = 10 or more) 28
    Tail shorter 29
28. Body short (L = 350 μm) . . . . *Achromadora longicauda*
    Body longer (L = 660–850 μm) . . . . *Achromadora gracilis*
29. Amphideal fovea at level of stoma 30
    Amphideal fovea posterior to stoma 31
30. Thermophile. Amphideal fovea small (one-sixth to one-seventh of head diameter), L = 700–800 μm . . . . *Achromadora chungsani*
    Not thermophile. Amphideal fovea larger (one-fourth of head diameter); L = 570–690 μm . . . . *Achromadora indica*
31. Pre-cloacal supplements present; distance V–anus 3–4 tail lengths . . . . *Achromadora micoletzkyi*
    Pre-cloacal supplements absent; distance V–anus smaller 32
32. Amphideal fovea at 2–2.5 head widths from anterior end; distance V–anus twice the tail length . . . . *Achromadora tenax*
    Amphideal fovea more anteriorly; distance V–anus 2.3–2.9 times the tail length . . . . *Achromadora sedata*

Order DESMODORIDA

Introduction

Desmodorida are characterized by a body cuticle with transverse striae in general without punctated ornamentation. Desmodorids are essentially marine animals except the genus *Prodesmodora,* with 40% of its species recorded from freshwater habitats. Within the order Desmodorida only five genera have species recorded from freshwater habitats.

Classification

In the classification by De Ley and Blaxter (2002) the superfamily Desmodoroidea Filipjev, 1922 *sensu* Lorenzen (1981, 1994) has been raised to order level. The following classification within the order largely follows Lorenzen (1981, 1994). The genera occurring in freshwater bodies are presented in bold. Within taxa without freshwater records, no lower ranking of taxa is presented.
Note: Asterisk indicates taxa reported from freshwater bodies.

Order DESMODORIDA De Coninck, 1965
- Suborder Desmodorina De Coninck, 1965
- Superfamily Desmodoroidea Filipjev, 1922
- Family Desmodoridae Filipjev, 1922
  - Subfamily Desmodorinae Filipjev, 1922*
    - Genus *Acanthopharyngoides* Chitwood, 1936
      - *Acanthopharynx* Marion, 1870
      - *Bolbonema* Cobb, 1920
      - *Croconema* Cobb, 1920
      - *Desmodora* de Man, 1989
      - *Desmodorella* Cobb, 1933*
      - *Echinodesmodora* Blome, 1982
      - *Metadesmodora* Schuurmans Stekhoven, 1942
      - *Paradesmodora* Schuurmans Stekhoven, 1950
      - *Psammonema* Verschelde & Vincx, 1995
      - *Pseudochromadora* Daday, 1899
      - *Pseudodesmodora* Boucher, 1975
      - *Sibayinema* Swart & Heyns, 1991*
      - *Stygodesmodora* Blome, 1982
      - *Zalonema* Cobb, 1920
  - Subfamily Spiriniinae Gerlach & Murphy, 1965
  - Subfamily Pseudonchinae Gerlach & Riemann, 1973
  - Subfamily Stilbonematinae Cobb, 1936
  - Subfamily Molgolaiminae Jensen, 1978
  - Subfamily Prodesmodorinae Lorenzen, 1981*
    - Genus *Prodesmodora* Micoletzky, 1923*

Family Epsilonematidae Steiner, 1927
Family Draconematidae Filipjev, 1918
Superfamily Microlaimoidea Micoletzky, 1922
Family Microlaimidae Micoletzky, 1922*
Genus *Aponema* Jensen, 1978
Bathynox Bussau & Vogel, 1999
Bolbolaimus Cobb, 1920
Caligocanna Bussau & Vopel, 1999
Calomicrolaimus Lorenzen, 1976
Cinctonema Cobb, 1920
Crassolaimus Kreis, 1929
Ixonema Lorezen, 1971
Microlaimus de Man, 1980*
Spirobolbolaimus Soetaert & Vincx, 1988
Family Aponchiidae Gerlach, 1963
Family Monoposthiidae Filipjev, 1934*
Genus *Monoposthia* de Man, 1889
Monoposthioides Hopper, 1963
Nudora Cobb, 1920*
Rhinema Cobb, 1920

Order DESMODORIDA De Coninck, 1965

Diagnosis. Chromadorea. Body appears yellowish to brownish in glycerin preparations. Body cuticle with transverse striae or annulated, rarely smooth; ornamentation present or absent, mostly without punctation. Anterior sensorial organs arranged in three circlets: the inner labial sensilla papilliform, the six outer labial sensilla and four cephalic sensilla usually setiform, with the four cephalic setae being usually longer than the six outer labial setae. Buccal cavity often with distinct dorsal tooth and smaller ventrosublateral teeth. Pharynx mostly with terminal bulb. Males mostly monorchic, but diorchic in some taxa of the Microlaimoidea. Pre-cloacal papillae or tubules may be present (based on Lorenzen (1981, 1994) for the Desmodoroidea).

A single suborder Desmodorina De Coninck, 1965 with same diagnosis as the order; two superfamilies Desmodoroidea Filipjev, 1922 and Microlaimoidea Micoletzky, 1922.

Key to superfamilies

Cuticle with transverse striae, never punctated; habitus brownish in fixed specimens.
Males monorchic; females didelphic . . . . .Desmodoroidea (Figs 17.8–17.10)

Cuticle with transverse striae, smooth or ornamented with dots, longitudinal bars or longitudinal ridges or rows of V-shaped structures; habitus brownish or greyish.
Males monorchic or diorchic; female didelphic or monodelphic . . . . .Microlaimoidea (Figs 17.11 and 17.12)

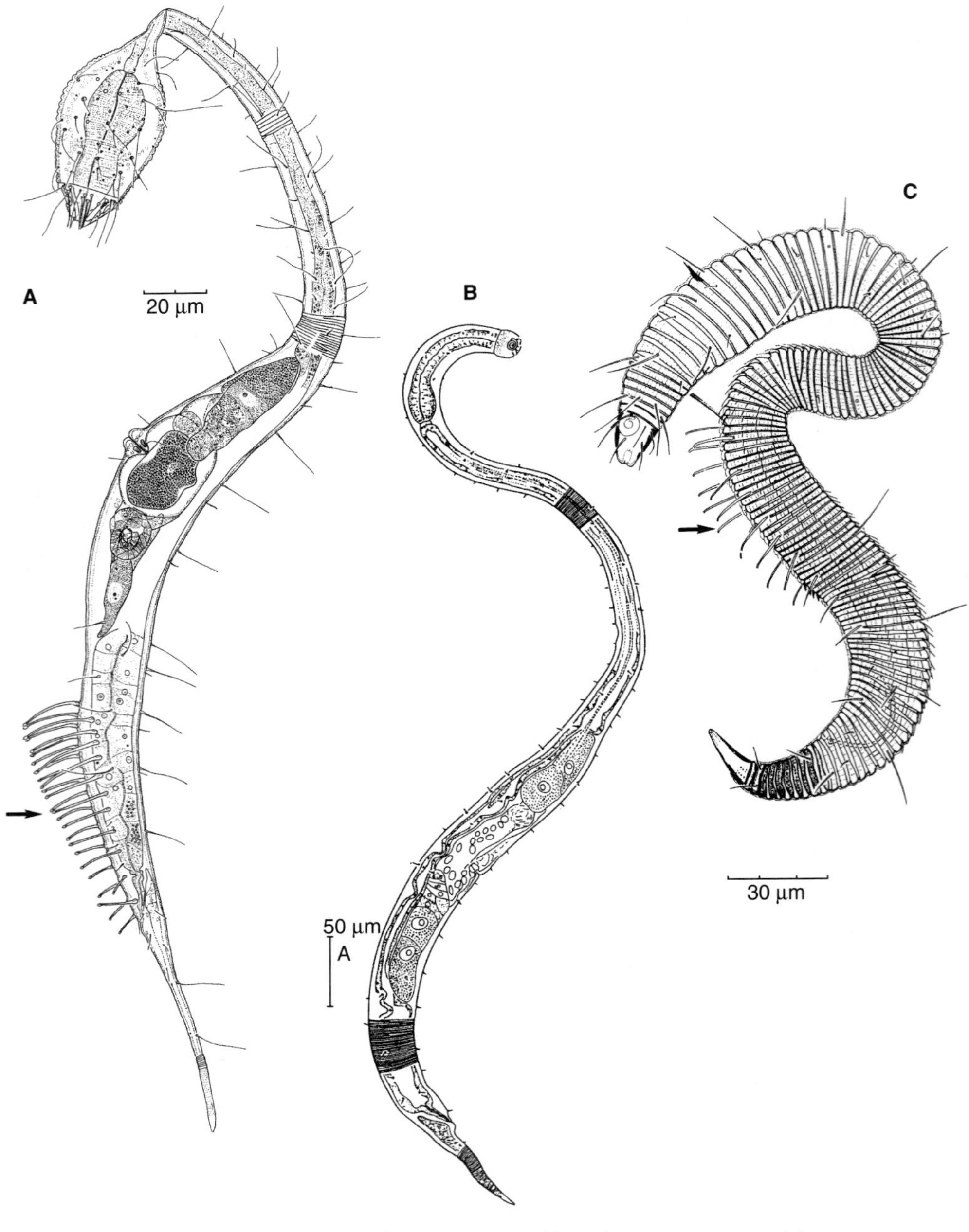

Fig. 17.8 Families of the Desmodoroidea, total view of female. A: Draconematidae, *Paradraconema maggentii*, arrows point to anterior and posterior adhesion tubes (PAT) (from Decraemer, 1989); B: Desmodoridae, *Desmodorella* (from Verschelde *et al.*, 1998); C: Epsilonematidae, *Epsilonema parvospina*, arrow points to ambulatory setae (from Decraemer, 1982). Courtesy *Bulletin of the Royal Belgian Institute of Natural Sciences* (A and C); *Nematologica* (B).

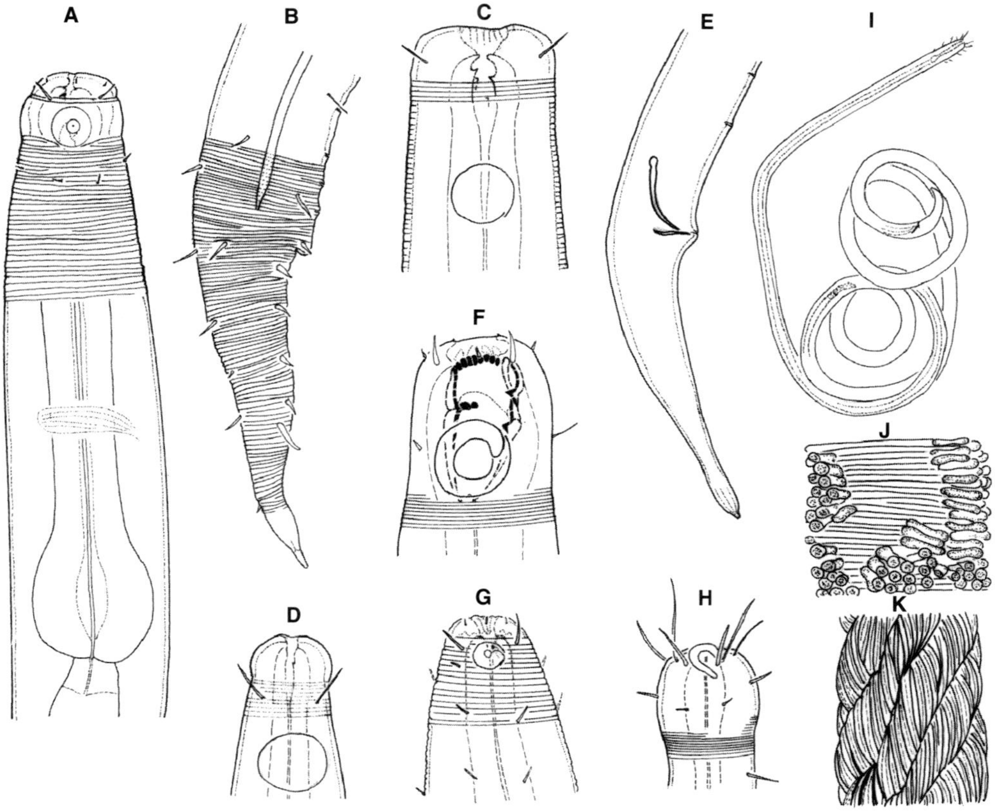

Fig. 17.9 Subfamilies of the family Desmodoridae. A and B: Desmodorinae, *Pseudochromadora*, anterior body and tail regions; C: Prodesmodorinae, *Prodesmodora*, head region; D and E: Molgolaiminae, *Molgolaimus*, head and posterior body regions; F: Pseudonchinae, *Pseudonchus*, head region (based on Platt and Warwick, 1988); G: Spiriniinae, *Spirinia*, head region; H–K: Stilbonematinae, H: *Leptonemella* head region, I: *Adelphus* male *in toto* (based on Ott, 1997), J and K: bacterial growth on body cuticle of *Catanema* and *Eubostrichus*, respectively (based on Ott *et al.*, 1991).

Superfamily DESMODOROIDEA Filipjev, 1922

Diagnosis. Desmodorina. Body cuticle with transverse striae, never punctated. Head region usually striated. Amphideal fovea spiral or cryptospiral. Male with a single testis. Females with two reflexed ovaries, vulva usually posterior to mid-body.
Three families: Desmodoridae Filipjev, 1922, Draconematidae Filipjev, 1918 and Epsilonematidae Steiner, 1927.

Key to families of the DESMODOROIDEA

1. Body cylindrical without specialized somatic setae such as adhesion tubes or ambulatory setae . . . . .Desmodoridae (Fig. 17.8B)

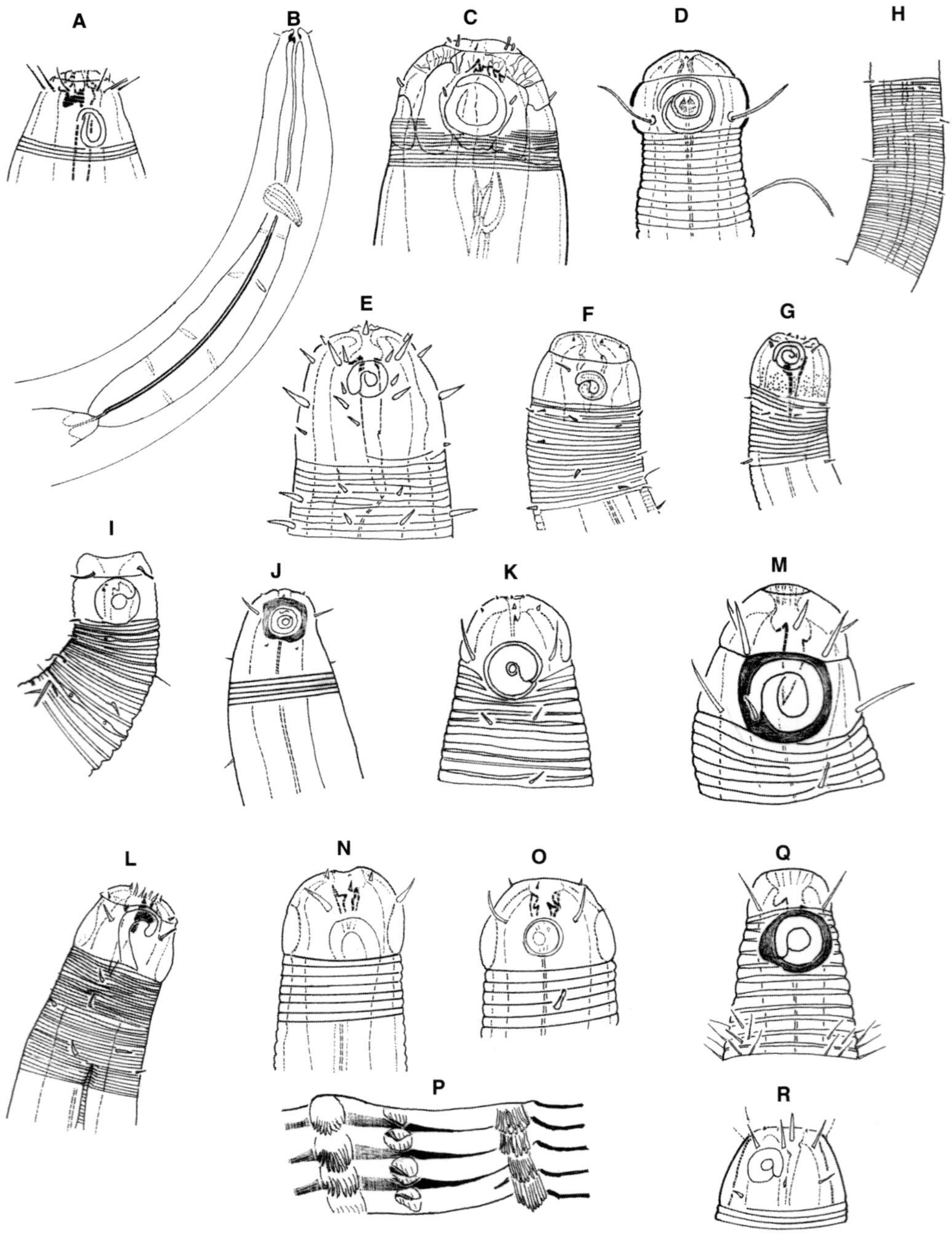

Fig. 17.10 Genera of the subfamily Desmodorinae. A and B, *Acanthopharynx*, head region and pharyngeal region; C: *Acanthopharyngoides*, head region (based on Riemann and Schrage, 1977); D: *Bolbonema*, head region; E: *Croconema*, head region; F: *Desmodora*, head region; G and H: *Desmodorella*, head region and detail of body cuticle at mid-body showing longitudinal differentiation (based on Verschelde *et al.*, 1998); I: *Echinodesmodora*, head region; J: *Metadesmodora*, head region; K: *Paradesmodora*, head region; L: *Psammonema*, anterior body region with beginning of lateral alae; M: *Pseudodesmodora*, head region; N–P: *Sibayinema*, head region of male and female and detail of body cuticle based on scanning electron microscopy (SEM) (based on Swart and Heyns, 1991); Q: *Stygodesmodora*, head region; R: *Zalonema*, head region.

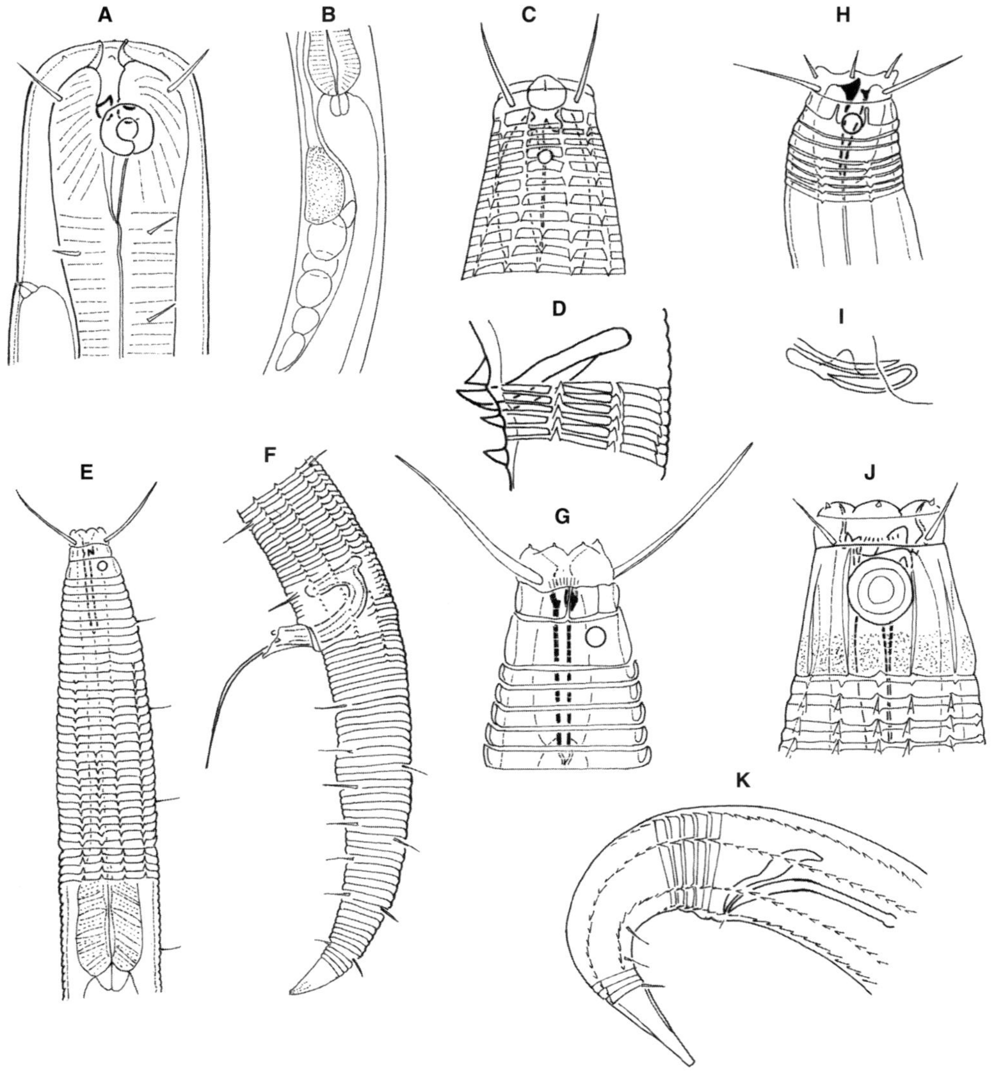

Fig. 17.11 Families of the superfamily Microlaimoidea. A and B: Family Aponchiidae, respectively head region of *Aponchium* and multicellular ventral gland of *Synonema* (based on Gerlach, 1962). C–K: Family Monoposthiidae, C and D: *Monoposthia*, head region and gubernaculum respectively (redrawn from Gerlach, 1963); E–G: *Monoposthioides*, anterior body region showing start of ridges, posterior body region, gubernaculum and head region, respectively (based on Hopper, 1963); H and I: *Nudora*, head region and copulatory apparatus respectively (redrawn from Inglis, 1963); J and K: *Rhinema*, head region and tail region with copulatory apparatus respectively (redrawn from Gerlach, 1963). Courtesy *Kieler Meeresforschung* (C and D); *Hydrobiologia* (H and I).

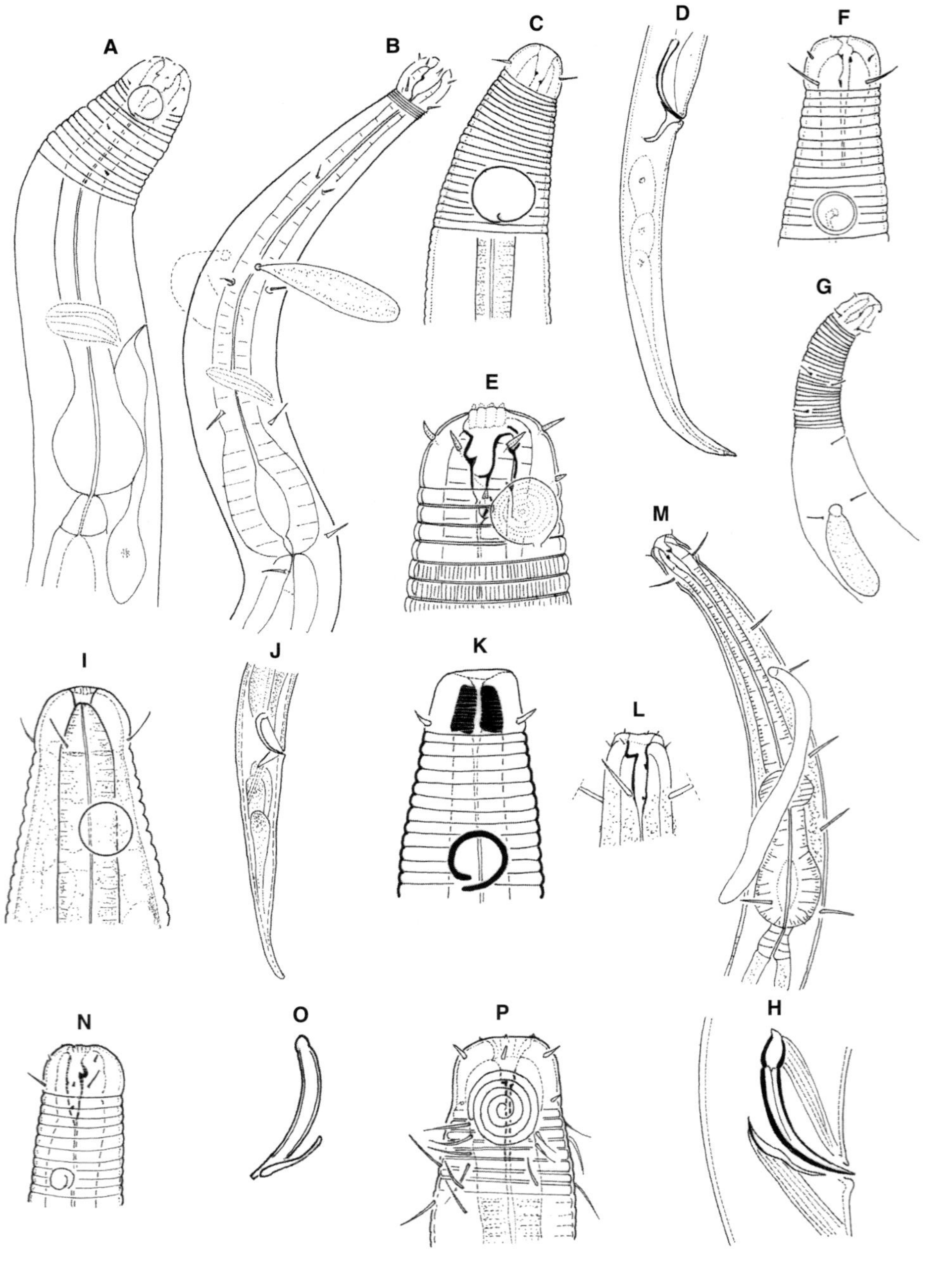

Fig. 17.12 Genera of the family Microlaimidae. A: *Bolbolaimus*, anterior body region with head region in surface view; B: *Bathynox*, anterior body region (redrawn from Bussau and Vopel, 1999); C and D: *Aponema*, head region and tail region with copulatory apparatus respectively; E: *Caligocanna*, head region, surface view with detail of cuticle ornamentation (redrawn from Bussau and Vopel, 1999); F–H: *Calomicrolaimus*, anterior body region with non-protruding and protruding *corpus gelatum* copulatory apparatus (based on Gourbault and Vincx, 1988); I and J: *Cinctonema*, anterior body region and tail region with copulatory apparatus; K: *Crassolaimus*, head region (redrawn from Wieser, 1951); L and M: *Ixonema*, detail of stoma and anterior body region with protruding *corpus gelatum* (redrawn from Muthumbi and Vincx, 1999); N and O: *Microlaimus*, head region and copulatory apparatus; P: *Spirobolbolaimus*, head region (based on Soetaert and Vincx, 1988). Courtesy *Annalen des Naturhistorischen Museums in Wien* (B and E); *Hydrobiologia* (M).

Body S- or epsilon-shaped with swollen pharyngeal region and mid-posterior body region with specialized somatic setae 2

2. Body S-shaped, provided with cephalic and posterior adhesion tubes (PAT). Female reproductive system anterior to region with PAT . . . . Draconematidae (Fig. 17.8A)

Body epsilon- or S-shaped, with ambulatory setae posterior to dorsal curvature; no adhesion tubes. Female reproductive system within region of ambulatory setae or posterior to it . . . . Epsilonematidae (Fig. 17.8C)

Family DESMODORIDAE Filipjev, 1922 (Fig. 17.9)

Diagnosis. Desmodoroidea. Amphideal fovea spiral usually of one or two turns, described as loop-shaped. Pharynx with muscular posterior bulb. Tail conical. Gubernaculum without dorsal apophysis.

Six subfamilies: Desmodorinae Filipjev, 1922; Spiriniinae Gerlach & Murphy, 1965; Pseudonchinae Gerlach & Riemann, 1973; Stilbonematinae Cobb, 1936; Molgolaiminae Jensen, 1978; and Prodesmodorinae Lorenzen, 1981.

Key to subfamilies of the Desmodoridae (Fig. 17.9)

1. Buccal cavity bilateral symmetrical, large tubular, subdivided with ventrosublateral teeth at junction of subdivision; pre-cloacal supplements when present at mid-ventral cuticular wing . . . . Pseudonchinae (Fig. 17.9F)

Buccal cavity different; supplements not at ventral cuticular wing 2

2. Habitus filliform, cuticle finely striated and mostly covered with Cyanophycea; numerous epidermal glands; marine, common in microxic sediment layers . . . . Stilbonematinae (Figs 17.9H–K)

Habitus variable, cuticle finely or more coarsely striated but not covered with Cyanobacteria 3

3. Head with cephalic capsule; amphideal fovea not surrounded by striation . . . . Desmodorinae (Figs 17.9A and B)

Head without cephalic capsule; amphid surrounded or not by striation 4

4. Amphideal fovea far anterior on body and surrounded by transverse striae of body cuticle; buccal cavity minute to medium-sized . . . . Spiriniinae (Fig. 17.9G)

Amphideal fovea more posteriorly located and surrounded by transverse striae; buccal cavity small to medium-sized 5

5. Secretory–excretory pore anterior to nerve ring; reproduction amphimictic; male common; marine . . . . Molgolaiminae (Figs 17.9D and E)

Secretory–excretory pore near nerve ring; reproduction parthenogenetic; male rare; limnic . . . . Prodesmodorinae (Fig. 17.9C)

Subfamily Desmodorinae Micoletzky, 1924
Diagnosis. Desmodoridae. Cuticle with transverse striae except in head region. Head region with thickened cuticle except in lip region and set off as a conspicuous cephalic capsule or helmet. Amphideal fovea in general not surrounded by striation of body cuticle; may be located on a cuticularized plate. Buccal cavity always with distinct teeth. Pharyngeal bulb round to elongated.

Type genus: *Desmodora* de Man, 1889 and 14 other genera (see classification).

Key to genera of the Desmodorinae (Fig. 17.10)

1. No cephalic helmet 2
   Cephalic helmet present 5
2. Amphideal fovea on a cuticularized plate 3
   Amphideal fovea not on a cuticularized plate 4
3. Striation of body cuticle starts posterior to the amphideal fovea . . . . *Metadesmodora* (Fig. 17.10J)
   Striation of body cuticle starts at level of amphideal fovea . . . . *Stygodesmodora* (Fig. 17.10Q)
4. Amphideal fovea completely surrounded by striation of body cuticle . . . . *Echinodesmodora* (Fig. 17.10I)
   Amphideal fovea partially surrounded by striation of body cuticle . . . . *Paradesmodora* (Fig. 17.10K)
5. Amphideal fovea on a cuticularized plate 6
   Amphideal fovea not on a cuticularized plate 7
6. Cephalic helmet with cuticularized plates; unispiral amphideal fovea; two or more ventrosublateral teeth . . . . *Acanthopharyngoides* (Fig. 17.10C)
   Cephalic helmet different; spiral amphideal fovea with 1.25 turns; a single small ventrosublateral tooth . . . . *Pseudodesmodora* (Fig. 17.10M)
7. Striated body cuticle with longitudinal ornamentation 8
   Striated body cuticle without longitudinal ornamentation 9
8. Body cuticle with longitudinal rows of ridges or spines; sub-cephalic setae present; amphideal fovea spiral with at least 1.5 turns . . . . *Desmoderella* (Figs 17.10G and H)
   Body cuticle with longitudinal rows of brush-like organs and lateral differentiation; sub-cephalic setae absent; amphideal fovea crescent-shaped in male, circular in female . . . . *Sibayinema* (Figs 17.10N–P)
9. Cephalic helmet longer than wide, with numerous sub-cephalic setae arranged in three or more rows . . . . *Croconema* (Figs 17.10E)
   Cephalic helmet different; sub-cephalic setae absent or when present less numerous and different arrangement 10
10. Body cuticle with differentiated lateral alae 11
    Lateral alae absent 12

11. Lateral alae thin, beginning at anterior third of pharynx; buccal cavity with denticles . . . . *Psammonema* (Fig. 17.10L)
    Lateral alae thicker, beginning posterior to pharynx; denticles absent . . . . *Pseudochromadora* (Figs 17.2D and F, and 17.9A and B)
12. Cephalic helmet globular; four cephalic setae at base of helmet. . . . . *Bolbonema* (Fig. 17.10D)
    Cephalic helmet with different shape; cephalic setae inserted more anteriorly 13
13. Pharyngeal bulb elongated, about half as long as the pharynx . . . . *Acanthopharynx* (Figs 17.10A and B)
    Pharyngeal bulb shorter 14
14. Amphideal fovea multispiral (at least two turns); sub-cephalic setae present . . . . *Zalonema* (Fig. 17.10R)
    Amphideal fovea cryptospiral or spiral with less turns; sub-cephalic setae present or absent . . . . *Desmodora* (Fig. 17.10F)

Genus *Acanthopharyngoides* Chitwood, 1936 (Fig. 17.10C)
Diagnosis. Desmodorinae. Body cuticle with fine transverse striae. Head helmet well developed, composed of cuticularized plates (joined). Sub-cephalic setae present. Amphideal fovea unispiral, located on a cuticularized plate. Buccal cavity cylindrical with a large dorsal tooth and well-developed ventrosublateral teeth (two or more). Pharyngeal bulb elongate, i.e. half as long as the pharynx and tripartite. Male with mid-ventral row of papilloid pre-cloacal supplements and a larger posteriormost supplement. Marine.

Type species: *A. scleratus* Chitwood, 1936.

Genus *Acanthopharynx* Marion, 1870 (Figs 17.10A and B)
Syn. *Xanthodora* Cobb, 1920.
Diagnosis. Desmodorinae. Similar to *Desmodora* except for elongate pharyngeal bulb (almost half the pharynx length). Marine.

Type species: *A. affinis* Marion, 1870.

Genus *Bolbonema* Cobb, 1920 (Fig. 17.10D)
Syn. *Desmodora* (*Bolbonema*) (Cobb, 1920) Gerlach, 1963
Diagnosis. Desmodorinae. Globular head capsule; cephalic setae inserted near base of helmet; long and shorter somatic setae (based on Verschelde *et al.*, 1998). Marine.

Type species: *B. brevicolle* Cobb, 1920.

Genus *Croconema* Cobb, 1920 (Fig. 17.10E)
Syn. *Aculconchus* Kreis, 1928; *Desmodora* (*Croconema*) (Cobb, 1920) Allgén, 1959.

Diagnosis. Desmodorinae. Thick body cuticle, coarse annuli; long head capsule; amphideal fovea located anteriorly on the head capsule; many sub-cephalic setae arranged in three (or more) circlets (Verschelde *et al*., 1998). Marine.

Type species: *C. cinctum* Cobb, 1920.

Genus *Desmodora* de Man, 1989 (Fig. 17.10F)
Syn. *Mastodex* Steiner, 1921; *Amphispira* Cobb, 1920.
Diagnosis. Desmodorinae. Body cuticle with transverse striae, no special ornamentation or appendages. Fine, short somatic setae arranged in six or eight longitudinal rows. Well-developed head capsule with differentiated lip region and a main part with thickened basal zone of cuticle and provided with amphideal fovea and subcephalic setae when present; main part of head capsule either smooth or ornamented with small vacuoles. Cephalic setae either on lip region or on main part of head capsule and inserted in front or at level of anterior edge of amphideal fovea. Amphideal fovea cryptospiral to multispiral (1–2 turns), seldom loop-shaped. Buccal cavity with large dorsal tooth and two (seldom one) ventrosublateral teeth. Cylindrical pharynx with end bulb. Tail short conical to long conico-cylindrical. Male monorchic; short arched spicules; pore-like pre-cloacal supplements may be present. Female didelphic–amphidelphic with reflexed ovaries (Verschelde *et al*., 1998). Marine.

Type species: *D. communis* (Buetschli, 1874) de Man, 1889.

Remarks: We follow Hope and Murphy (1972) who considered the genus *Amphispira* Cobb, 1920 as a junior synonym of *Desmodora*; Lorenzen (1981, 1994) listed the genus as valid but without comments. *Amphispira* is a monotypic, terrestrial genus based on a juvenile specimen (Cobb, 1920).

Genus *Desmodorella* Cobb, 1933 (Figs 17.10G and H)
Syn. *Desmodora* (*Desmodorella*) (Cobb, 1933) Wieser, 1954.
Diagnosis. Desmodorinae. Body cuticle with fine transverse striae, ornamented with regularly ordered longitudinal rows of ridges or spines. Somatic setae arranged in four, six or eight longitudinal rows. Well-developed conical to long conical cephalic capsule with lip region separated or not from the main part by a sutura; main part of capsule with thickened sclerotized cuticle, smooth or ornamented with vacuoles. Four cephalic setae inserted at level of anterior part of amphideal fovea; the latter large multispiral with at least 1.25 turns but mostly more than two. Up to 16 subcephalic setae may be present; when absent few additional setae are present. Buccal cavity with large dorsal tooth and two ventrosublateral teeth. Male either with slender slightly arched spicules with or without small capitulum, or very long filliform spicules without capitulum; gubernaculum may have crurae (Verschelde *et al*., 1998). Marine.

Type species: *D. cephalata* Cobb, 1933.

Genus *Echinodesmodora* Blome, 1982 (Fig. 17.10I)
Diagnosis. Desmodorinae. Cuticle with transverse striae surrounding the cryptospiral amphideal fovea. Body with stout spine-like somatic setae with, in between,

numerous irregularly arranged hair-like setae. No cephalic helmet. Buccal cavity with a small dorsal tooth and two minute ventrosublateral teeth. Pre-cloacal supplements tubiliform. Marine genus.

Type species: *E. axi* Blome, 1982.

Genus *Metadesmodora* Schuurmans Stekhoven, 1942 (Fig. 17.10J)
Diagnosis. Desmodorinae. Body with transverse striae beginning posterior to amphideal fovea. No cephalic helmet. Amphideal fovea located on a cuticularized plate. Marine.

Type species: *M. amphidiscata* Schuurmans Stekhoven, 1942.

Genus *Paradesmodora* Schuurmans Stekhoven, 1950 (Fig. 17.10K)
Diagnosis. Desmodorinae. Amphideal fovea partially surrounded by striated body cuticle and not situated on cuticularized plates. No cephalic helmet. Marine.

Type species: *P. cephalata* Schuurmans Stekhoven, 1950 but considered *species inquirenda* in Wieser and Hopper (1967).

Genus *Psammonema* Verschelde & Vincx, 1995 (Fig. 17.10L)
Diagnosis. Desmodorinae. Cuticle with numerous fine transverse striae without interdigitations; lateral alae present from anterior third of pharynx length to anal region. Different types of somatic setae in female. Cephalic capsule with well-offset labial region. No sub-cephalic setae but additional setae may be present. Amphideal fovea showing sexual dimorphism, open loop-shaped in male, cryptospiral to close loop-shaped in female. Buccal cavity with a crown of denticles, a strong dorsal tooth and two small ventrosublateral teeth. Pharynx with tripartite slightly prolonged end bulb. Brood protection may be present in female. Marine.

Type species: *P. ovisetosum* Verschelde & Vincx, 1995.

Genus *Pseudochromadora* Daday, 1899 (Figs 17.2D and E and 17.9 A and B)
Syn. *Desmodora* (*Pseudochromadora*) (Daday, 1899) Gerlach, 1967; *Micromicron* Cobb, 1920; *Bradylaimoides* Timm, 1961.
Diagnosis. Desmodorinae. Body cuticle with distinct lateral alae. Cephalic capsule wider than long with the amphideal fovea occupying most of its length, with large anterior labial capsule offset (but may be invaginated and therefore not always visible). Amphideal fovea unispiral (at least in female) located centrally on head capsule; no sub-cephalic setae; strong somatic setae. Buccal cavity without denticles. Terminal portion of tail smooth, i.e. without punctation (= small vacuoles) (adapted from Verschelde *et al.*, 1998). Mainly marine (six species); one freshwater species.

Type species: *P. quadripapillata* Daday, 1899.

Species from freshwater habitats:
P. quadripapillata Daday, 1899: a marine species (Costa Rica Pacific coast, New Guinea, Maryland (USA), Bay of Bengal, Nova Scotia); found in a small freshwater pool in a coral rock on Uipi Island, Marovo Lagoon, Solomon Islands.

Genus *Pseudodesmodora* Boucher, 1975 (Fig. 17.10M)
Diagnosis. Desmodorinae. Cephalic helmet present. Amphideal fovea spiral (1.25 turns), located on a large cuticularized plate extending to or into anteriormost striae of the body cuticle. Sub-cephalic setae present at base of helmet. Buccal cavity with a large dorsal tooth and a single small ventrosublateral tooth. Marine genus (male unknown).

Type species: *P. amphidiscata* Boucher, 1975.

Genus *Sibayinema* Swart & Heyns, 1991 (Figs 17.10N–P)
Diagnosis. Desmodorinae. Coarsely annulated body cuticle with longitudinal rows of brush-like ornamentation in adults and prominent lateral differentiation present. Cephalic capsule not subdivided, smooth; without sub-cephalic setae. Amphideal fovea crescent-shaped in male, circular in female. Pre-cloacal supplements represented by a row of mid-ventral setae. So far a monotypic genus from freshwater habitats.

Type species: *S. natalense* Swart & Heyns, 1991 from Lake Sibayi, Natal, South Africa.

Genus *Stygodesmodora* Blome, 1982 (Fig. 17.10Q)
Diagnosis. Desmodorinae. Cuticle with wide transverse striae surrounding the spiral amphideal fovea, which is located on a basal plate. No cephalic helmet. Buccal cavity with large dorsal tooth and a small ventrosublateral tooth. Pre-cloacal supplements setiform papillae. Marine.

Type species: *S. epixantha* Blome, 1982.

Genus *Zalonema* Cobb, 1920 (Fig. 17.10R)
Syn. *Desmodora* (*Zalonema*) (Cobb, 1920) Gerlach, 1963; *Heterodesmodora* Micoletzky, 1924.
Diagnosis. Body cuticle with transverse striae; rounded triangular cephalic capsule; sub-cephalic setae present; multispiral amphideal fovea (two turns or more). Pharynx slender cylindrical with pyriform terminal bulb (Verschelde *et al.*, 1998).

Type species: *Z. nudum* Cobb, 1920.

Subfamily Molgolaiminae Jensen, 1978 (Figs 17.9D and E)
Diagnosis. Desmodoridae. Body cuticle with weak transverse striae or smooth. Anterior two circlets of sensilla very small to papilloid, third circlet either setose or papilloid; no sub-cephalic setae. Amphideal fovea round, weakly sclerotized, at minimum one head diameter from anterior end. Buccal cavity weakly sclerotized with minute teeth. Secretory–excretory pore anterior to nerve ring. Copulatory apparatus slender, weakly sclerotized; gubernaculum without apophyses. Pre-cloacal supplements present and papilliform or absent. A marine taxon.

Type genus: *Molgolaimus* Ditlevsen, 1921.

Subfamily Prodesmodorinae Lorenzen, 1981 (Fig. 17.9C)
Diagnosis. Desmodoridae. Males are rare; reproduction parthenogenetic. Secretory–excretory pore posterior to nerve ring. Species purely limno-terrestrial. Only one genus known.

Type genus: *Prodesmodora* Micoletzky, 1923.

Genus *Prodesmodora* Micoletzky, 1923 (Fig. 17.9C)
Diagnosis. Prodesmodorinae. Amphideal fovea circular. Genus recorded from freshwater (three species) and terrestrial habitats (six species). Main characters used to differentiate are the position of the amphids and vulva from anterior end and morphometric features such as length of body, tail, cephalic setae. The genus is mainly terrestrial, with 30% of the species from freshwater habitats.

Type species: *P wolgensis* Micoletzky, 1923: junior synonym of *P. circulata* (Micoletzky, 1913) Micoletzky, 1923 by Micoletzky, 1925.

Species recorded from freshwater habitats:

P. circulata (Micoletzky, 1913) Micoletzky, 1923: freshwater, cosmopolitan, Russia, Ukraine, Uzbekistan, Far East, Sweden, Yugoslavia, Germany, Poland, River Danube, Bulgaria, Austria, Spain, Switzerland, Italy, Denmark, China, Japan, Michigan (USA), Colombia, Mexico.

P. loksai Andrássy, 1989: birch swamp, Hungary.

P. nurta Zullini, 1988: freshwater, Ethiopia: Lake Tana, Lake Ziway and River Abbay.

P. pantalicae Colomba & Vinciguerra, 1980: freshwater, Sicily.

Remarks: *P. lepturus* (De Cillis, 1917) Jensen, 1978: freshwater: Astronikrater Lake in Italy = *species inquirenda*. So far, only known by the original description without illustration and not precise enough to allow a reliable identification.

Subfamily Pseudonchinae Gerlach & Riemann, 1973 (Fig. 17.9F)
Diagnosis. Desmodoridae. Cuticle with fine transverse striae. Buccal cavity bilateral symmetric, large tubular, subdivided and with ventrosublateral teeth at level of junction. Sub-cephalic setae present. Amphideal fovea a single loop or round cryptospiral. Pre-cloacal supplements when present located at level of thin mid-ventral alae of the body cuticle. Marine.

Type genus: *Pseudonchus* Cobb, 1920.

Subfamily Spiriniinae Gerlach & Murphy, 1965 (Fig. 17.9G)
Diagnosis. Desmodoridae. Body cuticle finely striated. Head not demarcated as a cephalic capsule. Amphideal fovea a simple spiral, usually located far anterior on the body. Buccal cavity rather small, from minute to medium-sized, with a distinct or a minute dorsal tooth; two smaller ventrosublateral teeth may be present or absent. Marine.

Type genus: *Spirinia* Gerlach, 1963.

Subfamily Stilbonematinae Cobb, 1936 (Figs 17.9H–K)
Diagnosis. Desmodoridae. Habitus filliform; cuticle with faint transverse striae, mostly covered with Cyanophycea or bacteria. Numerous epidermal glands. Cephalic capsule absent or present but not subdivided. Sub-cephalic setae (4–8) present usually at anterior end of amphideal fovea. Pharynx with round end bulb.

Spicules with well-developed capitulum and velum. Marine, usually restricted to the microxic layers of the sediment.

Type genus: *Stilbonema* Cobb, 1920.

Family DRACONEMATIDAE Filipjev, 1918 (Fig. 17.8A)

Diagnosis. Desmodoroidea. Body S-shaped with more or less pronounced, enlarged pharyngeal and posterior body regions. Body cuticle with transverse striae except for cephalic helmet; usually not ornamented, rarely with spines or minute vacuoles. Cephalic adhesion tubes dorsally on helmet, posterior to helmet or in cephalic region. PAT arranged in four longitudinal rows located in posterior third of body, PAT with or without bell-shaped tip. Female reproductive system didelphic, located anterior to PAT region. Males monorchic. Marine.

Two subfamilies: Draconematinae Filipjev, 1918 and Prochaetosomatinae Allen & Noffsinger, 1978.

Family EPSILONEMATIDAE Steiner, 1927 (Fig. 17.8C)

Diagnosis. Desmodoroidea. Body epsilon- or S-shaped with more or less pronounced, enlarged pharyngeal and posterior body regions. Cuticle with coarse transverse striae (except for cephalic helmet); ornamented with vacuoles, longitudinal striae and/or spines or smooth. Presence of specialized somatic setae, the ambulatory setae, arranged in longitudinal rows from mid-ventral curvature to mid-posterior body region (rarely absent). Reproductive system posterior to dorsal curvature; Males monorchic, females didelphic–amphidelphic with ovaries usually reflexed. Marine.

Three subfamilies: Epsilonematinae Steiner, 1927, Glochinematinae Lorenzen, 1974 and Keratonematinae Gourbault & Decraemer, 1986.

Key to freshwater species of the Desmodoroidea

1. Head capsule present 2
   Head capsule absent 4
2. Cuticle with transverse striae ornamented with longitudinal rows of ridges, spines or brush-like structures 3
   Cuticle without such ornamentations .... *Pseudochromadora quadripapillata*
3. Cuticle with fine transverse striae with longitudinal ridges and spines; amphideal fovea usually large multispiral .... *Desmodorella aquaedulcis*
   Cuticle with strong transverse striae with longitudinal rows of brush-like ornamentation; amphidial fovea large crescent-shaped in male, small circular in female .... *Sibayinema natalense*
4. Amphideal fovea one head diameter or nearer to anterior end 5
   Amphideal fovea at greater distance from anterior end 6
5. Amphideal fovea 21–24% of corresponding body diameter (CBD) and less than one diameter form anterior end; two sub-ventral

teeth opposite dorsal tooth; body length (L) (female) = 470–960 µm .... *Prodesmodora circulata*
Amphideal fovea 25–29% of CBD, less than one diameter from anterior end
6. Body length (female) = 880–920 µm .... *Prodesmodora pantalicae*
Body length (female) = 390–500 µm .... *Prodesmodora nurta*

Superfamily MICROLAIMOIDEA Micoletzky, 1922 (Figs 17.11 and 17.12)

Diagnosis. Desmodorida. Cuticle with transverse striae ornamented with transverse rows of dots or longitudinal bars or smooth or strongly annulated with longitudinal ridges or V-shaped structures. Anterior sensorial organs arranged in three circlets. Amphideal fovea cryptospiral with circular to oval outline or non-spiral. Males monorchic or diorchic. Females monodelphic or didelphic, ovary (-ies) usually outstretched. Three families: Aponchiidae, Microlaimidae and Monoposthiidae; mainly marine.

Key to families of the Microlaimoidea (Figs 17.11 and 17.12)
1. Male monorchic; female monodelphic. Body cuticle smooth or with weak transverse striation .... Aponchiidae (Figs 17.11A and B)
Male diorchic; female mostly didelphic, rarely monodelphic. Body cuticle with transverse striae with or without ornamentation 2
2. Cuticle with fine or coarse transverse striae, smooth or ornamented with dots or bars. Female didelphic .... Microlaimidae (Fig. 17.12)
Cuticle with marked transverse striae and longitudinal ornamentation, mostly V-shaped and arranged on longitudinal lines. Female didelphic, exceptionally monodelphic .... Monoposthiidae (Figs 17.11C–K)

Family MICROLAIMIDAE Micoletzky, 1922 (Fig. 17.12)

Diagnosis. Microlaimoidea. Cuticle usually with transverse striae, ornamented with transverse rows of dots or longitudinal bars, rarely smooth. Anterior sensorial organs arranged in three circlets, the six outer labial and four cephalic sensilla usually setose with the four cephalic setae usually longer than the outer labial setae. Amphideal fovea cryptospiral with circular or oval outline. Buccal cavity sclerotized, with at least two compartments posteriorly and armed with small teeth; cheilorhabdia present. Pharynx sometimes enlarged at level of buccal cavity, posteriorly swollen to an oval or pyriform bulb. Secretory–excretory pore usually posterior to nerve ring. Male diorchic or rarely monorchic. Female didelphic–amphidelphic, branches outstretched, exceptionally reflexed. Caudal glands exceptionally with separate outlet (Jensen, 1978). Mainly marine.

Type genus: *Microlaimus* de Man, 1880 and eight other genera (see classification).

Only the genus *Microlaimus* has been recorded from freshwater habitats. Several of the genera need revision; the differentiation between the genera is often unclear because of overlapping of characters. The genus *Crassolaimus* has not been included in the key to genera of the family.

Key to genera of the MICROLAIMIDAE (Fig. 17.12)

1. Six outer labial sensilla longer than four cephalic setae 2
   Six outer labial sensilla shorter than four cephalic setae 3
2. Striated body cuticle ornamented with longitudinal bars. Amphideal fovea unispiral . . . . *Caligocanna* (Fig. 17.12E)
   Striated body cuticle not ornamented with longitudinal bars. Amphideal fovea multispiral . . . . *Spirobolbolaimus* (Fig. 17.12P)
3. Body cuticle smooth, covered with sediment. Amphids with protruding elongated *corpus gelatum*. Caudal glands with separate outlet . . . . *Ixonema* (Figs 17.12L and M)
   Body cuticle with transverse striae, not covered. Amphids mostly without protruding corpus gelatum (except for *Bathynox*, *Calomicrolaimus*). Caudal glands with common outlet 4
4. Amphideal fovea small round with club-shaped protruding *corpus gelatum*. Somatic setae inserted on peduncles . . . . *Bathynox* (Fig. 17.12B)
   Amphideal fovea variable in shape, without club-shaped protruding *corpus gelatum*. Somatic setae not inserted on peduncle 5
5. Anterior part of pharynx with more or less pronounced muscular peribuccal bulb . . . . *Bolbolaimus* (Fig. 17.12A)
   Anterior part of pharynx without muscular peribuccal bulb 6
6. Wall of anterior part of pharynx thickened and sclerotized . . . . *Crassonema* (Fig. 17.12K)
   Wall of anterior part of pharynx not thickened nor sclerotized 7
7. Cervical region narrow, elongate. Amphideal fovea small, round, located posteriorly (about mid-pharynx level); elongate protruding *corpus gelatum;* long cephalic setae . . . . *Calomicrolaimus* (Figs 17.12F–H)
   Cervical region different. Amphideal fovea with more anterior position 8

8. Male monorchic or diorchic; stoma small, funnel-shaped unarmed or with three minute teeth; gubernaculum with dorso-caudal apophyses . . . . *Aponema* (Figs 17.12C and D)
   Male diorchic; small, short, unarmed or larger funnel-shaped stoma with three teeth; gubernaculum with or without dorso-caudal apophyses 9
9. Stoma short, small, unarmed; gubernaculum with dorso-caudally orientated apophyses; amphideal fovea circular . . . . *Cinctonema* (Figs 17.12I and J)
   Stoma larger, with three teeth; gubernaculum without dorso-caudally orientated apophyses; amphideal fovea unispiral with circular outline . . . . *Microlaimus* (Figs 17.12N and O)

Genus *Aponema* Jensen, 1978 (Figs 17.12C and D)
Diagnosis. Microlaimidae. Body cuticle with fine transverse striae, without ornamentation. Amphideal fovea clearly sclerotized; may show sexual dimorphism in size and position. Buccal cavity small, funnel-shaped, with a minute dorsal and two ventrosublateral teeth, or unarmed. Pharynx with thick cuticular lining of posterior bulb. Male monorchic or diorchic, with strongly sclerotized copulatory apparatus, gubernaculum with dorso-caudally directed apophyses. Pre-cloacal supplements absent or a single supplement may be present. Female didelphic–amphidelphic with outstretched ovaries. Marine.

Type species: *A. torosum* (Lorenzen, 1973) Jensen, 1978.

Genus *Bathynox* Bussau & Vopel, 1999 (Fig. 17.12B)
Diagnosis. Microlaimidae. Cuticle with transverse striae without ornamentation; somatic setae inserted on peduncles. Sensilla of anterior second circlet with six short setae and shorter than the four cephalic setae of the third circlet. Amphideal fovea far behind anterior end (near mid-pharynx level); aperture small, with club-shaped protruding *corpus gelatum* with constant length and solid wall. Buccal cavity with one dorsal and two ventrosublateral teeth. Male monorchic; pre-cloacal supplements absent. Female with two outstretched ovaries. Marine.

Type species: *B. clavata* Bussau & Vogel, 1999.

Genus *Bolbolaimus* Cobb, 1920 (Fig. 17.12A)
Syn. *Pseudomicrolaimus* Sergeeva, 1976
Diagnosis. Microlaimidae. Cuticle with strong transverse striae. Amphideal fovea simple spiral or cryptospiral, located posterior to the insertion of the cephalic setae or partially in between; completely or partially surrounded by transverse striae. Buccal cavity well sclerotized and armed with a distinct dorsal tooth either opposite one or two pairs of small ventrosublateral teeth or numerous denticles on the ventrosublateral walls. Anterior part of pharynx with peribuccal bulb (swollen or not) and pharynx posteriorly with long oval end bulb. Male diorchic; copulatory apparatus well

sclerotized; pre-cloacal supplements tubular. Female didelphic–amphidelphic with reflexed ovaries. Marine.
Remarks: According to Platt and Warwick (1988), the main characteristics, i.e. presence of a muscular peribuccal bulb, strongly cuticularized buccal cavity and copulatory apparatus, may show different degrees of development, which makes the differentiation of *Bolbolaimus* from *Microlaimus* and *Calomicrolaimus* difficult; these three genera need revision.

Type species: *B. pellucidus* Cobb, 1920.

Genus *Caligocanna* Bussau & Vopel, 1999 (Fig. 17.12E)
Diagnosis. Microlaimidae. Cuticle with transverse striae, ornamented with longitudinal bars except far anteriorly. Six setiform sensilla of the second circlet longer than the four cephalic setae of the third circlet. Amphideal fovea unispiral, ventrally wound, located at level of striated cuticle without bars. Buccal cavity sclerotized and armed with two large teeth in anterior and two small teeth in posterior compartment. Male diorchic; pre-cloacal supplements absent. Female didelphic with outstretched ovaries. Marine, with assumed sediment tube-building ability.

Type species: *C. mirabilis* Bussau & Vopel, 1999.

Genus *Calomicrolaimus* Lorenzen, 1976 (Figs 17.12F–H)
Diagnosis. Microlaimidae. Cervical region more or less elongated. Body cuticle with fine transverse striae or coarse annules partially or over entire body. Cephalic setae rather long and slender. Amphideal fovea at some distance from anterior end; the *corpus gelatum* may be protruding. Somatic setae usually long, slender or reduced to papillae; specialized thorn- or spine-like setae and porids may be present. Male diorchic. Pre-cloacal supplements papilloid, additionally, a row of ventral thickenings of the body annules may be observed. Female didelphic–amphidelphic with outstretched ovaries but reflexed ovaries may occur. Tail slender, gradually tapering.
Remarks: The differentiation between *Microlaimus* and *Calomicrolaimus* is not always clear (Platt and Warwick, 1988). Marine.

Type species: *C. rugatus* Lorenzen, 1976.

Genus *Cinctonema* Cobb, 1920 (Figs 17.12I and J)
Diagnosis. Microlaimidae. Similar to *Microlaimus* but different from it in structure of gubernaculum with backwards-pointed apophyses (Timm, 1978a). Marine, rare.
Remarks: Unclear systematic position according to Jensen (1978).

Type species: *C. tenue* Cobb, 1920.

Genus *Crassolaimus* Kreis, 1929 (Fig. 17.12K)
Diagnosis. Microlaimidae. Cuticle with fine transverse striae. Of the anterior sensilla only the four setiform cephalic setae have been observed. Buccal cavity cylindrical, largely surrounded by the anterior part of the pharynx with thickened sclerotized wall; a small single ventral tooth present opposite a small indentation of the dorsal wall. Amphideal fovea with circular outline, unispiral. Male monorchic; pre-cloacal

supplements small papilliform; gubernaculum without dorso-caudal apophyses. Marine.

Remarks: According to Jensen (1978), a poorly understood genus with two species, of which only males are known.

Type species: *C. conicaudatus* Kreis, 1929.

Genus *Ixonema* Lorenzen, 1971 (Figs 17.12L and M)

Diagnosis. Microlaimidae. Body cuticle smooth under light microscopy, covered or not with sediment. Amphideal fovea circular, with protruding rod-like *corpus gelatum* and located relatively far from the anterior end (at least three times the distance of the cephalic setae from anterior end). Stoma small with three teeth, one larger dorsal tooth and two ventrosublateral teeth posterior to one another. Outer labial sensilla either papilliform as inner labial sensilla or short setiform. Outlet of caudal gland cells separate. Marine.

Type species: *I. sordidum* Lorenzen, 1971.

Genus *Microlaimus* de Man, 1880 (Figs 17.12N and O)

Syn. *Microlaimoides* Hoeppli, 1926; *Paracothonolaimus* Schulz, 1932.

Diagnosis. Microlaimidae. Cuticle with transverse striae, sometimes ornamented with dots or longitudinal bars; head often marked off. Somatic setae small, porids sometimes present. Amphideal fovea unispiral, circular to transverse or longitudinal oval in outline, mostly located far anteriorly, close to cephalic setae. Stoma small to medium-sized with mainly three minute to well-developed teeth, especially the dorsal one. Male diorchic; pre-cloacal supplements papilliform or absent. Female didelphic–amphidelphic with outstretched ovaries. Mainly marine (50 species), a few occasionally found in brackish habitats and two brackish species.

Type species: *M. globiceps* de Man, 1880.

Species recorded from freshwater habitats:

M. globiceps de Man, 1880: a marine species (North Sea, Baltic, Norway, Kattegat, Öresund (Scandinavia), Kieler Bay (Germany), Chile) also occurring in brackish waters (The Netherlands, southern Finnish archipelago, Poland, Germany) and freshwater (The Netherlands, Hungary, Uzbekistan, China, Chile).

M. setosus Hoeppli, 1926: freshwater, in thermal waters (40°C) of Yellowstone Park, Wyoming (USA).

Genus *Spirobolbolaimus* Soetaert & Vincx, 1988 (Fig. 17.12P)

Diagnosis. Microlaimidae. Cuticle with transverse striae, partially surrounding the amphideal fovea. Six external labial sensilla longer than four cephalic setae. Amphideal fovea distinctly sclerotized, multispiral, ventrally wound. Post-amphideal setae in six longitudinal rows. Pharynx with anterior bulb and oval posterior bulb. Male diorchic; no pre-cloacal supplements. Female didelphic–amphidelphic with outstretched ovaries. Marine.

Type species: *S. bathyalis* Soetaert & Vincx, 1988.

Family APONCHIIDAE Gerlach, 1963 (Figs 17.11A and B)

Diagnosis. Microlaimoidea. Body cuticle with slight transverse striae or smooth. Anterior sensilla arranged in three circlets; the posterior four cephalic setae are the most prominent sensilla. Amphideal fovea O-shaped loop or a spiral with a single turn. Buccal cavity with three nearly equal teeth. Ventral gland multicellular (4–7 cells). Male monorchic. Female monodelphic, only anterior ovary present (based on Lorenzen, 1994). Marine.

Type genus: *Aponchium* Cobb, 1920.

Family MONOPOSTHIIDAE Filipjev, 1934 (Figs 17.11C–K)

Diagnosis. Microlaimoidea. Body cuticle with strong transverse striae and longitudinal ornamentation. Anterior sensorial organs arranged in three circlets, the labial sensilla are papillae or setiform papillae; the posterior four cephalic setae are longer than those of the second circlet. Amphideal fovea circular, non-spiral. Male diorchic, testes opposed; spicules present or absent, gubernaculum present. Precloacal supplements present or absent. Females with one anterior ovary or two opposed ovaries; ovaries always reflexed. Anterior gonad left of intestine, posterior one to the right of the intestine (based on Lorenzen, 1994). Largely marine, common in coarse silt-free sediments.

Type genus: *Monoposthia* de Man, 1889 and three other genera (see classification).

Key to genera of the MONOPOSTHIIDAE (Fig. 17.11)

1. Females didelphic .... *Rhinema* (Figs 17.11J and K)
   Female with single anterior ovary 2
2. Spicules and gubernaculum present .... *Nudora* (Figs 17.11H and I)
   Spicules absent, gubernaculum present 3
3. Cuticle ornamentation of longitudinal V-shaped ridges absent in anterior body region .... *Monoposthioides* (Figs 17.11E–G)
   Cuticle ornamentation of longitudinal V-shaped ridges all over body .... *Monoposthia* (Figs 17.11C and D)

Genus *Monoposthia* de Man, 1889 (Figs 17.11C and D)

Diagnosis. Monoposthiidae. Body cuticle strongly annulated, ornamented with longitudinal ridges (alae) or V-shaped structures. Buccal cavity with strongly developed dorsal tooth opposed to two small ventrosublateral teeth; additional denticles may be present. Spicules absent, single gubernaculum heavily sclerotized; precloacal supplements papilliform or when absent a thickened pre-cloacal area of the ventral cuticle may be present. Female monodelphic, single anterior ovary reflexed. Marine.

Type species: *M. costata* (Bastian, 1865) de Man, 1889.

Genus *Monoposthioides* Hopper, 1963 (Figs 17.11E–G)

Diagnosis. Monoposthiidae. Similar to *Monoposthia* but with absence of longitudinal V-shaped ridges of body cuticle in the anterior part of the body; males with gubernaculum of different shapes , i.e. proximally with large ventral process. Marine.

Type species: *M. anonoposthia* Hopper, 1963.

Genus *Nudora* Cobb, 1920 (Figs 17.11H and I)
Diagnosis. Monoposthiidae. Strongly striated cuticle with longitudinal ornamentation usually in the form of several files of V-shaped markings. Buccal cavity with strongly developed dorsal tooth opposed to small ventrosublateral teeth. Tail conical. Single anterior reflexed ovary. Spicules present in addition to large unpaired gubernaculum; pre-cloacal papillae present. Genus largely similar to *Monoposthia* but with developed spicules. Mainly marine (13 species); one freshwater species.

Type species: *N. lineata* Cobb, 1920.

Species recorded from freshwater habitats:
N. omercooperi Inglis, 1965: from a freshwater spring below high tide mark, Claytons Rocks, South Africa.

Genus *Rhinema* Cobb, 1920 (Figs 17.11J and K)
Diagnosis. Monoposthiidae. Buccal cavity with well-developed dorsal tooth and small ventrosublateral teeth. Male copulatory apparatus with spicules and gubernaculum. Female didelphic, ovaries reflexed. Marine.

Type species: *R. retrorsum* Cobb, 1920.

Key to freshwater species of the Microlaimoidea

1. Cuticle marked transverse striae provided with longitudinal ornamentation; amphideal fovea circular; female reproductive system monodelphic . . . . *Nudora omercooperi*
   Cuticle with fine transverse striae without longitudinal ornamentation; amphideal fovea unispiral with almost circular outline; female reproductive system didelphic–amphidelphic 2
2. Amphideal fovea large, about half CBD; cephalic sensilla long (three-fourths of head width); distance vulva–anus = four times tail length; female L = 590–600 μm . . . . *Microlaimus globiceps*
   Amphideal fovea one-third of CBD; cephalic sensilla shorter; female L = 340 μm; distance vulva–anus = twice tail length . . . . *Microlaimus setosus*

Order DESMOSCOLECIDA

Introduction

In the new classification proposed by De Ley and Blaxter (2002), the Desmoscolecida are again recognized as a separate order within the Chromadoria (Chromadorea) i.e. different from the classification by Lorenzen (1981, 1994), who placed them as a suborder within the Chromadorida. Since no DNA sequence data of known species have as yet been obtained for desmoscolecids apart from an unidentified *Tricoma* species, the recently proposed classification is still based upon morphological features and phylogenetic reflections.

De Ley and Blaxter (2002) recognized within the order a single superfamily Desmoscolecoidea Shipley, 1896, composed of three families: the Desmoscolecidae Shipley, 1896, the Meyliidae De Coninck, 1965 and the Cyartonematidae Tchesunov, 1990.

Desmoscolecida are essentially marine nematodes, 255 out of 280 species. Records from non-marine habitats, e.g. salt marshes, brackish soils, freshwater and truly terrestrial habitats, are rare and, currently, mainly regard members of the subfamily Desmoscolecinae (*sensu* Decraemer, 1985) of the family Desmoscolecidae. The family Desmoscolecidae is a small group of free-living nematodes common in marine sublittoral or deep-sea habitats. They are unique because of their rather peculiar habitus of an 'annulated' body cuticle, i.e. a body cuticle with transverse striae covered with desmen,[2] which masks the internal morphology in the majority of individuals. Desmoscolecids show a widespread distribution and usually manifest high species diversity but low representation. They represent a higher percentage of the nematofauna in deep-sea habitats. Data on intraspecific variability are rare and may complicate unambiguous species identification.

Classification

The classification within the Desmoscolecida (adapted from De Ley and Blaxter, 2002; Decraemer, 1985; Malakhov *et al.*, 1982; Tchesunov, 1994).
Note: Asterisk indicates taxa reported from freshwater bodies.

Order DESMOSCOLECIDA Filipjev, 1929
- Superfamily Desmoscolecoidea Shipley, 1896
- Family Desmoscolecidae Shipley, 1896
  - Subfamily Desmoscolecinae Shipley, 1896
    - Tribe Desmoscolecini Shipley, 1896
      - Genus *Desmoscolex* Claparède, 1863*
        - Subgenera *Desmoscolex* Claparède, 1863*
          - *Desmolorenzenia* Freudenhammer, 1975*
          - *Pareudesmoscolex* Weischer, 1962
          - *Protricomoides* Timm, 1970
      - Genus *Prototricoma* Timm, 1970
          - *Spinodesmoscolex* Decraemer, 1983
    - Tribe Greeffiellini Decraemer, 1985
      - Genus *Calligyrus* Lorenzen, 1969
        - *Greeffiella* Cobb, 1922
        - *Greeffiellopsis* Schrage & Gerlach, 1975

[2]Desmos/desmen (plural) = rings of secretion and foreign particles, e.g. sand grains, shell debris (also referred to as main rings, concretion rings or *Hauptringe* in German); beneath the desmen the transverse striae may possess transverse row(s) of spines or pores. Main rings are separated by interzones (*Nebenringe* in German) of uncovered, transversely striated body cuticle, with or without ornamentation of spines or pores.

Hapalomus Lorenzen, 1969
Progreeffiella Timm, 1970
Subfamily Tricominae Lorenzen, 1969
Genus *Antarcticonema* Timm, 1969
Desmotimmia Freudenhammer, 1975
Desmotricoma Decraemer, 1984
Haptotricoma Lorenzen, 1977
Paratricoma Gerlach, 1964
Protricoma Timm, 1970
Quadricomoides Decraemer, 1976
Tricoma Cobb, 1894*
Subgenus *Quadricoma* Filipjev, 1922
Genus *Usarpnema* Timm, 1978.
Family MEYLIIDAE De Coninck, 1965
Genus *Boucherius* Decraemer & Jensen, 1982
Meylia Gerlach, 1956
Gerlachius Andrássy, 1974
Noffsingeria Decraemer & Jensen, 1982
Family Cyartonematidae Tchesunov, 1990
Genus *Cyartonema* Cobb, 1920
Paraterschellingia Kreis in Schuurmans Stekhoven, 1935

According to Tchesunov (1994), meyliids are profoundly different from 'true' desmoscolecids. In the past, the taxonomic position of the family Meyliidae has been subject to discussions and changes. The family Meyliidae was erected within the order Desmoscolecida by De Coninck (1965) for three rare species of the genus *Meylia* described by Gerlach (1956). Vitiello and De Coninck (1968) classified the genus *Peresiana* (junior synonym of *Manunema*) and the subfamily Peresianinae within the Meyliidae; they considered *Peresiana* as a link, via *Meylia* Gerlach, 1956, between the most primitive Desmoscolecida and the Araeolaimida (Decraemer, 1977). Because of the close relationship between the Meyliinae and the Peresianinae, several authors (Timm, 1970; Freudenhammer, 1975; Decraemer, 1977) considered the Meyliidae to belong to the order Monhysterida (including the Leptolaimidae with *Manunema*). Upon the discovery of seven new species by Decraemer and Jensen (1981, 1982), the Meyliidae were returned to the Desmoscolecida, a classification currently accepted.

Tchesunov (1989) studied the genus *Cyartonema* and in 1990 created the family Cyartonematidae. According to Tchesunov (1994) most of the cyartonematid characters are also common to meyliids. He considered the families Meyliidae and Cyartonematidae as sister taxa belonging to the superfamily Meyloidea, classified however, within the Leptolaimina; no synapomorphies were found for this superfamily. With the classification of the Cyartonematidae within the Desmoscolecida, De Ley and Blaxter (2002) only partially followed Tchesunov (1994), agreeing with the close relationship between Meyliidae and Cyartonematidae (Tchesunov, 1989, 1994).

The classification within the order Desmoscolecida is still open to further discussion. Phylogenetic analyses based solely on morphological features cannot solve the question of the relationships between the three families and the hypothesis that the Meyliidae are more closely related to the Cyartonematidae than to the Desmoscolecidae, and thus belong to a separate superfamily. This matter needs further investigation.

Order DESMOSCOLECIDA Filipjev, 1929 (Figs 17.13–17.15)

Diagnosis. Chromadoria. Body with transverse striae; cuticle smooth or ornamented with scales, warts, spines or desmen. Head with four setiform cephalic setae, mostly inserted on peduncles; amphideal fovea vesiculate, rounded or (elongate) oval in shape, rarely with spiral *corpus gelatum* (Meyliidae). Pharynx mostly cylindrical, with or without protruding pharyngeal glands, or pharynx with anterior part muscular, mid-part slender and widened glandular posterior part (Cyartonematidae). Intestinal cells anteriorly, mostly fine granular, posteriorly provided with large globular inclusions; rectum and anus present or absent. Pigment spots (ocelli) may be present, mostly just posterior to the pharyngo-intestinal junction. Female reproductive system didelphic–amphidelphic; branches outstretched or reflexed. Male repro-

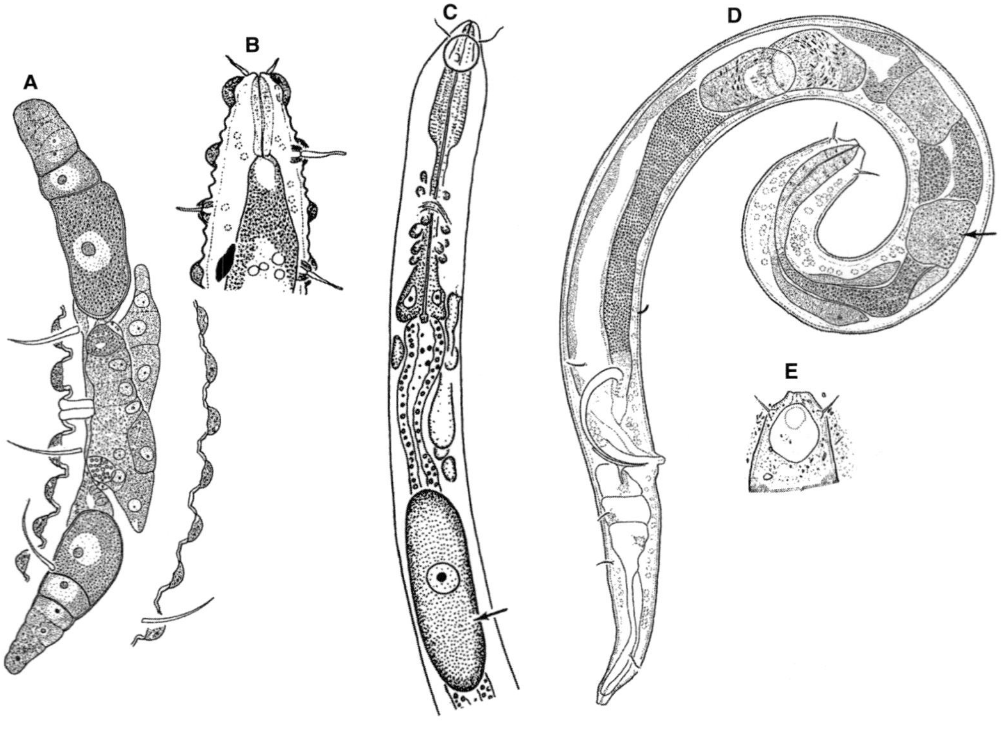

Fig. 17.13 Families of the Desmoscolecida. A and B: Family Desmoscolecidae, A: *Tricoma*, female reproductive system (from Decraemer, 1978); B: *Desmoscolex*, anterior body region (from Decraemer, 1984). C: Family Cyartonematidae, *Cyartonema*, anterior body region (from Tchesunov, 1994). D and E: Family Meyliidae, D: *Gerlachius*, total view of male; E: head region in surface view (from Decraemer and Jensen, 1981). Arrows indicate large pseudocoelomocytes cell along anterior intestine. Courtesy of *Australian Journal of Zoology* (A), *Hydrobiologia* (B), *Nematologica* (C), *Bulletin of the Royal Belgian Institute of Natural Sciences*, Brussels (D and E).

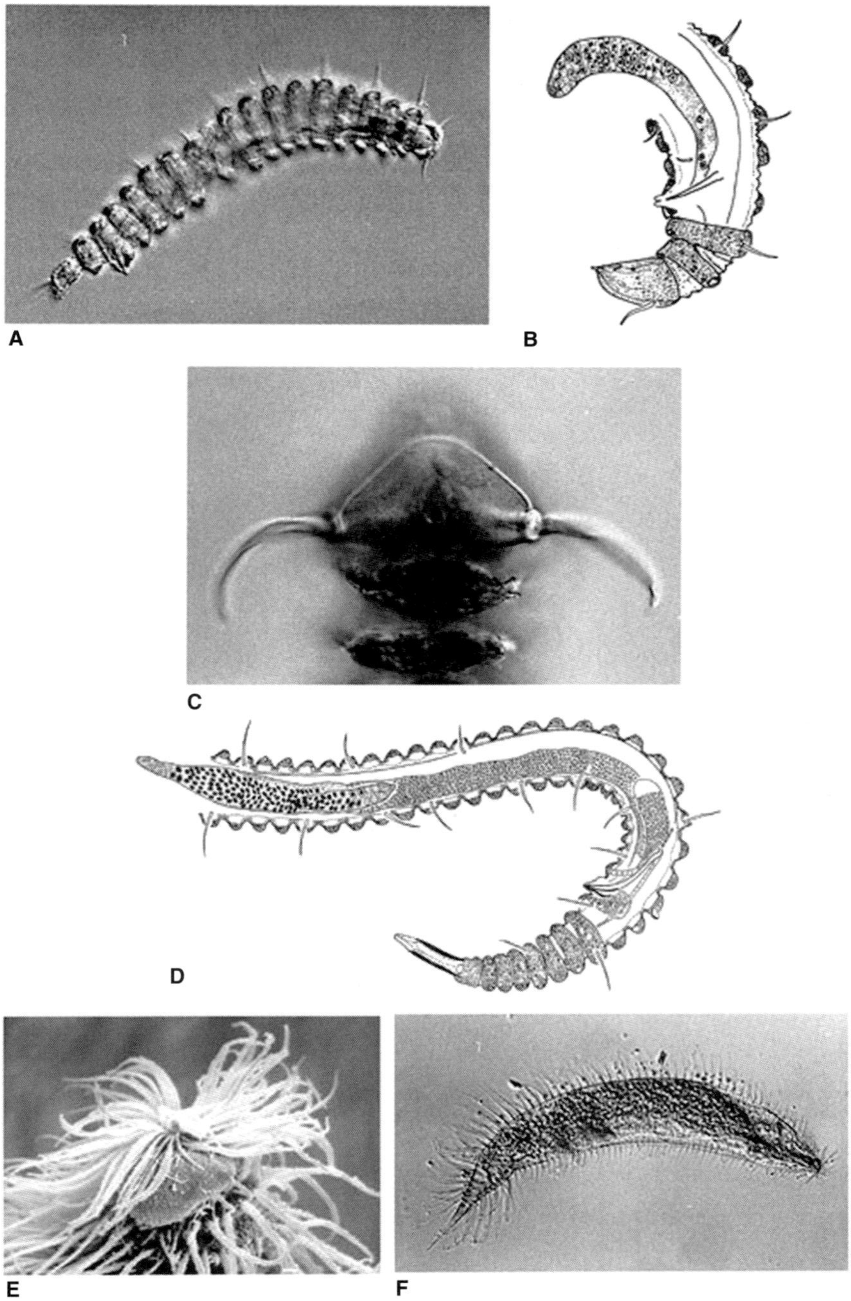

Fig. 17.14 Subfamilies of the Desmoscolecidae. A and B: Desmoscolecinae, Desmoscolecini, A: *Desmoscolex* (*Desmoscolex*) *paragranulatus*, total view of male (from Decraemer and Tchesunov, 1996); B: *D.* (*D.*) *parakoloensis*, male reproductive system (from Decraemer, 1984); C and D: Tricominae, C: *Tricoma septentrionalis*, surface view of head with vesicular amphid; D: *Tricoma riemanni* (from Decraemer, 1978). E and F: *Greeffiella* sp., scanning electron microscopy (SEM) of head region; *Greeffiella antarctica*, total view of male paratype. Courtesy of *Hydrobiologia* (B), *Australian Journal of Zoology* (C and D).

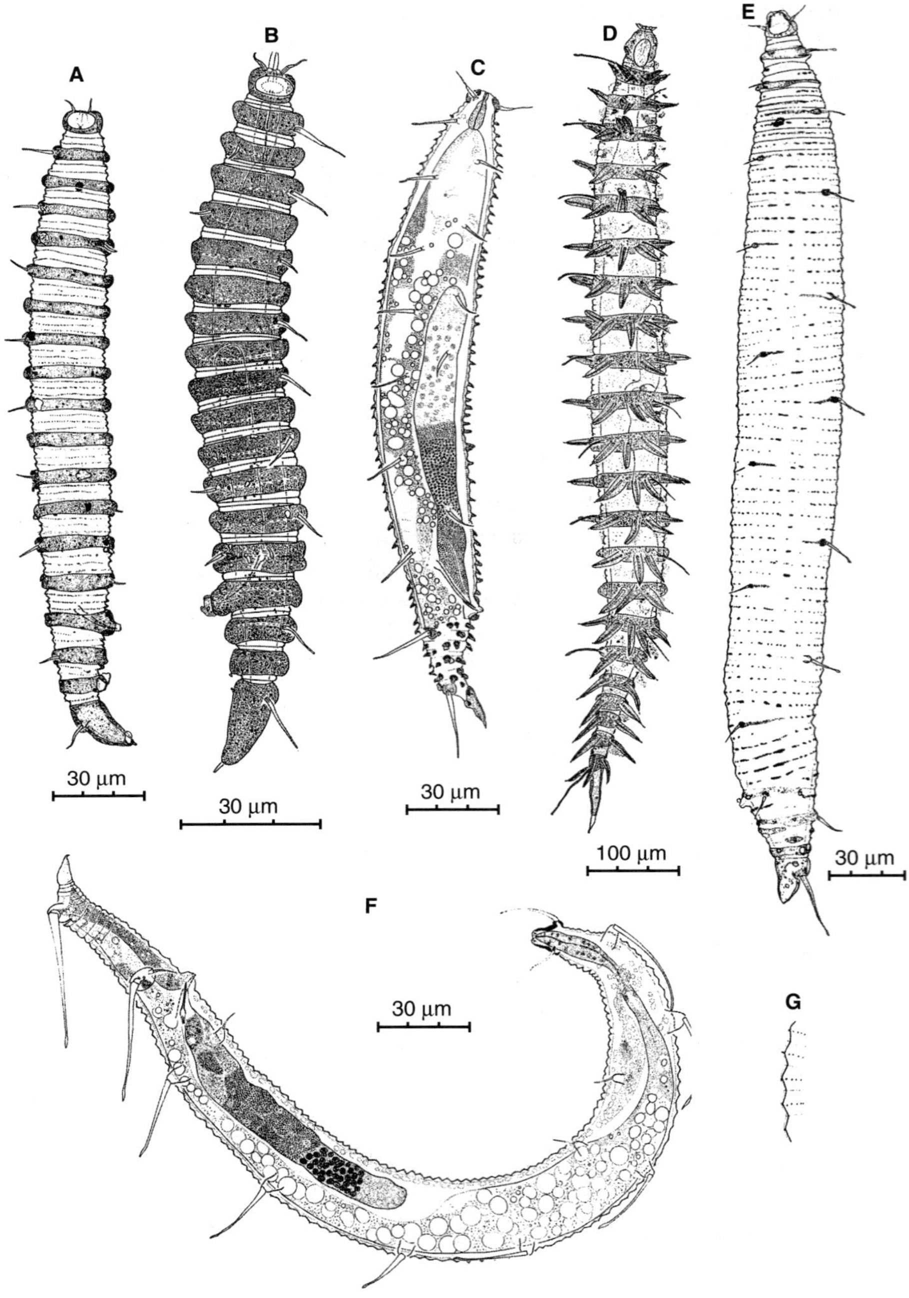

Fig. 17.15 Desmoscolecida, Desmoscolecinae. Genera of the Desmoscolecini, total views: A: *Desmoscolex* (*Desmoscolex*) *parakoloensis* (from Decraemer, 1984); B: *D.* (*D.*) *montana* (from Decraemer and Sturhan, 1982a); C: *Desmoscolex* (*Pareudesmoscolex*) *laciniosus* (from Decraemer and Sturhan, 1982b); D: *Spinodesmoscolex coronatus* (from Decraemer, 1983); E: *D.* (*P.*) *squamosus* (from Decraemer, 1985); F and G: *Prototricoma inaequalis*, total view of male (F), detail of body cuticle (G) (from Decraemer, 1984). Courtesy of *Hydrobiologia* (A and E), *Nematologica* (B and C), *Bulletin du Muséum National d'Histoire Naturelle*, Paris (D and F).

ductive system with one testis or two testes. Spicules paired, equal, usually cephalate. Gubernaculum usually present, with or without dorsal apophyses. Three caudal glands in tail region; terminal spinneret present. Most species marine, but some species inhabit freshwater, brackish water or soil. One superfamily Desmoscolecoidea Shipley, 1896.

The monophyly of the Desmoscolecida is based on the following two synapomorphies: (i) four cephalic setae inserted on peduncles; and (ii) vesicular amphideal fovea situated on the head to shortly behind it (Lorenzen, 1981, 1994; Decraemer, 1985).

Superfamily DESMOSCOLECOIDEA Shipley, 1896

Diagnosis. Similar as for order Desmoscolecida. Three families: Desmoscolecidae, Meyliidae and Cyartonematidae.

Key to families of the Desmoscolecoidea (Fig. 17.14)

1. Female reproductive system with two outstretched ovaries; male monorchic or diorchic; body short stout to long, more or less slender, with or without desmen; no large pseudocoelomocytes present along pre-gonadal midgut . . . . .Desmoscolecidae (Figs 17.13A and B, 17.14 and 17.15)
   Female reproductive system with two reflexed ovaries; male diorchic; body long slender, without desmen; two large pseudocoelomocytes present along pre-gonadal midgut 2
2. Cuticle smooth; renette cell present; pharynx with muscular anterior part, slender isthmus and posterior part with pharyngeal glands cells . . . . .Cyartonematidae (Figs 17.13C)
   Cuticle with transverse striae, ornamented or not with spines; pharynx different, i.e. with muscular anterior part and short slender posterior part; pharyngeal gland cells protruding; renette cell not observed . . . . .Meyliidae (Figs 17.13D and E)

Family DESMOSCOLECIDAE Shipley, 1896

Diagnosis. Desmoscolecoidea. Body with transverse striae, ornamented with spines, hairy spines, scales, warts or desmen; no desmen in juvenile stages. Head offset, its cuticle covered or not with concretion material except for the lip region and at the level of the amphideal fovea; in some species head cuticle is provided with cephalic tubercles (only visible with scanning electron microscope (SEM)). Anterior sensorial organs arranged in a circlet of six outer labial papillae (exceptionally small setiform) and a circlet of four sub-median cephalic setae. Cephalic setae mostly inserted on peduncles but may be inserted directly on head cuticle, especially when located near the anterior head border. Anteriormost circlet of inner labial papillae not visible, their dendrites ending in stoma wall. All anterior sensilla rarely absent (e.g. *Desmoscolex*

(*Desmoscolex*) *asetosus*). Amphideal fovea vesicular, rounded and confined to the head or (elongate) oval, extending beyond the posterior head border; amphideal fovea may appear bipartite; *corpus gelatum* never spiral. Opening of amphideal canal always on the head, mostly located near posterior head border. Two pigment spots (ocelli) usually present shortly posterior to the pharyngo-intestinal junction. Pharynx more or less cylindrical (exceptionally with asymmetric posterior bulb (*Quadricomoides*)). Intestinal cells anteriorly finely granular; cells in main part of intestine containing large globules. Female reproductive system didelphic–amphidelphic, with outstretched ovaries.

Key to subfamilies and tribes of the Desmocolecidae

1. Male with one testis; desmoscolecoid pattern of somatic setae (basic with nine pairs of sub-dorsal and eight pairs of sub-ventral setae) 2
   Male with two testes; somatic setal pattern different, usually with more sub-ventral than sub-dorsal setae . . . . .Tricominae (Fig. 17.16)
2. Desmen present or absent in adult; end ring desmoscolecoid in shape; sub-dorsal and sub-ventral somatic setae different in length and shape, glandular . . . . .Desmoscolecini (Fig. 17.15)
   Desmen absent; end ring different; somatic setae not differentiated; sub-ventral setae may be absent in female . . . . .Greeffiellini (Figs 17.14E and F)

Subfamily Desmoscolecinae Shipley, 1896

Diagnosis. Desmoscolecidae. Somatic setae arranged according to a desmoscolecoid pattern (Lorenzen, 1969), with nine pairs of sub-dorsal setae (terminal pair sub-dorsally on end ring) and eight pairs of sub-ventral setae arranged according to the following scheme:

Sub-dorsal: 1 2 3 4 5 6 7 8 9
Sub-ventral: 1 2 3 4 5 6 7 8

Juveniles without sub-ventral setae. Pharynx short cylindrical. Intestine anteriorly differentiated into a fine granular zone. Male reproductive system with one testis. Phasmata[3] and pigment spots present.

Tribe Desmoscolecini Shipley, 1896

Diagnosis. Desmoscolecinae. Adult body cuticle with transverse striae and ornamented with a band of desmos or with concretion scales (secretion and fine particle around a peg-like structure), or with a row of pores, hairy spines or minute concretion patches around pores. A thin layer of secretion and fine foreign particles may be spread irregularly over the body cuticle. Somatic setae, glandular; different in length and shape between sub-dorsal and sub-ventral body sides. Caudal glands often reduced. Consist of three genera: *Desmoscolex*, *Prototricoma* and *Spinodesmoscolex*; the last two are marine taxa. Records from freshwater habitats are restricted to species of the genus *Desmoscolex,* subgenus *Desmoscolex*; the other subgenera of this genus are also marine taxa.

---

[3]Name given by Chitwood (1951) to the phasmid-like structures located laterally on the end ring.

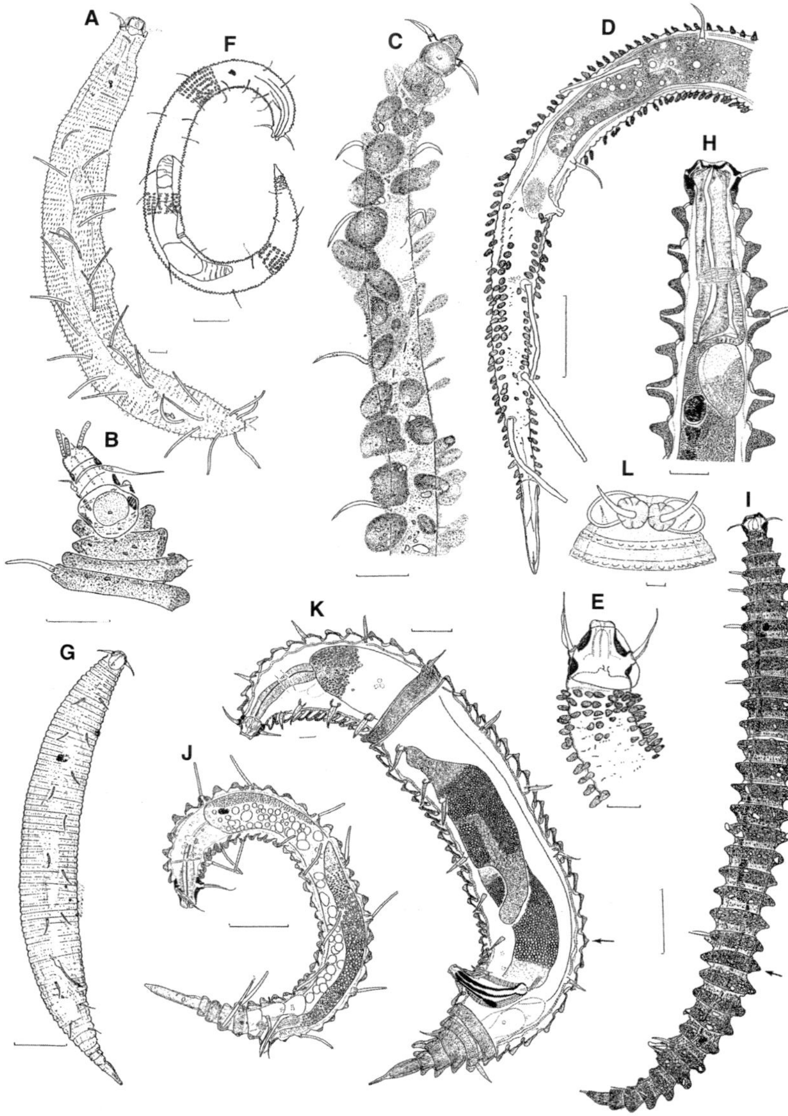

Fig. 17.16 Genera of the Tricominae. A: *Antarcticonema paracomecapitata*, female *in toto* (from Decraemer and Tchesunov, 1996); B: *Desmotimmia mirabilis*, head region (from Decraemer, 1975); C: *Desmotricoma spinicauda*, anterior body region (from Decraemer, 1983); D and E: *Haptotricoma boucheri*, posterior body region with partial surface view and head region in surface view (from Decraemer, 1979); F: *Paratricoma pratensis*, total view of female (from Lorenzen, 1969); G: *Protricoma squamosa*, male, surface view (from Decraemer, 1985); H and I: *Quadricomoides trilabiata*, pharyngeal region and male, surface view (from Decraemer, 1983); J: *Tricoma* (*Tricoma*) *gloriosa*, male with tail in surface view (from Decraemer, 1983); K: *Tricoma* (*Quadricoma*) *bahamaensis*, male with tail in surface view; arrow indicates ring with inversion in orientation; L: *Usarpnema*, male head, dorsoventral view (from Timm, 1978b). Scales of respectively 10 μm (B, D, E, K, L), 30 μm (A, C, F, G, H, J) and 100 μm (I). Courtesy of *Bulletin of Royal Belgian Institute of Natural Sciences* (C, J, K); *Bulletin du Muséum National d'Histoire Naturelle, Paris* (D, E, H, I); *Russian Journal of Nematology* (A); *Hydrobiologia* (G); *Veröffentlichungen des Instituts für Meeresforschung in Bremerhaven* (F); *Zeitschrift für Morphologie der Tiere* (B).

Key to genera of the Desmoscolecini (Figs 17.14 and 17.15)

1. Body cuticle covered with foreign particles and secretion either as bands (desmen), small patches, scales or warts 2
   Body cuticle without adhering foreign particles but ornamented with hairy spines or pores . . . .*Prototricoma* (Figs 17.15F and G)
2. Desmen provided with a transverse row of spine-like setae surrounded by concretion; head triradial symmetrical . . . .*Spinodesmoscolex* (Fig. 17.15D)
   Desmen without an additional transverse row of spine-like setae surrounded by concretion 3
3. Desmen present all over the body 4
   Desmen absent or when present restricted to anterior and posterior body region and often consisting of incomplete bands 5
4. Desmen mostly with rounded outline, when outline triangular never with abrupt reversal in direction . . . .*Desmoscolex* (*Desmoscolex*) (Figs 17.14A and B and 17.15A)
   Desmen with triangular outline with anteriorly or posteriorly directed slope and abrupt reversal in direction . . . .*Desmoscolex* (*Desmolorenzenia*) (Fig. 17.15B)
5. Desmen absent; homonomous striated body cuticle with transverse rows of warts or papillae (minute spines surrounded by fine concretion material) . . . .*Desmoscolex* (*Pareudesmoscolex*) (Fig. 17.15C)
   Desmen only present in anterior and posterior part of the body; on the rest of the body either scales or patches of concretion . . . .*Desmoscolex* (*Protricomoides*) (Fig. 17.15E)

Genus *Desmoscolex* Claparède, 1863

Syn. *Eudesmoscolex* Steiner, 1916; *Eutricoma* Allgén, 1939; *Heterodesmoscolex* Stammer, 1935; *Prodesmoscolex* Stauffer, 1924; *Protodesmoscolex* Timm, 1970.

Diagnosis. Desmoscolecini. Body cuticle with large main rings, separated by narrower to equally broad or broader interzones; transversely striated interzone naked or zone between two transverse striae[4] provided with a transverse row of hairy spines or pores. Desmoscolecoid setal pattern, sub-dorsal and sub-ventral setae usually differentiated in shape and length. Pharynx short cylindrical, extending to the second or third main ring in 17–18 ring species. Intestine posteriorly with (exceptionally without) post-rectal overlap. Mainly marine. Freshwater species restricted to the subgenus *Desmoscolex*.

Four subgenera: *Desmoscolex, Desmolorenzenia, Pareudesmoscolex* and *Protricomoides* (Decraemer, 1985).

[4]Will be henceforth indicated as 'annules'.

Type species: *D.* (*D.*) *minutus* Claparède, 1863.

Remarks:

- So far, juveniles of *Desmoscolex* (*Desmoscolex*) species from soil and freshwater habitats possess body 'annules' provided with warts; juvenile specimens of marine species have body 'annules' with hairy spines.
- The freshwater species of the subgenus *Pareudesmoscolex*, *Desmoscolex* (*Pareudesmoscolex*) *lacustris* Juget, 1969 are juvenile stages of *D.* (*D.*) *dussarti* Juget, 1969 and do not belong to the subgenus *Pareudesmoscolex*.

Freshwater desmoscolecids are small (body length of adult specimens smaller than 300 μm). They display a more or less similar habitus: 18 main body rings (except for 22 in *D.* (*D.*) *aquaedulcis*, females), a rounded head shape, cephalic setae inserted near anterior head border and orientated anteriorly (except for *D.* (*D.*) *lemani*) and show a pronounced sexual dimorphism in the setal pattern (1–3 sub-ventral setae in female *vs.* 8–9 in male). Freshwater desmoscolecids show a rather close phenetic resemblance with terrestrial desmoscolecids from circumtropical regions.

Records from freshwater habitats are restricted to four species of the genus *Desmoscolex* s. str., three species from temperate zones: *D.* (*D.*) *aquaedulcis* Stammer, 1935, *D.* (*D.*) *dussarti* Juget, 1969 and *D.* (*D.*) *lemani* Juget, 1969, and one species from the tropical zone, *D.* (*D.*) *algivorus* Coomans, Vincx & Decraemer, 1985.

Species recorded from freshwater habitats:

D. (*D.*) *algivorus* Coomans, Vincx & Decraemer, 1985: from a deep pool in a coral rock, Solomon Islands; pool characterized by the authors as freshwater but with rather more salts than usually present in waters considered fresh; its species composition consisting of 50% terrestrial, 45% marine and 5% freshwater species.

D. (*D.*) *aquaedulcis* Stammer, 1935: from freshwater springs in Croatia.

D. (*D.*) *dussarti* Juget, 1969: Lake Geneva, between 50 and 75 m depth at the east and west sides of the barrier between Grand Lac and Petit Lac.

D. (*D.*) *lemani* Juget, 1969: Lake Geneva, at 60 m depth, east of the barrier between Grand Lac and Petit Lac, build on fossil soil from remains of a submerged glacial moraine.

Three *Desmoscolex* species have been recorded from wet soils from river banks of the river Kolo, Solomon Islands (Decraemer, 1984). We consider them as terrestrial, but it might be possible to encounter them (rarely) in freshwater.

D. (*D.*) *koloensis* Decraemer (1984): terrestrial (freshwater: New Georgia, Solomon Islands, very wet soil from the bank of the River Kolo, at about 1.5 km from its mouth and 60 cm above watermark).

D. (*D.*) *parakoloensis* Decraemer, 1984: terrestrial, New Georgia, Solomon Islands, very wet soil from the bank of the river Kolo, at about 1.5 km from its mouth and 60 cm above watermark.

D. (*D.*) *montanus* Decraemer & Sturhan, 1982: terrestrial, from soil from forest or rainforest from tropical/subtropical zones worldwide (Nicaragua, Papua New Guinea, Guatemala, Philippines, Congo Republic; Surinam; New Georgia, Solomon Islands, very wet soil from the bank of the River Kolo at about 2 km from its mouth and 1 m above the watermark (see Decraemer, 1975; Decraemer and Sturhan, 1982a)).

Tribe Greeffiellini Decraemer, 1985

Diagnosis. Desmoscolecinae. Body with transverse striae, provided with short or long hairy spines or scale-like projections at posterior margin of 'annules'. Desmen absent. Somatic setal pattern desmoscolecoid with posteriormost pair subdorsally on end ring; their number may be reduced subdorsally or subventrally and sub-ventral setae may be absent in female.

Five genera, all described from marine habitats.

Key to the freshwater desmoscolecid species (measurements in μm): the most important features are the number of desmen and setal pattern (*D.* (*D.*) = *Demoscolex* (*Demoscolex*)).

Species	*D.* (*D.*) *algivorous*	*D.* (*D.*) *aquaedulcis*	*D.* (*D.*) *dussarti*	*D.* (*D.*) *lemani*	*D.* (*D.*) *koloensis*	*D.* (*D.*) *parakoloensis*	*D.* (*Desmolorenzenia*) *montana*
Shape of desmen	Rounded	Rounded	Rounded	Rounded	Rounded	Rounded	Triangular with abrupt change in orientation in ring 14 or 13
Number of desmen in male	18	18	–	–	18 (exc. 17)	18	18
Number of desmen in female	18	22	18 (exc. 19)	18	18 (exc. 17 or 19)	–	18
Number of sub-dorsal setae in male	9	9	–	–	9	9	9
Number of sub-dorsal setae in female	9	9	8	8	9	–	9
Number of sub-ventral setae in male	6	8	–	–	8	4/5	4/5
Number of sub-ventral setae in female	2	1	1	3	0	–	1/2
L in male	175	200–250	–	–	200–245	195–225	145–235
L in female	195	260–270	180–270	180–250	155–235	–	150–250
Spicule length	17	–	–	–	24–29	14	16–23
Length of end ring in male	23	30–56	–	–	27–32	21–23	24–38
Length of end ring in female	25	50–56	–	–	25–31	–	22–37
Insertion of cephalic setae	At front end	At front end	At front end	In anterior half of head	At front end	At front end	Subterminal

Subfamily Tricominae Lorenzen, 1969

Diagnosis. Desmoscolecidae. Body cuticle with transverse striae ornamented with desmen or transverse rows of short or long hairy spines or scales. Desmen, when present, separated by a narrow smooth interzone of a single 'annule'. Somatic setae

with tricomoid pattern (i.e. sub-ventral setae usually more numerous than sub-dorsal setae, paired or not, end ring without setae, posteriormost pair of somatic setae sub-ventral; Lorenzen, 1969). Male reproductive system with two testes.
Apart from a single record of an unidentified *Tricoma* species from a freshwater cavern in the French Pyrenees (Chodorowska, 1963 in Timm, 1970), the taxa of the Tricominae are marine and are not treated here in detail.

Key to genera of the Tricominae

1. Body cuticle with homogeneous transverse striae, without concretion material 2
   Body cuticle with homogeneous transverse striae, with concretion material 4
2. Somatic setae differentiated; body cuticle ornamented with warts and papillae .... *Usarpnema* (Fig. 17.16L)
   Somatic setae homogeneous; body cuticle ornamentation different 3
3. Body cuticle with differentiated lateral field; cuticle ornamentation of transverse rows of minute spines, absent at level of lateral fields .... *Protricoma* (Fig. 17.16G)
   Cuticle without differentiated lateral field; cuticular ornamentation of transverse rows of spines .... *Antarcticonema* (Fig. 17.16A)
4. Desmen or concretion bands present 5
   Desmen absent; concretion material differently arranged 8
5. Head annulated, consisting of three consecutive broader rings .... *Desmotimmia* (Fig. 17.16B)
   Head not annulated 6
6. Pharynx with asymmetric end bulb .... *Quadricomoides* (Figs 17.16H and I)
   Pharynx different 7
7. Desmen mostly with rounded outline, when outline triangular never with abrupt reversal in slope; end ring mainly cylindrical; head more or less triangular in longitudinal optical section .... *Tricoma* (*Tricoma*) (Fig. 17.16J)
   Desmen with triangular outline with anterior or posterior directed slope and abrupt reversal in direction; end ring often with fine offset spinneret; head shape more rounded with wide anterior end .... *Tricoma* (*Quadricoma*) (Fig. 17.16K)
8. Presence of large glandular setae posteriorly on the body, differentiated from the other somatic setae .... *Haptotricoma* (Figs 17.16D and E)
   Such glandular setae absent 9
9. Ornamentation of body cuticle with prominent protruding concretion masses around long hairy spines, more or less arranged in rows; tail tripartite .... *Desmotricoma* (Fig. 17.16C)

Ornamentation body cuticle with transverse rows of scales (= concretion material around a spine); scales more or less ordered in longitudinal rows . . . .*Paratricoma* (Fig. 17.16F)

Family MEYLIIDAE De Coninck, 1965 (Figs 17.13D and E)

Diagnosis. Desmoscolecoidea. Body slender or fusiform, usually with transverse striae and ornamented either with transverse rows of fine spines or longitudinal differentiations of ridges or spines; transverse striae may be interrupted by a narrow lateral field or cuticle rarely smooth, i.e. without striae. Somatic setae, not differentiated, sparse. Amphideal fovea large vesicular with inner spiral structure, mostly located on the head, rarely post-cephalic. Head cuticle, usually smooth, either thickened and sclerotized or thin. Pharynx anteriorly muscular, posteriorly tapering, usually with large overlapping dorsal gland. Pseudocoelomocytes usually present. Male reproductive system with two testes; spicules arched; gubernaculum with apophysis. Female reproductive system didelphic–amphidelphic with branches reflexed. Four genera, all marine.

Type genus: *Meylia* Gerlach, 1956

Family CYARTONEMATIDAE Tchesunov, 1989 (Fig. 17.13C)

Diagnosis. Desmoscolecoidea. Body usually slender, cuticle smooth. Pharynx with anterior muscular part, narrow isthmus and pear-like posterior part with pharyngeal gland cell bodies. Secretory–excretory pore posterior to nerve ring; renette cell present. Several pseudocoelomocytes of which two are prominent, present along pregonad midgut. Blind digestive system without rectum and anus in female. Female reproductive system with reflexed branches. Males with or without anteriorly directed dorsal processus of ejaculatory duct.

The taxa of the Cyartonematidae are marine.

Type genus: *Cyartonema* Cobb, 1920.

References

Andrássy, I. (1976) *Evolution and Systematization of Nematodes*. Pitman Publishing, London, 288 pp.

Andrássy, I. (1984) *Klasse Nematoda*. (*Ordnungen Monhysterida, Desmoscolecida, Araeolaimida, Chromadorida, Rhabditida*). Gustav Fischer Verlag, Stuttgart, Germany, 509 pp.

Blome, D. (1974) Zur Systematik von Nematoden aus dem Sandstrand der Nordseeinsel Sylt. *Mikrofauna des Meeresbodens* 33, 77–99.

Blome, D. (1982) Systematik der Nematoda eines Sandstrandes der Nordseeinsel Sylt. *Mikrofauna des Meeresbodens* 86, 1–194.

Bongers, T. (1988) *De nematoden van Nederland*. Stichting Uitgeverij Koninklijke Nederlandse Natuurhistorische Vereniging, Utrecht, The Netherlands. (Natuurhistorische bibliotheek nr 46), 408 pp.

Bussau, C. (1993) Taxonomische und ökologische Untersuchungen an Nematoden des Peru-Beckens. PhD thesis, Christian-Albrechts-Universität zu Kiel, Kiel, Germany, 621 pp.

Bussau, C. (1995) New deep-sea Nematoda (Enoplida, Thoracostomopsidae, Oncholaimidae, Enchelidiidae) from a manganese nodule area of the eastern South Pacific. *Zoological Scripta* 24, 1–12.

Bussau, C. and Vopel, K. (1999) New nematode species and genera (Chromadorida, Microlaimidae) from the deep sea of the eastern tropical South Pacific (Peru Basin). *Annalen des Naturhistorischen Museums in Wien* 101 B, 405–421.

Castillo-Fernandez, D. and Decraemer, W. (1993) *Cheironchus paravorax* n.sp. and *Cheironchus vorax* Cobb, 1917 from the Campeche Sound, an oil producing zone in the Gulf of Mexico (Nemata: Selachinematidae). *Bulletin van het Koninklijk Belgisch Instituut voor Natuurwetenschappen, Biologie* 63, 55–64.

Chitwood, B.G. (1951) North America Marine Nematodes. *The Texas Journal of Science* 3, 617–672.

Cobb, N.A. (1914) The North American free-living freshwater nematodes. *Transactions of the Microscopical Society* 33, 69–119.

Cobb, N.A. (1920) One hundred new nemas (type species of 100 new genera). *Contributions to a Science of Nematology (Baltimore)* 9, 217–343.

Dashchenko, O.I. (2002) Three new free-living marine nematode species of the subfamily Hypodontolaiminae (Chromadorida, Chromadoridae). *Zoologicheskii Zhurnal* 81, 771–778.

De Coninck, L. (1944) Les nématodes libres des eaux et des terres saumâtres. *Biologisch Jaarboek* 11, 165–220.

De Coninck, L.A. (1965) Classe des Nématodes – Généralités. In: Grassé, P. (ed.) *Traité de Zoologie. IV.*, Masson, Paris, 217 pp.

Decraemer, W. (1975) The aberrant structure of the head in the genus *Desmotimmia* Freudenhammer, 1975 (Nematodes, Desmoscolecida). *Zeitschrift für Morphologie der Tiere* 81, 191–194.

Decraemer, W. (1977) Origin and evolution of the Desmoscolecida, an aberrant group of nematodes. *Zeitschrift für Zoologie, Systematik und Evolutions-forsch*ung 15, 232–236.

Decraemer, W. (1978) Morphological and taxonomical study of the genus *Tricoma* Cobb (Nematoda: Desmoscolecida), with the description of new species from the Great Barrier Reef of Australia. *Australian Journal of Zoology, Supplementary Series* 55, 121 pp.

Decraemer, W. (1979) Desmoscolecids from sublittoral fine-sand of Pierre Noire (West Channel) (Nematoda, Desmoscolecida). *Bulletin du Muséum national d'Histoire naturelle (Paris)* 4è série 1, section A, 2, 299–321.

Decraemer, W. (1982) Draconematidae and Epsilonematidae (Nematoda) from Laing Island Papua New Guinea, with one new genus and three new species. *Bulletin van het Koninklijk Belgisch Instituut voor Natuurwetenschappen, Biologie* 54, 21 pp.

Decraemer, W. (1983) Tricominae (Nematoda – Desmoscolecida) from the northern part of the Mozambique Channel, with five new species and one new genus. *Bulletin van het Koninklijk Belgisch Instituut voor Natuurwetenschappen, Biologie* 55, 26 pp.

Decraemer, W. (1984) Desmoscolecids (Nematoda) from New Georgia, Solomon Islands. *Hydrobiologia* 119, 181–191.

Decraemer, W. (1985) Revision and phylogenetic systematics of the Desmoscolecida (Nematoda). *Hydrobiologia* 120, 259–283.

Decraemer, W. (1989) Three new draconematid species from Papua New Guinea. *Bulletin van het Koninklijk Belgisch Instituut voor Natuurwetenschappen, Biologie* 59, 5–24.

Decraemer, W. (1996) New information on the ultrastructure of the lip region in the genus *Desmoscolex* and description of *Desmoscolex (Demoscolex) parvospiculatus* sp.n. (Nemata: Desmoscolecida) from Papua New Guinea. *Nematologica* 42, 9–23.

Decraemer, W. and Coomans, A. (1978) Scientific report on the Belgian expedition to the Great Barrier Reef in 1967. Nematodes XIII. A description of four new species and a redescription of four known species from in and around mangroves on Lizard Island. *Australian Journal of Marine and Freshwater Research* 29, 509–541.

Decraemer, W. and Coomans, A. (1994) A compendium of our knowledge of the free-living nematofauna of ancient lakes. In: Martens, K., Goddeeris, B and Coulter, G. (eds) *Speciation in Ancient Lakes. Archiv für Hydrobiologie, Ergebnisse der Limnologie* 14, 508 pp.

Decraemer, W. and Jensen, P. (1981) Meyliidae (Nematoda-Desmoscolecida) with four new species and two new genera. *Bulletin van het Koninklijk Belgisch Instituut voor Natuurwetenschappen, Biologie* 53, 13 pp.

Decraemer, W. and Jensen, P. (1982) Revision of the subfamily Meyliinae De Coninck, 1965 (Desmoscolecoidea) with a discussion of its systematic position. *Zoological Journal of the Linnean Society* 75, 317–325.

Decraemer, W. and Sturhan, D. (1982a) A new terrestrial desmoscolecid, *Desmolorenzenia montana* (Nematoda, Desmoscolecida). *Nematologica* 27, 357–365.

Decraemer, W. and Sturhan, D. (1982b) Desmoscolecids from salt marshes in West Germany (Nematoda, Desmoscolecida). *Nematologica* 27, 385–396.

Decraemer, W. and Tchesunov, A.V. (1996) Some desmoscolecids from the White Sea (Nematoda: Desmoscolecida). *Russian Journal of Nematology* 4, 115–130.

De Ley, P. and Blaxter, M. (2002) Systematic position and phylogeny. In: Lee, D.L. (ed.) *The Biology of Nematodes*. Taylor & Francis, London, pp. 1–30.

de Man, J.G. (1880) Die einheimnischen, frei in der reinen Erde und in süszen Wasser lebende Nematoden monografisch bearbeitet. Vorläfiger Bericht und descriptive-systematischer Theil. *Tijdschrift der Nederlandsche Dierkundige Vereeniging* 5, 1–104.

Eyualem-Abebe and Coomans, A. (1996) Aquatic nematodes from Ethiopia V. Descriptions of *Achromadora inflata* n. sp., *Ethmolaimus zullinii* n.sp. and *Prodesmodora nurta* Zullini, 1988 (Chromadorida: Nematoda). *Hydrobiologia* 332, 27–39.

Filipjev, I. (1917) Un nématode libre nouveau de la mer Caspienne, *Chromadorissa* gen. nov. (Chromadoridae, Chromadorini). *Zoologicheskii Zhurnal* 2, 24–30.

Freudenhammer, I. (1975) Desmoscolecida aus der Iberischen Tiefsee, zugleich eine Revision dieser Nematoden-Ordnung. *Meteor-Forschungsergebnisse* D20, 1–65.

Gagarin, V.G. (1993) Free-living nematodes in the waters of Sakhalin Island. *Trudy Instituta Biologii Vnutrennikh Vod* 68, 53–77.

Gagarin, V.G. (2001) *Free-living Nematodes of the Fresh Waters of Russia and Adjacent Countries: Fauna and the Pathways of its Formation, Ecology, Taxonomy, Phylogeny*. Nauka Publisher, Moscow, 170 pp.

Gerlach, S. (1951) Nematoden aus der Familie der Chromadoridae von den deutschen Küsten. *Kieler Meeresforschung* 8, 106–132.

Gerlach, S.A. (1952) Nematoden aus dem Küstengrundwasser. *Abhandlungen. Akademie der Wissenschaften und der Literatur, Mainz. Mathematisch-Naturwissenschaftliche Klasse* 6, 315–372.

Gerlach, S.A. (1956) Diagnoses neuer Nematoden aus der Kieler Bucht. *Kieler Meeresforschung* 12, 85–109.

Gerlach, S.A. (1962) *Aponchium* Cobb, 1920, Typus einer neuen Familie freilebender Meeres – Nematoden. *Abhandlungen und Verhandlungen des Naturwissenschaftlichen Vereins in Hamburg* 7, 157–166.

Gerlach, S.A. (1963) Freilebende Meeresnematoden von den Malediven II. *Kieler Meeresforschung* 19, 67–103.

Gerlach, S.A. (1965) Freilebende Meeresnematoden aus der Gezeitenzone von Spitzbergen. *Veröffentlichungen des Instituts für Meeresforschung in Bremerhaven* 9, 109–172.

Gerlach, S. and Riemann, F. (1973) The Bremerhaven Checklist of Aquatic Nematodes. *Veröffentlichungen des Instituts für Meeresforschung in Bremerhaven*, Suppl. 4, part 1, 404 pp.

Gerlach, S. and Riemann, F. (1974) The Bremerhaven Checklist of Aquatic Nematodes. *Veröffentlichungen des Instituts für Meeresforschung in Bremerhaven*, Suppl. 4, part 2, 734 pp.

Gourbault, N. and Vincx, M. (1988) *Calomicrolaimus compridus* (Gerlach, 1956) n.comb., a marine nematode with a female producing a copulatory ring. *Revue de Nématologie* 11, 39–43.

Grimma, O.A. (1876) Das Kaspische Meer und seine Fauna. *Trudy aralo-kasp. Eksped* 2, 1–168.

Hirschmann, H. (1952) Die Nematoden der Wassergrenze mittelfränkischer Gewässer. *Zoologische Jarhbücher* 81, 313–407.

Hofmänner, B. (1913) Contribution à l'étude des nematodes libres du Lake Geneva. *Revue Suisse de Zoologie* 21, 589–658.

Hofmänner, B. and Menzel (1915) Die freilebenden Nematoden der Schweiz. *Revue Suisse de Zoologie* 23, 109–243.

Hope, W.D. and Murphy, D.G. (1972) A taxonomic hierarchy and checklist of the genera and higher taxa of marine nematodes. *Smithsonian Contributions to Zoology* 137, 101 pp.

Hopper, B.E. (1963) Marine nematodes from the coast line of the Gulf of Mexico. III. Additional species from Gulf shores, Alabama. *Canadian Journal of Zoology* 41, 841–863.

Inglis, W.G. (1963) *Nudora omercooperi* sp.nov. (Nematoda: Free-living) from an unusual locality in South Africa. *Hydrobiologia* 25, 176–180.

Jensen, P. (1978) Revision of Microlaimidae, erection of Molgolaimidae fam.n., and remarks on the systematic position of *Paramicrolaimus* (Nematoda, Desmodorida). *Zoologica Scripta* 7, 159–173.

Jensen, P. (1980) Description of the marine free-living nematode *Chromadora lorenzeni* n. sp. with notes on its microhabitat. *Zoologischer Anzeiger (Jena)* 205, 213–218.

Jensen, P. (1994) Revision of Ethmolaiminae Filipjev & Stekhoven, 1941 (Nematoda, Chromadorida), with descriptions of one new genus and three new species. *Hydrobiologia* 286, 1–15.

Khera, S. (1969) Nematodes from the banks of still and running waters. II *Kreisonema nandi* n.g., n. sp. Kreisoneminae n. Subfam. Family Leptolaimidae from India. *Labdev Journal of Science and Technology (India)* 7B, 64–66.

Kreis, H. (1928) Weitere Beitrag zur Kenntnis der freilebenden marinen und Süszwassernematoden. *Verhandlungen der Schweizerischen Naturforschenden Gessellschaft* 108, 196–197.

Litvaitis, M., Bates, J., Hope, W.D. and Moens, T. (2000) Inferring a classification of the Adenophorea (Nematoda) from nucleotide sequences of the D3 expansion segment (26/28S rDNA). *Canadian Journal of Zoology* 78, 911–922.

Loof, P.A.A. (1961) The nematode collection of Dr. J.G. De Man. *Mededelingen van het Laboratorium voor Fytopathologie* 190, 169–254.

Loof, P.A.A. (1971) Free-living and plant-parasitic nematodes from Spitzbergen, collected by Mr. H. Van Rossen. *Mededelingen Landbouwhogeschool Wageningen* 71, 1–86.

Lorenzen, S. (1969) Desmoscoleciden (eine Gruppe freilebender Meeresnematoden) aus Küstensalzwiesen. *Veröffentlichungen des Instituts für Meeresforschung in Bremerhaven* 12, 231–265.

Lorenzen, S. (1981) Entwurf eines phylogenetischen Systems der freilebenden Nematoden. *Veröffentlichungen des Instituts für Meeresforschung in Bremerhaven* Suppl. 7, 1–472.

Lorenzen, S. (1994) The phylogenetic systematics of free-living nematodes. *The Ray Society, London*, 162, 383 pp.

Malakhov, V.V., Ryzhikov, K.M. and Sonin, M.N. (1982) The classification of major taxa of nematodes: Subclasses, Orders, Suborders. *Zoologichesky Zhurnal* 64, 1125–1134.

Micoletzky, H. (1922) Die freilebenden Erdnematoden. *Archiv für Naturgeschichte* 87 A, 1–650.

Micoletzky, H. (1925) Die freilebenden Süszwasser- und Moornematoden Dänemarks. *Kongelige Dänske Videnskabers Selskabs Skrifter Math.-natur* 10, 55–310.

Mulvey, R.H. (1969) Soil-inhabiting nematodes of the orders Araeolaimida, Chromadorida, Enoplida and Monhysterida from the Canadian high Arctic. *Canadian Journal of Zoology* 47, 365–382.

Muthumbi, A.W. and Vincx, M. (1999) Microlaimidae (Microlamoidea: Nematoda) from the Indian Ocean: description of nine new and known species. *Hydrobiologia* 397, 39–58.

Ott, J.A. (1997) A new symbiotic marine nematode, *Adelphos rolandi* gen.n., sp.n. (Stilbonematinae), from the Caribbean Sea. *Annalen des Naturhistorischen Museums in Wien* 99B, 417–422.

Ott, J.A., Novak, R., Schiemer, F., Hentschel, U., Nebelsick, M. and Polz, M. (1991) Tackling the sulfide gradient: a novel strategy involving marine nematodes and chemoautotrophic ectosymbionts. *Marine Ecology* 12, 261–279.

Platonova, T.A. and Potin, V.V. (1972) On new genus *Harpagonchus* and *Harpagonchoides* (Nematoda, Chromadorida, Harpagonchidae fam.n.) living on the parapodia and gills of the Antarctic Polychaetes *Aglaophatus* Kinberg and *Hemipodes* Quatrefages, Islledovanija Fauny morjei 11 (8), *Resultatibilogitscheskich Issledovanii sovjetskich antarktitscheskich Ekspedizii* 5, 81–85 (Russian with English summary).

Platt, H.M. (1982) Revision of the Ethmolaimidae (Nematoda: Chromadorida). *Bulletin of the British Museum (Natural History)* 43, 185–252.

Platt, H.M. (1985) Further observations on the Ethmolaimidae (Nematoda: Chromadorida). *Journal of Natural History* 19, 139–149.

Platt, H.M. and Warwick, R. (1988) Free-living marine nematodes. Part II. British Chromadorids. In. *Synopses of the British Fauna (New Series)*, vol. 38. Cambridge University Press, Cambridge, Massachusetts, 502 pp.

Riemann, F. (1966) Die Interstitielle Fauna im Elbe-Aestuar. Verbreitung und Systematik. *Archive für Hydrobiologie* 31 (Suppl.), 1–279.

Riemann, F. (1970) Freilebende Nematoden aus dem Grenzbereich Meer-Süsz-Wasser in Kolumbien, Süszamerika. *Veröffentlichungen des Instituts für Meeresforschung in Bremerhaven* 12, 365–412.

Riemann, F. and Schrage, M. (1977) Zwei neue Nematoda Desmodorida aus der Iberischen Tiefsee. *'Meteor' Forschungsergebnisse*, Reihe D 25, 49–53.

Schneider, G. (1906) Süszwassernematodenaus Estland. *Zoologischer Anzeiger* 29, 679–683.

Schneider, W. (1937) Freilebende Nematoden der Deutschen Limnologischen Sundaexpedition nach Sumatra, Java und Bali. *Archiv für Hydrobiologie. Tropische Binnengewasser* 7 (Suppl. 15), 30–108.

Schulz, E. (1932) Beiträge zuur Kenntnis mariner Nematoden aus der Kieler Bucht. *Zoologische Jahrbücher* (Systematik) 62, 331–430.

Shoshin, A.V. (1998) Baikal nematodes of the genus *Ethmolaimus* (Nematoda: Chromadorida: Ethmolaimidae). *Zoosystematica Rossica* 7, 215–221.

Soetaert, K. and Vincx, M. (1988) *Spirobolbolaimus bathyalis* gen.n., sp.nov. (Nematoda, Microlaimidae) from the Mediterranean (Calvi). *Hydrobiologia* 164, 33–38.

Stoichev, S. (1996) On the free-living nematode fauna from Bulgarian inland waters. *Laterbornia* 25, 23–30.

Swart, A. and Heyns, J. (1991) *Desmodora (Sibayinema) natalensis* subg.nov., spec.nov. from Lake Sibayi, South Africa (Nematoda: Desmodorida). *Phytophylactica* 23, 273–278.

Tahseen, Q. (2001) A new and a known species of *Achromadora* Cobb, 1913 (Nematoda: Cyatholaimidae) from India. *Journal of Morphology and Systematics* 3, 101–108.

Tchesunov, A. (1989) The genus *Cyartonema* Cobb, 1920 (Nematoda, Chromadoria): morphological pecularity, new diagnosis and descriptions of three species from the White Sea. *Zoologicheskii Zhurnal* 68, 5–16.

Tchesunov, A. (1994) On the morphology and systematic position of the family Meyliidae (Nematoda: Chromadoria). *Nematologica*, 40, 369–378.

Timm, R.W. (1952) A survey of the marine nematodes of Chesapeake Bay, Maryland. *Contributions of the Chesapeake Biological Laboratory* 95, 3–70.

Timm, R.W. (1961) The marine nematodes of the Bay of Bengal. *Proceedings of the Pakistan Academy of Sciences* 1, 1–88.

Timm, R.W. (1969) *Megodontolaimus* new genus (Nematoda: Chromadoridae), with a description of two new species. *Proceedings of the Helminthological Society of Washington* 36, 36–39.

Timm, R.W. (1970) A revision of the nematode order Desmoscolecida Filipjev, 1929. *University of California Publications in Zoology* 93, 115 pp.

Timm, R.W. (1978a) A redescription of the marine nematodes of Shackleton's British Antarctic Expedition of 1907–1909. In: Pawson, D.L. (ed.) *Biology of the Antarctic Seas 6. Antarctic Research Series* 26, 237–255.

Timm, R.W. (1978b) Marine nematodes of the order Desmoscolecida from McMurdo Sound. Antarctica. In: Pawson, D.L. (ed.) *Biology of the Antarctic Seas 6. Antarctic Research Series* 26, 225–236.

Tsalolikhin, S.Y. (1985) *Nematodes of Fresh and Brackish Waters of Mongolia.* Nauka, Leningrad, Russia, 114 pp.

Verschelde, D. and Vincx, M. (1995) *Psammonema* gen.n. and Pseudochromadora Daday, 1889 (Nematoda, Desmodoridae) from sandy sediments of Gazi, Kenya. *Bulletin van het Koninklijk Belgisch Instituut voor Natuurwetenschappen, Biologie* 65, 114–139.

Verschelde, D., Gourbault, N. and Vincx, M. (1998) Revision of *Desmodora* with descriptions of new desmodorids (Nematoda) from hydrothermal vents of the Pacific. *Journal of the Marine Biological Association U.K.* 78, 75–112.

Vitiello, P. and De Coninck, L. (1968) *Peresiana annulata* n.gen., n.sp., type intéressant de Desmoscolecida. *Rapports et Procés-Verbaux des Réunions de la Commission Internationale pour l'Exploration Scientifique de la Mer Méditerranée* 19, 201–204.

Wieser, W. (1951) Untersuchungen über die algenbewhonende Mikrofauna mariner Hartböden I. Zur Ökologie und Systematik der Nematodenfauna von Plymouth. *Österreichische Zoologische Zeitschrift* 3, 425–480.

Wieser, W. (1954) Free-living marine nematodes. II. Chromadoroidea. Chile reports 17. *Lunds Universitets Arsskrift*, 16, 145 pp.

Wieser, W. (1959) *Free-living Nematodes and Other Small Invertebrates of Puget Sound Beaches*. University of Washington Press, Seattle, Washington, 179 pp.

Wieser, W. and Hopper, B. (1967) Marine nematodes of the east coast of North America. I. Florida. *Bulletin of the Museum of Comparative Zoology* 135, 239–344.

18 Order Monhysterida

A. Coomans[1] and Eyualem-Abebe[2]

[1]Department of Biology, Ghent University, Ledeganckstraat 35, B-9000 Ghent, Belgium; [2]Hubbard Centre for Genome Studies and Department of Zoology, University of New Hampshire, 35 Colovos Rd, Gregg Hall, Durham, NH 03824, USA

Order MONHYSTERIDA Filipjev, 1929

Morphology (Figs 18.1 and 18.2)

Body wall

Composed of cuticle, epidermis and somatic musculature.

Cuticle finely striated (sometimes appearing smooth under light microscope (LM)) or annulated, often bearing four rows of somatic setae.

Epidermis with four longitudinal chords and a thin interchordal part. Specialized epidermal cells form several glands (e.g. ventral gland, caudal glands).

Somatic musculature divided into four quadrants by the epidermal chords. The non-contractile part may contain crystalloids (genus *Monhystera*).

Body cavity and associated glands

The body cavity is a pseudocoelom, which is filled with fluid in the part posterior to the nerve ring or posterior to the cardia. Several gland cells may be present in the body cavity: (pseudo)coelomocytes, ventral gland, post-vulval gland, ejaculatory glands, rectal glands and caudal glands.

The unicellular ventral gland is the main part of the *secretory–excretory system*. It may be in (usually right) sub-ventral position. In some species or genera it is poorly developed and invisible under LM (then often described as absent), but it may be large in other species or genera, especially in epizoic forms. The cell leads to a long duct with a subterminal widening (ampulla) and a mid-ventral opening, the secretory–excretory pore, mostly called excretory pore. The latter opens behind the nerve ring (Sphaerolaimidae and some Xyalidae), more or less halfway between the anterior body end and the nerve ring (most Monhysteridae) or in the lip region (some Monhysteridae).

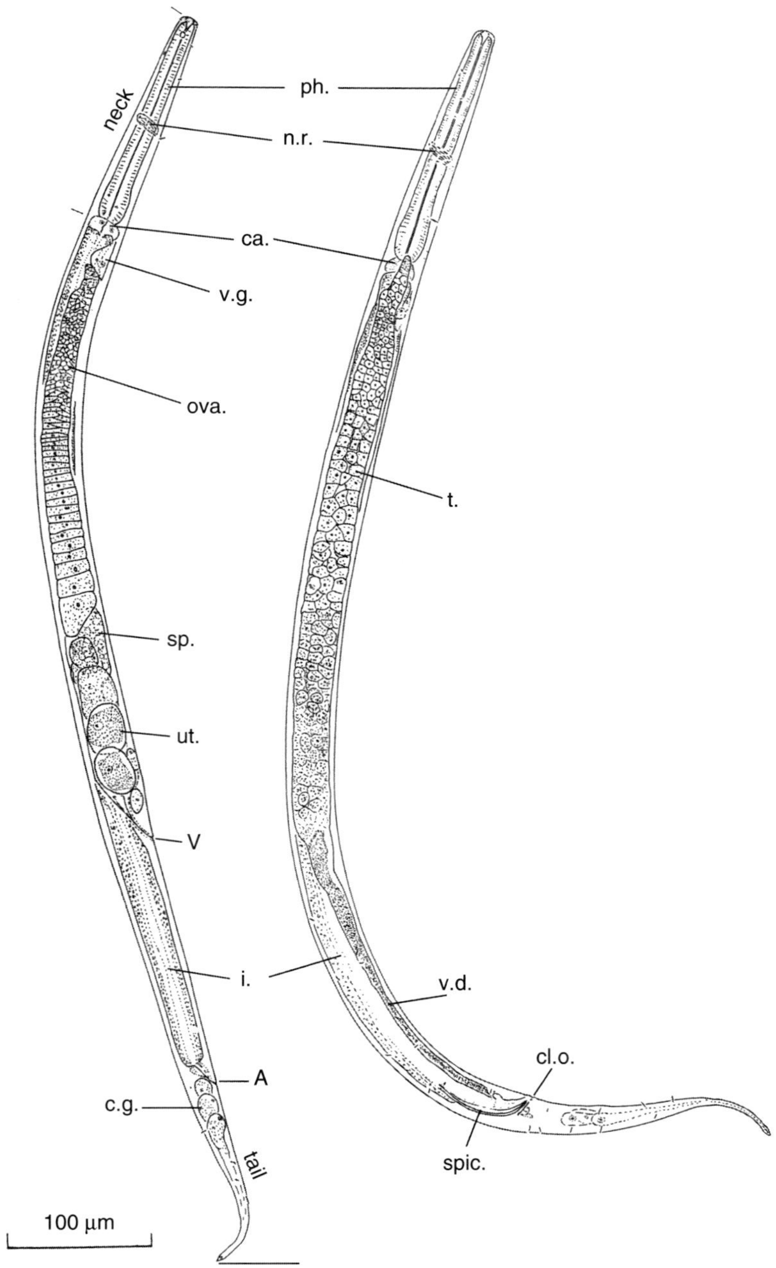

Fig. 18.1 General morphology of a monhysterid (based on *Monhystera*): Female left, male right. Abbreviations: A, anus; ca, cardia; c.g., caudal gland; cl.o., cloacal opening; i, intestine; n.r., nerve ring; ova., ovarium; ph., pharynx; sp., spermatheca; spic., spiculum; t., testis; ut., uterus; V, vulva; v.d., vas deferens; v.g., ventral gland. Neck: region between anterior end and cardia. Tail: region between anus (or cloacal opening) and tail tip.

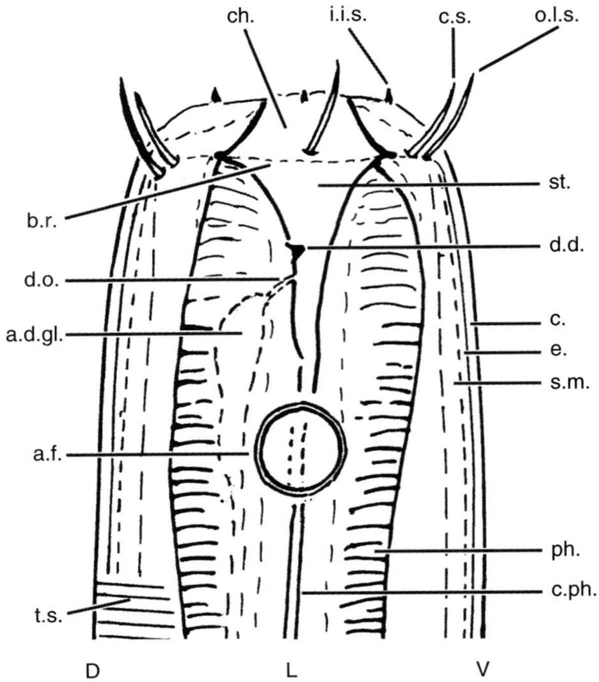

Fig. 18.2 Anterior body end of a monhysterid. Abbreviations: a.d.gl., ampulla of dorsal pharyngeal gland; a.f., amphidial fovea; b.r., buccal ring; c, cuticle; ch, cheilostom; c.ph, cuticular lining of pharynx; c.s., cephalic sensillum; D, dorsal; d.d., dorsal denticle; d.o., dorsal pharyngeal gland opening; e, epidermis; i.l.s., inner labial sensillum; L, lateral; o.l.s., outer labial sensillum; ph., pharynx; s.m., somatic musculature; st., stoma; t.s., transverse striae; V, ventral.

The caudal glands constitute the *adhesive system*. Typically there are three large caudal glands, which are more or less arranged in tandem, but sometimes the anterior one is poorly developed and possibly even absent. The glands may be contained in the tail proper or the anterior one may extend anteriad above the rectum. They produce an adhesive substance that allows temporary attachment to a substrate. Sometimes two smaller gland-like cells may be observed dorsally from the larger ones, these may represent 'release glands', with opposite action.

Digestive system

Composed of mouth opening, buccal cavity, pharynx, cardia, intestine and rectum.

Mouth (= *oral*) *opening* round or hexagonal, depending on degree of amalgamation of lips; when round, then surrounded by oral platelets. *Buccal cavity* variously structured according to genera; composed of a hexagonal *cheilostoma* (within the lip region) and a triradial *stoma*, the latter with or without one to three small teeth (denticles), uni- or bipartite and partly or completely surrounded by pharyngeal tissue. The junction between cheilostoma and stoma forms a 'buccal ring'.

Pharynx cylindrical or slightly enlarging posteriad, sometimes with a muscular basal bulb. *Pharyngo-intestinal junction* or *cardia* composed of an outer part between pharynx and intestine with three cardiac gland-like cells and a narrow inner part, which widens posteriorly where it penetrates into the lumen of the intestine.

Intestine tubular, with few cells (oligocytous). The cells have a large nucleus, a granular cytoplasm and a well-developed 'brush border'. The anteriormost ring of cells consists of small hyaline cells with flap-like projections into the lumen; the posteriormost cells are also hyaline.

Rectum usually about one anal body diameter (ABD) long in females and short in males, where it joins the vas deferens, thus forming a cloaca. Rectal (or cloacal) glands present along the dorsal wall of the rectum (or cloaca).

Sensory nervous system

The *nervous system* consists of a central *nerve ring* surrounding the pharynx about halfway along its length, longitudinal nerves, commissures and ganglia. Apart from the nerve ring and the ganglia around it, the nervous system cannot be accurately observed under LM without appropriate staining procedures.

The *sense organs* comprise anterior sensilla, amphids, ocelli, somatic setae and genital sensilla in the male.

The *anterior sensilla* are usually arranged in two (6 + 10), rarely in three (6 + 6 + 4) circlets. The most anterior circlet consists of the inner labial sensilla, which are papilliform or short setiform. The remaining anterior sensilla consist of six outer labial and four cephalic ones; typically their outer parts are short to medium-sized setae, but occasionally only the four cephalic sensilla are setiform. The outer labial setae are equal to or (mostly) longer than the cephalic setae in Monhysteridae and Xyalidae, but they are shorter in Sphaerolaimidae. Jointed setae can be found in Xyalidae.

Additional setae may occur in the head region, e.g. two extra lateral setae ventrally from the lateral outer labial setae (in many Xyalidae) or eight groups of 'subcephalic' (in fact 'post-labial') setae (in some Xyalidae and Sphaerolaimidae).

The outer part of the *amphid* or amphideal fovea (usually simply called 'amphid') is typically circular, but has in some cases been described as pseudospiral or unispiral.

When present, *ocelli* are located in the neck region between the anterior body end and the nerve ring, either paired and subdorsally very close to each other or united in one dorsal pigment pocket. In living specimens the pigment may be red (in some Monhysteridae) or yellow (in some Xyalidae).

Somatic setae are typically arranged in two laterodorsal and two lateroventral longitudinal rows. Those in the anterior body end region are called 'sub-cephalic' setae, those in the neck region 'cervical' setae and those in the tail region 'caudal' setae. In Sphaerolaimoidea there may be two or three terminal setae.

Genital sensilla are generally poorly developed or absent in freshwater species. In some brackish water (and marine) species one minute pre-cloacal, medioventral papilla or seta and peri-cloacal or post-cloacal setae or papillae may be present. Exceptionally, males possess complex supplements.

Sexual dimorphism in the size of some of the sense organs (amphids, setae) is not unusual.

Reproductive system

Female reproductive system

The *female reproductive system* consists of one or two outstretched branches. However, in *Sinanema* the ovary is reflexed and in gravid females of other genera the tip of the ovary may be sinuous or reflexed. A reproductive branch is composed of an ovary, an oviduct, a uterus and a spermatheca (*receptaculum seminis*) at the junction of oviduct and uterus. The uteri (or uterus) connect(s) to the vagina, which opens to the outside through a slit-like vulva.

The *ovary* has a distal germinal zone with germ cells packed in several rows leading to a growth zone where the cytoplasm of the germ cells increases and the number of cells decreases till they become arranged in a single row. Proximally the cytoplasm of the oocytes becomes more opaque due to yolk production.

The *oviduct* is long or short; in the latter case it is composed of few cells, and then is in fact more a constriction than a duct. It leads to the dorsal side of the uterus. The spermatheca is typically composed of a main part and a ventral blind sac extending anteriorly alongside the oviduct and the proximal part of the ovary; it originates from the distal uterus. In amphimictic species the spherical sperm are stored mainly either in the blind sac, in the distal uterus or in both.

The *uterus* is thin-walled and may be filled with a variable number of eggs (according to species) in gravid females. A post- and pre-vulval uterine sac may be present in some species (genus *Daptonema*).

The thin-walled *vagina* is usually shorter than the corresponding body width. Its inner cuticular wall may be sclerotized proximally in some genera (*Diplolaimella, Gammarinema, Odontobius*). It is surrounded by a variously developed (according to species) sphincter muscle.

The *vulva* is a transverse slit with posteriorly bent edges. Behind or partly alongside the vulva there may be several glandular cells, which form a single globular or a double V-shaped granular mass called the *post-vulvar* (or *post-vulval*) *gland.*

The female reproductive system may be accompanied by several coelomocytes.

Male reproductive system

The *male reproductive system* is composed of one *testis* or two *testes,* connected to a *vas deferens* which joins the rectum, thus forming a cloaca. The anterior testis is outstretched (occasionally the tip may be sinuous or reflexed), the posterior one is reflexed. The latter is mostly absent.

A *testis* comprises a germinal zone (with spermatogonia) and a growth zone (with spermatocytes), while the proximal part acts as a *vesicula seminalis* in which spermatids and spermatozoa are stored.

The *vas deferens,* or seminal duct, is a long tube with granular cells, which may be surrounded by a thin muscular sheath. It is flanked at both sides by a series of two to five (according to species) *ejaculatory gland cells,* which open posteriorly into the terminal part of the seminal duct.

The *cloaca* is lined with cuticle continuous with the body cuticle. Rectal (or cloacal) glands open in its dorsal wall. The distal part of the cloaca may be sclerotized to various degrees and forms a simple or more elaborate *gubernaculum*, which may be slipper-shaped with or without dorsal (caudal) *apophysis* or *apophyses*. The *gubernaculum* together with the *spicula* forms the *copulatory apparatus*. Each *spiculum* consists of a 'head', a shaft and often also a blade. The shape may vary from short and robust to long and slender, from slightly to strongly ventrally curved, according to species. The copulatory apparatus can be protruded or retracted by respectively protractor and retractor muscles.

Caudal alae (bursa) only rarely present (in *Diplolaimelloides, Monhystrium, Tripylium*).

Genital papillae mostly poorly developed or absent, genital setae present in some genera, complex supplements so far only found in one genus (*Sacrimarinema*).

Taxonomy (Figs 18.3–18.12)

Diagnosis. Small to medium-sized, more or less slender nematodes. Body slightly attenuated towards the rounded or truncated anterior body end, more strongly attenuated towards tail end. Cuticle finely striated (often appearing smooth under LM) or clearly annulated. Anterior sensilla 6 + 10 or 6 + 6 + 4; six inner labial sensilla papilli- or setiform; six outer labial sensilla and four cephalic sensilla usually setiform; setae may be jointed. Amphids typically with round fovea (sometimes described as pseudospiral or unispiral). Dorsal or sub-dorsal ocelli may be present.

Female reproductive system outstretched, with either two ovaries or only one (anterior) ovary.

Male reproductive system with either two testes or with only one (anterior) testis. Pre-anal genital papillae (supplements) rarely present.

Caudal glands mostly confined to tail, opening separately or through a common pore.

Mainly aquatic, some in wet terrestrial habitats.

Remarks:

1. Outstretched ovaries are considered to represent an (hol)apomorphy of the Monhysterida (Lorenzen, 1981, 1994), but other representatives of Chromadorea may possess outstretched ovaries (e.g. Microlaimidae, Aponchiidae, some Rhabditolaimidae). According to Lorenzen, this condition is derived independently from forms with reflexed ovaries within the respective families.

2. Lorenzen (1981, 1994) classified Axonolaimoidea under Monhysterida on the basis of their outstretched ovaries. Molecular analysis, however, demonstrated that this group belongs in Araeolaimida (De Ley and Blaxter, 2001, 2004), as was previously accepted in various classifications. It confirms that outstretched ovaries have originated several times independently during the evolution of the Chromadorea.

3. Reflexed ovaries occur in *Sinanema* and ovaries with sinuous or reflexed tip may occur in gravid females of some species.

Key to Superfamilies

Cuticle finely striated (often appearing smooth under LM);
anterior setae ten; groups of sub-cephalic setae lacking;

only one gonad present, at right side of intestine; tail
without long terminal setae . . . . .MONHYSTEROIDEA
Cuticle striated or annulated; anterior setae ten or more; eight groups
of sub-cephalic setae may be present; 1–2 testes, anterior one either left
or right of intestine, posterior one at opposite side; single ovary at right
or left side of intestine; tail often with 2–3 long
terminal setae . . . . .SPHAEROLAIMOIDEA
Cuticle striated or annulated; anterior setae four or more (ten in many marine
forms); female reproductive system primarily with two branches,
but posterior branch may be rudimentary; two testes; tail without
long terminal setae . . . . .SIPHONOLAIMOIDEA

Superfamily MONHYSTEROIDEA de Man, 1876

Diagnosis. Anterior sensilla in two circlets, 6 + 10. Inner labial sensilla papilliform or short setiform. Amphideal fovea round. Stoma usually funnel-shaped, sometimes tubular; with or without denticles. Pharynx without valved bulb.

Only one ovary or testis present, at right side of intestine.

One family: Monhysteridae (aquatic and terrestrial).

Family MONHYSTERIDAE de Man, 1876

Diagnosis. Small nematodes, i.e. usually less than 1.5 mm, exceptionally more than 2 mm, but then very slender. Cuticle finely striated (striae 1 μm or less apart), often appearing smooth under LM; usually with small sub-median setae along body length. Ocelli frequently present. Stoma surrounded by pharyngeal tissue; with or without denticles. Pharynx cylindrical, well muscularized, sometimes slightly swollen at its anterior end and in some genera (*Monhystrella*, *Sinanema* and partly in *Eumonhystera*) with more or less developed muscular basal bulb. Pharyngo-intestinal junction (cardia) with conoid outer part lying between pharynx and intestine, and oblong valve-like inner part penetrating into intestine. Intestine with few cells (oligocytous). Rectum usually one ABD[1] long, but up to three ABD in some *Geomonhystera* species. Ventral gland usually well developed, its cell body ventrally or right sub-ventrally opposite anterior end of intestine; secretory–excretory pore anterior to nerve ring, in neck region or on lip region. Female reproductive system monodelphic, prodelphic, with outstretched ovary at right side of intestine; post-vulvar uterine sac absent. Male reproductive system with single testis; spicules equal, very variable in size according to species; gubernaculum often with dorsal (= caudal) apophysis. Tail conoid to elongate-conoid, with caudal glands opening through a single pore at the terminal spinneret; without terminal setae.

Species reproducing amphimicticly or parthenogenetically. Non-selective deposit feeders or bacterivorous.

Thalassic and athalassic saline and freshwater bodies, some in wet soils. Three genera comprise species living in gill chambers of aquatic (*Gammarinema*) or terrestrial (*Monhystrium* and *Tripylium*) crustaceans.

[1] ABD = Anal Body Widths.

Remarks: The ovary in *Sinanema* is reflexed; this may also be the case over a short distance in gravid females of species belonging to other genera.

Type genus: *Monhystera* Bastian, 1865

Other genera:

Anguimonhystera Andrássy, 1981
Diplolaimella Allgén, 1929
Diplolaimelloides Meyl, 1954
Eumonhystera Andrássy, 1981
Gammarinema Kinne & Gerlach, 1953
Geomonhystera Andrássy, 1981
Monhystrella Cobb, 1918
Monhystrium Cobb, 1920
Odontobius Roussel de Vauzème, 1834
Sinanema Andrássy, 1960
Thalassomonhystera Jacobs, 1987
Tridentulus Eyualem-Abebe & Coomans, 1995
Tripylium Cobb, 1920
Genus inquirendum: *Sitadevinema* Khera, 1971

Key to genera of inland water Monhysteridae

1. Ovary rather short (= not reaching anterior part of intestine), reflexed; stoma with three small teeth just behind buccal ring (= margin between cheilostome and stoma proper); parthenogenetic; freshwater . . . . *Sinanema*
   Ovary outstretched (in gravid females tip of ovary may be reflexed or sinuous) 2
2. Small species (L usually < 0.5 mm); V halfway along the body; female genital system usually very short; cheilostome small, often well cuticularized; stoma narrow; pharynx with muscular basal bulb; spinneret typically a long cone or cylinder; thalassic and athalassic saline water, and freshwater . . . . *Monhystrella*
   Small to medium-sized species; v more posterior; female genital system long; cheilostome and stoma wider, funnel- or V-shaped; pharynx usually without basal bulb; spinneret short 3
3. Body long (1.5–2.4 mm), slender to very slender (a = 43–100) . . . . *Anguimonhystera*
   Body shorter and only rarely slender (a = up to 55 in *Eumonhystera gracilior*, but L = 0.9 mm) 4
4. V > 75%; V' (vulva as % of distance anterior end – anus) > 90%; tail rather short, conical to elongate-conical; mainly in moist soil (and marine) . . . . *Geomonhystera*
   V < 75%; V' < 90%; aquatic 5
5. Numerous crystalloids present in cell bodies of somatic muscles; ocelli often present; tail length < or = distance vulva to anus (or tail slightly longer than V–a distance in species with extended

uterus); amphimictic; mainly freshwater, sometimes in athalassic brackish water . . . . *Monhystera*
Crystalloids absent; fresh and brackish water (and marine) 6
6. Stoma subdivided into two chambers; thalassic brackish water or epizoic on gills of crustaceans 7
Stoma not or only obscurely subdivided; fresh and brackish water (and marine) 9
7. Cephalic setae > 20% of corresponding body diameter (CBD); secretory–excretory pore in lip region; tail rather short conoid to subcylindrical in both sexes; epizoic on gills of crustaceans (crayfish, freshwater crabs) . . . . *Gammarinema*
Cephalic setae < 20% of CBD; secretory–excretory pore behind lip region; tail long; free living 8
8. Fovea circular; tail filiform; caudal alae (bursa) absent; gubernaculum with strong dorsal apophysis . . . . *Diplolaimella*
Fovea ventrally unispiral; caudal alae present; gubernaculum without dorsal apophysis . . . . *Diplolaimelloides*
9. Tail < *V–a* distance; amphimictic, mostly marine . . . . *Thalassomonhystera*
Tail > or equal to *V–a* distance; amphimictic or parthenogenetic, mostly freshwater or brackish water 10
10. With 0–1 denticles in stoma . . . . *Eumonhystera*
With three denticles in stoma . . . . *Tridentulus*

Genus *Anguimonhystera* Andrássy, 1981 (Figs 18.3A and B)
Diagnosis. (adapted from Andrássy, 1981). Monhysteridae with long (L = 1.4–2.4 mm) and slender body (a = 43–100). Head wide, rounded, not offset. Cephalic setae comparatively long. Amphidial fovea at least one head diameter behind anterior body end. Ocelli absent. Stoma narrow, funnel-shaped, without denticles. Vulva at 60–80% of body length. Spicules short, gubernaculum slipper-shaped. Tail long, without terminal setae. Occurring in groundwater.

Type species: *Anguimonhystera stadleri* (Goffart, 1950) Andrássy, 1961

Other species:
A. ampliceps (Goffart, 1950) Andrássy, 1961
A. tenuissima (Goffart, 1950) Andrássy, 1961

Remarks: This genus has been treated as a *genus inquirendum* by Jacobs (1987a), who transferred *A. ampliceps* to the genus *Eumonhystera*, *A. stadleri* to *Thalassomonhystera* and *A. tenuissima* to *Geomonhystera.* Although we agree that this genus is doubtful, based mainly on the slender body and the occurrence in groundwater, we have retained it pending further information about these rare and incompletely known species.

Genus *Diplolaimella* Allgén, 1929 (Figs 18.4A–D)
Diagnosis. Monhysteridae with sub-median somatic setae less than 20% of CBD long. Anterior labial sensilla small, papilliform; posterior labial and cephalic sensilla small, setiform. Amphidial fovea circular, 1.5–3 CBDs from anterior end. Cheilostome inverted funnel- or cup-shaped. Second chamber of stoma may be denticulate.

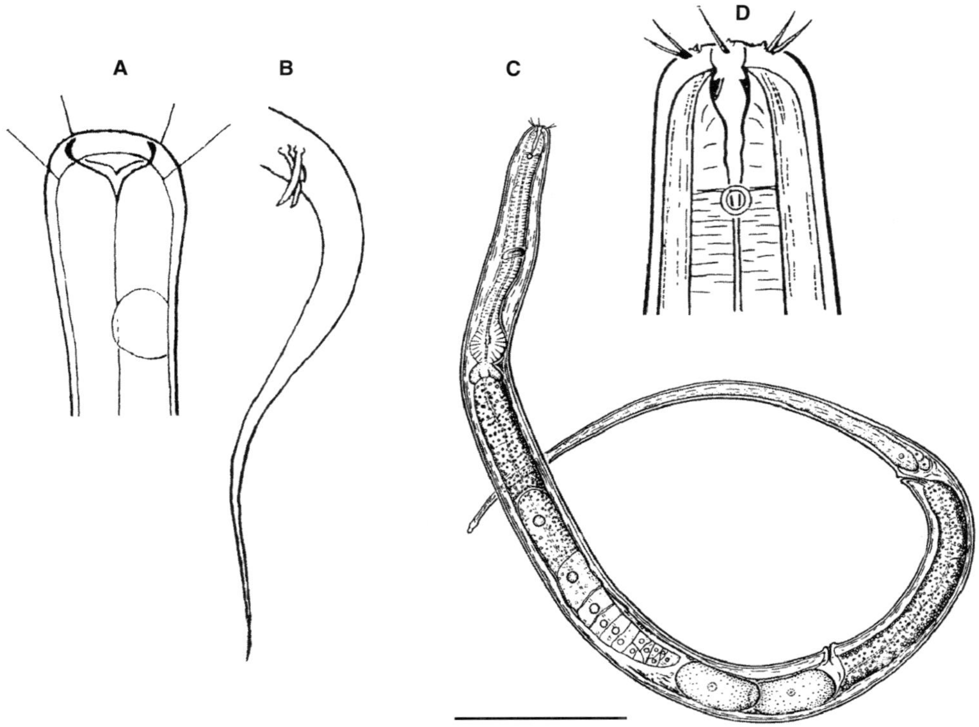

Fig. 18.3 A and B: *Anguimonhystera ampliceps* (Goffart, 1950). A: Male, anterior end; B: Male, tail region (A and B after Goffart, 1950, no scale given). C and D: *Sinanema ginlingensis* (Hoeppli & Chu, 1932). C: Female, entire; D: Female, anterior end (C and D after Hoeppli and Chu, 1932. Scale bars: 50 μm in C, 10 μm in D).

Ocelli usually present. Ventral gland obscure, opening in anterior neck region. Gravid females with one to two eggs in uterus. Vagina may be sclerotized. Vulva about halfway along the body. Spicules arcuate, up to 2 ABDs long. Gubernaculum well developed, usually with dorsal apophysis. Pre-cloacal papillae often present, post-cloacal papillae or setae always present. Tail very long, especially in the female. Amphimictic. Selective deposit feeders, mostly epibenthic in coastal water bodies, exceptionally in athalassic brackish waters.

Type species: *D. monhysteroides* Allgén, 1929

Species found in brackish water: *D. dievengatensis* Jacobs, Van de Velde, Geraert & Vranken, 1990; ? *D. ophthalmophora* Timm, 1952; *D. stagnosa* Lorenzen, 1966.

Genus *Diplolaimelloides* Meyl, 1954 (Figs 18.4E–H)
Diagnosis. Monhysteridae with very small (< 2 μm) somatic setae. Anterior labial sensilla are small papillae; posterior labial and cephalic sensilla are small setae. Fovea a ventrally wound spiral (may appear as circular), 1–2 CBD from anterior end. Cheilostome sub-cylindrical. Second chamber of stoma often denticulate. Ocelli usually present. Ventral gland sometimes well developed, opening in anterior region

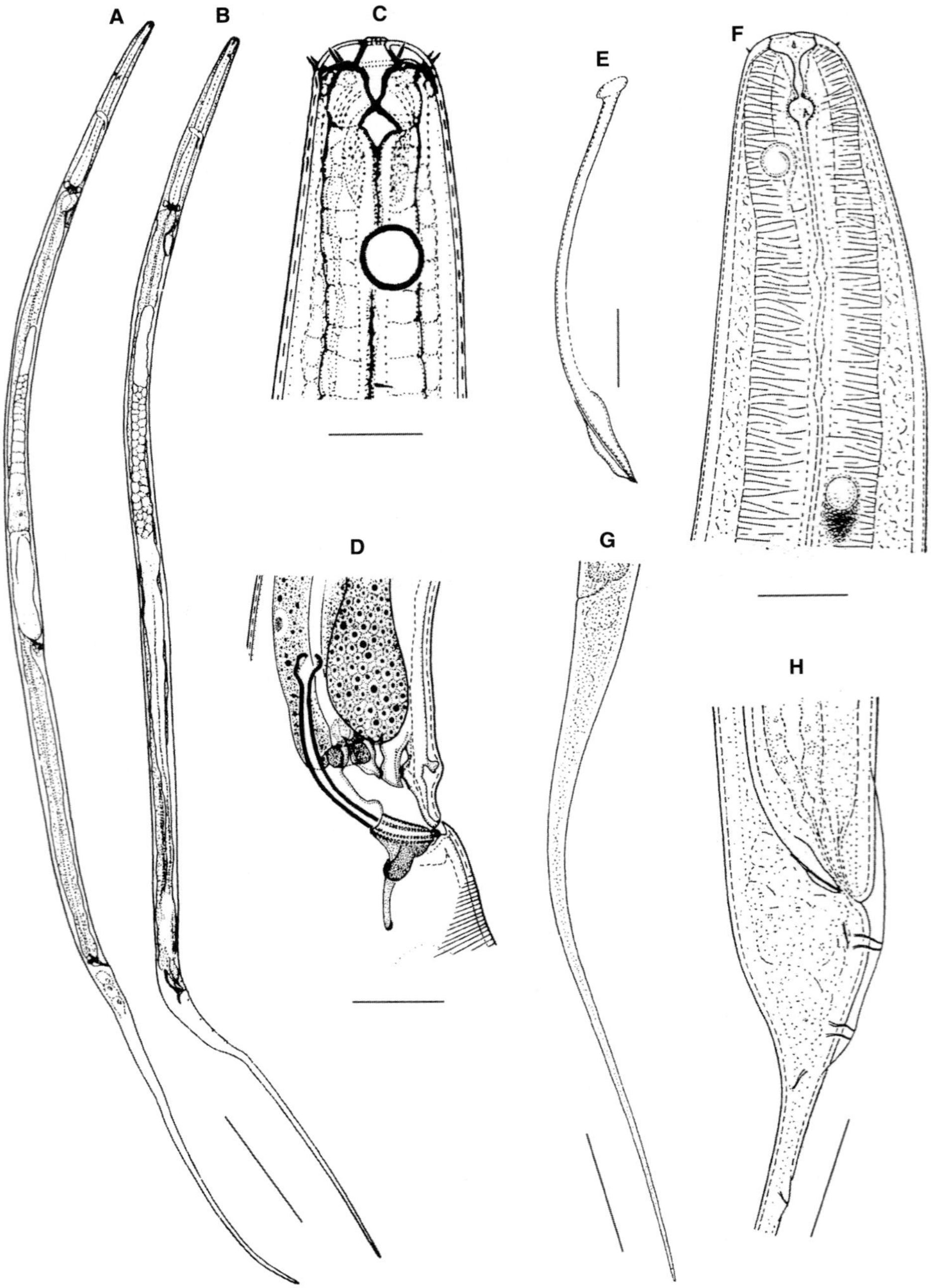

Fig. 18.4 A–D: *Diplolaimella dievengatensis* Jacobs, Van de Velde, Geraert, & Vranken 1990. A: Female, entire; B: Male, entire; C: Female, anterior end; D: Male, cloacal region (A–D after Jacobs *et al.,* 1990). E–H: *Diplolaimelloides delyi* Andrássy, 1958. E: Spicule and gubernaculum; F: Female, anterior end; G: Female, tail; H: Male, cloacal region (E–H after Andrássy, 1958). Scale bars: 100 µm in A and B; 50 µm in G; 20 µm in H; 10 µm in D–F; 5 µm in C.

of neck. Gravid females with one to two eggs in uterus. Vulva halfway along the body in long-tailed species, more posterior in shorter-tailed ones. Spicules slender, arcuate to almost straight, up to 4.4 ABD long. Gubernaculum poorly sclerotized, without dorsal apophysis. Caudal alae (bursa) present, supported by several pairs of post-cloacal papillae. Tail length equal in sexes, but variable according to species. Amphimictic. Microbivorous. Mainly epibenthic in coastal thalassic waters.

Type species: *D. islandicus* (De Coninck, 1943) Meyl, 1954

Species found in brackish water: *D. altherri* Meyl, 1954; *D. delyi* Andrássy, 1958; *D. oschei* Meyl, 1954; *D. palustris* Tsalolikhin, 1985.

Genus *Eumonhystera* Andrássy, 1981 (Figs 18.5A–C)
Diagnosis. Monhysteridae with submedian setae less than 30% of CBD. Crystalloids absent. Anterior margin of fovea usually more than 1 CBD from anterior end (exc. *E. geraerti*: 0.8, *E. barbata*: 1.0). Perioral platelets present. Stoma undivided, with or without a dorsal denticle. Pharynx with slightly swollen base (but more bulb-like in *E. hungarica* and *E. pseudobulbosa*). Ocelli (always?) absent. Ventral gland variously developed according to species (obscure to well developed). Uterus in gravid females with few eggs (usually 1–2). Post-vulvar gland usually absent or obscure, but clearly present in *E. andrássyi*, *E. filiformis* and *E. rustica*. Males only known in a few species (*E. andrássyi*, *E. filiformis*, *E. similis*). Spicula short (less than 2 ABD) and wide. Gubernaculum weakly developed. Tail usually slender with filiform posterior part often dorsally bent and spinneret slightly swollen. Tail longer than *V–a* distance, except in *E. dispar*, *E. sudanensis* and *E. tuporis*. Mostly parthenogenetic, oviparous. Non-selective deposit feeders. Freshwater.

Type species: *Eumonhystera vulgaris* (de Man, 1880) Andrássy, 1981

Other species:
E. andrássyi (Biro, 1969) Andrássy, 1981
E. barbata Andrássy, 1981
E. dispar (Bastian, 1865) Andrássy, 1981
E. elegans (Allgén, 1928) Jacobs, 1987
E. filiformis (Bastian, 1865) Andrássy, 1981
E. geraerti Eyualem-Abebe & Coomans, 1996
E. gracilior (Johnston, 1938) Andrássy, 1981
E. hungarica Andrássy, 1981
E. huruii Peng, Eyualem-Abebe & Coomans, 2002
E. longicaudatula (Gerlach & Riemann, 1973) Andrássy, 1981
E. media Hernández & Jordana, 1988
E. minuta (Filipjev, 1929) Jacobs, 1987
E. mwerazii (Meyl, 1957) Andrássy, 1981
E. papuana (Daday, 1899) Andrássy, 1981
E. patiens Armandáriz, Agudo & Hernández, 1991
E. pseudobulbosa (Daday, 1899) Andrássy, 1981
E. rustica (Bütschli, 1873) Andrássy, 1981
E. sibirica Gagarin, 2003
E. similis (Bütschli, 1873) Andrássy, 1981

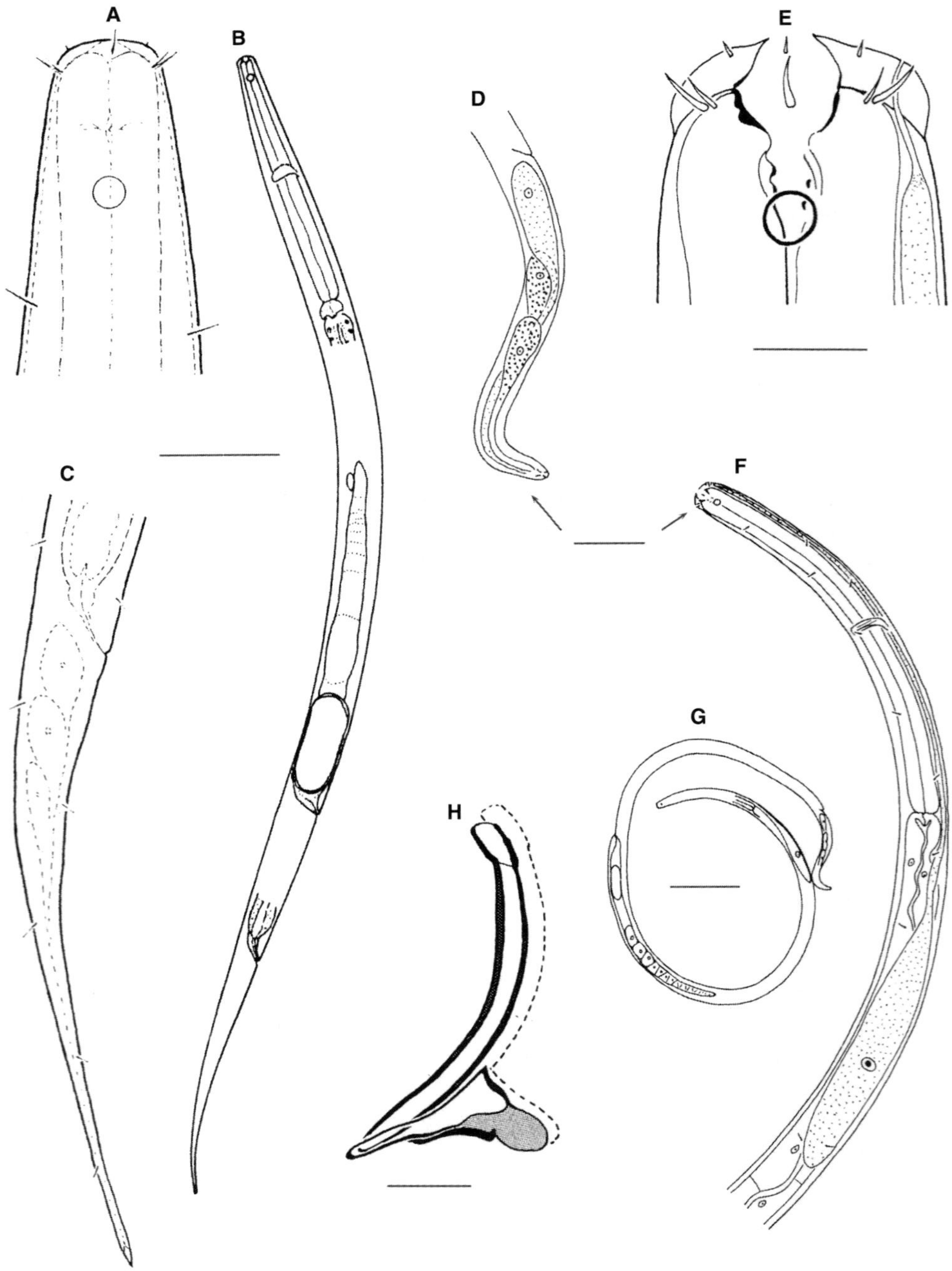

Fig. 18.5 A–C: *Eumonhystera vulgaris* (de Man, 1880). A: Female, anterior end; B: Female, entire; C: Female, tail (A, C after Andrássy, 1981; B after Eyualem-Abebe and Coomans, 1996b). D–H: *Gammarinema gammari* Kinne & Gerlach, 1983. D: Female, tail; E: Female, anterior end; F: Female, anterior body region; G: Female, entire; H: Spicules and gubernaculum (D–H after Lorenzen, 1986). Scale bars: 200 μm in G; 50 μm in B, D and F; 20 μm in C; 10 μm in A, E and H.

E. simplex (de Man, 1880) Andrássy, 1981
E. subfiliformis (Cobb, 1918) Andrássy, 1981
E. sudanensis Zeidan, Jacobs & Geraert, 1990
E. tatrica (Daday, 1897) Andrássy, 1981
E. tuporis Gagarin, 1991

Species inquirendae: *E. dadayi* (Goodey, 1963) Andrássy, 1981; *E. fuelleborni* (Daday, 1908) Andrássy, 1981; *E. paradoxa* (Allgén, 1926) Jacobs, 1987; *E. propinqua* (Daday, 1905) Andrássy, 1981.

Genus *Gammarinema* Kinne & Gerlach, 1953 (Figs 18.5D–H)
Diagnosis. Monhysteridae with somatic setae less than 20% of CBD long. Outer labial sensilla large papillae or small setae; outer labial and cephalic sensilla setiform. Fovea less than 1.5 CBD from anterior end. Stoma large, sclerotized, posterior part denticulate. Ocelli absent. Ventral gland well developed, opening in anterior region of neck or in lip region. Gravid females frequently with several eggs in uterus. Vagina more or less sclerotized. Spicules ventrally arcuate. Gubernaculum usually well developed and then with prominent dorsal apophyses. Pre-cloacal papillae may be present. Tail conoid to cylindroid with prominent spinneret. Amphimictic. Non-selective deposit feeders. Only known from gill chambers of various crustaceans.

Type species: *G. gammari* Kinne & Gerlach, 1953

Species found on freshwater crustaceans: *G. cambari* (Allgén, 1933) Osche, 1955 (on gills of freshwater crayfish *Cambarus*); *G. paratelphusae* (Farrouqui, 1967) Sudhaus, 1974 (on gills of freshwater crab *Paratelphusa*).

Genus *Geomonhystera* Andrássy, 1981 (Figs 18.6A–D)
Diagnosis. Monhysteridae with many somatic setae, less than 30% of CBD long. Outer labial and cephalic setae well developed in terrestrial species, less than 60% of CBD. Fovea 1–2 CBD from anterior end. Cheilostome wide, weakly cuticularized; stoma funnel-shaped. Rectum in terrestrial species longer than ABD. Ocelli absent. Ventral gland poorly developed in terrestrial species. Ovary and testis long, often reaching cardia region. Vulva posteriorly located ($V > 75\%$). Spicules ventrally arcuate and rather slender (1.5–2 times ABD). Gubernaculum without dorsal apophysis in terrestrial species. Median pre-cloacal papilla may be present. Tail relatively short and stout, conoid to elongate conoid, ventrally curved; longer than *V–a* distance; with short spinneret. Amphimictic or parthenogenetic. This genus comprises the only real terrestrial species of the family, mostly occurring in moist soils, in moss and under bark, but one species has also been found in thermal water from fumaroles.

Type species: *G. villosa* (Bütschli, 1873) Andrássy, 1981

Species found in thermal water: *G. aenariensis* (Meyl, 1953) Andrássy, 1981.

Genus *Monhystera* Bastian, 1865 (Fig. 18.7)
Diagnosis. Monhysteridae with sub-median somatic setae less than 50% of CBD long. Crystalloids present, mostly abundant, but sometimes sparse. Outer labial and cephalic setae less than 30% of CBD. Anterior margin of amphideal fovea less than

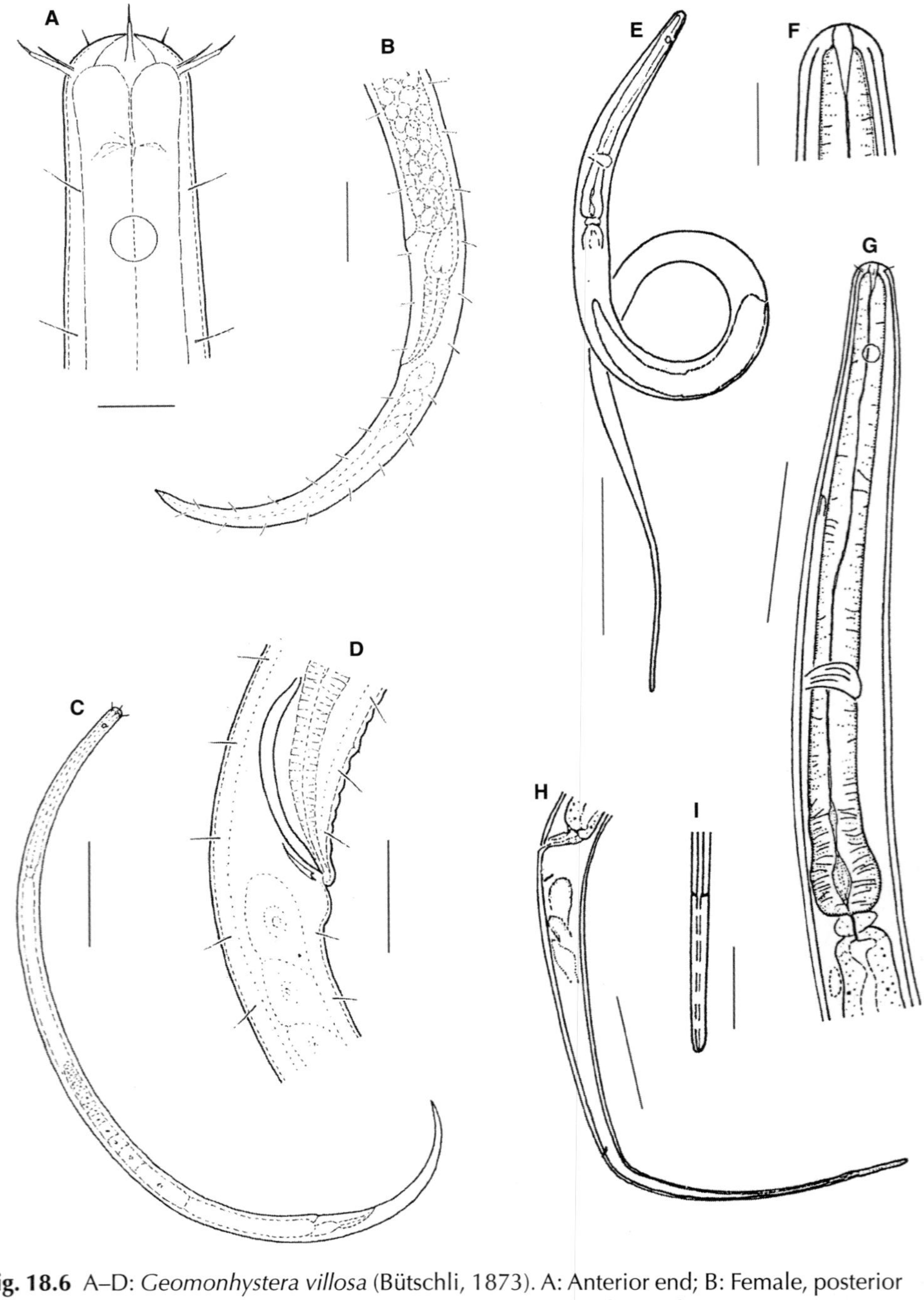

Fig. 18.6 A–D: *Geomonhystera villosa* (Bütschli, 1873). A: Anterior end; B: Female, posterior body region; C: Female, entire; D: Male, cloacal region (A–D after Andrássy, 1981). E–I: *Monhystrella woitorum* Eyualem-Abebe & Coomans, 1996. E: Female, entire; F: Anterior end; G: Neck region; H: Female, tail; I: Spinneret (E–I after Eyualem-Abebe and Coomans, 1996a). Scale bars: 100 µm in C; 50 µm in E; 30 µm in B; 20 µm in D, G and H; 10 µm in A; 5 µm in F and I.

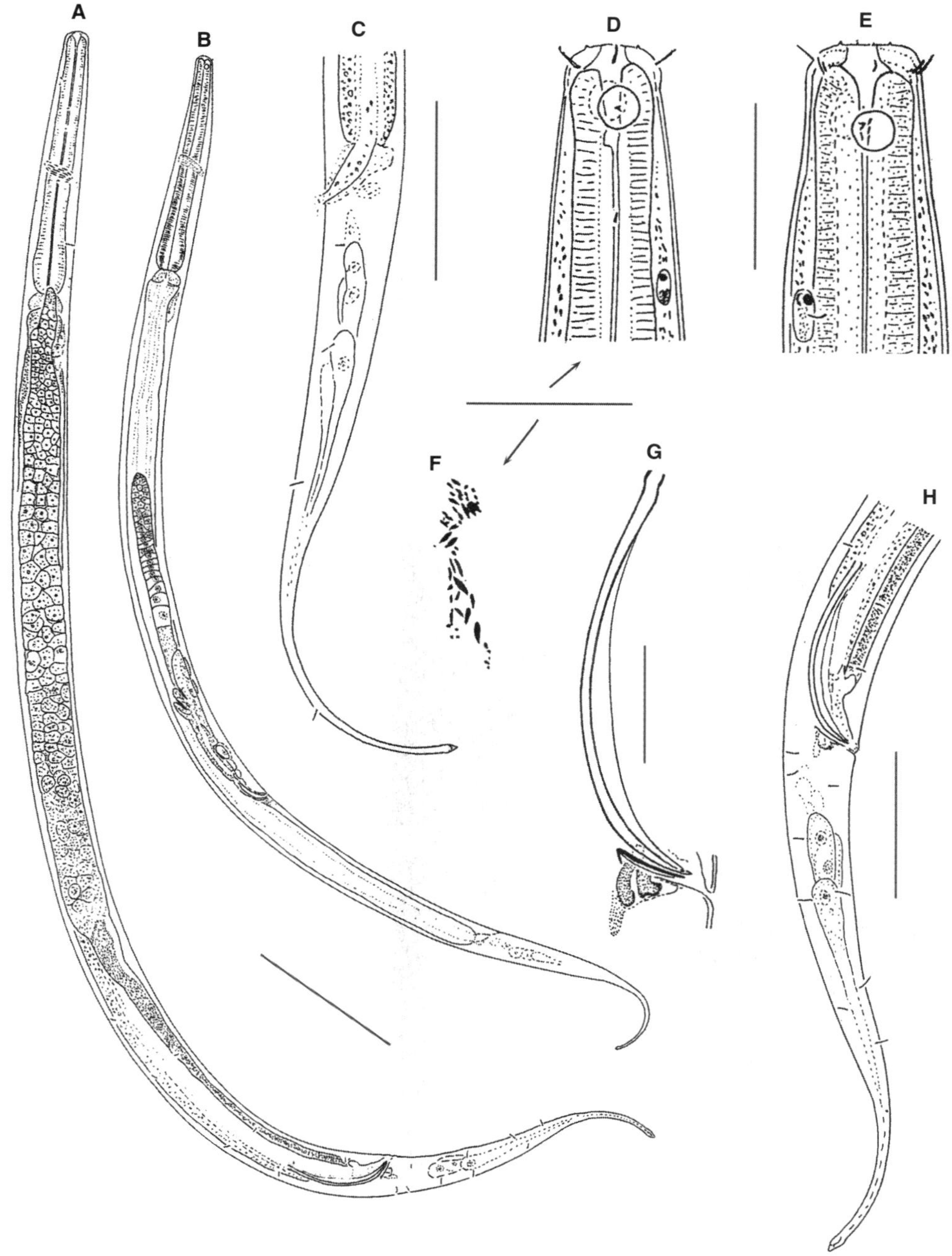

Fig. 18.7 A–H: *Monhystera stagnalis* Bastian, 1865. A: Male, entire; B: Young female, entire; C: Female, tail region; D: Female, anterior end; E: Male, anterior end; F: Crystalloids; G: Spicule and gubernaculum; H: Male, tail region (A–H original drawings by A. Coomans). Scale bars: 100 μm in A and B; 50 μm in C and H; 20 μm in D–G.

1.2 times CBD from anterior end. Perioral platelets present. Cheilostome wide, with weakly cuticularized wall. Stoma completely surrounded by pharyngeal tissue, wide funnel-shaped (when open), not subdivided, without or with one dorsal denticle near basis. Basis of the pharynx more or less widened. Dorsal to sub-dorsal ocelli frequently present (sometimes obscure in fixed specimens). Ventral gland present opposite anterior end of intestine, opening in neck region anterior to nerve ring. Gonad long, i.e. often reaching anterior part of intestine. Ovary sometimes sinuous or with reflexed tip in gravid females; in the latter uterus may contain many eggs and/or juveniles. Postvulvar gland often present. Spicules variable, from short, relatively wide and strongly curved to very long and slender. Gubernaculum well developed, usually with a prominent dorsocaudal apophysis. Female tail as long as or shorter than vulva–anus distance or only slightly longer in females with extended uterus. Male tail typically at first elongated conical and ventrally bent, then cylindrical and more or less dorsally bent. Tail tip slightly swollen. Mainly freshwater, but also terrestrial, estuarine and marine.

Amphimictic with variable sex ratio (depending on species and on environmental conditions); ovi-, ovovivi- or viviparous. Non-selective deposit feeders.

Type species: *M. stagnalis* Bastian, 1865

Other species:

M. africana Andrássy, 1964
M. altherri (Andrássy, 1981) Jacobs, 1987
M. coomansi Jacobs & Heyns, 1992
M. euromacramphis Jacobs, 1987
M. fasciculata Skwarra, 1921
M. gabaza Joubert & Heyns, 1980
M. lemani Juget, 1969
M. longicaudata Bastian, 1865
M. macramphis Filipjev, 1929
M. magnacephala Joubert & Heyns, 1980
M. nubiae Eyualem-Abebe & Coomans, 1996
M. paludicola de Man, 1880
M. paramacramphis Meyl, 1954
M. psammophila Juget, 1969
M. psilocephalus (Onorato de Cillis, 1917) Meyl, 1960
M. riemanni Jacobs & Heyns (in Jacobs, 1987a)
M. rivularis Bastian, 1865
M. robustospiculum Jacobs & Heyns (in Jacobs, 1987a)
M. somereni Allgén, 1952
M. taaiboschiensis Joubert & Heyns, 1980
M. uncispiculata Gagarin, 1979
M. wangi Wu & Hoeppli, 1929

Genus *Monhystrella* Cobb, 1918 (Figs 18.6E–I)
Diagnosis. Monhysteridae of small size (usually $L < 0.5$ mm, rarely up to 0.77 mm). Cuticle finely striated (striae < 0.25 μm apart), often with small somatic setae. Six inner labial sensilla papilliform; six outer labial and four cephalic setae usually in one circle. Amphidial fovea usually 1.5–2 or more CBD from anterior end, rarely

closer to anterior end (about 1 × CBD in *M. parvella*). Ocelli rarely present. Cheilostome small, tubiform, funnel-shaped or conoid, often with well cuticularized walls. Stoma narrow, tube-like to funnel-shaped; with dorsal denticle. Pharynx with muscular basal bulb, occasionally divided into two parts. Anterior part of intestine (*progaster*) often globular. Female reproductive system short with ovary comprising few oocytes and gravid females mostly with only one egg in uterus; vulva near mid-body. Males rare or unknown in freshwater species (except in *M. macrura*). Spicules rather short and slender. Tail long and filiform in both sexes. Spinneret typically a long and slender cone or cylinder. Marine and athalassic saline water, thermal and other freshwater; halophilic and thermophilic; some species euryhaline.

Type species: *M. plectoides* Cobb, 1918

Other species found in inland waters:

M. arsiensis Eyualem-Abebe & Coomans, 1996
M. atteae Eyualem-Abebe & Coomans, 1996
M. ethiopica Eyualem-Abebe & Coomans, 1996
M. fukiensis (Hoeppli & Chu, 1932) Andrássy, 1981
M. gracilis Khera, 1966
M. hastata Andrássy, 1968
M. hoogewijsi Eyualem-Abebe & Coomans, 1996
M. iranica Schiemer, 1965
M. jacobsi Eyualem-Abebe & Coomans, 1996
M. lepidura (Andrássy, 1963) Andrássy, 1968 (with four subspecies)
M. longistoma (Khera, 1971) Andrássy, 1981
M. macrura (de Man, 1880) Andrássy, 1981
M. paramacrura (Meyl, 1954) Andrássy, 1968
M. parelegantula (De Coninck, 1943) Andrássy, 1981
M. parvella (Filipjev, 1931) Jacobs, 1987 (with two subspecies)
M. spiralis (Wu & Hoeppli, 1929) Andrássy, 1981
M. stewarti (Khera, 1971) Andrássy, 1981
M. thermophila (Meyl, 1953) Andrássy, 1981
M. woitorum Eyualem-Abebe & Coomans, 1996

Species inquirenda: *M. bulbifera* (de Man, 1880) Steiner, 1920

Genus *Sinanema* Andrássy, 1960 (Figs 18.3C and D)
Diagnosis. Monhysteridae with small (L = 0.4–0.6 mm) body size. Inner labial sensilla papilliform; outer labial and cephalic sensilla setiform, in one circlet. Amphidial fovea at 1.4–2 CBD from anterior end. Cheilostome and stoma tube-like, latter with three small denticles (one dorsal, two ventrosublateral ones) in anterior part. Pharynx with muscular basal bulb. Cardia lobed (with three cardiac glands?). Nerve ring halfway along neck length. Secretory–excretory pore usually between anterior end and nerve ring. Ovary medium-sized and reflexed. Vulva at 50–66%. Males unknown. Tail long, elongate conoid at first, then filiform with a short spinneret at the end of slight terminal swelling. Freshwater and thermal springs.

Type species: *S. ginlingensis* (Hoeppli & Chu, 1932) Andrássy, 1960

Other species:

S. mongolicum Tsalolikhin, 1985
Species inquirendae: *S. godeti* (Steiner, 1920) Jacobs, 1987; *S. mysorensis* (Moorthy, 1938) Jacobs, 1987.
Note: Position of the genus uncertain due to presence of a reflexed ovary.

Genus *Thalassomonhystera* Jacobs, 1987 (Fig. 18.8)
Diagnosis. Monhysteridae with submedian somatic setae less than 30% of CBD. Inner labial sensilla usually papilliform, sometimes small setiform; cephalic setae sometimes equal to or slightly shorter than, but usually longer than outer labial setae; always shorter than CBD. Stoma undivided, funnel-shaped. Pharynx with slightly swollen base. Cardia variously shaped according to species. Ocelli mostly absent. Ventral gland usually well developed, opening in anterior neck region. Uterus of gravid females with up to two eggs. Spicules 1.2–2.2 ABD long, varying from arcuate and moderately wide to setaceous and slender. Gubernaculum either small and robust with dorsal apophysis or tube-like without apophysis. Tail elongate conoid with cylindrical terminal part, the former ventrally curved, the latter often dorsally bent. Tail of female shorter than *V–a* distance. Most species amphimictic. Non-selective deposit feeders. Most species are marine or occur in thalassic coastal waters; two species so far reported from freshwater.

Type species: *T. parva* (Bastian, 1865) Jacobs, 1987

Freshwater species:

T. amabilis (Gagarin, 1997) n.comb.
syn. *Monhystera amabilis* Gagarin, 1997
T. traesti Eyualem-Abebe, Peng & Coomans, 2001

Genus *Tridentulus* Eyualem-Abebe & Coomans, 1995 (Figs 18.9A–D)
Diagnosis. Monhysteridae with small to medium body size (L = 0.4–1.16 mm). Cuticle finely striated, with somatic setae. Inner labial sensilla papilliform; outer labial and cephalic sensilla setiform, in one circlet, the six labial setae longer than the four cephalic ones. Amphideal fovea at 0.8–1.8 CBD from anterior end. Cheilostome wide; stoma funnel-shaped, with three small denticles, one in each sector. Pharynx cylindrical with slightly wider basis. Ventral gland well developed. Males rare or unknown. Tail elongate conoid, tapering to a conoid (beak-shaped) spinneret. Tail length equal to or larger than *V–a* distance. Freshwater.

Type species: *T. floreanae* (Eyualem-Abebe & Coomans, 1995) Eyualem-Abebe & Coomans, 1995

Other species:

T. bidenticulatus (Gagarin, 1997) Gagarin, 2001
T. brzeskii Gagarin & Gusakov, 2000
T. minor Gagarin, 2001
T. obscurus Gagarin, 2001
T. palustris Gagarin & Holovachov, 2001
T. fluviatilis Gagarin, 2004.

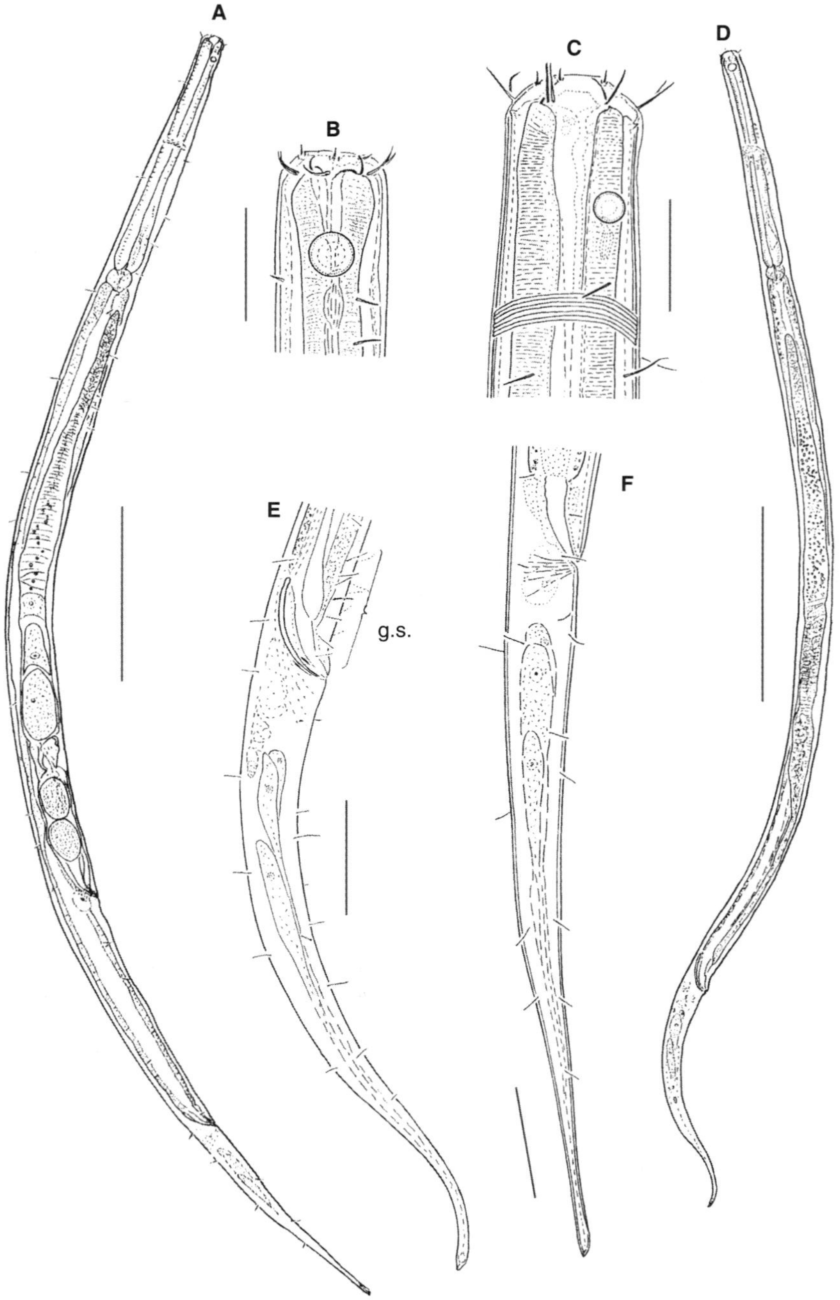

Fig. 18.8 A–F: *Thalassomonhystera traesti* Eyualem-Abebe, Peng Yun Liang & Coomans, 2001. A: Female, entire; B: Male, anterior end; C: Female, anterior end; D: Male, entire; E: Male, tail region; F: Female, tail region (A–F after Eyualem-Abebe *et al.*, 2001; g.s.: genital setae). Scale bars: 200 μm in A and D; 40 μm in E and F; 20 μm in B and C.

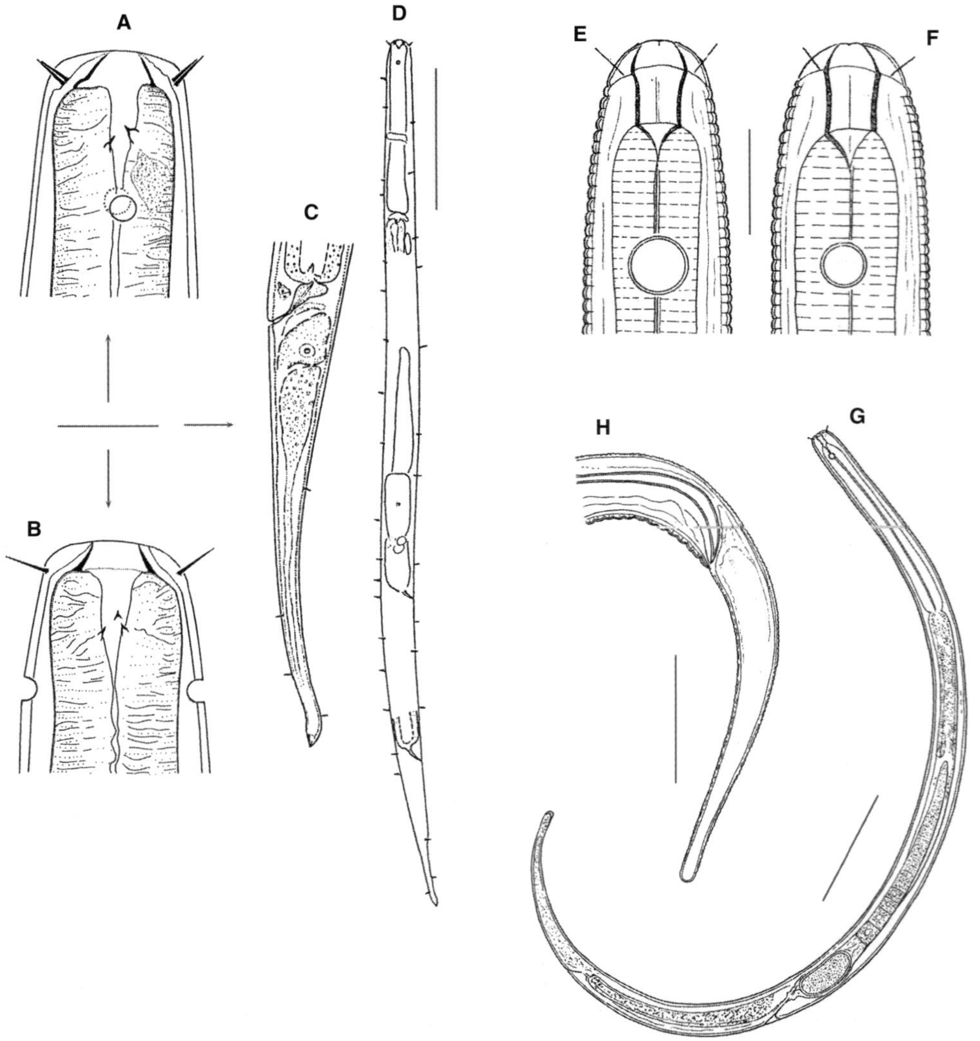

Fig. 18.9 A–D: *Tridentulus floreana* (Eyualem-Abebe & Coomans, 1995). A: Anterior end, lateral view; B: Anterior end, dorsoventral view; C: Female, tail; D: Female, entire (A and B after Eyualem-Abebe and Coomans, 1995a; C and D after Eyualem-Abebe and Coomans, 1995b). E–H: *Hofmaenneria brachystoma* (Hofmaenner in Hofmaenner & Menzel, 1914). E: Male, anterior end; F: Female, anterior end; G: Female, entire; H: Male, tail region (E–H after Hofmänner and Menzel, 1915). Scale bars: 100 µm in D and G; 25 µm in C and H; 10 µm in A, B, E and F.

Superfamily SPHAEROLAIMOIDEA Filipjev, 1918

Diagnosis. Cuticle transversely striated or clearly annulated; eight sub-cephalic setae may be present whereby the longest sub-cephalic setae are longer than the longest cephalic setae; usually two testes, anterior one at left or right side of intestine, posterior one at opposite side; single (anterior) ovary at left or right side of intestine; tail often with two to three terminal setae.

Two families: Sphaerolaimidae Filipjev, 1918 and Xyalidae Chitwood, 1951.

Key to families

1. The 4 cephalic setae > 6 outer labial setae; stoma barrel-shaped, only at basis surrounded by pharyngeal tissue; ventral gland present . . . . .Sphaerolaimidae
2. The 4 cephalic setae < or = 6 outer labial setae; stoma usually funnel-shaped, completely or only at basis surrounded by pharyngeal tissue; ventral gland absent (or invisible under LM) . . . . .Xyalidae

Family SPHAEROLAIMIDAE Filipjev, 1918

Diagnosis. Cuticle striated or annulated. Six outer labial setae and four cephalic setae in one circlet, with cephalic ones longer than labial ones. Eight groups of sub-cephalic setae present. Stoma wide, barrel-shaped with longitudinal ribs; only at basis surrounded by pharyngeal tissue. Inner wall of pharynx with thick cuticle. Ventral gland present, opening behind nerve ring. Males usually with two testes, anterior one either at left or at right side of intestine, posterior one at opposite side. Females with one (anterior) ovary at left or right side of intestine. Marine.

Type genus: *Sphaerolaimus* Bastian, 1865

Genus incerta sedis: *Hofmaenneria* Gerlach & Meyl, 1957 (Figs 18.9E–H)
This genus has been assigned to Sphaerolaimidae by Andrássy (1981) and to Xyalidae by Lorenzen (1981). It possesses characters of both these families. The species are small to medium-sized (0.4–2.0 mm) and have a transversally striated or finely annulated cuticle. Head with ten setae of which four cephalic setae equal to or longer than the six outer labial setae. Sub-cephalic setae absent. Amphideal fovea showing sexual dimorphism (larger in males than in females); about 0.8 to almost 2 CBD behind anterior end. Buccal cavity wide, cylindrical or barrel-shaped, with sclerotized walls. Stoma not completely surrounded by pharyngeal tissue. Small denticles may be present in anterior part of pharynx (or posterior part of stoma ?). Ventral gland present. Only one (anterior) gonad, located on right side of intestine (at least in *Hofmaenneria niddensis*). Spicules slender, ventrally arcuate; gubernaculum small, without dorsal apophysis. Tail elongate–conoid, with terminal pore (spinneret absent), without terminal setae. Freshwater or brackish water.

Type species: *H. brachystoma* (Hofmänner in Hofmänner & Menzel, 1914) Gerlach & Meyl, 1957

Other species:

H. elongata Gagarin, 1987
H. gratiosa Alekseev, 1983
H. hazanensis Mulvey, 1969
H. niddensis (Skwarra, 1921) Schneider, 1940
H. optata Alekseev, 1983

Family XYALIDAE Chitwood, 1951

Diagnosis. Cuticle annulated. Six outer labial setae and four cephalic setae in one circlet, with cephalic setae shorter than or at most equal to labial ones. Often additional cephalic setae and occasionally eight groups of sub-cephalic setae present. Stoma usually funnel-shaped, completely or only at basis surrounded by pharyngeal tissue. Ventral gland mostly absent (or invisible under LM). Males with usually two testes, anterior one to the left side of the intestine, posterior one (may be absent) to the right side. Females with one (anterior) ovary on left side of intestine. Majority of species marine, but some species occur in inland waters and terrestrial habitats.

Type genus: *Xyala* Cobb, 1920

Genera found in inland waters:

Daptonema Cobb, 1920
Sacrimarinema Shoshin, 2001
Theristus Bastian, 1865

Note: *Mesotheristus* Wieser, 1956 and *Mongolotheristus* Tsalolikhin, 1985 are here considered synonymous with *Daptonema* (see later).

Key to inland water genera of the family Xyalidae

1. Tail with two terminal setae; vulva at three-fourths of body length . . . .*Daptonema*
   Tail without terminal setae; vulva at two-thirds of body length 2
2. Lip region with 10–12 long setae; males without complex supplements . . . .*Theristus*
   Lip region with 14 long setae; males with complex supplements . . . .*Sacrimarinema*

Genus *Daptonema* Cobb, 1920 (Figs. 18.10F–J)

Diagnosis. Xyalidae with small to medium-sized (*L* in freshwater species: 0.8–1.6 mm) and stout species. Cuticle annulated, with lateral field. Somatic setae present. Lip region with 10 or 12 seta (these setae may be segmented when long). Cheilostome wide, dome-shaped; stoma funnel-shaped. Ventral gland and excretory pore absent or obscure. Usually two testes with anterior one on left and posterior one on right side of intestine. Spicules usually about one CBD long and strongly bent. Gubernaculum often with lateral guiding pieces and rarely also with dorsocaudal apophysis. Ovary always left of intestine. A post- and prevulval uterine sac may be present. Tail cylindrical in its posterior part, with two (rarely three or four) terminal setae. Most species marine, some in brackish water or freshwater.

Type species: *D. fissidens* Cobb, 1920

Inland species:

D. altaicus (Tsalolikhin, 1985) n. comb.
  syn. *Mongolotheristus altaicus* Tsalolikhin, 1985
D. aquaedulcis (Gagarin, 1987) Gagarin, 1993
  syn. *Cylindrotheristus aquadulcis* Gagarin, 1987
D. crassissimus (Ditlevsen, 1911) n. comb.
  syn. *Monhystera crassissima* Ditlevsen, 1911
    Theristus (*Mesotheristus*) *crassissimus* (Ditlevsen, 1911) Wieser, 1956
    Mesotheristus crassisimus (Ditlevsen, 1911) Wieser, 1956

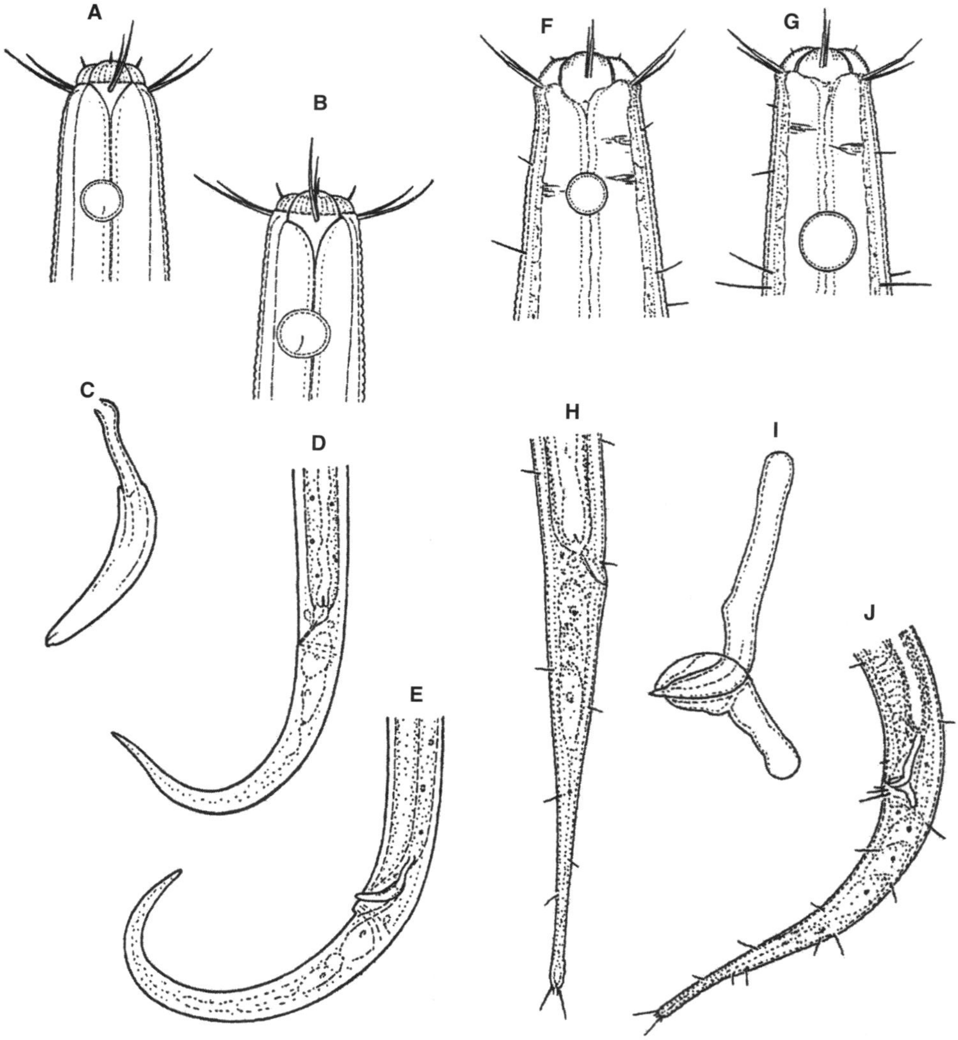

Fig. 18.10 A–E: *Theristus vesentinae* Andrássy, 1962. A: Female, anterior end; B: Male, anterior end; C: Spiculum; D: Female, tail region; E: Male, tail region. F–J: *Daptonema dubium* (Bütschli, 1873). F: Female, anterior end; G: Male, anterior end; H: Female, tail region; I: Spiculum and gubernaculum; J: Male, tail region (A–E after Andrássy, 1962; F–J after Andrássy, 1995; no scale given).

Monhystera crassoides Micoletzky, 1913
Theristus pseudesetosus Filipjev, 1918
Theristus subsetosus Schneider, 1943
D. dubium (Bütschli, 1873) Lorenzen, 1977
syn. *Monhystera dubia* Bütschli, 1873
Theristus dubius Micoletzky, 1922
Allomonhystera tripapillata Micoletzky, 1923
Mesotheristus dubius (Bütschli, 1873) Andrássy, 1995

D. fortis Gagarin, 1993
D. inversum Alekseev, 1984
D. limnobia Wu & Liang, 2000
D. osadchikhae (Tchesunov, 1980) Tchesunov, 1990
syn. *Mesotheristus osadchikhae* Tchesunov, 1980
D. sibiricum Gagarin, 2000
D. timoshkini (Gagarin, 2001) n. comb.
syn. *Mongolotheristus timoshkini* Gagarin, 2001

Remarks: Wieser (1956) regarded *Daptonema* as a subgenus of *Theristus* and proposed several other subgenera within that genus. One of these, *Mesotheristus*, was raised to genus level by Chitwood and Murphy (1964). Hopper (1969) considered the presence of a lateral field as typical for *Mesotheristus*. Cobb (1920) did indeed not mention the presence of a lateral field in his description of *Daptonema*, but he used the presence of a lateral field as a character to identify *Daptonema* in his key to genera. Lorenzen (1977) synonymized *Mesotheristus* with *Daptonema* because he could not find reliable characters to separate both genera. Andrássy (1995) did not agree with this action and considered *Mesotheristus* a valid genus with as most important characters: presence of lateral fields, nearly rectangularly curved spicules, gubernaculum with well-developed caudal apophysis and tail with long terminal setae. Some confusion may arise from the fact that Cobb (1920) could only observe the spicular apparatus in dorso-ventral position. The exact shape in lateral view is therefore unknown. From comparison with published illustrations of subsequently described species it is clear that the spicules of the type species are also strongly curved. The gubernaculum of *D. fissidens* has well-developed lateral pieces, but it cannot be ascertained whether it has a dorsal apophysis. However, it is not exceptional to find species with and without such an apophysis within one and the same genus. Long terminal setae are a character of *Daptonema*. In conclusion, we believe that not a single character can be found to separate *Mesotheristus* unambiguously from *Daptonema*. *Mongolotheristus* is also considered as a junior synonym of *Daptonema*. Indeed the so-called uniqueness of supplements in the male is not a valid character to separate these species from *Daptonema* since various types of supplements occur in the latter genus. Other characters are also in agreement with the generic diagnosis of *Daptonema*.

Genus *Sacrimarinema* Shoshin, 2001 (Fig. 18.11)
Diagnosis. Medium-sized (L = 1.3–2.6 mm). Cuticle finely annulated, with sparse somatic setae. Anterior sensilla all setiform with an anterior circlet of 6 small inner labial setae and a posterior circlet of 14 long setae (6 outer labial setae + 4 cephalic setae and 4 additional lateral setae). Outer labial setae about three-fourths of the lip region width, longer than the cephalic setae. Amphideal fovea round, slightly larger in males than in females; its anterior margin 0.7–1.4 CBD from anterior end. Cheilostome wide, dome-shaped; stoma funnel-shaped, without denticles. Pharynx cylindroid, without a basal bulb. Cardia elongate. Female reproductive system monodelphic, prodelphic. Vagina at two-thirds of body length. Vagina 1.0–2.5 CBD long. Uterus of gravid females with a single egg. Males with two testes, anterior one on left side, posterior one on right side of intestine. Spicules slender and long to very long (2.3–9.2 ×ABD). Gubernaculum long, slipper-shaped. 5–10 complex medioventral

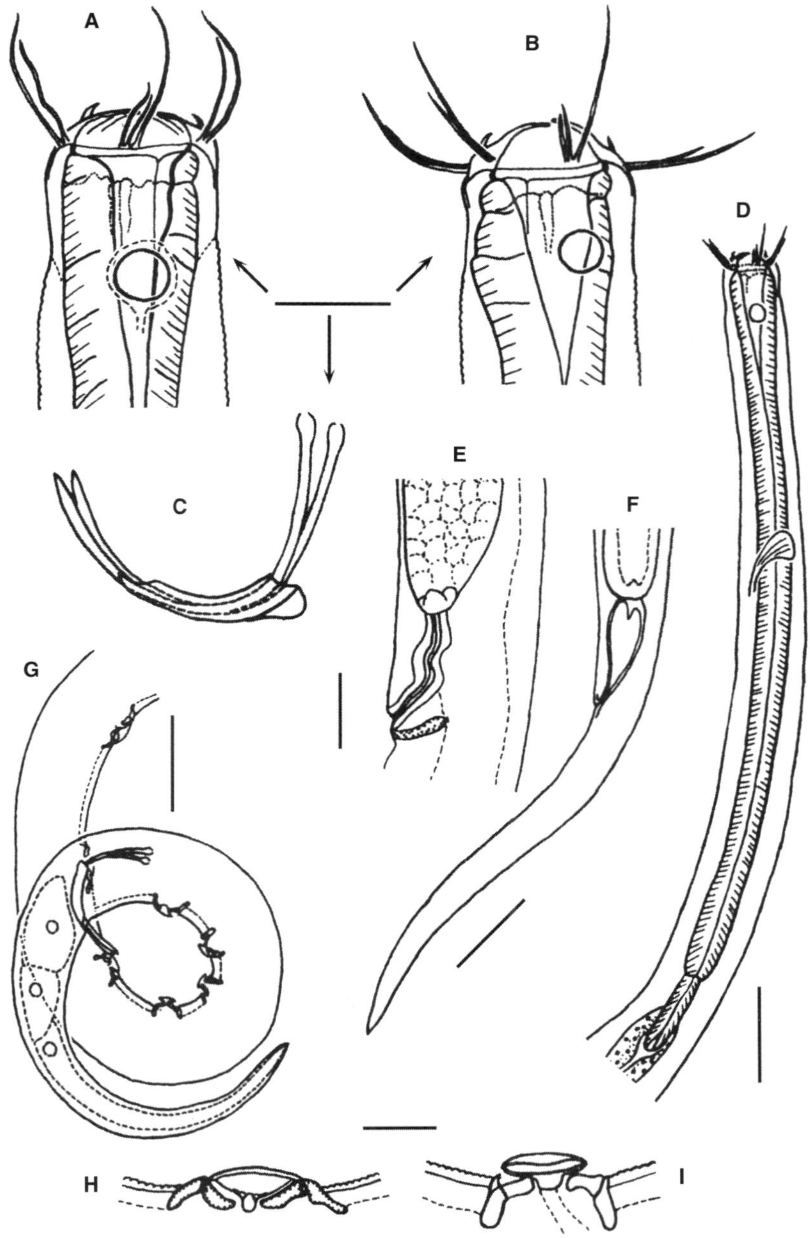

Fig. 18.11 *Sacrimarinema tatushae* Shoshin, 2001. A: Male, anterior end; B: Female, anterior end; C: Spicules and gubernaculum; D: Male, neck region; E: Vagina and vulva region; F: Female, tail; G: Male, posterior body region; H: Expanded supplement; I: Contracted supplement (A–I after Shoshin, 2001). Scale bars: 50 μm in D, F and G; 30 μm in E; 20 μm in A–C; 10 μm in H and I.

supplements present. Tail elongate conoid; terminal setae absent. Three species known, all from Lake Baikal.

Type species: *S. tatushae* Shoshin, 2001

Other species:
S. allae Shoshin, 2001
S. ljajiae Shoshin, 2001

Genus *Theristus* Bastian, 1865 (Figs 18.10A–E)
Diagnosis. Xyalidae with small to medium-sized species (*L* in freshwater species: 0.6–1.7 mm). Cuticle annulated, with somatic setae. Lip region with 10 or 12 setae (when 12, two additional setae occur ventrally from the lateral outer labial setae). Ocelli occasionally present. Cheilostome wide, dome-shaped; stoma narrow, funnel-shaped. Vulva at two-thirds of body length. Testes one or two. Spicules usually curved, sometimes very long, but mostly as long as ABD. Gubernaculum slipper-shaped, with or without dorsal apophysis. Tail elongate conoid, ventrally arcuate, without terminal setae. Mostly marine, a few species found in freshwater or brackish water, occasionally in groundwater and moist soil.

Type species: *T. acer* Bastian, 1865 (marine)

Inland water species:

T. agilis (de Man, 1880) Filipjev, 1918
T. athesinus Andrássy, 1962
T. borosi Andrássy, 1958 (in brackish water)
T. kaszabi Andrássy, 1977
T. ruffoi Andrássy, 1959 (groundwater and other freshwater)
T. pannonicus Andrássy, 1985 (in soil)
T. tessae Heyns & Coomans, 1989
T. vesentiniae Andrássy, 1962
T. wegelinae Andrássy, 1962 (in groundwater)

Species inquirendae: *T. helveticus* (Steiner, 1914) Andrássy, 1981; *T. lingi* (Hoeppli & Chu, 1932) Andrássy, 1960; *T. parasiticus* (Penso, 1938) Andrássy, 1981.

Superfamily SIPHONOLAIMOIDEA Filipjev, 1918

Diagnosis. Anterior sensilla in two or three circlets, either ten papillae and four setae or six papillae and ten setae. Amphidial fovea round. Stoma small; pharynx with often well-developed basal bulb. Gonads usually paired in the family Linhomoeidae (and the marine family Fusivermidae), but single in the exclusively marine family Siphonolaimidae. Spicules short; gubernaculum with dorsocaudal apophyses. Only one genus, belonging to Linhomoeidae, with species occurring in freshwater.

Genus *Terschellingia* de Man, 1888 (Fig. 18.12)
Diagnosis. Medium-sized Linhomoeidae (*L* = 1.0–3.5 mm), with transversely striated cuticle. Anterior sensilla with only four (cephalic) setae. Amphideal fovea close to

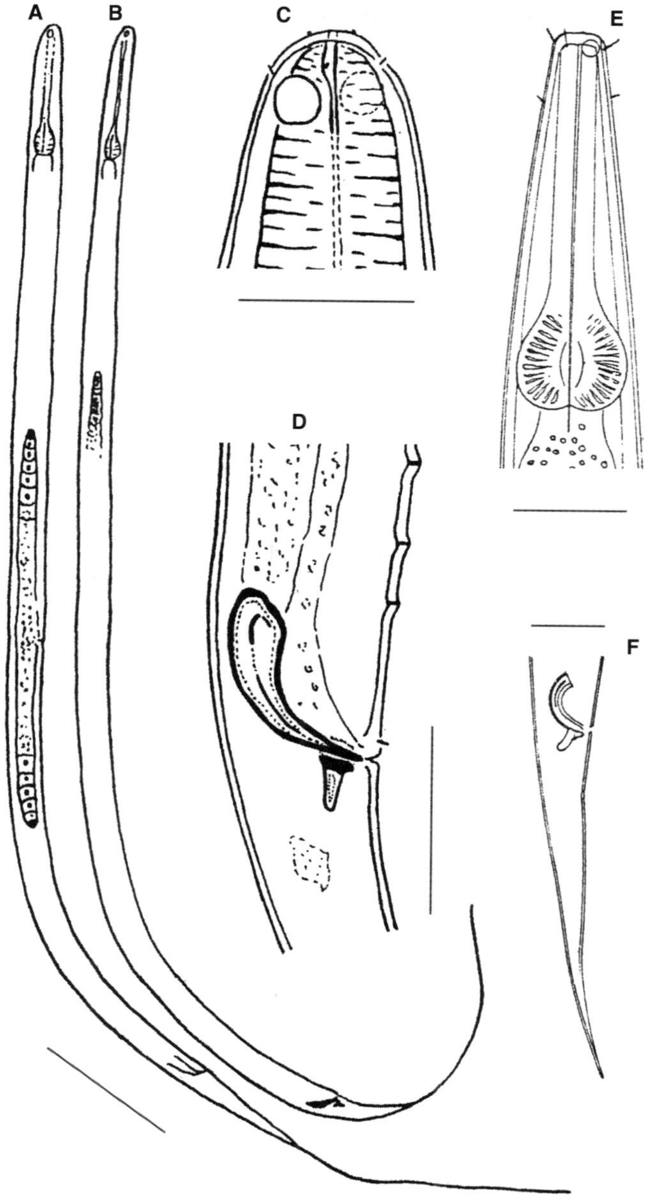

Fig. 18.12 A–D: *Terschellingia elegans* Gagarin & Nguyen Vu Thanh, 2003. A: Female, entire; B: Male, entire; C: Anterior end; D: Male, cloacal region (A–D after Gagarin and Nguyen Vu Thanh, 2003). E and F: *Terschellingia communis* de Man, 1888. E: Neck region; F: Male tail (E and F after de Man, 1888).

anterior end. Stoma small and narrow. Pharynx short, with well-developed terminal bulb. Female reproductive system with two ovaries. Testes paired. Spicules short and robust, ventrally arcuate. Gubernaculum with well-developed caudal apophyses. Pre-anal mid-ventral genital papillae present in some species. Tail elongate conical, with or without a long filiform extension. Mostly marine, some in freshwater.

Type species: *T. communis* de Man, 1888

Species found in inland waters:

T. elegans Gagarin & Nguyen Vu Thanh, 2003
T. supplementa Tchesunov, 1978

References

Andrássy, I. (1958) Ergebnisse der zoologischen Aufsammlungen des Ungarischen Naturwissenschaftlichen Museums in Aegypten im Jahre 1957. *Annales Historico-Naturales Musei Nationalis Hungarici L Series Nova* IX, 135–150.

Andrássy, I (1962) Nematoden aus dem Psammon des Adige-Flusses II. *Memorie del Museo Civico di Storia Naturale Verona* 10, 1–35.

Andrássy, I (1981) Revision of the order Monhysterida (Nematoda) inhabiting soil and inland waters. *Opuscula zoologica (Budapest)* 17–18, 13–47.

Andrássy, I (1995) A Balaton két Mesotheristus faja (Nematoda) (The two limnic species of the genus *Mesotheristus* Wieser, 1956 (Nematoda: Xyalidae)). *Állattani Közlemények* 79, 3–14.

Chitwood, B.G. and Murphy, D.G. (1964) Observations on two marine monhysterids – their classification, cultivation and behaviour. *Transactions of the American Microscopical Society* 83, 311–329.

Cobb, N.A. (1920) One hundred new nemas (type species of 100 genera). *Contributions to a Science of Nematology* 9, 217–343.

De Ley, P. and Blaxter, M.L. (2001) Systematic position and phylogeny. In: Lee, D.L. (ed.) *The Biology of Nematodes*. Taylor & Francis, London, pp. 1–30.

De Ley, P. and Blaxter, M.L. (2004) A new system for the Nematoda: combining morphological characters with molecular trees, and translating clades into ranks and taxa. *Nematology Monographs and Perspectives* 2, 633–653.

de Man, J.G. (1888) Sur quelques nématodes libres de la mer du Nord, nouveau ou peu connus. *Mémoires de la Société Zoologique de France* 1, 1–51.

Eyualem-Abebe and Coomans, A. (1995a) Freshwater nematodes of the Galápagos. *Hydrobiologia* 299, 1–51.

Eyualem-Abebe and Coomans, A. (1995b) *Tridentulus*, a new genus of Monhysteridae (Nematoda) from the Galápagos archipelago. *Bulletin de l'Institut Royal des Sciences Naturelles de Belgique* 65, 5–10.

Eyualem-Abebe and Coomans, A. (1996a) Aquatic nematodes from Ethiopia II. The genus *Monhystrella* Cobb, 1918 (Monhysteridae: Nematoda) with the description of six new species. *Hydrobiologia* 324, 53–77.

Eyualem-Abebe and Coomans, A. (1996b) Aquatic nematodes from Ethiopia III. The genus *Eumonhystera* Andrássy, 1981 (Monhysteridae: Nematoda) with the description of *E. geraerti* n. sp. *Hydrobiologia* 324, 79–97.

Eyualem-Abebe, Peng Yun Liang and Coomans, A. (2001) *Thalassomonhystera traesti* n. sp., *Eumonhystera andrássyi* and three *Monhystrella* species (Monhysteridae: Nematoda) from Li River, China. *Hydrobiologia* 462, 185–197.

Gagarin, V.G. and Nguyen Vu Thanh (2003) Three new species of free-living nematodes from Vietnamese water bodies. *Zoological Journal* 82, 1393–1401 (in Russian).

Goffart, H. (1950) Nematoden aus unterirdischen Gewässern. *Deutsche zoologische Zeitschrift* 1, 73–78.

Hoeppli, R. and Chu, H.J. (1932) Free-living nematodes from hot springs in China and Formosa. *The Hong Kong Naturalist* Suppl.1, 15–28.

Hofmänner, B. and Menzel, B. (1915) Die freilebenden Nematoden der Schweiz. *Revue Suisse de Zoologie* 23, 10–243.

Hopper, B.E. (1969) Marine nematodes of Canada II. Marine nematodes from the Minas Basin – Scots Bay area of the Bay of Fundy, Nova Scotia. *Canadian Journal of Zoology* 47, 671–690.

Jacobs, L.J. (1987a) *A Checklist of the Monhysteridae (Nematoda, Monhysterida)*. Rand Afrikaans University, Johannesburg, South Africa, 186 pp.

Jacobs, L.J., Van de Velde, M.C., Geraert, E. and Vranken, G. (1990) Description of *Diplolaimella dievengatensis* sp. n. (Nematoda: Monhysteridae). *Nematologica* 36, 1–21.

Lorenzen, S. (1977) Revision der Xylidae (freilebende Nematoden) auf der Grundlage einer kritischen Analyse von 56 Arten aus Nord- und Ostsee. *Veröffentlichungen des Instituts für Meeresforschung in Bremerhaven* 16, 197–261.

Lorenzen, S. (1981) Entwurf eines phylogenetischen Systems der freilebenden Nematoden. *Veröffentlichungen des Instituts für Meeresforschung in Bremerhaven*, Suppl. 7, 472 pp.

Lorenzen, S. (1986) *Odontobius* (Nematoda, Monhysteridae) from the baleen plates of whales and its relationship to *Gammarinema* living on crustaceans. *Zoologica Scripta* 15, 101–106.

Lorenzen, S. (1994) *The Phylogenetic Systematics of Free-living Nematodes*. The Ray Society, London, 383 pp.

Shoshin, A.V. (2001) *Sacrimarinema* gen. n. from Lake Baikal with descriptions of three species (Nematoda, Monhysterida: Xyalidae). *Zoological Institute* (St. Petersburg) 9, 253–257.

Wieser, W. (1956) Free-living marine nematodes III Axonolaimoidea and Monhysteroidea. Reports of the Lund University Chile Expedition 1948–49. *Lunds Universitets Arsskrift* 52, 1–115.

Further Reading

Gerlach, S.A. and Riemann, F. (1973/1974) The Bremerhaven Checklist of aquatic Nematodes. A catalogue of Nematoda Adenophorea excluding the Dorylaimida. *Veröffentlichungen des Instituts für Meeresforschung in Bremerhaven* Suppl. 4, 1–104 (1973) and 405–734 (1974). (Gives list of taxa up to species, synonyms and references.)

Jacobs, L.J. (1987b) A redefinition of the genus *Monhystrella* Cobb (Nematoda, Monhysteridae) with keys to the species. *Zoologica Scripta* 16, 191–197.

Jacobs, L.J. and Heyns, J. (1992) Morphology of *Monhystera coomansi* sp. n. from Algeria (Nematoda: Monhysteridae). *Nematologica* 38, 1–21. (Gives a detailed morphological description of a *Monhystera* species.)

Lorenzen, S. (1977) Revision der Xyalidae (freilebende Nematoden) auf der Grundlage einer kritischen Analyse van 56 Arten aus Nord- und Ostsee. *Veröffentlichungen des Instituts für Meeresforschung in Bremerhaven* 16, 197–261.

Lorenzen, S. (1978) The system of the Monhysteroidea (Nematodes) – a new approach. *Zoologischer Jahrbucher Systematik* 105, 515–536.

19 Order Araeolaimida

A. W. Muthumbi[1] and A. Vanreusel[2]

[1]Department of Zoology, University of Nairobi, P.O. Box 30197, Nairobi, Kenya;
[2]Marine Biology Section, Ghent University, Krijgslaan 281-S8 B-9000, Belgium

Introduction

In the most recent work on nematode systematics, De Ley and Blaxter (2004) gave Araeolaimida the rank of order under the superclass Chromadoria and the class Chromadorea. The order includes four families: Axonolaimidae, Comesomatidae, Dilopeltidae and Coninckiidae, originally classified in the superfamily Axonolaimoidea.

In the work of De Ley and Blaxter (2004), names of nematode clades higher than family-group ranks, whose rules concerning valid names are not specified in the International Code of Zoological Nomenclature, were based on chronological priority of the root word. Thus, the term 'Araeolaimida' was used based on the term 'Areolaimica' that was first used by De Coninck (1965). Araeolaimida was given the rank of order in the class Chromadoria. In earlier classifications (e.g. Lorenzen, 1994), the superfamily Axonolaimoidea was placed in the order Monhysterida and contained all the four families mentioned above.

Description

Araeolaimida De Coninck (1965) contains all the families in the superfamily Axonolaimoidea in the order Monhysterida by Lorenzen (1994). These families are different from the rest of the families in the order Monhysterida (Lorenzen, 1994) in that the amphids are spiral- or loop-shaped. The cuticle is striated or has punctations, and the stoma may be small or long, tubular and often without teeth. These are features unique to this superfamily, and in combination separate the group from the others.

Superfamily AXONOLAIMOIDEA, Filipjev 1918

Amphids are predominantly spiral- or bow-shaped and very seldom they may also be pore-shaped (Axonolaimidae). The cuticle is smooth or striated and in most cases in Comesomatidae it is covered with punctations. Labial sensilla are papilliform or not visible. The 6 + 4 sensilla are in most cases situated in two separate circles (except for some cases in Comesomatidae). The posterior four cephalic sensilla are longer than or at most equal in length to the six cephalic sensilla. The buccal cavity may have at most six (Axonolaimidae) or three (Comesomatidae partim) tooth-like structures on its front edge. Pharyngeal tubes occur frequently within the Axonolaimidae and Comesomatidae. The pharynx never widens at the end to form a muscular bulb. Females have two outstretched ovaries (seldom only one anterior ovary, Axonolaimidae). Males possess two testes or only one anterior testis. Males also have ventrally situated pre-anal papillae in Comesomatidae and only seldom in Axonolaimidae.

Family AXONOLAIMIDAE Filipjev, 1918

Buccal cavity has six movable tooth-like structures located at the front in the buccal cavity and referred to as odontia. However, they are insignificant in the genera *Axonolaimus*, *Ascolaimus* and *Apodontium*. Buccal cavity columnar or conical with six tooth-like structures. In the pharyngeal region and posterior to the buccal cavity, there are one ventral and two sub-dorsal pharyngeal tubules. The cuticle is smooth in most cases, seldom striated. The 6 + 4 cephalic sensilla stand in two separate circles, the posterior four being larger than the anterior six sensilla. Amphids are curved into a loop shape resembling either an inverted U or O (pore-shaped in *Apodontium*). In many species the ventral arm of the amphid is longer than the dorsal arm. In the juvenile stages, the amphids form only a spiral with one loop (Lorenzen, 1994 and references therein). There are two testes, the posterior one of which continues forwards at first and folds over in the anterior part. Most species have two outstretched ovaries, seldom only one posterior ovary (*Synodontium*, *Odontophoroides*). In most cases the anterior and posterior gonads each lie on either side of the intestine, except in *Ascolaimus*, where the location of the gonad varies for the whole length. Tail is always conical and caudal glands open through a common pore. The family Axonolaimidae is entirely marine.

Family COMESOMATIDAE Filipjev, 1918

The family is characterized by spiral amphids that have at least 2.5 loops except in *Kenyanema*, which has 1.5–2.0 turns of the amphid loop (Muthumbi *et al.*, 1997). Other features are: the cuticle is always striated and body annules are punctated except in *Cervonema*, *Laimella*, *Paracomesoma* and *Sabatieria hilarula* (Lorenzen, 1994). Males of most species have slightly arched, ventrally situated fine pre-cloacal supplements. Usually, the anterior gonad lies to the left and the posterior gonad to the right of the intestine. The buccal cavity is often cup-shaped and in Dorylaimopsiinae it has three or six tooth-like structures that are similar to those found in Axonolaimidae (except *Kenyanema*; Muthumbi *et al.*, 1997). Many species have pharyngeal tubules, one dorsal and two sub-ventral ones attached to the buccal cavity. Labial sensilla are usually short and the 6 + 4 cephalic ones are in two separate circles. The six anterior cephalic sensilla are shorter than the four posterior ones but are equal in length in *Metacomesoma* and *Pierrickia*. Gonads are two, outstretched, one anterior and the other posterior (except for *Kenyanema monorchis* Muthumbi *et al.*, 1997, which has a single anterior testis). Tail has a cylindrical end. The family Comesomatidae is entirely marine.

Family CONINCKIIDAE Lorenzen, 1981

The family is characterized by the following: amphids that are curved into an O-shape, 6 + 4 cephalic sensilla arranged in two separate circles and equal in length and a very tiny buccal cavity. Other features include smooth or striated cuticle and a pharynx that thickens slightly at the end. Two gonads lie in opposite directions. The family Coninckiidae is entirely marine.

Family DIPLOPELTIDAE Filipjev, 1918

The cuticle may be smooth or striated. The posterior four cephalic sensilla are longer than the anterior six, which may be very small or invisible, just like the six labial ones. Amphids are bow-shaped with only one loop. Ocelli may be present in Diplopeltinae. The buccal cavity is narrow, tubular-like without teeth. The pharynx may have one ventral and two sub-dorsal pharyngeal tubules (partly in Diplopeltinae). The pharynx never has a muscular end bulb. The females have two outstretched ovaries (only one ovary in *Cylindrolaimus monhystera*). Usually there are two testes and only occasionally the posterior one is absent. Pre-cloacal papillae may be present (Cylindrolaiminae) or absent. The tail is conical or has a cylindrical section. Caudal glands open together or separately. The subfamily Diplopeltinae is entirely marine, but the only and type genus of the subfamily Cylindrolaiminae, i.e. *Cylindrolaimus*, is limno-terrestrial.

Subfamily Diplopeltinae Filipjev, 1918

The cuticle is smooth or striated. Amphids are bow-shaped or have a spiral contour with one turn. The stoma lies terminally or is displaced to the dorsal side (*Pararaeolaimus nudus*, *Diplopeltis* and *Campylaimus*).

List of genera in Diplopeltinae:

Araeolaimus de Man, 1888
Campylaimus Cobb, 1920
Diplopeltis Cobb for Dipeltis Cobb, 1891
Diplopeltula Gerlach, 1950
Metaraeolaimoides de Coninck, 1936
Pararaeolaimus Timm, 1961
Pseudaraeolaimus Chitwood, 1951
Southerniella Allgen, 1932
Striatodora Timm, 1961

Subfamily Cylindrolaiminae Micoletzky, 1921

The cuticle has fine striations. Amphids circular or spiral with one turn. The buccal cavity is narrow and tubular and always terminal. The pharynx is cylindrical with a slightly enlarged terminal end but no muscular bulb. A single pre-cloacal papilla is present. Tail ventrally curved.

Type and only genus: *Cylindrolaimus* de Man, 1880.

Genus *Cylindrolaimus* de Man, 1880

Cylindrolaiminae. Body length varying from 0.4 to 1.4 mm. Cuticle with fine transverse striae. Head continuous with adjacent body, lips fused, papillae minute. Cephalic setae four, short. Amphids circular or spiroid, one-third to one-half as wide as corresponding body width. Buccal cavity cylindrical, 2–4 head widths long, without teeth. Pharynx simple, occasionally slightly swollen posteriorly. Vulva midway. Female gonads mostly amphidelphic, rarely prodelphic. Ovaries straight. Males extremely rare. Testes two. Spicula small, ventrally curved, gubernaculum inconspicuous. A single pre-anal papilla present. Tail plump, ventrally bent, terminus rounded; spinneret thin.

Type species: *C. communis* de Man, 1880.

List of species:
For each species locations are indicated as provided in Gerlach and Riemann (1973, 1974). If additional information is available on their habitats and geographical distribution, this is also summarized.

C. bambus Andrássy, 1968 (Paraguay). This species was also recorded from a *Pinus nigra* pinewood in Navarra (northern Spain) (Armendariz-Gonzalez and Hernandez-Minguillon, 1994). Furthermore, it was observed in naturally flooded meadows and forests and a thermophilous oak grove (www.sci.muni.cz).
C. baradlanus Andrássy, 1959 (Hungary)
C. communis de Man, 1880 (The Netherlands, Belgium, Germany, Switzerland, Denmark, England, Poland, Austria, Hungary, Bulgaria, Romania, Spain, Italy, Estonia, Moldavia, Japan, Egypt, USA). A common species found in humid prairies, sandy dunes (de Man, 1880) and in freshwater (Bongers, 1988). A morphological description is provided in Fig. 19.1.
C. melancholicus de Man, 1880 (The Netherlands, Belgium, Poland, Hungary, Arctic Canada). A morphological description is provided in Fig. 19.2. This species is found in coastal prairies (de Man, 1880) and oligotrophic freshwater (Bongers, 1988). It was also recorded from moist litter in Chongqing (China) (Wu-Jihua *et al.*, 1997).
C. monhystera Schneider, 1937 (First recorded from Indonesia. It was also found in Bulgaria (Andrássy, 1958)).
C. obtusus Cobb, 1916 (Washington DC (USA), Hungary)
C. ophis Andrássy, 1970 (Vietnam)
C. procerus Andrássy, 1968 (Paraguay)

The following species were indicated as *species inquirenda*, *species dubia* or *species incertae sedis* by Gerlach and Riemann (1973, 1974) and therefore not included in the original species list.

C. filicaudatus Allgén, 1933 (*species inquirenda* Central Africa)
C. lacustris Hofmänner, 1913 (*species incertae sedis*)
C. macrurus Daday, 1899 (*species dubia*)
C. politus Daday, 1905 (*species inquirenda*)

Molecular Information

Nearly full-length small subunit (SSU) rDNA sequences were determined from *Cylindrolaimus* sp. by Felix *et al.* (2000). The sequence was deposited with GenBank under the accession number 202149. In this study on the evolution of vulval development in the Cephalobina, *Cylindrolaimus* was used as an outgroup.

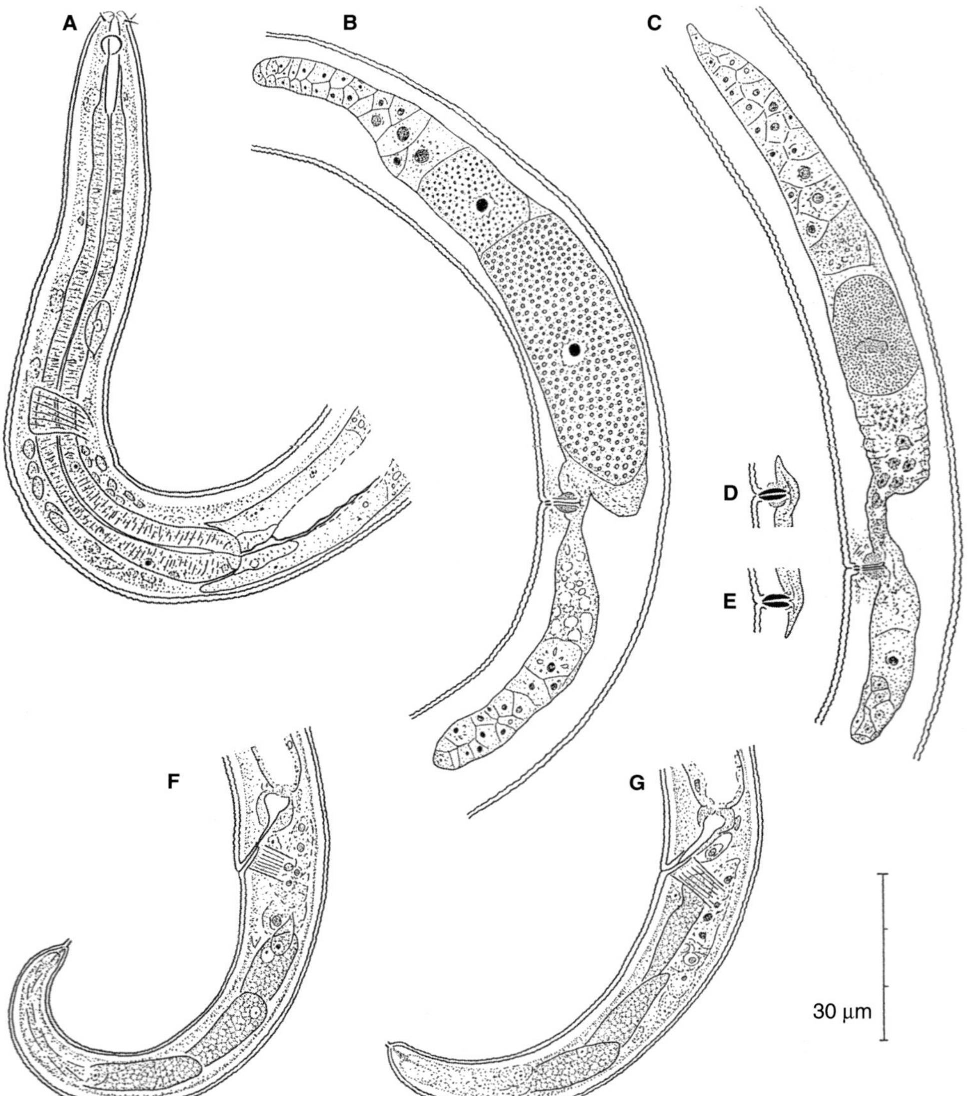

Fig. 19.1 *Cylindrolaimus communis*. A, Neck region; B and C, Reproductive system; D and E, Vagina of female shown in C with respect to left and right sub-median view; F and G, Tail. (Illustration courtesy of De Ley (1994)).

Summary

So far nearly all the families of Araeolaimida recorded are marine, except for the subfamily Cylindrolaiminae in the family Diplopeltidae. This subfamily is the only limno-terrestrial group in the order Araeolaimida and it is represented by only one genus, *Cylindrolaimus*.

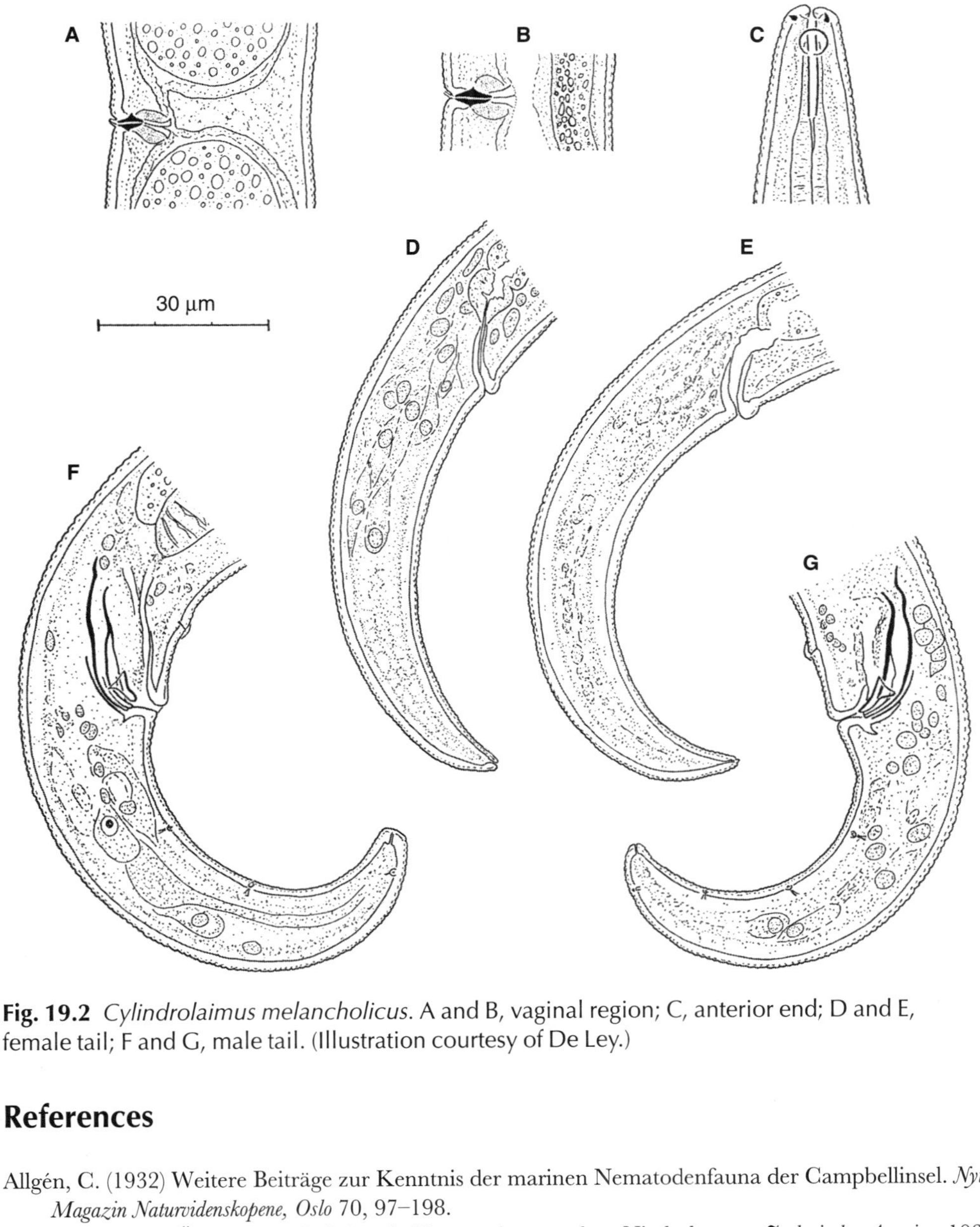

Fig. 19.2 *Cylindrolaimus melancholicus*. A and B, vaginal region; C, anterior end; D and E, female tail; F and G, male tail. (Illustration courtesy of De Ley.)

References

Allgén, C. (1932) Weitere Beiträge zur Kenntnis der marinen Nematodenfauna der Campbellinsel. *Nytt Magazin Naturvidenskopene, Oslo* 70, 97–198.

Allgén, C. (1933) Über einige freilebende Nematoden aus dem Niederkongo. *Zoologischer Anzeiger* 103, 312–320.

Andrássy, I. (1958) Erd- und Süβwassernematoden aus Bulgarien. *Acta Zoologica Hungarica* 4, 1–88.

Andrássy, I. (1959) Nematoden aus der Tropfsteinhöhle 'Baradla' bei Aggtelek (Ungarn), nebst einer Übersicht der bisher aus Höhlen bekannten freilebenden Nematoden-Arten. *Acta Zoologica Hungarica* 4, 235–277.

Andrássy, I. (1968) Fauna Paraguayensis 2. Nematoden aus den Galeriewäldern des Acaray-Flusses. *Opuscula Zoologica (Budapest)* 8, 167–315.

Andrássy, I. (1970) Freilebende Nematoden aus Vietnam. *Opuscula Zoologica (Budapest)* 10, 5–31.

Armendariz-Gonzalez, I and A. Hernandez-Minguillon, (1994) Species belonging to the Order Araeolaimida (Nematoda) found in pinewoods of *Pinus nigra* Arnold 1785 in Navarra

(Northern Spain). *Boletin de la real Sociedad Espanola de Historia Natural Seccion Biologica* 91, 135–141.

Bongers, T. (1988) De nematoden van Nederland. Stichting uitgeverij Koninklijke Nederlandse Natuurhistorische Vereniging Utrecht, 408 pp.

Chitwood, B.G. (1951) North American marine nematodes. *Texas Journal Science* 3, 617–672.

Cobb, N.A. (1891) Onyx and Dipeltis : new Nematoda genera, with a note on *Dorylaimus*. *Proceedings of the Linnean Society of New South Wales* 6 (2), 143–158.

Cobb, N.A. (1916) Revision of the genus *Cylindrolaimus*. *Journal of Parasitology* 2, 199–200.

Cobb, N.A. (1920) One hundred new nemas (type species of 100 new genera). *Contribution to a Science of Nematology (Baltimore)* 9, 217–343.

Daday, E.v. (1899) Uj-guineai szabadon élö nematodok. *Matematikai es termeszettudomanyi ertesito (Budapest)* 17, 557–572.

Daday, E.v. (1905) Untersuchungen über die Süβwasser-Mikrofauna Paraguays. *Zoologica (Stuttgart)* 18 (44), 374 .

De Coninck, L. (1936) *Metaraeolaimoides oxystoma* n.g. n.sp. (Nematoda) en zijne afleiding van Araeolaimoides de Man, 1893 door allometrie (avec résumé en Français). *Biologisch Jaarboek* 3, 182–204.

De Coninck, L. (1965) Classes des Nématodes – Systématique des Nématodes et sous-classe des Adonophorea. *Traité de Zoologie* (éd. Grasse) 4 (2), 586–681.

De Ley, P. (1994) The terrestrial nematofauna of the Galapagos. PhD Thesis, University of Ghent, Ghent, Belgium.

De Ley, P. and Blaxter, M.L. (2002) Systematic position and phylogeny. In: D.L. Lee (ed.) *The Biology of Nematodes*. Taylor & Francis, London, pp. 1–30.

De Ley, P. and Blaxter, M.L. (2004) A new system for Nematoda: combining morphological characters with molecular trees, and translating clades into ranks and taxa. *Nematology Monographs and Perspectives* 2, 633–653.

de Man, J.G. (1880) Die einheimischen, frei in der reinen Erde und im süβen Wasser lebenden Nematoden monographisch bearbeitet. Vorlaüfiger Bericht und descriptiv-systematischer Theil. *Tijdschrift der Nederladsche dierkundige Vereeniging* 5, 1–104.

de Man, J.G. (1888) Sur quelques Nématodes libres de la mer du Nord, nouveaux ou peu connus. *Mémoires de la Société Zoologique de France* 1, 1–51.

Felix, M-A., De Ley, P., Sommers, R.J., Frisse, L., Nadler, S.A., Thomas, W.K., Vanfleteren, J. and Strenberg, P.W. (2000) Evolution of vulva development in the Cephalobina (Nematoda) *Developmental Biology* 221, 68–86.

Filipjev, I. (1918) Free-living marine nematodes from the Sevastopol area. *Trudy osoboi zoologicheskoi Laboratorii Sebastopol'skoi biologischeskoi Stantsii* 4 (2), 1–350 (in Russian).

Gerlach, S.A. (1950) Die Diplopeltiden, eine Gruppe freilebender Nematoden. *Kieler Meeresforschung* 7, 138–156.

Gerlach, S.A. and Riemann, F. (1973) The Bremerhaven checklist of aquatic nematodes. *Veröffentlichungen des Instituts für Meeresforschung in Bremerhaven* Heft 1 (Suppl. 4), 1–404.

Gerlach, S.A. and Riemann, F. (1974) The Bremerhaven checklist of aquatic nematodes. *Veröffentlichungen des Instituts für Meeresforschung in Bremerhaven* Heft 2 (Suppl. 4), 405–734.

Hofmänner, B. (1913) Contribution à l'étude des nématodes libres du Lake Geneva. *Revue Suisse Zoologie* 21, 589–658.

Lorenzen, S. (1994) *The Phylogenetic Systematics of Free-living Marine Nematodes*. Ray Society, London.

Micoletzky, H. (1921) Die freilebenden Erdnematoden. *Arch.Naturgesch* 87A, 1–650.

Muthumbi, A.W.N., Soetart, K. and Vincx, M. (1997) Deep-sea nematodes from the Indian Ocean: Description of new and known species of the family Comesomatidae. *Hydrobiologia* 364, 25–57.

Schneider, W. (1937) Freilebende Nematoden der Deutschen Limnologischen Sundaexpedition nach Sumatra, Java und Bali. *Archives für Hydrobiologie* (Tropische binnengewässer) 7 (Suppl. 15), 30–108.

Timm, R.W. (1961) The marine nematodes of the Bay of Bengal. *Proceedings of the Pakistan Academy of Science* 1, 1–88.

Wu-Jihua, Liang-Yanling and Sun-Xida (1997) Newly recorded species of free-living nematodes from China (Chromadorida, Enoplida and Araeolaimida). *Acta Hydrobiologica Sinica* 21, 312–321.

20 Order Plectida

O. HOLOVACHOV[1] AND P. DE LEY[2]

[1]Department of Zoology, Ivan Franko National University of Lviv, Hrushevsky str. 4, Lviv 79005, Ukraine; [2]Department of Nematology, University of California, Riverside CA 92521, USA

Order PLECTIDA Malakhov, 1982

Morphology. Cuticle annulated, sometimes strongly refractive. Intracuticular sclerotization present only in Metateratocephalidae. Lateral differentiation of cuticle present or absent (often present in Leptolaimoidea and Plectoidea). Body pores, hypodermal glands and somatic sensilla usually present. Ocelli present only in several marine genera. Deirids present only in Plectidae, setiform, inside the lateral alae. Inner labial sensilla always papilliform, located on the anterior surface of lips or inside the cheilostom (in Plectoidea), sometimes not detectable. Outer labial sensilla usually papilliform, setiform only in *Setostephanolaimus*, Metateratocephalidae and some Ceramonematoidea. Cephalic sensilla setiform, with few exceptions (papilliform in *Pakira* and *Haliplectus*, transformed into cornua in Wilsonematinae). Amphid ventrally unispiral or derived therefrom. Secretory–excretory system usually present; renette cell located opposite intestine or posterior part of pharynx. Excretory ampulla present or absent; in latter case excretory duct is long, sometimes embedded into renette cell. Excretory pore usually opens to the exterior at about the level of nerve ring, sometimes between lips or inside the cheilostom (in Aphanolaimidae). Stoma cylindrical or funnel-shaped; cheilostom usually undifferentiated, bearing odontia in Camacolaimidae or six digitate projections in Rhadinematidae and *Anonchus*; gymnostom usually cylindrical; stegostom either undeveloped, funnel-shaped or cylindrical. Pharynx uniformly cylindrical, muscular; in some taxa with strongly developed glandular tissue in posterior section, or with basal bulb; it may also be subdivided into corpus, isthmus and basal bulb with strongly sclerotized lining or valvular apparatus. Dorsal gland orifice penetrates pharyngeal lumen somewhat posterior to stoma base or within stegostom. One pair of sub-ventral gland orifices is located somewhat posterior to stoma or at corpus–isthmus junction. Proximal part of rectum is surrounded by three gland-like cells: one dorsal and two ventrosublateral. Female reproductive system usually didelphic, amphidelphic; however, in some genera the anterior (*Procamacolaimus*) or posterior (*Chronogaster*, *Steratocephalus*, some *Anonchus*) genital branch may be absent. Ovarian branches reflexed antidromously. Spermathecae axial or offset (if present at all). Ventral tubuli or alveoli sometimes present in females. Male reproductive system usually diorchic, with opposed or outstretched (Camacolaimoidea) testes; rarely one of the testes is reduced (posterior in Metateratocephalidae) or non-functional (anterior in Aphanolaimidae). Pre-cloacal supplements often present, tubular and/or alveolar in Leptolaimoidea, Camacolaimoidea and Plectoidea; papilliform in Ceramonematoidea and Haliplectoidea. Mid-ventral pre-cloacal sensilla usually present in Leptolaimoidea, Camacolaimoidea and Plectoidea; sub-ventral post-cloacal sensilla usually present in Camacolaimoidea and Plectoidea. Spicules paired and symmetrical. Gubernaculum usually present. Caudal glands usually present, opening separately or via a spinneret on tail tip; reduced in some taxa (*Chronogaster*, Metateratocephalidae).

Taxonomy. The order Plectida was proposed by Malakhov (Malakhov *et al.*, 1982; Malakhov, 1986) for several families that were placed in orders Araeolaimida (Andrássy, 1976) or Chromadorida (Lorenzen, 1981) by other authors. It is consistent with recent molecular phylogenies, which place the family Plectidae as a separate clade compared to the type families of Araeolaimida and Chromadorida (Blaxter *et al.*, 1998). However, at the time of writing, many relevant taxa have not yet been analysed with these novel tools. Morphological analysis (Holovachov, 2004; Holovachov and Boström, 2004) suggests that five superfamilies should be recognized in the order Plectida, i.e. Leptolaimoidea, Plectoidea, Camacolaimoidea, Ceramonematoidea and Haliplectoidea. The latter three superfamilies include marine and brackish representatives, with the exception of the purely terrestrial family Aulolaimidae in Haliplectoidea. They are all omitted from the present review. Furthermore, according to the phylogenetic study of Holovachov and Boström (2004), the family Rhabdolaimidae should be placed within Microlaimoidea (see also Eyualem-Abebe and Coomans, 1996b), while families Prismatolaimidae, Bastianiidae and Odontolaimidae should be placed within Triplonchida. The monotypic family Ohridiidae could not be placed confidently in any of the three mentioned superfamilies of Plectida, due to its highly distinct morphology. We also omitted the type and only species *Prodomorganus tajmiris* Gagarin, 1993, which was described on the basis of two females from Lake Tajmyr on the Tajmyr peninsula, Northern Asia. Morphological data given in the original description are very limited and do not allow us to clarify its taxonomic position; it is considered here as *genus incertae sedis*. Type material was not available for restudy.

Biology. This group includes aquatic (marine, freshwater or brackish water) and terrestrial species. Some species of the genus *Domorganus* were found to be endosymbionts of annelid worms. *Smithsoninema* and several camacolaimid species are intracellular parasites or predators of marine Foraminifera; they were considered to be close to the marine invertebrate parasitic order Benthimermithida (Tchesunov, 1997).

Key to the order PLECTIDA

1. Pharynx with median and basal swellings; excretory pore is located posteriorly to cardia; male with only single pre-cloacal papilliform sensillum .... Ohridiidae (monotypic, *Domorganus*)
   Pharynx uniformly cylindrical or with basal bulb; excretory pore is located at about the nerve ring level, or between lips,

or in cheilostom; male with pre-cloacal sensillum, tubular supplements and caudal setae 2

2. Somatic sensilla are connected with prominent hypodermal glands; pharynx without pharyngeal tubes; each genital branch with two offset sac-like spermathecae; male usually has one pair of pre-cloacal setae and no post-cloacal papilliform genital sensilla . . . . .Leptolaimoidea

Somatic sensilla are separate from hypodermal glands; pharyngeal tubes originating at stoma base and extending posteriorly; each genital branch with single axial spermatheca or spermatheca absent; male usually has one pair of post-cloacal papilliform genital sensilla . . . . .Plectoidea

Family *incertae sedis* OHRIDIIDAE Andrássy, 1976

Genus *Domorganus* Goodey, 1947

Syn. = *Ohridia* Schneider, 1943 (homonym to a trematode genus) = *Ohridius* Gerlach & Riemann, 1973 = *Leoberginema* Tsalolikhin, 1977 = *Mikinema* Tchesunov, 1978

Morphology. Cuticle annulated; lateral alae consist of one to several smooth bands, extending posteriorly from the amphid level (Fig. 20.1A). Body pores, hypodermal glands, ocelli, deirids and somatic sensilla absent. Labial region modified into cephalic capsule. Labial sensilla indiscernible. Cephalic sensilla setiform, their bases located anterior to amphid. Amphid unispiral. Secretory–excretory system present; renette cell large, located opposite intestine, sub-lateral. Excretory ampulla absent, excretory duct long. Excretory pore posterior to cardia. Stoma cylindrical, cheilostom undifferentiated, gymnostom small barrel-shaped, sclerotized, stegostom undeveloped. Pharynx muscular, without pharyngeal tubes and valves, with two swellings: anterior one at mid-pharynx and posterior one at basal portion (Fig. 20.1C). Pharyngeal gland nuclei and orifices indistinct. Female reproductive system didelphic, amphidelphic, ovarian branches reflexed, anterior located on right-hand side and posterior located on left-hand side of intestine. Spermathecae absent. Vagina straight or bent posteriorly, in *D. macronephriticus* with epiptygmata and pre-vulval mid-ventral papilla. Male reproductive system diorchic, testes opposed (posterior reflexed dorsally). Spicules paired and symmetrical, arcuate. Gubernaculum present. Copulatory apparatus composed of only a single pre-cloacal mid-ventral papilliform sensillum, located about one corresponding body diameter anterior to cloaca (Fig. 20.1E). Three caudal glands present, their nuclei are caudal. Spinneret functional, weakly cuticularized.

Development. Embryonic development of *D. oligochaetophilus* was described by Valovaya and Malakhov (1990) and lasts 30–40 days at room temperature.

Biology. Amphimictic or thelytokous. *D. macronephriticus* was found in soils or associated with earthworms across Europe, located in the intestine (Tchesunov and Sturhan, 2004). *D. oligochaetophilus* and *D. beklemishevi* were found in the intestine of the littoral oligochaetes *Lumbricillus lineatus*, *L. pagenstecheri* and *Enchytraeus albidus*, in the Baltic and White seas. Other species were recorded as free living in soil in Spain,

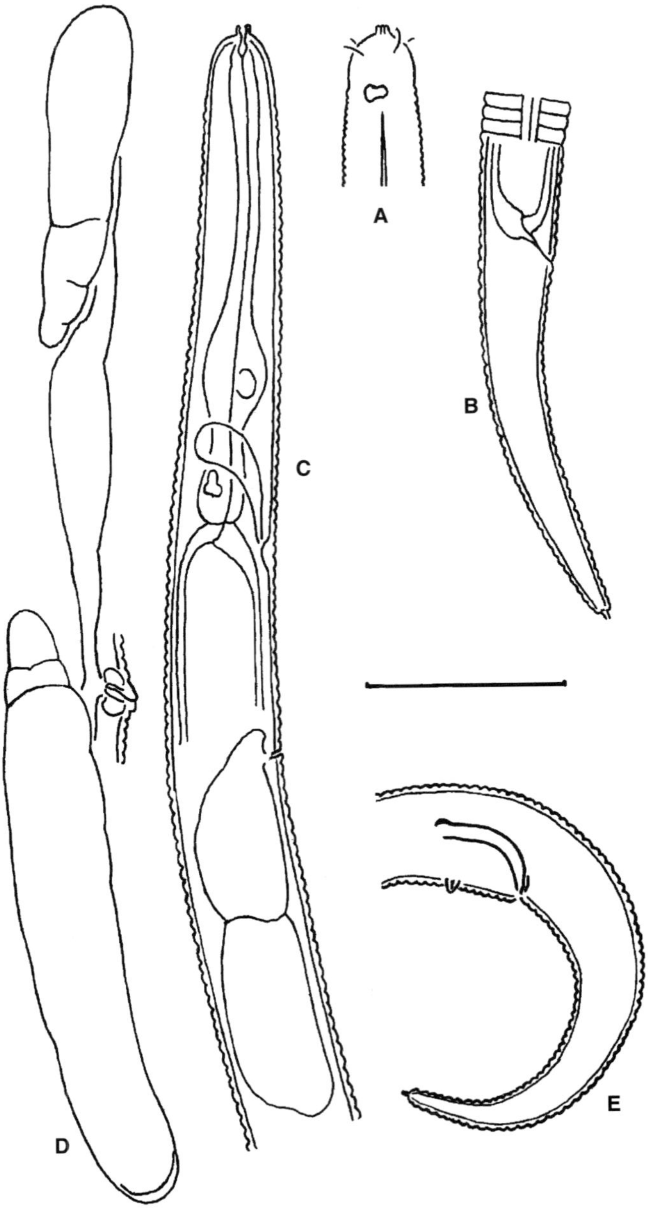

Fig. 20.1 Morphology of the genus *Domorganus* Goodey, 1947 as exemplified by *Domorganus* sp. from Costa Rica (A–D) and *D. macronephriticus* T. Goodey, 1947 (E). A: Anterior end, surface view; B: Female tail; C: Pharyngeal region; D: Female reproductive system; E: Male caudal region, redrawn from Goodey (1947), not to scale. (Scale bar: A–D = 20 μm.)

in sediments from Lake Baikal, Lake Ohrid, and the Caspian Sea. According to the personal communication of A. Shoshin, several undescribed species of this genus inhabit Baikal Lake. Two unidentifiable females were found in Costa Rica, Isla del Caco (Figs 20.1A–D). Secondary sexual dimorphism (in measurements) and adult growth were noted in *D. oligochaetophilus*, adult growth in *D. beklemishevi*.

Type species: *D. macronephriticus* T. Goodey, 1947

Other species:
D. acutus (Tsalolikhin, 1977) Lorenzen, 1981
D. bathybius (Schneider, 1943) Lorenzen, 1981
D. beklemishevi Valovaya, 1989
D. delgadoi Hernandez Minguillon & Jordana Butticaz, 1990
D. navarrensis Hernandez Minguillon & Jordana Butticaz, 1990
D. oligochaetophilus Thun, 1967
D. subtilis (Tchesunov, 1978) Lorenzen, 1981

Superfamily LEPTOLAIMOIDEA Örley, 1880

Morphology. Cuticle annulated. Lateral alae usually present, a narrow, smooth midlateral band of cuticle demarcated by two lines. Body pores arising from oval sublateral hypodermal gland cells. Somatic sensilla when present are connected with hypodermal glands and extending to the exterior through body pores. Labial region with six equally developed lips or cephalic capsule. Labial sensilla papilliform or invisible. Cephalic sensilla setiform, their bases located anteriorly to amphid. Amphid unispiral, or derived therefrom. Ocelli and deirids absent. Secretory–excretory system usually present; renette cell located opposite ventral side of cardia or anterior part of intestine. Excretory ampulla present in marine taxa, all freshwater representatives have a cuticularized excretory duct that originates from the renette cell, extends forward along ventral side of pharynx and opens to the exterior or in the cheilostom. Stoma cylindrical. Pharynx muscular, lacking pharyngeal tubes and valvular apparatus. Dorsal gland orifice penetrates pharyngeal lumen at stoma base. Cardia cylindrical, free, with posterior part surrounded by intestinal tissue. Female reproductive system didelphic, amphidelphic (except in a few *Anonchus* species), ovarian branches reflexed antidromously (except in *Manunema*). Oviduct short. Two offset, sac-like spermathecae located on right and left side of each gonoduct, in oviduct–uterus junction. Uterus an undifferentiated cylindrical tube. Vagina encircled by a single sphincter muscle, usually without sclerotizations (except in *Leptolaimus* and *Aphanonchus*). Females of some species have ventral alveoli or tubuli. Male reproductive system diorchic or monorchic, posterior testis reflexed dorsally. Alveolar and/or tubular supplements present, in a few species absent. Pre-cloacal sensillum papilliform in shape, located on anterior cloacal lip (except in *Manunema*). Setae in caudal region of males arranged in the following manner: usually one subventral pair located pre-cloacally at level or spicules and several pairs of setae on tail. Three caudal glands present, their nuclei are incaudal. Spinneret functional, weakly cuticularized.

Taxonomy. Two families are recognized here: Leptolaimidae and Aphanolaimidae, both present in freshwater (Holovachov and Boström, 2004).

Key to the families

1. Stoma long tubular, with strongly developed stegostom; pharynx usually with basal bulb; excretory pore opens to the exterior . . . . .Leptolaimidae (*Paraplectonema*)
   Stoma short, barrel- or pore-shaped, with weakly developed stegostom; pharynx uniformly cylindrical; excretory pore opens inside the cheilostom . . . . .Aphanolaimidae

Family LEPTOLAIMIDAE Örley, 1880

Morphology. Lateral alae usually present, demarcated by straight lines. Somatic sensilla present in both sexes or in males only. Inner labial sensilla never described. Outer labial sensilla papilliform, located on anterior surface of lips. Cephalic sensilla located at base of labial region. Amphid unispiral or derived therefrom. Excretory ampulla present in all genera except *Paraplectonema*, opening to the exterior. Stoma tubular, with developed stegostom (except in *Anomonema*). Pharynx muscular, cylindrical anteriorly, with distinct oval basal bulb. Anterior cylindrical part of pharynx subdivided by breaks in muscular pharyngeal tissue into a narrow procorpus, a somewhat wider metacorpus and a narrow isthmus. Basal bulb with thick and strongly cuticularized lumen. Female reproductive system didelphic. Ovarian branches reflexed antidromously or outstretched, on alternate sides of intestine. Females of some species of *Leptolaimus* have ventral alveoli or tubuli. Male reproductive system usually diorchic. Pre-cloacal sensillum usually papilliform, on the anterior cloacal lip (except in *Manunema*).

Taxonomy. The family is not subdivided into separate subfamilies. Seven genera are now recognized in the family by Holovachov and Boström (2004), five of these inhabit marine and brackish habitats. The monotypic genus *Leptoplectonema* was found only once in moist tundra soil on Tierra del Fuego, Chile. The genus *Paraplectonema* is the only typical freshwater inhabitant (Fig. 20.2).

Genus *Paraplectonema* Strand, 1934

Syn. = *Paraplectus* Filipjev, 1930 nec Raffay, 1989

Morphology. Cephalic sensilla setiform. Amphid unispiral, located anteriorly to cephalic sensillae bases (Figs 20.2D and 20.3A). Cuticularized excretory duct originates from renette cell, extends forward along ventral side of pharynx and opens to the exterior via a pore at mid-pharynx or on anterior surface of labial region. Stoma uniformly tubular: cheilostom short, supported by six fine longitudinal ribs; gymnostom short, with plate-like rhabdia; stegostom tubular, with uniformly thickened lining (Fig. 20.2A). Ovarian branches reflexed antidromously. Vagina straight, without sclerotizations. One sub-ventral and one sub-dorsal pair of caudal setae present in female (Fig. 20.2E). Male reproductive system diorchic or monorchic, with only anterior testis. Spicules with oval manubrium and gradually narrowing, arcuate shaft. Gubernaculum plate-like with strong caudal apophysis. Nine to 11 similar mid-ventral tubular supplements and a single mid-ventral pre-cloacal sensillum. Tubular supplements sigmoid, protrusible. Pre-cloacal sensillum papilliform, located on anterior cloacal lip (Fig. 20.2B).

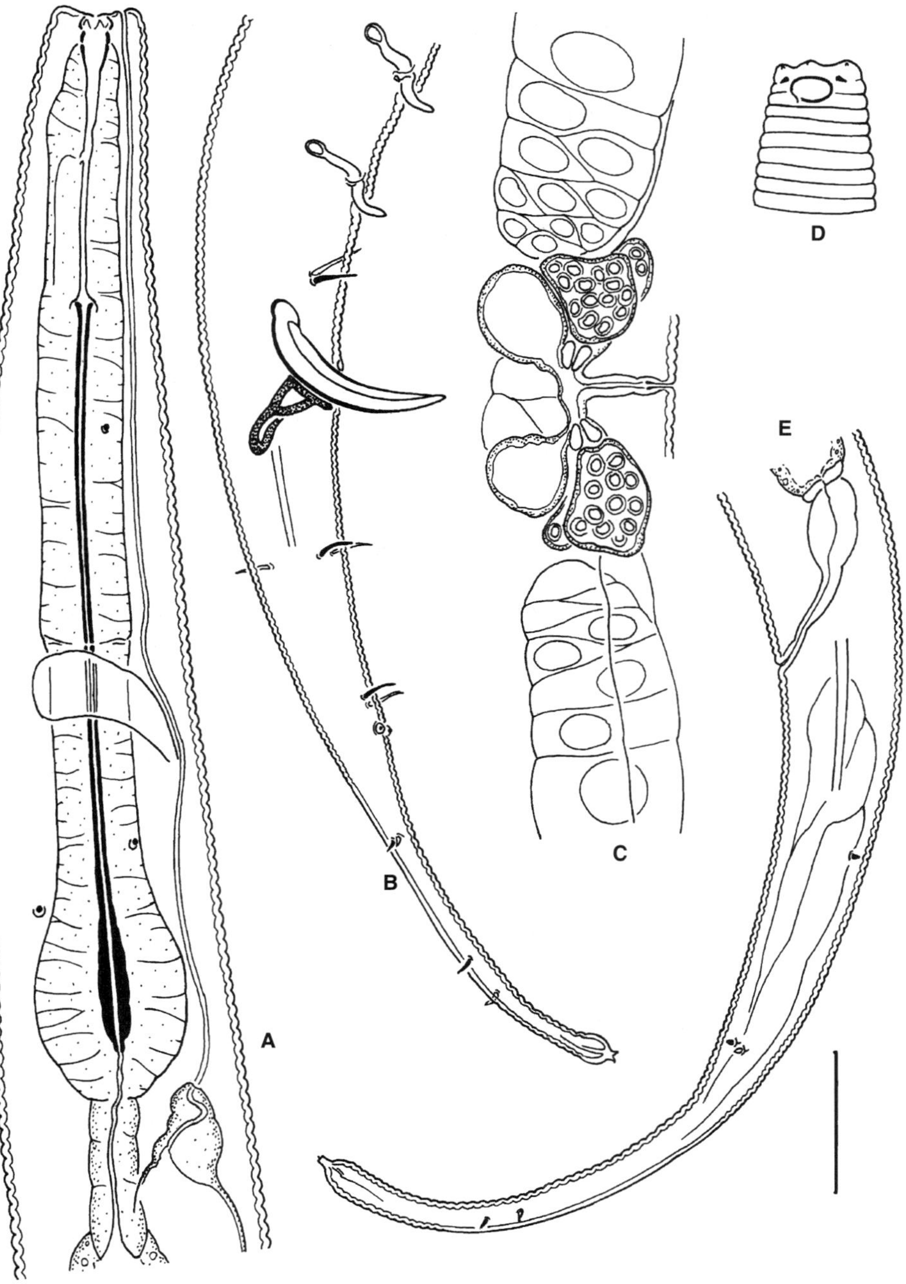

Fig. 20.2 Morphology of the genus *Paraplectonema* Strand, 1934 as exemplified by *Paraplectonema* sp. from Vietnam. A: Pharyngeal region; B: Male caudal region; C: Female reproductive system; D: Anterior end, surface view; E: Female tail. (Scale bar: A–E = 20 µm.)

Fig. 20.3 Scanning electron micrographs of the lip regions of Leptolaimidae, Metateratocephalidae and Chronogastridae. A: *Paraplectonema pedunculatum* (Hofmänner, 1913) Strand, 1934; B: *Metateratocephalus crassidens* (de Man, 1880) Eroshenko, 1973; C: *Chronogaster magnifica* Andrássy, 1956; D: *Chronogaster boettgeri* Kischke, 1956. Scale bars equal 2 μm. (A: After Coomans and Raski, 1991b; B: after Boström, 1989; C: after Eyualem-Abebe and Coomans, 1996a. A, B: Courtesy of *Revue de Nématologie;* C: with kind permission of Springer Science and Business Media; D: courtesy of Laboratorium voor Nematologie, Wageningen Universiteit.)

Biology. Amphimictic except for *P. pedunculatum* (thelytokous). All species inhabit freshwater or brackish water sediments. Recorded from many countries in Europe; from China, Vietnam, Japan, Far East and Caspian Sea in Asia; from Ethiopia in Africa; and from USA and Costa Rica in the Americas.
Bibliography. The only scanning electron micrograph (SEM) of *P. pedunculatum* (Fig. 20.3A) was published by Coomans and Raski (1991b). A diagnostic species compendium is given in Holovachov and Boström (2004).

Type species: *P. pedunculatum* (Hofmänner, 1913) Strand, 1934

Other species:
P. americanum Zullini, Loof & Bongers, 2002
P. canadianum Hopper, 1968
P. loofi Holovachov & Boström, 2004
P. multitubiferum (Imamura, 1931) Loof & Jairajpuri, 1965
P. primitivum (Gagarin, 2001) Holovachov & Boström, 2004
P. vietnamicum Gagarin, Thanh & Tu, 2003

Family APHANOLAIMIDAE Chitwood, 1936

Morphology. Lateral alae present. Somatic sensilla present in pharyngeal region only. Cephalic capsule developed, carrying cephalic sensilla and amphids. Labial sensilla absent on the outer surface of lips as seen under SEM. Cephalic sensilla setiform, their bases located anteriorly to amphid. Amphid unispiral, or derived therefrom. Renette cell oval or with anterior or posterior lobes. Excretory ampulla absent. Cuticularized excretory duct originates from renette cell and opens inside the cheilostom. Coomans and De Waele (1979) suggested a digestive function for the renette in Aphanolaimidae. Stoma cylindrical, stegostom undifferentiated. Pharynx muscular, uniformly cylindrical lacking radial tubules, bulbs and valvular apparatus. Female reproductive system didelphic, amphidelphic (except in a few *Anonchus* species), ovarian branches reflexed antidromously. There is a strong sphincter at the uterus–ovejector junction. Male reproductive system monorchic with outstretched glandular anterior part and dorsally reflexed functional posterior testis. Both alveolar and tubular, or only tubular supplements present. Pre-cloacal sensillum papilliform, located on anterior cloacal lip.
Taxonomy. Two subfamilies are recognized by Holovachov *et al.* (2002): Aphanolaiminae and Anonchinae (with only one genus, *Anonchus*).
Biology. Aquatic (inhabit marine and freshwater sediments, periphyton) and terrestrial (found in soil and litter).

Key to the subfamilies

1. Amphid unispiral, with or without central elevation; stoma cylindrical, small and narrow; cheilostom undifferentiated; pharynx weak; tubular supplements unequal (usually the last but one tubular supplement is located on a cuticular elevation) . . . . .Aphanolaiminae
   Amphid with oval aperture and larger sub-cuticular fovea; stoma large and broad; cheilostom with six digitate projections; pharynx strong; tubular supplements equal in structure . . . . .Anonchinae (monotypic, *Anonchus*)

Subfamily Aphanolaiminae Chitwood, 1936

Morphology. Bases of cephalic sensilla are located at anterior margin of amphid. Amphid unispiral, or derived therefrom. Oral aperture oval. Stoma uniformly cylindrical; gymnostom cylindrical, weakly or heavily sclerotized. Pharynx muscular, weak. Female reproductive system didelphic. Ventral alveoli present in females of two species of *Aphanonchus*. Male with tubular supplements, sometimes also with alveolar supplements. The last but one tubular supplement is located on a cuticular elevation (in two species the most posterior supplement is elevated instead of the second from cloaca).

Taxonomy. Three genera: *Aphanolaimus* (Figs 20.4B and 20.5A–C, H–M, O–Q, S), *Paraphanolaimus* (Fig. 20.5F, G and R) and *Aphanonchus* (Figs 20.4C and 20.5D, –E, N, T).

Biology. Aquatic (inhabit marine and freshwaters) and terrestrial.

Key to genera

1. Cheilostom is surrounded by peristomatal vestibulum; alveolar supplements usually present in male . . . . *Aphanonchus*

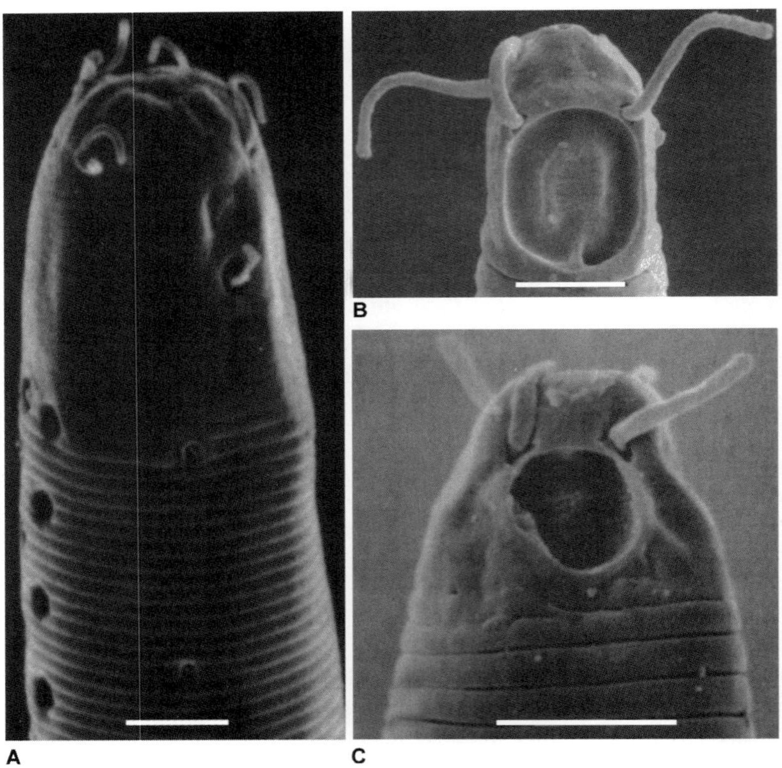

Fig. 20.4 Scanning electron micrographs of the lip regions of Aphanolaimidae. A: *Anonchus coomansi* Eyualem-Abebe, 1996; B: *Aphanolaimus elegans* Raski & Coomans, 1990; C: *Aphanonchus intermedius* Coomans & Raski, 1991. Scale bars equal 5 µm. (All after the respective original descriptions; A: with kind permission of Springer Science and Business Media; B, C: courtesy of *Nematologica*.)

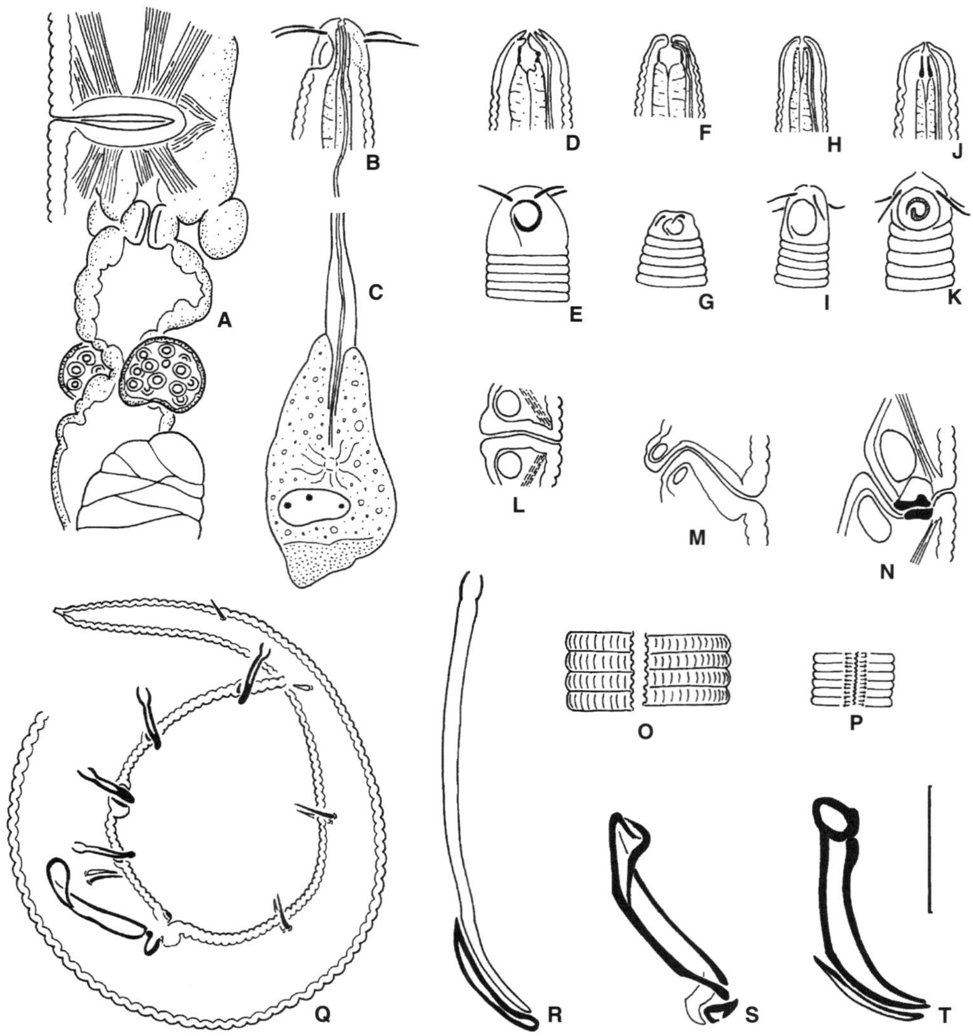

Fig. 20.5 Morphology of the genera *Aphanolaimus* de Man, 1880, *Aphanonchus* Coomans & Raski, 1991 and *Paraphanolaimus* Micoletzky, 1923 as exemplified by *Aphanolaimus aquaticus* von Daday, 1894 from Ukraine (A, H,I, L, S); *A. aymarae* De Waele & Coomans, 1993 from Costa Rica (B, C, M); *A.* cf. *microlaimus* De Coninck, 1935 from Dominica (J, K); *A. seshadrii* Raski & Coomans, 1990 from Vietnam (O); *A. costatus* Holovachov, 2004 from Cameroon (P); *A. pseudoattentus* Coomans & De Waele, 1983 from Ukraine (Q), *Aphanonchus bayensis* (Keppner, 1988) Holovachov & Sturhan, 2004 from Costa Rica (D, E, N, T); *Paraphanolaimus paraguayensis* Holovachov & Sturhan, 2004 from Paraguay (F, G); and *P. asiaticus* Gagarin, Thanh & Tu, 2003 from Vietnam (R). A: Part of posterior female genital branch; B: Anterior end showing excretory duct; C: Renette cell; D, F, H, J: Anterior end, median section; E, G, I, K: Anterior end, surface view; L–N: Vagina; O and P: Cuticle and lateral field; Q: Male caudal region; R–T: Spicula and gubernaculum. (Scale bar: A–T = 20 μm.)

Cheilostom narrow undifferentiated, peristomatal vestibulum absent; alveolar supplements absent 2

2. Gymnostom broad cylindrical, barrel-shaped; amphid without prominent central elevation . . . .*Paraphanolaimus*

Gymnostom short and narrow, with granule-shaped rhabdia; when gymnostom is barrel-shaped then amphid with prominent central elevation . . . .*Aphanolaimus*

Genus *Aphanolaimus* de Man, 1880

Morphology. Cuticular annulation smooth or with longitudinal incisures. Lateral alae demarcated by two crenate lines; in two species sub-lateral rows of dots are also present (Figs 20.5O and P). Somatic sensilla present in males of some species only. Amphid unispiral, with or without central elevation (Figs 20.4B and 20.5J and K), or reniform. Excretory duct opens inside the cheilostom (Fig. 20.5B). Stoma: cheilostom undifferentiated; gymnostom short and weak or cylindrical heavily sclerotized; stegostom narrow (Figs 20.5H and J). Spermathecae well developed (except in *A. aymarae*). Uterus short in oviparous species (Fig. 20.5A) and long in ovoviviparous species. Vagina straight or inverted V-shaped, without sclerotizations (Figs 20.5L and M). Spicules arcuate or with manubrium twisted along their axis. Gubernaculum rectangular or platelike. Mid-ventral tubular supplements 3–13, of which the last but one is located on a cuticular elevation (except in *A. camerunensis* and *A. aymarae,* where the most posterior supplement is elevated). One sub-ventral pair of pre-cloacal setae located at level of spicules (except in *A. costatus* and *A. camerunensis*). Two or three sub-ventral pairs and one or two sub-dorsal pairs of setae located on tail (Fig. 20.5Q).

Biology. Most species amphimictic. Three species (*A. aymarae, A. spiriferus* and *A. viviparus*) are ovoviviparous. Found on all continents except Antarctica. Inhabit freshwater sediments, moist soil, moss and litter.

Bibliography. The most recent identification key is given in Raski and Coomans (1990). Analysis of variability of some morphological features is given in Coomans and De Waele (1983) and Raski and Coomans (1990). SEM observations were made by Coomans and De Waele (1979), Raski and Coomans (1990) and Eyualem-Abebe (1996).

Type species: *A. attentus* de Man, 1880

Other species:

A. aquaticus von Daday, 1894
A. aymarae De Waele & Coomans, 1993
A. boliviensis De Waele & Coomans, 1993
A. brasiliensis De Waele & Coomans, 1993
A. camerunensis Holovachov, 2004
A. chilensis Raski & Coomans, 1990
A. coomansus Tsalolikhin, 1988
A. costatus Holovachov, 2004
A. deconincki Coomans & De Waele, 1983

A. elegans Raski & Coomans, 1990
A. fuegoensis Raski & Coomans, 1990
A. furcifer Andrássy, 1989
A. louisae Coomans & De Waele, 1979
A. microlaimus De Coninck, 1935
A. pseudoattentus Coomans & De Waele, 1983
A. quechuae De Waele & Coomans, 1993
A. seshadrii Raski & Coomans, 1990
A. solitudinis Andrássy, 1968
A. spiriferus Cobb, 1914
A. tudoranceai Zullini, 1988
A. viviparus Plotnikoff, 1899
A. yamani Raski & Coomans, 1990

Genus *Paraphanolaimus* Micoletzky, 1923
Syn. = *Bathyonchus* Kreis, 1936
Morphology. Cuticular annulation smooth. Somatic sensilla present in males. Amphid unispiral, without central elevation (Fig. 20.5G). Excretory duct opens inside the cheilostom. Stoma: cheilostom undifferentiated; gymnostom broad cylindrical, heavily sclerotized; stegostom undeveloped (Fig. 20.5F). Spermathecae well developed, in *P. behningi* and *P. terrestris* small non-functional. Uterus short in oviparous and long in ovoviviparous species, an undifferentiated cylindrical tube. Vagina straight or bent anteriorly, without sclerotizations. Spicules long arcuate (Fig. 20.5R) or with twisted along the axis manubrium. Gubernaculum rectangular. Mid-ventral tubular supplements 8–19, of which the last but one is located on cuticular elevation. One sub-ventral pair of pre-cloacal setae located at level of spicules. Two to three sub-ventral pairs and one to two sub-dorsal pairs of setae located on tail.
Biology. Most species amphimictic. Two species *P. behningi* and *P. asiaticus* are ovoviviparous. Found on all continents except Antarctica. Inhabit freshwater sediments and moist soil. Different populations of *P. behningi* were thelytokous (Holovachov and Sturhan, 2004a) or amphimictic (Eyualem-Abebe, 1996).
Bibliography. Identification keys are given in Raski and Coomans (1991) and Holovachov and Sturhan (2004a). SEM observations were made by Raski and Coomans (1991) and Eyualem-Abebe (1996).

Type species: *P. behningi* Micoletzky, 1923

Other species:
P. anisitsi (von Daday, 1905) Andrássy, 1968
P. asiaticus Gagarin, Thanh & Tu, 2003
P. embryonophorus (Alekseev & Naumova, 1977) Andrássy, 1984
P. latescens Raski & Coomans, 1991
P. paraguayensis Holovachov & Sturhan, 2004
P. tahoensis Raski & Coomans, 1991
P. terrestris Raski & Coomans, 1991

Genus *Aphanonchus* Coomans & Raski, 1991
Morphology. Cuticular annulation smooth. Somatic sensilla present in both sexes or in males only. Amphid unispiral, without central elevation (Figs 20.4C and 20.5E). Excretory duct opens inside the peristomatal vestibulum. Stoma: cheilostom is surrounded by peristomatal vestibulum, undifferentiated; gymnostom broad cylindrical, heavily sclerotized; stegostom undeveloped (Fig. 20.5D). Spermathecae well developed, in *A. europaeus* small non-functional. Uterus long. Vagina straight or sigmoid, with (Fig. 20.5N) or without sclerotizations. Ventral alveoli present in females of two species. Spicules arcuate (Fig. 20.5T). Gubernaculum arch-like. Males with 20–108 alveolar (none in *A. longiceras*) and 10–20 mid-ventral tubular supplements, of which the last but one is located on cuticular elevation. One sub-ventral pair of pre-cloacal setae located at level of spicules. Two sub-ventral pairs and two sub-dorsal pairs of setae located on tail.
Biology. Most species amphimictic. All but one species (*A. longiceras*) are ovoviviparous. Hitherto recorded from Germany in Europe; from China, Vietnam and Japan in Asia; Tanzania and Ivory Coast in Africa; from USA, Costa Rica, Colombia, Surinam and Paraguay in the Americas. Inhabit sediments in freshwater and brackish habitats.
Bibliography. Genus description and detailed morphological observations are presented in Coomans and Raski (1991a). The most recent identification key is given in Holovachov and Sturhan (2004b).

Type species: *A. intermedius* Coomans & Raski, 1991

Other species:
A. africanus Holovachov & Sturhan, 2004
A. europaeus Holovachov & Sturhan, 2004
A. bayensis (Keppner, 1988) Holovachov & Sturhan, 2004
A. longiceras (Tsalolikhin, 1989) Tsalolikhin, 2002
A. multipapillatus (von Daday, 1905) Coomans & Raski, 1991
A. obesus Gagarin & Thahn, 2003
A. orientalis Wu & Liang, 2000

Subfamily Anonchinae Andrássy, 1973

Genus *Anonchus* Cobb, 1913
Syn. = *Bathylaimus* Daday, 1905 nec Cobb, 1894 = *Pseudobathylaimus* Filipjev, 1918 = *Dadayia* Micoletzky, 1922 = *Assia* Gerlach, 1957 = *Haconnus* Andrássy, 1973
Morphology. Cuticular annulation smooth. Somatic sensilla present in male of *A. mirabilis* only. Bases of cephalic sensilla are located anteriorly to amphid. Four subcephalic setae present in male of *A. venezolanus*. Amphid unispiral, with rounded aperture and larger unispiral sub-cuticular fovea (Figs 20.4A and 20.6B). Excretory duct opens inside the cheilostom. Stoma cylindrical: cheilostom supported by six digitate projections; gymnostom short or long cylindrical, stegostom shallow. Pharynx muscular, strong (Figs 20.6A and C). Female reproductive system didelphic, amphidelphic or monodelphic, prodelphic. Spermathecae well developed (except in *A. millelacunatus* and *A. venezolanus*, where they are absent). Uterus short. Vagina straight or bent anteriorly, without sclerotiza-

tions (Fig. 20.6D). Spicules arcuate or with manubrium twisted along the axis. Gubernaculum rectangular. Pre-cloacal region with 51–816 alveolar supplements (none in *A. venezolanus* and *A. mirabilis*) and 1–72 mid-ventral tubular supplements. Usually one sub-ventral pair of pre-cloacal setae located at level of spicules (except in *A. venezolanus*, which has ten sub-lateral pairs). Five to 12 pairs of setae located on tail.

Biology. Amphimictic. Three species were found in mountainous areas or deep lakes and river estuaries of Europe, Asia (*A. mirabilis*, Fig. 20.6) and Africa (*A. coomansi*, *A. palaeotropicus*); other species inhabit brackish and freshwater sediments in South and Central America, with only *A. maculatus* being distributed in the whole of the New World. Several cases of sympatric species are known: *A. maculatus* + *A. winiszewskae* (Paraguay); *A. maculatus* + *A. winiszewskae* + *A.* sp. *apud* Holovachov *et al.* (2002) (Colombia); *A. winiszewskae* + *A.* sp. *apud* Holovachov *et al.* (2002) (Costa Rica); niche partitioning may apply here and could be reflected in differences in stoma size correlated with differences in feeding habitat/preferences (related, e.g. to maximum size of food particles). *A. mirabilis* from Italy was found with diatomaceous algae in the intestine, while a population from Germany apparently fed on testacean Rhizopoda (O. Holovachov, unpublished data). This species is therefore probably adapted to feeding on unicellular eukaryotes. Generation time in *Anonchus* sp. was estimated at 35–47 days depending on temperature (Laybourn, 1979); this author also studied effects of temperature on the respiration and reproduction of this species in laboratory conditions.

Bibliography. The most recent identification key is given in Holovachov *et al.* (2002). SEM observations (Fig. 20.4A) were made on *A. coomansi* by Eyualem-Abebe (1996), who found that the oral aperture is surrounded with five triangular liplets, including one dorsal liplet that is twice as wide as all others and presumably resulted from fusion of a sub-dorsal pair.

Type species: *A. maculatus* (Daday, 1905) T. Goodey, 1951

Other species:

A. coomansi Eyualem-Abebe, 1996
A. laureatus (Gerlach, 1957) Holovachov, Zullini, Loof & Bongers, 2002
A. mangrovi Gerlach, 1957
A. millelacunatus (Andrássy, 1973) Holovachov, Zullini, Loof & Bongers, 2002
A. mirabilis (Hofmänner in Hofmänner and Menzel, 1914) Chitwood & Chitwood, 1936
A. palaeotropicus Tsalolikhin, 1989
A. pulcher Zullini, Loof & Bongers, 2002
A. venezolanus Holovachov, Zullini, Loof & Bongers, 2002
A. winiszewskae Holovachov, Zullini, Loof & Bongers, 2002

Superfamily PLECTOIDEA Orley, 1880

Morphology. Cuticle weakly to coarsely annulated. Intracuticular ornamentation absent or present. Lateral alae absent or present. Body pores absent or arising from oval sub-lateral hypodermal gland cells. Somatic sensilla when present stand

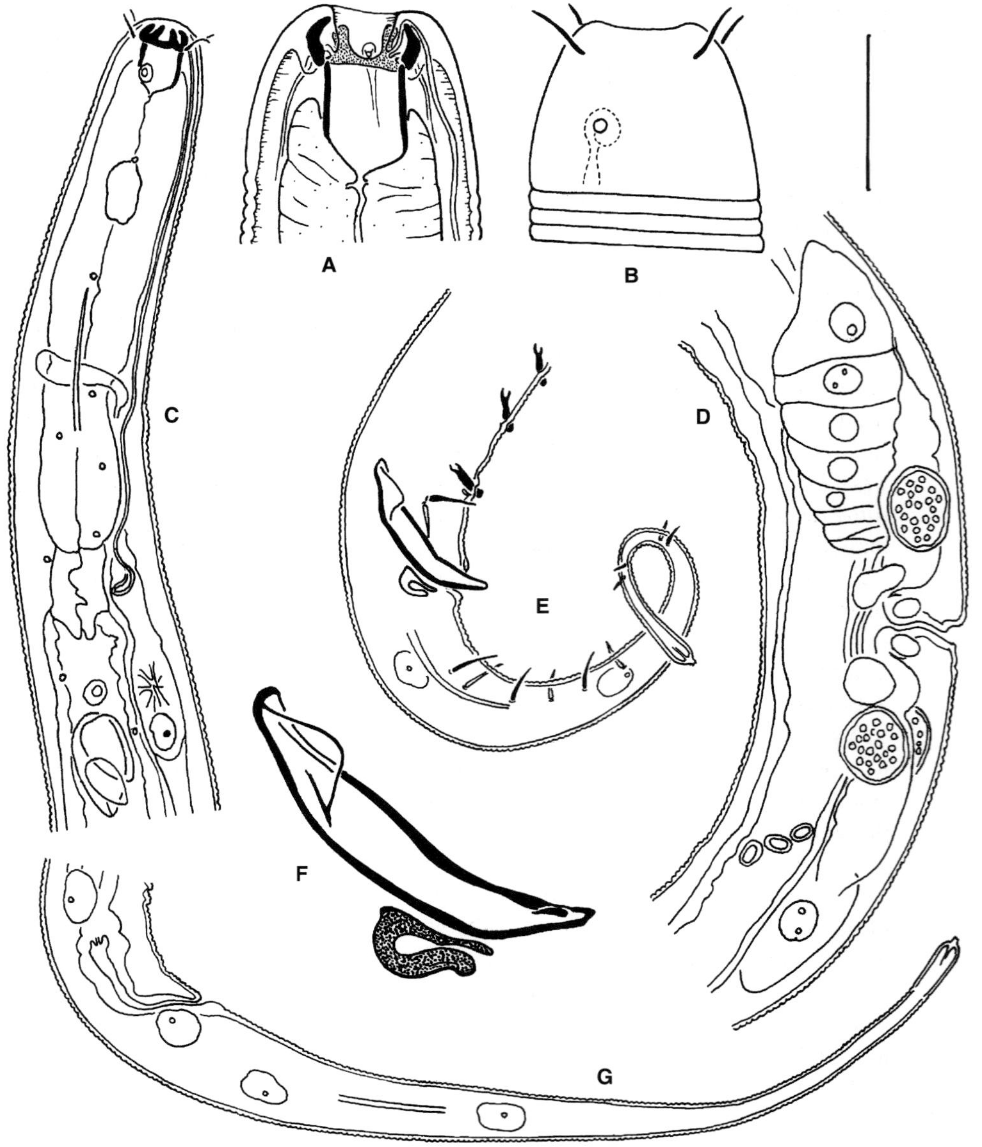

Fig. 20.6 Morphology of the genus *Anonchus* Cobb, 1913 as exemplified by *A. mirabilis* (Hofmänner in Hofmänner & Menzel, 1914) Chitwood & Chitwood, 1936 from Germany. A: Anterior end, median section; B: Anterior end, surface view; C: Pharyngeal region; D: Female reproductive system; E: Male caudal region; F: Spicule and gubernaculum; G: Female tail. (Scale bar: A–B and F = 20 μm, C–E and G = 50 μm.)

independently from the hypodermal glands. Setiform deirids absent or present. Labial region either with six equal lips, or with leaf-like lips with sclerotized edges, or with biradially arranged cuticular outgrowths. Inner labial sensilla open inside the cheilostom (in Metateratocephalidae their position is as yet unknown). Outer labial sensilla papilliform (setiform in Metateratocephalidae) or located in radial slits, on the outer surface of lips. Cephalic sensilla setiform or strongly modified; their bases located anteriorly to amphid. Amphid unispiral, or derived therefrom. Ocelli absent. Secretory–excretory system present; renette cell located opposite to ventral side of the posterior region of the isthmus (except *Pakira*). Excretory ampulla absent. Excretory duct with lateral loops around the pharynx, excretory pore located externally at about the nerve ring level. Stoma funnel-shaped: cheilostom narrow, with or without rhabdia; gymnostom convex or cylindrical, with sclerotized lining; stegostom subdivided into two sections: anterior section subcylindrical to conical and surrounded by thinner musculature, posterior section usually closed in fixed specimens. Pharynx muscular, with or without (in *Pakira*) distinct subdivision into corpus, isthmus and basal bulb. Pharyngeal tubes starting at stoma base and extending posteriorly for some distance (procorpus). Longitudinal or transverse valvular apparatus present in basal bulb. Dorsal gland orifice penetrates second section of the stegostom. Sub-ventral gland orifices penetrate pharyngeal lining at corpus–isthmus junction (at mid-pharynx in *Pakira*). Pharyngeal gland nuclei located in basal bulb. Cardia cylindrical, non-muscular, surrounded by intestinal tissue. Female reproductive system didelphic, amphidelphic (except in *Chronogaster* and *Steratocephalus*), ovarian branches reflexed antidromously, located on alternative sides of intestine with anterior branch usually on right side. Oviduct short. Spermathecae axial when present. Uterus cylindrical or funnel-shaped. Vagina encircled by one or two sphincter muscles. Male reproductive system diorchic or monorchic, posterior testis reflexed dorsally. Tubular supplements present or absent. Pre-cloacal sensillum papilliform or setiform, located at some distance anteriorly to cloaca, sometimes absent. Post-cloacal sensilla (papilliform in shape) absent or located sub-ventrally at the middle of tail. Setae in caudal region of males arranged in four rows, two sub-ventral and two sub-dorsal. Three caudal glands absent or present, with nuclei that are incaudal. Tail tip with mucro or with a spinneret that is weakly cuticularized.
Taxonomy. Four families are recognized here: Pakiridae, Chronogastridae, Plectidae and Metateratocephalidae.
Biology. Inhabit freshwater or brackish water as well as moist soils, sandy soils near beaches, mosses.

Key to families

1. Basal bulb clearly sub-terminal, equipped with longitudinal denticulate ridges; posterior stegostom section long; female reproductive system monoprodelphic . . . . .Chronogastridae (*Chronogaster*)
   Basal bulb terminal or slightly sub-terminal, equipped with transverse plates that are smooth, denticulate or corrugated; posterior stegostom section short; female reproductive system didelphic 2
2. Labial region crown-shaped; lips leaf-like with sclerotized edges; outer labial sensilla setiform; male reproductive system monorchic . . . . .Metateratocephalidae

Labial region with six equal lips or with biradially arranged cuticular outgrowths; outer labial sensilla papilliform; male reproductive system diorchic . . . . .Plectidae

Family CHRONOGASTRIDAE Gagarin, 1975

Morphology. Cuticle annulated. Sub-cuticular ornamentation absent. Lateral alae and deirids absent. Hypodermal glands present or absent. Somatic sensilla present in both sexes or in males only. Cephalic sensilla setiform. Labial region with six equal lips or cap-like. Amphid unispiral or derived therefrom (stirrup- or horseshoe-shaped). Excretory duct free, coiled around isthmus. Posterior section of the stegostom long. Basal bulb sub-terminal, weakly to strongly developed. Valves of the longitudinal, denticulate type (three denticulate ridges) or just plate-like thickenings of the pharyngeal lumen. Female reproductive system didelphic, amphidelphic or monodelphic, prodelphic. Spermatheca axial if present. Male reproductive system diorchic, with opposed testes. Tubular supplements present or absent. Pre-cloacal sensillum setiform or absent. Post-cloacal sensilla usually present. Caudal glands present or absent. Tail tip with mucro or spinneret.

Taxonomy. We include three genera, *Chronogaster* (Figs 20.3C and D and 20.7), *Cynura* and *Caribplectus*, in this family, not dividing them into separate subfamilies (Holovachov, 2004).

Biology. Inhabit freshwaters and moist soil (*Chronogaster*), brackish waters (*Caribplectus*) and marine sediments (*Cynura*).

Genus *Chronogaster* Cobb, 1913

Syn. = *Walcherenia* de Man, 1921 = *Keralanema* Siddiqi, 2003

Morphology. Cuticle annulated, sometimes with longitudinal incisures, ridges and spines. Hypodermal glands present or absent. Somatic sensilla present only in caudal region of males. Amphid unispiral (Figs 20.3D and 20.7A), stirrup- (Fig. 20.7B) or horseshoe- (Fig. 20.3C) shaped. Basal bulb strongly developed. Post-bulbal prolongation (cylindrical part of the pharyngeal muscular tissue) is present. Valves of the longitudinal, denticulate type, e.g. with three denticulate ridges (Fig. 20.7D). Female reproductive system monodelphic, prodelphic. Spermatheca absent. Tubular supplements present or absent. Pre-cloacal sensillum setiform or absent. Post-cloacal sensilla present (except in *C. spinicorpus*). Caudal glands present and functional in one species only (*C. boettgeri*) or absent. Tail tip with variously shaped mucro or spinneret (only in *C. boettgeri*). Sexual dimorphism in the tail terminus structure was noted in *C. andrássyi, C. boettgeri, C. indica* and *C. multispinatoides* (Fig. 20.7F); in these species the male tail is bluntly rounded and lacks the spinneret or mucro characteristic for the females.

Biology. Most species thelytokous. Only in the type population of *C. spinicorpus* were number of males and females almost equal, suggesting an amphimictic mode of reproduction in this species. Occur in freshwater, thermal springs, moist soil, fungal mats (*C. troglodytes*) and brackish habitats (*C. alata*). Ettema *et al.* (2000) studied population dynamics of five sympatric species of *Chronogaster* and concluded that long-term coexistence was maintained by unpredictable habitat fluctuations and differential recolonizations by the different species. Eyualem-Abebe and Coomans (1996a) discussed the external factors, which may play a role in the

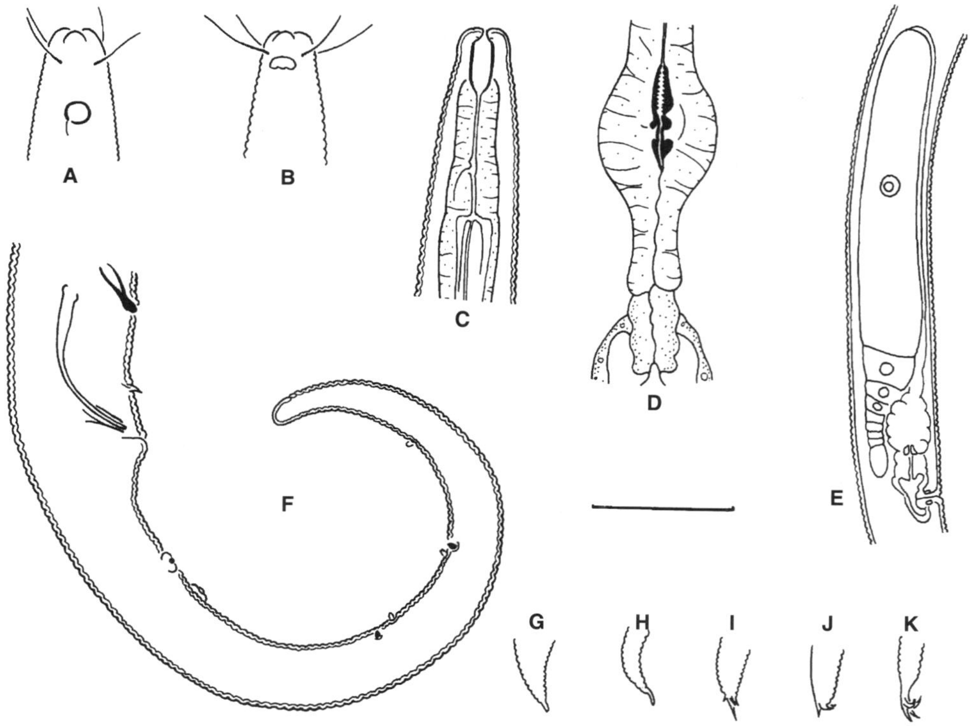

Fig. 20.7 Morphology of the genus *Chronogaster* Cobb, 1913 as exemplified by *C. boettgeri* Kischke, 1956 from Ukraine (A, G); *C. multispinatoides* Heyns & Coomans, 1984 from Costa Rica (B, C, I) and USA (F); *C. serrulata* Loof, 1973 from Costa Rica (D, E); *C. typica* (de Man, 1921) De Coninck, 1937 from Ukraine (H); *C. jankiewiczi* Winiszewska-Slipinska, 1997 from Costa Rica (J); and *C.* cf. *magnifica* Andrássy, 1956 from Benin (K). A, B: Anterior end, surface view; C: Anterior end, median section; D: Basal bulb; E: Female reproductive system; F: Male caudal region; G–K: Tail terminus. (Scale bar: A–D and F–K = 20 μm; E = 50 μm.)

presence/absence of crystalloid bodies in different species of this genus, and suggested that the finding of more than one species of *Chronogaster* from the same ecological niche may help in clarifying this question. Observations of the five species found by Ettema *et al.* (2000) revealed the following data: crystalloids are present along the entire body in *C. floridiensis* and *C.* cf. *tenuis*, only in the pharyngeal region in *C. multispinatoides* and indistinct in *C. daoi* and *C. parva*. In one sample of sandy soil collected on the shore of the freshwater Lake Pisotchne in Ukraine, we found three species: *C. typica*, *C. polonica* and *C. boettgeri*. The first two species have crystalloids in their body, while *C. boettgeri* does not. We may suppose that the appearance of crystalloids in the nematode body depends not only upon the environmental conditions, but also on the species-specific physiological peculiarities.

Taxonomy. Siddiqi (2003) proposed a separate genus *Keralanema* for *C. spinicorpus*. Phylogenetic analysis (Holovachov, 2004) suggests that *C. spinicorpus* is embedded within the genus *Chronogaster* and that separate generic status would render the genus *Chronogaster* paraphyletic.

Bibliography. Identification keys are given in Heyns and Coomans (1983) and Raski and Maggenti (1984), species compendia can be found in Heyns and Coomans (1980) and Eyualem-Abebe and Coomans (1996a).

Type species: *C. gracilis* Cobb, 1913

Other species:
C. africana Heyns & Coomans, 1980
C. alata Gerlach, 1956
C. andrássyi Loof & Jairajpuri, 1965
C. aspinata Raski & Maggenti, 1985
C. bengalensis Saha, Lal & Singh, 2001
C. bigubernacula Khera, 1972
C. boettgeri Kischke, 1956
C. brasiliensis Meyl, 1957
C. cameroonensis Heyns & Coomans, 1984
C. chilensis Raski & Maggenti, 1985
C. chilkensis Khera, 1972
C. citri Khan & Nanjappa, 1973
C. costaricae Zullini, Loof & Bongers, 2002
C. daoi Loof, 1964
C. elegans Raski & Maggenti, 1985
C. ethiopica Eyualem-Abebe & Coomans, 1996
C. floridensis Raski & Maggenti, 1985
C. getachevi Eyualem-Abebe & Coomans, 1996
C. glandifera Heyns & Coomans, 1980
C. indica Bajaj & Bhatti, 1979
C. jankiewiczi Winiszewska-Slipinska, 1997
C. lissa Loof, 1973
C. longicaudata Heyns & Coomans, 1980
C. longicollis (Daday, 1899) Andrássy, 1958
C. loofi Chaturvedi & Khera, 1979
C. magnifica Andrássy, 1956
C. mexicana Winiszewska-Slipinska, 1997
C. multispinata Heyns & Coomans, 1980
C. multispinatoides Heyns & Coomans, 1984
C. neoparva Tahseen & Siddiqi, 2003
C. neotypica Tahseen, Ahmad & Ahmad, 1994
C. parva Heyns & Coomans, 1984
C. polonica Winiszewska-Slipinska, 1997
C. pseudotypica Gagarin & Thanh, 2004
C. rotundicauda Heyns & Coomans, 1984
C. sclerostoma Tahseen & Siddiqi, 2003
C. serrulata Loof, 1973
C. spicata Heyns & Coomans, 1984
C. spinicauda Tahseen, Ahmad & Ahmad, 1994
C. spinicorpus Maggenti, Raski, Koshy & Sosamma, 1983
C. subtilis Andrássy, 1958

C. tenuis Loof & Jairajpuri, 1965
C. tesselata Mounport, 2005
C. troglodytes Poinar & Sarbu, 1994
C. typica (de Man, 1921) De Coninck, 1937
C. vacuoli Saha, Lal & Singh, 2001
C. zujarensis Ocaña & Coomans, 1991

Family PLECTIDAE Örley, 1880

Morphology. Cuticle annulated. Sub-cuticular ornamentation absent. Lateral alae consist of two sub-lateral wings usually separated by a distinct gap. Deirids setiform, located between wings of lateral field just posterior to nerve ring. Hypodermal glands and pores present or absent. Somatic sensilla present in both sexes. Labial region with six equal lips or biradially symmetrical, with outgrowths. Amphid unispiral or transverse slit. Excretory duct forming two loops on right- and left-hand sides of pharynx, enveloped by a renette cell. Stoma consisting of a relatively flexible cheilostom, a cylindrical or convex gymnostom with sclerotized lining and a stegostom that has a membranous lining and tapers posteriorly. Posterior section of the stegostom short. Basal bulb terminal or slightly sub-terminal, strongly developed, with rhabditid-like transverse valves or with *parietinus*-like, strongly arcuate longitudinal valves bearing numerous denticles (except in first-stage juveniles, which have linear denticulate valves). Female reproductive system didelphic, amphidelphic, ovarian branches reflexed antidromously, usually located on alternate sides of intestine with anterior branch on right side. Male reproductive system diorchic, anterior testis outstretched, posterior testis reflexed. Tubular supplements, pre-cloacal sensillum and post-cloacal sensilla usually present. Caudal glands and spinneret usually present (in a few species the spinneret has apparently become dysfunctional and lost its duct); spinneret sometimes surrounded by papillae.

Taxonomy. Three subfamilies are recognized: Plectinae, Anaplectinae and Wilsonematinae. Unpublished molecular data also suggest Anaplectinae, Plectinae and Wilsonematinae constitute a single clade. The latter include mostly terrestrial and moss-inhabiting species, only *Tylocephalus auriculatus* was often recorded from moist soil and freshwater, although it has also been recovered from dehydrated sand with shell fragments (De Ley and Coomans, 1997). Zell (1993) conducted an unusual phylogenetic analysis and relegated all wilsonematids to a single subgenus within *Plectus*. We consider this proposal premature until a more robust phylogenetic analysis is completed.

Biology. Inhabit freshwaters, soil, litter, moss, rotting wood etc. Functional morphology of the pharynx in *Plectus* and *Tylocephalus* was described by Fürst von Lieven (2003).

Key to subfamilies

1. Labial region with biradially arranged cuticular outgrowths; cervical expansions present; lateral outer labial sensilla elevated on the mid-lateral projections; cephalic sensilla leaf-shaped, sometimes with incised inner edge (cornua) . . . . .Wilsonematinae (*Tylocephalus*)
   Labial region with six equal lips; cuticular outgrowths and cervical expansions absent; outer labial sensilla papilliform or embedded in radial slits; cephalic sensilla setiform 2

2. Somatic sensilla restricted to tail only; papilliform cervical sensilla present posterior to amphid; spermatheca and crustaformeria present; several pairs of pre-cloacal papilliform sensilla present in male . . . . .Anaplectinae (*Anaplectus*)
   Somatic sensilla present over the whole body; cervical sensilla absent; spermatheca and crustaformeria absent . . . . .Plectinae (*Plectus*)

Subfamily Plectinae Örley, 1880

Morphology. Hypodermal glands present or absent. Somatic sensilla distributed over the whole body. Labial sensilla papilliform or embedded in radial slits, cephalic sensilla setiform. Labial region with six equal lips or lateral lips broader than sub-dorsal and sub-ventral ones. Amphid unispiral. Cervical sensilla absent. Gymnostom cylindrical. Spermatheca and crustaformeria absent. Tubular supplements, pre-cloacal and post-cloacal sensilla present or absent.

Taxonomy. This subfamily is considered to consist of a single large genus *Plectus* (Figs 20.8 and 20.9B and D) that is one of the most diverse and widespread genera of freshwater nematodes. New species are described relatively frequently, while the repertoire of useful identification characters has been expanded greatly in recent years by the application of SEM and by analysis of, e.g. the excretory system, setotaxy and vaginal structure. On the other hand, differences between species are often subtle and identification is often complicated by the limited information available for species described in the early years of nematology. Andrássy (1984, 1985) provided extensive keys and proposed the genera *Ceratoplectus* and *Chiloplectus* for two of the more distinctive species groups. In reality, both genera intergrade with other *Plectus* species. Phylogenetic analysis (Holovachov, 2004) suggests that the genus *Chiloplectus* (Fig. 20.9C) is deeply nested inside the genus *Plectus*, close to species of the '*parietinus*-group', while *Ceratoplectus* may either be close to *P. parvus* or, along with the subfamily Wilsonematinae, represent a sister taxon of *Plectus*. On the basis of present evidence, we agree with Zell (1993) in relegating *Ceratoplectus* to a subgenus within *Plectus*. Zell (1993) re-examined type material of many historically important *Plectus* species and provided a very different key from that of Andrássy (1984, 1985). In practice, identification of newly collected specimens often requires use of both keys and comparison of the outcome(s) with the relevant original description(s).

Biology. Inhabit freshwaters, soil, litter, moss, rotting wood etc.

Genus *Plectus* Bastian, 1865

Syn. = *Plectoides* de Man, 1904 = *Proteroplectus* Paramonov, 1964 = *Ceratoplectus* Andrássy, 1984 = *Chiloplectus* Andrássy, 1984

Morphology. Somatic setae distributed over the entire body. Hypodermal glands and body pores present or absent. Lips six, smooth or radially incised (Figs 20.9B and D). Outer labial sensilla papilliform or embedded in radial incisures. Amphid unispiral. Female reproductive system didelphic. Uterus short without clear differentiation of oviduct or ovejector. Vagina short and straight, rarely with cruciform folds between vaginal sphincters, with or without epiptygmata (Fig. 20.8B). Rectum short to long, with cuticularized lumen. Males with or without tubular supplements, usually with sub-ventral and sub-dorsal rows of setae in the pre-cloacal and caudal

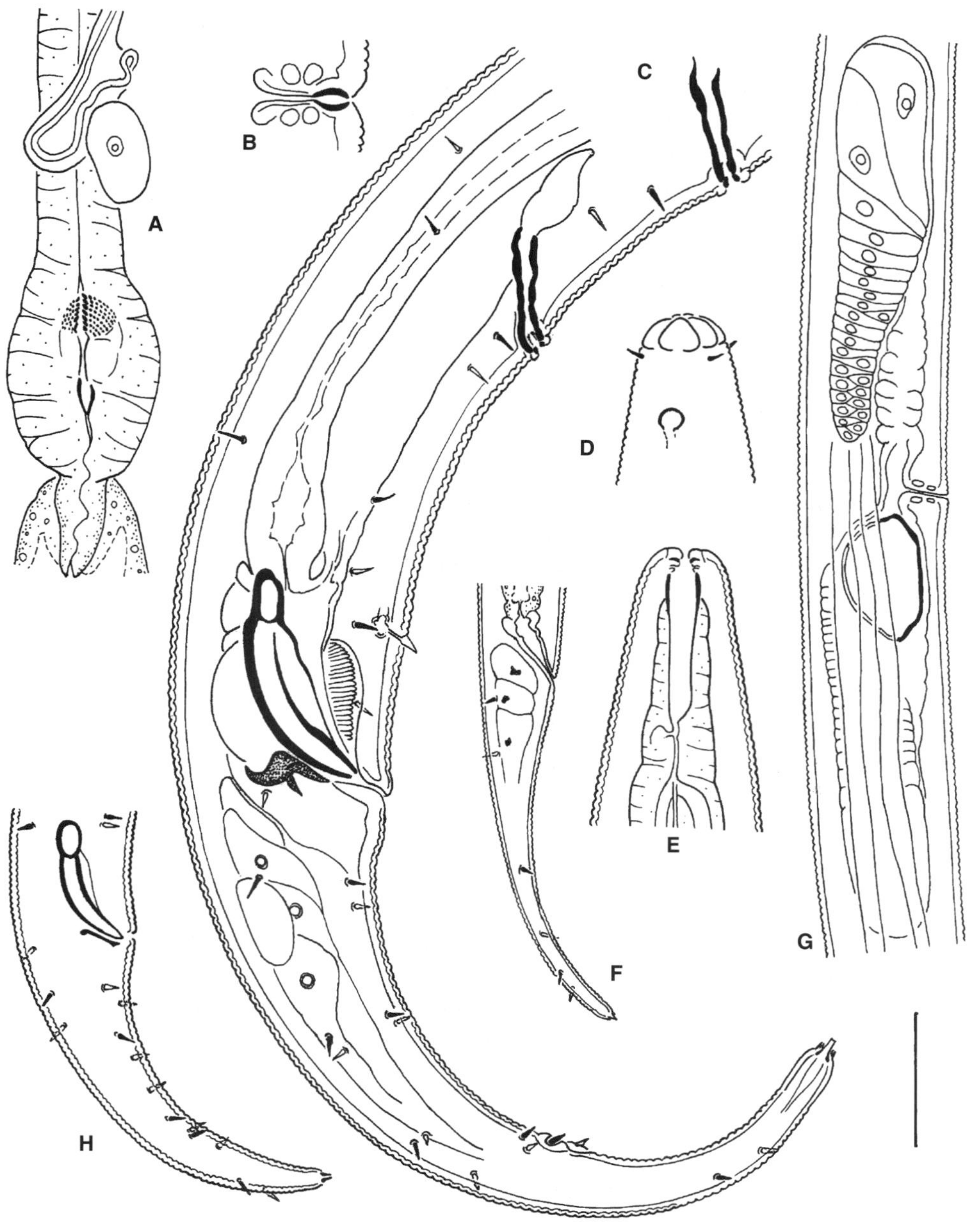

Fig. 20.8 Morphology of the genus *Plectus* Bastian, 1865 as exemplified by *P. aquatilis* Andrássy, 1985 (A, C–G); *P. velox* Bastian, 1865 (B); and *P. parvus* Bastian 1865 (H) from Ukraine. A: Basal bulb; B: Vagina with epiptygmata; C, H: Male caudal region; D: Anterior end, surface view; E: Anterior end, median section; F: Female tail; G: Female reproductive system with intrauterine egg. (Scale bar: A–E and H = 20 μm; F and G = 50 μm.)

Fig. 20.9 Scanning electron micrographs of the lip regions of Anaplectinae and Plectinae. A: *Anaplectus granulosus* (Bastian, 1865) De Coninck & Schuurmans Stekhoven, 1933; B: *Plectus lamproptychus* De Ley & Coomans, 1994; C: *Plectus cancellatus* Zullini, 1978; D: *Plectus tropicus* Zell, 1993. Scale bars equal 2 μm. (A: After Holovachov *et al.*, 2004b; B, D: after De Ley and Coomans, 1994; C: after Holovachov *et al.*, 2000. A: Courtesy of *Russian Journal of Nematology*; B, D: courtesy of the Royal Belgian Institute of Natural Sciences; C: courtesy of *Nematologica*.)

region (Figs 20.8C and H). Pre-cloacal sensillum setiform or absent. Post-cloacal sensilla papilliform or absent. Spicules arcuate, manubrium usually with large lateral fenestra. Gubernaculum plate-like, sometimes with caudal apophysis. Tail conoid to elongate subcylindrical, ventrally arcuate. Caudal glands present or absent. Spinneret usually functional, rarely without duct.

Biology. Parthenogenetic. Distributed all over the world in freshwater and terrestrial habitats. *Plectus aquatilis*, *P. indicus*, *P. palustris*, *P. sambesii*, *P. tenuis* are typical freshwater species, although other members of the genus may be found in aquatic habitats as well. Some species are capable of anhydrobiosis (Hendriksen, 1983; P. De Ley unpublished data) and the genus appears to contain the most easily cultured species within the order Plectida. Because of this relative ease, *Plectus* cultures have been used for research ranging from metabolic studies (Klekowski *et al.*, 1980) and environmental toxicology (e.g. Kammenga *et al.*, 2001) to phylogenetic (Saverimuttu *et al.*, 2000) and developmental analyses (see the following paragraph).

Development. In *P. palustris*, embryonic development lasts from 68 to 680 h depending on temperature; larval development takes 10–17 days depending on temperature and food supply; fecundity equals 4–49 eggs/day resulting in 900–1050 eggs produced during the life of one female; two females lived for about 50 days (Schiemer *et al.*, 1980). Embryonic development in *P. zelli* lasts 18–20 h, generation time is 7–9 days at 26–30°C (Tahseen *et al.*, 1992). Cleavage patterns in *Plectus* were described for *P. zelli* (Tahseen *et al.*, 1992), *P. acuminatus, P. aquatilis* and *P. minimus* (Lahl *et al.*, 2003). Post-embryonic development of *P. parietinus*, *P. communis* and *P. decens* was described by Holovachov (2004).

Key to subgenera

1. Labial setae conical, rod-shaped or filiform, not extending anterior to the lip region . . . .*Plectus*
   Labial setae flattened at their bases, protruding anterior to the lip region . . . .*Ceratoplectus*

Type species: *P. parietinus* Bastian, 1865

Other species:
P. acuminatus Bastian, 1865
P. acuticaudatus Truskova, 1976
P. americanus Zell, 1993
P. amorphotelus Ebsary, 1985
P. andrássyi Timm, 1971
P. antarcticus de Man, 1904
P. aquatilis Andrássy, 1985
P. araiensis Khan & Araki, 2001
P. araucanorum Andrássy, 2002
P. australis Cobb, 1893
P. belgicae de Man, 1904
P. cancellatus Zullini, 1976
P. capensis Khan & Araki, 2001
P. cirratus Bastian, 1865
P. cladinosus Holovachov & Susulovsky, 1999
P. coloradensis (Zell, 1993) comb. n. pro *Chiloplectus coloradensis* Zell, 1993
P. communis Bütschli, 1873
P. costatus Zell, 1993

P. cryptoptychus De Ley & Coomans, 1994
P. cylindricus Abdel-Rahman, 1993
P. decens Andrássy, 1985
P. elegans Zell, 1993
P. elongatus Maggenti, 1961
P. exinocaudatus Truskova, 1976
P. fragilis Zell, 1993
P. frigophilus Kirjanova, 1958
P. galapagensis De Ley & Coomans, 1994
P. geophilus de Man, 1880
P. glandulatus Tahseen, Baniyamuddin, Hussain & Ahmad, 2004
P. hyperboreus Zell, 1993
P. indicus Khera, 1972
P. infundibulifer Andrássy, 1985
P. inquirendus Andrássy, 1958
P. insolens Andrássy, 1998
P. intermedius Cobb, 1893
P. intorticaudatus Truskova, 1976
P. karachiensis Shahina, Tabassum & Maqbool, 2001
P. kyotoensis Khan & Araki, 2001
P. lamproptychus De Ley & Coomans, 1994
P. longicaudatus Bütschli, 1873
P. magadani Kuzmin, 1979
P. makrodemas Massey, 1964
P. masleni (Boström, 1997) comb. n. pro *Chiloplectus masleni*
P. meridianus Andrássy, 1998
P. minimus Cobb, 1893
P. montanus Zell, 1993
P. murrayi Yeates, 1970
P. neoindicus nom. n. pro *P. indicus* (Tahseen, Baniyamuddin, Hussain & Ahmad, 2004) comb. n. (= *Chiloplectus indicus*) nec *P. indicus* Khera, 1972
P. niaensis Khan & Araki, 2001
P. opisthocirculus Andrássy, 1952
P. palustris de Man, 1880
P. paracuminatus Zell, 1993
P. paratenuis Zell, 1993
P. parvus Bastian, 1865
P. patagonicus de Man, 1904
P. pseudoelongatus Zell, 1993
P. pulcher Zell, 1993
P. pusillus Cobb, 1893
P. pusteri Fuchs, 1930
P. refusus Tahseen, Ahmad & Jairajpuri, 1995
P. rhizophilus de Man, 1880
P. rotundilabiatus Zell, 1993
P. sabinae Hernandez, Imaz & Armendariz, 1998

P. sambesii Micoletzky, 1916
P. similis Zell, 1993
P. spicacaudatus Ebsary, 1985
P. tenuis Bastian, 1865
P. tolerans Andrássy, 1998
P. tropicus Zell, 1993
P. turricaudatus Truskova, 1976
P. varians Maggenti, 1961
P. velox Bastian, 1865
P. zelli Tahseen, Ahmad & Jairajpuri, 1992
P. (*Ceratoplectus*) *armatus* Bütschli, 1873 (Andrássy, 1984)
P. (*Ceratoplectus*) *assimilis* Bütschli, 1873 (Andrássy, 1984)
P. (*Ceratoplectus*) *brzeskii* Holovachov, 2001
P. (*Ceratoplectus*) *cornus* Maggenti, 1961 (Andrássy, 1984)
P. (*Ceratoplectus*) *lenis* Andrássy, 1985 (Zell, 1993)

Subfamily Anaplectinae Zell, 1993
Morphology. Hypodermal glands present, with more or less sclerotized openings. Somatic sensilla restricted to caudal region. Cephalic sensilla setiform. Labial region with six equal lips. Amphid unispiral or transverse slit. Cervical papilliform sensilla present at least in *Anaplectus* and *Perioplectus*. Gymnostom cylindrical to strongly convex. Axial spermatheca and crustaformeria present. Tubular supplements, pre-cloacal and post-cloacal sensilla present. Caudal sensilla distributed in the following way: 2–3 sub-ventral pairs pre-cloacal, one sub-ventral pair at level of cloaca, several sub-ventral and sub-dorsal pairs on tail.
Taxonomy. We include three genera in this subfamily: *Anaplectus*, *Perioplectus* and *Arctiplectus*.
Biology. Inhabit freshwaters, soil, litter and moss. Only a few species of the genus *Anaplectus* (Fig. 20.10) may be considered as true freshwater nematodes.

Genus *Anaplectus* De Coninck & Schuurmans Stekhoven, 1933
Syn. = *Marinoplectus* Kreis, 1963
Morphology. Hypodermal glands and body pores present. Lips six. Labial sensilla papilliform, cephalic sensilla setiform. Amphid a transverse slit (Figs 20.9A and 20.10A). Deirids setiform. Stegostom uniformly cylindrical (Fig. 20.10C). Uterus differentiated into spermatheca, crustaformeria and ovejector. Vagina short and straight, without epiptygmata. Rectum short. Males with tubular supplements, with sub-ventral and sub-dorsal rows of papillae in the pre-cloacal and caudal region. Pre-cloacal sensillum papilliform. Post-cloacal sensilla papilliform. Spicules arcuate, manubrium usually with large lateral fenestra. Gubernaculum plate-like, sometimes with caudal apophysis (Fig. 20.10D). Tail conoid, ventrally arcuate. Caudal glands present. Spinneret usually functional, rarely absent.
Biology. Amphimictic. *A. granulosus* (Fig. 20.9A) and *A. grandepapillatus* (Fig. 20.10) often inhabit freshwaters, while remaining species are mostly terrestrial.
Development. Post-embryonic development of *A. grandepapillatus* was described by Holovachov, 2004.
Bibliography. A first SEM study of *Anaplectus granulosus* (Fig. 20.9A) and the most recent identification key to the genus are given in Holovachov *et al.* (2004b).

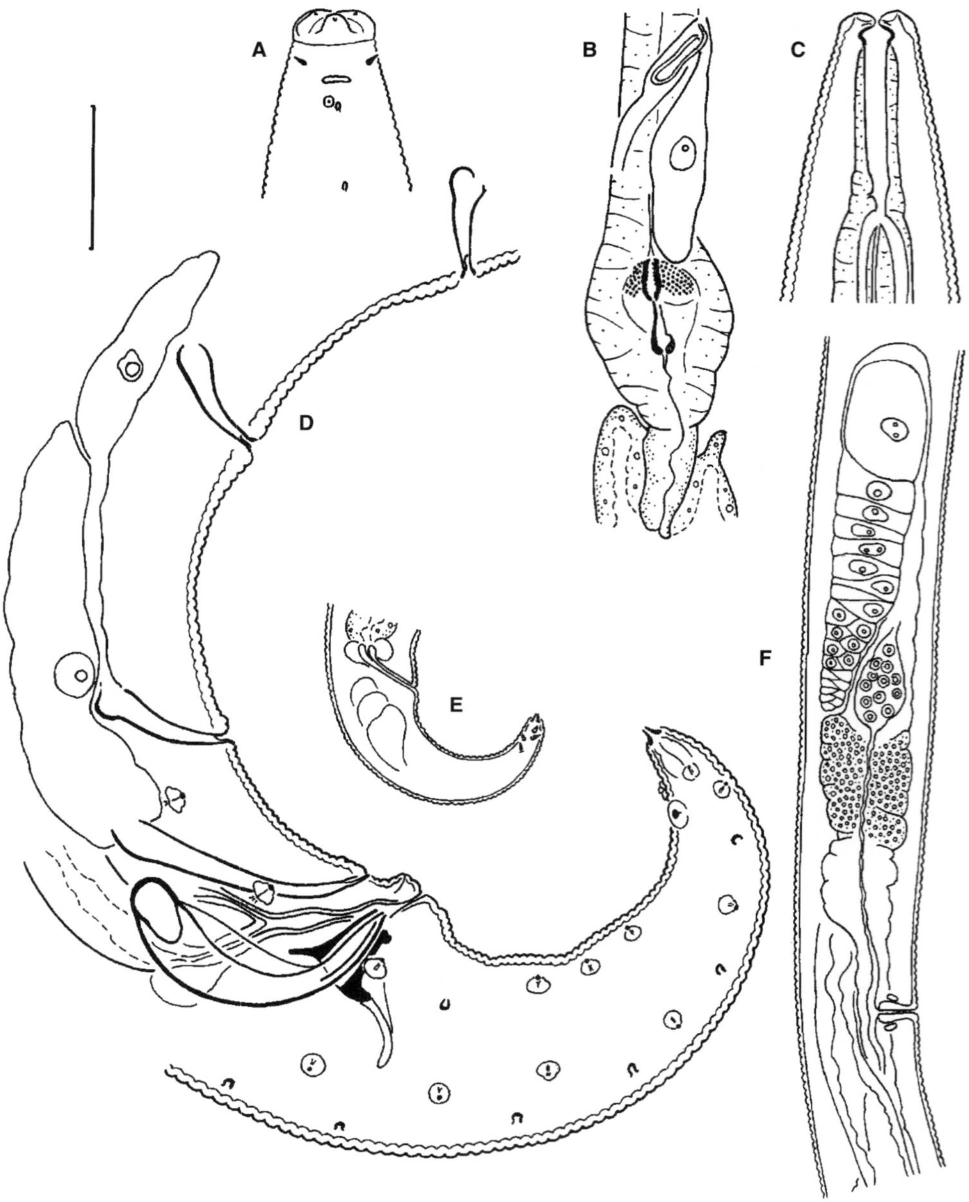

Fig. 20.10 Morphology of the genus *Anaplectus* De Coninck & Schuurmans Stekhoven, 1933 as exemplified by *A. grandepapillatus* (Ditlevsen, 1928) Andrássy, 1973 from Russia. A: Anterior end, surface view; B: Basal bulb; C: Anterior end, median section; D: Male caudal region (papillae and body pores are shown on the body side facing the viewer); E: Female tail; F: Anterior part of female reproductive system. (Scale bar: A–D = 20 μm; E and F = 50 μm.)

Type species: *A. granulosus* (Bastian, 1865) De Coninck & Schuurmans Stekhoven, 1933

Other species:
A. atubulatus Andrássy, 1987
A. brzeskii Holovachov, Boström, Winiszewska & Hánel, 2004
A. eurycerus (Massey, 1964) Andrássy, 1984
A. grandepapillatus (Ditlevsen, 1928) Andrássy, 1973
A. magnus Brzeski, 1963
A. octo Zullini, 1973
A. parasimilis Truskova, 1978
A. porosus Allen & Nofsinger, 1968
A. similis Allen & Noffsinger, 1968
A. subgranulosus Truskova, 1978
A. tortus Andrássy, 1986
A. varicaudatus Allen & Noffsinger, 1968

Subfamily Wilsonematinae Chitwood, 1951
Morphology. Hypodermal glands absent. Somatic sensilla distributed over the whole body. One (sub-dorsal) or two (sub-dorsal and sub-ventral) pairs of somatic setae located at level of stoma. Lateral outer labial sensilla elevated on mid-lateral projections. Cephalic sensilla cylindrical, leaf-shaped oval or with incised sub-lateral edge (cornua). Anterior end with bilateral and dorsoventral symmetry. Cuticle inflated on the dorsal and ventral sides of anterior end, forming bulbiform cervical expansions, which are annulated or smooth. Amphid unispiral. Gymnostom cylindrical. Spermatheca and crustaformeria absent. Tubular supplements, pre-cloacal and post-cloacal sensilla absent.
Development. Post-embryonic development was studied in *Tylocephalus auriculatus* and *Ereptonema arcticum* Loof, 1971 and is characterized by three (morphological types) juvenile stages (De Ley and Coomans, 1997; De Ley *et al.*, 2002).
Taxonomy. Following the revision of Holovachov *et al.* (2003), four genera are included here in the subfamily, e.g. *Tylocephalus*, *Neotylocephalus*, *Ereptonema* and *Wilsonema* (Fig. 20.12). Only members of *Tylocephalus* (Figs 20.11A and 20.12) were found in freshwater habitats.
Biology. De Ley and Coomans (1997) showed that the labial region in Wilsonematinae is probably specialized for sweeping up bacteria attached to more or less flat surfaces in the soil.

Genus *Tylocephalus* Crossman, 1933
Morphology. Anterior end with bilateral and dorsoventral symmetry. Sub-ventral and sub-dorsal pairs of lips fused, forming mid-dorsal and mid-ventral median ridges, respectively. Cervical expansions bulbiform, annulated under light microscope (LM) and SEM, each extends orally into one median ridge. Outer labial sensilla (sub-dorsal and sub-ventral pairs) are located at inner end of the median ridge. Four setiform or flat cornua (Figs 20.11A and B). Mid-lateral projection short oval. Lateral rims not fimbriate. Female reproductive system didelphic, amphidelphic, ovarian branches reflexed antidromously, usually located on alternate sides of intestine with

Fig. 20.11 Scanning electron micrographs of the lip regions of Wilsonematinae. A: *Tylocephalus auriculatus* (Bütschli, 1873) Anderson, 1966; B: *Ereptonema arcticum* Loof, 1971; C: *Neotylocephalus inflatus* (Yeates, 1967) Holovachov *et al.*, 2003; D: *Wilsonema otophorum* (de Man, 1880) Cobb, 1913. Scale bars equal 1 μm. (A: After De Ley and Coomans, 1997; B: kindly provided by Dr. I. Tandingan De Ley; C, D: after Holovachov *et al.*, 2003. A: Courtesy of *Nematologica*; C, D: courtesy of *Journal of Nematode Morphology and Systematics*.)

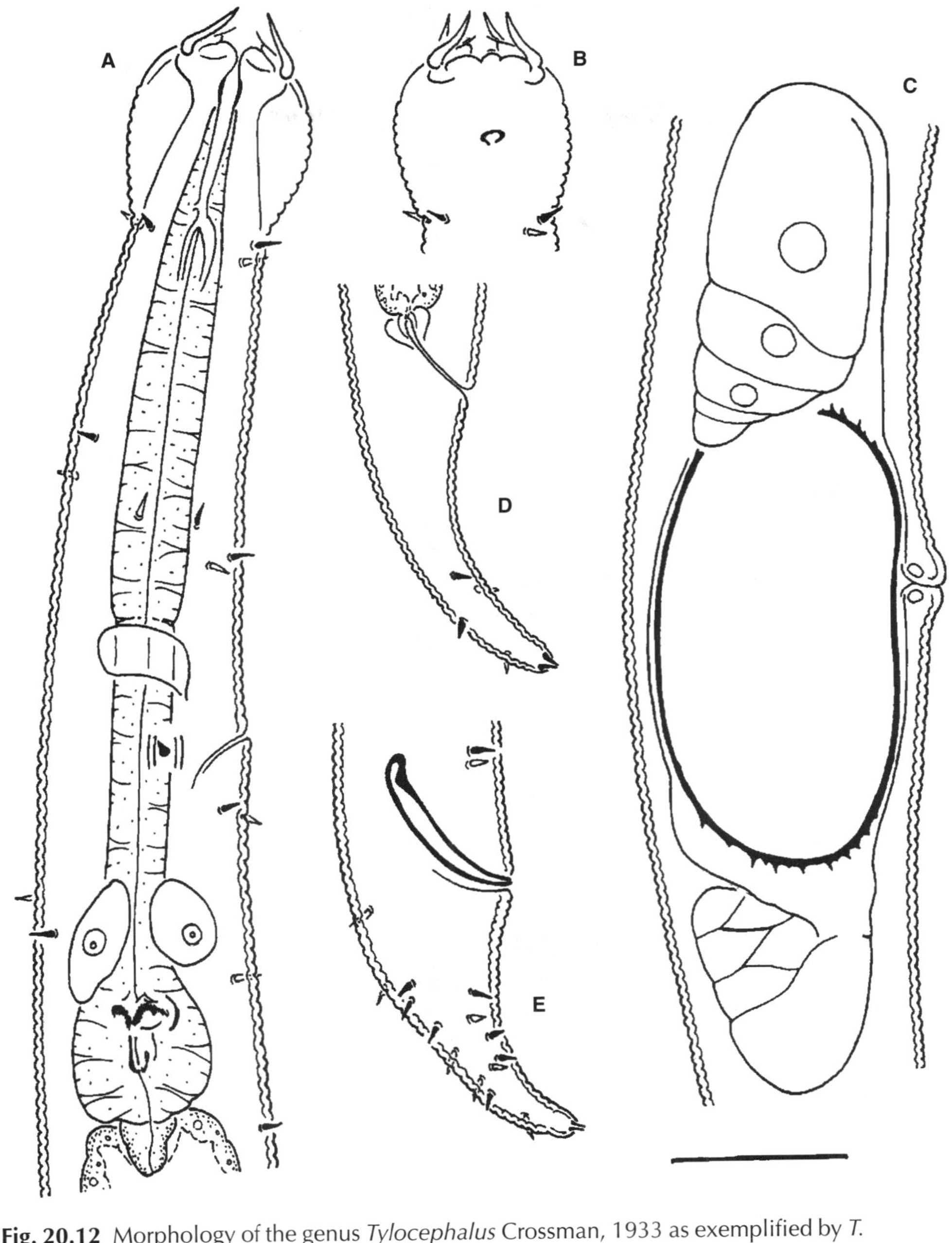

Fig. 20.12 Morphology of the genus *Tylocephalus* Crossman, 1933 as exemplified by *T. auriculatus* (Bütschli, 1873) Crossman, 1933 from Ukraine. A: Pharyngeal region; B: Anterior end, surface view; C: Female reproductive system with intrauterine egg; D: Female tail; E: Male caudal region. (Scale bar: A–E = 20 μm.)

anterior branch on right side. Vagina short and straight, without epiptygmata. Rectum short. Spicules arcuate, manubrium usually with large lateral fenestra. Gubernaculum absent (Fig. 20.12E). Tail conoid, ventrally arcuate. Caudal glands present. Spinneret functional.

Biology. *T. auriculatus* was found in a variety of habitats, ranging from aquatic plants to arid lands, mosses and rotting wood. *T. longicornis* was recovered from standing water in a bromeliad from Costa Rica.

Bibliography. The most recent identification key is given in Holovachov *et al.* (2004a).

Type species: *T. auriculatus* (Bütschli, 1873) Crossman, 1933

Other species:
T. andinus Zell, 1985
T. annulatus Zell, 1985
T. becki Zell, 1985
T. cephalatus (Cobb, 1893) Anderson, 1966
T. cornutus Zell, 1985
T. laticollis Zell, 1985
T. longicornis Holovachov, Boström & Mundo-Ocampo, 2004
T. nimius De Ley & Coomans, 1998
T. palmatus Tahseen, Ahmad & Jairajpuri, 1995
T. primitivus Holovachov, Boström & Mundo-Ocampo, 2004

Family METATERATOCEPHALIDAE Eroshenko, 1973

Morphology. Cuticle finely annulated. Subcuticular ornamentation present, appearing as punctations under LM. Lateral alae and deirids absent. Lateral differentiation of cuticle is formed by rows of enlarged punctations. Body pores present. Somatic sensilla present in both sexes. Lips with sclerotized edges, deeply divided. Outer labial sensilla and cephalic sensilla setiform. Amphid faint but relatively large with circular or oval subcuticular cavity and oval aperture. Excretory duct curving in a dense spiral on right-hand side of pharynx, enveloped by a renette cell. Stoma consisting of a relatively flexible or sclerotized cheilostom, a cylindrical gymnostom with sclerotized lining and a stegostom that has a sclerotized lining and tapers posteriorly. Posterior section of the stegostom short. Basal bulb strongly developed, with transverse valves. Female reproductive system didelphic, amphidelphic. Male reproductive system monorchic, posterior testis absent, anterior testis reflexed ventrally. Tubular supplements absent. Pre-cloacal sensillum papilliform. Post-cloacal sensilla absent. Caudal glands and spinneret absent. Tail with a pair of sub-lateral short setae (sometimes assumed to be phasmid homologues).

Taxonomy. We include three genera in this family: *Metateratocephalus* (Figs 20.3B and 20.13F), *Steratocephalus* and *Euteratocephalus* (Fig. 20.13A and E), not dividing them into separate subfamilies. The genus *Steratocephalus* and its only species *S. demani* (Stefanski, 1924) Andrássy, 1984 is known only from the original description, being found in mosses. It is similar morphologically to *Metateratocephalus*, but has a single anterior ovary. *Euteratocephalus minor* Mukhina, 1981 fits within the diagnosis of the genus *Metateratocephalus* by general morphology, differing only by the presence of a

single anterior ovary. Although this species was synonimized with *M. crassidens* by Andrássy (1984), it may actually belong to the genus *Steratocephalus*.
Biology. Inhabit freshwaters, wet soil, moss, litter and rotting wood.

Key to genera

1. Relaxed body and tail generally curved ventrad; labial region flat, continuous with body contour . . . . *Euteratocephalus*
   Relaxed body usually straight or curved dorsad, tail curved dorsad; labial region high, strongly offset from body contour . . . . *Metateratocephalus*

Genus *Metateratocephalus* Eroshenko, 1973
Morphology. Body straight or curved dorsad, tail curved towards the dorsal side. Labial region strongly offset from body contour. Lips leaf-like with sclerotized edges, deeply divided (Fig. 20.3B). Cheilostom sclerotized. Tail with a pair of sub-lateral short setae in female, with several sub-ventral pairs of setae in male (Fig. 20.13H).

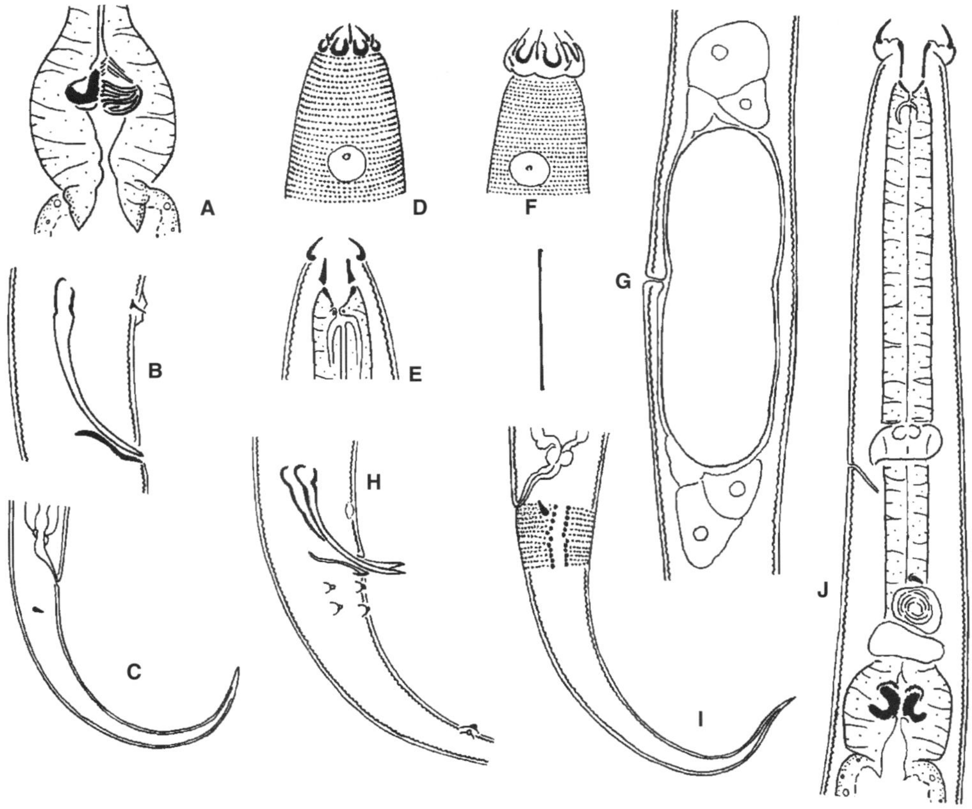

Fig. 20.13 Morphology of the genera *Euteratocephalus* Andrássy, 1958 and *Metateratocephalus* Eroshenko, 1973 as exemplified by *E. palustris* (de Man, 1880) Andrássy, 1968 (A–E) and *M. gracilicaudatus* Andrássy, 1985 (F–J) from Ukraine. A: Basal bulb; B, H: Male cloacal region; C, I: Female tail; D, F: Anterior end, surface view; E: Anterior end, median section; G: Female reproductive system with intrauterine egg; J: Pharyngeal region. (Scale bar: A,B and D–J = 20 μm; C = 50 μm.)

Biology. Thelytokous. *Metateratocephalus gracilicaudatus* (Figs 20.13F–J) is the most common in freshwaters, while *M. crassidens* occurs also in soil, litter, moss and rotting wood.
Bibliography. SEM observations were published by Boström (1989), Swart *et al.* (1989) and Karegar *et al.* (1997). Morphometric variability of species of this genus was studied by Holovachov (2003).

Type species: *M. crassidens* (de Man, 1880) Eroshenko, 1973

Other species:
M. bialowieziensis Holovachov, 2004
M. deconincki Andrássy, 1984
M. gracilicaudatus Andrássy, 1985
M. typicus Eroshenko, 1973

Genus *Euteratocephalus* Andrássy, 1968
Morphology. Body generally curved ventrad. Labial region continuous with body contour. Lips with sclerotized edges, deeply divided (Fig. 20.13D). Cheilostom flexible, not sclerotized (Fig. 20.13E). Tail with a pair of sub-lateral short setae in female, with several pairs of setae in male.
Biology. Thelytokous. All species inhabit aquatic habitats, including the very widely distributed *E. palustris* (Figs 20.13A–E).
Bibliography. The most recent identification key is given in Swart *et al.* (1991). SEM observations were published by Swart *et al.* (1989) and Karegar *et al.* (1997).

Type species: *E. palustris* (de Man, 1880) Andrássy, 1968

Other species:
E. capensis Heyns, 1977
E. punctatus Swart, De Waele & Heyns, 1991
E. spiraloides (Micoletzky, 1913) Heyns, 1977

References

Andrássy, I. (1976) *Evolution as a Basis for the Systematization of Nematodes*. Akadémiai Kiado, Budapest.

Andrássy, I. (1984) *Klasse Nematoda (Ordnungen Monhysterida, Desmoscolecida, Araeolaimida, Chromadorida, Rhabditida)*. Gustav Fischer Verlag, Stuttgart, Germany.

Andrássy, I. (1985) The genus *Plectus* Bastian, 1865 and its nearest relatives (Nematoda: Plectidae). *Acta Zoologica Hungarica* 31, 1–52.

Blaxter, M., De Ley, P., Garey, J.R., Liu, L.X., Scheldeman, P., Vierstraete, A., Vanfleteren, J.R., Mackey, L.Y., Dorris, M., Friesse, L.M., Vida, J.T. and Thomas W.K. (1998) A molecular evolutionary framework for the phylum Nematoda. *Nature* 392, 71–75.

Boström, S. (1989) The taxonomic position of some teratocephalid nematodes – a scanning electron micrograph study. *Revue de Nématologie* 12, 181–190.

Coomans, A. and De Waele, D. (1979) Species of *Aphanolaimus* (Nematoda, Araeolaimida) from Africa. *Zoologica Scripta* 8, 171–180.

Coomans, A. and De Waele, D. (1983) Species of *Aphanolaimus* (Nematoda: Araeolaimida) from Belgium. *Hydrobiologia* 101, 165–178.

Coomans, A.V. and Raski, D.J. (1991a) *Aphanonchus intermedius* n.g., n. sp. (Nemata: Araeolaimida). *Nematologica* 37, 8–19.

Coomans, A. and Raski, D.J. (1991b) Observations on *Paraplectonema* Strand, 1934 and description of *Leptoplectonema fuegoense* n.g., n.sp. (Nematoda: Leptolaimidae). *Revue de Nématologie* 14, 197–205.

De Ley, P. and Coomans, A. (1994) Terrestrial nematodes from the Galapagos Archipelago. IV. The genus *Plectus* Bastian, 1865 with description of three new species (Leptolaimina: Plectidae). *Bulletin van het Koninklijk Belgisch Institut voor Natuurwetenschappen Biologia* 64, 43–76.

De Ley, P. and Coomans A. (1997) Terrestrial nematodes from the Galapagos Archipelago. 7. Description of *Tylocephalus nimius* sp. n. and new data on the morphology, development and behaviour of *T. auriculatus* (Bütschli, 1873) Anderson, 1966 (Leptolaimina: Plectidae). *Fundamental and Applied Nematology* 20, 213–228.

De Ley, I.T., Holovachov, O. and De Ley, P. (2002) Scanning electron microscopy of the juvenile stages of some Wilsonematinae (Plectidae). *Nematology* 4, 244. (Abstract.)

Ettema, C.H., Rathbun, S.L. and Coleman, D.C. (2000) On spatiotemporal patchiness and the coexistence of five species of *Chronogaster* (Nematoda: Chronogasteridae) in a riparian wetland. *Oecologia* 125, 444–452.

Eyualem-Abebe (1996) Aquatic nematodes from Ethiopia. IV. The family Leptolaimidae Orley, 1880 *sensu* Lorenzen, 1981 (Chromadorida: Nematoda) with the description of *Anonchus coomansi* n. sp. *Hydrobiologia* 332, 1–26.

Eyualem-Abebe and Coomans, A. (1996a) Aquatic nematodes from Ethiopia. VI. The genera *Chronogaster* Cobb, 1913, *Plectus* Bastian, 1865 and *Prismatolaimus* de Man, 1880 with description of *C. ethiopica* n. sp. and *C. getachewi* n. sp. (Chromadorida: Nematoda). *Hydrobiologia* 332, 41–61.

Eyualem-Abebe and Coomans, A. (1996b) Aquatic nematodes from Ethiopia. VII. The family Rhabdolaimidae Chitwood, 1951 *sensu* Lorenzen, 1981 (Chromadorida: Nematoda) with the description of *Udonchus merhatibebi* n. sp. *Hydrobiologia* 341, 197–214.

Fürst von Lieven, A. (2003) Functional morphology and evolutionary origin of the three-part pharynx in nematodes. *Zoology* 106, 183–201.

Goodey, T. (1947) *Domorganus macronephriticus* n.g., n. sp., a new cylindrolaimid free-living soil nematode. *Journal of Helminthology* 21, 175–180.

Hendriksen, N.B. (1983) Anhydrobiosis in nematodes: studies on *Plectus* sp. In: Lebrun, P., André, H.M. and de Medts, A. (eds) *New Trends in Soil Biology*. Université Catholique de Louvain, Louvain-la-Neuve, Belgium, pp. 387–394.

Heyns, J. and Coomans, A. (1980) Freshwater nematodes from South Africa. 5. *Chronogaster* Cobb, 1913. *Nematologica* 26, 187–208.

Heyns, J. and Coomans, A. (1983) New and known species of *Chronogaster* Cobb, 1913 (Nematoda: Leptolaimidae). *Nematologica* 29, 245–265.

Holovachov, O. (2003) Morphometric variability of species of the genus *Metateratocephalus* Eroshenko, 1973 (Nematoda). *Visnyk of L'viv University, Series Biology* 33, 110–116. (In Ukrainian.)

Holovachov, O. (2004) Morphology, phylogeny and evolution of the superfamily Plectoidea Örley, 1880 (Nematoda: Plectida). *Annales Zoologici* 54, 631–672.

Holovachov, O. and Boström, S. (2004) Morphology and systematics of the superfamilies Leptolaimoidea Örley, 1880 and Camacolaimoidea Micoletzky, 1924 (Nematoda: Plectida). *Journal of Nematode Morphology and Systematics* 7, 1–49.

Holovachov, O. and Sturhan, D. (2004a) Studies on the genus *Paraphanolaimus* Micoletzky, 1923 (Nematoda: Aphanolaimidae) with description of *P. paraguayensis* sp. n. *Nematology* 5, 793–807.

Holovachov, O. and Sturhan, D. (2004b) Two new species of the genus *Aphanonchus* Coomans and Raski, 1991 (Nematoda: Aphanolaimidae) and a revised taxonomy of the genus. *Nematology* 6, 357–373.

Holovachov, O., Susulovsky, A. and Boström, S. (2000) Descriptions of two species of *Chiloplectus* Andrássy, 1984 (Nematoda: Plectidae) from Ukraine and a revised taxonomy of the genus. *Nematology* 2, 381–384.

Holovachov, O., Zullini, A., Loof, P.A.A. and Bongers, T. (2002) Morphology and systematics of the genus *Anonchus* Cobb, 1913 (Nematoda: Leptolaimina) and reappraisal of the family Aphanolaimidae Chitwood, 1936 n. rank. *Nematology* 4, 725–757.

Holovachov, O., Boström, S., Tandingan De Ley, I., De Ley, P. and Coomans, A. (2003) Morphology and systematics of the genera *Wilsonema* Cobb, 1913, *Ereptonema* Anderson, 1966 and *Neotylocephalus* Ali, Farooqui & Tejpal, 1969 (Leptolaimina: Wilsonematinae). *Journal of Nematode Morphology and Systematics* 5, 73–106.

Holovachov, O., Boström, S. and Mundo-Ocampo, M. (2004a) Description of three known and two new species of the genus *Tylocephalus* Crossmann, 1933 with a revised taxonomy of the genus and key to species of the subfamily Wilsonematinae (Plectida). *Russian Journal of Nematology* 12, 115–130.

Holovachov, O., Boström, S., Winiszewska, G. and Hanél, L. (2004b) Description of two known and one new species of the genus *Anaplectus* De Coninck and Schuurmans Stekhoven, 1933 from Europe, and a revised taxonomy of the genus (Nematoda: Plectida). *Russian Journal of Nematology* 12, 45–58.

Kammenga, J.E., van Gestel, C.A.M. and Hornung, E. (2001) Switching life-history sensitivities to stress in soil invertebrates. *Ecological Applications* 11, 226–238.

Karegar, A., De Ley, P. and Geraert, E. (1997) Three teratocephalid nematodes from Iran. *Fundamental and Applied Nematology* 20, 459–471.

Klekowski, R.Z., Schiemer, F. and Dunkan, A. (1980) A bioenergetic study of a benthic nematode, *Plectus palustris* de Man, 1880, throughout its life cycle. I. The respiratory metabolism at different densities of bacterial food. *Oecologia* 44, 119–124.

Lahl, V., Halama, C. and Schierenberg, E. (2003) Comparative and experimental embryogenesis of Plectidae (Nematoda). *Developmental Genes and Evolution* 213, 18–27.

Laybourn, J. (1979) The effects of temperature on the respiration and production of the freshwater nematode *Anonchus* sp. *Oecologia* 41, 329–337.

Lorenzen, S. (1981) Entwurf eines phylogenetischen Systems der freilebenden Nematoden. *Veröffentlichungen des Instituts für Meeresforschung Bremerhaven,* Supplement 7, 472 pp.

Malakhov, V.V. (1986) *Nematodes. Structure, Development, System and Phylogeny.* Nauka, Moskva. (In Russian.)

Malakhov, V.V., Ryzhikov, K.M. and Sonin, M.D. (1982) The classification of the major taxa of nematodes: subclasses, orders and suborders. *Zoologicheskii Zhurnal* 64, 1125–1134.

Raski, D.J. and Coomans, A.V. (1990) Five new species of *Aphanolaimus* (Nemata: Araeolaimida) with a key to species. *Nematologica* 36, 22–54.

Raski, D.J. and Coomans, A.V. (1991) Three new species of *Paraphanolaimus* (Nemata: Araeolaimida) with key to species. *Nematologica* 37, 44–62.

Raski, D.J. and Maggenti, A.R. (1984) Four new species of *Chronogaster* Cobb, 1913 (Nemata: Plectidae) with a key to species of the genus. *Nematologica* 30, 117–130.

Saverimuttu, J.K.C., Karunanayake, E.H., Chandrasekharan, N.V. and Jayasena, S.M.T. (2000) Molecular characterisation of the actin gene of the filarial parasite *Wuchereria bancrofti. International Journal for Parasitology* 30, 119–124.

Schiemer, F., Dunkan, A. and Klekowski, R.Z. (1980) A bioenergetic study of a benthic nematode, *Plectus palustris* de Man, 1880, throughout its life cycle. II. Growth, fecundity and energy budgets at different densities of bacterial food and general ecological considerations. *Oecologia* 44, 205–212.

Siddiqi, M.R. (2003) *Keralanema spinicorpus* (Maggenti *et al.,* 1983) gen. n., comb. n. (Nematoda: Chronogastridae). *International Journal of Nematology* 13, 236.

Swart, A., Meyer, A.J. and Heyns, J. (1989) Description of one new and two known species of Teratocephalidae (Nematoda) from South Africa. *Phylophylactica* 21, 367–377.

Swart, A., De Waele, D. and Heyns J. (1991) A review of the genus *Euteratocephalus* Andrássy, 1958, with a description of *E. punctatus* n. sp. *Revue de Nématologie* 14, 551–563.

Tahseen, Q., Ahmad, I. and Jairajpuri, M.S. (1992) Description and developmental biology of *Plectus zelli* n. sp. (Nematoda: Araeolaimida). *Fundamental and Applied Nematology* 15, 503–510.

Tchesunov, A.V. (1997) Marimermithid nematodes: anatomy, position in the nematode system, phylogeny. *Zoologicheskii Zhurnal* 76, 1283–1299.

Tchesunov, A.V. and Sturhan, D. (2004) Studies on *Domorganus macronephriticus* Goodey, 1947 (Nematoda: Ochridiidae). *Journal of Nematode Morphology and Systematics* 6, 139–150.

Valovaya, M.A. and Malakhov, V.V. (1990) Embryonic development of the nematode *Domorganus oligochaetophilus* from the White Sea supralittoral oligochaetous *Enchytraeus albidus*. *Parazitologiya* 24, 332–338.

Zell, H. (1993) Die Gattung *Plectus* Bastian, 1865 sensu lato (Nematoda, Plectidae) – Ein Beitrag zur Ökologie, Biogeographie, Phylogenie und Taxonomie der Plectidae. *Andrias* 11, 1–173

21 Order Rhabditida: Suborder Tylenchina

W. Bert and G. Borgonie

Department of Biology, Ghent University, Ledeganckstraat 35, 9000 Ghent, Belgium

Introduction

Strictly freshwater Tylenchina Thorne, 1949 are not known, nevertheless a relatively large number of genera can be found in freshwater habitats. Most of the Tylenchina recorded from freshwater are actually living in semiaquatic habitats, or their presence may be explained by passive transportation from surrounding terrestrial habitats (Faulkner and Bolander, 1970). Tylenchina recorded from (semi-) aquatic habitats (see Table 21.1) are typically common, eurytopic species or species able to withstand extreme conditions rather than species adapted to freshwater. Yet, regarding the existence of Tylenchina adapted to freshwater, their relative abundance, possible relation with aquatic plants and the tolerance of individual species to harsh environments, make them a significant group of nematodes in freshwater habitats.

Infraorder TYLENCHOMORPHA De Ley & Blaxter, 2002

The Tylenchomorpha comprises the largest and economically most important group of plant-parasitic nematodes. They have exploited all plant organs including flowers and seeds, but they mostly attack roots. All plants are susceptible to attack by one or more species of plant-parasitic nematodes, and the associations can range from transitory grazing by root-hair feeders to the highly complex host–pathogen interactions of gall-inducing nematodes and their hosts. Several Tylenchomorpha, particularly species of the Sphaerularioidea, are parasites of insects, mites, leeches or frogs. Feeding on fungi, algae, lichens or mosses is probably the ancestral feeding behaviour within the Tylenchomorpha and one family, the Halenchidae, has representatives exclusively parasitizing seaweeds. Tylenchomorpha occur in all possible habitats in soil, plants, freshwater and, exceptionally, in salt water (Siddiqi, 2000).

Table 21.1 Overview of Tylenchina recorded from freshwater.

	Major references[a]	Semi-aquatic habitat[b]	Ground-water	Swamp	Canal, river, stream	Pool, pond	Lake	Sub-merged aquatic plants	Periphyton	Aquatic, non-specified
TYLENCHOMORPHA De Ley & Blaxter, 2002										
Aglenchus agricola (de Man, 1884) Meyl, 1961	1, 2, 3, 5, 11, 12, 14	–			–	x	–			–
Amplimerlinius icarus (Wallace & Greet, 1964) Siddiqi, 1976	7					x				
A. macrurus (Goodey, 1932) Siddiqi, 1976	5	x								x
Aphelenchoides sp.	12, 14, 17, 20				–		–			–
Aphelenchoides arcticus Sanwal, 1965	5, 12	–						–	–	–
A. africanus Dassonville & Heyns, 1984	8	–								–
A. asteromucronates Eroshenko, 1967	5, 12	–						–	–	–
A. besseyi Christie, 1942	2	–								
A. bicaudatus (Imamura, 1931) Filipjev & Schuurmans Stekhoven, l941	2, 3, 5, 12	–			–			–	–	–
A. fluviatilis Andrássy, 1960	5								–	
A. fragariae (Ritzema Bos, 1890) Christie, l932	2, 5	–						–		–
A. helophilus (de Man, 1880) Goodey, 1933	2, 3, 5, 11	–			–					–
A. parietinus (Bastian, 1865) Steiner, 1932	1, 2, 3, 11, 12	–		x	–	x			–	–

Continued

Table 21.1 *Continued.* Overview of Tylenchina recorded from freshwater.

	Major references[a]	Semi-aquatic habitat[b]	Ground-water	Swamp	Canal, river, stream	Pool, pond	Lake	Sub-merged aquatic plants	Periphyton	Aquatic, non-specified
A. robustus Gagarin, 1997	12									x
A. sacchari Hooper, 1958	19				–					
A. saprophilus Franklin, 1957	12									–
A. subparietinis Sanwall, 1961	5, 12	–						–	–	–
Aphelenchus sp.										–
Aphelenchus avenae Bastian, 1865	2, 5, 12	–							x	–
Atylenchus sp.	20	x								–
Atylenchus decalineatus Cobb, 1913	5	–	x			x				
Basiria duplexa (Hagemeyer & Allen, 1952) Geraert, 1968	5	x								
B. flandriensis Geraert, 1968	5	x			x					
Belonolaimus sp.	20									–
Belonolaimus longicaudatus Rau, 1958	2	x								x
Boleodorus thylactus Thorne, 1941	2, 5	–								
Caloosia paradoxa (Luc, 1958) Brzeski, 1974	2	–								
Cephalenchus leptus Siddiqi, 1963	5				x					
C. hexalineatus (Geraert, 1962) Geraert & Goodey, 1964	5	–				x				
Coslenchus sp.	14						–			
Coslenchus aquaticus Geraert & Raski, 1989	5	–								x
C. costatus (De Man, 1921) Siddiqi, 1978	2, 5, 11, 12	x			x				x	–

C. polonicus Brzeski, 1982	5, 7, 14	–	x	x		x			
Criconema sp.	14, 17, 20			x		x			–
Criconema annuliferum (de Man, 1921) Micoletzky, 1925	5, 14	–				x			
C. demani Micoletzky, 1925	1, 5	–		x	x				
C. princeps (Andrássy, 1962) Raski & Luc, 1985	5	–							
Criconemoides sp.	20								–
Criconemoides informis (Micoletzky, 1922) Taylor, 1936	5, 12	–	x	x					
C. morgensis (Hofmänner in Hofmänner & Menzel, 1914) Taylor, 1936	5	–		x					
C. parvus Raski, 1952	5	–		x					
Ditylenchus sp.	12, 14, 19, 20			–		–			–
Ditylenchus angustus (Butler, 1913) Filipjev, 1936.	2	–							
D. brevicauda (Micoletzky, 1925) Filipjev, 1936	1								–
D. cyperi Husain & Khan, 1967	2	–							
D. danubialis (Andrássy, 1960) Fortuner & Maggenti, 1987	5							x	
D. dipsaci (Kühn, 1857) Filipjev, 1936	2	–					–		–
D. intermedius (De Man, 1880) Filipjev, 1936	1, 5, 11	–		x					–
D. loksai (Andrássy, 1959) Fortuner & Maggenti, 1987	12							x	x

Continued

Table 21.1 *Continued.* Overview of Tylenchina recorded from freshwater.

	Major references[a]	Semi-aquatic habitat[b]	Ground-water	Swamp	Canal, river, stream	Pool, pond	Lake	Sub-merged aquatic plants	Periphyton	Aquatic, non-specified
Dolichodorus sp.									–	
Dolichodorus heterocephalus Cobb, 1914	2	–			–	–	–	–		–
D. aquaticus Doucet, 1985	21	–								
D. longicaudatus Doucet, 1981	21				–					
D. miradvulvus Smart & Khuong, 1958	2	–								–
Ecphyadophora tenuissima de Man, 1921	5, 10	–		x	x					
Eutylenchus sp.	20									–
Eutylenchus africanus Sher, Corbett & Colbran, 1966	12									x
E. excretorius Ebsary & Eveleigh, 1981	5, 21	–			–		x			
E. gracilis Gagarin, 2003	16									
Filenchus sp.	14, 19				–		–			
Filenchus facultativus (Szczygiet, 1969) Raski & Geraert, 1987	14						x			
F. filiformis (Bütschli, 1873) Meyl, 1961	1, 2, 11, 12	–							x	–
F. polyhypnus (Steiner & Albin, 1946) Meyl, 1961	5	x				x				
F. thornei (Andrássy, 1954) Andrássy, 1963	5, 12, 14	–	x	x	–	x	x		x	–
F. uliginosus (Brzeski, 1977) Siddiqi, 1986	5	–								

F. vulgaris (Brzeski, 1963) Lownsbery & Lownsbery, 1985	5, 17	–	–	x			
Geocenamus brevidens (Allen, 1955) Brzeski, 1991	5	–		x			
G. quadrifer (Andrássy, 1954) Brzeski, 1991	5	–	x	x			
G. tesselatus (Goodey, 1952) Brzeski, 1991	5	–					–
Gracilancea graciloides (Micoletzky, 1925) Siddiqi, 1976	1, 5, 10	–					x
Helicotylenchus sp.	17, 20		–				–
Helicotylenchus africanus (Micoletzky, 1916) Andrássy, 1958	8						–
H. coomansi Ali & Loof, 1975	5, 10	–					
H. crenacauda Sher, 1966	2	–				x	x
H. dihystera (Cobb, 1893) Sher, 1961	2	–					
H. erythrinae (Zimmerman, 1904) Golden, 1956	2, 12	–					x
H. multicinctus (Cobb, 1893) Golden, 1956	12		x				
H. pseudorobustus (Steiner, 1914) Golden, 1956	5, 7, 14	–	x	–	x	–	–
H. varicaudatus Yuen, 1964	5, 7	–	x				
H. vulgaris Yuen, 1964	5	–	x				
Hemicriconemoides minutus Esser 1960	2	–					
H. pseudobrachyurus De Grisse, 1964	5	–	x				
H. wessoni Chitwood & Birchfield, 1957	2	–					
Hemicycliophora sp.	7, 12, 20	–	–				–

Continued

Table 21.1 *Continued.* Overview of Tylenchina recorded from freshwater.

	Major references[a]	Semi-aquatic habitat[b]	Ground-water	Swamp	Canal, river, stream	Pool, pond	Lake	Sub-merged aquatic plants	Periphyton	Aquatic, non-specified
Hemicycliophora aquatica (Micoletzky, 1913) Loos, 1948	5, 10, 11	–	–				x			
H. belemnis Germani & Luc, 1973	2	–								
H. conida Thorne, 1955	5, 7	–								–
H. diolaensis Germani & Luc, 1973	2	–								
H. floridensis (Chitwood & Birchfield, 1957) Goodey, 1963	2	–							2	–
H. fluvialis Bird, 1999	24									
H. labiata Colbran, 1960	2, 8	–								–
H. lutosa Loof & Heyns, 1969	8									–
H. micoletzky Goffart, 1951	5, 10, 11		–							–
H. nucleata Loof, 1968	5	–	x		x					
H. oostenbrinki Luc, 1958	2	–								
H. thienemanni (Schneider, 1925) Loos, 1948	5, 11	–	x		x	x	x			–
H. triangulum Loof, 1968	5	–			x	x				
H. typical de Man, 1921	2, 5	–								
Heterodera sp.	19, 20	–			x					–
Heterodera canadensis Mulvey, 1979	2	–								
H. cruciferae Franklin, 1945	2	–						x		x
H. elachista Oshima, 1974	2	–								

H. oryzae Luc & Berdon Brizuela, 1961	2, 12	–						–		–
H. schachtii Schmidt, 1871	2	–						x		x
Hirschmanniella anchoryzae Ebsary & Anderson, 1982	2	–								
H. behningi (Micoletzky, 1923) Luc & Goodey, 1963	2, 5, 12	–			–			–		–
H. belli Sher, 1968	2	–								
H. caudacrena Sher, 1968	2	–			–	–	–	–	–	–
H. diversa Sher, 1968	2	–								
H. gracilis (de Man 1880) Luc & Goodey, 1963	1,2, 5, 7, 11, 12	–	–	–	–	–	–	–		–
H. imamuri Sher, 1968	2	–								
H. loofi Sher, 1968	5, 7	–				–				
H. mucronata (Das, 1960) Luc & Goodey, 1963	2	–								
H. oryzae (van Breda de Haan, 1902) Luc & Goodey, 1963	2, 12	–			–			–		–
H. pisquidensis Ebsary & Pharoah, 1982	2	–								
H. spinicaudata (Schuurmans Stekhoven, 1944) Luc & Goodey, 1963	2	–								
Hoplolaimus sp.	20									–
Hoplolaimus columbus Sher, 1963	2	–								
H. galeatus (Cobb, 1913) Thorne, 1935	2	–								
H. indicus Sher, 1963	2	–								
H. tylenchiformis Daday, 1905	2	–						–		–
Lelenchus leptosoma (De Man, 1880) Andrássy, 1954	2, 5, 11, 14	–			–		–			–

Continued

Table 21.1 *Continued.* Overview of Tylenchina recorded from freshwater.

	Major references[a]	Semi-aquatic habitat[b]	Ground-water	Swamp	Canal, river, stream	Pool, pond	Lake	Sub-merged aquatic plants	Periphyton	Aquatic, non-specified
Malenchus acarayensis Andrássy, 1968	5, 14	–			x		–		x	x
M. andrássyi Merny, 1970	2, 5	–								
M. bryophilus (Steiner, 1914) Andrássy, 1980	2, 3, 5, 11, 14	–			x		–			–
M. gratiosus Andrássy, 1981	5	–		x						
M. neosulcus Geraert & Raski, 1986	5	–		x						
M. pachycephalus Andrássy, 1981	14						x			
M. pressulus (Kazachenko, 1975) Andrássy, 1981	14						x			
Meloidodera floridensis Chitwood, Hannon & Esser, 1956	2	–								
Meloidogyne sp.	20	–								–
Meloidogyne graminicola Golden & Birchfield, 1965	2	–								
M. incognita (Kofoid & White, 1919) Chitwood, 1949	2	–								
M. javanica (Treub, 1885) Chitwood, 1949	2	–								
M. kralli Jepson, 1984	2, 14	–					x			
M. oryzae Maas, Sanders & Dede, 1978	2	–								
M. sewelli Mulvey & Anderson, 1980	2	–								

Mesocriconema curvatum (Raski, 1952) Loof & De Grisse, 1989	2, 5	–	x			x	
M. involutum (Loof, 1978) Loof, 1989	5	–				x	
M. kirjanovae (Andrássy, (1962) Loof & De Grisse, 1989	2, 5	–	x				x
M. onoense (Luc, 1959) Loof & De Grisse, 1989	2	–					
M. pseudosolivagum (De Grisse, 1964) Andrássy, 1965	5	–					
M. raskiense (De Grisse, 1964) Andrássy, 1965	5	–					
M. rotundicauda (Loof, 1964) Loof, 1989	5	–					
M. rusticum (Micoletzky, 1915) Loof & De Grisse, 1989	5, 11	–					
M. sphaerocephala (Taylor, 1936) Loof, 1989	1, 2, 5	–					
M. vadense (Loof, 1964) Loof & De Grisse, 1989	5	–	x				
Miculenchus salvus Andrássy, 1959	14				–		
Nagelus leptus (Allen, 1955) Siddiqi, 1979	5	–			x		
Nagelus obscurus (Allen, 1955) Powers, Baldwin & Bell, 1983	5, 7	–	–	–			–
Neodolichodorus paralongicaudatus Rashid, Geraert & Heyns, 1990	8	–					
Neopsilenchus noctiscriptus (Andrássy, 1962) Khan, 1973	5			x			

Continued

Table 21.1 *Continued.* Overview of Tylenchina recorded from freshwater.

	Major references[a]	Semi-aquatic habitat[b]	Ground-water	Swamp	Canal, river, stream	Pool, pond	Lake	Sub-merged aquatic plants	Periphyton	Aquatic, non-specified
Ogma cobbi (Micoletzky, 1925) Siddiqi, 1986	2	–								
O. menzeli (Stefanski, 1924) Schuurmans Stekhoven & Teunissen, 1938	5	–								
O. multisquamata Kirjanova, 1948	5	–			x					
O. murrayi Southern, 1914	5	–								
O. octangulare (Cobb, 1914) Schuurmans Stekhoven & Teunissen, 1938	5	–		x						
Paraphelenchus pseudoparietinus (Micoletzky, 1922) Micoletzky, 1925	5, 12	–		x						x
Paratylenchus sp.	14						x			
Paratylenchus aquaticus Merni, 1966	2	–								
P. bukowinensis Micoletzky, 1922	1, 5	–								
P. goodeyi Oostenbrink, 1953	5	–								
P. microdorus Andrássy, 1959	2	–						x		
P. mimilus Raski, 1975	2	–								
P. nanus Cobb, 1923	5	–								–
P. projectus Jenkins, 1956	5	–								x
P. straeleni (de Coninck, 1931) Oostenbrink, 1960	2, 5	–								–

Peltamigratus christiei (Golden & Taylor 1956) Sher, 1954	2	–				
Pleurotylenchus sachsi (Hirschmann, 1952) Szczygiet, 1969	5	–	x	x		
Pratylenchoides crenicauda Winslow, 1958	5	–		x		x
P. magnicauda (Thorne, 1935) Baldwin, Luc & Bell, 1983	5, 7	–				
P. riparius (Andrássy, 1985) Luc, 1986	5	x				
P. ritteri Sher, 1970	5			x		
Pratylenchus sp.	20					–
Pratylenchus brachyurus (Godfrey, 1929) Filipjev & Schuurmans Stekhoven, 1941	2	–				
P. coffeae (Zimmermann, 1898) Filipjev & Schuurmans Stekhoven, 1941	2	–				
P. neglectus (Rensch, 1924) Filipjev & Schuurmans Stekhoven, 1941	2	–				
P. pratensis (de Man, 1880) Filipjev, 1936	2, 5	–				
P. thornei Sher & Allen, 1953	12	–				–
P. zeae Graham, 1951	2	–				
Prothallonema consobrinum (de Man, 1907) Siddiqi, 1986	1				x	
Psilenchus sp.	20					–

Continued

Table 21.1 *Continued.* Overview of Tylenchina recorded from freshwater.

	Major references[a]	Semi-aquatic habitat[b]	Groundwater	Swamp	Canal, river, stream	Pool, pond	Lake	Submerged aquatic plants	Periphyton	Aquatic, non-specified
Psilenchus aestuarius Andrássy, 1962	5, 7, 11	–				–				–
P. hilarulus de Man, 1921	2, 5, 6, 12	–			–					–
Radopholus sp.	20									–
Radopholus similis (Cobb, 1893) Thorne, 1949	2	–								
Rotylenchulus reniformis Linford & Oliveira, 1940	2	–								
R. buxophilus Golden, 1956	5				x					
R. capitatus Eroshenko, 1981	10	–								
R. quartus (Andrássy, 1958) Sher, 1961	5	–			x					
R. robustus (de Man, 1880) Filipjev, 1936	1, 5, 11, 12	–			–					–
R. uniformis (Thorne, 1949) Loof & Oostenbrink, 1958	5	–			x					
Scutellonema bradys (Steiner & LeHew, 1933) Andrássy, 1958	2	–								
Seinura demani (Goodey, 1928) Goodey, 1960	3, 5									x
S. tenuicaudata (de Man, 1895) Goodey, 1960	3, 5									x
Trophonema arenaria (Raski, 1956) Raski, 1957	10	–								
Trophotylenchulus floridensis Raski, 1957	2	–								
Tylenchorhynchus sp.	20	–								–

Tylenchorhynchus annulatus (Cassidy, 1930) Golden, 1971	2	–				x		x
T. bryobius Sturhan, 1966	5	–						x
T. dubius (Bütschli, 1873) Filipjev, 1936	1, 5, 11, 12	–		x	x			x
T. huesingi Paetzold, 1958	5, 11	x						
T. irregularis Wu, 1969	2	–				–		–
T. lamelliferus (de Man, 1880) Filipjev, 1936	5	–	x					
T. maximus Allen, 1955	5	–	x					x
Tylenchulus semipenetrans Cobb, 1913	2	–						
Tylenchus sp.	17, 20		–					–
Tylenchus davainei Bastian, 1865	1, 5, 11, 12	–	–	–	x			–
T. elegans de Man, 1876	2, 5	–					x	
T. rex Andrássy, 1979	14				–			
Verutus volvingentis Esser, 1981	2	–						
Zygotylenchus guevarai (Tobar, 1963) Braun & Loof, 1966	5	x						x
CEPHALOBOMORPHA De Ley & Blaxter, 2002								
Acrobeles ciliatus Linstow, 1877	3, 11, 12	–	–					–
A. (*Heterocephalobus*) *bisimilis* (Thorn, 1925)	19		–					
Acrobeloides sp. Andrássy, 1967	23, 12		–					
Acrobeloides butschlii (de Man, 1885) Steiner & Buhrer, 1933	3, 12	–						x
A. emarginatus (de Man, 1880) Thorne, 1937	3, 11	–						

Continued

Table 21.1 *Continued.* Overview of Tylenchina recorded from freshwater.

	Major references[a]	Semi-aquatic habitat[b]	Ground-water	Swamp	Canal, river, stream	Pool, pond	Lake	Sub-merged aquatic plants	Periphyton	Aquatic, non-specified
A. nanus (de Man, 1880) Anderson, 1968	3, 6, 11, 15, 17	–			–		–			x
A. tricornis (Thorne, 1925) Thorne, 1937	3	–								
Cephalobus sp.	20									–
Cephalobus (*Heterocephalobus*) *elongatus* de Man, 1880	1, 11, 12, 13, 15, 17				–					–
C. labiatus (Ivanova, 1968) Andrássy, 1984	19				x					
C. (*Heterocephalobus*) *longicaudatus* Bütschli, 1873	3	–								
C. persegnis Bastian, 1865	3, 11, 13, 15, 17, 18	–			–					x
Cervidellus cervus (Thorne, 1925) Thorne, 1937	6	–					–			
C. vexilliger (de Man, 1880) Thorne, 1937	3	–								x
Chiloplacus propinquus (de Man, 1921) Thorne, 1936	3	–								x
C. symmetricus (Thorne, 1925) Thorne, 1937	3	–								
Eucephalobus sp.	20, 23								x	–
Eucephalobus cornis (Thorne, 1925) Andrássy, 1967	14, 19				x					
E. oxyuroides (de Man, 1876) Steiner, 1936	3, 6, 14	–			x		x			

E. striatus (Bastian, 1865) Thorne, 1937	3, 6, 8, 12, 13, 15, 18, 19	–			–		–			x
Paracrobeles laterellus Heyns, 1968	8	–								
PANGROLAIMORPHA De Ley & Blaxter, 2002										
Panagrellus sp.	23									
Panagrellus filiformis (Sukul, 1971) Andrássy, 1984	4				x					
P. redivivus (Linnaeus, 1767) Goodey, 1945	13, 15, 17				–					–
Panagrolaimus sp.	20, 21, 23									–
Panagrolaimus hygrophilus Bassen, 1940	4, 12, 17	–			–		–			–
P. rigidus (Schneider, 1866) Thorne, 1937	1, 3, 11, 15, 17	–			–	–				–
P. subelongatus (Cobb, 1914) Thorne, 1937	6, 12	–		–						
P. thienemanni Hirschmann, 1952	3, 4	–								–
Turbatrix aceti (Müller, 1783) Peters, 1927	6					x				x

[a]Where possible, overview works are cited instead of the original references.

[b]Habitats that are wet or beside aquatic habitats but not permanently submerged, such as river banks, wet soil, swampy meadows, etc.

Note: X = Sporadically recorded; – = recorded multiple times or described as being generally present (discrimination between X and – can be vague).

References: 1, Micoletzky (1925); 2, Gerber and Smart (1987); 3, Mol (1984); 4, Andrássy (1984); 5, Loof (2001); 6, Esser and Buckingham (1987); 7, W. Bert (unpublished data); 8, Heyns (2002); 9, Micoletzky (1914); 10, Brzeski (1998); 11, Andrássy (1978); 12, Gagarin (2001); 13, Zullini and Ricci (1980); 14, Wanless and Hunter (2001); 15, Zullini (1982); 16, Gagarin (2003); 17, Beier and Traunspurger (2003); 18, Zullini (1988); 19, Ocaña and Picazo (1991); 20, Tarjan *et al.* (1977); 21, Brzeski (1996); 22, Doucet (1985); 23, Pennak (1978); and 24, Bird (1999).

Aquatic plant parasitism

Much information has been generated on plant-parasitic nematodes that live on terrestrial crops, but relatively little information is available on parasitic nematodes of aquatic vascular plants. The few studied are those nematode parasites of plants grown in aquatic environments as food crops, such as rice (Fortuner and Merny, 1979) or as commercial aquarium plants (Smart and Khoung, 1985). An extensive overview of plant-parasitic nematodes associated with aquatic vascular plants is given by Gerber and Smart (1987), who emphasize the possibilities of using nematodes for controlling harmful aquatic weeds. *Aphelenchoides fragariae* and *Hirschmanniella caudacrena* were forwarded as potential biocontrol candidates for submerged and floating devastating plants. Gerber and Smart's (1987) research was substantiated with experimental pathogenicity tests. A list of Tylenchomorpha associated with (semi-)aquatic plants is given in Table 21.2.

Table 21.2 Tylenchomorpha genera and their association with (semi-)aquatic plants.

Genus	References[a]	(Semi-)aquatic plants
Aglenchus	2	*Oryza sativa* L.
Aphelenchoides	2	*Cabomba* sp., *Carex* sp., *Ceratophyllum demersum* L., *Cyperus alternifolius*, *Cyperus iria* L., *Oryza sativa* L., *Panicum maximum* Jacq., *Phragmites* sp., *Potamogeton* spp., *Zizania* sp.
Atylenchus	5	*Oxycoccus* sp.
Belonolaimus	2	*Cyperus rotundus* L., *Typha* sp., *Sagittaria latifolia* Willd.
Boleodorus	2	*Panicum maximum* Jacq.
Coslenchus	2, 7	*Oryza sativa* L., *Phragmites* sp.
Criconema	2	*Ludwigia peruviana* (L.) Hara
Ditylenchus	2, 5	*Cyperus rotundus* L., *Leersia hexadra* Sw., *Myriophyllum spicatum* L., *Oryza sativa* L., *Potamogeton* spp.
Dolichodorus	2	*Anubias nana* Engler, *Elocharis dulcis* (Burm. F.), *Nymphae* sp., *Panicum hemitomon* Schult., *Scirpus validas* Vahl., etc.
Filenchus	2	*Oryza sativa* L.
Helicotylenchus	2	*Alternanthera philoxeroides* (Mart.), *Bacopa caroliniana* (Walt.), *Cyperus rotundus* L., *Eleocharis dulcis* (Burm. F.), *Eupatorium compostifolium* Walt, *Hydrochloa caroliniensis* Beauv., *Iris* sp., *Juncus effusus* L., *Ludwigia* spp., *Oryza sativa*, *Panicum* spp., *Polygonum hydropiperoides* Michx., *Xanthosoma sagittifolium* L., etc.
Hemicriconemoides	2	*Eupatorium compostifolium* Walt., *Juncus effusus* L., *Ludwigia* spp., etc.

Hemicycliophora	2	*Cyperus rotundus* L., *Eupatorium compostifolium* Walt., *Ludwigia* spp., *Oryza sativa* L., *Typha latifolia* L.
Heterodera	2	*Eleocharis acicularis* L., *Nasturtium* spp., *Oryza sativa* L.
Hirschmanniella	1, 2, 5, 7, 11	*Butomus umbellatus* L., *Cabomba* sp., *Ceratophyllum demersum, Cyperus* spp., *Elodea canadensis* Rich. in Michx., *Hydrilla verticillata* (L.f.), *Myriophyllum spicatum* L., *Nelumbium nelumbo* (L.), *Oryza sativa* L., *Phragmites communis* Trin., *Pistia stratiotes* L., *Potamogeton* spp. *Scirpus americanus* Pers., *Typha* sp., *Zizania aquatica* L., etc. (For *H. gracilis* alone, already more than 30 associated plant species are known)
Hoplolaimus	2	*Ceratophyllum demersum* L., *Cyperus rotundus* L., *Eupatorium compostifolium* Walt, *Panicum hemitomon* Schult, *Zizania aquatica* L.
Lelenchus	2, 5	*Oryza sativa* L., *Sphagnum* sp.
Malenchus	2, 5	*Oryza sativa* L., *Sphagnum* sp.
Meloidodera	2	*Eupatorium compostifolium* Walt., *Panicum hemitomon* Schult.
Meloidogyne	2	*Cyperus* spp., *Eleocharis acicularis* L., *Fimbristylis miliacea* L., *Glyceria striata* Lam., *Leersia hexadra* Sw., *Leersia oryzoides* L.,. *Oryza sativa* L., *Panicum maximum* Jacq., *Polygonum* spp., *Scirpus sylvaticus* L., *Vallisneria americana* Michx., *Xanthosoma sagittifolium* L.
Mesocriconema	1, 2, 5, 11	*Bacopa caroliniana* (Walt.), *Cyperus* spp., *Elodea canadensis, Eupatorium compostifolium* Walt., *Fimbristylis miliacea* L., *Iris* sp., *Juncus effusus* L., *Ludwigia erecta* L., *Oryza sativa* L., *Panicum hemitomon* Schult., *Paspalum hydrophilum* Henr., *Sphagnum* sp., *Xanthosoma sagittifolium* L.
Nagelus	5	*Iris* sp.
Ogma	2	*Panicum hemitomon* Schult, *Sphagnum* sp.
Paratylenchus	1, 2, 5	*Bacopa caroliniana* (Walt.), *Elocharis dulcis* (Burm. F.), *Eupatorium compostifolium* Walt., *Oryza sativa* L.
Peltamigratus	2	*Eupatorium compostifolium* Walt., *Ludwigia octovavis* (Jacq.), *Panicum hemitomon* Schult.
Pratylenchus	2, 5	*Cyperus iria* L., *Cyperus rotundus* L., *Eupatorium compositifolium* Walt., *Ludwigia octovavis* (Jacq.), *Oryza sativa* L., *Panicum maximum* Jacq., *Scirpus cernuus* Muhl., etc.

Continued

Table 21.2 *Continued.* Tylenchomorpha genera and their association with (semi-)aquatic plants.

Genus	References[a]	(Semi-)aquatic plants
Psilenchus	2	*Oryza sativa* L.
Radopholus	2	*Cyperus rotundus* L., *Panicum hemitomon* Schult., *Panicum maximum* Jacq., *Xanthosoma sagittifolium* L.
Rotylenchus	2	*Cyperus rotundus* L., *Panicum maximum* Jacq.
Sarisodera	2	*Panicum maximum* Jacq.
Scutellonema	2	*Eupatorium compositifolium* Walt.
Trophotylenchulus	2	*Panicum rigidulum* Nees
Tylenchorhynchus	2	*Sphagnum* sp.
		Hydrochloa spp., *Ludwigia octovavis* (Jacq.)
Tylenchulus	2	*Cyperus rotundus* L., *Panicum* spp.
Tylenchus	1, 5, 11	*Sphagnum* sp.
Verutus	2	*Ludwigia peruviana* L.

[a]Numbers refer to references given in Tale 21.1, main reference Gerber and Smart (1987) = 2.

Which Tylenchomorpha are adapted to the aquatic environment?

Classifying Tylenchomorpha into strict terrestrial species, or as those adapted to freshwater, is impossible. Large populations of plant-parasitic nematodes occur in waterways; in irrigation canals the density can reach up to 100 individuals/l, randomly distributed along the water column (Faulkner and Bolander, 1966). These plant-parasitic nematodes are able to survive extended aquatic transport and can also be extracted from sediments, but are most likely not able to live and reproduce in freshwater habitats. In the past, tylenchs were often simply omitted from freshwater studies, with the exception of *Psilenchus* and *Hirschmanniella* (e.g. Zullini and Ricci, 1980). Only a very limited number of Tylenchomorpha can be considered with certainty as being freshwater-adapted. Fortuner's (1977) experiments have shown that in Sénégal, species belonging to *Hirschmanniella* are the only ones that are perfectly adapted to constant flooding. Their possession of thorneian cells between the intestine and muscle layer, a unique feature within the Tylenchina, might be a morphological adaptation to aquatic life (Sher, 1968).

Several other genera and species are recorded repeatedly from freshwater habitats (Table 21.1). Most of them are common genera and consequently have a greater chance of being recorded. But their widespread occurrence alone is not sufficient to explain the differential presence of genera like *Coslenchus*, *Tylenchus*, *Lelenchus*, *Eutylenchus*, *Nagelus*, *Dolichodorus*, *Hemicycliophora*, *Ditylenchus* etc. in freshwater habitats. Several Tylenchomorpha species must have unknown adaptations to freshwater habitats. In this chapter we make the artificial choice to discuss the genera comprising species that are frequently recorded from permanently submerged freshwater habitats.

Infraorder CEPHALOBOMORPHA De Ley & Blaxter, 2002

Cephalobomorpha are bacterial-feeding, not closely associated with animal hosts or vectors, and include some of the most widespread opportunists, but also numerous specialists of sandy soils and extreme temperatures. In deserts and mountain soils Cephalobomorpha may constitute more than half of the total nematode density and diversity (e.g. De Ley, 1992). Many species are capable of anhydrobiosis, at most stages of development.

None of the Cephalobomorpha species is considered to have a preference for aquatic habitats. However, a very limited number of species is recorded repeatedly from freshwater habitats (Table 21.1). This may imply that only a few species are able to survive in freshwater habitats or that the actual species diversity is inappropriately evaluated (defining and identifying genera and species of cephalobs is problematic, see De Ley, 1997). The predominant freshwater recording of some genera compared to others is possibly more related to their general common occurrence (e.g. *Acrobeloides* and/or *Cephalobus* are found in most soils) rather than differential ecological adaptations to specific freshwater habitats. Perhaps the reverse is more correct, in that some genera have become so strongly adapted to survival in low-humidity soil conditions, at the expense of their ability to cope with complete inundation. The genera discussed below (*Acrobeles*, *Acrobeloides*, *Cephalobus* and *Eucephalobus*) have a relatively higher chance of being encountered in freshwater, but this list is not unequivocal in listing the 'freshwater genera' within the Cephalobomorpha.

Infraorder PANAGROLAIMOMORPHA De Ley & Blaxter, 2002

The Panagrolaimomorpha include the insect-parasitic Steinernematidae, the vertebrate-parasitic Strongyloididae, the amphibian/reptilian-parasitic Rhabdiasidae and the primarily free-living Panagrolaimidae. Only the latter group is commonly encountered in freshwater.

Panagrolaimidae are bacterial-feeding, occasionally associated with animal hosts or vectors, and include many opportunists (e.g. *Panagrolaimus* spp.), as well as several genera specifically associated with fermentation, decaying wood and wood-boring insects. Many species are capable of anhydrobiosis, and some appear to have the capacity to go through a dauer-like dispersal stage during development. Furthermore, many Panagrolaimidae are capable of withstanding a wide range of chemically harsh environments, e.g. *Turbatrix aceti* (the vinegar eelworm), *Panagrellus redivivus*, *P. silusiae*, *P. nepenthicola* and *Baujardia mirabilis* inhabit vinegar, bookbinder's paste, beer filters and pitcher plants, respectively. They are also commonly found in freshwater. The genera *Panagrellus* and *Panagrolaimus,* in particular, are generally considered to be partly aquatic (e.g. Tarjan *et al.*, 1977). *Panagrellus redivivus* is often associated with polluted freshwater and is used in aquatic toxicity bioassays (Sherry *et al.*, 1997) and as fish food in aquaculture (Santiago *et al*, 2004).

Infraorder DRILONEMATOMORPHA De Ley & Blaxter, 2002

Drilonematomorpha consist largely of parasites of terrestrial and aquatic annelids; free-living freshwater forms are not known, although it is conceivable that parasites of leeches might accidentally be recovered from freshwater samples.

Systematic Position and Diagnosis of Freshwater Tylenchina

The Tylenchomorpha (mainly plant-parasitic), the Cephalobomorpha and Panagrolaimomorpha (mainly free living) and the Drilonematomorpha (zooparasitic) were treated in the past as unrelated taxa based on their divergent ecological attributes and morphology. Combining molecular phylogenetic analysis with morphological data, De Ley and Blaxter (2002) established a drastically different classification and grouped the above taxa in the suborder Tylenchina.

The infraorder Tylenchomorpha is equivalent to the order Tylenchida Thorne, 1949 in the classification of Maggenti *et al.* (1987) or to Tylenchida and Aphelenchida Siddiqi, 1980 in the classification of Siddiqi (2000). The popular terms ‘tylenchs’ (= Criconematoidea, Sphaerularioidea and Tylenchoidea = Tylenchina *sensu* Geraert, 1966 = Tylenchida *sensu* Siddiqi, 1980) and ‘aphelenchs’ (= Aphelenchoidea = Aphelenchina *sensu* Maggenti *et al.*, 1987 = Aphelenchida *sensu* Siddiqi, 1980) are often used as less unambiguous terms, independent of changing classifications.

Suborder TYLENCHINA Thorne, 1949

Since Tylenchina encompasses the morphological disjunct Tylenchomorpha, Cephalobomorpha, Panagrolaimorpha and Drilonematomorpha; it is not possible to provide a morphology-based diagnosis for the Tylenchina. De Ley and Blaxter (2002) hypothesized a most recent common ancestor of the Tylenchina having the following important characteristics: a stegostom without epithelial radial cells or glottoid part, median bulb absent, monoprodelphy and bursa absent.

Infraorder TYLENCHOMORPHA De Ley & Blaxter, 2002

Vermiform nematodes, usually ranging from 0.2 to 1 mm long, occasionally over 3 mm. In some genera the female becomes globose. Cuticle usually clearly annulated. Amphids located on the head. Protrusible, narrow pointed stylet composed of a conus with a subterminal ventral aperture, a cylindrical shaft and, if present, three basal knobs. Pharynx generally divided into four parts: anterior procorpus, enlarged metacorpus (= median bulb) having typically crescentic valves to which attach radial muscles, a narrow isthmus encircled by a ring of nerves and a posterior bulb or set of lobes.

The book *Tylenchida, Parasites of Plants and Insects* (Siddiqi, 2000) is a standard reference work on the Tylenchomorpha (without Aphelenchoidea). It includes keys and detailed diagnosis to generic level and information on the biology, ecology and pathogenicity of the tylenchs. For the determination of Tylenchoidea, Criconematoidea and Anguinidae, Brzeski (1998) is especially convenient (emphasis on temperate Europe, but compendia are of worldwide use).

Another important work on the freshwater Tylenchomorpha of Central Europe is that of Loof (2001), published in the series *Süßwasserfauna von Mitteleuropa*. The book is arranged in key format, starting from order up to species level, and a short diagnosis is given for each taxonomic level. More than 100 species are presented as potential freshwater species. Precise information about their actual habitat preference is given, combined with notes about their distribution outside Central Europe.

The classification below family level followed in this chapter is largely based on 'a reappraisal of Tylenchina' (Fortuner, 1987; Fortuner and Luc, 1987; Geraert and Raski, 1987; Luc *et al.*, 1987; Maggenti *et al.*, 1987; Raski and Luc, 1987; Maggenti *et al.*, 1988). This approach emphasizes the reduction of the 'taxonomic inflation' to preserve a clear picture of the ordered relationships in a group of living organisms, rather than demonstrating the diversity above species level. For writing the taxa diagnoses below, Brzeski (1998) and Siddiqi (2000) were followed as major reference works. However, a few taxa diagnoses have to be revised in the light of the current systematics of De Ley and Blaxter (2002).

Superfamily TYLENCHOIDEA Örley, 1880

Cuticule usually finely annulated. Lip region generally hexaradiate and distinguished from general body contour. Labial region supported by a cuticularized skeleton that may or may not be well developed. Procorpus generally offset from metacorpus, usually slender and cylindrical. Isthmus narrow and leads to the expanded glandular region almost always wider than the metacorpus. Glandular region consists of three glands ending at the beginning of the intestine, or variously overlapping this structure.

Family TYLENCHIDAE Örley, 1880 (Fig. 21.1)

Small to medium-sized (0.33–1.3 mm), slender and vermiform nematodes. Lateral field mostly with two, three or four incisures. Deirids normally present. Phasmid-like structures, if present, usually advulval (occasional typical phasmid on tail). Amphids vary from small oblique slits to long sinuous clefts longitudinally directed. Lip region generally elevated, rounded and annulated. Cephalic framework weak, stylet delicate (6–21 μm, few exceptions). Pharynx divided into slender procorpus, elliptical metacorpus usually having valves, long slender isthmus followed by symmetrical pyriform glandular region. Female genital system prodelphic, seldom both branches developed. Columned uterus with four rows of cells. Length of post-vulval sac typically less than one vulval body diameter. Testis outstretched, tip rarely reflexed. Bursa not enveloping tail, rarely absent. Sperm cells with little cytoplasm. Tail elongated to filiform.

Genus *Tylenchus* Bastian, 1865
= *Aerotylenchus* Fotedar & Handoo, 1979; *Aerotylenchus* Fortuner, 1984

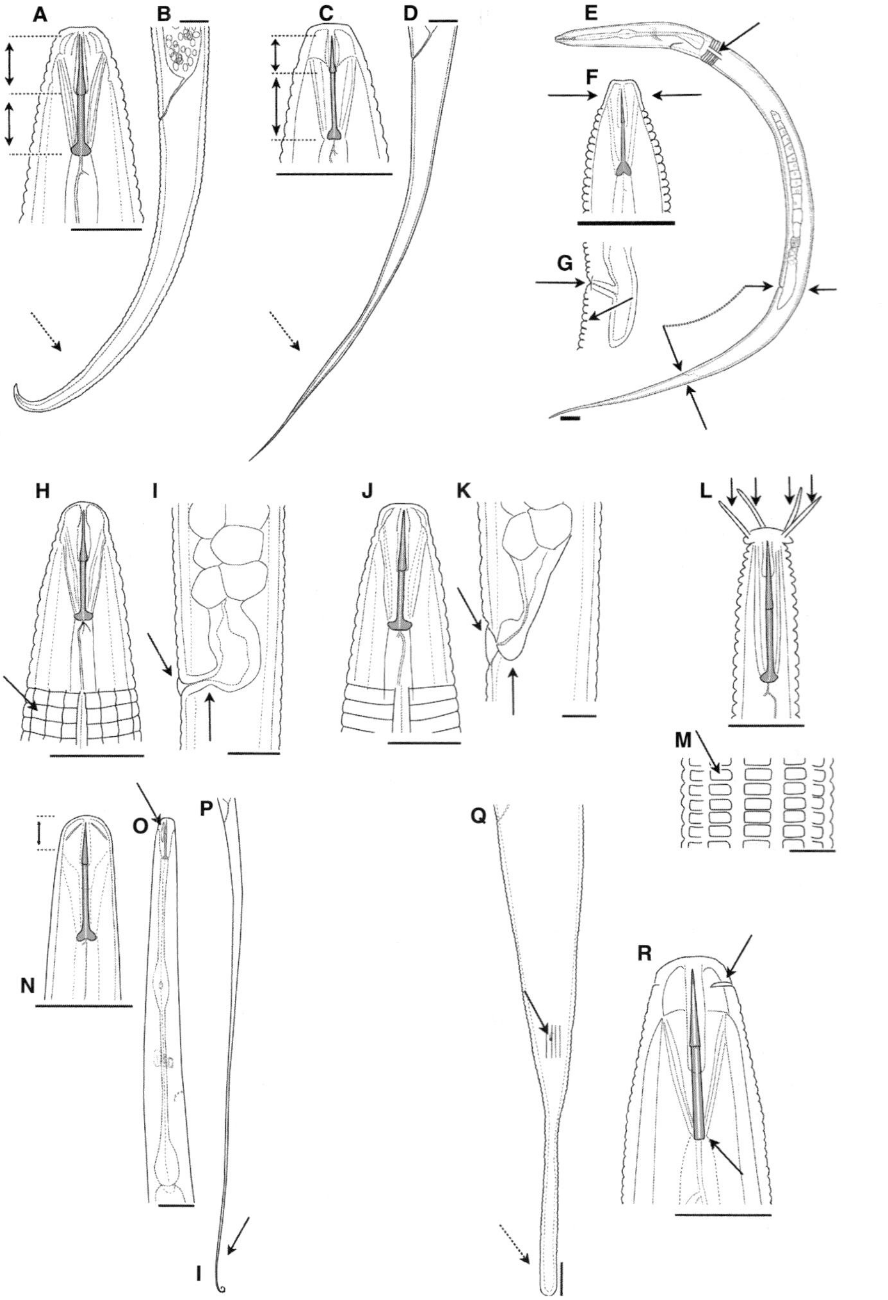

Fig. 21.1 Tylenchidae: *Tylenchus arcuatus* (A, B); *Filenchus vulgaris* (C, D); *Malenchus* sp. (E, F, G); *Coslenchus costatus* (H, I); *Aglenchus agricola* (J, K); *Eutylenchus setiferus* (L, M); *Lelenchus leptosoma* (N, O, P); and *Psilenchus aestuarius* (Q, R). Female head region (A, C, F, H, J, L, N); male head region (R); female tail (B, D, P, Q); female habitus (E); female vulva region (G, I, K); female midbody (M); and female neck region (O). *E. setiferus* based on Sher *et al.* (1966); *L. leptosoma* based on Raski and Geraert (1986). Scale bars: 10 μm.

Body ventrally curved, usually 0.6–1.2 mm long. Cuticle distinctly annulated. Lateral field with four lines. Deirids usually just behind excretory pore level. Stylet relatively robust, about 12–21 μm long, conus about half of stylet, knobs distinct. Dorsal gland orifice close to knobs. Median pharyngeal bulb well developed, muscular and oval. Glandular bulb offset from intestine, usually valvate and pyriform. Tail ventrally arcuate, often hooked, not filiform. Female reproductive system prodelphic, spermatheca mostly offset, short post-vulval sac.

Type species: *T. davainei* Bastian, 1865. Siddiqi (2000) lists 29 species (and 25 *species inquirendae* or *incertae sedis*). Keys to *Tylenchus* spp. are given by Andrássy (1979) and Maqbool and Shahina (1987).

Tylenchus spp. are often found in wet mosses and other moist habitats. The type species was described from moss covering a large boulder lying in a freshwater stream and is also recorded from some actual freshwater habitats (Table 21.1).

Genus *Filenchus* Andrássy, 1954
= *Dactylotylenchus* Wu, 1968; *Discotylenchus* Siddiqi, 1980; *Duosulcius* Siddiqi, 1979; *Lambertia* Brzeski, 1977; *Ottolenchus* Husain & Khan, 1965; *Zanenchus* Siddiqi, 1979
Body straight to arcuate, length seldom over 1 mm. Cuticle usually with fine annulations. Lateral field with two to four lines. Deirids at the level of excretory pore. Cephalic region broadly rounded or conoid rounded. Head wide at base, about half as high, not offset. Stylet delicate, about 6–15 μm long, conus about one-third of stylet length, knobs present. Dorsal gland orifice close to knobs. Median pharyngeal bulb muscular, oval to rounded, usually valvate. Glandular bulb offset from intestine, usually pyriform. Tail generally filiform and straight. Female reproductive system prodelphic, outstretched, spermatheca mostly offset, short post-vulval sac usually present.

Type species: *F. vulgaris* (Brzeski, 1963) Lownsbery & Lownsbery, 1983 (or *F. filiformis* (Bütschli, 1873) Meyl, 1960). A comment supporting the designation of *F. vulgaris* as type species (Fortuner *et al.*, 1988), and one opposing it (Siddiqi and Hunt, 1988), were published in the Bulletin of Zoological Nomenclature. Until now, the International Commission on Zoological Nomenclature (ICZN) has not proposed an opinion on this case. Raski and Geraert (1987) list 70 *Filenchus* spp. Siddiqi (2000) lists 94 species (*Discotylenchus*, *Duosulcius*, *Lambertia*, *Ottolenchus* and *Zanenchus* are considered synonyms here). Determination of *Filenchus* spp. is often problematic, but a key to *Filenchus* spp. is given by Raski and Geraert (1987).

Filenchus is an extremely common and diverse terrestrial genus and can be found also in freshwater habitats (Table 21.1).

Genus *Malenchus* Andrássy, 1968
= *Neomalenchus* Siddiqi, 1979; *Paramalenchus* Sumenkova, 1988
Small nematodes, usually not more than 0.6 mm. Body fusiform, tapering markedly towards both ends, width at anus about half that of vulva. Annules prominent, cuticular folds between them. Lateral field seen as single protruding ridge (larger number of lines visible only with SEM). Cephalic region high, flattened dorsoventrally. Stylet delicate, cone about one-third of stylet length, knobs small but elongated. Dorsal

gland orifice close to knobs. Median bulb weakly developed, with or without valve plates. Tail elongate conoid to a pointed or hooked tip. Vulva sunken in body cavity. Cloacal lips protruding.

Type species: *M. machadoi* (Andrássy, 1963) Andrássy, 1968. Siddiqi (2000) lists 32 species. A key to *Malenchus* spp. is given by Geraert and Raski (1986).

Several species are known to occur in wet riverbanks, mosses, peat and also in freshwater habitats (Table 21.1).

Genus *Aglenchus* Andrássy, 1954

Body small (0.4–0.7 mm), straight to slightly arcuate. Cuticle coarsely annulated. Lateral field composed of two protruding ridges, separated by a groove; in lateral view three or four lateral lines may appear depending on the width of the groove. Amphidial apertures restricted to labial plate. Stylet relatively strong, conus about half total length. Knobs rounded and well developed. Vulva sunken in body, larger outer and smaller inner lips, with prominent lateral membranes. Vaginal walls often markedly swollen. Post-vulval sac absent. Male with distinct bursa and large protruding pointed cloacal lips forming a tube.

Type species: *A. agricola* (de Man, 1884) Meyl, 1961. Siddiqi (2000) lists eight species (also two *species inquirendae*). Key to species is provided by Andrássy (1980).

A. agricola is an especially common species, known from several habitats, including those from freshwater (Table 21.1).

Genus *Coslenchus* Siddiqi, 1978

= *Cosaglenchus* Siddiqui & Khan, 1983; *Paktylenchus* Maqbool, 1983

Body small (0.3–1 mm), straight to slightly arcuate. Cuticle coarsely annulated with longitudinal ridges, showing minute squares outside lateral field. Lateral field composed of two to three ridges being more elevated than other body ridges. Pore-like amphidial apertures restricted to labial plate. Stylet relatively well developed, conus slightly shorter than shaft. Vulva sunken in body with thick lips and generally surrounded by lateral membranes. Vaginal walls mostly thickened. Male usually rare, cloacal lips forming a short tube.

Type species: *C. costatus* (de Man, 1921) Siddiqi, 1978. Siddiqi (2000) lists 37 species. Keys to *Coslenchus* spp. are given by Andrássy (1982) and Geraert and Raski (1988).

Coslenchus is a common genus, of which some species prefer moist habitats and are occasionally recorded in freshwater (Table 21.1).

Genus *Lelenchus* Andrássy, 1954

= *Tylenchus* (*Lelenchus*) Andrássy, 1954

Very slender and small nematodes, almost straight when relaxed. Cuticule smooth or indistinctly annulated. Lateral field obscure or with two incisures delimiting single ridge. Cephalic region high, dorsoventrally flattened. Amphideal apertures longitudinal slits on lateral side of head. Stylet delicate, with minute basal knobs. Dorsal gland orifice close to knobs. Median pharyngeal bulb spindle-shaped, weakly muscular. Tail very long, filiform, with pointed terminus. Spermatheca offset, vulva small, surrounded by lateral membranes. Vagina with thin walls. Narrow bursa.

Type species: *L. leptosoma* (de Man, 1880) Andrássy, 1954. Siddiqi (2000) lists only one additional species (*L. filicaudatus* Raski & Geraert, 1986) and transferred *L. elegans* Raski & Geraert, 1986 to *Chilenchus* Siddiqi, 2000.

Lelenchus is often found in wet soil (Loof, 2001), *L. leptosoma* is known from freshwater (Table 21.1).

Genus *Eutylenchus* Cobb, 1913

Small to medium-sized nematodes (0.5–1 mm), body actuate upon relaxation. Cuticle divided in blocks, formed by coarse annulus and 12 similar longitudinal ridges. Lateral ridges not different from other longitudinal ridges. Deirids reduced or absent. Distinct cephalic seta present, 9–12 μm long in females, male cephalic setae shorter. Cephalic region small, offset, four-lobed and each lobe with cephalic setae. Stylet long (18–32 μm) and slender. Median pharyngeal bulb with prominent refractive thickings, basal bulb large, saccate. Vulva with lateral cuticular membranes, post-vulval sac present. Tail elongate conoid to filiform, over ten anal body diameters long. Bursa lobed, with crenate margins. Gubernaculum about half the size of well-developed spicules.

Type species: *E. setiferus* (Cobb, 1893) Cobb, 1913. Siddiqi (2000) lists five species. Recently described species: *E. gracilis* Gagarin, 2003. Brzeski (1996) provides a key to five *Eutylenchus* species.

Species of this genus are seldom found; *Eutylenchus* spp. seem to have a preference for moist soil and actual freshwater habitats (Table 21.1). The pronounced head setae of *Eutylenchus* are noteworthy as a prominent morphological feature of aquatic nematodes.

Genus *Psilenchus* de Man, 1921

Medium-sized to large nematodes (0.7–1.70 mm), straight to curved when relaxed. Lateral field with four lines, inner lines often indistinct. Amphid apertures generally distinct, as transverse slits on lateral side. Distinct phasmids on tail, anterior to its middle. Cephalic region continuous or slightly narrower than body. Cephalic framework weakly developed. Stylet cylindrical, cone about one-third of stylet length, basal knobs absent but minute swellings may occur. Median bulb prominent with distinct thickenings of lumen wall. Female reproductive system paired, spermatheca axial and elongated. Tail elongated or shorter and thicker, with clavate or non-clavate rounded tip.

Type species: *P. hilarulus* de Man, 1921. Siddiqi (2000) lists 20 species. Keys to *Psilenchus* spp. are given by Kheiri (1970), Brzeski (1989) and Doucet (1996).

Psilenchus is often considered to be the only actual aquatic genus within the Tylenchidae (Zullini and Ricci, 1980). *P. hilarulus* and *P. aestuarius* Andrássy, 1962 are the most common *Psilenchus* species in freshwater habitats (Table 21.1).

Family BELONOLAIMIDAE Whitehead, 1959 and DOLICHODORIDAE Chitwood, 1950 (Fig. 21.2)

Small to large-sized (generally about 0.5–3 mm), habitus mostly arcuate. Cuticle usually strongly annulated. Lateral field with 1–6 incisures. Deirids present or absent.

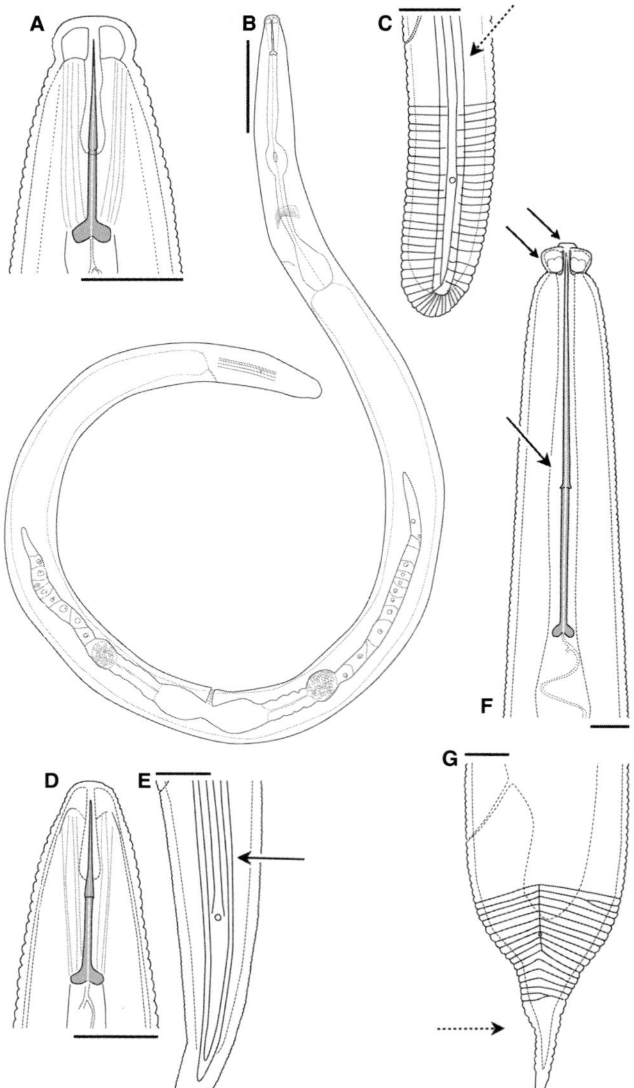

Fig. 21.2 Belonolaimidae and Dolichodoridae: *Tylenchorhynchus dubius* (A, B, C); *Nagelus obscurus* (D, E); *Dolichodorus heterocephalus* (F, G). Female head region (A, D, F); female habitus (B); female tail (C, E, G). *N. obscurus* based on Brzeski (1998); *D. heterocephalus* based on Siddiqi, 2000 and Orton Williams (1986). Scale bars: A, C, D, E, F and G = 10 μm; B = 50 μm.

Phasmids pore-like, caudal. Cephalic region annulated, hexagonal or 4-lobed. Stylet well developed, length especially variable. Pharynx with muscular median bulb, posterior bulb either offset or overlapping anterior part of intestine, sub-ventral glands not enlarged. Female genital system paired (except *Trophurus*, where posterior branch is reduced). Columned uterus with three rows of cells. Female tail variable in shape, typically 1.5–3.5 as long as anal body width. Bursa usually enveloping tail.

Ectoparasites of roots, sometimes also found inside host tissues.

Genus *Tylenchorhynchus* Cobb, 1913
= *Bitylenchus* Filipjev, 1934; *Telotylenchus* Siddiqi, 1960; *Quinisulcius* Siddiqi, 1971; *Dolichorhynchus* Mulk & Jairajpuri, 1974; *Triversus* Sher, 1974; *Meiodorus* Siddiqi, 1974; *Sauertylenchus* Sher, 1974; *Trilinellus* Lewis & Golden, 1981; *Divittus* Jairajpuri, 1984; *Mulveyotus* Anderson & Ebsary, 1982; *Morasinema* Javed, 1984; *Tesselus* Jairajpuri & Hunt, 1984; *Neodolichorhynchus* Jairajpuri & Hunt, 1984; *Prodolichorhynchus* Jairajpuri, 1985; *Mulkorhynchus* Jairajpuri, 1988; *Macrorhynchus* Sultan, Singh & Sakhuja, 1991; *Billineellus* Volkova, 1993.
Small to medium-sized nematodes (0.4–1.5 mm). Lateral field with two to five incisures. Longitudinal striation sometimes present on body. Deirids present or absent. Cephalic region continuous to slightly offset. Stylet 15–30 μm long, usually slender, with cone about as long as shaft, cone sometimes needle-like. Pharyngeal glands offset from intestine, or forming lobe extending over pharynx for a short distance. Cardia usually prominent. Vulva near middle of body. Tail conoid to subcylindroid, about three times as long as wide. Spicules with well-developed velum. Bursa rarely lobed, enveloping tail.

Tylenchorhynchus is considered here in a broad sense (following Fortuner and Luc, 1987 and Brzeski and Dolinski, 1998) and encompasses also the genera *Bitylenchus* Filipjev, 1934, *Neodolichorhynchus* Jairajpuri & Hunt, 1984, *Sauertylenchus* Sher, 1974, *Telotylenchus* Siddiqi, 1960, and *Quinisulcius* Siddiqi, 1971 pointed out by Siddiqi (2000).

Type species: *T. cylindricus* Cobb, 1913. Brzeski and Dolinski (1998) listed 177 species in their compendium of *Tylenchorhynchus sensu lato*.

Recently described species: *T. vishwanathensis* Pathak & Siddiqui, 1997; *T. iarius* Saha, Gaur & Lal, 1998; *T. paulettae* Bloemers & Wanless, 1998; *T. karnalensis* Saha, Singh, Lal & Kaushal, 2002; *T. shimizui* Talavera, Watanabe & Mizukubo, 2003. A key to *Tylenchorhynchus* spp. with four lateral lines is provided by Handoo (2000).

Tylenchorhynchus is a very common and diverse genus; some *Tylenchorhynchus* spp. are recorded from semiaquatic or aquatic habitats (Table 21.1).

Genus *Nagelus* Thorne & Malek, 1968
Body length 0.5–1.2 mm. Lateral field with six incisures. Deirids conspicuous, situated in six lines zone of lateral field. Cephalic region elevated, more or less conoid, usually slightly narrower than adjacent part of body. Stylet medium-sized to robust, conus about as long as shaft. Pharyngeal glands enclosed in posterior bulb offset from intestine. Vulva usually covered by epiptygma. Female tail elongate–conoid to subcylindrical, with hyaline distal part and striated terminus. Male with wide bursa enveloping tail.

Type species: *N. aberrans* Thorne & Malek, 1968. Siddiqi (2000) listed 26 species. Tabular information about *Nagelus* spp. is presented by Powers *et al.* (1983).

N. obscurus (Allen, 1955) Powers, Baldwin & Bell, 1983 (= *N. alpensis* Doucet & Luc, 1981) thrives in several types of freshwater habitats and is often associated with roots of *Iris* species.

Genus *Dolichodorus* Cobb, 1914
Body long, slender. Lateral field with three incisures, areolated. Amphidial apertures small, lateral directed slits. Cephalic region prominently four-lobed, labial disc conspicuous. Sclerotization of basal plate of framework massive. Stylet long (50–160 μm) and strong. Female tail usually tapering suddenly to form spicate end, rarely conoid. Amphids post-anal. Bursa large, trilobed. Spicules robust, with large flanges. Gubernaculum apparently protruding.

Type species: *D. heterocephalus* Cobb, 1914. Siddiqi (2000) listed 16 species. Keys to species have been provided by Lewis and Golden (1981) and Smart and Khoung (1985).

The type species of *Dolichodorus* was collected from a lake; it is usually found in the wet soil, around the edges of lakes, streams and ponds or in water (Smart and Nguyen, 1991). However, *D. heterocephalus* does not appear to be dependent on wet conditions. Smart and Nguyen (1991) hypothesize *Dolichodorus* to be in a state of evolution from aquatic to terrestrial.

Family PRATYLENCHIDAE Thorne, 1949 (Fig. 21.3)
Heterogeneous, supposedly not a monophyletic family. Vermiform nematodes (except *Nacobbus*), relatively straight when relaxed. Deirids absent (except *Pratylenchoides*). Phasmids pore-like, on tail well behind anus. Cephalic region low (except *Hirschmanniella*), annulated, continuous or slightly offset. Stylet strong, length not exceeding three cephalic region widths (except *Hirschmanniella*). Median bulb well developed, muscular with prominent refractive thickenings of luminal walls, glandular part of pharynx usually overlaps intestine and is variable in length (offset in a few *Pratylenchoides* spp.). Pharyngo-intestinal valve developed only in *Pratylenchoides*. Female genital system paired or prodelphic, columned uterus with three rows of cells, vulva without flaps or epiptygma. Male with bursa either enveloping or nearly reaching tail tip. Tail generally about 2–3 anal body-widths long.

Obligate migratory endoparasites of roots (except *Nacobbus*, a sedentary endoparasite). Several species are of great economic importance as pests of cultivated plants.

Genus *Hirschmanniella* Luc & Goodey, 1964
= *Hirschmannia* Luc & Goodey, 1962 nec Olofson, 1941
Very long vermiform nematodes (1–4 mm). Cuticle finely annulated, lateral field with four incisures, areolated towards extremities. Cephalic region continuous, anteriorly flattened or hemispherical, framework heavily sclerotized. Stylet massive, usually 3–5 times maximum width of cephalic region. Basal knobs large, rounded and close to shaft. Pharyngeal glands in line, overlap intestine ventrally in the form of a long lobe. Sub-ventral glands asymmetrical, larger and much longer than the dorsal gland; nuclei of the three glands lying in a row, well separated from each other. Intestine often with thorneian cells, a connective tissue-like formation with accumulated glycogen present between intestine and muscle layer of body wall. Female genital branches paired. Tail elongate-conoid, usually ending with small mucro. Phasmids in posterior part of tail. Crenate, subterminal bursa lacking phasmidial pseudoribs.

The genus *Hirschmanniella* is unique among Tylenchina in inhabiting soil, freshwater and marine habitats, as well as parasitizing plants in these environments;

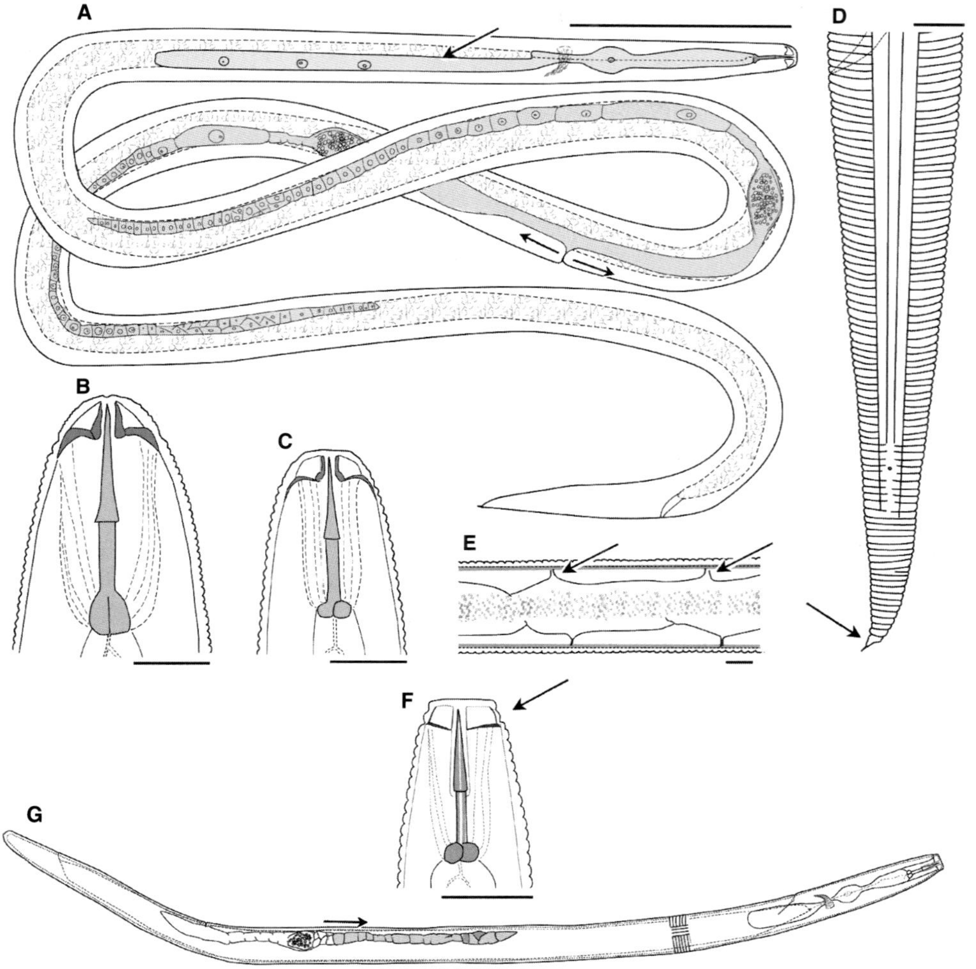

Fig. 21.3 Pratylenchidae: *Hirschmanniella oryzae* (A, E); *H. loofi* (B); *H. gracilis* (C, D); and *Pratylenchus flakkensis* (F, G). Female habitus (A); female head region (B, C, F); female tail (D); and midbody region showing 'thorneian cells' joining epidermis and intestine (E). Scale bars: B, C, D, E and F = 10 µm; A and G = 100 µm.

some species are of economic significance because of their impact on rice production. This genus is supposedly the most prevalent Tylenchina genus in freshwater and it can represent up to 56% of the total nematofauna in ponds (Messiaen, 2004).

Type species: *H. spinicaudata* (Schuurmans Stekhoven, 1944) Luc & Goodey, 1964. Siddiqi (2000) lists 34 species (and five *species inquirendae* or *nomina nuda*).

Recently described species: *H. caribbeana* Van den Berg & Quénéhervé, 2000. Keys are given by Ebsary and Anderson (1982), Loof (1991) and Sivakumar and Khan (1982).

Genus *Pratylenchus* Filipjev, 1936

Small vermiform nematodes. Lateral field with four to six incisures. Cephalic region low, usually flattened anteriorly; continuous with body outline, seldom slightly off-set. Cephalic framework well built. Stylet relatively short, less than 25 μm. Median bulb oval to round, very muscular. Pharyngeal glands overlapping intestine laterally and ventrally, the longest overlap on ventral side, overlap usually less than two body-widths long. Vulva at about 70–85% of body length, posterior genital branch reduced to post-vulval sac with or without rudiments of posterior ovary. Female tail subcylindrical to conoid, usually about 2–3 anal body-widths long. Phasmids at middle of tail. Bursa enclosing pointed male tail terminus.

Type species: *P. pratensis* (de Man, 1880) Filipjev, 1936. The genus is stenomorphic and identification of its species is difficult due to lack of diagnostic characters available and variability. Accurate identification requires examination of several specimens, which may be confused by the occurrence of more than one species (Brzeski, 1998). Siddiqi (2000) lists 89 species (and 16 *species inquirendae* or *nomina nuda*).

Recently described species: *P. brzeskii* Karssen, Waeyenberge & Moens, 2000; *P. arlingtoni* Handoo, Carta & Skantar, 2001; *P. jaehni* Inserra, Duncan, Troccoli, Dunn, dos Santos, Kaplan & Vovlas, 2001. The latest keys to species of *Pratylenchus* are given by Frederick and Tarjan (1989), Handoo and Golden (1989) and Café Filho and Huang (1989).

The root lesion nematodes are very common, with worldwide distribution, and damage several important crops. Several *Pratylenchus* species occur in wet meadows; few species are recorded from freshwater (Table 21.1).

Family HOPLOLAIMIDAE Filipjev, 1934 (Fig. 21.4)

Distinct sexual body shape dimorphism in sedentary forms. Endoparasitic nematodes: infective second-stage juvenile tail conical with long hyaline posterior part; male upon relaxation twisted along body axis; females globose (with exceptions), immobile, cuticle may be tanned and transformed into cysts. Habitus of ectoparasites often spiral. Head generally higher than half the diameter at basal ring, cephalic framework refractive. Sexual dimorphism in cephalic region often present. Deirids absent. Lateral field when present typical with four lines. Phasmids typically near anus level, exceptional migrated far anteriorly, generally small pore-like structures, sometimes migrated into scutella, rarely absent. Stylet robust, approximately 2–4 times head diameter. Pharyngeal gland usually overlapping anterior part of intestine. Tail absent or short, usually not longer than two anal diameters. Female reproductive system paired, posterior branch seldom reduced to sac. Columned uterus with three cell rows. Perineum (vulva–anus region in sedentary females) without fingerprint-like pattern. Bursa, when present, enclosing all or most of the tail.

Obligatory parasites of plant roots; migratory ectoparasites, sedentary semi-ectoparasites or sedentary endoparasites. Large family of economically important Tylenchomorpha. In former classifications the sedentary forms (except Rotylenchulinae) were classified within a separate family Heteroderidae Filipjev & Schuurmans Stekhoven, 1941.

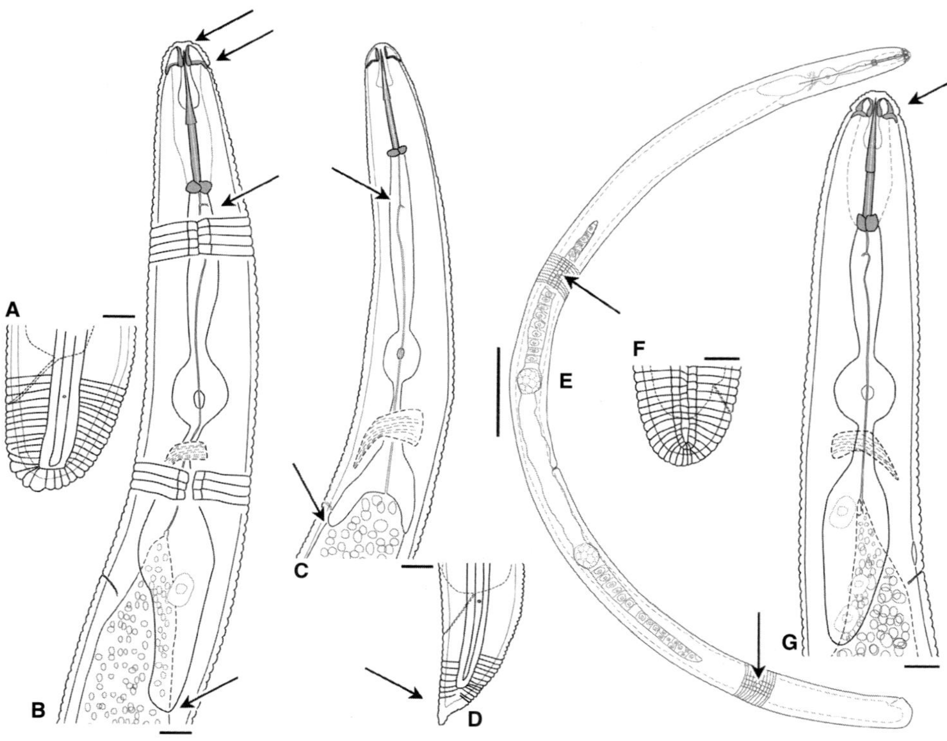

Fig. 21.4 Hoplolaimidae: *Rotylenchus uniformis* (A, B); *Helicotylenchus pseudorobustus* (C, D); and *Hoplolaimus galeatus* (E, F, G). Female neck region (B, C, G); female tail (A, D, F); and habitus (E). *H. galeatus* based on Sher (1961). Scale bars: A, B, C, D, F and G = 10 μm; E = 100 μm.

Genus *Hoplolaimus* von Daday, 1905
= *Nemonchus* Cobb, 1913; *Hoplolaimoides* Shakil, 1973; *Basirolaimus* Shamsi, 1979; *Ethiolaimus* Siddiqi, 2000.

Large-sized (1–2 mm). Female body relatively straight. Lateral field with two or four incisures. Cephalic region offset, rounded, marked by prominent transverse and longitudinal striae, labial disc distinct, basal annule marked by more than six longitudinal striae. Labial framework and stylet massive, stylet knobs tulip-shaped. Dorsal gland orifice usually 3–10 μm posterior to knobs. Pharyngeal glands overlapping intestine dorsally and laterally, intestine symmetrically arranged between the sub-ventral glands. Female reproductive system paired, branches opposite, outstretched and equally developed. Epiptygma present, usually indistinct. Tail short, rounded. Plasmids enlarged to scutella, not opposite each other, one pre-vulval, another post-vulval (exceptional both phasmids post-vulval). Spicules massive, somewhat cylindroid, may be dimorphic with distal flanges variable in size. Bursa surrounding tail terminus.

Type species: *H. tylenchiformis* von Daday, 1905. Siddiqi (2000) lists 32 species (and five *species inquirendae* or *nomina nuda*). Handoo and Golden (1992) provide a key to the species of *Hoplolaimus*.

H. tylenchiformis is found in association with several submerged aquatic plants (Gerber and Smart, 1987).

Genus *Rotylenchus* Filipjev, 1936
= *Anguillulina* (Rotylenchus) Filipjev, 1936; *Gottholdsteineria* Andrássy, 1958; *Orientylus* Jairaipuri & Siddiqi, 1977; *Calvatylus* Jairaipuri & Siddiqi, 1977; *Pararotylenchus* Baldwin & Bell, 1981; *Interrotylenchus* Eroshenko, 1984; *Scutellonemoides* Eroshenko, 1984; *Varotylus* Siddiqi, 1986.

Females C-shaped to spirally coiled. Lateral field with four lines. Cephalic framework strongly refractive. Labial disc usually distinct. Cephalic region with or without longitudinal striae. Stylet robust, conus and shaft approximately equal in length. Dorsal gland orifice principally less than one-fourth stylet length posterior to knobs. Pharyngeal glands overlapping intestine with longest overlap on dorsal side (exceptional enclosed in offset bulb). Female reproductive system with two opposed outstretched branches. Epiptygma single or double. Female tail short, rounded, annulation following its contour, seldom with projection. Phasmids small, near anus level. Spicules robust, distally flanged. Bursa surrounding tail terminus.

Type species: *R. robustus* (de Man, 1876) Filipjev, 1936. La Massése and Germani (2000) provide a tabular key to 103 *Rotylenchus* species.

Recently described species: *R. aquaticus* Germani & La Massése (2002); *R. incognitos* Germani & La Massése (2002); *R. landii* Germani & La Massése (2002) and *R. helenae* Germani & La Massése (2002).

Several *Rotylenchus* species are recorded from wet habitats, also exceptionally found in typical freshwater habitats (Table 21.4).

Genus *Helicotylenchus* Steiner, 1945
= *Rotylenchoides* Whitehead, 1958; *Zimmermannia* Shamsi, 1973

Small to medium-sized (0.4–1.2 mm). Females spirally coiled or rarely C-shaped. Lateral field with four lines. Cephalic region continuous or rarely offset, no longitudinal striae on head annules. Cephalic framework moderately developed, lateral ribs refractive. Stylet robust, average-sized, both parts of approximately equal length. Dorsal gland orifice usually one-third of stylet length posterior to knobs. Pharyngeal glands overlapping intestine laterally and ventrally, the longest overlap on ventral or lateral side of body, glands separate. Female reproductive system usually paired, branches opposite outstretched, sometimes posterior branch less developed to a variable degree. Epiptygma single or double. Tail 1–2 body diameters long, typically more curved dorsally, with or without terminal ventral process, sometimes rounded. Phasmids small, at level of anus. Bursa surrounding tail terminus.

Type species: *H. dihystera* (Cobb, 1893) Sher, 1961. Siddiqi (2000) lists 196 species (and three *species inquirendae*).

Recently described species: *H. jasminii* Jain, Siddiqui, Aruna, Parihar & Parihar (2000); *H. rajcolagri* Jain, Siddiqui, Aruna, Parihar & Parihar (2000); *H. samorensis* Jain, Siddiqui, Aruna, Parihar & Parihar (2000); *H. gulabi* Jain, Siddiqui, Aruna, Parihar & Parihar (2000); *H. crotonii* Jain, Siddiqui, Aruna, Parihar & Parihar (2000); *H. brevicaudatus* Jain, Siddiqui, Aruna, Parihar & Parihar (2000); *H. shervarayensis*

Giribabu & Saha (2002); *H. bajoriensis* Saha, Singh, Lal & Kaushal (2002); and *H. marethae* Marais, Queneherve, Tiedt & Meyer, 2003. *Helicotylenchus* contains many poorly differentiated species and their identification is often difficult (Brzeski, 1998). Keys to *Helicotylenchus* spp. are given by Fotedar and Kaul (1985) and Firoza and Maqbool (1994).

Helicotylenchus spp. are common nematodes in various soil types. Several *Helicotylenchus* species are recorded from freshwater habitats too (Table 21.1); some species are considered to live and reproduce in freshwater (Heyns, 2002) and to parasitize submerged aquatic plants (Gerber and Smart, 1987).

Superfamily SPHAERULARIOIDEA Lubbock, 1861

Group with remarkable heterogeneous morphology, sharing some ecological traits. Variable-shaped nematodes, vermiform to obese. Phasmid absent from lateral field. Amphidial apertures dorsosublateral. Cephalic framework usually low, smooth or finely striated. Stylet generally under 20 μm, with or without basal knobs. Female genital system monodelphic, prodelphic. Columned uterus with four or more rows of cells.

Primarily mycetophagous nematodes, many insect- or mite-parasitic, plant parasitism limited to above-ground parts (except *Subanguina*). Both free-living and entomoparasitic generations found in several genera.

Family ANGUINIDAE Nicoll, 1935 (Fig. 21.5)

Small to large-sized, in some genera adults may be obese and sedentary. Cuticle with fine transverse striations, often appearing smooth. Phasmid-like structures near vulva, dorsal from lateral field, not on tail. Cephalic region low, usually continuous with body contour, framework hexaradiate. Stylet delicate (under 15 μm), knobs small. Median bulb with or without valve plates, posterior bulb either offset or variable length overlapping intestine. Cardia absent, two anteriormost intestinal cells often acting as a valve. Female reproductive system prodelphic, ovary may be reflexed. Spermatheca, elongated, sac-like. Uterus with four or more cell rows. Postvulval sac regularly longer than body width. Sperm with large amount of cytoplasm (except *Pseudohalenchus*). Bursa usually subterminal. Tail typically elongate–conoid (seldom rounded or filiform).

Mainly fungal feeders and parasites of aerial parts of plants, also parasites of lower plants (mosses, seaweeds). Anguinidae are often considered to be more closely related to the Tylenchidae than to the remaining Sphaerularioidea (e.g. Maggenti *et al.*, 1987; Brzeski, 1998; Siddiqi, 2000).

Genus *Ditylenchus* Filipjev, 1936

= *Anguillulina* (*Ditylenchus*) Filipjev, 1936; *Nothotylenchus* Thorne, 1941; *Boleodoroides* Mathur, Khan & Prasad, 1966; *Diptenchus* Khan, Chawla & Seshadri, 1969; *Safianema* Siddiqi, 1980; *Orrina* Brzeski, 1981

Body length about 0.4–1.5 mm, all stages vermiform. Cephalic skeleton weak, stylet delicate. Median bulb muscular or non-muscular, with or without refractive

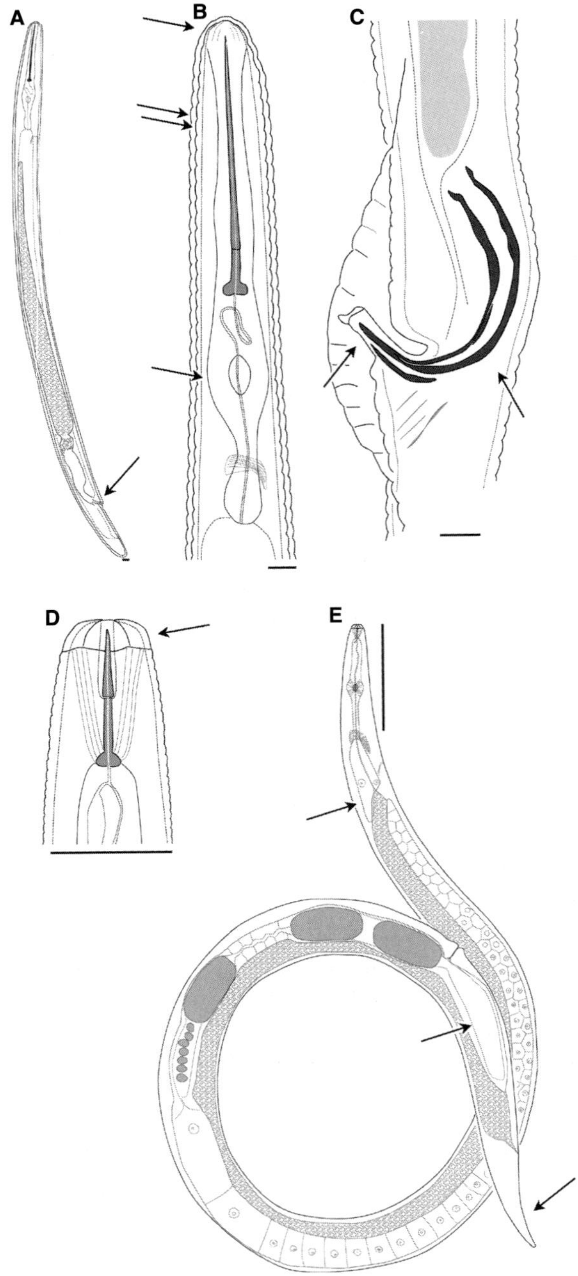

Fig. 21.5 *Hemicyclophora arenaria* (A, B); *H. typica* (C); and *Ditylenchus destructor* (D, E). Female head region (D); female neck region (B); male tail (C); and female habitus (A, E). *H. arenaria* based on Franklin and Stone (1974); *H. typica* based on Siddiqi (2000); and *D. destructor* based on Hooper (1972). Scale bars: A and E = 100 μm; B, C and D = 10 μm.

thickenings. Basal glandular bulb offset from intestine or forming lobe of variable length overlapping the intestine. Overlap usually short and nuclei of glands anterior to pharyngeal–intestinal junction. Ovary outstretched, with one or two rows of oocytes. Columned uterus with four rows of four cells. Spermatheca axial, elongated, 16-celled. Post-vulval sac usually large. Bursa adanal to subterminal, never enclosing tail end. Tail usually elongate conoid, sometimes filiform.

Type species: *D. dipsaci* (Kühn, 1857) Filipjev, 1936. Siddiqi (2000) lists 111 species (and 14 *species inquirendae, nomina nuda* or *incertae sedis*). Species identification of this species-rich group is difficult. A polytomal computerized key is available (Viscardi and Brzeski, 1993).

Principally terrestrial fungal feeders and parasites of higher plants. Also known from periphyton and as parasites of submerged plants (Table 21.5).

Superfamily CRICONEMATOIDEA Taylor, 1936 (Fig. 21.5)

Marked sexual dimorphism: male slender, female sausage-shaped, cylindrical or spheroidal; males and some juveniles lack a stylet or have a degenerated one and cannot feed on roots. Cuticle either thin and fine annulated or thick and coarsely annulated, in latter case provided with retrose annules, scales, crenation, spines or an extra cuticular sheet. Lateral field absent or present. Deirids absent except in some juveniles of Tylenchulidae. Phasmids absent. Cephalic region smooth or usually with one to three coarse annules, framework hexaradiate with light to strong sclerotization. Female and juveniles mostly with well-developed stylet, often very long, with cone markedly longer than shaft (shaft always 8–10 μm long); large knobs may be characteristically anchor-shaped. Pharynx with enormously developed, muscular corpus; procorpus and metacorpus merged; isthmus either slender and offset from glandular bulb or short and broad, being amalgamated with glandular bulb; basal bulb small, nearly always offset from intestine. Female genital system monodelphic, prodelphic. Post-vulval sac absent. Spermatheca small, usually offset, inclined laterally or ventrally. Bursa weakly developed (except Hemicycliophoridae), rarely enveloping tail tip, absent in several groups. Females and most juveniles obligate root ectoparasites, rarely endoparasitic.

Family HEMICYCLIOPHORIDAE Skarbilovich, 1959 (Fig. 21.5)

Moderate to large-sized (0.6–2mm), vermiform, habitus straight to arcuate. Thick cuticle with round, coarse, non-retrose annuli, usually numbering over 200. Juveniles and females with extra cuticle, except in *Caloosia*. Typical lateral field lacking, but often irregularities in body annuli or various longitudinal markings are present. Female cephalic region with one to three annules, with a labial disc but usually lacking sub-median lobes. Stylet in juveniles usually over 50 μm, basal knobs rounded, posteriorly sloping. Procorpus amalgamated with median bulb, valve plates long, isthmus short, expanding into offset basal bulb. Tail usually elongate-conoid, rarely filiform, cilindroid or hemispherical. Prominent bursa rarely covers more than half of the tail. Spicules setaceous, long, usually strongly curved to become semicircular, U- or hook-shaped, but may be arcuate or straight. Cloacal lips usually elongated to form a penial tube.

Genus *Hemicycliophora* de Man, 1921
= *Procriconema* Micoletzky, 1925; *Colbranium* Andrássy, 1979; *Aulosphora* Siddiqi, 1980; *Loofia* Siddiqi, 1980
Cuticular layer always appearing double, generally loose, never membranous. Typically two cephalic annuli (exceptionally three), usually not modified or separate. Vulva a transverse slit over half of body diameter long. Vagina straight or curved but not sigmoid. Labial region of male marked by a discontinuity in body annulation, usually offset, labial framework in lateral view appearing as 'spectacles mark'. Spicules arcuate, semicircular, U- or hook-shaped. Lips of cloaca forming a penial tube bearing a single hypothygma at its tip. Bursa covering less than one-third of tail (diagnosis following Raski and Luc, 1987).

Type species: *H. typica* de Man, 1921. Siddiqi (2000) lists 138 valid species.

Recently described new species: *H. hellenica* Vovlas, 2000 and *H. dulli* Van den Berg & Tiedt, 2000. Keys to *Hemicycliophora* spp. were provided by Loof (1968) and Van den Berg (1987).

Species of *Hemicycliophora* are found mainly in meadows and freshwater habitats, also groundwater and lakes (Table 21.5).

Superfamily APHELENCHOIDEA Fuchs, 1937

Mycetophagous, phytoparasitic, predaceous or entomoparasitic nematodes. Body 0.2–2.5 mm, vermiform, rarely obese except in some insect parasites. Amphidial apertures oval, pore-like, dorsosublateral on labial region. Cephalic region low rounded, continuous or offset, weak or moderate sclerotization. Basal swelling or knobs usually weakly developed or entirely absent. Pharynx comprising a narrow, cylindrical procorpus; a strongly developed, offset, ovoid to rounded rectangular median bulb with crescentic valve plates; and well-developed pharyngeal glands forming a dorsally overlapping lobe (except *Paraphelenchus*). Dorsal gland orifice located within the median bulb. Genital tract monoprodelphic, usually outstretched. Vulva posterior at 60–98%, usually in the form of a transverse slit. Spermatheca axial, post-vulval uterine sac usually present. Sperm large and rounded. Anus a broad transverse slit with an overhanging anterior lip (can be degenerate). Bursa usually absent, typically three pairs of caudal papillae present (diagnosis modified after Hunt, 1993).

Aphelenchs are predominantly free-living and mycetophagous; also phytoparasitic, predaceous or entomoparasitic. Hunt (1993) is a comprehensive reference work for the Aphelenchoidea and provides also keys and taxa diagnosis to generic level.

Family APHELENCHOIDIDAE Scarbilovich, 1947 (Fig. 21.6)
Lateral field with four or fewer incisures (very exceptionally six). Cephalic region usually high and offset from body contour. Stylet often with basal knobs or swellings. Pharynx with rudimentary isthmus or isthmus absent. Female tail conoid to a pointed or narrowly rounded terminus, which may be mucronate or adorned. Spicules

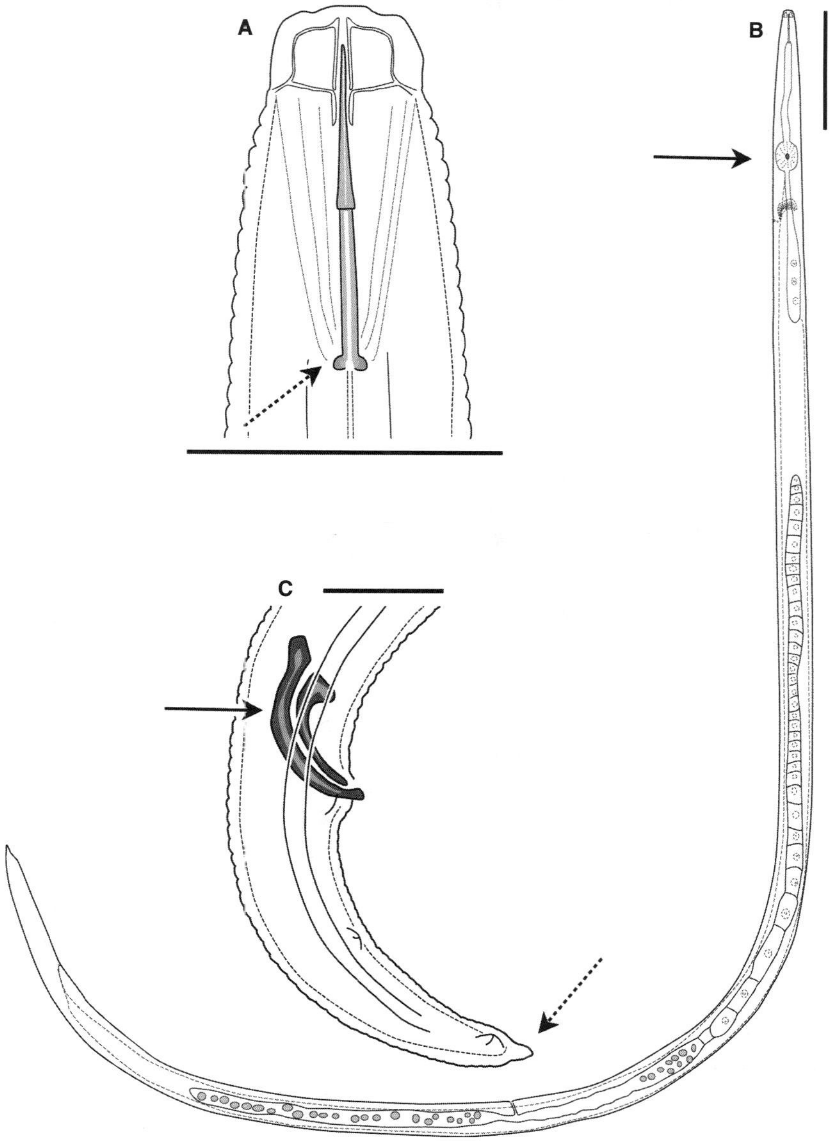

Fig. 21.6 *Aphelenchoides fragariae*. Female head region (A); female habitus (B); and male tail (C). Based on Siddiqi (1975). Scale bars: A and C = 10 µm; B = 50 µm.

rose-thorn-shaped or derived therefrom; usually with a prominent apex and rostrum. Gubernaculum absent or indistinct: if present it is a small structure, never elongated, linear or V-shaped in cross section (diagnosis modified after Hunt, 1993).

Genus *Aphelenchoides* Fischer, 1894
Habitus of females straight to ventrally arcuate. Habitus of males 'walking-stick'-shaped: tail region sharply curved ventrally. Cuticula finely annulated. Lateral field

often with four incisures, sometimes two or three. Cephalic region usually rounded and slightly offset, cephalic framework weak, six equal-sized lips. Stylet slender, usually with basal knobs or swellings, about 10–12 μm long. Vulva usually at 60–75% of the body length, only exceptionally more posterior. Tail conoid with a variable terminus, which may be bluntly or finely rounded, digitate or bifurcate or with a ventral projection. One or more mucrons of various shapes may be present. Spicules thorn-shaped. Typically three pairs of caudal papillae, one pair adanal, one pair subterminal and the other in between. Bursa absent (diagnosis modified after Hunt, 1993).

Type species: *A. kuehnii* Fischer, 1894. Hunt (1993) lists 139 species.

Recently described species: *A. involutus* Minagawa, 1992; *A. tsalolikhini* Ryss, 1993; *A. varicaudatus* Ibrahim & Hooper, 1994; *A. nechaleos* Hooper & Ibrahim, 1994; *A. paranechaleos* Hooper & Ibrahim, 1994; *A. resinosi* Kaisa, Harman & Harman, 1995; *A. robustus* Gagarin, 1996; *A. panaxofolia* Liu, Wu & Duan, 1999; *A. microstylus* Kaisa, 2000 and *A. ensete* Swart, Bogale & Tiedt, 2000. Sanwal (1961) and Shahina (1996) provide, respectively, a key and a compendium to the species of *Aphelenchoides*.

Several *Aphelenchoides* species are known from freshwater habitats (Table 21.6). *Aphelenchoides fragariae* lives in association with a wide diversity of submerged plants (Gerber and Smart, 1987; Table 21.2). Some *Aphelenchoides* spp. are accomplished swimmers (Hunt, 1993), which possibly facilitates terrestrial–aquatic transitions.

Infraorder CEPHALOBOMORPHA De Ley & Blaxter, 2002

Superfamily CEPHALOBOIDEA Fuchs, 1934 (only one superfamily)

Cuticle normally distinctly annulated. Lip region usually with complex processes. Stoma narrow but anterior part can be wide, unarmed, sometimes with small denticles. Stegostom comprises typically two-thirds or more of stoma length and its lining, often with distinct subdivisions. Pharynx tripartite; corpus cylindrical to fusiform without median bulb, isthmus thin, basal bulb rounded valvate. Female reproductive system monodelphic, prodelphic with offset spermatheca, ovary directed posteriad and having double flexure near germinal zone. Males without bursa, typically with two to three pairs of pre-cloacal papillae, five pairs of caudal papillae, phasmids and one minute mid-ventral pre-cloacal papilla. Life cycle without dauer stage, parasitism absent.

Family CEPHALOBIDAE Filipjev, 1931 (Fig. 21.7)

Cuticle distinctly annulated, annules at least 1 μm wide. Anterior end usually with triradiate symmetry, usually with more or less prominent cuticular processes or probolae. Amphids located dorsosublateraly on the lateral lips, minute slits or rarely circles. Cheilostom short and moderately wide, gymnostom and stegostom narrow, the latter longer than two-thirds of the total stoma length and lined by four

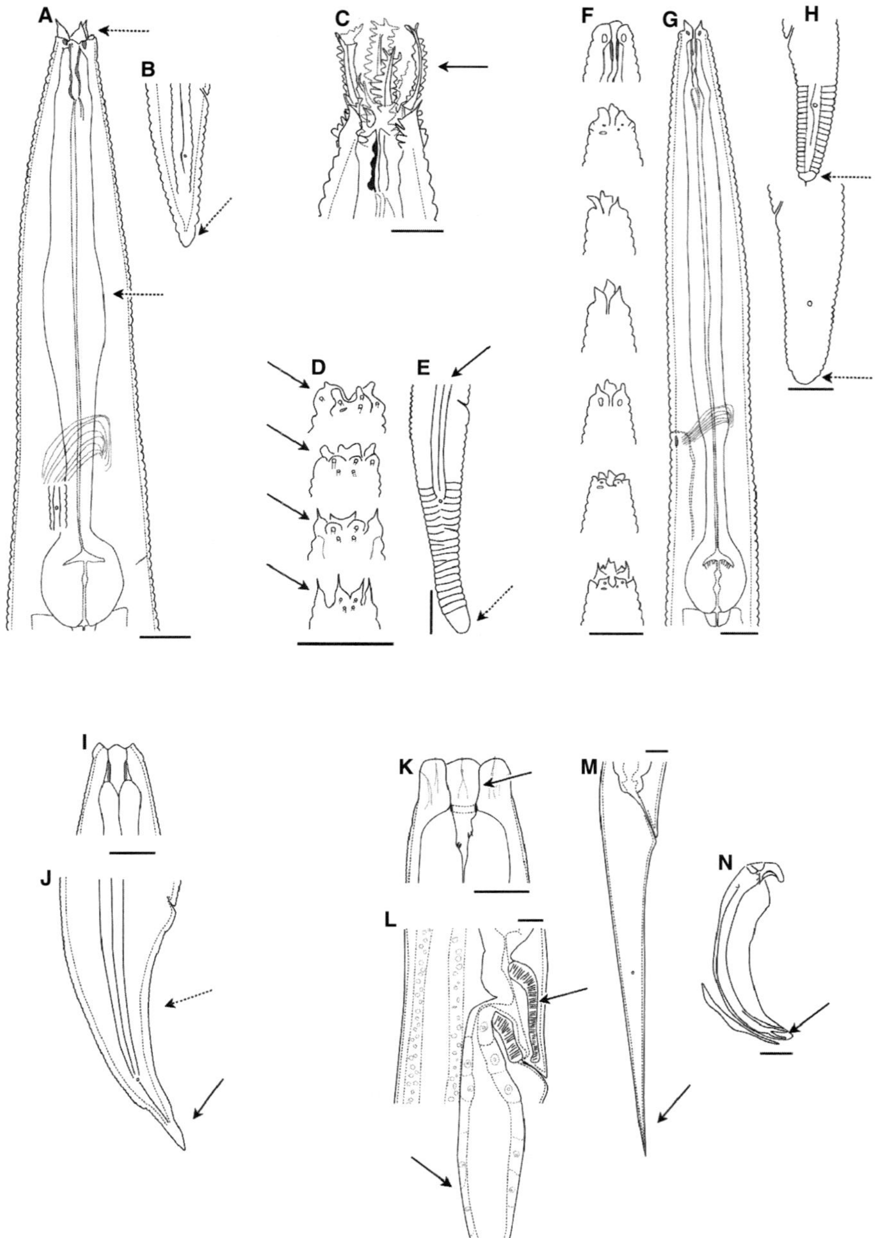

Fig. 21.7 Cephalobidae: *Acrobeloides elesmerensis* DWF 1110 (A, B); *Acrobeles zapatai* (C); *Cephalobus persegensis* (F–H), *Eucephalobus striatus* (D–E). Panagrolaimidae: *Pangrolaimus magnivulvatus* (I, J); *Panagrellus redivivus* (K, L, M, N). Female neck region (A, G); female tail (B, E, H, J, M); female head (C, I, K); variation of female head (D, F); vulval region (L); and spicule and gubernaculum (N). *C. persegensis* and *E. striatus* adapted from Anderson and Hooper (1970, 1971); *A. zapatai* modified from Mundo-Ocampo *et al.* (2003); *P. magnivulvatus* adapted from Boström (1995); and *P. redivivus* modified from Hechler (1971). Scale bars = 10 μm.

sets of muscle cells, often with serial cuticular thickenings of the stoma lining including one dorsal denticle in the metastegostom. Female tail short to long but never filiform. Rectum as long as anal body width. Male genital papillae typically showing following pattern: two or three ventrosublateral preanal pairs, one single midventral precloacal papilla, two caudal pairs near phasmid level and three caudal pairs near tail tip. Spicules with subapical dorsal fenestra, gubernaculum linear in lateral view.

Diagnosis of the family Cephalobidae, a key to genera, useful remarks on diagnostic features and an overview of taxonomic problems is provided by De Ley (1997).

The genera discussed below have been repeatedly reported from freshwater habitats (Table 21.1), but this list is not unequivocal in listing the 'freshwater genera' within the Cephalobomorpha.

Genus *Acrobeles* Linstow, 1877
= *Seleborca* Andrássy, 1985
Body small to large (0.3–1.1 mm). Cuticle 'single' or 'double', with large to very large annules, with or without longitudinal striae, punctuations and/or pores. Lateral field with two or three incisures. Amphids relatively distinct, circular. Labial probolae long, deeply bifurcated, each prong with at least seven tines. Cephalic probolae high, triangular, separate and fringed by numerous tines, demarcating cephalic axils that are equally deep. Primary axils with two indistinct guard processes. Cheilorhabdia large and spherical in cross-section. Pharyngeal corpus cylindrical to fusiform, not distinctly swollen posteriorly. Female tail convex-conoid, dorsally sigmoid-conoid or uniformly sigmoid-conoid, pointed. Diagnosis modified from Shahina & De Ley (1997).

Type species: *A. ciliatus* von Linstow, 1877.
Shahina & De Ley (1997) provide a tabular key of the 29 listed species and register also five species *inquirendae* or *incertae sedis*. Mundo-Ocampo *et al.* (2003) recently described *A. zapatai* Mundo-Ocampo, Baldwin, Dorado-Ramirez & Morales-Ruiz, 2003 and discussed the taxonomic limits of the genus *Acrobeles.*

Genus *Acrobeloides* Cobb, 1924
Body length highly variable (0.3–1.2 mm). Lateral field with three to five incisures. Lips varying from rounded to distinct lobe-like or apiculate plates. Labial probolae not furcate, of variable shape: low rounded, or conical rounded, or conical pointed. Pharyngeal corpus usually with elongate spindle-like swelling. Tail usually conoid-rounded. Males rare.

Abolafia & Peña-Santiago (2003) list 25 species and five *species inquirenda* and provide a key.

Genus *Cephalobus* Bastian, 1865
Body length between 0.4 and 0.9 mm. Lateral field with three to five incisures, reaching tail tip. Three to five lateral lines present. Labial probolae small, their shape exceptionally variable: low-rounded, appearing as three rounded lips, or conical, or bicornuate. Female tail subcylindrical or conoid, tip rounded with or without mucro. Male tail usually with mucro.

Type species: *C. persegnis* Bastian, 1865.
Andrássy (1984) provides a key to 16 species and lists 12 *species inquirendae*. Recently described species: *C. camberenensis* De Ley, Geraert & Coomans, 1990; *C. maximus* Lemzina & Gagarin, 1994; and *C. sacchari* Shahina & Tabassum, 2002.

Genus *Eucephalobus* Steiner, 1936
Body length between 0.4 and 0.9 mm. Lateral field with three incisures, fading out at or near the phasmid or reaching tail tip. Labial probolae low-bicornuate, appearing as six pointed lips; bifurcation of labial probolae can be more or less pronounced. Lips low-rounded without distinct axils or cephalic probolae. Female tail conoid-rounded, rounded or clavate; with or without mucro. Male tail usually with mucro.
Type species: *E. oxyuroides* (de Man, 1876) Steiner, 1936.
Andrássy (1984) provides a key to 11 species. Recently described species: *E. tribei* Swart & Heyns, 1997; and *E. panaxi* Mukhina, 1990.

Infraorder PANAGROLAIMORPHA De Ley & Blaxter, 2002

The assemblage of insect-parasitic Steinernematidae, vertebrate-parasitic Strongyloididae, amphibian–reptilian-parasitic Rhabdiasidae and the primarily free-living Panagrolaimidae is based on moderate molecular support (De Ley and Blaxter, 2002). Morphologically unifying characters are limited. Stoma unarmed or with small denticles and usually at least partially narrow. Pharynx typically comprising cylindrical corpus, valveless metacorpus and valvate basal bulb. Males usually without bursa.

Superfamily PANAGROLAIMOIDEA Thorne, 1937

Family PANAGROLAIMIDAE Thorne, 1937 (only one family) (Fig. 21.7)

Lip region without probolae or fimbriate processes. Labial sensilla papilliform, cephalic sensilla papilliform or rarely setiform. Cuticula finely annulated (annules usually < 1 μm wide). Stoma fairly wide anteriorly, stegostom usually tapering, comprising between half and two-thirds of stoma length, never with distinct subdivisions of its lining. Pharynx with cylindrical procorpus, metacorpus cylindrical or with valveless median bulb, terminal bulb valvate. Female reproductive system monodelphic, prodelphic. Spermatheca usually absent or axial, rarely offset as in Cephalobidae. Uterus–oviduct junction exceptionally with two offset spermatheca-like pouches. Post-vulval sac present or absent. Female tail conical with pointed tip. Males without bursa, with one unpaired papilla and five to seven pairs of genital papillae, arrangement variable. Spicules often with wide velum, gubernaculum often with posteriorly expanded outline. Diagnosis adapted from Bert *et al.* (2003).

Genus *Panagrolaimus* Fuchs, 1930
= *Neocephalobus* Steiner, 1929; *Cephalobus* (*Neocephalobus*) Steiner, 1929; *Procephalobus* Steiner, 1934; *Pseudorhabditis* Kreis, 1929 nec Perroncito, 1880; *Asymmetricus* Krejs,

1930; *Panagrodontus* Thorne, 1937; *Panagrolaimoides* Ivanova, 1958; *Panagromacra* Massey, 1964; *Eminensia* Mahajan, 1979.
Body length 0.3–1.9 mm. Cuticle finely annulated or almost smooth. Six lips with small papillae, amalgamated or more or less separate. Lining of cheilostom usually with distinct cheilorhabdia, gymnostom clearly longer than wide, sclerotized (often divided into two equal parts), stegostom regularly with a denticle on dorsal sector and sometimes denticles on sub-ventral sectors. Median bulb absent. Female reproductive system reflexed past the vulva, sometimes almost as far as the anus. Vulval lips can be markedly swollen. Post-vulval sac about a body-width long. Female tail conical, often with a characteristically sigmoid dorsal outline. Spicules relatively short and broad, their shaft with single or barely bifid tip.

Type species: *P. detritophagus* Fuchs, 1930. Andrássy (1984) provides a key to 35 *Panagrolaimus* species. Morphological species recognition within the genus *Panagrolaimus* is considered extremely difficult (Boström, 1995) and morphological data can not always discern biological species (Eyualem-Abebe and Blaxter, 2003).

Recently described species: *P. magnivulvatus* Böstrom, 1995
Panagrolaimus spp. are extremely widespread, occurring in many terrestrial as well as aquatic habitats. *P. hygrophilus* Bassen, 1940 and *P. rigidus* (Schneider, 1866) Thorne, 1937 are the most common species reported from freshwater (Table 21.1).

Genus *Panagrellus* Thorne, 1938
= *Anguillula* de Man, 1910 *pro parte*; *Neocephalobus* Steiner, 1936 *pro parte*; *Turbator* Goodey, 1943; *Tylorhabdus* Sukul, 1971.
Body length variable (0.5–3 mm). Cuticle finely annulated, annulations rarely visible with light microscope. Head continuous or only slightly offset from body. Six lips with small papillae, amalgamated or more or less separate. Cheilo- and gymnostoma combined about as wide as long. Lining of cheilostom not sclerotized. Stegostom funnel-shaped, half as long as stoma, usually with multiple small teeth or denticles. Corpus long, 55 to 65% of pharyngeal length. Median bulb absent. Vagina with muscular walls. Postvulval sac long. Mature females usually with many embryonated eggs. Spicules relatively long, their shaft with deeply bifid tip resembling a tuning fork.

Type species: *P. pycnus* Thorne, 1938. Andrássy (1984) provides a key to 12 *Panagrellus* species.

Panagrellus is reported from several freshwater habitats (Table 21.1); *P. redivivus* is often associated with polluted freshwater and is used in aquatic toxicity bioassays (Sherry *et al.*, 1997) and as fish food in aquaculture (Santiago *et al.*, 2004).

References

Abolatia, J. and Peña-Santiago, R. (2003) Nematodes of the order Rhabditina from Andalucia, Oriental, Spain. The genus *Acrobeloides* (Cobb, 1924) Thorn, 1937 with description of *A. arenicola* sp. n. and a key to its species. *Journal of Nematode Morphology and Systematics* 5, 107–130.

Anderson, R.V. and Hooper, D.J. (1970) A neotype of *Cephalobus persegnis* Bastian, 1865, redescription of the species, and observations on variability in taxonomic characters. *Canadian Journal of Zoology* 48, 457–469.

Anderson, R.V. and Hooper, D.J. (1971) A neotype for *Eucephalobus striatus* (Bastian, 1865) Thorne, 1937 (Nematoda) and redescription of the species from topotypes and their progeny. *Canadian Journal of Zoology* 49, 451–459.

Andrássy, I. (1978) Nematoda. In: Illies, J. (ed.) *Limnofauna Europea.* Gustav Fischer Verlag, Amsterdam, pp. 98–117.

Andrássy, I. (1979) The genera and species of the family Tylenchidae Örley, 1880 (Nematoda). The genus *Tylenchus* Bastian, 1865. *Acta Zoologica Academiae Scientiarum Hungaricae* 25, 1–33.

Andrássy, I. (1980) The genera and species of the family Tylenchidae Örley, 1880 (Nematoda). The genera *Aglenchus* (Andrássy, 1954) Meyl, 1961, *Miculenchus* Andrássy, 1959 and *Polenchus* gen. n. *Acta Zoologica Academiae Scientiarum Hungaricae* 26, 1–20.

Andrássy, I. (1982) The genera and species of the family Tylenchidae Orley, 1880 (Nematoda). The genus *Coslenchus* Siddiqi, 1978. *Acta Zoologica Academiae Scientiarum Hungaricae* 28, 193–232.

Andrássy, I. (1984) *Klasse Nematoda*, vol. 9. Bestimmungsbücher zur Bodenfauna Europas, Akademie-Verlag, Berlin.

Beier, S. and Traunspurger, W. (2003) Seasonal distribution of free-living nematodes in the stream Krähenbach: a fine-grained submountain carbonate stream in Southwest Germany. *Nematology* 5, 113–136.

Bert, W., Tandingan De Ley, I., Van Driessche, R., Segers, H. and De Ley, P. (2003) *Baujardia mirabilis* gen. n. sp. n. from pitcher plants and its phylogenetic position within Panagrolaimidae (Nematoda: Rhabditida). *Nematology* 5, 405–420.

Bird, A.F. (1999) Observations of some nematodes from Kangaroo Island, South Australia, including the description of a new species, *Hemicycliophora fluvialis* (Tylenchida: Hemicycliophoridae), from Rocky River. *Transactions of the Royal Society of South Australia* 123, 121.

Boström, S. (1995) Populations of *Plectus acuminatus* Bastian, 1865 and *Panagrolaimus magnivulvatus* n. sp. (Nematoda) from nunatakks in Dronning Maud Land, East Antarctica. *Fundamental and Applied Nematology* 18, 25–34.

Brzeski, M.W. (1989) Notes on the genus *Psilenchus* de Man, 1921, with description of *P. klingleri* sp. n. (Nematoda: Tylenchidae). *Annales Zoologici* 43, 51–69.

Brzeski, M.W. (1996) On the genus *Eutylenchus* Cobb, 1913 (Nematoda: Tylenchidae). *Nematologica* 42, 1–8.

Brzeski, M.W. (1998) *Nematodes of Tylenchina in Poland and Temperate Europe.* Muzeum i Instytut Zoologii Polska Akademia Nauk, Warszaw.

Brzeski, M.W. and Dolinski, C.M. (1998) Compendium of the genus *Tylenchorhynchus* Cobb, 1913 *sensu lato* (Nematoda: Belonolaimidae). *Russian Journal of Nematology* 6, 189–199.

Café Filho, A.C. and Huang, C.S. (1989) Description of *Pratylenchus pseudofallax* n. sp. with a key to the species of the genus *Pratylenchus* Filipjev, 1936 (Nematoda: Pratylenchidae). *Revue de Nématologie* 12, 7–15.

De Ley, P. (1992) The nematode community of a marginal soil at Cambérène, Senegal, with special attention to functional morphology and niche partitioning in the family Cephalobidae. *Mededelingen van de Koninklijke Academie voor Wetenschappen, Letteren en Schone Kunsten van België – Klasse der Wetenschappen* 53, 107–153.

De Ley, P. (1997) The current state of affairs in identification and diagnosis of the genera of the family Cephalobidae (Nematoda: Rhabditida). *Mededelingen Faculteit Landbouwkuadige en Toegepaste Biologische Wetenschappen Universiteit Gent* 62, 657–673.

De Ley, P. and Blaxter, M.L. (2002) Systematic position and phylogeny. In: Lee, D.L. (ed.) *The Biology of Nematodes.* Taylor & Francis, London, pp. 1–30.

Doucet, M.E. (1985) Description de *Dolichodorus aquaticus* n. sp. (Nematoda: Dolichodoridae). *Nematologica* 31, 143–150.

Doucet, M.E. (1996) New data on *Psilenchus hilarus* Siddiqi, 1963 and description of two new species of *Psilenchus* de Man, 1921 (Nematoda: Tylenchida) from Argentina. *Fundamental and Applied Nematology* 19, 449–461.

Ebsary, B.A. and Anderson, R.V. (1982) Two new species of *Hirschmanniella* Luc & Goodey, 1963 (Nematoda: Pratylenchidae) with a key to the nominal species. *Canadian Journal of Zoology* 60, 530–534.

Esser, R.P. and Buckingham, G.R. (1987) Genera and species of free-living nematodes occupying freshwater habitats in North America. In: Veech, J.A. and Dickson, D.W. (eds.) *Vistas on Nematology: a Commemoration of the Twenty-fifth Anniversary of the Society of Nematologists.* Society of Nematologists, Hyattsville, Maryland, pp. 477–487.

Eyualem-Abebe and Blaxter, M. (2003) Comparison of biological, molecular and morphological methods of species identification in a set of cultured *Panagrolaimus* isolates. *Journal of Nematology* 35, 119–128.

Faulkner, L.R. and Bolander, W.J. (1966) Occurrence of large nematode populations in irrigation canals of South Central Washington. *Nematologica* 12, 591–600.

Faulkner, L.R. and Bolander, W.J. (1970) Agriculturally polluted irrigation water as a source of plant-parasitic nematodes infestations. *Journal of Nematology* 2, 368–374.

Firoza, K. and Maqbool, M.A. (1994) A diagnostic compendium of the genus *Helicotylenchus* Steiner, 1945 (Nematoda: Hoplolaimidae). *Pakistan Journal of Nematology* 12, 11–50.

Fortuner, R. (1977) Fertilisation du riz et dégats causés par *Hirschmanniella oryzae* (Van Breda de Haan) Luc & Goodey. *Comptes Rendus des Séances de l'Académie d'Agriculture de France* 63, 624–630.

Fortuner, R. (1987) A reappraisal of Tylenchina (Nemata). 8. The family Hoplolaimidae Filipjev, 1934. *Revue de Nématologie* 10, 219–232.

Fortuner, R. and Luc, M. (1987) A reappraisal of Tylenchina (Nemata) 6. The family Belonolaimidae Whitehead, 1960. *Revue de Nématologie* 10, 183–202.

Fortuner, R. and Merny, G. (1979) Root-parasitic nematodes of rice. *Revue de Nématologie* 2, 79–102.

Fortuner, R., Maggenti, A.R. and Loof, P.A.A. (1988) Comment on the proposed designation of *Tylenchus vulgaris* Brzeski, 1963 as type species of *Filenchus* Andrássy, 1954 (Nematoda) Case 25. *Bulletin of Zoological Nomenclature* 45, 23–24.

Fotedar, D.N. and Kaul, V. (1985) A revised key to the species of the genus *Helicotylenchus* Steiner, 1945 (Nematoda: Rotylenchoidinae). *Indian Journal of Nematology* 15, 138–147.

Franklin, M.T. and Stone, A.R. (1974) *Hemicycliophora arenaria. C.I.H. Descriptions of Plant-parasitic Nematodes,* Set 3, No. 43. Commonwealth Institute of Parasitology. CAB International, Wallingford, UK, 3 pp.

Frederick, J.J. and Tarjan, A.C. (1989) A compendium of the genus *Pratylenchus* Filipjev, 1936 (Nemata: Pratylenchidae). *Revue de Nématologie* 12, 243–256.

Gagarin V.G. (2001) *Free-living Nematodes of the Freshwaters of Russia and Adjacent Countries: Fauna and the Pathways of its Formation, Ecology, Taxonomy, Phylogeny.* Pensoft publishers, Moscow.

Gagarin V.G. (2003) Some new and poorly known species of Tylenchidae and Monhysteridae from Siberia (Nematoda). *Zoosystematica-Rossica* 12, 1–6.

Geraert, E. (1966) The systematic position of the families Tylenchulidae and Criconematidae. *Nematologica* 12, 362–368.

Geraert, E. and Raski, D.J. (1986) Unusual *Malenchus* species (Nematoda: Tylenchidae). *Nematologica* 32, 27–55.

Geraert, E. and Raski, D.J. (1987) A reappraisal of Tylenchina (Nemata). 3. The family Tylenchidae Orley, 1880. *Revue de Nématologie* 10, 143–161.

Geraert, E. and Raski, D.J. (1988) Study of some *Aglenchus* and *Coslenchus* species (Nemata: Tylenchida). *Nematologica* 34, 6–46.

Gerber, K. and Smart, G.C. (1987) Plant-parasitic nematodes associated with aquatic vascular plants. In: Veech, J.A. and Dickson, D.W. (eds.) *Vistas on Nematology: a Commemoration of the Twenty-fifth Anniversary of the Society of Nématologists.* Society of Nematologists, Hyattsville, Maryland, pp. 488–501.

Handoo, Z.A. (2000) A key and diagnostic compendium to the species of the genus *Tylenchorhynchus* Cobb, 1913 (Nematoda: Belonolaimidae). *Journal of Nematology* 32, 20–34.

Handoo, Z.A. and Golden, A.M. (1989) A key and diagnostic compendium to the genus *Pratylenchus* Filipjev, 1936 (lesion nematodes). *Journal of Nematology* 21, 202–218.

Handoo, Z.A. and Golden, A.M. (1992) A key and diagnostic compendium to the species of the genus *Hoplolaimus* Daday, 1905 (Nematoda: Hoplolaimidae). *Journal of Nematology* 15, 45–53.

Hechler, H.C. (1971) Taxonomic notes on four species of *Panagrellus* Thorne (Nematoda: Cephalobidae). *Journal of Nematology* 3, 227–237.

Heyns, J. (2002) Checklist of free-living nematodes recorded from freshwater habitats in Southern Africa. *Water SA* 28, 449–456.

Hooper, D.J. (1972) *Ditylenchus dipsaci. C.I.H. Descriptions of Plant-parasitic Nematodes*, Set 1, No. 14. Commonwealth Institute of Parasitology, CAB International, Wallingford, UK, 3 pp.

Hunt, D.J. (1993) *Aphelenchida, Longidoridae, and Trichodoridae: Their Systematics and Bionomics*. CAB International, Wallingford, UK.

Kheiri, A. (1970) Two new species in the family Tylenchidae (Nematoda) from Iran, with a key to *Psilenchus* De Man, 1921. *Nematologica* 16, 359–368.

La Massése, C.S. and Germani, G. (2000) Description de quatre nouvelles espèces et de quatre populations de *Rotylenchus* (Nematoda: Hoplolaimidae). Proposition d'une clé tabulaire. *Nematology* 2, 699–718.

Lewis, S.A. and Golden, A.M. (1981) Description and SEM observation of *Dolichodorus marylandicus* n. sp. with a key to species of *Dolichodorus*. *Journal of Nematology* 13, 128–134.

Loof, P.A.A. (1968) Taxonomy of *Hemicycliophora* species from West and Central Europe (Nematoda: Criconematoidea). *Mededelingen Landbouwhogeschool Wageningen Nederland* 68, 12–15.

Loof, P.A.A. (1991) The family Pratylenchidae Thorn, 1949. In: Nickle, W.R. (ed.) *Manual of Agricultural Nematology*. Marcel Dekker, New York, pp. 363–421.

Loof, P.A.A. (2001) *Nematoda, Secernentea* (*Tylenchida, Aphelenchida*) *Süswasserfauna von Mitteleuropa*, 4/1–1. Spektrum Akademischer Verlag, Heidelberg, Berlin.

Luc, M., Maggenti, A.R., Fortuner, R., Raski, D.J. and Geraert, E. (1987) A reappraisal of Tylenchina (Nemata). 1. For a new approach to the taxonomy of Tylenchina. *Revue de Nématologie* 10, 127–134.

Maggenti, A.R., Luc, M., Raski, D.J., Fortuner, R. and Geraert, E. (1987) A reappraisal of Tylenchina (Nemata). 2. Classification of the suborder Tylenchina (Nemata: Diplogasteria). *Revue de Nématologie* 10, 135–142.

Maggenti, A.R., Luc, M., Raski, D.J., Fortuner, R. and Geraert, E. (1988) A reappraisal of Tylenchina (Nemata). 11. List of generic and supra-generic taxa, with their junior synonyms. *Revue de Nématologie* 11, 177–188.

Maqbool, M.A. and Shahina, F. (1987) Description of two new, and data on three, known species of the genus *Tylenchus* Bastian, 1985 (Nematoda: Tylenchidae) from Pakistan with a key to the species of *Tylenchus*. *Pakistan Journal of Nematology* 5, 53–67.

Messiaen, M. (2004) *Bijdrage tot de ecologie van zoetwaternematoden van kleine, stilstaande wateren in Vlaanderen*. MSc thesis, Ghent University, Ghent, Belgium.

Micoletzky, H. (1914) *Die freilebenden süsswasser-nematoden der Ost-Alpen mit besonderer Berücksichtigung des Lunzer Seengebietes*. Mitteiling aus der Biologischen Station Lunz und aus dem Zoologischen Institute der Universtät in Czernowitz, Vienna.

Micoletzky, H. (1925) Die freilebenden süsswasser- und moornematoden Dänemarks. Nebst anhang: über Amöbosporidien und andere parasiten bei freilebenden nematoden. *Det kongelige Danske videnskabernes selskab Biologiske skrifter* 8, 57–310.

Mol, A.W.M. (1984) Limnofauna Neerlandica, een lijst van meercellige ongewervelde dieren aangetroffen in binnenwateren van Nederland. *Nieuwsbrief European Invertebrate Survey Nederland* 15, 1–124.

Mundo-Ocanpo, M., Baldwin, J.G., Dorado-Ramirez, O. and Morales-Ruiz, M.C. (2003) *Acrobeloides zapata*. n. sp. (Rhabditida: Cephalobidae) from Biosphere Reserve 'Sierra de Huautla' (Mexico), with a discussion of the taxonomic limits of the genus. *Journal of Nematode Morphology and Systematics* 5, 21–33.

Ocaña, A. and Picazo, J.S. (1991) Study on nematode species encountered in the Monachil River (Granada, Spain): response to organic pollution. *Verhandlungen der Internationalen Vereinigung für Theoretische und Angewandte Limnologie* 24, 2729–2737.

Orton Williams, K.J. (1986) *Dolichodorus heterocephalus. C.I.H. Descriptions of Plant-parasitic Nematodes*, Set 4, No. 56. Commonwealth Institute of Parasitology, CAB International, Wallingford, UK, 3 pp.

Pennak, R.W. 1978 *Freshwater Invertebrates of the United States*, 2nd edn. John Wiley & Sons, New York.

Powers, T.O., Baldwin, J.G. and Bell, A.H. (1983) Taxonomic limits of the genus *Nagelus* (Thorne & Malek, 1968) Siddiqi, 1979 with a description of *Nagelus borealis* n. sp. from Alaska. *Journal of Nematology* 15, 582–593.

Raski, D.J. and Geraert, E. (1986) New species of *Lelenchus* Andrássy, 1954 and *Ecphyadophora* de Man, 1921 (Nemata: Tylenchidae) from southern Chile. *Nematologica* 31, 244–265.

Raski, D.J. and Geraert, E. (1987) Review of the genus *Filenchus* Andrássy, 1954 and descriptions of six new species (Nemata: Tylenchidae). *Nematologica* 32, 265–311.

Raski, D.J. and Luc, M. (1987) A reappraisal of Tylenchina (Nemata). 10. The superfamily Criconematoidea Taylor, 1936. *Revue de Nématologie* 10, 409–444.

Santiago, C.B., Ricciand, M. and Reyes-Lampa, A. (2004) Effect of nematode *Panagrellus redivivus* density on growth, survival, feed consumption and carcass composition of bighead carp *Aristichthys nobilis* (Richardson) larvae. *Journal of Applied Ichthyology* 20, 22.

Sanwal, K.C. (1961) A key to the species of the nematode genus *Aphelenchoides* Fischer, 1894. *Canadian Journal of Zoology* 39, 143–148.

Shahina, F. (1996) A diagnostic compendium of the genus *Aphelenchoides* Fischer, 1894 (Nematoda: Aphelenchida) with some new records of the group from Pakistan. *Pakistan Journal of Nematology* 14, 1–32.

Shahina, F. and De Ley, P. (1997) Two new species of Cephalobidae from Valle de la Luna, Argentina and observations on the genera *Acrobeles* and *Nothacrobeles* (Nematoda: Rhabditida). *Fundamental and Applied Nematology* 20, 329–347.

Sher, S.A. (1961) Revision of the Hoplolaiminae (Nematoda). I. Classification of nominal genera and nominal species. *Nematologica* 6, 155–169.

Sher, S.A. (1968) Revision of the genus *Hirschmanniella* Luc & Goodey, 1962 (Nematoda: Tylenchoidea). *Nematologica* 14, 243–257.

Sher, S.A., Corbett, D.C.M. and Colbran, R.C. (1966) Revision of the family Atylenchidae Skarbilovich, 1959 (Nematoda: Tylenchoidea). *Proceedings of the Helminthological Society of Washington* 33, 60–66.

Sherry, J. Scott, B. and Dutka, B. (1997) Use of various acute, sublethal and early life-stage tests to evaluate the toxicity of refinery effluents. *Environmental Toxicology and Chemistry* 16, 2249–2257.

Siddiqi, M.R. (1975) *Aphelenchoides fragariae. C.I.H. Descriptions of Plant-parasitic Nematodes*, Set 5, No. 74. Commonwealth Institute of Parasitology, CAB International, Wallingford, UK, 4 pp.

Siddiqi, M.R. (1980) The origin and phylogeny of the nematode orders Tylenchida Thorne,1949 and Aphelenchida n. ord. *Helminthologica Abstracts Series B* 49, 143–170.

Siddiqi, M.R. (2000) *Tylenchida, Parasites of Plants and Insects*, 2nd edn. CAB International, Wallingford, UK.

Siddiqi, M.R. and Hunt, D.J. (1988) Comment on the proposed designation of *Tylenchus vulgaris* Brzeski, 1963 as type species of *Filenchus* Andrássy, 1954 (Nematoda). *Bulletin of Zoological Nomenclature* 45, 55.

Sivakumar, C.V. and Khan, E. (1982) Description of *Hirschmanniella kaverni* sp. n. (Radopholidae: Nematoda) with a key for identification of *Hirschmanniella* spp. *Indian Journal of Nematology* 12, 86–90.

Smart, G.C. and Khoung, N.B. (1985) *Dolichodorus miradvulvus* n. sp. (Nematoda: Tylenchida) with a key to species. *Journal of Nematology* 17, 29–37.

Smart, G.C. and Nguyen, K.B. (1991) Sting and awl nematodes. In Nickle, W.R. (ed.) *Manual of Agricultural Nematology*. Marcel Dekker, New York, pp. 627–668.

Tarjan, A.C., Esser, R.P. and Chang, S.L. (1977) An illustrated key to nematodes found in freshwater. *Journal of the Water Pollution Control Federation* 49, 2318–2337.

Van den Berg, E. (1987) *Hemicycliophora* species from the western Cape Province with a key to the South African species (Hemicycliophoridae: Nematoda). *Phytophylactica* 19, 303–307.

Viscardi, T. and Brzeski, M.W. (1993) DITYL: computerized key for species identification of *Ditylenchus* (Nematoda: Anguinidae). *Fundamental and Applied Nematology* 16, 389–392.

Wanless, F.R. and Hunter, R. (2001) Freshwater nematodes from Loch Ness, Scotland, Part 1. The orders Tylenchida Thorne, 1949 and Rhabditida Chitwood, 1933 (Nematoda, Secernentea). *Bulletin of The Natural History Museum* 67, 1–24.

Zullini, A. (1982) *Guide per il reconoscimento delle specie animali delle acque interne italiane 17. Nematodi.* Consiglio Nazionale delle Ricerche, Verona, Italy.

Zullini, A. (1988) The ecology of the Lambro River. *Rivista di Idrobiologia* 27, 39–58.

Zullini, A. and Ricci, C. (1980) Bdelloid rotifers and nematodes in a small Italian stream. *Freshwater Biology* 10, 67–72.

22 Order Rhabditida: Suborder Rhabditina

J. Abolafia

Departamento de Biología Animal, Biología Vegetal y Ecología, Universidad de Jaén, Campus 'Las Lagunillas' s/n. 23071-Jaén, Spain

Introduction

The suborder Rhabditina *sensu* De Ley & Blaxter (2002) is an extensive group made up of species that occupy very diverse habitats. Although they are terrestrial in their habitat preference, some members have been recorded from freshwater or semiaquatic habitats. They are able to live and reproduce in these sites, where they occupy a variety of niches in the ecosystem (Timm, 1971; Loof, 1973; Gagarin, 2000; Traunspurger, 2000; Wanless and Hunter, 2001; Heyns, 2002a) and play an important role in nutrient cycling, being very useful as biological monitors of the environmental health of lakes and rivers. However, this group does not dominate the composition of species in aquatic habitats (see, e.g. Eyualem-Abebe *et al.*, 2001; Beier and Traunspurger, 2003b).

Unfortunately, or luckily for Rhabditina species, these habitats have frequent but slight inputs of anthropogenic pollution most probably due to agricultural, industrial and municipal activities and indirect discharges in the vicinity of these systems. These inputs add organic or inorganic compounds and change water physicochemical parameters (Adam *et al.*, 2001; Honnen *et al.*, 2001), favouring the development of bacteria and unicellular eukaryotes, which are an important food source for Rhabditina species (Beier and Traunspurger, 2001).

Of the different families of this suborder most of them belong to infraorder Diplogasteromorpha De Ley & Blaxter, 2002 (Diplogastroididae Filipjev & Schuurmans Stekhoven, 1941; Diplogastridae Micoletzky, 1922; Neodiplogastridae Paramonov, 1952 (Andrássy, 1984)) and infraorder Rhabditomorpha De Ley & Blaxter, 2002 (Rhabditidae Oerley, 1880) (Table 22.1). To obtain a better understanding of the functional significance of every aquatic Rhabditina, basic knowledge of populations, of habitat associations and of trophic dominance is required (Anderson, 1992).

Feeding types

Rhabditina nematodes represent different feeding types (Traunspurger, 1997). Predaceous aquatic species feed on other nematodes and small invertebrates

Table 22.1 List of genera cited in freshwater habitats.

Taxa	Feeding type	References
SUBORDER RHABDITINA Chitwood, 1933		
INFRAORDER BUNONEMATOMORPHA De Ley & Blaxter, 2002		
Family BUNONEMATIDAE Micoletzky, 1922		
Craspedonema Richters, 1908	d	Micoletzky (1922)
Rhodolaimus Fuchs, 1930	d	Zeidan and Geraert (1989)
INFRAORDER RHABDITOMORPHA De Ley & Blaxter, 2002		
Family DAUBAYLIIDAE Chitwood & Chitwood, 1934		
Daubaylia Chitwood & Chitwood, 1934	z	Anderson and Bartlett (1993)
Family MESORHABDITIDAE Andrássy, 1976		
Mesorhabditis Osche, 1952 (Dougherty, 1953)	d	Dassonville and Heyns (1984), Meyl (1953, 1954), Zeidan and Geraert (1989)
Family PELODERIDAE Andrássy, 1976		
Pelodera Schneider, 1866	d	Kozlowska (1962), Picazo *et al.* (1989, 1990), Santos-Lobatón *et al.*(2002), Volterra *et al.* (1999), Zullini (1982)
Family RHABDITIDAE Oerley, 1880		
Caenorhabditis (Osche, 1952) Dougherty, 1953	d	Volterra *et al.* (1999), Zullini (1982)
Cephaloboides (Rahm, 1928) Massey, 1974	d	Meyl (1953, 1954), Ocaña and Picazo (1991), Picazo *et al.* (1989, 1990), Volterra *et al.* (1999), Zullini (1982)
Cruznema Artigas, 1927	d	Ocaña and Picazo (1991), Picazo *et al.* (1989, 1990), Santos-Lobatón *et al.* (2002), Volterra *et al.* (1999), Zullini (1982)
Diploscapter Cobb, 1913	d	Eyualem-Abebe and Coomans (1997), Eyualem-Abebe *et al.* (2001), Beier and Traunspurger (2001, 2003a,b), Dassonville (1981), Meyl (1953), Ocaña and Picazo (1991), Picazo *et al.* (1989, 1990), Santos-Lobatón *et al.* (2002), Schneider (1937), Tahseen *et al.* (2002), Volterra *et al.* (1999), Zullini (1982)
Oscheius Andrássy, 1976	d	Kozlowska (1962), Volterra *et al.* (1999)

Continued

Table 22.1 *Continued.* List of genera cited in freshwater habitats.

Taxa	Feeding type	References
Pellioditis (Dougherty, 1953) Timm, 1960	d	Santos-Lobatón *et al.* (2002), Volterra *et al.* (1999), Zullini (1982)
Poikilolaimus Fuchs, 1930	d	Beier and Traunspurger (2003a,b), Dassonville (1981), Kozlowska (1962), Ocaña and Picazo (1991), Picazo *et al.* (1989, 1990)
Prodontorhabditis Timm, 1961	d	Schneider (1937)
Protorhabditis (Osche, 1952) Dougherty, 1953	d	Gadea (1963), Santos-Lobatón *et al.* (2002), Volterra *et al.* (1999), Zullini (1982)
Rhabditella (Cobb, 1929) Chitwood, 1933	d	Meyl (1954), Stefanski (1916, 1926), Volterra *et al.* (1999), Zullini (1981, 1982)
Rhabditis Dujardin, 1845	d	Beier and Traunspurger (2003b), Cobb (1914), Gadea (1963), Pott (1910), Santos-Lobatón *et al.* (2002) Vidakovic and Jakumetovic (1998)
Rhabditoides Goodey, 1929	d	Gagarin (2000), Volterra *et al.* (1999), Zullini (1982)
INFRAORDER DIPLOGASTEROMORPHA De Ley & Blaxter, 2002		
Family DIPLOGASTROIDIDAE Filipjev & Schuurmans Stekhoven, 1941		
Diplogasteroides de Man, 1912	n	De Cillis (1917)
Goffartia Hirschmann, 1952	d	Meyl (1953, 1954), Micoletzky (1916), Shoshin (1989), Volterra *et al.* (1999), Zullini (1982)
Rhabditidoides Rahm, 1928	d	Dassonville and Heyns (1984), Ocaña and Picazo (1991), Picazo *et al.* (1989, 1990), Volterra *et al.* (1999), Zullini (1982)
Family DIPLOGASTRIDAE Micoletzky, 1922		
Acrostichus Rahm, 1928	d	Dassonville and Heyns (1984), Gagarin (1977, 2001a,b), Ocaña and Picazo (1991), Picazo *et al.* (1989, 1990), Schuurmans Stekhoven (1943), Schuurmans Stekhoven and Teunissen (1938), Stefanski (1926), Volterra *et al.* (1999), Zullini (1982)
Butlerius Goodey, 1929	c	Khera (1969), Meyl (1953, 1954, 1957), Volterra *et al.* (1999), Zullini (1982)

Table 22.1 *Continued.* List of genera cited in freshwater habitats.

Taxa	Feeding type	References
Diplogaster Schultze in Carus, 1857	d	Beier and Traunspurger (2003a,b), Daday (1897), Gadea (1963), Kozlowska (1962), Tsalolichin (1980), Volterra *et al.* (1999), Zullini (1982)
Diplogastrellus Paramonov, 1952	d	Dassonville and Heyns (1984)
Paroigolaimella Paramonov, 1952	d	Beier and Traunspurger (2003b), Dassonville and Heyns (1984), Gagarin (1977, 2001b), Meyl (1953, 1954), Ocaña and Picazo (1991), Picazo *et al.* (1989, 1990), Santos-Lobatón *et al.* (2002), Steiner (1914), Volterra *et al.* (1999), Zullini (1982)
Family NEODIPLOGASTRIDAE Paramonov, 1952 (Andrássy, 1984)		
Fictor Paramonov, 1952	c	Beier and Traunspurger (2001, 2003b), Cobb (1914), Edmondson (1959), Gadea (1952), Kozlowska (1962), Picazo *et al.* (1990), Santos-Lobatón *et al.* (2002), Volterra *et al.* (1999), Zullini (1982)
Koerneria Meyl, 1960	c	Dassonville and Heyns (1984), Ocaña and Picazo (1991), Picazo *et al.* (1989, 1990), Volterra *et al.* (1999), Zullini (1982)
Mononchoides Rahm, 1928	c	Andrássy (1970), Daday (1897), Dassonville (1981), Dassonville and Heyns (1984), Meyl (1953, 1954), Picazo *et al.* (1989, 1990), Zullini (1981), Volterra *et al.* (1999), Zullini (1982)
Neodiplogaster Cobb, 1924	c	Santos-Lobatón *et al.* (2002)
Oigolaimella Paramonov, 1952	c	Meyl (1953, 1954)
Family TYLOPHARYNGIDAE Filipjev, 1934		
Tylopharynx de Man, 1876	f	Santos-Lobatón *et al.* (2002)

Feeding types: deposit-/detritus-feeders (d), chewers (c), fungal hypae-feeders (f), necromenic or cadaver-feeders (n), zooparasites (z).

(predators/omnivores), e.g. *Fictor* and *Mononchoides*; these feed on animal matter by ingestion through a broad stoma with teeth. Bacteria- or detritus-feeders (bacteriophages/detritophages/saprophages) are very frequent, e.g. *Rhabditis* and *Diplogaster*, and feed either on prokaryotic food resources using a tubular stoma, which is generally unarmed, or on organic substrate from which nutrients may be derived (Yeates, 1999). While ingestion of algae and diatoms has not been observed, some species of freshwater Rhabditina are characterized by bright yellow or green coloration associated with symbiotic algae in their guts. The dominance of deposit-feeders indicates that bacteria and unicellular eukaryotes should be an important food source in most running waters.

According to the Maturity Index, which is used as a measure of environmental disturbance (Bongers, 1990; Bongers *et al.*, 1991; Bongers and Bongers, 1998), nematode taxa are classified into five colonizer–persister (*c–p*) groups (from *c–p*1: colonizer species that need favourable conditions to *c–p*5: persister species that do not need favourable conditions), representing different life strategies and ecological requirements. Among Rhabditina, all free-living aquatic families belong to the *c–p*1 group, indicating the presence of food-enriched conditions.

They are enrichment opportunists (Bongers *et al.*, 1995), deposit-feeders and are assumed to feed mainly on bacteria, being the first to colonize organic deposits (Warwick, 1987; Bongers *et al.*, 1991). They also assimilate detrital compounds and, as in other deposit-feeding Rhabditina, may enhance the decomposition process of phytodetritus in aquatic systems by stimulating the activity of the microbial, especially the bacterial, communities (De Mesel *et al.*, 2003), although this was observed in marine systems.

Within infraorder Bunonematomorpha De Ley & Blaxter, 2002 there are few records; the genus *Craspedonema* Richters, 1908, belonging to family Bunonematidae Micoletzky, 1922, has semiaquatic species (Andrássy, 1984). Also, the genus *Bunonema* Jägerskiöld, 1905 may occur in freshwater habitats (Ward and Whipple, 1959; Volterra *et al.*, 1999). However, this genus is a typical bacterial-feeder from forest soils, moss and humus (Armendáriz and Hernández, 1992; Schouten *et al.*, 1998), and its presence in freshwaters could possibly be because of runoff after heavy rainfall (Coomans, 1992). Runoff can be a special problem in the study of freshwater Rhabditina and other aquatic nematodes. Terrestrial species end up in aquatic systems where they may survive for relatively long periods, but as they do not feed or reproduce here they cannot play any significant role in the ecosystem (Heyns, 1976, 2002a,b).

Species belonging to infraorder Rhabditomorpha De Ley & Blaxter, 2002 are also deposit-feeders of the *cp*1 group (Rhabditidae), having a small group that are parasitic. In this infraorder, the occurrence of members of the families Mesorhabditidae, Peloderidae and Rhabditidae[1] is indicative of organic enrichment (Zullini, 1976). The genus *Rhabditis* Dujardin, 1845 is frequently found in such environments. This genus has been reported where meiofauna nematode abundance and diversity were low (Vidakovic and Jakumetovic, 1998). Other species common in similar habitats

[1]Although, according to De Ley and Blaxter (2002, 2004), the genus *Diploscapter* is placed in a different family (Diploscapteridae); this genus is very close to *Protorhabditis*, this family being considered a junior synonym of family Rhabditidae (P. De Ley, personal communication).

are *Pelodera punctata* (Cobb, 1914) Dougherty, 1955, *Poikilolaimus oxycercus* (de Man, 1895) Sudhaus, 1980 and *Diploscapter coronatus* (Cobb, 1893) Cobb, 1913. These species are also widely reported from terrestrial habitats, and appear frequently in aquatic systems (Table 22.1).

The family Daubayliidae, represented by only one genus, *Daubaylia* Chitwood & Chitwood, 1934, is parasitic on freshwater snails, living in the lung cavity of species of family Planorbidae (Chitwood and Chitwood, 1934; Anderson and Bartlett, 1993).

The Diplogasteromorpha are deposit-feeders (Diplogastridae) and chewers as well as predators/omnivores (Diplogastridae and Neodiplogastridae). Widely recorded species are *Acrostichus nudicapitatus* (Steiner, 1914) Massey, 1962 and *Paroigolaimella bernensis* (Steiner, 1914) Andrássy, 1958. Other taxa that are reported from aquatic habitats but are characteristic of manure or compost (saprobiotic nematodes) include *Diplogastrellus* Paramonov, 1952 (Dassonville and Heyns, 1984; Volterra *et al.*, 1999). On the other hand, the genus *Diplogasteroides* de Man, 1912 was recorded from lake mud (De Cillis, 1917) and associated with beetle cadavers (necromany) (Manegold and Kiontke, 2001).

A representative predator species frequently found in aquatic habitats is *Fictor fictor* (Bastian, 1865) Paramonov, 1952. Other predator species, such as those belonging to genus *Butlerius* Goodey, 1929, are voracious predators on other nematodes and have been reported in superficial waters (see Volterra *et al.*, 1999). However, these nematodes typically inhabit debris or excrement produced by beetles (Taylor, 1964; Pillai and Taylor, 1968; Hunt, 1980; Andrássy, 1984).

In general, deposit-feeding Rhabditina in aquatic habitats crawl upon organic detritus looking for food and feed mainly on bacteria and unicellular eukaryotes, which are swallowed whole (Beier and Traunspurger, 2003c). They are capable of both taste and smell using an array of compounds to locate microhabitats in aquatic environments, but how these senses affect their capacity is not known (Höckelmann *et al.*, 2004). Here, bacterial biofilms may offer structure, shelter and food for nematodes and are known to produce a variety of odour compounds appearing to work through chemotaxis (Höckelmann *et al.*, 2004).

Interactions with Other Organisms

Members of the suborder of rhabditids share their habitats with other nematodes and other invertebrates (mainly microturbellarians, rotifers, oligochaetes, tardigrades and chironomids) with which they maintain different types of interactions. Most bacterial-feeding species may face competition, although they are not strongly affected by it (Michiels and Traunspurger, 2004). Also, predation by other nematodes and tardigrades takes place. Rhabditid nematodes are also susceptible to attack by species of the aquatic oomycete *Lagenidium* Schenk, 1857. Once the hyphae trap a nematode, they will invade the body cavity, resulting in death. Also, zoospores of this fungus encyst on nematode eggs. The zoospore is attracted to the egg by chemical secretions that leak from the egg. Zoospores encyst on the surface of the egg, penetrate the wall and then proliferate inside as they consume the contents (Kerry and Jaffee, 1997).

Indicators of pollution

Rhabditina and nematodes in general are affected by water quality, and their presence/absence in a particular habitat may be a bioindicator that a pollutant is present, or another toxic condition has developed. The degree of aquatic pollution affects the relative abundance of the different orders. In unpolluted sites species of the orders Chromadorida, Monhysterida, Enoplida and Dorylaimida dominate, whereas in polluted situations saprobic families of the suborder Rhabditina, such as Rhabditidae and Diplogastridae, become more abundant (Hirschmann, 1952; Zullini, 1976; Zullini and Ricci, 1980; Niemann *et al.*, 1996; Beier and Traunspurger, 2001). Rhabditina increase in number in polluted sites, and mainly Diplogastridae and Rhabditidae prefer some degree of pollution. Nevertheless, some species, such as the diplogastrid *Fictor fictor*, avoid polluted situations and prefer habitats that are rich in oxygen (Zullini, 1976; Niemann *et al.* 1996). An explanation for this exception might be that this species feeds more on diatoms than on bacteria (Bongers and Bongers, 1998) and that nematodes belonging to these families are only active when there is high microbial activity, forming dauer larvae when microbial activity decreases. Under food-rich conditions those nematodes which have a short generation time, mainly Diplogastridae and Rhabditidae, produce many small eggs resulting in an explosive population growth (Beier and Traunspurger, 2001, 2003c; Bongers and Bongers, 1998). As soon as food conditions become worse they form dauer larvae, waiting for better conditions. Due to this life strategy Diplogastridae and Rhabditidae are frequently found in eutrophic habitats (Hirschmann, 1952; Zullini, 1974, 1976, 1988; Ocaña and Picazo, 1991; Niemann *et al.*, 1996).

Although several authors (Zullini and Ricci, 1980; Niemann *et al.*, 1996; Beier and Traunspurger, 2001) regard the presence of the suborder Rhabditina as an indicator of polluted waters, these nematodes are not good indicators for monitoring pollution in small areas of shallow water because they are mainly terricolous and are easily transported into water from the surrounding land (Wu and Liang, 1999).

Also, these nematodes are influenced by the oxygen concentration in the water. They show significant positive correlations with biological oxygen demand (BOD_5) value (Ocaña and Picazo, 1991; Ocaña, 1992), i.e. BOD necessary to decompose organic material within 5 days (Gafny *et al.*, 2000), to an extent that the presence/absence of members of this suborder is used as an indicator of water quality (Bazzanti, 2000).

These variable conditions can modify the reproductive parameters of different species. It seems probable that certain saprozoic Rhabditina would present a still higher figure of egg production. The eggs, frequently, appear to represent rather distinct seasonal fluctuations in the population, so that at certain periods only larvae of a single stage are encountered (Ward and Whipple, 1959).

It appears that in lakes there is a variety of patterns with respect to these seasonal fluctuations in the population dynamics of Rhabditina. Some authors noted significant annual density fluctuations (Fisher, 1968; Tudorancea and Zullini, 1989), whereas another found no such distinct changes of nematode abundance (Strayer, 1985).

Other substances that can have varying effects on nematodes are humic substances that have hormone-like effects and which comprise the majority of dissolved organic matter in freshwater ecosystems (Steinberg, 2003; Steinberg *et al.*, 2004). Exposure to humic substances, such as fluvic acid, has been demonstrated as modulating the number of offspring in *Caenorhabditis elegans* (Maupas, 1899) Dougherty, 1953. At low concentrations it increases the number of offspring, and at elevated concentrations reduces them significantly.

This tolerance of Rhabditina to pollutants can be explained by the presence of a relatively impermeable cuticle (Bongers, 1999) that does not allow the passage of toxic substances. Also, the presence of voluminous gonads, which produce many small eggs and which mature often at the gonads (viviparity in Rhabditidae), results in explosive growth under conditions where the microbial activity is high. Thus, although rhabditid nematodes are not common in unpolluted freshwater habitats, they can be the dominant group in altered waters, being able to recolonize disturbed and azoic areas (Santos-Lobatón *et al.*, 2002).

On the other hand, a species of Diplogasteromorpha, *Tylopharynx phoetida* (Bütschli, 1874) Goffart, 1930 (Tylopharyngidae), has been recorded in putrid end-substances, feeding on fungal hyphae by ripping the cell wall with stegostomatal teeth (Fürst von Lieven, 2002).

Thus, according to the abovementioned literature, the presence of certain rhabditid groups in freshwaters, mainly Rhabditidae and different diplogastrids (see Volterra *et al.*, 1999; Gagarin, 2001b), is considered as an indication of pollution or faecalization of the aquatic system (Volterra *et al.*, 1999).

Survival Strategies

The cuticle may play an important role in controlling water loss during periods of desiccation. This includes anhydrobiosis, the production of dauer larvae and, in viviparous nematodes such as some members of Rhabditidae, larvae achieving a slow rate of water loss via the protection provided by the body of the mother (Lees, 1953; Ellenby, 1969).

In vernal pools, which only form during wet years and evaporate within a few weeks or months, Rhabditina species and other nematodes are capable of efficient anhydrobiosis and/or finding refuge in roots and plants which persist during the protracted dry phase awaiting the inundation period (De Ley *et al.*, 2004).

In the case of dauer larvae, they appear when environmental conditions are negative, such as during limited food supply or after desiccation of an aquatic system (Riddle, 1988; Riddle and Albert, 1997).

Diagnosis of Freshwater Taxa of suborder Rhabditina[2]

Freshwater Rhabditina belong to the three main groups: bunonematids, rhabditids *sensu stricto* and diplogastrids.

---

[2]The classification and differential diagnosis of freshwater taxa of suborder Rhabditina used here is according to Andrássy (1983, 1984), Sudhaus and Fitch (2001), De Ley and Blaxter (2002), De Ley *et al.* (2003) and Sudhaus and Fürst von Lieven (2003).

Infraorder BUNONEMATOMORPHA De Ley & Blaxter, 2002

Body asymmetrical, right side with fine lattice structure and usually also with warts. Anterior body region having bristle- and lamella-shaped appendages. Stoma tubular or prismatic. Pharynx corpus having an anterior part cylindrical (procorpus) and a posterior part swollen (metacorpus); isthmus narrow; pharyngeal bulb swollen, spherical. Genital system paired. Bursa weak, asymmetrical.

Family BUNONEMATIDAE Micoletzky, 1922

Cuticle not annulated, at the right side of body with fine warts or fins (large groups of warts). Labial region bilaterally symmetrical, with bristles and membranous appendages. Female tail usually longer than male tail. Bursa narrow, asymmetrical, leptoderan. Spicules very long and thin.

Genus *Craspedonema* Richters, 1908

= *Bunonema* (*Craspedonema* Richters, 1908) Sach, 1949.

Right side of body ornamented with a series of transverse saddles or crust-like warts, and left side with five longitudinal ridges. Lip region with setose appendages. Stoma prismatic having cheilostom cuticularized. Female gonads paired. Female tail conoid. Spicules long and slender. Bursa leptoderan, asymmetrical.

C. styriacum Micoletzky, 1922 has been reported from semiaquatic habitats associated with mosses in marshy areas (Micoletzky, 1922).

Genus *Rhodolaimus* Fuchs, 1930

= *Bunonema* (*Rhodolaimus* Fuchs, 1930) Sachs, 1949; *Bunonema* (*Stammeria* Sachs, 1949); *Stammeria* (Sachs, 1949) Andrássy, 1958.

Right side of body ornamented with papillae or warts composed of small papillae and arranged in one or two rows. Lip region with setose appendages. Stoma tubular, cheilostom cuticularized. Pharyngeal corpus swollen at posterior part. Female gonads paired, vulva at mid-body. Female tail conoid. Spicules very slender, separate. Bursa leptoderan, asymmetrical.

R. ritae Zeidan & Geraert, 1989 has been recorded at the border of a temporary water pool from Sudan (Zeidan and Geraert, 1989).

Infraorder RHABDITOMORPHA De Ley & Blaxter, 2002

Body bilaterally symmetrical. Anterior body region usually with six lips. Stoma tubular or prismatic, longer than wide. Cheilostom usually not cuticularized, gymnostom and promesostegostom fused to the buccal tube, metastegostom with or without glottoid apparatus and small denticles, telostegostom short. Pharyngeal corpus having cylindrical anterior part (procorpus) and often a swollen posterior part (metacorpus); isthmus narrow; pharyngeal bulb swollen, spherical. Female gonads paired or unpaired, prodelphic. Female tail of different shape, conoid or cupola-shaped. Spicules separate or fused at tip. Bursa present, leptoderan or peloderan.

Family MESORHABDITIDAE Andrássy, 1976

Lips well developed, separate, each with setiform papilla. Stoma wide, well developed. Cheilostom not cuticularized; gymnostom and promesostegostom with parallel walls; metastegostom with glottoid apparatus and small denticles. Pharyngeal corpus swollen bulb-like. Female genital system unpaired, prodelphic; vulva far back. Female tail conoid. Spicules often fused distally. Bursa peloderan, proximally open, with mid-bursa in phasmid position.

Genus *Mesorhabditis* Osche, 1952 (Dougherty, 1953)
= *Rhabditis* (*Mesorhabditis* Osche, 1952); *Rhabditis* (*Uniovaria* Khera, 1968); *Bursilla* Andrássy, 1976.
Lips well separated, rounded, each with one setiform papilla. Stoma with glottoid apparatus having setose denticles (Fig. 22.1A). Female gonad unpaired, prodelphic. Vulva far back. Female tail conoid. Phasmids near the anus. Spicules distally fused to a third or more. Bursa peloderan, anteriorly open.

M. inarimensis (Meyl, 1953) Dougherty, 1955 has been recorded in thermal waters from Italy (Meyl, 1953, 1954).

M. microbursaris (Steiner, 1926) in the borders of various water pools from Sudan (Zeidan and Geraert, 1989).

M. striatica Dassonville & Heyns, 1984 in freshwater habitats from South Africa (Dassonville and Heyns, 1984).

Family PELODERIDAE Andrássy, 1976

Lips hardly separate, papillae very small. Stoma well developed. Cheilostom not cuticularized; gymnostom and promesostegostom with parallel walls; metastegostom with glottoid apparatus and either with small warts or with bristle-like denticles. Pharyngeal corpus swollen, bulb-like. Female genital organ paired, amphidelphic; vulva at or near mid-body region. Tail of female conoid- to cupola-like. Spicules fused distally. Bursa well developed, peloderan, with mid-bursa phasmid position.

Genus *Pelodera* Schneider, 1866
= *Rhabditis* (*Pelodera* Schneider, 1866) Sudhaus, 1974; *Rhabditis* (*Rhabditis apud* Osche, 1952); *Pelodera* (*Coarctadera* Dougherty, 1953); *Pelodera* (*Cylindridera* Dougherty, 1953); *Coarctadera* (Dougherty, 1953) Andrássy, 1976; *Rhomborhabditis* Andrássy, 1983.
Cuticle annulated and also finely longitudinally striated. Lips separate. Stoma varying in length, metastegostom isoglottoid with setose denticles. Female gonads paired; vulva at mid-body. Tail of female conoid or cupola-like with or without tip. Spicules fused. Bursa peloderan, open or closed.

P. littoralis (Skwarra, 1921) Dougherty, 1955 was found in freshwater habitats from Poland (Kozlowska, 1962).

P. punctata (Cobb, 1914) Dougherty, 1955, an aquatic or semiaquatic species reported from the Monachil River, Spain (Picazo *et al.*, 1989, 1990) and in potable waters from Italy (Zullini, 1981, 1982; Volterra *et al.*, 1999).

P. serrata (Körner in Osche, 1952) Dougherty, 1955 and *P. teres* Schneider, 1866; both species recorded in sludge from waste water from Andalucía, Spain (Santos-Lobatón *et al.*, 2002).

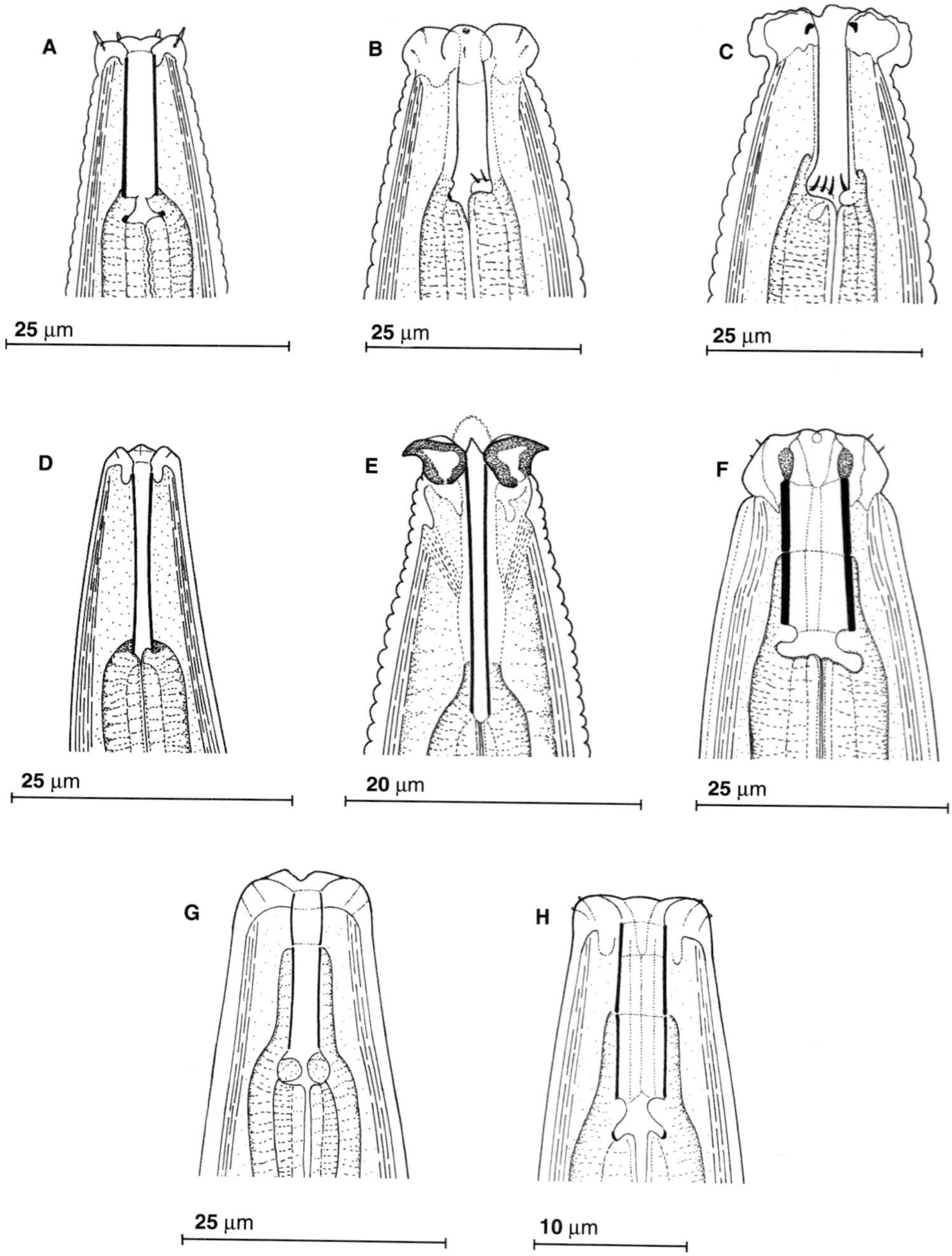

Fig. 22.1 Morphology of Rhabditomorpha: *Mesorhabditis* (A); *Cruznema* (B, C); *Protorhabditis* (D); *Diploscapter* (E); *Cephaloboides* (F); *Rhabditis* (G); *Oscheius* (H). All anterior region.

Family RHABDITIDAE Örley, 1880

= Diploscapteridae Micoletzky, 1922.

Lips closed or slightly separate. Stoma well developed, tubular, usually long. Cheilostom not cuticularized, gymnostom and promesostegostom with parallel walls; metastegostom with or without glottoid apparatus bearing minute warts or setiform denticles. Pharyngeal corpus often bulb-like. Female gonad didelphic, amphidelphic, or monodelphic, prodelphic. Vulva medial or posterior. Female tail mostly conical, sometimes cupola-shaped. Spicules usually separate. Bursa leptoderan or peloderan, open or closed, with posteriormost phasmid position.

Genus *Caenorhabditis* (Osche, 1952) Dougherty, 1953
= *Rhabditis* (*Caenorhabditis*) Osche, 1952.
Lips hardly separate, low. Stoma with metastegostom isoglottoid, each armed with bristle-like denticles. Female gonad paired, vulva at mid-body region. Female tail conoid. Spicules separate. Bursa peloderan, anteriorly closed, sucker-shaped.

C. elegans (Maupas, 1900) Dougherty, 1953 has been recorded in potable waters from Italy (Zullini, 1981, 1982; Volterra *et al.*, 1999).

Genus *Cephaloboides* (Rahm, 1928) Massey, 1974
= *Rhabditis* (*Cephaloboides* Rahm, 1928); *Odontorhabditis* Timm, 1959; *Curviditis* (Dougherty, 1953) Andrássy, 1983; *Rhabditis* (*Curviditis* Dougherty, 1953); *Flagicaudoides* Khan, Singh & Baird, 1999.
Lip region differentiated from neck. Stoma with cheilostom not or slightly cuticularized and metastegostom slightly iso- or anisoglottoid, without denticles or fine warts (Fig. 22.1F). Female gonad didelphic, amphidelphic; vulva medial to post-medial. Tail of both sexes cupola-shaped. Spicules separate. Bursa leptoderan, open.

C. boettgeri (Meyl, 1953) Andrássy, 1983 was reported in thermal waters from Italy (Meyl, 1953, 1954).

C. curvicaudata (Schneider, 1866) Dougherty, 1955 has been recorded in the Monachil river, Spain (Picazo *et al.*, 1989, 1990; Ocaña and Picazo, 1991), and in potable waters from Italy (Zullini, 1981, 1982; Volterra *et al.*, 1999).

Genus *Cruznema* Artigas, 1927
= *Pelodera* (*Cruznema* Artigas, 1927) Dougherty, 1953; *Rhabditis* (*Cruznema* Artigas, 1927) Sudhaus, 1974; *Epimenides* Gutiérrez, 1949.
Cuticle transversely annulated and longitudinally striated. Lips well separate and well differentiated from neck. Metastegostom isoglottoid, carrying small denticles (Fig. 22.1B and C). Female gonad unpaired, prodelphic; vulva in the posterior fifth of body length. Tails conoid. Spicules not fused distally. Bursa peloderan, anteriorly open. Phasmids near the anal opening.

C. tripartitum (Linstow, 1906) Zullini, 1982 has been recorded from the Monachil river, Spain (Picazo *et al.*, 1989, 1990; Ocaña and Picazo, 1991); in potable waters from Italy (Zullini, 1981, 1982; Volterra *et al.*, 1999); and in sludge from waste water from Andalucía, Spain (Santos-Lobatón *et al.*, 2002).

Genus *Diploscapter* Cobb, 1913
Lip region bilaterally symmetrical, ventral and dorsal with paired, cuticularized, hook-like appendages; lateral lips membrane-like (Fig. 22.1E). Metastegostom lacking glottoid apparatus. Ovaries paired, vulva medial or post-medial. Female tail conoid. Spicules free. Bursa peloderan, anteriorly open.

D. coronatus (Cobb, 1893) Cobb, 1913 from freshwater habitats in Indonesia (Schneider, 1937) and South Africa (Dassonville, 1981); in thermal waters from Italy (Meyl, 1953); in the Monachil river, Spain (Picazo *et al.*, 1989, 1990; Ocaña and Picazo, 1991); in potable waters from Italy (Zullini, 1981, 1982; Volterra *et al.*, 1999); and in Lake Tana, Ethiopia (Eyualem-Abebe *et al.*, 2001); in sludge from waste water from Andalucía, Spain (Santos-Lobatón *et al.*, 2002); or in the Körsch, a submountain stream in Germany (Beier and Traunspurger, 2003b).

D. indicus Tahseen *et al.*, 2002 found in sewage slurry from a drain in India.

Genus *Oscheius* Andrássy, 1976
= *Dolichorhabditis* Andrássy, 1983.
Lips separate. Stoma unusually short, buccal tube almost as long as wide, and metastegostom isoglottoid bearing minute or bristle-like denticles or warts (Fig. 22.1H). Pharyngeal corpus without an offset middle bulb. Female gonad paired, vulva at mid-body. Tail of both sexes conical. Phasmids at anal region. Spicules separate, with a knobbed or hooked terminus. Bursa pseudopeloderan or peloderan, open.

O. dolichurus (Schneider, 1866) Sudhaus & Hooper, 1994 has been recorded in freshwater from Poland (Kozlowska, 1962) and in potable waters from Italy (Zullini, 1981, 1982; Volterra *et al.*, 1999).

Genus *Pellioditis* (Dougherty, 1953) Timm, 1960
= *Rhabditis* (*Pellioditis* Dougherty, 1953); *Phasmarhabditis* Andrássy, 1976.
Cuticle more or less annulated or smooth and sometimes longitudinally striated. Lips separate. Stoma long, having metastegostom isoglottoid with minute warts. Female gonad paired, vulva at mid-body region. Female tail conoid, filiform or cupola-shaped. Phasmid near the anal opening. Spicules separate. Bursa peloderan, open.

P. friderici (Hirschmann, 1952) Andrássy, 1983 has been found in sludge from waste water from Andalucía, Spain (Santos-Lobatón *et al.*, 2002).

Genus *Poikilolaimus* Fuchs, 1930
= *Cuticularia* Van der Linde, 1938; *Peplorhabditis* Ivanova, 1960; *Praeputirhabditis* Khera, 1969.
Lips not separate. Stoma tubular having metastegostom slightly asymmetrical, anisoglottoid, with small or setose denticles. Female gonad paired, vulva postmedial. Female tail conical- or cupola-shaped. Spicules free. Bursa leptoderan, rudimentary.

P. oxycercus (De Man, 1895) Sudhaus, 1980 has been found in many different sites: in freshwater habitats from Poland (Kozlowska, 1962) and from South Africa (Dassonville, 1981); in the Monachil river, Spain (Picazo *et al.*, 1989, 1990; Ocaña and Picazo, 1991;); in potable waters from Italy (Volterra *et al.*, 1999); and in the Körsch, a submountain stream in Germany (Beier and Traunspurger, 2003b). Also, this genus has been recorded as *Cuticularia* sp. in the Krähenbach, a submountain stream in Germany (Beier and Traunspurger, 2003a).

Genus *Protorhabditis* (Osche, 1952) Dougherty, 1953
= *Rhabditis* (*Protorhabditis*) Osche, 1952.
Cuticle finely transversely annulated and sometimes longitudinally striated. Lips hardly separate. Stoma very long and narrow, having metastegostom lacking glottoid apparatus (Fig. 22.1D). Female gonad paired, vulva at mid-body. Female tail elongate conoid to filiform. Spicules free. Bursa peloderan, open or rarely closed.

P. filiformis (Bütschli, 1873) Sudhaus, 1976 has been found in the Carrión river, Spain (Gadea, 1963); in potable waters from Italy (Zullini, 1981, 1982; Volterra *et al.*, 1999); and in sludge from waste water from Andalucía, Spain (Santos-Lobatón *et al.*, 2002).

Genus *Prodontorhabditis* Timm, 1961
Cuticle finely transversely annulated and longitudinally striated. Stoma usually very long, having cheilostom cuticularized and metastegostom lacking glottoid apparatus. Female gonad paired; vulva at mid-body or lightly anterior. Female tail elongate conoid to filiform. Spicules separate. Bursa peloderan, anteriorly closed.

P. anthobia (Schneider, 1937) Andrássy, 1983 was reported from freshwater aquatic systems in Indonesia (Schneider, 1937).

Genus *Rhabditella* (Cobb, 1929) Chitwood, 1933
= *Rhabditis* (*Rhabditella* Cobb, 1929); *Rhabdibicauda* Sultan, Chhabra & Kaul, 1985.
Lips slightly separate. Stoma having metastegostom iso- or anisoglottoid, with very small denticles. Female gonad paired; vulva at mid-body or slightly anterior. Tail in both sexes long. Spicules separate with large dorsal projection. Bursa leptoderan.

R. macrospiculata (Stefanski, 1916) Andrássy, 1983 was found in sediment from Inn river (Ausina). (Stefanski, 1916, 1927).

R. pseudoelongata (Micoletzky, 1913) Andrássy, 1983 has been recorded from Italy in thermal waters (Meyl, 1954) and potable waters (Zullini, 1981, 1982; Volterra *et al.*, 1999).

Genus *Rhabditis* Dujardin, 1845
= *Tribactis* Dujardin, 1945; *Rhabditis* (*Choriorhabditis* Osche, 1952); *Paradoxorhabditis* Khera, 1971; *Ablechroiulus* Andrássy, 1974; *Rhabditis* (*Indorhabditis* Chaturvedi & Khera, 1979).
Lips closed, slightly or well separate. Stoma having metastegostom isoglottoid, rarely slightly developed, with fine warts (Fig. 22.1G). Female gonad paired; vulva near mid-body. Tail of both sexes conical, sometimes female tail cupola-shaped. Spicules free or distally fused. Bursa leptoderan, open or sometimes closed, or rarely pseudopeloderan, leaving a short and very fine tail filament free.

R. maupasi Seurat in Maupas, 1919 was found in sludge from waste water from Andalucía, Spain (Santos-Lobatón *et al.*, 2002).

R. producta (Schneider, 1866) Linstow, 1878 was found in sludge from waste water from Andalucía, Spain (Santos-Lobatón *et al.*, 2002); in Carrión river, Spain (Gadea, 1963); and in the Körsch, a submountain stream, in Germany (Beier and Traunspurger, 2003b).

R. sechellensis Pott, 1910 was found in several places such as the Seychelles Islands (Pott, 1910).

Genus *Rhabditoides* Goodey, 1929 (*incertae sedis*)
= *Rhabditis* (*Rhabditoides* Goodey, 1929) Sudhaus, 1974; *Rhabditis* (*Telorhabditis* Osche, 1952); *Telorhabditis* (Osche, 1952) Schuurmans Stekhoven, 1957; *Rhitis* Andrássy, 1983.
Lips slightly separate. Stoma having metastegostom iso- or anisoglottoid with small denticles. Female gonad paired; vulva at mid-body or near. Female tail elongate-conoid. Spicules separated. Bursa leptoderan, open, rudimentary, often hardly discernible.

R. eximius Gagarin, 2000 has been recorded in littoral of the Kurilskoe lake, Russia (Gagarin, 2000).
R. inermiformis (Osche, 1952) Dougherty, 1955 has been recorded in potable waters from Italy (Zullini, 1981, 1982; Volterra *et al.*, 1999).

Infraorder DIPLOGASTEROMORPHA De Ley & Blaxter, 2002

Cuticle smooth or longitudinally striated. Labial papillae often bristle-shaped; males in addition usually with another four cephalic bristles. Stoma symmetrically bilateral or asymmetrical, with parallel walls or developing different structures, more or less mobile, as claw-like teeth. Pharyngeal corpus bearing an anterior cylindrical part (procorpus) and a posterior swollen bulb-like part usually well developed (metacorpus); isthmus narrow; pharyngeal bulb swollen, spherical. Female gonad didelphic, amphidelphic, or monodelphic, prodelphic. Spicules free, rarely fused. Males usually lacking bursa. Genital papillae often bristle-shaped. Tail in both sexes long, filiform.

Family DIPLOGASTROIDIDAE Filipjev & Schuurmans Stekhoven, 1941[3]
Amphid large, wider than stoma. Stoma cylindrical having cheilostom as long as wide with curved walls; gymnostom and promesostegostom cylindrical; metastegostom with glottoid apparatus. Female gonad monodelphic, prodelphic or didelphic, amphidelphic. Males without bursa.

Genus *Diplogasteroides* de Man, 1912
= *Rhabditolaimus* Fuchs, 1914; *Rhabditolaimus* (*Rhabdontolaimus* Fuchs, 1931); *Rhabdontolaimus* (Fuchs, 1931) Filipjev & Schuurmans Sterkhoven, 1941; *Pseudodiplogaster* Takaki, 1941; *Dirhabdilaimus* Paramonov & Turlygina, 1955; *Diplogasteroides* (*Neodiplogasteroides* Rühm, 1956); *Masseyus* Paramonov, 1964.
Lips low, slightly separate. Stoma having cheilostom with or without per- and interradial flaps; gymnostom tube-shaped; stegostom with dorsal tooth formed by three rods. Female gonad paired, amphidelphic, or unpaired, prodelphic. Males without bursa. Tail in both sexes filiform.

D. bidentatus De Cillis, 1917 has been recorded from mud of a lake from Italy (De Cillis, 1917).

[3]Although, according to Sudhaus and Fürst von Lieven (2003), all traditional families of diplogastrids are synonyms of family Diplosgastridae; however, we use here the different ranks from De Ley and Blaxter (2002, 2004).

Genus *Goffartia* Hirschmann, 1952
= *Paramonovnema* Andrássy, 1968.
Lips very closed, with prominent papillae. Amphids very large. Stoma having cheilostom anteriorly converging; gymnostom barrel-shaped; stegostom without cuticularized projections. Female gonad paired, amphidelphic. Tail in both sexes filiform. Male lacking bursa.

G. africana (Micoletzky, 1916) Hirschmann, 1952 was found in freshwater algae from South Africa (Micoletzky, 1916).
G. filicaudata (Andrássy, 1968) Sudhaus & Fürst von Lieven, 2003 has appeared in roots at edge of a pond (Andrássy, 1968).
G. heteroceri Hirschmann, 1952 and *G. variabilis* (Micoletzky, 1922) Hirschmann, 1952 have been reported from Italy in thermal (Meyl, 1953, 1954) and potable waters (Zullini, 1981, 1982; Volterra *et al.*, 1999).
G. praepilata Shoshin, 1989 in Lake Baikal, Russia (Shoshin, 1989).

Genus *Rhabditidoides* Rahm, 1928
= *Anchidiplogasteroides* Paramonov & Turlygina, 1955.
Lips slightly separate. Stoma having gymnostom tube-shaped; stegostom with dorsal flap-like process. Female gonad paired, amphidelphic. Male without bursa. Tail in both sexes filiform.

R. stigmatus (Steiner, 1930) Andrássy, 1984 has been recorded in freshwater habitats from South Africa (Dassonville and Heyns, 1984); in the Monachil river, Spain (Picazo *et al.*, 1989, 1990; Ocaña and Picazo, 1991); and in potable waters from Italy (Zullini, 1981, 1982; Volterra *et al.*, 1999).

Family DIPLOGASTRIDAE Micoletzky, 1922
Cuticle more or less smooth or longitudinally striated. Lip region with bristle-shaped labial sensillae variable in length. Stoma having cheilostom with per- and interradial flaps; gymnostom and promesostegostom almost as wide as long; metastegostom often with very large teeth, dorsal tooth always larger than the sub-ventral, not mobile. Female gonad usually didelphic, amphidelphic, but also monodelphic, prodelphic. Bursa rudimentary.

Genus *Acrostichus* Rahm, 1928
= *Diplogasteritus* Paramonov, 1952; *Filipjevella* Lazarevskaja, 1965; *Aduncospiculum* Giblin & Kaya, 1984.
Cuticle longitudinally striated. Lips with short labial sensillae. Stoma having cheilostom with six adradial flaps and separated into six adradial plates; gymnostom almost as long as cheilostom; stegostom with thorn- or dagger-like dorsal cuticularized tooth (Figs 22.2A and B). Female gonad paired, amphidelphic. Male without bursa. Tail in both sexes filiform (Fig. 22.2C).

A. angustilaimus (Schuurmans Stekhoven & Teunissen, 1938) Massey, 1966 was recorded from Democratic Republic of Congo and *A. paramicrostoma* (Schuurmans Stekhoven, 1943) Sudhaus & Fürst von Lieven, 2003 was found in a spring in Belgium (Schuurmans Stekhoven, 1943).

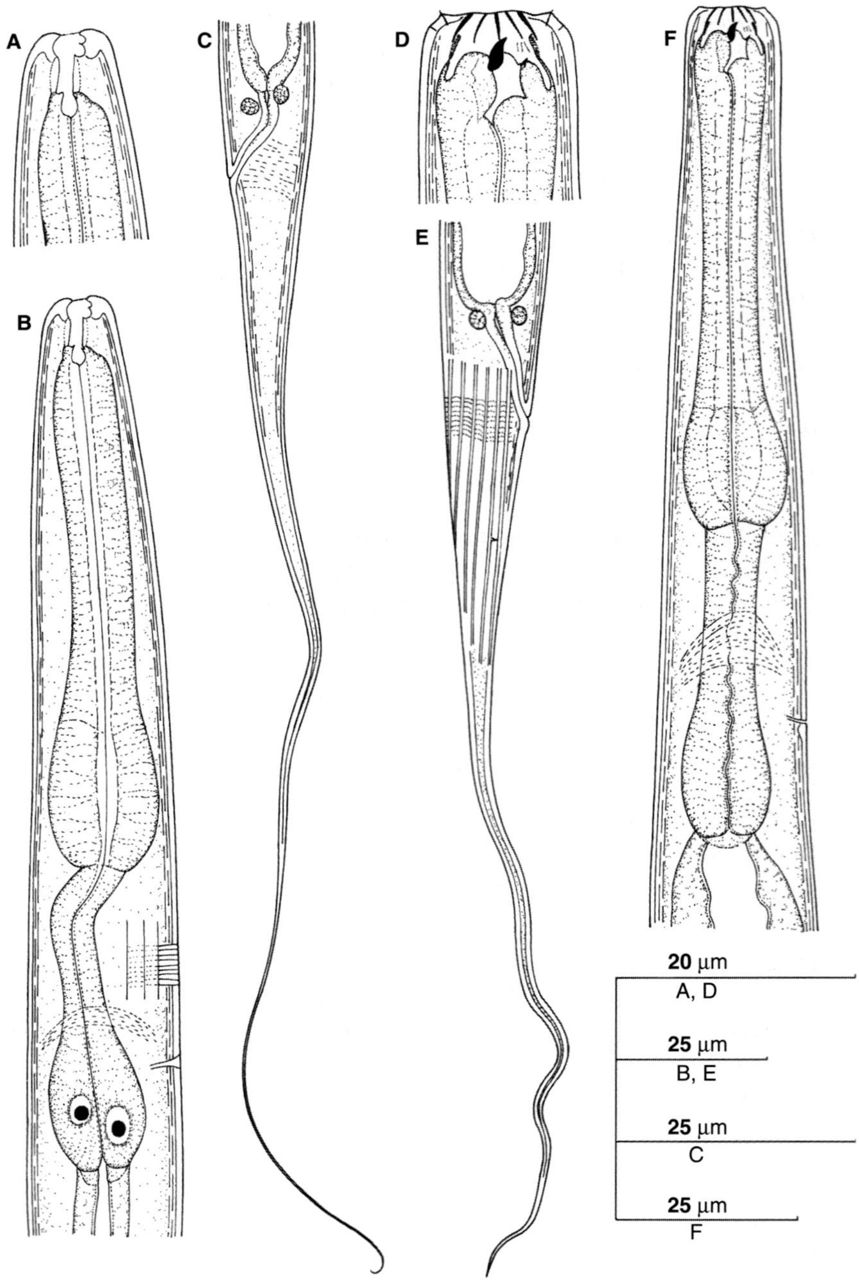

Fig. 22.2 Morphology of Diplogasteromorpha: *Acrostichus* (A: lip region; B: neck; C: tail); *Mononchoides* (D: lip region; E: tail; F: neck).

A. nudicapitatus (Steiner, 1914) Massey, 1962 was recorded in a stream in Prahova Valley, Romania (Stefanski, 1926, 1927); in sewage discharge from Russia (Gagarin, 1977); in aquatic habitats from South Africa (Dassonville and Heyns, 1984); in the Monachil River, Spain (Picazo *et al.*, 1989, 1990; Ocaña and Picazo, 1991); and from Italy in cold waters (Meyl, 1954) and in potable waters (Zullini, 1981, 1982; Volterra *et al.*, 1999).

Genus *Butlerius* Goodey, 1929
= *Butlerioides* Lordello & Zamith, 1959; *Butleriellus* Meyl, 1960; *Mesodiplogasteroides* Khera, 1969; *Monobutlerius* Andrássy, 1984; *Parabutlerius* Ebsary, 1986.
Lip region with long bristle-shaped labial sensillae. Stoma having cheilostom well developed forming well-developed flaps; gymnostom wider than long; stegostom with a dorsal flap or thorn-like tooth. Female gonad unpaired, prodelphic, or paired, amphidelphic. Male without bursa. Tail in both sexes filiform.

B. gerlachi Meyl, 1957 was found in standing water within bromeliads from Brazil (Meyl, 1957).

B. longipyge (Khera, 1969) Sudhaus & Fürst von Lieven, 2003 in banks of sewer (Khera, 1969) from Italy (Meyl, 1953, 1954).

B. micans Pillai & Taylor, 1968 has been reported in potable waters from Italy (Zullini, 1981, 1982; Volterra *et al.*, 1999).

B. spirifer (Skwarra, 1921) Zullini & Loof, 1980 in thermal water from Italy (Meyl, 1953, 1954).

Genus *Diplogaster* Schultze in Carus, 1857
Lip region with short bristle-shaped labial sensillae. Stoma having cheilostom with per- and interradial flaps; cheilo- and gymnostom wider than long; dorsal tooth dagger-like. Female gonad amphidelphic. Males without bursa. Female tail filiform.

Members of this genus have been found in the Krähenbach, a submountain stream in Germany (Beier and Traunspurger, 2003a).

D. gagarini (Tsalolichin, 1980) Sudhaus & Fürst von Lieven, 2003 has appeared in aquatic system, in muddy sand in Lake Baikal, Russia (Tsalolikhin, 1980).

D. rivalis (Leydig, 1854) Bütschli, 1873 has been recorded in freshwater habitats in Hungary (Daday, 1897) and in Poland (Kozlowska, 1962); in the Carrión river, Spain (Gadea, 1963); and in potable waters from Italy (Zullini, 1981, 1982; Volterra *et al.*, 1999).

Genus *Diplogastrellus* Paramonov, 1952
= *Metadiplogaster* Weingärtner, 1955 (Meyl, 1961); *Tawdenema* Suryawanshi, 1971.
Lip region with short bristle-shaped labial sensillae. Stoma having cheilostom as long as gymnostom, with six per- and interradial flaps; gymnostom with dorsal wall being shorter than ventral wall; stegostom with dorsal flap or tooth. Female gonad unpaired, prodelphic. Male lacking bursa. Tail in both sexes filiform.

D. heynsi (Kionktke & Sudhaus, 1996) Sudhaus & Fürst von Lieven, 2003 was found in sediment of river from South Africa (Dassonville and Heyns, 1984 described it as *Metadiplogaster secundus*).

Genus *Paroigolaimella* Paramonov, 1952
= *Diplogaster* (*Paradiplogaster apud* Hirschmann, 1952 and *apud* Meyl, 1954); *Diplogaster* (*Paradiplogaster* Weingärtner, 1955).
Lip region with short labial sensillae. Stoma having cheilostom with six adradial flaps and separated into six adradial plates; stegostom with thorn-like dorsal tooth and with sub-ventral warts. Female gonad paired, amphidelphic. Male without bursa. Tail in both sexes filiform.

P. anomala Gagarin, 1977 was found in polluted water and sewage (Gagarin, 1977).
P. bernensis (Steiner, 1914) Andrássy, 1958 was recorded in polluted water in Switzerland (Steiner, 1914); in freshwater habitats from South Africa (Dassonville and Heyns, 1984); in the Monachil River, Spain (Picazo *et al.*, 1989, 1990; Ocaña and Picazo, 1991); in the Körsch, a submountain stream in Germany (Beier and Traunspurger, 2003b); and in potable waters from Italy (Zullini, 1981, 1982; Volterra *et al.*, 1999).
P. coprophaga (de Man, 1876) Paramonov, 1952 was recorded in cold water in a stream from Italy (Meyl, 1953, 1954) and sludge from waste water in Andalucía, Spain (Santos-Lobatón *et al.*, 2002).

Family NEODIPLOGASTRIDAE Paramonov, 1952 (Andrássy, 1984)[3]
Lip region with bristle-shaped labial sensillae variable in length. Stoma having cheilostom longitudinally divided into plates; cheilostom and gymno-promesostegostom with similar length; metastegostom asymmetrical, dorsal with a quite large, claw-like bent, mobile tooth, at the right sub-ventral usually with a similarly formed, equal large, but more weakly cuticularized claw tooth, while at the left sub-ventral with two or more acute plates, with a very small tooth, or completely unarmed; telostegostom with variable length. Female gonad didelphic, amphidelphic. Males without bursa or thin rudimentary.

Genus *Koerneria* Meyl, 1960
= *Gobindonema* Khera, 1970; *Anchidiplogaster* Paramonov, 1952; *Allodiplogaster* Paramonov & Sobolev in Skrjabin, Shikhobalova, Sobolev, Paramonov & Sudarikov, 1954; *Glauxinemella* Gagarin, 1998.
Lips fused. Stoma having cheilostom divided into six per- and interradial plates or undivided, and vertically striated; stegostom with dorsal claw-like tooth, right sub-ventral tooth and left sub-ventral serrated plate; stegostom-cylinder short with a subventrally directed apodeme on each side. Female gonad paired, didelphic, or unpaired, prodelphic. Males lacking bursa. Tail in both sexes filiform.

K. agilis (Skwarra, 1921) Andrássy in Zullini, 1981 was reported in potable waters from Italy (Zullini, 1981, 1982; Volterra *et al.*, 1999).
K. angarensis (Gagarin, 1983) Ebsary, 1986 was found in running water from Russia (Gagarin, 1983); *K. aquatica* (Dassonville and Heyns, 1984) Sudhaus & Fürst von Lieven, 2003 also in running water from South Africa (Dassonville and Heyns, 1984) and in a lake from Russia (Gagarin and Lemsina, 1982).
K. baicalensis (Tsalolichin, 1972) Shoshin, 1989 was reported from a lake from Russia (Tsalolichin, 1972).

K. carinata (Zullini, 1981) Fürst von Lieven & Sudhaus, 2000 from polluted freshwater from Italy (Zullini, 1981); from the Monachil River, Spain (Picazo *et al.*, 1989, 1990; Ocaña and Picazo, 1991); and in potable waters from Italy (Zullini, 1981, 1982; Volterra *et al.*, 1999).

K. filicaudata (Khera, 1970) Andrássy, 1984 from the banks of a lake in India (Khera, 1970).

K. lupata Shoshin, 1989; *K. mordax* Shoshin, 1989; and *K. regia* Shoshin, 1989 in Lake Baikal, Russia (Shoshin, 1989).

K. pararmata (Schneider, 1938) Sudhaus & Fürst von Lieven, 2003 in brackish water rich with diatoms (Schneider, 1938) and *K. strenua* (Gagarin, 1983) Ebsary, 1986 in a lake from Russia (Gagarin, 1983).

Genus *Fictor* Paramonov, 1952
= *Paradiplogaster* Paramonov, 1952 nec Schuurmans Stekhoven & Teunissen, 1938; *Holodiplogaster* Meyl, 1960; *Eudiplogasterium* Meyl, 1960; *Paradoxogaster* Khera, 1970.
Lip region with short labial sensillae. Stoma having cheilostom forming rugae; stegostom with dorsal claw-like tooth, right sub-ventral tooth and left sub-ventral serrated plate. Female gonad paired, amphidelphic. Male without bursa. Tail in both sexes filiform.

F. fictor (Bastian, 1865) Paramonov, 1952 widely reported from the Noguera River, Spain (Gadea, 1952); in freshwater habitats from Poland (Kozlowska, 1962); in potable waters from Italy (Zullini, 1981, 1982; Volterra *et al.*, 1999); in the Körsch, a submountain stream in Germany (Beier and Traunspurger, 2003b).

F. similis (Bütschli, 1876) Goodey, 1963 recorded from sludge in Andalucía, Spain (Santos-Lobatón *et al.*, 2002).

Genus *Mononchoides* Rahm, 1928
= *Diplogaster* (*Loxolaimus* Rahm, 1928); *Glauxinema* Allgén, 1947; *Eudiplogaster* Paramonov, 1952; *Tridontus* Khera, 1965; *Syedella* Suryawanshi, 1971; *Diplenteron* Andrássy, 1964; *Pareudiplogaster* Paramonov, 1952; *Prosodontus* Paramonov & Sobolev in Skrjabin, Shikhobalova, Sobolev, Paramonov & Sudarikov, 1954.
Lip region with very short labial sensillae. Stoma having cheilostom with rugae; stegostom with dorsal claw-like tooth, right sub-ventral tooth and left sub-ventral serrated plate; post-dental part forming a stegostom-cylinder (Fig. 22.2D and F). Female gonad paired, amphidelphic. Male lacking bursa. Tail in both sexes filiform (Fig. 22.2E).

M. armatus (Hofmänner, 1913) Gagarin, 1998 has been found in sediment of lakes from European Russia (Gagarin 1998).

M. asiaticus Gagarin, 2001 in a stream from Japan (Gagarin, 2001a).

M. bollingeri Goodrich, Hechler & Taylor, 1968 in waste water (Goodrich *et al.*, 1968).

M. changi Goodrich, Hechler & Taylor, 1968 from a waste-treatment plant (Goodrich *et al.*, 1968).

M. intermedius Gagarin, 1993 from a river from Sachalin, Russia (Gagarin, 1993).

M. gracilis Dassonville & Heyns, 1984 found in sediment of river from South Africa (Andrássy, 1970; Dassonville, 1981; Dassonville and Heyns, 1984).

M. isolae (Meyl, 1953) Goodey in Goodey, 1963 associated with algae in thermal water from Italy (Meyl, 1953, 1954).

M. pulcher Zullini, 1981 from sewage and potable waters in Italy (Zullini, 1981, 1982; Volterra *et al.*, 1999).

M. ruffoi Zullini, 1981 is a species recorded in the Monachil river, Spain (Picazo *et al.*, 1989, 1990) and in sewage and potable waters from Italy (Zullini, 1981, 1982; Volterra *et al.*, 1999, respectively).

M. striatus (Bütschli, 1876) Goodey in Goodey, 1963 associated with algae in thermal water from Italy (Meyl, 1953, 1954) and freshwater from Hungary (Daday, 1897).

Genus *Neodiplogaster* Cobb, 1924
= *Tylenchodon* Fuchs, 1930.
Cuticle with transversal and longitudinal striations. Lips fused with short labial sensillae. Stoma having cheilostom with fine rugae; stegostom with dorsal claw-like and right sub-ventral tooth; long stegostom-cylinder with wing-like sub-dorsal apodemes at posterior end. Female gonad paired, amphidelphic. Male with bursa. Tail in both sexes elongate–conoid.

N. pissodis (Fuchs, 1930) Goodey, 1951 is a species found in waste water from Andalucía, Spain (Santos-Lobatón *et al.*, 2002).

Genus *Oigolaimella* Paramonov, 1952
= *Hemidiplogaster* Goodey in Goodey, 1963.
Cuticle lacking longitudinal striations. Lip region with long and acute bristle-shaped labial sensillae. Stoma having cheilostom with corona; gymnostom a very short wide ring; stegostom with dorsal claw-like and right sub-ventral tooth. Female gonad paired, amphidelphic. Male without bursa. Tail in both sexes filiform.

O. longicauda (Claus, 1862) Fürst von Lieven, 2003 was recorded in thermal waters from Italy (Meyl, 1953, 1954).

Family TYLOPHARYNGIDAE Filipjev, 1934[3]

Lip region with short labial sensillae. Cheilostom short, smooth; stegostom with dorsal claw-like and right sub-ventral tooth; stegostom cylindrical; telostegostom long and narrow with spherical sub-dorsal apodemes at posterior end. Female gonad didelphic, amphidelphic. Male lacking bursa.

Genus *Tylopharynx* de Man, 1876
= *Chitinotylenchoides* Arias & Jiménez-Millán, 1968.
Cuticle lacking striations. Lips fused with short labial sensillae. Cheilostom without rugae; stegostom with dorsal claw-like and right sub-ventral tooth; long and narrow stegostom-cylinder with spherical sub-dorsal apodemes at posterior end. Female gonad paired, amphidelphic. Male without bursa. Tail in both sexes filiform.

T. foetida (Bütschli, 1874) Goffart, 1930 is a species recorded in sludge from waste water from Andalucía, Spain (Santos-Lobatón *et al.*, 2002).

References

Adam, S., Pawert, M., Lehmann, R., Roth, B., Müller, E. and Triebskorn, R. (2001) Physicochemical and morphological characterisation of two small polluted streams in Southwest Germany. *Journal of Aquatic Ecosystem Stress and Recovery* 8, 179–194.

Anderson, R.V. (1992) Free-living nematode associations in Pool 19, Mississippi river. *Journal of Freshwater Ecology* 7, 243–250.

Anderson, R.C. and Bartlett, C.M. (1993) *Daubaylia pearsoni* n. sp. (Nematoda: Daubayliidae) in *Glyptophysa gibbosa* (Planorbidae) in Australia. *Journal of Parasitology* 79, 671–673.

Andrássy, I. (1968) Wissenschaftliche Ergebniss der ungarischen zoologischen Expedition nach Tasanien. 12. Bodennematoden aus der III. Expedition. *Acta Zoologica Academiae Scientiarum Hungaricae* 14, 239–257.

Andrássy, I. (1970) Nematoden aus einigen Fluss-Systemen Südafrikas. *Opuscula Zoologica (Budapest)* 10, 179–219.

Andrássy, I. (1983) *A Taxonomic Review of the Suborder Rhabditina (Nematoda: Secernentia)* ORSTOM, Paris, 241 pp.

Andrássy, I. (1984) *Klasse Nematoda: Ordnungen Monhysterida, Desmoscolecida, Araeolaimida, Chromodorida, Rhabditida.* Fischer, Akademie-Verlag, Stuttgart, Germany, pp. 1–509.

Armendáriz, I. and Hernández, M.A. (1992) Cuatro especies de la familia Bunonematidae Micoletzky, 1922 (Nematoda, Rhabditida) encontradas en pinares de *Pinus nigra* en Navarra (Norte de España) *Miscelánea Zoológica* 16, 7–12.

Bazzanti, M. (2000) Macrobenthic nematodes as biological indicators in a Mediterranean lowland river in Central Italy: a case study. *Archiv für Hydrobiologie* 148, 59–70.

Beier, S. and Traunspurger, W. (2001) The meiofauna community of two small German streams as indicator of pollution. *Journal of Aquatic Ecosystem Stress and Recovery* 8, 387–405.

Beier, S. and Traunspurger, W. (2003a) Seasonal distribution of free-living nematodes in the Krähenbach, a fine-grained submountain carbonate stream in southwest Germany. *Nematology* 5, 113–136.

Beier, S. and Traunspurger, W. (2003b) Seasonal distribution of free-living nematodes in the Körsch, a coarse-grained submountain carbonate stream in southwest Germany. *Nematology* 5, 481–504.

Beier, S. and Traunspurger, W. (2003c) Temporal dynamics of meiofauna communities in two small submountain carbonate streams with different grain size. *Hydrobiologia* 498, 107–131.

Bongers, T. (1990) The Maturity Index: an ecological measure of environmental disturbance based on nematode species composition. *Oecologia* 83, 14–19.

Bongers, T. (1999) The Maturity Index, the evolution of nematode life-history traits, adaptive radiation and *cp*-scaling. *Plant and Soil* 212, 13–22.

Bongers, T. and Bongers, M. (1998) Functional diversity of nematodes. *Applicated Soil Ecology* 10, 239–251.

Bongers, T., Alkemande, R., and Yeates, G.W. (1991) Interpretation of disturbance-induced maturity decrease in marine nematode assemblages by means of the Maturity Index. *Marine Ecology Progress Series* 76, 135–142.

Bongers, T., de Goede, R.G.M., Korthals, G.W. and Yeates, G.W. (1995) Proposed changes of *c–p* classification for nematodes. *Russian Journal of Nematology* 3, 61–62.

Chitwood, B.G. and Chitwood, M.B. (1934) *Daubaylia potomata* n. sp., a nematode parasite of snails with a note on other nemas associated with molluscs. *Proceedings of the Helminthological Society of Washington* 2, 8–9.

Cobb, N.A. (1914) North American freshwater nematodes. *Transactions of the American Microscopical Society* 33, 35–100.

Coomans, A. (1992) Aquatic Nematodes. *Nematologica* 38, 446–447.

Daday, J. 1897 Die freilebenden Sü(wasser-Nematoden Ungarns. *Zoologische Jahrbücher (Systematik)* 10, 91–134.

Dassonville, A.F. (1981) 'n Taksonomie en Ekologiese Studie van Varswater-nematode in Skinnerspruit, Pretoria. Unpublished PhD thesis, Rand Afrikaans University, Johannesburg, South Africa, 222 pp.

Dassonville, A.F. and Heyns, J. (1984) Freshwater nematodes from South Africa. 7. New and known species collected in Skinnerspruit, Pretoria. *Phytophylactica* 16, 15–32.

De Cillis, O.M. (1917) Nuovi generi e nuove specie di nematodi liberi d'acqua dolce. *Monitore Zoologico Italiano* 28, 57–62.

De Ley, P. and Blaxter, M. (2002) Systematic position and phylogeny. In: Lee, D.L. (ed) *The Biology of Nematodes.* Taylor & Francis, London, pp. 1–30 (635 pp.).

De Ley, P. and Blaxter, M. (2004) A new system for Nematoda: combining morphological characters with molecular trees, and translating clades into ranks and taxa. *Nematology Monographs and Perspectives* 2, 633–653.

De Ley, P., Mundo-Ocampo, M. and Tandingan De Ley, I. (2003) *Identification of Free-living Nematodes (Secernentea).* University of California, Riverside, California, 119 pp.

De Ley, P., De Ley, I.T., Esfahani, S. and Abolafia, J. (2004) Nematode communities from the rhizosphere of two vernal pools in the Santa Rosa Plateau Ecological Reserve, California. *International Congress 'Rhizosphere 2004'*, Munich, Germany, 72.

De Mesel, I., Derycke, S., Swings, J., Vincx, M. and Moens, T. (2003) Influence of bacterivorous nematodes on the decomposition of cordgrass. *Journal of Experimental Marine Biology and Ecology* 296, 227–242.

Edmondson, W.T. (1959) *Freshwater Biology*, 2nd edn. John Wiley & Sons, New York, 1248 pp.

Ellenby, C. (1969) Dormancy and survival in nematodes. *Symposium of the Society for Experimental Biology* 23, 83–97.

Eyualem-Abebe and Coomans, A. (1997) Aquatic nematodes from Ethiopia IX. One new and three known species of Dorylaimidae, *Mononchus truncatus* Bastian, 1865, and *Diploscapter coronatus* (Cobb, 1893) Cobb, 1913 (Nematoda). *Hydrobiologia* 353, 121–138.

Eyualem-Abebe, Mees, J. and Coomans, A. (2001) Nematode communities of Lake Tana and other inland water bodies of Ethiopia. *Hydrobiologia* 462, 41–73.

Fisher, K.D. (1968) Population patterns of nematodes in Lake Champlain. *Nematologica* 14, 1–19.

Fürst von Lieven, A. (2002) Functional morphology, origin and phylogenetic implications of the feeding mechanism of *Tylopharynx foetida* (Nematoda: Diplogastrina). *Russian Journal of Nematology* 10, 11–23.

Gadea, E. (1952) Contribución al estudio de los nematodos libres terrestres y dulceacuícolas de la fauna española. *Publicaciones del Instituto de Biología Aplicada, Serie Zoología* 5, 1–213.

Gadea, E. (1963) Sobre la nematofauna del río Carrión. *Publicaciones del Instituto de Biología Aplicada* 35, 141–147.

Gafny, S., Goren, M. and Gasith, A. (2000) Habitat condition and fish assemblage structure in a coastal Mediterranean stream (Yarqon, Israel) receiving domestic effluent. *Hydrobiologia* 422, 319–330.

Gagarin, V.G. (1977) Three new species of nematodes from the order Rhabditida from the regions of sewage discharge. *Zoologischeskii Zhurnal* 56, 1245–1248. (In Russian.)

Gagarin, V.G. (1983) New species of the genus *Mononchoides*, Nematoda, Diplogasterinae. *Zoologicheskii Zhurnal* 62, 25–31. (In Russian.)

Gagarin, V.G. (1993) Freilebende Nematoden asu den Gewässern der Insel Sachalin. In: *Presnovodnye bespozvonochnye: biologija, sistematika, evolyutsija. Trudy Instituta Biologii Vnutrennikh Vod im (St. Petersburg)* 68, 53–77. (In Russian.)

Gagarin, V.G. (1998) A review of the genus *Mononchoides* Rahm, 1928, with description of two new species from European Russia (Nematoda: Diplogasterida). *Zoosystematica Rossica* 7, 229–238.

Gagarin, V.G. (2000) Fauna of free-living nematodes from littoral of Kurilskoe Lake (Russia, Kamchatka Peninsula) and description of *Rhitis eximius* sp. n. (Nematoda, Rhabditida). *Biology of Inland Waters* 3.

Gagarin, V.G. (2001a) New species of free-living nematodes from Biwa Lake and inflowing stream (Honshu Island, Japan). *Zoologicheskii Zhurnal* 80, 12–25. (In Russian.)

Gagarin, V.G. (2001b) Some data on the fauna of saprobiotic nematodes in the Jaroslavi region (Central Russia). *Russian Journal of Nematology* 9, 148.

Gagarin, V.G. and Lemsina, L.V. (1982) A male of *Mononchus aquaticus* and two new species of free-living nematodes from Lake Issyk-kul. *Zoologicheskii Zhurnal* 61, 935–937. (In Russian.)

Goodrich, M., Hechler, H.C. and Taylor, D.P. (1968) *Mononchoides changi* n. sp. and *M. bollingeri* n. sp. (Nematoda: Diplogasterinae), from a waste treatment plant. *Nematologica* 14, 25–36.

Heyns, J. (1976) Preliminary results of a survey of freshwater nematodes in South Africa. *Journal of Limnological Society of South Africa* 2, 43–45.

Heyns, J. (2002a) Checklist of free-living nematodes recorded from freshwater habitats in Southern Africa. *Water SA* 28, 449–456.

Heyns, J. (2002b) Nematoda. In: Day, J.A. and De Moor, I.J. (eds.) *Guides to the Freshwater Invertebrates of Southern Africa, Vol. 5: Non-Arthropods.* Water Research Commission, Pretoria, South Africa.

Hirschmann, H. (1952) Die Nematoden der Wassergrenze mittel-fränkischer Gewässer. *Zoologische Jahrbücher Abteilung für Systematik* 81, 313–436.

Höckelmann, C., Moens, T. and Jüttner, F. (2004) Odor compounds from cyanobacterial biofilms acting as attractants and repellents for free-living nematodes. *Limnology and Oceanography* 49, 1809–1819.

Honnen, W., Rath, K., Schlegel, T., Schwinger, A. and Frahne, D. (2001) Chemical analyses of water, sediment and biota in two small streams in southwest Germany. *Journal of Aquatic Ecosystem Stress and Recovery* 8, 195–213.

Hunt, D. (1980) *Butlerius macrospiculum* n. sp. and *Cylindrocorpus walkeri* n. sp. (Nematoda: Diplogatroidea) from St. Lucia, West Indies. *Revue de Nématologie* 3, 155–160.

Kerry, B.R. and Jaffee, B.A. (1997) Fungi as biological control agents for plant parasitic nematodes. In: Esser, K. and Lemke, P.A. (eds.) *The Mycota*, vol. IV. Springer-Verlag, Heidelberg, Germany, pp. 204–218.

Khera, S. (1969) Nematodes from the banks of still and running waters. VI. Rhabditida from sewer. *Journal of Helminthology* 43, 347–363.

Khera, S. (1970) Nematodes from the banks of still and running waters. IX. Two new genera belonging to subfamily Diplogasterinae Micoletzky from India. *Revista Brasileira de Biologia (Rio de Janeiro)* 30, 405–409.

Kozlowska, J. (1962) Free-living aquatic nematodes (*Nematoda aquatica*) of the ód Upland. *Fragmenta Faunistica* 9, 281–305.

Lees, E. (1953) An investigation into the method of dispersal of *Panagrellus silusiae*, with particular reference to its desiccation resistance. *Journal of Helminthology* 27, 95–103.

Loof, P.A.A. (1973) Freshwater nematodes from Surinam collected by J. van der Land. *Zoologische-Verhandelingen* 129, 1–46.

Manegold, A. and Kiontke, K. (2001) The association of two *Diplogasteroides* species (Secernentea: Diplogastrina) and cockchafers (*Melolontha* spp., Scarabaeidae). *Nematology* 3, 603–606.

Meyl, A.H. (1953) Beiträge zur Kenntnis der Nematodenfauna vulkanish erhitzter Biotope. II. Mitteilung. Die in Thermalgewässern der Insel Ischia vorkommenden Nematoden. *Zeitschrift für Morphologie und Ökologie der Tiere* 42, 159–208.

Meyl, A.H. (1954) Die bisher in Italien gefundenen freilebenden Erd- und Süsswasser-Nematoden. *Archivo Zoologico Italiano* 39, 161–264.

Meyl, A.H. (1957) Two new free-living nematodes, found in the rainwater reserve of *Quesnelia arvensis* (Vell.) Mez. (Bromeliaceae) from Brazil. *Proceedings of the Helminthological Society of Washington* 24, 62–66.

Michiels, I.C. and Traunspurger, W. (2004) Impact of resource availability on species composition and diversity in freshwater nematodes. *Oecologia* 142, 98–103.

Micoletzky, H. (1916) Süßwasser-Nematoden aus Südafrika. Ergebnisse einer botanischen Forschungsreise nach Deutsch-Ostafrika una Südafrika (Kapland, Natal und Rhodesien). *Denkschriften der Kaiserlichen Akademie der Wissenschaften Wien* 92, 149–171.

Micoletzky, H. (1922) Die freilebenden Erd- Nematoden mit besonderer Berücksichtigung der Steiermark und der Bukowina, zugleich mit einer Revision sämtlicher nicht mariner, freilebender Nematoden in Form von Genus-Beschreibungen und Bestimmungsschlüsseln. *Archiv für Naturgeschichte*, 87, 1–650.

Niemann, R., Arens, M., Koczwana, K. and Sturhan, D. (1996) Untersuchung über die Eignung von Nematoden zur Güteberwertung von Fliessgewässem. *Mitteilungen aus des Biologischen Bundesanstal für Land- und Forstwirtschaft Berlin-Dahlem* 317, 195–208.

Ocaña, A. (1992) Relationship between nematode species and physicochemical characteristics of spring water. III. pH, oxygen and iron. *Nematologica Mediterranea* 20, 181–186.

Ocaña, A. and Picazo, J.S. (1991) Study on nematode species encountered in the Monachil River (Granada, Spain): response to organic pollution. *Verhandlungen – Internationale Vereinigung für Theoretische und Angewandte Limnologie* 24, 2729–2737.

Picazo, J.S., Ocaña, A. and Jiménez-Millán, F. (1989) Grupos tróficos de nematodos en el bentos del río Monachil (Granada). *Miscelánea Zoológica* 13, 1–11.

Picazo, J.S., Ocaña, A. and Jiménez-Millán, F. (1990) Estudio del orden Rhabditida (Nematoda) en un río con problemas de contaminación orgánica (río Monachil, Granada). *Zoologica Baetica* 1, 53–75.

Pillai, J.K. and Taylor, D.P. (1968) *Butlerius micans* n. sp. (Nematoda: Diplogasterinae) from Illinois, with observations on its feeding habits and a key to the species of *Butlerius* Goodey, 1929. *Nematologica* 14, 89–93.

Pott, F.A. (1910) Notes on the free-living nematodes. *Quarterly Journal of the Microbiological Society* 55, 433–484.

Riddle, D.L. (1988) The dauer larva. In: Wood, W.B. (ed.) *The Nematode* Caenorhabditis elegans. Cold Spring Harbor Laboratory Press, Cold Spring Harbor, New York, pp. 393–412 (667 pp.).

Riddle, D.L. and Albert, P.S. (1997) Genetic and environmental regulation of dauer larva development. In: Riddle, D.L., Blumenthal, T., Meyer, B.J. and Preiss, J.R. (eds.) C. elegans *II*. Cold Spring Harbor Laboratory Press, Cold Spring Harbor, New York, pp. 1–22 (1222 pp.).

Santos-Lobatón, C., Trujillo-Mena, E. and Casaus-Álvarez, M. (2002) Nematodes as indicators of waste water treatment. Fourth International Congress of Nematology, Tenerife (Spain). *Nematology* 4, 254–255.

Schneider, W. (1937) Freilebende Nematoden der Deutschen Limnologischen Sundaexpedition nach Sumatra, Java und Bali. *Archiv für Hydrobiologie* Suppl. 15, 30–108.

Schneider, W. (1938) *Diplogaster pararmatus* n. sp. nebst Bemerkungen über einige andere *Diplogaster*-Arten. *Zoologischer Anzeiger* 121, 37–43.

Schouten, A.J., van Esbroek, M.L.P. and Alkemade, J.R.M. (1998) Dynamics and stratification of functional groups of nematodes in the organic layer of a Scots pine forest in relation to temperature and moisture. *Biology and Fertility of Soils* 26, 293–304.

Schuurmans Stekhoven, J.H. (1943) Nématodes recueillis dans des grottes et des sources en Belgique. *Bulletin du Musée Royal d'Histoire Naturelle de Belgique* 19, 1–20.

Schuurmans Stekhoven, J.H. and Teunissen, R.J.H. (1938) Nématodes libres terrestres. *Exploration du Parc National Albert, Mission G. F. de Witte (1933–1935) Fasc.* 22, 1–229.

Shoshin, A.V. (1989) New namatode species of the families Diplogasteridae and Diplogasteroididae from the Baikal Lake. In: Ryss, A.Y. (ed.) *Tylenchida and Rhabditida (Nematoda) – Parasites of Plants and Insects*. Leningrad. *Trudy Zoologiceskogo Instituta Akademia Nauk SSSR* 194, 83–95.

Stefanski, W. (1916) Die freilebenden Nematoden des Inn, ihre Verbreitung und Systematik. *Zoologischer Anzeiger* 46, 363–384.

Stefanski, W. (1926) Sur les nématodes libres des torrents de Sinaia (Roumanie). *Sonderdruck aus Internationale Revue der gesamten Hydrobiologie und Hydrographie* 15, 347–357.

Stefanski, W. (1927) Les nématodes libres des torrents de Sinaia avec les considérations sur les nématodes des torrents en général. *Publicatiunile Societatii Naturalistilor din Romania* 9, 3–53.

Steinberg, C.E.W. (2003) *Ecology of Humic Substances in Freshwaters Determinants from Geochemistry to Ecological Niches*. Springer-Verlag, Heidelberg, Germany, 440 pp.

Steinberg, C.E.W., Höss, S., Kloas, W., Meinelt, T., Pflugmacher, S. and Wiegand, C. (2004) *Hormone-like Effects of Humic Substances on Fish, Amphibians, and Invertebrates.* Wiley InterScience (online), pp. 409–411.

Steiner, G. (1914) Freilebende Nematoden aus der Schweiz. 1.+2. Teil. *Archiv für Hydrobiologie und Planktonkunde* 9, 259–276, 420–438.

Strayer, D. (1985) The benthic micrometazoans of Mirror Lake, New Hampshire. *Archiv für Hydrobiologie* 3 (Suppl. 72), 287–426.

Sudhaus, W. and Fitch, D. (2001) Comparative studies on the phylogeny and systematics of the Rhabditidae (Nematoda). *Journal of Nematology* 33, 1–70.

Sudhaus, W. and Fürst von Lieven, A. (2003) A phylogenetic classification and catalogue of the Diplogastridae (Secernentea, Nematoda). *Journal of Nematode Morphology and Systematics* 6, 43–90.

Tahseen, Q., Siddiqi, M.R. and Rowe, J. (2002) Study on species of *Diploscapter* Cobb (Nematoda: Rhabditida) from India including *D. indicus* sp. n. *International Journal of Nematology* 12, 183–188.

Taylor, D.P. (1964) *Butlerius monhystera* (Nematoda: Diplogastrinae), a new species of predacious nematodes from Illinois. *Proceedings of the Helminthological Society of Washington* 31, 129–132.

Timm, R.W. (1971) *Antarctic soil and freshwater nematodes from McMurdo Sound region.* Proceedings of the Helminthological Society of Washington 38, 42–52.

Traunspurger, W. (1997) Bathymetric, seasonal and vertical distribution of feeding-types of nematodes in an oligotrophic lake. *Vie Milieu* 47, 1–7.

Traunspurger, W. (2000) The biology and ecology of lotic nematodes. *Freshwater Biology* 44, 29–45.

Tsalolikhin, S.Y. (1972) New species of free-living nematodes from the Baikal lake. *Zoologicheskii Zhurnal* 51, 1559–1563. (In Russian.)

Tsalolikhin, S.Y. (1980) *Free-living Nematodes of Baikal.* Nauka, Novosibirsk, Russia, 120 pp. (In Russian.)

Tudorancea, C. and Zullini, A. (1989) Associations and distribution of benthic nematodes in the Ethiopian Rift Valley lakes. *Hydrobiologia* 179, 81–96.

Vidakovic, J. and Jakumetovic, A. (1998) Meiofauna and free-living aquatic nematodes from some stations along the Drava River (Croatia). *Natura Croatica* 7, 349–358.

Volterra, L., Formichetti, P. and Mancini, L. (1999) Metodi per la ricerca e identificazione di nematodi in acque destinate al consumo umano. *Biologia Ambientale* 1, 3–18.

Wanless, F.R. and Hunter, R. (2001) Freshwater nematodes from Loch Ness, Scotland, Part I. The orders Tylenchida Thorne, 1949 and Rhabditida Chitwood, 1933 (Nematoda, Secernentea). *Bulletin-Natural History Museum London (Zoology)* 67, 1–24.

Ward, H.B. and Whipple, G.C. (1959) *Fresh-water Biology.* Chapman & Hall, London, 1248 pp.

Warwick, R.M. (1987) Meiofauna: their role in marine detrital systems. Detritus and microbial ecology in aquaculture. *ICLARM Conference Proceedings* 14, 282–295.

Wu, J. and Liang, Y. (1999) A comparative study of benthic nematodes in two Chinese lakes with contrasting sources of primary production. *Hydrobiologia* 411, 31–37.

Yeates, G.W. (1999) Effects of plants on nematode community structure. *Annual Review of Phytopathology* 37, 127–149.

Zeidan, A.B. and Geraert, E. (1989) Free-living nematodes from Sudan. *Nematologica* 35, 277–304.

Zullini, A. (1974) The nematological population of the Po River. *Bolletino di Zoologia* 41, 183–210.

Zullini, A. (1976) Nematodes as indicators of river pollution. *Nematologia Mediterranea* 4, 13–22.

Zullini, A. (1981) Three new species of Diplogastrids (Nematoda). *Bollettino del Museo Civico di Storia Naturale di Verona* 7, 395–404.

Zullini, A. (1982) Nematodi (Nematoda). En: *Guide per il riconoscimiento delle specie animali delle acque interne italiane* 17, 1–117.

Zullini, A. (1988) The ecology of the Lambro River. *Rivista di Idrobiologia* 27, 39–58.

Zullini, A. and Ricci, C. (1980) Bdelloid rotifers and nematodes in a small Italian stream. *Freshwater Biology* 10, 67–72.

Appendix
Order Mermithida and families *incertae sedis* (Bastianiidae and Rhabdolaimidae)

Oleksandr Holovachov[1] and Paul De Ley[2]

[1] *Department of Zoology, Biological Faculty, Ivan Franko National University of Lviv, Hrushevsky str. 4, Lviv 79005, Ukraine.* [2] *Department of Nematology, University of California, Riverside CA 92521, USA.*

Order Mermithida

The Mermithida are a biologically fascinating group of invertebrate parasites that show up relatively sporadically in nematode samples from freshwater habitats. However, their numbers can be overwhelming in those locations where they do occur. Their global diversity and local ecological importance are rarely fully appreciated by freshwater ecologists or by nematode taxonomists specializing in other orders. Mermithids can locally suppress the insect populations that are targeted by them, while the nematodes in turn provide food sources for other aquatic organisms such as e.g. copepods (Platzer and Mackenzie-Graham, 1980) or aquatic fungi (Platzer and Platzer, 1999). Although steinernematids have recently attracted greater commercial and scientific attention, mermithids continue to be studied intensively for potential or actual biocontrol applications. For example, depending on other factors, mermithid nematodes may (Buxton, 1989) or may not (Takaoka, 1994) succeed in naturally infecting a large fraction of the blackfly vectors of human or bovine onchocerciasis, thereby presumably significantly influencing the local prevalence and epidemiology of these diseases.

In view of this importance, the editors of this book had hoped to be able to include a full chapter on mermithids, but unfortunately circumstances beyond their control conspired to preclude at this time a suitably detailed and complete treatment. Instead, we will try to highlight in this part of the appendix some of the many special features of these organisms in a very short summary of their biology, and also to provide an undoubtedly very approximate list of genera reported from freshwater habitats. The detailed morphological and taxonomic study of 16 mermithid genera, both aquatic and terrestrial, followed by the brief discussion of life cycles and host specificity was published by Nickle (1972). The only detailed review of aquatic mermithids known to us, is the two-volume monograph by Rubzov published in Russian in 1972 and 1974 and in English translation in 1981. This work contains a treasure trove of biological and taxonomic detail for 152 valid aquatic species of the family Mermithidae, 138 of which were described from Eurasia, 11 from North America, one from South America and two from Africa (Rubzov, 1972, 1974). The author also gave a list of 139 recently described aquatic species of the family Mermithidae, most of which were also found in Eurasia but had not been covered in the two volumes (Rubzov, 1973). However, this important monograph on aquatic mermithids does not reflect the vast amount of more recent papers presenting e.g. taxonomic changes or physiological studies. Subsequent publications by scientists from the former Soviet Union deal mainly with the biology and systematics of Mermithida in general, or with terrestrial species only (Rubzov, 1977, 1978; Artyukhovsky, 1990; Gafurov, 1996; Kharchenko, 1999; Gubaidulin and Bekturganov, 2001). An identification key to North American mermithid genera was compiled by Poinar (2001). Although recently published, this again is a geographically limited work and furthermore it does not seek to provide monographic levels of detail.

A census of over 2000 abstracts from papers published on mermithids during the last 30 years yielded an estimated count of over 40 genera and more than 400 species reported from freshwater sites around the world, with more than 70 genera of Mermithida in total. No doubt our survey of the literature is very incomplete, while the overwhelming majority of the reported genera and species were described from temperate and subarctic locations in the northern hemisphere. It is certain that additional species were described in the literature, and that many hundreds of species remain to be discovered, especially in the tropics as well as further south. On the other hand, the specialized literature also contains repeated expressions of deep frustration with published descriptions, such as the estimate by Curran and Hominick (1981) that 70 of the then known 99 published species of *Gastromermis* had to be considered *species inquirendae* due to the inadequacy of their original descriptions. It is therefore quite likely that many proposed species are actually synonymous with one another, or at least that their diagnostic characters will never be redescribed in sufficient detail to be of practical use in future surveys.

Morphology. Long slender worms often attaining a length of 1–50 cm. Cuticle smooth, but may contain crisscross fibres. Head with four submedian and two lateral papillae. Amphids present, pouches may be very large in some aquatic forms. Mouth and oesophagus probably nonfunctional in adults. Oesophagus (stichosome) non-muscular, with (Mermithidae) or without four large cells (Tetradonematidae). Stoma of preparasitic and postparasitic larvae with a piercing tooth. Intestine in adult replaced by a pseudo-intestine or trophosome, serving as a storage organ; anus absent. Excretory system usually reduced. Gonads paired in both sexes. Vulva equatorial in position and leading to a muscular S- or barrel-shaped vagina. Uteri filled with eggs in sexually mature specimens. Eggs with or without byssi, variable in size depending on genus. Spicules paired or single. Male tail papillae numerous, arranged in three or more rows, rows often bifurcating around opening for spicule. Larval stages parasitic in terrestrial and aquatic invertebrates. Adults free living, non-feeding (Mermithidae) or persisting in host at least until mating (Tetradonematidae).

Biology. The free-living stages of many mermithids are aquatic, and can occur locally in substantial numbers in sandy sediments, swimming actively in the water column, or suspended on the water surface. Based on observations of a moult in the egg in *Romanomermis culicivorax* (Poinar and Otieno, 1974), it is assumed that mermithids hatch as second-stage juveniles. These preparasites emerge from their eggs in synchronized batches and swim upwards or crawl over the sediment surface, until they are passively ingested by larger invertebrates or until they actively penetrate the body wall of a suitable host. The host is typically an aquatic dipteran larva, although mermithids have been found in a much wider range of animals, including not only other aquatic insects but also other Arthropoda: many kinds of terrestrial insects; spiders and opilionids (Poinar, 2000); terrestrial, freshwater and even marine crustaceans (Poinar *et al.*, 2002); terrestrial molluscs (Grewal *et al.*, 2003); and marine nematodes (Tchesunov and Spiridonov, 1993). Species that parasitize predatory terrestrial arthropods, such as spiders, often use aquatic insects as paratenic hosts. Other mermithids have become entirely terrestrial, their free-living stages occurring in soil or even in aboveground vegetation, wet mulch and rotting wood (personal observation). In practice it can be as difficult to draw the line between freshwater and terrestrial habitats in these nematodes as in the many other groups presented in this book. A particular case in point is species like *Octmyomermis troglodytis*, which is a specialist of mosquitoes breeding in tree hollows rather than in surface waters (Washburn *et al.*, 1986). However, Rubzov (1977) stated that there is a generally clear taxonomic distinction between freshwater and terrestrial genera.

Once inside a suitable host that is unable to mount an effective immunological response, the infective juvenile will grow and moult repeatedly, feeding off host tissues and dissolved nutrients with the assistance of a more or less strongly modified resorptive body wall. In the family Mermithidae, a thicker cuticle will be formed after the third moult and usually, when prompted by external stimuli (e.g. contact with water), the fourth stage juvenile will erupt from its dying host and move down through water and into sediment. These preadult postparasitic juveniles can persist on the sediments for days or weeks, until a synchronized stimulus triggers moulting to adulthood, which is often immediately followed by copulation in large mating clusters. Eventually, fertilized females will deposit large numbers of eggs, and the cycle can then repeat itself. In the family Tetradonematidae, on the other hand, juveniles will remain inside the host to reach adulthood and mate, after which only the females will erupt from the host.

Mermithid biogeography has long puzzled mermithid experts, most species exhibiting an exceedingly patchy distribution pattern. Adjacent locations with similar conditions and identical hosts often differ in which species occur, or whether mermithids occur at all. The infective juveniles are relatively uncommon in sediments in terms of distribution, but they can occur in very substantial numbers when conditions are suitable, typically in sandy deposits of well-oxygenated oligotrophic waters that are free from pollutants and tannins (Rubzov, 1981).

Development. The life cycle of Mermithidae is rather reminiscent of the phylum Nematomorpha, and highly unusual compared to other metazoan endoparasites, in that the adults live outside the host. As with nematomorphs, adult mermithids can reach surprising body lengths compared to the size of their host, and in fact some of the larger species of mermithids could easily be mistaken for nematomorphs when seen moving through the water. Mermithid identification is traditionally based almost entirely on adult morphology, and the literature contains many examples where species are described from a single adult of a single sex, or where unidentifiable juveniles were found inside their hosts or in sediments. A few limited attempts have been made to develop identification keys for first-stage juveniles (Luo and Bao, 1991) but it is likely that the first real successes will only be achieved now that highly efficient techniques for molecular diagnostics have become available.

The second-stage juveniles of mermithids, although typically unidentifiable to species, are easily distinguished from other nematodes by the presence of a protruding tooth and a modified anterior intestinal structure known as the *stichosome* (Appendix Fig. 1). The remainder of the intestine typically lacks a lumen and is known as the *trophosome*. These juveniles are often similar in size to free-living nematodes and are readily extracted with the same variety of techniques as other nematodes. Mermithid preadults and adults, by contrast, are several orders of magnitude larger than any other nematodes sharing the same sediments; they are best collected by flotation and sieving sediments (Cagnolo *et al.*, 1999) or by collecting insect larvae and allowing the nematodes to emerge and moult in the laboratory.

Phylogeny. The earliest fossil representative of a putatively aquatic mermithid lineage is *Cretacimermis libani* from adult Chironomidae in 135 million-year-old Lebanese amber (Poinar *et al.*, 1994). Other more recent records of fossil Mermithidae, particularly of the genus *Heydenius*, are associated with various terrestrial arthropods. Molecular data indicate that Mermithida are the sister order to Mononchida (Blaxter *et al.*, 1998; Rusin *et al.*, 2003)

Taxonomy. We conclude this brief review with a list of nominal genera and species, those which are marked by the asterisk were reported at least once from freshwater habitats in the literature available to us. As noted above, our list is undoubtedly inaccurate and incomplete. Different authors have proposed different synonymies between various genera, further complicating matters. We therefore refer authors to the relevant existing taxonomic revisions.

Family MERMITHIDAE Braun, 1883

List of genera

Abathymermis Rubzov, 1971*
Agamermis Cobb, Steiner & Christie, 1923
Agamomermis Stiles, 1903*
Allomermis Steiner, 1925
Amphibiomermis Artyukhovskii, 1969*
Amphidomermis Filipjev, 1934
Amphimermis Kaburaki & Imamura 1932*
Aranimermis Poinar & Benton, 1986*
Aquamermis Rubzov, 1973*
Austromermis Poinar, 1990
Baikalomermis Rubzov, 1976*
Bathymermis Daday, 1911*

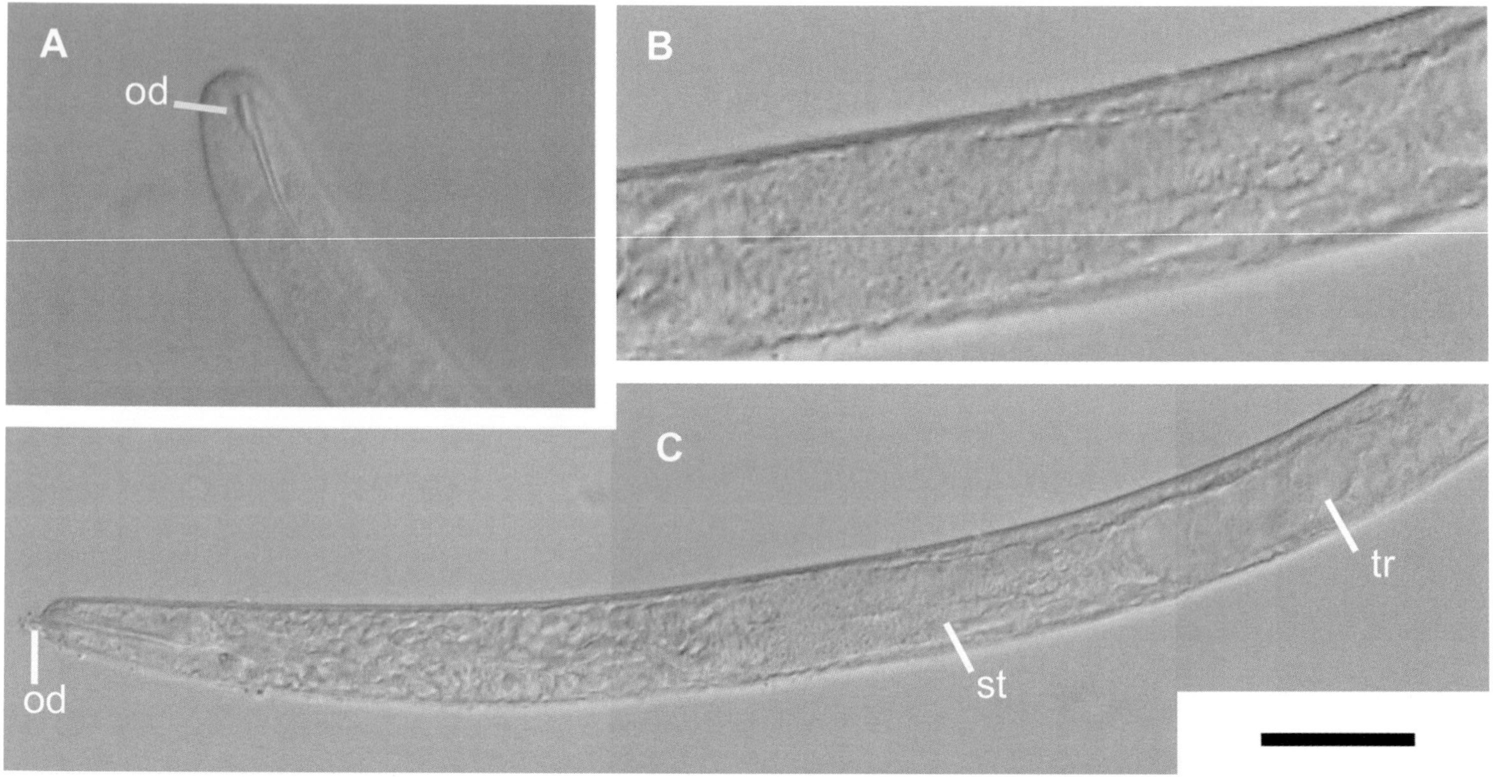

Appendix Fig. 1 Anterior morphology of infective stage juveniles of mermithids. *Strelkovimermis spiculatus* Poinar & Camino, 1986, a freshwater species from Argentina with free-swimming preparasites: A. Lip region with 'odontostyle' (od) in a live specimen. Unidentified mermithid from mulched avocado orchard in California; B. Stichosome of a specimen in glyerin mount; C. Anterior half of body with 'odontostyle' (od), stichosome (st) and part of trophosome (tr) of the same specimen as B. Scale bar is 6 μm for A, 10 μm for B and 20 μm for C. *S. spiculatus* kindly provided by Drs R. Perez Pacheco and E. Platzer.

Blepharomermis Poinar, 1990
Brevimermis Rubzov, 1972*
Capitomermis Rubzov, 1968*
Ceratomermis Rubzov, 1977
Coccinellimermis Rubzov, 1977
Comanimermis Artyukhovsky, 1969
Complexomermis Filipjev, 1934
Culicimermis Rubzov & Isaeva, 1975*
Decamermis Rubzov, 1977
Dendromermis Rubzov & Polishchuk, 1975*
Divisispiculimermis Doucet, 1986
Dipteromermis Rubzov, 1976
Discomermis Coman, 1969
Diximermis Nickle, 1972*
Drilomermis Poinar & Petersen, 1978*
Empidomermis Poinar, 1977*
Eumermis Daday, 1911*
Eurymermis Müller, 1931*
Filipjevimermis Pologentsev & Artyukhovsky, 1958
Gammaromermis Rubzov & Bekman, 1979*
Gastromermis Micoletzky, 1923*
Heleidomermis Rubzov, 1970*
Hexamermis Steiner, 1924*
Hydromermis Corti, 1902*
Ipatjevimermis Mukhamedzyanova, 1980*
Isomermis Coman, 1953*
Isthmusimermis Gafurov, 1980*
Kurshymermis Zahidov & Poinar, 1970
Lanceimermis Artyukhovskii, 1969*
Limnomermis Daday, 1911*
Linstovimemis Rubzov, 1977
Lugamermis Rubzov, 1977
Megalomermis Müller, 1931
Melolonthimermis Arthyukhovsky, 1963
Mermis Dujardin, 1845*
Mesomermis Daday, 1911*
Neolimnomermis Rubzov, 1977*
Neomermis Linstow, 1904*
Neomesomermis Rubzov, 1972*
Notonectomermis Rubzov, 1977*
Octomermis Steiner, 1929*
Octomyomermis Johnson, 1963*
Odontomermis Rubzov, 1973*
Oesophagomermis Artyukhovskii, 1969*
Onchiomermis Rubzov, 1976
Orthomermis Poinar, 1965
Ovomermis Rubzov, 1977

Paramermis Linstow, 1898*
Pentatomermis Rubzov, 1977
Perutilimermis Nickle, 1972*
Pheromermis Poinar, Lane & Thomas, 1976*
Phreatomermis Coman, 1953*
Pologenzevimermis Kirjanova, Karavajeva & Romanenko, 1959
Psammomermis Pologenzev, 1941
Pseudomermis de Man, 1903*
Quadrimermis Coman, 1961*
Reesimermis Tsai & Grundman, 1969*
Rhynchomermis Rubzov, 1977
Romanomermis Coman, 1961*
Schuurmanimermis Rubzov, 1977
Schmassmannimermis Rubzov, 1977
Scirtimermis Kaiser & Klausnitzer, 2001
Skrjabinomermis Pologenzev, 1952
Spiculimermis Artyukhovskii, 1963*
Strelkovimermis Rubzov, 1969*
Telomermis Johnson & Bowen, 1990*
Tetramermis Steiner, 1925*
Thalassomermis Tchesunov & Hope, 1992
Thaumamermis Poinar, 1981
Thraustomermis Song & Peng, 1995
Torsiomermis Kaiser & Schwank, 1985*
Trophomermis Johnson & Kleve, 1993
Tunicamermis Schuurmans Stekhoven, Mawson & Couturier, 1955
Unicamermis Rubzov, 1982*
Utriculimermis Kaiser & Deixelberger, 2001*
Welchimermis Rubzov, 1977

Family TETRADONEMATIDAE Cobb, 1919
List of genera
Aproctonema Keilin, 1917*
Bispiculum Zervos, 1980
Brevinema Rubzov, 1977
Brissonema Rubzov, 1977
Corethrellonema Nickle, 1969
Crassinema Rubzov, 1977
Didilia Tang, Hominick, Killick-Kendrick, Killick-Kendrick & Page, 1993
Heterogonema Waerebecke & Remillet, 1973
Mermithonema Goodey, 1941*
Nematimermis Tchesunov & Spiridonov, 1993
Paraproctonema Rubzov, 1977
Tetradomermis Rubzov, 1978*
Tetradonema Cobb, 1920
Trichonema Rubzov, 1977

Families *incertae sedis* (Bastianiidae and Rhabdolaimidae)

Taxonomy. In the classification scheme of the phylum Nematoda by De Ley and Blaxter (2002), the families Bastianiidae and Rhabdolaimidae, which also include freshwater representatives, were both placed within the superfamily Leptolaimoidea of the order Plectida. However, several recent morphological studies questioned the correctness of this taxonomic position: (i) Holovachov and Boström (2004) proposed to move families Bastianiidae and Odontolaimidae to the order Triplonchida, close to the family Prismatolaimidae (see also Coomans and Raski, 1988); (ii) based on the stoma morphology of the genus *Udonchus*, Eyualem and Coomans (1996) transferred the family Rhabdolaimidae to the superfamily Microlaimoidea of the order Desmodorida, while Holovachov and Boström (2004) pointed out the strong similarities between the genus *Tobrilia* and the family Tripylidae. At the time of writing, several molecular studies are underway that include the first DNA sequences from some of the relevant genera (see http://nematol.unh.edu/ for a provisional tree). In view of these various current opinions and forthcoming new data, we decided to omit the Bastianiidae and Rhabdolaimidae from Chapter 20 (Plectida) and instead to treat them separately for the time being as taxa of uncertain taxonomic position.

Family *incertae sedis* BASTIANIIDAE de Coninck, 1935

Morphology. Cuticle annulated, without any ornamentation (Appendix Fig. 2). Lateral differentiations of cuticle, deirids, body pores and hypodermal glands absent. Somatic sensilla present, setiform. Inner labial sensilla papilliform, located on the anterior surface of lips. Outer labial and cephalic sensilla setiform, arranged in single or in two separate circles, outer labial sensilla jointed and larger than cephalic ones. Amphid ventrally or dorsally spiral, unispiral or crescent-shaped. Secretory–excretory system absent. Stoma small, funnel-shaped, conisting of undifferentiated but visible cheilo- and gymnostomatal homologues. Pharynx uniformly cylindrical, muscular, not subdivided; without pharyngeal tubes or valves. Pharyngeal glands (nuclei and orifices) indistinct. Cardia free. Rectum short. Female reproductive system didelphic. Ovarian branches reflexed antidromously. Spermatheca axial. Male reproductive system diorchic, with opposed testes. Precloacal supplements present, papilliform or setiform in shape, in some species extending into pharyngeal region. Spicules paired and symmetrical, slender, enclosed in a muscular sheath. Gubernaculum present. Caudal glands and spinneret absent.

Taxonomy. The family is not subdivided into separate subfamilies. Two genera are currently recognized: the monotypic genus *Dintheria* inhabits soil in grassland habitats, while some species of the type genus *Bastiania* occur in freshwater sediments.

Genus *Bastiania* de Man, 1876

Morphology. Cuticle annulated, with scattered somatic setae. Outer labial and cephalic sensilla setiform, arranged in single or in two separate circles. Amphid crescent-shaped. *Organellum ovale* present in some species, opening on the dorsal body side, a few body diameters posterior to cardia. Stoma small, funnel-shaped. Female reproductive system didelphic. Vulva straight, without sclerotizations. Ventral papillae

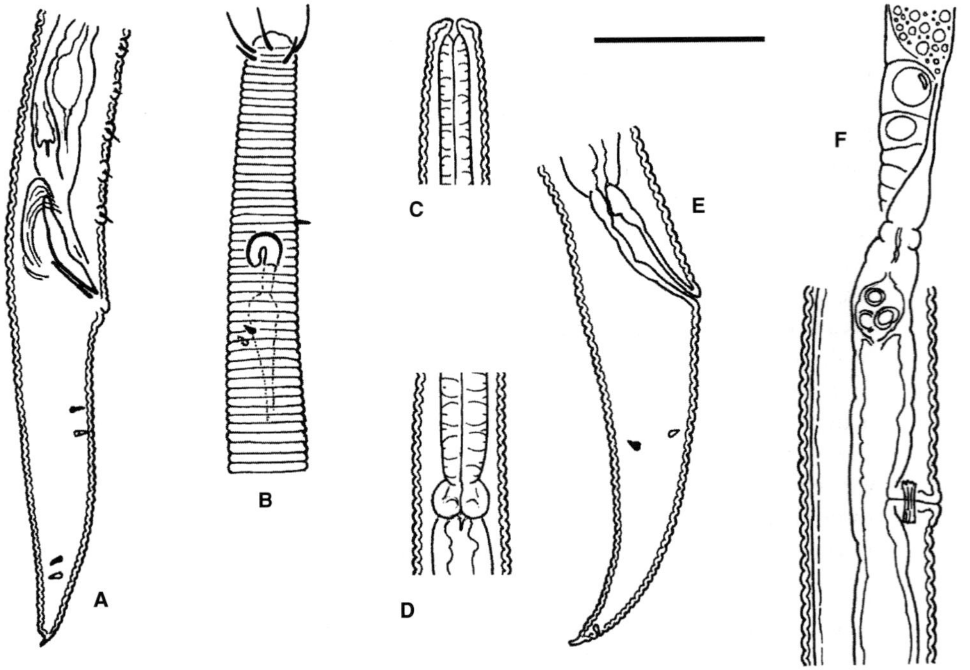

Appendix Fig. 2 Morphology of the genus *Bastiania* de Man, 1876 as exemplified by *B. gracilis* de Man, 1876 from Ukraine. A: Male caudal region; B: Anterior end, surface view; C: Anterior end, median section; D: Pharyngo-intestinal junction; E: Female tail; F: Female reproductive system. Scale bar: A–F = 20 µm.

present in females of some species. Male reproductive system diorchic. Precloacal supplements small papilliform or setiform, in some species extending into pharyngeal region. Spicules with partial capsular musculature sheath. Tail subcylindrical to elongate–conoid, with conical terminal part, usually with small terminal mucro.
Biology. Amphimictic. Inhabit freshwater sediments, soil, litter and moss. Among all species only *B. gracilis* was regularly found in aquatic habitats.
Bibliography. The most recent identification key is given in Andrássy (1991).

Type species: *B. gracilis* de Man, 1876

Other species:
B. acarayensis Andrássy, 1968
B. australis Cobb, 1893
B. eroshenkoi Andrássy, 1991
B. exilis Cobb, 1914
B. longicaudata de Man, 1880
B. octopapillata Meyl, 1954, 1961
B. papillata Meyer & Coomans, 1977
B. parexilis De Coninck, 1935
B. uncinata Andrássy, 1991
B. vesca Eroshenko, 1977

Family *incertae sedis* RHABDOLAIMIDAE Chitwood, 1951

Morphology. Cuticle finely annulated, without any ornamentation. Lateral differentiations of cuticle, deirids, somatic sensilla, body pores and hypodermal glands absent. Outer labial sensilla pore-like or papilliform, cephalic sensilla papilliform or setiform. Amphid oval or stirrup-shaped. Secretory–excretory system absent or present. Stoma tubular or funnel-shaped, with small cheilo- and gymnostom and developed stegostom, the latter with distinct or minute dorsal denticle and often with additional smaller denticles. Pharynx muscular, cylindrical anteriorly, with basal bulb or cylindrical posterior portion. Pharyngeal glands (nuclei and orifices) indistinct. Cardia small. Rectum short. Female reproductive system didelphic or monodelphic. Ovarian branches reflexed antidromously or outstretched. Spermatheca absent. Male reproductive system diorchic or monorchic. Supplements present or absent. Spicules paired and symmetrical, straight or arcuate. Gubernaculum usually absent. Tail elongate conoid. Caudal glands and spinneret present.

Taxonomy. The family is subdivided into two subfamilies: Monochromadorinae with a single aquatic genus, *Udonchus*, and Rhabdolaiminae, which covers the remaining three genera *Rhabdolaimus, Rogerus* (both freshwater) and *Tobrilia* (terrestrial).

Biology. Aquatic (inhabit saline and freshwater sediments) and terrestrial (soil inhabiting).

Key to subfamilies of Rhabdolaimidae

1. Cheilostom with twelve longitudinal rugae; stoma with three denticles, which are arranged in two separate groups at two levels; female reproductive system monodelphic .... Monochromadorinae (monotypic, *Udonchus*)
   Cheilostom without rugae; stoma with three denticles, which are arranged in one group, all at the same level; female reproductive system didelphic .... Rhabdolaiminae

Subfamily Monochromadorinae Andrássy, 1958
Syn. = *Udonchinae* Eyualem & Coomans, 1996 syn. n.

Genus *Udonchus* Cobb, 1937
Syn. = *Monochromadora* Goodey, 1951

Morphology. Outer labial and cephalic sensilla papilliform (Appendix Fig. 3). Amphid stirrup-shaped, with transverse slit-like aperture. Stoma tubular, with weakly developed cheilo- and gymnostom and sclerotized tubular stegostom. Three denticles arranged in stoma as following: one large dorsal and one smaller ventral located near the anterior edge of stegostom, and one small left lateral denticle near the middle of stegostom. Pharynx muscular, cylindrical anteriorly, with basal bulb. Cardia surrounded by gland-like cells, not embedded into intestine. Female reproductive system usually mono–prodelphic, but in some populations didelphic with asymmetrically developed branches (in *U. tenuicaudatus*, see Ocaña, 1991). Ovarian branches reflexed antidromously. Tail elongate conoid. Caudal glands and spinneret present.

Biology. Thelytokous. Inhabit freshwaters and thermal springs.

Bibliography. SEM observations were published by Eyualem & Coomans (1996).

Type species: *U. tenuicaudatus* Cobb, 1913

Other species:
U. crassicauda Schneider, 1937 and Schiemer, 1973
U. merhatibebi Eyualem & Coomans, 1996

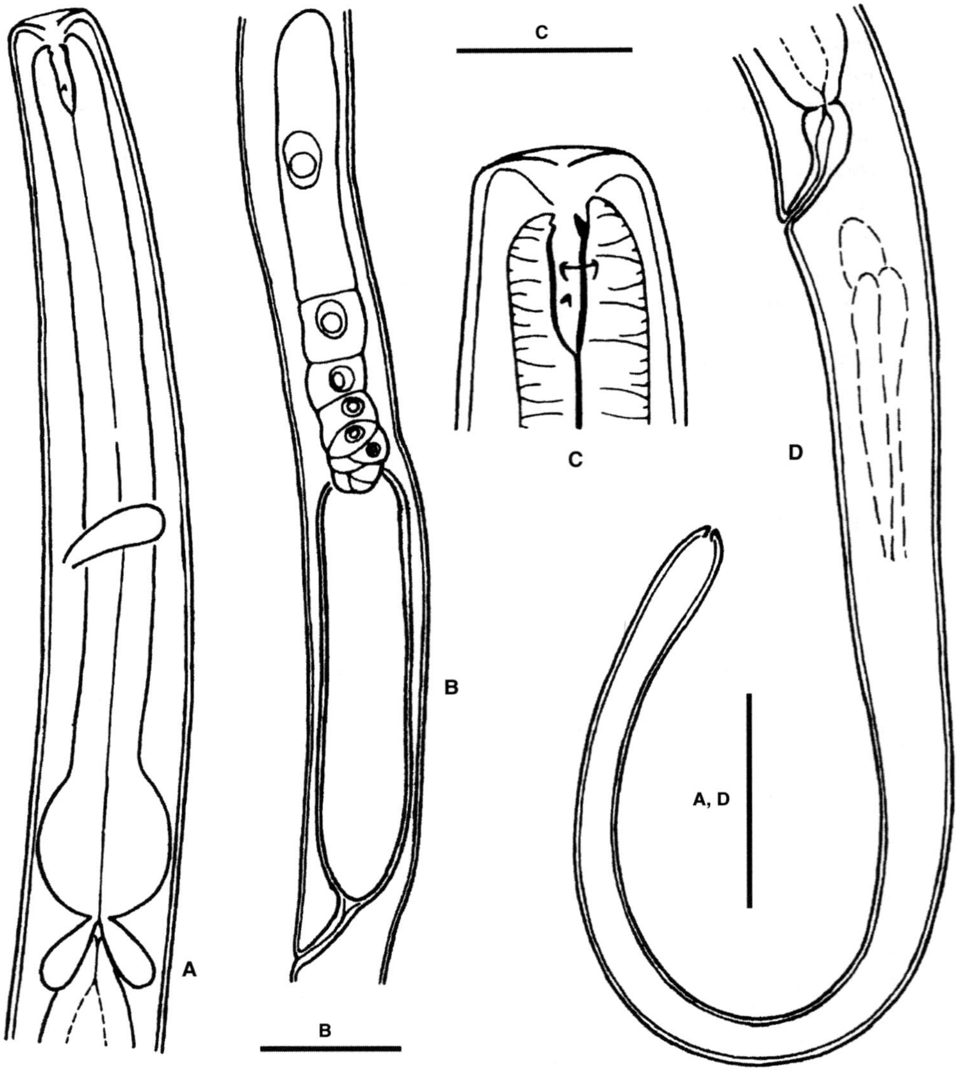

Appendix Fig. 3 Morphology of the genus *Udonchus* Cobb, 1937 as exemplified by *U. merhatibebi* Eyualem & Coomans, 1996 from Ethiopia (modified from Eyualem & Coomans, 1996). A: Pharyngeal region; B: Female reproductive system; C: Anterior end; D: Female tail. Scale bars: A, B, D = 20 μm, C = 10 μm.

Subfamily Rhabdolaiminae Chitwood, 1951

Syn. = *Rogerinae* Andrássy, 1976

Morphology. Outer labial sensilla pore-like or papilliform, cephalic sensilla papilliform or setiform. Amphid oval or stirrup-shaped. Secretory–excretory system absent or present. Stoma tubular or funnel-shaped, with small cheilo- and gymnostom and developed stegostom, the latter with distinct dorsal denticle and often with smaller denticles. Pharynx muscular, cylindrical anteriorly, with basal bulb or cylindrical posterior portion. Female reproductive system didelphic or monodelphic. Ovarian branches reflexed antidromously or outstretched. Male reproductive system diorchic or monorchic. Supplements present or absent. Spicules paired and symmetrical, straight or arcuate.

Taxonomy. Three genera: *Rhabdolaimus, Rogerus* and *Tobrilia.*

Biology. Aquatic (inhabit marine and freshwaters) and terrestrial.

Key to genera of Rhabdolaiminae

1. Cephalic sensilla papilliform; ovarian branches reflexed .... *Rhabdolaimus*
   Cephalic sensilla setiform; ovarian branches outstretched .... *Rogerus*

Genus *Rhabdolaimus* de Man, 1880

Syn. = *Pseudorhabdolaimus* Soós, 1937

Morphology. Outer labial sensilla pore-like, cephalic sensilla papilliform (Appendix Fig. 4). Amphid a transverse slit. Stoma tubular, with small cheilo- and gymnostom and long sclerotized tubular stegostom. Three small denticles located at anterior edge of stegostom. Pharynx muscular, cylindrical anteriorly, with basal bulb. Basal bulb with thickened lining. Cardia small. Rectum short. Female reproductive system didelphic. Ovary branches reflexed antidromously. Number and arrangement of testes was not explicitly stated in the description (Ocaña, 1991). Supplements and

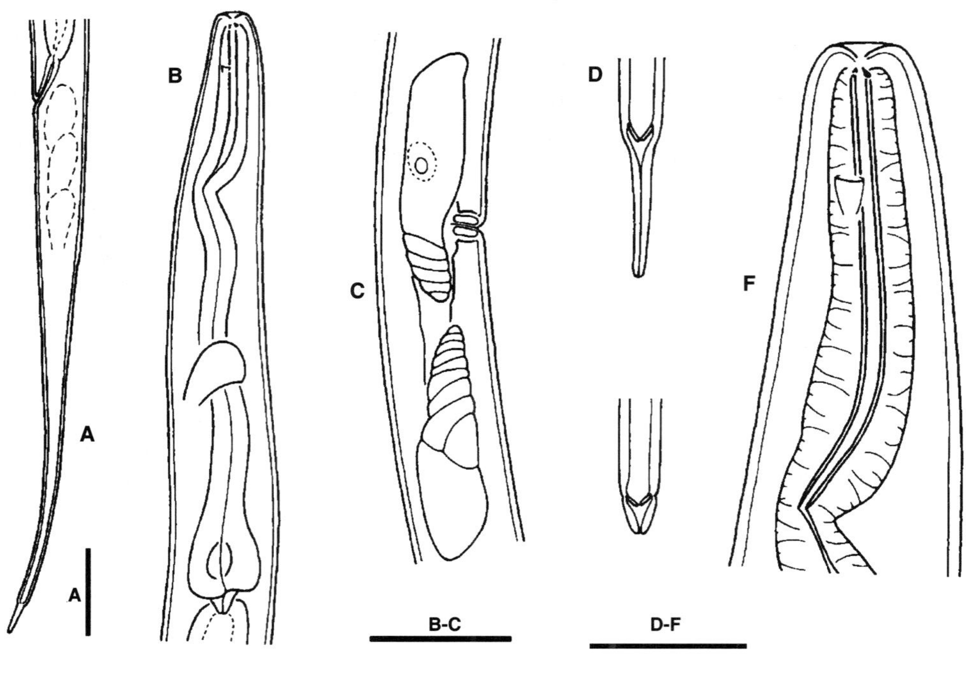

Appendix Fig. 4 Morphology of the genus *Rhabdolaimus* de Man, 1880 as exemplified by *R. terrestris* de Man, 1880 (A–D, F) and *R. aquaticus* de Man, 1880 (E) from Ethiopia (modified from Eyualem & Coomans, 1996). A: Female tail; B: Pharyngeal region; C: Female reproductive system; D-E: Spinneret; F: Anterior end. Scale bars: A–C = 20 µm, D–F = 10 µm.

gubernaculum absent. Spicules paired and symmetrical, arcuate, with oval manubrium and conoid shaft. Tail elongate–conoid. Caudal glands and spinneret present (the spinneret can easily be mistaken for a solid mucro).
Biology. Thelytokous. Commonly found in diverse inland habitats: fresh- and brackish water sediments, soil and moss.
Bibliography. SEM observations were published by Eyualem and Coomans (1996).

Type species: *R. terrestris* de Man, 1880

Other species:
R. aquaticus de Man, 1880
R. brachyuris Meyl, 1954
R. limnophilus (Soós, 1937) Andrássy, 1984
R. minor Cobb, 1914
R. nannus (Hoeppli, 1926) De Coninck, 1935

Genus *Rogerus* Hoeppli & Chu, 1934
Syn. = *Greenia* Hoeppli & Chu, 1932 nec Kirby, 1896, nec Oudemans, 1901 = *Greeninema* Andrássy, 1959
Morphology. Outer labial and cephalic sensilla setiform. Amphid oval. Stoma tubular, with small cheilo- and gymnostom and long sclerotized tubular stegostom. Three small denticles located at anterior edge of stegostom. Pharynx muscular, cylindrical anteriorly, with basal bulb. Basal bulb without thickenings. Cardia small. Rectum short. Female reproductive system didelphic. Ovarian branches outstretched. Male reproductive system monorchic. Two pairs of sub-ventral adcloacal papilla present. Spicules paired and symmetrical, arcuate. Tail elongate conoid. Caudal glands and spinneret present.
Biology. In thermal and saline waters.
Type and only species: *R. orientalis* (Hoeppli & Chu, 1932) Hoeppli & Chu, 1934

References

Andrássy, I. (1991) The free-living nematode fauna of the Bátorliget nature reserves. *The Bátorliget Nature Reserves – after forty years,* 129–197.

Artyukhovsky, A. K. (1990) *Soil Mermithids: Systematics, Biology and Utilization.* Voronezh, 158 pp. (In Russian).

Blaxter, M.L., De Ley, P., Garey, J.R., Liu, L.X., Scheldeman, P., Vierstraete, A., Vanfleteren, J.R., Mackey, L.Y., Dorris, M., Frisse, L.M., Vida, J.T. and Thomas, W.K. (1998) A molecular evolutionary framework for the phylum Nematoda. *Nature* 392, 71–75.

Buxton, J. (1989) Parasitism of *Simulium* species by mermithid nematodes in forest areas of Sierra Leone. *Antenna* 13, 17–23.

Cagnolo, S.R., de Doucet, M.M.A., Doucet, M.E. and Dirienzo J.A. (1999) A method for extraction of aquatic Mermithidae (Nematoda) from riverbed samples. *Nematology* 1, 757–758.

Coomans, A.V. and Raski, D.J. (1988) Two new species of *Prismatolaimus* de Man, 1880 (Nemata: Prismatolaimidae) in Southern Chile. *Journal of Nematology* 20, 288–303.

Curran, J. and Hominick, W.M. (1981) Description of *Gastromermis metae* sp. n. (Nematoda: Mermithidae) with an assessment of some diagnostic characters and species in *Gastromermis*. *Nematologica* 27, 258–273

De Ley, P. and Blaxter, M. (2002) Systematic position and phylogeny. In: Lee D.L. (ed.). *The Biology of Nematodes.* Taylor & Francis, London and New York, 1–30.

Eyualem, A. and Coomans, A. (1996) Aquatic nematodes from Ethiopia. VII. The family Rhabdolaimidae Chitwood, 1951 sensu Lorenzen, 1981 (Chromadorida: Nematoda) with the description of *Udonchus merhatibebi* n. sp. *Hydrobiologia* 341, 197–214.

Gafurov, A.K. (1996) *Mermithids: systematics, biology, phylogeny, practical importance*. Pensoft, Dushanbe, 382 pp. (In Russian).

Gubaidulin, N.A. and Bekturganov, B.B. (2001) Annotated list of the family Mermithidae Braun, 1883 – the parasites of invertebrates. *Tethys Entomological Research* 3, 155–176.

Holovachov, O. and Boström, S. (2004) Morphology and systematics of the superfamilies Leptolaimoidea Örley, 1880 and Camacolaimoidea Micoletzky, 1924 (Nematoda: Plectida). *Journal of Nematode Morphology and Systematics* 7, 1–49.

Kharchenko, N.A. (1999) *Ecology of Mermithids*. Voronezh, 278 pp. (In Russian).

Luo, D. and Bao, X. (1991) Studies on the taxonomy of pre-parasitic juveniles of *Mermithidae* (Nematoda). *Zoological Research* 12, 143–146.

Nickle, W.R. (1972) A contribution to our knowledge of the Mermithidae (Nematoda). *Journal of Nematology* 4, 113–146.

Ocaña, A. (1991) A redescription of two nematode species found in hot springs. *Nematologia Mediterranea* 19, 173–175.

Platzer, A.C. and Platzer, E.G. (1999) Interaction of *Catenaria anguillulae* with *Romanomermis culicivorax*. *Journal of Invertebrate Pathology* 73, 147–153.

Platzer, E.G. and Mackenzie-Graham, L.L. (1980) *Cyclops vernalis* as a predator of the preparasitic stages of *Romanomermis culicivorax*. *Mosquito News* 40, 252–257.

Poinar, G.O. Jr. (2001) Nematoda and Nematomorpha. In: Thorp, J.H. and Covich, A.P. *Ecology and Classification of North American Freshwater Invertebrates*. Academic Press, San Diego, California, pp. 255–295.

Poinar, G.O. Jr. and Otieno, W.A. (1974) Evidence of four molts in the Mermithidae. *Nematologica* 20, 370.

Rubzov, I.A. (1972) *Aquatic Mermithids*. Vol. 1. Nauka, Leningrad, 254 pp. (In Russian).

Rubzov, I.A. (1973) *Freshwater Mermithids from Estonia*. Tallin, Estonia, 172 pp. (In Russian).

Rubzov, I.A. (1974) *Aquatic Mermithids*. Vol. 2. Nauka, Leningrad, 222 pp. (In Russian).

Rubzov, I.A. (1977) *Mermithids: Origin, Biology, Distribution*. Nauka, Leningrad, 191 pp. (In Russian).

Rubzov, I.A. (1981) *Aquatic Mermithidae of the Fauna of the USSR*. 2 vols. & Amerind Publishing Co., New Delhi.

Rubzov, I.A. (1978) *Mermithids: Classification, Importance, Use*. Nauka, Leningrad, 207 pp. (In Russian).

Rusin, L.Y., Aleshin, V.V., Tchesunov, A.V. and Atrashkevich, G.I. (2002) The 18S ribosomal RNA gene of *Soboliphyme baturini* Petrow, 1930 (Nematoda: Dioctophymida) and its implications for phylogenetic relationships within Dorylaimia. *Nematology* 5, 615–628.

Takaoka, H. (1994) Natural vectors of three bovine *Onchocerca* species (Nematoda: Onchocercidae) and seasonal transmission by three blackfly species (Diptera: Simuliidae) in central Kyushu, Japan. *Journal of Medical Entomology* 31, 404–416.

Washburn, J.O., Anderson, J.R. and Egerter, D.E. (1986) Distribution and prevalence of *Octomyomermis troglodytis* (Nematoda: Mermithidae), a parasite of the western tree hole mosquito, *Aedes sierrensis*. *Journal of the American Mosquito Control Association* 2, 341–346.

Subject Index

abiotic factors 60
  alkalinity 85
  ammonia 63
  chloride 63, 171, 180, 183
  conductivity 62, 184
  granulometry 63
  nitrate 63, 82
  water chemistry 62
absorption 106, 114, 119, 120
abundance 60, 65, 79, 82, 133, 164, 191
  Eumonhystera 65, 66
  genera 62, 63
  island 185, 186
  lake 52–55, 56
  local 48
  production 100
  productivity 80, 83, 86
  river 57–59, 171
  seasonal 80, 81, 82, 172
  size range 48, 94
  spatial 79, 83
  spring 180, 182, 183, 184
  vertical 84
  water depth 68, 79, 83
acetate 111, 112
activity
  bacterial 102, 112, 115, 195
  enzymatic 117
  feeding 113, 120
  grazing 115
  metabolic 100, 101
  microbial 114, 212
adaptations 172–173, 201, 202
Africa 6, 47, 49, 190
algae 32, 84, 107, 109, 110, 111, 112, 132, 138, 171, 174
America
  North 5, 47, 190, 191, 193, 201
  South 6, 51, 191
anoxia 69
  anoxic condition 69
  anoxic sediment 146, 201
  anoxic zone 67
Antarctica 187
Arctic 185
  lakes 69, 83
Aufwuchs 98, 121
Australia 51, 170, 171, 185, 188, 190,191, 200

Bacteria 115, 174
  biofilm 112
  food 110, 111, 112, 113, 118, 119, 120
  grazing 118, 119
  production 101
  symbiotic 117
bacterivores 9, 117, 218, 219
behaviour 147
  feeding 105, 106–110
bioassay 145, 146, 147, 149, 153, 158
biodiversity 5, 48, 79, 86, 89, 113, 211, 219
biofilm 112, 113, 147
bioindicators 158
biomass 4, 52–55, 86, 87, 94, 95, 100, 101, 134, 140, 141, 173, 180
  distribution 139
  P/B ratio 94–98
  size spectra 135, 138
bioturbation 102, 117
bird 47, 121, 203
body mass 94, 95, 96, 100, 133, 134, 135, 137, 138, 139, 141
body size 48, 101, 132, 133–135, 139
bromeliad 179, 191, 192
bryophytes 171, 174

cadmium 147, 148
calcium 171, 183
canonical correspondence analysis 219
carbon 86, 94, 105, 119
  dissolved organic 105, 111, 152
carbonate 63, 171, 183
carnivores 4, 108
cave 193, 194, 203
chewer 110, 111, 154
chironomids 96, 121, 172
chloride 63, 171, 180, 183
ciliates 108, 118
circulation 78, 79
colonizer 152, 154, 157, 211, 212
competition 63, 65
consumption 106, 107, 114, 117, 118, 119, 120
contaminant 144, 145, 147, 154, 156, 158
contamination 38, 150, 154, 156
cosmopolitan 47, 48, 49, 51, 190
counting 37–38
crystalloid 194, 201

density 133, 134, 135, 136, 137
diatoms 109, 110, 118, 120
dispersal 47, 48, 79, 164, 173, 203
distribution 46, 47, 48, 50, 62, 63, 65, 67, 68, 69, 78, 79, 83, 84, 85, 86, 170, 171, 180
  biomass 139
  normalized size 139, 140
  size, 132, 135
  species 170
disturbance 67, 79, 153, 157, 164, 202, 212
diversity 40, 50, 56, 60, 79, 86, 88, 89, 90, 112, 113, 154, 170, 180, 213
  Brillouin 214
  Camargo 213, 214
  Hill 213, 218
  Margalef 214

egg index 151, 152, 153
energy 51, 56, 105, 106, 114, 121, 180
Europe 5, 47, 48, 164, 179, 185, 194
eutrophic lake 51, 64, 79, 80, 86, 88, 90
  abundance 54, 56, 80, 83
  feeding type 111
  genera 63
  species richness 51, 56, 60, 86
estuary 57, 115, 122, 188, 241
evenness 56, 60, 61, 62, 90, 150, 151, 213
extraction 32–37

feeders
  bacterial 107, 112, 113, 114, 116
  deposit 63, 107, 108, 109, 110, 111, 156, 171, 195
  epistrate 107, 108, 109, 110, 111, 154, 195
  hyphal 4, 63,107, 185, 186
  microbial 185, 203
  plant 107, 186, 188
  suction 63, 110, 111, 156
  unicellular eukaryote 108
feeding
  plant 164, 171
  rate 105, 106, 113, 114, 117, 118, 120
  type 85, 106, 107, 108, 109, 110, 111, 151, 152, 153, 156, 157
fixation 31, 32, 39
flotation 35, 36, 37, 44
food 63, 81, 105, 111, 112, 115, 118, 120
  availability 84, 85, 100, 112, 113, 150
  bacterial 112
  consumption 117, 118
  recognition 106, 109, 114
  source 105, 106, 107, 114, 119, 153
food web 105, 113, 115, 120, 138, 144, 145, 153, 156, 158, 173
functional response 114, 118

gene expression 147
generation time 5, 11, 47, 94, 95, 96, 97, 138, 146, 153, 183, 212
grain size 31, 60, 85, 100, 113,138–139, 170, 171, 172
grazing 105, 113, 115, 117, 118, 119, 120
groundwater 163, 164, 179, 191

heavy metal 156
humic substances 111
hyporheic 56, 63, 70, 164

indices 56, 60, 88, 150, 152, 153, 154, 157, 158, 211, 212, 213, 214–217, 218, 219
ingestion 11, 106, 107, 108, 114, 118, 120, 147

k-dominance curve 151
K-strategist 152, 153

lentic 88, 100, 139, 145, 157, 163
life history 108, 152
lotic habitats 56, 145, 165–169, 172

macrobenthos 100, 101
macrofauna 65, 94, 121, 138, 173
matter, dissolved organic *see* carbon
Maturity Index 150, 151, 157, 211, 220
meiobenthos 62, 65, 80, 101, 118
meiofauna 69, 86, 100, 111, 121, 138, 145
mesotrophic 56, 64, 78, 83
mounting 39–41
morphology
  buccal 4, 105, 106
mucous 117
multivariate method 157

niche 84, 105, 112, 138, 188, 195, 218
nitrate 63, 82
nutrient 78, 79, 81, 116, 117, 144, 172, 212

oligotrophic lake 79, 80, 83, 86, 88, 91
  abundance 52–55
  feeding type 111
  genera 63
  species richness 51–55, 56, 62, 86
omnivores 11, 108, 109, 110, 111, 156, 171, 185, 186, 195, 218
organic content 147, 148
organic matter 4, 69, 80, 85, 105, 111, 116, 152, 163
oxygen 63, 69, 81, 117, 201, 202
  spring 183, 184
  vertical 69, 79, 85
  water depth 67

P/*B* ratio 94–98
persister 152, 157, 211, 212
phytotelmata 192
planarian 121
plant-parasitic 164, 188, 198, 211, 212, 213
polar 179, 185, 186, 187, 188, 202
pollutant 144, 145, 147, 150, 219
pollution 145, 154, 157, 180
  feeding type 153
  indicator 152, 211
  sediment 149, 150, 158
predation 63, 108, 114, 118, 121, 174
predator 63, 65, 109, 110, 111, 118, 120–123, 153, 171, 186, 194, 195, 203
predator–prey 63, 65, 132, 138
predatory 63, 65, 121, 156, 185, 193
prey 108, 113, 114, 115, 118, 121, 173, 174
productivity 80, 219
  diversity 88, 89
  genera 63
  lake 56, 81, 83, 86, 139
  primary 62, 79, 88, 89, 219
  spring 183
production 94–101
  bacterial 101
  mucous 117
  primary 84, 194
  secondary 94, 116
protozoa 94, 101, 108, 109, 110, 118, 152

ratio (Secernentea) 157
refuge 121
reproduction 69, 97, 106, 112, 147, 148, 202
respiration 101, 106, 114, 115, 116
river 148, 164, 171, 173, 188
  abundance 57–59
  diversity 170
  evenness 61, 62
  pollution 152, 154, 157
  Shannon 61, 62
  species richness 57–59, 61, 62
r-strategist 152, 153

salinity 89, 150, 200, 201, 202
scale 48, 50, 65, 170
  global 81, 88, 89
  local 49, 51, 171
seasonality 77, 81, 82
sediment 83, 147
  grain size 85, 170, 171
  layer 68, 69, 82, 84, 85, 100, 121
  particle 108, 117, 170
  quality 148, 154, 211
  structure 117
  surface 115, 116, 121, 171
  test 148
  texture 113, 121
  toxicity 148, 149
Shannon–Wiener 56, 60, 61, 62, 88, 150, 151, 158, 213, 214, 218
similarity
  Bray-Curtis 156
  Jaccard 217
  Morisita 217
  Sörensen 217
size
  class 100, 132, 134, 135, 138, 139
  range 48, 94, 133, 138
  spectra 135, 138, 139
  structure 132, 133, 138
species
  composition 46–49, 60, 81, 82, 86, 95, 152, 153, 154, 155, 156, 157, 171, 191, 219
  number (see also species richness) 67, 68, 88, 133, 198, 199
  richness 48, 51–63, 65, 86, 89, 90, 145, 151, 182, 183, 184, 213
spring 179, 180, 183, 184, 185, 202
  hot 182, 203
  mineral 183
  thermal 180, 194
stream *see* river
substrate
  ingester 107, 186
sulphide 201, 202
swallower 110

temperature 77–79, 81, 84, 85, 113, 115, 116, 180, 182, 183, 202
toxicity 145, 146, 147, 148, 149, 150
trophic composition 185, 186
trophic status 51, 56, 63, 139, 171
tropic 77, 78, 79, 85, 86, 88, 89, 90

variability 48–50, 65, 88
variation 60
  seasonal 86
  spatial 83
  temporal 80, 82, 86, 90

variation *continued*
vertical 84
velocity 170, 171

waste water 158
water column 163, 164, 173
water depth 67, 68, 80, 81, 85, 134
abundance 65, 66, 83
biomass 86
production 97, 98
species richness 60, 65, 69, 86, 145

Taxonomic Index

Abunema 15, 294, 309, 310, 313, 318
Acanthonchus 501, 524, 526
Acanthopharyngoides 534, 538, 542, 543
Acanthopharynx 534, 538, 543
Achromadora 21, 64, 71, 181, 184, 190, 192, 195, 497 499, 500, 501, 520, 521, 523, 533
  key 531
  species list 522
Achromadoridae 21, 498, 501, 502, 503, 520, 521, 523
Achromadorinae 21, 498, 520
Acrobeles 661, 667, 688
Acrobeloides 193, 493, 661, 667, 688
Acrostichus 27, 698, 701, 711, 712
Actinca 18, 192, 416, 417, 424, 426, 427
Actinolaimidae 17, 212, 392, 393
  diagnosis of genera 416–429
  key to genera 416
Actinolaimoidinae 454, 455
Actinolaimus 18, 190, 192, 416, 417, 425
  diagnosis 426
Actus 470, 483, 491
  diagnosis 485
Aculonchus 357, 356
Acunemella 461
Adenolaimus 16, 388
  diagnosis 374
Adoncholaiminae 13, 248, 252, 254
Adoncholaimus 13, 111, 127, 129, 196, 228, 230, 248
  diagnosis 254
Adorus 270
Aetholaimidae 16, 374
  diagnosis 375
Aetholaiminae 16
  diagnosis 376
Aetholaimus 16
  diagnosis 376
Afractinca 18, 416, 417, 427
  diagnosis 428
Afractinolaimus 18, 192, 416, 417, 421
  diagnosis 422
Africanthion 236, 240
  diagnosis 241
Afrodorylaiminae 17, 394, 395
  diagnosis 410
  key to genera 412
Afrodorylaimus 17, 190, 411, 413
  diagnosis 412
Aglenchus 26, 649, 664, 670
  diagnosis 672
Agmodorus 368
Akrotonus 448
Alaimidae 14, 212, 232, 271, 277
  diagnosis 278
Alaimina 9, 14, 225, 227, 229, 230, 231, 232, 233, 278, 279
  diagnosis 278
Alaiminae 14, 277, 279
  diagnosis 278
Alaimoidea 14, 279
Alaimus 14, 181, 190, 276, 277
  diagnosis 279
Allodorylaimus 18, 127, 433, 435, 440
  diagnosis 443
Allomonhystera 597
Allotrichodorus 294, 297, 317
  diagnosis 298
Amblydorylaimus 434
Amphibelondira 339, 341
Amphidelinae 14, 277, 278, 281, 282, 283
  diagnosis 280
Amphidelus 14, 193, 276, 277, 283
  diagnosis 280
Amphidorylaiminae 394, 395
Amplimerlinius 646

Anaplectinae 23, 631, 632, 633, 634, 637
Anaplectus 23, 173, 193, 632, 634
    diagnosis 637
Anatonchidae 19, 469, 475
Anatonchinae 19, 469, 478, 495
Anatonchoidea 19, 469, 472, 474, 490
Anatonchus 19, 469, 477, 478, 491
Anchobelondira 341, 342
Anguimonhystera 22, 581, 582, 583
Anguinidae 26, 669
    diagnosis 681
Anomonema 616
Anonchinae 23, 619, 624
Anonchus 23, 611, 615, 619, 620, 625, 626
    diagnosis 624
Anoplostoma 13, 196, 237
    diagnosis 245
Anoplostomatidae 13, 237, 238, 246
    diagnosis 245
Anoplostomatinae 13, 237, 245
    diagnosis 245
Anticomidae 13, 232, 238
    diagnosis 246
Aphanolaimidae 23, 611, 612, 616, 620
    diagnosis 619
Aphanolaiminae 23, 619, 620
Aphanolaimus 23, 156, 173, 181, 183, 620, 621
    diagnosis 622
Aphanonchus 23, 615, 620, 621
    diagnosis 624
Aphelenchidae 26
Aphelenchoidea 26, 668
    diagnosis 684
Aphelenchoides 26, 112, 186, 188, 191, 192, 198, 200, 218, 649, 685, 686
Aphelenchoididae 26, 212
    diagnosis 684
Aphelenchus 191, 192, 198, 650
Apodontium 605
Apodorylaimus 411, 414
Apoleptonchus 17, 360, 409, 410, 412
Aponchiidae 21, 498, 535, 539, 549, 579
    diagnosis 554
Aponema 535, 540, 551
Aporcedorus 448
Aporcelaimellus 18, 115, 193, 327, 448, 449
    diagnosis 450
Aporcelaimidae 18, 212, 328, 392, 453
    diagnosis 447
Aporcelaiminae 18, 448, 449, 450
Aporcelaimoides 337, 379, 453
Aporcelaimus 18, 192, 448, 449, 466, 493
    diagnosis 449
Aporcella 448
Aquatides 17, 378, 380
    diagnosis 381
Araeolaimida 5, 9, 10, 22, 28, 29, 181, 182, 186, 557, 579, 604, 608, 610, 612
Araeolaimus 606
Archionchus 295
Arctidorylaiminae 18, 429
    diagnosis 430
Arctidorylaimus 430, 431
Arctiplectus 637
Arenasoma 273, 274
Aronema 260
Aronematinae 259
Ascolaimus 605
Asperotobrilus 15, 294, 304, 319, 321
    diagnosis 306
Assia 624
Athernema 16, 365
    diagnosis 366
Athernematinae 16, 366, 365
Atrochromadora 501, 509, 510
    diagnosis 508
Atylenchus 650, 664
Aulolaimidae 23, 612
Aulolaimoides 16, 365
    diagnosis 374
Aulolaimoididae 16, 349, 350, 367, 374
    diagnosis 373
Axonchium 324
Axonchoides 341, 342
Axonolaimidae 22, 604, 605
Axonolaimoidea 22, 579, 604
Axonolaimus 197, 605

Baicalobrilus 301
Baladorylaimus 401, 411
    diagnosis 410
Baqriella 434
Basiria 198, 200, 650
Basirotyleptus 16, 358, 359
Bastianiidae 612
Bathyodontus 471
Bathyeurystomina 250, 260, 261
Bathylaimus 273, 274, 275, 624
Bathynox 535, 540, 550, 551
Bathyodontidae 19, 471, 473
Bathyodontina *19, 469,* 471, 473, 489
Bathyonchus 623
Belbolla 230, 250, 259
    diagnosis 261
Belbollidae 259
Belbollinae 259
Belondira 15, 341, 342
    diagnosis 344
Belondirella 341
Belondiridae 15, 212, 340, 341, 343, 344, 345, 387
Belondirinae 15, 341, 342, 343, 344
Belondiroidea 15, 326, 331, 340, 343
Belonenchinae 16, 357, 358, 359
Belonolaimidae 26, 673, 674
Belonolaimus 198, 650, 664
Bertzuckermania 360, 361
Biarmifer 501, 526, 527
Bolbolaimus 535, 540, 550, 551
Bolbonema 534, 538, 543
Boleodorus 198, 650, 664
Boreolaimus 18, 434, 445, 447
Boucherius 557
Brachonchulus 488

Brasilaimus 18, 416, 417, 424, 427
diagnosis 428
Brevitobrilus 15, 294, 304, 308, 320, 321, 322
Bunonema 192, 700
Bunonematidae 27, 212, 697, 700, 704
Bunonematomorpha 27, 697, 700, 704
Butlerius 27, 698, 701
diagnosis 713

Cacolaimus 248
diagnosis 253
Caenorhabditis 28, 111, 112, 146, 148, 191, 192, 697, 707
Calcaridorylaimus 400, 401
Caligocanna 535, 540, 550, 552
Calligyrus 556
Calodorylaimus 17, 400, 401, 402, 403–404
Calolaiminae 16, 367, 371
Calolaimus 16, 198, 367, 371
Calomicrolaimus 535, 540, 550, 552
Caloosia 198, 650, 683
Calyptronema 13, 250, 259, 260, 261, 262
Camacolaimidae 611
Camacolaimoidea 612
Campydora 231, 232, 276
Campydoridae 14, 276
Campydorina 9, 14, 225, 227, 229, 231, 233, 234, 235, 275, 276
Campydoroidea 14, 276
Campylaimus 606
Capilonchus 363
Caprionchulus 15, 295, 315, 316, 320
Caputonchus 469, 472, 475, 490
Carcharolaiminae 429, 430
Caribplectus 197, 628
Caveonchus 360, 361
Caviputa 277, 280
Cephalenchus 650
Cephalobidae 26, 112, 212, 686, 687
Cephalobina 607
Cephaloboidea 26, 686
Cephaloboides 28, 697, 706, 707
Cephalobomorpha 26, 661, 667, 668, 686
Cephalobus 191, 192, 193, 494, 661, 667, 688, 689
Cephalodorylaiminae 454, 455
Ceratoplectus 632, 635, 637
Cervidellus 193, 662
Cervonema 605
Chaetonema 227, 245, 246
Chaetonematinae 245, 246
Chaolaimus 295
Chiloplacus 164, 186, 493, 662
Chiloplectus 632, 635, 636
Chitwoodielloides 352, 353
Chitwoodiellus 352, 353
Chitwoodius 16, 351, 352, 353, 354
Chromadora 499, 505, 508, 510, 518, 519, 520, 529
Chromadorea 8, 9, 502, 535, 555, 579, 604
Chromadorella 501, 508, 510
Chromadoria 8, 9, 10, 11, 497, 499, 555, 604
diagnosis 558
Chromadorida 5, 9, 10, 11, 20, 181, 182, 195, 497, 498, 499, 500, 502
Chromadoridae 497, 498, 502, 503, 504, 509, 521
Chromadorina 501, 505, 508, 510
Chromadorina 499, 502
Chromadorinae 499, 505, 507, 508
Chromadorissa 501, 515
Chromadorita 501, 515
Chromadoroidea 479, 502
Chronogaster 23, 113, 181, 183, 184, 191, 192, 193, 194, 203, 204, 611, 612, 618, 627, 628, 629
Chronogastridae 23, 618, 627, 628
Chrysodorus 17, 401, 405, 406
Chrysonema 18, 192, 446, 447
Chrysonematinae 18, 429, 430, 447
Chrysonemoides 198, 447
Cinctonema 535, 540, 551, 552
Cladocephalus 374
Clarkus 19, 192, 470, 483, 484, 492
Clavicauda 17, 244, 378, 380, 382
Clavicaudoides 17, 380, 382
Clavigula 358, 360, 361
Cobbonchinae 19, 470, 485–486
Cobbonchus 19, 470, 483, 485, 492
Comesomatidae 22, 604, 605
Comiconchus 19, 470, 483, 486, 492
Conilia 266
Coninckiidae 22, 604, 606
Coomansus 469, 482
Cosalaimus 277, 279
Coslenchus 26, 650, 664, 666, 670, 672
Craspedonema 27, 697, 700, 704
Crassibucca 470, 479, 480, 491
Crassogula 18, 434, 436, 438
Crassolaimus 535, 540, 550, 552
Crateronematidae 17, 393
Crestonchulus 470, 487, 489, 492
Cricodorylaimus 429
Cricohalalaimus 265, 271, 272
Criconema 193, 194, 198
Criconematidae 26
Criconematoidea 26, 668, 669, 683
Criconemoides 193, 194, 200, 651
Cristamphidelinae 227, 278, 283
Cristamphidelus 227, 283
Crocodorylaimus 17, 49, 64, 68, 400, 401, 402, 404
Croconema 534, 538, 542, 543
Cruznema 28, 697, 706, 707
Cryptenoplus 236, 240, 241
Cryptonchidae 19, 469, 471
Cryptonchoidea 19, 469, 471
Cryptonchulus 471
Cryptonchus 19, 469, 471, 472, 489
Curvidorylaimus 356
Curvolaimus 258
Cyartonema *556–558*, 561
Cyartonematidae 556–558, 561, 568
Cyatholaimidae 498, 501, 502, 503, 504, 521, 523, 524

Cyatholaiminae 501, 523, 525
Cylindrolaiminae 22, 606, 608
Cylindrolaimus 22, 181, 192, 193, 606, 607, 608, 609
Cylindrotheristus 596
Cynura 628
Cytolaimium 196, 230, 232

Dadayia 624
Daptonema 22, 40, 108, 110, 156, 181, 578, 596, 597, 598
Daubaylia 697, 701
Daubayliidae 697, 701
Deltanema 501, 514, 516
Denticulella 501, 513, 515, 516
Deontostoma 227
Desmodora 534, 538, 542, 543, 544, 545
Desmodorella 534, 538, 544, 548
Desmodorida 497, 498, 499, 534, 535
Desmodoridae 498, 534, 536, 537, 541
Desmodorina 534
Desmodorinae 534, 537, 542
Desmodoroidea 534, 537
Desmolorenzenia 556, 564, 566, 570
Desmoscolecida 497–9, 555–7, 556
Desmoscolecidae 557, 559, 561
Desmoscolecinae 556, 562
Desmoscolecini 556, 559, 560, 562
Desmoscolecoidea 556, 561
Desmoscolex 564, 569
Desmotricoma 557, 567
Dichromadora 498, 501, 513, 516
Diphtherophora 293, 295, 321
Diphtherophoridae 296
Diphtherophorina 295, 321
Diphtherophoroidea 295, 296
Diplogaster 27, 173, 699, 700, 713–715, 720
Diplogasteroides 27, 698, 701, 710, 719
Diplogasteromorpha 27, 696, 698, 719
Diplogastrellus 27, 699, 713
Diplogastridae 696, 698, 762, 711, 721
Diplogastroididae 696, 698, 710
Diplolaimella 22, 197, 578, 581, 582, 584, 603
Diplolaimelloides 22, 40, 190, 197, 579, 583
Diplopeltidae 606
Diplopeltinae 22, 606
Diplopeltis 606
Diplopeltula 606
Diploscapter 28, 153, 173, 697, 696, 701
  diagnosis 707
Discolaiminae 18, 429–431
Discolaimium 430
Discolaimoides 18, 430–432
Discolaimus 18, 430, 431
Discomyctus 16, 352, 354, 386
Ditlevsenella 247, 250, 260–262
Ditlevsenia 471
Ditylenchus 26, 112, 164, 186, 199, 218, 651, 661, 664, 679, 681, 682
Dolichodoridae 27
Dolicholaimus 266, 267
Domorganus 22, 612, 613
Dorella 368
Doronchus 470, 479, 480, 491
Dorydorella 19, 455, 456, 459, 460
Dorylaimellinae 15, 390
Dorylaimellus 15, 344, 387
Dorylaimida 5, 9, 10, 15, 180, 181, 186, 188, 191, 201, 203, 204, 231, 278, 293, 326–463
Dorylaimidae 17, 64, 189, 212, 267, 325, 392, 393, 394, 395, 398, 410, 415, 430, 466
Dorylaiminae 17, 394, 395, 398
Dorylaimoidea 17, 278, 340, 371, 392–467
Dorylaimoides 16, 367, 372
Dorylaimopsiinae 605
Dorylaimus 17, 50, 110, 156, 173, 186, 190, 192, 193, 330, 398, 399, 415, 467
Doryllium 16, 367, 368, 369
Draconematidae 21, 498, 535, 536, 541, 548
Drepanodorylaimus 17, 411, 412, 413
Duriella 346

Echinodesmodora 534, 538, 542
  diagnosis 544
Ecphyadophora 651
Ecuadorus 294, 297, 317
  diagnosis 298
Ecumenicus 18, 433, 434, 445
  diagnosis 445
Egtitus 17, 416–17, 421
  diagnosis 420
Enchelidiinae 259, 264
Enchelidium 259
  diagnosis 262
Enchodelus 18, 189, 327, 455, 456
  diagnosis 457
Enchodorella 461
Enoplacea 247
Enoplia 8–11, 13, 231, 293, 295
Enoplida 9, 13, 181–2, 186, 194, 225–292, 293, 702
  diagnosis 232, 238
Enoplidae 13, 227, 229, 238
  diagnosis 238
Enoplina 9, 13, 226–7, 229, 231, 233, 234, 235, 236, 237
  diagnosis 232
Enoplochilus 473
Enoploidea 13, 238, 245, 246, 267
  diagnosis 232
Enoploides 13, 108, 113, 185–6, 236, 239, 242, 243
  diagnosis 241
Enoplolaiminae 13, 236, 237, 241, 242, 243, 244, 245
  diagnosis 239
Enoplolaimus 236, 239, 240
  diagnosis 241
Enoplus 225, 227, 232, 236, 237, 238, 468
Epacanthion 236, 238, 240
  diagnosis 241

Epacrolaimus 18, 448, 452, 453
Epidorylaimus 18, 433, 435, 440, 444
Epitobrilini 15, 294, 309
Epitobrilus 15, 294, 309, 310, 320, 321, 322
Epsilonematidae 21, 498, 535, 536, 537, 541
 diagnosis 548
Ereptonema 639, 640
Etamphidelus 277
 diagnosis 281
Ethmolaimidae 21, 498, 501, 502, 503, 504, 521, 528
 diagnosis 527
Ethmolaimus 21, 40, 49, 64, 68, 83, 97, 181, 497, 499, 501, 521, 528, 530, 532
 diagnosis 528
Eucephalobus 186, 662, 667, 687
Euchromadorinae 501, 506, 509, 512
Eudorylaimus 18, 69, 173, 186, 187, 192, 201, 203, 433, 435, 440
 diagnosis 439
Eumonhystera 22, 48, 63, 64, 65, 66, 68, 95, 97, 98, 154, 155, 158, 170, 173, 181, 183, 184, 185, 186, 187, 203, 204, 580, 581, 582, 585
 diagnosis 585
Eurystomina 13, 250, 260
 diagnosis 262
Eurystomininae 259
Euteratocephalus 23, 192, 642, 643
 diagnosis 642
Eutobrilus 15, 170, 188, 189, 294, 304, 319, 322
 diagnosis 305
Eutylenchus 26, 652, 666, 670
 diagnosis 673

Falcihasta 15, 345, 346, 347
 diagnosis 347
Fenestrolaimus 236, 238, 240
 diagnosis 242
Feroxides 17, 380, 383
 diagnosis 383
Fictor 27, 173, 699, 700, 701
 diagnosis 715
Filenchus 26, 652, 664, 670
 diagnosis 671
Filidiscolaimus 431
Filipjevia 236, 240, 242
Fleuronema 236, 240
 diagnosis 242
Fotolaimus 249, 256
Funaria 16, 361, 362, 363
 diagnosis 362
Fuscheila 400, 401
Fusivermidae 22, 600
Fusonema 501, 505

Gairleanema 273, 274
Gammarinema 22, 578, 580, 581, 582, 586
 diagnosis 587
Geocenamus 652
Geomonhystera 22, 580, 581, 582, 587
 diagnosis 585
Gerlachius 557, 558
Gerthus 369
Glochidorella 358
Goferus 368
Goffartia 698, 711
Gopalus 18, 434, 435
Gracilancea 653
Granonchulus 489
Greeffiella 556, 557
Greeffiellini 22, 556, 562, 566
Greeffiellopsis 556
Gymnolaimus 471
Gymnotyleptus 364

Haconnus 624
Hadronchoides 478
Hadronchulus 478
Hadronchus 476
Halalaimus 14, 193, 196, 227, 265
 diagnosis 271
Haliplectoidea 23, 612
Haliplectus 40, 197, 611
Halodorylaimus 400, 401
Hapalomus 557
Haptotricoma 557, 567
Harpagonchinae 501, 505, 506, 509, 512
Helicotylenchus 26, 188, 189, 191, 192, 198, 200, 653, 664, 679, 680
 diagnosis 680
Helmabia 454
Helmabiinae 454
Hemicriconemoides 198, 200, 653, 664
Hemicycliophora 26, 198, 200, 653, 654, 665, 666
 diagnosis 684
Hemicycliophoridae 26, 683
Heterodera 188, 189, 198, 200, 203, 654, 665
Heynsaxonchium 341, 342
Hirschmanniella 26, 188, 189, 199, 654, 664, 665, 666, 676
 diagnosis 676
Hofmaenneria 22, 594, 595
Hoplolaimidae 26, 200, 678
 diagnosis 669
Hoplolaimus 26, 198, 200, 493, 655, 665
 diagnosis 679
Hypodontolaiminae 501, 506, 509, 512, 514, 515
Hypodontolaimus 109, 501, 513, 515, 517
Hyptiolaimus 239, 240

Idiodorylaimus 17, 400, 401, 405
 diagnosis 406
Immanigula 341, 342
Inbionema 434, 435
Incanema 360, 361
Indokochinema 445
Ingenia 273, 274
 diagnosis 275
Innocuonema 501, 513, 514
 diagnosis 515

Iotonchulus 469, 475
Iotonchus 19, 469, 472, 490, 493, 494
diagnosis 475
Ironella 266, 267
Ironidae 14, 227, 232, 265, 266, 268
Ironina 9, 14, 231, 233, 234, 235, 265, 266
diagnosis 264
Ironinae 14, 266, 267
Ironoidea 14, 266, 268, 269
Ironus 14, 64, 68, 85, 97, 153, 181, 183, 190, 191, 192, 265, 268
diagnosis 267
Ischiodorylaimus 17, 192, 398, 399
diagnosis 400
Ixonema 535, 540, 550, 553

Jensenonchus 475
Judonchulus 483

Kallidorylaimus 433, 434
Karkinochromadora 501, 513, 515, 517
Kenyanema 605
Keralanema 628, 629
Khantbhala 365
Kinonchulus 293
diagnosis 314
Kittydorylaimus 429
Koerneria 27, 699, 714
Kolodorylaimus 429
Krampia 258
Krampiinae 252
diagnosis 258
Kreisoncholaimus 248, 254
diagnosis 255
Kreisonematinae 501, 523
Kurikania 301, 321

Labronema 18, 108, 189, 330, 433, 434, 435
diagnosis 435
Labronemella 18, 433, 434, 436
diagnosis 437
Laevides 17, 378, 380, 381
diagnosis 383
Lagenidium 701
Laimella 605
Laimydorinae 17, 394, 395, 400, 403, 410
diagnosis 400
Laimydorus 17, 49, 64, 68, 190, 192, 328, 400, 402, 415
diagnosis 401
Lamuania 15, 294, 302, 318, 320
diagnosis 303
Latocephalus 428
Lauratonematidae 247
Lautus 267
Lawtonema 360
Laxamphidelus 277
diagnosis 281
Ledovitia 250, 260
diagnosis 262
Lelenchus 26, 652, 662, 663, 667, 670
diagnosis 672
Lenonchium 18, 453, 454, 458
diagnosis 459
Leoberginema 611
Leptolaimidae 22, 518, 521, 555, 616
diagnosis 616
Leptolaimoidea 22, 609, 610, 611
diagnosis 615
Leptolaimus 201, 300, 615, 616
Leptonchidae 16, 212, 347, 349, 350, 356, 357, 359, 360, 363, 364, 365
diagnosis 354
Leptonchinae 16, 357, 360, 361, 362
diagnosis 358
Leptonchus 16, 357, 360, 361
diagnosis 362
Leptoplectonema 616
Leptosomatidae 14, 225, 266
diagnosis 268
Leptosomatum 269
Limonchulus 15, 295, 309, 319
diagnosis 316
Lindseyus 15, 341, 344, 345, 347
Linhomoeidae 22, 398
Litinium 265, 269
diagnosis 270
Loncharionema 352, 354
Longibulbophora 294, 297, 316
diagnosis 296
Longidorella 18, 198, 460
diagnosis 461
Longidoridae 17, 331, 392, 393
diagnosis 394
Lordellonematidae 17
diagnosis 393
Lyranema 251, 259
diagnosis 263

Macrotobrilus 309
Mactinolaimus 18, 192, 416, 417, 423
diagnosis 424
Makatinus 18, 448, 452
diagnosis 451
Malenchus 26, 656, 665, 670, 671
diagnosis 671
Manunema 557, 615, 616
Margaronchulus 470, 487, 492
diagnosis 488
Margollus 368
Marimermithida 20, 225
Marinoplectus 637
Megamphidelus 277
diagnosis 281
Megaonchulus 469, 489
Megeurystomina 251, 260
diagnosis 263
Megodontolaimus 501, 513, 514
diagnosis 518

Meloidodera 656, 665
Meloidogyne 112, 188, 189, 198, 200, 203, 656, 665
Meloidogynidae 26
Mesacanthion 13, 196, 226, 236, 239, 240, 242
diagnosis 243
Mesacanthoides 237, 240
diagnosis 243
Mesocriconema 656, 665
Mesodorylaimus 17, 187, 190, 191, 192, 193, 200, 202, 203, 204, 218, 335, 401, 406, 415
diagnosis 406
Mesorhabditidae 28, 697, 700
diagnosis 705
Mesorhabditis 28, 173, 697, 705
diagnosis 705
Mesotheristus 596, 597, 598
Mesotobrilus 15, 294, 302, 318
diagnosis 301
Metacomesoma 605
Metactinolaimus 18, 414, 423
diagnosis 424
Metadesmodora 534, 538, 542, 545
Metadorylaiminae 352
Metadorylaimus 356
Metamphidelus 277
diagnosis 281
Metaparoncholaimus 249
diagnosis 256
Metaporcelaimus 448
Metaraeolaimoides 606
Metateratocephalidae 23, 611, 612, 618, 627
diagnosis 642
Metateratocephalus 23, 618, 642, 643, 644
diagnosis 643
Metenoploides 237, 240
diagnosis 243
Metoncholaimoides 248, 254
diagnosis 255
Metoncholaimus 107, 228, 249
diagnosis 256
Meyersia 228, 248, 254
diagnosis 255
Meylia 557
Meyliidae 556, 561, 568
Meylis 360, 361
Meylonema 352, 353
Micatonchus 478
Miconchoides 479
Miconchus 19, 189, 192, 193, 301, 468, 470, 479, 491
diagnosis 479
Microdorylaimus 18, 433, 435, 440
diagnosis 442
Microlaimidae 21, 498, 535, 540, 549, 550, 551, 552, 579
diagnosis 550
Microlaimoidea 21, 533, 535, 537, 539, 549, 552, 553, 610
diagnosis 549
key to freshwater species 555
Microlaimus 21, 197, 535, 540, 550, 551, 552, 553, 621, 623
diagnosis 553
Miculenchus 657, 691
Mikinema 613
Miodorylaimus 401
Mirolaimus 471
Molgolaiminae 534, 537, 541
Mongolotheristus 596, 598
Monhystera 22, 40, 49, 63, 64, 65, 67, 68, 83, 97, 98, 114, 146, 154, 156, 158, 170, 173, 181, 183, 185, 186, 187, 190, 192, 200, 201, 203, 204, 239, 494, 574, 575, 581, 582, 589, 592, 296, 597, 606, 607
diagnosis 587
list of freshwater species 590
Monhysterida 5, 8, 9, 10, 11, 22, 170, 181, 182, 183, 186, 195, 201, 202, 203, 204, 497, 557, 574–604, 644
Monhysteridae 22, 29, 115, 157, 158, 159, 212, 574, 577, 580, 581, 582, 583, 585, 587, 590, 591, 592
diagnosis 580
list of genera 579
Monhysteroidea 22, 580
Monhystrella 22, 181, 183, 184, 190, 193, 203, 204, 580, 581, 588, 590
diagnosis 590
Monhystrium 579, 580, 581
Monochromadora 193
Mononchida 9, 10, 19, 153, 181, 186, 191, 194, 203, 204, 468, 469, 470, 471, 472, 474, 475, 477, 479, 480, 481, 483, 485, 487, 489, 491, 493, 494
diagnosis 470, 471
Mononchidae 19, 108, 212, 470, 481
diagnosis 481
Mononchinae 19, 470, 481, 482
diagnosis 482, 483
Mononchoidea 19, 470, 481, 486, 490
diagnosis 481
Mononchoides 27, 173, 699, 700, 712, 715
Mononchulidae 19, 469, 473, 474
Mononchuloidea 19, 469
diagnosis 473
Mononchulus 19, 469, 473, 489
diagnosis 473
Mononchus 19, 108, 156, 164, 173, 186, 189, 190, 191, 192, 193, 203, 204, 468, 470, 480, 492, 493
diagnosis 481
Monoposthia 535, 539, 554
diagnosis 554
Monoposthiidae 21, 498, 535, 539, 549
diagnosis 554, 555
Monoposthioides 535, 539
diagnosis 554
Monotrichodorus 294, 297, 316
diagnosis 298
Morasia 16, 367, 372
diagnosis 373

Morenchus 478
Mulveyellus 469, 472, 490
diagnosis 475
Mumtazium 352, 353, 355
Mydonomidae 16, 349, 350, 351, 364, 367, 371, 372
diagnosis 371, 372
Mydonominae 16, 364, 367, 371
diagnosis 372, 373
Mydonomus 372
Mylodiscoides 431
Mylodiscus 430
Mylonchulidae 19, 470, 488
diagnosis 486
Mylonchulinae 19, 470, 487
Mylonchulus 19, 192, 193, 470, 487, 492, 493
diagnosis 486

Nagelus 26, 657, 665, 666, 674
diagnosis 675
Namaquanema 401, 411
diagnosis 410
Nanidorus 198, 298
Nemanema 265
diagnosis 270
Neoactinolaimus 18, 190, 192, 416, 417, 423
diagnosis 422
Neochromadora 20, 195, 197, 501, 510, 513, 514, 521, 531
diagnosis 518
Neodiplogaster 27, 699, 716
diagnosis 716
Neodiplogastridae 696, 699, 701
diagnosis 709
Neometadorylaimus 356
Neopsilenchus 657
Neotobrilinae 15, 306, 307, 310, 317
Neotobrilini 15, 294, 307, 318
Neotobrilus 15, 294, 302, 319, 322
diagnosis 307
Neotonchidae 20, 498, 501, 502, 503, 504, 521
diagnosis 530
Nigronchus 470, 480, 492
diagnosis 482
Nimigula 341
Noffsingeria 557
Nordia 40, 461
Nordiidae 18, 392, 393, 461
diagnosis 454, 455
Nordiinae 18, 454, 455, 461
Nudora 21, 535, 539, 554
diagnosis 555
Nullonchus 469, 472, 474, 475, 490
Nyctonematinae 501, 523, 524, 525
Nygellidae 16, 375, 376
diagnosis 376
Nygellinae 16, 376
diagnosis 376, 377
Nygellus 17, 377
diagnosis 377
Nygolaimellidae 17, 375, 377
diagnosis 377
Nygolaimellinae 17, 377
diagnosis 377, 379
Nygolaimellus 17, 379
diagnosis 379
Nygolaimidae 17, 379, 380
diagnosis 379–380
Nygolaimina 16, 17, 328, 339, 380
diagnosis 375
Nygolaiminae 17, 380
diagnosis 380
Nygolaimium 17, 379
diagnosis 379
Nygolaimoidea 16, 327, 330, 375
diagnosis 375
Nygolaimoides 19, 463, 464
diagnosis 463
Nygolaimus 17, 331, 378, 380, 381
diagnosis 384

Octonchinae 252, 258
Octonchus 258
Odontobius 578, 581
Odontolaimidae 23, 612
Odontophoroides 605
Ogma 658, 665
Ohridia 613
Ohridiidae 22, 23, 612, 613
Ohridius 613
Oigolaimella 27, 173, 699, 716
Oionchus 19, 192, 468, 469, 472, 473, 474, 489
diagnosis 473
Okranema 237, 240
diagnosis 243
Oligonchulus 487, 488, 493
Oncholaimacea 247
Oncholaimellinae 13, 248, 252, 253, 254
Oncholaimelloides 248, 252, 253
Oncholaimellus 196, 230, 246, 248, 253
Oncholaimidae 13, 108, 194, 212, 230, 247, 248, 254, 255, 258, 259
diagnosis 252
Oncholaimina 9, 13, 229, 231, 233, 234, 235, 247, 248, 249, 250, 251
Oncholaiminae 13, 249, 252, 255, 256, 257, 262
Oncholaimoidea 13, 245, 247, 259
diagnosis 252
Oncholaimoides 247, 248
diagnosis 253
Oncholaimus 13, 40, 190, 192, 194, 195, 196, 208, 228, 230, 231, 255, 256
diagnosis 257
Onchulidae 313, 318
Onchulus 15, 293, 295, 316, 320
diagnosis 314
Oonaguntus 393
Oostenbrinkella 16, 367, 368
diagnosis 369
Oostenbrinkia 374

Oscheius 28, 654
diagnosis 708
Oxybelondira 345, 388
Oxydirus 16, 387, 388, 389
diagnosis 347
Oxyonchus 226, 227, 237, 241
diagnosis 244
Oxystomina 14, 265, 269, 322
diagnosis 270
Oxystominidae 14, 230, 265, 266, 271, 272
diagnosis 269
Oxystomininae 14, 269, 270, 271

Pachydorylaimus 18, 433, 434, 445
diagnosis 445
Pakira 611, 627
Pakiridae 573
Pakmylonchulus 432
Panagrellus 690
Panagrolaimidae 25, 112, 212, 667, 687, 689
diagnosis 689
Panagrolaimoidea 25, 689
Panagrolaimomorpha 25, 665, 669
Panagrolaimus 25, 173, 187, 192, 194, 663, 667
diagnosis 689
Panduripharynx 501, 513, 514, 518
Pannagrellus 663
Paracanthonchinae 501, 523, 524, 525, 526
Paracanthonchus 501, 526, 527
Parachromadorita 501, 513, 515, 519
Paracomesoma 605
Paracrassibucca 481
Paracrobeles 662
Paractinolaimoides 17, 416, 418
diagnosis 420
Paractinolaimus 17, 190, 192, 416, 417, 418
diagnosis 419
Paracyatholaimoides 501, 526, 527
Paracyatholaimus 21, 501, 524, 525, 526, 527, 532, 533
Paradesmodora 534, 538, 542, 545
Paradorylaimus 17, 411, 412, 414, 415
diagnosis 415
Paraethmolaimus 501, 530
Parahadronchus 476, 477
Paraleptonchus 360, 361
Paramesacanthion 237, 240
diagnosis 244
Paramononchus 482
Paramphidelus 14, 277
diagnosis 282
Paramylonchulus 486
Paranticoma 246
Paranygolaimus 17, 380, 381
diagnosis 384
Paraoxybelondira 244
Paraoxydirus 16, 343, 346, 349
diagnosis 348
Paraphanolaimus 23, 156, 619, 623
diagnosis 623
Paraphelenchus 658, 684
Paraplectonema 23, 64, 181, 183, 616
diagnosis 616
Paraplectus 616
Pararaeolaimus 606
Parasaveljevia 237, 241
diagnosis 244
Parastomachoglossa 18, 416, 417, 424, 425
diagnosis 426
Paraterschellingia 557
Paratrichodorus 298
Paratricoma 557, 568
Paratrilobini 300
Paratrilobus 301, 320
Paratripyla 311
Paratripyloides 273, 274
diagnosis 275
Paratylenchus 198, 200, 658, 665
Paravulvus 17, 378, 380, 385
diagnosis 384
Paraxonchiinae 429, 430
Parenoplus 237, 240
diagnosis 244
Pareudesmoscolex 556, 560, 564, 565
Pareurystomina 251, 260
diagnosis 263
Pareurystominінae 259
Parironus 266
Parkellus 468, 469, 482
Paroigolaimella 27, 173, 609, 701
diagnosis 714
Paroxystomina 269, 272
Paroxystomininae 269, 272
Pelagonema 259
Pelagonematinae 252, 258
Pellioditis 28, 113, 698
diagnosis 708
Pelodera 28, 173, 697, 701
diagnosis 705
Peloderidae 28, 697, 700
diagnosis 705
Peltamigratus 198, 658, 665
Perioplectus 637
Peritobrilus 305
Phanodermatidae 13, 229, 238
diagnosis 246
Phellonema 368
Pheronus 266
Pierrickia 605
Planorbidae 701
Plectida 8, 9, 10, 22, 203, 204, 611, 612, 635
Plectidae 23, 112, 157, 212, 611, 612, 627, 628
diagnosis 631
Plectinae 23, 631, 632, 634
Plectoidea 23, 267, 611, 612, 613, 625
Plectoides 591, 632
Plectus 23, 49, 64, 68, 110, 114, 129, 147, 173, 181, 183, 184, 186, 187, 191, 192, 193, 203, 204, 631, 633, 634, 635
diagnosis 632
Pleurotylenchus 658

Poikilolaimus 28, 698, 701
diagnosis 708
Polygastrophora 14, 196, 230, 251, 259
diagnosis 263
Polyonchulus 469, 488
Pomponematinae 501, 524
Pontonema 69, 111, 226, 247, 259
Pontonematinae 252, 259
Porternema 341, 340
Postamphidelus 277
diagnosis 282
Practinocephalus 18, 416, 417, 428
diagnosis 429
Pratylenchidae 26, 199, 676, 686, 687
diagnosis 676
Pratylenchoides 189, 659, 676
Pratylenchus 15, 198, 200, 493, 659, 666, 677
diagnosis 678
Prionchulellus 469, 477, 490
diagnosis 476
Prionchuloides 19, 477, 490
diagnosis 476
Prionchulus 19, 110, 470, 483, 491, 493
diagnosis 484
Prismatolaimidae 14, 212, 293, 294, 299, 612
Prismatolaimoidea 14, 294, 295, 299, 321
Prismatolaimus 14, 299, 318
Procamacolaimus 14, 64, 181, 183, 186, 191, 192, 294, 297, 300, 316, 319, 320
diagnosis 299
Prochromadora 20, 501, 505, 508
diagnosis 511
Prochromadorella 508, 511
Prodesmodora 21, 48, 64, 154, 181, 192, 200, 497, 534, 537, 549
diagnosis 547
Prodesmodorinae 21, 534, 541, 546
Prodontorhabditis 28, 698
diagnosis 709
Prodorylaiminae 17, 394
diagnosis 395
Prodorylaimium 17, 395, 396
diagnosis 397
Prodorylaimus 17, 186, 192, 328, 395, 415
diagnosis 395
Progreeffiella 557
Proleptonchoides 16, 360, 361
diagnosis 362
Proleptonchus 16, 189, 360, 361, 362
diagnosis 363
Promiconchus 479
Promononchus 311
Promumtazium 352, 353
Prooncholaimus 249, 256, 257
Proteroplectus 632
Prothallonema 659
Protodorylaimus 17, 395, 396
Protorhabditis 28, 194, 698, 700, 706
diagnosis 709
Prototricoma 556, 560, 564
Protricoma 557, 567
Protricomoides 556, 564
Psammonema 534, 538, 543, 545
Pseudaraeolaimus 606
Pseudobathylaimus 621
Pseudochromadora 500, 534, 537, 543, 548
diagnosis 545
Pseudodesmodora 534, 538, 546
Pseudonchinae 534, 537, 541, 547
Pseudoncholaimus 13, 201, 249, 256
diagnosis 257
Pseudonchulus 15, 295, 309, 320
diagnosis 316
Pseudotylencholaimus, 352, 353
Psilenchus 26, 199, 659, 660, 666, 670, 688
diagnosis 673
Ptycholaimellus 40, 109, 501, 509, 514
diagnosis 519
Punctodora 20, 110, 497, 499, 501, 505, 508, 510, 531, 532
diagnosis 511
Pungentinae 15, 454, 455, 457, 459
Pungentus 18, 455, 456
diagnosis 457

Quadricoma 557, 567
Quadricomoides 557, 567
Quasibrilus 301
Qudsianema 439
Qudsianematidae 18, 212, 392, 393, 429, 433, 447
diagnosis 429
Qudsianematinae 18, 429, 430, 434, 435, 437, 438, 439, 442, 443, 444, 445
diagnosis 433
Qudsiella 346

Radopholus 199, 660
Rahmium 468, 474
Raritobrilus 305
Rhabditella 28, 698
diagnosis 709
Rhabditida 5, 6, 8, 9, 10, 11, 23, 152, 171, 181, 183, 186, 203, 204, 494, 648
Rhabditidae 28, 112, 157, 164, 212, 696, 697, 700, 702, 703
diagnosis 706
Rhabditidoides 27, 698
diagnosis 711
Rhabditina 9, 27, 153, 696–721
Rhabditis 9, 28, 181, 182, 186, 187, 189, 190, 192, 193, 203, 204, 698, 700, 705, 706, 707, 708, 709, 710
Rhabditoides 28, 173, 698
diagnosis 710
Rhabditolaimidae 579
Rhabditomorpha 27, 691, 697, 700, 706
diagnosis 704
Rhabdocoma 227
Rhabdodemania 227

Rhabdolaimidae 23, 164, 612
Rhabdolaimus 48, 49, 64, 68, 83, 97, 181, 182, 202; *see also* Appendix
Rhadinematidae 23, 611
Rhinema 535, 539, 554, 555
Rhodolaimus 27, 697, 704
Rhyssocolpus 19, 455, 456
 diagnosis 459
Roqueus 346
Rostrulium 352, 353
Rotylenchulus 199, 660
Rotylenchus 26, 187, 193, 199, 666, 679
 diagnosis 680

Sabatieria 40, 605
Sacrimarinema 579, 596, 599
 diagnosis 598
Saevadorella 461
Saveljevia 237, 244
 diagnosis 245
Scalpenchinae 16, 357, 363
Scalpenchus 364, 371
Scapidens 379
Scleralaimus 277
 diagnosis 282
Scleroactinolaimus 17, 416
 diagnosis 422
Sclerolaimus 352
Sclerostylus 358
Scutellonema 164, 199, 200, 660, 666
Sectonema 18, 450
 diagnosis 453
Sectonematinae 18, 448, 453
Seinura 108, 153, 199, 660
Selachinematidae 498, 501, 502, 503, 504, 521, 530
Semitobrilus 320, 322
 diagnosis 307
Setsalia 15, 294, 302, 319, 321
 diagnosis 306
Sibayinema 21, 534, 538, 542, 548
 diagnosis 546
Simpliconematidae 13, 247
Sinanema 22, 578, 579, 580, 581, 583, 591
 diagnosis 591
Siphonolaimoidea 22, 580
 diagnosis 600
Sitadevinema 579
Skibbenema 433, 434
Smithsoninema 612
Solididens 17, 378, 380, 381
 diagnosis 385
Southerniella 606
Sphaerolaimidae 22, 574, 577
 diagnosis 595
Sphaerolaimoidea 22, 577, 580, 594
Sphaerolaimus 40, 108
 diagnosis 595
Sphaerulariidae 26
Sphaerularioidea 26, 648, 668, 681
Spilipherinae 497, 506, 519
Spilophorella 501, 513, 514
 diagnosis 519
Spinodesmoscolex 556, 560, 562
Spiriniinae 534, 537, 541, 547
Spirobolbolaimus 535, 540, 550
 diagnosis 553
Sporonchulus 470, 483, 491
 diagnosis 485
Stenonchulus 15, 193, 194, 294, 309, 320, 323
 diagnosis 314
Stephanium 468, 474
Steratocephalus 611, 627, 642, 643
Stilbonematinae 534, 537, 541, 547
Stopractinca 18, 416, 417, 425
 diagnosis 424
Striatodora 606
Stygodesmodora 534, 538, 542
 diagnosis 546
Subclavatus 377
Swangeria 346
Swangeriinae 15, 341, 343, 347, 348
 diagnosis 345
Symplocostoma 251, 259, 260, 262
 diagnosis 264
Symplocostomella 251, 260
 diagnosis 264
Synodontium 605
Syringolaimus 230, 232, 267

Takakia 299
Takamangai 18, 433, 434, 436
 diagnosis 437
Talanema 434
Tantunema 16, 351, 352, 353
 diagnosis 355
Tectonchus 468
Terschellingia 22, 40, 197, 601
 diagnosis 600
Thalassironinae 266, 267, 268
Thalassironus 266, 267
Thalassoalaimus 14, 196, 265, 269
 diagnosis 270
Thalassomonhystera 22, 581, 582, 593
 diagnosis 592
Theristus 22, 40, 195, 197, 596, 597, 598
 diagnosis 600
Thonus 437
Thoonchinae 259
Thoonchus 251, 260
 diagnosis 264
Thoracostomopsidae 13, 227, 238, 239, 242
 diagnosis 238
Thoracostomopsinae 239
Thoracostomopsis 239
Thornedia 461
Thorneella 19, 464
 diagnosis 465
Thorneellinae 19, 462, 465
Thornenematidae 17, 212, 392, 393
 diagnosis 394

Thornia 19, 463, 464
diagnosis 463
Thorniidae 19, 392, 393, 465
diagnosis 462
Thorniinae 19, 462, 463
Tigronchoides 469, 478
Timmia 501, 505, 507
diagnosis 512
Timminema 463
Timmus 371
Tobrilia 15, 192, 294, 304, 318
diagnosis 313
Tobrilidae 11, 171, 188, 201, 202, 272, 293, 294, 307, 308, 315, 317, 321, 498
diagnosis 300
Tobrilinae 15, 294, 300, 303, 307, 308, 318
Tobrilina 9, 14, 272, 293, 294, 295, 299, 300, 321
ecology 320
Tobrilinae 15, 294, 300, 303, 318
Tobriloidea 14, 294, 295, 299, 300, 321
Tobriloides 295, 315, 316, 318
Tobrilus 10, 11, 15, 48, 63, 64, 67, 68, 83, 97, 98, 139, 154, 155, 156, 170, 173, 181, 183, 184, 189, 190, 191, 192, 193, 201, 202, 203, 204, 294, 299, 301, 302, 303, 304, 306, 307, 308, 309, 310, 315, 319, 321, 322
Torumanawa 433, 434
Trachactinolaimus 416, 417, 418
Trachypleurosum 416, 417, 418
Trefusia 231, 247
Trefusialaimus 247
Trefusiida 13, 247
Trefusiidae 13, 247
Trefusiina 9, 13, 231, 233, 234, 235
diagnosis 247
Trefusioidea 13, 247
Tricaenonchus 470, 486, 492
Trichodoridae 14, 200, 293, 294, 295, 298, 316, 319
diagnosis 296
Trichodorus 14, 199, 200, 294, 297, 316, 321, 385
diagnosis 296
Tricoma 22, 497, 555, 557, 558, 559, 563, 567
Tricominae 21, 557, 559, 562, 563, 566, 567
Tridentulus 22, 581, 582, 592, 594
Trileptiinae 239
Trilobus 15, 303, 305, 306, 307, 308, 309
Triplonchida 9, 10, 14, 170, 181, 186, 200, 201, 203, 204, 272, 293–323, 612
diagnosis 295
Triplonchium 293, 296
Tripyla 15, 48, 64, 67, 95, 153, 173, 181, 186, 293, 294, 304, 311, 313, 318, 322
Tripylella 294, 310, 318
diagnosis 312
Tripylidae 15, 154, 272, 293, 294, 295, 311, 312, 313, 322
diagnosis 309
Tripylina 13, 15, 294, 295, 309, 310, 313, 318, 320, 322
diagnosis 312
Tripylium 579, 580, 581
Tripyloidea 15, 267, 278, 294, 309, 313
Tripyloides 14, 40, 197, 273, 274
diagnosis 275
Tripyloididae 14, 227, 273, 274, 275
diagnosis 275
Tripyloidina 9, 14, 231, 233, 234, 235, 273
diagnosis 272
Tripyloidoidea 14, 272
Trischistoma 15, 181, 191, 192, 294, 304, 317, 321
diagnosis 312
Trissonchulus 266, 267
Trophonema 660
Trophotylenchulus 199, 660, 666
Truxonchus 469, 478
Tsukubanema 352
Tubixaba 448
Turbatrix 111, 663, 667
Tylenchidae 26, 164, 171, 213, 670, 673, 681
diagnosis 669
Tylenchina 9, 25, 648–695
diagnosis 668
Tylenchodoroides 354
Tylenchodorus 354
Tylenchoidea 26, 668
diagnosis 669
Tylencholaimellidae 16, 349, 350, 351, 366, 367
diagnosis 366–368
Tylencholaimellinae 16, 366, 367, 369, 370
Tylencholaimellus 16, 368, 370
diagnosis 370
Tylencholaimidae 16, 349, 351, 352, 354, 356, 366
diagnosis 350
Tylencholaiminae 16, 351, 352, 354, 355
Tylencholaimoidea 16, 324, 327, 331, 340, 349, 366, 371, 373
diagnosis 350
Tylencholaimus 16, 192, 218, 335, 352, 353, 493
diagnosis 355
Tylenchomorpha 26, 648, 649, 664, 665, 666, 668, 678
diagnosis 668
Tylenchorhynchus 26, 190, 199, 493, 660, 666, 674
diagnosis 675
Tylenchulidae 26, 683
Tylenchulus 199, 660, 666
Tylenchus 26, 186, 188, 189, 192, 193, 199, 218, 661, 666, 670, 671, 672
diagnosis 669, 672
Tyleptinae 16, 357, 359, 364
Tyleptus 16, 359
diagnosis 364
Tylocephalus 23, 631, 640, 641
diagnosis 639
Tylolaimophorus 239, 294, 297, 317
diagnosis 296

Tylopharyngidae 699, 703
diagnosis 716
Tylopharynx 699, 703
diagnosis 716

Udonchus 181, 182, 202
Usarpnema 557, 563, 567
Utahnema 364

Vanderlindia 350, 356
Vanderlindiinae 16, 352
diagnosis 356
Verutus 199, 661, 666
Viscosia 13, 196, 228, 230, 248, 253, 255
diagnosis 254

Walcherenia 628
Wasimellus 352, 354
Westindicus 17, 416, 417, 421
diagnosis 420
Wieseria 265, 270
diagnosis 271
Wiesoncholaimus 249, 256
diagnosis 258
Wilsonema 192, 639, 640
Wilsonematinae 23, 611, 613, 632, 640
diagnosis 639
Witoldinema 439

Xenellidae 13, 247
Xenocyatholaiminae 21, 501, 524
diagnosis 527
Xiphinemella 16, 199, 359
diagnosis 365
Xiphinemellinae 16, 357, 358, 359
diagnosis 365
Xyalidae 22, 574, 577, 600
diagnosis 595–596

Yunqueus 341

Zalonema 534, 537, 543
diagnosis 546
Zalophidera 365
Zetalaimus 358
Zingli 377
Zygotylenchus 661